a LANGE medical book

Harper's
Biochemistry

twenty-fourth edition

a LANGE medical book

Harper's
Biochemistry

twenty-fourth edition

Robert K. Murray, MD, PhD
Professor of Biochemistry
University of Toronto
Toronto, Ontario

Daryl K. Granner, MD
Professor and Chairman
Department of Molecular Physiology and Biophysics
Professor of Medicine
Vanderbilt University
Nashville, Tennessee

Peter A. Mayes, PhD, DSc
Reader in Biochemistry
Royal Veterinary College
University of London
London

Victor W. Rodwell, PhD
Professor of Biochemistry
Purdue University
West Lafayette, Indiana

APPLETON & LANGE
Stamford, Connecticut

Notice: The authors and the publisher of this volume have taken care to make
certain that the doses of drugs and schedules of treatment are correct and compatible
with the standards generally accepted at the time of publication. Nevertheless,
as new information becomes available, changes in treatment and in the use of drugs
become necessary. The reader is advised to carefully consult the instruction and
information material included in the package insert of each drug or therapeutic agent
before administration. This advice is especially important when using, administering,
or recommending new or infrequently used drugs. The authors and publisher disclaim
all responsibility for any liability, loss, injury, or damage incurred as a consequence,
directly or indirectly, of the use and application of any of the contents of this volume.

Copyright © 1996 by Appleton & Lange
A Simon & Schuster Company
Previous editions © 1993, 1990 by Appleton & Lange and © 1988 by Lange Medical Publications

All rights reserved. This book, or any parts thereof, may not be used or
reproduced in any manner without written permission. For information,
address Appleton & Lange, Four Stamford Plaza, PO Box 120041,
Stamford, Connecticut 06912–0041.

96 97 98 99 00 / 10 9 8 7 6 5 4 3 2 1

Prentice Hall International (UK) Limited, *London*
Prentice Hall of Australia Pty. Limited, *Sydney*
Prentice Hall Canada, Inc., *Toronto*
Prentice Hall Hispanoamericana, S.A., *Mexico*
Prentice Hall of India Private Limited, *New Delhi*
Prentice Hall of Japan, Inc., *Tokyo*
Simon & Schuster Asia Pte. Ltd., *Singapore*
Editora Prentice Hall do Brasil Ltda., *Rio de Janeiro*
Prentice Hall, *Upper Saddle River, New Jersey*

ISSN 0734-9866

Acquisitions Editor: John Dolan
Production Editor: Chris Langan
Senior Art Coordinator: Maggie Belis Darrow
Illustrators: Teshin Associates

PRINTED IN THE UNITED STATES OF AMERICA

ISBN 0-8385-3611-5

9 780838 536117

Table of Contents

SECTION III. METABOLISM OF PROTEINS & AMINO ACIDS

SECTION IV. STRUCTURE, FUNCTION, & REPLICATION OF INFORMATIONAL MACROMOLECULES

SECTION V. BIOCHEMISTRY OF EXTRACELLULAR & INTRACELLULAR COMMUNICATION

Preface

Harper's Biochemistry provides concise yet authoritative coverage of the principles of biochemistry and molecular biology. It also offers numerous examples of why a knowledge of biochemistry is imperative for understanding how health is maintained and for understanding the causes and rational treatment of many diseases, two major areas of interest to physicians and other health care workers.

CHANGES IN THE TWENTY-FOURTH EDITION

The two central goals guiding the preparation of this new edition have again been to reflect the latest advances in biochemistry that are important to medicine and to provide medical and other students of the health sciences with a book that is user-friendly and interesting. The changes described below reflect these objectives:

- The most striking feature of the present edition concerns the artwork. Attractive coloring has been introduced, many figures are new, and many have been redrawn.
- Every chapter has been updated and revised.
- The metabolism of lipids and carbohydrates is explained with the aid of figures designed to convey a sound grasp of the mechanisms that underlie metabolic diseases such as diabetes mellitus and atherosclerosis.
- The extensive coverage of lipoproteins has been updated.
- Many topics in metabolism have been updated, including the creatine phosphate shuttle, the role of glycogen in the synthesis of glycogen, glucose transporters, the role of peroxisomes in the synthesis of glyceryl either lipids, and the role of antioxidants in disease processes.
- Discussions of DNA replication and repair are updated.
- New information on ribozymes is included.
- Discussion of the regulation of protein synthesis has been updated.
- New aspects of the actions of insulin, growth hormone, and prolactin are described.
- Coverage of membranes has been expanded to include discussion of vesicular transport, chaperones, and diseases due to abnormalities of membrane constituents.
- Discussion of the role of glycoproteins in various diseases is expanded.
- The biochemistry of bone and cartilage is discussed, along with the bases of a number of diseases of these tissues.
- The role of nitric oxide in health and disease is described.
- The functions of various plasma proteins and the molecular bases of Wilson's disease and Menkes' disease are described.
- Discussion of the role of platelets in health and disease is updated.
- Information on cytochrome P450 is updated.
- Information on the Human Genome Project, gene therapy, and antisense therapy is expanded.
- The roles of apoE4 in Alzheimer's disease and of trinucleotide repeat expansions in various neurologic conditions are described.

ORGANIZATION OF THIS BOOK

The text is divided into three introductory chapters followed by six main sections.

Section I is devoted to proteins and enzymes, the workhorses of the body. Because most reactions in the human body are catalyzed by enzymes, it is vital to understand the properties of enzymes before moving on to other topics.

Section II sets out how various cellular reactions either utilize or produce energy and traces the pathways by which carbohydrates and lipids are synthesized and degraded. The functions of these two classes of molecules are also described.

Section III deals with the amino acids and their fates and shows how the metabolism of amino acids also uses and yields energy.

Section IV describes the structures and functions of the nucleic acids, covering the representation DNA→DNA→proteins. This section also describes the principles of recombinant DNA technology, a topic having tremendous implications for biomedical science.

Section V discusses hormones and their key roles in intercellular communication and metabolic regulation. In order to affect cells, hormones must first interact with the plasma membranes of cells; thus, membrane structure and function are addressed initially in this chapter.

Section VI consists of fourteen special topics, including the extracellular matrix, cancer, a description of the biochemical bases of certain neuropsychiatric disorders, and nine biochemical case histories.

The **Appendix** contains a list of the major biochemical laboratory tests and their reference ranges used in clinical medicine.

ACKNOWLEDGMENTS

The authors extend their appreciation to John Dolan for his ongoing interest, advice, and encouragement. They are particularly grateful to Jim Ransom for his highly professional editing and most helpful comments about many aspects of the text. It was a privilege to work with him. The authors wish to extend their gratitude to professional colleagues and friends throughout the world who have conveyed to us their suggestions for improvements and corrections. We wish to encourage their continued interest and effort. Particularly helpful were many constructive suggestions by Dr. Frank Vella of the University of Saskatchewan. Comments from many student readers were also invaluable and are warmly welcomed. RKM wishes to thank Drs. David Williams, Eric Degen, and Harry Schachter for access to materials that were invaluable in revising Chapters 43 and 56. He would also like to thank Drs. P. Hamel and J. R. Wherrett for very helpful suggestions regarding updating Chapters 62 and 64, respectively. The authors are grateful for the broad base of acceptance and support this book has received in many countries. Several editions of the English language version have been reprinted in Japan, Lebanon, Taiwan, the Philippines, and Korea. In addition, there are translations in Italian, Spanish, French, Portuguese, Japanese, Polish, German, Indonesian, Serbo-Croatian, and Greek.

RKM
DKG
PAM
VWR

March 1996

Biochemistry & Medicine

1

Robert K. Murray, MD, PhD

INTRODUCTION

Biochemistry is the science concerned with the various molecules that occur in living cells and organisms and with their chemical reactions. Anything more than an extremely superficial comprehension of life—in all its diverse manifestations—demands a knowledge of biochemistry. In addition, medical students who acquire a sound knowledge of biochemistry will be in a strong position to deal with two central concerns of the health sciences: (1) the understanding and maintenance of health and (2) the understanding and effective treatment of disease.

Biochemistry Is the Chemistry of Life

Biochemistry can be defined more formally as *the science concerned with the chemical basis of life* (Gk *bios* "life").

The **cell** is the structural unit of living systems. Consideration of this concept leads to a functional definition of biochemistry as *the science concerned with the chemical constituents of living cells and with the reactions and processes that they undergo.* By this definition, biochemistry encompasses large areas of **cell biology,** of **molecular biology,** and of **molecular genetics.**

The Aim of Biochemistry Is to Describe and Explain, in Molecular Terms, All Chemical Processes of Living Cells

The major objective of biochemistry is the complete understanding at the molecular level of all of the chemical processes associated with living cells. To achieve this objective, biochemists have sought to isolate the numerous molecules found in cells, determine their structures, and analyze how they function. To give one example, the efforts of many biochemists to understand the molecular basis of **contractility**—a process associated primarily, but not exclusively, with muscle cells—have entailed purification of many molecules, both simple and complex, followed by detailed structure-function studies. Through these efforts, some of the features of the molecular basis of **muscle contraction** have been revealed.

A further objective of biochemistry is to attempt to understand how life began. Knowledge of this fascinating subject is still embryonic.

The scope of biochemistry is as wide as life itself. Wherever there is life, chemical processes are occurring. Biochemists study the chemical processes that occur in microorganisms, plants, insects, fish, birds, mammals, and human beings. Students in the biomedical sciences will be particularly interested in the biochemistry of the two latter groups. However, an appreciation of the biochemistry of less complex forms of life is often of direct relevance to human biochemistry. For instance, contemporary theories on the regulation of the activities of genes and of enzymes in humans emanate from pioneering studies on bread molds and on bacteria. The field of **recombinant DNA** emerged from studies on bacteria and their viruses; their rapid multiplication times and the ease of extracting their genetic material make them suitable for genetic analyses and manipulations. Knowledge gained from the study of viral genes responsible for certain types of cancer in animals (**viral oncogenes**) has provided profound insights into how human cells become cancerous.

A Knowledge of Biochemistry Is Essential to All Life Sciences

The biochemistry of the nucleic acids lies at the heart of genetics; in turn, the use of genetic approaches has been critical for elucidating many areas of biochemistry. Physiology, the study of body function, overlaps with biochemistry almost completely. Immunology employs numerous biochemical techniques, and many immunologic approaches have found wide use by biochemists. Pharmacology and pharmacy rest on a sound knowledge of biochemistry and physiology; in particular, most drugs are metabolized by enzyme-catalyzed reactions, and the complex interactions among drugs are best understood biochemically. Poisons act on biochemical reactions or processes; this is the subject matter of toxicology. Biochemical approaches are being used increasingly to study basic aspects of pathology (the study of disease), such as inflammation, cell injury, and cancer.

Many workers in microbiology, zoology, and botany employ biochemical approaches almost exclusively. These relationships are not surprising, because life as we know it depends on biochemical reactions and processes. In fact, the old barriers among the life sciences are breaking down, and biochemistry is increasingly becoming their common language.

A Reciprocal Relationship Between Biochemistry and Medicine Has Stimulated Mutual Advances

As stated at the beginning of this chapter, the two major concerns for workers in the health sciences—and particularly physicians—are the understanding and maintenance of health and the understanding and effective treatment of diseases. Biochemistry impacts enormously on both of these fundamental concerns of medicine. In fact, the interrelationship of biochemistry and medicine is a wide, two-way street. Biochemical studies have illuminated many aspects of health and disease, and conversely, the study of various aspects of health and disease has opened up new areas of biochemistry. Some examples of this two-way street are shown in Figure 1–1. For instance, a knowledge of protein structure and function was necessary to elucidate the single biochemical difference between normal and sickle cell hemoglobin. On the other hand, analysis of sickle cell hemoglobin has contributed significantly to our understanding of the structure and function of both normal hemoglobin and other proteins. Analogous examples of reciprocal benefit between biochemistry and medicine could be cited for the other paired items shown in Figure 1–1. Another example is the pioneering work of Garrod, a physician in England during the early years of this century. He studied patients with a number of relatively rare disorders (alkaptonuria, albinism, cystinuria, and pentosuria; these are described in later chapters) and established that these conditions were genetically determined. Garrod designated these conditions as **inborn errors of metabolism.** His insights provided a major foundation for the development of the field of human biochemical genetics.

This relationship between medicine and biochemistry has important philosophical implications for the former. As long as medical treatment is firmly grounded in a knowledge of biochemistry and other relevant basic sciences (eg, physiology, microbiology, nutrition), the practice of medicine will have a rational basis that can be adapted to accommodate new knowledge. This contrasts with unorthodox health cults, which are often founded on little more than myth and wishful thinking and generally lack any intellectual basis.

NORMAL BIOCHEMICAL PROCESSES ARE THE BASIS OF HEALTH

The World Health Organization (WHO) defines health as a state of "complete physical, mental and social well-being and not merely the absence of disease and infirmity." From a strictly biochemical viewpoint, health may be considered that situation in which all of the many thousands of intra- and extracellular reactions that occur in the body are proceeding at rates commensurate with its maximal survival in the physiologic state. However, this is an extremely reductionist view, and it should be apparent that caring for the health of patients requires not only a wide knowledge of biologic principles but also of psychologic and social principles.

Biochemical Research Has Impact on Nutrition and Preventive Medicine

One major prerequisite for the maintenance of health is that there be optimal dietary intake of a number of chemicals; the chief of these are **vitamins,**

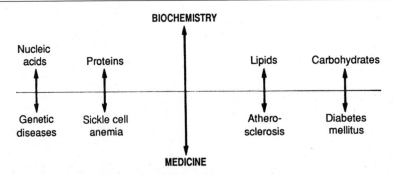

Figure 1–1. Examples of the two-way street connecting biochemistry and medicine. Knowledge of the biochemical compounds shown in the top part of the diagram has illuminated the diseases shown in the bottom half; conversely, analyses of the diseases shown below have illuminated many areas of biochemistry. Note that sickle cell anemia is a genetic disease, and both atherosclerosis and diabetes mellitus have genetic components.

certain **amino acids,** certain **fatty acids,** various **minerals,** and **water.** Because much of the subject matter of both biochemistry and nutrition is concerned with the study of various aspects of these chemicals, there is a close relationship between these two sciences. Moreover, as attempts are made to curb the rising costs of medical care, more emphasis is being placed on systematic attempts to maintain health and forestall disease, ie, on **preventive medicine.** Thus, nutritional approaches to—for example—the prevention of atherosclerosis and cancer are receiving increased emphasis. Understanding nutrition depends to a great extent on a knowledge of biochemistry.

All Disease Has a Biochemical Basis

All diseases are manifestations of abnormalities of molecules, chemical reactions, or processes. The major factors responsible for causing diseases in animals and humans are listed in Table 1–1. All of them affect one or more critical chemical reactions or molecules in the body.

Biochemical Studies Contribute to Diagnosis, Prognosis, and Treatment

There is a wealth of documentation of the uses of biochemistry in prevention, diagnosis, and treatment of disease; many examples will be cited throughout this text. Here, seven brief examples are given to illustrate the breadth of the subject and stimulate the reader's interest.

(1) Humans must ingest a number of complex organic molecules called **vitamins** in order to maintain health. If a particular vitamin is deficient in the diet, the reactions in which it is involved are compro-

Table 1–1. The major causes of diseases. All of the causes listed act by influencing the various biochemical mechanisms in the cell or in the body[1]

1. Physical agents: Mechanical trauma, extremes of temperature, sudden changes in atmospheric pressure, radiation, electric shock.
2. Chemical agents, including drugs: Certain toxic compounds, therapeutic drugs, etc.
3. Biologic agents: Viruses, rickettsiae, bacteria, fungi, higher forms of parasites.
4. Oxygen lack: Loss of blood supply, depletion of the oxygen-carrying capacity of the blood, poisoning of the oxidative enzymes.
5. Genetic disorders: Congenital, molecular.
6. Immunologic reactions: Anaphylaxis, autoimmune disease.
7. Nutritional imbalances: Deficiencies, excesses.
8. Endocrine imbalances: Hormonal deficiencies, excesses.

[1]Adapted, with permission, from Robbins SL, Cotram RS, Kumar V: *The Pathologic Basis of Disease,* 3rd ed. Saunders, 1984.

mised. This situation may be manifested as a deficiency disease such as scurvy or rickets (due to lack of intake of vitamin C and D, respectively). The elucidation of the roles played by the vitamins or their biologically active derivatives in animal and human cells has been a concern of biochemists and nutritionists since the turn of the century. Once a disease was established as resulting from a vitamin deficiency, it became rational to treat it by administration of the appropriate vitamin.

(2) The fact that many plants in Africa are deficient in one or more essential amino acids (ie, amino acids that must be supplied in the diet in order to maintain health) helps explain the debilitating malnutrition **(kwashiorkor)** suffered by those who depend on such plants as major dietary sources of protein. Treatment of deficiencies of essential amino acids is rational but, unfortunately, not always feasible. It consists of providing a well-balanced diet containing adequate amounts of all of the essential amino acids.

(3) Greenland Inuit consume large quantities of fish oils rich in certain **polyunsaturated fatty acids** and are known to have low plasma levels of cholesterol and a low incidence of atherosclerosis. These observations have stimulated interest in the use of polyunsaturated fatty acids to reduce plasma levels of cholesterol.

The vitamin deficiency diseases and the essential amino acid deficiencies are examples of nutritional imbalances (Table 1–1). Atherosclerosis may be considered as a nutritional imbalance, but other important factors (eg, genetic) are also involved.

(4) The condition known as **phenylketonuria,** if untreated, may lead to severe mental retardation in infancy. The biochemical basis of phenylketonuria has been known since 1953; the disorder is genetically determined and results from low or absent activity of the enzyme that converts the amino acid phenylalanine to the amino acid tyrosine. This in turn causes an elevation of the level of phenylalanine in the blood, resulting in damage to the developing central nervous system. When the nature of the biochemical lesion in phenylketonuria was revealed, it became rational to treat the disease by placing affected infants on a diet low in phenylalanine. Once biochemical screening tests for diagnosing phenylketonuria at birth became available, effective treatment could be started immediately.

(5) **Cystic fibrosis** is a common genetic disease of the exocrine glands and of the eccrine sweat glands. It is characterized by abnormally viscous secretions that plug up the secretory ducts of the pancreas and the bronchioles. In addition, patients with cystic fibrosis exhibit elevated amounts of chloride in their sweat. Victims often die at an early age from lung infections. The isolation and complete sequence of the gene responsible for this disease was reported in 1989. The normal gene codes for a transmembrane protein (the cystic fibrosis transmembrane conduc-

tance regulator), 1480 amino acids in length, which functions as a chloride channel. The abnormality in approximately 70% of patients with cystic fibrosis is a deletion of three bases in the gene, resulting in the transmembrane protein lacking amino acid number 508, a phenylalanine residue. How this deletion impairs the function of the transmembrane protein and results in the excessively thick mucus is being determined. This important work should facilitate the detection of carriers of the cystic fibrosis gene and, it is hoped, lead to more rational treatment of the disease than exists at present. For instance, it may be possible to design drugs that can correct the abnormality in the transmembrane protein; likewise, it may be possible to introduce the normal gene into lung cells by gene therapy. Phenylketonuria and cystic fibrosis are examples of genetic diseases (Table 1–1).

(6) Analysis of the mechanism of action of the bacterial toxin that causes **cholera** has provided important insights into how the clinical manifestations of this disease (copious diarrhea and loss of salt and water) are brought about.

(7) The finding that the mosquitoes which transmit the parasites (plasmodia) causing **malaria** can develop biochemically based resistance to the action of insecticides has important consequences for attempts to eradicate this serious disease. This example and the previous one are instances of diseases caused by biologic agents (Table 1–1).

Many Biochemical Studies Illuminate Disease Mechanisms, and Diseases Inspire Biochemical Research

The initial observations made by Garrod on a small group of inborn errors of metabolism in the early 1900s stimulated the investigation of the biochemical pathways affected in these conditions. Efforts to understand the basis of the genetic disease known as **familial hypercholesterolemia,** which results in severe atherosclerosis at an early age, have led to dramatic progress in knowledge of cell receptors and of mechanisms of uptake of cholesterol into cells. The ongoing studies of **oncogenes** in cancer cells have directed attention to the molecular mechanisms involved in the control of normal cell growth. These and many other possible examples illustrate how the study of disease can open up whole areas of cell function for basic biochemical research.

THIS TEXT WILL HELP RELATE BIOCHEMICAL KNOWLEDGE TO CLINICAL PROBLEMS

Brief descriptions of the biochemical mechanisms underlying many diseases are interspersed throughout the text. However, Chapters 63, 64, and 65 specifically describe the biochemical bases of a number of important diseases. The Appendix briefly discusses some basic considerations used in the interpretation of the results of the biochemical laboratory tests and lists the most widely used tests along with their ranges of normal values. The overall purpose of these final chapters and the Appendix is to assist and encourage the reader to translate knowledge of biochemistry into effective clinical use.

Some major uses of biochemical investigations and of laboratory tests in relation to diseases are summarized in Table 1–2. Additional examples of many of these uses are presented in various sections of this text.

SUMMARY

Biochemistry is the science concerned with studying the various molecules that occur in living cells and organisms and with their chemical reactions. Because life depends on biochemical reactions, biochemistry has become the basic language of all biologic sciences.

Biochemistry is concerned with the entire spectrum of life forms, from relatively simple viruses and bacteria to complex human beings.

Biochemistry and medicine are intimately related. Health depends on a harmonious balance of biochemical reactions occurring in the body, and disease reflects abnormalities in biomolecules, biochemical reactions, or biochemical processes.

Table 1–2. Some uses of biochemical investigations and laboratory tests in relation to diseases.

Use	Example
1. To reveal the fundamental causes and mechanisms of diseases	Demonstration of the nature of the genetic defects in cystic fibrosis.
2. To suggest rational treatments of diseases based on (1) above	Use of a diet low in phenylalanine for the treatment of phenylketonuria.
3. To assist in the diagnosis of specific diseases	Use of the plasma enzyme creatine kinase MB (CK-MB) in the diagnosis of myocardial infarction.
4. To act as screening tests for the early diagnosis of certain diseases	Use of measurement of blood thyroxine or thyroid-stimulating hormone (TSH) in the neonatal diagnosis of congenital hypothyroidism.
5. To assist in monitoring the progress (eg, recovery, worsening, remission, or relapse) of certain diseases	Use of the plasma enzyme alanine aminotransferase (ALT) in monitoring the progress of infectious hepatitis.
6. To assist in assessing the response of diseases to therapy	Use of measurement of blood carcinoembryonic antigen (CEA) in certain patients who have been treated for cancer of the colon.

Advances in biochemical knowledge have illuminated many areas of medicine. Conversely, the study of diseases has often revealed previously unsuspected aspects of biochemistry.

Biochemical approaches are often fundamental in illuminating the causes of diseases and in designing appropriate therapies.

The judicious use of various biochemical laboratory tests is an integral component of diagnosis and monitoring of treatment.

A sound knowledge of biochemistry and of other related basic disciplines is essential for the rational practice of medical and related health sciences.

REFERENCES

Garrod AE: Inborn errors of metabolism. (Croonian Lectures.) Lancet 1908;2:1, 73, 142, 214.

Kornberg A: Basic research: The lifeline of medicine. FASEB J 1992;6:3143.

Scriver CR et al (editors): *The Metabolic and Molecular Bases of Inherited Disease,* 7th ed. McGraw-Hill, 1995.

Williams DL, Marks V: *Scientific Foundations of Biochemistry in Clinical Practice,* 2nd ed. Butterworth-Heinemann, 1994.

Biomolecules & Biochemical Methods

Robert K. Murray, MD, PhD

INTRODUCTION

This chapter has five objectives. The first is to indicate **the composition of the body and the major classes of molecules** found in it. These molecules make up the principal subjects of this text.

The cell is the major structural and functional unit of biology. Most chemical reactions occurring in the body take place within cells. Thus, the second objective is to give a concise account of the **components of cells** and how they may be isolated; the details of how these components function form much of the fabric of this text.

The third objective concerns the fact that biochemistry is an experimental science. It is important to have an understanding and appreciation of the **experimental approach and methods** used in biochemistry, lest its study become an exercise in rote learning. Moreover, biochemistry is not an immutable corpus of knowledge but a constantly evolving field. Advances, like those in other biomedical areas, depend on the experimental approach and technologic innovation.

The fourth objective is to summarize briefly the **principal achievements** that have been made in biochemistry. The concise view of the science that is presented will assist in imparting to the reader a sense of the overall direction of the remainder of the text.

The fifth objective is to indicate **how little is known about certain areas,** eg, about development, differentiation, brain function, cancer, and many other human diseases. Perhaps this will serve as a stimulus to some readers to contribute to research in these areas.

THE HUMAN BODY IS COMPOSED OF A FEW ELEMENTS THAT COMBINE TO FORM A GREAT VARIETY OF MOLECULES

Carbon, Hydrogen, Oxygen, and Nitrogen Are the Major Elements

The elementary composition of the human body has been determined, and the major findings are listed in Table 2–1. Carbon, oxygen, hydrogen, and nitrogen are the major constituents of most biomolecules. **Phosphate** is a component of the nucleic acids and other molecules and is also widely distributed in its ionized form in the human body. **Calcium** plays a key role in innumerable biologic processes and is the focus of much current research. The elements listed in the third column of the table fulfill diverse roles. Most are encountered on an almost daily basis in medical practice in dealing with patients with electrolyte imbalances (K^+, Na^+, Cl^-, and Mg^{2+}), iron-deficiency anemia (Fe^{2+}), and thyroid diseases (I^-).

The Five Major Complex Biomolecules Are DNA, RNA, Proteins, Polysaccharides, and Complex Lipids

As shown in Table 2–2, the major complex biomolecules found in the cells and tissues of higher animals (including humans) are **DNA, RNA, proteins, polysaccharides,** and **lipids.** These complex molecules are constructed from simple biomolecules, which are also listed. The building blocks of DNA and RNA (collectively known as the nucleic acids) are **deoxynucleotides** and **ribonucleotides,** respectively. The building blocks of proteins are **amino acids.** Polysaccharides are built up from simple carbohydrates; in the case of **glycogen** (the principal polysaccharide found in human tissues), the carbohydrate is **glucose. Fatty acids** may be considered to be the building blocks of many lipids, although lipids are not polymers of fatty acids. DNA, RNA, proteins, and polysaccharides are referred to as **biopolymers** because they are composed of repeating units of their building blocks (the monomers). The above molecules essentially make up the "stuff of life"; most of this text will be largely concerned with descriptions of their various biochemical features and of their building blocks. The same complex molecules are also generally found in lower organisms, although the building blocks in certain cases may differ from those shown in Table 2–2. For instance, bacteria do not contain glycogen or triacylglycerols, but they do contain other polysaccharides and lipids.

Table 2–1. Approximate elementary composition of the human body (dry weight basis).[1]

Element	Percent	Element	Percent
Carbon	50	Potassium	1
Oxygen	20	Sulfur	0.8
Hydrogen	10	Sodium	0.4
Nitrogen	8.5	Chlorine	0.4
Calcium	4	Magnesium	0.1
Phosphorus	2.5	Iron	0.01
		Manganese	0.001
		Iodine	0.00005

[1]Reproduced, with permission, from West ES, Todd WR: *Textbook of Biochemistry,* 3rd ed. Macmillan, 1961.

Table 2–3. Normal chemical composition for a man weighing 65 kg.[1]

	Kg	Percent
Protein	11	17.0
Fat	9	13.8
Carbohydrate	1	1.5
Water[2]	40	61.6
Minerals	4	6.1

[1]Reproduced, with permission, from Davidson SD, Passmore R, Brock JF: *Human Nutrition and Dietetics,* 5th ed. Churchill Livingstone, 1973.
[2]The value of water can vary widely among different tissues, being as low as 22.5% for marrow-free bone. The percentage consisting of water also tends to diminish as body fat increases.

Protein, Fat, Carbohydrate, Water, and Minerals Are the Chief Components of the Human Body

The elementary composition of the human body is given above. Its chemical composition is shown in Table 2–3; protein, fat, carbohydrate, water, and minerals are the chief components. Water constitutes the major component, although its amount varies widely among different tissues. Its polar nature and ability to form hydrogen bonds render water ideally suited for its function as the solvent of the body. A detailed account of the properties of water is presented in Chapter 3.

THE CELL IS THE BASIC UNIT OF BIOLOGY

The cell was established as the fundamental unit of biologic activity by Schleiden and Schwann and other pioneers, such as Virchow, in the 19th century. However, in the years immediately after World War II, three developments helped usher in a period of unparalleled activity in biochemistry and cell biol-

Table 2–2. The major complex organic biomolecules of cells and tissues. The nucleic acids, proteins, and polysaccharides are biopolymers, constructed from the building blocks shown. The lipids are not generally biopolymers, and not all lipids have fatty acids as building blocks.

Biomolecule	Building Block	Major Functions
DNA	Deoxynucleotide	Genetic material
RNA	Ribonucleotide	Template for protein synthesis
Proteins	Amino acids	Numerous; usually they are the molecules of the cell that perform work (eg, enzymes, contractile elements)
Polysaccharide (glycogen)	Glucose	Short-term storage of energy as glucose
Lipids	Fatty acids	Numerous, eg, membrane components and long-term storage of energy as triacylglycerols

ogy. These were (1) the increasing availability of the **electron microscope;** (2) the introduction of methods permitting **disruption of cells** under relatively mild conditions that preserved function; and (3) the increasing availability of the high-speed, refrigerated **ultracentrifuge,** capable of generating centrifugal forces sufficient to separate the constituents of disrupted cells from one another without overheating them. Use of the electron microscope revealed many previously unknown or poorly observable cellular components, while disruption and ultracentrifugation permitted their isolation and analysis in vitro.

A Rat Hepatocyte Illustrates Features Common to Many Eukaryotic Cells

A diagram of the structure of a liver cell (hepatocyte) of a rat is shown in Figure 2–1; this cell is probably the most studied of all cells from a biochemical viewpoint, partly because of its availability in relatively large amounts, suitability for fractionation studies, and diversity of functions. The hepatocyte contains the major **organelles** found in eukaryotic cells (Table 2–4); these include the nucleus, mitochondria, endoplasmic reticulum, free ribosomes, Golgi apparatus, lysosomes, peroxisomes, plasma membrane, and certain cytoskeletal elements.

Physical Techniques Are Used to Disrupt Cells and to Isolate Intracellular Molecules and Subcellular Organelles

In order to study the function of any organelle in depth, it is first necessary to isolate it in relatively pure form, free of significant contamination by other organelles. The usual process by which this is achieved is called **subcellular fractionation** and generally entails three procedures: extraction, homogenization, and centrifugation. Much of the pioneering work in this area was done using rat liver.

A. Extraction: As a first step toward isolating a specific organelle (or molecule), it is necessary to extract it from the cells in which it is located. Most or-

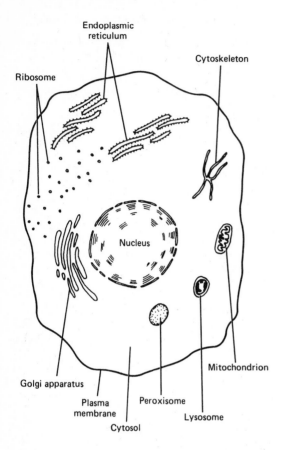

Figure 2–1. Schematic representation of a rat liver cell with its major organelles.

ganelles and many biomolecules are labile and subject to loss of biologic activities: they must be extracted using mild conditions (ie, employment of aqueous solutions and avoidance of extremes of pH and osmotic pressure and of high temperatures). In fact, most procedures for isolating organelles are performed at about 0–4 °C (eg, in a cold room or using material kept on ice). Significant losses of activity can occur at room temperature, partly owing to the action of various digestive enzymes (proteases, nucleases, etc) liberated when cells are disrupted. A common solution for extraction of organelles consists of sucrose, 0.25 mol/L (isosmotic), adjusted to pH 7.4 by TRIS (tris-[hydroxymethyl]aminomethane) hydrochloric acid buffer, 0.05 mol/L, containing K^+ and Mg^{2+} ions at near physiologic concentrations; this solution is conveniently called STKM. Not all solvents used for extraction are as mild as STKM; eg, organic solvents are used for the extraction of lipids and of nucleic acids.

B. Homogenization: To extract an organelle (or biomolecule) from cells, it is first necessary to disrupt the cells under mild conditions. Organs (eg,

liver, kidney, brain) and their contained cells may be conveniently disrupted by the process of homogenization, in which a manually operated or motor-driven pestle is rotated within a glass tube of suitable dimensions containing minced fragments of the organ under study and a suitable homogenizing medium, such as STKM. The controlled rotation of the pestle exerts mechanical shearing forces on cells and disrupts them, liberating their constituents into the sucrose. The resulting suspension, containing many intact organelles, is known as a **homogenate.**

C. Centrifugation: Subfractionation of the contents of a homogenate by differential centrifugation has been a technique of central importance in biochemistry. The classic method uses a series of three different centrifugation steps at successively greater speeds (Figure 2–2), each yielding a pellet and a supernatant. The supernatant from each step is subjected to centrifugation in the next step. This procedure provides three pellets, named the nuclear, mitochondrial, and microsomal fractions. None of these fractions are composed of absolutely pure organelles. However, it has been well established by the use of the electron microscope and by measurements of suitable "marker" enzymes and chemical components (eg, DNA and RNA) that the major constituents of each of these three fractions are nuclei, mitochondria, and microsomes, respectively. A "marker" enzyme or chemical is one that is almost exclusively confined to one particular organelle, eg, acid phosphatase to lysosomes and DNA to the nucleus (Table 2–4). The marker can thus serve to indicate the presence or absence in any particular fraction of the organelle in which it is contained. The **microsomal fraction (microsomes)** contains mostly a mixture of smooth endoplasmic reticulum, rough endoplasmic reticulum (ie, endoplasmic reticulum with attached ribosomes), and free ribosomes. The contents of the final supernatant correspond approximately to those of the **cell sap (cytosol).** Modifications of this basic approach, using different homogenization media or different protocols or methods of centrifugation (eg, the use of gradients—either continuous or discontinuous—of sucrose), have permitted the isolation in more or less pure form of all of the organelles illustrated in Figure 2–1 and listed in Table 2–4. The scheme described above is applicable in general terms to most organs and cells; however, cell fractionations of this type must be assessed by the use of measurements of marker enzymes and chemicals and by the electron microscope until the overall procedure can be considered to be standardized.

The importance of subcellular fractionation studies in the development of biochemistry and cell biology cannot be overemphasized. It has been one of the major components of the experimental approach (see below), and—largely because of its application—the functions of the organelles indicated in Table 2–4 have been elucidated. The information on function

Table 2–4. Major intracellular organelles and their functions. Only the major functions associated with each organelle are listed. In a number of instances, many other pathways, processes, or reactions occur in the organelle.

Organelle or Fraction[1]	Marker	Major Functions
Nucleus	DNA	Site of chromosomes Site of DNA-directed RNA synthesis (transcription)
Mitochondrion	Glutamic dehydrogenase	Citric acid cycle, oxidative phosphorylation
Ribosome[1]	High content of RNA	Site of protein synthesis (translation of mRNA into protein)
Endoplasmic reticulum	Glucose-6-phosphatase	Membrane-bound ribosomes are a major site of protein synthesis Synthesis of various lipids Oxidation of many xenobiotics (cytochrome P450)
Lysosome	Acid phosphatase	Site of many hydrolases (enzymes catalyzing degradative reactions)
Plasma membrane	Na^+ - K^+ ATPase 5′-Nucleotidase	Transport of molecules in and out of cells Intercellular adhesion and communication
Golgi apparatus	Galactosyl transferase	Intracellular sorting of proteins Glycosylation reactions Sulfation reactions
Peroxisome	Catalase Uric acid oxidase	Degradation of certain fatty acids and amino acids Production and degradation of hydrogen peroxide
Cytoskeleton[1]	No specific enzyme markers[2]	Microfilaments, microtubules, intermediate filaments
Cytosol[1]	Lactate dehydrogenase	Enzymes of glycolysis, fatty acid synthesis

[1]An organelle can be defined as a subcellular entity that is membrane-limited and is isolated by centrifugation at high speeds. According to this definition, ribosomes, the cytoskeleton, and the cytosol are not organelles. However, they are considered in this table along with the organelles because they also are usually isolated by centrifugation. They can be considered subcellular entities or fractions. An organelle as isolated by one cycle of differential centrifugation is rarely pure; to obtain a pure fraction usually requires at least several cycles.

[2]The cytoskeletal fractions can be recognized by electron microscopy or by analysis by electrophoresis of the characteristic proteins that they contain.

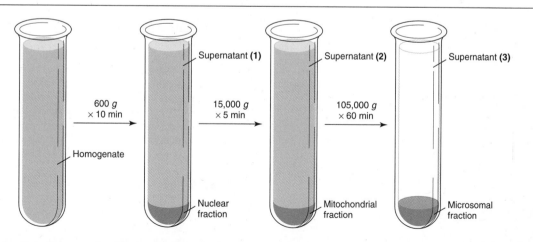

Figure 2–2. Scheme of separation of subcellular fractions by differential centrifugation. The homogenized tissue (eg, liver) is first subjected to low-speed centrifugation, which yields the nuclear fraction (containing both nuclei and unruptured cells) and supernatant *(1).* The latter is decanted and subjected to centrifugation at an intermediate speed, yielding the mitochondrial fraction (containing mitochondria, lysosomes, and peroxisomes) and supernatant *(2).* The latter is next decanted and subjected to high-speed centrifugation, yielding the microsomal fraction (containing a mixture of free ribosomes and smooth and rough endoplasmic reticulum) and the final clear solution, supernatant *(3).* The latter corresponds approximately to the cytosol or cell sap. By various modifications of this basic approach, it is generally possible to isolate each cell organelle in relatively pure form.

summarized in this table represents one of the major achievements of biochemical research (see below).

The Experimental Approach Has Three Components

There are three major components to the experimental approach used in biochemistry: (1) isolation of biomolecules and organelles (see Centrifugation, above) found in cells; (2) determination of the structures of biomolecules; and (3) analyses, using various preparations, of the function and metabolism (ie, synthesis and degradation) of biomolecules.

The Experimental Approach Requires Isolation of Biomolecules

As in the case of organelles, elucidation of the function of any biomolecule requires that it first be isolated in pure form. Table 2–5 lists the major methods that are used to separate and purify biomolecules. No details of these methods will be given here; certain of them are described briefly at various places in this text. A combination of the successive use of several of them is almost always needed to purify a biomolecule to **homogeneity** (freedom from contamination by any other biomolecule).

It is important to appreciate that advances in biochemistry are dependent upon the development of new methods of analysis, purification, and structural determination. For instance, the field of lipid biochemistry was turned on its heels by the introduction of **gas-liquid** and **thin-layer chromatography.** The analysis of membrane and many other proteins was extremely difficult until the introduction of **sodium dodecyl sulfate-polyacrylamide gel electrophoresis** (SDS-PAGE); the introduction of the detergent SDS

Table 2–5. Major methods used to separate and purify biomolecules. Most of these methods are suitable for analyzing the components present in cell extracts and other biochemical materials. The sequential use of several techniques will generally permit purification of most biomolecules. The reader is referred to texts on methods of biochemical research for details regarding each.

Salt fractionation (eg, precipitation with ammonium sulfate)
Chromatography
  Paper
  Ion exchange (anion and cation exchange)
  Affinity
  Thin-layer
  Gas-liquid
  High-pressure liquid
Gel filtration
Electrophoresis
  Paper
  High-voltage
  Agarose
  Cellulose acetate
  Starch gel
  Polyacrylamide
  SDS-polyacrylamide
Ultracentrifugation

permitted the "solubilization" for electrophoresis of many proteins that were previously rather insoluble. The development of methods for sequencing and cloning DNA has had a revolutionary impact on the study of the nucleic acids and of biology in general.

The Experimental Approach Requires Determination of the Structure of Biomolecules

Once a biomolecule has been purified, it is then necessary to determine its structure. This should allow detailed correlations to be made between structure and function. The major methods used to analyze the structures of biomolecules are listed in Table 2–6. They will be familiar to the reader who has some knowledge of organic chemistry. The known specificity of certain enzymes makes them powerful tools in the elucidation of the structural features of certain biomolecules. Improvements in resolution afforded by theoretic and technologic advances are increasingly making **mass spectrometry** and **nuclear magnetic resonance (NMR) spectroscopy** the routine methods of choice for structural determination. For instance, the structures of many proteins and of the extremely complex carbohydrate chains found in certain biomolecules, such as glycoproteins, can now often be elucidated by high-resolution NMR spectroscopy. The most detailed information about the structure of biomolecules is provided by **x-ray diffraction and crystallography.** Their use was critical in revealing the detailed structures of various proteins and enzymes and the double helical nature of DNA.

The Experimental Approach Requires Analysis, Using Various Preparations, of the Function and Metabolism of Biomolecules

Initial biochemical research on humans and animals was performed at the level of the whole animal. Examples were studies of respiration and of the fate of ingested compounds. It soon became apparent that the whole animal was too complex to permit definitive answers to be given to many questions. Accordingly, simpler in vitro preparations were developed that removed many of the complications experienced

Table 2–6. Principal methods used for determining the structures of biomolecules.

Elemental analysis
Ultraviolet, visible, infrared, and NMR spectroscopy
Use of acid or alkaline hydrolysis to degrade the biomolecule under study into its basic constituents
Use of a battery of enzymes of known specificity to degrade the biomolecule under study (eg, proteases, nucleases, glycosidases)
Mass spectrometry
Specific sequencing methods (eg, for proteins and nucleic acids)
X-ray crystallography

at the level of the whole animal. Table 2–7 summarizes the various types of preparations that are now available to study biochemical processes; most of the facts presented in this text have been obtained by their use. The items in the list are arranged in decreasing order of complexity. Just as the use of whole animals has its drawbacks, the other preparations also have limitations. Erroneous results (artifacts) can be derived from the use of in vitro approaches; eg, homogenization of cells can liberate enzymes that may partially digest cellular molecules.

STRATEGIES FOR STUDYING BIOCHEMICAL REACTIONS ARE COMPLEX AND MULTILEVEL

Much of this text is concerned with complex biochemical processes (eg, protein synthesis and muscle contraction), including metabolic pathways. A **metabolic pathway** is a series of reactions responsible for the synthesis of a more complex compound from one or more simple compounds, or for the degradation of a compound to its end product. The existence of a complex biochemical process or of certain metabolic pathways can be inferred from observations made at the level of the whole animal. For instance, direct observations on our fellow humans indicate that muscles contract. We know that glucose serves a source of energy for humans and other animals and can thus infer that it must be degraded (metabolized) in the body to yield energy. However, to understand fully how glucose is metabolized by human cells—and knowledge of this is still far from complete—requires analyses at a variety of levels. Figure 2–3 shows the various types of observations and analyses that are required in order to comprehend biochemical processes, such as the initial breakdown of glucose to yield energy (a process known as **glycolysis**). The scheme shown applies at a general level to all of the major biochemical processes discussed in this text and thus represents an overall strategy for elucidating biochemical processes; it should be borne in mind when each of the major biochemical processes described in this text (eg, glycolysis, fatty acid oxidation) is under consideration, although not every point listed will always be relevant.

A number of important points concerning Table 2–7 and Figure 2–3 merit discussion: (1) Despite the possibility of artifacts, it is absolutely necessary to isolate and identify each component of a biochemical process in pure form in order to understand the process at a molecular level. Numerous examples will be encountered subsequently. (2) It is also im-

Table 2–7. Hierarchy of preparations used to study biochemical processes.

Method	Comments
Studies at the whole-animal level	These can include (1) removal of an organ (eg, hepatectomy). (2) alterations of diet (eg, fasting-feeding). (3) administration of a drug (eg, phenobarbital). (4) administration of a toxin (eg, carbon tetrachloride). (5) use of an animal with a specific disease (eg, diabetes mellitus). (6) use of sophisticated techniques such as NMR spectroscopy and positron emission tomography. Studies at this level are often physiologic but can be difficult to interpret because of the interplay among organs mediated by the circulation and the nervous system.
Isolated perfused organ	Liver, heart, and kidney are particularly suitable. This method allows study of an organ removed from the influence of other organs or the nervous system. The use of perfusion generally ensures that organ function is maintained for at least several hours.
Tissue slice	Liver slices have been especially used. Removes the organ slice from other influences, but the preparations tend to deteriorate within a few hours, partly because of inadequate supply of nutrients.
Use of whole cells	(1) Particularly applicable to blood cells, which can be purified relatively easily. (2) Use of cells in tissue culture is indispensable in many areas of biology.
Homogenate	(1) Ensures a cell-free preparation. (2) Specific compounds can be added or removed (eg, by dialysis) and their effects studied. (3) Can be subfractionated by centrifugation to yield individual cell organelles.
Isolated cell organelles	Extensively used to study the function of mitochondria, the endoplasmic reticulum, ribosomes, etc.
Subfractionation of organelles	Extensively used, eg, in studies of mitochondrial function.
Isolation and characterization of metabolites and enzymes	A vital part of the analysis of any chemical reaction or pathway.
Cloning of genes for enzymes and proteins	Isolation of the cloned gene is vital for studying the details of its structure and regulation; it can also reveal the amino acid sequence of the enzyme or protein for which it codes.

Inference of existence of biochemical process or metabolic pathway from observations made at the whole-animal level

↓

Analyses of its control mechanisms in vivo

↓

Analyses of effects on it of specific diseases (eg, inborn errors of metabolism, cancer)

↓

Its localization to one or more organs

↓

Its localization to one or more cellular organelles or subcellular fractions

↓

Delineation of the number of reactions involved in it

↓

Purification of its individual substrates, products, enzymes, and cofactors or other components

↓

Analyses of its control mechanisms in vitro

↓

Establishment of reaction mechanisms involved in it

↓

Its reconstitution

↓

Studies of it at the gene level by the methods of recombinant DNA technology

Figure 2–3. Scheme of the general strategy used to analyze a biochemical process or metabolic pathway. The approaches used need not necessarily be performed in the precise sequence listed here. However, it is by their employment that the details of a biochemical process or pathway are usually elucidated. The scheme thus applies at a general level to all of the major metabolic pathways discussed in subsequent chapters of this text.

portant to be able to reconstitute the process under study in vitro by the systematic reassembly of its individual components. If the process does not function when its parts are reassembled, one explanation is that some critical component has escaped identification and has not been added back. (3) Recent technologic advances (eg, in NMR spectroscopy and positron emission tomography [PET] scanning) have permitted detection of certain biomolecules at the whole-organ level and monitoring of changes of their amounts with time. Such developments indicate that it is becoming possible to make sophisticated analyses of many biochemical processes at the in vivo level. (4) When the results obtained using many different levels of approach are consistent, then one is justified in concluding that real progress has been made in understanding the biochemical process under study. If major inconsistencies are obtained using the various approaches, then their causes must be investigated until rational explanations are obtained. (5) The preparations and levels of analysis outlined can be used to study biochemical alterations in ani-

mals with altered metabolic states (eg, fasting, feeding) or specific diseases (eg, diabetes mellitus, cancer). (6) Most of the methods and approaches indicated can be applied to studies of normal or diseased human cells or tissues. However, care must be taken to obtain such material in a fresh condition, and particular attention must be paid to those ethical considerations that apply to human experimentation.

Radioactive and Heavy Isotopes Have Contributed to Elucidation of Biochemical Processes

The introduction of the use of isotopes into biochemistry in the 1930s had a dramatic impact; consequently, their use deserves special mention. Prior to their employment, it was very difficult to "tag" biomolecules so that their metabolic fates could be monitored conveniently. Pioneering studies, particularly by Schoenheimer and colleagues, applied the use of certain stable isotopes (eg, ^2D, ^{15}N) combined with their detection by mass spectrometry to many biochemical problems. For instance, certain amino acids, sugars, and fatty acids could be synthesized containing a suitable stable isotope and then administered to an animal or added to an in vitro preparation to follow their metabolic fates (eg, half-lives, conversion to other biomolecules). Compounds labeled with suitable stable isotopes were used to investigate many aspects of the metabolism of proteins, carbohydrates, and lipids. From such studies, it has become apparent that metabolism is a very active process, with most of the compounds in the cell being continuously synthesized and degraded, although at widely differing rates. These findings were epitomized by Schoenheimer as "the dynamic nature of metabolism."

The subsequent introduction of **radioactive isotopes** and of instruments enabling their measurement was also extremely important. The principal stable and radioactive isotopes used in biologic systems are listed in Table 2–8. The use of isotopes, both stable and radioactive, is critical to the development of every area of biochemistry. Investigations of complex and simple biomolecules rely heavily on their use, either in vivo or in vitro. The tremendous progress made recently in **sequencing nucleic acids** and in measuring extremely small amounts of compounds found in biologic systems by **radioimmunoassay** has also depended on their employment.

Table 2–8. Principal isotopes used in biochemical research.

Stable Isotopes	Radioactive Isotopes
^2D	^3H
^{15}N	^{14}C
^{18}O	^{32}P
	^{35}S
	^{35}Ca
	^{125}I
	^{131}I

MAJOR ACHIEVEMENTS CHARACTERIZE THE CONTRIBUTIONS OF BIOCHEMISTRY TO CELL SCIENCE AND MEDICINE

The following paragraphs summarize the principal achievements made in the field of biochemistry, particularly in relation to human biochemistry. Much of this text expands on the topics listed here.

(1) The overall chemical composition of cells, tissues, and the body has been determined, and the principal compounds present have been isolated and their structures established.

(2) The functions of many of the simple biomolecules are understood, at least at a general level, and are described later in this text. The functions of the major complex biomolecules have also been established. Of central interest is that DNA is known to be the genetic material and to transmit its information to one type of RNA (messenger RNA, or mRNA), which in turn dictates the linear sequence of amino acids in proteins. The flow of information from DNA may be conveniently represented as DNA → RNA → protein. However, important exceptions to some of the above statements are known. RNA is the genetic material of certain viruses. In addition, in certain circumstances the information in RNA can be transcribed into DNA; this process is known as **reverse transcription** and is used, for instance, by the virus (human immunodeficiency virus-1, HIV-1) believed to cause AIDS.

(3) The development of recombinant DNA technology has been a pivotal achievement. This technology has revolutionized the study of gene structure and function and has had a revolutionary impact on all fields of biology, including medicine.

(4) The principal organelles of animal cells have been isolated and their major functions established.

(5) Almost all reactions occurring in cells have been found to be catalyzed by enzymes; many enzymes have been purified and studied, and the broad features of their mechanisms of action have been revealed. Although the great majority of enzymes are proteins, it is now firmly established that certain RNA molecules (ribozymes) also have biocatalytic activity.

(6) The metabolic pathways involved in the synthesis and degradation of the major simple and complex biomolecules have been delineated. In general, the pathway of synthesis of a compound has been found to be distinct from its pathway of degradation.

(7) A number of aspects of the regulation of metabolism have been clarified.

(8) The broad features of how cells conserve and utilize energy have been recognized.

(9) Many aspects of the structure and function of the various membranes found in cells are understood; proteins and lipids are their major components.

(10) Considerable information is available at a general level on how the major hormones act.

(11) Biochemical bases for a considerable number of diseases have been discovered.

MUCH REMAINS TO BE LEARNED

While it is important to know that much biochemical information has been accumulated, it is just as important to appreciate how little is known in many areas. Probably the two major problems to be solved concern establishing the biochemical bases of development and differentiation and of brain function. Although the chemical nature of the genetic material is now well established, almost nothing is known about the mechanisms that turn eukaryotic genes on and off during development. Understanding gene regulation is also a key area in learning how cells differentiate and how they become cancerous. Knowledge of cell division and growth—both normal and malignant—and of their regulation is very primitive. Virtually nothing is known concerning the biochemical bases of complex neural phenomena such as consciousness and memory. Only very limited information is available concerning mechanisms of cell secretion. Despite some progress, the molecular bases of most major genetic diseases are unknown, but approaches provided by recombinant DNA technology suggest that remarkable progress will be made in this area during the next few years. The human genome may be sequenced by the year 2005 or earlier; the information made available by this massive endeavor will have a tremendous impact on human biology and medicine.

SUMMARY

Carbon, oxygen, hydrogen, and nitrogen are the major constituents of most biomolecules. In addition, calcium, phosphorus, potassium, sodium, chlorine, magnesium, iron, manganese, iodine, and certain other elements are of great biologic and medical significance. Water, DNA, RNA, proteins, polysaccharides, and lipids are the major molecules found in cells and tissues.

Cells are the fundamental units of biology. They contain a number of organelles that have many specialized functions. These organelles can be separated by subcellular fractionation and their functions studied in detail.

Progress in biochemistry has depended on the isolation of cellular biomolecules, determination of their structures, and analyses of their function and metabolism. Many approaches, from the level of the whole animal to that of the isolated gene, are used to investigate the structure, function, and metabolism of biomolecules. In particular, the use of isotopes, both stable and radioactive, has been of tremendous importance in advancing biochemical knowledge.

The representation

Transcription   Translation
DNA————————————→RNA————————————→Protein

summarizes the thrust of much contemporary effort in biochemistry. However, many other advances have been made in biochemical knowledge, such as an appreciation of the composition of the body and partial understanding of the structures and functions of enzymes, hormones, and membranes. Biochemical and genetic bases for many diseases have been discovered, and the recent application of recombinant DNA technology has enormously accelerated progress in this area. Nevertheless, much remains unknown; major challenges for the future include mapping the human genome and providing molecular explanations of the mechanisms involved in development, differentiation, and brain function.

REFERENCES

Freifelder D: *Physical Biochemistry: Applications to Biochemistry and Molecular Biology.* Freeman, 1982.

Fruton JS: *Molecules and Life; Historical Essays on the Interplay of Chemistry and Biology.* Wiley-Interscience, 1972.

Green ED, Cox DR, Myers RM: The human genome project and its impact on the study of human disease. In: Scriver CR et al (editors): *The Metabolic and Molecular Bases of Inherited Disease,* 7th ed. McGraw-Hill, 1995.

Radda GK: Control, bioenergetics, and adaptation in health and disease: Non-invasive biochemistry from nuclear magnetic resonance. FASEB J 1992;6:3032.

Watson JD et al: *Recombinant DNA,* 2nd ed. Scientific American Books, 1992.

Wilson K, Walker J: *Principles and Techniques of Practical Biochemistry,* 4th ed. Cambridge Univ Press, 1994.

Water & pH

3

Victor W. Rodwell, PhD

INTRODUCTION

The polar organic and inorganic biomolecules of living cells exist and react primarily in an aqueous environment. Water, a remarkable molecule essential to life, solubilizes and modifies the properties of biomolecules such as nucleic acids, proteins, and carbohydrates by forming hydrogen bonds with their polar functional groups. These interactions modify the properties of the biomolecules and their conformations in solution. The accompanying changes impart properties to these biomolecules essential to the life process. Biomolecules—even relatively nonpolar biomolecules such as certain lipids—also alter the properties of water. An understanding of the homeostatic mechanisms by which organisms maintain a relatively constant intracellular environment must include consideration of pH and buffering in body fluids and subcellular compartments. Finally, the dissociation behavior of the functional groups of biomolecules in aqueous solution at various pH values is central to understanding their reactions and properties both in the living cells and in the laboratory.

BIOMEDICAL IMPORTANCE

Homeostasis, the maintenance of the composition of the internal environment that is essential for health, includes consideration of the distribution of water in the body and the maintenance of appropriate pH and electrolyte concentrations. Two-thirds of total body water (55–65% of body weight in men and about 10% less in women) is **intracellular fluid.** Of the remaining **extracellular fluid,** blood plasma constitutes approximately 25%.

Regulation of water balance depends on hypothalamic mechanisms for controlling thirst, on antidiuretic hormone, and on retention or excretion of water by the kidneys. States of water depletion and of excess body water, which are common, often are accompanied by sodium depletion or excess. **Water depletion** may result from decreased intake (eg, coma) or increased loss (eg, severe sweating; renal loss in diabetes mellitus; diarrhea of infants or from

cholera). Causes of **excess body water** include increased intake (eg, excessive administration of intravenous fluids) and decreased excretion (eg, severe renal failure). Osmotic and nonosmotic mechanisms safeguard water and osmotic homeostasis of the extracellular fluid. Two distinct responses, conservation of water by antidiuresis and acquisition of water by thirst, serve to maintain homeostasis. As little as a 2% increase in extracellular fluid osmolarity can trigger thirst and the release of the hypophysial antidiuretic hormone (ADH). A somewhat less sensitive mechanism triggers nonosmotic ADH release and thirst when circulating extracellular fluid volume decreases 10%. The genetic disorder nephrogenic diabetes insipidus, characterized by extreme thirst, high water intake, and an inability to concentrate urine or to respond to subtle changes in extracellular fluid osmolarity, results from the inability of renal tubular ADH osmoreceptors to respond to ADH.

Maintenance of the extracellular fluid between pH 7.35 and 7.45, in which the **bicarbonate buffer system** plays a key role, is essential for health. **Disturbances of acid-base balance** are diagnosed in the clinical laboratory by measuring the pH of arterial blood and the CO_2 content of venous blood. Causes of **acidosis** (blood pH < 7.35) include diabetic ketoacidosis and lactic acidosis, and causes of **alkalosis** (blood pH > 7.45) include the vomiting of acidic gastric contents or treatment with certain diuretics. Accurate diagnosis and prompt treatment of water imbalance and of acid-base disturbances rely in part upon the concepts considered in this chapter.

WATER IS AN IDEAL BIOLOGIC SOLVENT

Water Is a Slightly Skewed Tetrahedral Molecule

The water molecule is an irregular tetrahedron with oxygen at its center (Figure 3–1). The two bonds with hydrogen are directed toward two corners of the tetrahedron, while the unshared electrons on the two sp^3-hybridized orbitals occupy the two remaining corners. The angle between the two hydro-

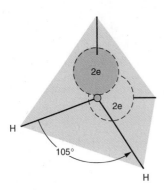

Figure 3–1. Tetrahedral structure of water.

gen atoms (105 degrees) is slightly less than the tetrahedral angle (109.5 degrees), forming a slightly skewed tetrahedron.

Water Molecules Form Dipoles

Because of its skewed tetrahedral structure, electrical charge is not uniformly distributed about the water molecule. The side of the oxygen opposite to the two hydrogens is relatively rich in electrons, while on the other side the relatively unshielded hydrogen nuclei form a region of local positive charge. The term "dipole" denotes molecules such as water that have electrical charge (electrons) unequally distributed about their structure.

Ammonia Is Also Tetrahedral and Dipolar

In ammonia, the bond angles between the hydrogens (107 degrees) approach the tetrahedral angle even more closely than in water (Figure 3–2). Many biochemicals are dipoles. Examples include alcohols, phospholipids, amino acids, and nucleic acids.

WATER MOLECULES FORM HYDROGEN BONDS

Hydrogen Bonds Confer Macromolecular Structure

Because of their dipolar character, water molecules can assume ordered arrangements (consider a snowflake). Like ice, liquid water also exhibits macromolecular structure that parallels the geometric disposition of water molecules in ice. The ability of water molecules to associate with one another in both solid and liquid states arises from the dipolar character of water. It remains a liquid rather than a solid because of the transient nature of these macromolecular complexes (the half-life for association-dissociation of water molecules is about 1 microsecond). In the solid state, each water molecule is associated with four other water molecules. In the liquid state, the number is somewhat less (about 3.5). With the exception of the transient nature of its intermolecular interactions, liquid water thus resembles ice in its macromolecular structure more closely than may at first be imagined.

The dipolar character of water molecules favors their mutual association in ordered arrays with a precise geometry dictated by the internal geometry of the water molecule (Figure 3–3).

The electrostatic interaction between the hydrogen nucleus of one water molecule and the unshared electron pair of another is termed a **hydrogen bond.** Compared with covalent bonds, hydrogen bonds are quite weak. To break a hydrogen bond in liquid water requires about 4.5 kcal of energy per mole—about 4% of the energy required to rupture the O—H bond in water (110 kcal/mol).

Hydrogen Bonds Stabilize Proteins and Nucleic Acids

While methane (molecular weight 16) and ammonia (molecular weight 17) are gases at room tempera-

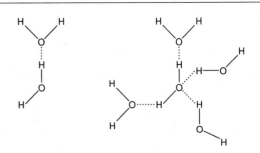

Figure 3–3. *Left:* Association of two dipolar water molecules. The dotted line represents a hydrogen bond. Note that a given water molecule can act as a hydrogen donor, or a hydrogen acceptor, or both simultaneously. *Right:* Association of a central water molecule with four other water molecules by hydrogen bonding. This structure is typical of ice and, to a lesser extent, of liquid water.

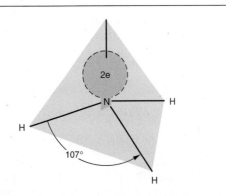

Figure 3–2. Tetrahedral structure of ammonia.

ture, water (molecular weight 18) is a liquid. Why is this so? The answer lies in water's ability to form hydrogen bonds, which also accounts for water's relatively high viscosity and surface tension.

The ability of water to serve as a solvent for ions and many organic molecules results from its dipolar character and its ability to form hydrogen bonds. Molecules that can form hydrogen bonds with water (eg, –OH or –SH compounds, amines, esters, aldehydes, and ketones) are readily solvated, and their water solubility is thereby increased. Macromolecules stabilized by intermolecular hydrogen bonds may exchange certain surface intermolecular hydrogen bonds for hydrogen bonds to water, again enhancing solubility. Thus, soluble proteins are coated with a water layer formed by exchange of surface intermolecular hydrogen bonds for intramolecular hydrogen bonds to water.

The dipolar character of water profoundly affects its interactions with biomolecules. In the aqueous environment of living cells, numerous charge interactions occur with the polar groups of biomolecules. DNA folds so as to expose its polar sugar and phosphate groups to water molecules. Similarly, the polar residues of proteins are present primarily at the biopolymer surface where they participate in extensive interactions with water molecules. Figure 3–4 illustrates the formation of hydrogen bonds between water and representative functional groups of biomolecules. Note that alcohols, like water, can participate both as hydrogen donors and as hydrogen acceptors for hydrogen bond formation with water or with other biomolecules.

Apolar groups such as those present in hydrocarbons lack the capacity to form hydrogen bonds and thus are water-insoluble. Such nonpolar groups can nevertheless affect the structure of water. When added to water, apolar molecules form spherical droplets with minimal water-exposed surface, a phenomenon illustrated by the tendency of olive oil in cold water to form a single large floating mass. Minimization of the water-exposed apolar surface area is an entropically driven process. The presence of apolar molecules decreases the number of possible orientations (degrees of freedom) of adjacent water molecules and thus is accompanied by an increase in entropy. Minimization of the exposed apolar surface area permits the maximum degrees of freedom (ie, maximizes disorder) of nearby water molecules and thus minimizes this increase in entropy. In water, hydrocarbons form rigid clathrate (cage-like) structures. Similarly, in the aqueous environment of living cells, the nonpolar portions of biopolymers tend to reside inside the structure of the biopolymer, minimizing contacts with water.

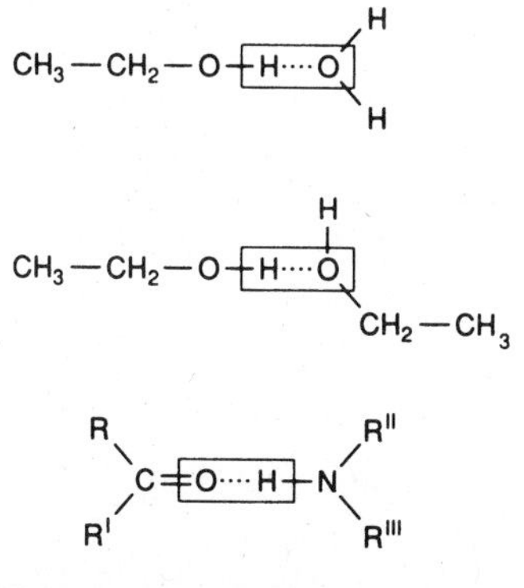

Water Molecules Exhibit a Slight But Physiologically Important Tendency to Dissociate

The ability of water to ionize, while slight, is of central importance for life on earth. Since water can act both as an acid and as a base, its ionization may be represented as an intermolecular proton transfer, forming a hydronium ion (H_3O^+) and a hydroxide ion (OH^-):

$$H_2O + H_2O \rightleftharpoons H_3O^+ + OH^-$$

The transferred proton is actually associated with a cluster of water molecules and exists in solution not even as H_3O^+ but as something like $H_5O_2^+$ or $H_7O_3^+$. While for practical purposes this apparently "naked" proton is almost always written as "H^+," it should not be forgotten that it is in fact highly hydrated.

Since ions continuously recombine to form water molecules, and vice versa, it cannot be stated whether an individual hydrogen or oxygen is present as an ion or as part of a water molecule. At one instant, it is an ion; an instant later, part of a molecule. Fortunately, individual ions or molecules need not be considered. Since 1 g of water contains 3.46×10^{22} molecules, the ionization of water can be described statistically. It is sufficient to know the *probability* that a hydrogen will be present as an ion or as part of a water molecule.

To state that the probability that a hydrogen exists as an ion is 0.01 means that a hydrogen atom has one chance in 100 of being an ion and 99 chances out of 100 of being in a water molecule. The actual probability of a hydrogen atom in pure water existing as a hydrogen ion is approximately 0.0000000018, or 1.8×10^{-9}. Consequently, the probability of its being part of a molecule is almost unity. Stated another way, for every hydrogen ion and hydroxyl ion in pure water, there are 1.8 billion or 1.8×10^9 water molecules. Hydrogen ions and hydroxyl ions nevertheless contribute significantly to the properties of water.

Figure 3–4. Formation of hydrogen bonds between an alcohol and water, between two molecules of ethanol, and between the peptide carbonyl oxygen and the hydrogen on the peptide nitrogen of an adjacent peptide.

For dissociation of water,

$$K = \frac{[H^+][OH^-]}{[H_2O]}$$

where the bracketed terms represent the molar concentrations of hydrogen ions, hydroxyl ions, and undissociated water molecules,* and K is termed the **dissociation constant.** To calculate the dissociation constant for water, recall that 1 mol of water weighs 18 g. One liter (L) (1000 g) of water therefore contains $1000 \div 18 = 55.56$ mol. Pure water is thus 55.56 molar. Since the probability that a hydrogen in pure water will exist as an H^+ ion is 1.8×10^{-9}, the molar concentration of H^+ ions (or of OH^- ions) in pure water is calculated by multiplying the probability, 1.8×10^{-9}, by the molar concentration of water, 55.56 mol/L. This result is 1.0×10^{-7} mol/L.

We can now calculate K for water:

$$K = \frac{[H^+][OH^-]}{[H_2O]} = \frac{[10^{-7}][10^{-7}]}{[55.56]}$$

$$= 0.018 \times 10^{-14} = 1.8 \times 10^{-16} \text{ mol/L}$$

The high concentration of molecular water (55.56 mol/L) is not significantly affected by dissociation. It is therefore convenient to consider it as essentially constant. This constant may then be incorporated into the dissociation constant, K, to provide a new constant, K_w, termed the **ion product** for water. The relationship between K_w and K is shown below:

$$K = \frac{[H^+][OH^-]}{[H_2O]} = 1.8 \times 10^{-16} \text{ mol/L}$$

$$K_w = (K)[H_2O] = [H^+][OH^-]$$

$$= (1.8 \times 10^{-16} \text{ mol/L})(55.56 \text{ mol/L})$$

$$= 1.00 \times 10^{-14} \text{ (mol/L)}^2$$

Note that the dimensions of K are moles per liter and of K_w moles^2 per liter^2. As its name suggests, the ion product, K_w, is numerically equal to the product of the molar concentrations of H^+ and OH^-:

$$K_w = [H^+][OH^-]$$

At 25 °C, $K_w = (10^{-7})^2 = 10^{-14}$ (mol/L)^2. At temperatures below 25 °C, K_w is less than 10^{-14}, and at temperatures above 25 °C, greater than 10^{-14}. For example, at the temperature of the human body (37 °C), the concentration of H^+ in pure water is slightly more than 10^{-7} mol/L. Within the stated limitations of the effect of temperature, $K_w = 10^{-14}$ **(mol/L)^2 for all aqueous solutions**—even those that contain acids or bases. We shall use this constant in the calculation of pH values for acidic and basic solutions.

---

*Strictly speaking, the bracketed terms represent molar activity rather than molar concentration.

pH IS THE NEGATIVE LOG OF THE HYDROGEN ION CONCENTRATION

The term **pH** was introduced in 1909 by Sorensen, who defined pH as **the negative log of the hydrogen ion concentration:**

$$pH = -\log[H^+]$$

This definition—while not rigorous*—is adequate for most biochemical purposes. To calculate the pH of a solution:

(1) Calculate hydrogen ion concentration, $[H^+]$.
(2) Calculate the base 10 logarithm of $[H^+]$.
(3) pH is the negative of the value found in step 2.

For example, for pure water at 25 °C,

$$pH = -\log[H^+] = -\log 10^{-7} = -(-7) = 7.0$$

Low pH values correspond to high concentrations of H^+ and high pH values to low concentrations of H^+.

Acids are **proton donors,** and bases are **proton acceptors.** A distinction is made, however, between **strong acids** (eg, HCl, H_2SO_4), which completely dissociate even in strongly acidic solutions (low pH), and **weak acids,** which dissociate only partially in acidic solutions. A similar distinction is made between **strong bases** (eg, KOH, NaOH) and **weak bases** (eg, Ca[OH]$_2$). Only strong bases are dissociated at high pH. Many biochemicals are weak acids. Exceptions include phosphorylated intermediates, which possess the strongly acidic primary phosphoric acid group.

The following examples illustrate how to calculate the pH of acidic and basic solutions.

Example: What is the pH of a solution whose hydrogen ion concentration is 3.2×10^{-4} mol/L?

$$pH = -\log[H^+]$$
$$= -\log(3.2 \times 10^{-4})$$
$$= -\log(3.2) - \log(10^{-4})$$
$$= -0.5 + 4.0$$
$$= 3.5$$

Example: What is the pH of a solution whose hydroxide ion concentration is 4.0×10^{-4} mol/L?

To approach this problem, we define a quantity **pOH** that is equal to $-\log[OH^-]$ and that may be derived from the definition of K_w:

$$K_W = [H^+][OH^-] = 10^{-14}$$

therefore:

$$\log[H^+] + \log[OH^-] = \log 10^{-14}$$

or:

$$pH + pOH = 14$$

---

*pH = $-\log$ (H^+ activity).

To solve the problem by this approach:

$$[OH^-] = 4.0 \times 10^{-4}$$
$$pOH = -\log [OH^-]$$
$$= -\log (4.0 \times 10^{-4})$$
$$= -\log (4.0) - \log (10^{-4})$$
$$= -0.60 + 4.0$$
$$= 3.4$$

Now:

$$pH = 14 - pOH = 14 - 3.40$$
$$= 10.6$$

Example: What are the pH values of (a) 2.0×10^{-2} mol/L KOH and of (b) 2.0×10^{-6} mol/L KOH? The OH^- arises from two sources: KOH and water. Since pH is determined by the *total* $[H^+]$ (and pOH by the *total* $[OH^-]$), both sources must be considered. In the first case, the contribution of water to the total $[OH^-]$ is negligible. The same cannot be said for the second case:

	Concentration (mol/L)	
	(a)	(b)
Molarity of KOH	2.0×10^{-2}	2.0×10^{-6}
$[OH^-]$ from KOH	2.0×10^{-2}	2.0×10^{-6}
$[OH^-]$ from water	1.0×10^{-7}	1.0×10^{-7}
Total $[OH^-]$	2.00001×10^{-2}	2.1×10^{-6}

Once a decision has been reached about the significance of the contribution by water, pH may be calculated as above.

In the above examples, it was assumed that the strong base KOH was completely dissociated in solution and that the molar concentration of OH^- ions was thus equal to the molar concentration of KOH. This assumption is valid for relatively dilute solutions of strong bases or acids but not for solutions of weak bases or acids. Since these weak electrolytes dissociate only slightly in solution, we must calculate the concentration of H^+ (or $[OH^-]$) produced by a given molarity of the acid (or base) using the **dissociation constant** before calculating total $[H^+]$ (or total $[OH^-]$), and subsequently calculating the pH.

Functional Groups That Are Weak Acids Have Great Physiologic Significance

Many biochemicals possess functional groups that are weak acids or bases. One or more of these functional groups—carboxyl groups, amino groups, or the secondary phosphate dissociation of phosphate esters—are present in all proteins and nucleic acids, most coenzymes, and most intermediary metabolites. The dissociation behavior (protonic equilibria) of weakly acidic and weakly basic functional groups is

therefore fundamental to understanding the influence of intracellular pH on the structure and biochemical activity of these compounds. Their separation and identification in research and clinical laboratories is also facilitated by knowledge of the dissociation behavior of their functional groups.

We term the protonated form of an acid (eg, HA or RNH_3^+) the **acid** and the unprotonated form (eg, A^- or RNH_2) its **conjugate base.** Similarly, we may refer to a **base** (eg, A^- or RNH_2) and its **conjugate acid** (eg, HA or RNH_3^+) (Latin *coniungere* "to join together"). Representative weak acids (left), their conjugate bases (center), and the pK values (right) include the following:

$R-CH_2-COOH$	$R-CH_2-COO^-$	$pK = 4-5$
$R-CH_2-NH_3^+$	$R-CH_2-NH_2$	$pK = 9-10$
H_2CO_3	HCO_3^-	$pK = 6.4$
$H_2PO_4^-$	$H_2PO_4^{-2}$	$pK = 7.2$

The relative strengths of weak acids and of weak bases are expressed quantitatively as their **dissociation constants,** which express their tendency to ionize. Shown below are the expressions for the dissociation constant (K) for two representative weak acids, R-COOH and R-NH_3^+.

$$R-COOH \rightleftharpoons R-COO^- + H^+$$
$$K = \frac{[R-COO^-][H^+]}{[R-COOH]}$$
$$R-NH_3^+ \rightleftharpoons R-NH_2 + H^+$$
$$K = \frac{[R-NH_2][H^+]}{[R-NH_3^+]}$$

Since the numerical values of K for weak acids are negative exponential numbers, it is convenient to express K as pK, where

$$pK = -\log K$$

Note that pK is related to K as pH is to H^+ concentration. Table 3–1 lists illustrative K and pK values for a monocarboxylic, a dicarboxylic, and a tricarboxylic acid. Observe that the stronger acid groups have lower pK values.

From the above equations that relate K to $[H^+]$ and

Table 3–1. Dissociation constants and pK values for representative carboxylic acids.

Acid	K		pK
Acetic		1.76×10^{-5}	4.75
Glutaric	(1st)	4.58×10^{-5}	4.34
	(2nd)	3.89×10^{-6}	5.41
Citric	(1st)	8.40×10^{-4}	3.08
	(2nd)	1.80×10^{-5}	4.74
	(3rd)	4.00×10^{-6}	5.40

to the concentrations of undissociated acid and its conjugate base, note that when

$$[R\!-\!COO^-] = [R\!-\!COOH]$$

or when

$$[R\!-\!NH_2] = [R\!-\!NH_3^+]$$

then

$$K = [H^+]$$

In words: When the associated (protonated) and dissociated (conjugate base) species are present in equal concentrations, the prevailing hydrogen ion concentration $[H^+]$ is numerically equal to the dissociation constant, K. If the logarithms of both sides of the above equation are taken and both sides are multiplied by -1, the expressions would be as follows:

$$K = [H^+]$$
$$-\log K = -\log [H^+]$$

Since $-\log K$ is defined as pK, and $-\log [H^+]$ defines pH, the equation may be rewritten as

$$pK = pH$$

ie, **the pK of an acid group is that pH at which the protonated and unprotonated species are present at equal concentrations.** The pK for an acid may be determined experimentally by adding 0.5 equivalent of alkali per equivalent of acid. The resulting pH will be equal to the pK of the acid.

The Behavior of Weak Acids and Buffers Is Described by the Henderson-Hasselbalch Equation

The pH of a solution containing a weak acid is related to its acid dissociation constant, as shown above for the weak acid water. The relationship can be stated in the convenient form of the **Henderson-Hasselbalch** equation, derived below.

A weak acid, HA, ionizes as follows:

$$HA \rightleftharpoons H^+ + A^-$$

The equilibrium constant for this dissociation is written as follows:

$$K = \frac{[H^+][A^-]}{[HA]}$$

Cross-multiply:

$$[H^+][A^-] = K\,[HA]$$

Divide both sides by $[A^-]$:

$$[H^+] = K\frac{[HA]}{[A^-]}$$

Take the log of both sides:

$$\log [H^+] = \log\!\left(K\,\frac{[HA]}{[A^-]} \right)$$
$$= \log K + \log \frac{[HA]}{[A^-]}$$

Multiply through by -1:

$$-\log [H^+] = -\log K - \log \frac{[HA]}{[A^-]}$$

Substitute pH and pK for $-\log [H^+]$ and $-\log K$, respectively; then:

$$pH = pK - \log \frac{[HA]}{[A^-]}$$

Then, to remove the minus sign, invert the last term:

$$\boxed{pH = pK + \log \frac{[A^-]}{[HA]}}$$

The Henderson-Hasselbalch equation is an expression of great predictive value in protonic equilibria. For example,

(1) When an acid is exactly half-neutralized, $[A^-] = [HA]$. Under these conditions,

$$pH = pK + \log \frac{[A^-]}{[HA]} = pK + \log \frac{1}{1} = pK + 0$$

Therefore, at half-neutralization, pH = pK.

(2) When the ratio $[A^-]/[HA] = 100{:}1$,

$$pH = pK + \log \frac{[A^-]}{[HA]}$$
$$pH = pK + \log 100/1 = pK + 2$$

(3) When the ratio $[A^-]/[HA] = 1{:}10$,

$$pH = pK + \log \tfrac{1}{10} = pK + (-1)$$

If the equation is evaluated at several ratios of $[A^-]/[HA]$ between the limits 10^3 and 10^{-3} and the calculated pH values plotted, the result obtained describes the titration curve for a weak acid (Figure 3–5).

Solutions of Weak Acids and Their Salts Buffer the pH

Solutions of weak acids and their conjugate bases (or of weak bases and their conjugate acids) exhibit **buffering**—the tendency of a solution to resist more effectively a change in pH following addition of a strong acid or base than does an equal volume of water. Important physiologic buffers include bicar-

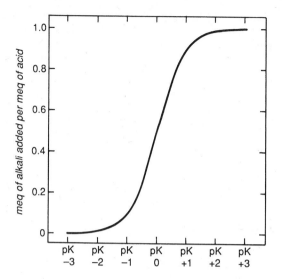

Figure 3–5. General form of a titration curve calculated from the Henderson-Hasselbalch equation.

bonate (HCO_3^-/H_2CO_3), inorganic orthophosphate ($H_2PO_4^{-1}/HPO_4^{-2}$), and intracellular proteins. Nonphysiologic buffers used in biochemical experimentation include TRIS (pK 8.2) and HEPES (pK 7.6). Buffering is best illustrated by titrating a weak acid or base using a pH meter. Alternatively, we may calculate the pH shift that accompanies addition of acid or base to a buffered solution. In the example, the buffered solution (a weak acid, pK = 5.0, and its conjugate base) is initially at one of four pH values. We will calculate the pH shift that results when 0.1 meq of KOH is added to 1 meq of each solution:

Initial pH	5.00	5.37	5.60	5.86
$[A^-]_{initial}$	0.50	0.70	0.80	0.88
$[HA]_{initial}$	0.50	0.30	0.20	0.12
$([A^-]/[HA])_{initial}$	1.00	2.33	4.00	7.33
Addition of 0.1 meq of KOH produces				
$[A^-]_{final}$	0.60	0.80	0.90	0.98
$[HA]_{final}$	0.40	0.20	0.10	0.02
$([A^-]/[HA])_{final}$	1.50	4.00	9.00	49.0
log $([A^-]/[HA])_{final}$	0.176	0.602	0.95	1.69
Final pH	5.18	5.60	5.95	6.69
ΔpH	0.18	0.60	0.95	1.69

Observe that the pH change per milliequivalent of OH^- added varies greatly depending on the pH. At pH values close to pK, the solution resists changes in pH most effectively, and it is said to exert a **buffering effect.** Solutions of weak acids and their conjugate bases buffer most effectively in the pH range pK ± 2.0 pH units. This means that to buffer a solution at

pH X, a weak acid or base whose pK is no more than one pH unit removed from pH X should be used.

Figure 3–6 illustrates the net charge on one molecule of the acid as a function of pH. A fractional charge of –0.5 means not that an individual molecule bears a fractional charge but that the statistical probability that a given molecule has a unit negative charge is 0.5. Consideration of the net charge on macromolecules as a function of pH provides the basis for many separatory techniques, including the electrophoretic separation of amino acids, plasma proteins, and abnormal hemoglobins.

SUMMARY

Water, which exists both in the liquid and solid state in the form of hydrogen-bonded molecular clusters, forms hydrogen bonds both with itself and with other proton donors or acceptors. Water's surface tension, viscosity, liquid state at room temperature, and solvent power result from its ability to form hydrogen bonds. Compounds containing O, N, or S are solvated by water through their ability to serve as hydrogen acceptors or hydrogen donors for hydrogen-bonding to water. Proteins and other macromolecules stabilized by intermolecular hydrogen bonds exchange surface intermolecular hydrogen bonds for hydrogen bonds to water. Water modifies the properties of biomolecules such as proteins and nucleic acids that possess both polar and nonpolar functional groups. Entropic forces dictate that macromolecules in aqueous solutions fold such that their polar portions contact the water interface while their nonpolar portions are buried in the interior of the biomolecule.

Water dissociates to form hydroxide ions and extensively hydrated protons, conventionally repre-

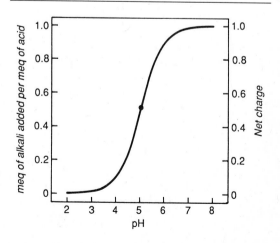

Figure 3–6. Titration curve for an acid of the type HA. The heavy dot in the center of the curve indicates the pK, 5.0.

sented as naked protons (H$^+$). pH, the negative log of [H$^+$], expresses relative acidity. Low pH denotes an acidic and high pH a basic solution.

Biochemists consider both carboxylic acids (eg, R—COOH) and amines (eg, R—NH$_3^+$) to be acids and term the corresponding proton acceptors (R—COO$^-$ and R—NH$_2$) the conjugate bases of these acids. These weak acids play key roles in metabolism. Their acid strength is expressed quantitatively by pK, the negative log of the acid dissociation constant. Relatively strong acids have low pKs and relatively weak acids high pKs.

Buffers resist a change in pH when protons are produced or consumed. Maximum buffering capacity occurs $\pm$ 1 pH unit on either side of pK. Important physiologic buffers include bicarbonate, orthophosphate, and proteins.

REFERENCE

Segel IM: *Biochemical Calculations*. Wiley, 1968.

Section I.
Structures & Fun[ctions of Protein]s & Enzymes

Amino Acids

4

Victor W. Rodwell, PhD

INTRODUCTION

Among the multiple functions that amino acids fulfill in living cells is to serve as the monomer units from which the polypeptide chains of proteins are constructed. Most proteins contain, in varying proportions, the same 20 L-α-amino acids. Many specific proteins contain, in addition, L-α-amino acids derived from some of the basic 20 by processes that occur after formation of the polypeptide backbone. These "unusual" amino acids fulfill highly specific functions for the protein in question. The kinds of amino acids, the order in which they are joined together, and their mutual spatial relationships dictate the three-dimensional structures and biologic properties of simple proteins and are major determinants of structure and function for the complex proteins that contain, in addition to amino acids, heme, carbohydrate, lipid, nucleic acids, etc. This chapter considers the structures, physical properties, stereochemistry, chemical reactions, and ionic equilibria of the L-α-amino acids present in proteins.

BIOMEDICAL IMPORTANCE

The human diet must contain adequate quantities of ten **essential L-α-amino acids,** since neither humans nor any other higher animal can synthesize these amino acids in amounts adequate to support infant growth or to maintain health in adults. In the form of proteins, amino acids perform a multitude of structural, hormonal, and catalytic functions essential to life. It thus is not surprising that genetic defects in the metabolism of amino acids can result in severe illness. Several comparatively rare genetic diseases of amino acid catabolism (eg, phenylketonuria and maple syrup urine disease) will, if left untreated, result in mental retardation and early death. Additional genetic diseases can result from an impaired ability to transport specific amino acids into cells. Since these transport defects typically result in the excretion in the urine of increased amounts of one or more amino acids, they often are referred to as **aminoacidurias.**

In addition to their roles in proteins, L-amino acids and their derivatives participate in intracellular functions as diverse as nerve transmission, regulation of cell growth, and the biosynthesis of porphyrins, purines, pyrimidines, and urea. L-α-Amino acids in low-molecular-weight peptides play additional roles as hormones, and both D- and L-α-amino acids are present in polypeptide antibiotics elaborated by microorganisms.

PROPERTIES OF AMINO ACIDS

The Genetic Code Specifies 20 L-α-Amino Acids

Although over 300 different amino acids occur in nature, a subset of only 20 constitute the monomer units from which the polypeptide backbones of proteins are constructed. While a nonredundant three-letter genetic code could accommodate over 20 amino acids, the redundancy of the universal genetic code limits available amino acid codons to the 20 L-α-amino acids listed in Table 4–1. Consequently, all proteins contain varying proportions of these 20 L-α-amino acids. In addition, however, certain proteins contain "unusual" L-α-amino acids which have arisen from some of the basic 20 by posttranslational processing (ie, processing that occurs after formation of the polypeptide backbone). A frequently encountered example is cystine, which arises by oxidation of the —SH groups of two cysteines to an —S—S— (disulfide) bond. Other less common modified or "unusual" amino acids also fulfill highly specific functions in the proteins in which they occur. Examples of types of posttranslational modifications include methylation, formylation, acetylation, prenylation, carboxylation, and phosphorylation (see Chapter 40).

Table 4–1. L-α-Amino acids present in proteins.*

Name	Symbol	Structural Formula	pK$_1$	pK$_2$	pK$_3$
With Aliphatic Side Chains			α-COOH	α-NH$_3^+$	R Group
Glycine	Gly [G]	H—CH—COO$^-$; $_+$NH$_3$	2.4	9.8	
Alanine	Ala [A]	CH$_3$—CH—COO$^-$; $_+$NH$_3$	2.4	9.9	
Valine	Val [V]	H$_3$C\CH—CH—COO$^-$/H$_3$C ; $_+$NH$_3$	2.2	9.7	
Leucine	Leu [L]	H$_3$C\CH—CH$_2$—CH—COO$^-$/H$_3$C ; $_+$NH$_3$	2.3	9.7	
Isoleucine	Ile [I]	CH$_3$—CH$_2$—CH—CH—COO$^-$; CH$_3$ $_+$NH$_3$	2.3	9.8	
With Side Chains Containing Hydroxylic (OH) Groups					
Serine	Ser [S]	CH$_2$—CH—COO$^-$; OH $_+$NH$_3$	2.2	9.2	about 13
Threonine	Thr [T]	CH$_3$—CH—CH—COO$^-$; OH $_+$NH$_3$	2.1	9.1	about 13
Tyrosine	Tyr [Y]	See below.			
With Side Chains Containing Sulfur Atoms					
Cysteine	Cys [C]	CH$_2$—CH—COO$^-$; SH $_+$NH$_3$	1.9	10.8	8.3
Methionine	Met [M]	CH$_2$—CH$_2$—CH—COO$^-$; S—CH$_3$ $_+$NH$_3$	2.1	9.3	
With Side Chains Containing Acidic Groups or Their Amides					
Aspartic acid	Asp [D]	$^-$OOC—CH$_2$—CH—COO$^-$; $_+$NH$_3$	2.0	9.9	3.9
Asparagine	Asn [N]	H$_2$N—C—CH$_2$—CH—COO$^-$; O $_+$NH$_3$	2.1	8.8	
Glutamic acid	Glu [E]	$^-$OOC—CH$_2$—CH$_2$—CH—COO$^-$; $_+$NH$_3$	2.1	9.5	4.1
Glutamine	Gln [Q]	H$_2$N—C—CH$_2$—CH$_2$—CH—COO$^-$; O $_+$NH$_3$	2.2	9.1	

(continued)

Table 4–1. L-α-Amino acids present in proteins.* (continued)

Name	Symbol	Structural Formula	pK_1	pK_2	pK_3
With Side Chains Containing Basic Groups			α-COOH	α-NH$_3^+$	R Group
Arginine	Arg [R]	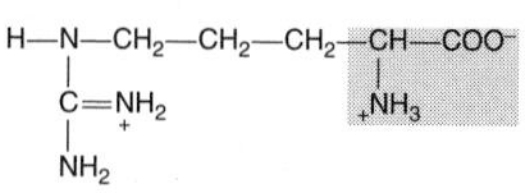 H—N—CH$_2$—CH$_2$—CH$_2$—CH—COO$^-$ C=NH$_2$ / NH$_3^+$ NH$_2$	1.8	9.0	12.5
Lysine	Lys [K]	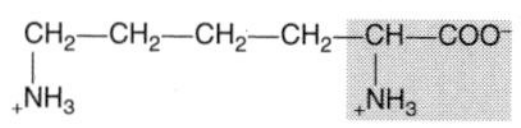 CH$_2$—CH$_2$—CH$_2$—CH$_2$—CH—COO$^-$ $_+$NH$_3$ $_+$NH$_3$	2.2	9.2	10.8
Histidine	His [H]	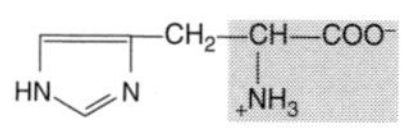 —CH$_2$—CH—COO$^-$ HN—N $_+$NH$_3$	1.8	9.3	6.0
Containing Aromatic Rings					
Histidine	His [H]	See above.			
Phenylala-nine	Phe [F]	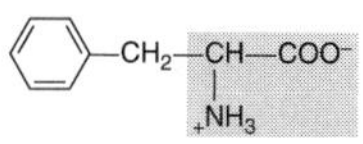 ⬡—CH$_2$—CH—COO$^-$ $_+$NH$_3$	2.2	9.2	
Tyrosine	Tyr [Y]	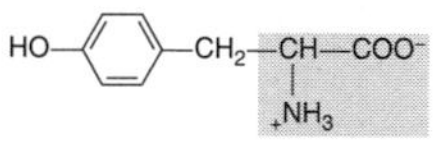 HO—⬡—CH$_2$—CH—COO$^-$ $_+$NH$_3$	2.2	9.1	10.1
Tryptophan	Trp [W]	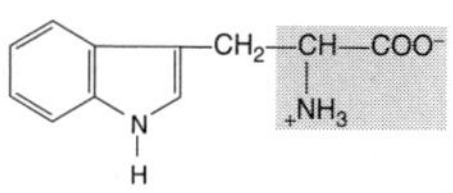 —CH$_2$—CH—COO$^-$ N-H $_+$NH$_3$	2.4	9.4	
Imino Acids					
Proline	Pro [P]	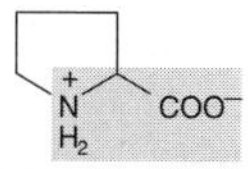 $_+$NH$_2$ COO$^-$	2.0	10.6	

Only L-α-Amino Acids Occur in Proteins

Amino acids contain both amino and carboxylic acid functional groups. In an α-amino acid, both are attached to the same carbon atom (Figure 4–1).

A carbon atom bearing four different substituents is said to be chiral. With the exception of glycine, for which R is a hydrogen atom (Figure 4–1), all four groups linked to the α-carbon atom of amino acids are different. This tetrahedral orientation of four different groups about the α-carbon atom confers optical activity (the ability to rotate the plane of plane-polarized light) on amino acids. Although some amino acids found in proteins are dextrorotatory and some levorotatory at pH 7.0, all have the absolute configurations of L-glyceraldehyde and hence are L-α-amino acids.

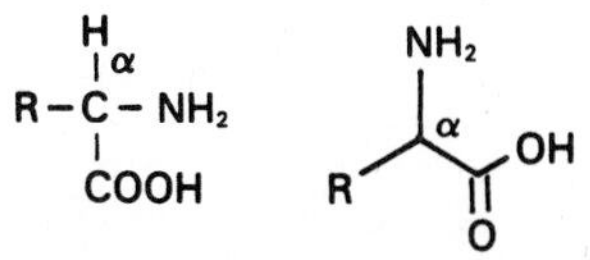

Since D-amino acids occur in nature and even in polypeptide antibiotics, why do only L-amino acids occur in proteins? We hypothesize that L-amino acids were selected by some fortuitous event that occurred extremely early in the evolution of life.

Amino Acids May Have Positive, Negative, or Zero Net Charge

Amino acids bear at least two ionizable weak acid groups, a —COOH and an —NH$_3^+$. In solution, two forms of these groups, one charged and one uncharged, exist in protonic equilibrium:

$$R—COOH \rightleftharpoons R—COO^- + H^+$$
$$R—NH_3^+ \rightleftharpoons R—NH_2 + H^+$$

R—COOH and R—NH$_3^+$ are the protonated, or acidic, partners in these equilibria. R—COO$^-$ and

Figure 4–1. Two representations of an α-amino acid.

R—NH$_2$ are the **conjugate bases** (proton acceptors) of the corresponding acids. Although both R—COOH and R—NH$_3^+$ are weak acids, R—COOH is a far stronger acid than is R—NH$_3^+$. At the pH of blood plasma or the intracellular space (7.4 and 7.1, respectively), carboxyl groups exist almost entirely as **carboxylate ions,** R—COO$^-$. At these pH values, most amino groups are predominantly in the protonated form, R—NH$_3^+$. The prevalent ionic species of amino acids in blood and most tissues should be represented as shown in Figure 4–2A. Structure B (Figure 4–2) cannot exist at *any* pH. At a pH sufficiently low to protonate the carboxyl group, the more weakly acidic amino group would also be protonated. At a pH 2 units below its pK$_a$, an acid is approximately 99% protonated. If the pH is gradually raised, the proton from the carboxylic acid will be lost long before that from the R—NH$_3^+$. At any pH sufficiently high for the uncharged conjugate base of the amino group to predominate, a carboxyl group is present as the carboxylate ion (R—COO$^-$). The B representation, however, is used for many equations not involving protonic equilibria.

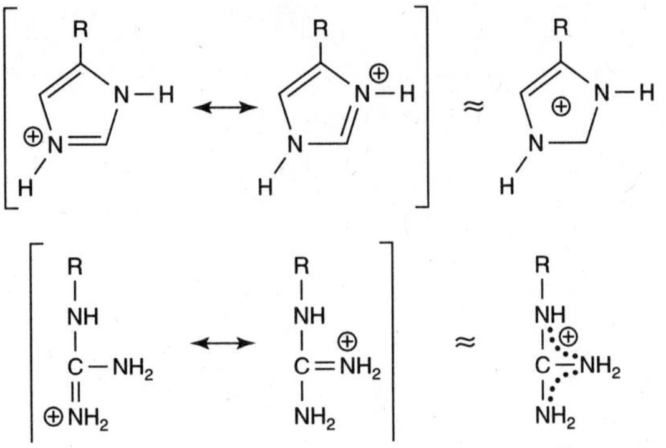

pK$_a$ Values Express the Strengths of Weak Acids

The relative acid strengths of weak acids are expressed by their acid dissociation constant, K$_a$, or by their pK$_a$, the negative log of the dissociation constant:

$$pK_a = -\log K_a$$

Table 4–1 lists, to the nearest tenth of a pH unit, the pK values for each of the acidic groups present on the 20 amino acids of proteins. For convenience, the *a* subscript in K$_a$ and pK$_a$ will be implied but dropped hereafter.

Figure 4–3 illustrates the protonated forms of the imidazole R group of histidine and the guanidino R group of arginine. Both exist as resonance hybrids and thus may be represented as shown on the right, with the positive charge distributed between both nitrogens (histidine) or all three nitrogens (arginine).

The **net charge** (the algebraic sum of all the positively and negatively charged groups present) of an

Figure 4–3. Resonance hybrids of the protonated forms of the R groups of histidine and arginine.

amino acid depends upon the pH, or proton concentration, of the surrounding solution. The ability to alter the charge on amino acids or their derivatives by manipulating the pH facilitates the physical separation of amino acids, peptides, and proteins.

At Its Isoelectric pH (pI), an Amino Acid Bears No Net Charge

For an aliphatic amino acid such as alanine, the isoelectric species is the form shown in Figure 4–4. The isoelectric pH is the pH midway between pK values on either side of the isoelectric species. For an amino acid with only two dissociating groups, there can be no possible ambiguity, as is shown below in the calculation of the pI for alanine. Since pK$_1$ (R—COOH) = 2.35 and pK$_2$ (R—NH$_3^+$) = 9.69, the isoelectric pH (pI) of alanine is as shown below:

$$pI = \frac{pK_1 + pK_2}{2} = \frac{2.35 + 9.69}{2} = 6.02$$

Calculation of pI for a compound with more than two dissociable groups carries more possibility for error. For example, from consideration of Figure 4–5, what would be the isoelectric pH (pI) for aspartic acid? First, write out all possible ionic structures for a compound in the order in which they occur, proceeding from strongly acidic to basic solution (eg,

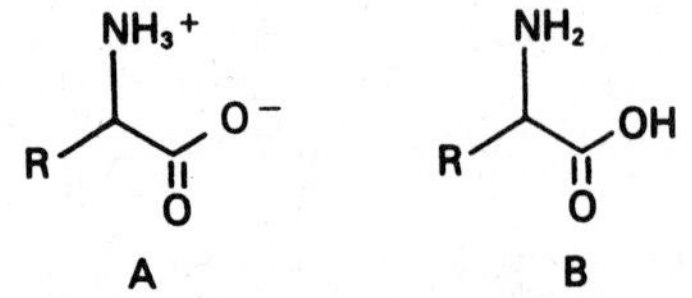

Figure 4–2. Ionically correct structure for an amino acid at or near physiologic pH **(A).** The uncharged structure **(B)** cannot exist at any pH but may be used as a convenience when discussing the chemistry of amino acids.

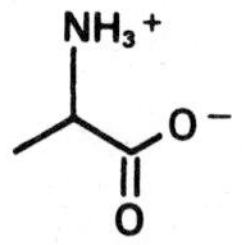

Figure 4–4. Isoelectric or "zwitterionic" structure of alanine. Although charged, the zwitterion bears no *net* charge and hence does not migrate in a direct current electric field.

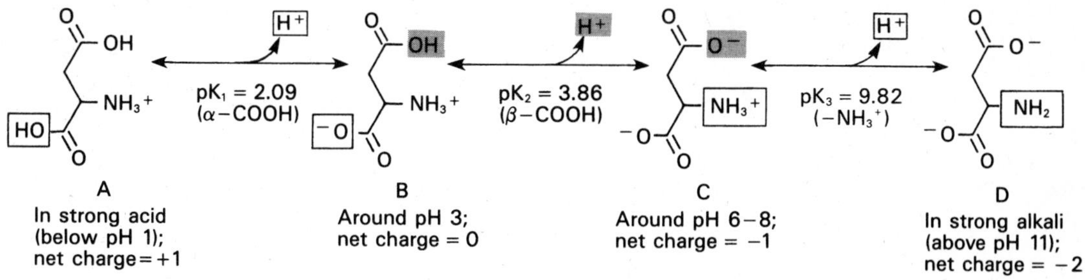

Figure 4–5. Protonic equilibria of aspartic acid.

as for aspartic acid in Figure 4–5). Next, identify the isoionic, zwitterionic, or neutral representation (as in Figure 4–5B). The pI is the pH at the midpoint between the pK values on either side of the isoionic species. In this example,

$$pI = \frac{2.09 + 3.86}{2} = 2.98$$

This approach works equally well for amino acids with additional dissociating groups, eg, lysine or histidine. After writing the formulas for all possible charged species of the basic amino acids lysine and arginine, observe that

$$pI = \frac{pK_2 + pK_3}{2}$$

For lysine, pI is 9.7; for arginine, pI is 10.8. The student should determine the pI for histidine.

Determining pK values on either side of the zwitterion by inspection of charged structures is not limited to amino acids. This approach may be applied to calculating the charge on a molecule with any number of dissociating groups. The ability to perform calculations of this type is of value in the clinical laboratory to predict the mobility of compounds in electrical fields and to select appropriate buffers for separations. For example, a buffer at pH 7.0 will separate two molecules with pI values of 6 and 8, respectively, because the molecule with pI = 6 will have a larger net negative charge at pH 7.0 than the molecule with pI = 8. Similar considerations apply to understanding separations on ionic supports such as positively or negatively charged polymers (eg, DEAE cellulose or Dowex 1 resin).

The Solubility and Melting Points of Amino Acids Reflect Their Ionic Character

Since multiple charged groups are present on amino acids, they are readily solvated by—and hence soluble in—polar solvents such as water and ethanol but are insoluble in nonpolar solvents such as benzene, hexane, or ether. Their high melting points (> 200 °C) reflect the high energy needed to disrupt the ionic forces that stabilize the crystal lattice.

AMINO ACIDS MAY BE CLASSIFIED BY THE POLARITIES OF THEIR R GROUPS

The amino acids in proteins may be divided into two broad groups on the basis of whether the R groups attached to the α-carbon atoms are polar or nonpolar (Table 4–2). The single-letter abbreviations (Table 4–1) are used to represent extremely long sequences of amino acids (eg, for listing the complete sequence of amino acids in a protein).

Amino acids that occur in free or combined states (but not in proteins) fulfill important roles in metabolic processes. For example, ornithine, citrulline, and argininosuccinate participate in the formation of urea, tyrosine in the formation of thyroid hormones, and glutamate in neurotransmitter biosynthesis. Over 20 D-amino acids occur naturally. These include the D-alanine and D-glutamate of certain bacterial cell walls and a variety of D-amino acids in antibiotics.

Table 4–2. Classification of the L-α-amino acids of proteins based on their relative hydrophilicity (tendency to associate with water) or hydrophobicity (tendency to avoid water in favor of a more nonpolar environment).

Hydrophobic	Hydrophilic	
Alanine	Arginine	Histidine
Isoleucine	Asparagine	Lysine
Leucine	Aspartic acid	Serine
Methionine	Cysteine	Threonine
Phenylalanine	Glutamic acid	
Proline	Glutamine	
Tryptophan	Glycine	
Tyrosine		
Valine		

THE α-R GROUPS DETERMINE THE PROPERTIES OF INDIVIDUAL AMINO ACIDS

Glycine, the smallest amino acid, can fit into regions of the three-dimensional structure of proteins inaccessible to other amino acids and occurs in regions where peptides bend sharply.

The aliphatic R groups of alanine, valine, leucine, and isoleucine and the aromatic R groups of phenylalanine, tyrosine, and tryptophan are hydrophobic, a property that has important consequences for the ordering of water molecules in proteins in their immediate neighborhood. These amino acids typically occur primarily in the interior of cytosolic proteins.

The charged R groups of basic and acidic amino acids stabilize specific protein conformations via formation of salt bonds. For example, rupture and reformation of salt bonds accompany oxygenation and deoxygenation of hemoglobin (see Chapter 7). In addition, amino acids with positively or negatively charged R groups function in "charge relay" systems that transmit charges across considerable distances during enzymatic catalysis. Finally, histidine plays unique roles in enzymatic catalysis, since the pK of its imidazole proton permits it, at pH 7.0, to function as either a base or an acid catalyst.

The primary alcohol group of serine and the primary thioalcohol (—SH) group of cysteine are excellent nucleophiles and may function as such during enzymatic catalysis. While the secondary alcohol group of threonine also is a good nucleophile, it is not known to perform this role in catalysis. In addition, the —OH of serine, tyrosine, and threonine also functions in regulation of the activity of certain enzymes whose catalytic activity depends on the phosphorylation state of specific hydroxylated aminoacyl residues.

Amino acids do not absorb visible light (ie, they are colorless) and, with the exceptions of the aromatic amino acids tryptophan, tyrosine, phenylalanine, and histidine, do not absorb ultraviolet light of a wavelength above 240 nm. Several amino acids, particularly tryptophan, absorb high-wavelength (250- to 290-nm) ultraviolet light (Figure 4–6). While relatively uncommon in most proteins, tryptophan therefore makes a major contribution to the ability of most proteins to absorb light in the region of 280 nm.

NINHYDRIN OR FLUORESCAMINE DETECTS AMINO ACIDS

Ninhydrin (Figure 4–7) oxidatively decarboxylates α-amino acids to CO_2, NH_3, and an aldehyde with one less carbon atom than the parent amino acid. The reduced ninhydrin then reacts with the liberated ammonia, forming a blue complex that maxi-

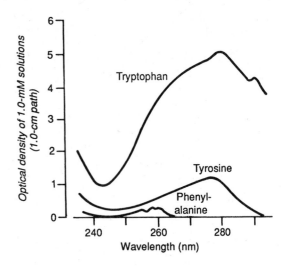

Figure 4–6. The ultraviolet absorption spectra of tryptophan, tyrosine, and phenylalanine.

mally absorbs light of wavelength 570 nm. This blue color forms the basis of a **quantitative test** for α-amino acids that can detect as little as 1 μg of amino acid. Amines other than α-amino acids also react with ninhydrin, forming a blue color, but without releasing CO_2. The evolution of CO_2 thus indicates an α-amino acid. NH_3 and peptides also react, but more slowly than α-amino acids. Proline and 4-hydroxyproline produce a yellow color with ninhydrin.

Fluorescamine (Figure 4–8), an even more sensitive reagent, can detect nanogram quantities of an amino acid. Like ninhydrin, fluorescamine forms a complex with amines other than amino acids.

VARIOUS TECHNIQUES SEPARATE AMINO ACIDS

Chromatography

In all chromatographic separations, molecules are partitioned between a stationary and a mobile phase. Separation depends on the relative tendencies of molecules in a mixture to associate more strongly with one or the other phase.

While these separatory techniques are discussed

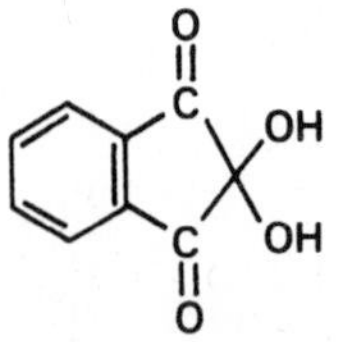

Figure 4–7. Ninhydrin.

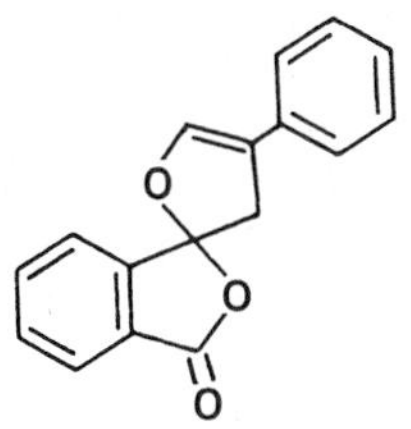

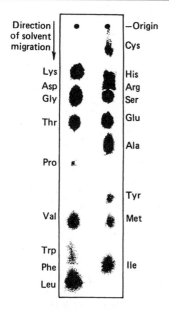

Figure 4–8. Fluorescamine.

principally with respect to amino acids, their use is by no means restricted to these molecules.

Paper Chromatography

For paper chromatography, a drop of solution containing one or more amino acids is applied about 5 cm from the end of a filter paper strip. The strip is placed in a closed vessel, where its end contacts a solvent, typically a water-saturated low-molecular-weight alcohol containing an acid or base. Following migration of the solvent through the paper, the strip is dried, treated with ninhydrin in acetone, and heated briefly, revealing the positions of the amino acids as purple spots (Figure 4–9).

Amino acids with large nonpolar R groups (Leu, Ile, Phe, Trp, Val, Met, Tyr) migrate farther than those with shorter nonpolar R groups (Pro, Ala) or with polar R groups (Thr, Glu, Ser, Arg, Asp, His, Lys, Cys) (Figure 4–9). For a nonpolar series (Gly, Ala, Val, Leu), increasing length of the nonpolar R group, which increases the nonpolar character, results in increased mobility.

The ratio of the distance traveled by an amino acid to that traveled by the solvent front, both measured from the point of application, is the **R_f value** (mobility relative to the solvent front). R_f values for a given amino acid vary with experimental conditions, eg, the solvent used. It is preferable to chromatograph amino acid standards simultaneously with unknowns so that mobility may be expressed relative to that of a standard (eg, as R_{ala} rather than as R_f), which varies less than R_f values from experiment to experiment.

Thin-Layer Chromatography

There are two distinct types of thin-layer chromatography (TLC): partition TLC (PTLC) and adsorption TLC (ATLC). In partition TLC, resolution involves partitioning of the components of a mixture between stationary and mobile liquid phases of differing polarity as the mobile phase passes over the stationary phase. Separation results from the differing solubilities of components in the stationary and mobile phases.

For normal PTLC, the stationary phase is the more polar component. The developing solvent contains both polar and weakly nonpolar components, typically water, a four- or five-carbon alcohol, and a weak acid or base such as acetic acid or ammonia. As it moves over the support, the solvent changes composition because the more polar components associate with the polar —OH groups of the cellulose. The advancing solvent front therefore becomes progressively more nonpolar. The more nonpolar components of a mixture therefore move farther than more polar components of equal size. For example, glutamate and lysine are retarded, whereas leucine and valine migrate with the solvent.

In "reversed-phase" PTLC, both the polarities of the phases and the mobilities of components of the mixture are reversed from those of normal PTLC. The mobile phase provides the polar phase. The stationary phase, typically cellulose coated with a nonpolar liquid such as a silicone oil, provides the more nonpolar phase. In contrast to normal PTLC, polar components migrate with the solvent while the mobility of less polar components is retarded.

The basis of separation in adsorption TLC (ATLC) is entirely different. Resolution involves adsorption of the components of a mixture to silica gel that has been heated to drive off all bound water. The mobile phase, typically a single or binary mixture of organic solvents, displaces adsorbed compounds by competing for binding sites on the silica gel. The less tightly bound components are displaced first, and resolution is achieved.

Figure 4–9. Identification of amino acids present in proteins. After descending paper chromatography in *n*-butanol:acetic acid:water, spots were visualized with ninhydrin.

Ion Exchange Chromatography

Analysis of amino acid residues after hydrolysis of a polypeptide generally involves **automated ion exchange chromatography.** Complete separation, identification, and quantitation require less than 3 hours. The procedure of Moore and Stein uses a short and a long column that contain the Na^+ form of a sulfonated polystyrene resin. When acid hydrolysate at pH 2 is applied to the columns, the amino acids bind via cation exchange with Na^+. The columns are then eluted with sodium citrate under preprogrammed conditions of pH and temperature. Eluted material is reacted with ninhydrin reagent, and color densities are monitored in a flow-through colorimeter. Data are displayed on a cathode ray tube with computer-linked integration of peak areas (Figure 4–10).

High-Voltage Electrophoresis (HVE)

Separations of amino acids, polypeptides, and other ampholytes (molecules whose net charge depends on the pH of the surrounding medium) in a direct current field have many applications in biochemistry. For amino acids, paper sheets or thin layers of powdered cellulose are most frequently used as supports. For large polypeptides or proteins, a cross-linked polyacrylamide gel is used. For nucleotide oligomers, both agarose and polyacrylamide supports are used.

Separations depend upon the strength of the direct current field, the net charge on the ampholyte, and the molecular weight of the compound being separated. For molecules with identical charge, those of lower molecular weight migrate farther. Net charge, however, is the more important factor in determining separation. Applications include amino acids, low-molecular-weight polypeptides, certain proteins, nucleotides, and phosphosugars. Samples are applied to the support, which is then moistened with buffer of an appropriate pH and connected to buffer reservoirs by paper wicks. When current is applied, molecules with a net negative charge at the selected pH migrate toward the anode and those with a net positive charge toward the cathode. For visualization, the **electropherogram** is treated with ninhydrin (amino acids, peptides) or with ethidium bromide and viewed under ultraviolet light (nucleotide oligomers). The choice of pH is dictated by the pK values of the dissociating groups on the molecules in the mixture.

FUNCTIONAL GROUPS DICTATE THE CHEMICAL REACTIONS OF AMINO ACIDS

The chemical reactions in which amino acids can participate are determined by the functional groups they possess, the α-amino and α-carboxylic acid functions, and the functional groups present in R groups. Each functional group can participate in all its characteristic chemical reactions. These reactions would include, for carboxylic acid groups, ionization and the formation of esters, amides, and acid anhydrides. These reactions include ionization, acylation, and esterification of amino groups, oxidation and alkylation of —SH groups, the esterification of —OH groups, etc.

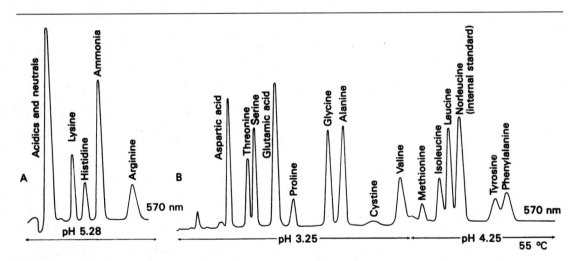

Figure 4–10. Automated analysis of amino acids present in a protein hydrolysate. The amino acids are adsorbed to the polystyrene-based cation exchanger Dowex 50, eluted with buffers of the indicated pH, reacted with ninhydrin, and the absorbancy at 570 nm and 440 nm (detects proline) recorded. Peak areas are proportionate to the quantity of each amino acid present. A short column **(A)** resolves basic amino acids at pH 5.28. Elution of a longer column **(B)** at pH 3.25 and pH 4.25 resolves the remaining amino acids. The nonprotein amino acid norleucine serves as an internal standard.

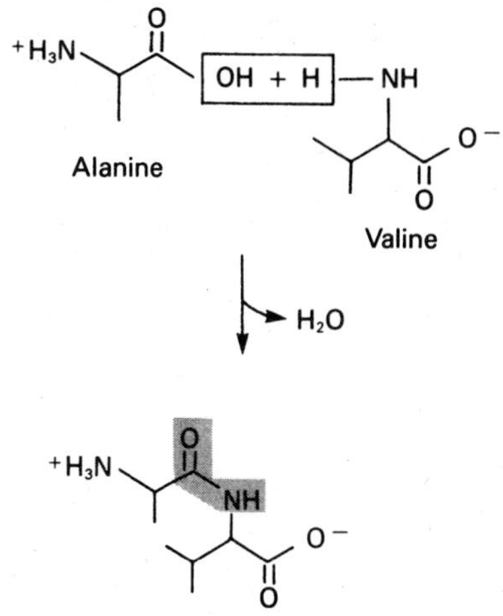

Figure 4–11. Amino acids united by a peptide bond (shaded).

THE MOST IMPORTANT REACTION OF AMINO ACIDS IS PEPTIDE BOND FORMATION

In principle, peptide bond formation involves removal of 1 mol of water between the α-amino group of one amino acid and the α-carboxyl group of a second amino acid (Figure 4–11). This reaction does not, however, proceed as written, since the equilibrium constant strongly favors peptide bond hydrolysis. To synthesize peptide bonds, the carboxyl group must first be activated. Chemically, this may involve prior conversion to an acid chloride. Biologically, activation involves initial condensation with ATP, forming an aminoacyladenylate.

SUMMARY

Only L-α-amino acids occur in proteins, though both D-amino acids and non-α-amino acids occur in nature. All amino acids possess at least two weakly acidic functional groups, $R—NH_3^+$ and $R—COOH$, which, for the amino acids of proteins, reside on the α-carbon atom. In addition, these and other weak acid functional groups (—OH, —SH, guanidino, imidazole) dictate that the net charge on an amino acid varies with pH. Amino acids thus are ampholytes whose net charge at a given pH depends on the pK_a values of their functional groups. pI is the pH at which an amino acid bears no net charge and hence does not move in a direct current electrical field.

In addition to determining charge relationships and the kinds of chemical reactions an amino acid will undergo (of which the formation of the peptide bond is most important), the R groups and their associated functionalities dictate unique biochemical functions for each amino acid. R group functionalities also provide a basis for classifying amino acids as basic, acidic, aromatic, aliphatic, or sulfur-containing. Following separation of mixtures of amino acids by partition or ion exchange chromatography, amino acids may be detected and quantified by reaction with ninhydrin.

REFERENCES

Barrett GC: *Chemistry and Biochemistry of the Amino Acids.* Chapman & Hall, 1985.

Davies JS: *Amino Acids and Peptides.* Chapman & Hall, 1985.

Gehrke CW et al: *Amino Acid Analysis by Gas Chromatography.* 3 vols. CRC Press, 1987.

Hancock WS: *Handbook of HPLC for the Separation of Amino Acids, Peptides, and Proteins.* 2 vols. CRC Press, 1984.

Hugli TE: *Techniques in Protein Chemistry.* Academic Press, 1989.

Rattenbury JM: *Amino Acid Analysis.* Halstead Press, 1981.

Peptides

Victor W. Rodwell, PhD

INTRODUCTION

The polymerization of L-α-amino acids by peptide bonds forms the structural framework for proteins. This chapter discusses characteristics of the peptide bond, properties of peptides, methods for determination of the primary structure of peptides and proteins, and methods for peptide synthesis.

BIOMEDICAL IMPORTANCE

Peptides are of immense biomedical interest, particularly in endocrinology. Many major hormones are peptides and may be given to patients to correct corresponding deficiency states (eg, administration of insulin to patients with diabetes mellitus). Some peptides act in the nervous system, either as neurotransmitters or as neuromodulators. Certain antibiotics are peptides (eg, valinomycin and gramicidin A), as are a few antitumor agents (eg, bleomycin). The dipeptide aspartame serves as a sweetening agent in many beverages. Rapid chemical synthesis and recombinant DNA technology have facilitated the manufacture of substantial amounts of peptide hormones, many of which are present in the body in relatively minute concentrations and thus difficult to isolate in quantities sufficient for therapy. The same technology allows the synthesis of other peptides, also available from natural sources in only small amounts (eg, certain viral peptides and proteins), for use in vaccines.

L-α-AMINO ACIDS LINKED BY PEPTIDE BONDS FORM PEPTIDES

Figure 5–1 shows a tripeptide made up of alanine, cysteine, and valine. Note that a tripeptide is one with three *residues,* not three peptide bonds. By convention, peptide structures are written with the amino terminal residue (the residue with a free α-amino group) at the left and with the carboxyl termi-nal residue (the residue with a free α-carboxyl group) at the right. This peptide has a single free α-amino group and a single free α-carboxyl group. However, in some peptides, the terminal amino or carboxyl group may be derivatized (eg, an *N*-formyl amine or an amide of the carboxyl group) and thus not free.

Peptide Structures Are Easy to Draw

By convention, peptides are drawn with their amino terminal on the left and carboxyl terminal on the right of the page. To illustrate peptide structures, first draw the main chain or backbone, then add appropriate R groups. Write a zigzag formed from the repeating sequence of main chain, or "backbone," atoms: α-nitrogen, α-carbon, carbonyl carbon.

Complete the amino and carboxyl terminals, add a hydrogen to each α-carbon and to each peptide nitrogen, and add oxygen to the carbonyl carbon.

Add appropriate R groups (shaded) to each α-carbon atom.

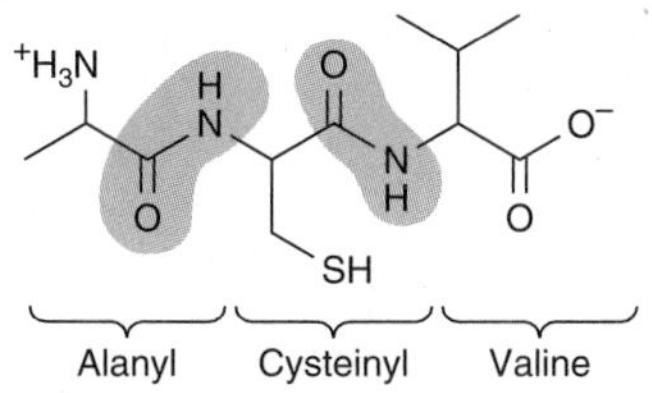

Alanyl   Cysteinyl   Valine

Figure 5–1. Structural formula for a tripeptide. Peptide bonds are shaded for emphasis.

Amino Acid Sequence Determines Primary Structure

When the number, structure, and order of all of the amino acid residues in a polypeptide are known, its **primary structure** has been determined. Amino acids whose α-carboxyl groups participate in the formation of peptide bonds are termed "aminoacyl residues." These residues are named by replacing the *-ate* or *-ine* endings of free amino acids by *-yl* (eg, alany*l,* aspart*yl,* tyros*yl*). Peptides are named as derivatives of the carboxyl terminal aminoacyl residue. For example, the tetrapeptide Lys-Leu-Tyr-Gln is named as a derivative of glutam*ine* and is called lysyl-leuc*yl*-tyros*yl*-glutam*ine*. The *-ine* ending on glutamine indicates that its α-carboxyl group is *not* involved in peptide bond formation.

Primary Structure Affects Biologic Activity

Mutation of DNA that alters codons may result in the insertion of inappropriate aminoacyl groups in a polypeptide or protein. While often without major biologic effect, even single substitutions may on occasion reduce or abolish biologic activity with accompanying minor effects (eg, alcohol intolerance in many Asians) or more serious consequences (eg, sickle cell disease). Many inherited metabolic errors involve a single change of this type. Powerful new methods to determine protein and DNA structure have greatly increased our understanding of the biochemical basis for many inherited metabolic diseases.

Abbreviations Are Used to Name the Amino Acids Present in Peptides

Both three- and one-letter abbreviations for the amino acids (Table 4–1) are used to represent primary structures (Figure 5–2). Three-letter abbreviations linked by straight lines represent a primary

Glu-Ala-Lys-Gly-Tyr-Ala

E   A   K   G   Y   A

Figure 5–2. Three- and one-letter representations of a hexapeptide.

Glu-Lys-(Ala,Gly,Tyr)-His-Ala

Figure 5–3. A heptapeptide containing a region of uncertain primary structure (in parentheses).

structure that is known and unambiguous. These lines are omitted for single-letter abbreviations. Where there is uncertainty about the precise order of a portion of a polypeptide, the questionable residues are enclosed in brackets and separated by commas (Figure 5–3).

Many Peptides Have Physiologic Activity

Animal, plant, and bacterial cells contain a variety of low-molecular-weight polypeptides (3–100 aminoacyl residues) with profound physiologic activity. Some, including most mammalian polypeptide hormones, contain only peptide bonds formed between α-amino and α-carboxyl groups of the L-α-amino acids of proteins. However, additional amino acids or derivatives of the protein amino acids may also be present in polypeptides.

Glutathione (Figure 5–4), in which the amino terminal glutamate is linked to cysteine via a non-α-peptidyl bond, is required by several enzymes. Glutathione and the enzyme glutathione reductase participate in the formation of the correct disulfide bonds of many proteins and polypeptide hormones and participate in the metabolism of xenobiotics (Chapter 61).

Polypeptide antibiotics elaborated by fungi contain both D- and L-amino acids and amino acids not present in proteins. Examples include tyrocidine and gramicidin S, cyclic polypeptides that contain D-phenylalanine, and the nonprotein amino acid ornithine.

Thyrotropin-releasing hormone (TRH) (Figure 5–5) illustrates yet another variant. The amino terminal glutamate is cyclized to pyroglutamic acid, and the carboxyl terminal prolyl carboxyl is amidated.

A mammalian polypeptide may contain more than one physiologically potent polypeptide. Within the

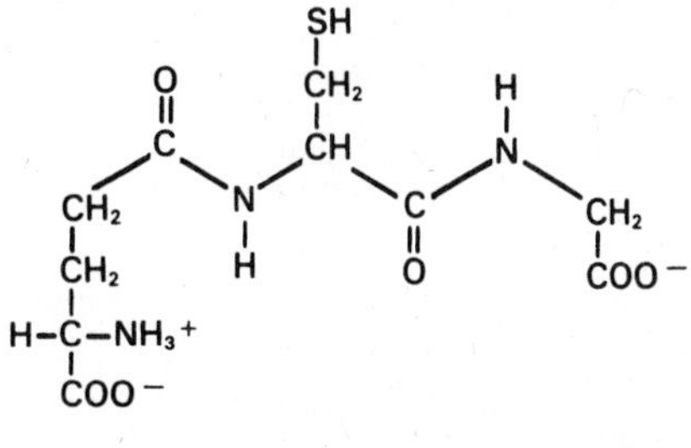

Figure 5–4. Glutathione (γ-glutamyl-cysteinyl-glycine). Note the non-α peptide bond that links Glu to Cys.

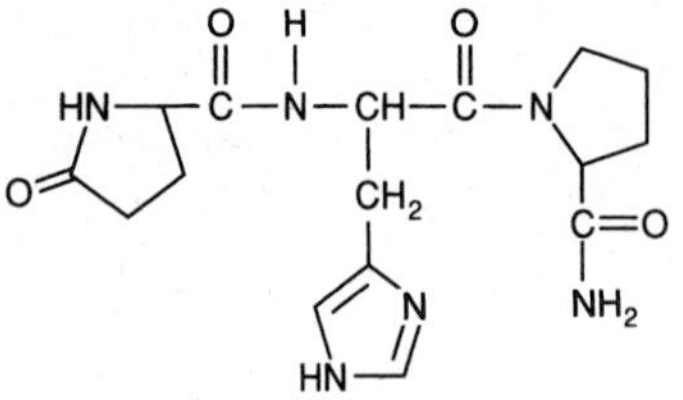

Figure 5–5. Pyroglutamyl-histidyl-prolinamide (TRH).

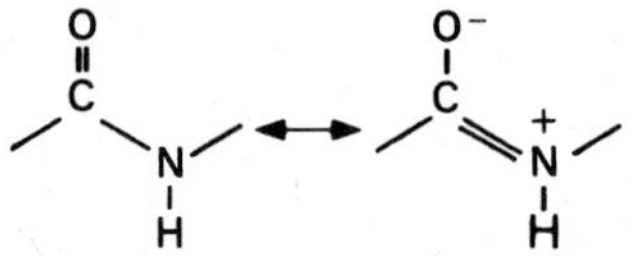

Figure 5–7. Resonance stabilization of the peptide bond confers partial double-bond character, and hence rigidity, on the C—N bond.

primary structure of **β-lipotropin**—a hypophysial hormone that stimulates the release of fatty acids from adipose tissue—are sequences of amino acids that are common to several other polypeptide hormones with diverse physiologic activities (Figure 5–6). The large polypeptide is a precursor of the smaller polypeptides.

Peptides Are Polyelectrolytes

The peptide (amide) bond is uncharged at any pH of physiologic interest. Formation of peptides from amino acids at pH 7.4 is therefore accompanied by a net loss of one positive and one negative charge per peptide bond formed. Peptides are, however, charged molecules at physiologic pH owing to their carboxyl and amino terminal groups and to acidic or basic R groups.

The Peptide Bond Has Partial Double-Bond Character

While peptides are written with a single bond connecting the α-carboxyl and α-nitrogen atoms, this bond in fact has partial double-bond character (Fig-

ure 5–7). There is no freedom of rotation about the bond that connects the C and N atoms. Consequently, all four of the atoms shown in Figure 5–7 lie in the same plane, or are **coplanar.** This semirigidity has important consequences for orders of protein structure above the primary level. There is, by contrast, ample freedom of rotation about the remaining bonds of the polypeptide backbone. In Figure 5–8, which summarizes these concepts, the bonds having free rotation are circled by arrows and the coplanar atoms are shaded.

Noncovalent Forces Constrain Peptide Conformations

While a large number of conformations (spatial arrangements) are possible for a polypeptide, in solution a narrow range of conformations tends to predominate. These favored conformations reflect factors such as steric hindrance, coulombic interactions, hydrogen bonding, and hydrophobic interactions. Both for peptides and proteins (Chapter 6), specific conformations are required for physiologic activity.

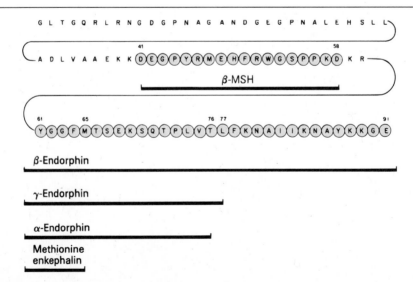

Figure 5–6. Primary structure of β-lipotropin. Residues 41–58 are melanocyte-stimulating hormone (β-MSH). Residues 61–91 contain the primary structures of the indicated endorphins.

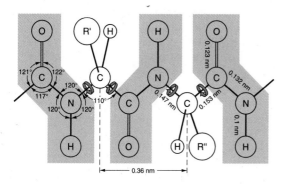

Figure 5–8. Dimensions of a fully extended polypeptide chain. The four atoms in shaded boxes, which are coplanar, comprise the polypeptide bond. The unshaded atoms are the α-carbon atom, the α-hydrogen atom, and the α-R group of the particular amino acid. Free rotation can occur about the bonds connecting the α-carbon with the α-nitrogen and α-carbonyl functions (white arrows). The extended polypeptide chain is thus a semirigid structure with two-thirds of the atoms of the backbone held in a fixed planar relationship one to another. The distance between adjacent α-carbon atoms is 0.36 nm (3.6 Å). The interatomic distances and bond angles, which are not equivalent, are also shown. (Redrawn and reproduced, with permission, from Pauling L, Corey LP, Branson HR: The structure of proteins: Two hydrogen-bonded helical configurations of the polypeptide chain. Proc Natl Acad Sci USA 1951;37:205.)

PROTEIN SEQUENCING: DETERMINATION OF PRIMARY STRUCTURE

Comparison of Peptide Sequences Reveals Key Residues and Phylogenetic Relationships

Comparison of the primary structures of a given protein from distantly related organisms reveals that a small number of amino acid residues are invariant between species. These conserved residues, which generally fulfill essential functions, provide vital clues concerning how a protein can serve as a cytoskeletal component, an electron carrier, or a catalyst.

The extent to which nonconserved residues of a given protein vary between species provides clues to species divergence during evolution. Close evolutionary relationships correlate with a high degree of sequence similarity; more distant relationships correlate with more sequence differences. First documented for the ubiquitous 104- to 111-residue electron carrier polypeptide cytochrome *c*, this relationship holds true for an ever-increasing number of proteins.

Peptides Are Purified Before Analysis

Peptide sequencing data will be uninterpretable unless the peptide homogeneity, assessed by SDS-

PAGE, exceeds 95%. Peptides must therefore first be purified by conventional protein purification techniques (Chapter 8).

Amino Acid Composition Usually Is First Determined

Determination of the amino acid composition of a peptide is usually undertaken prior to initiating sequencing in order to identify possible pitfalls and subsequently to verify sequencing data. The peptide bonds linking amino acid residues are first broken by hydrolysis. The liberated amino acids are then separated and identified by HPLC or ion exchange chromatography.

No procedure hydrolyzes peptides completely without some losses. The method of choice is hydrolysis of replicate samples in 6N HCl at 110 °C in sealed, evacuated glass tubes for 24, 48, 72, and 96 hours. This destroys all the Trp and Cys, and if metal ions are present some Met and Tyr are lost. Recovery of Ser and Thr is incomplete. Val-Val, Ile-Ile, Val-Ile, and Ile-Val bonds are extremely resistant to hydrolysis, while Gln and Asn are deamidated to Glu and Asp.

How are these difficulties overcome? Cys is oxidized to cysteic acid, an acid-stable derivative, prior to hydrolysis. Trp is estimated after basic hydrolysis, a process that destroys Ser, Thr, Arg, and Cys. Ser and Thr data are extrapolated back to zero time of hydrolysis. Val and Leu are estimated from 96-hour data. Dicarboxylic acids and their amides are reported collectively as "Glx" or "Asx." Finally, since the relative proportions of each amino acid must be integers, decimal fractions for proportions of each amino acid are rounded off to the nearest whole number.

Sanger Was First to Sequence a Polypeptide

Over four decades have elapsed since Sanger determined the complete primary structure of the polypeptide hormone insulin. Sanger's approach was first to separate the two polypeptide chains, A and B, of insulin and then to convert them by specific enzymatic cleavage into smaller peptides that contained regions of overlapping sequence. Using 1-fluoro-2,4-dinitrobenzene (Figure 5–9), he then removed and identified, one at a time, the amino terminal amino acid residues of these peptides. By comparing the sequences of overlapping peptides, he was able to deduce an unambiguous primary structure for both the A and the B chains.

Two later techniques have revolutionized determination of the primary structures of polypeptides (proteins). The first was the introduction by Edman in 1967 of a procedure for the automated sequential removal and identification of amino terminal amino acid residues as their phenylthiohydantoin derivatives. The second was the independent introduction

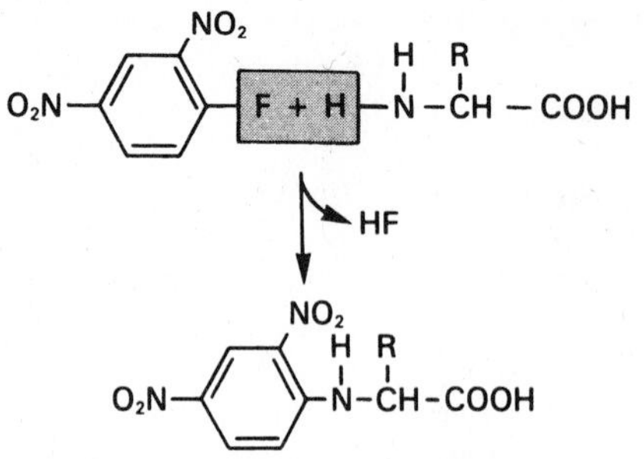

Figure 5–9. Reaction of an amino acid with 1-fluoro-2,4-dinitrobenzene (Sanger's reagent). The reagent is named for the Nobel laureate (1958) biochemist Frederick Sanger, who used it to determine the primary structure of insulin.

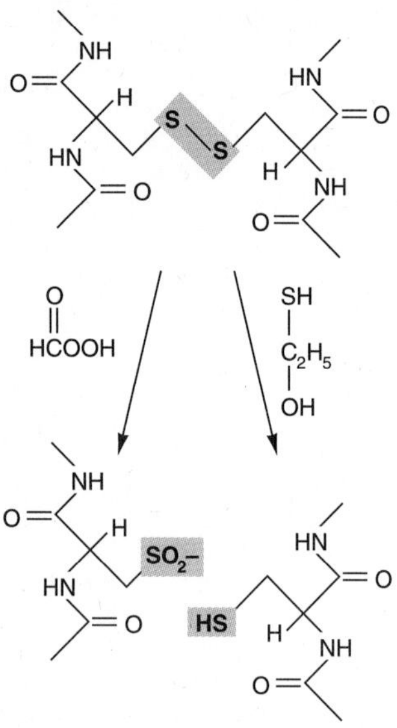

Figure 5–10. Oxidative cleavage of adjacent polypeptide chains linked by disulfide bonds (shaded) by performic acid (left) or reductive cleavage by β-mercaptoethanol (right) forms two peptides that incorporate cysteic acid residues or cysteinyl residues, respectively.

by Sanger and by Maxam and Gilbert of techniques for rapid sequencing of DNA. At present, optimal strategy is to utilize both approaches simultaneously.

Primary Structures Are Determined by the Automated Edman Technique

Since many proteins consist of more than one polypeptide chain associated by noncovalent forces or disulfide bridges, the first step may be to dissociate and separate individual polypeptide chains. Denaturing agents (urea, guanidine hydrochloride) disrupt hydrogen bonds and dissociate noncovalently associated polypeptides. Oxidizing and reducing agents disrupt disulfide bridges (Figure 5–10). Polypeptides are then separated by chromatography.

Large Polypeptides Are Cleaved Before Being Sequenced

Automated sequencing instruments (sequenators) operate most efficiently on polypeptides 20–60 residues long. Specific and complete cleavage at comparatively rare sites thus is desirable. The following reagents meet this requirement.

A. Cyanogen Bromide (CNBr): Cysteine residues are first modified with iodoacetic acid. CNBr then cleaves on the COOH side of Met. Since Met is comparatively rare in polypeptides, this usually generates peptide fragments of the desired size range.

B. Trypsin: Trypsin cleaves on the COOH side of Lys and Arg. If Lys residues are first derivatized with citraconic anhydride (a reversible reaction) to change their charge from positive to negative, trypsin now cleaves only after Arg. Derivatization of Arg residues is less useful because of the relative abundance of Lys residues. It is, however, useful for subsequent cleavage of CNBr fragments.

C. o-Iodosobenzene: o-Iodosobenzene cleaves the comparatively rare Trp-X residues. It requires no prior protection of other residues.

D. Hydroxylamine: Hydroxylamine cleaves the comparatively rare Asn-Gly bonds, although generally not in quantitative yield.

E. Protease V8: *Staphylococcus aureus* protease V8 cleaves Glu-X peptide residues with a preference for situations where X is hydrophobic. Glu-Lys resists cleavage. This reaction is useful for subsequent degradation of CNBr fragments.

F. Mild Acid Hydrolysis: This cleaves the rare Asp-Pro bond.

Sequencing Requires Multiple Digests

Two or three digests of the original polypeptide, normally at Met, Trp, Arg, and Asn-Gly, combined with appropriate subdigests of the resulting fragments, usually permit determination of the entire primary structure of the polypeptide. Barring unusual difficulties in purifying fragments, this can be accomplished with a few micromoles of polypeptide.

MIXTURES OF PEPTIDES MUST BE SEPARATED BEFORE SEQUENCING

Fragment purification is achieved, chiefly by gel filtration in acetic or formic acid, by reversed-phase high-performance liquid chromatography (HPLC) or by ion exchange chromatography on phosphocellulose or sulfophenyl Sephadex in solutions of phosphoric acid.

(A) Ion Exchange Chromatography and High-Voltage Electrophoresis (HVE): These techniques (Chapter 4), which separate on the basis of charge, are applicable to low-molecular-weight polypeptides as well as to amino acids. The pK value for the carboxyl terminal α-carboxylic acid group of a polypeptide is higher than that of the α-carboxyl group in the corresponding amino acid (ie, the peptide COOH is a weaker acid). Conversely, the amino terminal α-amino group is a stronger acid (has a lower pK) than the amino acid from which it has been derived.

(B) Gel Filtration: Automated sequencing utilizes small numbers of large (30- to 100-residue) polypeptides. However, many denatured, high-molecular-weight polypeptides may be insoluble owing to exposure during denaturation of previously buried hydrophobic residues. While insolubility can be overcome by urea, alcohols, organic acids, or bases, these restrict the subsequent use of ion exchange techniques for peptide purification. Gel filtration of large hydrophobic peptides, however, may be performed in between 1- and 4-molar formic or acetic acid. This technique separates molecules of different sizes on the basis of whether they are excluded from or included in the pores of a molecular sieve such as Sephadex.

(C) Reversed-Phase HPLC: A powerful technique for purification of high-molecular-weight, nonpolar peptides is high-performance liquid chromatography on a nonpolar support with elution by polar solvents (reversed-phase HPLC). Gel filtration and reversed-phase HPLC are used in conjunction to purify complex mixtures of peptides that result from partial digestion of proteins.

(D) HVE on Molecular Sieves: Molecular sieving may be used in conjunction with charge separation to facilitate separation. While starch and agarose are used, most commonly the support is a cross-linked polymer of acrylamide ($CH_2=CH—CONH_2$). For **polyacrylamide gel electrophoresis (PAGE),** protein solutions are applied to buffered tubes or slabs of polyacrylamide cross-linked 2–10% by inclusion of methylene bisacrylamide (bis) $(CH_2=CH—CONH_2)_2—CH_2$ or similar cross-linking reagents. Direct current is then applied. Visualization is by staining with Coomassie blue dye or Ag^+. A popular variant is PAGE under denaturing conditions. Proteins are boiled and subsequently electrophoresed in the presence of the denaturing agent sodium dodecyl sulfate (SDS). The negatively charged molecule $CH_3—(CH_2)_{11}—SO_3^{2-}$ coats peptides in the approximate proportion of one SDS per two peptide bonds. This "swamps" the native charge of the protein, making it strongly negative. Subsequent separations thus are based primarily on molecular size. SDS-PAGE is widely used to estimate the molecular weights of peptides by comparison of mo-

bilities with those of standards of known molecular weight.

Sequencing Employs the Edman Reaction

Automated sequencing employs the Edman reagent, phenyl isothiocyanate. The reaction sequence (Figure 5–11) releases the amino terminal amino acid as a phenylthiohydantoin derivative, which is then identified by HPLC. The next amino acid in sequence is then derivatized and removed, and the process is

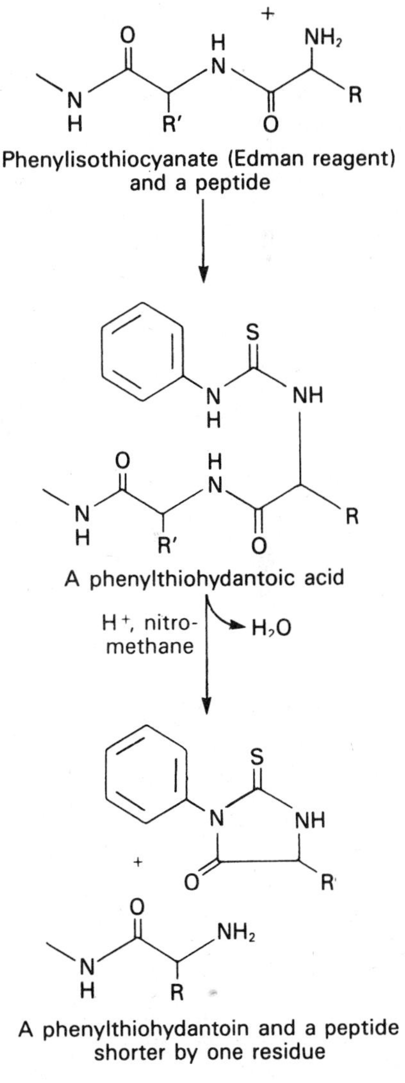

Phenylisothiocyanate (Edman reagent) and a peptide

A phenylthiohydantoic acid

H^+, nitromethane → H_2O

A phenylthiohydantoin and a peptide shorter by one residue

Figure 5–11. The Edman reaction. Phenylisothiocyanate derivatizes the amino terminal residue of a peptide as a phenylthiohydantoic acid. Treatment with acid in a nonhydroxylic solvent releases a phenylthiohydantoin, which is subsequently identified by its chromatographic mobility, and a peptide one residue shorter. The process is then repeated.

repeated. A sequence of 30–40 (or on occasion 60–80) aminoacyl residues is determined in one continuous operation. The Edman reactions take place either in a thin film on the wall of a spinning cup reaction chamber or on a solid matrix to which the carboxyl terminal of the peptide has been covalently coupled. Gas-phase sequencing, a solid-phase technique, employs gaseous reagents and facilitates removal of gaseous products. Solid-phase sequencing methods are applicable to extremely small quantities of peptide—as little as 1 μg in some instances.

Several Overlapping Peptides Must Be Sequenced

To have sequenced all the CNBr peptides from a polypeptide does not by itself provide its complete primary structure, since the *order* in which these peptides occur in the protein is not known. Additional peptides whose amino and carboxyl terminals overlap those of the CNBr peptides must also be prepared using techniques (eg, digestion with chymotrypsin) that cleave the protein at places other than Met residues, then sequenced. An unambiguous primary structure is then deduced by comparison of peptide sequences, a process analogous to the assembly of a jigsaw puzzle (Figure 5–12).

To locate disulfide bonds, peptides from untreated and from reduced or oxidized protein are separated by two-dimensional chromatography or by electrophoresis and chromatography (fingerprinting). Visualization with ninhydrin reveals fewer peptides in the digest from untreated protein and additional peptides in the digest from treated protein. With knowledge of the primary structure of these peptides, the positions of disulfide bonds can then be inferred.

Peptide and DNA Sequencing Are Complementary Techniques

While the simplicity and speed of DNA sequencing (Chapter 42) have revolutionized how peptide structures are determined, peptide and DNA sequencing are complementary rather than mutually exclusive techniques, each with certain advantages.

While the primary structures of proteins may be deduced by sequencing the genes that encode them, direct protein sequencing remains essential for structural analysis of proteins. DNA sequencing cannot reveal the position of disulfide bonds or the presence of numerous other posttranslational modifications. Triplet codons and appropriate tRNA molecules are available only for the 20 most common amino acids. DNA sequencing therefore cannot reveal the presence of prolyl, hydroxylysyl, or a host of methylated, isoprenylated, phosphorylated, esterified, or other posttranslationally modified aminoacyl residues that perform essential functions in specific proteins. The Edman technique thus will continue to prove essential for structural determination of peptides and proteins.

While peptide sequencing suffers from none of these disadvantages, many peptides undergo one or more posttranslational proteolytic processing events. DNA sequencing thus is of major assistance for detection of labile prepeptides and prepropeptides important in catalysis (Chapter 10) or in targeting of peptides to specific subcellular organelles.

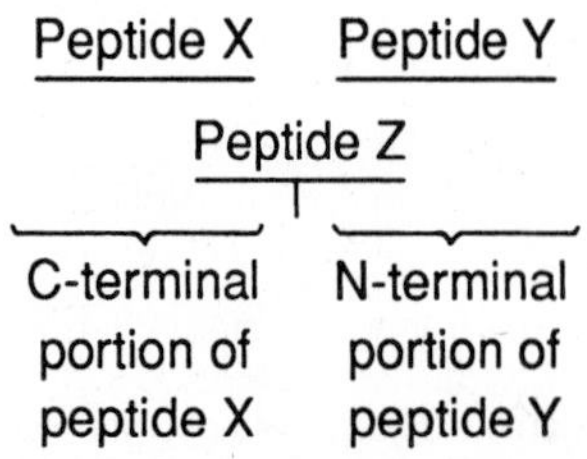

Mass Spectrometry Can Be Used to Sequence Peptides

Fast atom bombardment (FAB) coupled with mass spectrometry (MS) in two linked mass spectrometers provides an alternative way to sequence peptides of up to about 25 residues in length. Fast atom bombardment by argon or xenon atoms of a peptide dissolved in glycerol forms a positively charged peptide ion. The first mass spectrometer resolves the peptide ion from contaminating ions, then transfers it to a cell where it collides with helium atoms. Owing to the absorbed energy, the peptide ion is fragmented from both ends, forming multiple ions of decreasing size. The molecular masses of these ion fragments are then determined in the second mass spectrometer.

Comparison of ions of successively increasing mass permits identification of all aminoacyl fragments and of the sequence of the entire peptide. Computer-driven FAB-MS is extremely fast, does not require highly purified peptides, identifies posttranslationally modified residues, and, unlike the Edman technique, can sequence peptides with blocked (eg, acylated) amino terminals. In comparison with advanced Edman sequenators, the major drawbacks of FAB-MS are high cost and lower sensitivity.

PEPTIDES CAN BE SYNTHESIZED BY AUTOMATED TECHNIQUES

Chemically synthesized peptides serve as drugs or food additives. The ability to synthesize sets of similar peptides whose primary structures differ only slightly provides insight into the relationship of peptide structure to biologic activity.

The basic chemistry of peptide synthesis, developed in the late 1800s by the German chemist Emil

Figure 5–12. The overlapping peptide Z is used to deduce that peptides X and Y are present in the original protein in the order X → Y, not Y ← X.

Fisher, provides for activation of carboxyl groups and for the addition and removal of the blocking groups that prevent unwanted side reactions. Many years later, peptide synthesis techniques were revolutionized by Bruce Merrifield (Nobel Prize 1984), who originated solid phase peptide synthesis. The automated Merrifield synthesis technique that yields long synthetic peptides in short periods of time initiated a renaissance in peptide chemistry.

Solid-Phase Peptide Synthesis

Figure 5–13 illustrates synthesis of a representative dipeptide A-B (where A is the amino terminal and B the carboxyl terminal aminoacyl residue) by the Merrifield solid-phase technique and summarizes the reactions required to synthesize a peptide of any desired length. These steps are as follows:

(1) Block the amino terminals of amino acid A (open symbol) and amino acid B (shaded symbol) with the *t*-butyloxycarbonyl (*t*-BOC) group (■):

$$(CH_3)_3-\overset{\displaystyle O}{\overset{\|}{C}}-O-\overset{\|}{C}-$$

forming *t*-BOC-A and *t*-BOC-B.

(2) Activate the carboxyl group of *t*-BOC-B with dicyclohexyl carbodiimide (DCC) (▶):

$$C_6H_6-N=C=N-C_6H_6$$

(3) React the carboxyl group of amino acid A (which will become the carboxyl terminal residue of the peptide) with an activated, insoluble polystyrene resin (◉).

(4) Remove the blocking group from *t*-BOC-A with room temperature trifluoroacetic acid (TFA, $F_3C-COOH$).

Note: In practice, steps 3 and 4 may be omitted, since resins with any given *t*-BOC amino acid connected via an ester bond to a phenylacetamidomethyl (PAM) "linker" molecule attached to the polystyrene resin are commercially available.

(5) Condense the activated carboxyl group of *t*-BOC-B with the free amino group of immobilized A.

(6) Remove the *t*-BOC blocking group with TFA (see step [4]).

(7) Liberate the dipeptide A-B from the resin particle by treating at −2 °C with hydrofluoric acid (HF) in dichloromethane.

The initial achievements of the Merrifield technique were the synthesis of the A and B chains of insulin and of the enzyme pancreatic ribonuclease. Subsequent improvements have reduced the time for synthesis and have increased yields significantly. This has initiated new prospects for producing vaccines and polypeptide hormones and conceivably also for treating selected inborn errors of metabolism.

SUMMARY

Peptides, formed by loss of water between the carboxyl and amino groups of amino acids, are named for the number of amino acid residues present. Their primary structures (the order of the amino acid residues listed from the amino terminal) determine biologic activity and conformations. Small peptides may contain non-α peptidyl bonds and unusual amino acids. As polyelectrolytes, peptides are separable by ion exchange chromatographic and electrophoretic techniques.

The partial double-bond character of the bond that links the COOH carbon and the N of a peptide makes four atoms of the peptide bond coplanar. This limits the number of possible peptide conformations. Conformations in solution are further constrained by noncovalent forces.

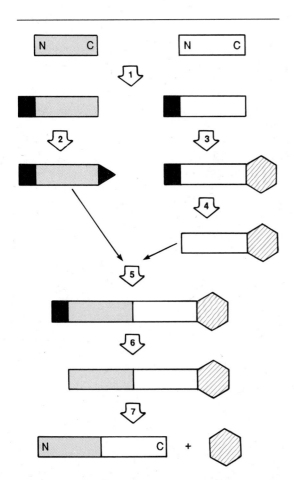

Figure 5–13. Representation of the synthesis of a dipeptide by the solid-phase synthesis technique pioneered by Merrifield. See accompanying text for explanations of symbols.

The amino acid composition of purified peptides is determined after acid hydrolysis and requires correction for losses of certain amino acids. Determination of primary structure employs automated Edman techniques, which can sequence a few micrograms of purified peptide. Disulfide bonds are first oxidized or reduced. Large peptides are cleaved at rare sites by reagents such as CNBr, and the resulting peptides are purified by gel filtration and HPLC. They may then be linked by their carboxyl terminal to a solid support to facilitate sequencing. The subsequent automated operations involve either liquid chemistry or gaseous reagents and products that facilitate sample recovery and purification. As many as 40 successive amino terminal residues can generally be sequenced. To determine the order of assembly of sequenced peptides in the original polypeptide, overlapping peptides generated by different cleavage techniques are sequenced. Alternatively, linkages between peptides may be inferred from DNA sequencing of the appropriate gene. DNA and peptide sequencing thus are complementary, not mutually exclusive, sequencing techniques. Automated techniques also permit the unambiguous synthesis of peptides of known primary structure and full biologic activity.

REFERENCES

Allen G: Sequencing of proteins and peptides. In: *Laboratory Techniques in Biochemistry and Molecular Biology,* 2nd ed. Burdon RH, Knippenberg PH (editors). Elsevier, 1989.

Bhwon AS: *Protein/Peptide Sequence Analysis: Current Methodologies.* CRC Press, 1988.

Biemann K: Mass spectrometry of peptides and proteins. Annu Rev Biochem 1992;61:977.

Deutscher MP (editor): *Guide to Protein Purification.* Methods in Enzymology, vol 182, 1990.

Doolittle RF (editor): *Molecular Evolution: Computer Analysis of Protein and Nucleic Acid Sequences.* Methods in Enzymology, vol 183, 1990.

Doolittle RF: Reconstructing history with amino acid sequences. Protein Sci 1992;1:191.

Fenselau C: Beyond gene sequencing: Analysis of protein structure with mass spectrometry. Annu Rev Biophys Biophys Chem 1991;20:205.

Hunkapiller MW, Strickler JE, Wilson KJ: Contemporary methodology for protein structure determination. Science 1984;226:304.

Karger BL, Chu YH, Foret F: Capillary electrophoresis of proteins and nucleic acids. Annu Rev Biophys Biomol Struct 1995;24:579.

Kent SBH: Chemical synthesis of peptides and proteins. Annu Rev Biochem 1988;57:957.

Merrifield RB: Solid phase synthesis. Science 1986;232:341.

Ozols J: Amino acid analysis. Methods Enzymol 1990;182:587.

Regnier FE, Gooding KM: High performance liquid chromatography of proteins. Anal Biochem 1980;103:1.

Sanger F: Sequences, sequences, and sequences. Annu Rev Biochem 1988;57:1.

Senko MW, McLafferty FW: Mass spectrometry of macromolecules: Has its time come now? Annu Rev Biophys Biomol Struct 1994;23:763.

Stellwagen E: Gel filtration. Methods Enzymol 1990;182:317.

Stoscheck CM: Quantitation of proteins. Methods Enzymol 1990;182:50.

Strickler JE, Hunkapiller MW, Wilson KJ: Utility of the gas-phase sequencer for both liquid- and solid-phase degradation of proteins and peptides at low picomole levels. Anal Biochem 1984;140:553.

Wilson KJ: Micro-level protein and peptide separations. Trends Biol Sci 1989;14:252.

Proteins: Structure & Function

6

Victor W. Rodwell, PhD

INTRODUCTION

Features common to all proteins include the restraints placed on their conformation by covalent and noncovalent bonds. This chapter discusses the secondary, tertiary, and quaternary structure of proteins, with emphasis on the forces that stabilize these higher orders of structure and the physical methods used to examine them. Major topics include the α-helix, β-sheet, β-turn, and loop; how these secondary structural features form domains and tertiary structure; and the assembly of polypeptides into the subunits of multimeric proteins. The paths by which proteins are thought to fold and the role of chaperonins and other accessory proteins that facilitate rapid and correct folding in vivo are examined. While the above features and comments are generally valid for all proteins, specific proteins may exhibit unique secondary-tertiary structures that confer properties which permit them to fulfill specific biologic functions. The close linkage between protein structure and biologic function is illustrated for two fibrous proteins that serve structural roles: silk fibroin and collagen. Subsequent chapters amplify the close linkage of structure and function for the globular proteins myoglobin and hemoglobin and for the catalytic proteins known as enzymes.

BIOMEDICAL IMPORTANCE

The thousands of proteins present in the human body perform functions too numerous to list. These functions include serving as carriers of vitamins, oxygen, and carbon dioxide plus structural, kinetic, catalytic, and signaling roles. It thus is not surprising that dire consequences can arise from mutations either in genes that encode proteins or in regions of DNA that control gene expression. Consequences equally dire can also result in deficiency of cofactors essential for maturation of a protein. Ehlers-Danlos syndrome illustrates a genetic defect in protein maturation and scurvy a deficiency of a cofactor essential for protein maturation.

PROTEINS ARE CLASSIFIED IN MANY WAYS

While no universally accepted classification system exists, proteins may be classified on the basis of their solubility, shape, biologic function, or three-dimensional structure. A system in limited use in clinical biochemistry distinguishes "albumins," "globulins," "histones," etc, based on their solubility in aqueous salt solutions. Proteins may also be classified based on their overall shape. Thus, **globular proteins** (eg, many enzymes) have compactly folded, coiled polypeptide chains and axial ratios (ratios of length to breadth) of less than 10 and generally not greater than 3–4. **Fibrous proteins** have axial ratios greater than 10.

Based on their biologic functions, proteins might be classified as enzymes (dehydrogenases, kinases), storage proteins (ferritin, myoglobin), regulatory proteins (DNA-binding proteins, peptide hormones), structural proteins (collagen, proteoglycans), protective proteins (blood clotting factors, immunoglobulins), transport proteins (hemoglobin, plasma lipoproteins), and contractile or motile proteins (actin, tubulin).

Specialized systems of classification distinguish certain complex proteins of high medical interest. Thus, **plasma lipoproteins** are termed "origin," α_1-, α_2-, or β-lipoproteins based on their electrophoretic mobility at pH 8.6; or as chylomicrons, VLDL, LDL, HDL, or VHDL based on their sedimentation behavior in an ultracentrifuge (Figure 6–1). Lipoproteins may also be classified by immunologic determination of which apoproteins (A, B, C, D, E, F) are present. Similarities in three-dimensional structure, revealed primarily by x-ray crystallography, provide a potentially valuable basis for protein classification. For instance, proteins that bind nucleotides share a nucleotide-binding domain of tertiary structure.

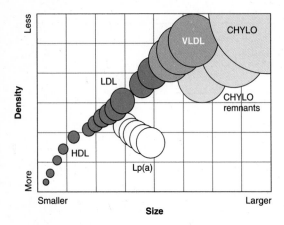

Figure 6–1. Size and density distribution of lipoproteins. (CHYLO, chylomicrons; VLDL, very low density lipoproteins; LDL, low-density lipoproteins; HDL, high-density lipoproteins; Lp[a], lipoprotein[a].) (Reproduced, with permission, from Segrest JP et al: The amphipathic α-helix: A multifunctional structural motif in plasma lipoproteins. Adv Protein Chem 1995;45:303.)

THE FOUR ORDERS OF PROTEIN STRUCTURE

PRIMARY STRUCTURE

The primary structure of the polypeptide chain of a protein is the order in which amino acids are joined together, and it includes the location of any disulfide bonds. Primary structure is determined by the methods described for polypeptides in Chapter 5.

The number of known protein sequences is so large and is increasing so rapidly that, rather than being available in printed form, sequence data are now deposited in electronic protein sequence databases that can be accessed via the Internet. Important continuously updated protein databases include EMBL (European Molecular Biology Laboratory Data Library), GenBank (Genetic Sequence Databank), and PIR (Protein Identification Resource Sequence Database).

SECONDARY STRUCTURE

Configuration and Conformation

The term "configuration" refers to the geometric relationship between a given set of atoms. Interconversion of configurational alternatives (eg, conversion of D- to L-alanine) can be achieved only by breaking and re-forming covalent bonds.

The similar sounding term "conformation" refers to the three-dimensional architecture of a protein, the spatial relationships of all the atoms to all the others. The interconversion of conformers involves not the rupture of covalent bonds but the rupture and re-formation of noncovalent forces (hydrogen bonds, salt bonds, hydrophobic interactions) that stabilize given conformations. Even after ruling out conformations precluded by steric interactions, the free rotation about two-thirds of the covalent bonds of the main chain of a polypeptide (Figure 5–8) allows for a staggeringly large number of possible conformations for a given protein. However, for a given protein, only a small number of possible conformations have biologic significance.

Various Forces Stabilize Protein Structures

Several individually weak but numerically formidable noncovalent interactions stabilize protein conformation. These forces include hydrogen bonds, hydrophobic interactions, electrostatic interactions, and van der Waals forces.

Hydrogen Bonds

Residues with polar R groups generally occur on the surface of globular proteins, where they form hydrogen bonds primarily to water molecules. Elsewhere, the aminoacyl residues of the backbone form hydrogen bonds with one another.

Hydrophobic Interactions

Hydrophobic interactions involve the nonpolar R groups of aminoacyl residues that in typical globular proteins reside in the interior of the protein. Formation of hydrophobic interactions is "entropically driven." A roughly spherical overall shape minimizes surface area. Concentration of nonpolar residues in the interior of the protein lowers the number of surface residues and maximizes the opportunity for the film of surface water molecules to form hydrogen bonds with one another, a process associated with an increase in entropy. By contrast, the nonpolar environment of biologic membranes favors hydrophobic surface residues whose nonpolar R groups participate in hydrophobic interactions with the alkyl side chains of the fatty acyl esters of membrane bilayers.

Electrostatic Interactions

Electrostatic interactions or salt bonds are formed between oppositely charged groups such as the amino terminal and carboxyl terminal groups of peptides and the charged R groups of polar aminoacyl residues. While all formally charged groups tend to be located on the surfaces of globular proteins, exceptions occur. Specific polar groups that perform essential biologic functions may reside in clefts that penetrate the interior of a protein. Since polar residues can also participate in ionic interactions, the presence of salts such as KCl can significantly decrease ionic interactions between surface residues.

Van der Waals Interactions

Van der Waals forces, which are extremely weak and act only over extremely short distances, include both an attractive and a repulsive component. The attractive force involves interaction between induced dipoles formed by momentary fluctuations in the electron distribution in nearby atoms. The repulsive force comes into play when two atoms come so close that their electron orbitals overlap. The distance at which the attractive force is maximal and the repulsive force is minimal is termed the **van der Waals contact distance.** Atoms have characteristic van der Waals radii, and the optimal contact distance between two atoms is the sum of their van der Waals radii.

Peptide Bonds Restrict Possible Secondary Conformations

Rotation is possible only about two of the three covalent bonds that form the polypeptide backbone of proteins. The partial double bond character of the bond that links the carbonyl carbon to the α-nitrogen of a peptide bond significantly restricts possible conformations of its associated atoms, which become coplanar (Figure 5–8). Free rotation is possible only about the bonds that link the α-carbon (C_α) to the carbonyl carbon (C_o) and the nitrogen. The angle about the C_α—N bond is termed the phi angle (Φ) and that about the C_α—C_o bond the psi angle (ψ). Combinations of phi-psi angles that result in steric hindrance between nonbonded atoms are not allowed. When all phi and psi angles are specified, the conformation of all atoms of the main chain or backbone of a polypeptide is known. The analysis of crystallographic data therefore begins with determination of the main chain phi and psi angles.

G.N. Ramachandran first used a plot of phi versus psi angles to illustrate both the allowed values and the numerous disallowed values (Figure 6–2). Sterically allowed combinations characterize the α-helix, β-pleated sheet, and collagen triple helix discussed below.

Both Regular and Irregular Conformations Characterize Secondary Structure

The conformation of the polypeptide backbones of proteins constitutes their secondary structure. Originally proposed on theoretical grounds, the existence of secondary structure has been amply confirmed by x-ray crystallographic analyses. The kinds of secondary structures are limited by the partial double bond character of the peptide bond (Figures 5–7 and 5–8) and by the size and shape of the aminoacyl R groups. Secondary structures include, in addition to

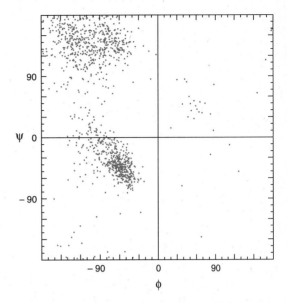

Figure 6–2. Ramachandran plot of the main chain phi and psi angles for approximately 1000 nonglycine residues in eight proteins whose structures were solved at high resolution. The dots represent allowable combinations and the spaces prohibited combinations of phi-psi angles. (Reproduced, with permission, from Richardson JS: The anatomy and taxonomy of protein structures. Adv Protein Chem 1981;34:167.)

the regular repeating units of α-helices, β-sheets, and β-bends, irregular conformations termed loops or coils.

The α-Helix

If the backbone of a polypeptide is twisted by equal amounts about each α-carbon, it forms a coil or helix. Different types of helices, formed by imparting differing extents and direction of twist, are described by the number (*n*) of aminoacyl residues per turn and the pitch (*p*) or distance per turn that the helix rises along its axis. Formed from chiral amino acids, polypeptide helices exhibit chirality—ie, they are either right- or left-handed. To visualize a right-handed helix, hold your right hand palm up with the fingers curled slightly and the thumb pointing away from you. Your thumb points in the direction of the carboxyl terminal of the polypeptide of a right-handed helix, which advances in the direction of your curled fingers. Although synthetic polypeptides of D-amino acids can form left-handed helices, steric interference by the R groups of L-aminoacyl residues dictates that left-handed helices do not occur in proteins. The types of polypeptide helices present in proteins are further limited by additional steric factors and by the number of possible hydrogen bonds and van der Waals interactions that can form and stabilize a given type of helix.

The α-helix has both favorable phi and psi angles and a hydrogen bonding pattern that confers maximum stability. The α-helices of proteins contain anywhere from 4 to 50 residues (average, about 12 residues). The relevant parameters of an α-helix (Figure 6–3) are $n = 3.6$ residues per turn and $p = 0.54$ nm (5.4 Å). The distance along the helix axis that separates equivalent main-chain atoms of adjacent residues is 0.15 nm (1.5 Å). The aminoacyl R groups are directed outward from the helix axis (Figures 6–4 and 6–5), minimizing mutual steric interference.

Hydrogen Bonds and Van der Waals Forces Stabilize α-Helices

Since the α-helix is the lowest-energy and most stable conformation for a polypeptide chain, it forms spontaneously. The stability of α-helices arises primarily from the formation of the maximum possible number of hydrogen bonds. Peptide nitrogens act as hydrogen donors, and the carbonyl oxygen of the residue fourth in line behind in a primary structural

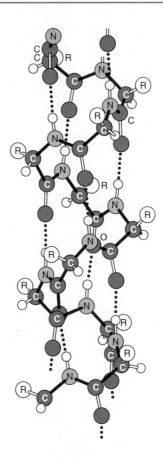

Figure 6–4. Hydrogen bonds (dots) formed between H and O atoms stabilize a polypeptide in an α-helical conformation. (Reprinted, with permission, from Haggis GH et al: *Introduction to Molecular Biology.* Wiley, 1964.)

sense acts as hydrogen acceptor (Figure 6–4). These hydrogen bonds have an essentially optimal N-to-O distance of 28 nm (2.8 Å). Van der Waals interactions also confer additional stability. The tightly packed atoms at the core of an α-helix are in van der Waals contact with one another across the axis of the α-helix.

Proline Can Bend an α-Helix

Ala, Glu, Leu, and Met residues are more common in α-helices than are Gly, Pro, Ser, or Tyr residues. This tendency is not useful, however, for structural predictions. Since the peptide nitrogen of a prolyl residue cannot form a hydrogen bond, proline fits only the first turn of an α-helix. Elsewhere it produces a bend. However, not all bends in α-helices are caused by prolyl residues. Bends often occur also at Gly residues.

α-Helices May Be Amphipathic

While commonly on the surface of proteins, α-helices may also be wholly or partially buried in the in-

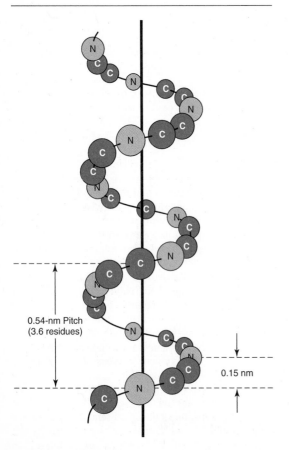

Figure 6–3. Orientation of the main chain atoms of a peptide about the axis of an α-helix.

0.54-nm Pitch (3.6 residues)

0.15 nm

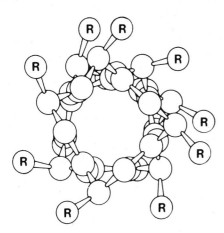

Figure 6–5. View down the axis of an α-helix. The side chains (R) are on the outside of the helix. The van der Waals radii of the atoms are larger than shown here; hence, there is almost no free space inside the helix. (Slightly modified and reproduced, with permission, from Stryer L: *Biochemistry,* 3rd ed. Freeman, 1995. Copyright © 1995 by W.H. Freeman and Co.)

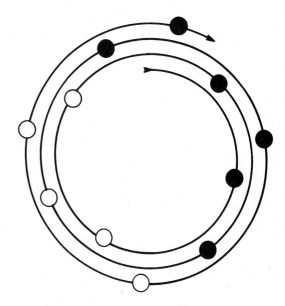

Figure 6–6. Spiral representation of amino acids in an amphipathic α-helix. Helices of plasma apolipoproteins are amphipathic, ie, one face is polar and the other nonpolar. Residues are plotted 100 degrees apart (360 degrees/3.6 residues = 100 degrees per residue) viewed down the helix axis. Solid circles represent nonpolar and open circles polar residues.

terior of a protein. The amphipathic helix, a special case in which residues switch between hydrophobic and hydrophilic about every three or four residues (Figure 6–6), occurs where α-helices interface with both a polar and a nonpolar environment. Amphipathic helices occur in plasma lipoproteins as well as in certain polypeptide hormones, venoms, antibiotics, human immunodeficiency virus glycoprotein, and calmodulin-regulated protein kinases.

Amino Acids From Different Regions Form β-Sheets

The second regular conformation, present in most proteins, is the β-pleated sheet. The "β" denotes that it was the second regular structure described. The term "pleated sheet" describes its appearance when viewed edge-on. The α-carbons and their associated R groups alternate between slightly above and slightly below the plane of the main chain of the polypeptide (Figure 6–7). Like α-helices, β-sheets have repeating phi and psi angles and are stabilized by the maximum possible number of hydrogen bonds. Polypeptides aligned alongside one another are stabilized by hydrogen bonds formed between the peptide nitrogen hydrogens and carbonyl oxygens of adjacent strands. However, while α-helices contain residues adjacent in a primary structural sense, β-sheets involve stretches of five to ten amino acids from different primary structural regions. Unlike the compact α-helix, the peptide backbone of a β-sheet is almost fully extended.

β-Pleated Sheets May Be Parallel or Antiparallel

Figure 6–7 illustrates antiparallel strands of β-pleated sheet. Adjacent polypeptide chains of antiparallel sheet proceed in opposite directions; those of parallel sheet, in the same direction. Except for the flanking strands, all possible hydrogen bonds are formed, and each sheet has a characteristic pattern of hydrogen bonding (Figure 6–8). Alternating pairs of narrowly and widely spaced hydrogen bonds stabilize antiparallel sheets. The hydrogen bonds that stabilize parallel strands are evenly spaced but angle across between strands. Almost all β-sheet strands are twisted in a right-handed sense. These twisted strands of β-sheet form the central core of many globular proteins. From 2 to 15 strands may be involved in a β-sheet, and regions of mixed parallel and antiparallel sheet are common. Perhaps because of somewhat lower stability, parallel β-strands of less than five strands are, however, rare.

Loop Regions Form Antigen-Binding Sites

Roughly half of the residues in a typical globular protein are present in α-helices or β-sheets. The remainder reside in "loop" or "coil" conformations which, while irregularly ordered, are no less biologically important than more regularly ordered sec-

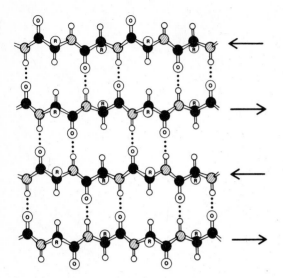

Figure 6–7. Model of an antiparallel β-pleated sheet. Adjacent strands run in opposite directions (see arrows). The R groups (R) are located above and below the plane of the sheet. Note that successive α-carbons are located just above and below this plane, giving a pleated effect. Hydrogen bonds grouped in pairs stabilize the β-sheet conformation. In parallel pleated sheets (not shown), adjacent chains run in the same direction.

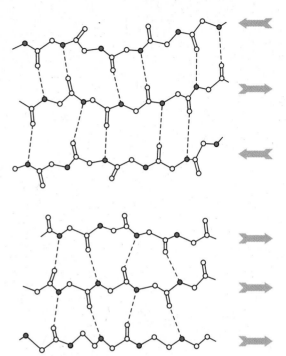

Figure 6–8. Spacing and bond angles of the hydrogen bonds of antiparallel and parallel pleated β-sheets. Arrows indicate the direction of each strand. The hydrogen-donating α-nitrogen atoms are shown as blue circles. For clarity in presentation, R groups and α-hydrogens are omitted. ***Top:*** Antiparallel β-sheet. Pairs of hydrogen bonds alternate between being close together and wide apart and are oriented approximately perpendicular to the polypeptide backbone. ***Bottom:*** Parallel β-sheet. The hydrogen bonds are evenly spaced but slant in alternate directions.

ondary structures. Loop or coil should not be confused with "random coil," a term that describes the disordered, biologically unimportant conformations of denatured proteins. Loop regions of varying size and shape form major surface features of proteins. Exposed to solvent, rich in charged and polar residues but lacking regular secondary structure, hairpin loops connect adjacent antiparallel β-sheets. Often the site for ligand interactions, loop regions of varying primary structure and length form the antigen-binding sites of antibodies.

Globular Proteins Contain β-Bends

The compact character of globular proteins arises from more or less straight stretches of 20 or so residues joined by regions of polypeptide that abruptly change direction. Strands of β-sheet are connected by regions of polypeptide which, if long, often contain α-helices. These reverse turns, or β-bends, involve four residues hydrogen-bonded in various ways and occur primarily at protein surfaces.

Proteins May Contain Disordered Regions

Not all residues are necessarily present as ordered secondary structures. Specific regions of many proteins exist in numerous conformations in solution and

thus are truly disordered. Examples include the long R groups of Lys residues and portions of the amino or carboxyl terminals of many polypeptides. This disorder conveys flexibility and can assume a vital biologic role. Many disordered regions become ordered when a specific ligand is bound. A common example is the stabilization of disordered regions of the catalytic centers of many enzymes when a ligand is bound.

TERTIARY STRUCTURE

Schematic Diagrams Simplify Protein Structures

The high information content of diagrams or models that include all atoms of a protein hinders the study of overall features of protein structure. Consequently, we employ simplified conventional representations to display structural features. The symbols used are cylinders for α-helices, broad arrows for β-strands, and ribbon-like strands for the remaining structures such as β-bends and loops.

Secondary Structures Can Form Supersecondary Motifs

In many globular proteins, the secondary structural motifs of α-helix or β-pleated sheet form recognizable "supersecondary" motifs. Figure 6–9 illustrates several supersecondary motifs: β-α-β, two strands of β-sheet connected by an α-helix; the β-hairpin, composed of antiparallel β-sheets connected by short regions of loop; and the "Greek key" motif, named for its resemblance to a decorative motif on ancient Greek vases. Repetitions of these supersecondary motifs can then form structures such as the regularly repeating β-α-β units of an entire protein (Figure 6–10).

The term "tertiary structure" refers to the spatial relationships between secondary structural elements. The secondary and supersecondary structures of large proteins often are organized as domains—compact units connected by the polypeptide backbone. Folding of a polypeptide within one domain usually occurs independently of folding in other domains. Tertiary structure describes the relationships of these domains, the ways in which protein folding can bring together amino acids far apart in a primary structural sense, and the bonds that stabilize these conformations.

Large Polypeptides Have Distinct Domains

Secondary-tertiary structure, particularly of large polypeptides, may be organized into connected but structurally relatively independent units termed **domains.** Domains often perform discrete functions such as binding specific ligands. Ligands may contact residues contributed exclusively by a single domain. Alternatively, a ligand may contact residues from more than one domain and thus reside at the interface between domains. In multimeric proteins, the residues of the domains of different subunits may

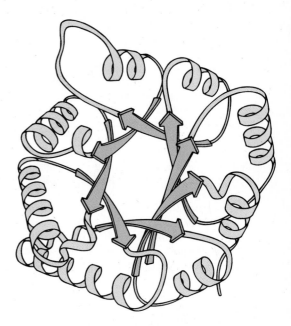

Figure 6–10. Tertiary structure. Triose phosphate isomerase, shown in an end-on view, is built up from four β-α-β motifs that are consecutive in both a primary and a tertiary structural sense. (Courtesy of J Richardson.)

bind ligands such as coenzymes at subunit interfaces whose contact surfaces derive from residues from different subunits.

Electrostatic Bonds Link Surface Residues

Salt (electrostatic) bonds link oppositely charged R groups of residues and the charged α-groups of carboxyl and amino terminal residues. For example, the R group of lysine (net charge +1 at physiologic pH) and aspartate or glutamate (net charge −1) can interact electrostatically to stabilize proteins.

Disulfide Bonds Confer Additional Stability

In addition to peptide bonds, covalent **disulfide bonds** can form between cysteine residues present in the same or different polypeptides. These disulfide bonds confer additional stability to specific conformations of proteins such as enzymes (eg, ribonuclease) and structural proteins (eg, keratin).

Hydrophobic Interactions Link Interior Residues

Nonpolar side chains of amino acids associate in the interior of globular proteins. These associations are individually extremely weak and are not stoichiometric. Their large number dictates, however, that hydrophobic interactions contribute significantly to maintaining protein structure.

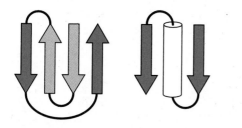

Figure 6–9. Supersecondary motifs. The α-helices and β-pleated sheets of many globular proteins are arranged in repeating units such as the Greek key (*left*) or β-α-β (*right*) supersecondary motifs.

X-Ray Crystallography Reveals Secondary and Tertiary Structure

X-rays directed at a crystal of a protein and of a derivative that contains an added heavy metal are scattered in patterns that depend on the electron densities in different parts of the protein. Data collected on photographic plates or computer-linked area detectors are analyzed mathematically, translated into electron density maps, and used to construct computer-generated three-dimensional models. Time-consuming, expensive, and requiring highly specialized training, x-ray crystallography has nevertheless yielded detailed three-dimensional views of over 1000 proteins. Its contributions to our understanding of protein structure can hardly be overemphasized. **Magnetic resonance imaging (MRI)** has been used to solve the three-dimensional structure of certain small proteins and holds promise for application to larger proteins.

Databases and Sophisticated Software Facilitate Study and Manipulation of Protein Structures

The coordinates of all the atoms of proteins whose structures have been determined by x-ray crystallography are accessible via the Internet from electronic databases. The coordinates can be downloaded and linked to sophisticated software programs—many in the public domain—that permit one to generate and manipulate on-screen models, rotate them in all directions, add or subtract residues, domains, or ligands, change bond angles, etc—in many instances using only a Pentium-based personal computer.

QUATERNARY STRUCTURE

Oligomeric Proteins Have Multiple Polypeptide Chains

Proteins that contain two or more polypeptide chains associated by noncovalent forces are said to exhibit quaternary structure. In these multimeric proteins, the individual polypeptide chains are termed protomers or subunits. Hydrogen bonds and electrostatic bonds formed between surface residues of adjacent subunits stabilize the association of subunits. Proteins composed of two or four subunits are termed dimeric or tetrameric proteins, respectively. Homodimers, homotetramers, etc, consist of identical subunits; hetero-oligomers, of dissimilar subunits. The different subunits of hetero-oligomeric proteins typically perform discrete functions. One subunit or set of identical subunits may perform a catalytic function; another subunit set, ligand recognition or a regulatory role. Different spatial orientations of subunits confer altered properties on the oligomer and permit multimeric proteins to play unique roles in intracellular regulation.

Protein Denaturants Disrupt Secondary, Tertiary, and Quaternary Structure

Reagents such as urea, sodium dodecyl sulfate (SDS), mild H^+, and mild OH^- rupture hydrogen bonds, hydrophobic bonds, and electrostatic bonds (but not peptide or disulfide bonds). They thus disrupt all orders of protein structure except primary structure and destroy biologic activity (Figure 6–11).

Physical Methods Reveal Molecular Weight and Quaternary Structure

Determination of the quaternary structure of oligomeric proteins involves determining the number and kind of protomers present, their mutual orientation, and the interactions that unite them. Providing that oligomers do not undergo denaturation during the procedure used to determine molecular weight, many methods can yield molecular weight data. These same techniques may be used to determine protomer molecular weight if the oligomer is first denatured.

A. Analytical Ultracentrifugation: Analytical ultracentrifugation, developed by Svedberg, measures the rate at which a protein sediments in a gravitational field of around $10^5 \times g$. An excellent physical technique for the determination of molecular weight, it has been supplanted in recent years by less complex methods.

B. Sucrose Density Gradient Centrifugation: This related technique is best applied to globular proteins. Protein standards and unknowns are layered on a gradient of buffered 5–20% sucrose, prepared by repeated freeze-thawing of 20% sucrose. Following overnight centrifugation at $10^5 \times g$, the tube contents are removed dropwise and analyzed for protein. The mobility of the unknown protein, expressed relative to standards of known molecular weight, is then computed and its molecular weight is calculated.

C. Gel Filtration: Columns of Sephadex or similar matrices that have "pores" of known size range are calibrated using proteins of known molecular weight. The molecular weight of an unknown protein is then calculated from its elution position relative to these standards. Large errors may result if the

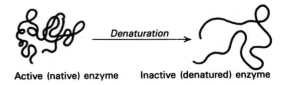

Active (native) enzyme   **Inactive (denatured) enzyme**

Figure 6–11. Representation of denaturation of a protomer.

protein is highly asymmetric or interacts strongly with the material from which the molecular sieve has been manufactured.

D. Polyacrylamide Gel Electrophoresis (PAGE): Protein unknowns and standards are separated by electrophoresis in 5–15% cross-linked acrylamide gels of varying porosity, which are then stained for protein using Coomassie blue stain or Ag^+. SDS-PAGE is used to determine the subunit size of oligomeric proteins. Proteins are first denatured by boiling them in the presence of β-mercaptoethanol and the negatively charged ionic detergent sodium dodecyl sulfate (SDS). The denatured proteins, all of which are coated with SDS and hence are negatively charged, are then separated on the basis of their size (not their charge) by SDS-PAGE on gels that contain SDS.

E. Electron Microscopy Visualizes Macromolecular Complexes: The electron microscope, which can magnify up to 100,000 diameters, visualizes virus particles, enzyme complexes, oligomeric proteins, and high-molecular-weight oligomers.

HOW PROTEINS FOLD

How, within a time frame of seconds, do polypeptides fold into their native physiologic states? No definitive answers are as yet available, and this question remains a subject of active investigation. Some general principles are, however, apparent. Folding cannot proceed via a "random search" approach, ie, one that involves exploration of all possible conformations en route to the native state. Even for a comparatively small polypeptide, folding via a random search path would require not seconds but a time longer than the estimated age of the universe! Folding clearly must be a more highly directed process that involves highly favored paths to intermediates.

General Principles

Based on present knowledge, the folding of polypeptides and of multimeric proteins appears to involve the following stages. First formed are short stretches of secondary structure—regions of α-helix, β-sheet, β-turn, etc. Diffusion-directed growth of these short stretches leads ultimately to a domain. In multidomain proteins, folded domains coalesce to form a so-called "molten globule" that has extensive secondary structure but little ordered tertiary structure. Successive conformational changes convert the molten globule to a compact form with ordered tertiary structure. Minor changes in conformation then lead to a native polypeptide structure. For monomeric proteins, the folding process now is complete. For proteins with quaternary structure, these ordered polypeptides serve as subunits that come together to form multimers.

Denatured Proteins Can Be Renatured

Many denatured proteins refold spontaneously in vitro with accompanying restoration of biologic activity. This led Anfinsen to conclude that the primary structure of polypeptides was itself sufficient to direct protein folding. However, spontaneous refolding takes hours—far too long to account for protein folding in vitro. Anfinsen's conclusion that primary structure directs in vivo folding remains valid but requires modification to incorporate the participation of accessory proteins that accelerate the folding process and direct it along specific paths. These accessory proteins include protein disulfide isomerase, prolyl *cis-trans* isomerase, and the chaperonins.

Protein Disulfide Isomerase

Disulfide (—S—S—, or cystine) bonds formed under oxidizing conditions within and between polypeptides stabilize both secondary and tertiary structure. A given cysteinyl residue can, however, form many different disulfide bond pairs, only one of which is appropriate for biologically correct folding. The enzyme protein disulfide isomerase facilitates "shuffling" of —S—S— bonds by accelerating the rate at which disulfides undergo mutual exchange, thereby accelerating the overall process of protein folding.

Prolyl *cis-trans* Isomerase

While the configuration of X-Pro peptide bonds of newly synthesized polypeptides is *trans,* up to 10% of the X-Pro bonds of mature proteins have the *cis* configuration. The enzyme prolyl *cis-trans* isomerase catalyzes this isomerization, thereby facilitating the folding process (Figure 6–12).

Molecular Chaperones

The molecular chaperones of bacteria, mitochondria, and chloroplasts are large, multisubunit proteins that accelerate the folding process by providing a protected environment where polypeptides fold into native conformations and form quaternary structures. Chaperonins include the heat-shock proteins that consist structurally of two rings of identical subunits

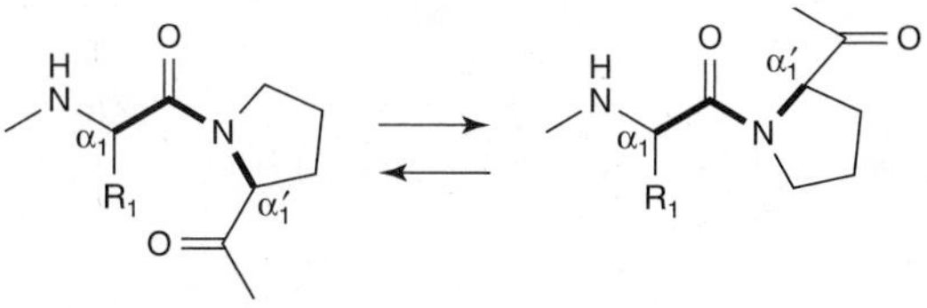

Figure 6–12. Isomerization of the N-α_1 prolyl peptide bond from a *cis* to a *trans* configuration relative to the backbone of the polypeptide.

arranged about a central space that can accommodate large polypeptides. Chaperonins favor paths that inhibit inappropriate interactions between complementary surfaces and facilitate appropriate interactions. The mechanism by which chaperonins achieve this is, however, imperfectly understood.

Chaperonins include the Hsp60 heat-shock proteins, elaborated by bacteria or chloroplasts in response to heat stress, which reverse the denaturation and aggregation of proteins that occur at elevated temperatures. The 14 identical subunits of the Hsp10 and Hsp60 proteins GroES and GroEL of *E coli*—and Cpn10 and Cpn60 of chloroplasts—form paired rings of seven subunits that surround a spacious central cavity which can accommodate large polypeptides. This protective environment counteracts the tendency of solvent-exposed hydrophobic regions of partially folded or denatured polypeptides to self-associate and form insoluble aggregates.

FORM DICTATES FUNCTION

Close examination of representative proteins illustrates not only the unique structures that have evolved to fulfill specific biologic functions but also the close linkage of protein structure and biologic function. Most fibrous proteins fulfill structural roles in skin, connective tissue, or fibers such as hair, silk, or wool. The atypical sequences of amino acids of fibrous proteins dictate specific secondary and tertiary structures that confer specific mechanical properties.

Hydrophobic Forces Stabilize the β-Sheet of Silk Fibroin

The insect protein silk fibroin illustrates stabilization of secondary structure by forces other than hydrogen bonds. The amino acid composition of fibroin consists almost entirely (85%) of Gly residues alternating with Ala or Ser residues. Fibroin polypeptide chains form arrays of extended, antiparallel β-sheet in which the R group of Gly residues (hydrogen) extends from one surface and the R groups of Ala or Ser from the other. These arrays are stabilized exclusively by hydrophobic interactions between Gly R groups on one surface and between Gly or Ser R groups on the other. Small quantities of residues such as Val or Tyr with bulky R groups occur periodically, interrupt β-sheet regions, and confer some flexibility. Fibroin's long regions of closely packed, almost fully extended antiparallel β-sheet resist stretching but remain highly flexible owing to the nondirectional character of the stabilizing hydrophobic forces.

Secondary Structure of Fibrous Proteins

The secondary structures of fibrous proteins, including those of skin, tendon, bone, and muscle that serve structural, protective, and motile functions, exhibit structural features that contrast sharply with those of globular proteins. Discussed below is the secondary structure of the most abundant mammalian protein, collagen.

COLLAGEN

Collagens Illustrate Diverse Structure-Function Relationships

The collagens, which comprise about 25% of total mammalian protein, strikingly illustrate the diverse structure-function relationships that characterize vertebrate fibrous proteins. The collagen of tendons forms highly asymmetric structures of high tensile strength. Skin collagen forms loosely woven, flexible fibers. The collagen of hard regions of teeth and bone contains hydroxyapatite, a calcium phosphate polymer. Finally, collagen in the cornea of the eye, ordered so as to be nearly crystalline, is transparent.

Tropocollagen Is a Triple Helical Molecule Rich in Gly, Pro, and Hyp

Tropocollagen consists of three approximately 1000-residue polypeptide chains organized as a left-handed non-α-helix that has approximately three residues per turn. Three left-handed helices entwine to form a right-handed triple helix, or supercoil stabilized by hydrogen bonds formed between (not within, as in the α-helix) individual polypeptide chains (Figure 57–1). This triple helical supercoil is extremely strong. The supercoil resists unwinding because it and its three polypeptides are coiled in opposite directions, a principle used also in the steel cables of suspension bridges. The highly asymmetric mature collagen fibers measure 1.5 nm by about 300 nm, reflecting their stable, extended conformation.

Every Third Residue Is Glycine

The primary structural motif of mature collagen is (Gly-X-Pro/Hyp)$_n$. Every third residue is glycine, the only residue with an R group small enough to fit within the central core of the superhelix. Each glycine is preceded by either a prolyl or a hydroxyprolyl residue. Mutual repulsion between prolyl residues forces the polypeptide into an extended left-handed helix. The conformationally inflexible prolyl and hydroxyprolyl residues confer rigidity.

Many Posttranslational Modifications Characterize Procollagen Maturation

Unlike the superhelix of mature collagen, the carboxyl and amino terminals of the collagen precursor procollagen have an amino acid composition typical of a globular protein. During maturation of procollagen, these carboxyl and amino terminal extensions are removed by selective proteolysis. Additional posttranslational modifications involve hydroxylation, oxidation, aldol condensation, reduction, and glycosylation. Hydroxylation of prolyl and lysyl residues is catalyzed by prolyl hydroxylase and lysyl hydroxylase, enzymes that require ascorbic acid (vitamin C). Specific hydroxylysyl residues are then glycosylated by galactosyl and glucosyl transferases.

Covalent Cross-Links Stabilize Collagen Fibers

Tropocollagen chains associate to form collagen microfibrils. Initially stabilized by intrachain hydrogen bonds, they subsequently are stabilized by covalent bonds formed within and between helices. This covalent cross-linking process involves the copper-requiring enzyme lysyl oxidase, which converts the non-α-amino groups of lysyl and hydroxyl residues to aldehydes. These aldehydes then either undergo an aldol condensation or condense with the non-α-amino groups of lysine or hydroxylysine, forming Schiff bases. Subsequent reductions form stable covalent cross-links that confer great tensile strength.

Nutritional and Genetic Disorders Can Involve Impaired Secondary Structure

The medical importance of stable secondary structures is amply documented by disorders of tropocollagen biosynthesis and maturation. In severe vitamin C deficiency, prolyl and lysyl hydroxylases are inactive, and tropocollagen cannot undergo the reactions that form covalent cross-links. The result is scurvy, a nutritional disorder characterized clinically by bleeding gums, poor wound healing, and ultimately death. Similarly, Menkes' disease, characterized by kinky hair and retardation of growth, reflects a dietary deficiency of the copper required by lysyl oxidase.

Genetic disorders of collagen biosynthesis include several forms of osteogenesis imperfecta, characterized clinically by fragile bones. Genetically distinct forms of the disease can result from defects in various genes of collagen biosynthesis. Similarly, several types of Ehlers-Danlos syndrome, characterized clinically by mobile joints and skin abnormalities, reflect

defects in the genes that encode α_1-procollagen, procollagen N-peptidase, or lysyl hydroxylase (Chapter 57).

SUMMARY

Structural features of proteins are considered under four orders: primary, secondary, tertiary, and (for oligomeric proteins only) quaternary. Primary structure, the sequence of amino acids and location of any disulfide bonds, is encoded in the genes. Secondary and tertiary structures, which concern the protein conformations permitted by peptide bonds, are dictated by the primary structure. Secondary structure describes the folding of polypeptide chains into multiply hydrogen-bonded motifs such as the α-helix and the β-pleated sheet. Combinations of these motifs can then form supersecondary motifs (eg, β-α-β). Tertiary structure concerns the relationships between secondary structural domains and between residues far apart in a primary structural sense. Quaternary structure, present only in proteins having two or more polypeptide chains (oligomeric proteins), describes contact points and other relationships between these polypeptides or subunits.

While primary structure involves covalent bonds, higher orders are stabilized only by weak forces that include multiple hydrogen bonds, salt (electrostatic) bonds between surface residues, and nonstoichiometric association of hydrophobic R groups in the interiors of proteins. Reagents that break noncovalent bonds (eg, urea or SDS) disrupt secondary, tertiary, and quaternary structure with attendant loss of biologic activity (denaturation). Physical techniques for study of higher orders of protein structure include x-ray crystallography (secondary and tertiary structure) plus ultracentrifugation, gel filtration, and gel electrophoresis (quaternary structure).

The close linkage of protein structure and biologic function is illustrated by two fibrous proteins: silk fibroin and collagen. Diseases associated with defects in collagen maturation include Ehlers-Danlos syndrome and the vitamin C deficiency disease scurvy.

REFERENCES

Advances in Protein Chemistry. Academic Press, 1944 to date. [Annual publication.]

Benner SA et al: Predicting the conformation of proteins from sequences: Progress and future progress. Adv Enzyme Regul 1994;34:269.

Bollag DM, Edelstein SJ: *Protein Methods.* Wiley-Liss, 1990.

Branden C, Tooze J: *Introduction to Protein Structure.* Garland, 1991.

Burley SK, Petsko GA: Weakly polar interactions in proteins. Adv Protein Chem 39;1989:125.

Copeland RA: Methods of protein analysis: A practical guide to laboratory protocols. Chapman & Hall, 1993.

Creighton TE: *Protein Structure: A Practical Approach.* Oxford, 1990.

Darby NJ, Creighton TE: *Protein Structure.* IRL Press, 1993.

Dunn MJ: *Gel Electrophoresis of Proteins.* Wright, 1986.

Georgopoulos C, Welch WJ: Role of major heat-shock proteins as molecular chaperones. Ann Rev Cell Biol 1993;9:601.

Gierasch LM, King J: *Protein Folding.* American Association for the Advancement of Science, 1990.

Hames BD, Rickwood D: *Gel Electrophoresis of Proteins: A Practical Approach.* IRL Press, 1990.

Hartl FU, Martin J, Neupert W: Protein folding in the cell: The role of molecular chaperones Hsp70 and Hsp60. Annu Rev Biophys Biomol Struct 1992;21:293.

Hendrick JP, Hartl FU: Molecular chaperone functions of heat-shock proteins. Annu Rev Biochem 1993;62:349.

Landry SJ, Giersach LM: Polypeptide interactions with molecular chaperones and their relationship to in vivo protein folding. Annu Rev Biophys Biomol Struct 1994;32:645.

Matthews CR: Pathways of protein folding. Annu Rev Biochem 1993;62:653.

Prockop DJ, Kivirikko KI: Collagens: Molecular biology, diseases, and potentials for therapy. Annu Rev Biochem 1995;64:403.

Richardson JS, Richardson DC: Principles and patterns of protein conformation. In: *Prediction of Protein Structure and the Principles of Protein Conformation.* Fasman GD (editor). Plenum Press, 1989.

Rose GD, Gierasch LM, Smith JA: Turns in proteins and peptides. Adv Protein Chem 1985;37:1

Schmid FX: Prolyl isomerase: Enzymatic catalysis of slow protein-folding reactions. Annu Rev Biophys Biomol Struct 1993;22:123.

Scholz JM, Baldwin RL: The mechanism of α-helix formation by peptides. Annu Rev Biophys Biomol Struct 1992;21:95.

Scopes RK: *Protein Purification: Principles and Practices.* Springer, 1993.

Segrest JP et al: The amphipathic α-helix: A multifunctional structural motif in plasma lipoproteins. Adv Protein Chem 45;1995:303.

Stein RL: Mechanism of enzymatic and nonenzymatic prolyl *cis-trans* isomerization. Adv Protein Chem 1993;44:1.

Wagner G, Hyberts SG, Havel TF: NMR structure determination and solution: A critique and comparison with x-ray crystallography. Annu Rev Biophys Biomol Struct 1992;21:167.

Weber PC: Physical principles of protein crystallization. Adv Protein Chem 1991;41:1.

Proteins: Myoglobin & Hemoglobin 7

Victor W. Rodwell, PhD

INTRODUCTION

As an example of protein structure-function relationships, this chapter discusses myoglobin and hemoglobin, proteins that are of significance both in their own right and for the insights they provide into the ways in which the structures of proteins conform to, or dictate, their biologic functions.

BIOMEDICAL IMPORTANCE

Heme proteins function in oxygen binding, oxygen transport, electron transport, and photosynthesis. Detailed study of hemoglobin and myoglobin illustrates structural themes common to many globular proteins. In a sense, their greatest medical significance is that this knowledge eloquently illustrates protein structure-function relationships. In addition, it provides insight into the molecular basis of genetic diseases such as sickle cell disease (a result of altered surface properties of the hemoglobin β-subunit), and the thalassemias (chronic, familial hemolytic diseases characterized by defective synthesis of hemoglobin). Cyanide and carbon monoxide kill because they disrupt the physiologic function of the heme proteins cytochrome oxidase and hemoglobin, respectively. Finally, stabilization of the quaternary structure of deoxyhemoglobin by 2,3-bisphosphoglycerate (BPG) is central to understanding the mechanisms of high-altitude sickness and of adaptation to high altitudes.

HEME AND FERROUS IRON CONFER THE ABILITY TO STORE AND TO TRANSPORT OXYGEN

Myoglobin and hemoglobin possess as their prosthetic group **heme,** a **cyclic tetrapyrrole.** Heme's extensive network of conjugated double bonds absorbs light at the low end of the visible spectrum, coloring it deep red. Tetrapyrroles consist of four molecules of pyrrole (Figure 7–1) linked in a planar ring by four α-methylene bridges. The β substituents determine whether a tetrapyrrole is heme or a related compound. In heme, these are methyl (M), vinyl (V),

and propionate (Pr) groups arranged in the order M, V, M, V, M, Pr, Pr, M (Figure 7–2). One atom of ferrous iron (Fe^{2+}) is at the center of this planar ring. Other proteins with tetrapyrrole prosthetic groups (and their associated metal ions) include cytochromes (Fe^{2+} and Fe^{3+}), the enzymes catalase and tryptophan pyrrolase, and chlorophyll (Mg^{2+}). In cytochromes, oxidation and reduction of the iron atom are essential to their biologic function. By contrast, oxidation of the Fe^{2+} of myoglobin or hemoglobin destroys biologic activity.

Oxymyoglobin Stores Oxygen

Myoglobin of red muscle tissue stores oxygen. Under conditions of oxygen deprivation (eg, severe exercise), this oxygen is released for use by muscle mitochondria for oxygen-dependent synthesis of ATP.

With Two Exceptions, Polar Residues Are on the Surface of Myoglobin

Myoglobin, a single polypeptide chain of molecular weight 17,000, is unexceptional with respect to its 153 aminoacyl residues. Clear differences, however, are apparent in their spatial distribution. The surface is polar and the interior nonpolar, a pattern characteristic of globular proteins. Residues with both polar and nonpolar regions (eg, Thr, Trp, and Tyr) orient their nonpolar regions inward. Apart from two His residues that function in oxygen binding, the interior of myoglobin contains only nonpolar residues (eg, Leu, Val, Phe, Met).

Myoglobin Is Rich in α-Helix

Myoglobin is a compact, roughly spherical molecule measuring 4.5 × 3.5 × 2.5 nm (Figure 7–3). Its conformation is atypical, however. To facilitate reference to particular regions of secondary and tertiary structure of a polypeptide, each α-helix, β-sheet, or loop is numbered or lettered, starting from the amino terminal of the polypeptide. Approximately 75% of the residues are present in eight right-handed α-helices from seven to 20 residues in length. Starting at the amino terminal, these are termed helices A through H. Interhelical regions are identified by the

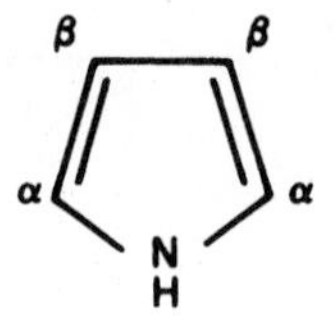

Figure 7–1. Pyrrole. The α carbons are linked by methylene bridges in a tetrapyrrole. The β carbons bear the substituents of a specific tetrapyrrole such as heme.

letters of the two helical regions they connect. Individual residues are designated by a letter for the helix in which they reside and a number that indicates their distance from the amino terminal of that helix. For example, "His F8" refers to the eighth residue in the F helix and identifies it as His. Residues far distant in a primary structural sense (eg, in different helices) may nevertheless be spatially close together, eg, the F8 (proximal) and E7 (distal) His residues (Figure 7–3).

The secondary and tertiary structure of myoglobin in solution closely resembles that of crystalline myoglobin. They exhibit virtually identical absorption spectra. Crystalline myoglobin binds oxygen, and the amount of α-helix in solution (estimated by optical rotatory dispersion and circular dichroism) closely approximates that revealed by x-ray crystallographic analysis.

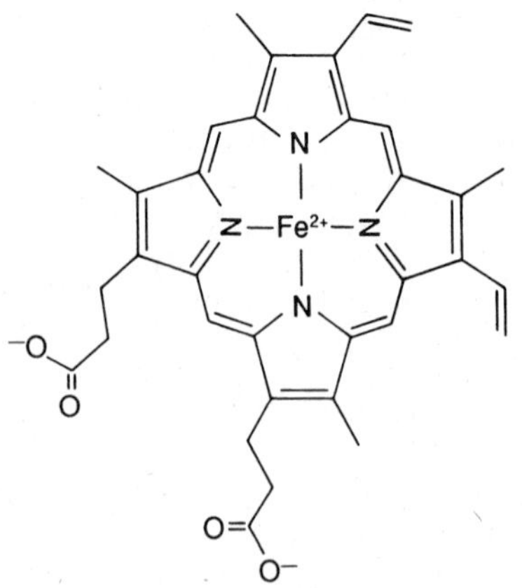

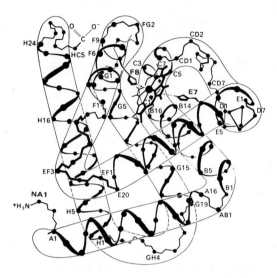

Figure 7–3. A model of myoglobin at low resolution. Only the α-carbon atoms are shown. The α-helical regions are named A through H. (Based on Dickerson RE in: *The Proteins,* 2nd ed. Vol. 2. Neurath H [editor]. Academic Press, 1964. Reproduced with permission.)

In the Presence of Heme, the Primary Structure of Apomyoglobin Dictates Correct Protein Folding

When apomyoglobin (myoglobin with the heme removed) is prepared by lowering the pH to 3.5, its α-helical content decreases dramatically. Subsequent addition of urea erases all α-helical content. If the urea is then removed by dialysis and heme is added, full α-helical content is restored, and addition of Fe^{2+} restores full biologic (oxygen-binding) activity. The primary structural information implicit in apomyoglobin therefore can, in the presence of heme, specify folding of the protein to its native, biologically active conformation. As discussed in Chapter 6, this important concept extends to other proteins: **The primary structure of a protein dictates its secondary and tertiary conformation.**

Histidines F8 and E7 Perform Unique Roles in Oxygen Binding

The heme of myoglobin, which resides in a crevice between helices E and F (Figure 7–3), is oriented with its polar propionate groups on the surface. The remainder projects into the interior of the myoglobin molecule, where, with the exception of His E7 and His F8, the surrounding residues are nonpolar. The fifth coordination position of the iron atom is linked to a ring nitrogen of the **proximal histidine,** His F8 (Figure 7–4). While not linked to the sixth coordination position of the iron, the **distal histidine**

Figure 7–2. Heme. The pyrrole rings and methylene bridge carbons are coplanar, and the iron atom (Fe^{2+}) resides in almost the same plane. The fifth and sixth coordination positions of Fe^{2+} are directed perpendicular to, and directly above and below, the plane of the heme ring. Observe the nature of the substituent groups on the β carbons of the pyrrole rings, the central iron atom, and the location of the polar side of the heme ring (at about 7 o'clock) that faces the surface of the myoglobin molecule.

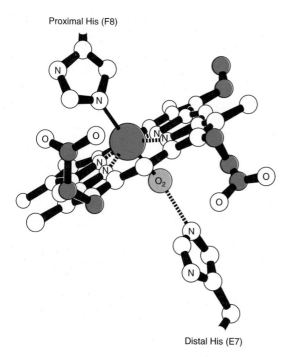

Proximal His (F8)

Distal His (E7)

Figure 7–4. Addition of oxygen to heme iron in oxygenation. Shown also are the imidazole side chains of the two important histidine residues of globin that attach to the heme iron. (Reproduced, with permission, from Harper HA et al: *Physiologische Chemie.* Springer, 1975.)

(His E7) lies on the side of the heme ring across from His F8 (Figure 7–4).

The Iron Moves Toward the Plane of the Heme When Oxygen Is Bound

In unoxygenated myoglobin, the heme iron resides about 0.03 nm (0.3 Å) outside the plane of the ring in the direction of His F8. In oxygenated myoglobin, an oxygen molecule occupies the sixth coordination position of the iron atom, which then lies only about 0.01 nm (0.1 Å) outside the plane of the heme. Oxygenation of myoglobin is therefore accompanied by movement of the iron atom, and consequently movement of His F8 and residues covalently linked to His F8, toward the plane of the ring. This motion brings about a new conformation of portions of the protein.

Apomyoglobin Provides a Hindered Environment for Heme Iron

When O_2 binds to myoglobin, the bond between one oxygen and the Fe^{2+} is perpendicular to the plane

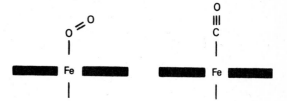

Figure 7–5. Preferred angles for bonding of oxygen and of carbon monoxide to the iron atom of heme (solid line).

of the heme ring. The second O_2 binds at an angle of 121 degrees to the plane of the heme and oriented away from the distal histidine (Figure 7–5).

Carbon monoxide (CO) binds to isolated heme about 25,000 times more strongly than does oxygen. The atmosphere contains traces of CO, and normal catabolism of heme itself forms small quantities of CO. Why then does not CO (rather than O_2) occupy the sixth coordination position of the heme iron of myoglobin? The answer lies in the **hindered environment** of heme in myoglobin. The preferred orientation for CO bound to heme iron is with all three atoms (Fe, C, O) perpendicular to the heme ring (Figure 7–5). While this orientation is possible for isolated heme, in myoglobin the distal histidine sterically hinders binding of CO at this angle (Figure 7–6). This forces CO to bind in a less favored configuration and reduces the strength of the heme-CO bond by over two orders of magnitude, to about 200 times that of the heme-O_2 bond. A small portion (about 1%) of myoglobin nevertheless normally is present in the form of myoglobin-CO.

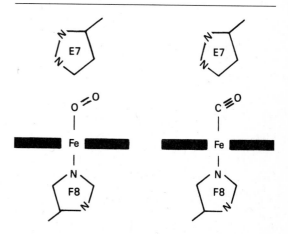

Figure 7–6. Angles for bonding of oxygen and carbon monoxide to the heme iron of myoglobin. The distal E7 histidine hinders bonding of CO at the preferred (180-degree) angle to the plane of the heme ring.

THE OXYGEN DISSOCIATION CURVES FOR MYOGLOBIN AND HEMOGLOBIN SUIT THEIR PHYSIOLOGIC ROLES

Why is myoglobin unsuitable as an oxygen transport protein but effective for oxygen storage? The quantity of oxygen bound to myoglobin (expressed as "percent saturation") depends upon the oxygen concentration (expressed as PO_2, the partial pressure of oxygen) in the immediate environment of the heme iron. The relationship between PO_2 and the quantity of oxygen bound may be expressed graphically as an oxygen saturation (oxygen dissociation) curve. For myoglobin, the shape of the oxygen adsorption isotherm is hyperbolic (Figure 7–7). Since PO_2 in the lung capillary bed is 100 mm Hg, myoglobin could effectively be loaded up with oxygen in the lungs. However, the PO_2 of venous blood is 40 mm Hg and that of active muscle about 20 mm Hg. Since myoglobin cannot deliver a large fraction of its bound oxygen even at 20 mm Hg, it cannot serve as an effective vehicle for delivery of oxygen from lungs to peripheral tissues. However, the oxygen deprivation that accompanies severe physical exercise can lower the PO_2 of muscle tissue to as little as 5 mm Hg. At 5 mm Hg, myoglobin readily releases its bound oxygen for oxidative synthesis of ATP by muscle mitochondria.

HEMOGLOBIN TRANSPORTS O_2, CO_2, AND PROTONS

Hemoglobins of vertebrate erythrocytes perform two major transport functions: (1) transport of O_2 from the respiratory organ to peripheral tissues and (2) transport of CO_2 and protons from peripheral tissues to the respiratory organ for subsequent excretion. While the comparative biochemistry of vertebrate hemoglobins provides fascinating insights, we shall here be concerned solely with human hemoglobins.

THE ALLOSTERIC PROPERTIES OF HEMOGLOBINS RESULT FROM THEIR QUATERNARY STRUCTURES

The properties of individual hemoglobins are consequences of their quaternary as well as of their secondary and tertiary structure. The quaternary structure of hemoglobin confers striking additional properties (absent from myoglobin) that adapt it to its unique biologic roles and permit precise regulation of its properties. The **allosteric** (Gk *allos* "other," *steros* "space") properties of hemoglobin provide, in addition, a model for understanding other allosteric proteins.

Unlike Myoglobin, Hemoglobin Is Tetrameric

Hemoglobins are tetrameric proteins composed of pairs of two different polypeptides (termed α, β, γ, δ, S, etc). While similar in overall length, the α (141-residue) and β (146-residue) polypeptides of hemoglobin A (HbA) are encoded by different genes and have different primary structures. By contrast, the primary structures of β, γ, and δ chains of human hemoglobins have highly conserved primary structures. The tetrameric structures of common hemoglobins follow: HbA (normal adult hemoglobin) = $\alpha_2\beta_2$, HbF (fetal hemoglobin) = $\alpha_2\gamma_2$, HbS (sickle cell hemoglobin) = α_2S_2, and HbA$_2$ (a minor adult hemoglobin) = $\alpha_2\delta_2$.

Myoglobin and the β Subunits of Hemoglobin Have Almost Identical Secondary and Tertiary Structures

Despite differences in the type and number of amino acids present, myoglobin and the β polypeptide of HbA exhibit almost identical secondary and tertiary structures. This similarity, which extends to the location of the heme and the eight helical regions, results in part from the presence of amino acids of similar properties at equivalent points in their primary structures. The β polypeptide also closely resembles myoglobin despite the presence of seven rather than eight helical regions. As for myoglobin, hydrophobic residues are internal and (again with the exception of two His residues per subunit) hydrophilic residues are surface features of both the α and β subunits of HbA.

Oxygenation of Hemoglobin Is Accompanied by Conformational Changes in the Apoprotein

Hemoglobins bind four oxygen molecules per tetramer (one per subunit heme), and the oxygen sat-

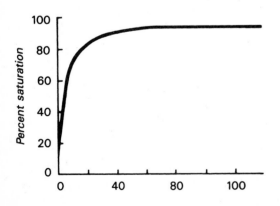

Figure 7–7. Oxygen-binding curve for myoglobin. Note the relationship between percent saturation and the partial pressures representative of lungs (100 mm Hg), tissues (20 mm Hg), and working muscle (5 mm Hg).

uration curves are sigmoidal (Figure 7–8). The facility with which O_2 binds to hemoglobin depends on whether other O_2 molecules are present on the same tetramer. If O_2 is already present, binding of subsequent O_2 molecules occurs more readily. Hemoglobin thus exhibits **cooperative binding kinetics,** a property that permits it to bind a maximal quantity of O_2 at the respiratory organ and to deliver a maximal quantity of O_2 at the Po_2 of the peripheral tissues. Compare, for example, these quantities at the Po_2 of human lungs (100 mm Hg) and tissues (20 mm Hg) with myoglobin (Figure 7–8).

P_{50} Compares the Relative Affinities of Different Hemoglobins for Oxygen

P_{50} is the partial pressure of oxygen that half-saturates a hemoglobin. Depending on the organism, P_{50} can vary widely but in all instances exceeds the peripheral tissue Po_2 in that organism. Human fetal hemoglobin (HbF) provides an illustrative example. For HbA, P_{50} = 26 mm Hg; for HbF, P_{50} = 20 mm Hg. This difference permits HbF to extract oxygen from the HbA of placental blood. Postpartum, however, HbF is unsuitable, since its high affinity for O_2 dictates that it can deliver less O_2 to the tissues.

The human fetus initially synthesizes not α and β but ζ and ε chains. By the end of the first trimester, α chains have replaced ζ chains and γ chains have replaced ε chains. Hemoglobin F, the hemoglobin of late fetal life, thus has the composition $α_2γ_2$. Beta

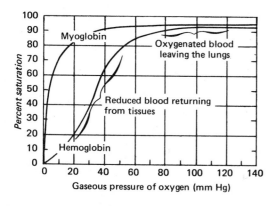

Figure 7–8. Oxygen-binding curves of both hemoglobin and myoglobin. Arterial oxygen tension is about 100 mm Hg; mixed venous oxygen tension is about 40 mm Hg; capillary (active muscle) oxygen tension is about 20 mm Hg; and the minimum oxygen tension required for cytochromes is about 5 mm Hg. Association of chains into a tetrameric structure (hemoglobin) results in much greater oxygen delivery than would be possible with single chains. (Modified, with permission, from Stanbury JB, Wyngaarden JB, Fredrickson DS [editors]: *The Metabolic Basis of Inherited Disease,* 4th ed. McGraw-Hill, 1978.)

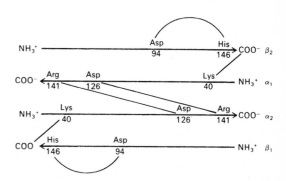

Figure 7–9. Salt links between and within subunits in deoxyhemoglobin. These noncovalent, electrostatic interactions are disrupted on oxygenation. (Slightly modified and reproduced, with permission, from Stryer L: *Biochemistry,* 2nd ed. Freeman, 1981.)

chains, whose synthesis starts in the third trimester, do not completely replace γ chains until some weeks postpartum.

Large Conformational Changes Accompany Oxygenation of Hemoglobin

Binding of O_2 is accompanied by the rupture of salt bonds between the carboxyl terminals of all four subunits (Figure 7–9). Subsequent O_2 binding is facilitated, since it involves rupture of fewer salt bonds. These changes also profoundly alter hemoglobin's secondary, tertiary, and quaternary structures. One pair of α/β subunits rotates with respect to the other α/β pair, compacting the tetramer and increasing the affinity of the hemes for O_2 (Figures 7–10 and 7–11).

The quaternary structure of partially oxygenated

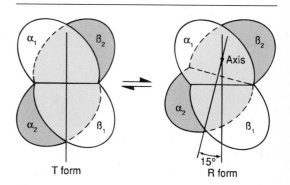

Figure 7–10. During the transition of the T form to the R form of hemoglobin, one pair of rigid subunits ($α_2/β_2$) rotates through 15 degrees relative to the other rigid pair ($α_1/β_1$). The axis of rotation is eccentric, and the $α_2/β_2$ pair also shifts toward the axis somewhat. In the diagram, the $α_1/β_1$ pair is unshaded and held fixed, while the shaded $α_2/β_2$ pair rotates and shifts.

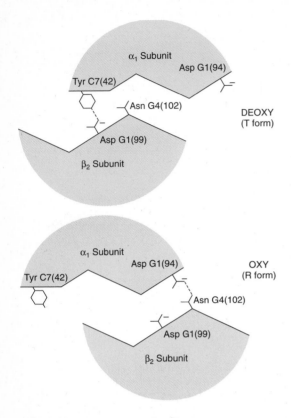

DEOXY
(T form)

OXY
(R form)

Figure 7–11. Changes at the α_1/β_2 contact on oxygenation. The contact "clicks" from one dovetailing area to another, involving a switch from one hydrogen bond to a second. The other bonds are nonpolar. (Reproduced, with permission, from Perutz MF: Molecular pathology of human hemoglobin: Stereochemical interpretation of abnormal oxygen affinities. Nature 1971;232:408.)

hemoglobin is described as the **T (taut) state** and that of oxygenated hemoglobin (HbO$_2$) as the **R (relaxed) state** (Figure 7–12). R and T are also used to characterize the quaternary structures of allosteric enzymes, where the T state has the lower substrate affinity.

Oxygenation of Hemoglobin Is Accompanied by Conformational Changes Near the Heme Group

On oxygenation, the iron atoms of deoxyhemoglobin (which lie about 0.06 nm beyond the plane of the heme ring) move into the plane of the heme ring (Figure 7–13). This motion is transmitted to the proximal (F8) histidine, which moves toward the plane of the ring, and to residues attached to His F8.

After Releasing O$_2$ at the Tissues, Hemoglobin Transports CO$_2$ and Protons to the Lungs

In addition to transporting oxygen from the lungs to peripheral tissues, hemoglobin facilitates the transport of CO$_2$ from tissues to the lungs for exhalation. Hemoglobin can bind CO$_2$ directly when oxygen is released, and about 15% of the CO$_2$ carried in blood is carried directly on the hemoglobin molecule. CO$_2$ reacts with the amino terminal α-amino groups of hemoglobin, forming a carbamate and releasing protons that contribute to the Bohr effect.

$$CO_2 + Hb-NH_3^+ = 2H^+ + Hb-\overset{H}{N}-COO^-$$

Conversion of the amino terminal from a positive to a negative charge favors salt bridge formation be-

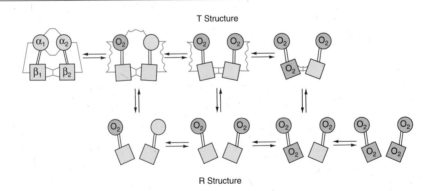

Figure 7–12. Transition from the T structure to the R structure becomes more probable as each heme group of a hemoglobin tetramer is oxygenated. In this model, salt bridges (thin lines) linking the subunits in the T structure break progressively as oxygen is added, and even those salt bridges that have not yet ruptured are progressively weakened (wavy lines). The transition from T to R does not take place after a fixed number of oxygen molecules have been bound, but it becomes more probable with each successive oxygen bound. The transition between the two structures is influenced by several factors, including protons, carbon dioxide, chloride, and BPG. The higher their concentration, the more oxygen must be bound to trigger the transition. Fully oxygenated molecules in the T structure and fully deoxygenated molecules in the R structure are not shown because they are too unstable to exist in significant numbers. (Modified and redrawn, with permission, from Perutz MF: Hemoglobin structure and respiratory transport. Sci Am [Dec] 1978;239:92.)

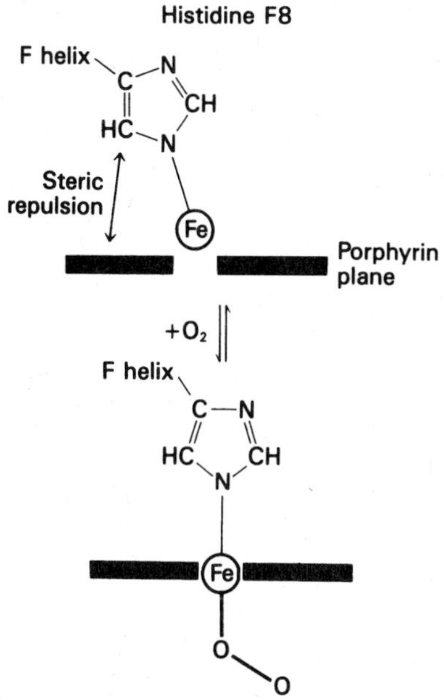

Histidine F8

F helix

Steric repulsion

Porphyrin plane

$+O_2$

F helix

Figure 7–13. The iron atom moves into the plane of the heme on oxygenation. Histidine F8 and its associated residues are pulled along with the iron atom. (Slightly modified and reproduced, with permission, from Stryer L: *Biochemistry,* 2nd ed. Freeman, 1981.)

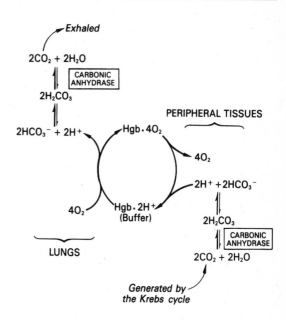

Figure 7–15. The Bohr effect. The carbon dioxide generated in peripheral tissues combines with water to form carbonic acid, which dissociates into protons and bicarbonate ions. The deoxygenated hemoglobin acts as a buffer by binding protons and delivering them to the lungs. In the lungs, the binding of oxygen by hemoglobin releases protons that combine with bicarbonate ion, forming carbonic acid, which when dehydrated by carbonic anhydrase becomes carbon dioxide which is then exhaled from the lungs.

tween the α and β chains, a situation characteristic of the deoxy state. At the lungs, oxygenation of hemoglobin is accompanied by expulsion and subsequent expiration of CO_2. As CO_2 is absorbed in blood, the carbonic anhydrase in erythrocytes catalyzes the formation of carbonic acid (Figure 7–14). Carbonic acid rapidly dissociates into bicarbonate and a proton. To avoid increasing the acidity of blood, a buffering system must absorb these excess protons. Hemoglobin binds two protons for every four oxygen molecules released and thus contributes significantly to the buffering capacity of blood (Figure 7–15). In the lungs, the process is reversed—ie, as oxygen binds to deoxygenated hemoglobin, protons are released and combine with bicarbonate, forming carbonic acid. With the aid of carbonic anhydrase, the carbonic acid forms CO_2, which is exhaled. Thus, the binding of oxygen forces the exhalation of CO_2. This reversible phenomenon is called

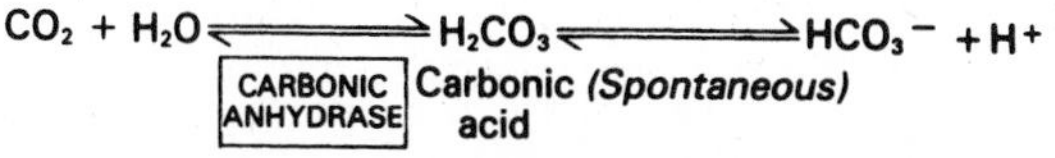

$$CO_2 + H_2O \rightleftharpoons H_2CO_3 \rightleftharpoons HCO_3^- + H^+$$

CARBONIC ANHYDRASE   Carbonic *(Spontaneous)* acid

Figure 7–14. The formation of carbonic acid, catalyzed by erythrocyte carbonic anhydrase, and the dissociation of carbonic acid to bicarbonate ion and a proton.

the **Bohr effect.** This effect is accompanied by a shift in the oxygenation curve to the right, ie, the hemoglobin is less saturated at a given partial pressure of oxygen. The Bohr effect is a property of tetrameric hemoglobin and is dependent upon its heme-heme interaction or cooperative effects. Myoglobin does not exhibit a Bohr effect.

The Protons Responsible for the Bohr Effect Arise by Rupture of Salt Bonds During Binding of O_2

The protons responsible for the Bohr effect are generated by the rupture of salt bridges during the binding of oxygen to the T structure. The protons, released from the nitrogen atoms of β-chain residue HC3 (His 146), drive bicarbonate toward carbonic acid, which is released as CO_2 in alveolar blood (Figure 7–15). Upon the release of oxygen, the T structure and its salt bridges re-form, and protons bind to the β-chain HC3 residues. Thus, protons favor the formation of salt bridges by protonating the terminal His residue of the β subunits. Re-formation of the salt bridges facilitates the release of oxygen from oxygenated (R form) hemoglobin. Overall, an increase in protons promotes oxygen release, while an increase in P_{O_2} promotes proton release. The former is represented in an oxygen dissociation curve by a rightward shift in the dissociation curve upon increasing the concentration of hydrogen ions (protons).

2,3-Bisphosphoglycerate (BPG) Stabilizes the T Structure of Hemoglobin

In peripheral tissues, an oxygen shortage causes an increased accumulation of 2,3-bisphosphoglycerate (BPG) (Figure 7–16). This compound is formed from the glycolytic intermediate 1,3-bisphosphoglycerate. One molecule of BPG is bound per hemoglobin tetramer in the central cavity formed by all four subunits. This cavity is of sufficient size for BPG only when the space between the H helices of the β chains is wide enough, ie, when hemoglobin is in the T form. BPG is bound by salt bridges between its oxygen atoms and both β chains via their amino terminal amino groups (Val NA1), and by Lys EF6 and His H21 (Figure 7–17). Thus, BPG stabilizes the T or deoxygenated form of hemoglobin by crosslinking the β chains and forming additional salt bridges that must be broken prior to formation of the R form.

BPG binds more weakly to fetal hemoglobin than to adult hemoglobin because the H21 residue of the γ chain of HbF is Ser rather than His and cannot form a salt bridge with BPG. Therefore, BPG has a less profound effect on the stabilization of the T form of HbF and is responsible for HbF appearing to have a higher affinity for oxygen than does HbA.

The trigger for the R to T transition of hemoglobin is the movement of the iron in and out of the plane of the porphyrin ring. Both steric and electrostatic factors mediate this trigger. Thus, a minimal change in the position of Fe^{2+} relative to the porphyrin ring induces significant switching of the conformations of hemoglobin and crucially affects its biologic function in response to an environmental signal.

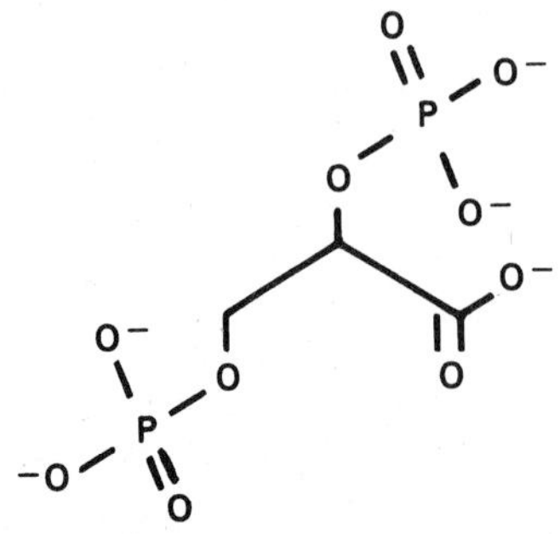

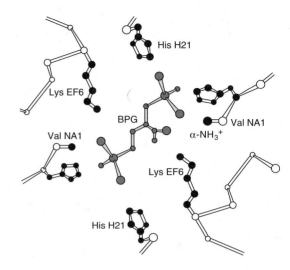

Figure 7–17. Mode of binding of 2,3-bisphosphoglycerate to human deoxyhemoglobin. BPG interacts with three positively charged groups on each β chain. (Based on Arnone A: X-ray diffraction study of binding of 2,3-diphosphoglycerate to human deoxyhemoglobin. Nature 1972; 237:146. Reproduced with permission.)

SEVERAL HUNDRED MUTANT HUMAN HEMOGLOBINS HAVE BEEN IDENTIFIED

Mutations in the genes that code for the α or β chains potentially can affect the biologic function of hemoglobin. Of the several hundred known mutant human hemoglobins (most of them extremely rare and benign) several in which biologic function is altered are described below. When biologic function is altered owing to a mutation in hemoglobin, the condition is known as a **hemoglobinopathy.**

In Hemoglobin M, Tyr Replaces His F8

Heme iron is stabilized in the Fe^{3+} state, since it forms a tight ionic complex with the phenolate anion of Tyr. In methemoglobinemia, the heme iron is ferric rather than ferrous. Fe^{3+} can be acquired (eg, by oxidation of Fe^{2+} to Fe^{3+} by agents such as sulfonamides), hereditary (owing to the presence of HbM), or a consequence of decreased activity of methemoglobin reductase, an enzyme that reduces the Fe^{3+} of MetHb to Fe^{2+}. Since methemoglobin does not bind O_2, it cannot participate in O_2 transport. In α-chain hemoglobin M variants, the R–T equilibrium favors the T form. Oxygen affinity is reduced, and a Bohr effect is absent. β-Chain hemoglobin M variants exhibit R–T switching, and a Bohr effect is therefore present.

Mutations (eg, hemoglobin Chesapeake) that favor the R form exhibit increased oxygen affinity. They

Figure 7–16. Structure of 2,3-bisphosphoglycerate (BPG).

therefore fail to deliver adequate oxygen to peripheral tissues. The resulting tissue hypoxia leads to **polycythemia** (increased concentration of erythrocytes).

In Hemoglobin S, a Valyl Residue Replaces Glutamate β-6

In hemoglobin S, Val replaces Glu A2(6)β, ie, residue 6 of the β chain located on the surface of the hemoglobin, exposed to water. This substitution replaces the polar Glu residue with a nonpolar one and generates a **"sticky patch"** on the surface of the β chain. The sticky patch is present on oxygenated and deoxygenated hemoglobin S but not on hemoglobin A. On the surface of deoxygenated hemoglobin, there exists a complement to the sticky patch, but in oxygenated hemoglobin this complementary site is masked (Figure 7–18).

Deoxyhemoglobin S Can Form Fibers That Distort Erythrocytes

When hemoglobin S is deoxygenated, the sticky patch can bind to the complementary patch on another deoxygenated HbS molecule. This binding causes a polymerization of deoxyhemoglobin S, forming long fibrous precipitates. These extend throughout the erythrocyte and mechanically distort it, causing lysis and multiple secondary clinical effects. Thus, if HbS can be maintained in an oxygenated state—or if the concentration of deoxygenated HbS can be minimized—formation of these polymers will not occur and "sickling" can be prevented. Clearly, it is the T form of HbS that polymerizes.

Although deoxyhemoglobin A contains the receptor sites for the sticky patch present on oxygenated or deoxygenated HbS (Figure 7–18), the binding of sticky hemoglobin S to deoxyhemoglobin A cannot extend the polymer, since the latter does not have a sticky patch to promote binding to still another hemoglobin molecule. Therefore, the binding of deoxyhemoglobin A to either the R or the T form of hemoglobin S will terminate polymerization.

The polymer forms a twisted helical fiber whose cross-section contains 14 HbS molecules (Figure 7–19). These tubular fibers distort the erythrocyte so that they take on the shape of a sickle (Figure 7–20) and are vulnerable to lysis as they penetrate the interstices of the splenic sinusoids.

BIOMEDICAL IMPLICATIONS

(1) Myoglobinuria: Following massive crush injury, myoglobin released from ruptured muscle fibers appears in the urine, coloring it dark red. While myoglobin can be detected in plasma following a myocardial infarct, assay of serum enzymes (see Chapter 8) provides a more sensitive index of myocardial injury.

(2) Anemias: Common anemias (reductions in the amount of red blood cells or of hemoglobin in the blood) result from impaired synthesis of hemoglobin (eg, in iron deficiency; Chapter 59) or impaired production of erythrocytes (eg, in folic acid or vitamin B_{12} deficiency; Chapter 52). Diagnosis of anemias begins with spectral measurement of blood hemoglobin levels. **Thalassemias** and the use of DNA probes for their diagnosis are considered in Chapters 8 and 42.

(3) Hemoglobinopathies: While mutation of certain critical residues of hemoglobin (eg, histidines E7 or F8) has serious consequences, mutation of many surface residues far removed from the heme-binding site may present no clinical abnormalities.

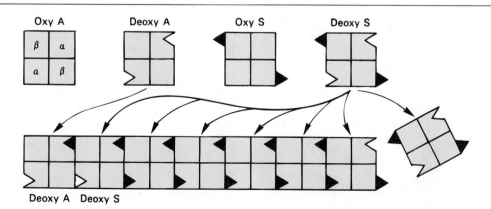

Figure 7–18. Representation of the sticky patch (▲) on hemoglobin S and its "receptor" (△) on deoxyhemoglobin A and deoxyhemoglobin S. The complementary surfaces allow deoxyhemoglobin S to polymerize into a fibrous structure, but the presence of deoxyhemoglobin A will terminate the polymerization by failing to provide sticky patches. (Modified and reproduced, with permission, from Stryer L: *Biochemistry,* 2nd ed. Freeman, 1981.)

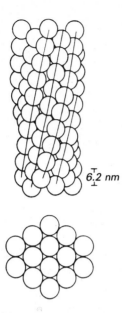

Figure 7–19. Representation of the twisted helical structure of a fiber of aggregated deoxyhemoglobin S. In this model, spheres represent individual molecules of HbS. (Reproduced, with permission, from Maugh T II: A new understanding of sickle cell emerges. Science 1981;211: 265. Copyright © 1981 by the American Association for the Advancement of Science.)

One notable exception is sickle cell anemia, in which all the signs and symptoms (eg, sickle cell crises, thromboses) emanate from the mutation of a single polar residue to a nonpolar residue.

(4) Glycosylated hemoglobin (HbA$_{1c}$): Hemoglobin is nonenzymatically glycosylated when blood glucose enters the erythrocytes and its anomeric hydroxyl derivatizes amino groups present on lysyl residues and at amino terminals. HbA$_{1c}$ may be separated from HbA by ion exchange chromatography or electrophoresis. The fraction of hemoglobin glycosylated, normally about 5%, is proportionate to blood glucose concentration. Measurement of HbA$_{1c}$ thus provides information useful for the management of diabetes mellitus. Since the mean half-life of an erythrocyte is 60 days, the HbA$_{1c}$ level reflects the average blood glucose concentration over the preceding 6–8 weeks. An elevated HbA$_{1c}$, which indicates poor control of blood glucose level, can guide the physician in selection of appropriate treatment (eg, more rigorous control of diet or increased insulin dosage).

Thalassemias Result From Reduced Synthesis of the α or β Polypeptides

In the thalassemias, the synthesis of either the α polypeptide chains (α-thalassemias) or β polypeptide chains (β-thalassemias) of hemoglobin is reduced. This results in anemia, which may be severe. (See Chapter 42.)

SUMMARY

The compact, globular proteins myoglobin (Mb) and hemoglobin (Hb) function in O$_2$ storage and O$_2$ transport, respectively. Mb is monomeric; Hb is a tetramer of two subunit types (α$_2$β$_2$ in adult hemoglobin, HbA). Mb and the β subunit of Hb, rich in α-helix, share a virtually identical secondary and tertiary (but not primary) structure. Amino acid residues

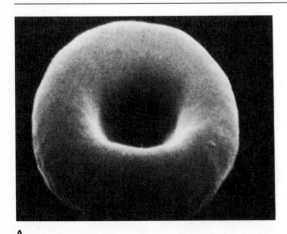

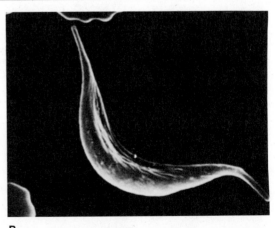

A **B**

Figure 7–20. Scanning electron micrograph of normal **(A)** and sickle **(B)** red blood cells. The change of the β-globin molecule that causes this structural alteration results from a single base mutation in DNA, T to A, which results in the substitution of valine for glutamate in the β-globin molecule.

of their 6 (A to H) helical regions are named (for example) "His F8"—the histidine that is the eighth residue in the F, or sixth, helix.

The heme prosthetic group of Mb and Hb is an essentially planar, slightly puckered, cyclic tetrapyrrole with a central Fe^{2+}. When heme is added to denatured apomyoglobin (Mb less heme), it refolds into an active conformation. The primary structural information implicit in apoMb (and by extension, in other proteins) thus suffices to dictate secondary and tertiary structure.

The heme Fe^{2+} is linked to all four nitrogen atoms of the heme and to two additional ligands located above and below the plane of the heme. One ligand is the F8 histidine. In oxyMb and oxyHb, the sixth ligand is O_2. In carbon monoxide poisoning, the sixth ligand is CO, which, despite steric hindrance by His F7, binds far more strongly than does O_2.

Saturation curves illustrate O_2 uptake and release. The curve for Mb is hyperbolic, whereas that for Hb is sigmoidal. Sigmoid shape is critical for an O_2 carrier, which must fully load O_2 at the lungs and deliver maximum O_2 at the tissues. Relative affinities of different hemoglobins are expressed as P_{50}, the partial pressure of O_2 which half-saturates them with O_2. Hemoglobins saturate at the partial pressures of their respective respiratory organ (eg, HbF is saturated at the P_{O_2} of the placenta). The proximal (F8)

and distal (E7) His residues lie on opposite sides of the heme ring. On oxygenation, the Fe^{2+}, His F8, and associated residues move toward the heme ring. Gross conformational changes thus accompany Hb oxygenation.

Addition of O_2 to deoxyHb ruptures inter- and intrasubunit salt bonds, loosening quaternary structure and facilitating binding of additional O_2 molecules. 2,3-Bisphosphoglycerate (BPG) in the central cavity of deoxyHb forms salt bonds with the β subunits that further stabilize deoxyHb. On oxygenation, the central cavity contracts, BPG is extruded, and the quaternary structure is again loosened.

Hb also functions in CO_2 and proton transport from tissues to lungs. Release of O_2 from oxyHb (HbO$_2$) at the tissues is accompanied by uptake of protons owing to a decrease in the pK_a of a His residue.

Prominent among the hundreds of (largely benign) Hb mutants is sickle cell Hb (HbS), in which Val replaces the $\beta6$ Glu of HbA. This creates a "sticky patch" that has a complement on deoxyHb (but not on oxyHb). At low oxygen concentrations, deoxyHbS polymerizes, forms fibers, and distorts erythrocytes into sickle shapes. α- and β-thalassemias are anemias that result from reduced production of α and β subunits of HbA, respectively.

REFERENCES

Bunn HF, Forget BG: *Hemoglobin: Molecular, Genetic, and Clinical Aspects.* Saunders, 1986.

Eaton WJ, Holfrichter J: Sickle cell hemoglobin polymerization. Adv Protein Chem 1990;40:63.

Embury SH et al: Rapid prenatal diagnosis of sickle cell anemia by a new method of DNA analysis. N Engl J Med 1987;316:656.

Embury SH: The clinical pathology of sickle-cell disease. Annu Rev Med 1986;37:361.

Everse J, Vandergriff KD, Winslow RM: Hemoglobins: Part B. Biochemical and analytical methods. Methods Enzymol 231:1994.

Everse J, Vandergriff KD, Winslow RM: Hemoglobins:

Part C. Biophysical methods. Methods Enzymol 232:1994.

Friedman JM: Structure, dynamics, and reactivity in hemoglobin. Science 1985;228:1273.

Saiki RK et al: Enzymatic amplification of beta-globin genomic sequences and restriction site analysis for diagnosis of sickle-cell anemia. Science 1985;230:1350.

Schacter L et al: Altered amount and activity of superoxide dismutase in sickle cell anemia. FASEB J 1988;2:237.

Weatherall DJ et al: The hemoglobinopathies. In: *The Metabolic and Molecular Bases of Inherited Disease,* 7th ed. Scriver CR et al (editors). McGraw-Hill, 1995.

8

Enzymes: General Properties

Victor W. Rodwell, PhD

INTRODUCTION

Topics considered in this chapter include the kinds of reactions catalyzed by enzymes, the participation of coenzymes in enzyme-catalyzed group transfer reactions, and aspects of enzyme specificity. Methods for the purification, analysis, and determination of the intracellular distribution of enzymes are introduced. Isoenzymes and the diagnostic and prognostic significance of serum enzymes are then discussed, concluding with the use of restriction endonucleases in the diagnosis of genetic diseases.

BIOMEDICAL IMPORTANCE

Without enzymes, life as we know it would not be possible. As the biocatalysts that regulate the rates at which all physiologic processes take place, enzymes occupy central roles in health and disease. While in health all physiologic processes occur in an ordered, regulated manner and homeostasis is maintained, homeostasis can be profoundly disturbed in pathologic states. For example, the severe tissue injury that characterizes liver cirrhosis can profoundly impair the ability of cells to form the enzymes which catalyze a key metabolic process such as urea synthesis. The resultant inability to convert toxic ammonia to nontoxic urea is then followed by ammonia intoxication and ultimately hepatic coma. A spectrum of rare but frequently debilitating and often fatal genetic diseases provides additional dramatic examples of the drastic physiologic consequences that can follow impairment of the activity of but a single enzyme.

Following severe tissue injury (eg, cardiac or lung infarct, crushed limb) or uncontrolled cell growth (eg, prostatic carcinoma), enzymes that may be unique to specific tissues are released into the blood. Measurement of these intracellular enzymes in blood serum therefore provides physicians with invaluable diagnostic and prognostic information.

ENZYMES ARE CLASSIFIED BY REACTION TYPE AND MECHANISM

A century ago, only a few enzymes were known, most of which catalyzed the hydrolysis of covalent bonds. These enzymes were identified by adding the suffix *-ase* to the name of the substance, or substrate, which they hydrolyzed. Thus, **lipases** hydrolyzed fat (Gk *lipos*), **amylases** hydrolyzed starch (Gk *amylon*), and **proteases** hydrolyzed proteins. Although numerous vestiges of this terminology persist to the present day, it proved inadequate when several enzymes were discovered that catalyzed different reactions of the same substrate, eg, oxidation or reduction of an alcohol function of a sugar. While the suffix *-ase* remains in use, present-day enzyme names emphasize the type of reaction catalyzed. For example, dehydrogenases catalyze the removal of hydrogen, while transferases catalyze group transfer reactions. As more and more enzymes were discovered, inevitable ambiguities arose, and it often was not clear which enzyme an investigator was discussing. To remedy this deficiency, the International Union of Biochemistry (IUB) adopted a complex but unambiguous system of enzyme nomenclature based on reaction mechanism. Although its clarity and lack of ambiguity commend the IUB nomenclature system for research purposes, more ambiguous but mercifully shorter names persist in use in textbooks and in the clinical laboratory. For this reason, only a skeletal outline of the IUB system is presented below.

(1) Reactions and the enzymes that catalyze them form six classes, each having 4–13 subclasses.

(2) The enzyme name has two parts. The first names the substrate or substrates. The second, ending in *-ase,* indicates the type of reaction catalyzed.

(3) Additional information, if needed to clarify the reaction, may follow in parentheses; eg, the enzyme catalyzing L-malate + NAD^+ = pyruvate + CO_2 + NADH + H^+ is designated 1.1.1.37 L-malate:NAD^+ oxidoreductase (decarboxylating).

(4) Each enzyme has a code number (EC) that characterizes the reaction type as to class (first digit),

subclass (second digit), and subsubclass (third digit). The fourth digit is for the specific enzyme. Thus, EC 2.7.1.1 denotes class 2 (a transferase), subclass 7 (transfer of phosphate), subsubclass 1 (an alcohol is the phosphate acceptor). The final digit denotes hexokinase, or ATP:D-hexose 6-phosphotransferase, an enzyme catalyzing phosphate transfer from ATP to the hydroxyl group on carbon 6 of glucose.

MANY ENZYMES REQUIRE A COENZYME

Many enzymes that catalyze group transfer and other reactions require, in addition to their substrate, a second organic molecule known as a coenzyme, without which they are inactive. To distinguish them from metal ion activators and from enzymes themselves, coenzymes are defined as heat-stable, low-molecular-weight organic compounds required for the activity of enzymes. Most coenzymes are linked to enzymes by noncovalent forces. Those which form covalent bonds to enzymes may also be termed **prosthetic groups.**

Enzymes that require coenzymes include those which catalyze oxidoreductions, group transfer and isomerization reactions, and reactions that form covalent bonds (IUB classes 1, 2, 5, and 6). Lytic reactions, including the hydrolytic reactions catalyzed by digestive enzymes, do not require coenzymes.

Coenzymes May Be Regarded as Second Substrates

For two major reasons, it often is helpful to regard a coenzyme as a second substrate. First, the chemical changes in the coenzyme exactly counterbalance those taking place in the substrate. For example, in oxidoreduction reactions, when one molecule of substrate is oxidized, one molecule of coenzyme is reduced (Figure 8–1).

A second reason to accord equal emphasis to the coenzyme is that this aspect of the reaction may be of greater fundamental physiologic significance. For example, the importance of the ability of muscle working anaerobically to convert pyruvate to lactate does not reside in pyruvate or lactate. The reaction serves merely to oxidize the reduced coenzyme NADH to NAD^+. Without NAD^+, glycolysis cannot continue and anaerobic ATP synthesis (and hence work) ceases. Under anaerobic conditions, reduction of pyruvate to lactate reoxidizes NADH and permits synthesis of ATP. Other reactions can serve this function equally well. For example, in bacteria or yeast growing anaerobically, metabolites derived from pyruvate serve as oxidants for NADH and are themselves reduced (Table 8–1).

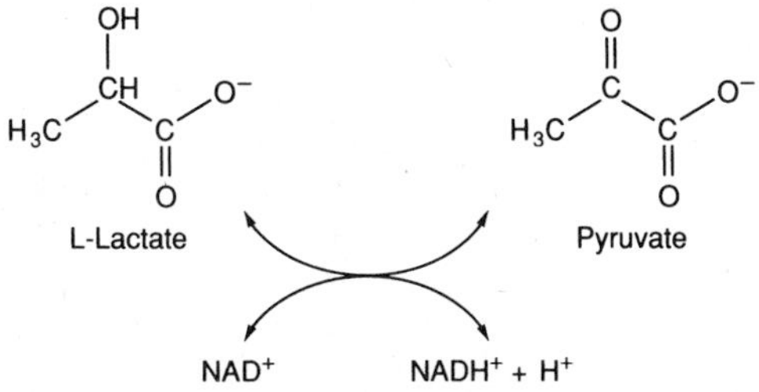

Coenzymes Function as Group Transfer Reagents

Biochemical group transfer reactions of the type

$$D—G + A = A—G + D$$

in which a functional group, G, is transferred from a donor molecule, D—G, to an acceptor molecule, A, usually involve a coenzyme either as the ultimate acceptor (eg, dehydrogenation reactions) or as an intermediate group carrier (eg, transamination reactions). The following display illustrates the latter concept.

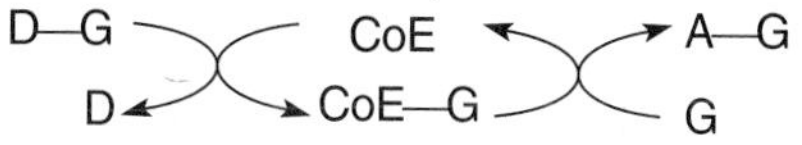

While this suggests formation of a single CoE—G complex during the course of the overall reaction, several intermediate CoE—G complexes may be involved in a particular reaction (eg, transamination).

When the group transferred is hydrogen, it is customary to represent only the left "half reaction":

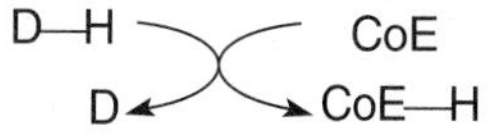

That this actually represents only a special case of general group transfer can best be appreciated in terms of the reactions that occur in intact cells (Table 8–1). These can be represented as follows:

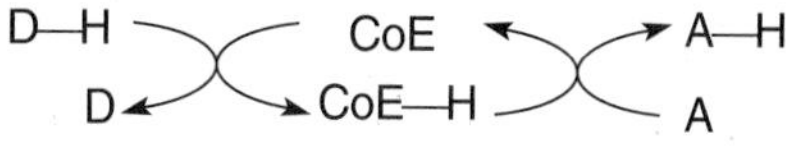

Figure 8–1. NAD$^+$ acts as a cosubstrate in the lactate dehydrogenase reaction.

Table 8–1. Mechanisms for anaerobic regeneration of NAD$^+$.

Oxidant	Reduced Product	Life Form
Pyruvate	Lactate	Muscle, lactic bacteria
Acetaldehyde	Ethanol	Yeast
Dihydroxyacetone phosphate	α-Glycerophosphate	*Escherichia coli*
Fructose	Mannitol	Heterolactic bacteria

Coenzymes Can Be Classified According to the Group Whose Transfer They Facilitate

Based on the above concept, we might classify coenzymes as follows:

For transfer of groups other than hydrogen–
  Sugar phosphates
  CoA-SH
  Thiamin pyrophosphate
  Pyridoxal phosphate
  Folate coenzymes
  Biotin
  Cobamide (B_{12}) coenzymes
  Lipoic acid
For transfer of hydrogen–
  NAD^+, $NADP^+$
  FMN, FAD
  Lipoic acid
  Coenzyme Q

Many Coenzymes Are Derivatives of B Vitamins and of Adenosine Monophosphate

B vitamins form part of the structure of many coenzymes. The B vitamins **nicotinamide, thiamin, riboflavin,** and **pantothenic acid** are essential constituents of coenzymes for biologic oxidations and reductions, and **folic acid** and **cobamide** coenzymes function in one-carbon metabolism. Many coenzymes contain adenine, ribose, and phosphate and are derivatives of adenosine monophosphate (AMP). Examples include NAD^+ and $NADP^+$ (Figure 8–2).

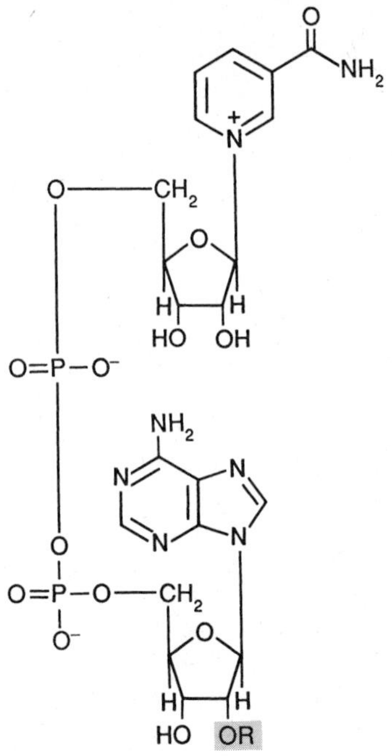

Enzymes Are Stereospecific Catalysts

Most substrates form at least three bonds with enzymes. This "three-point attachment" can confer asymmetry on an otherwise symmetric molecule. Figure 8–3 shows a substrate molecule, represented as a carbon atom bearing three different groups, about to attach at three points to an enzyme site. If the site can be approached only from one side and only complementary atoms and sites can interact (valid assumptions for actual enzymes), the molecule can bind in only one way. The reaction may be confined to the atoms bound at sites 1 and 2, even though atoms 1 and 4 are identical. By mentally turning the substrate molecule in space, note that it can attach at three points to one side of the planar site with only one orientation. Consequently, atoms 1 and 4, although identical, become distinct when the substrate is attached to the enzyme. A chemical change thus can involve atom 1 but not atom 4, or vice versa. While the active sites of enzymes are not planar surfaces such as that shown in Figure 8–3, this figure nevertheless illustrates why, for example, the enzyme-catalyzed reduction of optically inactive pyruvate forms L-lactate and not D,L-lactate.

Figure 8–2. NAD(P)$^+$. In NAD^+, R = H; in $NADP^+$, R = —OPO_3^{2-}.

ENZYMES CATALYZE A SPECIFIC REACTION OR REACTION TYPE

The ability of an enzyme to catalyze one specific reaction and essentially no others is perhaps its most significant property. Rates of metabolic processes may thus be regulated by changes in the catalytic efficiency of specific enzymes. However, most enzymes catalyze the same type of reaction (phosphate transfer, oxidation-reduction, etc) with a small number of structurally related substrates, although often at significantly lower rates. Reactions with alternative substrates tend to take place if these are present in high concentration. Which of the possible reac-

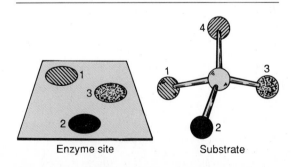

Enzyme site   Substrate

Figure 8–3. Representation of three-point attachment of a substrate to a planar active site of an enzyme.

tions will occur in living organisms depends on the relative concentration of alternative substrates in the cell and the relative affinity of the enzyme for those substrates.

Enzymes Exhibit Optical Specificity

Except for epimerases (racemases), which catalyze interconversion of optical isomers, enzymes exhibit absolute optical specificity for at least a portion of a substrate molecule. Thus, enzymes of the glycolytic and direct oxidative pathways catalyze the interconversion of D- but not L-phosphosugars. With few exceptions (eg, kidney D-amino acid oxidase), most mammalian enzymes act on the L-isomers of amino acids.

Optical specificity may extend to a portion of the substrate molecule or to its entirety. Glycosidases illustrate both extremes. These catalyze hydrolysis of glycosidic bonds between sugars and alcohols, are highly specific for the sugar portion and for the linkage (α or β), but are relatively nonspecific for the aglycone (alcohol portion).

Enzymes Are Specific for the Type of Reaction Catalyzed

Lytic enzymes act on specific chemical groupings, eg, glycosidases on glycosides, pepsin and trypsin on peptide bonds, and esterases on esters. Many different peptide substrates may be attacked, lessening the number of digestive enzymes otherwise required. Proteases may also catalyze hydrolysis of esters. While this is of limited physiologic importance, the use of esters as synthetic substrates has contributed significantly to the study of the mechanism of action of proteases.

Certain lytic enzymes exhibit higher specificity. Chymotrypsin hydrolyzes peptide bonds in which the carboxyl group is contributed by the aromatic amino acids phenylalanine, tyrosine, or tryptophan. Carboxypeptidases and aminopeptidases remove amino acids one at a time from the carboxyl or amino terminal end of polypeptide chains, respectively.

Although a few oxidoreductases utilize either NAD$^+$ or NADP$^+$ as electron acceptor, most use exclusively one or the other. In general, oxidoreductases functional in biosynthetic processes in mammalian systems (eg, fatty acid or sterol synthesis) use NADPH as reductant, while those functional in degradative processes (eg, glycolysis, fatty acid oxidation) use NAD$^+$ as oxidant.

THE CATALYTIC ACTIVITY OF AN ENZYME FACILITATES ITS DETECTION

The small quantities of enzymes present in cells complicate measuring the amount of an enzyme in tissue extracts or fluids. Fortunately, the catalytic ac-

tivity of an enzyme provides a sensitive and specific probe for its own measurement.

To measure the amount of an enzyme in a sample of tissue extract or other biologic fluid, the rate of the reaction catalyzed by the enzyme in the sample is measured. Under appropriate conditions, the measured rate of the reaction is proportionate to the quantity of enzyme present. Since it is difficult to determine the number of molecules or mass of enzyme present, results are expressed in **enzyme units.** Relative amounts of enzyme in different extracts may then be compared. Enzyme units are best expressed in micromoles (μmol; 10^{-6} mol), nanomoles (nmol; 10^{-9} mol), or picomoles (pmol; 10^{-12} mol) of substrate reacting or product produced per minute. The corresponding International Enzyme Units are μU, nU, and pU.

NAD$^+$-Dependent Dehydrogenases Are Assayed at 340 nm

In reactions involving NAD$^+$ or NADP$^+$ (dehydrogenases), advantage is taken of the property of NADH or NADPH (but not NAD$^+$ or NADP$^+$) to absorb light of wavelength 340 nm (Figure 8–4). The oxidation of NADH to NAD$^+$ is accompanied by a decrease in optical density (OD) at 340 nm proportionate to the quantity of NADH oxidized. Similarly, when NAD$^+$ is reduced, the OD at 340 nm rises in proportion to the quantity of NADH formed. This change in OD at 340 nm can be exploited for the quantitative analysis of any NAD$^+$- or NADP$^+$-dependent dehydrogenase as follows. For a dehydro-

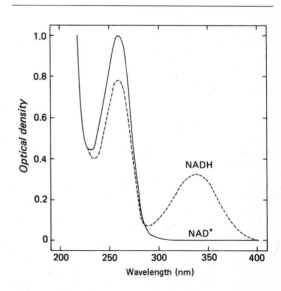

Figure 8–4. Absorption spectra of NAD$^+$ and NADH. Densities are for a 44 mg/L solution in a cell of 1 cm light path. NADP$^+$ and NADPH have spectra analogous to those of NAD$^+$ and NADH, respectively.

genase that catalyzes the oxidation of NADH by its oxidized substrate, the rate of decrease in OD at 340 nm is directly proportionate to the enzyme concentration. Measurement of the rate of decrease in OD at 340 nm therefore permits one to deduce the quantity of enzyme, expressed in activity units, present in a given biologic sample such as serum or a tissue extract.

Many Enzymes May Be Assayed by Coupling to a Dehydrogenase

In the above example, the rate of formation of a product (NADH) was measured to determine enzyme activity. Enzymes other than dehydrogenases may also be assayed by measuring the rate of appearance of a product (or, less commonly, the rate of disappearance of a substrate). The physicochemical properties of the product or substrate determine the specific method for quantitation. It often is convenient to "couple" the product of a reaction to a dehydrogenase for which this product is a substrate (Figure 8–5).

PURE ENZYMES ARE ESSENTIAL FOR UNDERSTANDING THEIR STRUCTURE, FUNCTION, REACTION MECHANISM, AND REGULATION

Knowledge about the reactions and chemical intermediates in metabolic pathways and of the regulatory mechanisms that operate at the level of catalysis derives to a great extent from studies of purified enzymes. Reliable information concerning the kinetics,

cofactors, active sites, structure, and mechanism of action also requires highly purified enzymes.

Purification Increases Specific Activity

The objective of enzyme purification is to isolate a specific enzyme from a crude cell extract containing many other components. Small molecules may be removed by dialysis or gel filtration, nucleic acids by precipitation with the antibiotic streptomycin, and so on. The problem is to separate the desired enzyme from hundreds of chemically and physically similar proteins.

The progress of a typical enzyme purification for a liver enzyme with good recovery and 490-fold overall purification is shown in Table 8–2. Note how specific activity and recovery of initial activity are calculated. The aim is to achieve the maximum specific activity (enzyme units per milligram of protein) with the best possible recovery of initial activity.

Purification Employs Chromatography on Ion Exchange or on Size Exclusion Supports

Useful classic purification procedures include precipitation with varying salt concentrations (generally ammonium or sodium sulfate) or solvents (acetone or ethanol), differential heat or pH denaturation, differential centrifugation, gel filtration, and electrophoresis. Selective adsorption and elution of proteins from the cellulose anion exchanger diethylaminoethyl (DEAE) cellulose and the cation exchanger carboxymethylcellulose (CMC) have also been extremely successful for extensive and rapid purification of proteins.

For example, an aqueous extract of liver tissue is adjusted to pH 7.5, at which pH most of the proteins bear a net negative charge. The mixture of soluble proteins is then applied to a column of DEAE cellulose buffered at pH 7.5, at which pH DEAE is positively charged. Negatively charged proteins bind to the DEAE by opposite charge interaction, while uncharged or positively charged proteins flow directly through the column. The column is then eluted, typically using a gradient of NaCl, extending from low to high concentration, dissolved in pH 7.5 buffer. Since Cl^- competes with proteins for binding to the positively charged support, proteins are selectively eluted, the most weakly bound first and the most strongly bound last. Analogous protein separations employ a negatively charged support such as carboxymethylcellulose or phosphocellulose at a somewhat lower pH to ensure that proteins are more positively charged.

Protein separations may also be achieved on porous materials known as "molecular sieves." When a mixture of proteins flows through a column containing a molecular sieve, small proteins distribute themselves both in the spaces between particles and in the internal spaces or pores of the support. As a

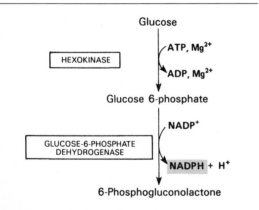

Figure 8–5. Coupled assay for hexokinase activity. The reaction is coupled to that catalyzed by glucose-6-phosphate dehydrogenase. Glucose-6-phosphate dehydrogenase, glucose, ATP, Mg^{2+}, and $NADP^+$ all are added in excess. The quantity of hexokinase present then determines the rate of the overall coupled reaction and therefore the rate of formation of NADPH, which can be measured at 340 nm.

Table 8–2. Summary of a typical enzyme purification scheme.

Enzyme Fraction	Total Activity (mU)[1]	Total Protein (mg)	Specific Activity (mU/mg)	Overall Recovery (%)
Crude liver homogenate	100,000	10,000	10	(100)
100,000 × g supernatant liquid	98,000	8,000	12.2	98
40–50% $(NH_4)_2SO_4$ precipitate	90,000	1,500	60	90
20–35% acetone precipitate	60,000	250	240	60
DEAE column fractions 80–110	58,000	29	2,000	58
43–48% $(NH_4)_2SO_4$ precipitate	52,000	20	2,600	52
First crystals	50,000	12	4,160	50
Recrystallization	49,000	10	4,900	49

[1]mU = millienzyme units; millimoles of substrate turned over per minute.

mixture of proteins of different sizes flows through the column, the mobility of small proteins is retarded relative to proteins too large to enter these pores. Large proteins therefore emerge from the column before smaller proteins. The elution profile presents a graded pattern from largest to smallest proteins. However, molecular sieves have limited capacity and generally are used only after prior purification has been achieved by other techniques.

Affinity Chromatographic Supports Recognize Specific Regions of Enzymes

The salient feature of affinity chromatography is its ability to selectively remove one particular protein, or at most a small number of particular proteins, from a complex protein mixture. The technique employs an immobilized ligand that interacts specifically with the enzyme whose purification is desired. When the protein mixture is exposed to this immobilized ligand, the only proteins that bind are those which interact strongly with the ligand. Unwanted protein flows through the column and is discarded. The desired protein is then eluted from the immobilized ligand, generally with a high concentration of salt or of the soluble form of the ligand. Purifications achieved by affinity chromatographic techniques are impressive, often surpassing that possible by successive application of numerous classic techniques.

Favored ligands are substrate and coenzyme derivatives covalently attached to a support such as Sephadex, generally via a linker molecule three to eight carbon atoms in length. The hydrophobic "linker," however, may complicate the separation by introducing an element of hydrophobic ligand chromatography (see below). Examples of successful affinity chromatography include purification of many different dehydrogenases on NAD^+ affinity supports. While many dehydrogenases may be bound and may be eluted together when the column is treated with soluble NAD^+, subsequent use of substrate (rather than coenzyme) affinity supports or elution with an "abortive ternary mixture" of one product and one substrate have proved successful in many instances.

Dye-Ligand and Hydrophobic Chromatography Are Analogous Affinity Techniques

Dye-ligand chromatography on supports such as blue, green, or red Sepharose and hydrophobic ligand chromatography on supports such as octyl- or phenyl-Sepharose are techniques closely related to affinity chromatography. The former employs as the immobilized ligand an organic dye that serves as an analog of substrate, coenzyme, or allosteric effector. Elution generally is achieved using salt gradients.

In hydrophobic ligand chromatography, an alkyl or aryl hydrocarbon is attached to a support such as Sephadex. Retention of proteins on these supports involves hydrophobic interactions between the alkyl chain and hydrophobic regions on the protein. Proteins are applied in solutions that contain a high concentration of a salt (eg, $[NH_4]_2SO_4$) and are eluted with *decreasing* gradients of the same salt.

Recombinant DNA Technology Can Provide a Rich Source of Enzymes

Until comparatively recently, the only available starting materials for preparing purified enzymes were the animal, plant, or bacterial cells in which they reside. The quantity of a given enzyme present was in many instances extremely small, imposing an upper limit on the quantity of purified enzyme that could be obtained at reasonable cost. This situation has been profoundly altered by recombinant DNA technology. The genes that encode many enzymes have been cloned and sequenced and can be placed in a plasmid- or phage-derived expression vector under the control of a powerful promoter. These vectors are transformed into *Escherichia coli,* yeast, or cultured mammalian cells. In the presence of an appropriate inducer, the protein synthetic machinery of the host cell is diverted to the synthesis of the desired enzyme. Subsequent purification is facilitated both by the relatively large quantities of the recombinant enzyme and by the fact that, being derived from a life form different from that of the host, it may differ significantly from host proteins in properties such as heat stability. Yields range from a few milligrams to hundreds of milligrams of homogeneous protein per liter of *E coli* cells, quantities adequate for extensive x-ray crystallographic and other physical studies needed to determine three-dimensional structure.

Polyacrylamide Gel Electrophoresis Detects Contaminants

Protein homogeneity is best assessed by polyacrylamide gel electrophoresis (PAGE) under several conditions. One-dimensional PAGE of the native protein will, if sufficient sample is applied, reveal major and minor protein contaminants. In two-dimensional (O'Farrell) PAGE, the first dimension separates denatured proteins on the basis of their pI values by equilibrating them in an electrical field that contains urea and a pH gradient maintained by polymerized ampholytes. The second dimension then separates proteins, after treatment with SDS, on the basis of the molecular sizes of their protomer units.

ENZYMES MAY BE PRESENT IN SPECIFIC ORGANELLES

The spatial arrangement and compartmentalization of enzymes, substrates, and cofactors within the cell are of cardinal significance. In liver cells, for example, the enzymes of glycolysis are located in the cytoplasm, whereas enzymes of the citric acid cycle are in the mitochondria. The distribution of enzymes among subcellular organelles may be studied following fractionation of cell homogenates by high-speed centrifugation. The enzyme content of each fraction is then examined.

Localization of a particular enzyme in a tissue or cell in a relatively unaltered state is frequently accomplished by histochemical procedures ("histoenzymology"). Thin (2- to 10-μm) frozen sections of tissue are treated with a substrate for a particular enzyme. Where the enzyme is present, the product of the enzyme-catalyzed reaction is formed. If the product is colored and insoluble, it remains at the site of formation and localizes the enzyme. Histoenzymology provides a graphic and relatively physiologic picture of patterns of enzyme distribution.

ISOZYMES ARE PHYSICALLY DISTINCT FORMS OF THE SAME CATALYTIC ACTIVITY

When techniques for purification of enzymes were applied, for example, to malate dehydrogenase from different sources (eg, rat liver and *Escherichia coli*), it became apparent that while rat liver and *E coli* malate dehydrogenase both catalyze the same reaction, their physical and chemical properties exhibited many significant differences. Physically distinct forms of the same catalytic activity may also be present in different tissues of the same organism, in different cell types, in subcellular compartments, or within a prokaryote such as *E coli*. This discovery followed from the application of electrophoretic separation procedures to separation of electrophoretically distinct forms of a particular enzymatic activity.

While the term "isozyme" (isoenzyme) encompasses all the above examples of physically distinct forms of a given catalytic activity, in practice and particularly in clinical medicine, "isozyme" has a more restricted meaning, namely, the physically distinct and separable forms of a given enzyme present in different cell types or subcellular compartments of a human subject. Isozymes are common in sera and tissues of all vertebrates, insects, plants, and unicellular organisms. Both the kind and the number of enzymes involved are equally diverse. Isozymes of numerous dehydrogenases, oxidases, transaminases, phosphatases, and proteolytic enzymes are known. Different tissues may contain different isozymes, and these isozymes may differ in their affinity for substrates.

Separation and Identification of Isozymes Is of Diagnostic Value

Medical interest in isozymes was stimulated by the discovery that human sera contained several lactate dehydrogenase isozymes and that their relative proportions changed significantly in certain pathologic conditions. Subsequently, many additional examples of changes in isozyme proportions as a result of disease have been described.

Serum lactate dehydrogenase isozymes may be visualized by subjecting a serum sample to electrophoresis, usually at pH 8.6, on a starch, agar, or polyacrylamide gel support. The isozymes have different charges at this pH and migrate to five distinct regions of the electropherogram. Isozymes are then detected by their ability to catalyze reduction of a colorless dye to an insoluble, colored form.

A typical dehydrogenase assay mixture contains NAD^+, a reduced substrate, the oxidized form of a redox dye such as nitro blue tetrazolium (NBT), an intermediate electron carrier required for transfer of electrons from NADH to NBT, and buffer and activating ions as required. Figure 8–6 illustrates application of such an assay to the visualization and quantitation of serum lactate dehydrogenase (LDH) isozymes in the clinical laboratory. Lactate dehydrogenase catalyzes transfer of two electrons and one H^+ from lactate to NAD^+ (Figure 8–1). The reaction proceeds at a measurable rate only in the presence of lactate dehydrogenase. When the assay mixture is spread on the electropherogram and incubated at 37 °C, concerted electron transfer reactions take place only in those regions where lactate dehydrogenase is present (Figure 8–6). The relative intensities of the colored bands may then be quantitated by a

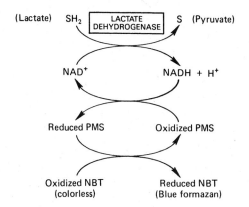

Figure 8–6. Coupled reactions in detection of lactate dehydrogenase activity on an electropherogram. (NBT, nitroblue tetrazolium; PMS, phenazine methosulfate.)

scanning photometer (Figure 8–7). The most negative isoenzyme is termed I_1.

Isozymes Are Products of Closely Related Genes

Oligomeric enzymes with dissimilar protomers can exist in several forms. Frequently, one tissue produces one protomer predominantly and another tissue a different protomer. If these can combine in various

ways to construct an active enzyme (eg, a tetramer), isozymes of that enzymatic activity are formed. This phenomenon is illustrated below for an enzyme of clinical diagnostic significance, lactate dehydrogenase.

Lactate dehydrogenase isozymes differ at the level of quaternary structure. The oligomeric lactate dehydrogenase molecule (molecular weight 130,000) consists of four protomers of two types, H and M (molecular weight about 34,000). Only the tetrameric molecule possesses catalytic activity. If order is unimportant, these protomers might be combined in the following five ways:

HHHH
HHHM
HHMM
HMMM
MMMM

Markert used conditions known to disrupt and reform quaternary structure to clarify the relationships between the lactate dehydrogenase isozymes. Disruption and reconstitution of the quaternary structure of homogeneous lactate dehydrogenase-I_1 or homogeneous lactate dehydrogenase-I_5 produced no new isozymes. These therefore consist of a single type of protomer. When a mixture of lactate dehydrogenase-I_1 and lactate dehydrogenase-I_5 was subjected to the same treatment, lactate dehydrogenase-I_2, -I_3, and -I_4

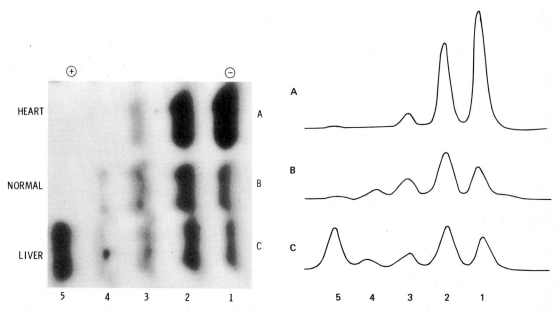

Figure 8–7. Normal and pathologic patterns of lactate dehydrogenase (LDH) isozymes in human serum. LDH isozymes of serum were separated on cellulose acetate at pH 8.6 and stained for enzyme. The photometer scan shows the relative proportion of the isozymes. Pattern A is serum from a patient with a myocardial infarct, B is normal serum, and C is serum from a patient with liver disease. (Courtesy of Dr Melvin Black and Mr Hugh Miller, St Luke's Hospital, San Francisco.)

were generated. The proportions of the isozymes found are those which would result if the relationship were–

Lactate Dehydrogenase Isozyme	Subunits
I_1	HHHH
I_2	HHHM
I_3	HHMM
I_4	HMMM
I_5	MMMM

Syntheses of H and M subunits are controlled by distinct genetic loci that are differentially expressed in different tissues, eg, the heart and skeletal muscle.

THE QUANTITATIVE ANALYSIS OF CERTAIN PLASMA ENZYMES IS OF DIAGNOSTIC SIGNIFICANCE

Certain enzymes, proenzymes, and their substrates are present at all times in the circulation of normal individuals and perform a physiologic function in blood. Examples of **functional plasma enzymes** include lipoprotein lipase, pseudocholinesterase, and the proenzymes of blood coagulation and of blood clot dissolution. They generally are synthesized in the liver but are present in blood in equivalent or higher concentrations than in tissues.

As the name implies, **nonfunctional plasma enzymes** perform no known physiologic function in blood. Their substrates frequently are absent from plasma, and the enzymes themselves are present in the blood of normal individuals at levels up to a millionfold lower than in tissues. Their presence in plasma at levels elevated above normal values suggests an increased rate of tissue destruction. Measurement of these nonfunctional plasma enzyme levels can thus provide the physician with valuable diagnostic and prognostic information.

Nonfunctional plasma enzymes include those in exocrine secretions and true intracellular enzymes. Exocrine enzymes—pancreatic amylase, lipase, bile alkaline phosphatase, and prostatic acid phosphatase—diffuse into the plasma. True intracellular enzymes normally are absent from the circulation.

Low Levels of Nonfunctional Plasma Enzymes Result From Normal Destruction of Cells

Low levels of nonfunctional enzymes found ordinarily in plasma apparently arise from the routine, normal destruction of erythrocytes, leukocytes, and other cells. With accelerated cell death, soluble enzymes enter the circulation. Although elevated plasma enzyme levels are generally interpreted as ev-

idence of cellular necrosis, vigorous exercise releases significant quantities of muscle enzymes.

Nonfunctional Plasma Enzymes Aid Diagnosis and Prognosis

Practicing physicians have long made use of quantitation of the levels of certain nonfunctional plasma enzymes. This valuable diagnostic and prognostic information is in most instances obtained on fully automated equipment. Table 8–3 contains a list of the principal enzymes employed in the field of diagnostic enzymology. The range of normal values for the principal enzymes used for diagnostic purposes is given in the Appendix.

RESTRICTION ENDONUCLEASES FACILITATE DIAGNOSIS OF GENETIC DISEASES

The diagnosis of genetic diseases received tremendous impetus from recombinant DNA technology. While all molecular diseases have long been known to be a consequence of altered DNA, techniques for direct examination of DNA sequences have become available comparatively recently. Development of hybridization probes for DNA fragments has led to techniques of sensitivity sufficient for prenatal screening for hereditary disorders by restriction enzyme mapping of DNA derived from fetal cells in the amniotic fluid.

In principle, DNA probes can be constructed for the diagnosis of most genetic diseases. For example, for prenatal detection of thalassemias (characterized by defects in the synthesis of hemoglobin subunits;

Table 8–3. Principal serum enzymes used in clinical diagnosis. Many of the enzymes are not specific for the disease listed.

Serum Enzyme	Major Diagnostic Use
Aminotransferases Aspartate aminotransferase (AST, or SGOT) Alanine aminotransferase (ALT, or SGPT)	Myocardial infarction Viral hepatitis
Amylase	Acute pancreatitis
Ceruloplasmin	Hepatolenticular degeneration (Wilson's disease)
Creatine kinase	Muscle disorders and myocardial infarction
γ-Glutamyl transpeptidase	Various liver diseases
Lactate dehydrogenase (isozymes)	Myocardial infarction
Lipase	Acute pancreatitis
Phosphatase, acid	Metastatic carcinoma of the prostate
Phosphatase, alkaline (isozymes)	Various bone disorders, obstructive liver diseases

see Chapter 7), a DNA probe constructed against a portion of the gene for a normal hemoglobin subunit can detect a shortened or absent restriction fragment arising from a deletion in that gene, as occurs in some α-thalassemias and certain rare types of β- and β, δ-thalassemias. Alternatively, a synthetic cDNA probe has been constructed that hybridizes to a β-globin sequence that contains a nonsense mutation present in certain β-thalassemias, but not to the normal β-globin gene. Absence of the plasma protease inhibitor α_1-antitrypsin is associated with emphysema and with infantile liver cirrhosis. The presence of inactive α_1-antitrypsin has been detected by a probe constructed against the inactive allele that contains a point mutation in the α_1-antitrypsin gene.

Hybridization probes may also be used to detect genetic alterations that lead to the loss of a restriction endonuclease site (see Chapter 42). For example, the point mutation of the Glu codon (GAG) to the Val codon (GTG) characteristic of sickle cell disease can be detected in the β-globin gene from cells in as little as 10 mL of amniotic fluid using the restriction endonuclease MstII or SauI.

PROBED RESTRICTION DIGESTS CAN REVEAL HOMOLOGOUS GENES WITH DIFFERENT BASE SEQUENCES

DNA probes may also be used to detect DNA sequences tightly linked to, but not actually within, the gene of interest. The analyses may be extended to detection of chromosome-specific variations (differences in sequence between homologous chromosomes). Digestion of the DNA with a restriction endonuclease generates different restriction maps (patterns of DNA fragments) from homologous genes that contain different base sequences. This phenomenon is termed **"restriction fragment length polymorphism" (RFLP).** For a genetic disease linked to a restriction length polymorphism, a human carrier of the disease will bear one chromosome with a normal gene and one with the defective gene. When the restriction fragments are resolved and probed, two hydridizing bands are detected (as distinct from a single band where both genes are identical). Offspring who inherit the disease-bearing chromosome exhibit only a single hybridizing band that differs from the band produced by normal chromosomes. The phenomenon of associated RFLP has been applied to the analysis of sickle cell trait (based on an associated HpaI RFLP) and of β-thalassemia (linked to a HindIII and a BamHI RFLP).

A screen based on RFLPs has been developed for the detection of infant phenylketonuria (see Chapter 32). Note, however, that since the RFLP does not cause the disease but is merely located near the defective gene, this general approach is not infallible.

RFLPs are points of departure for identification of the gene responsible for linked diseases. This approach has been applied to a screen for the mutational events involved in formation of retinoblastoma tumors and Huntington's disease.

Further examples of the uses of restriction enzymes for diagnosis are given in Chapter 42.

CATALYTIC RNAs

While clearly not proteins, certain ribonucleic acids (RNAs) exhibit highly substrate-specific catalytic activity. These RNAs, which meet all the classic criteria for definition as enzymes, are termed **ribozymes.** Although the substrates acted on by ribozymes are limited to the phosphodiester bonds of RNAs, the specificity of their action is fully comparable to that of classic enzymes. Ribozymes catalyze *trans*-esterification, and ultimately hydrolysis, of phosphodiester bonds in RNA molecules. These reactions are facilitated by free —OH groups, for example, on guanosyl residues. Ribozymes play key roles in the **intron splicing events** essential for the conversion of pre-mRNAs to mature mRNAs (see Chapter 39).

SUMMARY

Enzymes are protein catalysts that regulate the rates at which physiologic processes take place. Consequently, defects in enzyme function frequently cause disease. Enzymes that catalyze reactions involving group transfer, isomerization, oxidoreduction, or synthesis of covalent bonds require a cosubstrate known as a coenzyme. Since many coenzymes are B vitamin derivatives, vitamin deficiency can adversely affect enzyme function and therefore homeostasis. Many coenzymes also contain the nucleotide AMP. Most enzymes are highly specific for their substrates, coenzymes, and type of reaction catalyzed. However, some proteases also cleave esters. For enzymes acting on low-molecular-weight substrates, substrate analogs may also react, but generally at lower rates.

Measurement of enzyme activity is central to enzyme quantitation in research or clinical laboratories. The activity of NAD(P)$^+$-dependent dehydrogenases is assayed spectrophotometrically by measuring the change in absorption at 340 nm that accompanies oxidation or reduction of NAD(P)$^+$/NAD(P)H. Coupling other enzymes to dehydrogenases can facilitate their analysis. For investigation of their structure, mechanism of action, and regulation of their activity, enzymes must be purified to over 95% homogeneity. Techniques for enzyme purification include selective precipitation by salts or organic solvents and chromatography on ion exchange, gel filtration, substrate affinity, dye-ligand, or hydrophobic interaction supports. The ability to exploit recombinant DNA techniques to express enzymes in hosts of convenience has revolutionized enzyme purification by providing large quantities of enzymes that can in most instances be readily purified to homogeneity. The progress of a purification is assessed by measuring the increase in an enzyme's specific activity (activity per unit mass) and its ultimate homogeneity by polyacrylamide gel electrophoresis (PAGE). Precise intracellular localization of enzymes is inferred by histochemical and cell fractionation techniques coupled to enzymatic analysis of tissue slices or fractions of cell homogenates. Isozymes, physically distinct forms of the same catalytic activity, are present in all forms of life and tissues. Distinctive isozyme patterns of nonfunctional serum enzymes reveal damage to specific human tissues and provide valuable diagnostic and prognostic information. Finally, the ability of restriction endonucleases to detect subtle changes in gene structure permits physicians to diagnose genetic diseases in which mutations result in defective or nonfunctional enzymes.

While almost all enzymes are proteins, catalytic RNAs known as ribozymes catalyze highly specific hydrolysis of phosphodiester bonds in RNAs. These reactions are important in the processing events involved in maturation of pre-mRNAs.

REFERENCES

Advances in Enzymology. Issued annually. Academic Press.

Aitken A: *Identification of Protein Consensus Sequences: Active Site Motifs, Phosphorylation, and Other Posttranslational Modifications.* Horwood, 1990.

Bergmeyer H, Bergmeyer J, Grass M: *Methods of Enzymic Analysis.* Vol 11: *Antigens and Antibodies.* VCH Publishers, 1986.

Fersht A: *Enzyme Structure and Mechanism,* 2nd ed. Freeman, 1985.

Freifelder D: *Physical Biochemistry: Applications to Biochemistry and Molecular Biology.* Freeman, 1982.

Kaiser T, Lawrence DS, Rokita SZ: The chemical modification of enzyme specificity. Annu Rev Biochem 1985;54:597.

Methods in Enzymology. Over 130 volumes, 1955–present. Academic Press.

Naqui A: Where are the asymptotes of Michaelis-Menten? Trends Biochem Sci 1986;11:64.

Scopes R: *Protein Purification: Principles and Practice.* Springer, 1982.

Suckling CJ: *Enzyme Chemistry.* Chapman & Hall, 1990.

Tijssen P: *Practice and Theory of Enzyme Immunoassays.* Elsevier, 1985.

Uhlenbeck OC: Catalytic RNAs. Curr Opin Structural Biol 1991;1:459.

Enzymes: Kinetics

9

Victor W. Rodwell, PhD

INTRODUCTION

Many features of enzyme kinetics derive logically from concepts inherent in the kinetics of noncatalyzed chemical reactions. This chapter therefore first reviews aspects of reaction rate theory valid for chemical reactions in general, then considers factors unique to enzyme-catalyzed reactions.

BIOMEDICAL IMPORTANCE

All major factors that affect the rates of enzyme-catalyzed reactions (enzyme concentration, substrate concentration, temperature, pH, and the presence of inhibitors) are of clinical interest. The cardinal biologic principle of homeostasis states that good health requires that the composition of the internal milieu of the body be maintained within relatively narrow limits. Good health thus requires not only that hundreds of enzyme-catalyzed reactions take place but that they proceed at appropriate rates. Failure to achieve these objectives disturbs the homeostatic balance of our tissues, with potentially profound consequences. A physician thus must understand how pH, enzyme and substrate concentration, and inhibitors influence the rates of enzyme-catalyzed reactions. Some selected examples illustrate this point.

The rates of certain enzyme-catalyzed reactions respond to the subtle shifts in intracellular pH that characterize metabolic acidosis or alkalosis. Because the rates of enzymatic catalysis rise and fall in response to corresponding fluctuations in temperature, fever and hypothermia disturb homeostasis by altering the rates of many enzyme-catalyzed reactions. These subtle changes may, however, be turned to the physician's advantage. For example, the decrease in activity of all enzymes that accompanies lowered body temperature (hypothermia) may be exploited to reduce overall metabolic demand during open heart surgery or the transportation of organs for transplantation surgery.

Toxicology and pharmacology exploit an understanding of factors that affect the rates of enzyme-catalyzed reactions. Metabolic poisons such as mercurials, curare, and nerve gases exert their toxicity by inhibiting enzymes and consequently slowing down or abolishing essential metabolic reactions. Finally, many therapeutically important drugs act by lowering the rates of metabolic reactions by competing with the natural substrate for a key metabolic enzyme. These drugs often resemble natural substrates. Examples include lovastatin and zidovudine (AZT), drugs used to treat hypercholesterolemia and AIDS, respectively. Both exert their therapeutic effects by altering the rates of enzyme-catalyzed reactions.

REACTIONS PROCEED VIA TRANSITION STATES

The concept of **transition states** is integral to an understanding of catalysis of any kind, including catalysis by enzymes. For example, consider a displacement reaction in which a leaving group **L**, attached initially to **R**, is displaced by an entering group **E:**

$$E + R{-}L = E{-}R + L$$

This displacement actually involves a **transition state, $E{\cdot\cdot}R{\cdot\cdot}L$,** whose formation and subsequent decay may be represented by two "partial reactions," each having a characteristic change in free energy:

$$E + R{-}L = E{\cdot\cdot}R{\cdot\cdot}L \quad \Delta G_F$$
$$E{\cdot\cdot}R{\cdot\cdot}L = E{-}R + L \quad \Delta G_D$$

ΔG_F and ΔG_D are the changes in free energy associated with the formation and decay of the transition state, respectively. ΔG, the change in free energy associated with the *overall* reaction, is the sum of the free energies of formation and decay of the transition state:

$$\Delta G = \Delta G_F + \Delta G_D$$

As for any equation with two terms, it is not possible to infer the sign or magnitude of either ΔG_F or ΔG_D by inspecting the algebraic sign and magnitude of ΔG. Stated another way, we are unable simply from consideration of the change in free energy for the overall reaction, ΔG, to infer anything whatever concerning the free energy changes associated with formation and decay of transition states. Since catalysis is intimately associated with ΔG_F and ΔG_D, it follows that the thermodynamics of the overall reaction (ΔG) can tell us nothing of the path a reaction follows (ie, its mechanism). This is the task of kinetics.

More complex reactions involve several successive transition states and proceed via a series of partial reactions, each with an associated change in free energy. It is important to note that the change in free energy for the *overall* reaction, ΔG, is **independent of the number or kind of transition states.** This inference derives from the fact that ΔG is the algebraic sum of the free energy changes for each participating partial reaction.

NUMEROUS FACTORS AFFECT THE REACTION RATE

The **kinetic** or **collision theory** of chemical kinetics incorporates two key concepts: (1) Only molecules that collide, ie, come within bond-forming distance of one another, can react; and (2) for each chemical reaction, there is an **energy barrier** that must be overcome in order for a reaction to occur. For a collision to result in a reaction, the reacting molecules must possess sufficient energy to overcome this energy barrier. Thus, factors that (a) raise the kinetic energy of reacting molecules, (b) lower the energy barrier for reaction, or (c) increase collision frequency, should increase the rate of reaction.

Temperature

Raising temperature increases the number of molecules that can react both by elevating their kinetic energy and by increasing their frequency of collision. As shown in Figure 9–1, the number of molecules whose kinetic energy exceeds the energy barrier for reaction (vertical bar) increases as the temperature rises from low (A) through intermediate (B) to high (C) temperatures. In addition, any rise in temperature increases molecular motion and thus increases collision frequency. Both factors contribute to the increase in reaction rate that accompanies a rise in reaction temperature. This increase in reaction rate does not, however, continue indefinitely, since eventually a temperature is reached at which the reacting molecules are no longer stable. This limiting temperature tends to be extremely high for inorganic reactants, moderately high for most organic molecules, but well below 100 °C for most reactions of biologic interest.

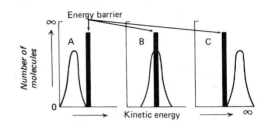

Figure 9–1. The energy barrier for chemical reactions.

Reactant Concentration

At high reactant concentrations, both the number of molecules with sufficient energy to react and their frequency of collision are high. This is true whether all or only a fraction of the molecules have sufficient energy to react. Consider reactions involving two different molecules, A and B:

$$A + B \rightarrow AB$$

Doubling the concentration either of A or of B will double the reaction rate. Doubling the concentration of both A and B will increase the probability of collision fourfold. The reaction rate therefore increases fourfold. The reaction rate is proportionate to the concentrations of the reacting molecules. Square brackets are used to denote molar concentrations;* $\propto$ means "proportionate to." The rate expression is

$$\text{Rate} \propto \text{[reacting molecules]}$$

or

$$\text{Rate} \propto [A][B]$$

For the situation represented by

$$A + 2B \rightarrow AB_2$$

the rate expression is

$$\text{Rate} \propto [A][B][B]$$

or

$$\text{Rate} \propto [A][B]^2$$

For the general case when n molecules of A react with m molecules of B

$$nA + mB \rightarrow A_nB_m$$

---

*Strictly speaking, molar activities rather than concentrations should be used.

the rate expression is

$$\text{Rate} \propto [A]^n [B]^m$$

K_{eq} Is a Ratio of Rate Constants

Since all chemical reactions are reversible, for the reverse reaction where A_nB_m forms n molecules of A and m molecules of B

$$A_nB_m \rightarrow nA + mB$$

the appropriate rate expression is

$$\text{Rate} \propto [A_nB_m]$$

Reversibility is represented by double arrows,

$$nA + mB \rightleftharpoons A_nB_m$$

This expression reads as follows: "n molecules of A and m molecules of B are in equilibrium with A_nB_m." The "proportionate to" symbol ($\propto$) may be replaced with an equal sign by inserting a proportionality constant, k, characteristic of the reaction under study. For the general case

$$nA + mB \rightleftharpoons A_nB_m$$

expressions for the rates of the forward reaction (Rate_1) and back reaction (Rate_{-1}) are

$$\text{Rate}_1 = k_1[A]^n[B]^m$$

and

$$\text{Rate}_{-1} = k_{-1}[A_nB_m]$$

When the rates of the forward and back reactions are equal, the system is said to be at equilibrium, ie,

$$\text{Rate}_1 = \text{Rate}_{-1}$$

Then

$$k_1[A]^n[B]^m = k_{-1}[A_nB_m]$$

and

$$\frac{k_1}{k_{-1}} = \frac{[A_nB_m]}{[A]^n[B]^m} = K_{eq}$$

The ratio of k_1 to k_{-1} is termed the **equilibrium constant, K_{eq}.** The following important properties of a system at equilibrium should be kept in mind.

(1) The equilibrium constant is the ratio of the reaction rate *constants* k_1/k_{-1}.

(2) At equilibrium, the reaction *rates* (not the reaction rate constants) of the forward and back reactions are equal.

(3) Equilibrium is a **dynamic state.** Although no *net* change in concentration of reactant or product molecules occurs at equilibrium, A and B are continually being converted to A_nB_m and vice versa.

(4) The equilibrium constant may be given a numerical value if the concentrations of A, B, and A_nB_m at equilibrium are known.

ΔG^0 May Be Calculated From K_{eq}

The equilibrium constant is related to ΔG^0 as follows:

$$\Delta G^0 = -RT \ln K_{eq}$$

R is the gas constant and T the absolute temperature. Since these are known, knowledge of the numerical value of K_{eq} permits one to calculate a value for ΔG^0. If the equilibrium constant is greater than 1, the reaction is spontaneous; ie, the reaction as written (from left to right) is favored. If it is less than 1, the opposite is true; ie, the reaction is more likely to proceed from right to left. Note, however, that although the equilibrium constant for a reaction indicates the direction in which a reaction is *spontaneous,* it does not indicate whether it will take place *rapidly.* That is, it does not tell us anything about the magnitude of the energy barrier for the reaction (ie, ΔG_F; see above). This follows because K_{eq} determines ΔG^0, previously shown to concern only initial and final states. **Reaction rates depend on the magnitude of the energy barrier, not on the magnitude of ΔG^0.**

Most factors that affect the velocity of enzyme-catalyzed reactions do so by changing local reactant concentration.

THE KINETICS OF ENZYMATIC CATALYSIS

Enzymes Provide Alternative Transition States

Like all true catalysts, enzymes are recovered unchanged following completion of the reaction. Nevertheless, during catalysis enzymes participate directly in the formation of transition states. Enzyme-substrate transition states typically are at significantly lower energy levels than the transition states formed when the same reaction proceeds in the absence of enzymatic catalysis. A major factor contributing to the ability of an enzyme to accelerate the rate of a given reaction thus is that ΔG_F for formation of an enzyme-substrate transition state is significantly lower than ΔG_F for the corresponding uncatalyzed reaction. Stated another way, enzymes lower the energy barrier for a reaction. Consequently, a greater proportion of substrate molecules are able to react. Note, however, that while enzymes affect ΔG_F, they do *not* affect ΔG. ΔG^0 for the *overall* reaction is the same whether or not the reaction is enzyme-catalyzed. Since the equilibrium constant for a chemical reaction is a function of the standard free energy change for a reaction

$$\Delta G^0 = -RT \text{ in } K_{eq}$$

it follows that enzymes and other catalysts have no effect on the equilibrium constant for a reaction.

Enzymes Catalyze Formation or Rupture of Covalent Bonds

For the group transfer reaction

$$D\text{—}G + A \rightleftharpoons A\text{—}G + D$$

a group, G, is transferred from a donor, D—G, to an acceptor, A. The overall reaction involves both rupture of the D—G bond and formation of a new A—G bond.

Enzyme-catalyzed group transfer reactions may be represented as follows:

This emphasizes three important features of enzyme-catalyzed group transfer reactions:

(1) Each partial reaction involves both the rupture and formation of a covalent bond.

(2) The enzyme is a reactant coequal with D—G and with A.

(3) Whereas in the overall reaction the enzyme acts catalytically (ie, is required only in trace quantities and may be recovered unchanged when the reaction is complete), for each partial reaction, the enzyme is a stoichiometric reactant (ie, it is required in a 1:1 molar ratio with the other reactants).

Many additional biochemical reactions may be considered as special cases of group transfer in which D, A, or both may be absent. For example, isomerization reactions (eg, the interconversion of glucose 6-phosphate and glucose 1-phosphate) might be represented as reactions in which both D and A are absent:

EnzS Complexes Participate in Catalysis

The above representations fail to emphasize yet another key feature of enzyme-catalyzed reactions—participation in the overall reaction of two or more intermediate forms of EnzS complex and the conse-

quent participation of a set of several sequential partial reactions. A representation of a group transfer reaction that emphasizes these features might be

in which EnzG, EnzG*, and EnzG** represent successive EnzS complexes formed during the course of the overall reaction.

Enzymes Enhance Reactant Proximity and Local Concentration

For any of the above reactions to occur, all of the reactants must come within bond-forming (or bond-breaking) distance of one another. For homogeneous solution chemistry in the absence of catalysts, the concentrations of the reacting molecules are constant throughout the solution. This condition no longer obtains following introduction of a catalyst. Catalysts have surface domains that bind the reacting molecules. Although this binding is a reversible process, the overall equilibrium constant for binding strongly favors the bound rather than the free forms of the reacting molecules. Qualitatively, we might represent this as follows:

$$\text{Reactant} + \text{Catalyst} \rightleftharpoons \text{Reactant-catalyst complex}$$

Quantitatively, we might express the tightness of association between a reactant, R, and a catalyst, C, in terms of the dissociation constant, K_d, for the R—C complex, or the equilibrium constant for the reaction

$$R\text{–}C \rightleftharpoons R + C$$

$$K_d = \frac{[R][C]}{[R\text{–}C]}$$

A low value for K_d thus represents a tight R—C complex.

One important consequence is that when a reactant binds to a catalyst this raises the concentration of the reactant in a localized area of the solution well above that of its concentration in free solution. Thus, we are no longer dealing with homogeneous but with heterogeneous solution chemistry.

If the catalyst for a bimolecular (two-reactant) reaction binds both reactants, the local concentration of each reactant is increased by a factor that depends on its individual affinity (K_d value) for the catalyst. Since the rate of the overall bimolecular reaction

$$A + B \rightarrow A\text{—}B$$

is, as we have seen above, proportionate to the concentrations of both A and B, binding of both A and B by the catalyst can result in an enormous (several thousand-fold) increase in overall reaction rate.

CATALYSIS OCCURS AT THE ACTIVE SITE

A key property of enzymes is their ability to bind reactants with an accompanying increase in local reactant concentration, and hence in local reaction rate. Enzymes are both extremely efficient and highly selective catalysts. To understand these distinctive properties of enzymes, we must introduce the concept of the "active" or "catalytic" site.*

The large size of proteins relative to substrates led to the concept that a restricted region of the enzyme, the "active site," was concerned with catalysis. Initially, it was puzzling why enzymes were so large when only a portion of their structure appeared to be required for substrate binding and catalysis. Today, we recognize that a far greater portion of the protein interacts with the substrate than was formerly supposed. When the need for allosteric sites of equal size also arises, the size of enzymes should no longer be surprising.

A Rigid Catalytic Site Model Explains Many Properties of Enzymes

The model of a catalytic site proposed by Emil Fischer visualized interaction between substrate and enzyme in terms of a "lock and key" analogy. This lock and key or rigid template model (Figure 9–2) is still useful for understanding certain properties of enzymes—eg, the ordered binding of two or more substrates (Figure 9–3) or the kinetics of a simple substrate saturation curve.

Substrates Induce Conformational Changes in Enzymes

An unfortunate feature of the Fischer model is the implied rigidity of the catalytic site. A more general model is the **"induced fit"** model of Koshland. This model has considerable experimental support. In the Fischer model, the catalytic site is presumed to be preshaped to fit the substrate. In the induced fit model, the substrate induces a conformational change in the enzyme. This aligns amino acid residues or other groups on the enzyme in the correct spatial orientation for substrate binding, catalysis, or both.

In the example (Figure 9–4), hydrophobic groups (hatched) and charged groups (stippled) both are in-

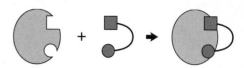

Figure 9–2. Representation of formation of an EnzS complex according to the Fischer template hypothesis.

volved in substrate binding. A phosphoserine (—P) and the —SH of a cysteine residue are involved in catalysis. Other residues involved in neither process are represented by Lys and Met residues. In the absence of substrate, the catalytic and the substrate-binding groups are several bond distances apart. Approach of the substrate induces a conformational change in the enzyme protein, aligning the groups correctly for substrate binding and for catalysis. At the same time, the spatial orientations of other regions are also altered—the Lys and Met are now closer together (Figure 9–4).

Substrate analogs may cause some but not all of the correct conformational changes (Figure 9–5). On attachment of the true substrate (A), all groups (shown as closed circles) are brought into correct alignment. Attachment of a substrate analog that is too "bulky" (Figure 9–5B) or too "slim" (Figure 9–5C) induces incorrect alignment. One final feature is the site shown as a small notch on the right. One may visualize a regulatory molecule attaching at this point and "holding down" one of the polypeptide arms bearing a catalytic group. Substrate binding, but not catalysis, might then occur. As illustrated by the induced fit model, the residues that comprise the catalytic site may be distant one from another in the primary structure but spatially close in the three-dimensional (tertiary) structure.

Many Aminoacyl Residues Contribute to the Active Site

Which portions of an enzyme contribute to—and form portions of—its active site? This question is best addressed by examining the three-dimensional structures of the ternary (three-component) complex formed by an enzyme and both substrates. However, even in the crystalline state, if all substrates are present, catalysis will occur, complicating data interpre-

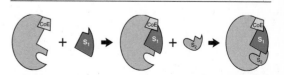

Figure 9–3. Representation of sequential adsorption of a coenzyme (CoE) and of two substrates (S_1 and S_2) to an enzyme in terms of the template hypothesis. The coenzyme is assumed to bear a group essential for binding the first substrate (S_1), which in turn facilitates binding of S_2.

---

*While many texts equate the active and catalytic sites of enzymes, there are on enzymes other "active" sites concerned with regulation of enzyme activity rather than with the intermediary enzymology of the catalytic process.

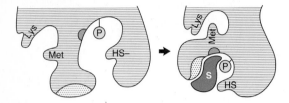

Figure 9–4. Two-dimensional representation of an induced fit by a conformational change in the protein structure. Note the relative positions of key residues before and after the substrate is bound.

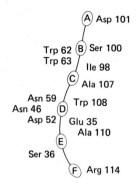

Figure 9–6. Two-dimensional representation of the catalytic site in the cleft region of lysozyme. A to F represent the glycosyl moieties of a hexasaccharide. Some residues in the cleft region are shown with their numbers in the lysozyme sequence. (Adapted from Koshland.)

tation. It therefore has proved helpful to examine the structure of so-called abortive ternary complexes composed of one substrate and one product, or of enzyme-substrate analog inhibitor complexes. These complexes, which cannot undergo turnover, provide a close approximation of the active site during the initial step in catalysis, substrate binding. These studies reveal that many aminoacyl residues are in contact with bound substrates or coenzymes and thus constitute part of the active site.

The extensive size and three-dimensional character of active sites dictates that they are optimally visualized on computer screens using software that permits one to turn the structure in all directions, zoom in on portions of interest, and view structures in three dimensions. Limitations of the printed page permit only an imperfect depiction of active sites. Some generalizations can, however, be stated.

Active Sites Often Are Located in Clefts

For enzymes that consist of a single subunit, the active site often resides in a cleft in the enzyme. One such example is lysozyme, an enzyme present in tears and nasal mucus that lyses the cell walls of many airborne gram-positive bacteria. A single polypeptide chain of 129 residues, lysozyme has a deep central cleft which harbors a catalytic site with six subsites (Figure 9–6) that bind various substrates. The residues responsible for bond cleavage lie between sites D and E close to the carboxyl groups of Asp 52 and Glu 35. Glu 35 apparently protonates the

acetal bond of the substrate, while the negatively charged Asp 52 stabilizes the resulting carbonium ion from the back side.

The catalytic site of ribonuclease also lies within a cleft similar to that of lysozyme, across which lie His 12 and His 119, residues implicated by chemical evidence as being at the catalytic site.

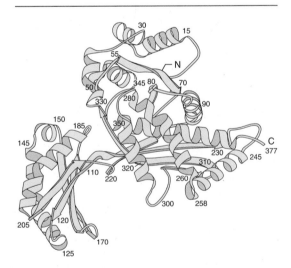

Figure 9–7. Structure of a single subunit of the HMG-CoA reductase of the bacterium *Pseudomonas mevalonii*. The subunit consists of two domains, the smaller of which (lower left) contains a nucleotide binding fold composed of β-sheet and α-helices. Aminoacyl residues are numbered to facilitate tracing the course of the polypeptide from its amino terminal (N) to its carboxyl terminal (C). Note that the polypeptide starts in the large domain, enters the small domain, and ultimately ends in the large domain. (Redrawn, with permission, from Lawrence CM, Rodwell VW, Stauffacher CV: Crystal structure of *Pseudomonas mevalonii* HMG-CoA reductase at 3.0 angstrom resolution. Science 1995;268:1758.)

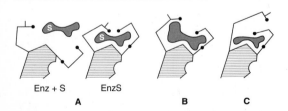

Figure 9–5. Representation of conformational changes in an enzyme protein when binding substrate **(A)** or inactive substrate analogs **(B, C)**. (After Koshland.)

Table 9–1. Amino acid sequences in the neighborhood of the catalytic sites of several bovine proteases. Regions shown are those on either side of the catalytic site seryl (S) and histidyl (H) residues.[1]

Enzyme	Sequence Around Serine Ⓢ																	Sequence Around Histidine Ⓗ									
Trypsin	D	S	C	Q	D	G	Ⓢ	G	G	P	V	V	C	S	G	K		V	V	S	A	A	Ⓗ	C	Y	K	S	G
Chymotrypsin A	S	S	C	M	G	D	Ⓢ	G	G	P	L	V	C	K	K	N		V	V	T	A	A	Ⓗ	G	G	V	T	T
Chymotrypsin B	S	S	C	M	G	D	Ⓢ	G	G	P	L	V	C	Q	K	N		V	V	T	A	A	Ⓗ	C	G	V	T	T
Thrombin	D	A	C	E	G	D	Ⓢ	G	G	P	F	V	M	K	S	P		V	L	T	A	A	Ⓗ	C	L	L	Y	P

[1]Reproduced, with permission, from Dayhoff MO [editor]: *Atlas of Protein Sequence and Structure,* Vol. 5. National Biomedical Research Foundation, 1972.

Active Sites of Multimeric Enzymes May Reside at Subunit Interfaces

For certain enzymes that contain multiple polypeptides or subunits, the active site resides at the interface between subunits. Aminoacyl residues from both substrates contribute to the active site, serving to bind substrates or to function in catalysis. One such example is the enzyme HMG-CoA reductase, the rate-limiting enzyme of cholesterologenesis (Figure 9–7).

Catalytic Residues Are Highly Conserved

Certain amino acids, notably cysteine and hydroxylic, acidic, or basic amino acids, perform key roles in catalysis. These residues serve as nucleophiles, as general base or general acid catalysts, or as acceptors of a group undergoing transfer. For all forms of an enzyme that catalyze a given reaction or reaction type, the chemical reaction mechanism is likely to be identical irrespective of the tissue or life form of origin. Consequently, the amino acids that perform highly specific roles in catalysis, and to some extent adjacent amino acids, tend to be highly conserved, even between genera. Table 9–1 illustrates this primary structural conservation at the catalytic sites of related but different hydrolytic enzymes. Table 9–2 illustrates conservation of primary structural motifs across gen-

era for the biosynthetic enzyme HMG-CoA reductase. However, since a given tertiary structural motif can be constructed from a variety of different primary structures, overall primary structure of enzymes is far less highly conserved than is tertiary structure.

MULTIPLE FACTORS AFFECT THE RATES OF ENZYME-CATALYZED REACTIONS

Temperature

While raising temperature increases the rate of an enzyme-catalyzed reaction, this holds only over a strictly limited range of temperatures (Figure 9–8). The reaction rate initially increases as temperature rises owing to increased kinetic energy of the reacting molecules. Eventually, however, the kinetic energy of the enzyme exceeds the energy barrier for breaking the weak hydrogen and hydrophobic bonds that maintain its secondary-tertiary structure. At this temperature, denaturation, with an accompanying precipitate loss of catalytic activity, predominates. Enzymes therefore exhibit an optimal temperature. This is not, however, a fundamental parameter, since the optimal temperature depends upon the duration of

Table 9–2. The primary structure of some regions of a catalytic site may be conserved across genera. Shown are the amino acids that surround the glutamate residue Ⓔ thought to function in catalysis by the enzyme HMG-CoA reductase. Underlined residues deviate from the consensus sequence for the mammalian enzyme.

Genus	Amino Acids										
Human	M	G	A	C	C	Ⓔ	N	V	I	G	Y
Hamster	M	G	A	C	C	Ⓔ	N	V	I	G	Y
Frog	M	G	A	C	C	Ⓔ	N	V	I	G	Y
Sea urchin	<u>S</u>	G	A	C	C	Ⓔ	N	V	I	G	Y
Fruit fly	<u>L</u>	N	A	C	C	Ⓔ	N	V	<u>L</u>	G	Y
Schistosome	<u>Y</u>	G	Q	C	C	Ⓔ	E	<u>V</u>	<u>I</u>	G	Y
Yeast	<u>F</u>	G	A	C	C	Ⓔ	N	V	I	G	Y

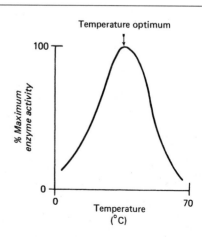

Figure 9–8. Effect of temperature on the velocity of a hypothetical enzyme-catalyzed reaction.

the assay used to determine it; ie, the longer an enzyme is maintained at a temperature at which its structure is marginally stable, the more likely it is to be denatured.

The factor by which the rate of a biologic process increases for a 10 °C temperature rise is the Q_{10}, or **temperature coefficient.** The rate of many biologic processes—eg, the rate of contraction of an excised heart—approximately doubles with a 10 °C rise in temperature ($Q_{10} = 2$). The alteration in the rates of many enzyme-catalyzed reactions that accompanies a rise or fall in body temperature constitutes an essential survival feature for life forms such as lizards that do not maintain a constant body temperature. By contrast, homeothermic organisms such as humans tolerate only strictly limited alterations in body temperature. For human subjects, changes in reaction rate due to a change in temperature thus are of limited physiologic significance except in fever or hypothermia.

Since raising the number of molecules having sufficient kinetic energy to overcome the energy barrier for reaction offers meager physiologic options for homeothermic organisms, how then do enzymes increase reaction rates? The answer lies in the ability of enzymes to lower the energy barriers for reactions and elevate local reactant concentrations.

For most enzymes, optimal temperatures are at or above those of the cells in which the enzymes occur. For example, enzymes from microorganisms adapted to growth in natural hot springs or deep oceanic thermal areas may exhibit optimal temperatures close to the boiling point of water.

pH

When enzyme activity is measured at several pH values, optimal activity typically is observed between pH values of 5 and 9. However, a few enzymes (eg, pepsin) are active at pH values well outside this range.

The shape of pH activity curves is determined by the following:

(1) Enzyme denaturation at high or low pH.

(2) Alterations in the charged state of the enzyme or substrates (either or both). For the enzyme, pH can affect activity by changing the structure or by changing the charge on a residue functional in substrate binding or catalysis. To illustrate, consider a negatively charged enzyme (Enz^-) reacting with a positively charged substrate (SH^+):

$$Enz^- + SH^+ \rightarrow EnzSH$$

At low pH, Enz^- protonates and loses its negative charge:

$$Enz^- + H^+ \rightarrow EnzH$$

At high pH, SH^+ ionizes and loses its positive charge:

$$SH^+ \rightarrow S + H^+$$

Since, by definition for this example, the only forms that will interact are SH^+ and Enz^-, extreme pH values will lower the effective concentration of Enz^- and SH^+, thus lowering the reaction velocity (Figure 9–9). Only in the cross-hatched area are both Enz and S in the appropriate ionic state, and the maximal concentrations of Enz and S are correctly charged at pH X.

Enzymes may also undergo changes in conformation when the pH is varied. A charged group distal to the region where the substrate is bound may be necessary to maintain an active tertiary or quaternary structure. As the charge on this group is changed, the protein may unravel, become more compact, or dissociate into protomers—all with resulting loss of activity. Depending upon the severity of these changes, activity may or may not be restored when the enzyme is returned to its optimal pH.

Enzymes Do Not Affect Equilibrium Constants

The enzyme is a reactant that combines with substrate to form an **enzyme-substrate complex, EnzS,** which decomposes to form a product, P, and free enzyme. In its simplest form, this may be represented as

$$Enz + S \underset{k_{-1}}{\overset{k_1}{\rightleftharpoons}} Enz + P$$

While the rate expressions for the forward, back, and *overall* reactions include the term [Enz],

$$Enz + S \underset{k_{-1}}{\overset{k_1}{\rightleftharpoons}} Enz + P$$
$$Rate_1 = k_1[Enz][S]$$
$$Rate_{-1} = k_{-1}[Enz][P]$$

in the expression for the *overall* equilibrium constant, [Enz] cancels out.

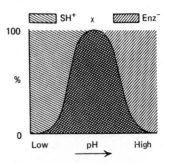

Figure 9–9. Effect of pH on enzyme activity.

$$K_{eq} = \frac{k_1}{k_{-1}} = \frac{[Enz][P]}{[Enz][S]} = \frac{[P]}{[S]}$$

The enzyme concentration thus has no effect on the equilibrium constant. Stated another way, since enzymes affect rates, not rate constants, they cannot affect K_{eq}, which is a ratio of rate constants. **The K_{eq} of a reaction is the same regardless of whether equilibrium is approached with or without enzymatic catalysis (recall ΔG^0).** Enzymes change the reaction path but do not affect the initial and final equilibrium concentrations of the reactants and products, the factors that determine K_{eq} and ΔG^0.

INITIAL RATE IS PROPORTIONATE TO ENZYME CONCENTRATION

The initial rate of a reaction is the rate measured before sufficient product has been formed to permit the reverse reaction to occur. The initial rate of an enzyme-catalyzed reaction is always proportionate to the concentration of enzyme. Note, however, that this statement holds only for *initial* rates.

SUBSTRATE CONCENTRATION AFFECTS REACTION RATE

In the following discussion, enzyme reactions are treated as if they had a single substrate and a single product. While this is the case for some enzyme-catalyzed reactions, most enzyme-catalyzed reactions have two or more substrates and products. This consideration does not, however, invalidate the discussion.

If the concentration of a substrate [S] is increased while all other conditions are kept constant, the measured initial velocity, v_i (the velocity measured when very little substrate has reacted), increases to a maximum value, V_{max}, and no further (Figure 9–10).

The velocity increases as the substrate concentration is increased up to a point where the enzyme is said to be "saturated" with substrate. The measured initial velocity reaches a maximal value and is unaffected by further increases in substrate concentration, because substrate is present in large molar excess over the enzyme. For example, if an enzyme with a molecular weight of 100,000 acts on a substrate with a molecular weight of 100 and both are present at a concentration of 1 mg/mL, there are 1000 mol of substrate for every mole of enzyme. More realistic figures might be

$$[Enz] = 0.1 \ \mu g/mL = 10^{-9} \ molar$$
$$[S] = 0.1 \ mg/mL = 10^{-3} \ molar$$

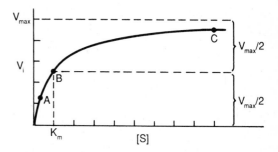

Figure 9–10. Effect of substrate concentration on the velocity of an enzyme-catalyzed reaction.

giving a 10^6 molar excess of substrate over enzyme. Even if [S] is decreased 100-fold, substrate is still present in 10,000-fold molar excess over enzyme.

The situations at points A, B, and C in Figure 9–10 are illustrated in Figure 9–11. At points A and B only a portion of the enzyme present is combined with substrate, even though there are many more molecules of substrate than of enzyme. This is because the equilibrium constant for the reaction Enz + S ⇌ EnzS (formation of the EnzS complex) is not infinitely large. At point A or B, increasing or decreasing [S] will therefore increase or decrease the amount of Enz associated with S as EnzS, and v_i will thus depend on [S].

At C, essentially all the enzyme is combined with substrate, so that a further increase in [S] cannot result in increased rates of reaction since no free enzyme is available to react.

Case B depicts a situation of major theoretical interest where exactly half the enzyme molecules are "saturated" with substrate. The velocity is accordingly **half the maximal velocity** ($V_{max}/2$) attainable at that particular enzyme concentration.

THE MICHAELIS-MENTEN AND HILL EQUATIONS MODEL THE EFFECTS OF SUBSTRATE CONCENTRATION

The Michaelis-Menten Equation

The substrate concentration that produces half-maximal velocity, termed the **K_m value** or **Michaelis constant,** may be determined experimentally by graphing v_i as a function of [S] (Figure 9–10). K_m has the dimensions of molar concentration.

When [S] is approximately equal to the K_m, v_i is very responsive to changes in [S], and the enzyme is working at half-maximal velocity. In fact, many enzymes possess K_m values that approximate the physiologic concentration of their substrates.

The Michaelis-Menten expression

$$v_i = \frac{V_{max}[S]}{K_m + [S]}$$

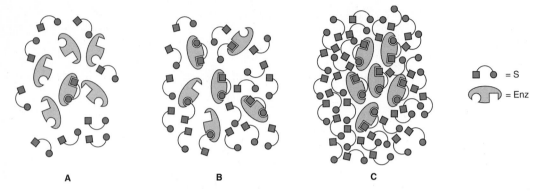

Figure 9–11. Representation of an enzyme at low *(A)*, at high *(C)*, and at the K_m concentration of substrate *(B)*. Points A, B, and C correspond to those of Figure 9–10.

describes the behavior of many enzymes as substrate concentration is varied. The dependence of the initial velocity of an enzyme-catalyzed reaction on [S] and on K_m may be illustrated by evaluating the Michaelis-Menten equation as follows:

(1) **When [S] is very much less than K_m** (point A in Figures 9–10 and 9–11). Adding [S] to K_m in the denominator now changes its value very little, so that the [S] term can be dropped from the denominator. Since V_{max} and K_m are both constants, we can replace their ratio by a new constant, K:

$$v_i = \frac{V_{max}[S]}{K_m + [S]}; \quad v_i \approx \frac{V_{max}[S]}{K_m} \approx \frac{V_{max}}{K_m}[S] \approx K[S]$$

[≈ means "approximately equal to."]

In other words, when the substrate concentration is considerably below that required to produce half-maximal velocity (the K_m value), the initial velocity, v_i, depends upon the substrate concentration, [S].

(2) **When [S] is very much greater than K_m** (point C in Figures 9–10 and 9–11). Now adding K_m to [S] in the denominator changes the value of the denominator very little, so the term K_m can be dropped from the denominator:

$$v_i = \frac{V_{max}[S]}{K_m + [S]}; \quad v_i \approx \frac{V_{max}[S]}{[S]} \approx V_{max}$$

This states that when the substrate concentration [S] far exceeds the K_m value, the initial velocity, v_i, is maximal, V_{max}.

(3) **When [S] = K_m** (point B in Figures 9–10 and 9–11).

$$v_i = \frac{V_{max}[S]}{K_m + [S]}; \quad v_i = \frac{V_{max}[S]}{[S] + [S]} = \frac{V_{max}[S]}{2[S]} = \frac{V_{max}}{2}$$

This states that when the substrate concentration

is equal to the K_m value, the initial velocity, v_i, is half-maximal. It also tells how to evaluate K_m, namely, to determine experimentally the substrate concentration at which the initial velocity is half-maximal.

A Linear Form of the Michaelis-Menten Equation Is Used to Determine K_m and V_{max}

Since many enzymes give saturation curves that do not readily permit evaluation of V_{max} (and hence of K_m) when v_i is plotted versus [S], it is convenient to rearrange the Michaelis-Menten expression to simplify evaluation of K_m and V_{max}. The Michaelis-Menten equation may be inverted and factored as follows:

$$v_i = \frac{V_{max}[S]}{K_m + [S]}$$

Invert:

$$\frac{1}{v_i} = \frac{K_m + [S]}{V_{max}[S]}$$

Factor:

$$\frac{1}{v_i} = \frac{K_m}{V_{max}} \cdot \frac{1}{[S]} + \frac{[S]}{V_{max}[S]}$$

Simplify:

$$\frac{1}{v_i} = \frac{K_m}{V_{max}} \cdot \frac{1}{[S]} + \frac{1}{V_{max}}$$

This is the equation for a straight line

$$y = a \cdot x + b$$

where

$$y = \frac{1}{v_i} \text{ and } x = \frac{1}{[S]}$$

If y, or $1/v_i$, is plotted as a function of x, or $1/[S]$, the y intercept, b, is $1/V_{max}$, and the slope, a, is K_m/V_{max}. The negative x intercept may be evaluated by setting $y = 0$. Then

$$x = -\frac{b}{a} = -\frac{1}{K_m}$$

Such a plot is called a double-reciprocal plot; ie, the reciprocal of v_i ($1/v_i$) is plotted versus the reciprocal of $[S]$($1/[S]$).

K_m may be estimated from the **double-reciprocal or Lineweaver-Burk plot** (Figure 9–12) using either the slope and y intercept or the negative x intercept. Since $[S]$ is expressed in molarity, **the dimensions of K_m are molarity or moles per liter.** Velocity, v_i, may be expressed in any units, since **K_m is independent of [Enz].** The double-reciprocal treatment requires relatively few points to define K_m and is the method most often used to determine K_m.

Experimentally, use of the Lineweaver-Burk approach to evaluate K_m can give rise to unwarranted emphasis on data gathered at low substrate concentrations. This is the case if the substrate concentrations selected for study differ by a constant increment. This shortcoming may be circumvented by selecting substrate concentrations whose *reciprocals* differ by constant increments.

An alternative approach to the experimental evaluation of K_m and of V_{max} is that of Eadie and Hofstee. The Michaelis-Menten equation may be rearranged as follows:

$$\frac{v_i}{[S]} = -v_i \cdot \frac{1}{K_m} + \frac{V_{max}}{K_m}$$

To evaluate K_m and V_{max}, plot $v_i/[S]$ (y axis) versus v_i (x axis). The y intercept is then V_{max}/K_m, and the x intercept is V_{max}. The slope is $-1/K_m$.

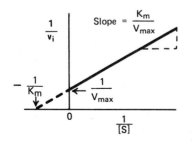

Figure 9–12. Double reciprocal or Lineweaver-Burk plot of $1/v_i$ versus $1/[S]$ used to evaluate K_m and V_{max}.

While both the Lineweaver-Burk and Eadie-Hofstee approaches are useful in selected instances, rigorous determination of K_m and of V_{max} requires statistical treatment.

K_m values are of considerable practical value. At a substrate concentration of 100 times K_m, an enzyme will act at essentially maximum rate, and therefore the maximal velocity (V_{max}) will reflect the amount of active enzyme present. This situation is generally desirable in the assay of enzymes. The K_m value tells how much substrate to use in order to measure V_{max}. Double reciprocal treatments also find extensive application in the evaluation of enzyme inhibitors.

K_m May Approximate a Binding Constant

The affinity of an enzyme for its substrate is equal to the inverse of the dissociation constant, K_d, for the enzyme-substrate complex, EnzS.

$$Enz + S \underset{k_{-1}}{\overset{k_1}{\rightleftharpoons}} EnzS$$

$$K_d = \frac{k_{-1}}{k_1}$$

Stated another way, the smaller the tendency of the substrate and enzyme to dissociate, the greater is the affinity of the enzyme for the substrate.

The K_m value of an enzyme for its substrate may also serve as a measure of its K_d. However, for this to be true, an assumption included in the derivation of the Michaelis-Menten expression must be valid. The derivation assumed that the first step of the enzyme-catalyzed reaction

$$Enz + S \underset{k_{-1}}{\overset{k_1}{\rightleftharpoons}} EnzS$$

is fast and always at equilibrium. In other words, the rate of dissociation of EnzS to Enz + S must be much faster than its dissociation to enzyme + product:

$$EnzS \underset{k_{-2}}{\overset{k_2}{\rightleftharpoons}} Enz + P$$

In the Michaelis-Menten expression, the $[S]$ that gives $v_i = V_{max}/2$ is

$$[S] = \frac{k_2 + k_{-1}}{k_1} = K_m$$

But when

$$k_{-1} \gg k_2$$

then

$$k_2 + k_{-1} \approx k_{-1}$$

and

$$[S] = \frac{k_{-1}}{k_1} \sim K_d$$

Under these conditions, $1/K_m = 1/K_d$ = affinity. If k_2 is not approximately equal to k_{-1}, then $1/K_m$ underestimates the affinity, $1/K_d$.

The Hill Equation

Certain enzymes and other ligand-binding proteins such as hemoglobin do not exhibit classic Michaelis-Menten saturation kinetics. When [S] is plotted versus v_i, the saturation curve is sigmoid (Figure 9–13). This generally indicates cooperative binding of substrate to multiple sites. Binding at one site affects binding at the others, as described in Chapter 7 for hemoglobin.

For sigmoid substrate saturation kinetics, the methods of graphic evaluation of the substrate concentration that produces half-maximal velocity discussed above are invalid (straight lines are not produced). To evaluate sigmoid saturation kinetics, we employ a graphic representation of the Hill equation, an equation originally derived to describe the cooperative binding of O_2 to hemoglobin. Written in the form of a straight line, the Hill equation is

$$\log \frac{v_i}{V_{max} - v_i} = n\log [S] - \log k'$$

where k' is a complex constant. The equation states that, when [S] is low compared to k', the reaction velocity increases as the nth power of [S]. Figure 9–14 illustrates a Hill plot of kinetic data for an enzyme with cooperative binding kinetics. A plot of $v_i/(V_{max} - v_i)$ versus log [S] yields a straight line with slope = n, where n is an empirical parameter whose value depends on the number of substrate-binding sites and

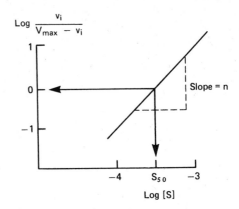

Figure 9–14. Graphic evaluation of the Hill equation to determine the substrate concentration that produces half-maximal velocity. This method is employed when substrate saturation kinetics are sigmoid.

the number and type of interactions between these binding sites. When $n = 1$, the binding sites act independently of one another. If $n > 1$, the sites are cooperative; and the greater the value of n, the stronger is the cooperativity and thus the more "sigmoid" are the saturation kinetics. If $n < 1$, the sites are said to exhibit negative cooperativity.

At half-maximal velocity ($v_i = V_{max}/2$), $v_i/(V_{max} - v_i) = 1$, and hence log $v_i/(V_{max} - v_i) = 0$. Thus, to determine S_{50}, the concentration of substrate that produces half-maximal velocity, drop a perpendicular line to the x axis from the point where log $v_i/(V_{max} - v_i) = 0$.

KINETIC ANALYSIS DISTINGUISHES COMPETITIVE FROM NONCOMPETITIVE INHIBITION

While we distinguish inhibitors of enzyme activity on the basis of whether the inhibition is or is not relieved by increasing the substrate concentration, many inhibitors do not exhibit the idealized properties of pure competitive or noncompetitive inhibition discussed below. An alternative way to classify inhibitors is by their site of action. Some bind to the enzyme at the same site as does the substrate (the catalytic site); others bind at some site (an allosteric site) removed from the catalytic site.

Competitive Inhibitors Typically Resemble the Substrate

Classic competitive inhibition occurs at the substrate-binding (catalytic) site. The chemical structure of a substrate analog inhibitor (I) generally resembles that of the substrate (S). It therefore combines reversibly with the enzyme, forming an enzyme-inhibitor (EnzI) complex rather than an EnzS complex.

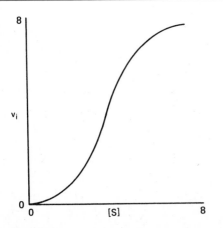

Figure 9–13. Sigmoid saturation kinetics.

When both the substrate and this type of inhibitor are present, they compete for the same binding sites on the enzyme. A classic case of competitive inhibition is that of malonate (I) with succinate (S) for succinate dehydrogenase.

Succinate dehydrogenase catalyzes formation of fumarate by removal of one hydrogen atom from each α-carbon atom of succinate (Figure 9–15).

Malonate ($^-$OOC—CH$_2$—COO$^-$) can combine with the dehydrogenase, forming an EnzI complex. This cannot be dehydrogenated, since there is no way to remove even one hydrogen atom from the single α-carbon atom of malonate without forming a pentavalent carbon atom. The only reaction the EnzI complex can undergo is decomposition back to free enzyme plus inhibitor. For the reversible reaction,

$$EnzI \underset{k_{-1}}{\overset{k_1}{\rightleftharpoons}} Enz + I$$

the equilibrium constant, K_i, is

$$K_i = \frac{[Enz][I]}{[EnzI]} = \frac{k_1}{k_{-1}}$$

The action of competitive inhibitors may be understood in terms of the following reactions:

$$Enz \underset{\pm S}{\overset{\pm I}{\rightleftharpoons}} \begin{array}{l} EnzI \text{ (inactive)} \xrightarrow{\times\times} Enz + P \\ EnzS \text{ (active)} \rightarrow Enz + P \end{array}$$

The rate of product formation depends solely on the concentration of EnzS. Suppose that the inhibitor binds very tightly to the enzyme (K_i = a small number). There now is little free enzyme (Enz) available to combine with S to form EnzS and eventually Enz + P. The reaction rate (formation of P) will therefore be slow. For analogous reasons, an equal concentration of a less tightly bound inhibitor (K_i = a larger number) will not decrease the rate of the catalyzed reaction so markedly.

Suppose that, at a fixed concentration of I, more S is added. This increases the probability that Enz will combine with S rather than with I. The ratio of EnzS to EnzI and thus the reaction rate also rise. At a suffi-

ciently high concentration of S, the concentration of EnzI should be vanishingly small. If so, the rate of the catalyzed reaction will be the same as in the absence of the inhibitor (Figure 9–16).

Double Reciprocal Plots Facilitate Evaluation of Inhibitors

Figure 9–16 represents a typical case of competitive inhibition shown graphically in the form of a Lineweaver-Burk plot. The reaction velocity (v_i) at a fixed concentration of inhibitor was measured at various concentrations of S. The lines drawn through the experimental points coincide at the y axis. Since the y intercept is $1/V_{max}$, this states that **at an infinitely high concentration of S ($1/[S] = 0$), v_i is the same as in the absence of inhibitor.** However, the intercept on the x axis (which is related to K_m) varies with inhibitor concentration and becomes a larger number ($-1/K'_m$ is smaller than $-1/K_m$) in the presence of the inhibitor. Thus, **a competitive inhibitor raises the apparent K_m (K'_m) for the substrate.** Since K_m is the substrate concentration at which the concentration of free enzyme is equal to the concentration of enzyme present as EnzS, substantial free enzyme is available to combine with inhibitor. For simple competitive inhibition, the intercept on the x axis is

$$x = \frac{1}{K_m\left(1 + \frac{[I]}{K_i}\right)}$$

K_m may be evaluated in the absence of I, and K_i may be evaluated using the above equation. If the number of moles of I added is much greater than the number of moles of enzyme present, [I] may generally be taken as the added (known) concentration of inhibitor.

The K_i values for a series of substrate analog (competitive) inhibitors indicate which are most effective. At a low concentration, those with the lowest K_i values will cause the greatest degree of inhibition.

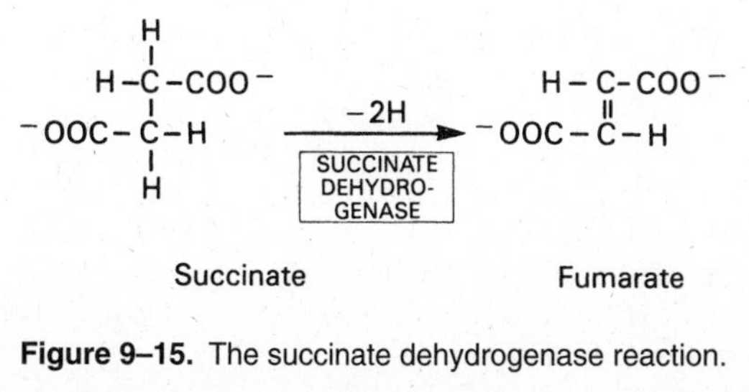

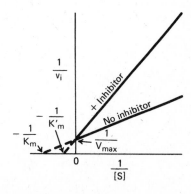

Figure 9–16. Lineweaver-Burk plot of classic competitive inhibition. Note the complete relief of inhibition at high [S] (low 1/[S]).

Figure 9–15. The succinate dehydrogenase reaction.

Many clinically efficacious drugs act as competitive inhibitors of important enzyme activities in microbial and animal cells.

Reversible Noncompetitive Inhibitors Lower V_{max} But Do Not Affect K_m

In noncompetitive inhibition, no competition occurs between S and I. The inhibitor usually bears little or no structural resemblance to S and may be assumed to bind to a different domain on the enzyme. Reversible noncompetitive inhibitors lower the maximum velocity attainable with a given amount of enzyme (lower V_{max}) but usually do not affect K_m. Since I and S may combine at different sites, formation of both EnzI and EnzIS complexes is possible. Since EnzIS may break down to form product at a slower rate than does EnzS, the reaction may be slowed but not halted. The following competing reactions may occur:

$$\text{Enz} \underset{\pm S}{\overset{\pm I}{\rightleftharpoons}} \begin{matrix} \text{EnzI} \underset{\pm I}{\overset{\pm S}{\rightleftharpoons}} \\ \text{EnzS} \end{matrix} \text{EnzIS} \to \text{Enz} + \text{P}$$
$$\text{Enz} + \text{P}$$

If S has equal affinity both for Enz and for EnzI (I does not affect the affinity of Enz for S), the results shown in Figure 9–17 are obtained when $1/v_i$ is plotted against $1/[S]$ in the presence and absence of inhibitor. It is assumed that there has been no significant alteration of the conformation of the active site when I is bound.

Irreversible Inhibitors "Poison" Enzymes

A variety of enzyme "poisons," eg, iodo-acetamide, heavy metal ions (Ag^+, Hg^{2+}), oxidizing agents, etc, reduce enzyme activity. Since these inhibitors bear no structural resemblance to the substrate, an increase in substrate concentration generally does not relieve this inhibition. The presence of one or more substrates or products, however, may protect the enzyme against inactivation. Kinetic analysis of the type discussed above may not distinguish between enzyme poisons and true reversible noncompetitive inhibitors. Reversible noncompetitive inhibition is, in any case, rare. Unfortunately this is not always appreciated, since both reversible and irreversible noncompetitive inhibition exhibit similar kinetics.

Most Reactions Involve Two or More Substrates

The preceding treatment of enzyme-catalyzed reactions considers only reactions that involve single substrates and products. Most biochemical reactions, however, involve two or more substrates and products. While the detailed kinetic analysis of multisubstrate reaction mechanisms lies beyond the scope of this chapter, the two-substrate, two-product reactions termed "Bi Bi" reactions are considered below.

Special Terms Describe Complex Enzymatic Reactions

The following conventions are used to characterize complex enzyme-catalyzed reactions. Substrates are termed A, B, C, and D in the order in which they add to the enzyme. Similarly, products are termed P, Q, R, and S in the order in which they leave the enzyme. Different forms of the enzyme are termed E, F, and G, where E is the free enzyme.

The terms **Uni** (1), **Bi** (2), **Ter** (3), and **Quad** (4) denote the number of reactants and products. While many enzyme-catalyzed reactions involve more reactants and products (eg, Bi Ter reactions), only **Bi Bi reactions**—those having two substrates and two products—are considered here.

Sequential or Single Displacement Reactions

In **sequential reactions,** all substrates must combine with the enzyme before any product is released. Reactions of this type are also termed single-displacement reactions because the group undergoing transfer (G) is passed directly from the donor substrate A—G to the acceptor substrate B.

Depending on how substrates add to the enzyme, we distinguish **random order** reactions and **compulsory order** reactions. In random order reactions, either substrate may be added first, forming either E—A or E—B. In compulsory order reactions, only A can react with E, while B can react only with the E—A complex (Figure 9–18). One explanation for a compulsory order of substrate addition is that the addition of A induces a conformational change that aligns residues essential for binding B.

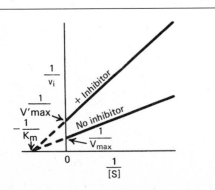

Figure 9–17. Lineweaver-Burk plot for reversible noncompetitive inhibition.

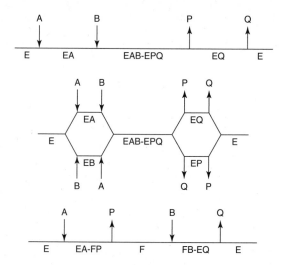

Figure 9–18. Representation of three classes of Bi Bi reaction mechanisms. Lines represent the enzyme and arrows the addition of substrates and departure of products. *Top:* An ordered Bi Bi mechanism, characteristic of many NAD(P)-dependent oxidoreductases. *Center:* A random Bi Bi reaction, characteristic of some dehydrogenases and kinases. *Bottom:* A Ping-Pong reaction, characteristic of aminotransferases (transaminases) and of chymotrypsin.

Ping-Pong Reactions

The term **Ping-Pong** applies to mechanisms in which one or more products are released from the enzyme before all the substrates have been added. Ping-Pong Bi Bi reactions are **double displacement reactions** because the group undergoing transfer (G) is first displaced from A—G by E and subsequently from E—G by B, forming the product Q—G.

Kinetic Data Distinguish Between Types of Mechanisms

Steady state kinetic data and appropriate rate equations are used to distinguish between different mechanisms. The terms in these kinetic equations, while more numerous, are analogous to those in the single-substrate Michaelis-Menten equation. These terms include (1) V_{max}, the reaction rate measured when both A and B are present at saturating concentrations; (2) the concentrations of A and B that result in half-maximal velocity when the other substrate is present at a saturating concentration; and (3) the dissociation constants for the release of A and B from the enzyme.

Double-Reciprocal Plots Distinguish Between Ping-Pong and Sequential Mechanisms

To distinguish between different types of mechanisms, the dependence of initial rate on the concentration of substrate A is measured at several sets of different (but constant) concentrations of substrate B. Data are then gathered for B as the variable substrate and for A as the constant substrate. The patterns of lines when data are graphed as double-reciprocal plots reveal the type of mechanism. For example, parallel lines are diagnostic of a Ping-Pong Bi Bi mechanism, and lines that intersect in the negative quadrant are diagnostic of an ordered Bi Bi mechanism.

Product inhibition studies are used to complement these kinetic analyses and to distinguish between ordered Bi Bi and random Bi Bi mechanisms. For example, in a rapid-equilibrium Bi Bi reaction, either product is a competitive inhibitor of A at constant [B] and of B at constant [A]. The more complex inhibition patterns of ordered Bi Bi reactions involve both competitive and mixed patterns of inhibition depending on which product is studied and on which substrate is varied and which is held constant.

SUMMARY

Temperature, pH, enzyme concentration, substrate concentration, and inhibitors alter the rates of enzyme-catalyzed reactions with important implications for health and disease. Enzymes alter reaction rates by (1) lowering the activation energy for formation of transition states, (2) serving as templates that increase local substrate concentrations and maintain substrates in conformations that favor specific chemical reactions, and (3) providing amino acid residues whose functional groups perform specific roles in catalysis. While enzymes profoundly alter reaction rates, they do not affect equilibrium constants or overall free energy changes for reactions. Enzymatic catalysis involves transition states of lower energy than those for the corresponding noncatalyzed reaction. This reduces the energy barrier for reaction.

Substrate binding and catalysis occur at the catalytic (active) site, a three-dimensional region of an enzyme that may contain, in addition to amino acid residues, coenzymes or metal ions. Active sites often reside in clefts in enzymes or at the interface between subunits of multimeric enzymes. When substrates approach, catalytic sites may undergo conformational changes that can be transmitted well beyond the catalytic site (recall O_2 binding to hemoglobin). These substrate-induced conformational changes may "create" substrate binding pockets and align key catalytic residues for catalysis.

As the temperature of an enzyme is raised, the reaction rate increases but only until the kinetic energy level of the enzyme exceeds that for rupture of the weak, noncovalent forces that maintain its native secondary and tertiary structure. At this point, activity decreases precipitously. More moderate changes in temperature also affect the rates of enzyme-catalyzed reactions in vivo but to a strictly limited extent in homeothermic organisms such as humans.

Moderate changes in pH (ie, in the region pH 5–9) affect enzyme activity by altering the charged state of the weakly acidic or basic R groups of amino acids that function in catalysis, substrate binding, or conformation of the catalytic site. pH changes can also alter the charged state of substrates. Enzymes thus exhibit pH values at which they are optimally active (pH optima). Extremes of pH denature enzymes by protonating or deprotonating acidic or basic amino acid residues so that they can no longer form the salt bonds that maintain secondary and tertiary structure—and in some cases also quaternary structure.

In almost all situations of physiologic interest, the molar concentration of an enzyme [E] is orders of magnitude below the molar concentration of its substrate [S]. Since ample substrate is available to react with free enzyme, any increase or decrease in enzyme concentration thus is accompanied by a corresponding increase or decrease in reaction rate. However, changes in substrate concentration [S] affect reaction rates only when there is sufficient free enzyme around to react. When all the enzyme is bound up as EnzS complex (V_{max} conditions), further increases in [S] no longer increase the reaction rate. The Michaelis-Menten equation describes the effects of substrate concentration on the activity of many enzymes. The Michaelis constant (K_m) is the substrate concentration that results in a half-maximal reaction rate ($V_{max}/2$; half the enzyme is present as EnzS). K_m values have the dimensions of molarity and are determined by graphing. For a double-reciprocal plot of $1/v_i$ versus $1/[S]$, the x intercept is $-1/K_m$ and the y intercept is $1/V_{max}$. The effect of [S] on the rates of reaction catalyzed by allosteric enzymes is described by the Hill equation. Most enzyme-catalyzed reactions involve two or more substrates or products. Different classes of Bi Bi reactions—reactions that involve two substrates and two products—are distinguished based on whether substrates bind sequentially or randomly and whether products leave before all substrates are bound (Ping-Pong Bi Bi reactions).

Substrate analogs that bind reversibly to the catalytic site and act as competitive inhibitors of enzymes include many drugs with medical value. Most chemicals that bear little or no structural resemblance to substrates and bind at the catalytic site or elsewhere act as noncompetitive inhibitors of enzyme activity. Briefly stated, competitive inhibitors decrease the apparent K_m but have no effect on V_{max}, while noncompetitive inhibitors decrease V_{max} without affecting K_m. These inhibitors may be distinguished by measuring enzyme activity at several concentrations of substrate in the presence and absence of inhibitor. Both data sets are then plotted as $1/v_i$ versus $1/[S]$. For classic competitive inhibition, the two lines intersect on the y axis (ie, V_{max} is unchanged), but the x intercept ($-1/K_m$) is decreased by the presence of inhibitor (ie, apparent K_m has increased). For noncompetitive inhibition, the y intercept ($1/V_{max}$) is increased in the presence of the inhibitor (ie, V_{max} is reduced), but the x intercept is unaffected.

REFERENCES

Bender ML, Bergeron RJ, Komiyama M: *The Bioorganic Chemistry of Enzymatic Catalysis.* Wiley-Interscience, 1984.

Freeman RB, Hawkins HC (editors): *The Enzymology of Posttranslational Modification of Proteins.* Academic Press, 1985.

Lawrence CM, Rodwell VW, Stauffacher CV: Crystal structure of *Pseudomonas mevalonii* HMG-CoA reductase at 3.0 angstrom resolution. Science 1995;268:1758.

Suckling CJ: *Enzyme Chemistry.* Chapman & Hall, 1990.

Symons RH: Small catalytic RNAs. Annu Rev Biochem 1992;61:641.

Walsh CT: Suicide substrates, mechanism-based enzyme inactivators: Recent developments. Annu Rev Biochem 1984;53:493.

See also references in Chapter 7.

Enzymes: Mechanisms of Action

10

Victor W. Rodwell, PhD

INTRODUCTION

The mechanisms of enzymatic catalysis closely mirror chemical reaction mechanisms. However, in contrast to inorganic or simpler organic catalysts, enzymes exhibit an extremely high order of substrate specificity and enormous catalytic efficiency—consequences of the protracted evolutionary development of active sites exquisitely suited to rapid and selective catalysis of individual reactions.

The pathways or routes by which enzymes convert substrates to products involve, rather than a single enzyme-substrate intermediate, a succession of enzyme-bound intermediates. The ultimate goal of mechanistic studies is to understand, at a molecular level, why enzymes are efficient and specific catalysts. This goal, which is far from having been reached, requires identification of the rate-limiting step in the overall reaction, characterization of all enzyme-bound intermediates, and identification of which metal ions or functional groups of amino acid residues, coenzymes, or prosthetic groups participate in binding substrates, products, and intermediates at the active site and which participate directly in catalysis. By analogy to intermediary metabolism, the study of the intermediates in metabolic pathways, study of the enzyme-bound intermediates formed during catalysis has been aptly termed "intermediary enzymology." In this chapter, major features of enzymatic catalysis are illustrated by the enzyme chymotrypsin, a comparatively simple protein that contains no coenzymes, prosthetic groups, or metal ions.

BIOMEDICAL IMPORTANCE

The molecular events that accompany conversion of substrates to products constitute the subject matter of the mechanism of enzyme action. These studies lead to a rational approach to therapy and drug design—areas of great potential for development in the immediate future. High-resolution structural information obtained by x-ray crystallography, combined with mechanistic information, now facilitates the design of drugs to inhibit specific enzymes such as HMG-CoA reductase, the pacemaker enzyme of cho-

lesterol biosynthesis. Mechanistic studies also suggest ways in which recombinant DNA technology and site-directed mutagenesis may be used to modify enzyme specificity or catalytic efficiency. These techniques ultimately will facilitate the design and introduction into human subjects of enzymes with specific desired properties.

THE "INTERMEDIARY ENZYMOLOGY" OF CHYMOTRYPSIN ILLUSTRATES GENERAL FEATURES OF ENZYMATIC CATALYSIS

Chymotrypsin catalyzes hydrolysis of peptide bonds in which the carboxyl group is contributed by an aromatic amino acid (Phe, Tyr, or Trp) or by one with a bulky nonpolar R group (Met). Like many other proteases, chymotrypsin also catalyzes the hydrolysis of certain esters. While the ability of chymotrypsin to catalyze ester hydrolysis is of no physiologic significance, it facilitates study of the catalytic mechanism.

The synthetic substrate *p*-nitrophenylacetate (Figure 10–1) facilitates colorimetric analysis of chymotrypsin activity because hydrolysis of *p*-nitrophenylacetate releases *p*-nitrophenol. In alkali, this converts to the yellow *p*-nitrophenylate anion.

STOP-FLOW KINETICS REVEALS THE "INTERMEDIARY ENZYMOLOGY" OF CHYMOTRYPSIN

The kinetics of chymotrypsin hydrolysis of *p*-nitrophenylacetate can be studied in a "stop-flow" apparatus. Stop-flow experiments use substrate quantities of

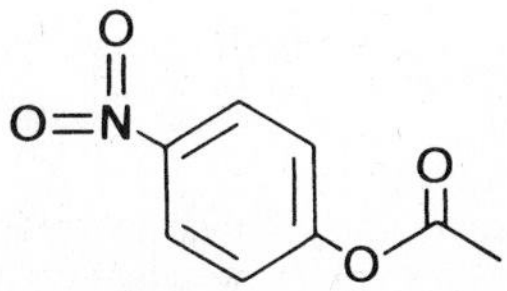

Figure 10–1. *p*-Nitrophenylacetate (PNPA).

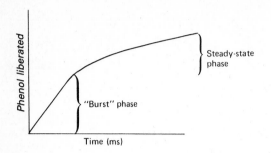

Figure 10–2. Kinetics of release of *p*-nitrophenylate anion when chymotrypsin hydrolyzes *p*-nitrophenylacetate in a stop-flow apparatus. "Phenol liberated" was calculated from the optical density of *p*-nitrophenylate anion.

enzyme (roughly equimolar quantities of enzyme and of substrate) and measure events that occur in the first few milliseconds after enzyme and substrate are mixed. The stop-flow apparatus has two syringes: one for chymotrypsin and the other for *p*-nitrophenylacetate. Instantaneous mixing of enzyme and substrate is achieved by a mechanical device that rapidly and simultaneously expels the contents of both syringes into a single narrow tube that passes through a spectrophotometer. The optical density as a function of time after mixing is displayed on a cathode ray tube.

Release of *p*-Nitrophenylate Is Biphasic

Release of *p*-nitrophenylate anion takes place in two distinct phases (Figure 10–2): (1) a "burst" phase, characterized by rapid liberation of *p*-nitrophenylate anion; and (2) a subsequent, slower release of additional *p*-nitrophenylate anion. The biphasic character of the release of *p*-nitrophenylate anion is comprehensible in terms of the successive steps in catalysis shown in Figure 10–3.

The Slow Step Is Hydrolysis of the Chymotrypsin-Acetate (CT-Ac) Complex

Once all of the available chymotrypsin has been converted to CT-Ac, no further release of *p*-nitrophenylate anion can occur until more free chymotrypsin is liberated by the slow, hydrolytic removal of acetate anion from the CT-Ac complex (Figure 10–3). The "burst" phase of *p*-nitrophenylate anion release (Figure 10–2) corresponds to the conversion of all of the available free chymotrypsin to the CT-Ac complex with simultaneous release of *p*-nitrophenylate anion. The subsequent release of *p*-nitrophenylate anion (phenol; PNPA) that follows the burst phase results from the slow liberation of free chymotrypsin by hydrolysis of the CT-Ac complex. This free chymotrypsin then is available for further formation of the CT-PNPA and CT-Ac complexes with attendant release of PNPA. Indeed, the magnitude of the "burst" phase (ie, moles of *p*-nitrophenylate anion released) is directly proportionate to the number of moles of chymotrypsin present initially.

Serine 195 Plays a Key Role in Catalysis

The acyl group of the acyl-CT intermediate is linked to a highly reactive seryl residue—serine 195—of chymotrypsin. The high reactivity of Ser 195 is shown by its ability (but not by the ability of the remaining 27 seryl residues of chymotrypsin) to react with diisopropylphosphofluoridate (DIPF) (Figure 10–4). Analogous reactions occur with other serine proteases.

Derivatization of Ser 195 inactivates chymotrypsin. Many other proteases are inactivated by DIPF by an analogous mechanism. These are termed "serine proteases."

A Charge Relay Network Provides a Proton Shuttle During Catalysis

The charge relay network of chymotrypsin involves three aminoacyl residues that are far apart in a primary structural sense but within bond-forming distance of one another in a tertiary structural sense. These residues are Asp 102, His 57, and Ser 195. While most of the charged residues of chymotrypsin are at the surface of the molecule, those of the charge relay network are "buried" in the otherwise nonpolar interior of the molecule. The three residues are aligned in the order Asp 102–His 57–Ser 195.

Recall that Ser 195 is the residue that is acylated

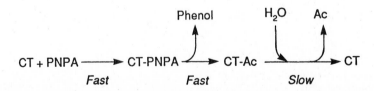

Figure 10–3. Intermediate steps in catalysis of the hydrolysis of *p*-nitrophenylacetate by chymotrypsin. (CT, chymotrypsin; PNPA, *p*-nitrophenylacetate; CT-PNPA, chymotrypsin-*p*-nitrophenylacetate complex; CT-Ac, chymotrypsin-acetate complex; phenol, *p*-nitrophenylate anion; Ac⁻, acetate anion.) Formation of the CT-PNPA and CT-Ac complexes is fast relative to the hydrolysis of the CT-Ac complex.

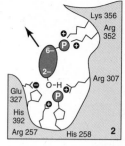

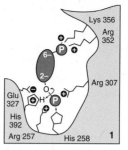

E • Fru-2,6-P₂ — E-P • Fru-6-P

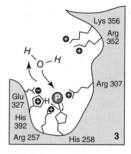

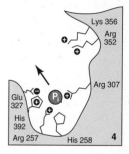

E-P • H₂O   E • Pᵢ

Figure 10–4. Reaction of the primary hydroxyl of Ser 195 of chymotrypsin with diisopropylphosphofluoridate (DIPF).

during catalysis by chymotrypsin. The approach of the acetate anion (derived from *p*-nitrophenylacetate) to the oxygen atom on the R group of Ser 195 triggers sequential proton shifts that "shuttle" protons from Ser 195 through His 57 to Asp 102 (Figure 10–5).

During deacylation of the acyl–Ser 195 intermediate, protons shuttle in the reverse direction. An analogous series of proton shifts is believed to accompany hydrolysis of a physiologic chymotrypsin substrate such as a peptide.

Like many proteases, chymotrypsin is released from ribosomes as a catalytically inactive enzyme precursor called a preenzyme or zymogen. As will be discussed in Chapter 11, conversion of the preenzyme chymotrypsinogen to the active enzyme chymotrypsin involves a series of selective proteolytic events that ultimately result in formation of the active site and alignment of the charge relay complex responsible for catalysis.

Catalysis by Fructose-2,6-bisphosphatase

The phosphohydrolase fructose-2,6-bisphosphatase (Chapter 21) catalyzes the hydrolytic removal of the phosphate on carbon 2 of fructose 2,6-bisphosphate. Figure 10–6 illustrates the roles of seven active site residues in catalysis by this phosphohydro-

Figure 10–6. Catalysis by fructose-2,6-bisphosphatase. *(1)* The quadruple negative charge of the bound substrate is stabilized by charge-charge interactions with Arg 257, Arg 307, Arg 352, and Lys 356. One member of the catalytic triad, Glu 327, stabilizes the positive charge on His 392. *(2)* The nucleophile His 392 attacks the C-2 phosphoryl group, transferring the phosphate to His 258 and forming an enzyme-phosphate intermediate. Fructose 6-phosphate leaves the enzyme. *(3)* A second nucleophilic attack by a water molecule, possibly assisted by Glu 327 acting as a base, forms inorganic phosphate. *(4)* Inorganic orthophosphate is released from Arg 257 and Arg 307. (Reproduced, with permission, from Pilkis SJ et al: 6-Phosphofructo-2-kinase/fructose-2,6-bisphosphatase: A metabolic signaling enzyme. Annu Rev Biochem 1995;64:799.)

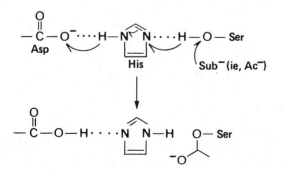

Figure 10–5. Operation of the proton shuttle of chymotrypsin during acylation of Ser 195 by the substrate (Sub⁻).

lase. As for catalysis by serine proteases, catalysis involves a "catalytic triad" but one which consists of two His and one Glu residue. Also illustrated is the stabilization by positively charged Arg and Lys residues of the highly negatively charged substrate.

COENZYMES PARTICIPATE DIRECTLY IN ENZYMATIC CATALYSIS

The direct participation of coenzymes in catalysis is illustrated below for the class of enzymes known as aminotransferases, or more commonly as transaminases. Transaminases catalyze the exchange of the α-amino groups of α-amino acids with the α-oxo groups of the α-oxo acids pyruvate, oxaloacetate, or α-ketoglutarate, reactions central to amino acid biosynthesis and catabolism (Chapters 30 and 31). As is the case for many reactions that require coenzymes, catalysis by transaminases involves the Ping-Pong mechanisms discussed in Chapter 9, with release of one product before the second substrate is added (Figure 10–7).

RESIDUES AT THE CATALYTIC SITE CAN ACT AS ACID-BASE CATALYSTS

Once substrate has bound at the catalytic site, the charged (or chargeable) functional groups of the side chains of nearby aminoacyl residues may participate in catalysis by functioning as acidic or basic catalysts.

There are two broad categories of acid-base catalysis by enzymes: **general** and **specific** acid (or base) catalysis. Reactions whose rates vary in response to changes in H+ or H3O+ concentration but are independent of the concentrations of other acids or bases present in the solution are said to be subject to **specific acid** or **specific base catalysis.** Reactions whose rates are responsive to *all* the acids (proton donors) or bases (proton acceptors) present in solution are said to be subject to **general acid** or **general base catalysis.**

Varying pH & Buffer Concentration Distinguishes General From Specific Acid-Base Catalysis

If the rate of the reaction changes as a function of pH at constant buffer concentration, the reaction is said to be *specific base-catalyzed* (if the pH is above 7) or *specific acid-catalyzed* (if the pH is below 7). If the reaction rate at constant pH increases as the buffer concentration increases, the reaction is said to be subject to *general base catalysis* (if the pH is above 7) or *general acid catalysis* (if the pH is be-low 7).

As an example of specific acid catalysis, consider the conversion of a substrate (S) to a product (P) that occurs in two steps—a rapid, reversible proton transfer step,

$$S + H_3O^+ \rightleftharpoons SH^+ + H_2O$$

followed by a slower, and therefore rate-determining, step of rearrangement of the protonated substrate to product

$$SH^+ + H_2O \rightarrow P + H_3O^+$$

Increasing the concentration of hydronium ion [H3O+] increases the reaction rate by elevating the concentration of SH+, the conjugate acid of the substrate, which is the substrate for the rate-determining step in the overall reaction. Stated mathematically,

$$Rate = \frac{d[P]}{dt} = k[SH^+]$$

where P = the product, t = time, k = the specific rate constant, and [SH+] = the concentration of the conjugate acid of the substrate.

Since the concentration of SH+ depends upon both the concentration of S and the concentration of H3O+, the general rate expression for specific acid-catalyzed reactions is

$$\frac{d[P]}{dt} = k'[S][H_3O^+]$$

Note that it is a requirement of specific acid catalysis that the rate expression contain *only* terms for S and for H3O+.

Next consider that, in addition to the specific acid catalysis described above, there is also catalysis by imidazolium ion of an imidazole buffer. Since imidazole is a weak acid (pKa about 7), it is a poor proton donor; hence the reaction

$$S + Imidazole\text{---}H^+ \rightarrow SH^+ + Imidazole$$

is slow and is rate-determining for the overall reaction. Note that the fast and slow steps are reversed when the mechanism changes from specific to general acid catalysis. The rate expressions for general acid catalysis frequently are complex and for this reason are not discussed here.

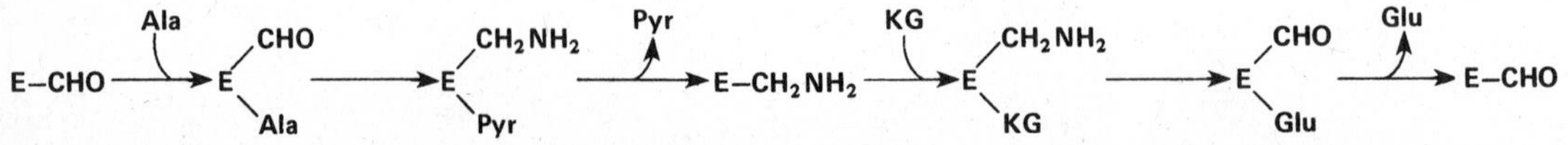

Figure 10–7. Ping-Pong mechanism for transamination. E—CHO and E—CH2NH2 represent the enzyme-pyridoxal phosphate and enzyme-pyridoxamine phosphate complexes, respectively. (Ala, alanine; Pyr, pyruvate; KG, α-ketoglutarate; Glu, glutamate.)

SITE-DIRECTED MUTAGENESIS PROVIDES MECHANISTIC INSIGHTS

Molecular biologic techniques provide powerful tools for investigation of the mechanism of action of enzymes. Prominent among these are the ability to alter at will the nucleotide base sequence of genes and to express proteins in unicellular hosts such as cultured mammalian cells or the bacterium *Escherichia coli*. In combination with x-ray crystallographic solution of three-dimensional structure and classic kinetic investigations, these molecular techniques provide new insights into the mechanisms by which enzymes act.

Where physical and kinetic data implicate a given amino acid residue as performing a specific function in catalysis (ie, serving as a general base, or as a group transfer reagent), support for this inference may be provided by changing the amino acid to one incapable of performing the postulated function. In a few instances, this can be achieved by chemical modification of the enzyme. However, a more general approach is to employ **site-directed mutagenesis** to mutate the gene that encodes the enzyme, then to express and characterize the mutant enzyme. By altering the base sequence of a specific codon, site-directed mutagenesis can replace a given amino acid by any desired protein amino acid. Site-directed mutagenesis can also generate multiple point mutants. A short (ie, a 19-mer) oligonucleotide, produced by automated synthesis, that encodes a mutant amino acid is annealed in vitro to the isolated wild-type gene. A mutant gene is then produced by adding DNA polymerase, DNA ligase, and deoxyribonucleoside triphosphates. This mutant gene is then moved into an expression vector behind a strong promoter, expressed, and the resulting mutant enzyme is purified and its properties examined.

METAL IONS MAY FACILITATE SUBSTRATE BINDING AND CATALYSIS

Over 25% of all enzymes contain tightly bound metal ions or require them for activity. The functions of these metal ions are studied by x-ray crystallography, magnetic resonance imaging (MRI), and electron spin resonance (ESR). Coupled with knowledge of the formation and decay of metal complexes and of reactions within the coordination spheres of metal ions, this provides insight into the roles of metal ions in enzymatic catalysis. These roles are considered below.

Metalloenzymes contain a definite quantity of functional metal ion that is retained throughout purification. **Metal-activated enzymes** bind metals less tightly but require added metals. The distinction between metalloenzymes and metal-activated enzymes thus rests on the affinity of a particular enzyme for its metal ion. The mechanisms whereby metal ions perform their functions appear to be similar in metalloenzymes and metal-activated enzymes.

Ternary Complexes With Metals Function in Catalysis

For ternary (three-component) complexes of the catalytic site (Enz), a metal ion (M), and substrate (S) that exhibit 1:1:1 stoichiometry, four schemes are possible:

Enz—S—M
Substrate-bridge complex

M—Enz—S
Enzyme-bridge complex

Enz—S—M
Simple metal-bridge complex

$$\text{Enz}\diagdown_{\diagdown}^{M}$$

Cyclic metal-bridge complex

All four are possible for metal-activated enzymes. Metalloenzymes cannot form the EnzSM complex, because they retain the metal throughout purification (ie, are already as EnzM). Three generalizations can be stated: (1) Most but not all kinases (ATP:phosphotransferases) form substrate-bridge complexes of the type Enz-nucleotide-M. (2) Phosphotransferases using pyruvate or phosphoenolpyruvate as substrate, enzymes catalyzing other reactions of phosphoenolpyruvate, and carboxylases form metal-bridge complexes. (3) A given enzyme may form one type of bridge complex with one substrate and a different type with another.

A. Enzyme-Bridge Complexes (MEnzS): The metals in enzyme-bridge complexes are presumed to perform structural roles maintaining an active conformation (eg, glutamine synthase) or to form a metal bridge to a substrate (eg, pyruvate kinase). In addition to its structural role, the metal ion in pyruvate kinase appears to hold one substrate (ATP) in place and to activate it.

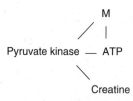

B. Substrate-Bridge Complexes (EnzSM): The formation of ternary substrate-bridge complexes of nucleoside triphosphates with enzyme, metal, and substrate appears attributable to displacement of H_2O from the coordination sphere of the metal by ATP:

$$ATP^{4-} + M(H_2O)_6^{2+} \rightleftharpoons ATP\text{---}M(H_2O)_3^{2-} + 3H_2O$$

Enzyme then binds, forming the ternary complex:

$$ATP\text{---}M(H_2O)_3^{2-} + Enz \rightleftharpoons Enz\text{---}ATP\text{---}M(H_2O)_3^{2-}$$

In phosphotransferase reactions, metal ions are thought to activate the phosphorus atoms and form a rigid, polyphosphate-adenine complex of appropriate conformation in the active, quaternary complex.

C. Metal-Bridge Complexes:

$$Enz\text{---}M\text{---}S \text{ or } Enz \begin{array}{c} / M \\ | \\ \backslash S \end{array}$$

Crystallographic and sequencing data have established that a His residue is concerned with metal binding at the active site of many proteins (eg, carboxypeptidase A, cytochrome c, rubredoxin, metmyoglobin, and methemoglobin; see Chapter 7). For binary (two-component) EnzM complexes, the rate-limiting step is in many cases the departure of water from the coordination sphere of the metal ion. For many peptidases, activation by metal ions is a slow process requiring many hours. The slow reaction probably is conformational rearrangement of the binary EnzM complex to an active conformation, eg,

Metal binding:

$$\overset{Rapid}{Enz + M(H_2O)_6 \rightarrow Enz\text{---}M(H_2O)_{6-n} + nH_2O}$$

Rearrangement to active conformation (Enz^*):

$$\overset{Slow}{Enz\text{---}M(H_2O)_{6-n} \rightarrow Enz^*\text{---}M(H_2O)_{6-n}}$$

For metalloenzymes, however, the ternary metal-bridge complex must be formed by combination of the substrate (S) with the binary EnzM complex:

$$Enz\text{---}M + S \rightleftharpoons Enz\text{---}M\text{---}S \text{ or } Enz \begin{array}{c} / M \\ | \\ \backslash S \end{array}$$

Metal Ions Perform Multiple Functions in Catalysis

Metal ions may participate in each of the four mechanisms by which enzymes are known to accelerate the rates of chemical reactions: (1) general acid-base catalysis, (2) covalent catalysis, (3) approximation of reactants, and (4) induction of strain in the enzyme or substrate. Other than iron and manganese,

which function in heme proteins, the metal ions most commonly concerned in enzymatic catalysis are Mg^{2+}, Mn^{2+}, and Ca^{2+}, although other metal ions (eg, K^+) are important for the activity of certain enzymes.

Metal ions, like protons, are Lewis acids (electrophiles) and can share an electron pair, forming a sigma bond. Metal ions may also be considered "super acids," since they exist in neutral solution, frequently have a positive charge of > 1, and may form π bonds. In addition (and unlike protons), metals can serve as three-dimensional templates for orientation of basic groups on the enzyme or substrate.

Metal ions can also accept electrons via σ or π bonds to activate electrophiles or nucleophiles (general acid-base catalysis). By donating electrons, metals can activate nucleophiles or act as nucleophiles themselves. The coordination sphere of a metal may bring together enzyme and substrate (approximation) or form chelate-producing distortion in either the enzyme or substrate (strain). A metal ion may also "mask" a nucleophile and thus prevent an otherwise likely side reaction. Finally, stereochemical control of the course of an enzyme-catalyzed reaction may be achieved by the ability of the metal coordination sphere to act as a three-dimensional template to hold reactive groups in a specific steric orientation (Table 10–1).

SUMMARY

The protease chymotrypsin (CT) illustrates many general features of enzymatic catalysis. These features include the formation of enzyme-bound intermediates (CT-PNPA and CT-Ac), the stepwise nature of catalysis (three partial reactions), the rate-limiting character of a single partial reaction

Table 10–1. Selected examples of the roles of metal ions in the mechanism of action of enzymes.[1]

Enzyme	Role of Metal Ion
Histidine deaminase	Masking a nucleophile
Kinases, lyases, pyruvate decarboxylase	Activation of an electrophile
Carbonic anhydrase	Activation of a nucleophile
Cobamide enzymes	Metal acts as a nucleophile
Pyruvate carboxylase, carboxypeptidase, alcohol dehydrogenase	π-Electron withdrawal
Nonheme iron proteins	π-Electron donation
Pyruvate kinase, pyruvate carboxylase, adenylyl kinase	Metal ion gathers and orients ligands
Phosphotransferase, D-xylose isomerase, hemoproteins	Strain effects

[1]Adapted from Mildvan AS: Metals in enzyme catalysis. Vol 2. Page 456 in: *The Enzymes*. Boyer PD, Lardy H, Myrbäck K (editors). Academic Press, 1970.

(hydrolysis of CT-Ac), and the sequential release of products (phenol followed by acetate). Functions of particular amino acids at the catalytic site were illustrated by the role of a Ser residue (Ser 195) as the acceptor of a group (Ac^-) undergoing transfer to water, and by roles of specific Ser, His, and Asp residues in forming a charge-relay network. This relay network illustrates, in addition, that amino acid residues far apart in a primary structural sense (Ser 195, His 57, and Asp 102) can form contiguous portions of an enzyme's catalytic site. Selective proteolysis of the inactive preprotein chymotrypsinogen creates the active site of chymotrypsin and brings the three residues that constitute the charge-relay complex within bond-forming distance of one another.

Catalytic site residues can act as general acids (Glu, Asp) or general bases (His, Arg, Lys) during catalysis. Addition of substrates and release of products may proceed in an ordered or in a random manner. In a Ping-Pong mechanism (eg, transamination), the reaction proceeds via alternate substrate addition and product release events.

Metal ions facilitate substrate binding and catalysis by forming several kinds of bridge complexes of enzyme, metal, and substrate. Metal ions may function as general acid catalysts or may facilitate approximation of reactants.

REFERENCES

Coleman JE: Structure and mechanism of alkaline phosphatase. Annu Rev Biophys Biomol Struct 1992;21:441.

Fersht A: *Enzyme Structure and Mechanism,* 2nd ed. Freeman, 1985.

Freeman RB, Hawkins HC (editors): *The Enzymology of Post-translational Modification of Proteins.* Academic Press, 1985.

Purich DL (editor): Enzyme kinetics and mechanisms. Parts A and B in: Methods in Enzymology. Vol 63, 1979; Vol 64, 1980. Academic Press.

Regan L: The design of metal binding sites in proteins. Annu Rev Biophys Biomol Struct 1993;22:257.

See also references in Chapters 7 and 8.

11 Enzymes: Regulation of Activities

Victor W. Rodwell, PhD

INTRODUCTION

In this chapter, mechanisms by which metabolic processes are controlled by altering the quantity or the catalytic efficiency of enzymes are illustrated by selected examples. The intent is to characterize overall patterns of regulation. Throughout this book, reference is made to many additional specific examples that illustrate these diverse features of metabolic regulation.

BIOMEDICAL IMPORTANCE

How do cells and intact organisms regulate and coordinate overall metabolism? This question is of concern to workers in areas of biomedical science as diverse as cancer, heart disease, aging, microbial physiology, differentiation, metamorphosis, hormone action, and drug action. Each of these areas provides important examples of normal or abnormal regulation of enzymes. For example, many cancer cells exhibit abnormalities in the regulation of their enzyme complement, illustrating that alterations of gene control are fundamental events in cancer cells. Again, certain oncogenic viruses contain a gene that codes for a tyrosine protein kinase. When this kinase is expressed in host cells, it can phosphorylate proteins and enzymes that are normally not phosphorylated and thus lead to dramatic changes in cell phenotype. A change of this nature appears to lie at the heart of certain types of viral oncogenic transformation. Drug action provides another important example involving enzyme regulation. Enzyme induction is one important biochemical cause of drug interactions, the situation in which the administration of one drug results in a significant change in the metabolism of another (Chapter 61).

REGULATION OF METABOLISM ACHIEVES HOMEOSTASIS

The concept of homeostatic regulation of the internal milieu advanced by Bernard in the late 19th century stressed the ability of animals to maintain the constancy of their intracellular environments despite changes in the external environment. This concept implies that enzyme-catalyzed reactions proceed at rates responsive to changes in the internal and external environment. A cell or organism might be defined as diseased when it responds inadequately or incorrectly to an internal signal or to an external stress. Knowledge of factors affecting the rates of enzyme-catalyzed reactions is essential both to understand the mechanism of homeostasis in normal cells and to comprehend the molecular basis of disease.

Metabolite Flow Tends to Be Unidirectional

All chemical reactions, including enzyme-catalyzed reactions, are to some extent reversible.[*] Within living cells, however, reversibility may not occur, because reaction products are promptly removed by additional enzyme-catalyzed reactions. Metabolite flow in living cells is analogous to the flow of water in a pipe. Although the pipe can transfer water in either direction, in practice the flow is unidirectional. Metabolite flow in living cells also is largely unidirectional. True equilibrium, far from being characteristic of life, is approached only when cells die. The living cell is a dynamic steady-state system maintained by a unidirectional flow of metabolites (Figure 11–1). In mature cells, the mean concentrations of metabolites remain relatively constant over considerable periods of time.[†] The flexibility of this steady-state system is illustrated by the delicate shifts and balances by which organisms maintain the constancy of the internal environment despite wide variations in food, water, and mineral intake, work output, or external temperature.

---

[*]A readily reversible reaction has a small numerical value of ΔG. One with a large negative value for ΔG might be termed "effectively irreversible" in most biochemical situations.
[†]Short-term oscillations of metabolite concentrations and of enzyme levels do occur, however, and are of profound physiologic importance.

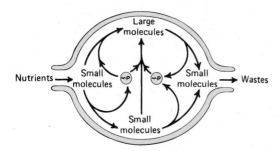

Figure 11–1. An idealized cell in steady state. Note that metabolite flow is unidirectional.

Figure 11–2. Enzyme quantity is determined by the net balance between enzyme synthesis and enzyme degradation. K_s and K_{deg} are the rate constants for the overall processes of synthesis and degradation, respectively.

Reaction Rates Respond to Changing Physiologic Need

For life to proceed in orderly fashion, metabolite flux through anabolic and catabolic pathways must be regulated. All requisite chemical events must proceed at rates consistent with the requirements of an organism in relation to its environment. ATP production, synthesis of macromolecular precursors, transport, secretion, and tubular reabsorption all must respond to subtle changes in the environment of the cell, organ, or intact animal. These processes must be coordinated and must respond to short-term changes in the external environment (eg, addition or removal of a nutrient) as well as to periodic intracellular events (eg, DNA replication). Until recently, the molecular details of regulation were best understood in bacteria, which lack the complexities of hormonal or neural control and in which genetic studies facilitate study of molecular events. Our understanding of molecular regulation in animal cells is in a state of rapid expansion. While metabolic regulation in mammals differs significantly from superficially similar phenomena in bacteria, regulation of metabolic processes in bacteria provides a conceptual framework for considering regulation in humans.

THREE GENERAL MECHANISMS REGULATE ENZYME ACTIVITY

Net flow of carbon through any enzyme-catalyzed reaction may be influenced (1) by changing the absolute *quantity* of enzyme present, (2) by altering the pool size of *reactants* other than enzyme, and (3) by altering the *catalytic efficiency* of the enzyme. All three options are exploited in most forms of life.

Rates of Synthesis and Degradation Determine Enzyme Quantity

The absolute quantity of an enzyme present is determined by its rate of synthesis (k_s) and rate of degradation (k_{deg}) (Figure 11–2). The quantity of an enzyme in a cell may be increased either by elevating its rate of synthesis (increase in k_s), by decreasing its rate of degradation (decrease in k_{deg}), or by both. Similarly, a lower quantity of enzyme can result from a decrease in k_s, an increase in k_{deg}, or both processes. Examples of changes in both k_s and k_{deg} occur in human subjects. In all forms of life, enzyme synthesis from amino acids and enzyme degradation to amino acids are distinct processes catalyzed by entirely different sets of enzymes. Independent regulation of enzyme synthesis and enzyme degradation is thus readily achieved (Figure 11–2).

Inducers Stimulate Enzyme Synthesis

Cells can synthesize specific enzymes in response to specific low-molecular-weight inducers. For example, *Escherichia coli* grown on glucose will not catabolize lactose owing to the absence of the enzyme β-galactosidase, which hydrolyzes lactose to galactose and glucose. If lactose or certain other β-galactosides are added to the growth medium, synthesis of the β-galactosidase and of a galactoside permease is **induced,** and the culture then can catabolize lactose.

Although many inducers are substrates for the enzymes they induce, compounds structurally similar to the substrate may be inducers but not substrates. These are termed **gratuitous inducers.** Conversely, a compound may be a substrate but not an inducer. Inducers often induce several enzymes of a catabolic pathway (eg, β-galactoside permease and β-galactosidase are both induced by lactose).

Enzymes whose concentration in a cell is independent of added inducer are termed **constitutive.** A particular enzyme may be constitutive in one strain, inducible in another, and absent in a third. Cells capable of being induced for a particular enzyme usually contain a small measurable basal level of that enzyme even in the absence of added inducer. The genetic heritage of the cell determines both the nature and the magnitude of the response to an inducer. "Constitutive" and "inducible" thus are relative terms, like "hot" and "cold," that represent the extremes of a spectrum of responses.

Enzyme induction also occurs in eukaryotes. Examples of inducible enzymes in animals include tryptophan pyrrolase, threonine dehydrase, tyrosine-α-ketoglutarate transaminase, invertase, enzymes of the urea cycle, HMG-CoA reductase, and cytochrome P450.

End Products Repress Enzyme Synthesis

The presence in the growth medium of bacteria of a metabolite being biosynthesized may curtail new synthesis of that metabolite via **repression.** Both induction and repression involve *cis* elements, specific DNA sequences located upstream of genes that encode a given enzyme, and *trans*-acting regulatory proteins. In addition, specific metabolites serve as corepressors or coinducers that, when bound to a *trans*-acting protein, either enhance or weaken its association with a *cis* element. High intracellular levels of a metabolite such as a purine or amino acid thus can block synthesis of the enzymes involved in its own biosynthesis. For example, in *Salmonella typhimurium,* addition of histidine represses synthesis of all the enzymes of His biosynthesis, and leucine represses synthesis of the first three enzymes unique to Leu biosynthesis. Following removal or exhaustion of an essential biosynthetic intermediate from the medium, enzyme biosynthesis again occurs. This constitutes **derepression.**

The above examples illustrate **product feedback repression** characteristic of biosynthetic pathways in bacteria. **Catabolite repression,** a related phenomenon, is the ability of an intermediate in a sequence of catabolic enzyme-catalyzed reactions to repress synthesis of catabolic enzymes. This effect was first noted in *E coli* growing on a carbon source (X) other than glucose. Addition of glucose repressed synthesis of the enzymes concerned with catabolism of X. This phenomenon was initially termed the "glucose effect." Since oxidizable nutrients other than glucose produce similar effects, the term "catabolite repression" was adopted. Catabolite repression is mediated by cAMP. The molecular mechanisms of induction, repression, and derepression are discussed in Chapter 41.

PROTEIN TURNOVER OCCURS IN ALL FORMS OF LIFE

The combined processes of enzyme synthesis and degradation constitute **enzyme turnover,** recognized as a characteristic property of all mammalian cells long before it was shown also to occur in bacteria. The existence of protein turnover in humans, deduced from dietary experiments well over a century ago, was conclusively shown by Schoenheimer's classic work just prior to and during World War II. By measuring the rates of incorporation of ^{15}N-labeled amino acids into protein and the rates of loss of ^{15}N from protein, Schoenheimer deduced that body proteins are in a state of "dynamic equilibrium," a

concept subsequently extended to other body constituents, including lipids and nucleic acids.

Rates of Degradation of Specific Enzymes Are Regulated

The degradation of mammalian proteins by ATP and ubiquitin-dependent pathways and by ATP-independent pathways is discussed in Chapter 31. The susceptibility of an enzyme to proteolytic degradation depends upon its conformation. The presence or absence of substrates, coenzymes, or metal ions, which can alter protein conformation, alters proteolytic susceptibility. The concentrations of substrates, coenzymes, and possibly ions in cells thus may influence the rates at which specific enzymes are degraded. Arginase and tryptophan oxygenase (tryptophan pyrrolase) illustrate these concepts. Regulation of liver arginase levels can involve a change either in k_s or in k_{deg}. After a protein-rich diet is ingested, liver arginase levels rise owing to an increased rate of arginase synthesis. Liver arginase levels also rise in starved animals. Here, however, it is arginase degradation that is decreased, while k_s remains unchanged.

In a second example, injection of glucocorticoids and ingestion of tryptophan both elevate levels of tryptophan oxygenase in mammals. The hormone raises the rate of oxygenase synthesis (raises k_s). Tryptophan, however, has no effect on k_s but lowers k_{deg} by stabilizing the oxygenase against proteolytic digestion. Contrast these two examples with enzyme induction in bacteria. For arginase, the increased intake of nitrogen on a high-protein diet may elevate liver arginase levels (Chapter 31). The increased rate of arginase synthesis thus superficially resembles that of substrate induction in bacteria. For tryptophan pyrrolase, however, even though tryptophan may act as an inducer in bacteria (affects k_s), its effect in mammals is solely on the enzyme degradative process (lowers k_{deg}).

Enzyme levels in mammalian tissues may be altered by a wide range of physiologic, hormonal, or dietary factors. Examples are known for a variety of tissues and metabolic pathways, but knowledge of the molecular details that account for these changes is fragmentary.

Glucocorticoids increase the concentration of tyrosine transaminase by stimulating k_s. This was the first clear case of a hormone regulating the synthesis of a mammalian enzyme. Insulin and glucagon, despite mutually antagonistic physiologic effects, both independently increase k_s four- to fivefold. The effect of glucagon is mediated via cAMP.

Regulatory Advantages Result From Synthesis as Inactive Precursors

Enzyme activity can be regulated by converting an inactive proenzyme to a catalytically active form. To become catalytically active, the proenzyme must undergo limited proteolysis, a process accompanied by

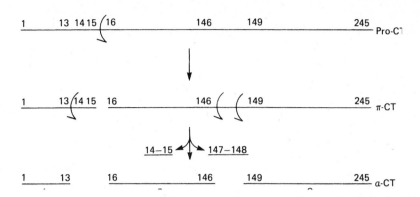

Figure 11–3. Conversion of prochymotrypsin (pro-CT) to π-chymotrypsin (π-CT) and subsequently to the mature catalytically active enzyme α-chymotrypsin (α-CT).

conformational changes that either reveal or "create" the catalytic site. This is illustrated below for the enzyme chymotrypsin.

MANY PROTEASES ARE SECRETED AS CATALYTICALLY INACTIVE PROENZYMES

Certain proteins are manufactured and secreted in the form of inactive precursor proteins known as **proproteins.** When the proteins are enzymes, the proproteins are termed **proenzymes** or **zymogens.** Conversion of a proprotein to the mature protein involves selective proteolysis, a process that converts the proprotein by one or more successive proteolytic "clips" to a form having the characteristic activity of the mature protein (its enzymatic activity). Examples of proteins manufactured as proproteins include the hormone insulin (proprotein = proinsulin), the digestive enzymes pepsin, trypsin, and chymotrypsin (proproteins = pepsinogen, trypsinogen, and chymotrypsinogen, respectively), several factors of the blood clotting and of the blood clot dissolution cascades (see Chapter 59), and the connective tissue protein collagen (proprotein = procollagen).

The conversion of prochymotrypsin (pro-CT), a 245-aminoacyl residue polypeptide, to the active enzyme α-chymotrypsin involves three proteolytic clips and the formation of an active intermediate known as π-chymotrypsin (π-CT) (Figure 11–3).

In α-chymotrypsin, the A, B, and C chains remain associated owing to the presence in α-CT of two interchain disulfide bonds (Figure 11–4).

Proenzymes Facilitate Rapid Mobilization of an Activity in Response to Physiologic Demand

Why are certain proteins secreted in an inactive form? Certain proteins are needed at essentially all times. Others (for example, the enzymes of blood clot formation and dissolution) are needed only intermittently. Furthermore, when these intermittently needed enzymes are required, they frequently are needed rapidly. Certain physiologic processes such as digestion are intermittent but fairly regular and predictable (although this may not have been the case for primitive humans). Others, such as blood clot formation, clot dissolution, and tissue repair, need only to be brought "on line" in response to pressing physiologic or pathophysiologic need.

It may readily be appreciated that the processes of blood clot formation and dissolution must be temporally coordinated to achieve homeostasis. In addition, the synthesis of proteases as catalytically inactive precursor proteins serves to protect the tissue of origin (eg, the pancreas) from autodigestion. Autodigestion can occur in pancreatitis. De novo synthesis of the required proteins might not be sufficiently rapid to respond to a pressing pathophysiologic demand such as the loss of blood. Moreover, an adequate and

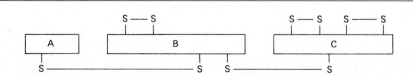

Figure 11–4. Intra- and interchain disulfide bonds of α-chymotrypsin (α-CT).

complete pool of the precursor amino acids must be available. Furthermore, the secretion process may be slow relative to the physiologic demand.

Activation of Prochymotrypsin Requires Selective Proteolysis

The example of conversion of a proprotein to its mature, physiologically active form illustrates the following general principles of proprotein to protein conversions:

(1) The process involves selective proteolysis, which in some instances requires only a single proteolytic clip.

(2) The polypeptide products may separate or may remain associated in the mature protein.

(3) The process may (or may not) be attended by a significant change in molecular weight.

(4) A major consequence of selective proteolysis is the attainment of a new conformation.

(5) If the proprotein is an enzyme, the above conformational change generates the catalytic site of the enzyme. Indeed, selective proteolysis of a proenzyme may be viewed as a process that triggers essential conformational changes that "create" the catalytic site.

Note that while His 57 and Asp 102 reside on the B peptide of α-chymotrypsin, Ser 195 resides on the C peptide (Figure 11–3). The selective proteolysis of prochymotrypsin (chymotrypsinogen) facilitates approximation of the three residues of the charge relay network. This illustrates how selective proteolysis can give rise to the catalytic site. Note also that contact and catalytic residues can be located on different peptide chains but still be within bond-forming distance of bound substrate.

MULTIPLE OPTIONS ARE AVAILABLE FOR REGULATING CATALYTIC ACTIVITY

If a physiologic manipulation alters the level of enzyme activity, we must inquire whether the quantity of enzyme has changed or whether the enzyme is a more efficient or less efficient catalyst. We refer to all changes in enzyme activity that occur without change in the quantity of enzyme present as "effects on catalytic efficiency."

ENZYME COMPARTMENTATION FACILITATES REGULATION OF METABOLISM

The importance of compartmentation of metabolic processes in eukaryotic cells, including those of mammals, cannot be overemphasized. Localization of specific metabolic processes in the cytosol or in subcellular organelles facilitates regulation of these processes independent of processes proceeding elsewhere. The extensive compartmentation of metabolic processes characteristic of higher forms of life thus confers the potential for finely tuned regulation of metabolism. At the same time, it poses problems with respect to translocation of metabolites across compartmental barriers. This is achieved via **shuttle mechanisms** that convert the metabolite to a form permeable to the compartmental barrier followed by transport and conversion back to the original form on the other side of the barrier. These interconversions often require cytosolic and compartmented forms of the same catalytic activity. Since these two forms of the enzyme are physically separated, their independent regulation is facilitated. The role of shuttle mechanisms in achieving regulation of metabolic pools of reducing equivalents, of citric acid cycle intermediates, and of other amphibolic intermediates is discussed in Chapter 18.

ENZYMES CAN FORM MACROMOLECULAR COMPLEXES

Organization of a set of enzymes that catalyze a sequence of metabolic reactions into a macromolecular complex channels intermediates along a metabolic path. Appropriate alignment of the enzymes can facilitate transfer of product between enzymes without prior equilibration with metabolic pools. This permits a finer level of metabolic control than is possible with the isolated components of the complex. In addition, conformational changes in one component of the complex may be transmitted by protein-protein interactions to other enzymes of the complex. Amplification of regulatory effects thus is possible.

LOCAL CONCENTRATIONS OF SUBSTRATES, COENZYMES, AND CATIONS CAN REGULATE ENZYMES

The mean intracellular concentration of a substrate, coenzyme, or metal ion may have little meaning for the in vivo behavior of an enzyme. Information on the concentrations of essential metabolites in the immediate neighborhood of the enzyme in question is needed. However, even measuring metabolite concentrations in different subcellular compartments does not take into account local discontinuities in metabolite concentrations within compartments brought about by factors such as proximity to the site of entry or production of a metabolite. Finally, little consideration generally is given to the discrepancy between total and free metabolite concentrations. For example, while the total concentration of 2,3-bisphosphoglycerate in erythrocytes is extremely high, the concentration of free (ie, not bound to hemoglobin) bisphosphoglycerate is comparable to that of other tissues. Similar considerations apply to other metabolites in the presence of proteins that bind them effectively and reduce their concentrations in the free state.

An assumption of the Michaelis-Menten kinetic approach is that the concentration of total substrate is essentially equal to the concentration of free substrate. As noted above, this assumption may be invalid in vivo, where concentrations of free substrates may approach those of enzyme concentrations.

Metal ions, which perform catalytic and structural roles in over one-fourth of all known enzymes (Chapter 10), may also fulfill regulatory roles, particularly for reactions where ATP and other polyanions are substrates. Maximal activity typically is observed at a molar ratio of ATP to metal of about unity. Excess metal or excess ATP is inhibitory. Since nucleoside di- and triphosphates form stable complexes with divalent cations, intracellular concentrations of the nucleotides can influence intracellular concentrations of free metal ions and hence the activity of certain enzymes.

CERTAIN ENZYMES ARE REGULATED BY ALLOSTERIC EFFECTORS

The catalytic activity of certain **regulatory enzymes** is modulated by low-molecular-weight allosteric effectors that generally have little or no structural similarity to the substrates or coenzymes for the regulated enzyme. **Feedback inhibition** refers to the inhibition of the activity of an enzyme in a biosynthetic pathway by an end product of that pathway. For example, for the biosynthesis of D from A catalyzed by enzymes Enz_1 through Enz_3:

$$\overset{Enz_1}{} \overset{Enz_2}{} \overset{Enz_3}{}$$
$$A \rightarrow B \rightarrow C \rightarrow D$$

a high concentration of D typically inhibits conversion of A to B. This involves not simple "backing up" of intermediates but the ability of D to bind to and inhibit Enz_1. D thus acts as a **negative allosteric effector** or **feedback inhibitor** of Enz_1. Feedback inhibition of Enz_1 by D therefore regulates the synthesis of D. Typically, D binds to the sensitive enzyme at an **allosteric site** spatially remote from the catalytic site.

The kinetics of feedback inhibition may be competitive, noncompetitive, partially competitive, uncoupled, or mixed. Feedback inhibition is commonest in biosynthetic pathways. Frequently the feedback inhibitor is the last small molecule before a macromolecule (eg, amino acids before proteins, nucleotides before nucleic acids). **Feedback regulation generally occurs at the earliest functionally irreversible* step unique to a particular biosynthetic**

*One strongly favored (in thermodynamic terms) in a single direction, ie, one with a large negative ΔG.

sequence. A much-studied example is inhibition by CTP of bacterial aspartate transcarbamoylase (see below and Chapter 36).

Frequently, a biosynthetic pathway is branched, with the initial portion serving for synthesis of two or more essential metabolites. Figure 11–5 shows probable sites of simple feedback inhibition in a branched biosynthetic pathway (eg, for amino acids, purines or pyrimidines). S_1, S_2, and S_3 are precursors of all four end products (A, B, C, and D), S_4 is a precursor of B and C, and S_5 is a precursor solely of D. The sequences

$$S_3 \rightarrow A$$
$$S_4 \rightarrow B$$
$$S_4 \rightarrow C$$
$$S_3 \rightarrow S_5 \rightarrow D$$

thus constitute linear reaction sequences that might be expected to be feedback-inhibited by their end products. Again, nucleotide biosynthesis (Chapter 36) provides specific examples.

Multiple Feedback Loops Regulate Branched Biosynthetic Pathways

Additional fine control is provided by multiple feedback loops (Figure 11–6). For example, if B is present in excess, the requirement for S_2 decreases. The ability of B to decrease production of S_2 thus confers a biologic advantage. However, if excess B inhibits not only the portion of the pathway unique to its own synthesis but also portions common to that for synthesis of A, C, or D, excess B should curtail synthesis of all four end products. Clearly, this is undesirable. Mechanisms have, however, evolved to circumvent this difficulty.

In **cumulative feedback inhibition,** the inhibitory effect of two or more end products on a single regulatory enzyme is strictly additive.

In **concerted** or **multivalent feedback inhibition,**

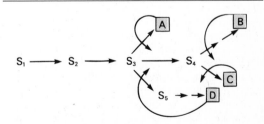

Figure 11–5. Sites of feedback inhibition in a branched biosynthetic pathway. S_1–S_5 are intermediates in the biosynthesis of end products A through D. Straight arrows represent enzymes catalyzing the indicated conversions. Curved arrows represent feedback loops and indicate probable sites of feedback inhibition by specific end products.

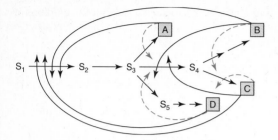

Figure 11–6. Multiple feedback inhibition in a branched biosynthetic pathway. Superimposed on simple feedback loops (dashed, curved arrows) are multiple feedback loops (solid, curved arrows) that regulate enzymes common to biosynthesis of several end products.

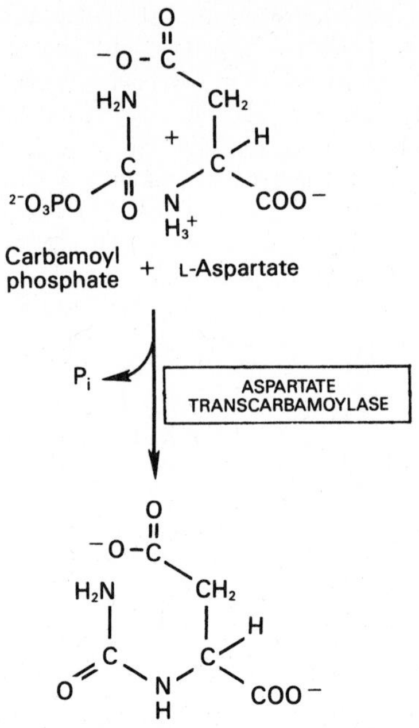

complete inhibition occurs only when two or more end products both are present in excess.

In **cooperative feedback inhibition,** a single end product present in excess inhibits the regulatory enzyme, but the inhibition when two or more end products are present far exceeds the additive effects of cumulative feedback inhibition.

Aspartate Transcarbamoylase Is a Model Allosteric Enzyme

Aspartate transcarbamoylase (ATCase) catalyzes the first reaction unique to pyrimidine biosynthesis (Figure 11–7). ATCase is feedback-inhibited by cytidine triphosphate (CTP). Following treatment with mercurials, ATCase loses its sensitivity to inhibition by CTP but retains its full activity for carbamoyl aspartate synthesis. This suggests that CTP is bound at a different (allosteric) site from either substrate. ATCase consists of multiple catalytic and regulatory protomers. Each catalytic protomer contains four aspartate (substrate) sites and each regulatory protomer at least two CTP (regulatory) sites.

Allosteric and Catalytic Sites Are Spatially Distinct

Monod noted the lack of structural similarity between a feedback inhibitor and the substrate for the enzyme whose activity it regulated. Since the effectors are not **isosteric** with a substrate but **allosteric** ("occupy another space"), he proposed that enzymes whose activity is regulated by **allosteric effectors** (eg, feedback inhibitors) bind the effector at an *allosteric site* that is physically distinct from the catalytic site. **Allosteric enzymes** thus are enzymes whose activity at the catalytic site may be modulated by the presence of allosteric effectors at an allosteric site. Lines of evidence that support the existence of physically distinct allosteric sites on regulated enzymes include the following:

(1) Regulated enzymes modified by chemical or physical techniques frequently become insensitive to

Figure 11–7. The aspartate transcarbamoylase (ATCase) reaction.

their allosteric effectors without alteration of their catalytic activity. Selective denaturation of allosteric sites has been achieved by treatment with mercurials, urea, x-rays, proteolytic enzymes, extremes of ionic strength or pH, by aging at 0–5 °C, by freezing, or by heating.

(2) Allosteric effectors frequently protect the catalytic site from denaturation under conditions where the substrates themselves do not protect. Since it seems unlikely that an effector bound at the catalytic site would protect when substrates do not, this suggests a second, allosteric site elsewhere on the enzyme molecule.

(3) In certain bacterial and mammalian cell mutants, the regulated enzymes have altered regulatory properties but wild-type catalytic properties. The structures of the allosteric and catalytic sites thus are genetically distinct.

(4) Binding studies of substrates and of allosteric effectors to regulated enzymes show that each may bind independently of the other.

(5) In certain cases (eg, ATCase), the allosteric site is present on a different protomer from the catalytic site.

Allosteric Enzymes Typically Exhibit Sigmoidal Substrate Saturation Kinetics

Figure 11–8 illustrates the rate of a reaction catalyzed by a representative allosteric enzyme measured at several concentrations of substrate in the presence and absence of an allosteric inhibitor. In the

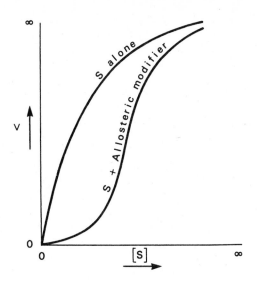

Figure 11–8. Sigmoid saturation curve for substrate in the presence of an allosteric inhibitor.

absence of the allosteric inhibitor, normal hyperbolic saturation kinetics are observed. In its presence, the substrate saturation curve is distorted from a hyperbola into a sigmoid, which at high substrate concentrations may merge with the hyperbola. Note the analogy to the relationship between the O_2 saturation curves for myoglobin and hemoglobin (Chapter 7).

Kinetic analysis of feedback inhibition shows that it can be competitive, noncompetitive, partially competitive, or of other types. If, at high concentrations of substrate, comparable activity is observed in the presence or absence of the allosteric inhibitor, the kinetics superficially resemble those of competitive inhibition. However, since the substrate saturation curve is sigmoid rather than hyperbolic, it is not possible to obtain meaningful results by graphing data for allosteric inhibition by the double-reciprocal technique. This method of analysis was developed for substrate competitive inhibition at the catalytic site. Since allosteric inhibitors act at a different (allosteric) site, that kinetic model is invalid.

The sigmoid character of the v versus [S] curve in the presence of an allosteric inhibitor reflects the phenomenon of **cooperativity.** At low concentrations of substrate, the activity in the presence of the inhibitor is low relative to that in its absence. However, as [S] is increased, the extent of inhibition becomes less severe. The kinetics are consistent with the presence of two or more interacting substrate-binding sites, where the presence of a substrate molecule at one catalytic site facilitates binding of a second substrate molecule at a second site. Cooperativity of substrate binding has been described in Chapters 7 and 9: the sigmoid O_2 saturation curve that results from cooperative interactions between four O_2 binding

sites located on different protomers of hemoglobin and the use of the Hill plot to quantify cooperativity.

Allosteric Effects May Be on K_m or V_{max}

Reference to the kinetics of allosteric inhibition as "competitive" or "noncompetitive" with substrate carries mechanistic implications that are misleading. We refer instead to two classes of regulated enzymes: K-series and V-series enzymes. For K-series allosteric enzymes, the substrate saturation kinetics are competitive in the sense that K_m is raised without effect on V_{max}. For V-series allosteric enzymes, the allosteric inhibitor lowers V_{max} without affecting the apparent K_m. Alterations in K_m or V_{max} probably result from conformational changes at the catalytic site induced by binding of the allosteric effector at the allosteric site. For a K-series allosteric enzyme, this conformational change may weaken the bonds between substrate and substrate-binding residues. For a V-series allosteric enzyme, the primary effect may be to alter the orientation or charge of catalytic residues so as to lower V_{max}. Intermediate effects on K_m and V_{max}, however, may be observed consequent to these conformational changes.

Cooperative Binding of a Substrate Confers Physiologic Advantage

The advantages of cooperative substrate-binding kinetics are analogous to those resulting from the cooperative binding of O_2 to hemoglobin. At low substrate concentrations, the allosteric effector is an effective inhibitor. It thus regulates most effectively at the time of greatest need, ie, when intracellular concentrations of substrates are low. As more substrate becomes available, stringent regulation is less necessary. As substrate concentration rises, the degree of inhibition lessens, and more product is formed. As with hemoglobin, the sigmoid substrate saturation curve in the presence of inhibitor also ensures that relatively small changes in substrate concentration result in large changes in activity. Sensitive control of catalytic activity thus is achieved by *small* changes in substrate concentration. Finally, by analogy with the differing O_2 saturation curves of hemoglobins from different species, regulatory enzymes from different sources may have sigmoid saturation curves shifted to the left or right to accommodate to the range of in vivo concentrations of substrate.

FEEDBACK REGULATION IS NOT SYNONYMOUS WITH FEEDBACK INHIBITION

In both mammalian and bacterial cells, end products "feed back" and control their own synthesis. In many instances, this involves feedback inhibition of an early biosynthetic enzyme. We must, however, distinguish between **feedback regulation,** a phenomenologic term devoid of mechanistic implications, and **feedback inhibition,** a mechanism for regulation of many bacterial and mammalian enzymes. For example, dietary cholesterol restricts the synthesis of cholesterol from acetate in mammalian tissues. This feedback regulation, however, does not appear to involve feedback inhibition of an early enzyme of cholesterol biosynthesis. An early enzyme (HMG-CoA reductase) is affected, but the mechanism involves curtailment by cholesterol or a cholesterol metabolite of the expression of the gene that encodes HMG-CoA reductase (ie, enzyme repression). Cholesterol added directly to HMG-CoA reductase has no effect on its catalytic activity.

REVERSIBLE, COVALENT MODIFICATION REGULATES KEY MAMMALIAN ENZYMES

Reversible modulation of the catalytic activity of enzymes can occur by covalent attachment of a phosphate group to one or more Ser, Thr, Tyr, or His residues. Enzymes that undergo covalent modification with attendant modulation of their activity are termed "interconvertible enzymes." Interconvertible enzymes exist in two activity states, one of high and the other of low catalytic efficiency. Depending on the enzyme concerned, the phospho- or the dephosphoenzyme may be the more active catalyst (Table 11–1).

Enzymes May Have Multiple Phosphorylation Sites

A specific seryl residue is phosphorylated, forming an *O*-phosphoseryl residue, or a tyrosyl residue is phosphorylated to form an *O*-phosphotyrosyl residue. While an interconvertible enzyme may contain many Ser or Tyr residues, phosphorylation is highly selective and occurs at only a small number of sites. These sites probably do not form part of the catalytic site, at least in a primary structural sense, and thus constitute another example of an allosteric site.

Table 11–1. Examples of mammalian enzymes whose catalytic activity is altered by covalent phosphorylation-dephosphorylation.

Enzyme	Activity State	
	Low	High
Acetyl-CoA carboxylase	EP	E
Glycogen synthase	EP	E
Pyruvate dehydrogenase	EP	E
HMG-CoA reductase	EP	E
Glycogen phosphorylase	E	EP
Citrate lyase	E	EP
Phosphorylase b kinase	E	EP
HMG-CoA reductase kinase	E	EP

E, dephosphoenzyme; EP, phosphoenzyme.

Protein Kinases and Phosphatases Are Converter Proteins

Phosphorylation and dephosphorylation of proteins are catalyzed by a variety of protein kinases and protein phosphatases (converter proteins), respectively (Figure 11–9). The ability of these protein kinases to recognize distinctive motifs, or patterns, of primary structure in part accounts for their high specificity. The converter proteins themselves may be interconvertible enzymes (Table 11–1). Thus, there are protein kinase kinases and protein kinase phosphatases that catalyze the interconversion of these converter proteins. The activity of protein phosphatases is regulated, and the activity of both protein kinases and protein phosphatases is under hormonal and neural control, although the precise details by which these agents act are in most instances far from clear.

Phosphorylation-Dephosphorylation Consumes ATP

The reactions shown in Figure 11–9 resemble those for interconversion of glucose and glucose 6-phosphate or of fructose 6-phosphate and fructose 1,6-bisphosphate (Chapter 19). The net result of phosphorylating and then dephosphorylating 1 mol of substrate (enzyme or sugar) is the hydrolysis of 1 mol of ATP. The activities of the kinases that catalyze reactions 1 and 3 (below) and of the phosphatases that catalyze reactions 2 and 4 must be regulated, for otherwise they would act together to catalyze uncontrolled hydrolysis of ATP.

1. Glucose + ATP → ADP + Glucose 6-P
2. H_2O + Glucose 6-P → P_i + Glucose

Net: H_2O + ATP → ADP + P_i

3. Enz—Ser—OH + ATP → ADP + Enz—Ser—O—P
4. H_2O + Enz—Ser—O—P → P_i + Enz—Ser—OH

Net: H_2O + ATP → ADP + P_i

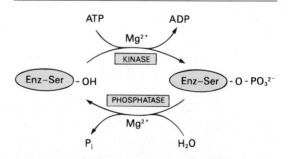

Figure 11–9. Covalent modification of a regulated enzyme by phosphorylation-dephosphorylation of a seryl residue.

Covalent Modification Regulates Metabolite Flow

Regulation of enzyme activity by phosphorylation-dephosphorylation has analogies to regulation by feedback inhibition. Both provide for short-term, readily reversible regulation of metabolite flow in response to specific physiologic signals. Both act without altering gene expression. Both act on early enzymes of a protracted (often biosynthetic) metabolic sequence, and both act at allosteric rather than catalytic sites. Feedback inhibition, however, involves a single protein and lacks hormonal and neural features. By contrast, regulation of mammalian enzymes by phosphorylation-dephosphorylation involves several proteins and ATP and is under direct neural and hormonal control.

SUMMARY

Regulation of enzyme activity contributes in a major way to preserving homeostasis: maintaining a relatively constant intracellular and intraorganismal environment in the face of wide fluctuations in the external environment (such as changes in temperature, the presence or absence of water or specific foods). To achieve homeostasis, the rates of numerous biochemical reactions must respond to physiologic need. How is this achieved?

Local substrate concentrations, enzyme compartmentation, and secretion as catalytically inactive proenzymes or zymogens (eg, chymotrypsin) all contribute to the regulation of metabolic processes. Many proteases and other exported proteins are secreted as biologically inert proproteins that must undergo selective proteolytic cleavage to form a biologically active enzyme or hormone. Examples include chymotrypsin, insulin, and the proteases of the blood clotting and blood clot dissolution cascades. Secretion as an inactive precursor provides protection against their action until need arises and facilitates rapid mobilization of an activity without requiring new protein synthesis. One or more selective proteolytic events trigger conformational changes that approximate and align formerly distant residues, forming the catalytic site. The peptides formed by selective proteolysis may be discarded (eg, the two dipeptides of CT) or may remain linked by disulfide bonds (the A, B, and C peptides of CT and the A and B chains of insulin).

Rapid and subtle alterations in the catalytic activity of key regulated enzymes play major roles in the selective channeling of metabolites toward one or another metabolic process. Regulated enzymes tend to be those which catalyze an early—frequently the earliest—reaction unique to a given sequence of metabolic reactions. The catalytic activity of regulated enzymes can be modulated (eg, switched repeatedly between low and high states of catalytic activity) in the absence of new protein synthesis or protein degradation. Modulation is achieved when specific metabolites bind to the regulated enzyme, generally at allosteric sites far removed from the catalytic site. Modulation of enzyme activity via conformational changes at the catalytic site may involve altering K_m for a substrate, changing V_{max} for the overall reaction, or effects on both K_m and V_{max}. The substrate saturation curves for allosteric inhibition often are sigmoidal, hence the Michaelis-Menten equation does not apply. Quantitative evaluation of cooperativity of allosteric enzymes employs the Hill equation. For biosynthetic processes, feedback inhibition involves regulatory metabolites "biosynthetically upstream" of the enzyme in question (eg, inhibition of aspartate transcarbamoylase by CTP). For catabolic reaction sequences, metabolites "biodegradatively downstream" of the regulated enzyme act as regulators (eg, inhibition of phosphofructokinase by ATP or citrate). Where more than one metabolite can feedback-regulate a given enzyme, the action of multiple regulatory metabolites may be cumulative, cooperative, or multivalent. In addition, a given metabolite may regulate the activity of several enzymes, each unique to a particular sequence of metabolic reactions (multiple feedback loops).

In humans and other eukaryotes, the activity of many enzymes is regulated by covalent modification, most commonly by ATP-dependent phosphorylation, with subsequent hydrolytic removal of phosphate as inorganic orthophosphate. Selective phosphorylation, most commonly of specific Ser or Tyr residues, and subsequent dephosphorylation, are catalyzed by protein kinases and protein phosphatases (converter enzymes). The activities of these converter enzymes are themselves regulated. The target enzyme and its converter enzymes thus may form a portion of a regulatory cascade that responds to a signal triggered by a hormone or by a second messenger such as cAMP.

REFERENCES

Crabtree B, Newsholme EA: A systematic approach to describing and analyzing metabolic control systems. Trends Biochem Sci 1987;12:4.

Falfvey E, Schibler U: How are the regulators regulated? Fed Am Soc Exp Biol J 1990;5:309.

Harris RA et al: Molecular cloning of the branched-chain α-ketoacid dehydrogenase kinase and the CoA-dependent methylmalonate semialdehyde dehydrogenase. Adv Enzyme Regul 1993;33:255.

Johnson LN, Barford D: The effect of phosphorylation on

the structure and function of proteins. Annu Rev Biophys Biomol Struct 1993;22:199.

Kacser H, Porteus JW: Control of metabolism: What have we to measure? Trends Biochem Sci 1987;12:5.

Kennelly PJ, Krebs EG: Consensus sequences as substrate specificity determinants for protein kinases and protein phosphatases. J Biol Chem 1991;266:15555.

Pilkis SJ et al: 6-Phosphofructo-2-kinase/fructose-2,6-bisphosphatase: A metabolic signaling enzyme. Annu Rev Biochem 1995;64:799.

Roach PJ: Multisite and hierarchical protein phosphorylation. J Biol Chem 1991;266:14139.

Schlesinger MJ, Hershko A: *The Ubiquitin System.* Cold Spring Harbour Press, 1988.

Scriver CR et al (editors): *The Metabolic and Molecular Bases of Inherited Disease,* 7th ed. McGraw-Hill, 1995.

Soderling TR: Role of hormones and protein phosphorylation in metabolic regulation. Fed Proc 1982;41:2615.

Stadtman ER, Chock PB (Editors): *Current Topics in Cellular Regulation.* Academic, 1969 to the present.

Weber G (editor): *Advances in Enzyme Regulation.* Pergamon Press, 1963–present.

Section II.
Bioenergetics & the Metabolism of Carbohydrates & Lipids

Bioenergetics: The Role of ATP

12

Peter A. Mayes, PhD, DSc

INTRODUCTION

Bioenergetics, or biochemical thermodynamics, is the study of the energy changes accompanying biochemical reactions. It provides the underlying principles to explain why some reactions may occur while others do not. Nonbiologic systems may utilize heat energy to perform work, but biologic systems are essentially **isothermic** and use chemical energy to power the living processes.

BIOMEDICAL IMPORTANCE

Suitable fuel is required to provide the energy that enables the animal to carry out its normal processes. How the organism obtains this energy from its food is basic to the understanding of normal nutrition and metabolism. Death from **starvation** occurs when available energy reserves are depleted, and certain forms of malnutrition are associated with energy imbalance **(marasmus).** The rate of energy release, measured by the metabolic rate, is controlled by the thyroid hormones, whose malfunction is a cause of disease. Excess storage of surplus energy results in **obesity,** one of the most common diseases of Western society.

FREE ENERGY IS THE USEFUL ENERGY IN A SYSTEM

Change in free energy (ΔG) is that portion of the total energy change in a system that is available for doing work; ie, it is the useful energy, also known in chemical systems as the chemical potential.

Biologic Systems Conform to the General Laws of Thermodynamics

The first law of thermodynamics states that **the total energy of a system, including its surroundings, remains constant.** This is also the law of conserva-

tion of energy. It implies that within the total system, energy is neither lost nor gained during any change. However, within that total system, energy may be transferred from one part to another or may be transformed into another form of energy. For example, in living systems, chemical energy may be transformed into heat, electrical energy, radiant energy, or mechanical energy.

The second law of thermodynamics states that **the total entropy of a system must increase if a process is to occur spontaneously. Entropy** represents the extent of disorder or randomness of the system and becomes maximum in a system as it approaches true equilibrium. Under conditions of constant temperature and pressure, the relationship between the free energy change (ΔG) of a reacting system and the change in entropy (ΔS) is given by the following equation, which combines the two laws of thermodynamics:

$$\Delta G = \Delta H - T\Delta S$$

where ΔH is the change in **enthalpy** (heat) and T is the absolute temperature.

Under the conditions of biochemical reactions, because ΔH is approximately equal to ΔE, the total change in internal energy of the reaction, the above relationship may be expressed in the following way:

$$\Delta G = \Delta E - T\Delta S$$

If ΔG is negative in sign, the reaction proceeds spontaneously with loss of free energy; ie, it is **exergonic.** If, in addition, ΔG is of great magnitude, the reaction goes virtually to completion and is essentially irreversible. On the other hand, if ΔG is positive, the reaction proceeds only if free energy can be gained; ie, it is **endergonic.** If, in addition, the magnitude of ΔG is great, the system is stable, with little or no tendency for a reaction to occur. If ΔG is zero, the system is at equilibrium and no net change takes place.

When the reactants are present in concentrations of 1.0 mol/L, ΔG^0 is the standard free energy change. For biochemical reactions, a standard state is defined as having a pH of 7.0. The standard free energy change at this standard state is denoted by $\Delta G^{0'}$.

The standard free energy change can be calculated from the equilibrium constant K'_{eq}.

$$\Delta G^{0'} = -2.303\ RT \log K'_{eq}$$

where R is the gas constant and T is the absolute temperature (Chapter 9). It is important to note that the actual ΔG may be larger or smaller than $\Delta G^{0'}$ depending on the concentrations of the various reactants, including the solvent, various ions, and proteins.

In a biochemical reaction system, it must be appreciated that an enzyme only speeds up the attainment of equilibrium; it never alters the final concentrations of the reactants at equilibrium after they are detached from the enzyme.

ENDERGONIC PROCESSES PROCEED BY COUPLING TO EXERGONIC PROCESSES

The vital processes—eg, synthetic reactions, muscular contraction, nerve impulse conduction, and active transport—obtain energy by chemical linkage, or **coupling**, to oxidative reactions. In its simplest form, this type of coupling may be represented as shown in Figure 12–1. The conversion of metabolite A to metabolite B occurs with release of free energy. It is coupled to another reaction, in which free energy is required to convert metabolite C to metabolite D. As some of the energy liberated in the degradative reaction is transferred to the synthetic reaction in a form other than heat, the normal chemical terms "exother-

mic" and "endothermic" cannot be applied to these reactions. Rather, the terms "exergonic" and "endergonic" are used to indicate that a process is accompanied by loss or gain, respectively, of free energy, regardless of the form of energy involved. In practice, an endergonic process cannot exist independently but must be a component of a coupled exergonic-endergonic system where the overall net change is exergonic. The exergonic reactions are termed **catabolism** (generally, the breakdown or oxidation of fuel molecules), whereas the synthetic reactions that build up substances are termed **anabolism.** The combined catabolic and anabolic processes constitute **metabolism.**

If the reaction shown in Figure 12–1 is to go from left to right, then the overall process must be accompanied by loss of free energy as heat. One possible mechanism of coupling could be envisaged if a common obligatory intermediate (I) took part in both reactions, ie,

$$A + C \rightarrow I \rightarrow B + D$$

Some exergonic and endergonic reactions in biologic systems are coupled in this way. It should be appreciated that this type of system has a built-in mechanism for biologic control of the rate at which oxidative processes are allowed to occur, since the existence of a common obligatory intermediate allows the rate of utilization of the product of the synthetic path (D) to determine by mass action the rate at which A is oxidized. Indeed, these relationships supply a basis for the concept of **respiratory control,** the process that prevents an organism from burning out of control. An extension of the coupling concept is provided by dehydrogenation reactions, which are coupled to hydrogenations by an intermediate carrier (Figure 12–2).

An alternative method of coupling an exergonic to an endergonic process is to synthesize a compound of high-energy potential in the exergonic reaction and to incorporate this new compound into the endergonic reaction, thus effecting a transference of free energy from the exergonic to the endergonic pathway. In Figure 12–3, ~Ⓔ is a compound of high potential energy and ~Ⓔ is the corresponding compound of low potential energy. The biologic advantage of this mechanism is that ~Ⓔ, unlike I in the previous system, need not be structurally related to

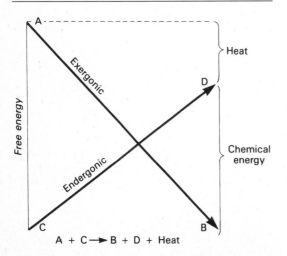

Figure 12–1. Coupling of an exergonic to an endergonic reaction.

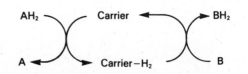

Figure 12–2. Coupling of dehydrogenation and hydrogenation reactions by an intermediate carrier.

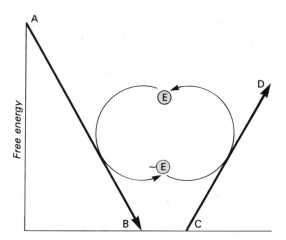

Figure 12–3. Transfer of free energy from an exergonic to an endergonic reaction via a high-energy intermediate compound.

A, B, C, or D. This would allow Ⓔ to serve as a transducer of energy from a wide range of exergonic reactions to an equally wide range of endergonic reactions or processes, as shown in Figure 12–4. In the living cell, the principal high-energy intermediate or carrier compound (designated ~Ⓔ) is **adenosine triphosphate (ATP).**

HIGH-ENERGY PHOSPHATES PLAY A CENTRAL ROLE IN ENERGY CAPTURE AND TRANSFER

In order to maintain living processes, all organisms must obtain supplies of free energy from their environment. **Autotrophic** organisms couple their metabolism to some simple exergonic process in their surroundings; eg, green plants utilize the energy of sunlight, and some autotrophic bacteria utilize the reaction $Fe^{2+} \rightarrow Fe^{3+}$. On the other hand, **heterotrophic** organisms obtain free energy by coupling their metabolism to the breakdown of complex organic molecules in their environment. In all these organisms, ATP plays a central role in the transference of free energy from the exergonic to the endergonic processes (Figures 12–3 and 12–4). ATP is a nucleoside triphosphate containing adenine, ribose, and three phosphate groups. In its reactions in the cell, it functions as the Mg^{2+} complex (Figure 12–5).

The importance of phosphates in intermediary metabolism became evident with the discovery of the chemical details of glycolysis and of the role of ATP, adenosine diphosphate (ADP), and inorganic phosphate (P_i) in this process (Chapter 19). ATP was considered a means of transferring phosphate radicals in the process of phosphorylation. The role of ATP in biochemical energetics was indicated in experiments in the 1940s demonstrating that ATP and creatine phosphate were broken down during muscular contraction and that their resynthesis depended on supplying energy from oxidative processes in the muscle. It was not until Lipmann introduced the concept of "high-energy phosphates" and the "high-energy phosphate bond" that the role of these compounds in bioenergetics was clearly appreciated.

The Intermediate Value for the Free Energy of Hydrolysis of ATP Compared With That of Other Organophosphates Has Important Bioenergetic Significance

The standard free energy of hydrolysis of a number of biochemically important phosphates is shown in Table 12–1. An estimate of the comparative tendency of each of the phosphate groups to transfer to a suitable acceptor may be obtained from the $\Delta G^{0'}$ of hy-

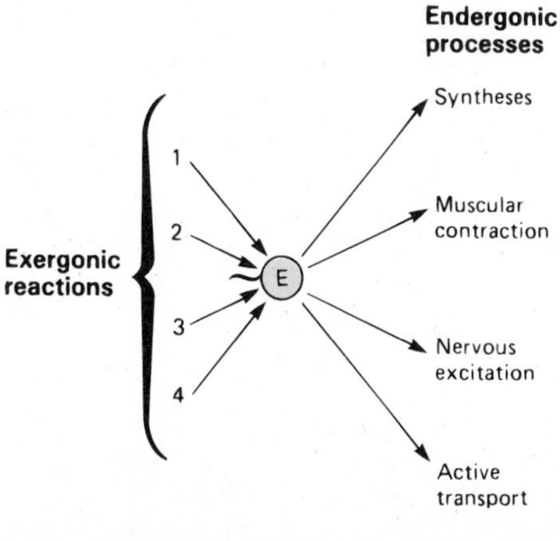

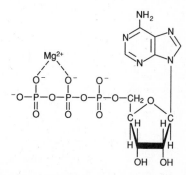

Figure 12–4. Transduction of energy through a common high-energy compound to energy-requiring (endergonic) biologic processes.

Figure 12–5. Adenosine triphosphate (ATP) shown as the magnesium complex. ADP forms a similar complex with Mg^{2+}.

Table 12–1. Standard free energy of hydrolysis of some organophosphates of biochemical importance.[1,2]

Compound	ΔG⁰′	
	kJ/mol	kcal/mol
Phosphoenolpyruvate	–61.9	–14.8
Carbamoyl phosphate	–51.4	–12.3
1,3-Bisphosphoglycerate (to 3-phosphoglycerate)	–49.3	–11.8
Creatine phosphate	–43.1	–10.3
ATP → ADP + P$_i$	–30.5	–7.3
ADP → AMP + P$_i$	–27.6	–6.6
Pyrophosphate	–27.6	–6.6
Glucose 1-phosphate	–20.9	–5.0
Fructose 6-phosphate	–15.9	–3.8
AMP	–14.2	–3.4
Glucose 6-phosphate	–13.8	–3.3
Glycerol 3-phosphate	–9.2	–2.2

[1]P$_i$, inorganic orthophosphate.
[2]Values for ATP and most others taken from Krebs and Kornberg (1957). They differ between investigators depending on the precise conditions under which the measurements are made.

drolysis (measured at 37 °C). It may be seen from the table that the value for the hydrolysis of the terminal phosphate of ATP divides the list into two groups. One group of **low-energy phosphates,** exemplified by the ester phosphates found in the intermediates of glycolysis, has $\Delta G^{0'}$ values smaller than that of ATP, while in the other group, designated **high-energy phosphates,** the value is higher than that of ATP. The components of this latter group, including ATP, are usually anhydrides (eg, the 1-phosphate of 1,3-bis-phosphoglycerate), enolphosphates (eg, phospho-enolpyruvate), and phosphoguanidines (eg, creatine phosphate, arginine phosphate). The intermediate position of ATP allows it to play an important role in energy transfer. The high free energy change on hydrolysis of ATP is due to charge repulsion of adjacent negatively charged oxygen atoms and to stabilization of the reaction products, especially phosphate, as resonance hybrids. Other biologically important compounds that are classed as "high-energy compounds" are thiol esters involving coenzyme A (eg, acetyl-CoA), acyl carrier protein, amino acid esters involved in protein synthesis, S-adenosylmethionine (active methionine), UDPGlc (uridine diphosphate glucose), and PRPP (5-phosphoribosyl-1-pyrophosphate).

High-Energy Phosphates Are Designated by ~Ⓟ

To indicate the presence of the high-energy phosphate group, Lipmann introduced the symbol ~ Ⓟ. The symbol indicates that the group attached to the bond, on transfer to an appropriate acceptor, results in transfer of the larger quantity of free energy. For this reason, the term **group transfer potential** is preferred by some to "high-energy bond." Thus, ATP contains two high-energy phosphate groups and ADP

contains one, whereas the phosphate in AMP (adenosine monophosphate) is of the low-energy type, since it is a normal ester link (Figure 12–6).

HIGH-ENERGY PHOSPHATES ACT AS THE "ENERGY CURRENCY" OF THE CELL

As a result of its position midway down the list of standard free energies of hydrolysis (Table 12–1), ATP is able to act as a donor of high-energy phosphate to form those compounds below it in the table. Likewise, provided the necessary enzymatic machinery is available, ADP can accept high-energy phosphate to form ATP from those compounds above ATP in the table. In effect, an **ATP/ADP cycle** connects those processes that generate ~ Ⓟ to those processes that utilize ~Ⓟ (Figure 12–7). Thus, ATP is continuously consumed and regenerated. This occurs at a very rapid rate, since the total ATP/ADP pool is extremely small and sufficient to maintain an active tissue only for a few seconds.

There are three major sources of ~Ⓟ taking part in **energy conservation** or **energy capture:**

(1) Oxidative phosphorylation: Oxidative phosphorylation is the greatest quantitative source of ~Ⓟ in aerobic organisms. The free energy to drive this process comes from respiratory chain oxidation using molecular O$_2$ within mitochondria (Chapter 13).

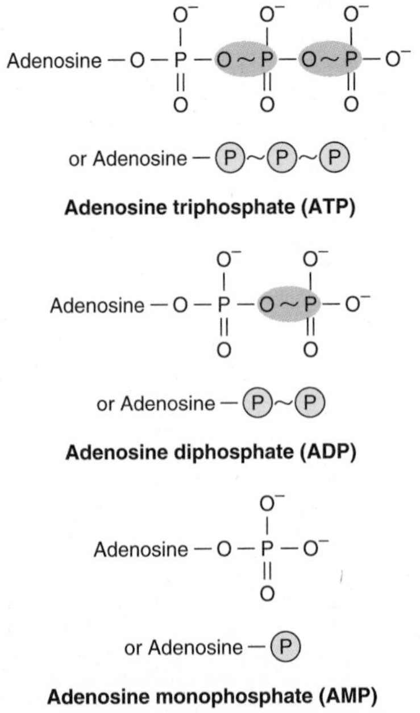

Figure 12–6. Structure of ATP, ADP, and AMP showing the position and the number of high-energy phosphates (~Ⓟ).

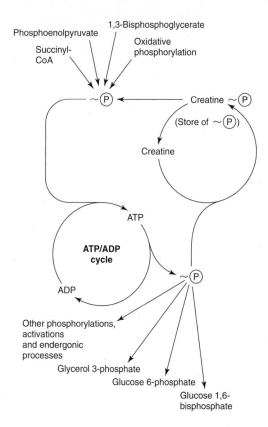

Figure 12–7. Role of ATP/ADP cycle in transfer of high-energy phosphate. Note that ~Ⓟ does not exist in a free state but is transferred in the reactions shown.

(2) Glycolysis: A net formation of two ~Ⓟ results from the formation of lactate from one molecule of glucose generated in two reactions catalyzed by phosphoglycerate kinase and pyruvate kinase, respectively (Figure 19–2).

(3) The citric acid cycle: One ~Ⓟ is generated directly in the cycle at the succinyl thiokinase step (Figure 18–3).

Another group of compounds, **phosphagens,** act as storage forms of high-energy phosphate. These include creatine phosphate, occurring in vertebrate skeletal muscle, heart, spermatozoa, and brain, and arginine phosphate, occurring in invertebrate muscle. Under physiologic conditions, phosphagens permit ATP concentrations to be maintained in muscle when ATP is rapidly being utilized as a source of energy for muscular contraction. On the other hand, when ATP is plentiful and the ATP/ADP ratio is high, their concentration can build up to act as a store of high-energy phosphate (Figure 12–8). In muscle, a **creatine phosphate shuttle** has been described that transports high-energy phosphate from mitochondria to the sarcolemma and acts as a high-energy phosphate buffer (Figure 14–16). In the myocardium, this buffer

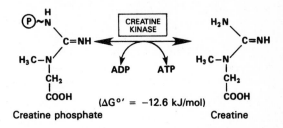

Figure 12–8. Transfer of high-energy phosphate between ATP and creatine.

may be of significance in affording immediate protection against the effects of infarction.

When ATP acts as a phosphate donor to form those compounds of lower free energy of hydrolysis (Table 12–1), the phosphate group is invariably converted to one of low energy, eg,

$$\text{Glycerol + Adenosine} - Ⓟ \sim Ⓟ \sim Ⓟ \xrightarrow{\boxed{\text{GLYCEROL KINASE}}}$$
$$\text{Glycerol} - Ⓟ + \text{Adenosine} - Ⓟ \sim Ⓟ$$

ATP Allows the Coupling of Thermodynamically Unfavorable Reactions to Favorable Ones

The energetics of coupled reactions are depicted in Figures 12–1 and 12–3. Such a reaction is the first in the glycolytic pathway (Figure 19–2), the phosphorylation of glucose to glucose 6-phosphate, which is highly endergonic and cannot proceed as such under physiologic conditions.

(1) Glucose + P_i → Glucose 6-phosphate + H_2O
$(\Delta G^{0\prime} = + 13.8 \text{ kJ/mol})$

To take place, the reaction must be coupled with another reaction that is more exergonic than the phosphorylation of glucose is endergonic. Such a reaction is the hydrolysis of the terminal phosphate of ATP.

(2) ATP → ADP + P_i $(\Delta G^{0\prime} = - 30.5 \text{ kJ/mol})$

When (1) and (2) are coupled in a reaction catalyzed by hexokinase, phosphorylation of glucose readily proceeds in a highly exergonic reaction that under physiologic conditions is far from equilibrium and thus irreversible for practical purposes.

$$\text{Glucose + ATP} \xrightarrow{\boxed{\text{HEXOKINASE}}}$$
$$\text{Glucose 6-phosphate + ADP}$$
$$(\Delta G^{0\prime} = - 16.7 \text{ kJ/mol})$$

Many "activation" reactions follow this pattern.

Adenylyl Kinase Interconverts Adenine Nucleotides

The enzyme adenylyl kinase (myokinase) is present in most cells. It catalyzes the interconversion of ATP and AMP on the one hand and ADP on the other:

$$\text{ATP + AMP} \xrightarrow{\text{ADENYLYL KINASE}} \text{2ADP}$$

This reaction has three functions:

(1) It allows high-energy phosphate in ADP to be used in the synthesis of ATP.

(2) It allows AMP, formed as a consequence of several activating reactions involving ATP, to be recovered by rephosphorylation to ADP.

(3) It allows AMP to increase in concentration when ATP becomes depleted and act as a metabolic (allosteric) signal to increase the rate of catabolic reactions, which in turn lead to the generation of more ATP (Chapter 21).

When ATP Forms AMP, Inorganic Pyrophosphate (PP$_i$) Is Produced

This occurs, for example, in the activation of long-chain fatty acids:

$$\text{ATP + CoA} \cdot \text{SH + R} \cdot \text{COOH} \xrightarrow{\text{ACYL-COA SYNTHETASE}} \text{AMP + PP}_i + \text{R} \cdot \text{CO—SCoA}$$

This reaction is accompanied by loss of free energy as heat, which ensures that the activation reaction will go to the right; this is further aided by the hydrolytic splitting of PP$_i$, catalyzed by **inorganic pyrophosphatase,** a reaction that itself has a large $\Delta G^{0'}$ of -27.6 kJ/mol. Note that activations via the pyrophosphate pathway result in the loss of two ~ⓅP rather than one ~Ⓟ as occurs when ADP and P$_i$ are formed.

$$\text{PP}_i + \text{H}_2\text{O} \xrightarrow{\text{INORGANIC PYROPHOSPHATASE}} 2 \text{ P}_i$$

A combination of the above reactions makes it possible for phosphate to be recycled and the adenine nucleotides to interchange (Figure 12–9).

Other Nucleoside Triphosphates Participate in the Transfer of High-Energy Phosphate

By means of the enzyme nucleoside diphosphate kinase, nucleoside triphosphates similar to ATP but containing an alternative base to adenine can be synthesized from their diphosphates, eg,

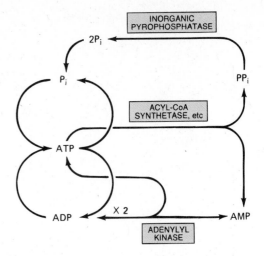

Figure 12–9. Phosphate cycles and interchange of adenine nucleotides.

$$\text{ATP + UDP} \xleftrightarrow{\text{NUCLEOSIDE DIPHOSPHATE KINASE}} \text{ADP + UTP}$$
(uridine triphosphate)

$$\text{ATP + GDP} \longleftrightarrow \text{ADP + GTP}$$
(guanosine triphosphate)

$$\text{ATP + CDP} \longleftrightarrow \text{ADP + CTP}$$
(cytidine triphosphate)

All of these triphosphates take part in phosphorylations in the cell. Similarly, nucleoside monophosphate kinases, specific for each purine or pyrimidine nucleoside, catalyze the formation of nucleoside diphosphates from the corresponding monophosphates.

$$\text{ATP + Nucleoside} - Ⓟ \xleftrightarrow{\text{SPECIFIC NUCLEOSIDE MONOPHOSPHATE KINASE}} \text{ADP + Nucleoside} - Ⓟ \sim Ⓟ$$

Thus, adenylyl kinase is a specialized monophosphate kinase.

SUMMARY

1. Biologic systems are essentially isothermic and use chemical energy to power the living processes.

2. Reactions take place spontaneously when there is loss of free energy (ΔG is negative), ie, they are exergonic. If ΔG is positive, the reaction occurs only if free energy can be gained, ie, it is endergonic.

3. Endergonic processes occur only when coupled to exergonic processes.

4. ATP acts as the "energy currency" of the cell, transferring free energy derived from substances of higher energy potential to those of lower energy potential.

REFERENCES

de Meis L: The concept of energy-rich phosphate compounds: Water, transport ATPases, and entropy energy. Arch Biochem Biophys 1993;306:287.

Ernster L (editor): *Bioenergetics.* Elsevier, 1984.

Harold FM: *The Vital Force: A Study of Bioenergetics.* Freeman, 1986.

Klotz IM: *Introduction to Biomolecular Energetics.* Academic Press, 1986.

Krebs HA, Kornberg HL: *Energy Transformations in Living Matter.* Springer, 1957.

Lehninger AL: *Bioenergetics: The Molecular Basis of Biological Energy Transformations,* 2nd ed. Benjamin, 1971.

13

Biologic Oxidation

Peter A. Mayes, PhD, DSc

INTRODUCTION

Chemically, oxidation is defined as the removal of electrons and reduction as the gain of electrons, as illustrated by the oxidation of ferrous to ferric ion.

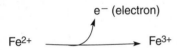

It follows that oxidation is always accompanied by reduction of an electron acceptor. This principle of oxidation-reduction applies equally to biochemical systems and is an important concept underlying understanding of the nature of biologic oxidation. It will be appreciated that many biologic oxidations can take place without the participation of molecular oxygen, eg, dehydrogenations.

BIOMEDICAL IMPORTANCE

Although certain bacteria (anaerobes) survive in the absence of oxygen, the life of higher animals is absolutely dependent upon a supply of oxygen. The principal use of oxygen is in **respiration,** which may be defined as the process by which cells derive energy in the form of ATP from the controlled reaction of hydrogen with oxygen to form water. In addition, molecular oxygen is incorporated into a variety of substrates by enzymes designated as **oxygenases;** many drugs, pollutants, and chemical carcinogens (xenobiotics) are metabolized by enzymes of this class, known as the **cytochrome P450 system.** Administration of oxygen can be lifesaving in the treatment of patients with respiratory or circulatory failure and, occasionally, administration of oxygen at high pressure (hyperbaric oxygen therapy) has proved of value, although this can result in oxygen toxicity.

FREE ENERGY CHANGES CAN BE EXPRESSED IN TERMS OF REDOX POTENTIAL

In reactions involving oxidation and reduction, the free energy change is proportionate to the tendency of reactants to donate or accept electrons. Thus, in addition to expressing free energy change in terms of $\Delta G^{0'}$ (Chapter 12), it is possible, in an analogous manner, to express it numerically as an **oxidation-reduction** or **redox potential** (E_0'). It is usual to compare the redox potential of a system (E_0) against the potential of the hydrogen electrode, which at pH 0 is designated as 0.0 volts. However, for biologic systems, it is normal to express the redox potential (E_0') at pH 7.0, at which pH the electrode potential of the hydrogen electrode is −0.42 volts. The redox potentials of some redox systems of special interest in mammalian biochemistry are shown in Table 13–1. The relative positions of redox systems in the table allows prediction of the direction of flow of electrons from one redox couple to another.

ENZYMES INVOLVED IN OXIDATION AND REDUCTION ARE DESIGNATED OXIDOREDUCTASES

In the following account, oxidoreductases are classified into four groups: **oxidases, dehydrogenases, hydroperoxidases,** and **oxygenases.**

OXIDASES USE OXYGEN AS A HYDROGEN ACCEPTOR

Oxidases catalyze the removal of hydrogen from a substrate using oxygen as a hydrogen acceptor.* They form water or hydrogen peroxide as a reaction product (Figure 13–1).

---

*Sometimes the term "oxidase" is used collectively to denote all enzymes that catalyze reactions involving molecular oxygen.

Table 13–1. Some redox potentials of special interest in mammalian oxidation systems.

System	E_0' volts
H^+/H_2	−0.42
$NAD^+/NADH$	−0.32
Lipoate; ox/red	−0.29
Acetoacetate/3-hydroxybutyrate	−0.27
Pyruvate/lactate	−0.19
Oxaloacetate/malate	−0.17
Fumarate/succinate	+0.03
Cytochrome b; Fe^{3+}/Fe^{2+}	+0.08
Ubiquinone; ox/red	+0.10
Cytochrome c_1; Fe^{3+}/Fe^{2+}	+0.22
Cytochrome a; Fe^{3+}/Fe^{2+}	+0.29
Oxygen/water	+0.82

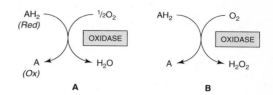

Figure 13–1. Oxidation of a metabolite catalyzed by an oxidase **(A)** forming H_2O, **(B)** forming H_2O_2.

Some Oxidases Contain Copper

Cytochrome oxidase is a hemoprotein widely distributed in many tissues, having the typical heme prosthetic group present in myoglobin, hemoglobin, and other cytochromes (Chapter 7). It is the terminal component of the chain of respiratory carriers found in mitochondria and is therefore responsible for the reaction whereby electrons resulting from the oxidation of substrate molecules by dehydrogenases are transferred to their final acceptor, oxygen. The enzyme is poisoned by carbon monoxide, cyanide, and hydrogen sulfide. It has also been termed cytochrome a_3. It was formerly assumed that cytochrome a and cytochrome a_3 were separate compounds, since each has a distinct spectrum and different properties with respect to the effects of carbon monoxide and cyanide. More recent studies show that the two cytochromes are combined with a single protein, and the complex is known as **cytochrome aa_3.** It contains two molecules of heme, each having one Fe atom that oscillates between Fe^{3+} and Fe^{2+} during oxidation and reduction. Furthermore, two atoms of Cu are present, each associated with a heme unit.

Other Oxidases Are Flavoproteins

Flavoprotein enzymes contain **flavin mononucleotide (FMN)** or **flavin adenine dinucleotide (FAD)** as prosthetic groups. FMN (Figure 52–2) and FAD (Figure 52–3) are formed in the body from the vitamin **riboflavin** (Chapter 52).

FMN and FAD are usually tightly—but not covalently—bound to their respective apoenzyme protein.

Many flavoprotein enzymes contain one or more metals as essential cofactors and are known as **metalloflavoproteins.**

Enzymes belonging to this group of oxidases include **L-amino acid oxidase,** an FMN-linked enzyme found in kidney with general specificity for the oxidative deamination of the naturally occurring L-amino acids. **Xanthine oxidase** has a wide distribution, occurring in milk, small intestine, kidney, and liver. It contains molybdenum and plays an important role in the conversion of purine bases to uric acid (Chapter 36). It is of particular significance in the liver and kidney of birds, which excrete uric acid as the main nitrogenous end product, not only of purine metabolism but also of protein and amino acid catabolism.

Aldehyde dehydrogenase is an FAD-linked enzyme present in mammalian livers. It is a metalloflavoprotein containing molybdenum and nonheme iron and acts upon aldehydes and N-heterocyclic substrates.

Of interest because of its use in estimating glucose is **glucose oxidase,** an FAD-specific enzyme prepared from fungi.

The mechanisms of oxidation and reduction of these enzymes are complex. However, evidence points to reduction of the isoalloxazine ring taking place in two steps via a semiquinone (free radical) intermediate (Figure 13–2).

DEHYDROGENASES CANNOT USE OXYGEN AS A HYDROGEN ACCEPTOR

There are a large number of enzymes in this class. They perform two main functions:

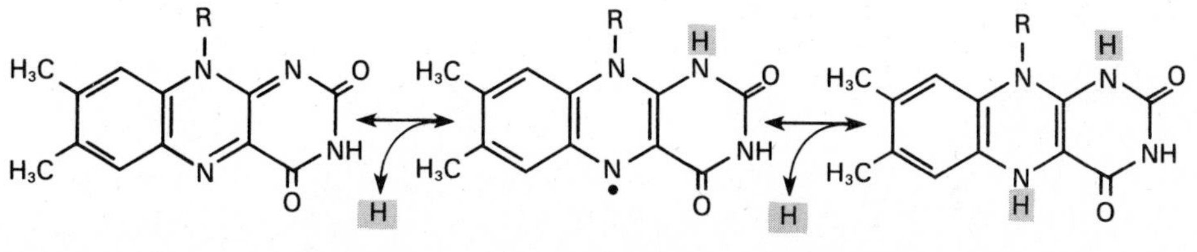

Figure 13–2. Oxidoreduction of isoalloxazine ring in flavin nucleotides.

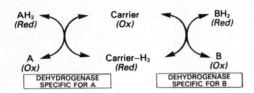

Figure 13–3. Oxidation of a metabolite catalyzed by coupled dehydrogenases.

(1) Transfer of hydrogen from one substrate to another in a coupled oxidation-reduction reaction (Figure 13–3). These dehydrogenases are specific for their substrates but often utilize the same coenzyme or hydrogen carrier as other dehydrogenases, eg, NAD$^+$. Since the reactions are reversible, these properties enable reducing equivalents to be freely transferred within the cell. This type of reaction, which enables a substrate to be oxidized at the expense of another, is particularly useful in enabling oxidative processes to occur in the absence of oxygen, such as during the anaerobic phase of glycolysis (Figure 19–2).

(2) As components in a **respiratory chain** of electron transport from substrate to oxygen (Figure 13–4).

Many Dehydrogenases Depend on Nicotinamide Coenzymes

Many dehydrogenases are specific for either **nicotinamide adenine dinucleotide (NAD$^+$)** or **nicotinamide adenine dinucleotide phosphate (NADP$^+$)** as coenzymes. However, some dehydrogenases can use either NAD$^+$ or NADP$^+$. NAD$^+$ and NADP$^+$ are formed in the body from the vitamin **niacin** (Figure 52–4). The coenzymes are reduced by the specific substrate of the dehydrogenase and reoxidized by a suitable electron acceptor (Figure 13–5). Unlike FMN and FAD$^+$, they may freely and reversibly dissociate from their respective apoenzymes.

Generally, NAD-linked dehydrogenases catalyze oxidoreduction reactions in the oxidative pathways of metabolism, particularly in glycolysis, in the citric acid cycle, and in the respiratory chain of mitochondria. NADP-linked dehydrogenases are found characteristically in reductive syntheses, as in the ex-

tramitochondrial pathway of fatty acid synthesis and steroid synthesis. They are also to be found as coenzymes to the dehydrogenases of the pentose phosphate pathway. Some nicotinamide coenzyme-dependent dehydrogenases have been found to contain zinc, notably alcohol dehydrogenase from liver and glyceraldehyde-3-phosphate dehydrogenase from skeletal muscle. The zinc ions are not considered to take part in oxidation and reduction.

Other Dehydrogenases Depend on Riboflavin

The flavin groups associated with these dehydrogenases are similar to FMN and FAD occurring in oxidases. They are generally more tightly bound to their apoenzymes than are the nicotinamide coenzymes. Most of the riboflavin-linked dehydrogenases are concerned with electron transport in (or to) the respiratory chain (Chapter 14). **NADH dehydrogenase** is a member of the respiratory chain acting as a carrier of electrons between NADH and the more electropositive components (Figure 14–3). Other dehydrogenases such as **succinate dehydrogenase, acyl-CoA dehydrogenase,** and **mitochondrial glycerol-3-phosphate dehydrogenase** transfer reducing equivalents directly from the substrate to the respiratory chain (Figure 14–4). Another role of the flavin-dependent dehydrogenases is in the dehydrogenation (by dihydrolipoyl dehydrogenase) of reduced lipoate, an intermediate in the oxidative decarboxylation of pyruvate and α-ketoglutarate (Figure 14–4). In this particular instance, owing to the low redox potential, the flavoprotein (FAD) acts as a hydrogen carrier from reduced lipoate to NAD (Figure 19–5). The **electron-transferring flavoprotein** is an intermediary carrier of electrons between acyl-CoA dehydrogenase and the respiratory chain (Figure 14–4).

Cytochromes May Also Be Regarded as Dehydrogenases

Except for cytochrome oxidase (previously described), the cytochromes are also classified as dehydrogenases. Their identification and study are facilitated by the presence in the reduced state of characteristic absorption bands that disappear on oxidation. In the respiratory chain, they are involved as

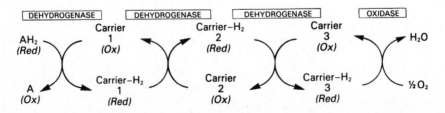

Figure 13–4. Oxidation of a metabolite by dehydrogenases and finally by an oxidase in a respiratory chain.

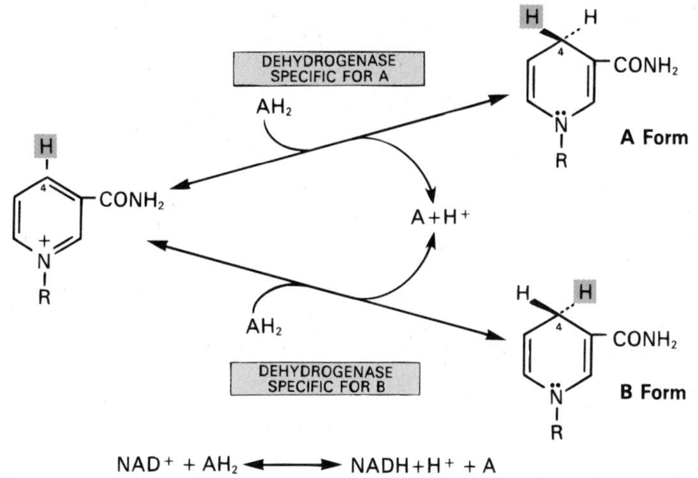

Figure 13–5. Mechanism of oxidation and reduction of nicotinamide coenzymes. There is stereospecificity about position 4 of nicotinamide when it is reduced by a substrate AH_2. One of the hydrogen atoms is removed from the substrate as a hydrogen nucleus with two electrons (hydride ion, H^-) and is transferred to the 4 position, where it may be attached in either the A or B position according to the specificity determined by the particular dehydrogenase catalyzing the reaction. The remaining hydrogen of the hydrogen pair removed from the substrate remains free as a hydrogen ion.

carriers of electrons from flavoproteins on the one hand to cytochrome oxidase on the other (Figure 14–4). The cytochromes are iron-containing hemoproteins in which the iron atom oscillates between Fe^{3+} and Fe^{2+} during oxidation and reduction. Several identifiable cytochromes occur in the respiratory chain, ie, cytochromes b, c_1, c, a, and a_3 (cytochrome oxidase). Of these, only cytochrome c is soluble. Besides the respiratory chain, cytochromes are found in other locations, eg, the endoplasmic reticulum (cytochromes P450 and b_5), plant cells, bacteria, and yeasts.

HYDROPEROXIDASES USE HYDROGEN PEROXIDE OR AN ORGANIC PEROXIDE AS SUBSTRATE

Two type of enzymes fall into this category: **peroxidases** and **catalase.** These two types are found both in animals and in plants.

Hydroperoxidases protect the body against harmful peroxides. Accumulation of peroxides can lead to generation of free radicals, which in turn can disrupt membranes and perhaps cause cancer and atherosclerosis (Chapters 16 and 53 for a discussion and summary of the mechanisms of defense against free radicals).

Peroxidases Reduce Peroxides Using Various Electron Acceptors

Although originally considered to be plant enzymes, peroxidases are found in milk and in leukocytes, platelets, and other tissues involved in eicosanoid metabolism (Chapter 25). The prosthetic group is protoheme, which, unlike the situation in most hemoproteins, is only loosely bound to the apoprotein. In the reaction catalyzed by peroxidase, hydrogen peroxide is reduced at the expense of several substances that will act as electron acceptors, such as ascorbate, quinones, and cytochrome c. The reaction catalyzed by peroxidase is complex, but the overall reaction is as follows:

$$H_2O_2 + AH_2 \xrightarrow{\text{PEROXIDASE}} 2H_2O + A$$

In erythrocytes and other tissues, the enzyme **glutathione peroxidase,** containing **selenium** as a prosthetic group, catalyzes the destruction of H_2O_2 and lipid hydroperoxides by reduced glutathione, protecting membrane lipids and hemoglobin against oxidation by peroxides (Chapter 22).

Catalase Uses Hydrogen Peroxide as Electron Donor and Electron Acceptor

Catalase is a hemoprotein containing four heme groups. In addition to possessing peroxidase activity, it is able to use one molecule of H_2O_2 as a substrate electron donor and another molecule of H_2O_2 as oxidant or electron acceptor.

$$2H_2O_2 \xrightarrow{\text{CATALASE}} 2H_2O + O_2$$

Under most conditions in vivo, the peroxidase activity of catalase seems to be favored. Catalase is found in blood, bone marrow, mucous membranes, kidney, and liver. Its function is assumed to be the destruction of hydrogen peroxide formed by the action of oxidases. Microbodies or **peroxisomes** are found in many tissues, including liver. They are rich in oxidases and in catalase, which suggests that there may be a biologic advantage in grouping the enzymes that produce H_2O_2 with the enzyme that destroys it (Figure 13–6). In addition to the peroxisomal enzymes, mitochondrial and microsomal electron transport systems as well as xanthine oxidase must be considered as sources of H_2O_2.

OXYGENASES CATALYZE THE DIRECT TRANSFER AND INCORPORATION OF OXYGEN INTO A SUBSTRATE MOLECULE

Oxygenases are concerned with the synthesis or degradation of many different types of metabolites rather than taking part in reactions that have as their purpose the provision of energy to the cell. Enzymes in this group catalyze the incorporation of oxygen into a substrate molecule. This takes place in two steps: (1) oxygen binding to the enzyme at the active site and (2) the reaction in which the bound oxygen is reduced or transferred to the substrate. Oxygenases may be divided into two subgroups:

Dioxygenases Incorporate Both Atoms of Molecular Oxygen Into the Substrate

The basic reaction is shown below:

$$A + O_2 \rightarrow AO_2$$

Examples of this type include enzymes that contain iron, such as **homogentisate dioxygenase** (oxidase) and **3-hydroxyanthranilate dioxygenase** (oxidase) from the supernatant fraction of the liver, and enzymes

utilizing heme, such as **L-tryptophan dioxygenase** (tryptophan pyrrolase) from the liver (Chapter 32).

Monooxygenases (Mixed-Function Oxidases, Hydroxylases) Incorporate Only One Atom of Molecular Oxygen Into the Substrate

The other oxygen atom is reduced to water, an additional electron donor or cosubstrate being necessary for this purpose.

$$A{-}H + O_2 + ZH_2 \rightarrow A{-}OH + H_2O + Z$$

Microsomal Cytochrome P450 Monooxygenase Systems Are Important for the Hydroxylation of Many Drugs

These monooxygenases are found in the microsomes of the liver together with cytochrome P450 and **cytochrome b_5**. Both NADH and NADPH donate reducing equivalents for the reduction of these cytochromes (Figure 13–7), which in turn are oxidized by substrates in a series of enzymatic reactions collectively known as the **hydroxylase cycle** (Figure 13–8).

$$\text{DRUG{-}H} + O_2 + 2Fe^{2+} + 2H^+ \xrightarrow{\text{Hydroxylase}}$$
$$\text{(P450)}$$
$$\text{DRUG{-}OH} + H_2O + 2Fe^{3+}$$
$$\text{(P450)}$$

Among the drugs metabolized by this system are benzpyrene, aminopyrine, aniline, morphine, and benzphetamine. Many drugs such as phenobarbital have the ability to induce the formation of microsomal enzymes and of cytochrome P450.

Mitochondrial Cytochrome P450 Monooxygenase Systems Catalyze Steroidal Hydroxylations

These systems are found in steroidogenic tissues such as adrenal cortex, testis, ovary, and placenta and are concerned with the biosynthesis of steroid hormones from cholesterol (hydroxylation at C_{22} and C_{20} in side-chain cleavage and at the 11β and 18 positions). Renal systems catalyze 1α- and 24-hydroxylations of 25-hydroxycholecalciferol, and the liver catalyzes 26-hydroxylation in bile acid biosynthesis. In the adrenal cortex, mitochrondrial cytochrome P450 is six times more abundant than cytochromes of the respiratory chain.

THE SUPEROXIDE FREE RADICAL MAY ACCOUNT FOR OXYGEN TOXICITY

The potential toxicity of oxygen has hitherto been attributed to the formation of H_2O_2. Recently, how-

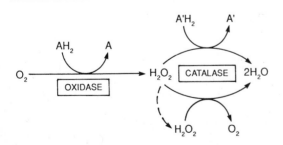

Figure 13–6. Role of catalase in the destruction of hydrogen peroxide.

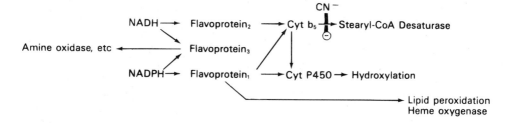

Figure 13–7. Electron transport chain in microsomes. Cyanide (CN^-) inhibits the indicated step.

ever, the ease with which oxygen can be reduced in tissues to the superoxide anion free radical (O_2^-) and the occurrence of **superoxide dismutase** in aerobic organisms (although not in obligate anaerobes) have suggested that the toxicity of oxygen is due to its conversion to superoxide.

Superoxide is formed when reduced flavins—present, for example, in xanthine oxidase—are reoxidized univalently by molecular oxygen.

$$Enz—Flavin—H_2 + O_2 \rightarrow Ezn—Flavin—H + O_2^- + H^+$$

Superoxide can reduce oxidized cytochrome c

$$O_2^- + Cyt\ c\ (Fe^{3+}) \rightarrow O_2 + Cyt\ c\ (Fe^{2+})$$

or be removed by the presence of the specific enzyme superoxide dismutase

$$O_2^- + O_2^- + 2H^+ \xrightarrow{\text{SUPEROXIDE DISMUTASE}} H_2O_2 + O_2$$

In this reaction, superoxide acts as both oxidant and reductant. The chemical effects of the superoxide free radical in the tissues are amplified by its giving rise to free radical chain reactions (Chapter 16). It has been proposed that O_2^- bound to cytochrome P450 is an intermediate in the activation of oxygen in hydroxylation reactions (Figure 13–8).

The function of superoxide dismutase seems to be that of protecting aerobic organisms against the potential deleterious effects of superoxide. The enzyme occurs in several different compartments of the cell. The cytosolic enzyme is composed of two similar subunits, each one containing one equivalent of Cu^{2+} and Zn^{2+}, whereas the mitochondrial enzyme contains Mn^{2+}, similar to the enzyme found in bacteria.

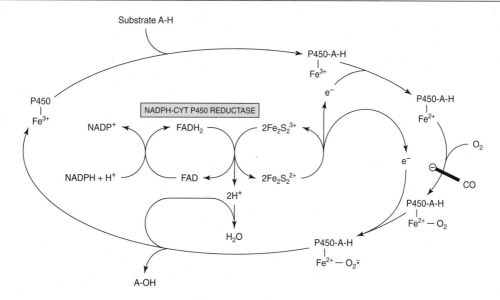

Figure 13–8. Cytochrome P450 hydroxylase cycle in microsomes. The system shown is typical of steroid hydroxylases of the adrenal cortex. Liver microsomal cytochrome P450 hydroxylase does not require the iron-sulfur protein Fe_2S_2. Carbon monoxide (CO) inhibits the indicated step.

This finding supports the hypothesis that mitochondria have evolved from a prokaryote that entered into symbiosis with a protoeukaryote. The dismutase is present in all major aerobic tissues. Although exposure of animals to an atmosphere of 100% oxygen causes an adaptive increase of the enzyme, particularly in the lungs, prolonged exposure leads to lung damage and death. Antioxidants, eg, α-tocopherol (vitamin E), act as scavengers of free radicals and reduce the toxicity of oxygen (Chapter 53).

SUMMARY

1. In biologic systems, as in chemical systems, oxidation (loss of electrons) is always accompanied by reduction of an electron acceptor.

2. Oxidoreductases are classified into four groups: oxidases, dehydrogenases, hydroperoxidases, and oxygenases.

3. Oxidases and dehydrogenases have a variety of roles in metabolism, but both classes of enzymes play major roles in respiration.

4. Hydroperoxidases protect the body against damage by free radicals, and oxygenases mediate the hydroxylation of drugs and steroids.

5. Oxygen toxicity may be caused by the superoxide free radical. Tissues are protected from superoxide by the specific enzyme superoxide dismutase.

REFERENCES

Bonnett R: Oxygen activation and tetrapyrroles. Essays Biochem 1981;17:1.
Coon MJ et al: Cytochrome P450: Progress and predictions. FASEB J 1992;6:669.
Ernster L (editor): *Bioenergetics.* Elsevier, 1984.
Fleischer S, Packer L (editors): Biological oxidations, microsomal cytochrome P450, and other hemoprotein systems. In: *Enzymology.* Vol. 52. *Biomembranes,* part C. Academic Press, 1978.
Friedovich I: Superoxide dismutases. Annu Rev Biochem 1975;44:147.
Nicholls DG: *Cytochromes and Cell Respiration.* Carolina Biological Supply Company, 1984.
Tolbert NE: Metabolic pathways in peroxisomes and glyoxysomes. Annu Rev Biochem 1981;50:133.
Tyler DD: *The Mitochondrion in Health and Disease.* VCH Publishers, 1992.
Tyler DD, Sutton CM: Respiratory enzyme systems in mitochondrial membranes. In: *Membrane Structure and Function.* Vol. 5. Bittar EE (editor). Wiley, 1984.
Yang CS, Brady JF, Hong JY: Dietary effects on cytochromes P450, xenobiotic metabolism, and toxicity. FASEB J 1992;6:737.

The Respiratory Chain & Oxidative Phosphorylation

14

Peter A. Mayes, PhD, DSc

INTRODUCTION

The mitochondrion has appropriately been termed the "powerhouse" of the cell, since it is within this organelle that most of the capture of energy derived from respiratory oxidation takes place. The system in mitochondria that couples respiration to the generation of the high-energy intermediate, ATP, is termed **oxidative phosphorylation.**

BIOMEDICAL IMPORTANCE

Oxidative phosphorylation enables aerobic organisms to capture a far greater proportion of the available free energy of respiratory substrates compared with anaerobic organisms. The **chemiosmotic theory** offers an insight into how this is accomplished. A number of drugs (eg, **amobarbital**) and poisons (eg, **cyanide, carbon monoxide**) inhibit oxidative phosphorylation, usually with fatal consequences. A number of inherited defects of mitochondria involving components of the respiratory chain and oxidative phosphorylation have been reported. Patients present with **myopathy** and **encephalopathy** and often have **lactic acidosis.**

SPECIFIC ENZYMES ACT AS MARKERS OF COMPARTMENTS SEPARATED BY THE MITOCHONDRIAL MEMBRANES

Mitochondria have an outer membrane that is permeable to most metabolites, an inner membrane which is selectively permeable and which is thrown into folds or cristae, and a matrix within the inner membrane (Figure 14–1). The outer membrane may be removed by treatment with digitonin and is characterized by the presence of monoamine oxidase, acyl-CoA synthetase, glycerolphosphate acyltransferase, monoacyl glycerolphosphate acyltransferase, and phospholipase A_2. Adenylyl kinase and creatine kinase are found in the intermembrane space. The phospholipid cardiolipin is concentrated in the inner membrane.

The soluble enzymes of the citric acid cycle and the enzymes of β-oxidation of fatty acids are found in the matrix, necessitating mechanisms for transporting metabolites and nucleotides across the inner membrane. Succinate dehydrogenase is found on the inner surface of the inner mitochondrial membrane, where it transports reducing equivalents to the respiratory chain enzymes, major constituents of the inner membrane. 3-Hydroxybutyrate dehydrogenase is also bound to the matrix side of the inner mitochondrial membrane. Glycerol-3-phosphate dehydrogenase is found on the outer surface of the inner membrane, where it is suitably located to participate in the glycerophosphate shuttle (Figure 14–15).

THE RESPIRATORY CHAIN COLLECTS AND OXIDIZES REDUCING EQUIVALENTS

All of the useful energy liberated during the oxidation of fatty acids and amino acids and nearly all of that released from the oxidation of carbohydrate is made available within the mitochondria as reducing equivalents (—H or electrons). The mitochondria contain the series of catalysts known as the respiratory chain that collect and transport reducing equivalents and direct them to their final reaction with oxygen to form water. Also present is the machinery for trapping the liberated free energy as high-energy phosphate. Mitochondria also contain the enzyme systems responsible for producing most of the reducing equivalents in the first place, ie, the enzymes of β-oxidation and of the citric acid cycle. The latter is the final common metabolic pathway for the oxidation of all the major foodstuffs. These relationships are shown in Figure 14–2.

Components of the Respiratory Chain Are Arranged in Order of Increasing Redox Potential

The major components of the respiratory chain are shown in Figure 14–3. Hydrogen and electrons flow through the chain in steps from the more electronega-

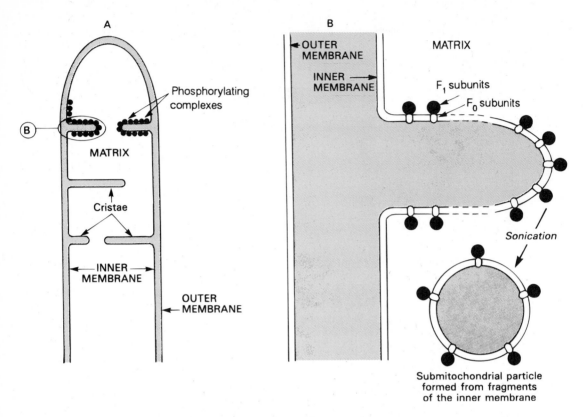

Figure 14–1. Structure of the mitochondrial membranes. Submitochondrial particles are "inside out" and allow study of an enclosed membrane system where the phosphorylating subunits are on the outside.

tive components to the more electropositive oxygen through a redox span of 1.1 V from $NAD^+/NADH$ to $O_2/2H_2O$ (Table 13–1).

The respiratory chain in mitochondria consists of a number of redox carriers that proceed from the NAD-linked dehydrogenase systems, through flavoproteins and cytochromes, to molecular oxygen. Not all substrates are linked to the respiratory chain through NAD-specific dehydrogenases; some, because their redox potentials are more positive (eg, fu-

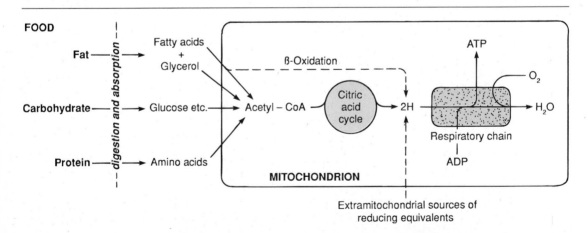

Figure 14–2. Role of the respiratory chain of mitochondria in the conversion of food energy to ATP. Oxidation of the major foodstuffs leads to the generation of reducing equivalents (2H) that are collected by the respiratory chain for oxidation and coupled generation of ATP.

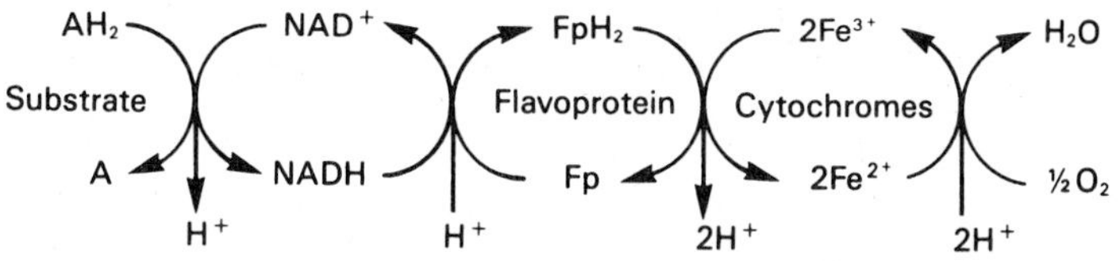

$$\text{AH}_2 \rightleftharpoons \text{NAD}^+ \rightleftharpoons \text{FpH}_2 \rightleftharpoons 2\text{Fe}^{3+} \rightleftharpoons \text{H}_2\text{O}$$

Substrate   Flavoprotein   Cytochromes

$$\text{A} \rightleftharpoons \text{NADH} \qquad \text{Fp} \rightleftharpoons 2\text{Fe}^{2+} \qquad \tfrac{1}{2}\text{O}_2$$

H⁺   H⁺   2H⁺   2H⁺

Figure 14–3. Transport of reducing equivalents through the respiratory chain.

marate/succinate; Table 13–1), are linked directly to flavoprotein dehydrogenases, which in turn are linked to the cytochromes of the respiratory chain (Figure 14–4).

It has become clear that an additional carrier is present in the respiratory chain linking the flavoproteins to cytochrome *b*, the member of the cytochrome chain of lowest redox potential. This substance, which has been named **ubiquinone** or **Q (coenzyme Q;** see Figure 14–5), exists in mitochondria in the oxidized quinone form under aerobic conditions and in the reduced quinol form under anaerobic conditions. Q is a constituent of the mitochondrial lipids; the other lipids are predominantly phospholipids that constitute part of the mitochondrial membrane. The structure of Q is very similar to vitamin K and vitamin E (Chapter 53). It is also similar to plastoquinone, found in chloroplasts. All these substances are characterized by the possession of a polyisoprenoid side chain. In mitochondria, there is a large stoichiometric excess of Q compared with other members of the respiratory chain; this is compatible with Q acting on a mobile component of the respiratory chain that collects reducing equivalents from the more fixed flavoprotein complexes and passes them on to the cytochromes.

An additional component found in respiratory chain preparations is the **iron-sulfur protein (FeS;** nonheme iron) (Figure 14–6). It is associated with the flavoproteins (metalloflavoproteins) and with cytochrome *b*. The sulfur and iron are thought to take part in the oxidoreduction mechanism between flavin and Q, which involves only a single e⁻ change, the iron atom undergoing oxidoreduction between Fe^{2+} and Fe^{3+}.

A current view of the sequence of the principal components of the respiratory chain is shown in Figure 14–4. At the electronegative end of the chain, dehydrogenase enzymes catalyze the transfer of electrons from substrates to NAD of the chain. Several differences exist in the manner in which this is carried out. The α-keto acids pyruvate and ketoglutarate

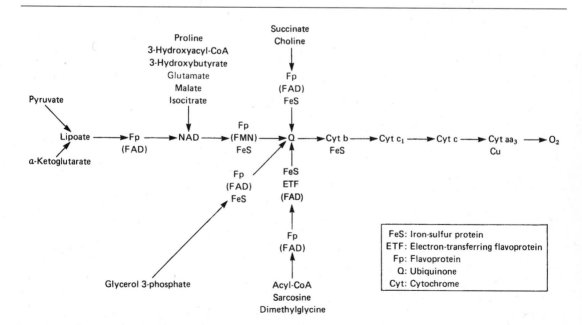

Figure 14–4. Components of the respiratory chain in mitochondria showing the collecting points for reducing equivalents from important substrates. FeS occurs in the sequences on the O_2 side of Fp or Cyt *b*.

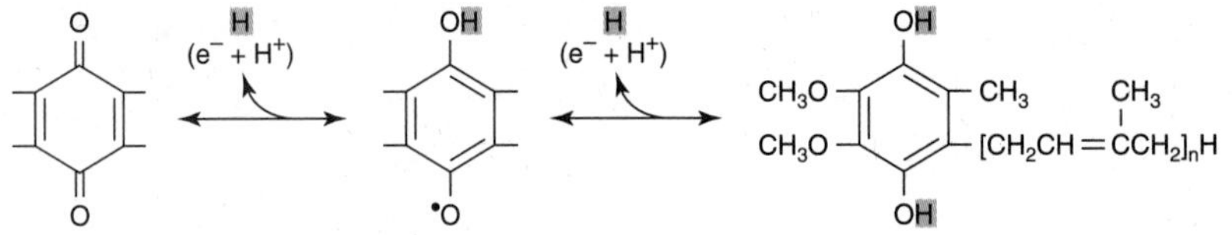

Figure 14–5. Structure of ubiquinone (Q). n = Number of isoprenoid units, which is 10 in higher animals, ie, Q_{10}.

have complex dehydrogenase systems involving lipoate and FAD prior to the passage of electrons to NAD of the respiratory chain. Electron transfers from other dehydrogenases such as L(+)-3-hydroxy-acyl-CoA, D(−)-3-hydroxybutyrate, proline, glutamate, malate, and isocitrate dehydrogenases couple directly with NAD of the respiratory chain.

The reduced NADH of the respiratory chain is in turn oxidized by a metalloflavoprotein enzyme—**NADH dehydrogenase.** This enzyme contains FeS and FMN, is tightly bound to the respiratory chain, and passes reducing equivalents onto Q. Q is also the collecting point in the respiratory chain for reducing equivalents derived from other substrates that are linked directly to the respiratory chain through flavo-protein dehydrogenases. These substrates include succinate, choline, glycerol 3-phosphate, sarcosine, dimethylglycine, and acyl-CoA (Figure 14–4). The flavin moiety of all these dehydrogenases is FAD.

Electrons flow from Q, through the series of cytochromes shown in Figure 14–4, to molecular oxygen. The cytochromes are arranged in order of increasing redox potential. The terminal cytochrome aa_3 (cytochrome oxidase) is responsible for the final combination of reducing equivalents with molecular oxygen. It has been noted that this enzyme system contains copper, a component of several oxidase enzymes. Cytochrome oxidase has a very high affinity for oxygen, which allows the respiratory chain to function at the maximum rate until the tissue has become virtually depleted of O_2. Since this is an irreversible reaction (the only one in the chain), it gives direction to the movement of reducing equivalents in the respiratory chain and to the production of ATP, to which it is coupled.

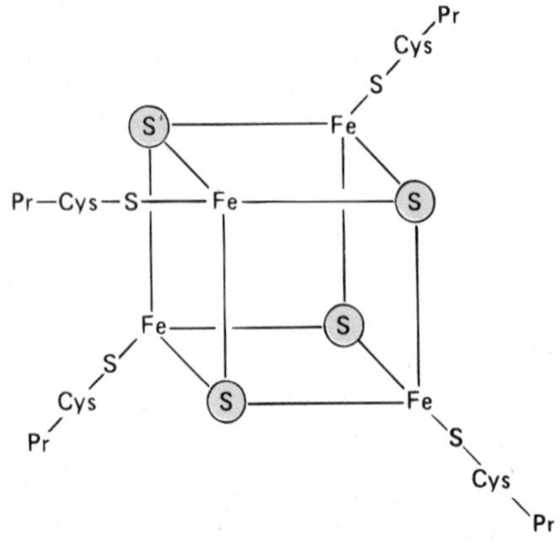

The structural organization of the respiratory chain has been the subject of considerable study. Of significance is the finding of nearly constant molar proportions between the components. Functionally and structurally, the components of the respiratory chain are present in the inner mitochondrial membrane as four **protein-lipid respiratory chain complexes** that span the membrane. Cytochrome c is the only soluble cytochrome and, together with Q, seems to be a more mobile component of the respiratory chain connecting the fixed complexes (Figures 14–7 and 14–10).

THE RESPIRATORY CHAIN PROVIDES MOST OF THE ENERGY CAPTURED IN METABOLISM

ADP is a molecule that captures, in the form of high-energy phosphate, a significant proportion of the free energy released by catabolic processes. The resulting ATP passes on this free energy to drive those processes requiring energy. Thus, ATP has been called the energy "currency" of the cell (Figure 12–7).

There is a net direct capture of two high-energy phosphate groups in the glycolytic reactions (Table 19–1), equivalent to approximately 103.2 kJ/mol of glucose. (In vivo, ΔG for the synthesis of ATP from ADP has been calculated as approximately 51.6 kJ/mol, allowing for the actual concentration of reactants present in cells. It is greater than $\Delta G^{0'}$ for the hydrolysis of ATP as given in Table 12–1, which is obtained under standard concentrations of 1.0 mol/L.) Since 1 mol of glucose yields approximately 2870 kJ on complete combustion, the energy captured by phosphorylation in glycolysis is small. The reactions of the citric acid cycle, the final pathway for the com-

Figure 14–6. Iron-sulfur-protein complex (Fe_4S_4). Ⓢ, acid-labile sulfur; Pr, apoprotein; Cys, cysteine residue. Some iron-sulfur proteins contain two iron atoms and two sulfur atoms (Fe_2S_2).

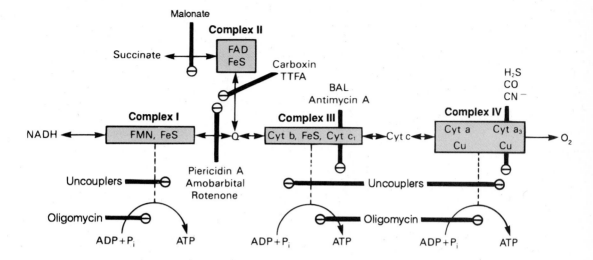

Figure 14–7. Proposed sites of inhibition $\ominus$ of the respiratory chain by specific drugs, chemicals, and antibiotics. The sites that appear to support phosphorylation are indicated. BAL, dimercaprol. TTFA is an Fe-chelating agent. Complex I, NADH:ubiquinone oxidoreductase; complex II, succinate:ubiquinone oxidoreductase; complex III, ubiquinol: ferricytochrome c oxidoreductase; complex IV, ferrocytochrome c:oxygen oxidoreductase. Other abbreviations as in Figure 14–4.

plete oxidation of glucose, include one phosphorylation step, the conversion of succinyl-CoA to succinate, which allows the capture of only two more high-energy phosphates per mole of glucose. All of the phosphorylations described so far occur at the **substrate level.** Examination of intact respiring mitochondria reveals that when substrates are oxidized via an NAD-linked dehydrogenase and the respiratory chain, approximately 3 mol of inorganic phosphate are incorporated into 3 mol of ADP to form 3 mol of ATP per $1/2$ mol of O_2 consumed; ie, the P:O ratio = 3 (Figure 14–7). On the other hand, when a substrate is oxidized via a flavoprotein-linked dehydrogenase, only 2 mol of ATP are formed; ie, P:O = 2. These reactions are known as **oxidative phosphorylation at the respiratory chain level.** Dehydrogenations in the pathway of catabolism of glucose in both glycolysis and the citric acid cycle, plus phosphorylations at the substrate level, can now account for 68% of the free energy resulting from the combustion of glucose, captured in the form of high-energy phosphate. It is evident that the respiratory chain is responsible for a large proportion of total ATP formation.

(Table 14–1). Generally, most cells in the resting state are in state 4, and respiration is controlled by the availability of ADP. When work is performed, ATP is converted to ADP, allowing more respiration to occur, which in turn replenishes the store of ATP (Figure 14–8). It would appear that under certain conditions the concentration of inorganic phosphate could also affect the rate of functioning of the respiratory chain. As respiration increases (as in exercise), the cell approaches state 3 or state 5 when either the capacity of the respiratory chain becomes saturated or the PO_2 decreases below the K_m for cytochrome a_3. There is also the possibility that the ADP/ATP transporter (Figure 14–2), which facilitates entry of cytosolic ADP into the mitochondrion, becomes rate-limiting.

Thus, the manner in which biologic oxidative processes allow the free energy resulting from the oxidation of foodstuffs to become available and to be captured is stepwise, efficient (approximately 68%), and controlled—rather than explosive, inefficient, and uncontrolled, as in many nonbiologic processes. The remaining free energy that is not captured as

Respiratory Control Ensures a Constant Supply of ATP

The rate of respiration of mitochondria can be controlled by the concentration of ADP. This is because oxidation and phosphorylation are tightly coupled; ie, oxidation cannot proceed via the respiratory chain without concomitant phosphorylation of ADP. Chance and Williams have defined five conditions that can control the rate of respiration in mitochondria

Table 14–1. States of respiratory control.

	Conditions Limiting the Rate of Respiration
State 1	Availability of ADP and substrate
State 2	Availability of substrate only
State 3	The capacity of the respiratory chain itself, when all substrates and components are present in saturating amounts
State 4	Availability of ADP only
State 5	Availability of oxygen only

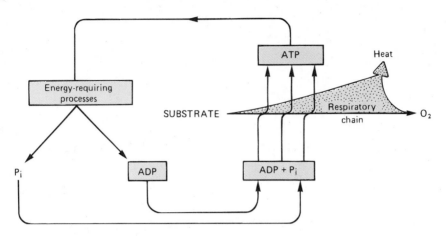

Figure 14–8. The role of ADP in respiratory control.

high-energy phosphate is liberated as **heat.** This need not be considered "wasted," since it ensures the respiratory system as a whole is sufficiently exergonic to be removed from equilibrium, allowing continuous unidirectional flow and constant provision of ATP. In the warm-blooded animal it contributes to maintenance of body temperature.

MANY POISONS INHIBIT THE RESPIRATORY CHAIN

Much information about the respiratory chain has been obtained by the use of inhibitors, and, conversely, this has provided knowledge about the mechanism of action of several poisons. The proposed loci of action are shown in Figure 14–7. For descriptive purposes, they may be divided into inhibitors of the respiratory chain proper, inhibitors of oxidative phosphorylation, and uncouplers of oxidative phosphorylation.

Inhibitors that arrest respiration by blocking the respiratory chain act at three sites. The first is inhibited by **barbiturates** such as amobarbital, by the antibiotic **piericidin A,** and by the insecticide and fish poison **rotenone.** These inhibitors prevent the oxidation of substrates that communicate directly with the respiratory chain via an NAD-linked dehydrogenase by blocking the transfer from FeS to Q. At sufficient dosage, they are fatal in vivo.

Dimercaprol and **antimycin A** inhibit the respiratory chain between cytochrome b and cytochrome c. The classic poisons **H$_2$S, carbon monoxide,** and **cyanide** inhibit cytochrome oxidase and can therefore totally arrest respiration. **Carboxin** and **TTFA** specifically inhibit transfer of reducing equivalents from succinate dehydrogenase to Q, whereas **malonate** is a competitive inhibitor of succinate dehydrogenase.

The antibiotic **oligomycin** completely blocks oxidation and phosphorylation in intact mitochondria. However, in the additional presence of the uncoupler **dinitrophenol,** oxidation proceeds without phosphorylation, indicating that oligomycin does not act directly on the respiratory chain but subsequently on a step in phosphorylation (Figure 14–9).

Atractyloside inhibits oxidative phosphorylation that is dependent on the transport of adenine nucleotides across the inner mitochondrial membrane.

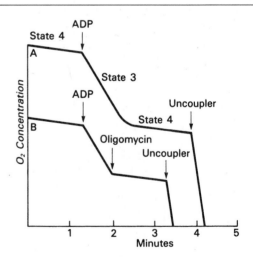

Figure 14–9. Respiratory control in mitochondria. Experiment A shows the basic state of respiration in state 4 that is accelerated upon addition of ADP. When the exogenous ADP has been phosphorylated to ATP, respiration reverts to state 4. The addition of uncoupler, eg, dinitrophenol, releases respiration from phosphorylation. In experiment B, addition of oligomycin blocks phosphorylation of added ADP and therefore of respiration as well. Addition of uncoupler again releases respiration from phosphorylation.

It is considered to inhibit the transporter of ADP into the mitochondrion and of ATP out of the mitochondrion (Figure 14–12).

The action of **uncouplers** is to dissociate oxidation in the respiratory chain from phosphorylation, and this action can explain the toxic action of these compounds in vivo. This results in respiration becoming uncontrolled, since the concentration of ADP or P_i no longer limits the rate of respiration. The uncoupler that has been used most frequently is 2,4-dinitrophenol, but other compounds act in a similar manner, including dinitrocresol, pentachlorophenol, and CCCP (*m*-chlorocarbonyl cyanide phenylhydrazone). The latter is about 100 times as active as dinitrophenol.

THE CHEMIOSMOTIC THEORY EXPLAINS THE MECHANISM OF OXIDATIVE PHOSPHORYLATION

Two principal hypotheses, the chemical and the chemiosmotic, have been advanced to account for the coupling of oxidation and phosphorylation. The **chemical hypothesis** postulated direct chemical coupling at all stages of the process, as in the reactions that generate ATP in glycolysis. However, energy-rich intermediates linking oxidation with phosphorylation were never isolated and the hypothesis has become discredited.

The Respiratory Chain Is a Proton Pump

Mitchell's **chemiosmotic theory** postulates that the energy from oxidation of components in the respiratory chain generates hydrogen ions that are ejected to the outside of a coupling membrane in the mitochondrion, ie, the inner membrane. The electrochemical potential difference resulting from the asymmetric distribution of the hydrogen ions (protons, H^+) is used to drive the mechanism responsible for the formation of ATP (Figure 14–10).

Each of the respiratory chain complexes I, III, and IV (Figures 14–7 and 14–10) acts as a **proton pump.** The inner membrane is impermeable to ions in general but particularly to protons, which accumulate outside the membrane, creating an **electrochemical**

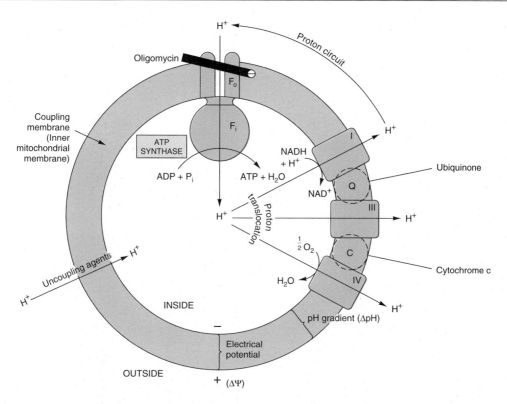

Figure 14–10. Principles of the chemiosmotic theory of oxidative phosphorylation. The main proton circuit is created by the coupling of oxidation to proton translocation from the inside to the outside of the membrane, driven by the respiratory chain complexes I, III, and IV, each of which acts as a *proton pump.* F_1, F_0, protein subunits which utilize energy from the proton gradient to promote phosphorylation. Uncoupling agents such as dinitrophenol allow leakage of H^+ across the membrane, thus collapsing the electrochemical proton gradient. Oligomycin specifically blocks conduction of H^+ through F_0.

potential difference across the membrane ($\Delta\mu_H^+$). This consists of a chemical potential (difference in pH) and an electrical potential. The precise number of protons pumped by each complex for each NADH oxidized is not known with certainty, but current estimates suggest that complex I translocates 3–4, complex III translocates 4, and complex IV translocates 2. Therefore, the P:O ratio would not be a complete integer, ie, 3, but possibly 2.5. For simplicity, a value of 3 for the oxidation of NADH + H$^+$ and 2 for the oxidation of FADH$_2$ will continue to be used throughout this text. A possible mechanism to account for proton pumping by complex III, the Q cycle, is shown in Figure 14–11.

A Membrane-Located ATP Synthase Forms ATP

The electrochemical potential difference is used to drive a membrane-located ATP synthase which in the presence of P$_i$ + ADP forms ATP (Figure 14–10). Thus, there is no high-energy intermediate that is common to both oxidation and phosphorylation as in the chemical hypothesis.

Scattered over the surface of the inner membrane are the phosphorylating complexes responsible for the production of ATP (Figure 14–1). These consist of several protein subunits collectively known as F$_1$, which project into the matrix and which contain the ATP synthase (Figure 14–10). These subunits are attached, possibly by a stalk, to a membrane protein complex known as F$_0$, which probably extends through the membrane and also consists of several protein subunits (Figure 14–10). Protons pass through the F$_0$-F$_1$ complex, leading to the formation of ATP from ADP and P$_i$. It is of interest that similar phosphorylating units are found inside the plasma membrane of bacteria but outside the thylakoid membrane of chloroplasts. It is significant that the proton gradient is from outside to inside in mitochondria and bacteria but in the reverse direction in chloroplasts.

The mechanism of coupling of proton translocation to the ATP synthase system is conjectural. Studies have suggested that ATP synthesis, which may take place while attached to the enzyme, is not the main energy-requiring step—rather it is the release of ATP from the active site. This may involve conformational changes in the F$_1$ subunit brought about by the proton gradient.

Experimental Findings Support the Chemiosmotic Theory

(1) Addition of protons (acid) to the external medium of intact mitochondria leads to the generation of ATP.

(2) Oxidative phosphorylation does not occur in soluble systems where there is no possibility of a

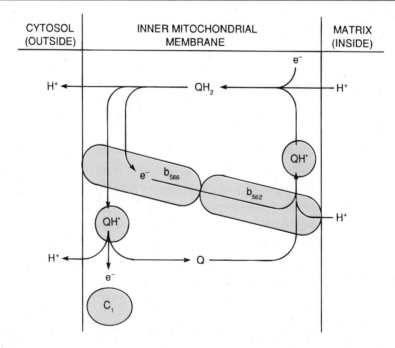

Figure 14–11. Proton-motive "Q" cycle. The figure shows how the components of complex III are organized as a proton pump. QH· is anchored on each side of the membrane by attachment to a Q-binding protein, whereas QH$_2$ and Q are mobile. Cytochromes are shown respectively as b, c_1.

vectorial ATP synthase. A closed membrane must be present in order to obtain oxidative phosphorylation (Figure 14–10).

(3) The respiratory chain contains components organized in a sided manner (transverse asymmetry) as required by the chemiosmotic theory.

The Chemiosmotic Theory Can Account for Respiratory Control

The electrochemical potential difference across the membrane, once established as a result of proton translocation, inhibits further transport of reducing equivalents through the respiratory chain unless discharged by back-translocation of protons across the membrane through the vectorial ATP synthase. This in turn depends on availability of ADP and P_i.

The Chemiosmotic Theory Explains the Action of Uncouplers

These compounds (eg, dinitrophenol) are amphipathic (Chapter 16) and increase the permeability of mitochondria to protons (Figure 14–10), thus reducing the electrochemical potential and short-circuiting the ATP synthase. In this way, oxidation can proceed without phosphorylation.

The Chemiosmotic Theory Explains the Existence of Mitochondrial Exchange Transporter Systems

These are a consequence of the coupling membrane, which must be impermeable to protons and other ions in order to maintain the electrochemical gradient (see below).

THE RELATIVE IMPERMEABILITY OF THE INNER MITOCHONDRIAL MEMBRANE NECESSITATES EXCHANGE TRANSPORTERS

Exchange diffusion systems are present in the membrane for exchange of anions against OH^- ions and cations against H^+ ions. Such systems are necessary for uptake and output of ionized metabolites while preserving electrical and osmotic neutrality.

The inner bilipoid mitochondrial membrane is freely permeable to uncharged small molecules, such as oxygen, water, CO_2, and NH_3, and to monocarboxylic acids, such as 3-hydroxybutyric, acetoacetic, and acetic. Long-chain fatty acids are transported into mitochondria via the carnitine system (Figure 24–1), and there is also a special carrier for pyruvate involving a symport that utilizes the H^+ gradient from outside to inside the mitochondrion (Figure 14–12). However, dicarboxylate and tricarboxylate anions and amino acids require specific transporter or carrier systems to facilitate their transport across the membrane. It appears that monocarboxylic acids pen-

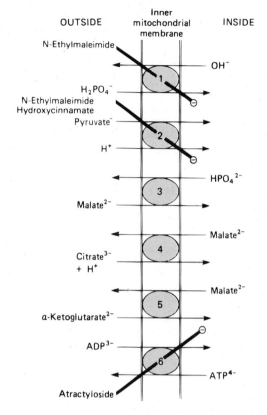

Figure 14–12. Transporter systems in the inner mitochondrial membrane. ①, phosphate transporter; ②, pyruvate symport; ③, dicarboxylate transporter; ④, tricarboxylate transporter; ⑤, α-ketoglutarate transporter; ⑥, adenine nucleotide transporter. N-Ethylmaleimide, hydroxycinnamate, and atractyloside inhibit (⊖) the indicated systems. Also present (but not shown) are transporter systems for glutamate/aspartate (Figure 14–15), glutamine, ornithine, neutral amino acids, and carnitine (Figure 24–1).

etrate more readily, because of their lesser degree of dissociation. It is the undissociated and more lipid-soluble acid that is thought to be the molecular species that penetrates the lipoid membrane.

The transport of di- and tricarboxylate anions is closely linked to that of inorganic phosphate, which penetrates readily as the $H_2PO_4^-$ ion in exchange for OH^-. The net uptake of malate by the dicarboxylate transporter requires inorganic phosphate for exchange in the opposite direction. The net uptake of citrate, isocitrate, or cis-aconitate by the tricarboxylate transporter requires malate in exchange. α-Ketoglutarate transport also requires an exchange with malate. Thus, by the use of exchange mechanisms, osmotic balance is maintained. It will be appreciated that citrate transport across the mitochondrial membrane depends not only on malate transport but on the transport of inorganic phosphate as well. The adenine nucleotide transporter allows the exchange of ATP and ADP but not AMP. It is vital in allowing ATP exit from mitochondria to the sites of extramitochondrial utilization and in allowing the return of ADP for ATP production within the mitochondrion (Figure 14–13). Na^+ can be exchanged for H^+, driven by the proton gradient. It is believed that active uptake of Ca^{2+} by mitochondria occurs with a net charge transfer of 1 (Ca^+ uniport), possibly through a Ca^{2+}/H^+ antiport. Calcium release from mitochondria is facilitated by exchange with Na^+.

Ionophores Permit Specific Cations to Penetrate Membranes

Ionophores are so termed because of their ability to complex specific cations and facilitate their transport through biologic membranes. This property of ionophoresis is due to their lipophilic character, which allows penetration of lipoid membranes such as the mitochondrial membrane. An example is the antibiotic **valinomycin,** which allows penetration of K^+ through the mitochondrial membrane discharging the membrane potential between the inside and the outside of the mitochondrion. **Nigericin** also acts as an ionophore for K^+ but in exchange for H^+. It therefore abolishes the pH gradient across the membrane. In the presence of both valinomycin and nigericin, both the membrane potential and the pH gradient are eliminated, and phosphorylation is therefore completely inhibited. The classic uncouplers such as dinitrophenol are, in fact, proton ionophores.

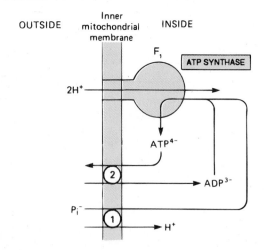

Figure 14–13. Combination of phosphate transporter (①) with the adenine nucleotide transporter (②) in ATP synthesis. The H^+/P_i, symport shown is equivalent to the P_i/OH^- antiport shown in Figure 14–12. Three or possibly four protons are taken into the mitochondrion for each ATP exported. However, one less proton would be taken in when ATP is used inside the mitochondrion.

A Proton-Translocating Transhydrogenase Is a Source of Intramitochondrial NADPH

This energy-linked transhydrogenase, a protein in the inner mitochondrial membrane, couples the passage of protons down the electrochemical gradient from outside to inside the mitochondrion with the transfer of H from intramitochondrial NADH to form NADPH. It appears to function as an energy-linked redox buffer and as a source of NADPH for intramitochondrial enzymes, such as glutamate dehydrogenase and hydroxylases, involved in steroid synthesis.

Oxidation of Extramitochondrial NADH Is Mediated by Substrate Shuttles

NADH cannot penetrate the mitochondrial membrane, but it is produced continuously in the cytosol by 3-phosphoglyceraldehyde dehydrogenase, an enzyme in the glycolysis sequence (Figure 19–2). However, under aerobic conditions, extramitochondrial NADH does not accumulate and is presumed to be oxidized by the respiratory chain in mitochondria. Several possible mechanisms have been considered to permit this process. These involve transfer of reducing equivalents through the mitochondrial membrane via substrate pairs, linked by suitable dehydrogenases. It is necessary that the specific dehydrogenase be present on both sides of the mitochondrial membrane. The mechanism of transfer using the **glycerophosphate shuttle** is shown in Figure 14–14). It is to be noted that since the mitochondrial enzyme is linked to the respiratory chain via a flavoprotein rather than NAD, only 2 rather than 3 mol of ATP are formed per atom of oxygen consumed. In some species, the activity of the FAD-linked enzyme decreases after thyroidectomy and increases after administration of thyroxine. Although this shuttle is present in insect flight muscle, brain, brown adipose tissue, and white muscle and might be important in liver, in other tissues (eg, heart muscle) the mitochondrial glycerol-3-phosphate dehydrogenase is deficient. It is therefore believed that a transport system involving malate and cytosolic and mitochondrial malate dehydrogenase is of more universal utility. The **malate shuttle** system is shown in Figure 14–15. The complexity of this system is due to the impermeability of the mitochondrial membrane to oxaloacetate, which must react with glutamate and transaminate to aspartate and α-ketoglutarate before transport through the mitochondrial membrane and reconstitution to oxaloacetate in the cytosol.

Ion Transport in Mitochondria Is Energy-Linked

Actively respiring mitochondria in which oxidative phosphorylation is taking place maintain or accumulate cations such as K^+, Na^+, Ca^{2+}, and Mg^{2+}, and P_i. Uncoupling with dinitrophenol leads to loss of ions from the mitochondria, but the ion uptake is

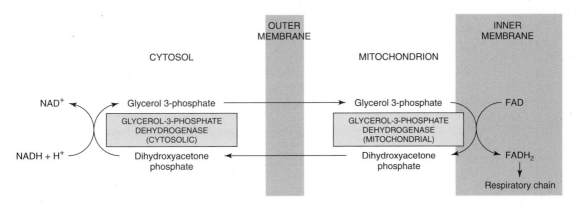

Figure 14–14. Glycerophosphate shuttle for transfer of reducing equivalents from the cytosol into the mitochondrion.

not inhibited by oligomycin, suggesting that the energy need not be supplied by phosphorylation of ADP. It is envisaged that a primary proton pump drives cation exchange.

The Creatine Phosphate Shuttle Facilitates Transport of High-Energy Phosphate From Mitochondria

This shuttle (Figure 14–16) augments the functions of creatine phosphate as an energy buffer by acting as a dynamic system for transfer of high-energy phosphate from mitochondria, in active tissues such as heart and skeletal muscle. An isoenzyme of creatine kinase (CK_m) is found in the mitochondrial intermembrane space, catalyzing the transfer of high-

energy phosphate to creatine from ATP emerging from the adenine nucleotide transporter. In turn, the creatine phosphate is transported into the cytosol via protein pores in the outer mitochondrial membrane, becoming available for generation of extramitochondrial ATP. Different isoenzymes of creatine kinase mediate transfer of high-energy phosphate to and from the various systems that utilize or generate it, eg, muscle contraction, glycolysis (Figure 14–16).

CLINICAL ASPECTS

The condition known as **fatal infantile mitochondrial myopathy and renal dysfunction** involves severe diminution or absence of most oxidoreductases

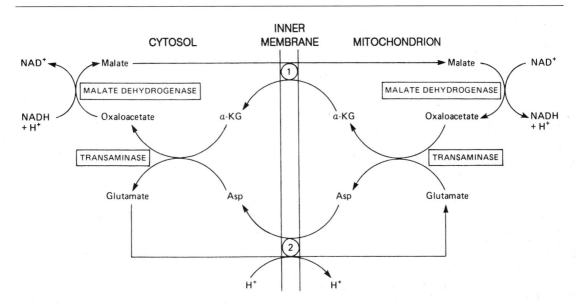

Figure 14–15. Malate shuttle for transfer of reducing equivalents from the cytosol into the mitochondrion. ①, Ketoglutarate transporter; ②, glutamate-aspartate transporter (note the proton symport with glutamate).

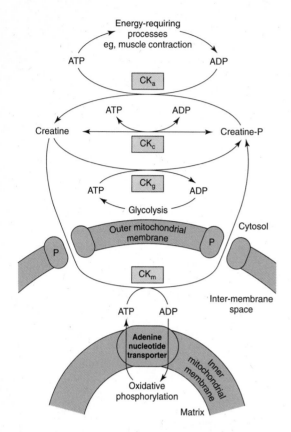

Figure 14–16. The creatine phosphate shuttle of heart and skeletal muscle. The shuttle allows rapid transport of high-energy phosphate from the mitochondrial matrix into the cytosol. (CK$_a$, creatine kinase concerned with large requirements for ATP, eg, muscular contraction; CK$_c$, creatine kinase for maintaining equilibrium between creatine and creatine phosphate and ATP/ADP; CK$_g$, creatine kinase coupling glycolysis to creatine phosphate synthesis; CK$_m$, mitochondrial creatine kinase mediating creatine phosphate production from ATP formed in oxidative phosphorylation; P, pore protein in outer mitochondrial membrane.)

of the respiratory chain. **MELAS** (mitochondrial myopathy, encephalopathy, lactic acidosis, and stroke) is an inherited condition due to NADH: ubiquinone oxidoreductase (complex I) or cytochrome oxidase deficiency. Diseases involving deficiencies of most enzymes of oxidative phosphorylation have been described.

A number of drugs and poisons act by inhibition of oxidative phosphorylation (see above).

SUMMARY

1. Virtually all energy released from the oxidation of carbohydrate, fat, and protein is made available in mitochondria as reducing equivalents (—H or e⁻). These are funneled into the respiratory chain, where they are passed down a redox gradient of carriers to their final reaction with oxygen to form water.

2. The redox carriers are grouped into respiratory chain complexes in the inner mitochondrial membrane. These use the energy released in the redox gradient to pump protons to the outside of the membrane, creating an electrochemical potential across the membrane.

3. Spanning the membrane are ATP synthase complexes that use the potential energy of the proton gradient to synthesize ATP from ADP and P$_i$. In this way, oxidation is closely coupled to phosphorylation to meet the energy needs of the cell.

4. Because the inner mitochondrial membrane is impermeable to protons and other ions, special exchange transporters span the membrane to allow passage of ions such as OH⁻, P$_i$⁻, ATP⁴⁻, and ADP³⁻, and metabolites, without discharging the electrochemical gradient across the membrane.

5. Many well-known poisons such as cyanide arrest respiration by inhibition of the respiratory chain.

REFERENCES

Balaban RS: Regulation of oxidative phosphorylation in the mammalian cell. Am J Physiol 1990;258:C377.
Boyer PD: The unusual enzymology of ATP synthase. Biochemistry 1987;26:8503.
Cross RL: The mechanism and regulation of ATP synthesis by F$_1$-ATPases. Annu Rev Biochem 1981;50:681.
Harold FM: *The Vital Force: A Study of Bioenergetics.* Freeman, 1986.
Hatefi Y: The mitochondrial electron transport and oxidative phosphorylation system. Annu Rev Biochem 1985; 54:1015.
Mitchell P: Keilin's respiratory chain concept and its chemiosmotic consequences. Science 1979;206:1148.
Morgan-Hughes JA et al: Mitochondrial myopathies: Clinical defects. Biochem Soc Trans 1990;18:523.

Nicholls DG: *Bioenergetics: An Introduction to the Chemiosmotic Theory.* Academic Press, 1982.
Prince RC: The proton pump of cytochrome oxidase. Trends Biochem Sci 1988;13:159.
Scholte HR, et al: Defects in oxidative phosphorylation. Biochemical investigations in skeletal muscle and expression of the lesion in other cells. J Inher Metab Dis 1987;10, Suppl 1:81.
Tyler DD: *The Mitochondrion in Health and Disease.* VCH Publishers, 1992.
Wallimann T et al: Intracellular compartmentation, structure and function of creatine kinase isoenzymes in tissues with high and fluctuating energy demands. Biochem J 1992;281:21.

Carbohydrates of Physiologic Significance

15

Peter A. Mayes, PhD, DSc

INTRODUCTION

Carbohydrates are widely distributed in plants and animals, where they fulfill both structural and metabolic roles. In plants, glucose is synthesized from carbon dioxide and water by photosynthesis and stored as starch or is converted to the cellulose of the plant framework. Animals can synthesize some carbohydrate from fat and protein, but the bulk of animal carbohydrate is derived ultimately from plants.

BIOMEDICAL IMPORTANCE

Knowledge of the structure and properties of the physiologically significant carbohydrates is essential to understanding their role in the economy of the mammalian organism. The sugar **glucose** is the most important carbohydrate. It is as glucose that the bulk of dietary carbohydrate is absorbed into the bloodstream or into which it is converted in the liver, and it is from glucose that all other carbohydrates in the body can be formed. Glucose is a major fuel of the tissues of mammals (except ruminants) and a universal fuel of the fetus. It is converted to other carbohydrates having highly specific functions, eg, **glycogen** for storage; **ribose** in nucleic acids; **galactose** in lactose of milk, in certain complex lipids, and in combination with protein in glycoproteins and proteoglycans. Diseases associated with carbohydrates include **diabetes mellitus, galactosemia, glycogen storage diseases,** and **lactose intolerance.**

CARBOHYDRATES ARE ALDEHYDE OR KETONE DERIVATIVES OF POLYHYDRIC ALCOHOLS

They are classified as follows:

(1) **Monosaccharides** are those carbohydrates that cannot be hydrolyzed into simpler carbohydrates: They may be subdivided into **trioses, tetroses, pentoses, hexoses, heptoses,** or **octoses,** depending upon the number of carbon atoms they possess; and as **aldoses** or **ketoses** depending upon whether the aldehyde or ketone group is present. Examples are listed in Table 15–1.

(2) **Disaccharides** yield two molecules of monosaccharide when hydrolyzed: Examples are maltose, yielding two molecules of glucose, and sucrose, yielding one molecule of glucose and one of fructose.

(3) **Oligosaccharides** yield two to ten monosaccharide units on hydrolysis: Maltotriose* is an example.

(4) **Polysaccharides** yield more than ten molecules of monosaccharides on hydrolysis: Examples of polysaccharides, which may be linear or branched, are the starches and dextrins. These are sometimes designated as hexosans or pentosans, depending upon the identity of the monosaccharides they yield on hydrolysis.

BIOMEDICALLY, GLUCOSE IS THE MOST IMPORTANT MONOSACCHARIDE

The Structure of Glucose Can Be Represented in Three Ways

The straight-chain structural formula (aldohexose; Figure 15–1A) can account for some of the properties of glucose, but a cyclic structure is favored on thermodynamic grounds and accounts for the remainder of its chemical properties. For most pur-

Table 15–1. Classification of important sugars.

	Aldoses	Ketoses
Trioses ($C_3H_6O_3$)	Glycerose	Dihydroxyacetone
Tetroses ($C_4H_8O_4$)	Erythrose	Erythrulose
Pentoses ($C_5H_{10}O_5$)	Ribose	Ribulose
Hexoses ($C_6H_{12}O_6$)	Glucose	Fructose

*Note that this is not a true triose but a trisaccharide containing three α-glucose residues.

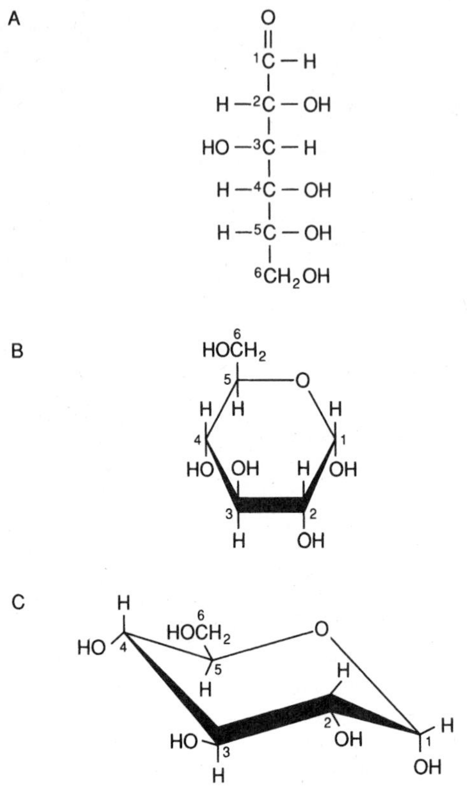

A

B

C

Figure 15–1. D-Glucose. *A:* straight chain form. *B:* α-D-glucose; Haworth projection. *C:* α-D-glucose; chair form.

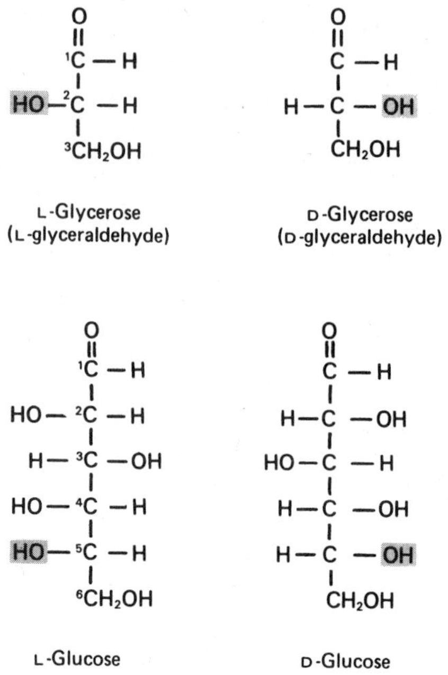

L-Glycerose
(L-glyceraldehyde)

D-Glycerose
(D-glyceraldehyde)

L-Glucose

D-Glucose

Figure 15–2. D- and L-isomerism of glycerose and glucose.

poses, the structural formula may be represented as a simple ring in perspective as proposed by Haworth (Figure 15–1B). X-ray diffraction analysis shows that the six-membered ring containing one oxygen atom is actually in the form of a chair (Figure 15–1C).

Sugars Exhibit Various Forms of Isomerism

Compounds that have the same structural formula but differ in spatial configuration are known as **stereoisomers.** The presence of asymmetric carbon atoms (carbon atoms attached to four different atoms or groups) allows the formation of isomers. The number of possible isomers of a compound depends on the number of asymmetric carbon atoms (n) and is equal to 2^n. Glucose, with four asymmetric carbon atoms, therefore has 16 isomers. The more important types of isomerism found with glucose are as follows:

(1) D and L isomerism: The designation of a sugar isomer as the D form or of its mirror image as the L form is determined by its spatial relationship to

the parent compound of the carbohydrate family, the three-carbon sugar glycerose (glyceraldehyde). The L and D forms of this sugar are shown in Figure 15–2 together with the corresponding isomers of glucose. The orientation of the —H and —OH groups around the carbon atom adjacent to the terminal primary alcohol carbon (eg, carbon atom 5 in glucose) determines whether the sugar belongs to the D or L series. When the —OH group on this carbon is on the right (as seen in Figure 15–2), the sugar is a member of the D series; when it is on the left, it is a member of the L series. Most of the monosaccharides occurring in mammals are of the D configuration, and enzymes responsible for their metabolism are specific for this configuration.

The presence of asymmetric carbon atoms also confers **optical activity** on the compound. When a beam of plane-polarized light is passed through a solution of an **optical isomer,** it will be rotated either to the right, dextrorotatory (+), or to the left, levorotatory (−). A compound may be designated D(−), D(+), L(−), or L(+), indicating structural relationship to D or L glycerose but not necessarily exhibiting the same optical rotation. For example, the naturally occurring form of fructose is the D(−) isomer.

When equal amounts of D and L isomers are present, the resulting mixture has no optical activity, since the activities of each isomer cancel one an-

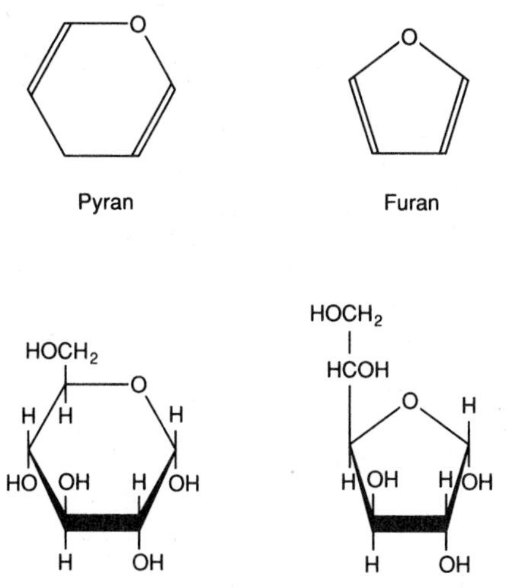

Pyran   Furan

α-D-Glucopyranose
(α-anomer)

β-D-Glucopyranose
(β-anomer)

Acyclic
aldehyde
form

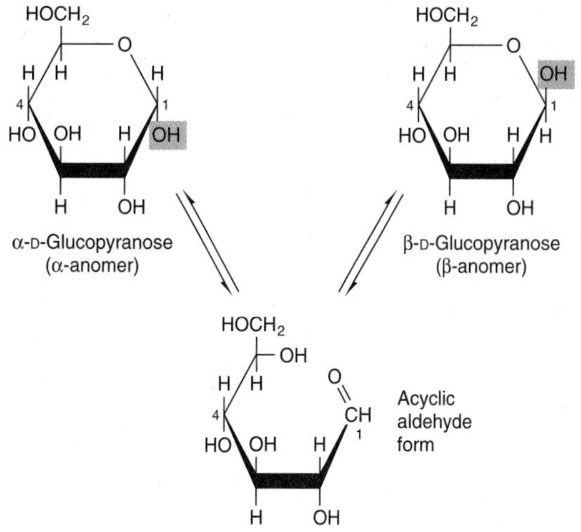

Figure 15–5. Mutarotation of glucose.

Figure 15–3. Pyranose and furanose forms of glucose.

α- D -Fructopyranose

α- D -Fructofuranose

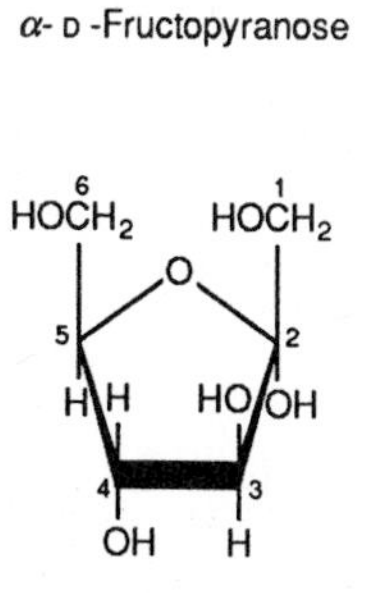

Figure 15–4. Pyranose and furanose forms of fructose.

other. Such a mixture is said to be a **racemic**—or DL—mixture. Synthetically produced compounds are necessarily racemic because the opportunities for the formation of each optical isomer are identical.

(2) Pyranose and furanose ring structures: This terminology is based on the fact that the stable ring structures of monosaccharides are similar to the ring structures of either pyran or furan (Figure 15–3).

Ketoses may also show ring formation, eg, D-fructofuranose or D-fructopyranose (Figure 15–4). In the case of glucose in solution, more than 99% is in the pyranose form.

(3) Alpha and beta anomers: The ring structure of an aldose is a hemiacetal, since it is formed by combination of an aldehyde and an alcohol group (Figure 15–5). Similarly, the ring structure of a ketose is a hemiketal. Crystalline glucose is α-D-glucopyranose. The cyclic structure is retained in solution, but isomerism takes place about position 1, the carbonyl or **anomeric carbon atom,** to give a mixture of α-glucopyranose (38%) and β-glucopyranose (62%). Less than 0.3% is represented by α and β anomers of glucofuranose. This equilibration is accompanied by optical rotation (**mutarotation**) as the hemiacetal ring opens and re-forms with change of position of the —H and —OH groups on carbon 1. The change probably takes place via a hydrated straight-chain acyclic molecule, although polarography has indicated that glucose exists only to the extent of 0.0025% in the acyclic form. The optical rotation of glucose in solution is dextrorotatory; hence, the alternative name of **dextrose,** often used in clinical practice.

(4) Epimers: Isomers differing as a result of variations in configuration of the —OH and —H on carbon atoms 2, 3, and 4 of glucose are known as epimers. Biologically, the most important epimers of glucose are mannose and galactose, formerly by epimerization at carbons 2 and 4, respectively (Figure 15–6).

(5) Aldose-ketose isomerism: Fructose has the same molecular formula as glucose but differs in its structural formula, since there is a potential keto

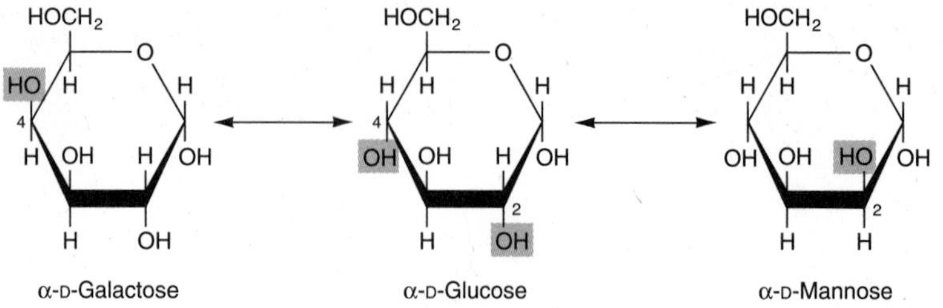

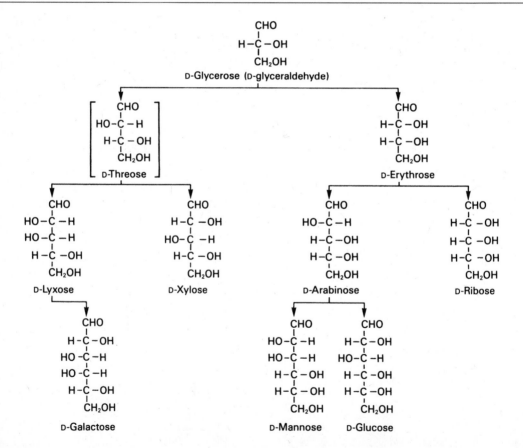

Figure 15–6. Epimerization of glucose.

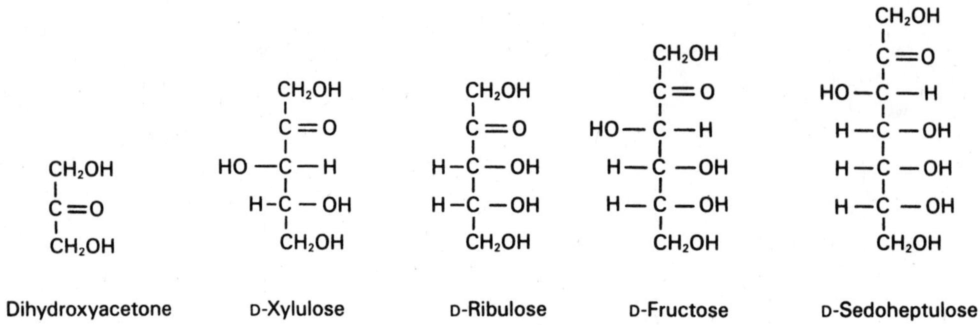

Figure 15–7. Examples of ketoses of physiologic significance.

Figure 15–8. The structural relations of the aldoses, D series, of physiologic significance. D-Threose is not of physiologic significance. The series is built up by the theoretical addition of a CH_2O unit to the —CHO group of the sugar.

Table 15–2. Pentoses of physiologic importance.

Sugar	Where Found	Biochemical Importance	Clinical Significance
D-Ribose	Nucleic acids.	Structural elements of nucleic acids and coenzymes, eg, ATP, NAD, NADP, flavoproteins. Ribose phosphates are intermediates in pentose phosphate pathway.	
D-Ribulose	Formed in metabolic processes.	Ribulose phosphate is an intermediate in pentose phosphate pathway.	
D-Arabinose	Gum arabic. Plum and cherry gums.	Constituent of glycoproteins.	
D-Xylose	Wood gums, proteoglycans, glycosaminoglycans.	Constituent of glycoproteins.	
D-Lyxose	Heart muscle.	A constituent of a lyxoflavin isolated from human heart muscle.	
L-Xylulose	Intermediate in uronic acid pathway.		Found in urine in essential pentosuria.

group in position 2 of fructose (Figure 15–7), whereas there is a potential aldehyde group in position 1 of glucose (Figure 15–8).

Many Monosaccharides Are Physiologically Important

Triose derivatives are formed in the course of the metabolic breakdown of glucose by the glycolysis pathway. Derivatives of trioses, tetroses, and pentoses and of a seven-carbon sugar (sedoheptulose) are formed in the breakdown of glucose via the pentose phosphate pathway. Pentose sugars are important constituents of nucleotides, nucleic acids, and many coenzymes (Table 15–2). Of the hexoses, glucose, galactose, fructose, and mannose are physiologically the most important (Table 15–3).

The structures of the aldo sugars of biochemical significance are shown in Figure 15–8. Five keto sugars which are important in metabolism are shown in Figure 15–7. Of additional significance are carboxylic acid derivatives of glucose such as D-glu-

curonate (important in glucuronide formation and present in glycosaminoglycans) and its metabolic derivatives, L-iduronate (present in glycosaminoglycans) (Figure 15–9) and L-gulonate (a member of the uronic acid pathway; see Figure 22–4).

Sugars Form Glycosides With Other Compounds and With Each Other

Glycosides are compounds formed from a condensation between the hydroxyl group of the anomeric carbon of a monosaccharide, or monosaccharide residue, and a second compound that may, or may not (in the case of an **aglycone**), be another monosaccharide. If the second group is a hydroxyl, the *O*-glycosidic bond is an **acetal** link because it results from a reaction between a hemiacetal group (formed from an aldehyde and an —OH group) and another —OH group. If the hemiacetal portion is glucose, the resulting compound is a **glucoside;** if galactose, a **galactoside;** and so on. If the second group is an

Table 15–3. Hexoses of physiologic importance.

Sugar	Source	Importance	Clinical Significance
D-Glucose	Fruit juices. Hydrolysis of starch, cane sugar, maltose, and lactose.	The "sugar" of the body. The sugar carried by the blood, and the principal one used by the tissues.	Present in the urine (glycosuria) in diabetes mellitus owing to raised blood glucose (hyperglycemia).
D-Fructose	Fruit juices. Honey. Hydrolysis of cane sugar and of inulin (from the Jerusalem artichoke).	Can be changed to glucose in the liver and so used in the body.	Hereditary fructose intolerance leads to fructose accumulation and hypoglycemia.
D-Galactose	Hydrolysis of lactose.	Can be changed to glucose in the liver and metabolized. Synthesized in the mammary gland to make the lactose of milk. A constituent of glycolipids and glycoproteins.	Failure to metabolize leads to galactosemia and cataract.
D-Mannose	Hydrolysis of plant mannans and gums.	A constituent of many glycoproteins.	

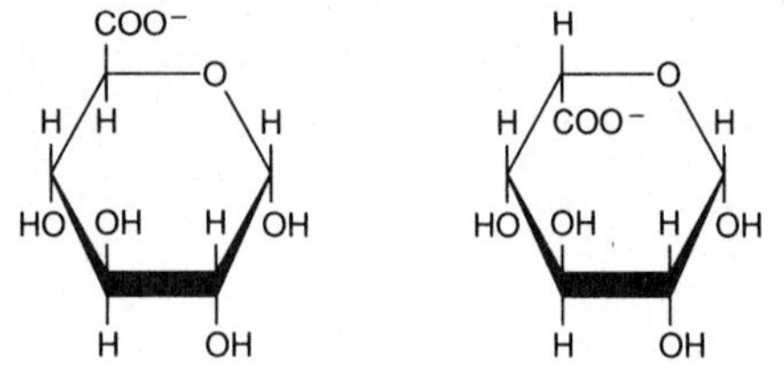

Figure 15–9. α-D-Glucuronate (*left*) and β-L-iduronate (*right*).

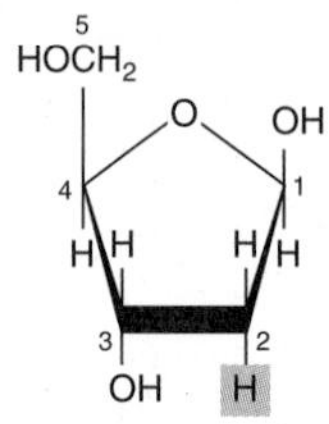

Figure 15–11. 2-Deoxy-D-ribofuranose (β form).

amine, an *N*-glycosidic bond is formed, eg, between adenine and ribose in nucleotides such as ATP (Figure 12–5).

Glycosides are found in many drugs and spices and in the constituents of animal tissues. The aglycone may be methanol, glycerol, a sterol, a phenol, or a base such as adenine. The glycosides that are important in medicine because of their action on the heart (**cardiac glycosides**) all contain steroids as the aglycone component. These include derivatives of digitalis and strophanthus such as **ouabain,** an inhibitor of the Na^+-K^+ ATPase of cell membranes. Other glycosides include antibiotics such as **streptomycin** (Figure 15–10).

Deoxy Sugars Lack an Oxygen Atom

Deoxy sugars are those in which a hydroxyl group attached to the ring structure has been replaced by a hydrogen atom. An example is **deoxyribose** (Figure 15–11) occurring in nucleic acids (DNA). Also found as a carbohydrate of glycoproteins is the deoxy sugar L-fucose (Figure 15–17), and of importance as an inhibitor of glucose metabolism is 2-deoxyglucose.

Amino Sugars (Hexosamines) Are Components of Glycoproteins, Gangliosides, & Glycosaminoglycans

Examples of amino sugars are D-glucosamine (Figure 15–12), D-galactosamine, and D-mannosamine, all of which have been identified in nature. Glucosamine is a constituent of hyaluronic acid. Galactosamine (chondrosamine) is a constituent of chondroitin (Chapter 57).

Several **antibiotics** (eg, erythromycin) contain amino sugars. The amino sugars are believed to be related to the antibiotic activity of these drugs.

MALTOSE, SUCROSE, AND LACTOSE ARE IMPORTANT DISACCHARIDES

The disaccharides are sugars composed of two monosaccharide residues united by a glycosidic linkage (Figure 15–13). Their chemical name reflects their component monosaccharides. The physiologically important disaccharides are maltose, sucrose, and lactose (Table 15–4). Hydrolysis of sucrose yields a crude mixture called "invert sugar" because the strongly levorotatory fructose thus produced

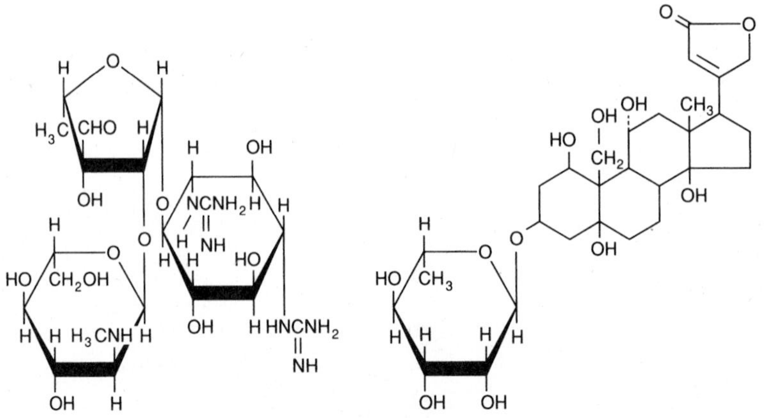

Figure 15–10. Streptomycin (*left*) and ouabain (*right*).

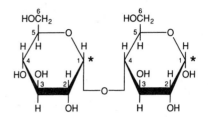

Figure 15–12. Glucosamine (2-amino-D-glucopyranose) (α form). Galactosamine is 2-amino-D-galactopyranose. Both glucosamine and galactosamine occur as *N*-acetyl derivatives in more complex carbohydrates, eg, glycoproteins.

Maltose

O-α-D-Glucopyranosyl-(1→4)-α-D-glucopyranose

Sucrose

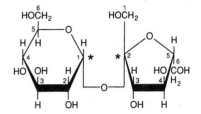

O-α-D-Glucopyranosyl-(1→2)-β-D-fructofuranoside

Lactose

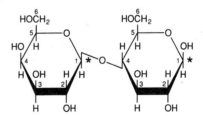

O-β-D-Galactopyranosyl-(1→4)-β-D-glucopyranose

Figure 15–13. Structures of important disaccharides. The α and β refer to the configuration at the anomeric carbon atom (*). When the anomeric carbon of the second residue takes part in the formation of the glycosidic bond, as in sucrose, the residue becomes a glycoside known as a furanoside or pyranoside. As the disaccharide no longer has an anomeric carbon with a free potential aldehyde or ketone group, it no longer exhibits reducing properties.

changes (inverts) the previous dextrorotatory action of the sucrose.

POLYSACCHARIDES SERVE STORAGE AND STRUCTURAL FUNCTIONS

Polysaccharides include the following physiologically important carbohydrates.

Starch is formed of an α-glucosidic chain. Such a compound, yielding only glucose on hydrolysis, is a homopolymer called a **glucosan** or **glucan.** It is the most important food source of carbohydrate and is found in cereals, potatoes, legumes, and other vegetables. The two chief constituents are **amylose** (15–20%), which has a nonbranching helical structure (Figure 15–14), and **amylopectin** (80–85%), which consists of branched chains composed of 24–30 glucose residues united by 1 → 4 linkages in the chains and by 1 → 6 linkages at the branch points.

Glycogen (Figure 15–15) is the storage polysaccharide of the animal body. It is often called animal starch. It is a more highly branched structure than amylopectin and has chains of 12–14-α-D-glucopyranose residues (in α[1 → 4]-glucosidic linkage) with branching by means of α(1 → 6)-glucosidic bonds.

Inulin is a starch found in tubers and roots of dahlias, artichokes, and dandelions. It is hydrolyzable to fructose, and hence it is a fructosan. This starch, unlike potato starch, is readily soluble in warm water and has been used in physiologic investigations for determination of the rate of glomerular filtration.

Dextrins are substances formed in the course of the hydrolytic breakdown of starch. Limit dextrins are the first formed products as hydrolysis reaches a certain degree of branching.

Cellulose is the chief constituent of the framework of plants. It is insoluble and consists of β-D-glucopyranose units linked by β(1 → 4) bonds to form long, straight chains strengthened by cross-linked hydrogen bonds. Cellulose cannot be digested by many

Table 15–4. Disaccharides.

Sugar	Source	Clinical Significance
Maltose	Digestion by amylase or hydrolysis of starch. Germinating cereals and malt.	
Lactose	Milk. May occur in urine during pregnancy.	In lactase deficiency, malabsorption leads to diarrhea and flatulence.
Sucrose	Cane and beet sugar. Sorghum. Pineapple. Carrot roots.	In sucrase deficiency, malabsorption leads to diarrhea and flatulence.
Trehalose[1]	Fungi and yeasts. The major sugar of insect hemolymph.	

[1]o-α-D-Glucopyranosyl-(1 → 1)-α-D-glucopyranoside.

mammals, including humans, because of the absence of a hydrolase that attacks the β linkage. Thus, it is an important source of "bulk" in the diet. In the gut of ruminants and other herbivores, there are microorganisms that can attack the β linkage, making cellulose available as a major calorigenic source. This process can also take place to a limited extent in the human colon.

Chitin is an important structural polysaccharide of invertebrates. It is found, for example, in the exoskeletons of crustaceans and insects. Structurally, chitin consists of N-acetyl-D-glucosamine units joined by β(1 → 4)-glycosidic linkages (Figure 15–16).

Glycosaminoglycans (mucopolysaccharides) consist of chains of complex carbohydrates characterized by their content of **amino sugars** and **uronic acids.** When these chains are attached to a protein molecule, the compound is known as a **proteoglycan.** As the ground or packing substance, they are associated with the structural elements of the tissues such as bone, elastin, and collagen. Their property of

holding large quantities of water and occupying space, thus cushioning or lubricating other structures, is assisted by the large number of —OH groups and negative charges on the molecules, which, by repulsion, keep the carbohydrate chains apart. Examples are **hyaluronic acid, chondroitin sulfate,** and **heparin** (Figure 15–16), discussed in detail in Chapter 57.

Glycoproteins (mucoproteins) occur in many different situations in fluids and tissues, including the cell membranes (Chapters 43 and 56). They are proteins containing carbohydrates in varying amounts attached as short or long (up to 15 units) branched or unbranched chains. Such chains are usually called oligosaccharide chains (even though they may sometimes exceed 10 units). Constituent carbohydrates are listed in Table 15–5.

Glucose is not found in mature glycoproteins apart from collagen, and, in contrast to the glycosaminoglycans and proteoglycans, uronic acids are absent.

The **sialic acids** are N- or O-acyl derivatives of

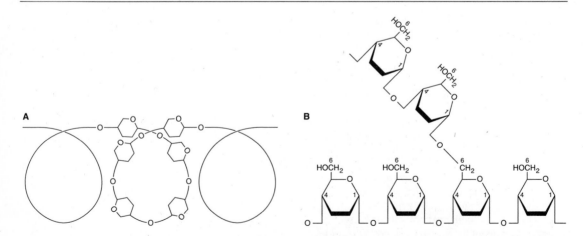

Figure 15–14. Structure of starch. **A:** Amylose, showing helical coil structure. **B:** Amylopectin, showing 1 → 6 branch point.

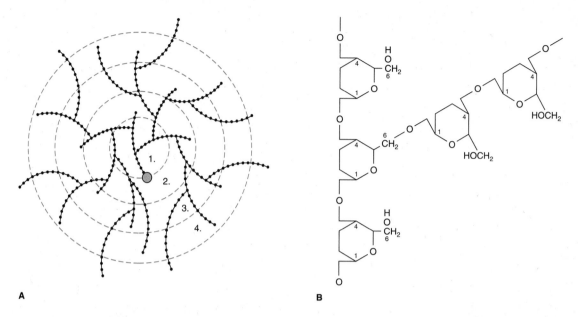

Figure 15–15. The glycogen molecule. **A:** General structure. **B:** Enlargement of structure at a branch point. The molecule is a sphere approximately 21 nm in diameter that can be visualized in electron micrographs. It has a molecular mass of 10^7 Da and consists of polysaccharide chains each containing about 13 glucose residues. The chains are either branched or unbranched and are arranged in 12 concentric layers (only four are shown in the figure). The branched chains (each has two branches) are found in the inner layers and the unbranched chains in the outer layer. (G, glycogenin, the primer molecule for glycogen synthesis.)

neuraminic acid (Figure 15–18). **Neuraminic acid** is a nine-carbon sugar derived from mannosamine (an epimer of glucosamine) and pyruvate. Sialic acids are constituents of both **glycoproteins** and **gangliosides** (Chapters 16 and 56) Gangliosides are also glycolipids.

CARBOHYDRATES OCCUR IN CELL MEMBRANES AND IN LIPOPROTEINS

The lipid structure of the cell membrane is described in Chapters 16 and 43. However, analysis of

Table 15–5. Carbohydrates found in glycoproteins.

Hexoses	Mannose (Man) Galactose (Gal)
Acetyl hexosamines	N-Acetylglucosamine (GlcNAc) N-Acetylgalactosamine (GalNAc)
Pentoses	Arabinose (Ara) Xylose (Xyl)
Methyl pentose	L-Fucose (Fuc; see Figure 15–17)
Sialic acids	N-Acyl derivatives of neuraminic acid, eg, N-acetylneuraminic acid (NeuAc; see Figure 15–18), the predominant sialic acid.

mammalian cell membrane components indicates that approximately 5% are carbohydrates, present in glycoproteins and glycolipids. Carbohydrates are also present in some lipoproteins, eg, low-density lipoproteins (LDL). Their presence on the outer surface of the plasma membrane (the **glycocalyx**) has been shown with the use of plant **lectins,** protein agglutinins that bind specifically with certain glycosyl residues. For example, **concanavalin A** has a specificity toward α-glucosyl and α-mannosyl residues. **Glycophorin** is a major integral membrane glycoprotein of human erythrocytes. It has 130 amino acid residues and spans the lipid membrane, having free polypeptide portions outside both the external and internal (cytoplasmic) surfaces. Carbohydrate chains are only attached to the amino terminal portion outside the external surface (Chapter 43).

SUMMARY

1. Carbohydrates are major constituents of animal food and animal tissues. They may be characterized by the type and number of monosaccharide residues in their molecules.

2. Glucose is the most important carbohydrate in mammalian biochemistry because nearly all carbohydrate in food is converted to glucose for further metabolism.

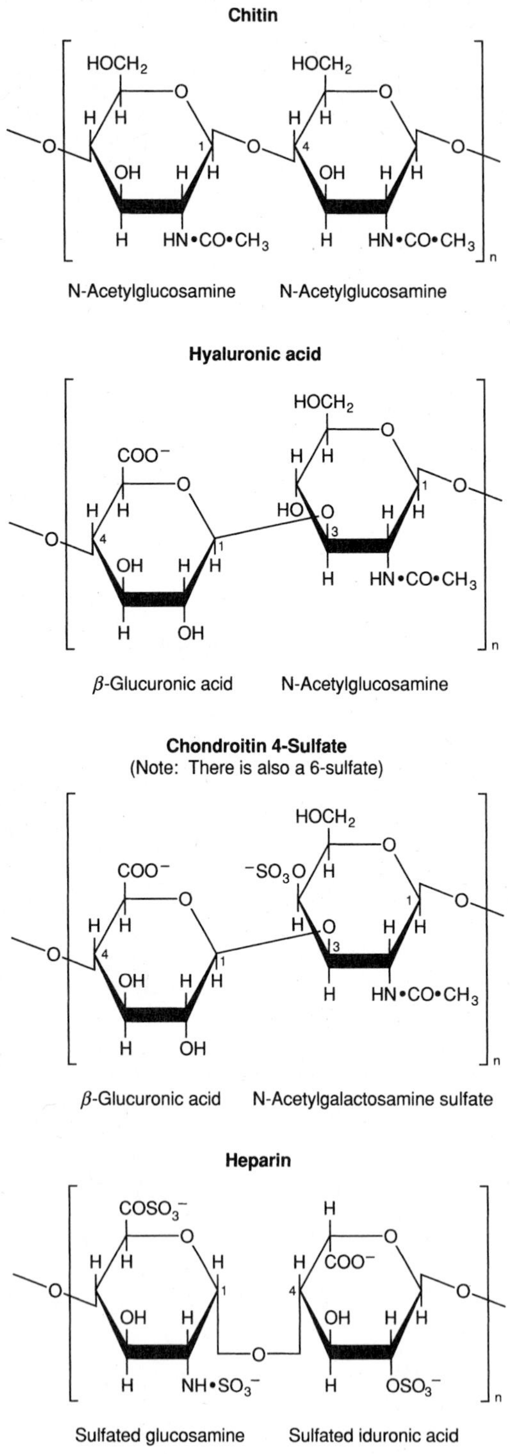

Chitin

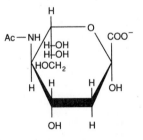

N-Acetylglucosamine   N-Acetylglucosamine

Hyaluronic acid

β-Glucuronic acid   N-Acetylglucosamine

Chondroitin 4-Sulfate
(Note: There is also a 6-sulfate)

β-Glucuronic acid   N-Acetylgalactosamine sulfate

Heparin

Sulfated glucosamine   Sulfated iduronic acid

Figure 15–16. Structure of some complex polysaccharides and glycosaminoglycans.

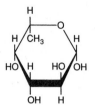

Figure 15–17. β-L-Fucose (6-deoxy-β-L-galactose).

3. Sugars have large numbers of stereoisomers because they contain several asymmetric carbon atoms.

4. The monosaccharides of physiologic importance include glucose, the "blood sugar," and ribose, an important constituent of nucleotides and nucleic acids.

5. The disaccharides of physiologic importance include maltose, an important intermediate in the digestion of starch and glycogen; sucrose, important as a dietary constituent containing fructose as well as glucose; and lactose, the unique sugar found in milk that contains galactose and glucose.

6. Starch and glycogen are storage polymers of glucose in plants and animals, respectively. They are major sources of energy in the diet.

7. Complex carbohydrates contain other sugar derivatives such as amino sugars, uronic acids, and sialic acids. They include proteoglycans and glycosaminoglycans, which are associated with structural elements of the tissues, and glycoproteins, which are proteins containing attached oligosaccharide chains; they are found in many situations including the cell membrane.

Figure 15–18. Structure of N-acetylneuraminic acid, a sialic acid (Ac = CH_3—CO—).

REFERENCES

Collins PM (editor): *Carbohydrates.* Chapman & Hall, 1987.

El-Khadem HS: *Carbohydrate Chemistry: Monosaccharides and Their Oligomers.* Academic Press, 1988.

Ferrier RJ, Collins PM: *Monosaccharide Chemistry.* Penguin Books, 1972.

Hughes RC: The complex carbohydrates of mammalian cell surfaces and their biological roles. Essays Biochem 1975;11:1.

Lindahl U, Höök M: Glycosaminoglycans and their binding to biological macromolecules. Annu Rev Biochem 1978;47:385.

Melendes-Hevia E, Waddell TG, Shelton ED: Optimization of molecular design in the evolution of metabolism: The glycogen molecule. Biochem J 1993;295:477.

Pigman WW, Horton D (editors): *The Carbohydrates,* vols 1A (1972) and 1B (1980). Academic Press.

Rees DA: *Polysaccharide Shapes.* Wiley, 1977.

Sharon N: Carbohydrates. Sci Am 1980;245:90

16

Lipids of Physiologic Significance

Peter A. Mayes, PhD, DSc

INTRODUCTION

The lipids are a heterogeneous group of compounds related more by their physical rather than by their chemical properties. They have the common property of being (1) relatively **insoluble in water** and (2) **soluble in nonpolar solvents** such as ether, chloroform, and benzene. Thus, the lipids include fats, oils, steroids, waxes, and related compounds.

Lipids are important dietary constituents not only because of their high energy value but also because of the fat-soluble vitamins and the essential fatty acids contained in the fat of natural foods.

BIOMEDICAL IMPORTANCE

In the body, fat serves as an efficient source of energy, both directly and potentially when stored in **adipose tissue.** It serves as a thermal insulator in the subcutaneous tissues and around certain organs, and nonpolar lipids act as **electrical insulators** allowing rapid propagation of depolarization waves along myelinated nerves. The fat content of **nerve tissue** is particularly high. Combinations of fat and protein (lipoproteins) are important cellular constituents, occurring both in the cell **membrane** and in the mitochondria within the cytoplasm, and serving also as the means of **transporting lipids** in the blood. A knowledge of lipid biochemistry is important in understanding many current biomedical areas of interest, eg, **obesity, atherosclerosis,** and the role of various **polyunsaturated fatty acids** in nutrition and health.

LIPIDS ARE CLASSIFIED AS SIMPLE OR COMPLEX

The following classification of lipids is modified from Bloor:

1. **Simple lipids:** Esters of fatty acids with various alcohols.
   a. **Fats:** Esters of fatty acids with glycerol. A fat in the liquid state is known as an oil.
   b. **Waxes:** Esters of fatty acids with higher molecular weight monohydric alcohols.

2. **Complex lipids:** Esters of fatty acids containing groups in addition to an alcohol and a fatty acid.
   a. **Phospholipids:** Lipids containing, in addition to fatty acids and an alcohol, a phosphoric acid residue. They frequently have nitrogen-containing bases and other substituents, eg, in **glycerophospholipids** the alcohol is glycerol and in **sphingophospholipids** the alcohol is sphingosine.
   b. **Glycolipids (glycosphingolipids):** Lipids containing a fatty acid, sphingosine, and carbohydrate.
   c. **Other complex lipids:** Lipids such as sulfolipids and aminolipids. Lipoproteins may also be placed in this category.

3. **Precursor and derived lipids:** These include fatty acids, glycerol, steroids, alcohols in addition to glycerol and sterols, fatty aldehydes, and ketone bodies (Chapter 24), hydrocarbons, lipid-soluble vitamins, and hormones.

Because they are uncharged, acylglycerols (glycerides), cholesterol, and cholesteryl esters are termed **neutral lipids.**

FATTY ACIDS ARE ALIPHATIC CARBOXYLIC ACIDS

Fatty acids occur mainly as esters in natural fats and oils but do occur in the unesterified form as **free fatty acids,** a transport form found in the plasma. Fatty acids that occur in natural fats are usually straight-chain derivatives and contain an even number of carbon atoms because they are synthesized from two-carbon units. The chain may be **saturated** (containing no double bonds) or **unsaturated** (containing one or more double bonds).

Fatty Acids Are Named After Corresponding Hydrocarbons

The most frequently used systematic nomenclature is based on naming the fatty acid after the hydrocarbon with the same number and arrangement of carbon atoms, **-oic** being substituted for the final **-e** in the name of the hydrocarbon (Genevan system). Thus, saturated acids end in **-anoic,** eg, octanoic acid, and unsaturated acids with double bonds end in **-enoic,** eg, octadecenoic acid (oleic acid).

Carbon atoms are numbered from the carboxyl carbon (carbon No. 1). The carbon atom adjacent to the carboxyl carbon (No. 2) is also known as the α-carbon. Carbon atoms No. 3 and No. 4 are the β and γ carbons, respectively, and the terminal methyl carbon is known as the ω-carbon or n-carbon atom.

Various conventions are in use for indicating the number and position of the double bonds; eg, Δ^9 indicates a double bond between carbon atoms 9 and 10 of the fatty acid; $\omega9$ indicates a double bond on the ninth carbon counting from the ω-carbon atom. Widely used conventions to indicate the number of carbon atoms, the number of double bonds, and the positions of the double bonds are shown in Figure 16–1. In animals, additional double bonds are introduced only between the existing double bond (eg, $\omega9$, $\omega6$, or $\omega3$) and the carboxyl carbon, leading to three series of fatty acids known as the $\omega9$, $\omega6$, and $\omega3$ families, respectively.

Saturated Fatty Acids Contain No Double Bonds

Saturated fatty acids may be envisaged as based on acetic acid (CH_3—COOH) as the first member of the series in which —CH_2— is progressively added between the terminal CH_3— and —COOH groups. Examples are shown in Table 16–1. Other higher members of the series are known to occur, particularly in waxes. A few branched-chain fatty acids have also been isolated from both plant and animal sources.

Unsaturated Fatty Acids Contain One or More Double Bonds (Table 16–2)

Fatty acids may be further subdivided as follows:

(1) Monounsaturated (monoethenoid, mono-enoic) acids, containing one double bond.

(2) Polyunsaturated (polyethenoid, polyenoic) acids, containing two or more double bonds.

Table 16–1. Saturated fatty acids.

Common Name	Number of C Atoms	
Formic[1]	1	Takes part in the metabolism of "C_1" units (formate)
Acetic	2	Major end product of carbohydrate fermentation by rumen organisms[2]
Propionic	3	An end product of carbohydrate fermentation by rumen organisms[2]
Butyric	4	In certain fats in small amounts (especially butter). An end product of carbohydrate fermentation by rumen organisms[2]
Valeric	5	
Caproic	6	
Caprylic (octanoic)	8	In small amounts in many fats (including butter), especially those of plant origin
Capric (decanoic)	10	
Lauric	12	Spermaceti, cinnamon, palm kernel, coconut oils, laurels, butter
Myristic	14	Nutmeg, palm kernel, coconut oils, myrtles, butter
Palmitic	16	Common in all animal and plant fats
Stearic	18	
Arachidic	20	Peanut (arachis) oil
Behenic	22	Seeds
Lignoceric	24	Cerebrosides, peanut oil

[1]Strictly, not an alkyl derivative.
[2]Also formed in the cecum of herbivores and to a lesser extent in the colon of humans.

(3) Eicosanoids: These compounds, derived from eicosa- (20-carbon) polyenoic fatty acids, comprise the **prostanoids** and **leukotrienes** (LTs), **and lipoxins** (LXs). Prostanoids include **prostaglandins** (PGs), **prostacyclins** (PGIs), and **thromboxanes** (TXs).

Prostaglandins were originally discovered in seminal plasma but are now known to exist in virtually every mammalian tissue, acting as local hormones; they have important physiologic and pharmacologic activities. They are synthesized in vivo by cyclization of the center of the carbon chain of 20-carbon (eicosanoic) polyunsaturated fatty acids (eg, arachidonic acid) to form a cyclopentane ring (Figure 16–2). A related series of compounds, the **thromboxanes,** discovered in platelets, have the cyclopentane ring interrupted with an oxygen atom (oxane ring) (Figure 16–3). Three different eicosanoic fatty acids give rise to three groups of eicosanoids characterized by the number of double bonds in the side chains, eg, PG_1, PG_2, PG_3. Variations in the substituent groups attached to the rings give rise to different types in each series of prostaglandins and thromboxanes, labeled A, B, etc. For example, the "E" type of prostaglandin (as in PGE_2) has a keto group in posi-

18:1;9 *or* Δ^9 18:1

$$\overset{18}{C}H_3(CH_2)_7\overset{10}{C}H = \overset{9}{C}H(CH_2)_7\overset{1}{C}OOH$$

or

$\omega9,C18:1$ *or* n–9, 18:1

$$\overset{\omega}{C}H_3\overset{2}{C}H_2\overset{3}{C}H_2\overset{4}{C}H_2\overset{5}{C}H_2\overset{6}{C}H_2\overset{7}{C}H_2\overset{8}{C}H_2\overset{9}{C}H = \overset{10}{C}H(CH_2)_7\overset{18}{C}OOH$$

Figure 16–1. Oleic acid. n – 9 (n minus 9) is equivalent to $\omega9$.

Table 16–2. Unsaturated fatty acids of physiologic and nutritional significance.

Number of C Atoms and Number and Position of Double Bonds	Series	Common Name	Systematic Name	Occurrence
Monoenoic acids (one double bond)				
16:1;9	ω7	Palmitoleic	*cis*-9-Hexadecenoic	In nearly all fats.
18:1;9	ω9	Oleic	*cis*-9-Octadecenoic	Possibly the most common fatty acid in natural fats.
18:1;9	ω9	Elaidic	*trans*-9-Octadecenoic	Hydrogenated and ruminant fats.
22:1;13	ω9	Erucic	*cis*-13-Docosenoic	Rape and mustard seed oils.
24:1;15	ω9	Nervonic	*cis*-15-Tetracosenoic	In cerebrosides.
Dienoic acids (two double bonds)				
18:2;9,12	ω6	Linoleic	all-*cis*-9,12-Octadecadienoic	Corn, peanut, cottonseed, soybean, and many plant oils.
Trienoic acids (three double bonds)				
18:3;6,9,12	ω6	γ-Linolenic	all-*cis*-6,9,12-Octadecatrienoic	Some plants, eg, oil of evening primrose, borage oil; minor fatty acid in animals.
18:3;9,12,15	ω3	α-Linolenic	all-*cis*-9,12,15-Octadecatrienoic	Frequently found with linoleic acid but particularly in linseed oil.
Tetraenoic acids (four double bonds)				
20:4;5,8,11,14	ω6	Arachidonic	all-*cis*-5,8,11,14-Eicosatetraenoic	Found in animal fats and in peanut oil; important component of phospholipids in animals.
Pentaenoic acids (five double bonds)				
20:5;5,8,11,14,17	ω3	Timnodonic	all-*cis*-5,8,11,14,17-Eicosapentaenoic	Important component of fish oils, eg, cod liver, mackerel, menhaden, salmon oils.
22:5;7,10,13,16,19	ω3	Clupanodonic	all-*cis*-7,10,13,16,19-Docosapentaenoic	Fish oils, phospholipids in brain.
Hexaenoic acids (six double bonds)				
22:6;4,7,10,13,16,19	ω3	Cervonic	all-*cis*-4,7,10,13,16,19-Docosahexaenoic	Fish oils, phospholipids in brain.

tion 9, whereas the "F" type has a hydroxyl group in this position. The **leukotrienes** and **lipoxins** are a third group of eicosanoid derivatives formed via the lipoxygenase pathway rather than cyclization of the fatty acid chain (Figure 16–4). First described in leukocytes, they are characterized by the presence of three or four conjugated double bonds, respectively.

Most Naturally Occurring Unsaturated Fatty Acids Have *cis* Double Bonds

The carbon chains of saturated fatty acids form a zigzag pattern when extended, as at low temperatures. At higher temperatures, some bonds rotate, causing chain shortening, which explains why bio-

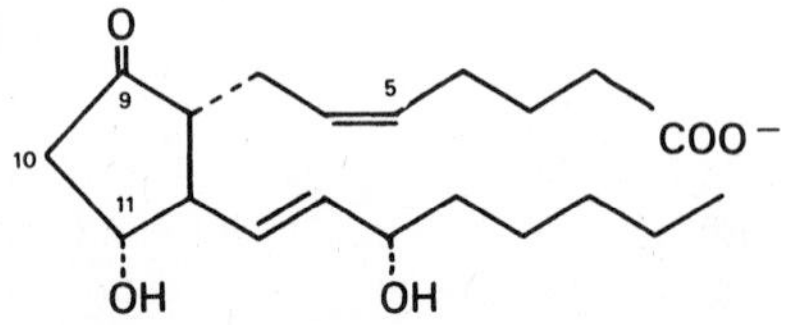

Figure 16–2. Prostaglandin E_2 (PGE$_2$).

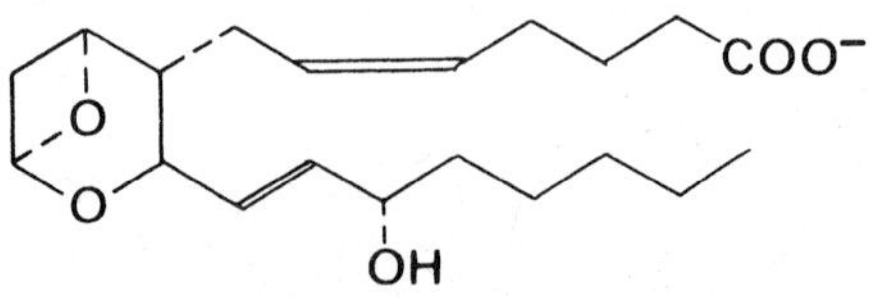

Figure 16–3. Thromboxane A_2 (TXA$_2$).

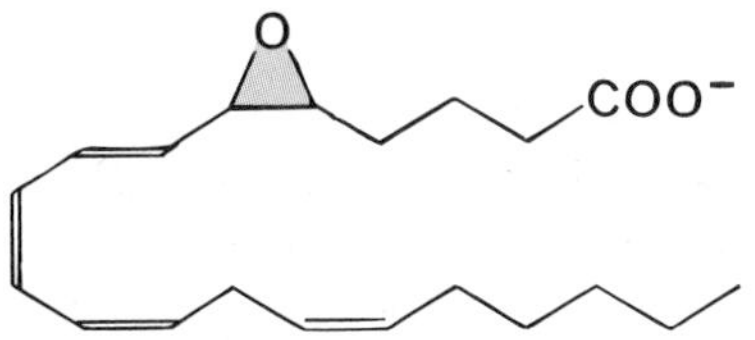

Figure 16–4. Leukotriene A_4 (LTA$_4$).

membranes become thinner with increase in temperature. A type of **geometric isomerism** occurs in unsaturated fatty acids, depending on the orientation of atoms or groups around the axes of double bonds. If the acyl chains are on the same side of the bond, it is *cis-*, as in oleic acid; if on opposite sides, it is *trans-*, as in elaidic acid, the unnatural isomer of oleic acid (Figure 16–5). Naturally occurring unsaturated long-chain fatty acids are nearly all of the *cis* configuration, the molecules being "bent" 120 degrees at the double bond. Thus, oleic acid has an L shape, whereas elaidic acid remains "straight" at its *trans* double bond. Increase in the number of *cis* double bonds in a fatty acid leads to a variety of possible spatial configurations of the molecule, eg, arachidonic acid, with four double *cis* bonds, may have "kinks" or a U shape. This may have profound significance on molecular packing in membranes and on the positions occupied by fatty acids in more complex molecules such as phospholipids. The presence of *trans* double bonds will alter these spatial relationships. *Trans* fatty acids are present in certain foods. Most arise as a by-product during the saturation of fatty acids in the process of hydrogenation, or "hardening," of natural oils in the manufacture of margarine. An additional small contribution comes from the ingestion of ruminant fat that contains *trans* fatty acids arising from the action of microorganisms in the rumen.

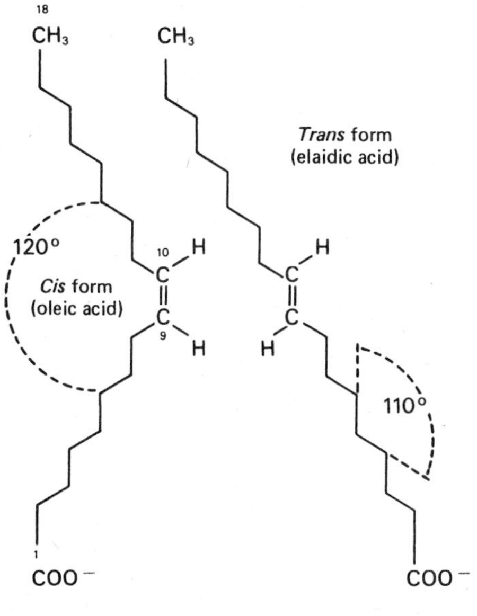

Physical and Physiologic Properties of Fatty Acids Reflect Chain Length and Degree of Unsaturation

Thus, the melting points of even-numbered-carbon fatty acids increase with chain length and decrease according to unsaturation. A triacylglycerol containing all saturated fatty acids of 12 carbons or more is solid at body temperature, whereas if all three fatty acid residues are 18:2, it is liquid to below 0 °C. In practice, natural acylglycerols contain a mixture of fatty acids tailored to suit their functional roles. The membrane lipids, which must be fluid at all environmental temperatures, are more unsaturated than storage lipids. Lipids in tissues that are subject to cooling, eg, in hibernators or in the extremities of animals, are more unsaturated.

Certain Alcohols Are Found in Natural Lipids

Alcohols associated with lipids include glycerol, cholesterol, and higher alcohols (eg, cetyl alcohol, $C_{16}H_{33}OH$), usually found in the waxes, and the polyisoprenoid alcohol dolichol (Figure 16–26).

TRIACYLGLYCEROLS (TRIGLYCERIDES)* ARE THE MAIN STORAGE FORMS OF FATTY ACIDS

The triacylglycerols are esters of the alcohol glycerol and fatty acids. In naturally occurring fats, the proportion of triacylglycerol molecules containing the same fatty acid residue in all three ester positions is very small. They are nearly all **mixed acylglycerols.** In Figure 16–6, if all three fatty acids represented by R were stearic acid, the fat would be known as tristearin. An example of a mixed acylglycerol is shown in Figure 16–7.

*According to the standardized terminology of the International Union of Pure and Applied Chemistry (IUPAC) and the International Union of Biochemistry (IUB), the monoglycerides, diglycerides, and triglycerides should be designated monoacylglycerols, diacylglycerols, and triacylglycerols, respectively. However, the older terminology is still widely used, particularly in clinical medicine.

Figure 16–5. Geometric isomerism of Δ^9, 18:1 fatty acids (oleic and elaidic acids).

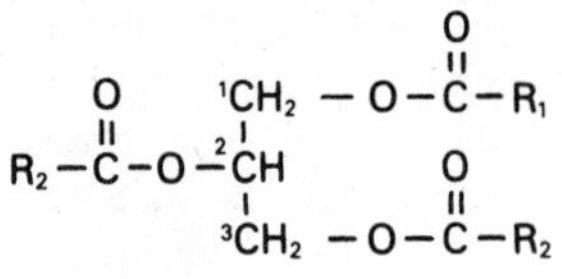

Figure 16–6. Triacylglycerol.

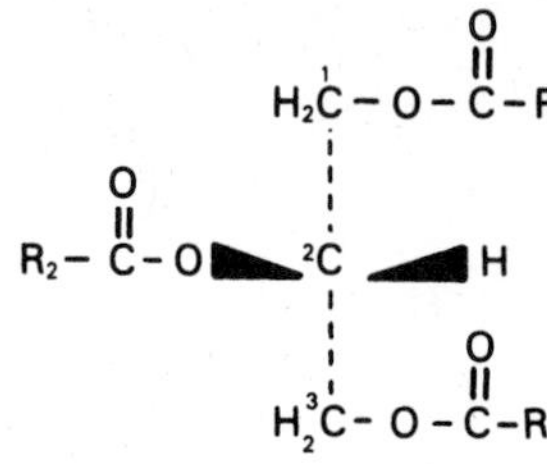

Figure 16–8. Triacyl-sn-glycerol.

Carbons 1 and 3 of Glycerol Are Not Identical

When it is required to number the carbon atoms of glycerol unambiguously, the -sn- (stereochemical numbering) system is used, eg, 1,3-distearyl-2-palmityl-sn-glycerol (shown as a projection formula also in Figure 16–8). It is important to realize that carbons 1 and 3 of glycerol are not identical when viewed in three dimensions. Enzymes readily distinguish between them and are nearly always specific for one or the other carbon; eg, glycerol is always phosphorylated on sn-3 by glycerol kinase to give glycerol 3-phosphate and not glycerol 1-phosphate.

Partial acylglycerols consisting of mono- and diacylglycerols wherein a single fatty acid or two fatty acids are esterified with glycerol are also found in the tissues. These are of particular significance in the synthesis and hydrolysis of triacylglycerols.

PHOSPHOLIPIDS ARE THE MAIN LIPID CONSTITUENTS OF MEMBRANES

The phospholipids include the following: (1) phosphatidic acid and phosphatidylglycerol, (2) phosphatidylcholine, (3) phosphatidylethanolamine, (4) phosphatidylinositol, (5) phosphatidylserine, (6) lysophospholipids, (7) plasmalogens, and (8) sphingomyelins. All of these are **phosphoacylglycerols** apart from the sphingomyelins, which do not contain glycerol. They may be regarded as derivatives of **phosphatidic acid** (Figure 16–9), in which the phosphate is esterified with the —OH of a suitable alcohol.

Phosphatidic acid is important as an intermediate in the synthesis of triacylglycerols as well as phosphoglycerols but is not found in any great quantity in tissues.

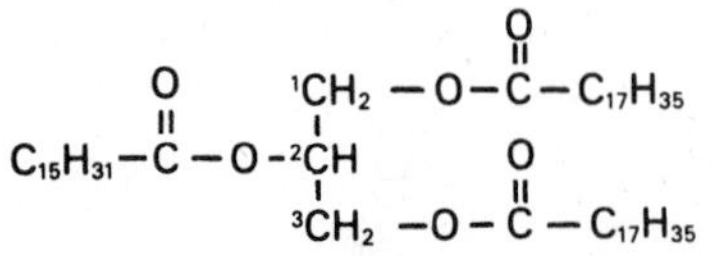

Cardiolipin Is a Major Lipid of Mitochondrial Membranes

Phosphatidic acid is a precursor of **phosphatidylglycerol** which, in turn, gives rise to **cardiolipin** (Figure 16–10) in mitochondria.

Phosphatidylcholines (Lecithins) Occur in Cell Membranes

These are phosphoacylglycerols containing choline (Figure 16–11). They are the most abundant phospholipids of the cell membrane and represent a large proportion of the body's store of choline. Choline is important in nervous transmission, as acetylcholine, and as a store of labile methyl groups. **Dipalmitoyl lecithin** is a very effective surface-active agent and a major constituent of the **surfactant** preventing adherence, due to surface tension, of the inner surfaces of the lungs. Its absence from the lungs of premature infants causes **respiratory distress syndrome.** It is to be noted, however, that most phospholipids have a saturated acyl radical in the sn-1 position but an unsaturated radical in the sn-2 position of glycerol.

Phosphatidylethanolamine (cephalin) differs from phosphatidylcholine only in that ethanolamine replaces choline (Figure 16–12).

Phosphatidylserine contains the amino acid serine rather than choline and is found in most tissues (Figure 16–13). Phospholipids containing threonine have also been isolated.

Phosphatidylinositol Is a Precursor of Second Messengers

The inositol is present as the stereoisomer, my-oinositol (Figure 16–14). **Phosphatidylinositol 4,5-**

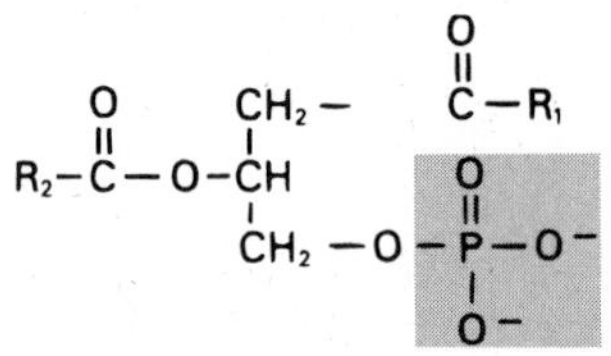

Figure 16–7. 1,3-Distearopalmitin.

Figure 16–9. Phosphatidic acid.

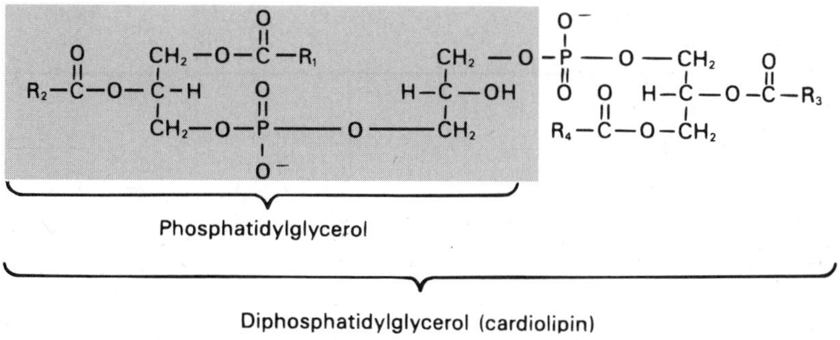

Phosphatidylglycerol

Diphosphatidylglycerol (cardiolipin)

Figure 16–10. Cardiolipin (diphosphatidylglycerol).

Choline

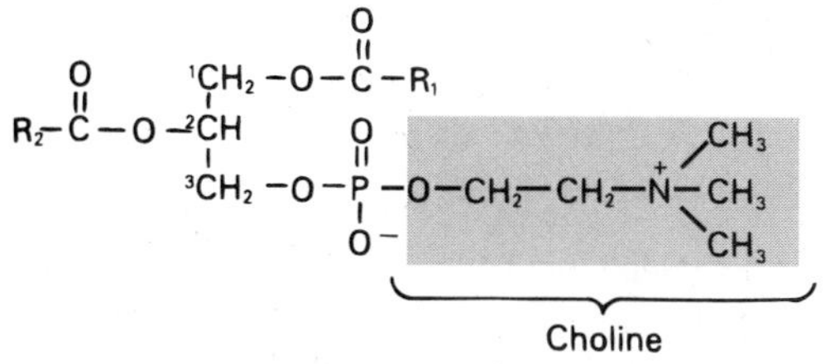

Figure 16–11. 3-Phosphatidylcholine.

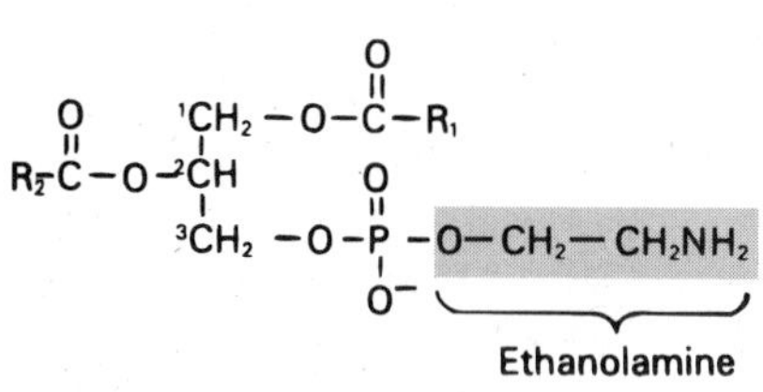

Ethanolamine

Figure 16–12. 3-Phosphatidylethanolamine.

Serine

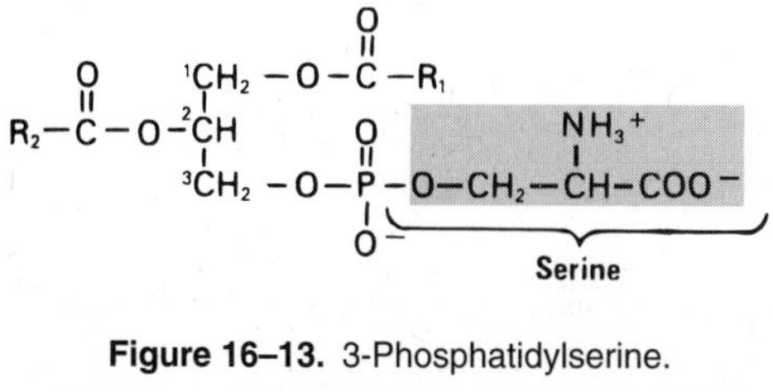

Figure 16–13. 3-Phosphatidylserine.

bisphosphate is an important constituent of cell membrane phospholipids; upon stimulation by a suitable hormone agonist, it is cleaved into **diacylglycerol** and **inositol trisphosphate,** both of which act as internal signals or second messengers (Chapter 44).

Lysophospholipids Are Intermediates in the Metabolism of Phosphoglycerols

These are phosphoacylglycerols containing only one acyl radical, eg, **lysolecithin,** important in the metabolism and interconversion of phospholipids (Figure 16–15).

Plasmalogens Occur in Brain and Muscle

These compounds constitute as much as 10% of the phospholipids of brain and muscle. Structurally, the plasmalogens resemble phosphatidylethanolamine but possess an ether link on the *sn*-1 carbon instead of the normal ester link found in acylglycerols. Typically, the alkyl radical is an unsaturated

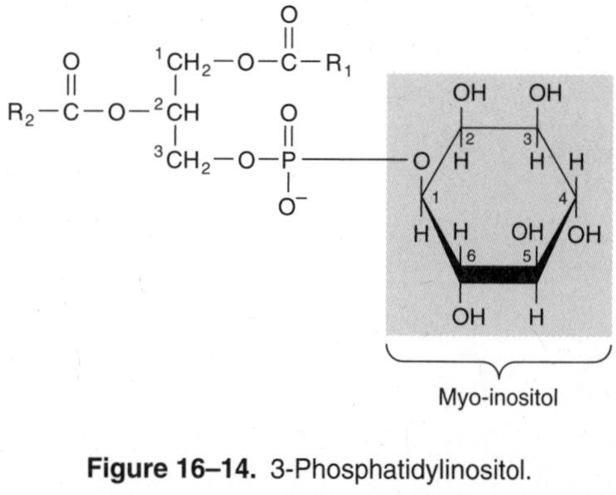

Myo-inositol

Figure 16–14. 3-Phosphatidylinositol.

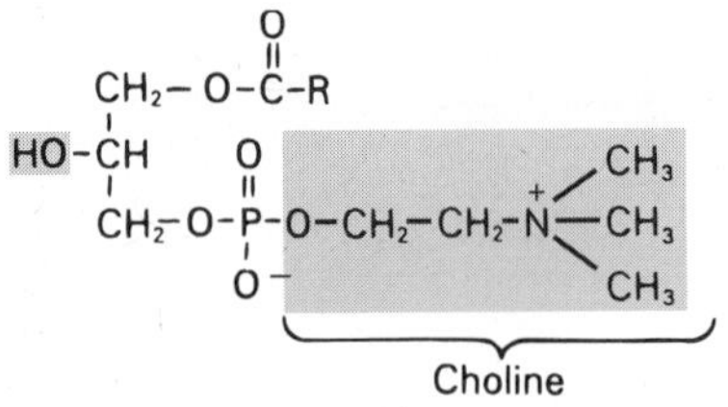

Figure 16–15. Lysolecithin.

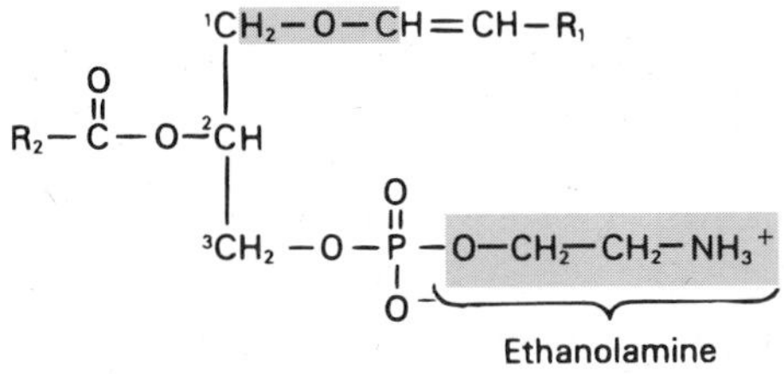

Figure 16–16. Plasmalogen (phosphatidal ethanolamine).

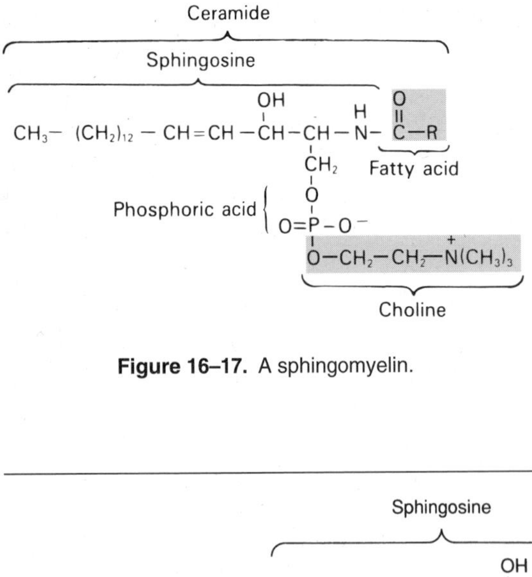

Figure 16–17. A sphingomyelin.

alcohol (Figure 16–16). In some instances, choline, serine, or inositol may be substituted for ethanolamine.

Sphingomyelins Are Found in the Nervous System

Sphingomyelins are found in large quantities in brain and nerve tissue. On hydrolysis, the sphingomyelins yield a fatty acid, phosphoric acid, choline, and a complex amino alcohol, **sphingosine** (Figure 16–17). No glycerol is present. The combination of sphingosine plus fatty acid is known as **ceramide**, a structure also found in the glycosphingolipids (see below).

GLYCOLIPIDS (GLYCOSPHINGOLIPIDS) ARE IMPORTANT IN NERVE TISSUES AND IN THE CELL MEMBRANE

Glycolipids are widely distributed in every tissue of the body, particularly in nervous tissue such as brain. They occur particularly in the outer leaflet of the plasma membrane, where they contribute to **cell surface carbohydrates.**

The major glycolipids found in animal tissues are glycosphingolipids. They contain ceramide and one or more sugars. The two simplest are **galactosylceramide** and **glucosylceramide.** Galactosylceramide is a major glycosphingolipid of brain and other nervous tissue, but it is found in relatively low amounts elsewhere. It contains a number of characteristic C_{24} fatty acids, eg, cerebronic acid. Galactosylceramide (Figure 16–18) can be converted to sulfogalactosylceramide (classic **sulfatide**), which is present in high amounts in myelin. Glucosylceramide is the predominant simple glycosphingolipid of extraneural tissues, but it also occurs in the brain in small amounts. Gangliosides are complex glycosphingolipids derived

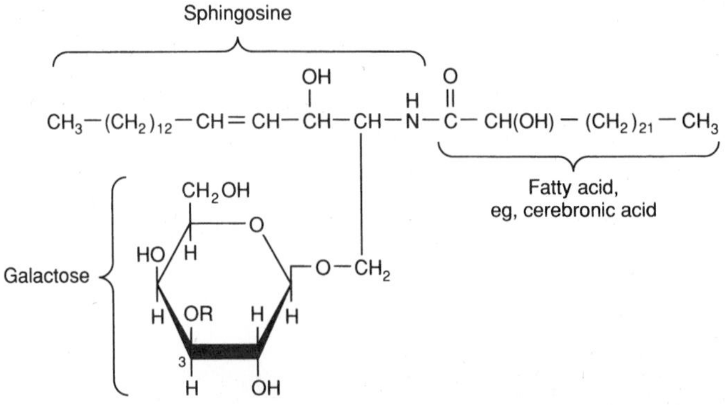

Figure 16–18. Structure of galactosylceramide (galactocerebroside, R = H), and sulfogalactosylceramide (a sulfatide, R = SO_4^{2-}).

Ceramide–Glucose–Galactose–N-Acetylgalactosamine–Galactose
(Acyl-
sphingo- |
sine)   NeuAc

or

Cer–Glc–Gal–GalNAc–Gal
|
NeuAc

Figure 16–19. G_{M1} ganglioside, a monosialoganglioside, the receptor in human intestine for cholera toxin.

from glucosylceramide that contain in addition one or more molecules of a **sialic acid.** Neuraminic acid (NeuAc; see Chapter 15) is the principal sialic acid found in human tissues. Gangliosides are also present in nervous tissues in high concentration. They appear to have receptor and other functions. The simplest ganglioside found in tissues is G_{M3}, which contains ceramide, one molecule of glucose, one molecule of galactose, and one molecule of NeuAc. In the shorthand nomenclature used, G represents ganglioside; M is a monosialo-containing species; and the subscript 3 is a number assigned on the basis of chromatographic migration. The structure of a more complex ganglioside derived from G_{M3}, named G_{M1}, is shown in Figure 16–19. G_{M1} is a compound of considerable biologic interest as it is known to be the receptor in human intestine for cholera toxin. Other gangliosides can contain anywhere from one to five molecules of sialic acid, giving rise to di-, trisialogangliosides, etc.

STEROIDS PLAY MANY PHYSIOLOGICALLY IMPORTANT ROLES

Cholesterol is probably the best known steroid because of its association with **atherosclerosis.** However, biochemically it is also of significance because it is the precursor of a large number of equally important steroids which include the bile acids, adrenocortical hormones, sex hormones, D vitamins, cardiac glycosides, sitosterols of the plant kingdom, and some alkaloids.

All of the steroids have a similar cyclic nucleus resembling phenanthrene (rings A, B, and C) to which

a cyclopentane ring (D) is attached. The carbon positions on the steroid nucleus are numbered as shown in Figure 16–20. It is important to realize that in structural formulas of steroids, a simple hexagonal ring denotes a completely saturated six-carbon ring with all valences satisfied by hydrogen bonds unless shown otherwise; ie, it is not a benzene ring. All double bonds are shown as such. Methyl side chains are shown as single bonds unattached at the farther (methyl) end. These occur typically at positions 10 and 13 (constituting C atoms 19 and 18). A side chain at position 17 is usual (as in cholesterol). If the compound has one or more hydroxyl groups and no carbonyl or carboxyl groups, it is a **sterol,** and the name terminates in -ol.

Because of Asymmetry in the Steroid Molecule, Many Stereoisomers Are Possible

Each of the six-carbon rings of the steroid nucleus is capable of existing in the three-dimensional conformation either of a "chair" or a "boat" (Figure 16–21).

In naturally occurring steroids, virtually all the rings are in the "chair" form, which is the more stable conformation. With respect to each other, the rings can be either *cis* or *trans* (Figure 16–22).

The junction between the A and B rings can be *cis*

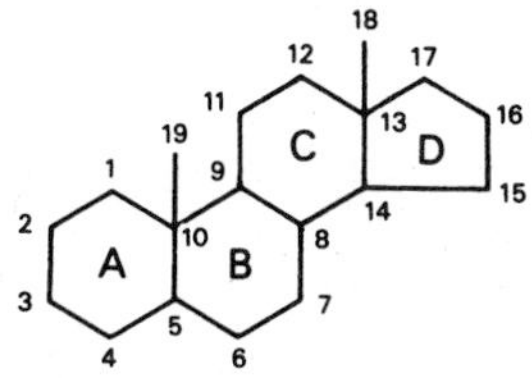

Figure 16–20. The steroid nucleus.

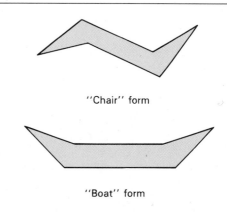

"Chair" form

"Boat" form

Figure 16–21. Conformations of stereoisomers of the steroid nucleus.

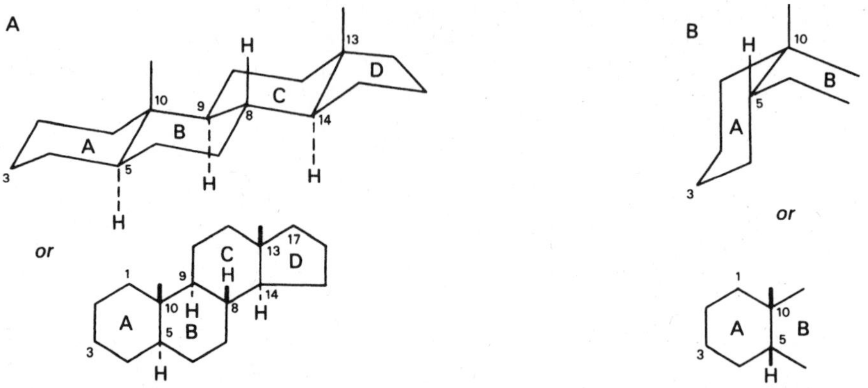

Figure 16–22. Generalized steroid nucleus, showing *(A)* an all-*trans* configuration between adjacent rings and *(B)* a *cis* configuration between rings A and B.

or *trans* in naturally occurring steroids. That between B and C is *trans* and the C/D junction is *trans* except in cardiac glycosides and toad poisons. Bonds attaching substituent groups above the plane of the rings are shown with bold solid lines (β), whereas those bonds attaching groups below are indicated with broken lines (α). The A ring of a 5α steroid is always *trans* to the B ring, whereas it is *cis* in a 5β steroid. The methyl groups attached to C_{10} and C_{13} are invariably in the β configuration.

Cholesterol Is a Significant Constituent of Many Tissues

Cholesterol (Figure 16–23) is widely distributed in all cells of the body, but particularly in nervous tissue. It is a major constituent of the plasma membrane and of plasma lipoproteins. It is often found as **cholesteryl ester,** where the hydroxyl group on position 3 is esterified with a long-chain fatty acid. It occurs in animal fats but not in plant fats.

Ergosterol Is a Precursor of Vitamin D

Ergosterol occurs in plants and yeast and is important as a precursor of vitamin D (Figure 16–24).

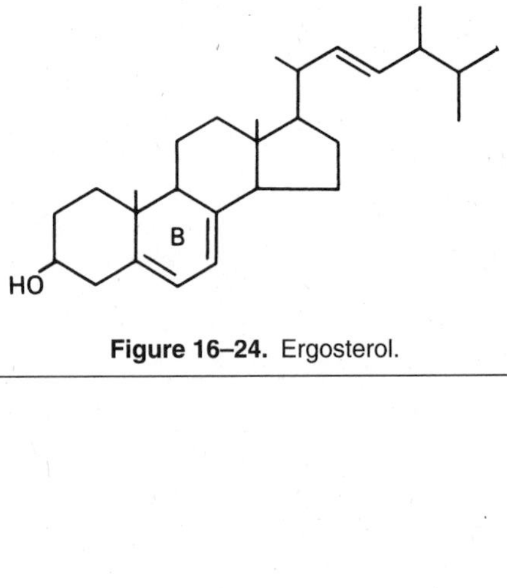

Figure 16–24. Ergosterol.

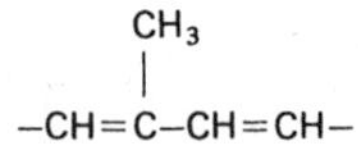

Figure 16–25. Isoprene unit.

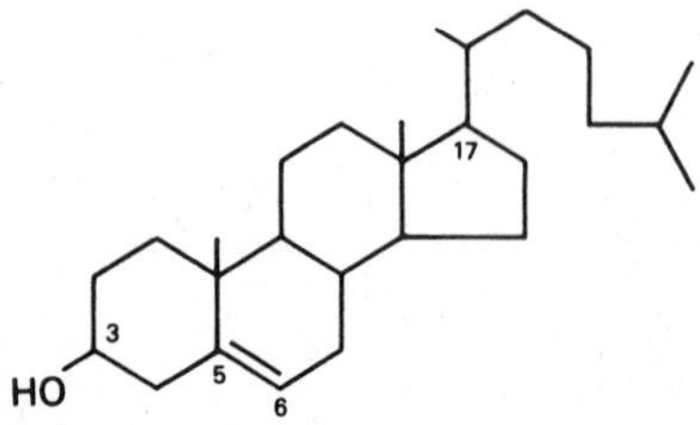

Figure 16–23. Cholesterol, 3-hydroxy-5,6-cholestene.

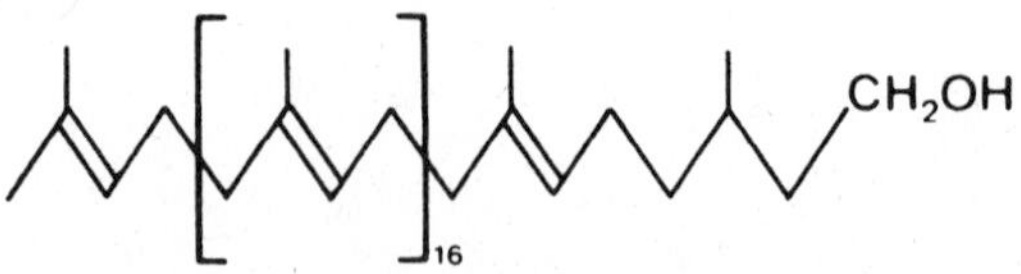

Figure 16–26. Dolichol—a C_{95} alcohol.

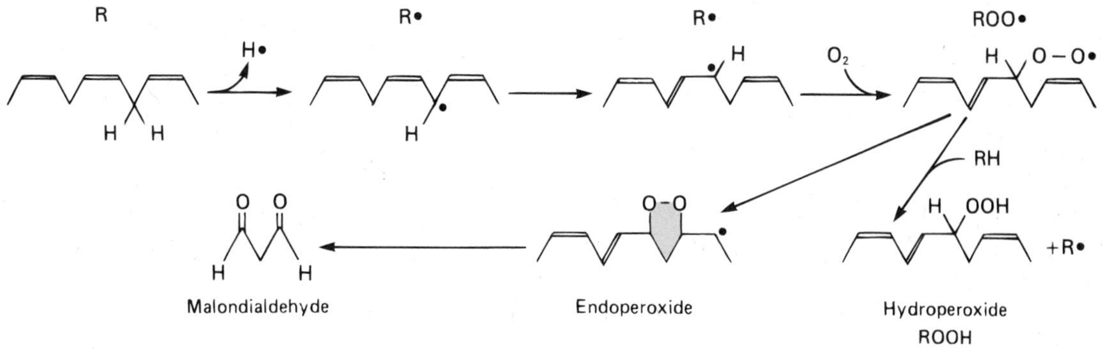

Figure 16–27. Lipid peroxidation. The reaction is initiated by light or by metal ions. Malondialdehyde is only formed by fatty acids with three or more double bonds and is used as a measure of lipid peroxidation together with ethane from the terminal two carbons of ω3 fatty acids and pentane from the terminal five carbons of ω6 fatty acids.

When irradiated with ultraviolet light, it acquires antirachitic properties consequent to the opening of ring B (Figure 53–6).

Coprosterol Occurs in Feces

Coprosterol (coprostanol) occurs in feces as a result of the reduction of the double bond of cholesterol between C_5 and C_6 by bacteria in the intestine.

Polyprenoids Share the Same Parent Compound as Cholesterol

Although not steroids, these compounds are related because they are synthesized, like cholesterol (Figure 28–2), from five-carbon isoprene units (Figure 16–25). They include **ubiquinone** (Chapter 14), a member of the respiratory chain in mitochondria, and the long-chain alcohol **dolichol** (Figure 16–26), which takes part in glycoprotein synthesis by transferring carbohydrate residues to asparagine residues of the polypeptide (Chapter 56). Plant-derived isoprenoid compounds include rubber, camphor, the fat-soluble vitamins A, D, E, and K, and β-carotene (provitamin A).

LIPID PEROXIDATION IS A SOURCE OF FREE RADICALS

Peroxidation (**auto-oxidation**) of lipids exposed to oxygen is responsible not only for deterioration of foods (**rancidity**) but also for damage to tissues in vivo, where it may be a cause of cancer, inflammatory diseases, atherosclerosis, aging, etc. The deleterious effects are initiated by free radicals (ROO˙, RO˙, OH˙) produced during peroxide formation from fatty acids containing methylene-interrupted double bonds, ie, those found in the naturally occurring polyunsaturated fatty acids (Figure 16–27). Lipid peroxidation is a chain reaction providing a continuous supply of free radicals that initiate further peroxidation. The whole process can be depicted as follows:

(1) Initiation:

$$ROOH + metal^{(n)+} \rightarrow ROO˙ + metal^{(n-1)+} + H^+$$
$$X˙ + RH \rightarrow R˙ + XH$$

(2) Propagation:

$$R˙ + O_2 \rightarrow ROO˙$$
$$ROO˙ + RH \rightarrow ROOH + R˙, \text{ etc.}$$

(3) Termination:

$$ROO˙ + ROO˙ \rightarrow ROOR + O_2$$
$$ROO˙ + R˙ \rightarrow ROOR$$
$$R˙ + R˙ \rightarrow RR$$

Since the molecular precursor for the initiation process is generally the hydroperoxide product ROOH, lipid peroxidation is a chain reaction with potentially devastating effects. To control and reduce lipid peroxidation, both humans in their activities and nature invoke the use of **antioxidants.** Propyl gallate, butylated hydroxyanisole (BHA), and butylated hydroxytoluene (BHT) are antioxidants used as food additives. Naturally occurring antioxidants include vitamin E (tocopherol), which is lipid-soluble, and urate and vitamin C, which are water-soluble. Beta-carotene is an antioxidant at low Po_2. Antioxidants fall into two classes: (1) preventive antioxidants, which reduce the rate of chain initiation; and (2) chain-breaking antioxidants, which interfere with chain propagation. Preventive antioxidants include catalase and other peroxidases that react with ROOH and chelators of metal ions such as DTPA (diethylenetriaminepentaacetate) and EDTA (ethylenediaminetetraacetate). Chain-breaking antioxidants are often phenols or aromatic amines. In vivo, the principal chain-breaking antioxidants are superoxide dismutase, which acts in the aqueous phase to trap superoxide free radicals (O_2^-); perhaps urate; and vitamin E, which acts in the lipid phase to trap ROO$^\bullet$ radicals (Figure 53–9).

Peroxidation is also catalyzed in vivo by heme compounds and by **lipoxygenases** found in platelets and leukocytes, etc.

CHROMATOGRAPHIC METHODS SEPARATE AND IDENTIFY LIPIDS

The older methods of separation and identification of lipids, based on classic chemical procedures of crystallization, distillation, and solvent extraction, have now been largely supplanted by chromatographic procedures. Particularly useful for the separation of the various lipid classes is **thin-layer chromatography** (TLC) (Figure 16–28) and for the separation of the individual fatty acids, **gas-liquid chromatography** (GLC). Before these techniques are applied to wet tissues, the lipids are extracted by a solvent system based usually on a mixture of chloroform and methanol (2:1).

AMPHIPATHIC LIPIDS SELF-ORIENT AT OIL:WATER INTERFACES

They Form Membranes, Micelles, Liposomes, and Emulsions

In general, lipids are insoluble in water, since they contain a predominance of nonpolar (hydrocarbon) groups. However, fatty acids, phospholipids, sphingolipids, bile salts, and, to a lesser extent, cholesterol contain polar groups. Therefore, part of the molecule is **hydrophobic,** or water-insoluble, and part is **hydrophilic,** or water-soluble. Such molecules are described as **amphipathic** (Figure 16–29). They become oriented at oil-water interfaces with the polar group in the water phase and the nonpolar group in the oil phase. A bilayer of such amphipathic lipids has been regarded as a basic structure in biologic membranes (Chapter 43). When a critical concentration of these lipids is present in an aqueous medium, they form **micelles.** Aggregations of bile salts into micelles and liposomes, and the formation of mixed micelles with the products of fat digestion are important in facilitating absorption of lipids from the intestine. Liposomes may be formed by sonicating an amphipathic lipid in an aqueous medium. They consist of spheres of lipid bilayers that enclose part of the aqueous medium. They are of potential clinical use, particularly when combined with tissue-specific antibodies, as carriers of drugs in the circulation, targeted to specific organs, eg, in cancer therapy. In addition, they are being used for gene transfer into vascular cells and as carriers for topical and transdermal delivery of drugs and cosmetics. **Emulsions** are much larger particles, formed usually by nonpolar lipids in an aqueous medium. These are stabilized by emulsifying agents such as amphipathic lipids (eg, lecithin), which form a surface layer separating the main bulk of the nonpolar material from the aqueous phase (Figure 16–29).

SUMMARY

1. Lipids have the common property of being relatively insoluble in water (hydrophobic) but soluble in nonpolar solvents. However, the amphipathic lipids contain, in addition, one or more polar groups, making them particularly suitable as constituents of membranes at lipid/water interfaces.

2. The lipids of major physiologic significance are fatty acids and their esters, together with cholesterol and other steroids.

3. Long-chain fatty acids may be saturated, monounsaturated, or polyunsaturated, according to the number of double bonds present. Their fluidity decreases with chain length and increases according to degree of unsaturation.

4. Eicosanoids are formed from 20-carbon polyunsaturated fatty acids and make up an important group of physiologically and pharmacologically active compounds known as prostaglandins, thromboxanes, leukotrienes, and lipoxins.

5. The esters of glycerol are quantitatively the most significant lipids, represented by triacylglycerol ("fat"), important as a major constituent of lipoproteins and as the storage form of lipid in adipose tis-

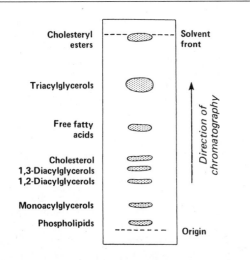

Figure 16–28. Separation of major lipid classes by thin-layer chromatography. A suitable solvent system for the above would be hexane-diethyl ether-formic acid (80:20:2 v/v/v).

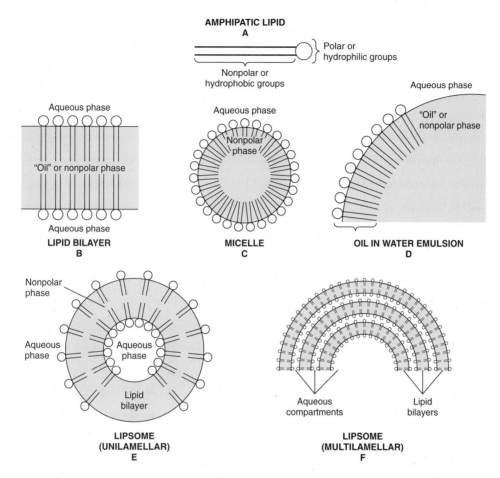

Figure 16–29. Formation of lipid membranes, micelles, emulsions, and liposomes from amphipathic lipids, eg, phospholipids.

sue. Phosphoacylglycerols are amphipathic lipids and fulfill many important roles, eg, as major constituents of membranes and the outer layer of lipoproteins, as surfactant in the lung, as precursors of second messengers, and as important constituents of nervous tissue.

6. Glycolipids are also important constituents of nervous tissue such as brain and the outer leaflet of the cell membrane, where they contribute to the carbohydrates on the cell surface.

7. Cholesterol, as an amphipathic lipid, is an important component of membranes. It is the parent molecule from which all other steroids in the body are synthesized. These include major hormones such as the adrenocortical and sex hormones, D vitamins, and bile acids.

REFERENCES

Christie WW: *Lipid Analysis,* 2nd ed. Pergamon Press, 1982.

Cotgreave IA, Moldeus P, Orrenius S: Host biochemical defense mechanisms against prooxidants. Annu Rev Pharmacol Toxicol 1988;28:189.

Frankel EN: Chemistry of free radical and singlet oxidation of lipids. Prog Lipid Res 1985;23:197.

Gunstone FD, Harwood JL, Padley FB: *The Lipid Handbook.* Chapman & Hall, 1986.

Gurr MI, Harwood JL: *Lipid Biochemistry: An Introduction,* 4th ed. Chapman & Hall, 1991.

Hawthorne JN, Ansell GB (editors): *Phospholipids.* Elsevier, 1982.

Small DM: Lateral chainpacking in lipids and membranes. J Lipid Res 1984;25:1490.

Vance DE, Vance JE (editors): *Biochemistry of Lipids, Lipoproteins and Membranes.* Elsevier, 1991.

17 Overview of Intermediary Metabolism

Peter A. Mayes, PhD, DSc

INTRODUCTION

The fate of dietary components after digestion and absorption constitutes intermediary metabolism. Thus, it encompasses a wide field that not only seeks to describe the metabolic pathways taken by individual molecules but also attempts to understand their interrelationships and the mechanisms that regulate the flow of metabolites through the pathways. Metabolic pathways fall into three categories (Figure 17–1): (1) **Anabolic pathways** are those involved in the synthesis of the compounds constituting the body's structure and machinery. Protein synthesis is such a pathway. The free energy required for these processes comes from the next category. (2) **Catabolic pathways** involve oxidative processes that release free energy, usually in the form of high-energy phosphate or reducing equivalents, eg, the respiratory chain and oxidative phosphorylation. (3) **Amphi-**

bolic pathways have more than one function and occur at the "crossroads" of metabolism, acting as links between the anabolic and catabolic pathways, eg, the citric acid cycle.

BIOMEDICAL IMPORTANCE

A knowledge of metabolism in the normal animal is a prerequisite to a sound understanding of abnormal metabolism underlying many diseases. Normal metabolism includes the variations and adaptation in metabolism due to periods of starvation, exercise, pregnancy, and lactation. Abnormal metabolism results from, for example, nutritional deficiency, enzyme deficiency, or abnormal secretion of hormones. An important example of a disease resulting from abnormal metabolism (a "metabolic disease") is **diabetes mellitus.**

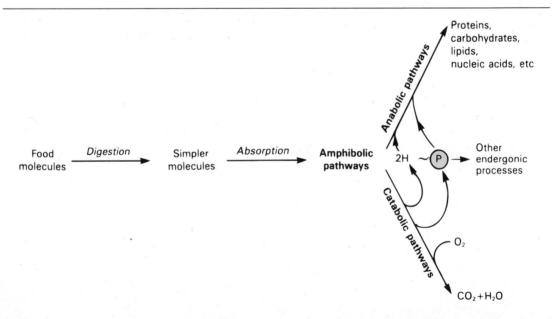

Figure 17–1. The three major categories of metabolic pathways. Catabolic pathways release free energy in the form of reducing equivalents (2H) or high-energy phosphate (~$\circledP$) to power the anabolic pathways. Amphibolic pathways act as links between the other two categories of pathways.

THE BASIC METABOLIC PATHWAYS PROCESS THE MAJOR PRODUCTS OF DIGESTION

The nature of the diet sets the basic pattern of metabolism in the tissues. Mammals such as humans need to process the absorbed products of digestion of dietary carbohydrate, lipid, and protein. These are mainly glucose, fatty acids and glycerol, and amino acids, respectively. In ruminants (and to a lesser extent other herbivores), cellulose in the diet is digested by symbiotic microorganisms to lower fatty acids (acetic, propionic, butyric), and tissue metabolism in these animals is adapted to utilize lower fatty acids as major substrates. All these products of digestion are processed by their respective metabolic pathways to a **common product, acetyl-CoA,** which is then completely oxidized by the **citric acid cycle** (Figure 17–2).

Carbohydrate Metabolism Is Centered on the Provision and Fate of Glucose (Figure 17–3)

Glucose is metabolized to pyruvate and lactate in all mammalian cells by the pathway of **glycolysis.** Glucose is a unique substrate because glycolysis can

occur in the absence of oxygen (anaerobic), when the end product is lactate only. However, tissues that can utilize oxygen (aerobic) are able to metabolize pyruvate to **acetyl-CoA,** which can enter the **citric acid cycle** for complete oxidation to CO_2 and H_2O, with liberation of much free energy as ATP in the process of **oxidative phosphorylation** (Figure 18–2). Thus, glucose is a major fuel of many tissues. But it (and certain of its metabolites) also takes part in other processes, eg, (1) conversion to its storage polymer, **glycogen,** particularly in skeletal muscle and liver. (2) The **pentose phosphate pathway,** which arises from intermediates of glycolysis. It is a source of reducing equivalents (2H) for biosynthesis—eg, of fatty acids—and it is also the source of **ribose,** which is important for nucleotide and nucleic acid formation. (3) Triose phosphate gives rise to the **glycerol moiety** of acylglycerols (fat). (4) Pyruvate and intermediates of the citric acid cycle provide the carbon skeletons for the synthesis of **amino acids,** and acetyl-CoA is the building block for long-chain **fatty acids** and **cholesterol,** the precursor of all steroids synthesized in the body. **Gluconeogenesis** is the process that produces glucose from noncarbohydrate precursors, eg, lactate, amino acids, and glycerol.

Lipid Metabolism Is Concerned Mainly With Fatty Acids and Cholesterol (Figure 17–4)

The source of long-chain fatty acids is either dietary lipid or de novo synthesis from acetyl-CoA derived from carbohydrate. In the tissues, fatty acids may be oxidized to **acetyl-CoA** (β-oxidation) or esterified to acylglycerols, where, as **triacylglycerol** (fat) they constitute the body's main caloric reserve. Acetyl-CoA formed by β-oxidation has several important fates.

(1) As in the case of acetyl-CoA derived from carbohydrate, it is **oxidized completely** to $CO_2 + H_2O$ via the **citric acid cycle.** Fatty acids yield considerable energy both in β-oxidation and in the citric acid cycle and are therefore very effective tissue fuels.

(2) It is a source of the carbon atoms in **cholesterol** and other **steroids.**

(3) In the liver, it forms **ketone bodies,** alternative water-soluble tissue fuels that become important sources of energy under certain conditions (eg, starvation).

Much of Amino Acid Metabolism Involves Transamination (Figure 17–5)

The amino acids are necessary for protein synthesis. Some must be supplied specifically in the diet (the **essential amino acids**), since the tissues are unable to synthesize them. The remainder, or **nonessential amino acids,** are also supplied in the diet, but they also can be formed from intermediates

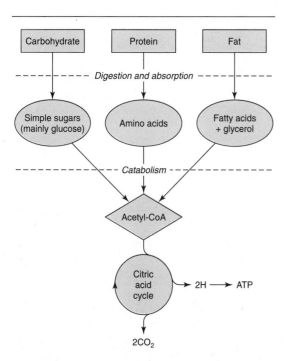

Figure 17–2. Outline of the pathways for the catabolism of dietary carbohydrate, protein, and fat. All the pathways lead to the production of acetyl-CoA, which is oxidized in the citric acid cycle, ultimately yielding ATP in the process of oxidative phosphorylation.

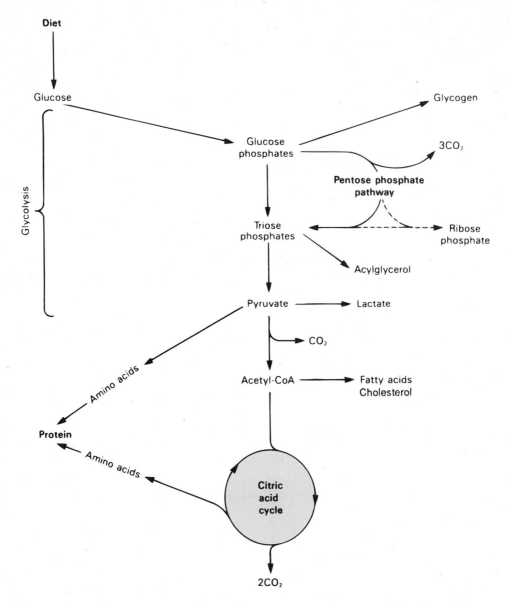

Figure 17–3. Overview of carbohydrate metabolism showing the major pathways and end products. Gluconeogenesis is not shown.

by **transamination** using the amino nitrogen from other surplus amino acids. After **deamination,** excess amino nitrogen is removed as **urea,** and the carbon skeletons that remain after transamination (1) are oxidized to CO_2 via the citric acid cycle, (2) form glucose (gluconeogenesis), or (3) form ketone bodies.

In addition to their requirement for protein synthesis, the amino acids are also the precursors of many other important compounds, eg, purines, pyrimidines, and hormones such as epinephrine and thyroxine.

METABOLIC PATHWAYS MAY BE STUDIED AT DIFFERENT LEVELS OF ORGANIZATION

So far we have looked at metabolism as it occurs in the whole organism. The location and integration of metabolic pathways is revealed by studies at lower levels of organization, namely: (1) At the **tissue and organ level**—the nature of the substrates entering and metabolites leaving tissues and organs is defined, and their overall fate is described. (2) At the **subcel-**

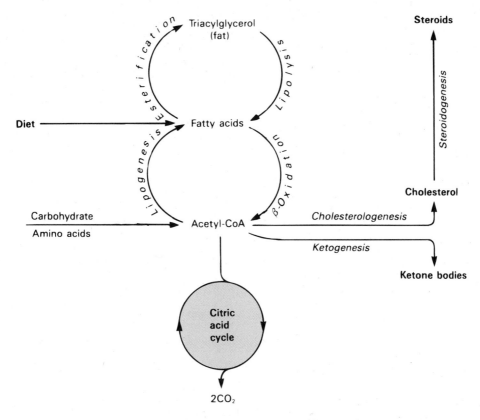

Figure 17–4. Overview of fatty acid metabolism showing the major pathways and end products. Ketone bodies comprise the substances acetoacetate, 3-hydroxybutyrate, and acetone.

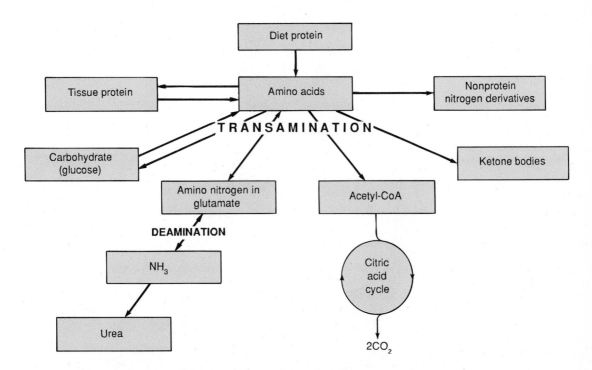

Figure 17–5. Overview of amino acid metabolism showing the major pathways and end products.

lular level—each cell organelle (eg, the mitochondrion) or compartment (eg, the cytosol) carries out specific biochemical roles that form part of a subcellular pattern of metabolic pathways.

At the Tissue and Organ Level, the Blood Circulation Integrates Metabolism

Amino acids resulting from the digestion of dietary protein and **glucose** resulting from the digestion of carbohydrate share a common route of absorption via the **hepatic portal vein.** This ensures that both types of metabolites and other water-soluble products of digestion are initially directed to the **liver** (Figure 17–6). The liver has the primary metabolic function of regulating the blood concentration of most metabolites, particularly glucose and amino acids. In the case of glucose, this is achieved by taking up excess glucose and converting it to glycogen (**glycogenesis**) or to fat (**lipogenesis**). Between meals, the liver can draw upon its glycogen stores to replenish glucose in the blood (**glycogenolysis**) or, in company with the kidney, to convert noncarbohydrate metabolites such as lactate, glycerol, and amino acids to glucose (**glu-coneogenesis**). The maintenance of an adequate concentration of blood glucose is vital for certain tissues in which it is an obligatory fuel, eg, brain and erythrocytes. The liver also has the task of **synthesizing the major plasma proteins** (eg, albumin) and of **deaminating amino acids** that are in excess of requirements, with formation of urea, which is transported via the blood to the kidney and excreted.

Skeletal muscle utilizes glucose as a fuel, forming both lactate and CO_2. It stores glycogen as a fuel for its use in muscular contraction and synthesizes muscle protein from plasma amino acids. Muscle accounts for approximately 50% of body mass and consequently represents a considerable store of protein that can be drawn upon to supply plasma amino acids, particularly during dietary shortage.

Lipids in the diet (Figure 17–7), represented mainly by triacylglycerol, upon digestion form monoacylglycerols and fatty acids. These are recombined in the intestinal cells, combined with protein, and secreted initially into the lymphatic system and then into the circulation as a **lipoprotein** known as a **chylomicron.** All hydrophobic, lipid-soluble products of digestion form lipoproteins, which facilitates

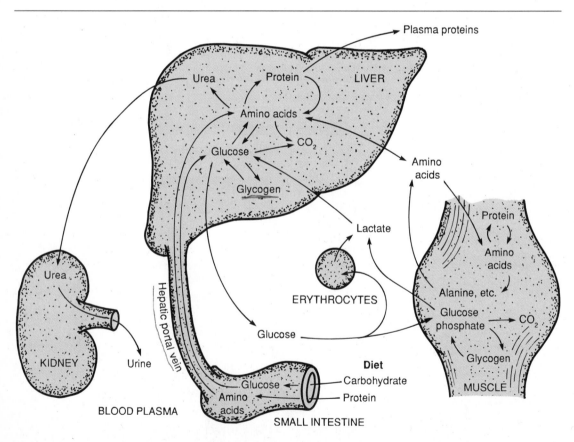

Figure 17–6. Transport and fate of major carbohydrate and amino acid substrates and metabolites. Note that there is little free glucose in muscle, since it is rapidly phosphorylated upon entry.

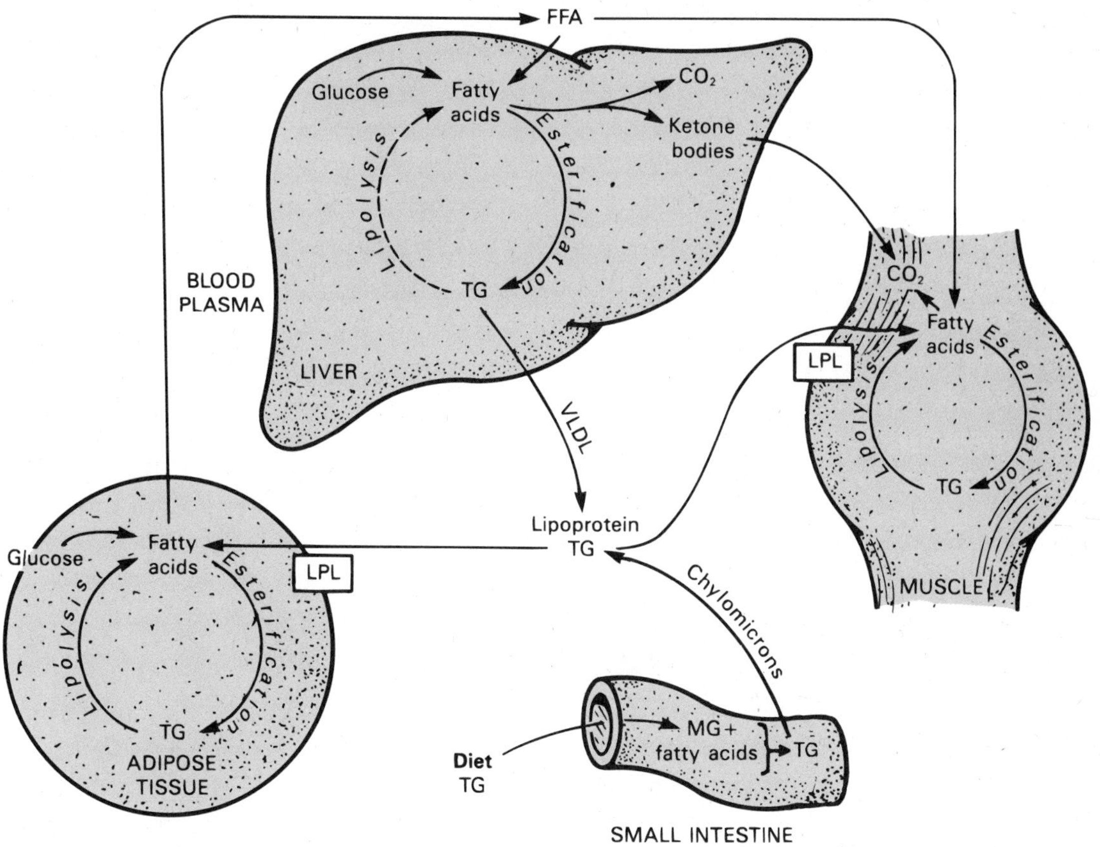

Figure 17–7. Transport and fate of major lipid substrates and metabolites. (FFA, free fatty acids; LPL, lipoprotein lipase; MG, monoacylglycerol; TG, triacylglycerol; VLDL, very low density lipoprotein.)

their transport between tissues in an aqueous environment—the plasma. Unlike glucose and amino acids, chylomicron triacylglycerol is not taken up directly by the liver. It is first metabolized by extrahepatic tissues possessing the enzyme **lipoprotein lipase,** which hydrolyzes the triacylglycerol, releasing fatty acids that are incorporated into tissue lipids or oxidized as fuel. The other major source of long-chain fatty acid is synthesis **(lipogenesis)** from carbohydrate, mainly in adipose tissue and the liver.

Adipose tissue triacylglycerol is the main fuel reserve of the body. Subsequent to its hydrolysis **(lipolysis),** fatty acids are released into the circulation as free fatty acids. These are taken up by most tissues (but not brain or erythrocytes) and esterified to acylglycerols or oxidized as a major fuel to CO_2. Two pathways of additional importance occur in liver: (1) Surplus triacylglycerol, arising from both lipogenesis and free fatty acids, is secreted into the circulation as **very low density lipoprotein** (VLDL). This triacylglycerol undergoes a fate similar to that of chylomicrons. (2) Partial oxidation of free fatty acid leads to ketone body production **(ketogenesis).** Ketone bodies are transported to extrahepatic tissues, where they act as another major fuel source.

At the Subcellular Level, Glycolysis Occurs in the Cytosol and the Citric Acid Cycle in the Mitochondria

A summary of the main biochemical functions of the subcellular components and organelles of the cell is given in Table 2–4. However, most cells are specialized in their functions and tend to emphasize certain metabolic pathways and relegate others. Figure 17–8 depicts the major metabolic pathways and their integration in a hepatic parenchymal cell, with special emphasis on their intracellular location.

The central role of the **mitochondrion** is immediately apparent, since it acts as the focus and crossroads of carbohydrate, lipid, and amino acid metabolism. In particular, it houses the enzymes of the citric acid cycle, of the respiratory chain and ATP synthase, of β-oxidation of fatty acids, and of ketone body production. In addition, it is the collecting point for the carbon skeletons of amino acids after

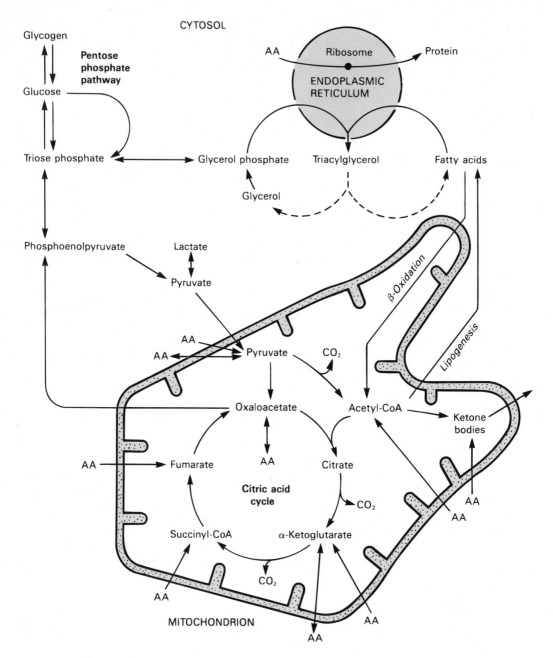

Figure 17–8. Intracellular location and overview of major metabolic pathways in a liver parenchymal cell. (AA →, metabolism of one or more essential amino acids; AA ↔, metabolism of one or more nonessential amino acids.)

transamination and for providing these skeletons for the synthesis of the nonessential amino acids.

Glycolysis, the pentose phosphate pathway, and fatty acid synthesis all occur in the cytosol. It will be noticed that in gluconeogenesis even substances such as lactate and pyruvate that are formed in the cytosol must enter the mitochondrion and form **oxaloacetate** before conversion to glucose.

The membranes of the **endoplasmic reticulum**

contain the enzyme system for **acylglycerol synthesis,** and the **ribosomes** are responsible for **protein synthesis.**

It will be appreciated that the transport of metabolites of varying size, charge, and solubility through the membranes separating organelles involves complex mechanisms. Some have been discussed in relation to the mitochondrial membranes (Chapter 14), and others will be discussed in succeeding chapters.

THE FLUX OF METABOLITES IN METABOLIC PATHWAYS MUST BE REGULATED IN A CONCERTED MANNER

Regulation of the overall flux along a metabolic pathway is often concerned with the control of only one or perhaps two key reactions in the pathway, catalyzed by **"regulatory enzymes."** The physicochemical factors that control the rate of an enzyme-catalyzed reaction, eg, substrate concentration, are of primary importance in the control of the overall rate of a metabolic pathway (Chapter 9). However, temperature and pH, factors that can influence enzyme activity, are held constant in warm-blooded vertebrates and have little regulatory significance. (Note, however, the variation in pH in the gastrointestinal tract and its effects on digestion [Chapter 55].)

"Nonequilibrium" Reactions Are Potential Control Points

In a reaction at equilibrium, the forward and reverse reactions take place at equal rates, and there is therefore no net flux in either direction. Many reactions in metabolic pathways are of this type, ie, **equilibrium reactions:**

$$A \leftrightarrow B \leftrightarrow C \leftrightarrow D$$

In vivo, under "steady-state" conditions, there would probably be a net flux from left to right owing to continuous supply of A and continuous removal of D. Such a pathway could function, but there would be little scope for control of the flux via regulation of enzyme activity, since an increase in activity would only serve to speed up attainment of the equilibrium.

In practice, there are invariably one or more **nonequilibrium** type reactions in a metabolic pathway, where the reactants are present in concentrations that are far from equilibrium. In attempting to reach equilibrium, large losses of free energy occur as heat, which cannot be reutilized, making this type of reaction essentially nonreversible, eg,

$$\text{Heat}$$
$$A \longleftrightarrow B \overset{\nearrow}{\longrightarrow} C \longleftrightarrow D$$
Nonequilibrium reaction

Such a pathway has both flow and direction but would exhaust itself if control were not exerted. The enzymes catalyzing nonequilibrium reactions are usually low in concentration and are subject to other controlling mechanisms. This is similar to the opening and shutting of a "one-way" valve, making it possible to control the net flow.

The Flux-Generating Reaction Is the First Reaction in a Pathway That Is Saturated With Substrate

It may be identified as a nonequilibrium reaction in which the K_m of the enzyme is considerably lower than the normal substrate concentration. The first reaction in glycolysis catalyzed by **hexokinase** (Figure 19–2) is such a flux-generating step because its K_m for glucose of 0.05 mmol/L is well below the normal blood glucose concentration of 5 mmol/L.

ALLOSTERIC AND HORMONAL MECHANISMS ARE IMPORTANT IN THE METABOLIC CONTROL OF ENZYME-CATALYZED REACTIONS

A hypothetical metabolic pathway, A, B, C, D, is shown in Figure 17–9, in which reactions A $\leftrightarrow$ B and C $\leftrightarrow$ D are equilibrium reactions and B $\rightarrow$ C is a nonequilibrium reaction. The flux through such a pathway can be regulated by the availability of substrate A. This depends on its supply from the blood, which in turn depends on adequate food intake to the gut or on certain key reactions that maintain and release major substrates to the blood, eg, the flux-generating reactions catalyzed by phosphorylase in liver (Figure 20–1), which attacks glycogen stores and provides blood glucose, and hormone-sensitive lipase in adipose tissue (Figure 27–8), which supplies free fatty acids, a major fuel for the tissues. The flux also depends on the ability of substrate A to permeate the cell membrane. It will also be determined by the efficiency of removal of the end product D and on the availability of cosubstrate or cofactors represented by X and Y.

Enzymes catalyzing nonequilibrium reactions are often allosteric proteins subject to the rapid actions of "feed-back" or "feed-forward" control by **allosteric modifiers,** often in immediate response to the needs of the cell (Chapter 11). Frequently, the product of a biosynthetic pathway, eg, long-chain acyl-CoA, will inhibit the enzyme catalyzing the first reaction in the pathway, eg, acetyl-CoA carboxylase. Other control mechanisms depend on the action of **hormones** responding to the needs of the body as a whole. These act by several different mechanisms (Chapter 44). One is **covalent modification** of the enzyme by **phosphorylation** and **dephosphorylation.** This is rapid and is often mediated through the formation of the second messenger **cAMP,** which in turn causes the conversion of an inactive enzyme into an active enzyme. This change is brought about via the activity of a **cAMP-dependent protein kinase** that phosphorylates the enzyme or of specific phosphatases that dephosphorylate the enzyme. The active form of the enzyme can be either the phosphorylated enzyme, as in enzymes catalyzing **degradative pathways** (eg, phosphorylase a), or the dephosphory-

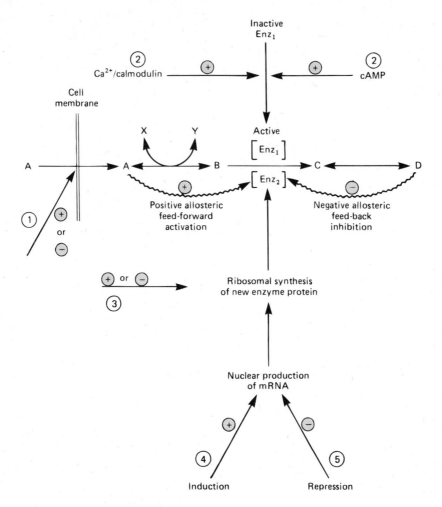

Figure 17–9. Mechanisms of control of an enzyme-catalyzed reaction. Circled numbers indicate possible sites of action of hormones. ①, Alteration of membrane permeability; ②, conversion of an inactive to an active enzyme, usually involving phosphorylation/dephosphorylation reactions; ③, alteration of the rate of translation of mRNA at the ribosomal level; ④, induction of new mRNA formation; and ⑤, repression of mRNA formation. ① and ② are rapid, whereas ③–⑤ are slower ways of regulating enzyme activity.

lated enzyme, as in enzymes catalyzing **synthetic processes** (eg, glycogen synthase a).

Some regulatory enzymes can be phosphorylated without the mediation of cAMP and cAMP-dependent protein kinase. These enzymes respond to other metabolic signals such as the [ATP]/[ADP] ratio, eg, pyruvate dehydrogenase (Figure 19–6), or Ca^{2+}/calmodulin-dependent protein kinase, eg, phosphorylase kinase (Figure 20–6).

The synthesis of rate-controlling enzymes can be affected by hormones. Because this involves new protein synthesis, it is not a rapid change but is often a response to a change in nutritional state. Hormones can act as inducers or repressors of mRNA formation in the nucleus or as stimulators of the translation stage of protein synthesis at the ribosomal level (Chapters 41 and 44).

A significant feature that aids metabolic control is that pathways catalyzing degradation of a substance are not the simple reversal of synthesis. Usually two entirely separate pathways are involved, allowing separate control of each, eg, glycogen synthesis and breakdown (Figure 20–1).

SUMMARY

1. The products of digestion provide the tissues with the building blocks for the biosynthesis of complex molecules and also with the fuel to power the living processes.

2. Nearly all products of digestion of carbohydrate, fat, and protein are metabolized to a common

metabolite, acetyl-CoA, before final oxidation to CO_2 in the citric acid cycle.

3. Acetyl-CoA is also used as the building block for the biosynthesis of long-chain fatty acids, cholesterol, and other steroids from carbohydrate and of cholesterol and ketone bodies from fatty acids.

4. Glucose provides carbon skeletons for the glycerol moiety of fat and of several nonessential amino acids.

5. All water-soluble products of digestion are transported directly to the liver via the hepatic portal vein for processing. This often involves oxidation or synthesis of molecules, some of which are exported to the rest of the body, eg, plasma proteins. The liver has a direct role in regulating the concentration of many blood constituents, including glucose and amino acids, since its primary function is to serve the extrahepatic tissues.

6. In addition to the nucleus, there are three primary subcellular metabolic compartments. The cytosol contains the pathways of glycolysis, glycogenesis, glycogenolysis, the pentose phosphate pathway, and lipogenesis. The mitochondrion contains the principal enzymes of oxidation including those of the citric acid cycle, β-oxidation of fatty acids, and the respiratory chain. Amino acid metabolism takes place, not only in the cytosol and mitochondria but also in the endoplasmic reticulum, where at the ribosomal site, amino acids are converted into proteins. The membranes of the endoplasmic reticulum also contain the enzymes for many other processes including glycerolipid formation and drug metabolism.

7. Metabolic pathways are regulated by rapid mechanisms affecting the activity of existing enzymes, eg, allosteric and covalent modification. The latter is often initiated by the action of hormones. Hormones also regulate by longer term mechanisms, through promotion or inhibition of the synthesis of enzyme protein through gene expression.

REFERENCES

Cohen P: *Control of Enzyme Activity,* 2nd ed. Chapman & Hall, 1983.

Hue L, Van de Werve G (editors): *Short-Term Regulation of Liver Metabolism.* Elsevier/North Holland, 1981.

Newsholme EA, Crabtree B: Flux-generating and regulatory steps in metabolic control. Trends Biochem Sci 1981;6:53.

Newsholme EA, Start C: *Regulation in Metabolism.* Wiley, 1973.

18

The Citric Acid Cycle: The Catabolism of Acetyl-CoA

Peter A. Mayes, PhD, DSc

INTRODUCTION

The citric acid cycle (Krebs cycle, tricarboxylic acid cycle) is a series of reactions in mitochondria that bring about the catabolism of acetyl residues, liberating hydrogen equivalents, which, upon oxidation, lead to the release and capture as ATP of most of the available energy of tissue fuels. The acetyl residues are in the form of **acetyl-CoA** (CH$_3$—CO~S—CoA, active acetate), an ester of coenzyme A. CoA contains the vitamin pantothenic acid.

BIOMEDICAL IMPORTANCE

The major function of the citric acid cycle is to act as the final common pathway for the oxidation of carbohydrate, lipids, and protein. This is because glucose, fatty acids, and many amino acids are all metabolized to acetyl-CoA or intermediates of the cycle. It also plays a major role in gluconeogenesis, transamination, deamination, and lipogenesis. Several of these processes are carried out in many tissues, but the liver is the only tissue in which all occur to a significant extent. The repercussions are therefore profound when, for example, large numbers of hepatic cells are damaged or replaced by connective tissue, as in acute **hepatitis** and **cirrhosis,** respectively. A mute testimony to the vital importance of the citric acid cycle is the fact that very few if any genetic abnormalities of its enzymes have been reported in humans; such abnormalities are presumably incompatible with normal development.

THE CITRIC ACID CYCLE PROVIDES SUBSTRATE FOR THE RESPIRATORY CHAIN

Essentially, the cycle comprises the combination of a molecule of acetyl-CoA with the four-carbon dicarboxylic acid oxaloacetate, resulting in the formation of a six-carbon tricarboxylic acid, citrate. There follows a series of reactions in the course of which two molecules of CO_2 are released and oxaloacetate is regenerated (Figure 18–1). Since only a small quantity of oxaloacetate is needed to facilitate the conversion of a large quantity of acetyl units to CO_2, oxaloacetate may be considered to play a **catalytic role.**

The citric acid cycle is an integral part of the process by which much of the free energy liberated during the oxidation of carbohydrate, lipids, and amino acids is made available. During the course of oxidation of acetyl-CoA in the cycle, **reducing equivalents** in the form of hydrogen or of electrons are formed as a result of the activity of specific dehydrogenases. These reducing equivalents then enter the respiratory chain, where large amounts of ATP are generated in the process of oxidative phosphorylation (Figure 18–2; see also Chapter 14). This process is **aerobic,** requiring oxygen as the final oxidant of the reducing equivalents. Therefore, absence

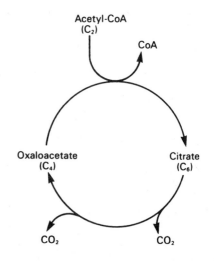

Figure 18–1. Citric acid cycle, illustrating the catalytic role of oxaloacetate.

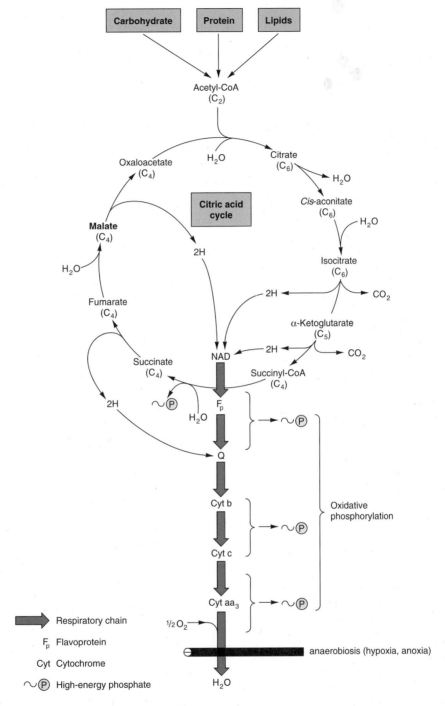

Figure 18–2. The citric acid cycle: the major catabolic pathway for acetyl-CoA in aerobic organisms. Acetyl-CoA, the product of carbohydrate, protein, and lipid catabolism, is taken into the cycle, together with H_2O, and oxidized to CO_2 with the release of reducing equivalents (2H). Subsequent oxidation of 2H in the respiratory chain leads to coupled phosphorylation of ADP to ATP. For one turn of the cycle, 11 ~ⓅⒶ are generated via oxidative phosphorylation and one ~Ⓟ arises at substrate level from the conversion of succinyl-CoA to succinate.

(anoxia) or partial deficiency (hypoxia) of O_2 causes total or partial inhibition of the cycle.

The enzymes of the citric acid cycle are located in the **mitochondrial matrix,** either free or attached to the inner surface of the inner mitochondrial membrane, which facilitates the transfer of reducing equivalents to the adjacent enzymes of the respiratory chain, situated also in the inner mitochondrial membrane.

REACTIONS OF THE CITRIC ACID CYCLE LIBERATE REDUCING EQUIVALENTS AND CO$_2$ (Figure 18–3)*

The initial condensation of acetyl-CoA with oxaloacetate to form citrate is catalyzed by a condensing enzyme, **citrate synthase,** which effects synthesis of a carbon-to-carbon bond between the methyl carbon of acetyl-CoA and the carbonyl carbon of oxaloacetate. The condensation reaction, which forms citryl-CoA, is followed by hydrolysis of the thioester bond of CoA, accompanied by considerable loss of free energy as heat, ensuring that the reaction goes to completion.

$$\text{Acetyl-CoA} + \text{Oxaloacetate} + H_2O \rightarrow \text{Citrate} + \text{CoA}$$

Citrate is converted to isocitrate by the enzyme **aconitase** (aconitate hydratase), which contains iron in the Fe^{2+} state in the form of an iron-sulfur protein (Fe:S). This conversion takes place in two steps: dehydration to *cis*-aconitate, some of which remains bound to the enzyme, and rehydration to isocitrate.

$$\text{Citrate} \rightleftharpoons \underset{H_2O}{\overset{\text{*Cis*-aconitate}}{\text{(enzyme bound)}}} \rightleftharpoons \underset{H_2O}{\text{Isocitrate}}$$

The reaction is inhibited by **fluoroacetate,** which, in the form of fluoroacetyl-CoA, condenses with oxaloacetate to form fluorocitrate. The latter inhibits aconitase, causing citrate to accumulate.

Experiments using ^{14}C-labeled intermediates indicate that aconitase reacts with citrate in an asymmetric manner, with the result that aconitase always acts on that part of the citrate molecule that is derived from oxaloacetate. This was puzzling, since citric acid appeared to be a symmetric compound. However, it is now realized (when the molecule is viewed in three dimensions) that the two —CH_2COO^-

*From Circular No. 200 of the Committee of Editors of Biochemical Journals Recommendations (1975): "According to standard biochemical convention, the ending *ate* in, eg, palmitate, denotes any mixture of free acid and the ionized form(s) (according to pH) in which the cations are not specified." The same convention is adopted in this text for all carboxylic acids.

groups are not identical in space with respect to the —OH and —COO^- groups. The consequences of the asymmetric action of aconitase, which is brought about by the three-point attachment of the enzyme to the substrate (Figure 8–3), may be appreciated by reference to the fate of labeled acetyl-CoA in the citric acid cycle as shown in Figure 18–3. It is possible that *cis*-aconitate may not be an obligatory intermediate between citrate and isocitrate but may in fact be a side branch from the main pathway.

Isocitrate undergoes dehydrogenation in the presence of **isocitrate dehydrogenase** to form oxalosuccinate. Three different isocitrate dehydrogenases have been described. One, which is NAD$^+$-specific, is found only in mitochondria. The other two enzymes are NADP$^+$-specific and are found in the mitochondria and the cytosol, respectively. Respiratory chain-linked oxidation of isocitrate proceeds almost completely through the NAD$^+$-dependent enzyme.

$$\text{Isocitrate} + NAD^+ \leftrightarrow \text{Oxalosuccinate} \leftrightarrow$$
$$\text{(enzyme bound)}$$

$$\alpha\text{-Ketoglutarate} + CO_2 + NADH + H^+$$

There follows a decarboxylation to α-ketoglutarate, also catalyzed by isocitrate dehydrogenase. Mn^{2+} (or Mg^{2+}) is an important component of the decarboxylation reaction. It would appear that oxalosuccinate remains bound to the enzyme as an intermediate in the overall action.

Next, α-ketoglutarate undergoes **oxidative decarboxylation** in a manner analogous to the oxidative decarboxylation of pyruvate (Figure 19-5), both substrates being α-keto acids.

$$\alpha\text{-Ketoglutarate} + NAD^+ + \text{CoA} \rightarrow$$
$$\text{Succinyl-CoA} + CO_2 + NADH + H^+$$

The reaction, catalyzed by an α-**ketoglutarate dehydrogenase complex,** also requires identical cofactors to the pyruvate dehydrogenase complex—eg, thiamin diphosphate, lipoate, NAD$^+$, FAD, and CoA—and results in the formation of succinyl-CoA, a high-energy thioester. The equilibrium of this reaction is so much in favor of succinyl-CoA formation that the reaction must be considered as physiologically unidirectional. As in the case of pyruvate oxidation (Chapter 19), arsenite inhibits the reaction, causing the substrate, α-**ketoglutarate,** to accumulate.

To continue the cycle, succinyl-CoA is converted to succinate by the enzyme **succinate thiokinase (succinyl-CoA synthetase).**

$$\text{Succinyl-CoA} + P_i + \text{ADP} \leftrightarrow \text{Succinate} + \text{ATP} + \text{CoA}$$

This is the only example in the citric acid cycle of the generation of a high-energy phosphate at the sub-

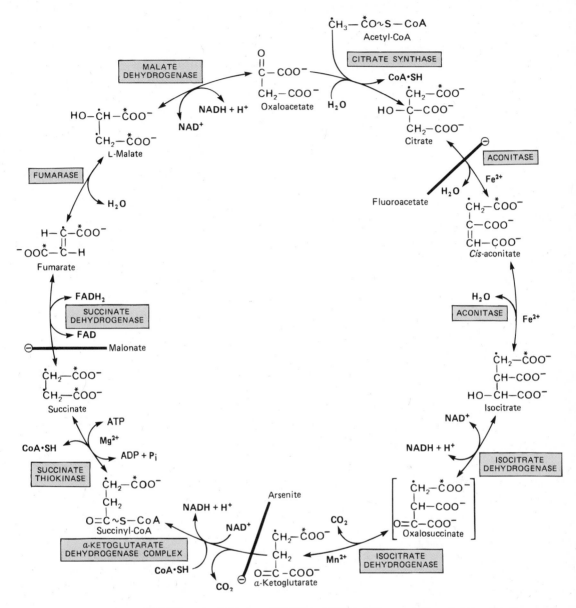

Figure 18–3. The citric acid (Krebs) cycle. Oxidation of NADH and $FADH_2$ in the respiratory chain leads to the generation of ATP via oxidative phosphorylation. In order to follow the passage of acetyl-CoA through the cycle, the two carbon
atoms of the acetyl radical are shown labeled on the carboxyl carbon (using the designation *) and on the methyl carbon
(using the designation •). Although two carbon atoms are lost as CO_2 in one revolution of the cycle, these atoms are not
derived from the acetyl-CoA that has immediately entered the cycle but from that portion of the citrate molecule which
derived from oxaloacetate. However, on completion of a single turn of the cycle, the oxaloacetate that is regenerated is
now labeled, which leads to labeled CO_2 being evolved during the second turn of the cycle. Because succinate is a symmetric compound and because succinate dehydrogenase does not differentiate between its two carboxyl groups, "ran-
domization" of label occurs at this step such that all four carbon atoms of oxaloacetate appear to be labeled after one
turn of the cycle. During gluconeogenesis, some of the label in oxaloacetate is incorporated into glucose and glycogen
(Figure 21–1). For a discussion of the stereochemical aspects of the citric acid cycle, see Greville (1968). The sites of inhibition ($\ominus$) by fluoroacetate, malonate, and arsenite are indicated.

strate level and arises because the release of free energy from the oxidative decarboxylation of α-ketoglutarate is sufficient to generate a high-energy phosphate in addition to the formation of NADH (equivalent to 3 ~(P)).

The mitochondrial matrix also contains a second succinyl-CoA synthetase, specific for guanine nucleotides, but this does not take part in the citric acid cycle. An alternative reaction in extrahepatic tissues, which is catalyzed by **succinyl-CoA-acetoacetate-CoA transferase (thiophorase),** is the conversion of succinyl-CoA to succinate coupled with the conversion of acetoacetate to acetoacetyl-CoA (Chapter 24).

Succinate is metabolized further by undergoing a dehydrogenation followed by the addition of water and subsequently by a further dehydrogenation that regenerates oxaloacetate.

$$\text{Succinate} + \text{FAD} \leftrightarrow \text{Fumarate} + \text{FADH}_2$$

The first dehydrogenation reaction is catalyzed by **succinate dehydrogenase,** which is bound to the inner surface of the inner mitochondrial membrane—unlike the other enzymes of the cycle, which are found in the matrix. It is the only dehydrogenation in the citric acid cycle that involves the direct transfer of hydrogen from the substrate to a flavoprotein without the participation of NAD^+. The enzyme contains FAD and iron-sulfur (Fe:S) protein. Fumarate is formed as a result of the dehydrogenation. Isotopic experiments have shown that the enzyme is stereospecific for the *trans* hydrogen atoms of the methylene carbons of succinate. Addition of malonate or oxaloacetate to suitable preparations inhibits succinate dehydrogenase competitively, resulting in succinate accumulation.

Fumarase (fumarate hydratase) catalyzes the addition of water to fumarate to give malate.

$$\text{Fumarate} + \text{H}_2\text{O} \leftrightarrow \text{L-Malate}$$

In addition to being specific for the L-isomer of malate, fumarase catalyzes the addition of the elements of water to the double bond of fumarate in the *trans* configuration. Malate is converted to oxaloacetate by **malate dehydrogenase,** a reaction requiring NAD^+.

$$\text{L-Malate} + \text{NAD}^+ \leftrightarrow \text{Oxaloacetate} + \text{NADH} + \text{H}^+$$

Although the equilibrium of this reaction strongly favors malate, the net flux is toward the direction of oxaloacetate because this compound, together with the other product of the reaction (NADH), is removed continuously in further reactions.

The enzymes of the citric acid cycle, except for the α-ketoglutarate and succinate dehydrogenase, are also found outside the mitochondria. While they may catalyze similar reactions, some of the enzymes, eg, malate dehydrogenase, may not in fact be the same proteins as the mitochondrial enzymes of the same name, ie, they are isoenzymes.

TWELVE ATP ARE FORMED PER TURN OF THE CITRIC ACID CYCLE

As a result of oxidations catalyzed by dehydrogenase enzymes of the citric acid cycle, three molecules of NADH and one of $FADH_2$ are produced for each molecule of acetyl-CoA catabolized in one revolution of the cycle. These reducing equivalents are transferred to the respiratory chain in the inner mitochondrial membrane (Figure 18–2). During passage along the chain, reducing equivalents from NADH generate three high-energy phosphate bonds by the esterification of ADP to ATP in the process of oxidative phosphorylation (see Chapter 14). However, $FADH_2$ produces only two high-energy phosphate bonds because it transfers its reducing power to Q, bypassing the first site for oxidative phosphorylation in the respiratory chain (Figure 14–7). A further high-energy phosphate is generated at the level of the cycle itself (ie, at substrate level) during the conversion of succinyl-CoA to succinate. Thus, 12 ATP molecules are generated for each turn of the cycle (Table 18–1).

VITAMINS PLAY KEY ROLES IN THE CITRIC ACID CYCLE

Four of the soluble vitamins of the B complex have precise roles in the functioning of the citric acid cycle. They are (1) **riboflavin,** in the form of **flavin adenine dinucleotide (FAD),** a cofactor in the α-ketoglutarate dehydrogenase complex and in succinate dehydrogenase; (2) **niacin,** in the form of **nicotinamide adenine dinucleotide (NAD),** the coenzyme for three dehydrogenases in the cycle, **iso-**

Table 18–1. Generation of ATP by the citric acid cycle.

Reaction Catalyzed By	Method of ~(P) Production	ATP Molecules Formed
Isocitrate dehydrogenase	Respiratory chain oxidation of NADH	3
α-Ketoglutarate dehydrogenase	Respiratory chain oxidation of NADH	3
Succinate thiokinase	Phosphorylation at substrate level	1
Succinate dehydrogenase	Respiratory chain oxidation of $FADH_2$	2
Malate dehydrogenase	Respiratory chain oxidation of NADH	3
		Net 12

citrate dehydrogenase, α-ketoglutarate dehydro-genase, and malate dehydrogenase; (3) thiamin (vitamin B$_1$), as thiamin diphosphate, the coenzyme for decarboxylation in the α-ketoglutarate dehydro-genase reaction; and (4) pantothenic acid, as part of coenzyme A, the cofactor attached to "active" car-boxylic acid residues such as acetyl-CoA and suc-cinyl-CoA.

THE CITRIC ACID CYCLE PLAYS A PIVOTAL ROLE IN METABOLISM

Some metabolic pathways end in a constituent of the citric acid cycle, while other pathways originate from the cycle. These pathways concern the processes of gluconeogenesis, transamination, deamination, and fatty acid synthesis. Therefore, the citric acid cycle plays roles in both oxidative and synthetic processes; ie, it is amphibolic. These roles are summarized below.

The Citric Acid Cycle Takes Part in Gluconeogenesis, Transamination, and Deamination

All major members of the cycle, from citrate to oxaloacetate, are potentially glucogenic, since they can give rise to a net production of glucose in the liver or kidney, the organs that contain a complete set of enzymes necessary for gluconeogenesis (Chap-ter 21). The key enzyme that facilitates the net trans-fer out of the cycle into the main pathway of gluco-neogenesis is phosphoenolpyruvate carboxykinase, which catalyzes the decarboxylation of oxaloacetate to phosphoenolpyruvate, GTP acting as the source of high-energy phosphate (Figure 18–4).

Oxaloacetate + GTP →

Phosphoenolpyruvate + CO$_2$ + GDP

Net transfer into the cycle occurs as a result of sev-eral different reactions. Among the most significant

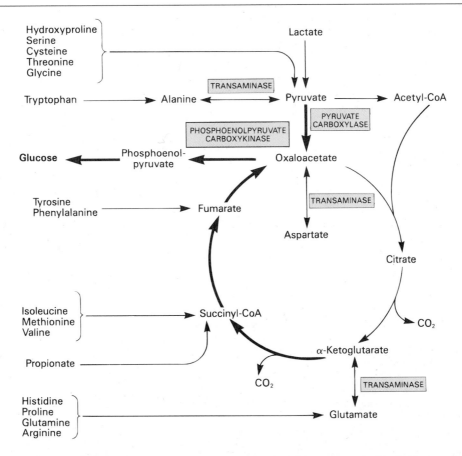

Figure 18–4. Involvement of the citric acid cycle in transamination and gluconeogenesis. The bold arrows indicate the main pathway of gluconeogenesis.

is the formation of oxaloacetate by the carboxylation of pyruvate, catalyzed by **pyruvate carboxylase.**

ATP + CO_2 + H_2O + Pyruvate →
$$\text{Oxaloacetate + ADP + } P_i$$

This reaction is considered important in maintaining adequate concentrations of oxaloacetate for the condensation reaction with acetyl-CoA. If acetyl-CoA accumulates, it acts as an allosteric activator of pyruvate carboxylase, thereby ensuring a supply of oxaloacetate. Lactate, an important substrate for gluconeogenesis, enters the cycle via conversion to pyruvate and then oxaloacetate.

Aminotransferase (transaminase) reactions produce pyruvate from alanine, oxaloacetate from aspartate, and α-ketoglutarate from glutamate. Because these reactions are reversible, the cycle also serves as a source of carbon skeletons for the synthesis of nonessential amino acids, eg,

Aspartate + Pyruvate ↔ Oxaloacetate + Alanine

Glutamate + Pyruvate ↔ α-Ketoglutarate + Alanine

Other amino acids contribute to gluconeogenesis because all or part of their carbon skeletons enter the citric acid cycle after deamination or transamination. Examples are alanine, cysteine, glycine, hydroxyproline, serine, threonine, and tryptophan, which form pyruvate; arginine, histidine, glutamine, and proline, which form α-ketoglutarate via glutamate; isoleucine, methionine, and valine, which form succinyl-CoA; and tyrosine and phenylalanine, which form fumarate (Figure 18–4). Substances forming pyruvate have the option of complete oxidation to CO_2 if they follow the pyruvate dehydrogenase pathway to acetyl-CoA, or they may follow the gluconeogenic pathway via carboxylation to oxaloacetate.

Of particular significance to ruminants is the conversion of propionate, the major glucogenic product of rumen fermentation, to succinyl-CoA via the methylmalonyl-CoA pathway (Figure 21–2).

The Citric Acid Cycle Takes Part in Fatty Acid Synthesis (Figure 18–5)

Acetyl-CoA, formed from pyruvate by the action of pyruvate dehydrogenase, is the major building block for long-chain fatty acid synthesis in nonruminants. (In ruminants, acetyl-CoA is derived directly from acetate.) Since pyruvate dehydrogenase is a mitochondrial enzyme and the enzymes responsible for fatty acid synthesis are extramitochondrial, the cell needs to transport acetyl-CoA through the mitochondrial membrane, which is impermeable to acetyl-CoA. This is accomplished by allowing acetyl-CoA to form citrate in the citric acid cycle, transporting citrate out of the mitochondria, and finally making acetyl-CoA available in the cytosol by cleaving citrate in a reaction catalyzed by the enzyme **ATP-citrate lyase.**

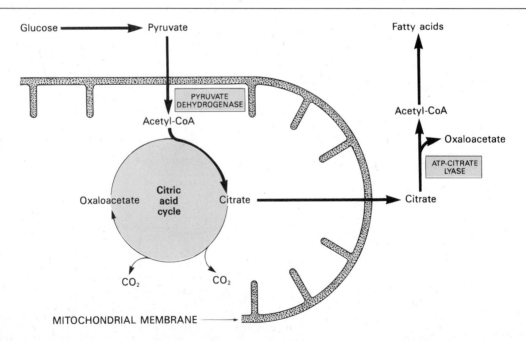

Figure 18–5. Participation of the citric acid cycle in fatty acid synthesis from glucose. See also Figure 23–5.

Citrate + ATP + CoA →
$$\text{Acetyl-CoA} + \text{Oxaloacetate} + \text{ADP} + P_i$$

Regulation of the Citric Acid Cycle Depends Primarily on a Supply of Oxidized Cofactors

In most tissues, where the primary function of the citric acid cycle is to provide energy, there is little doubt that **respiratory control** via the respiratory chain and oxidative phosphorylation is the overriding control on citric acid cycle activity. Thus, activity is immediately dependent on the supply of oxidized dehydrogenase cofactors (eg, NAD), which in turn, because of the tight coupling between oxidation and phosphorylation, is dependent on the availability of ADP and ultimately, therefore, on the rate of utilization of ATP. Therefore, provided there is adequate O_2, the rate of doing work via utilization of ATP determines both the rate of respiration and the activity of the citric acid cycle. In addition to this overall or coarse control, the properties of some enzymes of the cycle indicate that control might also be exerted at the level of the cycle itself to reinforce the coarse control.

The most likely sites for regulation are the nonequilibrium reactions. These are catalyzed by pyruvate dehydrogenase, citrate synthase, NAD-linked isocitrate dehydrogenase, and α-ketoglutarate dehydrogenase. All these dehydrogenases are activated by Ca^{2+}, which increases in concentration during muscular contraction and secretion, when there is increased energy demand. In a tissue such as brain, which is largely dependent on carbohydrate to supply acetyl-CoA, control of the citric acid cycle may occur at the pyruvate dehydrogenase step. In the cycle proper, several enzymes are responsive to the energy status as expressed by the [ATP]/[ADP] and [NADH]/[NAD$^+$] ratios. Thus, there is allosteric inhibition of citrate synthase by ATP and long-chain fatty acyl-CoA. Allosteric activation of mitochondrial NAD-dependent isocitrate dehydrogenase by ADP is counteracted by ATP and NADH. The α-ketoglutarate dehydrogenase complex appears to be under control analogous to that of pyruvate dehydrogenase (Figure 19–6). Succinate dehydrogenase is inhibited by oxaloacetate, and the availability of oxaloacetate, as controlled by malate dehydrogenase, depends on the [NADH]/[NAD$^+$] ratio. Since the K_m for oxaloacetate of citrate synthase is of the same order of magnitude as the intramitochondrial concentration, it would appear that the concentration of oxaloacetate could play a part in controlling the rate of citrate formation. Which (if any) of these mechanisms operates in vivo has still to be resolved.

SUMMARY

1. The citric acid cycle is the final pathway for the oxidation of carbohydrate, lipid, and protein. It catalyzes the combination of their common metabolite acetyl-CoA with oxaloacetate to form citrate. By a series of dehydrogenations and decarboxylations, citrate is degraded, releasing reducing equivalents and $2CO_2$ and regenerating oxaloacetate.

2. The reducing equivalents are oxidized by the respiratory chain with release of ATP. Thus, the cycle is the major route for the generation of ATP and is located in the matrix of mitochondria adjacent to the enzymes of the respiratory chain and oxidative phosphorylation.

3. The citric acid cycle is amphibolic, since it has other metabolic roles in addition to oxidation. It takes part in gluconeogenesis, transamination, deamination, and the synthesis of fatty acids.

REFERENCES

Baldwin JE, Krebs HA: The evolution of metabolic cycles. Nature 1981;291:381.

Boyer PD (editor): *The Enzymes,* 3rd ed. Academic Press, 1971.

Goodwin TW (editor): *The Metabolic Roles of Citrate.* Academic Press, 1968.

Greville GD: Vol 1, p 297, in: *Carbohydrate Metabolism and Its Disorders.* Dickens F, Randle PJ, Whelan WJ (editors). Academic Press, 1968.

Kay J, Weitzman PDJ (editors): *Krebs' Citric Acid Cycle—Half a Century and Still Turning.* Biochemical Society, London, 1987.

Lowenstein JM (editor): *Citric Acid Cycle: Control and Compartmentation.* Dekker, 1969.

Lowenstein JM (editor): *Citric Acid Cycle.* Vol 13 in: *Methods in Enzymology.* Academic Press, 1969.

Srere PA: The enzymology of the formation and breakdown of citrate. Adv Enzymol 1975;43:57.

Tyler DD: *The Mitochondrion in Health and Disease.* VCH Publishers, 1992.

19

Glycolysis & the Oxidation of Pyruvate

Peter A. Mayes, PhD, DSc

INTRODUCTION

Most tissues have at least a minimal requirement for glucose. In some cases, eg, brain, the requirement is substantial, while in others, eg, erythrocytes, it is nearly total. Glycolysis is the major pathway for the utilization of glucose and is found in the cytosol of all cells. It is a unique pathway, since it can utilize oxygen if available **(aerobic),** or it can function in the total absence of oxygen **(anaerobic).** However, to oxidize glucose beyond the pyruvate end stage of glycolysis requires not only molecular oxygen but also mitochondrial enzyme systems such as the pyruvate dehydrogenase complex, the citric acid cycle, and the respiratory chain.

BIOMEDICAL IMPORTANCE

Glycolysis is not only the principal route for glucose metabolism leading to the production of acetyl-CoA and oxidation in the citric acid cycle, but it also provides the main pathway for the metabolism of fructose and galactose derived from the diet. Of crucial biomedical significance is the ability of glycolysis to provide ATP in the absence of oxygen, because this allows skeletal muscle to perform at very high levels when aerobic oxidation becomes insufficient and it allows tissues with significant glycolytic ability to survive anoxic episodes. Conversely, heart muscle, which is adapted for aerobic performance, has relatively poor glycolytic ability and poor survival under conditions of **ischemia.** A small number of diseases occur in which enzymes of glycolysis (eg, pyruvate kinase) are deficient in activity; these conditions are mainly manifested as **hemolytic anemias** or, if they occur in skeletal muscle (eg, phosphofructokinase), as **fatigue.** In fast-growing cancer cells, glycolysis proceeds at a much higher rate than is required by the citric acid cycle. Thus, more pyruvate is produced than can be metabolized. This in turn results in excessive production of lactate, which favors a relatively acid local environment in the tumor, a situation that may have implications for certain types of cancer therapy. **Lactic acidosis** results from several causes, including pyruvate dehydrogenase deficiency.

GLYCOLYSIS CAN FUNCTION UNDER ANAEROBIC CONDITIONS

At an early period in the course of investigations on glycolysis it was realized that the process of fermentation in yeast was similar to the breakdown of glycogen in muscle. It was noted that when a muscle contracts in an anaerobic medium, ie, one from which oxygen is excluded, **glycogen disappears** and **lactate appears** as the principal end product. When oxygen is admitted, aerobic recovery takes place and glycogen reappears, while lactate disappears. However, if contraction takes place under aerobic conditions, lactate does not accumulate and pyruvate becomes the major end product of glycolysis. Pyruvate is oxidized further to CO_2 and water (Figure 19–1). As a result of these observations, it has been customary to separate carbohydrate metabolism into anaerobic and aerobic phases. However, this distinction is arbitrary, since the reactions in glycolysis are the same in the presence of oxygen as in its absence, except in extent and end products. When oxygen is in short supply, reoxidation of NADH formed from NAD^+ during glycolysis is impaired. Under these circumstances, NADH is reoxidized by coupling to the reduction of pyruvate to lactate, and the NAD^+ so formed allows further glycolysis to proceed (Figure 19–1). Thus, glycolysis can take place under anaerobic conditions, but this has a price, for it limits the amount of energy liberated per mole of glucose oxidized. Consequently, to provide a given amount of energy, much more glucose must undergo glycolysis under anaerobic as compared with aerobic conditions.

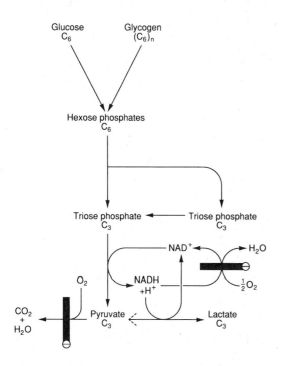

Figure 19–1. Summary of glycolysis. $\ominus$, blocked by anaerobic conditions or by absence of mitochondria containing key respiratory enzymes, eg, as in erythrocytes.

THE REACTIONS OF GLYCOLYSIS CONSTITUTE THE MAIN PATHWAY OF GLUCOSE UTILIZATION

The overall equation for glycolysis from glucose to lactate is as follows:

Glucose + 2ADP + 2P$_i$ $\rightarrow$
$$2\text{L}(+)\text{-Lactate} + 2\text{ATP} + 2\text{H}_2\text{O}$$

All of the enzymes of the glycolysis pathway (Figure 19–2) are found in the extramitochondrial soluble fraction of the cell, the cytosol, although evidence is accumulating to indicate that some of the enzymes may be associated with subcellular structures in the cell as well as the cytosol. They catalyze the reactions involved in the glycolysis of glucose to pyruvate and lactate, as follows.

Glucose enters into the glycolytic pathway by phosphorylation to glucose 6-phosphate, accomplished by the enzyme **hexokinase.** However, in liver parenchymal cells and in pancreatic islet cells, this function is carried out by **glucokinase,** whose activity in the liver is inducible and affected by changes in the nutritional state. ATP is required as phosphate donor, and as in many reactions involving phosphorylation, it reacts as the Mg-ATP complex. The terminal high-energy phosphate of ATP is utilized, and ADP is produced. The reaction is accompanied by

considerable loss of free energy as heat and therefore, under physiologic conditions, may be regarded as irreversible. Hexokinase is inhibited in an allosteric manner by the product, glucose 6-phosphate.

$$\text{Glucose} + \text{ATP} \xrightarrow{\text{Mg}^{2+}} \text{Glucose 6-phosphate} + \text{ADP}$$

Hexokinase, present in virtually all extrahepatic cells, has a high affinity (low K_m) for its substrate, glucose. Its function is to ensure a supply of glucose for the tissues, even in the presence of low blood glucose concentrations, by phosphorylating all the glucose that enters the cell, thereby maintaining a large glucose concentration gradient between the blood and the intracellular environment. It acts on both the α- and β-anomers of glucose and will also catalyze the phosphorylation of other hexoses but at a much slower rate than glucose.

The function of glucokinase is to remove glucose from the blood following a meal. In contrast to hexokinase, it has a high K_m for glucose and operates optimally at blood glucose concentrations above 5 mmol/L (Figure 21–5). It is specific for glucose.

Glucose 6-phosphate is an important compound at the junction of several metabolic pathways (glycolysis, gluconeogenesis, the pentose phosphate pathway, glycogenesis, and glycogenolysis). In glycolysis it is converted to fructose 6-phosphate by **phosphohexoseisomerase,** which involves an aldose-ketose isomerization. Only the α anomer of glucose 6-phosphate is acted upon.

α-D-Glucose 6-phosphate $\leftrightarrow$
α-D-Fructose 6-phosphate

This reaction is followed by another phosphorylation with ATP catalyzed by the enzyme **phosphofructokinase (phosphofructokinase-1)** to produce fructose 1,6-bisphosphate. Phosphofructokinase is both an allosteric and an inducible enzyme whose activity is considered to play a major role in the regulation of the rate of glycolysis. The phosphofructokinase reaction is another that may be considered to be functionally irreversible under physiologic conditions.

D-Fructose 6-phosphate + ATP $\rightarrow$
D-Fructose 1,6-bisphosphate

Fructose 1,6-bisphosphate is split by **aldolase** (fructose 1,6-bisphosphate aldolase) into two triose phosphates, glyceraldehyde 3-phosphate and dihydroxyacetone phosphate.

D-Fructose 1,6-bisphosphate $\leftrightarrow$
D-Glyceraldehyde 3-phosphate +
Dihydroxyacetone phosphate

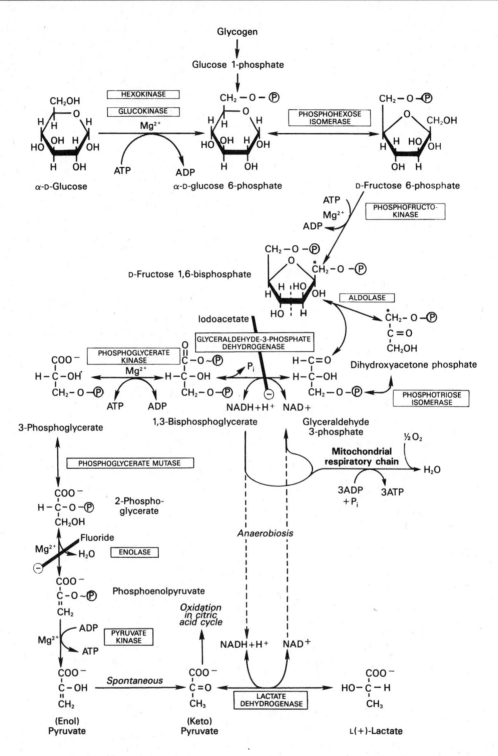

Figure 19–2. The pathway of glycolysis. ⓟ, —PO₃²⁻; Pᵢ, HOPO₃²⁻; ⊖, inhibition. *Carbon atoms 1–3 of fructose bisphosphate form dihydroxyacetone phosphate, whereas carbons 4–6 form glyceraldehyde 3-phosphate. The term "bis-," as in bisphosphate, indicates that the phosphate groups are separated, whereas diphosphate, as in adenosine diphosphate, indicates that they are joined.

Several different aldolases have been described, all of which contain 4 subunits. Aldolase A occurs in most tissues, and, in addition, aldolase B occurs in liver and kidney. The fructose phosphates exist in the cell mainly in the furanose form, but they react with phosphohexose isomerase, phosphofructokinase, and aldolase in the open-chain configuration.

Glyceraldehyde 3-phosphate and dihydroxyacetone phosphate are interconverted by the enzyme **phosphotriose isomerase.**

D-Glyceraldehyde 3-phosphate $\leftrightarrow$

Dihydroxyacetone phosphate

Glycolysis proceeds by the oxidation of glyceraldehyde 3-phosphate to 1,3-bisphosphoglycerate, and because of the activity of phosphotriose isomerase, the dihydroxyacetone phosphate is also oxidized to 1,3-diphosphoglycerate via glyceraldehyde 3-phosphate.

D-Glyceraldehyde 3-phosphate + NAD⁺ + Pᵢ $\leftrightarrow$

1,3-Bisphosphoglycerate + NADH + H⁺

The enzyme responsible for the oxidation, **glyceraldehyde 3-phosphate dehydrogenase,** is NAD-dependent. Structurally, it consists of four identical polypeptides (monomers) forming a tetramer. Four —SH groups are present on each polypeptide, derived from cysteine residues within the polypeptide chain. One of the —SH groups is found at the active site of the enzyme (Figure 19–3). The substrate initially combines with this —SH group, forming a thiohemiacetal that is converted to a high-energy thiol ester by oxidation; the hydrogens removed in this oxidation are transferred to NAD⁺ bound to the enzyme. The NADH produced on the enzyme is not so firmly bound to the enzyme as is NAD⁺. Consequently, NADH is easily displaced by another molecule of NAD⁺. Finally, by phosphorolysis, inorganic phosphate (Pᵢ) is added, forming 1,3-bisphosphoglycerate, and the free enzyme with a reconstituted —SH group is liberated. Energy released during the oxidation is conserved by the formation of a high-energy sulfur group that becomes, after phosphorolysis, a high-energy phosphate group in position 1 of 1,3-bisphosphoglycerate. This high-energy phosphate is captured as ATP in a further reaction with ADP cat-

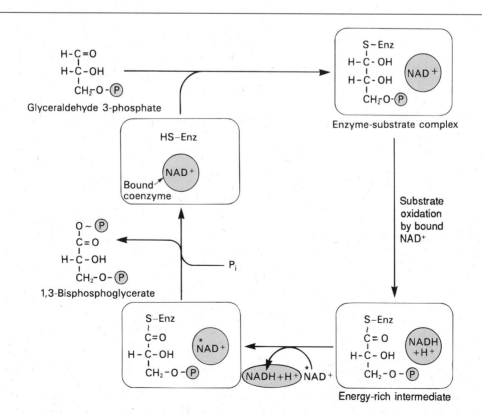

Figure 19–3. Mechanism of oxidation of glyceraldehyde 3-phosphate. (Enz, glyceraldehyde-3-phosphate dehydrogenase.) The enzyme is inhibited by the —SH poison iodoacetate, which is thus able to inhibit glycolysis.

alyzed by **phosphoglycerate kinase,** leaving 3-phosphoglycerate.

$$1,3\text{-Bisphosphoglycerate} + ADP \leftrightarrow$$
$$3\text{-Phosphoglycerate} + ATP$$

Since two molecules of triose phosphate are formed per molecule of glucose undergoing glycolysis, two molecules of ATP are generated at this stage per molecule of glucose, an example of phosphorylation "at the substrate level."

If arsenate is present, it will compete with inorganic phosphate (P_i) in the above reactions to give 1-arseno-3-phosphoglycerate, which hydrolyzes spontaneously to give 3-phosphoglycerate plus heat, without generating ATP. This is an important example of the ability of arsenate to accomplish uncoupling of oxidation and phosphorylation.

3-Phosphoglycerate arising from the above reactions is converted to 2-phosphoglycerate by the enzyme **phosphoglycerate mutase.** It is likely that 2,3-bisphosphoglycerate (diphosphoglycerate, DPG) is an intermediate in this reaction.

$$3\text{-Phosphoglycerate} \leftrightarrow 2\text{-Phosphoglycerate}$$

The subsequent step is catalyzed by **enolase** and involves a dehydration and redistribution of energy within the molecule, raising the phosphate on position 2 to the high-energy state, thus forming phosphoenolpyruvate. Enolase is inhibited by **fluoride,** a property that can be made use of when it is required to prevent glycolysis in blood prior to the estimation of glucose. The enzyme is also dependent on the presence of either Mg^{2+} or Mn^{2+}.

$$2\text{-Phosphoglycerate} \leftrightarrow \text{Phosphoenolpyruvate} + H_2O$$

The high-energy phosphate of phosphoenolpyruvate is transferred to ADP by the enzyme **pyruvate kinase** to generate, at this stage, two molecules of ATP per molecule of glucose oxidized. Enolpyruvate formed in this reaction is converted spontaneously to the keto form of pyruvate. This is another nonequilibrium reaction that is accompanied by considerable loss of free energy as heat and must be regarded as physiologically irreversible.

$$\text{Phosphoenolpyruvate} + ADP \rightarrow \text{Pyruvate} + ATP$$

The redox state of the tissue now determines which of two pathways is followed. If **anaerobic** conditions prevail, the reoxidation of NADH by transfer of reducing equivalents through the respiratory chain to oxygen is prevented. Pyruvate is reduced by the NADH to lactate, the reaction being catalyzed by **lactate dehydrogenase.** Several isoenzymes of this enzyme have been described and have clinical significance (Chapter 8).

$$\text{Pyruvate} + NADH + H^+ \leftrightarrow \text{L(+)-Lactate} + NAD^+$$

The reoxidation of NADH via lactate formation allows glycolysis to proceed in the absence of oxygen by regenerating sufficient NAD^+ for another cycle of the reaction catalyzed by glyceraldehyde-3-phosphate dehydrogenase. Under **aerobic conditions,** pyruvate is taken up into mitochondria, and after conversion to acetyl-CoA is oxidized to CO_2 by the citric acid cycle. The reducing equivalents from the $NADH + H^+$ formed in glycolysis are taken up into mitochondria for oxidation via one of the two shuttles described in Chapter 14.

Tissues That Function Under Hypoxic Circumstances Tend to Produce Lactate (Figure 19–2)

This is true of skeletal muscle, particularly the white fibers, where the rate at which the organ performs work is not limited by its capacity for oxygenation. The additional quantities of lactate produced may be detected in the tissues and in the blood and urine. Glycolysis in erythrocytes, even under aerobic conditions, always terminates in lactate, because mitochondria that contain the enzymatic machinery for the aerobic oxidation of pyruvate are absent. The mammalian erythrocyte is unique in that at least 90% of its total energy requirement is provided by glycolysis. Besides skeletal muscle white fibers, smooth muscle, and erythrocytes, other tissues that normally derive most of their energy from glycolysis and produce lactate include brain, gastrointestinal tract, renal medulla, retina, and skin. The liver, kidneys, and heart usually take up lactate and oxidize it but will produce it under hypoxic conditions.

Glycolysis Is Regulated at Three Steps Involving Nonequilibrium Reactions

Although most of the glycolytic reactions are reversible, three of them are markedly exergonic and must therefore be considered physiologically irreversible. These reactions are catalyzed by **hexokinase** (and glucokinase), **phosphofructokinase,** and **pyruvate kinase** and are the major sites of regulation of glycolysis. Cells that are capable of effecting a net movement of metabolites in the synthetic direction of the glycolytic pathway (gluconeogenesis) do so because of the presence of different enzyme systems that provide alternative routes around the irreversible reactions catalyzed by the above-mentioned enzymes. These, together with the regulation of glycolysis, are discussed with the regulation of gluconeogenesis in Chapter 21.

In Erythrocytes, the Second Site in Glycolysis for ATP Generation May Be Bypassed

In the erythrocytes of many mammalian species, the step catalyzed by **phosphoglycerate kinase** may be bypassed by a process that effectively dissipates as heat the free energy associated with the high-energy phosphate of 1,3-bisphosphoglycerate (Figure 19–4). An additional enzyme, **bisphosphoglycerate mutase,** catalyzes the conversion of 1,3-bisphosphoglycerate to 2,3-bisphosphoglycerate. The latter is converted to 3-phosphoglycerate by **2,3-bisphosphoglycerate phosphatase,** an activity also attributed to phosphoglycerate mutase. The loss of a high-energy phosphate, which means that there is no net production of ATP when glycolysis takes this route, may be of advantage to the economy of the red cell, since it would allow glycolysis to proceed when the need for ATP was minimal. However, 2,3-bisphosphoglycerate, which is present in high concentration, combines with hemoglobin, causing a decrease in affinity for oxygen and a displacement of the oxyhemoglobin dissociation curve to the right. Thus, its presence in the red cells helps oxyhemoglobin to unload oxygen (Chapter 7).

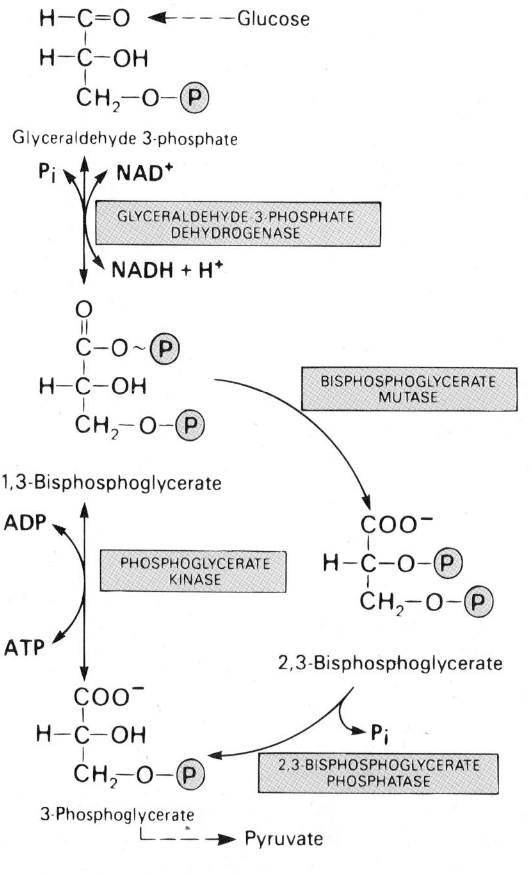

THE OXIDATION OF PYRUVATE TO ACETYL-CoA IS THE IRREVERSIBLE ROUTE FROM GLYCOLYSIS TO THE CITRIC ACID CYCLE

Before pyruvate can enter the citric acid cycle, it must be transported into the mitochondrion via a special **pyruvate transporter** that aids its passage across the inner mitochondrial membrane. This involves a symport mechanism whereby one proton is cotransported (Figure 14–12). Within the mitochondrion, pyruvate is oxidatively decarboxylated to acetyl-CoA. This reaction is catalyzed by several different enzymes working sequentially in a multienzyme complex that is associated with the inner mitochondrial membrane. They are collectively designated as the **pyruvate dehydrogenase complex** and are analogous to the α-ketoglutarate dehydrogenase complex of the citric acid cycle (Figure 18–3). Pyruvate is decarboxylated by the **pyruvate dehydrogenase** component of the enzyme complex to a hydroxyethyl derivative of the thiazole ring of enzyme-bound **thiamin diphosphate,** which in turn reacts with oxidized lipoamide, the prosthetic group of **dihydrolipoyl transacetylase,** to form acetyl lipoamide (Figure 19–5). Thiamin is an important member of the vitamin B complex (Chapter 52). Acetyl lipoamide reacts with coenzyme A to form acetyl-CoA and reduced lipoamide. The cycle of reaction is completed when the latter is reoxidized by a flavoprotein, containing FAD, in the presence of **dihydrolipoyl dehydrogenase.** Finally, the reduced flavoprotein is oxidized by NAD^+, which in turn transfers reducing equivalents to the respiratory chain.

Pyruvate + NAD^+ + CoA →

Acetyl-CoA + NADH + H^+ + CO_2

The pyruvate hydrogenase complex consists of a number of polypeptide chains of each of the three component enzymes, all organized in a regular spatial configuration. Movement of the individual enzymes appears to be restricted, and the metabolic intermediates do not dissociate freely but remain bound to the enzymes. Such a complex of enzymes, in which the substrates are handed on from one enzyme to the next, increases the reaction rate and eliminates side reactions, increasing overall efficiency.

It is to be noted that the pyruvate dehydrogenase system is sufficiently electronegative with respect to the respiratory chain that, in addition to generating a reduced coenzyme (NADH), it also generates a high-energy thio ester group in acetyl-CoA.

Figure 19–4. 2,3-Bisphosphoglycerate pathway in erythrocytes.

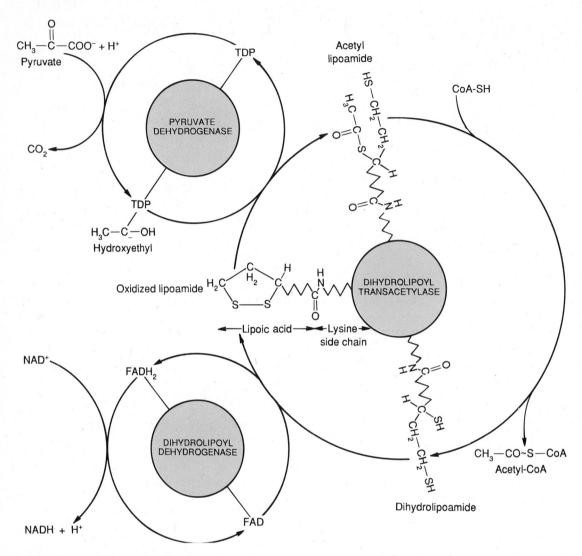

Figure 19–5. Oxidative decarboxylation of pyruvate by the pyruvate dehydrogenase complex. Lipoic acid is joined by an amide link to a lysine residue of the transacetylase component of the enzyme complex. It forms a long flexible arm, allowing the lipoic acid prosthetic group to rotate sequentially between the active sites of each of the enzymes of the complex. (NAD⁺, nicotinamide adenine dinucleotide; FAD, flavin adenine dinucleotide; TDP, thiamin diphosphate.)

Pyruvate Dehydrogenase Is Regulated by End-Product Inhibition and Covalent Modification

Pyruvate dehydrogenase is inhibited by its products, acetyl-CoA and NADH (Figure 19–6). It is also regulated by phosphorylation of three serine residues on the pyruvate dehydrogenase component of the multienzyme complex involving an ATP-specific kinase that causes a decrease in activity, and by dephosphorylation by a phosphatase that causes an increase in activity of the dehydrogenase. The kinase is activated by increases in the [acetyl-CoA]/[CoA], [NADH]/[NAD⁺], or [ATP]/[ADP] ratios. Thus, pyruvate dehydrogenase—and therefore glycolysis—is inhibited not only by a high energy potential, but

also under conditions of fatty acid oxidation, which leads to increases in these ratios. Thus, in starvation, when free fatty acid concentrations increase, there is a decrease in the proportion of the enzyme in the active form, leading to a sparing of carbohydrate. An increase in activity in adipose tissue occurs after administration of insulin but not in the liver.

Oxidation of Glucose Yields Up to 38 Mol of ATP Under Aerobic Conditions But Only 2 Mol When O₂ Is Absent

When 1 mol of glucose is combusted in a calorimeter to CO₂ and water, approximately 2870 kJ are liberated as heat. When oxidation occurs in the

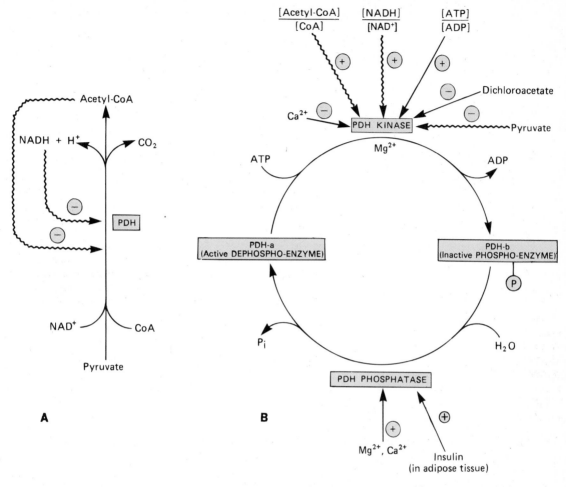

Figure 19–6. Regulation of pyruvate dehydrogenase (PDH). Arrows with wavy shafts indicate allosteric effects. *A:* Regulation by end-product inhibition. *B:* Regulation by interconversion of active and inactive forms.

tissues, some of this energy is not lost immediately as heat but is "captured" as high-energy phosphate. On the order of 38 mol of ATP are generated per molecule of glucose oxidized to CO_2 and water. In vivo, ΔG for the ATP synthase reaction has been calculated as approximately 51.6 kJ. It follows that the total energy captured in ATP per mole of glucose oxidized is 1961 kJ, or approximately 68% of the energy of combustion. Most of the ATP is formed as a consequence of oxidative phosphorylation resulting from the reoxidation of reduced coenzymes by the respiratory chain. The remainder is generated by phosphorylation at the "substrate level" (Chapter 14). Table 19–1 indicates the reactions responsible for the generation of high-energy phosphate during oxidation of glucose and the net production under aerobic and anaerobic conditions.

CLINICAL ASPECTS

Inhibition of Pyruvate Metabolism Leads to Lactic Acidosis

Arsenite or mercuric ions complex the —SH groups of lipoic acid and inhibit pyruvate dehydrogenase, as does a **dietary deficiency of thiamin,** allowing pyruvate to accumulate. Nutritionally deprived alcoholics are thiamin-deficient and if given glucose exhibit rapid accumulation of pyruvate and lactic acidosis, which is frequently lethal. Patients with **inherited pyruvate dehydrogenase deficiency,** which can be due to defects in one or more of the components of the enzyme complex, present with a similar lactic acidosis, particularly after a glucose load. Because of its dependence on glucose as a fuel, brain is a prominent tissue where these metabolic defects manifest themselves in neurologic disturbances.

Mutations have been reported for virtually all of the enzymes of carbohydrate metabolism, each asso-

Table 19–1. Generation of high-energy phosphate in the catabolism of glucose.

Pathway	Reaction Catalyzed by	Method of ~ⓟ Production	Number of ~ⓟ Formed per Mole of Glucose
Glycolysis	Glyceraldehyde-3-phosphate dehydrogenase	Respiratory chain oxidation of 2 NADH	6*
	Phosphoglycerate kinase	Phosphorylation at substrate level	2
	Pyruvate kinase	Phosphorylation at substrate level	2
			10
	Allow for consumption of ATP by reactions catalyzed by hexokinase and phosphofructokinase		−2
			Net 8
	Pyruvate dehydrogenase	Respiratory chain oxidation of 2 NADH	6
	Isocitrate dehydrogenase	Respiratory chain oxidation of 2 NADH	6
	α-Ketoglutarate dehydrogenase	Respiratory chain oxidation of 2 NADH	6
Citric acid cycle	Succinate thiokinase	Phosphorylation at substrate level	2
	Succinate dehydrogenase	Respiratory chain oxidation of 2 $FADH_2$	4
	Malate dehydrogenase	Respiratory chain oxidation of 2 NADH	6
			Net 30
	Total per mole of glucose under aerobic conditions		38
	Total per mole of glucose under anaerobic conditions		2

*It is assumed that NADH formed in glycolysis is transported into mitochondria via the malate shuttle (see Fig 14–16). If the glycerophosphate shuttle is used, only 2 ~ⓟ would be formed per mole of NADH, the total net production being 36 instead of 38. The calculation ignores the small loss of ATP due to a transport of H^+ into the mitochondrion with pyruvate and a similar transport of H^+ in the operation of the malate shuttle, totaling about 1 mol of ATP. Note that there is a substantial benefit under anaerobic conditions if glycogen is the starting point, since the net production of high-energy phosphate in glycolysis is increased from 2 to 3, as ATP is no longer required by the hexokinase reaction.

ciated with human disease. Inherited aldolase A deficiency and pyruvate kinase deficiency in erythrocytes cause **hemolytic anemia.** The exercise capacity of patients with **muscle phosphofructokinase deficiency** is low, particularly on high-carbohydrate diets. By providing an alternative lipid fuel, eg, during starvation, when blood free fatty acid and ketone bodies are increased, work capacity is improved.

SUMMARY

1. Glycolysis is the pathway found in the cytosol of all mammalian cells for the metabolism of glucose (or glycogen) to pyruvate and lactate.

2. It can function anaerobically by regenerating oxidized NAD^+ required in the glyceraldehyde-3-phosphate dehydrogenase reaction, by coupling this reaction to the reduction of pyruvate to lactate.

3. Lactate is the end product of glycolysis under anaerobic conditions (eg, in exercising muscle) or when the metabolic machinery is absent for the further oxidation of pyruvate (eg, in erythrocytes).

4. Glycolysis is regulated by three enzymes catalyzing nonequilibrium reactions, namely, hexokinase (or glucokinase), phosphofructokinase, and pyruvate kinase.

5. In erythrocytes, the second site in glycolysis for generation of ATP may be bypassed, leading to the formation of 2,3-bisphosphoglycerate, important in decreasing the affinity of hemoglobin for O_2.

6. Pyruvate is oxidized to acetyl-CoA by a multienzyme complex known as pyruvate dehydrogenase that is dependent on the vitamin cofactor thiamin diphosphate.

7. Conditions that involve an inability to metabolize pyruvate frequently lead to lactic acidosis.

REFERENCES

Behal RH et al: Regulation of the pyruvate dehydrogenase multienzyme complex. Annu Rev Nutr 1993;13:497.
Boiteux, A, Hess B: Design of glycolysis. Phil Trans R Soc London B 1981;293:5.
Boyer PD (editor): *The Enzymes,* 3rd ed, vols 5–9. Academic Press, 1972.
Fothergill-Gilmore LA: The evolution of the glycolytic pathway. Trends Biochem Sci 1986;11:47.
Randle PJ, Steiner DF, Whelan WJ (editors): *Carbohydrate Metabolism and Its Disorders,* vol 3. Academic Press, 1981.

Scriver CR et al (editors): *The Metabolic and Molecular Bases of Inherited Disease,* 7th ed. McGraw-Hill, 1995.
Sols A: Multimodulation of enzyme activity. Curr Top Cell Reg 1981;19:77.
Srere PA: Complexes of sequential metabolic enzymes. Annu Rev Biochem 1987;56:89.
Veneziale CM (editor): *The Regulation of Carbohydrate Formation and Utilization in Mammals.* University Park Press, 1981.

Metabolism of Glycogen

20

Peter A. Mayes, PhD, DSc

INTRODUCTION

Glycogen is the major storage form of carbohydrate in animals and corresponds to starch in plants. It occurs mainly in liver (up to 6%) and muscle, where it rarely exceeds 1%. However, because of its greater mass, muscle represents some three to four times as much glycogen store as liver (Table 20–1). Like starch, it is a branched polymer of α-D-glucose (Figure 15–15).

BIOMEDICAL IMPORTANCE

The function of muscle glycogen is to act as a readily available source of hexose units for glycolysis within the muscle itself. Liver glycogen is largely concerned with storage and export of hexose units for maintenance of the **blood glucose,** particularly between meals. After 12–18 hours of fasting, the liver becomes almost totally depleted of glycogen, whereas muscle glycogen is only depleted significantly after prolonged vigorous exercise. **Glycogen storage diseases** are a group of inherited disorders characterized by deficient mobilization of glycogen or deposition of abnormal forms of glycogen, leading to muscular weakness or even death.

GLYCOGENESIS OCCURS MAINLY IN MUSCLE AND LIVER

The Pathway of Glycogen Biosynthesis Involves a Special Nucleotide of Glucose (Figure 20–1)

Glucose is phosphorylated to glucose 6-phosphate, a reaction that is common to the first reaction in the pathway of glycolysis from glucose. This reaction is catalyzed by **hexokinase** in muscle and **glucokinase** in liver. Glucose 6-phosphate is converted to glucose 1-phosphate in a reaction catalyzed by the enzyme **phosphoglucomutase.** The enzyme itself is phosphorylated, and the phospho- group takes part in a reversible reaction in which glucose 1,6-bisphosphate is an intermediate.

Enz-P + Glucose 6-phosphate ↔

Enz + Glucose 1,6-bisphosphate ↔

Enz-P + Glucose 1-phosphate

Next, glucose 1-phosphate reacts with uridine triphosphate (UTP) to form the active nucleotide **uridine diphosphate glucose (UDPGlc)*** (Figure 20–2).

The reaction between glucose 1-phosphate and uridine triphosphate is catalyzed by the enzyme **UDPGlc pyrophosphorylase.**

UTP + Glucose 1-phosphate ↔ UDPGlc + PP_i

The subsequent hydrolysis of inorganic pyrophosphate by **inorganic pyrophosphatase** pulls the reaction to the right of the equation.

By the action of the enzyme **glycogen synthase,** the C_1 of the activated glucose of UDPGlc forms a glycosidic bond with the C_4 of a terminal glucose residue of glycogen, liberating uridine diphosphate (UDP). A preexisting glycogen molecule, or "glycogen primer," must be present to initiate this reaction. The glycogen primer may in turn be formed on a protein primer known as **glycogenin.**

$$UDPGlc + (C_6)_n \rightarrow UDP + (C_6)_{n+1}$$
$$\text{glycogen} \qquad \text{glycogen}$$

Glycogenin is a protein of 37 kDa which becomes glucosylated on a specific tyrosine residue by UDPGlc. Further glucose residues are attached in the 1→4 position to make a short chain that is acted upon by glycogen synthase. In skeletal muscle, glycogenin remains attached in the center of the glycogen molecule (Figure 15–15), whereas in liver the number of glycogen molecules is in excess of the number of glycogenin molecules.

---

*Other nucleoside diphosphate sugar compounds are known, eg, UDPGal. In addition, the same sugar may be linked to different nucleotides. For example, glucose may be linked to uridine (as shown above) as well as to guanosine, thymidine, adenosine, or cytidine nucleotides.

Table 20–1. Storage of carbohydrate in postabsorptive normal adult humans (70 kg).

Liver glycogen	4.0% = 72 g[1]
Muscle glycogen	0.7% = 245 g[2]
Extracellular glucose	0.1% = 10 g[3]
	327 g

[1] Liver weight 1800 g
[2] Muscle mass 35 kg
[3] Total volume 10 L

Branching Involves Detachment of Existing Glycogen Chains

The addition of a glucose residue to a preexisting glycogen chain, or "primer," occurs at the nonreducing, outer end of the molecule so that the "branches" of the glycogen "tree" become elongated as successive 1→4 linkages are formed (Figure 20–3). When the chain has been lengthened to a minimum of 11 glucose residues, a second enzyme, the **branching enzyme** (amylo[1→4]→[1→6]-transglucosidase), transfers a part of the 1→4-chain (minimum length 6 glucose residues) to a neighboring chain to form a 1→6 linkage, thus establishing a **branch point** in the

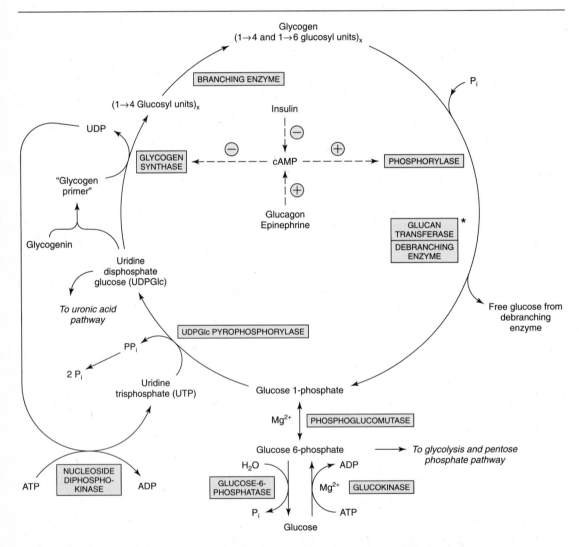

Figure 20–1. Pathway of glycogenesis and of glycogenolysis in the liver. Two high-energy phosphates are used in the incorporation of 1 mol of glucose into glycogen. ⊕, Stimulation; ⊖, inhibition. Insulin decreases the level of cAMP only after it has been raised by glucagon or epinephrine: ie, it antagonizes their action. Glucagon is active in heart muscle but not in skeletal muscle. *Glucan transferase and debranching enzyme appear to be two separate activities of the same enzyme.

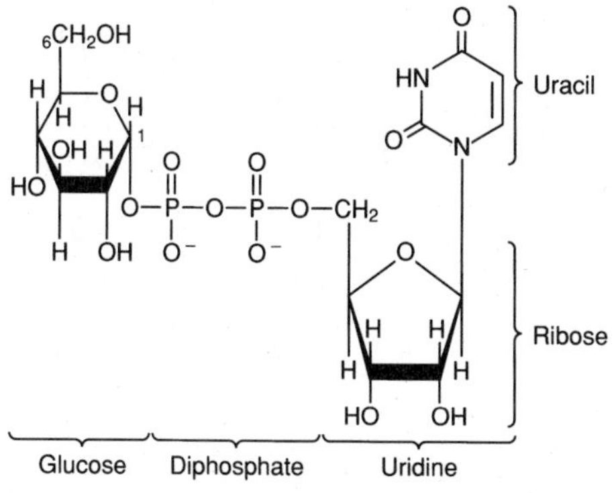

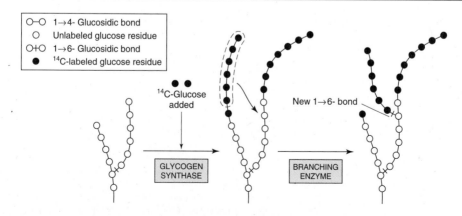

Figure 20–2. Uridine diphosphate glucose (UDPGlc).

molecule. The branches grow by further additions of 1→4-glucosyl units and further branching. As the number of nonreducing terminal residues increases, the total number of reactive sites in the molecule increases, speeding up both glycogenesis and glycogenolysis.

GLYCOGENOLYSIS IS NOT THE REVERSE OF GLYCOGENESIS BUT IS A SEPARATE PATHWAY

Degradation Involves a Debranching Mechanism (Figure 20–1)

It is the step catalyzed by **phosphorylase** that is rate-limiting in glycogenolysis.

$$(C_6)_n + P_i \rightarrow (C_6)_{n-1} + \text{Glucose 1-phosphate}$$
glycogen   glycogen

This enzyme is specific for the phosphorylytic breaking (phosphorolysis; cf hydrolysis) of the 1→4 linkages of glycogen to yield glucose 1-phosphate. The terminal glucosyl residues from the outermost chains of the glycogen molecule are removed sequentially until approximately four glucose residues remain on either side of a 1→6 branch (Figure 20–4). Another enzyme (**α-[1→4]→α-[1→4] glucan transferase**) transfers a trisaccharide unit from one branch to the other, exposing the 1→6 branch point. The **hydrolytic** splitting of the 1→6 linkages requires the action of a specific **debranching enzyme (amylo[1→6]glucosidase).** With the removal of the branch, further action by phosphorylase can proceed. The combined action of phosphorylase and these other enzymes leads to the complete breakdown of glycogen. The reaction catalyzed by phosphoglucomutase is reversible, so that glucose 6-phosphate can be formed from glucose 1-phosphate. In **liver** (and **kidney**), but not in muscle, there is a specific enzyme, **glucose-6-phosphatase,** that removes phosphate from glucose 6-phosphate, enabling glucose to diffuse from the cell into the blood. This is the final step in hepatic glycogenolysis, which is reflected by an increase in the blood glucose.

CYCLIC AMP INTEGRATES THE REGULATION OF GLYCOGENOLYSIS AND GLYCOGENESIS

The principal enzymes controlling glycogen metabolism—glycogen phosphorylase and glycogen synthase—are regulated by a complex series of reactions involving both allosteric mechanisms and covalent modifications due to reversible phosphorylation and dephosphorylation of enzyme protein (Chapter 11).

Figure 20–3. The biosynthesis of glycogen. The mechanism of branching as revealed by adding ^{14}C-labeled glucose to the diet in the living animal and examining the liver glycogen at further intervals.

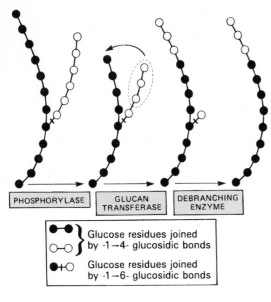

Figure 20–4. Steps in glycogenolysis.

Many covalent modifications are due to the action of cAMP (3′,5′-cyclic adenylic acid; cyclic AMP) (Figure 20–5). cAMP is the intracellular intermediate compound or **second messenger** through which many hormones act. It is formed from ATP by an enzyme, **adenylyl cyclase,** occurring in the inner surface of cell membranes. Adenylyl cyclase is activated by hormones such as **epinephrine** and **norepinephrine** acting through β-adrenergic receptors on the cell membrane and additionally in liver by **glucagon** acting through an independent **glucagon receptor.** cAMP is destroyed by a **phosphodiesterase,** and it is the activity of this enzyme that maintains the normally low level of cAMP. Insulin has been reported to increase its activity in liver, thereby lowering the concentration of cAMP.

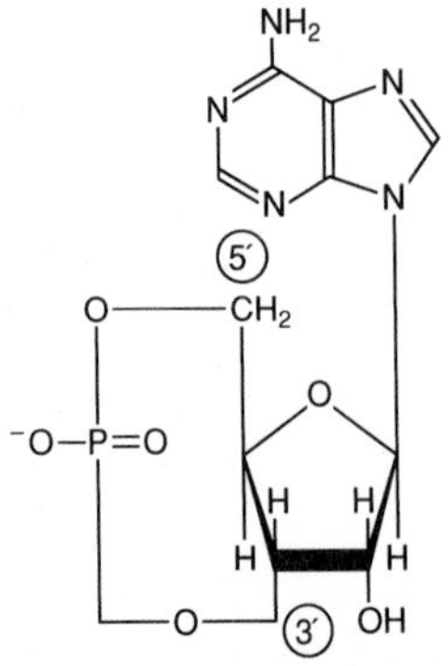

Figure 20–5. 3′,5′-Adenylic acid (cyclic AMP; cAMP).

Phosphorylase Differs Between Liver and Muscle

In liver, the enzyme exists in both an active and an inactive form. Active phosphorylase (**phosphorylase a**) has one of its serine hydroxyl groups phosphorylated in an ester linkage. By the action of a specific phosphatase, **protein phosphatase-1,** the enzyme is inactivated to **phosphorylase b** in a reaction that involves hydrolytic removal of the phosphate from the serine residue. Reactivation requires rephosphorylation with ATP and a specific enzyme, **phosphorylase kinase.**

Muscle phosphorylase is immunologically and genetically distinct from that of liver. It is a dimer, each monomer containing 1 mol of pyridoxal phosphate. It is present in two forms: **phosphorylase a,** which is phosphorylated and active in either the presence or absence of AMP (its allosteric modifier), and **phosphorylase b,** which is dephosphorylated and active only in the presence of AMP. This occurs during exercise when the level of AMP rises, providing, by this mechanism, fuel for the muscle. Phosphorylase a is the normal physiologically active form of the enzyme.

cAMP Activates Muscle Phosphorylase

Phosphorylase in muscle is activated by epinephrine (Figure 20–6). However, this occurs not as a direct effect but rather by way of the action of cAMP. Increasing the concentration of cAMP activates an enzyme of rather wide specificity, **cAMP-dependent protein kinase.** This kinase catalyzes the phosphorylation by ATP of inactive **phosphorylase kinase b** to active **phosphorylase kinase a,** which in turn, by means of a further phosphorylation, activates phosphorylase b to phosphorylase a.

Inactive cAMP-dependent protein kinase comprises two pairs of subunits, each pair consisting of a regulatory subunit (R), which binds 2 mol of cAMP, and a catalytic subunit (C), which contains the active site. Combination with cAMP causes the R_2C_2 complex to dissociate, releasing active C monomers.

$$R_2C_2 + 4cAMP \leftrightarrow 2C + 2(R—cAMP_2)$$

Inactive enzyme	Active enzyme

Ca²⁺ Synchronizes the Activation of Phosphorylase With Muscle Contraction

Glycogenolysis increases in muscle several hundred-fold immediately after the onset of contraction. This involves the rapid activation of phosphorylase owing to activation of phosphorylase kinase by Ca^{2+}, the same signal that initiates contraction. Muscle phosphorylase kinase has four types of subunits—α, β, γ, and δ—in a structure represented as $(\alpha\beta\gamma\delta)_4$.

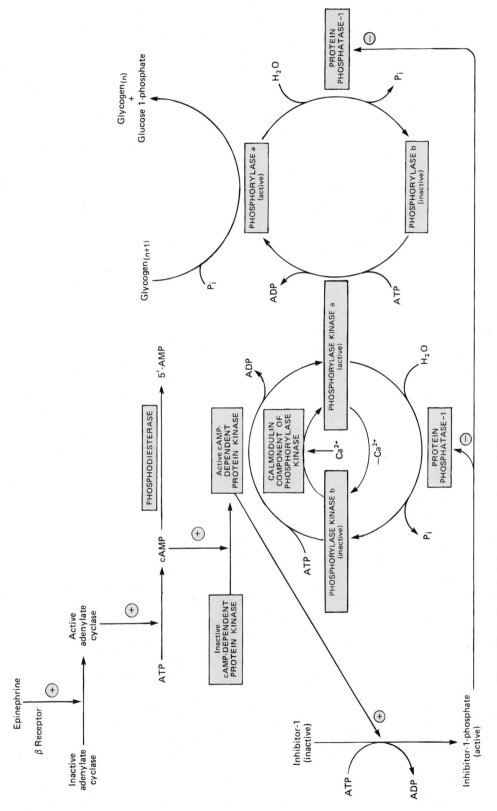

Figure 20–6. Control of phosphorylase in muscle. The sequence of reactions arranged as a cascade allows amplification of the hormonal signal at each step. (n = number of glucose residues.)

The α and β subunits contain serine residues that are phosphorylated by cAMP-dependent protein kinase. The β subunit binds four Ca^{2+} and is identical to the Ca^{2+}-binding protein **calmodulin** (Chapter 44). The binding of Ca^{2+} activates the catalytic site of the γ subunit while the molecule remains in the dephosphorylated b configuration. However, the phosphorylated a form is only fully activated in the presence of Ca^{2+}. It is of significance that calmodulin is similar in structure to TpC, the Ca^{2+}-binding protein in muscle. A second molecule of calmodulin or TpC can interact with the phosphorylase kinase, causing further activation. Thus, activation of muscle contraction and glycogenolysis are carried out by the same Ca^{2+}-binding protein, ensuring their synchronization.

Glycogenolysis in Liver Can Be cAMP Independent

In addition to the major action of **glucagon** in causing formation of cAMP and activation of phosphorylase in liver, studies have shown that **α₁ receptors** are the major mediators of epinephrine and norepinephrine stimulation of glycogenolysis. This involves a **cAMP-independent** mobilization of Ca^{2+} from mitochondria into the cytosol, followed by the stimulation of a **Ca^{2+}/calmodulin-sensitive phosphorylase kinase.** cAMP-independent glycogenolysis is also caused by vasopressin, oxytocin, and angiotensin II acting through calcium or the phosphatidylinositol bisphosphate pathway (Figure 44–7).

Protein Phosphatase-1 Inactivates Phosphorylase

Both phosphorylase a and phosphorylase kinase a are dephosphorylated and inactivated by **protein phosphatase-1.** Protein phosphatase-1 is inhibited by a protein called **inhibitor-1,** which is active only after it has been phosphorylated by cAMP-dependent protein kinase. Thus, cAMP controls both the activation and inactivation of phosphorylase (Figure 20–6).

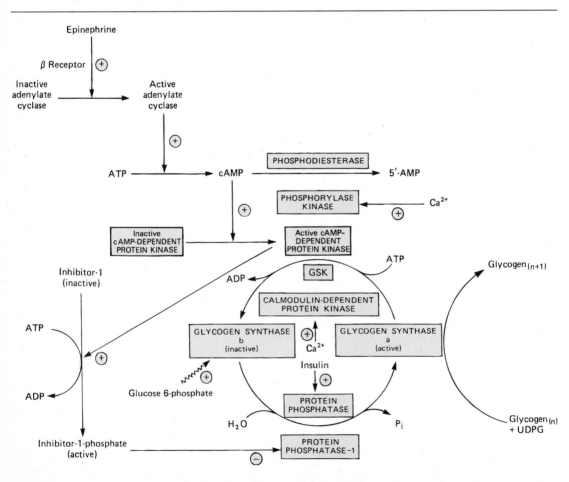

Figure 20–7. Control of glycogen synthase in muscle (n = number of glucose residues). The sequence of reactions arranged in a cascade causes amplification at each step, allowing only nanomole quantities of hormone to cause major changes in glycogen concentration. (GSK, glycogen synthase kinase-3, -4, and -5; wavy arrow, allosteric activation.)

Glycogen Synthase and Phosphorylase Activity Are Reciprocally Regulated (Figure 20–7)

Like phosphorylase, glycogen synthase exists in either a phosphorylated or nonphosphorylated state. However, unlike phosphorylase, the active form is dephosphorylated (**glycogen synthase a**) and may be inactivated to **glycogen synthase b** by phosphorylation on seven serine residues by no fewer than six different protein kinases. All seven phosphorylation sites are contained on each of four identical subunits. Two of the protein kinases are Ca^{2+}/calmodulin-dependent (one of these is phosphorylase kinase). Another kinase is cAMP-dependent protein kinase, which allows cAMP-mediated hormonal action to inhibit glycogen synthesis synchronously with the activation of glycogenolysis. The remaining kinases are known as glycogen synthase kinase-3, -4, and -5. Glucose 6-phosphate is an allosteric activator of glycogen synthase b, causing a decrease in K_m for UDP-glucose and allowing glycogen synthesis by the phosphorylated enzyme. Glycogen also exerts an inhibition on its own formation, and **insulin** also stimulates glycogen synthesis in muscle by promoting dephosphorylation and activation of glycogen synthase b. Normally, dephosphorylation of glycogen synthase b is carried out by protein phosphatase-1, which is under the control of cAMP-dependent protein kinase (Figure 20–7).

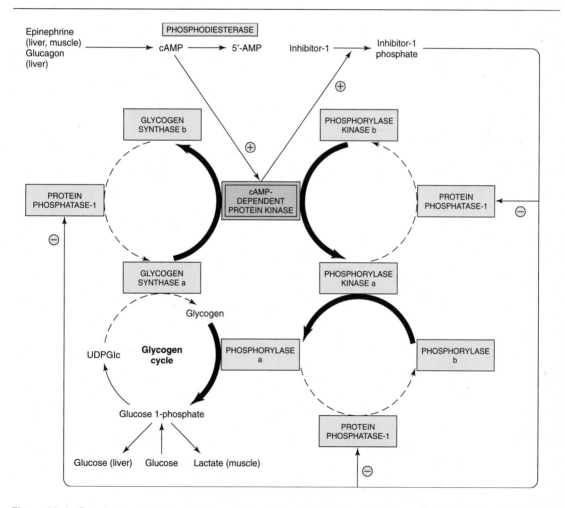

Figure 20–8. Coordinated control of glycogenolysis and glycogenesis by cAMP-dependent protein kinase. The reactions that lead to glycogenolysis as a result of an increase in cAMP concentrations are shown with bold arrows, and those that are inhibited by activation of protein phosphatase-1 are shown as broken arrows. The reverse occurs when cAMP concentrations decrease as a result of phosphodiesterase activity, leading to glycogenesis.

REGULATION OF GLYCOGEN METABOLISM IS EFFECTED BY A BALANCE IN ACTIVITIES BETWEEN GLYCOGEN SYNTHASE AND PHOSPHORYLASE
(Figure 20–8)

Glycogen synthase and phosphorylase are under substrate control (through allostery) as well as hormonal control. Not only is phosphorylase activated by a rise in concentration of cAMP (via phosphorylase kinase), but glycogen synthase is at the same time converted to the inactive form; both effects are mediated via **cAMP-dependent protein kinase.** Thus, inhibition of glycogenolysis enhances net glycogenesis, and inhibition of glycogenesis enhances net glycogenolysis. Of further significance in the regulation of glycogen metabolism is the finding that the dephosphorylation of phosphorylase a, phosphorylase kinase, and glycogen synthase b is accomplished by a single enzyme of wide specificity—**protein phosphatase-1.** In turn, protein phosphatase-1 is inhibited by cAMP-dependent protein kinase via inhibitor-1 (Figure 20–8). Thus, glycogenolysis can be terminated and glycogenesis can be stimulated synchronously, or vice versa, because both processes are keyed to the activity of cAMP-dependent protein kinase. Both phosphorylase kinase and glycogen synthase may be reversibly phosphorylated in more than one site by separate kinases and phosphatases. These secondary phosphorylations modify the sensitivity of the primary sites to phosphorylation and dephosphorylation (**multisite phosphorylation).**

The Major Factor That Controls Glycogen Metabolism in the Liver Is the Concentration of Phosphorylase a

Not only does this enzyme control the rate-limiting step in glycogenolysis, but it also inhibits the activity of protein phosphatase-1 and thereby controls glycogen synthesis (Figure 20–8). Inactivation of phosphorylase occurs as a result of allosteric inhibition by glucose as it rises in concentration after a meal. Activation is caused by 5′-AMP responding to depletion of ATP. Administration of insulin causes an immediate inactivation of phosphorylase followed by activation of glycogen synthase. The effects of insulin require the presence of glucose.

Regulation of the branching and debranching enzymes does not occur.

CLINICAL ASPECTS

Glycogen Storage Diseases Are Inherited

"Glycogen storage disease" is a generic term intended to describe a group of inherited disorders characterized by deposition of an abnormal type or quantity of glycogen in the tissues. The principal glycogenoses are summarized in Table 20–2. Deficiencies of **adenylyl kinase** and **cAMP-dependent protein kinase** have also been reported. Some of the conditions described have benefited from liver transplantation.

Table 20–2. Glycogen storage diseases.

Glycogenosis	Name	Cause of Disorder	Characteristics
Type I	von Gierke's disease	Deficiency of glucose-6-phosphatase	Liver cells and renal tubule cells loaded with glycogen. Hypoglycemia, lacticacidemia, ketosis, hyperlipemia.
Type II	Pompe's disease	Deficiency of lysosomal α-1→4- and 1→6-glucosidase (acid maltase)	Fatal, accumulation of glycogen in lysosomes in heart failure.
Type III	Limit dextrinosis, Forbes' or Cori's disease	Absence of debranching enzyme	Accumulation of a characteristic branched polysaccharide.
Type IV	Amylopectinosis, Andersen's disease	Absence of branching enzyme	Accumulation of a polysaccharide having few branch points. Death due to cardiac or liver failure in first year of life.
Type V	Myophosphorylase deficiency, McArdle's syndrome	Absence of muscle phosphorylase	Diminished exercise tolerance; muscles have abnormally high glycogen content (2.5–4.1%). Little or no lactate in blood after exercise.
Type VI	Hers' disease	Deficiency of liver phosphorylase	High glycogen content in liver, tendency toward hypoglycemia.
Type VII	Tarui's disease	Deficiency of phosphofructokinase in muscle and erythrocytes	As for type V but also possibility of hemolytic anemia.
Type VIII		Deficiency of liver phosphorylase kinase	As for type VI.

SUMMARY

1. Glycogen represents the principal storage form of carbohydrate in the mammalian body, present mainly in the liver and muscle.

2. In the liver, its major function is to service the other tissues via formation of blood glucose. In muscle, it serves the needs of that organ only, as a ready source of metabolic fuel.

3. Glycogen is synthesized from glucose and other precursors by the pathway of glycogenesis. It is broken down by a separate pathway known as glycogenolysis. Glycogenolysis leads to glucose formation in liver and lactate formation in muscle owing to the respective presence or absence of glucose-6-phosphatase.

4. Cyclic AMP integrates the regulation of glycogenolysis and glycogenesis in a reciprocal fashion by promoting the activation of phosphorylase and inhibition of glycogen synthase.

5. Inherited deficiencies in specific enzymes of glycogen metabolism in both liver and muscle are the causes of glycogen storage diseases.

REFERENCES

Cohen P: *Control of Enzyme Activity,* 2nd ed. Chapman & Hall, 1983.

Cohen P: The role of protein phosphorylation in the hormonal control of enzyme activity. Eur J Biochem 1985;151:439.

Ercan N, Gannon MC, Nuttall FQ: Incorporation of glycogenin into a hepatic proteoglycogen after oral glucose administration. J Biol Chem 1994;269:22328.

Exton JH: Molecular mechanisms involved in α-adrenergic responses. Mol Cell Endocrinol 1981;23:233.

Geddes R: Glycogen: A metabolic viewpoint. Bioscience Rep 1986;6:415.

Hers HG: The control of glycogen metabolism in the liver. Annu Rev Biochem 1976;45:167.

Randle PJ, Steiner DF, Whelan WJ (editors): *Carbohydrate Metabolism and Its Disorders,* vol 3. Academic Press, 1981.

Raz I, Katz A, Spencer MK: Epinephrine inhibits insulin-mediated glycogenesis but enhances glycolysis in human skeletal muscle. Am J Physiol 1991;260:E430.

Scriver CR et al (editors): *The Metabolic and Molecular Bases of Inherited Disease,* 7th ed. McGraw-Hill, 1995.

Selby R et al: Liver transplantation for type IV glycogen storage disease. N Engl J Med 1991;324:39.

21

Gluconeogenesis & Control of the Blood Glucose

Peter A. Mayes, PhD, DSc

INTRODUCTION

Gluconeogenesis is the term used to include all mechanisms and pathways responsible for converting noncarbohydrates to glucose or glycogen. The major substrates for gluconeogenesis are the glucogenic amino acids, lactate, glycerol, and propionate. Liver and kidney are the major tissues involved, since they contain a full complement of the necessary enzymes.

BIOMEDICAL IMPORTANCE

Gluconeogenesis meets the needs of the body for glucose when carbohydrate is not available in sufficient amounts from the diet. A continual supply of glucose is necessary as a source of energy, especially for the nervous system and the erythrocytes. Failure of gluconeogenesis is usually fatal. Below a critical blood glucose concentration, there is brain dysfunction, which can lead to coma and death. Glucose is also required in adipose tissue as a source of glyceride-glycerol, and it probably plays a role in maintaining the level of intermediates of the citric acid cycle in many tissues. It is clear that even under conditions where fat may be supplying most of the caloric requirement of the organism, there is always a certain basal requirement for glucose. Glucose is the only fuel that will supply energy to skeletal muscle under anaerobic conditions. It is the precursor of milk sugar (lactose) in the mammary gland, and it is taken up actively by the fetus. In addition, gluconeogenic mechanisms are used to clear the products of the metabolism of other tissues from the blood, eg, lactate, produced by muscle and erythrocytes, and glycerol, which is continuously produced by adipose tissue. Propionate, the principal glucogenic fatty acid produced in the digestion of carbohydrates by ruminants, is a major substrate for gluconeogenesis in these species.

GLUCONEOGENESIS INVOLVES GLYCOLYSIS, THE CITRIC ACID CYCLE, PLUS SOME SPECIAL REACTIONS (Figure 21–1)

Thermodynamic Barriers Prevent a Simple Reversal of Glycolysis

Krebs pointed out that energy barriers obstruct a simple reversal of glycolysis between pyruvate and phosphoenolpyruvate, between fructose 1,6-bisphosphate and fructose 6-phosphate, between glucose 6-phosphate and glucose, and between glucose 1-phosphate and glycogen. These reactions are all nonequilibrium, releasing much free energy as heat and therefore physiologically irreversible. They are circumvented by special reactions.

A. Pyruvate and Phosphoenolpyruvate: Present in mitochondria is an enzyme, **pyruvate carboxylase,** which in the presence of ATP, the B vitamin biotin, and CO_2 converts pyruvate to oxaloacetate. The function of the biotin is to bind CO_2 from bicarbonate onto the enzyme prior to the addition of the CO_2 to pyruvate (Figure 52–13). A second enzyme, **phosphoenolpyruvate carboxykinase,** catalyzes the conversion of oxaloacetate to phosphoenolpyruvate. High-energy phosphate in the form of GTP or ITP is required in this reaction, and CO_2 is liberated. Thus, with the help of these two enzymes catalyzing endergonic transformations and lactate dehydrogenase, lactate can be converted to phosphoenolpyruvate, overcoming the energy barrier between pyruvate and phosphoenolpyruvate.

In pigeon, chicken, and rabbit liver, phosphoenolpyruvate carboxykinase is a mitochondrial enzyme, and phosphoenolpyruvate is transported into the cytosol for conversion into fructose 1,6-bisphosphate by reversal of glycolysis. In the rat and the mouse, the enzyme is in the cytosol, which creates a problem because oxaloacetate does not diffuse through the mitochondrial inner membrane. This is overcome by conversion to malate, which can be transported into the cytosol followed by reconversion to oxaloacetate by extramitochondrial malate dehy-

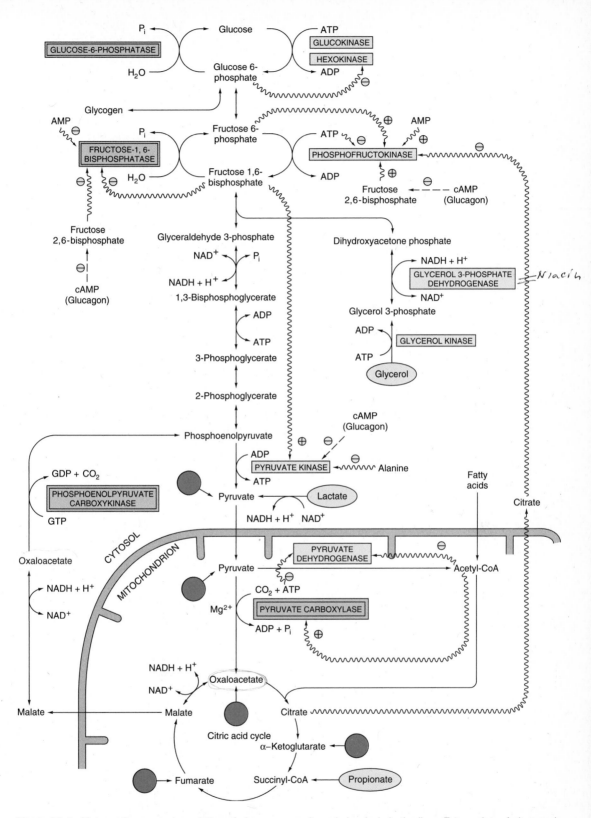

Figure 21–1. Major pathways and regulation of gluconeogenesis and glycolysis in the liver. Entry points of glucogenic amino acids after transamination are indicated by arrows extended from circles. (See also Figure 18–4.) The key gluconeogenic enzymes are enclosed in double-bordered boxes. The ATP required for gluconeogenesis is supplied by the oxidation of long-chain fatty acids. Propionate is of quantitative importance only in ruminants. Arrows with wavy shafts signify allosteric effects; dash-shafted arrows, covalent modification by reversible phosphorylation. High concentrations of alanine act as a "gluconeogenic signal" by inhibiting glycolysis at the pyruvate kinase step.

drogenase. In humans, the guinea pig, and the cow, the enzyme is equally distributed between mitochondria and cytosol.

B. Fructose 1,6-Bisphosphate and Fructose 6-Phosphate: The conversion of fructose 1,6-bisphosphate to fructose 6-phosphate, necessary to achieve a reversal of glycolysis, is catalyzed by a specific enzyme, **fructose-1,6-bisphosphatase.** This is a key enzyme from another point of view in that its presence determines whether or not a tissue is capable of synthesizing glycogen not only from pyruvate but also from triosephosphates. It is present in liver and kidney and has been demonstrated in striated muscle. It is held to be absent from heart muscle and smooth muscle.

C. Glucose 6-Phosphate and Glucose: The conversion of glucose 6-phosphate to glucose is catalyzed by another specific phosphatase, **glucose-6-phosphatase.** It is present in liver and kidney but absent from muscle and adipose tissue. Its presence allows a tissue to add glucose to the blood.

D. Glucose 1-Phosphate and Glucogen: The breakdown of glycogen to glucose 1-phosphate is carried out by phosphorylase. The synthesis of glycogen involves an entirely different pathway through the formation of uridine diphosphate glucose and the activity of **glycogen synthase** (Figure 20–1).

These key enzymes allow reversal of glycolysis to play a major role in gluconeogenesis. The relationships between gluconeogenesis and the glycolytic pathway are shown in Figure 21–1. After transamination or deamination, glucogenic amino acids form either pyruvate or members of the citric acid cycle. Therefore, the reactions described above can account for the conversion of both glucogenic amino acids and lactate to glucose or glycogen. Thus, lactate forms pyruvate and must enter the mitochondria before conversion to oxaloacetate and ultimate conversion to glucose.

Propionate is a major source of glucose in ruminants and enters the main gluconeogenic pathway via the citric acid cycle after conversion to succinyl-CoA. Propionate is first activated with ATP and CoA by an appropriate **acyl-CoA synthetase.** Propionyl-CoA, the product of this reaction, undergoes a CO_2 fixation reaction to form D-methylmalonyl-CoA, catalyzed by **propionyl-CoA carboxylase** (Figure 21–2). This reaction is analogous to the fixation of CO_2 in acetyl-CoA by acetyl-CoA carboxylase (Chapter 23) in that it forms a malonyl derivative and requires the vitamin **biotin** as a coenzyme. D-Methylmalonyl-CoA must be converted to its stereoisomer, L-methylmalonyl-CoA, by **methylmalonyl-CoA racemase** before its final isomerization to succinyl-CoA by the enzyme **methylmalonyl-CoA isomerase,** which requires vitamin B_{12} as a coenzyme. Vitamin B_{12} deficiency in humans and animals results in the excretion of large amounts of methylmalonate (**methylmalonic aciduria**).

Although the pathway to succinate is its main route of metabolism, propionate may also be used as the priming molecule for the synthesis—in adipose tissue and mammary gland—of fatty acids that have an odd number of carbon atoms in the molecule. C_{15} and C_{17} fatty acids are found particularly in the lipids of ruminants. As such they form an important source of these fatty acids in human diets and are ultimately broken down to propionate in the tissues (Chapter 24).

Glycerol is a product of the metabolism of adipose tissue, and only tissues that possess the activating enzyme, **glycerol kinase,** can utilize it. This enzyme, which requires ATP, is found in liver and kidney, among other tissues. Glycerol kinase catalyzes the

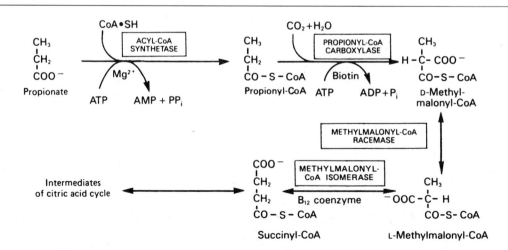

Figure 21–2. Metabolism of propionate.

conversion of glycerol to glycerol 3-phosphate. This pathway connects with the triosephosphate stages of the glycolysis pathway, because glycerol 3-phosphate may be oxidized to dihydroxyacetone phosphate by NAD^+ in the presence of **glycerol-3-phosphate dehydrogenase.** Liver and kidney are able to convert glycerol to blood glucose by making use of the above enzymes, some enzymes of glycolysis, and the specific enzymes of the gluconeogenic pathway, fructose-1,6-bisphosphatase and glucose-6-phosphatase (Figure 21–1).

SINCE GLYCOLYSIS AND GLUCONEOGENESIS SHARE THE SAME PATHWAY BUT IN OPPOSITE DIRECTIONS, THEY MUST BE REGULATED RECIPROCALLY

Changes in availability of substrates are either directly or indirectly responsible for most changes in metabolism. Fluctuations in their blood concentrations due to changes in dietary availability may alter the rate of secretion of hormones that influence, in turn, the pattern of metabolism in metabolic pathways—often by affecting the activity of key enzymes that attempt to compensate for the original change in substrate availability. Three types of mechanism can be identified as responsible for regulating the activity of enzymes concerned in carbohydrate metabolism and may be identified in Table 21–1: (1) changes in the rate of enzyme synthesis, (2) covalent modification by reversible phosphorylation, and (3) allosteric effects.

Induction and Repression of Key Enzyme Synthesis Requires Several Hours

Some of the better-documented changes in enzyme activity that are considered to occur under various metabolic conditions are listed in Table 21–1. The information in this table applies mainly to the liver. The enzymes involved catalyze nonequilibrium reactions that may be regarded physiologically as "one-way" rather than balanced reactions. Often the effects are reinforced because the activity of the enzymes catalyzing the changes in the opposite direction varies reciprocally (Figure 21–1). It is of importance that the key enzymes involved in a metabolic pathway are all activated or depressed in a coordinated manner. Table 21–1 shows that this is clearly the case. The enzymes involved in the utilization of glucose (ie, those of glycolysis and lipogenesis) all become more active when there is a superfluity of glucose, and under these conditions the enzymes responsible for producing glucose by the pathway of gluconeogenesis are all low in activity. The secretion of insulin, which is responsive to increases in the blood glucose concentration, enhances the synthesis of the key enzymes in glycolysis. Likewise, it antagonizes the effect of the glucocorticoids and glucagon-stimulated cAMP, which induce synthesis of the key enzymes responsible for gluconeogenesis. All these effects can be prevented by agents that block the synthesis of protein, such as puromycin and ethionine. The regulation of the mRNA species of these enzymes and modulation of the expression of their genes have been demonstrated.

Both dehydrogenases of the pentose phosphate pathway can be classified as adaptive enzymes, since they increase in activity in the well-fed animal and when insulin is given to a diabetic animal. Activity is low in diabetes or fasting. "Malic enzyme" and ATP-citrate lyase behave similarly, indicating that these two enzymes are involved in lipogenesis rather than gluconeogenesis (Chapter 23).

Covalent Modification by Reversible Phosphorylation Is Rapid

Glucagon, and to a lesser extent **epinephrine,** hormones that are responsive to decreases in blood glucose, inhibit glycolysis and stimulate gluconeogenesis in the liver by increasing the concentration of cAMP. This in turn activates cAMP-dependent protein kinase, leading to the phosphorylation and inactivation of **pyruvate kinase.** They also affect the concentration of fructose 2,6-bisphosphate and therefore glycolysis and gluconeogenesis, as explained below.

Allosteric Modification Is Also Rapid

Several examples are available from carbohydrate metabolism to illustrate allosteric control of the activity of an enzyme. In gluconeogenesis, the synthesis of oxaloacetate from bicarbonate and pyruvate, which is catalyzed by the enzyme pyruvate carboxylase, requires the presence of acetyl-CoA as an **allosteric activator.** The addition of acetyl-CoA results in a change in the tertiary structure of the protein, lowering the K_m value for bicarbonate. This effect has important implications for the self-regulation of intermediary metabolism, for, as acetyl-CoA is formed from pyruvate, it automatically ensures the provision of oxaloacetate and, therefore, its further oxidation in the citric acid cycle, by activating pyruvate carboxylase. The activation of pyruvate carboxylase and the reciprocal inhibition of pyruvate dehydrogenase by acetyl-CoA derived from the oxidation of fatty acids help to explain the action of fatty acid oxidation in sparing the oxidation of pyruvate and in stimulating gluconeogenesis in the liver. The reciprocal relationship between the activity of pyruvate dehydrogenase and pyruvate carboxylase in both liver and kidney alters the metabolic fate of pyruvate as the tissue changes from carbohydrate oxidation, via glycolysis, to gluconeogenesis during

Table 21–1. Regulatory and adaptive enzymes of the rat (mainly liver).

	Activity in		Inducer	Repressor	Activator	Inhibitor
	Carbo-hydrate Feeding	Starva-tion and Diabetes				
Enzymes of glycogenesis, glycolysis and pyruvate oxidation						
Glycogen synthase system	↑	↓	Insulin		Insulin	Glucagon (cAMP), phosphorylase, gly-cogen
Hexokinase						Glucose 6-phosphate[1]
Glucokinase	↑	↓	Insulin	Glucagon (cAMP)		
Phosphofructokinase-1	↑	↓	Insulin		AMP, fructose 6-phosphate, P_i, fruc-tose 2,6-bisphos-phate[1]	Citrate (fatty acids, ketone bodies),[1] ATP,[1] glucagon (cAMP)
Pyruvate kinase	↑	↓	Insulin, fructose	Glucagon (cAMP)	Fructose 1,6-bisphosphate[1], in-sulin	ATP, alanine, glucagon (cAMP), epinephrine
Pyruvate dehydro-genase	↑	↓			CoA, NAD^+, insu-lin,[2] ADP, pyruvate	Acetyl-CoA, NADH, ATP, (fatty acids, ketone bodies)
Enzymes of gluconeogenesis						
Pyruvate carboxylase	↓	↑	Glucocorticoids, glucagon, epi-nephrine (cAMP)	Insulin	Acetyl-CoA[1]	ADP[1]
Phosphoenolpyruvate carboxykinase	↓	↑	Glucocorticoids, glucagon, epi-nephrine (cAMP)	Insulin	Glucagon?	
Fructose-1,6-bisphosphatase	↓	↑	Glucocorticoids, glucagon, epi-nephrine (cAMP)	Insulin	Glucagon (cAMP)	Fructose 1,6-bisphosphate, AMP, fructose 2,6-bisphosphate[1]
Glucose-6-phosphatase	↓	↑	Glucocorticoids, glucagon, epi-nephrine (cAMP)	Insulin		
Enzymes of the pentose phosphate pathway and lipogenesis						
Glucose-6-phosphate dehydrogenase	↑	↓	Insulin			
6-Phosphogluconate dehydrogenase	↑	↓	Insulin			
"Malic enzyme"	↑	↓	Insulin			
ATP-citrate lyase	↑	↓	Insulin			ADP
Acetyl-CoA carboxylase	↑	↓	Insulin?		Citrate,[1] insulin	Long-chain acyl-CoA, cAMP, glucagon
Fatty acid synthase	↑	↓	Insulin?			

[1]Allosteric.
[2]In adipose tissue but not in liver.

transition from a fed to a starved state (Figure 21–1). A major role of fatty acid oxidation in promoting gluconeogenesis is to supply ATP required in the pyruvate carboxylase and phosphoenolpyruvate car-boxykinase reactions as well as reversing the phos-phoglycerate kinase reaction in glycolysis.

Another enzyme that is subject to feedback control is **phosphofructokinase (phosphofructokinase-1).** It occupies a key position in regulating glycolysis.

Phosphofructokinase-1 is inhibited by citrate and by ATP and is activated by AMP. AMP acts as an indi-cator of the energy status of the cell. The presence of **adenylyl kinase** in liver and many other tissues al-lows rapid equilibration of the reaction:

$$ATP + AMP \leftrightarrow 2ADP$$

Thus, when ATP is used in energy-requiring processes resulting in formation of ADP, [AMP] rises. As [ATP] may be 50 times [AMP] at equilibrium, a small fractional decrease in [ATP] will cause a severalfold increase in [AMP]. Thus, a large change in [AMP] acts as a metabolic amplifier of a small change in [ATP]. This mechanism allows the activity of phosphofructokinase-1 to be highly sensitive to even small changes in energy status of the cell and to control the quantity of carbohydrate undergoing glycolysis prior to its entry into the citric acid cycle. The increase in [AMP] can also explain why glycolysis is increased during hypoxia when [ATP] decreases. Simultaneously, AMP activates phosphorylase, increasing glycogenolysis. The inhibition of phosphofructokinase-1 by citrate and ATP is another explanation of the sparing action of fatty acid oxidation on glucose oxidation and also of the **Pasteur effect,** whereby aerobic oxidation (via the citric acid cycle) inhibits the anaerobic degradation of glucose. A consequence of the inhibition of phosphofructokinase-1 is an accumulation of glucose 6-phosphate that, in turn, inhibits further uptake of glucose in extrahepatic tissues by allosteric inhibition of hexokinase.

Fructose 2,6-Bisphosphate Plays a Unique Role in the Regulation of Glycolysis and Gluconeogenesis in Liver

The most potent positive allosteric effector of phosphofructokinase-1 and inhibitor of fructose-1,6-bisphosphatase in liver is **fructose 2,6-bisphosphate.** It relieves inhibition of phosphofructokinase-1 by ATP and increases affinity for fructose 6-phosphate. It inhibits fructose-1,6-bisphosphatase by increasing the K_m for fructose 1,6-bisphosphate. Its concentration is under both substrate (allosteric) and hormonal control (covalent modification) (Figure 21–3).

Fructose 2,6-bisphosphate is formed by phosphorylation of fructose 6-phosphate by **phosphofructokinase-2.** The same enzyme protein is also responsible for its breakdown, since it contains **fructose-2,6-bisphosphatase** activity. This **bifunctional enzyme** is under the allosteric control of fructose 6-phosphate, which when raised in concentration owing to an abundance of glucose, ie, in the well-fed state, stimulates the kinase and inhibits the phosphatase. On the other hand, when glucose is short, glucagon stimulates the production of cAMP, activating cAMP-dependent protein kinase, which in turn inactivates phosphofructokinase-2 and activates fructose-2,6-bisphosphatase by phosphorylation.

Thus, under a superfluity of glucose, fructose 2,6-bisphosphate increases in concentration, stimulating glycolysis by activating phosphofructokinase-1 and inhibiting fructose-1,6-bisphosphatase. Under conditions of glucose shortage, gluconeogenesis is stimulated by a decrease in the concentration of fructose 2,6-bisphosphate, which deactivates phosphofructokinase-1 and deinhibits fructose-1,6-bisphosphatase.

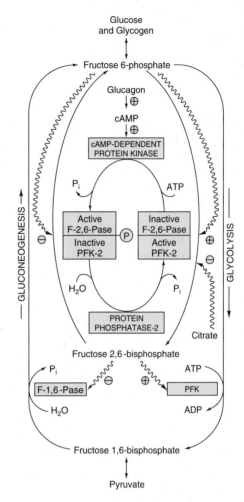

Figure 21–3. Control of glycolysis and gluconeogenesis in the liver by fructose 2,6-bisphosphate and the bifunctional enzyme PFK-2/F-2,6-Pase (6-phosphofructo-2-kinase/fructose 2,6-bisphosphatase). (PFK, phosphofructokinase [6-phosphofructo-1-kinase]; F-1,6-Pase, fructose-1, 6-bisphosphatase. Arrows with wavy shafts indicate allosteric effects.)

This mechanism also ensures that glucagon stimulation of glycogenolysis in liver results in glucose release rather than glycolysis.

Recent investigations indicate that glucose 1,6-bisphosphate plays a similar role in some extrahepatic tissues.

Substrate (Futile) Cycles Allow Fine Tuning

It will be apparent that many of the control points in glycolysis and glycogen metabolism involve a cycle of phosphorylation and dephosphorylation catalyzed by the following enzymes: glucokinase and glucose-6-phosphatase; phosphofructokinase-1 and fructose-1,6-bisphosphatase; pyruvate kinase, pyruvate carboxylase, and phosphoenolpyruvate carboxykinase; and glycogen synthase and phosphorylase. If these were allowed to cycle unchecked, they would amount to futile cycles whose net result was hydrolysis of ATP. That this does not occur extensively is due to the various control mechanisms, which ensure that one limb of the cycle is inhibited as the other is stimulated, according to the needs of the tissue and of the body. However, there may be a physiologic advantage in allowing some cycling. For example, in the phosphofructokinase and fructose-1,6-bisphosphatase cycle, an amplification of the effect of an allosteric modifier, eg, fructose 2,6-bisphosphate, would occur, causing a larger change in net flux of metabolites in either direction than would occur in the absence of substrate cycling. This "fine tuning" of metabolic control occurs only at the expense of some loss of ATP.

THE CONCENTRATION OF BLOOD GLUCOSE IS REGULATED WITHIN NARROW LIMITS

In the postabsorptive state, the concentration of blood glucose in individual humans and many mammals is set within the range 4.5–5.5 mmol/L. After the ingestion of a carbohydrate meal, it may rise to 6.5–7.2 mmol/L. During fasting, the levels fall to around 3.3–3.9 mmol/L. The blood glucose level in birds is considerably higher (14.0 mmol/L) and in ruminants considerably lower (approximately 2.2 mmol/L in sheep and 3.3 mmol/L in cattle). These lower normal levels appear to be associated with the fact that ruminants ferment virtually all dietary carbohydrate to lower (volatile) fatty acids, and these largely replace glucose as the main metabolic fuel of the tissues in the fed condition.

A sudden decrease in blood glucose will cause convulsions, as in insulin overdose, owing to the immediate dependence of the brain on a supply of glucose. However, much lower concentrations can be tolerated, provided progressive adaptation is allowed; eg, rats adapted to high-fat diets behave normally with a blood glucose concentration as low as 1.1 mmol/L.

BLOOD GLUCOSE IS DERIVED FROM THE DIET, GLUCONEOGENESIS, AND GLYCOGENOLYSIS

Most digestible carbohydrates in the diet ultimately form glucose. The dietary carbohydrates that are actively digested contain glucose, galactose, and fructose residues that are released in the intestine. These are transported to the liver via the **hepatic portal vein.** Galactose and fructose are readily converted to glucose in the liver (Chapter 22).

Glucose is formed from glucogenic compounds that undergo gluconeogenesis (Figures 18–4 and 21–1). These compounds fall into two categories: (1) those which involve a direct net conversion to glucose without significant recycling, such as some **amino acids** and **propionate;** and (2) those which are the products of the partial metabolism of glucose in certain tissues and are conveyed to the liver and kidney to be resynthesized to glucose. Thus, **lactate,** formed by the oxidation of glucose in skeletal muscle and by erythrocytes, is transported to the liver and kidney where it re-forms glucose, which again becomes available via the circulation for oxidation in the tissues. This process is known as the **Cori cycle,** or **lactic acid cycle** (Figure 21–4). Glycerol 3-phosphate for the synthesis of triacylglycerols in adipose tissue is derived from blood glucose. Acylglycerols of adipose tissue are continuously undergoing hydrolysis to form free **glycerol,** which cannot be utilized by adipose tissue and therefore diffuses out into the blood. It is converted back to glucose by gluconeogenic mechanisms in the liver and kidney (Figure 21–1).

Of the amino acids transported from muscle to the liver during starvation, alanine predominates. This has led to the postulation of a **glucose-alanine cycle** (Figure 21–4) that has the effect of cycling glucose from liver to muscle with formation of pyruvate, followed by transamination to alanine, then transport of alanine to liver, followed by gluconeogenesis back to glucose. A net transfer of amino nitrogen from muscle to liver and of free energy from liver to muscle is effected. The energy required for the hepatic synthesis of glucose from pyruvate is derived from the oxidation of fatty acids.

Glucose is also formed from liver glycogen by glycogenolysis (Chapter 20).

Metabolic and Hormonal Mechanisms Regulate the Concentration of the Blood Glucose

The maintenance of stable levels of glucose in the blood is one of the most finely regulated of all homeostatic mechanisms and one in which the liver, the extrahepatic tissues, and several hormones play a part. Liver cells appear to be freely permeable to glucose (via the GLUT 2 transporter), whereas cells of extrahepatic tissues (apart from pancreatic islets) are relatively impermeable. As a result, the passage through the cell membrane is the rate-limiting step in the uptake of glucose in extrahepatic tissues, and glucose is rapidly phosphorylated by hexokinase on entry into the cells. On the other hand, it is probable that the activity of certain enzymes and the concentration of key intermediates exert a much more direct effect on the uptake or output of glucose from liver.

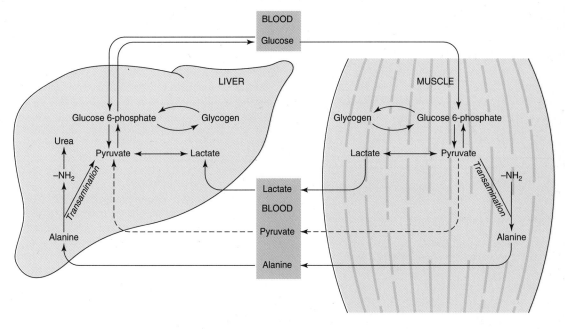

Figure 21–4. The lactic acid (Cori) cycle and glucose-alanine cycle.

Nevertheless, the concentration of glucose in the blood is an important factor controlling the rate of uptake of glucose in both liver and extrahepatic tissues.

The role of various glucose transporter proteins found in cell membranes, each having 12 transmembrane domains, is shown in Table 21–2.

Glucokinase Is Important in Regulating Blood Glucose After a Meal

It is to be noted that hexokinase is inhibited by glucose 6-phosphate, so that some feedback control may be exerted on glucose uptake in extrahepatic tissues dependent on hexokinase for glucose phosphorylation. The liver is not subject to this constraint be-

cause glucokinase is not affected by glucose 6-phosphate. Glucokinase, which has a higher K_m (lower affinity) for glucose than does hexokinase, increases in activity over the physiologic range of glucose concentrations (Figure 21–5) and seems to be specifically concerned with glucose uptake into the liver at the higher concentrations found in the hepatic portal vein after a carbohydrate meal. Its absence from the liver of ruminants, which have little glucose entering the portal circulation from the intestines, is compatible with this function.

At normal systemic-blood glucose concentrations (4.5–5.5 mmol/L), the liver appears to be a net producer of glucose. However, as the glucose level rises, the output of glucose ceases, so that at high levels there is a net uptake. In the rat, it has been estimated

Table 21–2. Glucose transporters.

	Tissue Location	Functions
Facilitative bidirectional transporters		
GLUT 1	Brain, kidney, colon, placenta, erythrocyte	Uptake of glucose
GLUT 2	Liver, pancreatic B cell, small intestine, kidney	Rapid uptake and release of glucose
GLUT 3	Brain, kidney, placenta	Uptake of glucose
GLUT 4	Heart and skeletal muscle, adipose tissue	Insulin-stimulated uptake of glucose
GLUT 5	Small intestine	Absorption of glucose
Sodium-dependent unidirectional transporter		
SGLT 1	Small intestine and kidney	Active uptake of glucose from lumen of intestine and reabsorption of glucose in proximal tubule of kidney against a concentration gradient

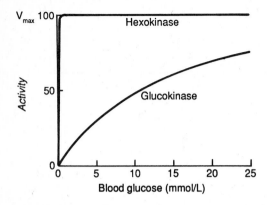

Figure 21–5. Variation in glucose phosphorylating activity of hexokinase and glucokinase with increase of blood glucose concentration. The K_m for glucose of hexokinase is 0.05 mmol/L and of glucokinase is 10 mmol/L.

that the rate of uptake of glucose and the rate of output are equal at a hepatic portal vein blood glucose concentration of 8.3 mmol/L.

Insulin Plays a Central Role in Regulating Blood Glucose

In addition to the direct effects of hyperglycemia in enhancing the uptake of glucose into both the liver and peripheral tissues, the hormone insulin plays a central role in regulating the blood glucose concentration. It is produced by the B cells of the islets of Langerhans in the pancreas as a direct response to the degree of hyperglycemia. The islet cell is freely permeable to glucose via the GLUT 2 transporter, and the glucose is phosphorylated by the high-K_m glucokinase. Therefore, the blood glucose concentration determines the flux through glycolysis, the citric acid cycle, and the generation of ATP. Increase in ATP concentration inhibits the ATP-sensitive K^+ channels causing depolarization of the B-cell membrane, which increases Ca_{2+} influx via voltage-sensitive Ca^{2+} channels, stimulating exocytosis of insulin. It is noteworthy that sulfonylurea drugs used to stimulate insulin secretion in type II diabetes mellitus (non-insulin-dependent diabetes mellitus; NIDDM), do so by inhibition of the ATP-sensitive K^+ channels. Thus, the concentration of insulin in the blood parallels that of the blood glucose. Its administration results in prompt **hypoglycemia.** Other substances causing release of insulin from the pancreas include amino acids, free fatty acids, ketone bodies, glucagon, secretin, and the drug tolbutamide. Epinephrine and norepinephrine block the release of insulin. Insulin has an immediate effect of increasing glucose uptake in tissues such as adipose tissue and muscle. This action is due to an enhancement of glucose transport through the cell membrane by recruitment of glucose transporters (GLUT 4) from the interior of the cell to the plasma membrane. In contrast, there is no direct effect of insulin on glucose penetration of hepatic cells; this finding agrees with the fact that glucose metabolism by liver cells is not rate-limited by their permeability to glucose. However, insulin does indirectly enhance long-term uptake of glucose by the liver as a result of its actions on the synthesis of enzymes controlling glycolysis, glycogenesis, and gluconeogenesis. Insulin has an immediate effect in activating glycogen synthase (Chapter 20).

Glucagon Opposes the Actions of Insulin

Glucagon is the hormone produced by the A cells of the islets of Langerhans of the pancreas. Its secretion is stimulated by hypoglycemia. When it reaches the liver (via the portal vein), it causes glycogenolysis by activating phosphorylase. Most of the endogenous glucagon (and insulin) is cleared from the circulation by the liver. Unlike epinephrine, glucagon does not have an effect on muscle phosphorylase. Glucagon also enhances gluconeogenesis from amino acids and lactate. Both hepatic glycogenolysis and gluconeogenesis contribute to the **hyperglycemic effect** of glucagon, whose actions oppose those of insulin.

Other Hormones Affect the Blood Glucose

The **anterior pituitary gland** secretes hormones that tend to elevate the blood glucose and therefore antagonize the action of insulin. These are growth hormone, ACTH (corticotropin), and possibly other "diabetogenic" principles. Growth hormone secretion is stimulated by hypoglycemia. Growth hormone decreases glucose uptake in certain tissues, eg, muscle. Some of this effect may not be direct, since it mobilizes free fatty acids from adipose tissue which themselves inhibit glucose utilization. Chronic administration of growth hormone leads to diabetes. By producing hyperglycemia, it stimulates secretion of insulin, eventually causing B cell exhaustion.

The **glucocorticoids** (11-oxysteroids) are secreted by the adrenal cortex and are important in carbohydrate metabolism. Administration of these steroids causes increased gluconeogenesis. This is a result of increased protein catabolism in the tissues, increased hepatic uptake of amino acids, and increased activity of aminotransferases and other enzymes concerned with gluconeogenesis in the liver. In addition, glucocorticoids inhibit the utilization of glucose in extrahepatic tissues. In all these actions, glucocorticoids act in a manner antagonistic to insulin.

Epinephrine is secreted by the adrenal medulla as a result of stressful stimuli (fear, excitement, hemorrhage, hypoxia, hypoglycemia, etc) and leads to glycogenolysis in liver and muscle owing to stimulation of phosphorylase. In muscle, as a result of the absence of glucose-6-phosphatase, glycogenolysis ensues with the formation of lactate, whereas in liver, glucose is the main product leading to increase in blood glucose.

Thyroid hormone should also be considered as affecting the blood glucose. There is experimental evidence that thyroxine has a diabetogenic action and that thyroidectomy inhibits the development of diabetes. It has also been noted that there is a complete absence of glycogen from the livers of thyrotoxic animals. In humans, the fasting blood glucose is elevated in hyperthyroid patients and decreased in hypothyroid patients. However, hyperthyroid patients apparently utilize glucose at a normal or increased rate, whereas hypothyroid patients have a decreased ability to utilize glucose. In addition, hypothyroid patients are much less sensitive to insulin than are normal or hyperthyroid individuals.

FURTHER CLINICAL ASPECTS

Glycosuria Occurs When the Renal Threshold for Glucose Is Exceeded

When the blood glucose rises to relatively high levels, the kidney also exerts a regulatory effect. Glucose is continuously filtered by the glomeruli but is ordinarily returned completely to the blood by the reabsorptive system of the renal tubules. The reabsorption of glucose against its concentration gradient is linked to the provision of ATP in the tubular cells. The capacity of the tubular system to reabsorb glucose is limited to a rate of about 350 mg/min. When the blood levels of glucose are elevated, the glomerular filtrate may contain more glucose than can be reabsorbed; the excess passes into the urine to produce **glycosuria.** In normal individuals, glycosuria occurs when the venous blood glucose concentration exceeds 9.5–10.0 mmol/L. This is termed the **renal threshold** for glucose.

Glycosuria may be produced in experimental animals with **phlorhizin,** which inhibits the glucose reabsorptive system in the tubule. This is known as renal glycosuria. Glycosuria of renal origin may result from inherited defects in the kidney, or it may be acquired as a result of disease processes. The presence of glycosuria is frequently an indication of diabetes mellitus.

Fructose-1,6-Bisphosphatase Deficiency Causes Lactic Acidosis and Hypoglycemia

Blockage of gluconeogenesis by deficiency of this enzyme prevents lactate and other glucogenic substrates from being converted to glucose in the liver. The condition may be controlled by feeding high-carbohydrate diets low in fructose and sucrose and with avoidance of fasting.

Impairment of Fatty Acid Oxidation Is a Cause of Hypoglycemia

Several conditions in which fatty acid oxidation is defective are characterized by hypoglycemia. This is due to the dependence of gluconeogenesis on active fatty acid oxidation (see above). These conditions are discussed in Chapter 24.

Hypoglycemia May Occur During Pregnancy and in the Neonate

During pregnancy, fetal glucose consumption increases and there is a risk of maternal and possibly fetal hypoglycemia, particularly if there are long intervals between meals or at night. Furthermore, premature and low-birth-weight babies are more susceptible to hypoglycemia, since they have little adipose tissue to generate alternative fuels such as free fatty acids or ketone bodies during the transition from fetal dependency to the free-living state. The enzymes of gluconeogenesis may not be completely functional at this time, and the process is dependent on a supply of free fatty acids for energy. Glycerol, which would normally be released from adipose tissue, is less available for gluconeogenesis.

The Body's Ability to Utilize Glucose May Be Ascertained by Measuring Its Glucose Tolerance

Glucose tolerance is indicated by the nature of the blood glucose curve following the administration of a test amount of glucose (Figure 21–6). **Diabetes mellitus** (type I, or insulin-dependent diabetes mellitus, IDDM) is characterized by decreased glucose tolerance due to decreased secretion of insulin in response to the glucose challenge. This is manifested by elevated blood glucose levels (hyperglycemia) and glycosuria and may be accompanied by changes in fat metabolism. Tolerance to glucose declines not only in type I diabetes but also in conditions where the liver is damaged; in some infections; in type II diabetes mellitus (non-insulin-dependent diabetes mellitus, NIDDM), which is often associated with obesity and raised levels of plasma free fatty acids; under the influence of some drugs; and sometimes in atherosclerosis. It can also be expected to occur in the presence of hyperactivity of the pituitary or adrenal cortex because of the antagonism of the hormones of these endocrine glands to the action of insulin.

Insulin increases glucose tolerance. Injection of insulin lowers the content of the glucose in the blood and increases its utilization and its storage in the liver and muscle as glycogen. An excess of insulin may cause severe **hypoglycemia,** resulting in convulsions and even in death unless glucose is administered promptly. Increased tolerance to glucose is observed in pituitary or adrenocortical insufficiency, attribut-

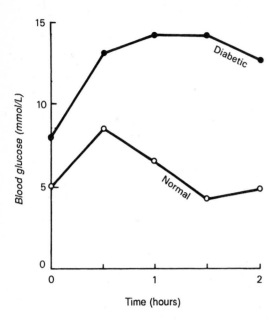

Figure 21–6. Glucose tolerance test. Blood glucose curves of a normal and a diabetic individual after oral administration of 50 g of glucose. Note the initial raised concentration in the diabetic. A criterion of normality is the return of the curve to the initial value within 2 hours.

able to a decrease in the antagonism to insulin by the hormones normally secreted by these glands.

SUMMARY

1. Gluconeogenesis is the mechanism for converting noncarbohydrates to glucose or glycogen. It provides the body with glucose when carbohydrate is not available from the diet. Important substrates are glucogenic amino acids, lactate, glycerol, and propionate.

2. The pathway of gluconeogenesis, found in the liver and kidney, utilizes those reactions in glycolysis which are reversible plus 4 additional reactions that circumvent the irreversible nonequilibrium reactions. The enzymes catalyzing the additional reactions are pyruvate carboxylase, phosphoenolpyruvate carboxykinase, fructose-1,6-bisphosphatase, and glucose-6-phosphatase.

3. Lactate forms pyruvate, which enters the mitochondrion for carboxylation to oxaloacetate, before conversion to phosphoenolpyruvate followed by biosynthesis of glucose in the cytosol.

4. Since glycolysis and gluconeogenesis share the same pathway but operate in opposite directions, their activities must be regulated reciprocally. This is achieved by 3 main mechanisms affecting the activity of key enzymes: (1) induction or repression of enzyme synthesis, (2) covalent modification by reversible phosphorylation, and (3) allosteric effects.

5. The liver cell, which is freely permeable to glucose, is a principal means of regulating the blood glucose concentration because it contains the high-K_m glucokinase that is specifically adapted to disposing of glucose after a meal. Insulin is secreted as a direct response to hyperglycemia; it assists the liver to store glucose as glycogen and facilitates uptake of glucose into extrahepatic tissues. Glucagon is secreted as a response to hypoglycemia and activates both glycogenolysis and gluconeogenesis in the liver, causing release of glucose into the blood.

6. Defective enzymes of gluconeogenesis lead to hypoglycemia and lactic acidosis. Blocks in fatty acid oxidation are an additional cause of impaired gluconeogenesis and hypoglycemia.

7. A deficiency in insulin secretion results in type I diabetes mellitus.

REFERENCES

Boyle PJ, Shar SD, Cryer PE: Insulin, glucagon, and catecholamines in prevention of hypoglycemia during fasting. Am J Physiol 1989;256:E651.

Buchalter SE, Crain MR, Kreisberg R: Regulation of lactate metabolism in vivo. Diabetes Metab Rev 1989;5:379.

Burant CF et al: Mammalian glucose transporters: Structure and molecular regulation. Recent Prog Horm Res 1991;47:349.

Krebs HA: Gluconeogenesis. Proc R Soc London (Biol) 1964;159:545.

Lenzen S: Hexose recognition mechanisms in pancreatic B-cells. Biochem Soc Trans 1990;18:105.

Newsholme EA, Chaliss RAJ, Crabtree B: Substrate cycles: Their role in improving sensitivity in metabolic control. Trends Biochem Sci 1984;9:277.

Newsholme EA, Start C: *Regulation in Metabolism.* Wiley, 1973.

Pilkis SJ, El-Maghrabi MR, Claus TH: Hormonal regulation of hepatic gluconeogenesis and glycolysis. Annu Rev Biochem 1988;57:755.

Watford M: What is the metabolic fate of dietary glucose? Trends Biochem Sci 1988;13:329.

Yki-Jarvinen H: Action of insulin on glucose metabolism in vivo. Baillieres Clin Endocrinol Metab 1993;7:903.

The Pentose Phosphate Pathway & Other Pathways of Hexose Metabolism

22

Peter A. Mayes, PhD, DSc

INTRODUCTION

The pentose phosphate pathway is an alternative route for the metabolism of glucose. It does not generate ATP but has two major functions: (1) The generation of **NADPH** for reductive syntheses such as fatty acid and steroid biosynthesis, and (2) the provision of **ribose** residues for nucleotide and nucleic acid biosynthesis.

Glucose, fructose, and galactose are quantitatively the most important hexoses absorbed from the gastrointestinal tract. They are derived from dietary starch, sucrose, and lactose, respectively. Specialized pathways have been developed, particularly in the liver, for the conversion of fructose and galactose to glucose.

BIOMEDICAL IMPORTANCE

The major metabolic routes for the utilization of glucose are glycolysis and the pentose phosphate pathway. Deficiencies of certain enzymes of the pentose phosphate pathway are major causes of hemolysis of red blood cells, resulting in one type of **hemolytic anemia.** The principal enzyme involved is glucose-6-phosphate dehydrogenase. More than 100 million people in the world may have genetically determined low levels of this enzyme.

Of minor quantitative importance but of major significance for the excretion of metabolites and foreign chemicals (xenobiotics) as **glucuronides** is the elaboration of glucuronic acid from glucose via the **uronic acid pathway.** A deficiency in the pathway leads to the condition of **essential pentosuria.** The total absence of one particular enzyme of the pathway in all primates accounts for the fact that **ascorbic acid** (vitamin C) is required in the diet of humans but not in that of most other mammals. Deficiencies in the enzymes of fructose and galactose metabolism lead to metabolic diseases such as **essential fructosuria** and the **galactosemias.** Fructose has been used for parenteral nutrition, but at high concentration it can cause depletion of adenine nucleotides in liver and hepatic necrosis.

THE PENTOSE PHOSPHATE PATHWAY GENERATES NADPH AND RIBOSE PHOSPHATE
(Figure 22–1)

The pentose phosphate pathway (hexose monophosphate shunt) is a more complex pathway than glycolysis. It is a multicyclic process in which three molecules of glucose 6-phosphate give rise to three molecules of CO_2 and three five-carbon residues. The latter are rearranged to regenerate two molecules of glucose 6-phosphate and one molecule of the glycolytic intermediate, glyceraldehyde 3-phosphate. Since two molecules of glyceraldehyde 3-phosphate can regenerate glucose 6-phosphate, the pathway can account for the complete oxidation of glucose.

3 Glucose 6-phosphate + 6NADP$^+$ →

3CO$_2$ + 2 Glucose 6-phosphate +

Glyceraldehyde 3-phosphate + 6NADPH + 6H$^+$

REACTIONS OF THE PENTOSE PHOSPHATE PATHWAY OCCUR IN THE CYTOSOL

The enzymes of the pentose phosphate pathway, as in glycolysis, are found in the cytosol. As for glycolysis, oxidation is achieved by dehydrogenation; but in the case of the pentose phosphate pathway, **NADP$^+$** and not NAD$^+$ is used as a hydrogen acceptor.

The sequence of reactions of the pathway may be divided into two phases: an **oxidative nonreversible phase** and a **nonoxidative reversible phase.** In the first, glucose 6-phosphate undergoes dehydrogenation and decarboxylation to give a pentose, ribulose 5-phosphate. In the second phase, ribulose 5-phos-

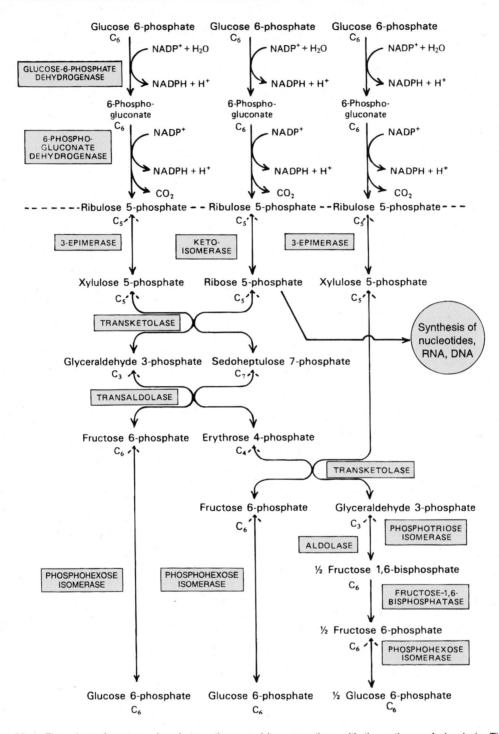

Figure 22–1. Flow chart of pentose phosphate pathway and its connections with the pathway of glycolysis. The full pathway, as indicated, consists of three interconnected cycles in which glucose 6-phosphate is both substrate and end product. The reactions above the broken line are nonreversible, whereas all reactions under that line are freely reversible.

phate is converted back to glucose 6-phosphate by a series of reactions involving mainly two enzymes: **transketolase** and **transaldolase** (Figure 22–1).

The Oxidative Phase
Generates NADPH
(Figures 22–1 and 22–2)

Dehydrogenation of glucose 6-phosphate to 6-phosphogluconate occurs via the formation of 6-phosphogluconolactone catalyzed by **glucose-6-phosphate dehydrogenase,** an NADP-dependent enzyme. The hydrolysis of 6-phosphogluconolactone is accomplished by the enzyme **gluconolactone hydrolase.** A second oxidative step is catalyzed by **6-phosphogluconate dehydrogenase,** which also requires NADP⁺ as hydrogen acceptor. Decarboxylation follows with the formation of the ketopentose, ribulose 5-phosphate. The reaction probably takes place in two steps through the intermediate 3-keto-6-phosphogluconate.

The Nonoxidative Phase
Generates Ribose Precursors

Ribulose 5-phosphate now serves as substrate for two different enzymes. **Ribulose 5-phosphate 3-epimerase** alters the configuration about carbon 3, forming the epimer xylulose 5-phosphate, another ketopentose. **Ribose 5-phosphate ketoisomerase** converts ribulose 5-phosphate to the corresponding aldopentose, ribose 5-phosphate, which is the precursor for the ribose residues required in nucleotide and nucleic acid synthesis.

Transketolase transfers the two-carbon unit comprising carbons 1 and 2 of a ketose to the aldehyde carbon of an aldose sugar. It therefore effects the conversion of a ketose sugar into an aldose with two carbons less and simultaneously converts an aldose sugar into a ketose with two carbons more. The reaction requires the B vitamin **thiamin** as the coenzyme thiamin diphosphate in addition to Mg^{2+} ions. The two-carbon moiety transferred is probably glycolaldehyde bound to thiamin diphosphate. Thus, transketolase catalyzes the transfer of the two-carbon unit from xylulose 5-phosphate to ribose 5-phosphate, producing the seven-carbon ketose sedoheptulose 7-phosphate and the aldose glyceraldehyde 3-phosphate. These two products then enter another reaction known as transaldolation. **Transaldolase** allows the transfer of a three-carbon dihydroxyacetone moiety (carbons 1–3) from the ketose sedoheptulose 7-phosphate to the aldose glyceraldehyde 3-phosphate to form the ketose fructose 6-phosphate and the four-carbon aldose erythrose 4-phosphate.

A further reaction takes place, again involving **transketolase,** in which xylulose 5-phosphate serves as a donor of glycolaldehyde. In this case the erythrose 4-phosphate formed above acts as acceptor, and the products of the reaction are fructose 6-phosphate and glyceraldehyde 3-phosphate.

In order to oxidize glucose completely to CO_2 via the pentose phosphate pathway, it is necessary that enzymes be present in the tissue to convert glyceraldehyde 3-phosphate to glucose 6-phosphate. This involves enzymes of the glycolysis pathway working in a reverse direction and, in addition, the gluconeogenic enzyme **fructose-1,6-bisphosphatase.** If this enzyme is absent, glyceraldehyde 3-phosphate follows the normal pathway of glycolysis to pyruvate.

The Two Major Pathways
for the Catabolism of Glucose
Have Little in Common

Although some metabolites are common to both, eg, glucose 6-phosphate, the pentose phosphate pathway is markedly different from glycolysis. Oxidation occurs in the first reactions utilizing NADP rather than NAD, and CO_2, which is not produced at all in the glycolysis pathway, is a characteristic product. ATP is not generated in the pentose phosphate pathway, whereas it is a major function of glycolysis. Ribose phosphates are generated in the pentose phosphate pathway but not in glycolysis.

Reducing Equivalents Are
Generated in Those Tissues
Specializing in Reductive
Syntheses

Estimates of the activity of the pentose phosphate pathway in various tissues indicate its metabolic significance. It is active in liver, adipose tissue, adrenal cortex, thyroid, erythrocytes, testis, and lactating mammary gland. It is not active in nonlactating mammary gland, and its activity is low in skeletal muscle. All of the tissues in which the pathway is active use NADPH in reductive syntheses, eg, synthesis of fatty acids, steroids, amino acids via glutamate dehydrogenase, or reduced glutathione in erythrocytes. It is probable that the presence of active lipogenesis or of a system which utilizes NADPH to produce NADP⁺ stimulates an active degradation of glucose via the pentose phosphate pathway. The synthesis of glucose-6-phosphate dehydrogenase and 6-phosphogluconate dehydrogenase may also be induced by insulin during conditions associated with the "fed state" (Table 21–1).

Ribose Can Be Synthesized
in Virtually All Tissues

The pentose phosphate pathway provides ribose residues for nucleotide and nucleic acid synthesis (Figure 22–2). The source is the ribose 5-phosphate intermediate that reacts with ATP to form PRPP used in nucleotide biosynthesis (Chapter 36). Muscle tissue contains only very small amounts of glucose-6-phosphate dehydrogenase and 6-phosphogluconate dehydrogenase. Nevertheless, skeletal muscle, like most other tissues, is capable of synthesizing ribose

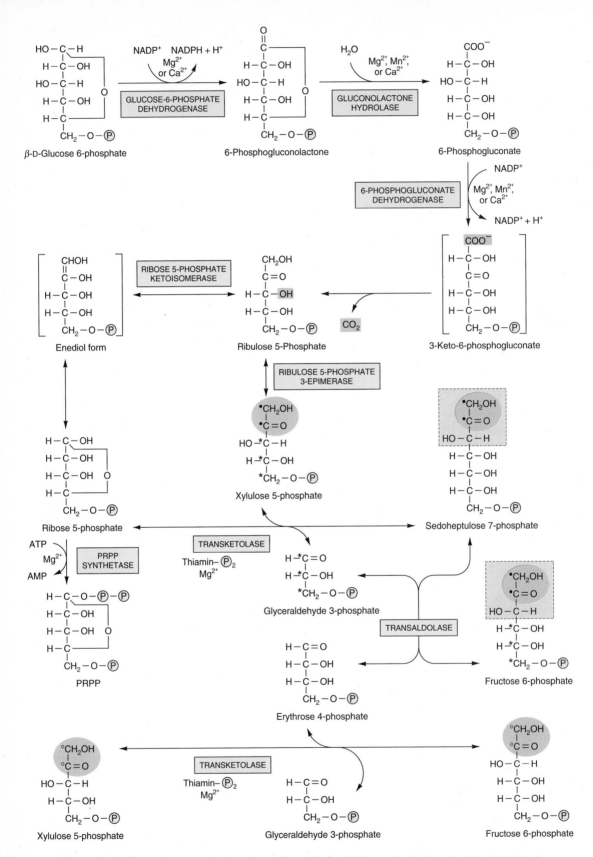

Figure 22–2. The pentose phosphate pathway. (Ⓟ, —PO$_3^{2-}$; PRPP, 5-phosphoribosyl-1-pyrophosphate.)

5-phosphate for nucleotide synthesis. This is accomplished by a reversal of the nonoxidative phase of the pentose phosphate pathway utilizing fructose 6-phosphate. Thus, it is not necessary to have a completely functioning pentose phosphate pathway for a tissue to synthesize ribose phosphates.

Ribose is not a significant constituent of systemic blood. Therefore, tissues must satisfy their own requirement for this important precursor of nucleotides.

THE PENTOSE PHOSPHATE PATHWAY ASSISTS GLUTATHIONE PEROXIDASE IN PROTECTING ERYTHROCYTES AGAINST HEMOLYSIS

The pentose phosphate pathway in the erythrocyte provides NADPH for the reduction of oxidized glutathione to reduced glutathione catalyzed by **glutathione reductase,** a flavoprotein enzyme containing FAD. In turn, reduced glutathione removes H_2O_2 from the erythrocyte in a reaction catalyzed by **glutathione peroxidase,** an enzyme that contains the trace element **selenium** (Figure 22–3). This reaction is important, since accumulation of H_2O_2 may decrease the life span of the erythrocyte by increasing the rate of oxidation of hemoglobin to methemoglobin.

GLUCURONATE, A PRECURSOR OF PROTEOGLYCANS AND CONJUGATED GLUCURONIDES, IS A PRODUCT OF THE URONIC ACID PATHWAY

Besides the major pathways of metabolism of glucose 6-phosphate that have been described, there exists a pathway for the conversion of glucose to glucuronic acid, ascorbic acid, and pentoses that is referred to as the **uronic acid pathway.** It is also an alternative oxidative pathway for glucose, but like the pentose phosphate pathway, it does not lead to the generation of ATP. In the uronic acid pathway, glucuronate is formed from glucose by the reactions shown in Figure 22–4. Glucose 6-phosphate is converted to glucose 1-phosphate, which then reacts with uridine triphosphate (UTP) to form the active nucleotide, uridine diphosphate glucose (UDPGlc). This latter reaction is catalyzed by the enzyme **UDPGlc pyrophosphorylase.** All of the steps up to this point are those previously described in the pathway of glycogenesis in the liver (Chapter 20). UDPGlc is oxidized at carbon 6 by a two-step process to glucuronate. The product of the oxidation, which is catalyzed by an NAD-dependent **UDPGlc dehydrogenase,** is UDP-glucuronate.

UDP-glucuronate is the "active" form of glucuronate for reactions involving incorporation of glucuronic acid into proteoglycans or for reactions in which glucuronate is conjugated to such substrates as steroid hormones, certain drugs, or bilirubin (Figure 34–13).

In an NADPH-dependent reaction, glucuronate is reduced to L-gulonate. This latter compound is the direct precursor of **ascorbate** in those animals capable of synthesizing this vitamin. In humans and other primates, as well as in guinea pigs, ascorbic acid cannot be synthesized because of the absence of the enzyme **L-gulonolactone oxidase.**

L-Gulonate is oxidized to 3-keto-L-gulonate, which is then decarboxylated to the pentose L-xylulose. L-Xylulose is converted to the D isomer by an NADPH-dependent reduction to xylitol, which is then oxidized in an NAD-dependent reaction to D-xylulose. After conversion to D-xylulose 5-phosphate, it is further metabolized in the pentose phosphate pathway.

INGESTION OF LARGE QUANTITIES OF FRUCTOSE HAS PROFOUND METABOLIC CONSEQUENCES

Diets high in sucrose or in high-fructose syrups (HFS) used in manufactured foods and beverages lead to large amounts of fructose (and glucose) entering the hepatic portal vein.

Fructose is more rapidly glycolyzed by the liver

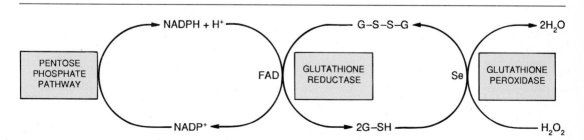

Figure 22–3. Role of the pentose phosphate pathway in the glutathione peroxidase reaction of erythrocytes. (G-S-S-G, oxidized glutathione; G-SH, reduced glutathione; Se, selenium cofactor.)

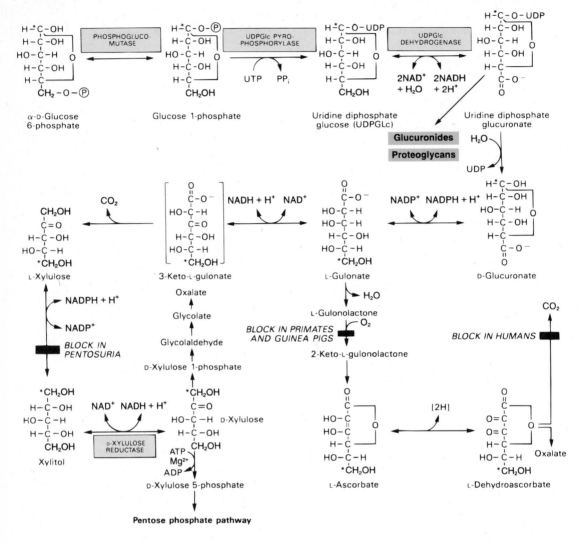

Figure 22–4. Uronic acid pathway. (*Indicates the fate of carbon 1 of glucose; Ⓟ, —PO_3^{2-}.)

than glucose. This is because it bypasses the step in glucose metabolism catalyzed by phosphofructokinase, at which point metabolic control is exerted on the rate of catabolism of glucose (Figure 22–5). This allows fructose to flood the pathways in the liver, leading to enhanced fatty acid synthesis, increased esterification of fatty acids, and increased VLDL secretion, which may raise serum triacylglycerols and ultimately raise LDL cholesterol concentrations (Figure 27–7). The extra glucose taken into the blood stream stimulates more insulin secretion, which enhances all these effects.

A specific kinase, **fructokinase,** is present in liver that effects the transfer of phosphate from ATP to fructose, forming fructose 1-phosphate. It has also been demonstrated in kidney and intestine. This enzyme will not phosphorylate glucose, and, unlike

glucokinase, its activity is not affected by fasting or by insulin, which may explain why fructose disappears from the blood of diabetic patients at a normal rate. The K_m for fructose of the enzyme in liver is very low, indicating a very high affinity of the enzyme for its substrate. It seems probable that this is the major route for the phosphorylation of fructose.

Fructose 1-phosphate is split into D-glyceraldehyde and dihydroxyacetone phosphate by **aldolase B,** an enzyme found in the liver. This enzyme also functions in glycolysis in the liver by attacking fructose 1,6-bisphosphate. D-Glyceraldehyde gains entry to the glycolysis sequence of reactions via another enzyme present in liver, **triokinase,** which catalyzes its phosphorylation to glyceraldehyde 3-phosphate. The two triose phosphates, dihydroxyacetone phosphate and glyceraldehyde 3-phosphate, may be degraded

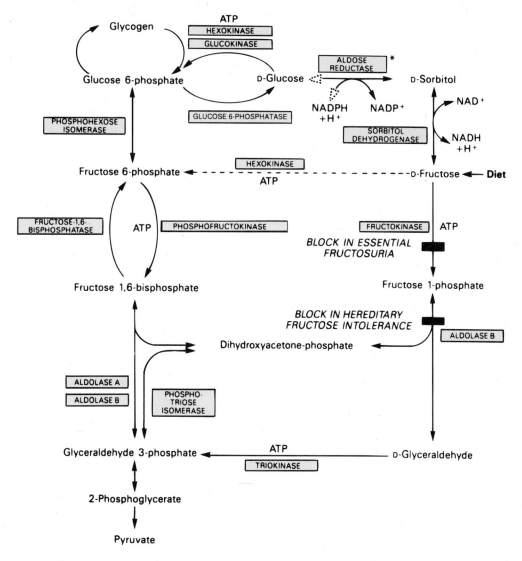

Figure 22–5. Metabolism of fructose. Aldolase A is found in all tissues except the liver, where only aldolase B is present. (*, not found in liver.)

via the glycolysis pathway or they may combine under the influence of aldolase and be converted to glucose. The latter is the fate of much of the fructose metabolized in the liver.

Hexokinase will catalyze the phosphorylation of most hexose sugars, including fructose. However, when fructose is present with glucose, its phosphorylation is largely inhibited by glucose. Nevertheless, it is probably by this route that some fructose can be metabolized in adipose tissue and muscle. Free fructose is found in seminal plasma and is secreted in quantity into the fetal circulation of ungulates and whales, where it accumulates in the amniotic and allantoic fluids. In all these situations, it represents a potential fuel source.

GALACTOSE IS NEEDED FOR THE SYNTHESIS OF LACTOSE, GLYCOLIPIDS, PROTEOGLYCANS, AND GLYCOPROTEINS

Galactose is derived from intestinal hydrolysis of the disaccharide **lactose,** the sugar of milk. It is readily converted in the liver to glucose. The ability of the liver to accomplish this conversion may be used as a test of hepatic function in the **galactose tolerance test.** The pathway by which galactose is converted to glucose is shown in Figure 22–6. Galactose is phosphorylated with the aid of **galactokinase,** using ATP as phosphate donor. The product, galactose 1-phosphate, reacts with uridine diphosphate glucose

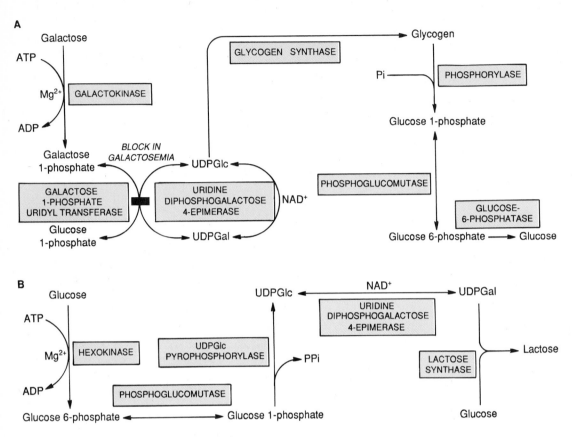

Figure 22–6. Pathway of conversion of **(A)** galactose to glucose in the liver and **(B)** glucose to lactose in the lactating mammary gland.

(UDPGlc) to form uridine diphosphate galactose (UDPGal) and glucose 1-phosphate. In this step, which is catalyzed by an enzyme called **galactose 1-phosphate uridyl transferase,** galactose is transferred to a position on UDPGlc, replacing glucose. The conversion of galactose to glucose takes place in a reaction of the galactose-containing nucleotide that is catalyzed by an **epimerase.** The product is UDPGlc. Epimerization probably involves an oxidation and reduction at carbon 4 with NAD as coenzyme. Finally, glucose is liberated from UDPGlc as glucose 1-phosphate, probably after incorporation into glycogen followed by phosphorolysis.

Since the epimerase reaction is freely reversible, glucose can be converted to galactose, so that preformed galactose is not essential in the diet. Galactose is required in the body not only in the formation of lactose but also as a constituent of glycolipids (cerebrosides), proteoglycans, and glycoproteins.

In the synthesis of lactose in the mammary gland, glucose is converted to UDPGal by the enzymes described above. UDPGal condenses with glucose to yield lactose, catalyzed by **lactose synthase** (Figure 22–6).

Glucose Is the Precursor of All Amino Sugars (Hexosamines)

Amino sugars are important components of **glycoproteins** (Chapter 56), of certain **glycosphingolipids** (eg, gangliosides) (Chapter 16), and of **glycosaminoglycans** (Chapter 57). The major amino sugars are **glucosamine, galactosamine,** and **mannosamine** (these are all hexosamines) and the nine-carbon compound **sialic acid.** The principal sialic acid found in human tissues is *N*-acetylneuraminic acid (NeuAc). A summary of the interrelationships among the amino sugars is shown in Figure 22–7; the following are its important features: (1) **Glucosamine** is the major amino sugar. It is formed as glucosamine 6-phosphate from fructose 6-phosphate, using glutamine as the donor of the amino group. (2) The amino sugars occur mainly in the *N*-acetylated form. The acetyl donor is acetyl-CoA. (3) *N*-Acetylmannosamine 6-phosphate is formed by epimerization of *N*-Acetylglucosamine 6-phosphate. (4) **NeuAc** is formed by the condensation of *N*-Acetylmannosamine 6-phosphate with phosphoenolpyruvate. (5) **Galactosamine** is formed by the epimerization of UDP-*N*-acetylglucosamine (UDPGlcNAc) to UDP-*N*-acetyl-

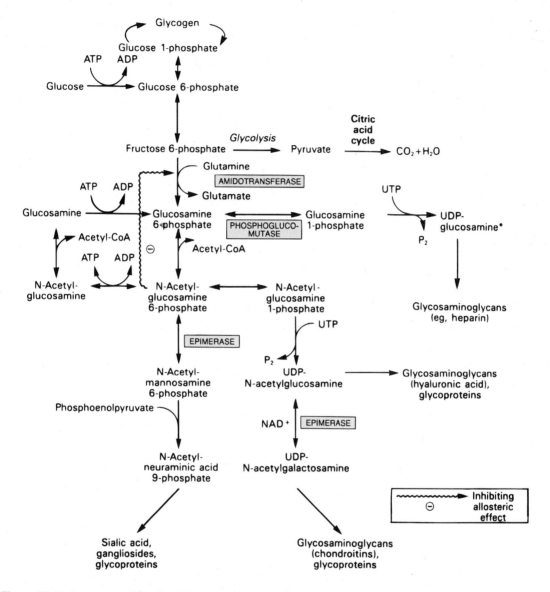

Figure 22–7. A summary of the interrelationships in metabolism of amino sugars. (*, Analogous to UDPGlc.) Other purine or pyrimidine nucleotides may be similarly linked to sugars or amino sugars. Examples are thymidine diphosphate (TDP)-glucosamine and TDP-*N*-acetylglucosamine.

galactosamine (UDPGalNAc). (6) **Nucleotide sugars** are the forms in which amino sugars are used for the biosynthesis of glycoproteins and other complex compounds; the important amino sugar-containing nucleotides are UDPGlcNAc, UDPGalNAc, and CMP-NeuAc.

CLINICAL ASPECTS

Impairment of the Pentose Phosphate Pathway Leads to Erythrocyte Hemolysis

A mutation present in some populations causes a deficiency in glucose-6-phosphate dehydrogenase, with consequent impairment of the generation of NADPH. This impairment is manifested as red cell hemolysis when the susceptible individual is subjected to oxidants, such as the antimalarial primaquine, aspirin, or sulfonamides, or when the susceptible individual has eaten fava beans (*Vicia fava*—favism). Glutathione peroxidase is a natural antioxidant found in many tissues dependent upon a supply of NADPH. It will attack organic peroxides in addition to H_2O_2. Together with vitamin E, it is part of the body's defense against lipid peroxidation (Figure 16–27). An association between the incidence of some cancers and low levels of blood selenium and glutathione peroxidase activity has been reported.

Measurement of **transketolase** activity in blood reflects the degree of thiamin deficiency. The only condition in which its activity is raised is pernicious anemia.

Disruption of the Uronic Acid Pathway Is Caused by Enzyme Defects and Some Drugs

In the rare hereditary disease **essential pentosuria,** considerable quantities of **L-xylulose** appear in the urine. This finding may be explained by the absence in pentosuric patients of the enzyme necessary to accomplish reduction of L-xylulose to xylitol. Parenteral administration of xylitol may lead to **oxalosis** involving calcium oxalate deposition in brain and kidneys. This results from the conversion of D-xylulose to oxalate via xylulose 1-phosphate, glycolaldehyde, and glycolate formation (Figure 22–4).

Various drugs markedly increase the rate at which glucose enters the uronic acid pathway. For example, administration of barbital or of chlorobutanol to rats results in a significant increase in the conversion of glucose to glucuronate, L-gulonate, and ascorbate. Aminopyrine and antipyrine have been reported to increase the excretion of L-xylulose in pentosuric subjects.

Loading of the Liver With Fructose May Potentiate Hypertryacylglycerolemia, Hypercholesterolemia, and Hyperuricemia

Reference has already been made to the effects of fructose on liver metabolism in promoting triacylglycerol synthesis and VLDL secretion and hypertriacylglycerolemia—and that this can lead to increased LDL cholesterol concentrations—all of which can be regarded as potentially atherogenic (Chapter 28). In addition, acute loading of the liver with fructose, as can occur with intravenous infusion or even very high fructose intakes, causes sequestration of inorganic phosphate in fructose 1-phosphate and diminished ATP synthesis. Consequently, the inhibition by ATP of the enzymes of adenine nucleotide degradation is removed and uric acid formation accelerates, causing hyperuricemia, which is a cause of gout (Chapter 36). These effects are of particular significance to individuals who already have a predisposition toward hypertriacylglycerolemia or hyperuricemia.

Defects in Fructose Metabolism Cause Disease (Figure 22–5)

A lack of hepatic fructokinase causes **essential fructosuria,** and absence of hepatic aldolase B, which attacks fructose 1-phosphate, leads to **hereditary fructose intolerance.** Diets low in fructose and sucrose are beneficial for both conditions. The defect in aldolase B does not affect as badly the activity toward fructose 1,6-bisphosphate, so gluconeogenesis is only mildly impaired.

One consequence of hereditary fructose intolerance and of another condition due to **fructose-1,6-bisphosphatase deficiency** is a fructose-induced **hypoglycemia** despite the presence of high glycogen reserves. Apparently, the accumulation of fructose 1-phosphate and fructose 1,6-bisphosphate inhibits the activity of liver phosphorylase by allosteric mechanisms.

Fructose and Sorbitol in the Lens Are Associated With Diabetic Cataract

Both fructose and **sorbitol** are found in the human lens, where they increase in concentration in diabetes and may be involved in the pathogenesis of **diabetic cataract.** The **sorbitol (polyol) pathway** (not found in liver) is responsible for fructose formation from glucose (Figure 22–5) and increases in activity as the glucose concentration rises in diabetes in those tissues that are not insulin-sensitive, ie, the lens, peripheral nerves, and renal glomeruli. Glucose undergoes reduction by NADPH to sorbitol catalyzed by **aldose reductase,** followed by oxidation of sorbitol to fructose in the presence of NAD^+ and sorbitol dehydrogenase (polyol dehydrogenase). Sorbitol does not diffuse through cell membranes easily and therefore accumulates, causing osmotic damage. Simultaneously, myoinositol levels fall. Sorbitol accumulation, myoinositol depletion, and diabetic cataract can be prevented by aldose reductase inhibitors in diabetic rats, and promising results have been obtained in clinical practice.

Aldose reductase is found in the placenta of the ewe and is responsible for the secretion of sorbitol into the fetal blood. The presence of sorbitol dehydrogenase in the liver, including the fetal liver, is responsible for the conversion of sorbitol into fructose. The pathway is also responsible for the occurrence of fructose in seminal fluid. When sorbitol is administered intravenously, it is converted to fructose rather than to glucose, though if it is given by mouth, much escapes absorption from the gut and is fermented in the colon by bacteria to products such as acetate and H_2. Abdominal pain **(sorbitol intolerance)** may be caused by "sugar-free" sweeteners containing sorbitol.

Enzyme Deficiencies in the Galactose Pathway Cause Galactosemia

Inability to metabolize galactose occurs in the **galactosemias,** which may be caused by inherited defects in galactokinase, uridyl transferase, or 4-epimerase (Figure 22–6A), though a deficiency in **uridyl transferase** is the best known. Galactose, which increases in concentration in the blood, is reduced by aldose reductase in the eye to the corresponding polyol (galactitol), which accumulates, causing cataract. The general condition is more severe if it is due to a defect in the uridyl transferase, since galactose 1-phosphate accumulates and depletes the liver of inorganic phosphate. Ultimately, liver failure and mental deterioration result. In uridyl transferase deficiency, the epimerase is, however, present in adequate amounts, so that the galactosemic individual can still form UDPGal from glucose. This explains how it is possible for normal growth and development of affected children to occur regardless of the galactose-free diets used to control the symptoms of the disease. Several different genetic defects have been described that cause reduced rather than total transferase deficiency. As the enzyme is normally present in excess, a reduction in activity to 50% or even less does not cause clinical disease, which manifests itself only in homozygotes. The epimerase has been found deficient in erythrocytes but present in liver and elsewhere, and this third condition appears to be symptom-free.

SUMMARY

1. The pentose phosphate pathway, which is present in the cytosol, can account for the complete oxidation of glucose, the main products being NADPH and CO_2. The pathway does not generate ATP.

2. The pathway has an oxidative phase, which is nonreversible and which generates NADPH, and a nonoxidative phase, which is reversible and which provides ribose precursors for nucleotide synthesis, including RNA and DNA. The complete pathway is present only in those tissues having a requirement for NADPH for reductive syntheses, eg, lipogenesis or steroidogenesis, whereas the nonoxidative phase is present in all cells requiring ribose.

3. In erythrocytes, the pathway has a major function in preventing hemolysis by providing NADPH for maintaining glutathione in the reduced state. In turn, glutathione is a substrate for glutathione peroxidase, which is instrumental in removing harmful H_2O_2 from the cell.

4. The uronic acid pathway is the source of glucuronic acid used for conjugation with many endogenous and exogenous substances before elimination in urine as glucuronides.

5. Fructose gains ready access to the metabolic pathways, bypassing the main control step in glycolysis catalyzed by phosphofructokinase. Thus, it stimulates fatty acid synthesis and hepatic triacylglycerol secretion. Like galactose, it readily forms glucose in the liver.

6. Galactose is synthesized from glucose in the lactating mammary gland and in other tissues where it is required for the synthesis of glycolipids, proteoglycans, and glycoproteins.

REFERENCES

Couet C, Jan P, Debry G: Lactose and cataract in humans: A review. J Am Coll Nutr 1991;10:79.

Cross NCP, Cox TM: Hereditary fructose intolerance. Int J Biochem 1990;22:685.

James HM et al: Models for the metabolic production of oxalate from xylitol in humans. Aust J Exp Biol Med Sci 1982;60:117.

Kador PF: The role of aldose reductase in the development of diabetic complications. Med Res Rev 1988;8:325.

Macdonald I, Vrana A (editors): *Metabolic Effects of Dietary Carbohydrates.* Karger, 1986.

Mayes PA: Intermediary metabolism of fructose. Am J Clin Nutr 1993;58:754S.

Randle PJ, Steiner DF, Whelan WJ (editors): *Carbohydrate Metabolism and Its Disorders,* vol 3. Academic Press, 1981.

Scriver CR et al (editors): *The Metabolic and Molecular Bases of Inherited Disease,* 7th ed. McGraw-Hill, 1995.

Van den Berghe G: Inborn errors of fructose metabolism. Annu Rev Nutr 1994;14:41.

Wood T: *The Pentose Phosphate Pathway.* Academic Press, 1985.

23

Biosynthesis of Fatty Acids

Peter A. Mayes, PhD, DSc

INTRODUCTION

Like many other degradative and synthetic processes (eg, glycogenolysis and glycogenesis), fatty acid synthesis (lipogenesis) was formerly considered to be merely the reversal of oxidation within the mitochondria. However, it is now apparent that a highly active **extramitochondrial** system is responsible for the complete synthesis of palmitate from acetyl-CoA in the cytosol. Another system for **fatty acid chain elongation** is also present in liver endoplasmic reticulum.

BIOMEDICAL IMPORTANCE

There are wide variations among species both in disposition of the principal lipogenic pathways between the tissues and in the main substrates for fatty acid synthesis. In the rat, a species that has provided most information about lipogenesis, the pathway is well represented in adipose tissue and liver, whereas in humans adipose tissue may not be an important site, and liver has only low activity. In birds, lipogenesis is confined to the liver, where it is particularly important in providing lipids for egg formation. In most mammals, glucose is the primary substrate for lipogenesis, but in ruminants acetate, which is the main fuel molecule produced by the diet, takes over this role. Since the lipogenesis pathway may be of reduced importance in humans, it is not surprising that critical diseases of the pathway have not been reported. However, variations in its activity between individuals may have a bearing on the nature and extent of **obesity,** and one of the lesions in type I, insulin-dependent **diabetes mellitus** is inhibition of lipogenesis.

THE MAIN PATHWAY FOR DE NOVO SYNTHESIS OF FATTY ACIDS (LIPOGENESIS) OCCURS IN THE CYTOSOL

This system is present in many tissues, including liver, kidney, brain, lung, mammary gland, and adipose tissue. Its cofactor requirements include NADPH, ATP, Mn^{2+}, biotin, and HCO_3^- (as a source of CO_2). **Acetyl-CoA** is the immediate substrate, and **free palmitate** is the end product.

Production of Malonyl-CoA Is the Initial and Controlling Step in Fatty Acid Synthesis

Bicarbonate as a source of CO_2 is required in the initial reaction for the carboxylation of acetyl-CoA to **malonyl-CoA** in the presence of ATP and **acetyl-CoA carboxylase.** Acetyl-CoA carboxylase has a requirement for the vitamin **biotin** (Figure 23–1). The enzyme contains a variable number of identical subunits, each containing biotin, biotin carboxylase, biotin carboxyl carrier protein, and transcarboxylase, as well as a regulatory allosteric site. It is therefore a **multienzyme protein.** The reaction takes place in two steps: (1) carboxylation of biotin (involving ATP; see Figure 52–13) and (2) transfer of the carboxyl to acetyl-CoA to form malonyl-CoA.

The Fatty Acid Synthase Complex Is a Polypeptide Containing Seven Enzyme Activities

There appear to be two types of **fatty acid synthase.** In bacteria, plants, and lower forms, the individual enzymes of the system are separate, and the acyl radicals are found in combination with a protein called the **acyl carrier protein (ACP).** However, in yeast, mammals, and birds, the synthase system is a multienzyme complex that may not be subdivided without loss of activity, and ACP is part of this complex. ACP of both bacteria and the multienzyme complex contains the vitamin **pantothenic acid** in the form of 4′-phosphopantetheine (Figure 52–6). In this system, ACP takes over the role of CoA. The aggregation of all the enzymes of a particular pathway into one multienzyme functional unit offers great efficiency and freedom from interference by competing reactions, thus achieving the effect of compartmentalization of the process within the cell without the erection of permeability barriers. Another advantage of a single multienzyme polypeptide is that synthesis

$$CH_3 - CO \sim S - CoA \longrightarrow {}^-OO\overset{*}{C} - CH_2 - CO \sim S - CoA$$

Acetyl-CoA   Malonyl-CoA

$$Enz-biotin\overset{*}{-}COO^- \qquad Enz-biotin$$

$$ADP + P_i$$

$$ATP + H\overset{*}{C}O_3^- + Enz-biotin$$

Figure 23–1. Biosynthesis of malonyl-CoA. (Enz, acetyl-CoA carboxylase.)

of all enzymes in the complex is coordinated, since it is encoded by a single gene.

The fatty acid synthase complex is a dimer (Figure 23–2). In mammals, each monomer is identical, consisting of one remarkable polypeptide chain containing all seven enzyme activities of fatty acid synthase and an ACP with a 4'-phosphopantetheine —SH group. In close proximity is another thiol of a cysteine residue of **3-ketoacyl synthase (condensing**

enzyme) of the other monomer (Figure 23–2). This is brought about because the two monomers lie in a "head-to-tail" configuration. Since both thiols participate in the synthase activity, **only the dimer is active.**

Initially, a priming molecule of acetyl-CoA combines with the cysteine —SH group catalyzed by **acetyl transacylase** (Figure 23–3). Malonyl-CoA combines with the adjacent —SH on the 4'-phospho-

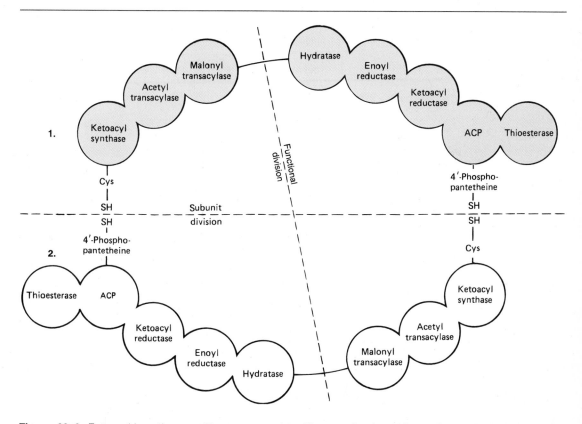

Figure 23–2. Fatty acid synthase multienzyme complex. The complex is a dimer of two identical polypeptide monomers, 1 and 2, each consisting of seven enzyme activities and the acyl carrier protein (ACP). (Cys—SH, cysteine thiol.) The —SH of the 4'-phosphopantetheine of one monomer is in close proximity to the —SH of the cysteine residue of the ketoacyl synthase of the other monomer, suggesting a "head-to-tail" arrangement of the two monomers. The detailed sequence of the enzymes in each monomer is tentative (based on Wakil). Though each monomer contains all the partial activities of the reaction sequence, the actual functional unit consists of one-half of one monomer interacting with the complementary half of the other. Thus, two acyl chains are produced simultaneously.

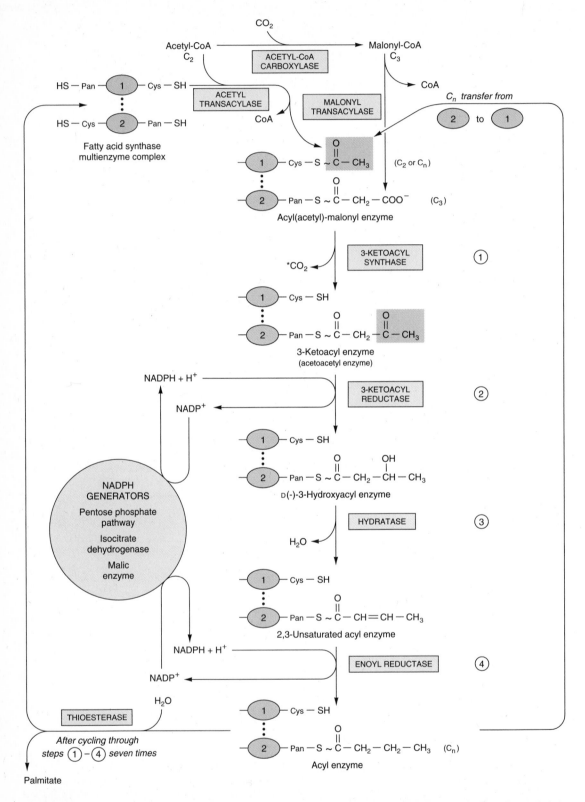

Figure 23–3. Biosynthesis of long-chain fatty acids. Details of how addition of a malonyl residue causes the acyl chain to grow by two carbon atoms. (Cys-, cysteine residue; Pan, 4'-phosphopantetheine.) Details of the fatty acid synthase dimer are shown in Figure 23–2. ① and ② represent the individual monomers of fatty acid synthase.

pantetheine of ACP of the other monomer, catalyzed by **malonyl transacylase,** to form **acetyl (acyl)-malonyl enzyme.** The acetyl group attacks the methylene group of the malonyl residue, catalyzed by **3-ketoacyl synthase,** and liberates CO_2, forming 3-ketoacyl enzyme (acetoacetyl enzyme). This frees the cysteine —SH group, hitherto occupied by the acetyl group. Decarboxylation allows the reaction to go to completion, acting as a pulling force for the whole sequence of reactions. The 3-ketoacyl group is reduced, dehydrated, and reduced again to form the corresponding saturated acyl-S-enzyme. A new malonyl-CoA molecule combines with the —SH of 4′-phosphopantetheine, displacing the saturated acyl residue onto the free cysteine —SH group. The sequence of reactions is repeated 6 more times, a new malonyl residue being incorporated during each sequence, until a saturated 16-carbon acyl radical (palmityl) has been assembled. It is liberated from the enzyme complex by the activity of a seventh enzyme in the complex, **thioesterase** (deacylase). The free palmitate must be activated to acyl-CoA before it can proceed via any other metabolic pathway. Its usual fate is esterification into acylglycerols (Figure 23–4).

In mammary gland, there is a separate thioesterase specific for acyl residues of C_8, C_{10}, or C_{12}, which are subsequently found in milk lipids. In ruminant mammary gland, this enzyme is part of the fatty acid synthase complex.

There would appear to be two centers of activity in one dimer complex that function independently and simultaneously to form two molecules of palmitate. The equation for the overall synthesis of palmitate from acetyl-CoA and malonyl-CoA is shown below:

$$CH_2CO \cdot S \cdot CoA + 7HOOC \cdot CH_2CO \cdot S \cdot CoA$$
$$+ 14NADPH + 14H^+ \rightarrow$$
$$CH_3(CH_2)_{14}COOH + 7CO_2 + 6H_2O$$
$$+ 8CoA \cdot SH + 14NADP^+$$

The acetyl-CoA used as a primer forms carbon atoms 15 and 16 of palmitate. The addition of all the subsequent C_2 units is via malonyl-CoA formation.

Butyryl-CoA may act as a primer molecule in mammalian liver and mammary gland. If propionyl-CoA acts as primer, long-chain fatty acids having an odd number of carbon atoms result. These are found particularly in ruminants, where propionate is formed by microbial action in the rumen.

The Main Source of NADPH for Lipogenesis Is the Pentose Phosphate Pathway

NADPH is involved as donor of reducing equivalents in both the reduction of the 3-ketoacyl and of the 2,3-unsaturated acyl derivatives. The oxidative reactions of the pentose phosphate pathway (see p 201) are the chief source of the hydrogen required for the reductive synthesis of fatty acids. It is significant that tissues specializing in active lipogenesis, ie, liver, adipose tissue, and the lactating mammary gland also possess an active pentose phosphate pathway. Moreover, both metabolic pathways are found in the cytosol of the cell, so there are no membranes or permeability barriers for the transfer of NADPH from one pathway to the other. Other sources of NADPH include the reaction that converts malate to pyruvate catalyzed by the **"malic enzyme"** (NADP malate dehydrogenase) (Figure 23–5) and the extramitochondrial **isocitrate dehydrogenase** reaction (probably not a substantial source).

Acetyl-CoA Is the Principal Building Block of Fatty Acids

It is formed from carbohydrate via the oxidation of pyruvate within the mitochondria. However, acetyl-CoA does not diffuse readily into the extramitochondrial cytosol, the principal site of fatty acid synthesis. The activity of the extramitochondrial **ATP-citrate lyase,** like the "malic enzyme," increases in activity in the well-fed state, closely paralleling the activity of the fatty-acid synthesizing system (see Table 21–1). It is now believed that utilization of glucose for lipogenesis is by way of citrate. The pathway involves glycolysis followed by the oxidative decarboxylation of pyruvate to acetyl-CoA within the mitochondria and subsequent condensation with ox-

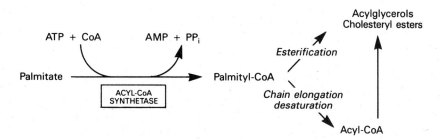

Figure 23–4. Fate of palmitate after biosynthesis.

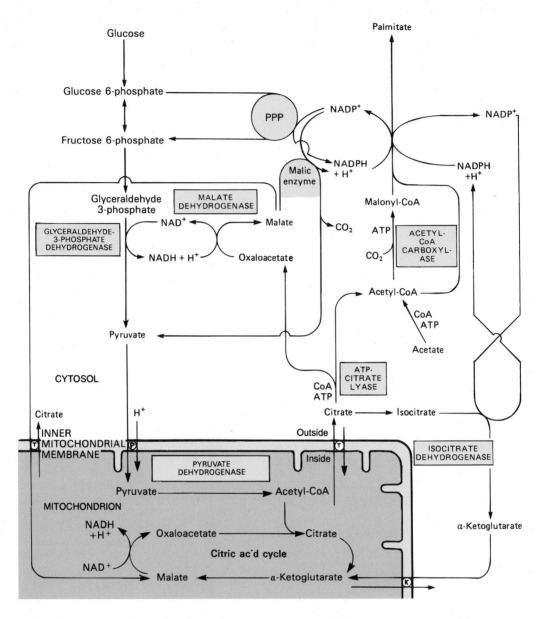

Figure 23–5. The provision of acetyl-CoA and NADPH for lipogenesis. (PPP, pentose phosphate pathway; T, tricarboxylate transporter; K, α-ketoglutarate transporter; P, pyruvate transporter.)

aloacetate to form citrate, as part of the citric acid cycle. This is followed by the translocation of citrate into the extramitochondrial compartment via the tricarboxylate transporter, where in the presence of CoA and ATP, it undergoes cleavage to acetyl-CoA and oxaloacetate catalyzed by ATP-citrate lyase. The acetyl-CoA is then available for malonyl-CoA formation and synthesis to palmitate (Figure 23–5). The resulting oxaloacetate can form malate via NADH-linked malate dehydrogenase, followed by the generation of NADPH via the malic enzyme. In turn, the

NADPH becomes available for lipogenesis. This pathway is a means of transferring reducing equivalents from extramitochondrial NADH to NADP. Alternatively, malate can be transported into the mitochondrion, where it is able to re-form oxaloacetate. It is to be noted that the citrate (tricarboxylate) transporter in the mitochondrial membrane requires malate to exchange with citrate (see Figure 14–13).

There is little ATP-citrate lyase or malic enzyme in ruminants, probably because in these species acetate (derived from the rumen) is the main source of acetyl-CoA. Since the acetate arises and is activated to acetyl-CoA extramitochondrially, there is no necessity for citrate transport out of mitochondria prior to incorporation of acetyl-CoA into long-chain fatty acids. Generation of NADPH via extramitochondrial isocitrate dehydrogenase is more important in these species because of the deficiency in malic enzyme.

Elongation of Fatty Acid Chains Occurs in the Endoplasmic Reticulum

This pathway (the "microsomal system") converts fatty acyl-CoA to an acyl-CoA derivative having two carbons more, using malonyl-CoA as acetyl donor and NADPH as reductant catalyzed by the microsomal **fatty acid elongase** system of enzymes (Figure 23–6). The acyl groups that may act as a primer molecule include the saturated series from C_{10} upward, as well as unsaturated fatty acids. Fasting largely abolishes chain elongation. Elongation of stearyl-CoA in brain increases rapidly during myelination in order to provide C_{22} and C_{24} fatty acids that are present in sphingolipids.

THE NUTRITIONAL STATE REGULATES LIPOGENESIS

Many animals, including humans, take their food as spaced meals and must therefore store much of the energy of their diet for use between meals. The process of lipogenesis is concerned with the conversion of glucose and intermediates such as pyruvate, lactate, and acetyl-CoA to fat, which constitutes the anabolic phase of this cycle. The nutritional state of the organism is the **main factor controlling the rate of lipogenesis.** Thus, the rate is higher in the well-fed animal whose diet contains a high proportion of carbohydrate. It is depressed under conditions of restricted caloric intake, on a high-fat diet, or when there is a deficiency of insulin, as in diabetes mellitus. All these conditions are associated with increased concentrations of plasma free fatty acids. The regulation of the mobilization of free fatty acids from adipose tissue is described in Chapter 27.

There is an inverse relationship between hepatic lipogenesis and the concentration of serum-free fatty acids (Figure 23–7). The greatest inhibition of lipogenesis occurs over the range of free fatty acids (0.3–0.8 μmol/mL of plasma) through which the plasma free fatty acids increase during transition from the fed to the starved state. Fat in the diet also causes depression of lipogenesis in the liver, and when there is more than 10% of fat in the diet, there is little conversion of dietary carbohydrate to fat. Lipogenesis is higher when sucrose is fed instead of glucose because fructose bypasses the phosphofructokinase control point in glycolysis and floods the lipogenic pathway (Figure 22–5).

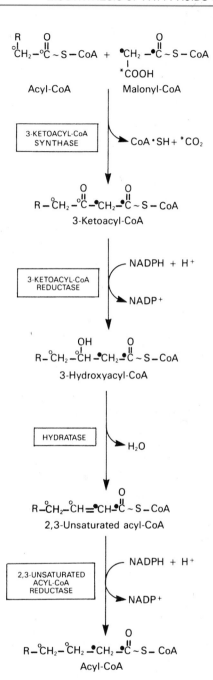

Figure 23–6. Microsomal elongase system for fatty acid chain elongation.

SHORT- AND LONG-TERM MECHANISMS REGULATE LIPOGENESIS

Long-chain fatty acid synthesis is controlled in the short term by allosteric and covalent modification of enzymes and in the long term by changes in gene expression governing rates of synthesis of enzymes.

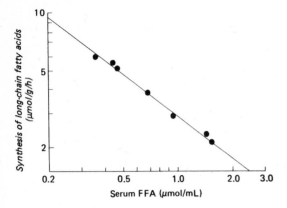

Figure 23–7. Direct inhibition of hepatic lipogenesis by free fatty acids. Lipogenesis was determined from the incorporation of ^3H_2O into long-chain fatty acids in the perfused rat liver. (FFA, free fatty acids.) (Experiments from the author's laboratory with DL Topping.)

Citrate and Acyl-CoA Regulate Acetyl-CoA Carboxylase

The rate-limiting reaction in the lipogenic pathway is at the **acetyl-CoA carboxylase step.** Acetyl-CoA carboxylase is an allosteric enzyme and is activated by **citrate,** which increases in concentration in the well fed state and is an indicator of a plentiful supply of acetyl-CoA. However, it is inhibited by long-chain acyl-CoA molecules, an example of metabolic negative feedback inhibition by a product of a reaction sequence. Thus, if acyl-CoA accumulates because it is not esterified quickly enough, it will automatically reduce the synthesis of new fatty acid. Likewise, if acyl-CoA accumulates as a result of increased lipolysis or an influx of free fatty acids into the tissue, this will also inhibit synthesis of new fatty acid (Figure 23–7). Allosteric activation of the enzyme involves aggregation from a dimeric to a polymeric configuration of several million in molecular mass. Acyl-CoA may also inhibit the mitochondrial **tricarboxylate transporter,** thus preventing activation of the enzyme by egress of citrate from the mitochondria into the cytosol.

Pyruvate Dehydrogenase Is Also Regulated by Acyl-CoA

There is also an inverse relationship between free fatty acid concentration and the proportion of active to inactive pyruvate dehydrogenase that regulates the availability of acetyl-CoA for lipogenesis. Acyl-CoA causes an inhibition of pyruvate dehydrogenase by inhibiting the ATP-ADP exchange transporter of the inner mitochondrial membrane, which leads to increased intramitochondrial [ATP]/[ADP] ratios and therefore to conversion of active to inactive pyruvate

dehydrogenase (see Figure 19–6). Also, oxidation of acyl-CoA due to increased levels of free fatty acids may increase the ratio of [acetyl-CoA]/[CoA] and [NADH]/[NAD$^+$] in mitochondria, inhibiting pyruvate dehydrogenase.

Hormones Also Regulate Lipogenesis

Insulin stimulates lipogenesis by several mechanisms. It increases the transport of glucose into the cell (eg, in adipose tissue) and thereby increases the availability of both pyruvate for fatty acid synthesis and glycerol 3-phosphate for esterification of the newly formed fatty acids. Insulin converts the inactive form of pyruvate dehydrogenase to the active form in adipose tissue but not in liver. In addition, acetyl-CoA carboxylase is an enzyme that can be regulated by reversible phosphorylation. **Insulin activates acetyl-CoA carboxylase.** This activation involves dephosphorylation by a protein phosphatase. Also, insulin, by its ability to depress the level of intracellular cAMP, **inhibits lipolysis** in adipose tissue and thereby reduces the concentration of plasma free fatty acids and therefore long-chain acyl-CoA, an inhibitor of lipogenesis. By this same mechanism insulin antagonizes the actions of **glucagon** and **epinephrine,** which inhibit acetyl-CoA carboxylase, and therefore lipogenesis, by increasing **cAMP,** allowing cAMP-dependent protein kinase to inactivate the enzyme by phosphorylation. Another protein kinase dependent on 5′-AMP will also inactivate the enzyme.

In ruminants, **acetate**—not glucose—is the starting material for lipogenesis. It follows that, in these species, many of the control mechanisms involving mitochondria are bypassed and thus do not apply.

The Fatty Acid Synthase Complex and Acetyl-CoA Carboxylase Are Adaptive Enzymes

These enzymes adapt to the body's physiologic needs by increasing in total amount in the fed state and by decreasing in fasting, feeding of fat, and diabetes. **Insulin** is an important hormone causing gene expression and induction of enzyme biosynthesis, and glucagon antagonizes this effect. Feeding fats containing polyunsaturated fatty acids coordinately regulates the inhibition of expression of key enzymes of glycolysis and lipogenesis. These mechanisms for longer-term control of lipogenesis take several days to become fully manifested and augment the direct and immediate effect of free fatty acids and hormones such as insulin and glucagon.

SUMMARY

1. The synthesis of long-chain fatty acids (lipogenesis) is carried out by two enzyme systems present in

the cytosol of the cell: acetyl-CoA carboxylase and fatty acid synthase.

2. The pathway converts acetyl-CoA to palmitate and requires NADPH, ATP, Mn^{2+}, biotin, pantothenic acid, and HCO_3^- as cofactors.

3. Acetyl-CoA carboxylase is required to convert acetyl-CoA to malonyl-CoA. In turn, fatty acid synthase, a multienzyme complex of one polypeptide chain with seven separate enzymatic activities, catalyzes the assembly of palmitate from one acetyl-CoA and seven malonyl-CoA molecules.

4. Lipogenesis is regulated at the acetyl-CoA carboxylase step by allosteric modifiers, covalent modification, and induction and repression of enzyme synthesis. Citrate activates the enzyme, and long-chain acyl-CoA inhibits its activity. Insulin activates acetyl-CoA carboxylase in the short term by dephosphorylation and in the long term by induction of synthesis. Glucagon and epinephrine have opposite actions to insulin.

5. Lengthening of long-chain fatty acids takes place in the endoplasmic reticulum catalyzed by a microsomal elongase enzyme system.

REFERENCES

Goodridge AG: Fatty acid synthesis in eukaryotes. In: *Biochemistry of Lipids and Membranes.* Vance DE, Vance JE (editors). Benjamin/Cummings, 1985.

Goodridge AG: Dietary regulation of gene expression: Enzymes involved in carbohydrate and lipid metabolism. Annu Rev Nutr 1987;7:157.

Haystead TAJ et al: Roles of the AMP-activated and cyclic-AMP-dependent protein kinases in the adrenaline-induced inactivation of acetyl-CoA carboxylase in rat adipocytes. Eur J Biochem 1990;187:199.

Jump DB et al: Coordinate regulation of glycolytic and lipogenic gene expression by polyunsaturated fatty acids. J Lipid Res 1994;35:1076.

Mabrouk GM et al: Acute hormonal control of acetyl-CoA carboxylase. J Biol Chem 1990;265:6330.

Wakil SJ: Fatty acid synthase, a proficient multifunctional enzyme. Biochemistry 1989;28:4523.

24

Oxidation of Fatty Acids: Ketogenesis

Peter A. Mayes, PhD, DSc

INTRODUCTION

Fatty acids are both oxidized to acetyl-CoA and synthesized from acetyl-CoA. Although the starting material of one process is identical to the product of the other and the chemical stages involved are comparable, fatty acid oxidation is not the simple reverse of fatty acid biosynthesis but an entirely different process taking place in a separate compartment of the cell. The separation of fatty acid oxidation from biosynthesis allows each process to be individually controlled and integrated with tissue requirements.

Fatty acid oxidation takes place in mitochondria; each step involves acyl-CoA derivatives catalyzed by separate enzymes, utilizes NAD^+ and FAD as coenzymes, and generates ATP. In contrast, fatty acid biosynthesis (lipogenesis) takes place in the cytosol, involves acyl derivatives continuously attached to a multienzyme complex, utilizes $NADP^+$ as coenzyme, and requires both ATP and bicarbonate ion. Fatty acid oxidation is an aerobic process, requiring the presence of oxygen.

BIOMEDICAL IMPORTANCE

Increased fatty acid oxidation is characteristic of starvation and of diabetes mellitus, leading to **ketone body** production by the liver **(ketosis).** Ketone bodies are acidic and when produced in excess over long periods, as in diabetes, cause **ketoacidosis,** which is ultimately fatal. Because gluconeogenesis is dependent upon fatty acid oxidation, any impairment in fatty acid oxidation leads to **hypoglycemia.** This occurs in various states of **carnitine deficiency** or deficiency of essential enzymes in fatty acid oxidation, eg, **carnitine palmitoyltransferase,** or inhibition of fatty acid oxidation by poisons, eg, **hypoglycin.**

OXIDATION OF FATTY ACIDS OCCURS IN MITOCHONDRIA

Fatty Acids Are Transported in the Blood As Free Fatty Acids (FFA)

The term "free fatty acid" refers to fatty acids that are in the **unesterified state.** Alternative nomenclature is UFA (unesterified fatty acids) or NEFA (nonesterified fatty acids). In plasma, FFA of longer-chain fatty acids are combined with **albumin,** and in the cell they are attached to a **fatty acid binding protein,** or Z-protein, so that in fact they are never really "free." Shorter-chain fatty acids are more water soluble and exist as the un-ionized acid or as a fatty acid anion.

Fatty Acids Are Activated Before Being Catabolized

As in the metabolism of glucose, fatty acids must first be converted in a reaction with ATP to an active intermediate before they will react with the enzymes responsible for their further metabolism. This is the only step in the complete degradation of a fatty acid that requires energy from ATP. In the presence of ATP and coenzyme A, the enzyme **acyl-CoA synthetase (thiokinase)** catalyzes the conversion of a fatty acid (or free fatty acid) to an "active fatty acid" or acyl-CoA, accompanied by the expenditure of one high-energy phosphate.

$$\text{Fatty acid} + \text{ATP} + \text{CoA} \rightarrow \text{Acyl-CoA} + PP_i + \text{AMP}$$

ACYL-CoA
SYNTHETASE

The presence of **inorganic pyrophosphatase** ensures that activation goes to completion by facilitating the loss of the additional high-energy phosphate associated with pyrophosphate. Thus, in effect, two high-energy phosphates are expended during the activation of each fatty acid molecule.

$$PP_i + H_2O \rightarrow 2\ P_i$$

INORGANIC
PYROPHOSPHATASE

Acyl-CoA synthetases are found in the endoplasmic reticulum and inside and on the outer membrane of mitochondria. Several acyl-CoA synthetases have been described, each specific for fatty acids of different chain length.

Long-Chain Fatty Acids Penetrate the Inner Mitochondrial Membrane as Carnitine Derivatives

Carnitine (β-hydroxy-γ-trimethylammonium butyrate), $(CH_3)_3N^+$—CH_2—$CH(OH)$—CH_2—COO^-, is widely distributed and is particularly abundant in muscle. It is synthesized from lysine and methionine in liver and kidney. Activation of lower fatty acids, and their oxidation within the mitochondria, may occur independently of carnitine, but long-chain acyl-CoA (or FFA) will not penetrate the inner membrane of mitochondria and become oxidized unless they form acylcarnitines. An enzyme, **carnitine palmitoyltransferase I,** present in the outer mitochondrial membrane, converts long-chain acyl CoA to acylcarnitine, which is able to penetrate the inner membrane of mitochondria and gain access to the β-oxidation system of enzymes (see reaction below). **Carnitine-acylcarnitine translocase** acts as an inner membrane carnitine exchange transporter. Acylcarnitine is transported in, coupled with the transport out of one molecule of carnitine. The acylcarnitine then reacts with CoA, catalyzed by **carnitine palmitoyltransferase II,** located on the inside of the inner membrane. Acyl-CoA is re-formed in the mitochondrial matrix, and carnitine is liberated (Figure 24–1).

$$\text{Acyl-CoA + Carnitine} \leftrightarrow \text{Acylcarnitine + CoA}$$

CARNITINE PALMITOYLTRANSFERASE

Another enzyme, **carnitine acetyltransferase,** is present within mitochondria and catalyzes the transfer of short-chain acyl groups between CoA and carnitine. The function of this enzyme is obscure, but it may facilitate transport of acetyl groups through the mitochondrial membrane. Together with fructose and lactate, acetylcarnitine is an important fuel source for sperm, supporting motility.

$$\text{Acetyl-CoA + Carnitine} \leftrightarrow \text{Acetylcarnitine + CoA}$$

CARNITINE ACETYLTRANSFERASE

β-OXIDATION OF FATTY ACIDS INVOLVES SUCCESSIVE CLEAVAGE WITH RELEASE OF ACETYL-CoA

In β-oxidation (Figure 24–2), two carbons are cleaved at a time from acyl-CoA molecules, starting at the carboxyl end. The chain is broken between the

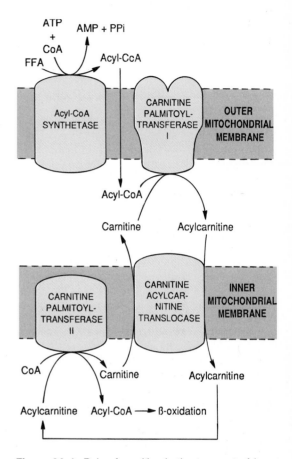

Figure 24–1. Role of carnitine in the transport of long-chain fatty acids through the inner mitochondrial membrane. Long-chain acyl-CoA cannot pass through the inner mitochondrial membrane, but its metabolic product, acylcarnitine, can.

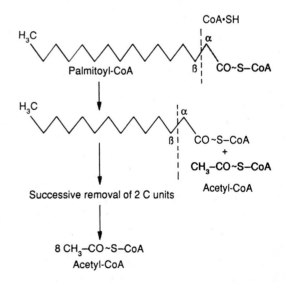

Figure 24–2. Overview of β-oxidation of fatty acids.

α(2)- and β(3)-carbon atoms, hence the name β-oxidation. The two-carbon units formed are acetyl-CoA; thus, palmitoyl-CoA forms eight acetyl-CoA molecules.

The Cyclic Reaction Sequence Generates NADH and FADH$_2$

Several enzymes, known collectively as "fatty acid oxidase," are found in the mitochondrial matrix adjacent to the respiratory chain (which is found in the inner membrane). These catalyze the oxidation of acyl-CoA to acetyl-CoA, the system being coupled with the phosphorylation of ADP to ATP (Figure 24–3).

After the penetration of the acyl moiety through the mitochondrial membrane via the carnitine transporter system and the re-formation of acyl-CoA, there follows the removal of two hydrogen atoms from the 2(α)- and 3(β)-carbon atoms, catalyzed by **acyl-CoA dehydrogenase.** This results in the formation of Δ^2-*trans*-enoyl-CoA. The coenzyme for the dehydrogenase is a flavoprotein, containing FAD as prosthetic group, whose reoxidation by the respiratory chain requires the mediation of another flavoprotein, termed **electron-transferring flavoprotein** (Chapter 13). Water is added to saturate the double bond and form 3-hydroxyacyl-CoA, catalyzed by the enzyme **Δ^2-enoyl-CoA hydratase.** The 3-hydroxy derivative undergoes further dehydrogenation on the 3-carbon (**L(+)-3-hydroxyacyl-CoA dehydrogenase**) to form the corresponding 3-ketoacyl-CoA compound. In this case, NAD$^+$ is the coenzyme involved in the dehydrogenation. Finally, 3-ketoacyl-CoA is split at the 2,3-position by **thiolase** (3-ketoacyl-CoA-thiolase), which catalyzes a thiolytic cleavage involving another molecule of CoA. The products of this reaction are acetyl-CoA and an acyl-CoA derivative containing two carbons less than the original acyl-CoA molecule that underwent oxidation. The acyl-CoA formed in the cleavage reaction reenters the oxidative pathway at reaction 2 (Figure 24–3). In this way, a long-chain fatty acid may be degraded completely to acetyl-CoA (C$_2$ units). Since acetyl-CoA can be oxidized to CO$_2$ and water via the citric acid cycle (which is also found within the mitochondria), the complete oxidation of fatty acids is achieved.

Oxidation of a Fatty Acid With an Odd Number of Carbon Atoms Yields Acetyl-CoA Plus a Molecule of Propionyl-CoA

Fatty acids with an odd number of carbon atoms are oxidized by the pathway of β-oxidation, producing acetyl-CoA until a three-carbon (propionyl-CoA) residue remains. This compound is converted to succinyl-CoA, a constituent of the citric acid cycle (Figure 21–2). Hence, **the propionyl residue from an odd-chain fatty acid is the only part of a fatty acid that is glucogenic.**

Oxidation of Fatty Acids Produces a Large Quantity of ATP

Transport in the respiratory chain of electrons from FADH$_2$ and NADH will lead to the synthesis of five high-energy phosphates (Chapter 14) for each of the first seven acetyl-CoA molecules formed by β-oxidation of palmitate ($7 \times 5 = 35$). A total of 8 mol of acetyl-CoA is formed, and each will give rise to 12 mol of ATP on oxidation in the citric acid cycle, making $8 \times 12 = 96$ mol of ATP derived from the acetyl-CoA formed from palmitate. Two must be subtracted for the initial activation of the fatty acid, yielding a net gain of 129 mol of ATP per mole of palmitate, or $129 \times 51.6^* = 6656$ kJ. As the free energy of combustion of palmitic acid is 9791 kJ/mol, the process captures as high-energy phosphate on the order of 68% of the total energy of combustion of the fatty acid.

Peroxisomes Oxidize Very Long Chain Fatty Acids

A modified form of β-oxidation is found in peroxisomes and leads to the formation of acetyl-CoA and H$_2$O$_2$ (from the flavoprotein-linked dehydrogenase step), which is broken down by catalase. Thus, this first dehydrogenation is not linked directly to phosphorylation and the generation of ATP but, with the initial activation by a very long chain acyl-CoA synthetase, facilitates the oxidation of very long chain fatty acids (eg, C$_{20}$, C$_{22}$). These enzymes are induced by high-fat diets and in some species by hypolipidemic drugs such as clofibrate.

The enzymes in peroxisomes do not attack shorter-chain fatty acids; the β-oxidation sequence ends at octanoyl-CoA. Octanoyl and acetyl groups are subsequently removed from the peroxisomes in the forms of octanoyl and acetylcarnitine, and both are further oxidized in mitochondria. A further role of peroxisomal β-oxidation is to shorten the side chain of cholesterol in bile acid formation (Chapter 28). Peroxisomes also take part in the synthesis of ether glycerolipids (Chapter 26), cholesterol, and dolichol (Figure 28–2). Peroxisomes do not contain carnitine palmitoyltransferase.

α- AND ω-OXIDATION OF FATTY ACIDS ARE SPECIALIZED PATHWAYS

Quantitatively, β-oxidation in mitochondria is the most important pathway for fatty acid oxidation. However, α-oxidation, ie, the removal of one carbon at a time from the carboxyl end of the molecule, has been detected in brain tissue. It does not require CoA intermediates and does not generate high-energy phosphates.

---

*ΔG for the ATP reaction, as explained in Chapter 19.

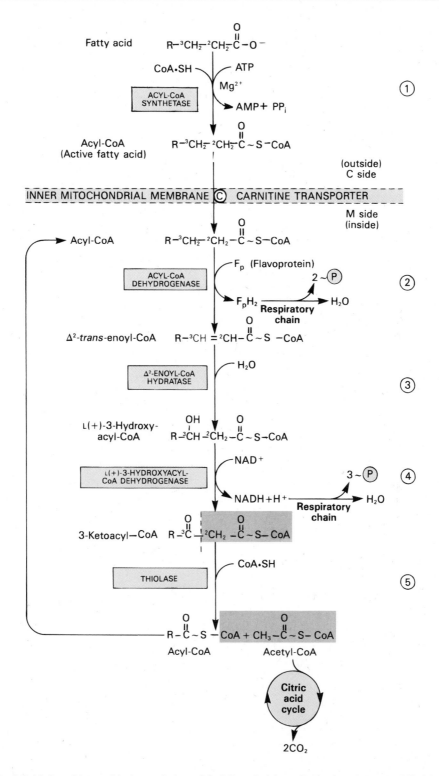

Figure 24–3. β-Oxidation of fatty acids. Long-chain acyl-CoA is cycled through reactions 2–5, acetyl-CoA being split off, each cycle, by thiolase (reaction 5). When the acyl radical is only four carbon atoms in length, two acetyl-CoA molecules are formed in reaction 5.

ω-Oxidation is normally a very minor pathway and is brought about by hydroxylase enzymes involving cytochrome P450 in the endoplasmic reticulum (Chapter 13). The —CH$_3$ group is converted to a —CH$_2$OH group that subsequently is oxidized to —COOH, thus forming a dicarboxylic acid. This is β-oxidized usually to adipic (C$_6$) and suberic (C8) acids, which are excreted in the urine.

OXIDATION OF UNSATURATED FATTY ACIDS OCCURS BY A MODIFIED β-OXIDATION PATHWAY

The CoA esters of these acids are degraded by the enzymes normally responsible for β-oxidation until either a Δ^3-*cis*-acyl-CoA compound or a Δ^4-*cis*-acyl-CoA compound is formed, depending upon the position of the double bonds (Figure 24–4). The former compound is isomerized (**Δ^3*cis*→Δ^2-*trans*-enoyl-CoA isomerase**) to the corresponding Δ^2-*trans*-CoA stage of β-oxidation for subsequent hydration and oxidation. Any Δ^4-*cis*-acyl-CoA either remaining, as in the case of linoleic acid (Figure 24–4), or entering the pathway at this point is converted by acyl-CoA dehydrogenase to Δ^2-*trans*-Δ^4-*cis*-dienoyl-CoA. This is converted to Δ^3-*trans*-enoyl-CoA by an NADP-dependent enzyme, **Δ^2-*trans*-Δ^4-*cis*-dienoyl-CoA reductase. Δ^3-*cis* (or *trans*-)→Δ^2-*trans*-enoyl-CoA isomerase** will attack the Δ^3-*trans* double bond to produce Δ^2-*trans*-enoyl-CoA, an intermediate in β-oxidation.

KETOGENESIS OCCURS WHEN THERE IS A HIGH RATE OF FATTY ACID OXIDATION IN THE LIVER

Under certain metabolic conditions associated with a high rate of fatty acid oxidation, the liver produces considerable quantities of **acetoacetate** and **D(−)-3-hydroxybutyrate** (β-hydroxybutyrate). Acetoacetate continually undergoes spontaneous decarboxylation to yield **acetone.** These three substances are collectively known as the **ketone bodies** (also called acetone bodies or [incorrectly*] "ketones") (Figure 24–5). Acetoacetate and 3-hydroxybutyrate are interconverted by the mitochondrial enzyme **D(−)-3-hydroxybutyrate dehydrogenase;** the equilibrium is controlled by the mitochondrial ratio of [NAD$^+$] to [NADH], ie, the **redox state.** The ratio [3-hydroxybutyrate]/[acetoacetate] in blood varies between 1:1 and 10:1. The concentration of total ketone bodies in the blood of well-fed mammals does

*The term "ketones" should not be used because 3-hydroxybutyrate is not a ketone and there are ketones in blood that are not ketone bodies, eg, pyruvate, fructose.

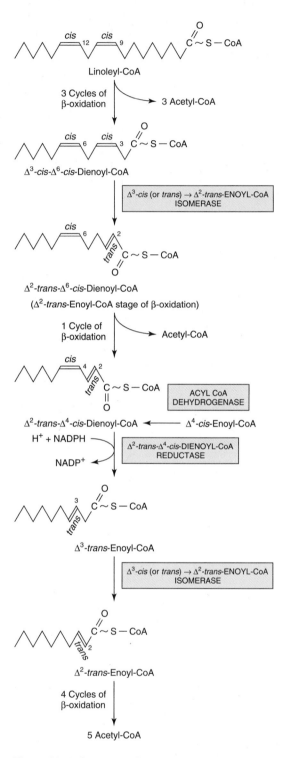

Figure 24–4. Sequence of reactions in the oxidation of unsaturated fatty acids, eg, linoleic acid. Δ^4-*cis*-fatty acids or fatty acids forming Δ^4-*cis*-enoyl-CoA enter the pathway at the position shown. NADPH for the dienoyl-CoA reductase step is supplied by intramitochondrial sources such as glutamate dehydrogenase, isocitrate dehydrogenase, and NAD(P)H transhydrogenase.

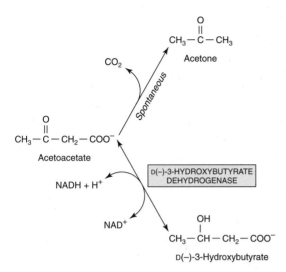

Figure 24–5. Interrelationships of the ketone bodies. D(–)-3-Hydroxybutyrate dehydrogenase is a mitochondrial enzyme.

not normally exceed 0.2 mmol/L. It is somewhat higher in ruminants owing to 3-hydroxybutyrate formation from butyric acid (a product of ruminal fermentation) in the rumen wall. Loss via the urine is usually less than 1 mg/24 h in humans.

In vivo, the liver appears to be the only organ in nonruminants to add significant quantities of ketone bodies to the blood. Extrahepatic tissues utilize them as respiratory substrates. Extrahepatic sources of ketone bodies, eg, the rumen epithelium in fed ruminants, do not contribute significantly to the occurrence of ketosis in these species.

The net flow of ketone bodies from the liver to the extrahepatic tissues results from an active enzymatic mechanism in the liver for the production of ketone bodies coupled with very low activity of enzymes responsible for their utilization. The reverse situation occurs in extrahepatic tissues (Figure 24–6).

3-Hydroxy-3-Methylglutaryl-CoA (HMG-CoA) Is an Intermediate in the Pathway of Ketogenesis

Enzymes responsible for ketone body formation are associated mainly with the mitochondria. Originally it was thought that only one molecule of acetoacetate was formed from the terminal four carbons of a fatty acid upon oxidation. Later, to explain both the production of more than one equivalent of acetoacetate from a long-chain fatty acid and the formation of ketone bodies from acetic acid, it was proposed that C_2 units formed in β-oxidation condensed with one another to form acetoacetate. This may occur by a reversal of the **thiolase** reaction whereby two molecules of acetyl-CoA condense to form acetoacetyl-CoA. Thus, acetoacetyl-CoA, which is the starting material for ketogenesis, arises either directly

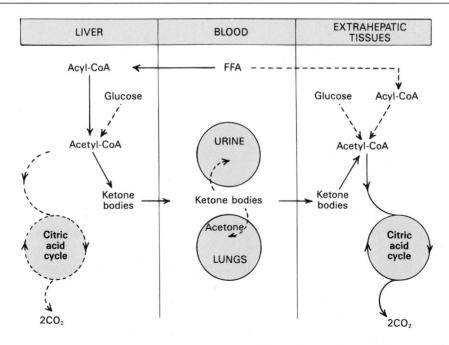

Figure 24–6. Formation, utilization, and excretion of ketone bodies. (The main pathway is indicated by the solid arrows.)

during the course of β-oxidation or as a result of the condensation of acetyl-CoA (Figure 24–7). The pathway involves the condensation of acetoacetyl-CoA with another molecule of acetyl-CoA to form **HMG-CoA,** catalyzed by **3-hydroxy-3-methylglutaryl-CoA synthase.** The presence of another enzyme in the mitochondria, **3-hydroxy-3-methylglutaryl-CoA lyase,** causes acetyl-CoA to split off from the HMG-CoA, leaving free acetoacetate. The carbon atoms split off in the acetyl-CoA molecule are derived from the original acetoacetyl-CoA molecule (Figure 24–7). **Both enzymes must be present in mitochondria for ketogenesis to take place.** This occurs solely in liver and rumen epithelium. Although there is an increase in activity of HMG-CoA lyase in fasting, evidence does not suggest that this enzyme is rate-limiting in ketogenesis.

Acetoacetate is in equilibrium with D(–)-3-hy-droxybutyrate catalyzed by **D(–)-3-hydroxybutyrate dehydrogenase,** which is present in mitochondria of many tissues, including the liver (Figure 24–5). D(–)-3-Hydroxybutyrate is quantitatively the predominant ketone body present in the blood and urine in ketosis.

Ketone Bodies Serve as a Fuel for Extrahepatic Tissues

While the liver is equipped with an active enzymatic mechanism for the production of acetoacetate from acetoacetyl-CoA, acetoacetate once formed cannot be reactivated directly in the liver except in the cytosol, where it is a precursor in cholesterol synthesis, a much less active pathway. This accounts for the net production of ketone bodies by the liver.

The main pathway of utilization in extrahepatic

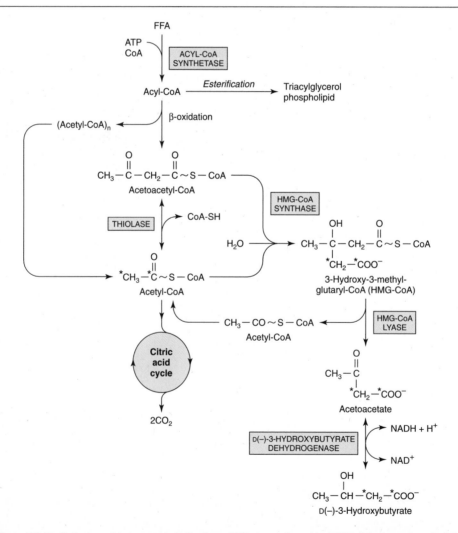

Figure 24–7. Pathways of ketogenesis in the liver. (FFA, free fatty acids; HMG, 3-hydroxy-3-methylglutaryl.)

tissues for the activation of acetoacetate to acetoacetyl-CoA involves succinyl-CoA and the enzyme **succinyl-CoA-acetoacetate CoA transferase.** Acetoacetate reacts with succinyl-CoA, and the CoA is transferred to form acetoacetyl-CoA, leaving free succinate (Figure 24–8). The acetoacetyl-CoA formed by these reactions is split to acetyl-CoA by thiolase and oxidized in the citric acid cycle as shown in Figure 24–8. Ketone bodies are oxidized in extrahepatic tissues proportionately to their concentration in the blood. If the blood level is raised, oxidation of ketone bodies increases until, at a concentration of approximately 12 mmol/L, they saturate the oxidative machinery. When this occurs, a large proportion of the oxygen consumption may be accounted for by the oxidation of ketone bodies.

Most of the evidence suggests that **ketonemia is due to increased production of ketone bodies** by the liver rather than to a deficiency in their utilization by extrahepatic tissues. However, the results of experiments on depancreatized rats support the possibility that ketosis in the severe diabetic may be enhanced by a reduced ability to catabolize ketone bodies. While acetoacetate and D(−)-3-hydroxybutyrate are readily oxidized by extrahepatic tissues, acetone is difficult to oxidize in vivo and to a large extent is volatilized in the lungs.

In moderate ketonemia, the loss of ketone bodies via the urine is only a few percent of the total ketone body production and utilization. Since there are renal threshold-like effects (there is not a true threshold) that vary between species and individuals, measurement of the ketonemia, not the ketonuria, is the preferred method of assessing the severity of ketosis.

KETOGENESIS IS REGULATED AT THREE CRUCIAL STEPS

(1) Control is exercised initially in adipose tissue. Ketosis does not occur in vivo unless there is an increase in the level of circulating free fatty acids that arise from lipolysis of triacylglycerol in adipose tissue. **Free fatty acids are the precursors of ketone bodies in the liver.** The liver, both in fed and in fasting conditions, has the ability to extract about 30% or more of the free fatty acids passing through it, so that at high concentrations of free fatty acids the flux passing into the liver is substantial. **Therefore, the factors regulating mobilization of free fatty acids from adipose tissue are important in controlling ketogenesis** (Figures 24–9 and 27–9).

(2) One of two fates awaits the free fatty acids upon uptake by the liver and after they are activated

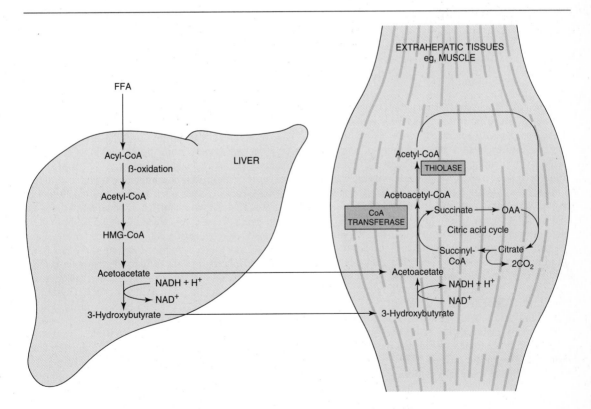

Figure 24–8. Transport of ketone bodies from the liver and mechanism of utilization and oxidation in extrahepatic tissues.

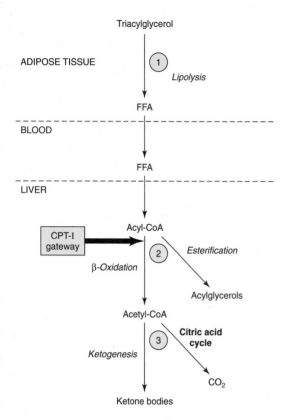

Figure 24–9. Regulation of ketogenesis. ①–③ show three crucial steps in the pathway of metabolism of free fatty acids (FFA) that determine the magnitude of ketogenesis. (CPT-I, carnitine palmitoyltransferase-I.)

to acyl-CoA: They are β-oxidized to CO_2 or ketone bodies or **esterified** to triacylglycerol and phospholipid. The capacity for esterification as an antiketogenic factor depends on the availability of precursors in the liver to supply sufficient glycerol 3-phosphate (Figure 26–2). However, the availability of glycerol 3-phosphate does not limit esterification in fasting livers. Whether its availability in the liver is ever rate-limiting on esterification is not clear; neither is there critical information on whether the activities in vivo of the enzymes involved in esterification are rate-limiting. It does not seem that they are, since neither free fatty acids nor any intermediates in their pathway of esterification to triacylglycerol (Figure 26–2) ever accumulate in the liver.

Perfused livers from starved rats oxidize considerably more ^{14}C-labeled free fatty acids to [14]CO_2 and ^{14}C-ketone bodies and esterify less as ^{14}C-acylglycerol than livers from fed rats. These results may be explained by the fact that **carnitine palmitoyltransferase I** activity in the outer mitochondrial membrane regulates the entry of long-chain acyl groups into mitochondria prior to β-oxidation (Figures 24–1 and 24–10). Its activity is low in the fed state, when

fatty acid oxidation is depressed, and high in starvation, when fatty acid oxidation increases. **Malonyl-CoA,** the initial intermediate in fatty acid biosynthesis (Figure 23-1), which increases in concentration in the fed state, inhibits this enzyme, thereby switching off β-oxidation. Thus, in the fed condition there is active lipogenesis and high [malonyl-CoA], which inhibits carnitine palmitoyltransferase I (Figure 24–10). Free fatty acids, in the fed state, enter the liver cell in low concentrations and are nearly all esterified to acylglycerols and transported out of the liver in very low density lipoproteins (VLDL). However, as the concentration of free fatty acids increases with the onset of starvation, acetyl-CoA carboxylase is inhibited directly by acyl-CoA, and [malonyl-CoA] decreases, releasing the inhibition of carnitine palmitoyltransferase I and allowing more acyl-CoA to be β-oxidized. These events are reinforced in starvation by decrease in the **[insulin]/[glucagon] ratio.** The direct result of this decrease is inhibition of acetyl-CoA carboxylase in the liver by covalent phosphorylation; indirectly, the result is to increase lipolysis in adipose tissue and release free fatty acids, which, after uptake, form acyl-CoA in the liver, inhibiting acetyl-CoA carboxylase (Figure 24–10). Thus, β-oxidation from free fatty acids is controlled by the carnitine palmitoyltransferase-I gateway into the mitochondria, and the balance of the free fatty acid uptake not oxidized is esterified.

(3) In turn, the acetyl-CoA formed in β-oxidation is oxidized in the citric acid cycle, or it enters the pathway of ketogenesis to form ketone bodies. As the level of serum free fatty acids is raised, proportionately more free fatty acid is converted to ketone bodies and less is oxidized via the citric acid cycle to CO_2. The partition of acetyl-CoA between the ketogenic pathway and the pathway of oxidation to CO_2 is so regulated that the total free energy captured in ATP which results from the oxidation of free fatty acids remains constant. This may be appreciated when it is realized that complete oxidation of 1 mol of palmitate involves a net production of 129 mol of ATP via β-oxidation and CO_2 production in the citric acid cycle (see above), whereas only 33 mol of ATP is produced when acetoacetate is the end product and only 21 mol when 3-hydroxybutyrate is the end product. Thus, ketogenesis may be regarded as a mechanism that allows the liver to oxidize increasing quantities of fatty acids within a tightly coupled system of oxidative phosphorylation, without increasing its total energy expenditure.

Several other hypotheses have been advanced to account for the diversion of fatty acid oxidation from CO_2 formation to ketogenesis. Theoretically, a fall in concentration of oxaloacetate, particularly within the mitochondria, could impair the ability of the citric acid cycle to metabolize acetyl-CoA. This fall may occur because of an increase in the [NADH]/NAD[+] ratio caused by increased β-oxidation. This would af-

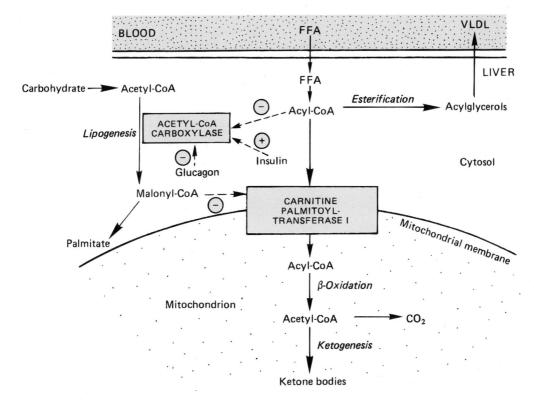

Figure 24–10. Regulation of long-chain fatty acid oxidation in the liver. (FFA, free fatty acids; VLDL, very low density lipoprotein.) Positive (⊕) and negative (⊖) regulatory effects are represented by broken lines and substrate flow by solid lines.

fect the equilibrium between oxaloacetate and malate, decreasing the concentration of oxaloacetate. Krebs has suggested that since oxaloacetate is also on the main pathway of gluconeogenesis, enhanced gluconeogenesis leading to a fall in the level of oxaloacetate may be the cause of the severe forms of ketosis found in diabetes and the ketosis of cattle. However, Utter and Keech have shown that pyruvate carboxylase, which catalyzes the conversion of pyruvate to oxaloacetate, is activated by acetyl-CoA. Consequently, when there are significant amounts of acetyl-CoA, there should be sufficient oxaloacetate to initiate the condensing reaction of the citric acid cycle.

CLINICAL ASPECTS

Impaired Oxidation of Fatty Acids Gives Rise to Diseases Often Associated With Hypoglycemia

Carnitine deficiency can occur particularly in the newborn—and especially in preterm infants—owing to inadequate biosynthesis or renal leakage. Losses can also occur in hemodialysis; patients with organic aciduria have large losses of carnitine, which is excreted conjugated to the organic acids. This indicates a vitamin-like dietary requirement for carnitine in some individuals. Signs and symptoms of deficiency include episodic periods of hypoglycemia owing to reduced gluconeogenesis resulting from impaired fatty acid oxidation and ketogenesis in the presence of raised plasma FFA, leading to a lipid accumulation with muscular weakness. Treatment is by oral supplementation of carnitine. The symptoms are similar to **Reye's syndrome** (encephalopathy with fatty degeneration of the viscera), in which carnitine is adequate, but the cause of Reye's syndrome is unknown.

Carnitine palmitoyltransferase I deficiency affects only the liver, resulting in reduced fatty acid oxidation and ketogenesis with hypoglycemia. On the other hand, **carnitine palmitoyltransferase II deficiency** affects primarily skeletal muscle (weakness and necrosis with myoglobinuria) and, in its most severe form, the liver. In a similar manner, the hypoglycemic sulfonylureas (**glyburide [glibenclamide]** and **tolbutamide**) reduce fatty acid oxidation by inhibiting carnitine palmitoyltransferase.

Inherited defects in the enzymes of β-oxidation

also lead to hypoglycemia, coma, and fatty liver, eg, deficiency in long-chain β-hydroxyacyl-CoA dehydrogenase may be a cause of **acute fatty liver of pregnancy.** Defects in the genes for 3-ketoacyl-CoA thiolase are also known. Also inborn errors of ketogenesis have been reported, such as **HMG-CoA lyase deficiency,** which also affects the degradation of leucine, a ketogenic amino acid (Chapter 32).

Jamaican vomiting sickness is caused by eating the unripe fruit of the akee tree, which contains a toxin, **hypoglycin,** that inactivates medium- and short-chain acyl-CoA dehydrogenase, inhibiting β-oxidation and causing hypoglycemia with excretion of medium- and short-chain mono- and dicarboxylic acids.

Dicarboxylic aciduria is characterized by the excretion of C_6–C_{10} ω-dicarboxylic acids and by nonketotic hypoglycemia. It is caused by a lack of mitochondrial **medium-chain acyl-CoA dehydrogenase.** This impairs β-oxidation but increases ω-oxidation of long- and medium-chain fatty acids, which are then shortened by β-oxidation to medium-chain dicarboxylic acids, which are excreted. Deficiencies of long- and short-chain acyl-CoA dehydrogenases have now been reported, together with deficiencies of hydroxyacyl-CoA dehydrogenase and 2,4-dienoyl-CoA reductase, which leads to urinary excretion of 2-*trans*-4-*cis*-decadienoylcarnitine as a result of a failure in β-oxidation of polyunsaturated fatty acids.

Refsum's disease is a rare neurologic disorder caused by accumulation of phytanic acid, formed from phytol, a constituent of chlorophyll found in plant foodstuffs. Phytanic acid contains a methyl group on carbon 3 that blocks β-oxidation. Normally, an initial α-oxidation removes the methyl group, but persons with Refsum's disease have an inherited defect in α-oxidation that allows accumulation of phytanic acid.

Zellweger's (cerebrohepatorenal) syndrome occurs in individuals with a rare inherited absence of peroxisomes in all tissues. They accumulate C_{26}-C_{38} polyenoic acids in brain tissue owing to inability to oxidize long-chain fatty acids in peroxisomes but also exhibit a generalized loss of peroxisomal functions, eg, impaired bile acid and ether lipid synthesis.

Ketoacidosis Results From Prolonged Ketosis

Higher than normal quantities of ketone bodies present in the blood or urine constitute **ketonemia** (hyperketonemia) or **ketonuria,** respectively. The overall condition is called **ketosis.** Acetoacetic and 3-hydroxybutyric acids are both moderately strong acids and are buffered when present in blood or other tissues. However, their continual excretion in quantity entails some loss of buffer cation (in spite of ammonia production by the kidney) that progressively depletes the alkali reserve, causing **ketoacidosis.** This may be fatal in uncontrolled **diabetes mellitus.**

The simplest form of ketosis occurs in **starvation** and involves depletion of available carbohydrate coupled with mobilization of free fatty acids. No other condition in which ketosis occurs seems to differ qualitatively from this general pattern of metabolism, but quantitatively it may be exaggerated to produce the pathologic states found in **diabetes mellitus, pregnancy toxemia in sheep,** and **ketosis in lactating cattle.** Nonpathologic forms of ketosis are found under conditions of high-fat feeding and after severe exercise in the postabsorptive state.

SUMMARY

1. Fatty acid oxidation in mitochondria leads to the generation of large quantities of ATP by a process called β-oxidation that cleaves acetyl-CoA units sequentially from fatty acyl chains. The acetyl-CoA is oxidized in the citric acid cycle, generating further ATP.

2. Oxidation of odd-numbered-carbon fatty acids yields acetyl-CoA plus one molecule of propionyl-CoA, which is glucogenic.

3. Peroxisomes are capable of oxidizing very long chain fatty acids but only as far as octanoyl-CoA, which must then be transferred to mitochondria for further oxidation.

4. The ketone bodies (acetoacetate, 3-hydroxybutyrate, and acetone) are formed in hepatic mitochondria when there is a high rate of fatty acid oxidation. The pathway of ketogenesis involves synthesis and breakdown of 3-hydroxy-3-methylglutaryl-CoA (HMG-CoA) by two key ketogenic enzymes, HMG-CoA synthase and HMG-CoA lyase.

5. Ketone bodies are important fuels in extrahepatic tissues.

6. Ketogenesis is regulated at three crucial steps: (1) control of free fatty acid mobilization from adipose tissue; (2) the activity of carnitine palmitoyltransferase-I in liver, which determines the proportion of the fatty acid flux that is oxidized rather than esterified; and (3) partition of acetyl-CoA between the pathway of ketogenesis and the citric acid cycle.

7. Diseases associated with impairment of fatty acid oxidation lead to hypoglycemia, fatty infiltration of organs, and hypoketonemia.

8. Ketosis is mild in starvation but severe in diabetes mellitus and ruminant ketosis.

REFERENCES

Boyer PD (editor): *The Enzymes,* 3rd ed. Vol 16 of *Lipid Enzymology.* Academic Press, 1983.

Mayes PA, Laker ME: Regulation of ketogenesis in the liver. Biochem Soc Trans 1981;9:339.

McGarry JD, Foster DW: Regulation of hepatic fatty acid oxidation and ketone body production. Annu Rev Biochem 1980;49:395.

Osmundsen H, Hovik R: β-Oxidation of polyunsaturated fatty acids. Biochem Soc Trans 1988;16:420.

Poulos A: Lipid metabolism in Zellweger's syndrome. Prog Lipid Res 1989;28:35.

Reddy JK, Mannaerts GP: Peroxisomal lipid metabolism. Annu Rev Nutr 1994;14:343.

Scholte HR et al: Primary carnitine deficiency. J Clin Chem Biochem 1990;28:351.

Schulz H: Beta oxidation of fatty acids. Biochim Biophys Acta 1991;1081:109.

Scriver CR et al (editors): *The Metabolic and Molecular Bases of Inherited Disease,* 7th ed. McGraw-Hill, 1995.

Treem WR et al: Acute fatty liver of pregnancy and long-chain 3-hydroxyacyl-coenzyme A dehydrogenase deficiency. Hepatology 1994;19:339.

25

Metabolism of Unsaturated Fatty Acids & Eicosanoids

Peter A. Mayes, PhD, DSc

INTRODUCTION

Compared with plants, animal tissues have limited ability in desaturating fatty acids. This necessitates dietary intake of certain polyunsaturated fatty acids derived ultimately from a plant source. These **essential fatty acids** give rise to eicosanoic (C_{20}) fatty acids, from which are derived families of compounds known as **eicosanoids.** These make up the prostaglandins, thromboxanes, leukotrienes, and lipoxins.

BIOMEDICAL IMPORTANCE

The content of unsaturated fatty acids in a natural fat is a major determinant of its melting point and therefore its fluidity. Similarly, phospholipids of the cell membrane contain unsaturated fatty acids important in maintaining membrane fluidity. A high ratio of polyunsaturated fatty acids to saturated fatty acids (P:S ratio) in the diet is a major factor in lowering plasma cholesterol concentrations by dietary means and is considered to be beneficial in preventing coronary heart disease. The prostaglandins and thromboxanes are local hormones that are synthesized rapidly when required and act near their sites of synthesis. The major physiologic roles played by prostaglandins are as modulators of adenylyl cyclase activity, eg, (1) in controlling platelet aggregation, and (2) in inhibiting the effect of antidiuretic hormone in the kidney. Nonsteroidal anti-inflammatory drugs, such as **aspirin,** act by inhibiting prostaglandin synthesis. Leukotrienes have muscle contractant and chemotactic properties, suggesting that they are important in allergic reactions and inflammation. A mixture of leukotrienes has been identified as the slow-reacting substance of anaphylaxis (SRS-A). By varying the proportions of the different polyunsaturated fatty acids in the diet, it is possible to influence the type of eicosanoids synthesized, indicating that it might be possible to influence the course of disease by dietary means.

SOME POLYUNSATURATED FATTY ACIDS CANNOT BE SYNTHESIZED BY MAMMALS AND ARE THEREFORE NUTRITIONALLY ESSENTIAL

Some long-chain unsaturated fatty acids of metabolic significance in mammals are shown in Figure 25–1. (For a review of the nomenclature of fatty acids, see Chapter 16.)

Other C_{20}, C_{22}, and C_{24} polyenoic fatty acids may be detected in the tissues. These may be derived from oleic, linoleic, and α-linolenic acids by chain elongation. It is to be noted that all double bonds present in naturally occurring unsaturated fatty acids of mammals are of the *cis* **configuration.**

Palmitoleic and oleic acids are not essential in the diet, because the tissues are capable of introducing a double bond at the Δ^9 position into the corresponding saturated fatty acid. Experiments with labeled palmitate have demonstrated that the label enters freely into palmitoleic and oleic acids but is absent from **linoleic and α-linolenic acids.** These are the only fatty acids known to be essential for the complete nutrition of many species of animals, including humans, and must therefore be supplied in the diet; as a consequence, they are known as the **nutritionally essential fatty acids. Arachidonic acid** can be formed from linoleic acid in most mammals (Figure 25–4) but not in the cat family, where it must be classified as an essential fatty acid. In most animals, double bonds can be introduced at the Δ^4, Δ^5, Δ^6, and Δ^9 positions (counting from the carboxyl terminal; see Chapter 16) but never beyond the Δ^9 position. In contrast, plants are able to introduce additional double bonds at the Δ^{12} and Δ^{15} positions and can therefore synthesize the nutritionally essential fatty acids.

MONOUNSATURATED FATTY ACIDS ARE SYNTHESIZED BY A Δ^9 DESATURASE SYSTEM

As far as the nonessential monounsaturated fatty acids are concerned, several tissues including the

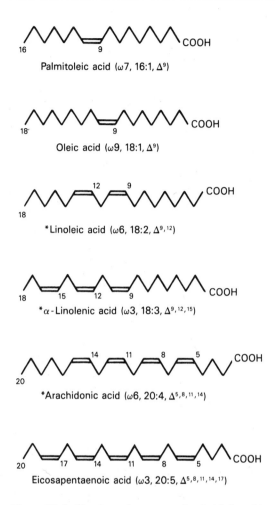

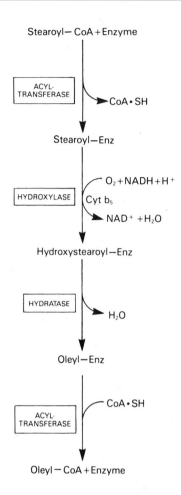

Figure 25–1. Structure of some unsaturated fatty acids. Although the carbon atoms in the molecules are conventionally numbered, ie, numbered from the carboxyl terminal, the ω numbers (eg, ω7 in palmitoleic acid) are calculated from the reverse end (the methyl terminal) of the molecules. The information in parentheses shows, for instance, that α-linolenic acid contains double bonds starting at the third carbon from the methyl terminal, has 18 carbons and 3 double bonds, and has these double bonds at the 9th, 12th, and 15th carbons from the carboxyl terminal. (*, Classified as "essential fatty acids.")

Figure 25–2. Microsomal Δ^9 desaturase system.

SYNTHESIS OF POLYUNSATURATED FATTY ACIDS INVOLVES DESATURASE AND ELONGASE ENZYME SYSTEMS

Additional double bonds introduced into existing monounsaturated fatty acids are always separated from each other by a methylene group (methylene interrupted) except in bacteria. In animals, the additional double bonds are all introduced between the existing double bond and the carboxyl group, but in plants they may also be introduced between the existing double bond and the ω (methyl terminal) carbon. Thus, since animals have a Δ^9 desaturase, they are able to synthesize the ω9 (oleic acid) family of unsaturated fatty acids completely by a combination of chain elongation and desaturation (Figure 25–3). However, since they are unable to synthesize either linoleic (ω6) or α-linolenic (ω3) acids because the required desaturases are absent, these acids must be supplied in the diet to accomplish the synthesis of the other members of the

liver are considered to be responsible for their formation from saturated fatty acids. The first double bond introduced into a saturated fatty acid is nearly always in the Δ^9 position. An enzyme system, **Δ^9 desaturase** (Figure 25–2), in the endoplasmic reticulum will catalyze the conversion of palmitoyl-CoA or stearoyl-CoA to palmitoleyl-CoA or oleyl-CoA, respectively. Oxygen and either NADH or NADPH are necessary for the reaction. The enzymes appear to be those of a typical monooxygenase system involving cytochrome b_5 (hydroxylase) (Chapter 13).

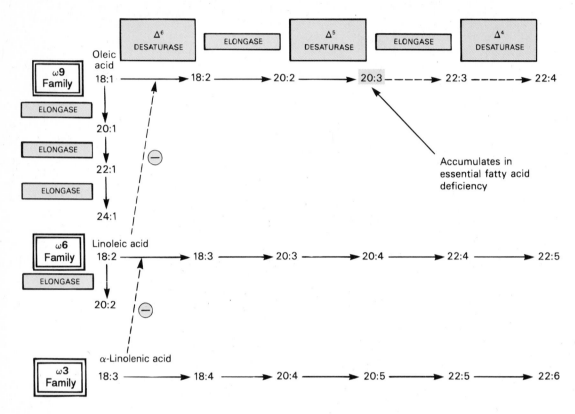

Figure 25–3. Biosynthesis of the ω9, ω6, and ω3 families of polyunsaturated fatty acids. Each step is catalyzed by the microsomal chain elongation or desaturase system. ω9 Polyunsaturated fatty acids only become quantitatively significant when linoleic and α-linolenic acids are withheld from the diet. This is because each series competes for the same enzyme systems, and affinities decrease from the ω3 to ω9 series. (⊖, Inhibition.)

ω6 and ω3 families of polyunsaturated fatty acids. Linoleate may be converted to arachidonate (Figure 25–4). The pathway is first by dehydrogenation of the CoA ester through **γ-linolenate,** followed by the addition of a two-carbon unit via malonyl-CoA in the microsomal system for chain elongation (Figure 23–6), to give eicosatrienoate (dihomo γ-linolenate). The latter forms arachidonate by a further dehydrogenation. The dehydrogenating system is similar to that described above for saturated fatty acids. The nutritional requirement for arachidonate may thus be dispensed with if there is adequate linoleate in the diet.

The desaturation and chain elongation system is greatly diminished in the fasting state, upon glucagon and epinephrine administration, and in the absence of insulin as in type I diabetes mellitus.

DEFICIENCY SYMPTOMS ARE PRODUCED WHEN THE ESSENTIAL FATTY ACIDS (EFA) ARE ABSENT FROM THE DIET

In 1928, Evans and Burr noticed that rats fed on a purified nonlipid diet to which vitamins A and D

were added exhibited a reduced growth rate and a reproductive deficiency. Later work showed that the deficiency syndrome was cured by the addition of **linoleic, α-linolenic,** and **arachidonic acids** to the diet. Further diagnostic features of the syndrome include scaly skin, necrosis of the tail, and lesions in the urinary system, but the condition is not fatal. These fatty acids are found in high concentrations in various vegetable oils (Table 16–2) and in small amounts in animal carcasses.

The functions of the essential fatty acids appear to be various, though not well defined, apart from prostaglandin and leukotriene formation (see below). Essential fatty acids are found in the structural lipids of the cell and are concerned with the structural integrity of the mitochondrial membrane.

Arachidonic acid is present in membranes and accounts for 5–15% of the fatty acids in phospholipids. Docosahexaenoic acid (DHA; ω3,2:6), which is synthesized from α-linolenic acid or obtained directly from fish oils, is present in high concentrations in retina, cerebral cortex, testis, and sperm. DHA is particularly needed for development of the brain and retina development and is supplied via the placenta and milk. The outer segments of the retinal rods con-

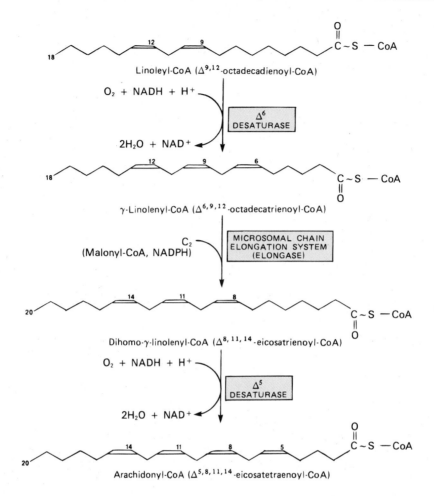

Figure 25–4. Conversion of linoleate to arachidonate. Cats cannot carry out this conversion owing to absence of Δ^6 desaturase and must obtain arachidonate in their diet.

tain very high concentrations of DHA, with most of the phospholipids containing at least one molecule. It appears that the resulting high fluidity is necessary for the functioning of rhodopsin, activation by a photon causing lateral and rotational movement within the membrane. Patients with **retinitis pigmentosa** are reported to have low blood levels of DHA. Premature infants have a low Δ^4 desaturase activity, reducing their potential for synthesizing DHA from n-3 fatty acid precursors.

In many of their structural functions, essential fatty acids are present in phospholipids, mainly in the 2 position. **In essential fatty acid deficiency,** nonessential polyenoic acids of the ω9 family replace the essential fatty acids in phospholipids, other complex lipids, and membranes, particularly $\Delta^{5,8,11}$-eicosatrienoic acid (Figure 25–3). The triene:tetraene ratio in plasma lipids can be used to diagnose the extent of essential fatty acid deficiency.

Trans-Fatty Acids May Compete With Cis-Fatty Acids

Traces of *trans*-unsaturated fatty acids are found in ruminant fat, where they arise from the action of microorganisms in the rumen, but the presence of large amounts of *trans*-unsaturated fatty acids in partially hydrogenated vegetable oils (eg, margarine) raises the question of their safety as food additives. Their long-term effects in humans are difficult to assess but they have been in the human diet for many years. Up to 15% of tissue fatty acids have been found at autopsy to be in the *trans* configuration. To date, no serious effects have been substantiated. They are metabolized more like saturated than like the *cis*-unsaturated fatty acids. This may be due to their similar straight-chain conformation (Chapter 16). In this respect, they tend to raise LDL levels and lower HDL levels and are thus contraindicated with respect to the enhancement of atherosclerosis and coronary heart disease. *Trans*-polyunsaturated fatty acids do not possess essential fatty acid activity and may antagonize the metabolism of essential fatty acids and exacerbate essential fatty acid deficiency.

EICOSANOIDS ARE FORMED FROM C_{20} POLYUNSATURATED FATTY ACIDS

Arachidonate and some other C_{20} fatty acids with methylene-interrupted bonds give rise to **eicosanoids,** physiologically and pharmacologically active compounds known as **prostaglandins (PG), thromboxanes (TX), leukotrienes (LT),** and **lipoxins (LX)** (Chapter 16). Physiologically, they are considered to act as local hormones functioning through G-protein linked receptors to elicit their biochemical effects.

Arachidonate, usually derived from the 2-position of phospholipids in the plasma membrane, as a result of phospholipase A_2 activity (Figure 26–6), is the substrate for the synthesis of the PG_2, TX_2, LT_4, and LX_4 compounds. The pathways of metabolism are divergent, the synthesis of the PG_2 and TX_2 series **(prostanoids)** competing with the synthesis of LT_4 and LX_4 for the arachidonate substrate. These two pathways are known as the **cyclooxygenase** and **lipoxygenase pathways,** respectively (Figure 25–5).

There are three groups of eicosanoids (each comprising PG, TX, LT, and possibly LX) that are synthesized from C_{20} eicosanoic acids derived from the essential fatty acids **linoleate** and **α-linolenate,** or directly from arachidonate and eicosapentaenoate in the diet (Figure 25–6).

THE CYCLOOXYGENASE PATHWAY IS RESPONSIBLE FOR PROSTANOID SYNTHESIS

Prostanoid synthesis (Figure 25–7) involves the consumption of two molecules of O_2 catalyzed by **prostaglandin endoperoxide synthase,** which possesses two separate enzyme activities, **cyclooxygenase** and **peroxidase.** The product of the cyclooxygenase pathway, an endoperoxide (PGH), is converted to prostaglandins D, E, and F as well as to the thromboxane (TXA_2) and prostacyclin (PGI_2). Each cell type produces only one type of prostanoid. **Aspirin** inhibits the cyclooxygenase, as do indomethacin and ibuprofen.

Essential Fatty Acid Activity and Production of Prostaglandins Are Correlated

Although there is marked correlation between essential fatty acid activity of various fatty acids and their ability to be converted to prostaglandins, it does not seem that essential fatty acids exert all their physiologic effects via prostaglandin synthesis. The role of essential fatty acids in membrane formation is unrelated to prostaglandin formation. Prostaglandins do not relieve symptoms of essential fatty acid deficiency, and an essential fatty acid deficiency syndrome is not caused by chronic inhibition of prostaglandin synthesis.

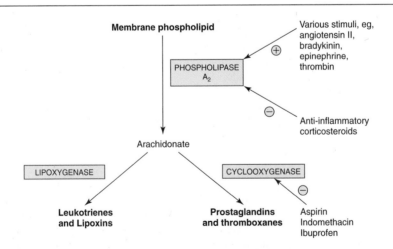

Figure 25–5. Conversion of arachidonic acid to prostaglandins and thromboxanes via the cyclooxygenase pathway and to leukotrienes and lipoxins via the lipoxygenase pathway. The figure indicates why steroids, which inhibit total eicosanoid production, are better anti-inflammatory agents than aspirin-like drugs, which only inhibit the cyclooxygenase pathway. Anti-inflammatory steroids are thought to inhibit phospholipase A_2 through induction of an inhibitory protein named lipocortin.

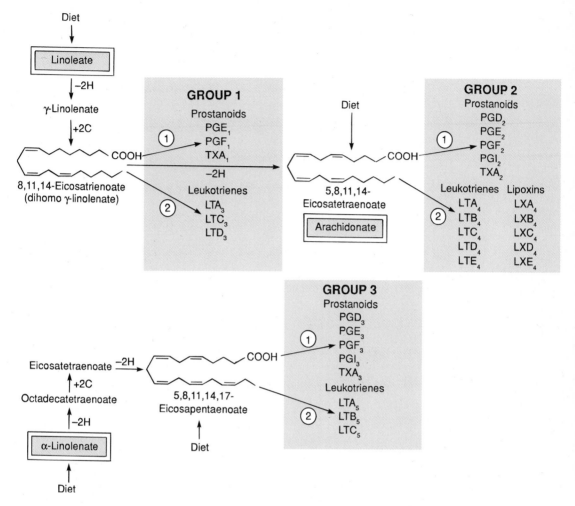

Figure 25–6. The three groups of eicosanoids and their biosynthetic origins. (PG, prostaglandin; PGI, prostacyclin; TX, thromboxane; LT, leukotriene; LX, lipoxin; ①, cyclooxygenase pathway; ②, lipoxygenase pathway.) The subscript denotes the total number of double bonds in the molecule and the series to which the compound belongs.

Cyclooxygenase Is a "Suicide Enzyme"

"Switching off" of prostaglandin formation is partly achieved by a remarkable property of cyclooxygenase—that of self-catalyzed destruction; ie, it is a "suicide enzyme." The inactivation of prostaglandins, once formed, is rapid. The presence of the enzyme **15-hydroxyprostaglandin dehydrogenase** in most mammalian tissues is probably the principal cause. Blocking the action of this enzyme with sulfasalazine or indomethacin can prolong the half-life of prostaglandins in the body.

LEUKOTRIENES AND LIPOXINS ARE FORMED BY THE LIPOXYGENASE PATHWAY

The leukotrienes are a family of conjugated trienes formed from eicosanoic acids in leukocytes, mastocytoma cells, platelets, and macrophages by the **lipoxygenase pathway,** in response to both immunologic and nonimmunologic stimuli (Figure 25–8). Three different lipoxygenases (dioxygenases) insert oxygen into the 5, 12, and 15 positions of arachidonic acid, giving rise to hydroperoxides (HPETE). Only **5-lipoxygenase** forms leukotrienes. The first formed is leukotriene A_4, which in turn is metabolized to either leukotriene B_4 or leukotriene C_4. Leukotriene C_4 is formed by the addition of the peptide glutathione via a thioether bond. The subsequent

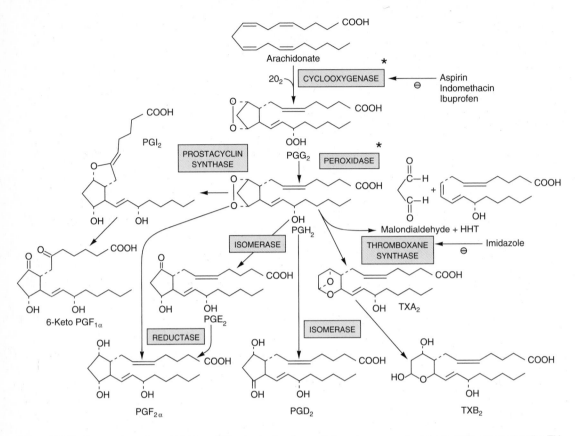

Figure 25–7. Conversion of arachidonic acid to prostaglandins and thromboxanes of series 2. (PG, prostaglandin; TX, thromboxane; PGI, prostacyclin; HHT, hydroxyheptadecatrienoate.) (*, Both of these starred activities are attributed to one enzyme: prostaglandin endoperoxide synthase. Similar conversions occur in prostaglandins and thromboxanes of series 1 and 3.)

removal of glutamate and glycine generates leukotriene D_4 and leukotriene E_4, sequentially.

Lipoxins are a family of conjugated tetraenes also arising in leukocytes. They are formed by the combined action of more than one lipoxygenase, introducing more oxygen into the molecule. Several lipoxins are formed (LXA_4 to LXE_4) in a manner similar to that described above for leukotrienes.

CLINICAL ASPECTS

Humans Also Show Symptoms When Deficient in Essential Fatty Acids

Skin symptoms and impairment of lipid transport have been noted in human subjects ingesting a diet lacking in essential fatty acids. In adults subsisting on ordinary diets, no signs of essential fatty acid deficiencies have been reported. However, infants receiving formula diets low in fat developed skin symptoms that were cured by giving linoleate. Deficiencies attributable to a lack of essential fatty acids, including α-linolenic acid, also occur among patients

maintained for long periods exclusively by intravenous nutrition low in essential fatty acids. Deficiency can be prevented by an essential fatty acid intake of 1–2% of the total caloric requirement.

Abnormal Metabolism of Essential Fatty Acids Occurs in Several Diseases

Apart from essential fatty acid deficiency and changes in unsaturated fatty acid patterns in chronic malnutrition, abnormal metabolism of essential fatty acids, which may be connected with dietary insufficiency, has been noted in cystic fibrosis, acrodermatitis enteropathica, hepatorenal syndrome, Sjögren-Larsson syndrome, multisystem neuronal degeneration, Crohn's disease, cirrhosis and alcoholism, and Reye's syndrome. Elevated levels of very long chain polyenoic acids have been found in the brains of patients with Zellweger's syndrome (Chapter 24). Diets with a high P:S (polyunsaturated:saturated fatty acid) ratio lower serum cholesterol levels, particularly in low-density lipoproteins. This is considered to be beneficial in view of the re-

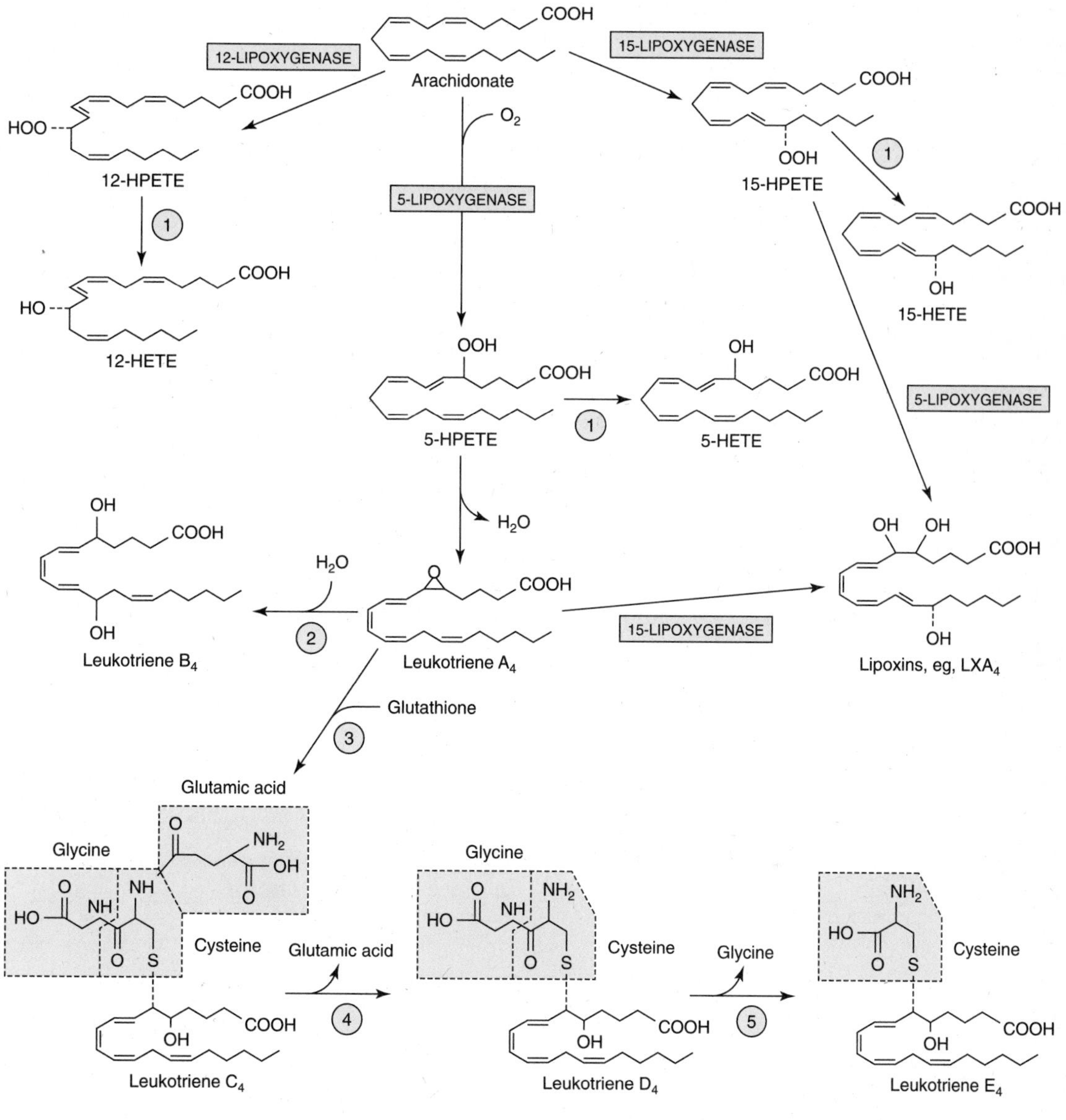

Figure 25–8. Conversion of arachidonic acid to leukotrienes and lipoxins of series 4 via the lipoxygenase pathway. Some similar conversions occur in series 3 and 5 leukotrienes. (HPETE, hydroperoxyeicosatetraenoate; HETE, hydroxyeicosatetraenoate.) (①, peroxidase; ②, leukotriene A_4 epoxide hydrolase; ③, glutathione S-transferase; ④, γ-glutamyltransferase; ⑤, cysteinyl-glycine dipeptidase.)

lationship between serum cholesterol level and coronary heart disease.

Prostanoids Are Potent Biologically Active Substances

Thromboxanes are synthesized in platelets and upon release cause vasoconstriction and platelet aggregation. **Prostacyclins (PGI$_2$)** are produced by blood vessel walls and are potent inhibitors of platelet aggregation. Thus, thromboxanes and prostacyclins are antagonistic. The low incidence of heart disease, diminished platelet aggregation, and prolonged clotting times in Greenland Eskimos have been attributed to their high intake of fish oils containing 20:5 ω3 (EPA, or eicosapentaenoic acid), which gives rise to the series 3 prostaglandins (PG$_3$) and thromboxane TX$_3$ (Figure 25–6). PG$_3$ and TX$_3$ inhibit the release of arachidonate from phospholipids and the formation of PG$_2$ and TX$_2$. PGI$_3$ is as potent an antiaggregator of platelets as PGI$_2$, but TXA$_3$ is a weaker aggregator than TXA$_2$; thus, the balance of activity is shifted toward nonaggregation.

In addition, the plasma concentrations of cholesterol, triacylglycerol, and low-density and very low density lipoproteins are all low in Eskimos, whereas the high-density lipoprotein concentration is raised—all factors considered to militate against atherosclerosis and myocardial infarction.

As little as 1 ng/mL of prostaglandins causes contraction of smooth muscle in animals. Potential therapeutic uses include prevention of conception, induction of labor at term, termination of pregnancy, prevention or alleviation of gastric ulcers, control of inflammation and of blood pressure, and relief of asthma and nasal congestion.

Prostaglandins increase cAMP in platelets, thyroid, corpus luteum, fetal bone, adenohypophysis, and lung but lower cAMP in renal tubule cells and adipose tissue (Chapter 27).

Leukotrienes and Lipoxins Are Potent Regulators of Many Disease Processes

The slow-reacting substance of anaphylaxis (**SRS-A**) is a mixture of leukotrienes C_4, D_4, and E_4. This mixture of leukotrienes is 100–1000 times more potent than histamine or prostaglandins as a constrictor of the bronchial airway musculature. These leukotrienes together with leukotriene B_4 also cause vascular permeability and attraction and activation of leukocytes and seem to be important regulators in many diseases involving inflammatory or immediate hypersensitivity reactions, such as asthma. Leukotrienes are vasoactive, and 5-lipoxygenase has been found in arterial walls.

Evidence supports a role for lipoxins in vasoactive and immunoregulatory function, eg, as counter-regulatory compounds (chalones) of the immune response.

SUMMARY

1. Biosynthesis of unsaturated long-chain fatty acids is achieved by a combination of desaturase enzymes, which introduce double bonds, and elongase enzymes, which lengthen existing acyl chains by 2 carbons at a time.

2. Animals are restricted to Δ^4, Δ^5, Δ^6, and Δ^9 desaturases, which does not allow for insertion of new double bonds beyond the 9 position of fatty acids. As a result, linoleic ($\omega6$) and α-linolenic ($\omega3$) acids cannot be synthesized and must be provided in the diet. They are called the essential fatty acids.

3. The essential fatty acids give rise to C_{20} (eicosanoic) fatty acids from which are synthesized very important groups of physiologically and pharmacologically active compounds, the eicosanoids. They comprise the prostaglandins, thromboxanes, leukotrienes, and lipoxins. Prostaglandins and thromboxanes are synthesized via the cyclooxygenase pathway (inhibited by aspirin); leukotrienes and lipoxins are synthesized via the lipoxygenase pathway.

4. Since different groups of eicosanoids are synthesized from the essential fatty acids, the balance between the physiologic effects of the various eicosanoids may be manipulated by changing the fatty acid composition of the diet.

REFERENCES

Brenner RR: Endocrine control of fatty acid desaturation. Biochem Soc Trans 1990;18:773.

Fischer S: Dietary polyunsaturated fatty acids and eicosanoid formation in humans. Adv Lipid Res 1989;23:169.

Holman RT: Control of polyunsaturated acids in tissue lipids. J Am Coll Nutr 1986;5:183.

Kinsella JE, Lokesh B, Stone RA: Dietary n-3 polyunsaturated fatty acids and amelioration of cardiovascular disease: Possible mechanisms. Am J Clin Nutr 1990;52:1.

Lagarde M, Gualde N, Rigaud M: Metabolic interactions between eicosanoids in blood and vascular cells. Biochem J 1989;257:313.

Neuringer M, Anderson GJ, Connor WE: The essentiality of n-3 fatty acids for the development and function of the retina and brain. Annu Rev Nutr 1988;8:517.

Serhan CN: Lipoxin biosynthesis and its impact in inflammatory and vascular events. Biochim Biophys Acta 1994;1212:1.

Smith WL: The eicosanoids and their biochemical mechanisms of action. Biochem J 1989;259:315.

Smith WL, Borgeat P: The eicosanoids. In: *Biochemistry of Lipids and Membranes*. Vance DE, Vance JE (editors). Benjamin/Cummings, 1985.

Metabolism of Acylglycerols & Sphingolipids

26

Peter A. Mayes, PhD, DSc

INTRODUCTION

Acylglycerols constitute the majority of lipids in the body. Triacylglycerols are the major lipids in fat deposits and in food. In addition, acylglycerols, particularly phospholipids, are major components of the plasma and other membranes. Phospholipids also take part in the metabolism of many other lipids. Glycosphingolipids, which contain sphingosine and sugar residues as well as fatty acids, account for 5–10% of the lipids of the plasma membrane.

BIOMEDICAL IMPORTANCE

The role of triacylglycerol in lipid transport and storage and in various diseases such as obesity, diabetes, and hyperlipoproteinemia will be described in detail in subsequent chapters. Phosphoglycerols, phosphosphingolipids, and glycosphingolipids are all amphipathic and consequently ideally suited as the main lipid constituents of biologic membranes. Some phospholipids have specialized functions; eg, dipalmitoyl lecithin is a major component of **lung surfactant,** the lack of which in premature infants is responsible for respiratory distress syndrome of the newborn. Inositol phospholipids in the cell membrane act as precursors of **hormone second messengers,** and **platelet-activating factor** is an alkylphospholipid. Glycosphingolipids, found in the outer leaflet of the plasma membrane with their oligosaccharide chains facing outward, form part of the glycocalyx of the cell surface and are considered to be important (1) in intercellular communication and contact; (2) as receptors for bacterial toxins (eg, the toxin that causes cholera); and (3) as ABO blood group substances. A dozen or so **glycolipid storage diseases** have been described (eg, Gaucher's disease, Tay-Sachs disease), each due to a specific deficiency in a hydrolase enzyme in the pathway of glycolipid breakdown in lysosomes.

CATABOLISM OF ACYLGLYCEROLS IS NOT THE REVERSAL OF BIOSYNTHESIS

Hydrolysis Initiates Catabolism of Triacylglycerols

Triacylglycerols must be hydrolyzed by a suitable **lipase** to their constituent fatty acids and glycerol before further catabolism can proceed. Much of this hydrolysis (lipolysis) occurs in adipose tissue with release of free fatty acids into the plasma, where they are found combined with serum albumin. This is followed by free fatty acid uptake into tissues and subsequent oxidation or reesterification. Many tissues (including liver, heart, kidney, muscle, lung, testis, brain, and adipose tissue) have the ability to oxidize long-chain fatty acids, although brain cannot extract them readily from the blood. The utilization of glycerol depends upon whether such tissues possess the necessary activating enzyme, **glycerol kinase.** The enzyme has been found in significant amounts in liver, kidney, intestine, brown adipose tissue, and lactating mammary gland.

TRIACYLGLYCEROLS AND PHOSPHOGLYCEROLS ARE FORMED BY ACYLATION OF TRIOSE PHOSPHATES

The major pathways of triacylglycerol and phosphoglycerol biosynthesis are outlined in Figure 26–1. From glycerol 3-phosphate are formed many significant substances, each playing vital roles in cell metabolism. These range from the major triacylglycerol stores to phosphatidyl derivatives of choline, ethanolamine, inositol, and cardiolipin, a constituent of mitochondrial membranes. Two important branch points in the pathway occur at the phosphatidate and diacylglycerol intermediate steps. From dihydroxyacetone phosphate are derived phosphoglycerols con-

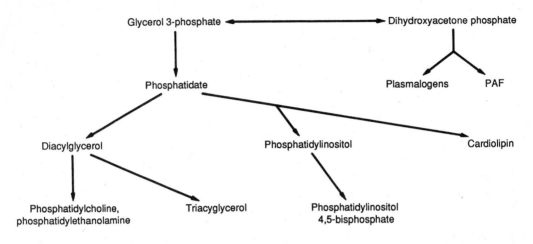

Figure 26–1. Overview of acylglycerol and phosphoglycerol biosynthesis. (PAF, platelet-activating factor.)

taining an ether link (—C—O—C—), the best-known of which are plasmalogens and platelet-activating factor (PAF). It will be noted that glycerol 3-phosphate or dihydroxyacetone phosphate is derived from or is a member of the glycolysis pathway, pointing to a very important connection between carbohydrate and lipid metabolism.

Phosphatidate Is the Common Precursor in the Biosynthesis of Triacylglycerols, Many Phosphoglycerols, and Cardiolipin

Although reactions involving the hydrolysis of triacylglycerols by lipase can be reversed in the laboratory, this is not the mechanism by which acylglycerols are synthesized in tissues. Both glycerol and fatty acids must be activated by ATP before they can be incorporated into acylglycerols. Glycerol kinase will catalyze the activation of glycerol to *sn*-glycerol 3-phosphate. If this enzyme is absent—or low in activity, as it is in muscle or adipose tissue—most of the glycerol 3-phosphate must be derived from an intermediate of the glycolytic system, dihydroxyacetone phosphate, which forms glycerol 3-phosphate by reduction with NADH catalyzed by **glycerol-3-phosphate dehydrogenase** (Figure 26–2).

A. Biosynthesis of Triacylglycerols: Fatty acids are activated to acyl-CoA by the enzyme **acyl-CoA synthetase,** utilizing ATP and CoA (Chapter 24). Two molecules of acyl-CoA combine with glycerol 3-phosphate to form **phosphatidate** (1,2-di-

acylglycerol phosphate). This takes place in two stages via lysophosphatidate, catalyzed first by **glycerol-3-phosphate acyltransferase** and then by **1-acylglycerol-3-phosphate acyltransferase.** Phosphatidate is converted by **phosphatidate phosphohydrolase** to a 1,2-diacylglycerol. A further molecule of acyl-CoA is esterified with the diacylglycerol to form a triacylglycerol, catalyzed by **diacylglycerol acyltransferase.** In intestinal mucosa, a **monoacylglycerol pathway** exists whereby monoacylglycerol is converted to 1,2-diacylglycerol as a result of the presence of **monoacylglycerol acyltransferase.** Most of the activity of these enzymes resides in the endoplasmic reticulum of the cell, but some is found in mitochondria, eg, glycerol-3-phosphate acyltransferase. Phosphatidate phosphohydrolase activity is found mainly in the particle-free supernatant fraction but also is membrane-bound.

B. Biosynthesis of Phosphoglycerols: These phospholipids are synthesized either from phosphatidate, eg, phosphatidylinositol, or from 1,2-diacylglycerol, eg, phosphatidylcholine and phosphatidylethanolamine. In the synthesis of phosphatidylinositol, cytidine triphosphate (CTP), a high-energy phosphate formed from ATP (Chapter 12), reacts with phosphatidate to form a cytidine-diphosphate-diacylglycerol (CDP-diacylglycerol). Finally, this compound reacts with inositol, catalyzed by the enzyme **CDP-diacylglycerol inositol transferase,** to form phosphatidylinositol (Figure 26–2). By successive phosphorylations, phosphatidylinositol is transformed first to phosphatidylinositol 4-phosphate and

---

Figure 26–2 (opposite). Biosynthesis of triacylglycerol and phospholipids. ①, Monoacylglycerol pathway; ②, glycerol phosphate pathway. Phosphatidylethanolamine may be formed from ethanolamine by a pathway similar to that shown for the formation of phosphatidylcholine from choline.

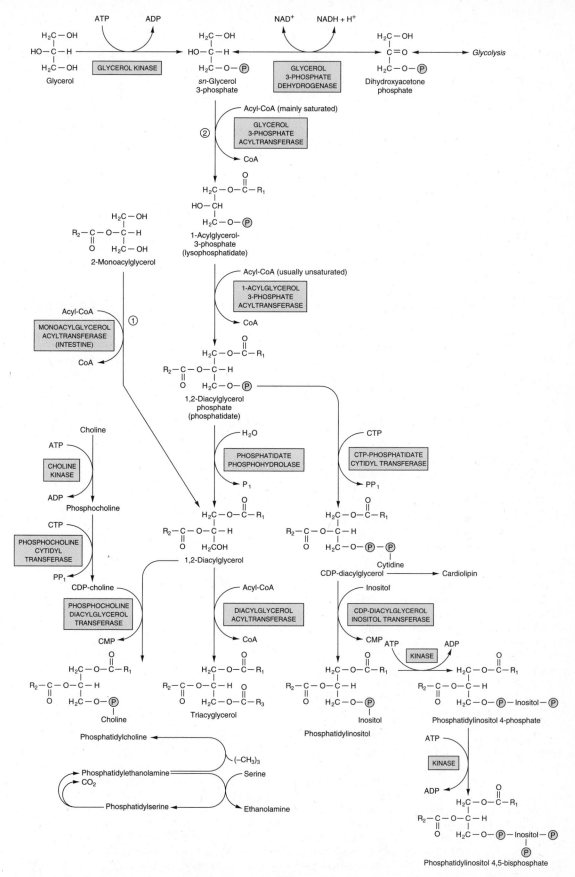

then to phosphatidylinositol 4,5-bisphosphate. The latter is broken down into **diacylglycerol** and **inositol trisphosphate** by hormones that increase [Ca^{2+}], eg, vasopressin. These two products act as **second messengers** in the action of the hormone (Chapter 44).

In the biosynthesis of phosphatidylcholine and phosphatidylethanolamine (lecithins and cephalins), choline or ethanolamine must first be converted to "active choline" or "active ethanolamine," respectively. This is a two-stage process involving, first, a reaction with ATP to form the corresponding monophosphate, followed by a further reaction with CTP to form either cytidine diphosphocholine (CDP-choline) or cytidine diphosphoethanolamine (CDP-ethanolamine). In this form, choline or ethanolamine reacts with 1,2-diacylglycerol so that a phosphorylated base (either phosphocholine or phospho-ethanolamine) is transferred to the diacylglycerol to form either phosphatidylcholine or phosphatidyl-ethanolamine, respectively. The cytidyl transferase appears to be the regulatory enzyme of the phosphatidylcholine pathway.

Phosphatidylserine is formed from phosphatidylethanolamine directly by reaction with serine (Figure 26–2). Phosphatidylserine may re-form phosphatidylethanolamine by decarboxylation. An alternative pathway in liver enables phosphatidyletha-nolamine to give rise directly to phosphatidylcholine by progressive methylation of the ethanolamine residue utilizing S-adenosylmethionine as the methyl donor. In turn, the methyl group in methionine can be derived from methyl-H_4 folate (Figure 52–15). In spite of these sources of choline, it is considered to be an essential nutrient in many mammalian species, but this has not been established in humans.

A phospholipid present in mitochondria is **cardiolipin** (diphosphatidylglycerol; Figure 16–10). It is formed from phosphatidylglycerol, which in turn is synthesized from CDP-diacylglycerol (Figure 26–2) and glycerol 3-phosphate according to the scheme shown in Figure 26–3. Cardiolipin, found in the inner membrane of mitochondria, is specifically required for the functioning of the phosphate transporter and for cytochrome oxidase activity.

C. Biosynthesis of Glycerol Ether Phospholipids: This pathway appears to be exclusively located in peroxisomes. Dihydroxyacetone phosphate is the precursor of the glycerol moiety of glycerol ether phospholipids (Figure 26–4). This compound combines with acyl-CoA to give 1-acyldihydroxy-acetone phosphate. An exchange reaction takes place between the acyl group and a long-chain alcohol to give a 1-alkyldihydroxyacetone phosphate (containing the ether link), which in the presence of NADPH is converted to 1-alkylglycerol 3-phosphate. After further acylation in the 2 position, the resulting 1-alkyl-2-acylglycerol 3-phosphate (analogous to phosphatidate in Figure 26–2) is hydrolyzed to give

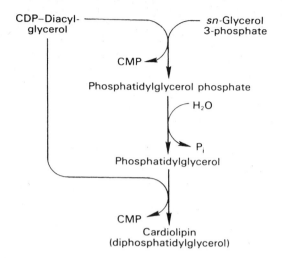

Figure 26–3. Biosynthesis of cardiolipin.

the free glycerol derivative. **Plasmalogens** are formed by desaturation of the analogous 3-phosphoethanolamine derivative (Figure 26–4). Much of the phospholipid in mitochondria consists of plasmalogens. **Platelet-activating factor (PAF)** is synthesized from the corresponding 3-phosphocholine derivative and has been identified as 1-alkyl-2-acetyl-sn-glycerol-3-phosphocholine. It is formed by many blood cells and other tissues and aggregates platelets at concentrations as low as 10^{-11} mol/L. It also has hypotensive and ulcerogenic properties and is involved in a variety of biologic responses including inflammation, chemotaxis, and protein phosphorylation.

Phospholipases Allow Degradation and Remodeling of Phosphoglycerols

Degradation of many complex molecules in tissues proceeds to completion, eg, proteins to amino acids. Thus, a turnover time can be determined for such a molecule. Although phospholipids are actively degraded, each portion of the molecule turns over at a different rate; eg, the turnover time of the phosphate group is different from that of the 1-acyl group. This is due to the presence of enzymes that allow partial degradation followed by resynthesis (Figure 26–5). **Phospholipase A_2** catalyzes the hydrolysis of the ester bond in position 2 of glycerophospholipids to form a free fatty acid and lysophospholipid, which in turn may be reacylated by acyl-CoA in the presence of an acyltransferase. Alternatively, lysophospholipid (eg, lysolecithin) is attacked by **lysophospholipase** (phospholipase B), removing the remaining 1-acyl group and forming the corresponding glyceryl phosphoryl base, which in turn may be split by a hydrolase liberating glycerol 3-phosphate plus base. **Phospholipase A_1** attacks the ester bond

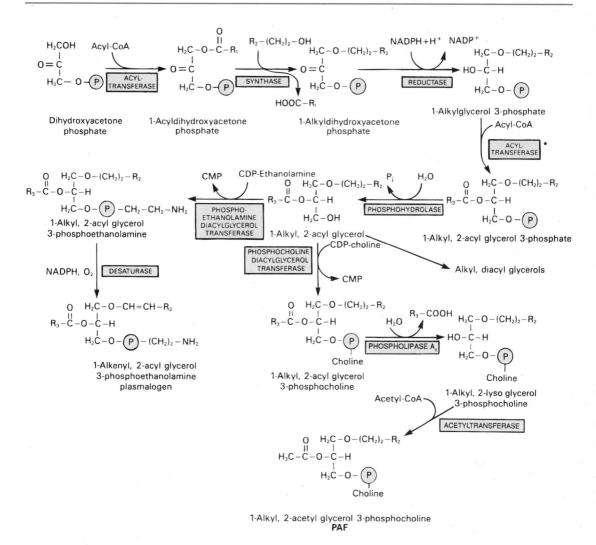

Figure 26–4. Biosynthesis of ether lipids, plasmalogens, and platelet-activating factor (PAF). In the de novo pathway for PAF synthesis, acetyl-CoA is incorporated at stage *, avoiding the last two steps in the pathway shown here.

in position 1 while phospholipase A_2 attacks the bond in position 2 of phospholipids (Figure 26–6). **Phospholipase C** attacks the ester bond in position 3, liberating 1,2-diacylglycerol plus a phosphoryl base. It is one of the major toxins secreted by bacteria. **Phospholipase D** is an enzyme, described mainly in plants, that hydrolyzes the nitrogenous base from phospholipids.

Lysolecithin may be formed by an alternative route that involves **lecithin:cholesterol acyltransferase (LCAT).** This enzyme, found in plasma and synthesized in liver, catalyzes the transfer of a fatty acid residue from the 2 position of lecithin to cholesterol to form cholesteryl ester and is considered to be responsible for much of the cholesteryl ester in plasma lipoproteins. The consequences of **LCAT deficiency** are shown in Table 28–1.

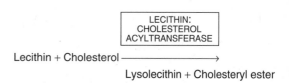

Long-chain saturated fatty acids are found predominantly in the 1 position of phospholipids, whereas the polyunsaturated acids (eg, the precursors of prostaglandins) are incorporated more into the 2 position. The incorporation of fatty acids into lecithin occurs by complete synthesis of the phospholipid, by transacylation between cholesteryl ester and lysolecithin, and by direct acylation of lysolecithin by acyl-CoA. Thus, a continuous exchange of the fatty acids is possible, particularly with regard to in-

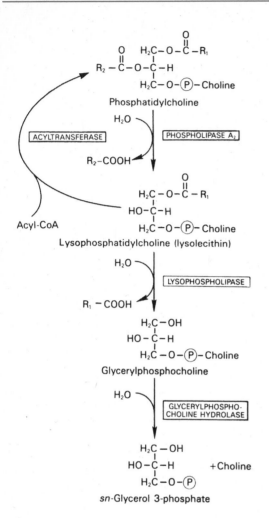

Figure 26–5. Metabolism of phosphatidylcholine (lecithin).

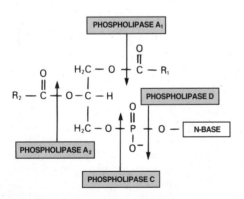

Figure 26–6. Sites of the hydrolytic activity of phospholipases on a phospholipid substrate.

troducing essential fatty acids into phospholipid molecules.

ALL SPHINGOLIPIDS ARE FORMED FROM CERAMIDE

Ceramide (Figure 26–7) is synthesized in the endoplasmic reticulum. First, the amino acid serine, following activation by combination with pyridoxal phosphate, combines with palmitoyl-CoA to form 3-ketosphinganine. This is converted to dihydrosphingosine in a reductive step utilizing NADPH. Dihydroceramide is formed by a combination with acyl-CoA followed by desaturation to form ceramide.

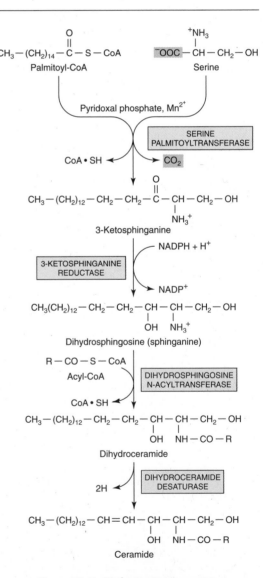

Figure 26–7. Biosynthesis of ceramide.

There is evidence that ceramide may act as a lipid mediator (second messenger), activating a protein kinase and opposing some of the actions of diacylglycerol (Chapter 44). The acyl group is represented frequently by long-chain saturated or monoenoic acids.

Sphingomyelins (Figure 16–17) are phospholipids and are formed when ceramide reacts with phosphatidylcholine to form sphingomyelin plus diacylglycerol (Figure 26–8). This occurs mainly in the Golgi apparatus and to a lesser extent in the plasma membrane. In organelles involved in secretory and endocytic processes, sphingomyelin is restricted to the luminal aspect.

Glycosphingolipids Are a Combination of Ceramide With One or More Sugar Residues

Characteristically, C_{24} fatty acids occur in many glycosphingolipids, particularly those in brain (lignoceric, cerebronic, and nervonic acids). Lignoceric acid ($C_{23}H_{47}COOH$) is completely synthesized from acetyl-CoA. Cerebronic acid, the 2-hydroxy derivative of lignoceric acid, is formed from it. Nervonic acid ($C_{23}H_{45}COOH$), a monounsaturated acid, is formed by elongation of oleic acid.

The simplest glycosphingolipids (**cerebrosides**) are **galactosylceramide (GalCer)** and **glucosylceramide (GlcCer).** GalCer is a major lipid of myelin, whereas GlcCer is the major glycosphingolipid of extraneural tissues and a precursor of most of the more complex glycosphingolipids. **Uridine diphosphogalactose epimerase** (Figure 26–9) utilizes uridine diphosphate glucose (UDPGlc) as substrate and accomplishes epimerization of the glucose moiety to galactose, thus forming uridine diphosphate galactose (UDPGal). The reaction in brain is similar to that described in Figure 22–6 for the liver and mammary gland. Galactosylceramide is formed in a reaction between ceramide and UDPGal. **Sulfogalactosylceramide** is formed after further reaction with 3′-phosphoadenosine-5′-phosphosulfate (PAPS; "active sulfate"). PAPS is also involved in the biosynthesis of the other sulfolipids, ie, the **sulfo-(galacto)glycerolipids** and the **steroid sulfates.**

Gangliosides are synthesized from ceramide by the stepwise addition of activated sugars (eg, UDPGlc and UDPGal) and a **sialic acid,** usually *N*-acetylneuraminic acid (Figure 26–10). A large number of gangliosides of increasing molecular weight may be formed. Most of the enzymes transferring sugars from nucleotide sugars (glycosyl transferases) are found in the Golgi apparatus.

Glycosphingolipids are constituents of the outer leaflet of plasma membranes and as such may be important in **intercellular communication and contact.** Some are antigens, eg, the Forssman antigen and ABO blood group substances. Similar oligosaccharide chains are found in glycoproteins in the plasma membrane. Certain gangliosides function as receptors for bacterial toxins (eg, for cholera toxin, which subsequently activates adenylyl cyclase).

CLINICAL ASPECTS

Deficiency of Lung Surfactant Causes Respiratory Distress Syndrome

Lung surfactant is a secretion with marked surface-active properties, composed mainly of lipid with some proteins and carbohydrate, that prevent the alveoli from collapsing. Surfactant activity is largely attributed to the presence of a phospholipid, **dipalmitoylphosphatidylcholine,** which is synthesized shortly before parturition in full-term infants. Deficiency of lung surfactant in the lungs of many preterm newborns gives rise to the **respiratory distress syndrome.** Administration of either natural or artificial surfactant has been of therapeutic benefit.

Phospholipids and Sphingolipids Are Involved in Multiple Sclerosis and Lipidoses

Certain diseases are characterized by abnormal quantities of these lipids in the tissues, often in the nervous system. They may be classified into two groups: (1) true demyelinating diseases and (2) sphingolipidoses.

In **multiple sclerosis,** which is a demyelinating disease, there is loss of both phospholipids (particularly ethanolamine plasmalogen) and of sphingolipids from white matter. Thus, the lipid composition of white matter resembles that of gray matter. Cholesteryl esters may be found in white matter, though normally they are absent. The cerebrospinal fluid shows raised phospholipid levels.

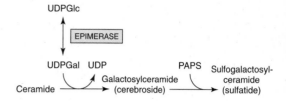

Figure 26–9. Biosynthesis of galactosylceramide and its sulfo derivative. (PAPS, "active sulfate," phosphoadenosine-5′-phosphosulfate.)

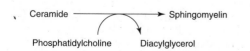

Figure 26–8. Biosynthesis of sphingomyelin.

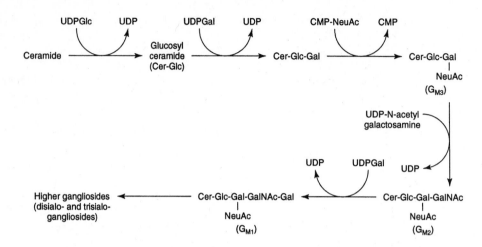

Figure 26–10. Biosynthesis of gangliosides. (NeuAc, *N*-acetylneuraminic acid.)

The **sphingolipidoses** are a group of inherited diseases that are often manifested in childhood. These diseases are part of a larger group of lysosomal disorders.

Lipid storage diseases exhibit several constant features: (1) In various tissues, there is an accumulation of complex lipids that have a portion of their structure in common—**ceramide.** (2) The rate of **synthesis** of the stored lipid is comparable to that in normal humans. (3) The enzymatic defect in each of these diseases is **a deficiency due to gene mutation of a specific lysosomal hydrolytic enzyme necessary to break down the lipid or of a key activator protein of the enzyme.** (4) The extent to which the activity of the affected enzyme is decreased is similar in all of the tissues of the affected individual. As a result of these unifying basic considerations, procedures for the diagnosis of patients with these disorders have been developed. It has also become possible to detect heterozygous carriers of the genetic abnormalities responsible for these diseases as well as to discover in the unborn fetus the fact that a sphingolipodystrophy is present. Enzyme replacement therapy has been tried for many years with little success. More recently, successful treatments have been obtained with enzymes that have been chemically modified to ensure binding to receptors of target cells, prior to receptor-mediated endocytosis, eg, to macrophages in the liver in order to deliver β-glucosidase (glucocerebrosidase) in the treatment of Gaucher's disease. Gene therapy for lysosomal disorders is currently under investigation. A summary of the more important lipidoses is shown in Table 26–1.

Multiple sulfatase deficiency results in accumulation of sulfogalactosylceramide, steroid sulfates, and proteoglycans, owing to a combined deficiency of arylsulfatases A, B, and C and steroid sulfatase (Table 57–6).

SUMMARY

1. Triacylglycerols are the major energy-storing lipids, whereas phosphoglycerols, sphingomyelin, and glycosphingolipids are amphipathic and fulfill many roles, ranging from structural functions in cell membranes to specialized functions, eg, precursors for hormone second messengers, lung surfactant, and platelet-activating factor (PAF).

2. Triacylglycerols and some phosphoglycerols are synthesized by progressive acylation of glycerol 3-phosphate. The pathway bifurcates at phosphatidate, forming inositol phospholipids and cardiolipin on the one hand and triacylglycerol and choline and ethanolamine phospholipids on the other.

3. Plasmalogens and PAF are ether phospholipids formed by acylation and alkylation of dihydroxyacetone phosphate.

4. All sphingolipids are formed from ceramide (*N*-acylsphingosine). Sphingomyelin is a phospholipid present characteristically in membranes of organelles involved in secretory processes (eg, Golgi apparatus). The simplest glycosphingolipids are a combination of ceramide plus a sugar residue (eg, GalCer in myelin). Gangliosides are more complex glycosphingolipids containing more sugar residues plus sialic acid. They are present in the outer layer of the plasma membrane, where they contribute to the glycocalyx and are important as antigens and cell receptors.

5. Phospholipids and sphingolipids are involved in several disease processes including respiratory distress syndrome (lack of lung surfactant), multiple sclerosis (demyelination), and sphingolipidoses (inability to break down sphingolipids in lysosomes due to inherited defects in hydrolase enzymes).

Table 26–1. Examples of sphingolipidoses.

Disease	Enzyme Deficiency	Lipid Accumulating	Clinical Symptoms
Fucosidosis	α-Fucosidase	Cer—Glc—Gal—GalNAc—Gal$\div$Fuc H-Isoantigen	Cerebral degeneration, muscle spasticity, thick skin.
Generalized gangliosidosis	G_{M1}-β-galactosidase	Cer—Glc—Gal(NeuAc)—GalNAc$\div$Gal G_{M1} Ganglioside	Mental retardation, liver enlargement, skeletal deformation.
Tay-Sachs disease	Hexosaminidase A	Cer—Glc—Gal(NeuAc)$\div$GalNAc G_{M2} Ganglioside	Mental retardation, blindness, muscular weakness.
Tay-Sachs variant or Sandhoff's disease	Hexosaminidase A and B	Cer—Glc—Gal—Gal$\div$GalNAc Globoside plus G_{M2} ganglioside	Same as Tay-Sachs, but progressing more rapidly.
Fabry's disease	α-Galactosidase	Cer—Glc—Gal$\div$Gal Globotriaosylceramide	Skin rash, kidney failure (full symptoms only in males; X-linked recessive).
Ceramide lactoside lipidosis	Ceramide lactosidase (β-galactosidase)	Cer—Glc$\div$Gal Ceramide lactoside	Progressing brain damage, liver and spleen enlargement.
Metachromatic leukodystrophy	Arylsulfatase A	Cer—Gal$\div$OSO$_3$ 3-Sulfogalactosylceramide	Mental retardation and psychologic disturbances in adults; demyelination.
Krabbe's disease	β-Galactosidase	Cer$\div$Gal Galactosylceramide	Mental retardation; myelin almost absent.
Gaucher's disease	β-Glucosidase	Cer$\div$Glc Glucosylceramide	Enlarged liver and spleen, erosion of long bones, mental retardation in infants.
Niemann-Pick disease	Spingomyelinase	Cer$\div$P—choline Sphingomyelin	Enlarged liver and spleen, mental retardation; fatal in early life.
Farber's disease	Ceramidase	Acyl$\div$Sphingosine Ceramide	Hoarseness, dermatitis, skeletal deformation, mental retardation; fatal in early life.

NeuAc, *N*-acetylneuraminic acid; Cer, ceramide; Glc, glucose; Gal, galactose; Fuc, fucose. $\div$, site of deficient enzyme reaction.

REFERENCES

Hannun YA: The sphingomyelin cycle and second messenger function of ceramide. J Biol Chem 1994;269:3125.

Hawthorne JN, Ansell GB (editors): *Phospholipids*. Elsevier, 1982.

Koval M, Pagano RE: Intracellular transport and metabolism of sphingomyelin. Biochim Biophys Acta 1991; 1082:113.

Neufeld EF: Lysosomal storage diseases. Annu Rev Biochem 1991;60:257.

Scriver CR et al (editors): *The Metabolic and Molecular Bases of Inherited Disease,* 7th ed. McGraw-Hill, 1995.

Snyder F: Platelet-activating factor and related acetylated lipids as potent biologically active cellular mediators. Am J Physiol 1990;259:C697.

Tijburg LBM, Geelen MJH, van Golde LMG: Regulation of the biosynthesis of triacylglycerol, phosphatidylcholine and phosphatidylethanolamine in the liver. Biochim Biophys Acta 1989;1004:1.

Vance DE, Vance JE (editors): Metabolism of triacylglycerols; phospholipid metabolism; ether-linked glycerolipids; sphingolipids. In: *Biochemistry of Lipids and Hormones.* Benjamin/Cummings, 1985.

van Echten G, Sandhoff K: Ganglioside metabolism. J Biol Chem 1993;268:5341.

VanGolde LMG, Batenburg JJ, Robertson B: The pulmonary surfactant system: biochemical aspects and functional significance. Physiol Rev 1988;68:374.

27

Lipid Transport & Storage

Peter A. Mayes, PhD, DSc

INTRODUCTION

Fat absorbed from the diet and lipids synthesized by the liver and adipose tissue must be transported between the various tissues and organs for utilization and storage. Since lipids are insoluble in water, the problem arises of how to transport them in an aqueous environment—the blood plasma. This is solved by associating nonpolar lipids (triacylglycerol and cholesteryl esters) with amphipathic lipids (phospholipids and cholesterol) and proteins to make watermiscible lipoproteins.

BIOMEDICAL IMPORTANCE

In a meal-eating omnivore such as the human, excess calories are ingested in the anabolic phase of the feeding cycle, followed by a period of negative caloric balance when the organism draws upon its carbohydrate and fat stores. Lipoproteins mediate this cycle by transporting lipids from the intestines as chylomicrons, and from the liver as very low density lipoproteins (VLDL), to most tissues for oxidation and to adipose tissue for storage. Lipid is mobilized from adipose tissue as free fatty acids (FFA) attached to serum albumin. Abnormalities of lipoprotein metabolism occur at the sites of production or utilization of lipoproteins, causing various **hypo-** or **hyperlipoproteinemias.** The most common of these is **diabetes mellitus,** where insulin deficiency causes excessive mobilization of FFA and underutilization of chylomicrons and VLDL, leading to **hypertriacylglycerolemia.** Most other pathologic conditions affecting lipid transport are due primarily to inherited defects in synthesis of the apoprotein portion of the lipoprotein, of key enzymes, or of lipoprotein receptors. Some of these defects cause **hypercholesterolemia** and premature **atherosclerosis.** Excessive fat deposits constitute **obesity.** Abdominal obesity, in particular, is a risk factor for increased mortality, hypertension, non-insulin-dependent diabetes mellitus (NIDDM), hyperlipidemia, hyperglycemia, and various endocrine dysfunctions.

LIPIDS ARE TRANSPORTED IN THE PLASMA AS LIPOPROTEINS

Four Major Lipid Classes Are Present in Lipoproteins

Extraction of the plasma lipids with a suitable lipid solvent and subsequent separation of the extract into various classes of lipids show the presence of **triacylglycerols, phospholipids, cholesterol,** and **cholesteryl esters** and, in addition, the existence of a much smaller fraction of unesterified long-chain fatty acids (free fatty acids) that accounts for less than 5% of the total fatty acid present in the plasma (Table 27–1). This latter fraction, the **free fatty acids (FFA),** is now known to be metabolically the most active of the plasma lipids.

Four Major Groups of Plasma Lipoproteins Have Been Identified

Pure fat is less dense than water; it follows that as the proportion of lipid to protein in a lipoprotein increases, the density decreases (Table 27–2). Use is made of this property in separating the various lipoproteins in plasma by **ultracentrifugation.** The composition of the different lipoprotein fractions obtained by centrifugation is shown in Table 27–2. The various chemical classes of lipids are seen to occur in varying amounts in most of the lipoprotein fractions. Since the density fractions represent the physiologic entities present in the plasma, mere chemical analysis of the plasma lipids (apart from FFA) yields little information on their physiology.

In addition to FFA, four major groups of lipoproteins have been identified that are important physiologically and in clinical diagnosis. These are (1) **chylomicrons,** derived from intestinal absorption of triacylglycerol; (2) **very low density lipoproteins** (VLDL, or pre-β-lipoproteins), derived from the liver for the export of triacylglycerol; (3) **low-density lipoproteins** (LDL, or β-lipoproteins), representing a final stage in the catabolism of VLDL; and (4) **high-density lipoproteins** (HDL, or α-lipoproteins), involved in VLDL and chylomicron metabolism and also in cholesterol transport. Triacylglycerol is the predominant lipid in chylomicrons and VLDL, whereas cholesterol and phospholipid are the pre-

Table 27–1. Lipids of the blood plasma in humans.

Lipid	mmol/L	
	Mean	Range
Triacylglycerol	1.6	0.9–2.0[2]
Total phospholipid[1]	3.1	1.8–5.8
Total cholesterol	5.2	2.8–8.3
Free cholesterol (nonesterified)	1.4	0.7–2.7
Free fatty acids (nonesterified)	0.4	0.2–0.6[2]

Of total fatty acids, 45% are triacylglycerols, 35% phospholipids, 15% cholesteryl ester, and less than 5% free fatty acids. The ranges can be exceeded under abnormal or pathologic conditions.

[1]Analyzed as lipid phosphorus.
[2]Varies with nutritional state.

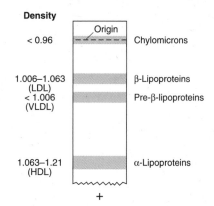

Figure 27–1. Separation of plasma lipoproteins by electrophoresis on agarose gel.

dominant lipids in LDL and HDL, respectively (Table 27–2).

In addition to the use of techniques depending on their density, lipoproteins may be separated according to their electrophoretic properties into α-, β-, and **pre-β-lipoproteins** (Figure 27–1) and may be identified more accurately by means of immunoelectrophoresis.

Amphipathic Lipids Are Essential Components of Lipoproteins

A typical lipoprotein—such as a chylomicron or VLDL—consists of a **lipid core** of mainly **nonpolar** **triacylglycerol and cholesteryl ester** surrounded by a **single surface layer** of **amphipathic phospholipid and cholesterol** molecules. These are oriented so that their polar groups face outward to the aqueous medium, as in the cell membrane (Chapter 16).

The protein moiety of a lipoprotein is known as an **apolipoprotein** or **apoprotein**, constituting nearly 60% of some HDL and as little as 1% of chylomicrons. Some apolipoproteins are integral and cannot be removed, whereas others are free to transfer to other lipoproteins (Figure 27–2).

Table 27–2. Composition of the lipoproteins in plasma of humans.

							Composition				
								Percentages of Total Lipid			
Fraction	Source	Diameter (nm)	Density	Sf[1]	Protein (%)	Total Lipid (%)	Triacylglycerol	Phospholipid	Cholesteryl Ester	Cholesterol (Free)	Free Fatty Acids
Chylomicrons	Intestine	90–1000	<0.95	>400	1–2	98–99	88	8	3	1	. . .
Very low density lipoproteins (VLDL)	Liver (intestine)	30–90	0.95–1.006	20–400	7–10	90–93	56	20	15	8	1
Intermediate-density lipoproteins (IDL)	VLDL	25–30	1.006–1.019	12–20	11	89	29	26	34	9	1
Low-density lipoproteins (LDL)	VLDL	20–25	1.019–1.063	2–12	21	79	13	28	48	10	1
High-density proteins HDL₂	Liver and intestine. VLDL, chylomicrons	10–20	1.063–1.125		33	67	16	43	31	10	. . .
HDL₃		7.5–10	1.125–1.210		57	43	13	46	29	6	6
Albumin-FFA	Adipose tissue		>1.281		99	1	0	0	0	0	100

[1]One Sf (Svedberg) unit is equal to 10^{-13} cm/s/dyne/g at 26°C.
FFA, free fatty acids. VHDL (very high density lipoprotein) is a minor fraction occurring at density 1.21–1.25.

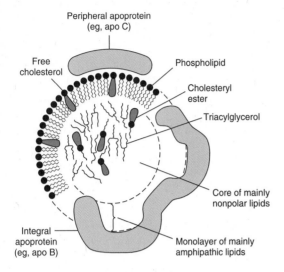

Figure 27–2. Generalized structure of a plasma lipoprotein. The similarities with the structure of the plasma membrane are to be noted. A small amount of cholesteryl ester and triacylglycerol is to be found in the surface layer and a little free cholesterol in the core.

The Distribution of Apolipoproteins Characterizes the Lipoprotein

One or more apolipoproteins (proteins or polypeptides) are present in each lipoprotein. According to the ABC nomenclature, the major apolipoprotein of HDL (α-lipoprotein) is designated A. The main apolipoprotein of LDL (β-lipoprotein) is apolipoprotein B and is found also in VLDL and chylomicrons. However, apo B of chylomicrons (B-48) is smaller than apo B-100 of LDL or VLDL. B-48 is synthesized in the intestine and B-100 in the liver. (In the rat, the liver forms B-48 in addition to B-100.)

Apo B-100 is one of the longest single polypeptide chains known, having 4536 amino acids. Apo B-48 (48% as large as B-100) is formed from the same mRNA as apo B-100. Apparently, in intestine, a stop codon that is not present in genomic DNA is introduced by an RNA-editing mechanism that stops translation at amino acid residue 2153 to liberate apo B-48. Apolipoproteins C-I, C-II, and C-III are smaller polypeptides freely transferable between several different lipoproteins (Table 27–3). Carbohydrates account for approximately 5% of apo B and include mannose, galactose, fucose, glucose, glucosamine, and sialic acid. Thus, some lipoproteins are also glycoproteins. Several other apolipoproteins have also been found in plasma lipoproteins. One is the arginine-rich apolipoprotein E isolated from VLDL and HDL; it contains arginine to the extent of 10% of the total amino acids and accounts for 5–10% of total VLDL apolipoproteins in normal subjects but is present in excess in the broad β-VLDL of patients with type III hyperlipoproteinemia.

Apolipoproteins carry out several roles: (1) they are enzyme cofactors, eg, C-II for lipoprotein lipase, A-I for lecithin:cholesterol acyltransferase; (2) they can act as lipid transfer proteins; and (3) they act as ligands for interaction with lipoprotein receptors in tissues, eg, apo B-100 and apo E for the LDL recep-

Table 27–3. Apolipoproteins of human plasma lipoproteins.

Apolipoprotein	Lipoprotein	Molecular Mass (Da)	Comments
Apo A-I	HDL, chylomicrons	28,000	Activator of lecithin:cholesterol acyltransferase (LCAT). Ligand for HDL receptor.
Apo A-II	HDL, chylomicrons	17,000	Structure is two identical monomers joined by a disulfide bridge. Inhibitor of LCAT?
Apo A-IV	Secreted with chylomicrons but transfers to HDL	46,000	Associated with the formation of triacylglycerol-rich lipoproteins. Function unknown. Synthesized by intestine.
Apo B-100	LDL, VLDL, IDL	550,000	Synthesized in liver. Ligand for LDL receptor.
Apo B-48	Chylomicrons, chylomicron remnants	260,000	Synthesized in intestine.
Apo C-I	VLDL, HDL, chylomicrons	7,600	Possible activator of LCAT.
Apo C-II	VLDL, HDL, chylomicrons	8,916	Activator of lipoprotein lipase.
Apo C-III	VLDL, HDL, chylomicrons	8,750	Several polymorphic forms depending on content of sialic acids.
Apo D	Subfraction of HDL	19,300	May act as lipid transfer protein.
Apo E	VLDL, HDL, chylomicrons, chylomicron remnants	34,000	Present in excess in the β-VLDL of patients with type III hyperlipoproteinemia. The sole apoprotein found in HDL$_c$ of diet-induced hypercholesterolemic animals. Ligand for chylomicron remnant receptor in liver and LDL receptor.

tor, apo E for the remnant receptor, and apo A-I for the HDL receptor.

FREE FATTY ACIDS ARE RAPIDLY METABOLIZED

The free fatty acids (FFA, nonesterified fatty acids, unesterified fatty acids) arise in the plasma from lipolysis of triacylglycerol in adipose tissue or as a result of the action of lipoprotein lipase during uptake of plasma triacylglycerols into tissues. They are found **in combination with albumin,** a very effective solubilizer, in concentrations varying between 0.1 and 2.0 μeq/mL of plasma. They comprise the long-chain fatty acids found in adipose tissue, ie, palmitic, stearic, oleic, palmitoleic, linoleic, and other polyunsaturated acids, and smaller quantities of other long-chain fatty acids. Binding sites on albumin of varying affinity for the fatty acids have been described. Low levels of free fatty acids are recorded in the fully fed condition, rising to about 0.5 μeq/mL in the postabsorptive and between 0.7 and 0.8 μeq/mL in the fully fasting state. In uncontrolled **diabetes mellitus,** the level may rise to as much as 2 μeq/mL. The level falls just after eating and rises again prior to the next meal; however, in such continuous feeders as ruminants—where there is uninterrupted influx of nutrient from the intestine—the free fatty acids remain at a very low level.

The rate of removal of free fatty acids from the blood is extremely rapid. Some of the uptake is oxidized and supplies about 25–50% of the energy requirements in fasting. The remainder of the uptake is esterified. In starvation, considerably more fat is oxidized than can be traced to the oxidation of free fatty acids. This difference is accounted for by the oxidation of esterified lipids from the circulation or of those present in tissues. The latter is thought to occur particularly in heart and skeletal muscle, where considerable stores of lipid are to be found in the muscle cells.

The free fatty acid turnover is related directly to the free fatty acid concentration. Thus, the rate of free fatty acid production by adipose tissue controls the free fatty acid concentration in plasma, which in turn determines the free fatty acid uptake by other tissues. The nutritional condition does not appear to have a great effect on the fractional uptake of free fatty acids by tissues. It does, however, alter the proportion of the uptake which is oxidized compared to the fraction which is esterified, more being oxidized in the fasting than in the fed state. After dissociation of the fatty acid-albumin complex at the plasma membrane, fatty acids bind to a **membrane fatty acid-binding protein** that acts as a transmembrane cotransporter with Na^+. On entering the cytosol, free fatty acids are bound by a **fatty acid-binding protein,** or **Z-protein.** The role of this protein in intracellular transport is thought to be similar to the role of serum albumin in extracellular transport of long-chain fatty acids.

TRIACYLGLYCEROL IS TRANSPORTED FROM THE INTESTINES IN CHYLO-MICRONS AND FROM THE LIVER IN VERY LOW DENSITY LIPOPROTEINS

By definition, **chylomicrons** are found in **chyle** formed only by the lymphatic system **draining the intestine.** They are responsible for the transport of all dietary lipids into the circulation. Smaller and denser particles having the physical characteristics of VLDL are also to be found in chyle. Their formation occurs even in the fasting state, their lipids originating mainly from bile and intestinal secretions. On the other hand, chylomicron formation increases with the load of triacylglycerol absorbed. Most of the plasma **VLDL** are of hepatic origin. **They are the vehicles of transport of triacylglycerol from the liver to the extrahepatic tissues.**

There are many similarities in the mechanism of formation of chylomicrons by intestinal cells and of VLDL by hepatic parenchymal cells (Figure 27–3). Apolipoprotein B is synthesized by ribosomes in the rough endoplasmic reticulum and is incorporated into lipoproteins in the smooth endoplasmic reticulum, which is the main site of synthesis of triacylglycerol. Lipoproteins pass through the Golgi apparatus, where more lipid and carbohydrate residues are added to the lipoprotein. The chylomicrons and VLDL are released from either the intestinal or hepatic cell by fusion of the secretory vacuole with the cell membrane (reverse pinocytosis). Chylomicrons pass into the spaces between the intestinal cells, eventually making their way into the lymphatic system (lacteals) draining the intestine. VLDL are secreted by hepatic parenchymal cells into the space of Disse and then into the hepatic sinusoids through fenestrae in the endothelial lining. The similarities between the two processes and the anatomic mechanisms are striking, for—apart from the mammary gland—the intestine and liver are the only tissues from which particulate lipid is secreted. The inability of particulate lipid of the size of chylomicrons and VLDL to pass through endothelial cells of the capillaries without prior hydrolysis is probably the reason dietary fat enters the circulation via the lymphatics (thoracic duct) and not via the hepatic portal system.

Although both chylomicrons and VLDL isolated from blood contain apolipoproteins C and E, the newly secreted or "nascent" lipoproteins contain little or none, and it would appear that the full complement of apo C and apo E polypeptides is taken up by transfer from HDL once the chylomicrons and VLDL have entered the circulation (Figures 27–4 and 27–5).

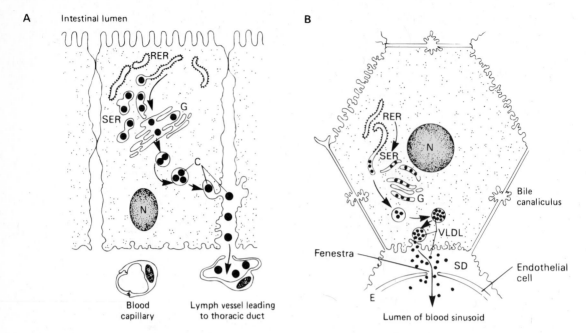

Figure 27–3. The formation and secretion of **(A)** chylomicrons by an intestinal cell and **(B)** very low density lipoproteins by a hepatic cell. (RER, rough endoplasmic reticulum; SER, smooth endoplasmic reticulum; G, Golgi apparatus; N, nucleus; C, chylomicrons; VLDL, very low density lipoproteins; E, endothelium; SD, space of Disse, containing blood plasma.) The figure is a diagrammatic representation of events that can be observed with electron microscopy.

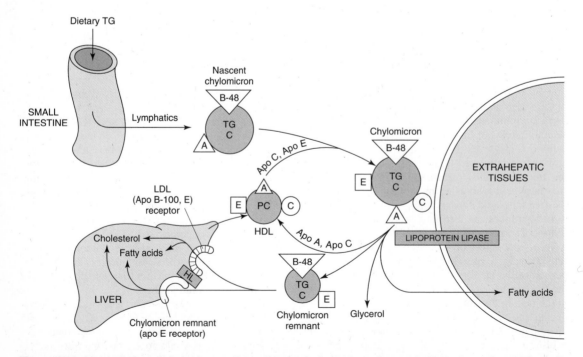

Figure 27–4. Metabolic fate of chylomicrons. (A, apolipoprotein A; B-48, apolipoprotein B-48; ©, apolipoprotein C; E, apolipoprotein E; HDL, high-density lipoprotein; TG, triacylglycerol; C, cholesterol and cholesteryl ester; P, phospholipid; HL, hepatic lipase.) Only the predominant lipids are shown.

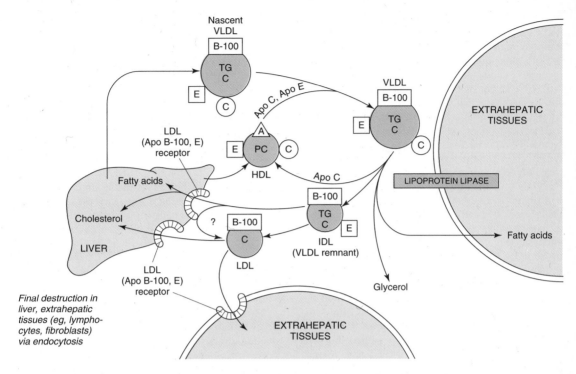

Figure 27–5. Metabolic fate of very low density lipoproteins (VLDL) and production of low-density lipoproteins (LDL). (A, apolipoprotein A; B-100, apolipoprotein B-100; ©, apolipoprotein C; E, apolipoprotein E; HDL, high-density lipoprotein; TG, triacylglycerol; IDL, intermediate-density lipoprotein; C, cholesterol and cholesteryl ester; P, phospholipid.) Only the predominant lipids are shown. It is possible that some IDL is also metabolized via the chylomicron remnant (apo E) receptor.

A more detailed account of the factors controlling hepatic VLDL secretion is given below.

Apo B is essential for chylomicron and VLDL formation. In **abetalipoproteinemia** (a rare disease), apo B is not able to function because of a defect in a triacylglycerol transfer protein which prevents loading of the apo B with lipid; therefore, lipoproteins containing this apolipoprotein are not formed, and lipid droplets accumulate in the intestine and liver.

CHYLOMICRONS AND VERY LOW DENSITY LIPOPROTEINS ARE RAPIDLY CATABOLIZED

The clearance of labeled chylomicrons from the blood is rapid, the half-time of disappearance being on the order of minutes in small animals (eg, rats) but longer in larger animals (eg, humans), in whom it is still under 1 hour. Larger particles are catabolized more quickly than smaller ones. When chylomicrons labeled in the triacylglycerol fatty acids are administered intravenously, some 80% of the label is found in adipose tissue, heart, and muscle and approximately 20% in the liver. As experiments with the perfused organ have shown that **the liver does not metabolize native chylomicrons or VLDL significantly,** the label in the liver must result secondarily from their metabolism in extrahepatic tissues.

Triacylglycerols of Chylomicrons and VLDL Are Hydrolyzed by Lipoprotein Lipase

There is a significant correlation between the ability of a tissue to incorporate the fatty acids of the triacylglycerols of lipoproteins and the activity of the enzyme **lipoprotein lipase.** It is located on the walls of blood capillaries, anchored to the endothelium by proteoglycan chains of heparan sulfate. It has been found in heart, adipose tissue, spleen, lung, renal medulla, aorta, diaphragm, lactating mammary gland, and neonatal liver. It is not active in adult liver. Normal blood does not contain significant quantities of the enzyme; however, following injection of **heparin,** lipoprotein lipase is released from its heparan sulfate binding into the circulation and is accompanied by the clearing of lipemia. A lipase, **hepatic lipase,** is also released from the liver by large quantities of heparin, but this enzyme has properties different from those of lipoprotein lipase and does not react readily with chylomicrons. It is found on the endothelial cells of the liver and is concerned with chylomicron remnant and HDL metabolism.

Both **phospholipids** and **apolipoprotein C-II** are required as cofactors for lipoprotein lipase activity. Apo C-II contains a specific phospholipid binding site through which it is attached to the lipoprotein. Thus, chylomicrons and VLDL provide the enzyme for their metabolism with both its substrate and cofactors. Hydrolysis takes place while the lipoproteins are attached to the enzyme on the endothelium. The triacylglycerol is hydrolyzed progressively through a diacylglycerol to a monoacylglycerol that is finally hydrolyzed to free fatty acid plus glycerol. Some of the released free fatty acids return to the circulation, attached to albumin, but the bulk is transported into the tissue (Figures 27–4 and 27–5). Heart lipoprotein lipase has a low K_m for triacylglycerol, whereas the K_m of the enzyme in adipose tissue is 10 times greater. As the concentration of plasma triacylglycerol decreases in the transition from the fed to the starved condition, the heart enzyme remains saturated with substrate but the saturation of the enzyme in adipose tissue diminishes, **thus redirecting uptake from adipose tissue toward the heart.** A similar redirection occurs during lactation, where adipose tissue activity diminishes and mammary gland activity increases, allowing uptake of lipoprotein triacylglycerol long-chain fatty acid for milk fat synthesis.

In adipose tissue, insulin enhances lipoprotein lipase synthesis in adipocytes and its translocation to the luminal surface of the capillary endothelium.

The Action of Lipoprotein Lipase Forms Remnant Lipoproteins

Reaction with lipoprotein lipase results in the loss of approximately 90% of the triacylglycerol of chylomicrons and in the loss of the apo C (which returns to HDL) but not apo E (which is retained). The resulting lipoprotein or **chylomicron remnant** is about half the diameter of the parent chylomicron and in terms of the percentage composition becomes relatively enriched in cholesterol and cholesteryl esters because of the loss of triacylglycerol (Figure 27–4). Similar changes occur to VLDL, with the formation of VLDL remnants or IDL (intermediate-density lipoprotein) (Figure 27–5).

The Liver Is Responsible for the Uptake of Remnant Lipoproteins

Chylomicron remnants are taken up by the liver by receptor-mediated endocytosis, and the cholesteryl esters and triacylglycerols are hydrolyzed and metabolized. Uptake appears to be mediated by a **receptor specific for apo E** (Figure 27–4). Present evidence suggests that both the LDL (apo B-100, E) receptor and a remnant receptor specific for apo E take part in remnant uptake. Hepatic lipase has a dual role (1) in

acting as a ligand to the lipoprotein and (2) in hydrolyzing its triacylglycerol and phospholipid.

Studies using apo B-100-labeled VLDL have shown that VLDL is the precursor of IDL and that IDL is the precursor of LDL. Only one molecule of apo B-100 is present in each of these lipoprotein particles, and this is conserved during the transformations. Thus, each LDL particle is derived from only one VLDL particle (Figure 27–5). Two possible fates await IDL. It can be taken up by the liver directly via the LDL (apo B-100, E) receptor, or it is converted to LDL. In the rat, most of the IDL is taken up by the liver, whereas in humans a much larger proportion forms LDL, accounting for the increased concentrations of LDL in humans compared with the rat (and many other mammals).

LDL IS METABOLIZED VIA THE LDL RECEPTOR

Most LDL appears to be formed from VLDL, as described above, but there is evidence for some production directly by the liver. The half-time of disappearance from the circulation of apo B-100 in LDL is approximately 2 days.

Studies on cultured fibroblasts, lymphocytes, arterial smooth muscle cells, and liver have shown the existence of specific binding sites, or **receptors,** for LDL, the LDL (B-100, E) receptor. It is so designated because it is specific for apo B-100 but not B-48, and under some circumstances it will take up lipoproteins rich in apo E. Apo B-48 lacks the carboxyl terminal domain of B-100 that contains the ligand for the LDL receptor. These receptors are defective in **familial hypercholesterolemia.** Approximately 30% of LDL is degraded in extrahepatic tissues and 70% in the liver. A positive correlation exists between the incidence of **coronary atherosclerosis** and the plasma concentration of LDL cholesterol. For further discussion of the regulation of the LDL receptor, see Chapter 28.

HDL TAKES PART IN BOTH LIPOPROTEIN TRIACYLGLYCEROL AND CHOLESTEROL METABOLISM

HDL is synthesized and secreted from both liver and intestine (Figure 27–6). However, nascent (newly secreted) HDL from intestine does not contain apo C or E but only apo A. Thus, apo C and apo E are synthesized in the liver and transferred from liver HDL to intestinal HDL when the latter enters the plasma. A major function of HDL is to act as a repository for apo C and apo E that are required in the metabolism of chylomicrons and VLDL (Figures 27–4 and 27–5).

Nascent HDL consists of discoid phospholipid bi-

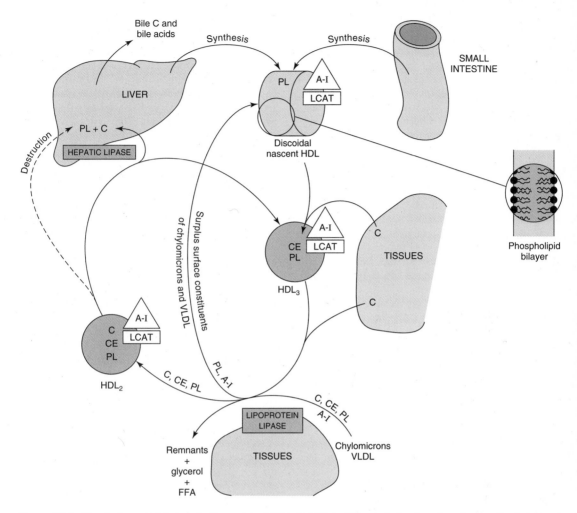

Figure 27–6. Metabolism of high-density lipoprotein (HDL). (LCAT, lecithin:cholesterol acyltransferase; C, cholesterol; CE, cholesteryl ester; PL, phospholipid; FFA, free fatty acids; A-I, apoprotein A-I.) The figure illustrates the role of the three enzymes hepatic lipase, LCAT, and lipoprotein lipase in the postulated HDL cycle for the transport of cholesterol from the tissues to the liver. HDL_2, HDL_3—see Table 27–2. In addition to triacylglycerol, hepatic lipase hydrolyzes phospholipid on the surface of HDL_2, releasing cholesterol for uptake into the liver, allowing formation of smaller and more dense HDL_3. Hepatic lipase activity is increased by androgens and decreased by estrogens, which may account for higher concentrations of plasma HDL_2 in women.

layers containing apolipoprotein and free cholesterol. These lipoproteins are similar to the particles found in the plasma of patients with a deficiency of the plasma enzyme **lecithin:cholesterol acyltransferase (LCAT)** and in the plasma of patients with obstructive jaundice. LCAT—and the LCAT activator apo A-I—bind to the disk. Catalysis by LCAT converts surface phospholipid and free cholesterol into cholesteryl esters and lysolecithin. The nonpolar cholesteryl esters move into the hydrophobic interior of the bilayer, whereas lysolecithin is transferred to plasma albumin. The reaction continues, generating a nonpolar core that pushes the bilayer apart until a spherical, pseudomicellar HDL is formed, covered by a surface film of polar lipids and apolipoproteins. Thus, the

LCAT system is involved in the removal of excess unesterified cholesterol from lipoproteins and from the tissues. The liver is the final site of degradation of HDL cholesteryl ester. Whether there is a true HDL or apo A-I receptor is not clear. It does not appear that HDL containing apo A-I is taken up significantly by the liver. Apo A-I that dissociates from HDL is finally destroyed by the kidney.

An HDL cycle has been proposed to account for the transport of cholesterol from the tissues to the liver, a process known as **reverse cholesterol transport** (Figure 27–6). The cycle involves uptake and esterification of cholesterol by HDL_3, which becomes less dense, forming HDL_2. Hepatic lipase hydrolyses HDL phospholipid and triacylglycerol, allowing the particle to release its cholesteryl ester cargo to the liver, whereupon the particle becomes more dense, re-forming HDL_3, which reenters the cycle. HDL concentrations vary reciprocally with plasma triacylglycerol concentrations and directly with the activity of lipoprotein lipase. This may be due to surplus surface constituents, eg, phospholipid and apo A-I being released during hydrolysis of chylomicrons and contributing toward the formation of nascent HDL. HDL (HDL_2) concentrations are **inversely related to the incidence of coronary atherosclerosis,** possibly because they reflect the efficiency of cholesterol scavenging from the tissues. HDL_c (HDL_1) is found in the blood of diet-induced hypercholesterolemic animals. It is rich in cholesterol, and its sole apolipoprotein is apo E. It is taken up by the liver via the apo E remnant receptor but also by LDL receptors. It is for this reason that the latter are sometimes designated apo B-100, E receptors.

It appears that all plasma lipoproteins are interrelated components of one or more metabolic cycles that together are responsible for the complex process of plasma lipid transport.

THE LIVER PLAYS A CENTRAL ROLE IN LIPID TRANSPORT AND METABOLISM

Much of the lipid metabolism of the body was formerly thought to be the prerogative of the liver. The discovery that most tissues have the ability to oxidize fatty acids completely and the knowledge which has accumulated showing that adipose tissue is extremely active metabolically have tended to modify the former emphasis on the role of the liver. Nonetheless, the concept of a central and unique role for the liver in lipid metabolism is still an important one. The liver carries out the following major functions in lipid metabolism: (1) It facilitates the digestion and absorption of lipids by the production of bile, which contains cholesterol and bile salts synthesized within the liver de novo or from uptake of lipoprotein cholesterol (Chapter 28). (2) The liver has active enzyme systems for synthesizing and oxidizing fatty acids (Chapters 23 and 24) and for synthesizing triacylglycerols and phospholipids (Chapter 26). (3) It converts fatty acids to ketone bodies (ketogenesis) (Chapter 24). (4) It plays an integral part in the synthesis and metabolism of plasma lipoproteins (this chapter).

Hepatic VLDL Secretion Is Related to Dietary and Hormonal Status

The cellular events involved in VLDL formation and secretion have been described above. Hepatic triacylglycerols are the immediate precursors of triacylglycerols contained in plasma VLDL (Figure 27–7).

The synthesis of triacylglycerol provides the immediate stimulus for the formation and secretion of VLDL. The fatty acids used in the synthesis of hepatic triacylglycerols are derived from two possible sources: (1) synthesis within the liver from acetyl-CoA derived mainly from carbohydrate (perhaps not important in humans) and (2) uptake of free fatty acids from the circulation. The first source is predominant in the well-fed condition, when fatty acid synthesis is high and the level of circulating free fatty acids is low. As triacylglycerol does not normally accumulate in the liver under this condition, it must be inferred that it is transported from the liver in VLDL as rapidly as it is synthesized and that the synthesis of apo B-100 is not rate-limiting. On the other hand, during fasting, during the feeding of high-fat diets, or in diabetes mellitus, the level of circulating free fatty acids is raised and more is abstracted into the liver. Under these conditions, lipogenesis is inhibited and free fatty acids are the main source of triacylglycerol fatty acids in the liver and in VLDL. The enzyme mechanisms responsible for the synthesis of triacylglycerols and phospholipids have been described in Chapter 26. Factors that enhance both the synthesis of triacylglycerol and the secretion of VLDL by the liver include (1) the fed state rather than the fasting state; (2) the feeding of diets high in carbohydrate (particularly if they contain sucrose or fructose), leading to high rates of lipogenesis and esterification of fatty acids; (3) high levels of circulating free fatty acids; (4) ingestion of ethanol; and (5) the presence of high concentrations of insulin and low concentrations of glucagon, which enhance fatty acid synthesis and esterification and inhibit their oxidation.

CLINICAL ASPECTS

Imbalance in the Rate of Triacylglycerol Formation and Export Causes Fatty Liver

For a variety of reasons, lipid—mainly as triacylglycerol—can accumulate in the liver (Figure 27–7). Extensive accumulation is regarded as a pathologic condition. When accumulation of lipid in the liver becomes chronic, fibrotic changes occur in the cells that progress to **cirrhosis** and impaired liver function.

Fatty livers fall into two main categories. The first type is associated with **raised levels of plasma free fatty acids** resulting from mobilization of fat from adipose tissue or from the hydrolysis of lipoprotein triacylglycerol by lipoprotein lipase in extrahepatic tissues. Increasing amounts of free fatty acids are taken up by the liver and esterified. The production of VLDL does not keep pace with the influx of free fatty acids, allowing triacylglycerol to accumulate, causing a fatty liver. The quantity of triacylglycerol present in the liver is significantly increased during

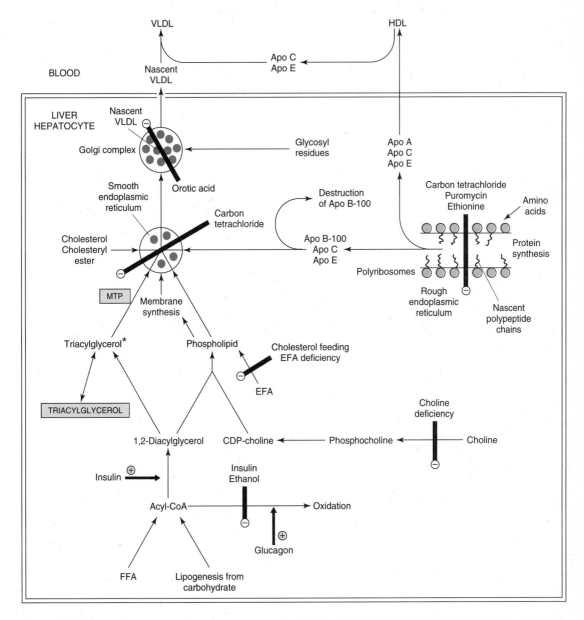

Figure 27–7. The synthesis of very low density lipoprotein (VLDL) and the possible loci of action of factors causing accumulation of triacylglycerol and a fatty liver. (EFA, essential fatty acids; FFA, free fatty acids; HDL, high-density lipoproteins; apo, apolipoprotein; MTP, microsomal triacylglycerol transfer protein.) The pathways indicated form a basis for events depicted in Figure 27–3B. The main triacylglycerol pool in liver is not on the direct pathway of VLDL synthesis from acyl-CoA. Thus, FFA, insulin, and glucagon have immediate effects on VLDL secretion as their effects impinge directly on the small triacylglycerol* precursor pool. In the fully fed state, apo B-100 is synthesized in excess of requirements for VLDL secretion and the surplus is destroyed in the liver.

starvation and the feeding of **high-fat diets.** In many instances (eg, in starvation), the ability to secrete VLDL is also impaired. This may be due to low levels of insulin and impaired protein synthesis. In uncontrolled **diabetes mellitus, pregnancy toxemia of ewes,** and **ketosis in cattle,** fatty infiltration is sufficiently severe to cause visible pallor (fatty appear-

ance) and enlargement of the liver with possible liver dysfunction.

The second type of fatty liver is usually due to a **metabolic block in the production of plasma lipoproteins,** thus allowing triacylglycerol to accumulate. Theoretically, the lesion may be due to (1) a block in apolipoprotein synthesis, (2) a block in the synthesis of the lipoprotein from lipid and apolipo-protein, (3) a failure in provision of phospholipids that are found in lipoproteins, or (4) a failure in the secretory mechanism itself.

One type of fatty liver that has been studied extensively in rats is due to a deficiency of **choline,** which has therefore been called a **lipotropic factor.** As choline may be synthesized using labile methyl groups donated by methionine in the process of **transmethylation** (Chapters 32 and 33), the deficiency is basically due to a shortage of the type of methyl group donated by methionine. Several mechanisms have been suggested to explain the role of choline as a lipotropic agent, including its absence, causing an impairment in synthesis of lipoprotein phospholipids.

The antibiotic puromycin inhibits protein synthesis and causes a fatty liver and a marked reduction in concentration of VLDL in rats. Other substances that act similarly include ethionine (α-amino-γ-mercaptobutyric acid), carbon tetrachloride, chloroform, phosphorus, lead, and arsenic. Choline will not protect the organism against these agents but appears to aid in recovery. It is very likely that carbon tetrachloride also affects the secretory mechanism itself or the conjugation of the lipid with apolipoprotein. Its effect is not direct but depends on further transformation of the molecule. This probably involves formation of free radicals that disrupt lipid membranes in the endoplasmic reticulum by formation of lipid peroxides. Some protection against carbon tetrachloride-induced lipid peroxidation is provided by the antioxidant action of vitamin E-supplemented diets. The action of ethionine is thought to be due to a reduction in availability of ATP. This results when ethionine, replacing methionine in S-adenosylmethionine, traps available adenine and prevents synthesis of ATP. Orotic acid also causes fatty livers; as VLDL accumulate in the Golgi apparatus, it is considered that orotic acid interferes with glycosylation of the lipoprotein, thus inhibiting its release and accounting for the marked decrease in plasma lipoproteins containing apo B.

A deficiency of vitamin E enhances the hepatic necrosis of the choline deficiency type of fatty liver. Added vitamin E or a source of selenium has a protective effect by combating lipid peroxidation. In addition to protein deficiency, essential fatty acid and vitamin deficiencies (eg, linoleic acid, pyridoxine and pantothenic acid) can cause fatty infiltration of the liver. A deficiency of essential fatty acids is thought to depress the synthesis of phospholipids; therefore, other substances such as cholesterol that compete for available essential fatty acids for esterification can also cause fatty livers.

Ethanol Also Causes Fatty Liver

Alcoholism leads to fat accumulation in the liver, hyperlipidemia, and ultimately **cirrhosis.** The exact mechanism of action of ethanol in the long term is still uncertain. Whether or not extra free fatty acid mobilization plays some part in causing the accumulation of fat is not clear, but several studies have demonstrated elevated levels of free fatty acids in the rat after administration of a single intoxicating dose of ethanol. However, ethanol consumption over a long period leads to the accumulation of fatty acids in the liver that are derived from endogenous synthesis rather than from adipose tissue. There is no impairment of hepatic synthesis of protein after ethanol ingestion. There is good evidence of increased hepatic triacylglycerol synthesis, decreased fatty acid oxidation, and decreased citric acid cycle activity, caused by oxidation of ethanol in the hepatic cytosol by **alcohol dehydrogenase,** leading to excess production of NADH.

$$CH_3-CH_2-OH \xrightarrow[NAD^+ \quad NADH + H^+]{\boxed{\text{ALCOHOL DEHYDROGENASE}}} CH_3-CHO$$
Ethanol → Acetaldehyde

The NADH generated competes with reducing equivalents from other substrates for the respiratory chain, inhibiting their oxidation. The increased [NADH]/[NAD$^+$] ratio causes a shift to the left in the equilibrium malate ⇌ oxaloacetate, which may reduce activity of the citric acid cycle. The net effect of inhibiting fatty acid oxidation is to cause increased esterification of fatty acids in triacylglycerol, which appears to be the cause of the fatty liver. Oxidation of ethanol leads to the formation of acetaldehyde, which is oxidized by **aldehyde dehydrogenase,** predominantly in mitochondria, acetate being the end product. Other effects of ethanol may include increased lipogenesis and cholesterol synthesis from acetyl-CoA. The increased [NADH]/[NAD$^+$] ratio also causes an increased [lactate]/[pyruvate] ratio that results in hyperlacticacidemia, which in turn decreases the capacity of the kidney to excrete uric acid. The latter is probably the cause of aggravation of gout by drinking alcohol. Although the major route for ethanol metabolism is via the alcohol dehydrogenase pathway, some metabolism takes place via a cytochrome P450-dependent microsomal ethanol oxidizing system (MEOS) involving NADPH and O$_2$. This system increases in activity in **chronic alcoholism** and may account for the increased metabolic clearance in this condition as indicated by increased blood levels of both acetaldehyde and acetate. Ethanol will also inhibit the metabolism of some drugs, eg, barbiturates, by competing for cytochrome P450-dependent enzymes (Chapter 61).

$$CH_2-CH_2-OH + NADPH + H^+ + O_2 \xrightarrow{\boxed{\text{MEOS}}}$$
Ethanol
$$CH_3-CHO + NADP^+ + 2H_2O$$
Acetaldehyde

Alcohol dehydrogenase is also present in gastric mucosa but is of the order of 60% less in activity in women than in men. The increased availability of ethanol resulting from the decreased gastric utilization may account for the increased susceptibility of women to the effects of alcohol consumption. Some Asian populations and Native Americans, following alcohol consumption, are prone to increased adverse reactions to acetaldehyde owing to a genetic defect of mitochondrial aldehyde dehydrogenase.

ADIPOSE TISSUE IS THE MAIN STORE OF TRIACYLGLYCEROL IN THE BODY

The triacylglycerol stores in adipose tissue are continually undergoing lipolysis (hydrolysis) and reesterification (Figure 27–8). These two processes are not the forward and reverse phases of the same reaction. Rather, they are entirely different pathways involving different reactants and enzymes. This al-

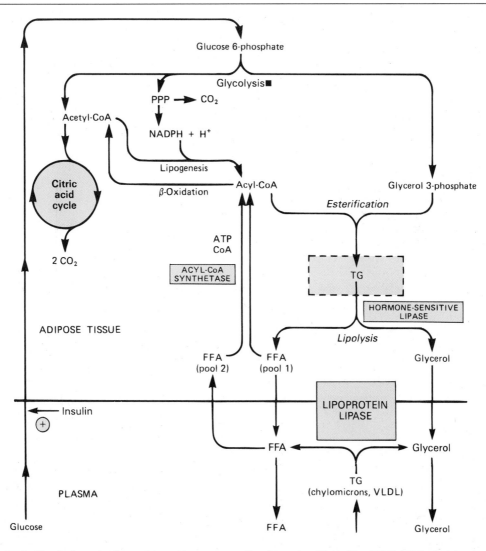

Figure 27–8. Metabolism of adipose tissue. Hormone-sensitive lipase is activated by ACTH, TSH, glucagon, epinephrine, norepinephrine, and vasopressin and inhibited by insulin, prostaglandin E_1, and nicotinic acid. Details of the formation of glycerol 3-phosphate from intermediates of glycolysis are shown in Figure 26–2. (PPP, pentose phosphate pathway; TG, triacylglycerol; FFA, free fatty acids; VLDL, very low density lipoprotein.)

lows many of the nutritional, metabolic, and hormonal factors that regulate the metabolism of adipose tissue to act either upon the process of esterification or on lipolysis. The resultant of these two processes determines the magnitude of the free fatty acid pool in adipose tissue, which in turn is the source and determinant of the level of free fatty acids circulating in the plasma. Since the level of plasma free fatty acids has most profound effects upon the metabolism of other tissues, particularly liver and muscle, the factors operating in adipose tissue that regulate the outflow of free fatty acids exert an influence far beyond the tissue itself.

The Provision of Glycerol 3-Phosphate Regulates Esterification: Lipolysis Is Controlled by Hormone-Sensitive Lipase

In adipose tissue, triacylglycerol is synthesized from acyl-CoA and glycerol 3-phosphate according to the mechanism shown in Figure 26–2. Because the enzyme **glycerol kinase** is low in activity in adipose tissue, glycerol cannot be utilized to any great extent in the esterification of acyl-CoA. For the provision of glycerol 3-phosphate, the tissue is dependent on glycolysis and a supply of glucose.

Triacylglycerol undergoes hydrolysis by a **hormone-sensitive lipase** to form free fatty acids and glycerol. This lipase is distinct from lipoprotein lipase that catalyzes lipoprotein triacylglycerol hydrolysis before its uptake into extrahepatic tissues (see above). Since glycerol cannot be utilized readily in this tissue, it diffuses out into the plasma, whence it is utilized by such tissues as liver and kidney, which possess an active glycerol kinase. The free fatty acids formed by lipolysis can be reconverted in the tissue to acyl-CoA by **acyl-CoA synthetase** and reesterified with glycerol 3-phosphate to form triacylglycerol. Thus, **there is a continuous cycle of lipolysis and reesterification within the tissue.** However, when the rate of reesterification is not sufficient to match the rate of lipolysis, free fatty acids accumulate and diffuse into the plasma, where they bind to albumin and raise the concentration of plasma free fatty acids: These are a most important source of fuel for many tissues.

Increased Glucose Metabolism Reduces the Output of Free Fatty Acids

When the utilization of glucose by adipose tissue is increased, the free fatty acid outflow decreases. However, the release of glycerol continues, demonstrating that the effect of glucose is not mediated by reducing the rate of lipolysis. It is believed that the effect is due to the provision of glycerol 3-phosphate, which enhances esterification of free fatty acids via acyl-CoA.

Glucose can take several pathways in adipose tissue, including oxidation to CO_2 via the citric acid cycle, oxidation in the pentose phosphate pathway, conversion to long-chain fatty acids, and formation of acylglycerol via glycerol 3-phosphate. When glucose utilization is high, a larger proportion of the uptake is oxidized to CO_2 and converted to fatty acids. However, as total glucose utilization decreases, the greater proportion of the glucose is directed to the formation of glycerol 3-phosphate for the esterification of acyl-CoA, which helps to minimize the efflux of free fatty acids.

Free Fatty Acids Are Taken Up as a Result of Lipoprotein Lipase Activity

There is more than one free fatty acid pool within adipose tissue. It has been shown that the free fatty acid pool (Figure 27–8, pool 1) formed by lipolysis of triacylglycerol is the same pool that supplies fatty acids for reesterification; also, it releases them into the external medium (plasma). Fatty acids taken up from the external medium as a result of the action of lipoprotein lipase on the triacylglycerol of chylomicrons and VLDL do not label pool 1 before they are incorporated into triacylglycerol but travel through a small pool 2 of high turnover.

HORMONES REGULATE FAT MOBILIZATION

Insulin Reduces the Output of Free Fatty Acids

The rate of release of free fatty acids from adipose tissue is affected by many hormones that influence either the rate of esterification or the rate of lipolysis. Insulin inhibits the release of free fatty acids from adipose tissue, which is followed by a fall in circulating plasma free fatty acids. It enhances lipogenesis and the synthesis of acylglycerol and increases the oxidation of glucose to CO_2 via the pentose phosphate pathway. All these effects are dependent on the presence of glucose and can be explained, to a large extent, on the basis of the ability of insulin to enhance the uptake of glucose into adipose cells. This is achieved by insulin causing the translocation of glucose transporters from the Golgi apparatus to the plasma membrane (Figure 51–9). Insulin has also been shown to increase the activity of **pyruvate dehydrogenase, acetyl-CoA carboxylase, and glycerol phosphate acyltransferase,** which would reinforce the effects arising from increased glucose uptake on the enhancement of fatty acid and acylglycerol synthesis. These three enzymes are now known to be regulated in a coordinate manner by covalent modification, ie, by phosphorylation-dephosphorylation mechanisms.

A principal action of insulin in adipose tissue is to

inhibit the activity of the **hormone-sensitive lipase,** reducing the release not only of free fatty acids but of glycerol as well. Adipose tissue is much more sensitive to insulin than are many other tissues, which points to adipose tissue as a major site of insulin action in vivo.

Several Hormones Promote Lipolysis

Other hormones accelerate the release of free fatty acids from adipose tissue and raise the plasma free fatty acid concentration by increasing the rate of lipolysis of the triacylglycerol stores (Figure 27–9). These include epinephrine, norepinephrine, glucagon, adrenocorticotropic hormone (ACTH), α- and β-melanocyte-stimulating hormones (MSH), thyroid-stimulating hormone (TSH), growth hormone (GH), and vasopressin. Many of these activate the hormone-sensitive lipase. For an optimal effect, most of these lipolytic processes require the presence of **glu-cocorticoids** and **thyroid hormones.** On their own, these particular hormones do not increase lipolysis markedly but act in a **facilitatory** or **permissive** capacity with respect to other lipolytic endocrine factors.

The hormones that act rapidly in promoting lipolysis, ie, catecholamines, do so by stimulating the activity of **adenylyl cyclase,** the enzyme that converts ATP to cAMP. The mechanism is analogous to that responsible for hormonal stimulation of glycogenolysis (Chapter 20). It appears that cAMP, by stimulating **cAMP-dependent protein kinase,** converts inactive hormone-sensitive triacylglycerol lipase into active lipase. Lipolysis is controlled largely by the amount of cAMP present in the tissue. It follows that processes which destroy or preserve cAMP have an effect on lipolysis. cAMP is degraded to $5'$-AMP by the enzyme **cyclic $3',5'$-nucleotide phosphodi-esterase.** This enzyme is inhibited by methylxan-

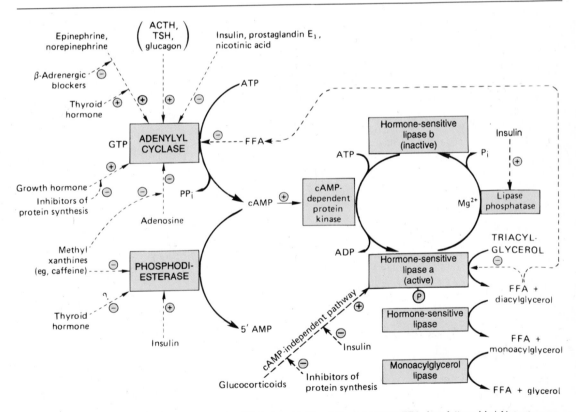

Figure 27–9. Control of adipose tissue lipolysis. (TSH, thyroid-stimulating hormone; FFA, free fatty acids.) Note the cascade sequence of reactions affording amplification at each step. The lipolytic stimulus is "switched off" by removal of the stimulating hormone; the action of lipase phosphatase; the inhibition of the lipase and adenylyl cyclase by high concentrations of FFA; the inhibition of adenylyl cyclase by adenosine; and the removal of cAMP by the action of phosphodiesterase. ACTH, TSH, and glucagon may not activate adenylyl cyclase in vivo, since the concentration of each hormone required in vitro is much higher than is found in the circulation. Positive ($\oplus$) and negative ($\ominus$) regulatory effects are represented by broken lines and substrate flow by solid lines.

∍ines such as **caffeine** and **theophylline.** It is significant that the drinking of coffee, containing caffeine, causes elevation of plasma FFA in humans.

Insulin antagonizes the effect of the lipolytic hormones. It is now considered that lipolysis may be more sensitive to changes in concentration of insulin than are glucose utilization and esterification. The antilipolytic effects of insulin, nicotinic acid, and prostaglandin E_1 may be accounted for by inhibition of the synthesis of cAMP at the adenylyl cyclase site, acting through a G_i protein. Insulin also stimulates phosphodiesterase and the lipase phosphatase that inactivates hormone-sensitive lipase. Possible mechanisms for the action of thyroid hormones include an augmentation of the level of cAMP by facilitation of the passage of the stimulus from the receptor site on the outside of the cell membrane to the adenylyl cyclase site on the inside of the membrane and an inhibition of phosphodiesterase activity. The effect of growth hormone in promoting lipolysis is slow. It is dependent on synthesis of proteins involved in the formation of cAMP. Glucocorticoids promote lipolysis via synthesis of new lipase protein by a cAMP-independent pathway, which may be inhibited by insulin. These findings help to explain the role of the pituitary gland and the adrenal cortex in enhancing fat mobilization.

The sympathetic nervous system, through liberation of norepinephrine in adipose tissue, plays a central role in the mobilization of free fatty acids by exerting a tonic influence even in the absence of augmented nervous activity. Thus, the increased lipolysis caused by many of the factors described above can be reduced or abolished by denervation of adipose tissue, by ganglionic blockade with hexamethonium, or by depleting norepinephrine stores with reserpine.

A Variety of Mechanisms Have Evolved for Fine Control of Adipose Tissue Metabolism

Human adipose tissue may not be an important site of lipogenesis. This is indicated by the observation that there is not significant incorporation of label into long-chain fatty acids from labeled glucose or pyruvate. ATP-citrate lyase, a key enzyme in lipogenesis, does not appear to be present and has extremely low activity in liver. Other enzymes—eg, glucose-6-phosphate dehydrogenase and the malic enzyme—which in the rat undergo adaptive changes coincident with increased lipogenesis, do not undergo similar changes in human adipose tissue. Indeed, it has been suggested that in humans there is a "carbohydrate excess syndrome" due to a unique limitation in ability to dispose of excess carbohydrate by lipogenesis. In birds, lipogenesis is confined to the liver, where it is particularly important in providing lipids for egg formation, stimulated by estrogens.

Human adipose tissue is unresponsive to most of

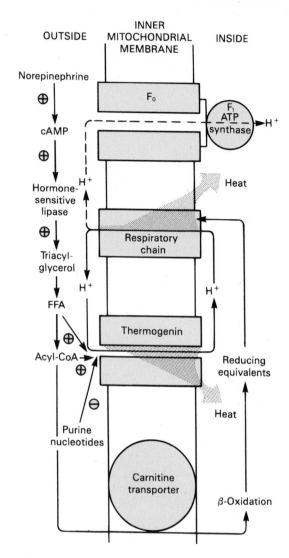

Figure 27–10. Thermogenesis in brown adipose tissue. Activity of the respiratory chain produces heat in addition to translocating protons (Chapter 14). These protons dissipate more heat when returned to the inner mitochondrial compartment via thermogenin instead of generating ATP when returning via the F_1 ATP synthase. The passage of H^+ via thermogenin is inhibited by purine nucleotides when brown adipose tissue is unstimulated. Under the influence of norepinephrine, the inhibition is removed by the production of free fatty acids (FFA) and acyl-CoA. Note the dual role of acyl-CoA in both facilitating the action of thermogenin and supplying reducing equivalents for the respiratory chain. $(\oplus)$ and $(\ominus)$ signify positive or negative regulatory effects.

the lipolytic hormones apart from the catecholamines. Of further interest is the lack of lipolytic response to epinephrine in the rabbit, guinea pig, pig, and chicken; the pronounced lipolytic effect of glucagon in birds, together with an absence of any antilipolytic effect of insulin; and the lack of acylglycerol-glycerol synthesis from glucose in the pigeon.

On consideration of the profound derangement of metabolism in diabetes mellitus (which is due mainly to increased release of free fatty acids from the depots) and the fact that insulin to a large extent corrects the condition, it must be concluded that insulin plays a prominent role in the regulation of adipose tissue metabolism.

BROWN ADIPOSE TISSUE PROMOTES THERMOGENESIS

Brown adipose tissue is involved in metabolism particularly at times when heat generation is necessary. Thus, the tissue is extremely active in some species in arousal from hibernation, in animals exposed to cold (nonshivering thermogenesis), and in heat production in the newborn animal. Though not a prominent tissue in humans, it is present in normal individuals, where it appears to be responsible for **"diet-induced thermogenesis,"** which may account for how some persons can "eat and not get fat." It is noteworthy that brown adipose tissue is reduced or absent in obese persons. Brown adipose tissue is characterized by a well-developed blood supply and a high content of mitochondria and cytochromes but low activity of ATP synthase. Metabolic emphasis is placed on oxidation of both glucose and fatty acids.

Norepinephrine liberated from sympathetic nerve endings is important in increasing lipolysis in the tissue. Oxidation and phosphorylation are not coupled in mitochondria of this tissue, since dinitrophenol has no effect and there is no respiratory control by ADP. The phosphorylation that does occur is at the substrate level, eg, at the succinate thiokinase step and in glycolysis. Thus, **oxidation produces much heat, and little free energy is trapped in ATP.** In terms of the chemiosmotic theory (Chapter 14), it would appear that the proton gradient, normally present across the inner mitochondrial membrane of coupled mitochondria, is continually dissipated in brown adipose tissue by a thermogenic uncoupling protein, **thermogenin,** which acts as a proton conductance pathway through the membrane. This would explain the apparent lack of effect of uncouplers (Figure 27–10).

SUMMARY

1. Since lipids are insoluble in water, nonpolar lipids must be combined with amphipathic lipids and proteins to make water-miscible lipoproteins for transport between the tissues in the aqueous blood plasma.

2. Four major groups of lipoproteins are recognized: Chylomicrons transport lipids resulting from digestion and absorption. Very low density lipoproteins (VLDL) transport triacylglycerol from the liver. Low-density lipoproteins (LDL) are cholesterol-rich lipoproteins resulting from the metabolism of VLDL, and high-density lipoproteins (HDL) are also cholesterol rich but are involved in removing cholesterol from the tissues and in the metabolism of other lipoproteins.

3. Chylomicrons and VLDL are first metabolized by hydrolysis with lipoprotein lipase in extrahepatic tissues. Most of the triacylglycerol is removed, and a lipoprotein remnant is left in the circulation. These remnants are taken up into the liver by receptor-mediated endocytosis, but some of the remnants (IDL) resulting from VLDL form LDL and are finally taken up by the liver and other tissues via the LDL receptor.

4. Apolipoproteins constitute the protein moiety of lipoproteins. They act as enzyme activators (eg, apo C-II and apo A-I) or as ligands for cell receptors (eg, apo A-I, apo E, and apo B-100).

5. Imbalance in the rate of triacylglycerol formation in the liver and in VLDL secretion leads to fatty liver. This is of clinical importance in alcoholism, where it proceeds to cirrhosis and liver dysfunction.

6. Triacylglycerol is the main storage lipid in adipose tissue. It is released after hydrolysis by hormone-sensitive lipase to free fatty acids and glycerol. Free fatty acids are bound to serum albumin for transport to the tissues, where they are used as an important fuel source. Hormone-sensitive lipase is stimulated by epinephrine and norepinephrine and inhibited by insulin.

7. Brown adipose tissue is the site of "nonshivering thermogenesis." It is found in hibernating and newborn animals and is present in small quantity in humans, where it may be responsible for "diet-induced thermogenesis." Thermogenesis results from the presence of a protein, thermogenin, which acts as a proton conductance pathway through the inner mitochondrial membrane, uncoupling oxidation from phosphorylation.

REFERENCES

Borensztajn J (editor): *Lipoprotein Lipase.* Evener Publishers, 1987.

Brewer HB et al: Apolipoproteins and lipoproteins in human plasma: An overview. Clin Chem 1988;34:B4.

Coppack SW et al: Effects of insulin on human adipose tissue metabolism in vivo. Clin Sci 1989;77:663.

Eisenberg S: High density lipoprotein metabolism. J Lipid Res 1984;25:1017.

Fielding CJ, Fielding PE: Metabolism of cholesterol and lipoproteins. In: *Biochemistry of Lipids and Membranes.*

Vance DE, Vance JE (editors). Benjamin/Cummings, 1985.

Kaikans RM, Bass NM, Ockner RK: Functions of fatty acid binding proteins. Experientia 1990;46:617.

Lardy H, Shrago E: Biochemical aspects of obesity. Annu Rev Biochem 1990;59:689.

Lieber CS: Alcohol, liver, and nutrition. J Am Coll Nutr 1991;10:602.

Various authors: Brown adipose tissue—role in nutritional energetics. (Symposium.) Proc Nutr Soc 1989;48:165.

Cholesterol Synthesis, Transport, & Excretion

28

Peter A. Mayes, PhD, DSc

INTRODUCTION

Cholesterol is present in tissues and in plasma lipoproteins either as free cholesterol or, combined with a long-chain fatty acid, as cholesteryl ester. It is synthesized in many tissues from acetyl-CoA and is ultimately eliminated from the body in the bile as cholesterol or bile salts. Cholesterol is the precursor of all other steroids in the body such as corticosteroids, sex hormones, bile acids, and vitamin D. It is typically a product of animal metabolism and therefore occurs in foods of animal origin such as egg yolk, meat, liver, and brain.

BIOMEDICAL IMPORTANCE

Cholesterol is an amphipathic lipid and as such is an essential structural component of membranes and of the outer layer of plasma lipoproteins. Lipoproteins transport free cholesterol in the circulation, where it readily equilibrates with cholesterol in other lipoproteins and in membranes. Cholesteryl ester is a storage form of cholesterol found in most tissues. It is transported as cargo in the hydrophobic core of lipoproteins. LDL is the mediator of cholesterol and cholesteryl ester uptake into many tissues. Free cholesterol is removed from tissues by HDL and transported to the liver for conversion to bile acids in the process known as **reverse cholesterol transport.** Cholesterol is a major constituent of **gallstones.** However, its chief role in pathologic processes is as a factor in the genesis of **atherosclerosis** of vital arteries, causing cerebrovascular, coronary, and peripheral vascular disease. Coronary atherosclerosis correlates with a high plasma LDL:HDL cholesterol ratio.

CHOLESTEROL IS DERIVED ABOUT EQUALLY FROM THE DIET AND FROM BIOSYNTHESIS

A little more than half the cholesterol of the body arises by synthesis (about 700 mg/d), and the remain-der is provided by the average diet. The liver accounts for approximately 10% of total synthesis in humans, the intestines for about another 10%.

Virtually all tissues containing nucleated cells are capable of synthesizing cholesterol. The microsomal (endoplasmic reticulum) and cytosol fraction of the cell is mainly responsible for cholesterol synthesis.

Acetyl-CoA Is the Source of All Carbon Atoms in Cholesterol

The biosynthesis of cholesterol may be divided into five stages. (1) Mevalonate, a six-carbon compound, is synthesized from acetyl-CoA (Figure 28–1). (2) Isoprenoid units are formed from mevalonate by loss of CO_2 (Figure 28–2). (3) Six isoprenoid units condense to form the intermediate, squalene. (4) Squalene cyclizes to give rise to the parent steroid, lanosterol. (5) Cholesterol is formed from lanosterol after several further steps, including the loss of three methyl groups (Figure 28–3).

Step 1—Acetyl-CoA Forms HMG-CoA and Mevalonate: The pathway through HMG-CoA (3-hydroxy-3-methylglutaryl-CoA) follows the same sequence of reactions described in Chapter 24 for the synthesis in mitochondria of ketone bodies. However, since cholesterol synthesis is extramitochondrial, the two pathways are distinct. Initially, two molecules of acetyl-CoA condense to form acetoacetyl-CoA catalyzed by a cytosolic **thiolase** enzyme. Alternatively, in liver, acetoacetate made inside the mitochondrion in the pathway of ketogenesis (Chapter 24) diffuses into the cytosol and may be activated to acetoacetyl-CoA by **acetoacetyl-CoA synthase,** requiring ATP and CoA. Acetoacetyl-CoA condenses with a further molecule of acetyl-CoA catalyzed by **HMG-CoA synthase** to form HMG-CoA.

HMG-CoA is converted to **mevalonate** in a two-stage reduction by NADPH catalyzed by **HMG-CoA reductase,** a microsomal enzyme considered to catalyze the rate-limiting step in the pathway of cholesterol synthesis and is the site of action of the most effective class of cholesterol-lowering drugs, the HMG-CoA reductase inhibitors (statins) (Figure 28–1).

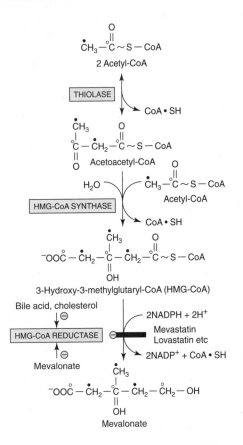

Figure 28–1. Biosynthesis of mevalonate. (HMG, 3-hydroxy-3-methylglutaryl.) The synthesis of HMG-CoA reductase is inhibited by the fungal metabolites mevastatin (compactin), lovastatin (mevinolin), provastatin, and simvastatin.

Step 2—Mevalonate Forms Active Isoprenoid Units: Mevalonate is phosphorylated by ATP to form several active phosphorylated intermediates (Figure 28–2). By means of a decarboxylation, the active isoprenoid unit, **isopentenyl pyrophosphate,** is formed.

Step 3—Six Isoprenoid Units Form Squalene: This stage involves the condensation of three molecules of isopentenylpyrophosphate to form **farnesyl pyrophosphate.** This occurs via an isomerization of isopentenyl pyrophosphate involving a shift of the double bond to form **dimethylallyl pyrophosphate,** followed by condensation with another molecule of isopentenyl pyrophosphate to form the ten-carbon intermediate, **geranyl pyrophosphate** (Figure 28–2). A further condensation with isopentenyl pyrophosphate forms farnesyl pyrophosphate. Two molecules of farnesyl pyrophosphate condense at the pyrophosphate end in a reaction involving first an elimination of pyrophosphate to form presqualene pyrophosphate, followed by a reduction with NADPH with

elimination of the remaining pyrophosphate radical. The resulting compound is **squalene.** An alternative pathway known as the "*trans*-methylglutaconate shunt" may be present. This pathway removes a significant proportion (5% in fed livers, rising to 33% in fasted livers) of the dimethylallyl pyrophosphate and returns it, via *trans*-3-methylglutaconate-CoA, to HMG-CoA. This pathway may have regulatory potential with respect to the overall rate of cholesterol synthesis.

Step 4—Squalene Is Converted to Lanosterol: Squalene has a structure that closely resembles the steroid nucleus (Figure 28–3). Before ring closure occurs, squalene is converted to squalene 2,3-oxide by a mixed-function oxidase in the endoplasmic reticulum, **squalene epoxidase.** The methyl group on C_{14} is transferred to C_{13} and that on C_8 to C_{14} as cyclization occurs, catalyzed by **oxidosqualene:lanosterol cyclase.**

Step 5—Lanosterol Is Converted to Cholesterol: In this last stage (Figure 28–3), the formation of cholesterol from **lanosterol** takes place in the membranes of the endoplasmic reticulum and involves changes in the steroid nucleus and side chain. The methyl group on C_{14} is oxidized to CO_2 to form 14-desmethyl lanosterol. Likewise, two more methyl groups on C_4 are removed to produce zymosterol. $\Delta^{7,24}$-Cholestadienol is formed from zymosterol by the double bond between C_8 and C_9 moving to a position between C_8 and C_7. **Desmosterol** is formed at this point by a further shift in the double bond in ring B to take up a position between C_5 and C_6, as in cholesterol. Finally, cholesterol is produced when the double bond of the side chain is reduced, although this can occur at any stage of the overall conversion to cholesterol. The exact order in which the steps described actually take place is not known with certainty.

It is probable that the intermediates from squalene to cholesterol are attached to a special carrier protein known as the **squalene and sterol carrier protein.** This protein binds sterols and other insoluble lipids, allowing them to react in the aqueous phase of the cell. In addition, it seems likely that it is in the form of cholesterol-sterol carrier protein that cholesterol is converted to steroid hormones and bile acids and participates in the formation of membranes and of lipoproteins.

Farnesyl Pyrophosphate Gives Rise to Other Important Isoprenoid Compounds

Farnesyl pyrophosphate is the branch point for the synthesis of the other polyisoprenoids, **dolichol** and **ubiquinone.** The polyisoprenyl alcohol dolichol (Figure 16–26 and Chapter 56) is formed by the further addition of up to 16 isopentenyl pyrophosphate residues, whereas the side chain of ubiquinone (Figure 14–5) is formed by the addition of a further three

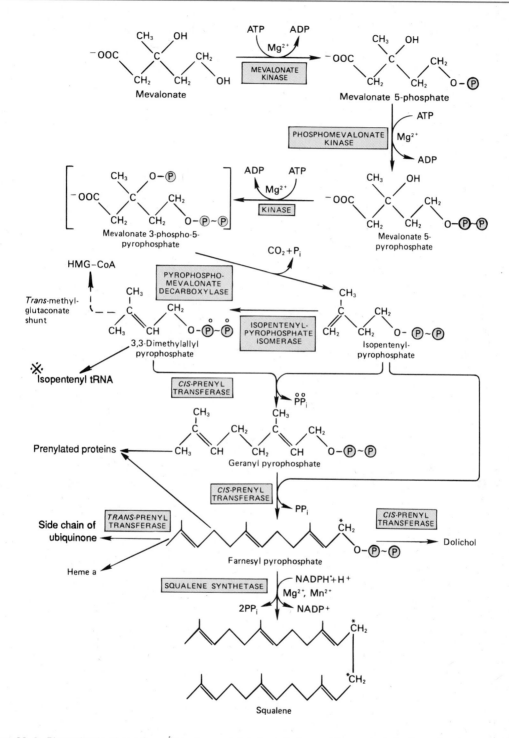

Figure 28–2. Biosynthesis of squalene, ubiquinone, dolichol, and other polyisoprene derivatives. (HMG, 3-hydroxy-3-methylglutaryl; ✳, cytokinin.) A farnesyl residue is present in heme a of cytochrome oxidase. The carbon marked * becomes C_{11} or C_{12} in squalene. Squalene synthetase is a microsomal enzyme; all other enzymes indicated are soluble cytosolic proteins.

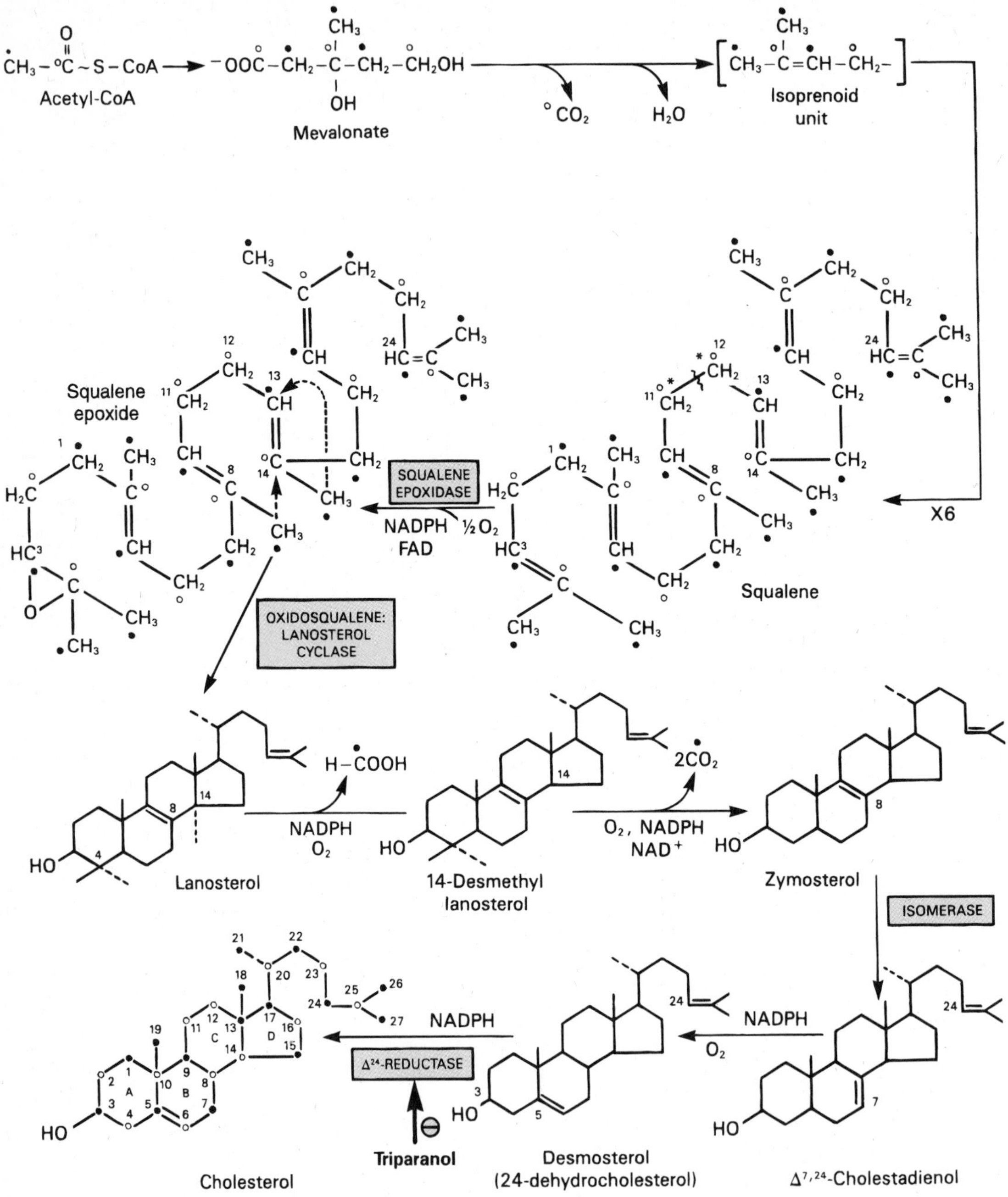

Figure 28–3. Biosynthesis of cholesterol. The numbered positions are those of the steroid nucleus and the open and solid circles indicate the fate of each of the carbons in the acetyl moiety of acetyl-CoA. *Refers to labeling of squalene in Figure 28–2.

to seven isoprenoid units. Some GTP-binding proteins in the cell membrane may undergo prenylation by combining with geranyl and farnesyl residues. This may facilitate the anchoring of prenylated proteins into lipoid membranes.

CHOLESTEROL SYNTHESIS IS CONTROLLED BY REGULATION OF HMG-CoA REDUCTASE

Regulation of cholesterol synthesis is exerted near the beginning of the pathway, at the HMG-CoA reductase step. There is a marked decrease in the activity of HMG-CoA reductase in fasting animals, which explains the reduced synthesis of cholesterol during fasting. There is a feedback mechanism whereby HMG-CoA reductase in liver is inhibited by mevalonate, the immediate product, and by cholesterol, the main product of the pathway. Since a direct inhibition of the enzyme by cholesterol cannot be demonstrated, cholesterol (or a metabolite, eg, oxygenated sterol) may act either by repression of the synthesis

of new reductase or by inducing the synthesis of enzymes that degrade existing reductase. Cholesterol synthesis is also inhibited by LDL-cholesterol taken up via LDL receptors (apo B-100, E receptors). A **diurnal variation** occurs in both cholesterol synthesis and reductase activity. However, there are more rapid effects on reductase activity than can be explained solely by changes in the rate of protein synthesis. Administration of insulin or thyroid hormone increases HMG-CoA reductase activity, whereas glucagon or glucocorticoids decrease it. The enzyme exists in both active and inactive forms that may be reversibly modified by phosphorylation-dephosphorylation mechanisms, some of which may be cAMP dependent and therefore immediately responsive to glucagon (Figure 28–4).

The effect of variations in the amount of cholesterol in the diet on the endogenous production of cholesterol has been studied in rats. When there was only 0.05% cholesterol in the diet, 70–80% of the cholesterol of the liver, small intestine, and adrenal gland was synthesized within the body, whereas when the dietary intake was raised to 2%, the endogenous production fell. It appears that it is only he-

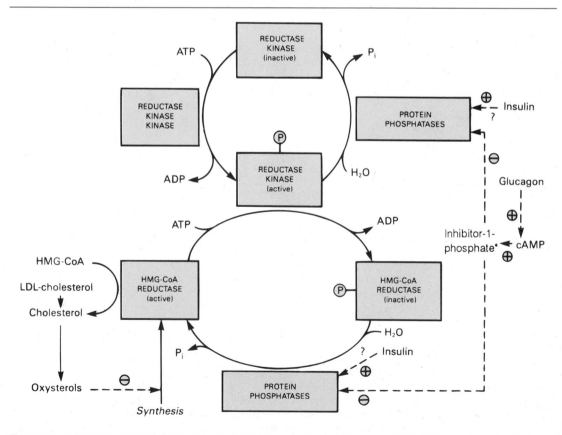

Figure 28–4. Possible mechanisms in the regulation of cholesterol synthesis by HMG-CoA reductase. Insulin has a dominant role compared with glucagon. *See Figure 20–6.

patic synthesis which is inhibited. Experiments with the perfused liver have demonstrated that cholesterol-rich chylomicron remnants, which are taken up by the liver (Chapter 27), inhibit sterol synthesis.

Attempts to lower plasma cholesterol in humans by reducing the amount of cholesterol in the diet produce variable results. Generally, a decrease of 100 mg in dietary cholesterol causes a decrease of approximately 0.13 mmol/L of serum.

MANY FACTORS INFLUENCE THE CHOLESTEROL BALANCE IN TISSUES

At the tissue level, the following processes are considered to govern the cholesterol balance of cells (Figure 28–5).

Increase is due to (1) uptake of cholesterol-containing lipoproteins by receptors, eg, the LDL receptor or the scavenger receptor; (2) uptake of cholesterol-containing lipoproteins by a non-receptor-mediated pathway; (3) uptake of free cholesterol from cholesterol-rich lipoproteins to the cell membrane; (4) cholesterol synthesis; and (5) hydrolysis of cholesteryl esters by the enzyme **cholesteryl ester hydrolase.**

Decrease is due to (1) efflux of cholesterol from the membrane to lipoproteins of low cholesterol potential, particularly to HDL_3 or nascent HDL, promoted by **LCAT** (lecithin:cholesterol acyltransferase); (2) esterification of cholesterol by **ACAT** (acyl-CoA:cholesterol acyltransferase); and (3) utilization of cholesterol for synthesis of other steroids, such as hormones, or bile acids in the liver.

The LDL Receptor Is Highly Regulated

The LDL (apo B-100, E) receptors occur on the cell surface in pits that are coated on the cytosolic side of the cell membrane with a protein called clathrin. The receptor is a glycoprotein that spans the

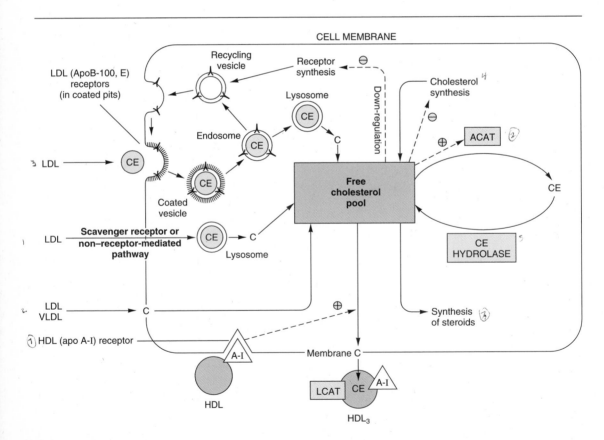

Figure 28–5. Factors affecting cholesterol balance at the cellular level. Reverse cholesterol transport may be initiated by HDL binding to an HDL (apo A-I) receptor, which, via protein kinase C (Chapter 44), stimulates translocation of cholesterol to the plasma membrane. (C, cholesterol; CE, cholesteryl ester; ACAT, acyl-CoA:cholesterol acyltransferase; LCAT, lecithin:cholesterol acyltransferase; A-I, apoprotein A-I; LDL, low-density lipoprotein; VLDL, very low density lipoprotein.) LDL and HDL are not shown to scale.

membrane, the B-100 binding region being at the exposed amino terminal end. After binding with the receptor, the LDL is taken up intact by endocytosis. It is broken down in the lysosomes, which involves hydrolysis of the apoprotein and cholesteryl ester followed by translocation of cholesterol into the cell. The receptors are not destroyed but return to the cell surface. This influx of cholesterol inhibits HMG-CoA reductase and cholesterol synthesis and stimulates ACAT activity. The number of LDL receptors on the cell surface is regulated by the cholesterol requirement for membranes, steroid hormones, or bile acid synthesis. Thus, influx of cholesterol down-regulates the number of LDL receptors (Figure 28–5).

The apo B-100, E receptor is a "high-affinity" LDL receptor, which may be saturated under most circumstances. Other "low-affinity" LDL receptors also appear to be present in addition to a scavenger pathway, which is not regulated.

CHOLESTEROL IS TRANSPORTED BETWEEN TISSUES IN PLASMA LIPOPROTEINS
(Figure 28–6)

In humans subsisting on westernized diets, the total plasma cholesterol is about 5.2 mmol/L, rising with

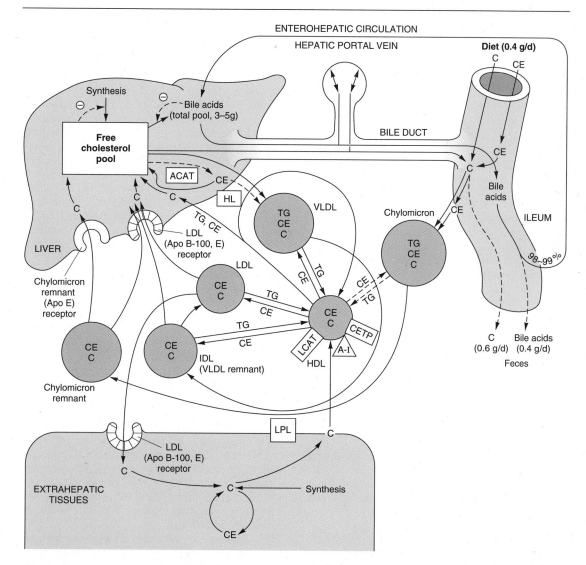

Figure 28–6. Transport of cholesterol between the tissues in humans. (C, free cholesterol; CE, cholesteryl ester; TG, triacylglycerol; VLDL, very low density lipoprotein; IDL, intermediate-density lipoprotein; LDL, low-density lipoprotein; HDL, high-density lipoprotein; ACAT, acyl-CoA:cholesterol acyltransferase; LCAT, lecithin:cholesterol acyltransferase; A-I, apoprotein A-I; CETP, cholesteryl ester transfer protein; LPL, lipoprotein lipase; HL, hepatic lipase.)

age, although there are wide variations between individuals. The greater part is found in the esterified form. It is transported in lipoproteins of the plasma, and the highest proportion of cholesterol is found in the LDL. However, under conditions where the VLDL are quantitatively more prominent, an increased proportion of the plasma cholesterol will reside in this fraction.

Dietary cholesterol takes several days to equilibrate with cholesterol in the plasma and several weeks to equilibrate with cholesterol of the tissues. The turnover of cholesterol in the liver is relatively fast compared with the half-life of the total body cholesterol. Free cholesterol in plasma and liver equilibrates in a matter of hours, since cholesterol exchanges and transfers readily between cell membranes, plasma lipoproteins, and erythrocyte membranes.

Cholesteryl ester in the diet is hydrolyzed to free cholesterol, which mixes with dietary free cholesterol and biliary cholesterol before absorption from the intestine in company with other lipids. It mixes with cholesterol synthesized in the intestines and is incorporated into chylomicrons. Of the cholesterol absorbed, 80–90% is esterified with long-chain fatty acids in the intestinal mucosa. The plant sterols (sitosterols) are poorly absorbed. When chylomicrons react with lipoprotein lipase to form chylomicron remnants, only about 5% of the cholesteryl ester is lost. The rest is taken up by the liver when the remnant reacts with the apo E or LDL receptor and is hydrolyzed to free cholesterol. VLDL formed in the liver transports cholesterol into the plasma. Most of the cholesterol in VLDL is retained in the VLDL remnant (IDL) that is taken up by the liver or converted to LDL, which in turn is taken up by the LDL receptor in liver and extrahepatic tissues.

LCAT Is Responsible for Most of the Plasma Cholesteryl Ester

The activity of plasma LCAT is responsible for virtually all plasma cholesteryl ester in humans. (This is not so in other species such as the rat, where there is appreciable ACAT activity in liver, allowing significant export of cholesteryl ester in nascent VLDL.) LCAT activity is associated with HDL containing apo A-I. As cholesterol in HDL becomes esterified, it creates a concentration gradient and draws in cholesterol from tissues and from other lipoproteins (Figures 28–5 and 28–6). It becomes less dense, forming HDL_2, which is thought to deliver cholesterol to the liver. Thus, HDL features prominently in **reverse cholesterol transport,** the process whereby tissue cholesterol is transported to the liver.

Cholesteryl Ester Transfer Protein Facilitates Transfer of Cholesteryl Ester From HDL to Other Lipoproteins

This protein, found in plasma of the human but not the rat, is also associated with HDL. It facilitates transfer of cholesteryl ester from HDL to VLDL, IDL, LDL, and, to a lesser extent, chylomicrons, and it allows triacylglycerol to transfer in the opposite direction. It therefore relieves product inhibition of LCAT activity in HDL. Thus, in humans, much of the cholesteryl ester formed by LCAT in HDL finds its way to the liver via VLDL remnants (IDL) or LDL (Figure 28–6). Simultaneously, the HDL_2, which has been enriched with triacylglycerol, unloads this cargo in the liver after reaction with hepatic lipase and is recycled as HDL_3.

ULTIMATELY, CHOLESTEROL MUST ENTER THE LIVER AND BE EXCRETED IN THE BILE AS CHOLESTEROL OR BILE ACIDS (SALTS)

About 1 g of cholesterol is eliminated from the body per day. Approximately half is excreted in the feces after conversion to bile acids. The remainder is excreted as cholesterol. Much of the cholesterol secreted in the bile is reabsorbed, and it is believed that at least some of the cholesterol that serves as precursor for the fecal sterols is derived from the intestinal mucosa. **Coprostanol** is the principal sterol in the feces; it is formed from cholesterol by the bacteria in the lower intestine. A large proportion of the biliary excretion of bile salts is reabsorbed into the portal circulation, taken up by the liver, and reexcreted in the bile. This is known as the **enterohepatic circulation.** The bile salts not reabsorbed, or their derivatives, are excreted in the feces.

Bile Acids Are Formed From Cholesterol

The **primary bile acids** are synthesized in the liver from cholesterol. These are **cholic acid** (found in the largest amount) and **chenodeoxycholic acid,** both formed from a common precursor, itself derived from cholesterol (Figure 28–7).

The 7α-hydroxylation of cholesterol is the first committed step in the biosynthesis of bile acids, and it is this reaction that is rate-limiting in the pathway for synthesis of the acids. The reaction is catalyzed by **7α-hydroxylase,** a microsomal enzyme. It requires oxygen, NADPH, and cytochrome P450 and appears to be a typical monooxygenase. Subsequent hydroxylation steps are also catalyzed by monooxygenases. Vitamin C deficiency interferes with bile acid formation at the 7α-hydroxylation step and leads to cholesterol accumulation and atherosclerosis in scorbutic guinea pigs.

The pathway of bile acid biosynthesis divides early into one subpathway leading to **cholyl CoA,** characterized by an extra α-OH group on position 12, and another pathway leading to **chenodeoxycholyl CoA.** Apart from this difference, both pathways involve similar hydroxylation reactions and

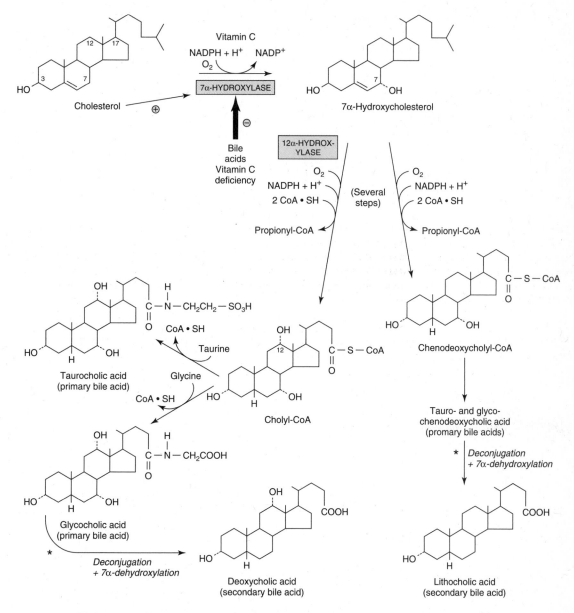

Figure 28–7. Biosynthesis and degradation of bile acids. *Catalyzed by microbial enzymes.

shortening of the side chain (Figure 28–7) to give the typical bile acid structures of α-OH groups on positions 3 and 7 and full saturation of the steroid nucleus. These primary bile acids enter the bile as glycine or taurine conjugates. In humans, the ratio of the glycine to the taurine conjugates is normally 3:1. Since bile contains significant quantities of sodium and potassium and the pH is alkaline, it is assumed that the bile acids and their conjugates are actually in a salt form—hence the term "bile salts."

A portion of the primary bile acids in the intestine is subjected to further changes by the activity of the intestinal bacteria. These include deconjugation and

7α-dehydroxylation, which produce the **secondary bile acids,** deoxycholic acid from cholic acid, and lithocholic acid from chenodeoxycholic acid (Figure 28–7).

Most Bile Acids Return to the Liver in the Enterohepatic Circulation

Although products of fat digestion, including cholesterol, are absorbed in the first 100 cm of small intestine, the primary and secondary bile acids are absorbed almost exclusively in the ileum, returning to the liver by way of the portal circulation about 98–

99% of the bile acids secreted into the intestine. This is known as the **enterohepatic circulation.** However, lithocholic acid, because of its insolubility, is not reabsorbed to any significant extent.

A small fraction of the bile salts—perhaps only as little as 400 mg/d—escapes absorption and is therefore eliminated in the feces. Even though this is a very small amount, it nonetheless represents a major pathway for the elimination of cholesterol. The enterohepatic circulation of the bile salts is so efficient that each day the relatively small pool of bile acids (about 3–5 g) can be cycled through the intestine six to ten times with only a small amount lost in the feces; ie, approximately 1–2% per pass through the enterohepatic circulation. However, each day, an amount of bile acid equivalent to that lost in the feces is synthesized from cholesterol by the liver, so that a pool of bile acids of constant size is maintained. This is accomplished by a system of feedback controls.

Bile Acid Synthesis Is Regulated at the 7α-Hydroxylase Step

The principal rate-limiting step in the biosynthesis of bile acids is at the **7α**-hydroxylase reaction, and in the biosynthesis of cholesterol it is at the HMG-CoA reductase step (Figure 28–1). The activities of these two enzymes often change in parallel, and consequently it has been difficult to ascertain whether inhibition of bile acid synthesis takes place primarily at the HMG-CoA reductase step or at the 7α-hydroxylase reaction. Both enzymes undergo similar diurnal variation in activity. Induction of the gene for 7α-hydroxylase by dietary cholesterol and suppression by bile acids has been demonstrated. In this regard, the return of bile salts to the liver via the enterohepatic circulation is an important control that, if interrupted, leads to activation of 7α-hydroxylase. 7α-Hydroxylase (as well as HMG-CoA reductase) can be controlled by covalent phosphorylation-dephosphorylation. In contrast to HMG-CoA reductase, it is the phosphorylated form that results in increased activity of 7α-hydroxylase.

CLINICAL ASPECTS

The Serum Cholesterol Is Correlated With the Incidence of Atherosclerosis and Coronary Heart Disease

Of the serum lipids, cholesterol is most often singled out as being chiefly concerned in the relationship. However, other parameters—such as serum triacylglycerol concentration—show similar correlations. Patients with arterial disease can have any one of the following abnormalities: (1) elevated concentrations of VLDL with normal concentrations of LDL; (2) elevated LDL with normal VLDL; or (3) elevation of both lipoprotein fractions. There is also an inverse relationship between HDL (HDL$_2$) concentrations and coronary heart disease, and some consider that the most predictive relationship is the **LDL:HDL cholesterol ratio.** This relationship is explainable in terms of the proposed roles of LDL in transporting cholesterol to the tissues and of HDL acting as the scavenger of cholesterol in reverse cholesterol transport.

Atherosclerosis is characterized by the deposition of cholesterol and cholesteryl ester of lipoproteins containing apo B-100 in the connective tissue of the arterial walls. Diseases in which prolonged elevated levels of VLDL, IDL and chylomicron remnants, or LDL occur in the blood (eg, diabetes mellitus, lipid nephrosis, hypothyroidism, and other conditions of hyperlipidemia) are often accompanied by premature or more severe atherosclerosis.

Experiments on the induction of atherosclerosis in animals indicate a wide species variation in susceptibility. The rabbit, pig, monkey, and humans are species in which atherosclerosis can be induced by feeding cholesterol. The rat, dog, and cat are resistant. Thyroidectomy or treatment with thiouracil drugs will allow induction of atherosclerosis in the dog and rat. Low blood cholesterol is a characteristic of hyperthyroidism.

Changes in Diet Play an Important Role in Reducing Serum Cholesterol

Hereditary factors play the greatest role in determining individual blood cholesterol concentrations, but of the dietary and environmental factors that lower blood cholesterol, the substitution in the diet of **polyunsaturated and monounsaturated fatty acids** for some of the saturated fatty acids is most beneficial. Naturally occurring oils that contain a high proportion of polyunsaturated fatty acids include sunflower, cottonseed, corn, and soybean oil, and olive oil contains a high concentration of monounsaturated fatty acids. On the other hand, butterfat, beef fat, and palm oil contain a high proportion of saturated fatty acids. Sucrose and fructose have a greater effect in raising blood lipids, particularly triacylglycerols, than do other carbohydrates.

The reason for the cholesterol-lowering effect of polyunsaturated fatty acids is still not clear. However, several hypotheses have been advanced, including the stimulation of cholesterol excretion into the intestine and the stimulation of the oxidation of cholesterol to bile acids. Diets rich in palmitate inhibit the conversion of cholesterol to bile acids. There is other evidence that the effect is due to a shift in distribution of cholesterol from the plasma into the tissues because of increased catabolic rate of LDL due to up-regulation of the LDL receptor by poly- and monounsaturated fatty acids and down-regulation by saturated fatty acids. Saturated fatty acids cause the formation of smaller VLDL particles that contain relatively more cholesterol, and they are utilized by extrahepatic tissues at a slower rate than are larger particles, tendencies that may be regarded as atherogenic.

Lifestyle Affects the Serum Cholesterol Level

Additional factors considered to play a part in coronary heart disease include high blood pressure, smoking, male gender, obesity (particularly abdominal obesity), lack of exercise, and drinking soft as opposed to hard water. Elevation of plasma free fatty acids will also lead to increased VLDL secretion by the liver, involving extra triacylglycerol and cholesterol output into the circulation. Factors leading to higher or fluctuating levels of free fatty acids include emotional stress, nicotine from cigarette smoking, coffee drinking, and partaking of a few large meals rather than more continuous feeding. Premenopausal women appear to be protected against many of these deleterious factors, possibly because they have higher concentrations of HDL than do men and postmenopausal women. It is of interest that studies have shown an association between moderate alcohol consumption and a lower incidence of coronary heart disease. This may be due to elevation of HDL concentrations, but it has been claimed that red wine is particularly beneficial, perhaps because of its content of antioxidants.

When Diet Changes Fail, Hypolipidemic Drugs Will Reduce Serum Cholesterol and Triacylglycerol

Hypercholesterolemia may be treated by interrupting the enterohepatic circulation of bile acids. Significant reductions of plasma cholesterol can be effected by the use of **cholestyramine resin,** or surgically by the ileal exclusion operations. Both procedures cause a block in the reabsorption of bile acids. Then, because of release from feedback regulation normally exerted by bile acids, the conversion of cholesterol to bile acids is greatly enhanced in an effort to maintain the pool of bile acids. Consequently, LDL receptors in the liver are up-regulated, causing increased uptake of LDL with consequent lowering of plasma cholesterol. **Sitosterol** is a hypocholesterolemic agent that acts by blocking the absorption of cholesterol from the gastrointestinal tract.

Several drugs are known to block the formation of cholesterol at various stages in the biosynthetic pathway. The fungal inhibitors of HMG-CoA reductase, **mevastatin** and **lovastatin,** reduce LDL cholesterol levels by up-regulation of the LDL receptors. **Clofibrate** and **gemfibrozil** exert at least part of their hypolipidemic effect (mainly on triacylglycerol) by diverting the hepatic inflow of free fatty acids from the pathways of esterification into those of oxidation, thus decreasing the secretion of triacylglycerol and cholesterol containing VLDL by the liver. In addition, they stimulate hydrolysis of VLDL triacylglycerols by lipoprotein lipase. **Probucol** appears to increase LDL catabolism via receptor-independent pathways, but its antioxidant properties may be more important in preventing accumulation of oxidized LDL in arterial walls. Oxidized LDL may be a prime cause of atherosclerosis. **Nicotinic acid** reduces the flux of FFA by inhibiting adipose tissue lipolysis, thereby inhibiting VLDL production by the liver.

Primary Disorders of the Plasma Lipoproteins (Dyslipoproteinemias) Are Inherited

A few individuals in the population exhibit inherited defects in lipoprotein metabolism, leading to the primary condition of either **hypo-** or **hyperlipoproteinemia** (Table 28–1). Many others having diseases such as diabetes mellitus, hypothyroidism, kidney disease (nephrotic syndrome), and atherosclerosis show secondary abnormal lipoprotein patterns that are very similar to one or another of the primary inherited conditions. Virtually all of these primary conditions are due to a defect at a stage in lipoprotein formation, transport, or destruction (see Figures 27–4, 28–5, and 28–6). Not all of the abnormalities are harmful.

SUMMARY

1. Cholesterol is the precursor of all other steroids in the body, eg, corticosteroids, sex hormones, bile acids, and vitamin D. It is also an important amphipathic lipid, which allows it to play a structural role in membranes and in the outer layer of lipoproteins.

2. Cholesterol is synthesized in the body entirely from acetyl-CoA via a complex pathway. Three molecules of acetyl-CoA form mevalonate via the important rate-limiting reaction for the pathway, catalyzed by HMG-CoA reductase. A five-carbon isoprenoid unit is formed from mevalonate, and six isoprenoid units condense to form squalene. Squalene undergoes cyclization to form the parent steroid lanosterol, which, after the loss of three methyl groups, forms cholesterol.

3. Cholesterol synthesis in the liver is regulated partly by the influx of dietary cholesterol in cholesterol-rich chylomicron remnants. In tissues, in general, a cholesterol balance is maintained between the factors causing gain of cholesterol (eg, synthesis, uptake via the LDL or scavenger receptors, hydrolysis of cholesteryl ester) and the factors causing loss of cholesterol (eg, steroid synthesis, cholesteryl ester formation, and reverse cholesterol transport via HDL). The activity of the LDL receptor is down-regulated by a high level of cell cholesterol and is up-regulated when cholesterol is depleted.

4. In reverse cholesterol transport, HDL attaches

Table 28–1. Primary disorders of plasma lipoproteins (dyslipoproteinemias).[1]

Name	Defect	Remarks
Hypolipoproteinemias Abetalipoproteinemia	No chylomicrons, VLDL, or LDL are formed because of defect in triacylglycerol transfer protein (MTP), which prevents the loading of apo B with lipid.	Rare; blood acylglycerols low; intestine and liver accumulate acyglycerols.
Familial hypobetalipoproteinemia	LDL concentration is 10–60% of normal.	Chylomicron formation still occurs; most individuals healthy and long-lived.
Familial alpha-lipoprotein deficiency Tangier disease Fish-eye disease Apo-A-I deficiencies	All have low or near absence of HDL.	No impairment of chylomicron or VLDL formation. No pre-β-lipoprotein but broad β-band on agarose electrophoresis; tendency toward hypertriacylglycerolemia as a result of absence of apo C-II, which activates lipoprotein lipase. Low LDL levels. Atherosclerosis in the elderly.
Hyperlipoproteinemias Familial lipoprotein lipase deficiency (type I)	(a) Deficiency of LPL, or (b) production of abnormal LPL, or (c) apo-C-II deficiency.	Slow clearance of chylomicrons and VLDL. Low levels of LDL and HDL. Treat by reducing fat and increasing complex carbohydrates in diet. No increased risk of coronary disease.
Familial hypercholesterolemia (type II)	Type IIa: Defective LDL receptors or mutation in ligand region of apo B-100. Type IIb: Tendency for VLDL to be elevated in addition.	Reduced rate of LDL clearance leads to elevated LDL levels and hypercholesterolemia, resulting in atherosclerosis and coronary disease.
Wolman's disease (cholesteryl ester storage disease)	Deficiency of cholesteryl ester hydrolase in lysosomes of cells such as fibroblasts that normally metabolize LDL.	Reduced rate of LDL clearance and consequences as above.
Familial type III hyperlipoproteinemia (broad beta disease, remnant removal disease, familial dysbetalipoproteinemia)	Deficiency in remnant clearance by the liver is due to abnormality in apo E, which is normally present in 3 isoforms: E2, E3, and E4. Patients have only E2, which does not react with the E receptor. Truncated apo B species present.	Increase in chylomicron and VLDL remnants of density <1.019, which appear as a broad β-band on electrophoresis (β-VLDL). Causes hypercholesterolemia, xanthomas, and atherosclerosis in peripheral and coronary arteries.
Familial hypertriacylglycerolemia (type IV)	Overproduction of VLDL often associated with glucose intolerance and hyperinsulinemia, which may be a cause of the overproduction.	Cholesterol levels rise with the VLDL concentration. LDL and HDL tend to be subnormal. This type of pattern is commonly associated with coronary heart disease, type II non–insulin-dependent diabetes mellitus, obesity, alcoholism, and administration of progestational hormones.
Familial type V hyperlipoproteinemia	Unknown cause leading to elevated chylomicrons and VLDL.	Hypertriacylglycerolemia and hypercholesterolemia with low LDL and HDL. Increased coronary heart disease risk in some patients.
Familial hyperalphalipoproteinemia	Increased concentrations of HDL.	A rare condition apparently beneficial to health and longevity.
Hepatic lipase deficiency	Deficiency of the enzyme leads to accumulation of large triacylglycerol-rich HDL and VLDL remnants.	Patients have anthomas and coronary heart disease.
Familial lecithin: cholesterol acyltransferase (LCAT) deficiency	Absence of LCAT leads to block in reverse cholesterol transport. HDL remains as nascent disks in stacks or rouleaux, incapable of taking up and esterifying cholesterol.	Plasma concentrations of cholesteryl esters and lysolecithin are low. Present is an abnormal LDL fraction, lipoprotein X, found also in patients with cholestasis. VLDL is abnormal (β-VLDL).
Familial lipoprotein(a) excess	Lp(a) consists of 1 mol of LDL attached to 1 mol of apo(a). Apo(a) shows structural homologies to plasminogen.	Premature coronary heart disease due to atherosclerosis, plus thrombosis due to inhibitioin of fibrinolysis.

[1]There is an association between patients possessing the apo E4 allele and the incidence of Alzheimer's disease. Apparently, apo E4 binds more avidly to β-amyloid found in neuritic plaques.

to the apo A-I receptor, causing cholesterol to translocate to the cell membrane, where it is taken up by HDL. A concentration gradient is maintained by the activity of LCAT, which allows the cholesterol to be esterified and deposited in the core of HDL. The cholesteryl ester in HDL is taken up by the liver, either directly, or after transfer to VLDL, IDL, or LDL via the cholesteryl ester transfer protein.

5. Excess cholesterol is excreted from the liver in the bile as cholesterol or bile salts. A large proportion of bile salts is absorbed into the portal circulation and returned to the liver as part of the enterohepatic circulation.

6. Elevated levels of cholesterol present in VLDL, IDL, or LDL are associated with atherosclerosis, whereas high levels of HDL have a protective effect.

7. A few individuals in the population have inherited defects in lipoprotein metabolism leading to a primary condition of hypo- or hyperlipoproteinemia. Many others having conditions such as diabetes mellitus, hypothyroidism, kidney disease, and atherosclerosis exhibit secondary abnormal lipoprotein patterns that are similar to one or another of the primary conditions.

REFERENCES

Fears R, Sabine JR (editors): *Cholesterol 7α-Hydroxylase (7α-Monooxygenase)*. CRC Press, 1986.

Fielding CJ, Fielding PE: Metabolism of cholesterol and lipoproteins. In: *Biochemistry of Lipids and Membranes.* Vance DE, Vance JE (editors). Benjamin/Cummings, 1985.

Kane JB, Havel RJ: Treatment of hypercholesterolemia. Annu Rev Med 1986;37:427.

Mahley RW, Innerarity TL: Lipoprotein receptors and cholesterol homeostasis. Biochim Biophys Acta 1983;737: 197.

Mendez AJ, Oram JF, Bierman EL: Protein kinase C as a mediator of high density lipoprotein receptor dependent efflux of intracellular cholesterol. J Biol Chem 1991;266:10104.

Report of the Expert Panel on Detection. Evaluation and Treatment of High Blood Cholesterol in Adults. NIH Publication No. 88-2925, January 1988.

Russell DW: Cholesterol biosynthesis and metabolism. Cardiovascular Drugs Therap 1992;6:103.

Russell DW, Setchell KDR: Bile acid biosynthesis. Biochemistry 1992;31:4737.

Spady DK, Woollett LA, Dietschy JM: Regulation of plasma LDL-cholesterol levels by dietary cholesterol and fatty acids. Annu Rev Nutr 1993;13:355.

Various authors: The cholesterol facts—A summary of the evidence relating dietary fats, serum cholesterol, and coronary heart disease. Circulation 1990;81:1721.

Various authors: The hypertriglyceridemias: Risk and management. Am J Cardiol 1991;68:1A.

29

Integration of Metabolism & the Provision of Tissue Fuels

Peter A. Mayes, PhD, DSc

INTRODUCTION

Carbohydrates and lipids play many structural and metabolic roles, but it is as the provider of a large proportion of dietary calories that they have their greatest impact on metabolism and health. The regulation of this fuel influx and the manner in which it is integrated with other tissue fuels are of central interest, since they impinge on many other metabolic processes and are concerned with metabolic disease.

BIOMEDICAL IMPORTANCE

Under positive caloric balance, a significant proportion of the food energy intake is stored as either glycogen or fat. If the diet is mainly carbohydrate, glucose will be the principal fuel of the tissues. However, in some tissues, even under fed conditions, fatty acids are oxidized in preference to glucose, but particularly under conditions of caloric deficit or starvation. The purpose is to spare glucose for those tissues (eg, brain and erythrocytes) that require it under all conditions. Thus, regulatory mechanisms, often hormone-mediated, ensure a supply of suitable fuel for all tissues, at all times, from the fully fed to the totally starved state. Breakdown of these mechanisms occurs owing to hormone imbalance (eg, insulin deficiency in diabetes mellitus), to metabolic imbalance due to heavy lactation (eg, ketosis of cattle), or to high metabolic demands in pregnancy (eg, pregnancy toxemia in sheep). All these conditions are pathologic aberrations of the **starvation syndrome,** which is a complication of many medical situations when appetite is diminished.

NOT ALL MAJOR FOODSTUFFS ARE INTERCONVERTIBLE
(Figure 29–1)

That animals may be fattened on a predominantly carbohydrate diet demonstrates the ease of conversion of carbohydrate into fat. However, as has been pointed out, humans may be limited in the extent to which glucose can be converted to fatty acids, particularly in adipose tissue. A most significant reaction in this respect is the conversion of pyruvate to acetyl-CoA, as acetyl-CoA is the starting material for the synthesis of long-chain fatty acids. With respect to the reverse process, the conversion of fatty acids to glucose, **the pyruvate dehydrogenase** reaction is essentially nonreversible, which prevents the direct conversion of acetyl-CoA to pyruvate. In addition, there cannot be a net conversion of acetyl-CoA to oxaloacetate via the citric acid cycle, since one molecule of oxaloacetate is required to condense with acetyl-CoA and only one molecule of oxaloacetate is regenerated. For similar reasons, there cannot be a net conversion of fatty acids having an even number of carbon atoms (which form acetyl-CoA) to glucose or glycogen. Only the terminal three-carbon portion of a fatty acid having an odd number of carbon atoms is glucogenic, as this portion of the molecule will ultimately form **propionyl-CoA** upon β-oxidation. Nevertheless, it is possible for labeled carbon atoms of fatty acids to be found ultimately in glycogen after traversing the citric acid cycle. This is because oxaloacetate is an intermediate both in the citric acid cycle and in the pathway of gluconeogenesis. The glycerol moiety of triacylglycerol will form glucose after activation to glycerol 3-phosphate, and this is an important source of glucose in starvation.

Many carbon skeletons of the nonessential amino acids can be produced from carbohydrate via the citric acid cycle and transamination. By reversal of these processes, glucogenic amino acids yield carbon skeletons that are either members or precursors of the citric acid cycle. They are therefore readily converted by gluconeogenic pathways to glucose and glycogen. The ketogenic amino acids give rise to acetoacetate, which will in turn be metabolized as ketone bodies, forming acetyl-CoA in extrahepatic tissues.

For the same reasons that it is not possible for a net conversion of fatty acids to carbohydrate to occur, it is not possible for a net conversion of fatty acids to glucogenic amino acids to take place. Neither is it possible to reverse the pathways of breakdown of ketogenic and other amino acids, which fall

284

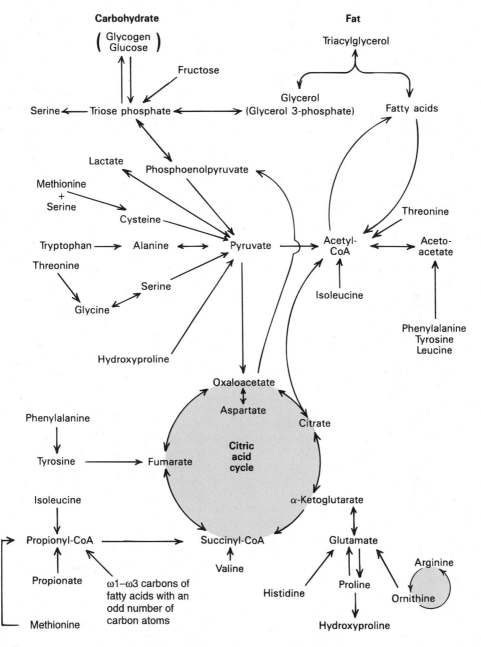

Figure 29–1. Interconversion of the major foodstuffs.

into the category of **nutritionally essential amino acids.** Conversion of the carbon skeletons of glucogenic amino acids to fatty acids is possible either by formation of pyruvate and acetyl-CoA or by reversal of nonmitochondrial reactions of the citric acid cycle from α-ketoglutarate to citrate followed by the action of ATP-citrate lyase to give acetyl-CoA (Chapter 23). Generally, however, the net conversion of amino acids to fat is not a significant process. Even in carnivores on a high-protein diet, the fat intake is also high, which would inhibit lipogenesis.

THE ECONOMICS OF CARBOHYDRATE AND LIPID METABOLISM ENCOMPASS THE WHOLE BODY

Glucose Is a Metabolic Necessity for the Brain and Erythrocytes in All Nutritional States

Many details of the interplay between carbohydrate and lipid metabolism in various tissues have been described, particularly the ready conversion of many glucogenic substances to glucose and glycogen by **gluconeogenesis.** Gluconeogenesis is particularly important because certain tissues and cell types, including the central nervous system and the erythrocytes, are dependent on a continual supply of glucose. A minimal supply of glucose is probably necessary in extrahepatic tissues to maintain oxaloacetate concentrations and the integrity of the citric acid cycle. In addition, glucose appears to be the main source of glycerol 3-phosphate in tissues devoid of glycerol kinase such as adipose tissue. There is, therefore, **a minimal and obligatory rate of glucose oxidation under all conditions.** Large quantities of glucose are also a necessity for fetal nutrition and the synthesis of lactose in milk. Certain mechanisms, in addition to gluconeogenesis, safeguard essential supplies of glucose in times of shortage by allowing other substrates to spare its oxidation.

The Preferential Utilization of Ketone Bodies and Free Fatty Acids Spares Glucose for Its Essential Functions

Ketone bodies and free fatty acids spare the oxidation of glucose in muscle by impairing its entry into the cell, its phosphorylation by hexokinase and phosphofructokinase, and its oxidative decarboxylation to pyruvate. Oxidation of free fatty acids and ketone bodies causes an increase in the intracellular concentration of citrate that in turn inhibits phosphofructokinase allosterically. Oxidation of these substrates also causes [acetyl-CoA]/[CoA] and [ATP]/[ADP] ratios to increase, inhibiting pyruvate dehydrogenase activity (Figure 19–6). These observations and others, demonstrating that acetoacetate is oxidized in the perfused heart preferentially to free fatty acids, justify the conclusion that under conditions of carbohydrate shortage available fuels are oxidized in the following order of preference: (1) **ketone bodies** (and probably other short-chain fatty acids, eg, acetate), (2) **free fatty acids,** and (3) **glucose.** This does not imply that any particular fuel is oxidized to the total exclusion of any other, as a mixture of different fuels is usually used (Figure 29–2). These mechanisms are more important in tissues having a high capacity for aerobic oxidation of fatty acids, eg, heart and slow-twitch muscle, than in tissues with a low capacity, eg, fast-twitch muscle.

The combination of the effects of free fatty acids in sparing glucose utilization in muscle and heart and the feedback effect of the spared glucose in inhibiting free fatty acid mobilization in adipose tissue has been called the **glucose–fatty acid cycle.**

In moderate endurance exercise, lipid is the main fuel, but in intense endurance exercise it becomes less adequate and carbohydrate assumes the role of major fuel until muscle glycogen is exhausted. Of importance are the triacylglycerol stores in muscle cells themselves.

DURING STARVATION, A CONTINUAL SUPPLY OF FUEL FOR THE TISSUES IS PROVIDED

In animals fed high-carbohydrate diets, fatty acid oxidation is spared. This is because lipolysis in adipose tissue is inhibited owing to high blood glucose and insulin concentrations, and therefore free fatty acid levels remain low (Figure 29–3). In a normal fed human, the proportions of the various calorific nutrients oxidized are set by their relative proportions in the diet. As the animal passes from the fed to the fasting condition, glucose availability from food becomes less, and liver glycogen is drawn upon in an attempt to maintain the blood glucose. The concentration of insulin in the blood decreases, and glucagon increases. As glucose utilization diminishes in adipose tissue and the inhibitory effect of insulin on lipolysis becomes less, fat is mobilized as free fatty acids and glycerol. The free fatty acids are transported to the tissues, where they are either oxidized or esterified. Glycerol joins the carbohydrate pool after activation to glycerol 3-phosphate, mainly in the liver and kidney. During this transition phase from the fully fed to the fully fasting state, endogenous glucose production (from amino acids and glycerol) does not keep pace with its utilization and oxidation, since the liver glycogen stores become depleted and blood glucose tends to fall. Thus, fat is mobilized at an ever-increasing rate, and in several hours the plasma free fatty acids and blood glucose stabilize at the fasting level (0.7–0.8 and 3.3–3.9 mmol/L, respectively). At this point, it must be presumed that in the whole animal the supply of glucose balances the obligatory demands for glucose utilization and oxidation. This is achieved by the increased oxidation of free fatty acids and ketone bodies, sparing the nonobligatory oxidation of glucose. This fine balance is disturbed in conditions that demand more glucose (eg, pregnancy and lactation) or in which glucose utilization is impaired (eg, diabetes mellitus) and which therefore lead to further mobilization of fat. The provision of carbohydrate by adipose tissue, in the form of **glycerol,** is an important function, for it is only this source of carbohydrate together with that provided by **gluconeogenesis from protein** that can supply the starving organism with the glucose needed for those processes which must utilize glucose. In prolonged starvation in humans, **gluconeogenesis from protein is diminished** owing to reduced release of amino acids, particularly alanine, from muscle. This coincides with adaptation of the brain to **replace approximately half of the glucose oxidized with ketone bodies.** In prolonged starvation, glucose contributes less than 5% of the total substrate oxidized in the whole body.

After refeeding of glucose to starved animals, glycogenesis in the liver occurs via a three-carbon intermediate such as lactate, indicating that gluconeo-genesis continues for some time after refeeding.

Ketosis Is a Metabolic Adaptation to Starvation

The primary function of ketogenesis is to remove excess fatty acid carbon from the liver in a form that is readily oxidized by extrahepatic tissues in place of glucose. Ketosis arises as a result of a deficiency in available carbohydrate. This has the following actions in fostering ketogenesis (Figures 24–9 and 24–10). (1) It causes an imbalance between esterifi-

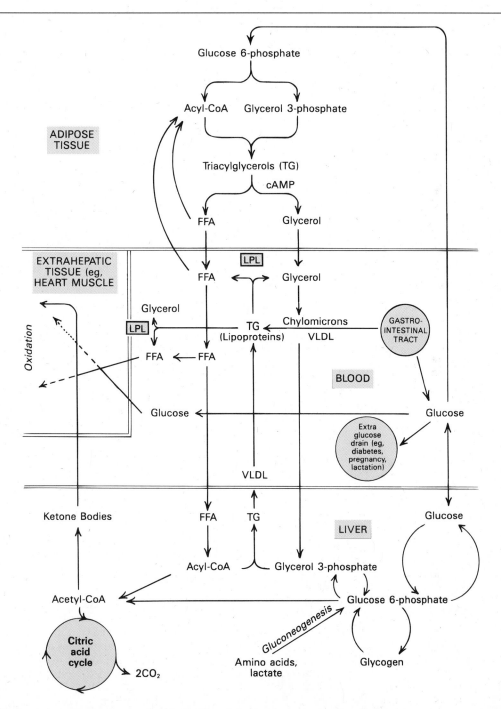

Figure 29–2. Metabolic interrelationships between adipose tissue, the liver, and extrahepatic tissues. (LPL, lipoprotein lipase; FFA, free fatty acids, VLDL; very low density lipoproteins.)

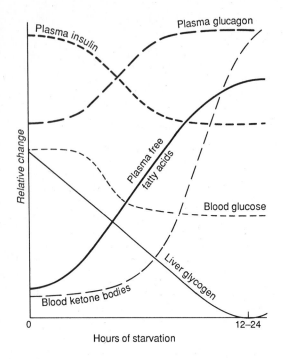

Figure 29–3. Relative changes in metabolic parameters during the onset of starvation.

cation and lipolysis in adipose tissue as a result of lower insulin levels, with consequent release of free fatty acids into the circulation. Free fatty acids are the principal substrates for ketone body formation in the liver, and therefore all factors, metabolic or endocrine, affecting the release of free fatty acids from adipose tissue influence ketogenesis. (2) Upon entry of free fatty acids into the liver, the balance between their esterification and oxidation is governed by the activity of carnitine palmitoyltransferase I, which is increased by the concentration of free fatty acids and the increased glucagon:insulin ratio. (3) As more fatty acid is oxidized, more forms ketone bodies and less forms CO_2 regulated in such a manner that the total ATP production of the liver remains constant (Figure 24–9).

A feedback mechanism for controlling free fatty acid output from adipose tissue in starvation may operate as a result of the action of ketone bodies and free fatty acids to directly stimulate the pancreas to produce insulin.

Under most conditions, free fatty acids are mobilized in excess of oxidative requirements, since a large proportion are esterified, even during fasting. As the liver takes up and esterifies a considerable proportion of the free fatty acid output, it plays a regulatory role in removing excess free fatty acids from the circulation. When carbohydrate supplies are adequate, most of the influx is esterified and ultimately retransported from the liver as VLDL to be utilized by other tissues. However, in the face of an increased influx of free fatty acids, an alternative route, **ketogenesis,** is available that enables the liver to continue to retransport much of the influx of free fatty acids in a form readily utilized by extrahepatic tissues under all nutritional conditions.

Most of these principles are depicted in Figure 29–2. It will be noted that there is a carbohydrate cycle involving release of glycerol from adipose tissue and its conversion in the liver to glucose, followed by its transport back to adipose tissue to complete the cycle. The other cycle, a lipid cycle, involves release of free fatty acid by adipose tissue, its transport to and esterification in the liver, and retransport as VLDL back to adipose tissue.

THE MAJOR METABOLIC PATHWAYS ARE REGULATED BY ONE OR TWO KEY ENZYMES CATALYZING NONEQUILIBRIUM REACTIONS

A summary of the major regulators of the main metabolic pathways is given in Table 29–1. As noted in previous chapters, these pathways are controlled at one or two nonequilibrium reactions occurring usually at the beginning of the pathway.

THE MAJOR PATTERNS OF METABOLISM IN INDIVIDUAL ORGANS OR TISSUES ARE DETERMINED BY THE PRESENCE OR ABSENCE OF KEY ENZYMES

A summary of the major and unique metabolic features of some of the principal organs is presented in Table 29–2. These metabolic patterns are determined by the distribution of key enzymes between the organs and tissues, which is the main factor affecting the type of substrates taken up and products formed, all of which determines the flux and direction of metabolites in the blood.

CLINICAL ASPECTS

Pathologic Ketosis Is Caused by an Amplification of the Factors Causing Starvation Ketosis

The ketosis that occurs in starvation and fat feeding is relatively mild compared with the condition encountered in uncontrolled **diabetes mellitus, pregnancy toxemia of ewes,** or **ketosis of lactating cattle.** The main reason appears to be that in the severe conditions carbohydrate is even less available to the tissues than in the mild conditions. Thus, in the milder forms of diabetes mellitus, in fat feeding, and in chronic starvation, glycogen is present in the liver

Table 29–1. Summary of the major regulators of metabolic pathways.

Pathway	Major Regulatory Enzymes	Activator	Inhibitor	Effector Hormone	Remarks
Citric acid cycle	Citrate synthase		ATP, long-chain acyl-CoA		Regulated mainly by the need for ATP and therefore by the supply of NAD$^+$
Glycolysis	Phosphofructokinase	AMP, fructose 2,6-bisphosphate in liver, fructose 1,6-biphosphate in muscle	Citrate (fatty acids, ketone bodies), ATP, cAMP	Glucagon ↓	Induced by insulin
Pyruvate oxidation	Pyruvate dehydrogenase	CoA, NAD, ADP, pyruvate	Acetyl-CoA, NADH, ATP (fatty acids, ketone bodies)	Insulin ↑ (in adipose tissue)	Also important in regulating the citric acid cycle
Gluconeogenesis	Pyruvate carboxylase,	Acetyl-CoA	ADP	Glucagon?	Induced by glucocorticoids, glucagon, cAMP
	Phosphoenolpyruvate carboxykinase	cAMP?			
	Fructose-1,6-bisphosphatase	cAMP	AMP, fructose 2,6-bisphosphate in liver, fructose 1,6-bisphosphate in muscle	Glucagon	Repressed by insulin
Glycogenesis	Glycogen synthase		Phosphorylase (in liver). cAMP, Ca^{2+} (in muscle)	Insulin ↑ Glucagon (liver) ↓ Epinephrine ↓	Induced by insulin
Glycogenolysis	Phosphorylase	cAMP, Ca^{2+} (muscle)		Insulin ↓ Glucagon (liver) ↑ Epinephrine ↑	
Pentose phosphate pathway	Glucose-6-phosphate dehydrogenase	NADP$^+$	NADPH		Induced by insulin
Lipogenesis	Acetyl-CoA carboxylase	Citrate	Long-chain acyl-CoA, cAMP	Insulin ↑ Glucagon (liver) ↓	Induced by insulin
Cholesterol synthesis	HMG-CoA reductase		Cholesterol, cAMP, mevalonate, bile acids	Insulin ↑ Glucagon (liver) ↓	Inhibited by certain drugs, eg, lovastatin

289

Table 29–2. Summary of the major and unique features of metabolism of the principal organs.

Organ	Major Function	Major Pathways	Main Substrates	Major Products	Specialist Enzymes
Liver	Service for the other organs and tissues	Most represented, including gluconeogenesis; β-oxidation; ketogenesis; lipoprotein formation; urea, uric acid, and bile acid formation; cholesterol synthesis; lipogenesis[1]	Free fatty acids, glucose (well fed), lactate, glycerol, fructose, amino acids	Glucose, VLDL (triacylglycerol), HDL, ketone bodies, urea, uric acid, bile acids, plasma proteins	Glucokinase, glucose-6-phosphatase, glycerol kinase, phosphoenolpyruvate carboxykinase, fructokinase, arginase, HMG-CoA synthase and lyase, 7α-hydroxylase
			(Ethanol)	(Acetate)	(Alcohol dehydrogenase)
Brain	Coordination of the nervous system	Glycolysis, amino acid metabolism	Glucose, amino acid, ketone bodies (in starvation) Polyunsaturated fatty acids in neonate	Lactate	
Heart	Pumping of blood	Aerobic pathways, eg, β-oxidation and citric acid cycle	Free fatty acids, lactate, ketone bodies, VLDL and chylomicron triacylglycerol, some glucose		Lipoprotein lipase. Respiratory chain well developed.
Adipose tissue	Storage and breakdown of triacylglycerol	Esterification of fatty acids and lipolysis; lipogenesis[1]	Glucose, lipoprotein triacylglycerol	Free fatty acids, glycerol	Lipoprotein lipase, hormone-sensitive lipase
Muscle Fast twitch	Rapid movement	Glycolysis	Glucose	Lactate	Lipoprotein lipase.
Slow twitch	Sustained movement	Aerobic pathways, eg, β-oxidation and citric acid cycle	Ketone bodies, triacylglycerol in VLDL and chylomicrons, free fatty acids		Respiratory chain well developed.
Kidney	Excretion and gluconeogenesis	Gluconeogenesis	Free fatty acids, lactate, glycerol	Glucose	Glycerol kinase, phosphoenolpyruvate carboxykinase
Erythrocytes	Transport of O_2	Glycolysis, pentose phosphate pathway. No mitochondria and therefore no β-oxidation or citric acid cycle.	Glucose	Lactate	(Hemoglobin)

[1]In many species but not in humans.

in variable amounts, and free fatty acid levels are lower, which probably accounts for the less severe ketosis associated with these conditions.

In type I diabetes mellitus, the lack (or relative lack) of insulin probably affects adipose tissue more than any other tissue, because of its extreme sensitivity to this hormone. As a result, free fatty acids are released in quantities that give rise to plasma free fatty acid levels more than twice those in fasting normal subjects, with correspondingly higher concentrations of ketone bodies. Many changes also occur in the activity of enzymes within the liver, and these changes enhance both the rate of gluconeogenesis and the rate of transfer of glucose to the blood despite high levels of circulating glucose.

In ketosis of ruminants, there is a severe drain of glucose from the blood owing to excessive fetal demands of twins or the demands of heavy lactation (Figure 29–2). Extreme hypoglycemia results, coupled with negligible amounts of glycogen in the liver. Ketosis in these conditions tends to be severe. As hypoglycemia develops, the secretion of insulin diminishes, allowing not only less glucose utilization but also enhancement of lipolysis in adipose tissue. Pregnant women often exhibit mild ketosis.

In untreated type I diabetes mellitus, death occurs as a result of complications of acidosis caused by long-term depletion of base needed to neutralize acidic ketone bodies excreted in the urine (Chapter 65, Case No. 9: Diabetes Mellitus Type I With Ketoacidosis). In pregnancy toxemia of ewes, coma and death occur rapidly owing to severe hypoglycemia.

SUMMARY

1. Many of the major foodstuffs are interconvertible. Carbohydrate is converted to fatty acids via the pyruvate dehydrogenase reaction. Since this reaction is essentially nonreversible, the opposite process cannot take place. Nor can there be any net conversion of acetyl-CoA (or acetyl-CoA–forming substances) to glucose via the citric acid cycle, since one molecule of oxaloacetate is consumed for every molecule of oxaloacetate converted to glucose.

2. Many carbon skeletons of nonessential amino acids can be produced from carbohydrate via the citric acid cycle and transamination. Reversal of this process allows glucogenic amino acids to enter the pathway of gluconeogenesis.

3. During starvation, free fatty acids and ketone bodies are oxidized in preference to glucose, which is spared for those tissues, such as the brain, that require glucose at all times. This is achieved through inhibition of phosphofructokinase and pyruvate dehydrogenase. This effect, coupled with the inhibition by the spared glucose of free fatty acid mobilization in adipose tissue, is called the glucose–fatty acid cycle.

4. Ketosis is a metabolic adaptation to starvation and is exacerbated in pathologic conditions such as diabetes mellitus and ruminant ketosis.

REFERENCES

Brooks GA, Mercier J: Balance of carbohydrate and lipid utilization during exercise: The "crossover" concept. J Appl Physiol 1994;76:2253.

Caprio S et al: Oxidative fuel metabolism during mild hypoglycemia: Critical role of free fatty acids. lAm J Physiol 1989;256:E413.

Cohen P: *Control of Enzyme Activity,* 2nd ed. Chapman & Hall, 1983.

Hue L, Van de Werve G (editors): *Short-Term Regulation of Liver Metabolism.* Elsevier/North Holland, 1981.

Knopp RH et al: Lipoprotein metabolism in pregnancy, fat transport to the fetus and the effects of diabetes. Biol Neonate 1986;50:297.

Randle PJ: The glucose-fatty acid cycle—biochemical aspects. Atherosclerosis Rev 1991;22:183.

Zorzano A et al: Effects of starvation and exercise on concentrations of citrate, hexose phosphates and glycogen in skeletal muscle and heart: Evidence for selective operation of the glucose–fatty acid cycle. Biochem J 1985;232:585.

Section III.
Metabolism of Proteins & Amino Acids

Biosynthesis of the Nutritionally Nonessential Amino Acids

30

Victor W. Rodwell, PhD

INTRODUCTION

To refer to nutritionally essential amino acids as "essential" or "indispensable" and to nutritionally nonessential amino acids as "nonessential" or "dispensable" is misleading (Table 30–1). While in a nutritional context these terms are correct, they obscure the biologically essential nature of all 20 amino acids. It might be argued that the nutritionally nonessential amino acids are more important to the cell than the nutritionally essential ones, since organisms (eg, humans) have evolved that lack the ability to manufacture the latter but not the former group.

Since this book emphasizes metabolic processes of human tissues, we will discuss only biosynthesis of the nutritionally nonessential amino acids, not of the nutritionally essential amino acids by plants and microorganisms.

BIOMEDICAL IMPORTANCE

Medical implications of the material in this chapter relate to amino acid deficiency states that can result if any of the nutritionally essential amino acids are omitted from the diet or are present in inadequate amounts. Since certain grains are relatively poor sources of tryptophan and lysine, in regions where the diet relies heavily on these grains for total protein and is unsupplemented by protein sources such as milk, fish, or meat, dramatic deficiency states may be observed. Kwashiorkor and marasmus are endemic in certain regions of West Africa. Kwashiorkor results when a child is weaned onto a starchy diet poor in protein. In marasmus, both caloric intake and specific amino acids are deficient.

NUTRITIONALLY ESSENTIAL AMINO ACIDS HAVE PROTRACTED BIOSYNTHETIC PATHWAYS

The existence of nutritional requirements suggests that dependence on an external supply of a required intermediate can be of greater survival value than the ability to biosynthesize it. If a specific intermediate is present in the food, an organism that can synthesize it is reproducing and transferring to future generations genetic information of negative survival value. The survival value is negative rather than nil because ATP and nutrients are used to synthesize "unnecessary" DNA. The number of enzymes required by prokaryotic cells to synthesize the nutritionally essential amino acids is large relative to the number of enzymes required to synthesize the nutritionally nonessential amino acids (Table 30–2). This suggests that there is a survival advantage in retaining the

Table 30–1. Amino acid requirements of humans.

Nutritionally Essential	Nutritionally Nonessential
Arginine[1]	Alanine
Histidine[1]	Asparagine
Isoleucine	Aspartate
Leucine	Cysteine
Lysine	Glutamate
Methionine	Glutamine
Phenylalanine	Glycine
Threonine	Hydroxyproline[2]
Tryptophan	Hydroxylysine[2]
Valine	Proline
	Serine
	Tyrosine

[1]"Nutritionally semiessential." Synthesized at rates inadequate to support growth of children.
[2]Not necessary for protein synthesis but formed during posttranslational processing of collagen.

Table 30–2. Enzymes required for the synthesis of amino acids from amphibolic intermediates.

Number of Enzymes Required to Synthesize			
Nutritionally Essential		Nutritionally Nonessential	
Arg[1]	7	Ala	1
His	6	Asp	1
Thr	6	Asn[2]	1
Met	5 (4 shared)	Glu	1
Lys	8	Gln[1]	1
Ile	8 (6 shared)	Hyl[3]	1
Val	1 (7 shared)	Hyp[4]	1
Leu	3 (7 shared)	Pro[1]	3
Phe	10	Ser	3
Trp	5 (8 shared)	Gly[5]	1
	59	Cys[6]	2
		Tyr[7]	1
			17

[1]From Glu. [2]From Asp. [3]From Lys. [4]From Pro. [5]From Ser. [6]From Ser plus S^{2-}. [7]From Phe.

ability to manufacture "easy" amino acids while losing the ability to make "difficult" amino acids.

NUTRITIONALLY NONESSENTIAL AMINO ACIDS HAVE SHORT BIOSYNTHETIC PATHWAYS

Of the 12 nutritionally nonessential amino acids (Table 30–1), nine are formed from amphibolic intermediates. The remaining three (Cys, Tyr, Hyl) are formed from nutritionally essential amino acids.

Glutamate dehydrogenase, glutamine synthetase, and transaminases occupy central positions in amino acid biosynthesis. Their combined effect is to catalyze transformation of inorganic ammonium ion into the organic α-amino nitrogen of various amino acids.

(1) Glutamate: Reductive amination of α-ketoglutarate is catalyzed by glutamate dehydrogenase (Figure 30–1). In addition to forming L-glutamate from the amphibolic intermediate α-ketoglutarate, this reaction constitutes a key first step in the biosynthesis of many additional amino acids.

(2) Glutamine: Biosynthesis of glutamine from

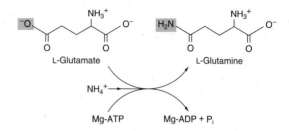

Figure 30–2. The glutamine synthetase reaction.

glutamate is catalyzed by glutamine synthetase (Figure 30–2). The reaction exhibits both similarities to and differences from the glutamate dehydrogenase reaction. Both "fix" inorganic nitrogen—one into amino and the other into amide linkage. Both reactions are coupled to highly exergonic reactions—for glutamate dehydrogenase the oxidation of NAD(P)H and for glutamine synthetase the hydrolysis of ATP.

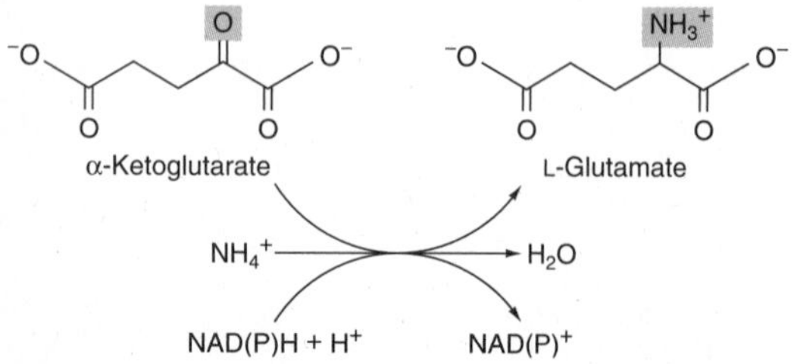

(3) Alanine and aspartate: Transamination of pyruvate forms L-alanine (Figure 30–3), and transamination of oxaloacetate forms L-aspartate. Transfer of the α-amino group of glutamate to the amphibolic intermediates pyruvate and oxaloacetate illustrates the ability of a transaminase to channel ammonium ion, via glutamate, into the α-amino nitrogen of amino acids.

(4) Asparagine: Formation of asparagine from aspartate, catalyzed by asparagine synthetase (Figure 30–4), resembles glutamine synthesis (Figure 30–2). However, since the mammalian enzyme uses glutamine rather than ammonium ion as the nitrogen source, mammalian asparagine synthetase does not "fix" inorganic nitrogen. By contrast, bacterial asparagine synthetases do use ammonium ion and hence do "fix" nitrogen. As for other reactions in which pyrophosphate (PP_i) is formed, hydrolysis of PP_i to P_i by pyrophosphatase ensures that the reaction is strongly favored energetically.

(5) Serine: Serine is formed from the glycolytic intermediate D-3-phosphoglycerate (Figure 30–5).

Figure 30–1. The glutamate dehydrogenase reaction. Reductive amination of α-ketoglutarate by NH_4^+ proceeds at the expense of NAD(P)H.

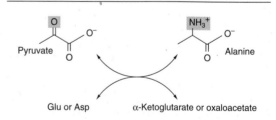

Figure 30–3. Formation of alanine by transamination of pyruvate. The amino donor may be glutamate or aspartate. The other product thus is α-ketoglutarate or oxaloacetate. If oxaloacetate is the oxo acid rather than pyruvate, transfer to the amino group of glutamate forms aspartate.

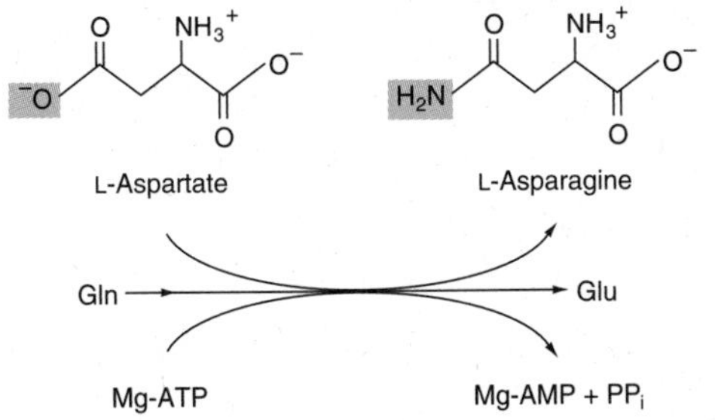

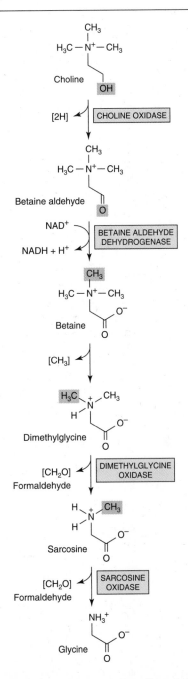

tissues can occur in several ways. Liver cytosol contains glycine transaminases that catalyze the synthesis of glycine from glyoxylate and glutamate or alanine. Unlike most transaminase reactions, this strongly favors glycine synthesis. Two additional important mammalian routes for glycine formation are from choline (Figure 30–6) and from serine via the

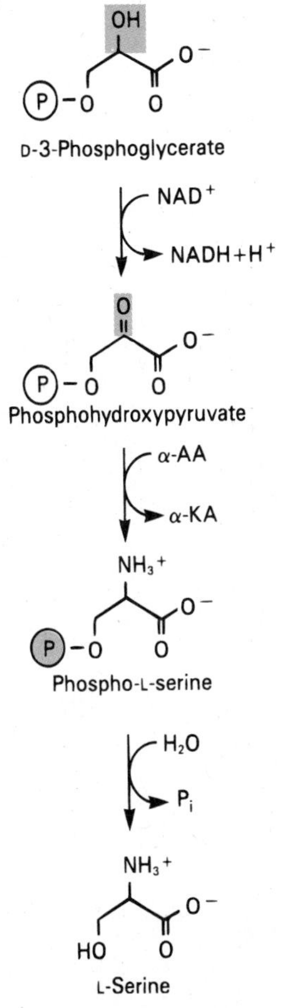

Figure 30–4. The asparagine synthetase reaction. Note similarities to and differences from the glutamine synthetase reaction (Figure 30–2).

The α-hydroxyl group is oxidized to an oxo group by NAD^+, then transaminated, forming phosphoserine. This is then dephosphorylated, forming serine.

(6) Glycine: Synthesis of glycine in mammalian

Figure 30–5. Serine biosynthesis. (α-AA, α-amino acids; α-KA, α-keto acids.)

Figure 30–6. Formation of glycine from choline.

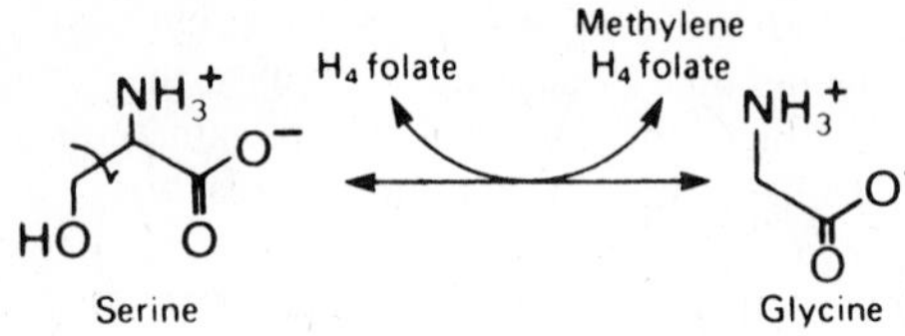

Figure 30-7. The serine hydroxymethyltransferase reaction. The reaction is freely reversible. (H₄ folate, tetrahydrofolate.)

serine hydroxymethyltransferase reaction (Figure 30-7).

(7) Proline: In mammals and some other life forms, proline is formed from glutamate by reversal of the reactions of proline catabolism (Figure 30-8).

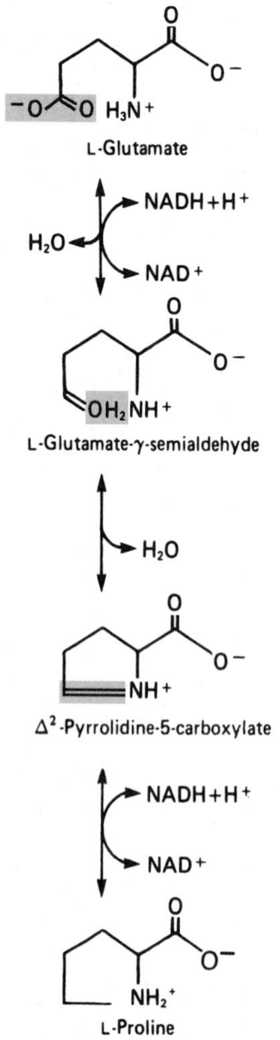

(8) Cysteine: Cysteine, while not itself nutritionally essential, is formed from methionine (nutritionally essential) and serine (nutritionally nonessential). Methionine is first converted to homocysteine via *S*-adenosylmethionine and *S*-adenosylhomocysteine (Chapter 32). Conversion of homocysteine and serine to cysteine and homoserine is shown in Figure 30-9.

(9) Tyrosine: Tyrosine is formed from phenylalanine by the reaction catalyzed by phenylalanine hydroxylase (Figure 30-10). Thus, whereas phenylalanine is a nutritionally essential amino acid, tyrosine is not—provided the diet contains adequate quantities of phenylalanine. The reaction is not reversible, so tyrosine cannot replace the nutritional requirement for phenylalanine. The phenylalanine hydroxylase complex is a mixed-function oxygenase present in mammalian liver but absent from other tissues. The reaction involves incorporation of one atom of molecular oxygen into the *para* position of

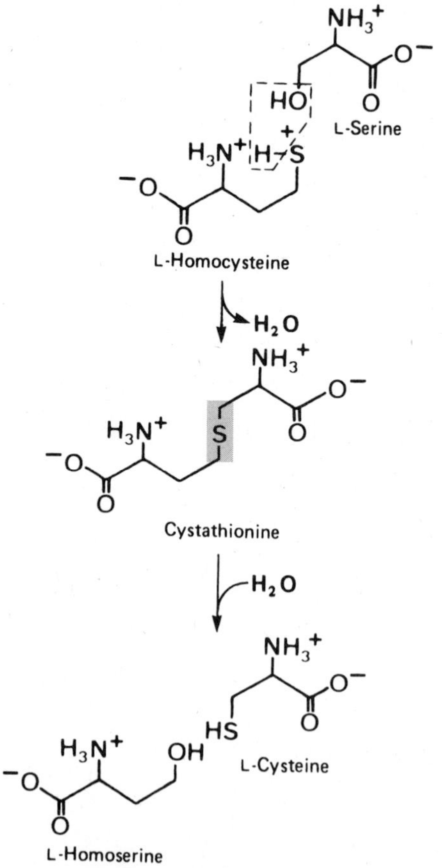

Figure 30-8. Biosynthesis of proline from glutamate by reversal of the reactions of proline catabolism.

Figure 30-9. Conversion of homocysteine and serine to homoserine and cysteine. Note that while the sulfur of cysteine derives from methionine by transsulfuration, the carbon skeleton is provided by serine.

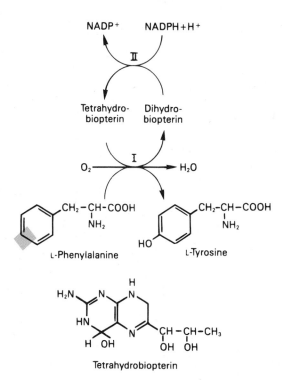

Figure 30–10. The phenylalanine hydroxylase reaction. Two distinct enzymatic activities are involved. Activity II catalyzes reduction of dihydrobiopterin by NADPH, and activity I the reduction of O_2 to H_2O and of phenylalanine to tyrosine. This reaction is associated with several defects of phenylalanine metabolism discussed in Chapter 32.

phenylalanine while the other atom is reduced, forming water (Figure 30–11). The reducing power, supplied ultimately by NADPH, is immediately provided as tetrahydrobiopterin, a pteridine which resembles folic acid.

(10) Hydroxyproline: Since proline serves as a precursor of hydroxyproline, proline and hydroxy-

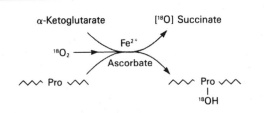

Figure 30–11. The prolyl hydroxylase reaction. The substrate is a proline-rich peptide. During the course of the reaction, molecular oxygen is incorporated into both succinate and proline (shown by the use of heavy oxygen, $^{18}O_2$). Lysyl hydroxylase catalyzes an analogous reaction.

proline belong to the glutamate family of amino acids. Although both 3- and 4-hydroxyprolines occur in mammalian tissues, what follows refers solely to *trans*-4-hydroxyproline.

Hydroxyproline, like hydroxylysine, is present principally in collagen, the most abundant protein of mammalian tissues. Collagen contains about one-third glycine and one-third proline plus hydroxyproline. Hydroxyproline, which accounts for many of the amino acid residues of collagen, stabilizes the collagen triple helix to digestion by proteases. Unlike the hydroxyl groups of hydroxylysine, which serve as sites for attachment of galactosyl and glucosyl residues, the hydroxyl groups of collagen hydroxyproline are unsubstituted.

An unusual feature of hydroxyproline and hydroxylysine metabolism is that the preformed amino acids of ingested food protein are not incorporated into collagen. There is no tRNA capable of accepting hydroxyproline or hydroxylysine and inserting them into an elongating polypeptide chain. Dietary proline, however, is a precursor of collagen hydroxyproline, and dietary lysine is a precursor of collagen hydroxylysine. Hydroxylation of peptide-bound proline or lysine is catalyzed by prolyl hydroxylase (Figure 30–11) or by lysyl hydroxylase, enzymes associated with the microsomal fraction of many tissues (skin, liver, lung, heart, skeletal muscle, and granulating wounds). These enzymes are peptidyl hydroxylases, since hydroxylation only occurs subsequent to incorporation of proline or lysine into polypeptide linkage.

Both hydroxylases are mixed-function oxygenases that require, in addition to substrate, molecular O_2, ascorbate, Fe^{2+}, and α-ketoglutarate. Prolyl hydroxylase has been more extensively studied, but lysyl hydroxylase appears to be an entirely analogous enzyme. For every mole of proline hydroxylated, 1 mol of α-ketoglutarate is decarboxylated to succinate. During this process, one atom of molecular O_2 is incorporated into proline and one into succinate (Figure 30–11).

(11) Hydroxylysine: 5-Hydroxylysine (α,ε-diamino-δ-hydroxycaproate) is present in collagen but absent from most other mammalian proteins. Collagen hydroxylysine arises directly from dietary lysine, not dietary hydroxylysine. Before lysine is hydroxylated, it must first be incorporated into peptide linkage. Hydroxylation of the lysyl peptide is then catalyzed by lysyl hydroxylase, a mixed-function oxidase analogous to prolyl hydroxylase (Figure 30–11).

The Keto Acids of Valine, Leucine, and Isoleucine Can Replace the Amino Acids in the Diet

While leucine, valine, and isoleucine are all nutritionally essential amino acids for humans and other higher animals, mammalian tissue transaminases reversibly interconvert all 3 amino acids and their corresponding α-keto acids. These α-keto acids thus can replace their amino acids in the diet.

Histidine and Arginine Are Nutritionally Semiessential

Arginine, a nutritionally essential amino acid for growing humans, can be synthesized by rats but not in quantities sufficient to permit normal growth.

Histidine, like arginine, is nutritionally semiessential. Adult humans and adult rats have been maintained in nitrogen balance for short periods in the absence of histidine. The growing animal does, however, require histidine in the diet. If studies were to be carried on for longer periods, it is probable that a requirement for histidine in adult human subjects would also be apparent.

SUMMARY

All vertebrates, including humans, can form the 12 nutritionally nonessential amino acids from amphibolic intermediates or from other dietary amino acids. Vertebrates cannot, however, biosynthesize the 10 nutritionally essential amino acids. Only the nutritionally nonessential group are considered here. Vertebrates biosynthesize amino acids from amphibolic intermediates via metabolic pathways that involve 5 or fewer enzyme-catalyzed reactions. The parent amphibolic intermediates and the amino acids to which they give rise are the citric acid cycle intermediates α-ketoglutarate (Glu, Gln, Pro, Hyp) and oxaloacetate (Asp, Asn) and the glycolytic intermediate 3-phosphoglycerate (Ser, Gly). Three other amino acids (Cys, Tyr, Hyl) are formed from nutritionally essential amino acids. Serine provides the carbon skeleton and homocysteine the sulfur for cysteine biosynthesis, while phenylalanine hydroxylase converts phenylalanine to tyrosine. Since no codon or tRNA dictates the insertion of Hyp or Hyl into peptides, neither dietary hydroxyproline nor dietary hydroxylysine is incorporated into proteins. These hydroxylated amino acids arise via posttranslational hydroxylation by mixed-function oxidases of peptidyl Pro or Lys.

REFERENCES

Mercer LP, Dodds SJ, Smith DI: Dispensable, indispensable, and conditionally indispensable amino acid ratios in the diet. In: *Absorption and Utilization of Amino Acids.* Friedman M (editor). CRC Press, 1989.

Scriver CR et al (editors): *The Metabolic and Molecular Bases of Inherited Disease,* 7th ed. McGraw-Hill, 1995.

Catabolism of Proteins & of Amino Acid Nitrogen

31

Victor W. Rodwell, PhD

INTRODUCTION

We here consider how nitrogen is removed from amino acids and converted to urea, and medical problems that arise when there are defects in these processes.

BIOMEDICAL IMPORTANCE

Nitrogen balance refers to the difference between total nitrogen intake and total nitrogen loss in feces, urine, and perspiration. Positive nitrogen balance, ingestion of more nitrogen than is excreted, characterizes growing infants and pregnant women. Normal adult subjects typically are in nitrogen equilibrium, ie, nitrogen intake matches output. Negative nitrogen balance, where nitrogen output exceeds intake, may occur following surgery, in advanced cancer, and following failure to ingest adequate or sufficiently high-quality protein (eg, kwashiorkor, marasmus).

Ammonia, derived mainly from deamination of the α-amino nitrogen of amino acids, is toxic to all animals. Human tissues therefore initially detoxify ammonia by converting it to glutamine for transport to the liver. Deamination of glutamine in the liver releases ammonia, which then is efficiently converted to the nontoxic, nitrogen-rich compound urea. Efficient biosynthesis of urea is essential for health. Where liver function is seriously compromised, eg, in individuals with massive cirrhosis or severe hepatitis, ammonia accumulates in the blood and generates clinical signs and symptoms. Rare but injurious metabolic disorders of all five urea cycle enzymes have been reported. Appropriate management of those few infants born with a deficiency in the activity of an enzyme of the urea cycle requires an understanding of the biochemistry of urea synthesis.

PROTEIN TURNOVER CHARACTERIZES ALL FORMS OF LIFE

Protein turnover, the continuous degradation and resynthesis of all cellular proteins, is a key physio-logic process in all forms of life. While turnover involves both synthesis and degradation of proteins, this chapter discusses only protein and amino acid catabolism. Protein synthesis is discussed in Chapter 40.

Adults Degrade 1–2% of Their Body Protein Daily

Each day, humans turn over 1–2% of their total body protein, principally muscle protein. Of the liberated amino acids, 75–80% are reutilized for new protein synthesis. The nitrogen of the remaining 20–25% forms urea. The carbon skeletons are then degraded to amphibolic intermediates (Figure 31–1).

Proteins Are Degraded at Varying Rates

Individual proteins are degraded at vastly different rates, and these rates vary in response to physiologic demand. High mean rates of protein degradation characterize tissues undergoing major structural rearrangement (eg, uterine tissue during pregnancy; tadpole tail tissue during metamorphosis; degradation of skeletal muscle proteins in severe starvation).

The susceptibility of a protein to degradation is expressed as its half-life, $t_{1/2}$, the time required to reduce its concentration to 50% of its initial value. Half-lives for liver proteins range from under 30

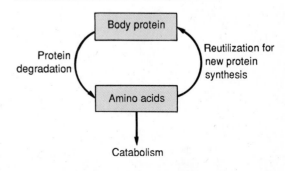

Figure 31–1. Protein and amino acid turnover.

minutes to over 150 hours. Proteins with short half-lives have PEST sequences, regions rich in the amino acids proline (P), glutamate (E), serine (S) and threonine (T), which target them for rapid degradation. Many key regulatory enzymes have short half-lives. For tryptophan oxygenase, tyrosine transaminase, and HMG-CoA reductase, $t_{1/2} = 0.5$–2 hours. These values contrast sharply with half-lives of over 100 hours for aldolase, lactate dehydrogenase, and cytochromes. In response to physiologic demand, degradation of key regulated enzymes may be accelerated or retarded, altering enzyme levels, and hence altering metabolite flux and partitioning metabolites between different metabolic pathways.

Excess Amino Acids Are Degraded, Not Stored

To maintain health, typical Western adults require 30–60 g of protein per day, or its equivalent in free amino acids. Protein quality, the proportion of essential amino acids in a food relative to their proportion in proteins undergoing synthesis, is, however, of critical importance. Excess amino acids are not stored. Regardless of source, those not immediately incorporated into new protein are rapidly degraded. Consumption of excess amino acids thus serves no purpose that cannot equally well be served at a lower cost by carbohydrates and lipids.

PROTEASES AND PEPTIDASES DEGRADE PROTEINS TO AMINO ACIDS

Intracellular proteases hydrolyze internal peptide bonds, releasing peptides, which are then degraded to free amino acids by peptidases. Endopeptidases cleave internal bonds, forming shorter peptides. Aminopeptidases and carboxypeptidases remove amino acids sequentially from the amino and carboxyl terminals, respectively. The ultimate products are free amino acids.

Proteins Are Degraded by ATP-Dependent and ATP-Independent Pathways

Two major pathways degrade intracellular proteins of eukaryotic cells. Extracellular, membrane-associated, and long-lived intracellular proteins are degraded in cellular organelles termed lysosomes by ATP-independent processes. By contrast, degradation of abnormal and other short-lived proteins requires ATP and ubiquitin and occurs in the cytosol.

Asialoglycoprotein Receptors Bind Glycoproteins Destined for Degradation

For proteins in the circulation such as peptide hormones, loss of a sialic acid moiety from the nonreducing ends of their oligosaccharide chains targets them for degradation. These asialated glycoproteins are recognized and internalized by liver cell asialoglycoprotein receptors, then degraded in lysosomes by proteases termed "cathepsins."

Ubiquitin Targets Many Intracellular Proteins for Degradation

Ubiquitin, a small (8.5 kDa) protein present in all eukaryotic cells, targets many intracellular proteins for degradation. The primary structure of ubiquitin is highly conserved; only 3 of 76 residues differ between yeast and human ubiquitin. Proteins destined for degradation via ubiquitin-dependent reactions are derivatized by several molecules of ubiquitin. These are attached by reactions that form non-α-peptide bonds between the carboxyl terminus of ubiquitin and ε-amino groups of lysyl residues in a protein (Figure 31–2). Whether or not a given protein is derivatized by ubiquitin depends on which aminoacyl residue is present at its amino terminal. Reaction with ubiquitin is retarded by amino terminal methionyl or seryl residues and is accelerated by amino terminal aspartyl or arginyl residues.

1. $UB-\overset{\displaystyle O}{\overset{\|}{C}}-O^- + E_1 -SH + ATP \rightarrow AMP + PP_i + UB-\overset{\displaystyle O}{\overset{\|}{C}}-S-E_1$

2. $UB-\overset{\displaystyle O}{\overset{\|}{C}}-S-E_1 + E_2 -SH \rightarrow E_1 -SH + UB-\overset{\displaystyle O}{\overset{\|}{C}}-S-E_2$

3. $UB-\overset{\displaystyle O}{\overset{\|}{C}}-S-E_2 + H_2N-\epsilon-Protein \overset{E_3}{\rightarrow} E_2 -SH + UB-\overset{\displaystyle O}{\overset{\|}{C}}-\overset{\displaystyle H}{\overset{|}{N}}-\epsilon-Protein$

Figure 31–2. Partial reactions in the attachment of ubiquitin (UB) to proteins. (1) The terminal COOH of ubiquitin forms a thioester bond with an —SH of E_1 in a reaction driven by conversion of ATP to AMP and PP_i. Subsequent hydrolysis of PP_i by pyrophosphatase ensures that reaction 1 will proceed readily in the direction shown. (2) A thioester exchange reaction transfers activated ubiquitin to E_2. (3) E_3 catalyzes transfer of ubiquitin to ε-lysyl groups of the target protein.

ANIMALS CONVERT α-AMINO NITROGEN TO VARIED END PRODUCTS

Animals excrete nitrogen from amino acids and other sources as one of three end products: ammonia, uric acid, or urea. Which product predominates depends on the availability of water in the ecologic niche occupied by each animal. Teleostean fish, which are ammonotelic, excrete nitrogen as ammonia. Their aqueous niche, which compels them to excrete water continuously, facilitates continuous excretion of highly toxic ammonia. Land animals convert nitrogen either to uric acid (uricotelic organisms) or to urea (ureotelic organisms). Birds, which must conserve water and maintain low weight, are uricotelic. The relatively insoluble end product uric acid is then excreted as semisolid guano. Many land animals, including humans, are ureotelic and excrete the highly water soluble compound urea. Urea is nontoxic. The high blood urea levels in patients with renal disease are a consequence, not a cause, of impaired renal function.

BIOSYNTHESIS OF UREA

Urea biosynthesis is divided for discussion into 4 stages: (1) transamination, (2) oxidative deamination of glutamate, (3) ammonia transport, and (4) reactions of the urea cycle. Figure 31–3 relates these areas to the overall catabolism of amino acid nitrogen.

α-Amino Groups Are Removed by Transamination

Free amino acids released from dietary or intracellular proteins are metabolized in identical ways. Following removal of the α-amino nitrogen by transamination, the resulting carbon "skeleton" is then degraded by pathways discussed in Chapter 32.

α-Amino Acid Nitrogen Is Channeled Into Glutamate

Transamination interconverts a pair of amino acids and a pair of keto acids, generally an α-amino acid and an α-keto acid (Figure 31–4). While most amino acids undergo transamination, exceptions include lysine, threonine, and the cyclic imino acids proline and hydroxyproline. Since transaminations are freely reversible, **transaminases** (aminotransferases) can function both in amino acid catabolism and biosynthesis. Pyridoxal phosphate resides at the catalytic site of all transaminases and of many other enzymes with amino acid substrates. For all pyridoxal phos-

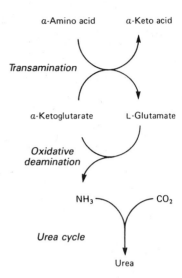

Figure 31–3. Overall flow of nitrogen in amino acid catabolism.

phate-dependent reactions of amino acids, the initial step is formation of an enzyme-bound Schiff base intermediate that is stabilized by interaction with a cationic region of the active site. This intermediate can rearrange in various ways. During transamination, bound coenzyme serves as a carrier of amino groups. Rearrangement forms a keto acid and enzyme-bound pyridoxamine phosphate. Bound pyridoxamine phosphate then forms a Schiff base with a second keto acid.

Alanine-pyruvate transaminase (alanine transaminase) and glutamate-α-ketoglutarate transaminase (glutamate transaminase), present in most mammalian tissues, catalyze transfer of amino groups from most amino acids to form alanine (from pyruvate) or glutamate (from α-ketoglutarate) (Figure 31–5). Serum

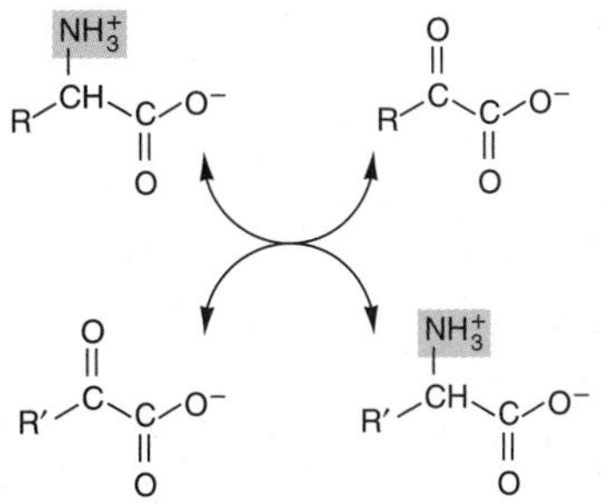

Figure 31–4. Transamination. The reaction is shown for two α-amino and two α-keto acids. Non-α-amino and non-α-oxo groups also participate in transamination, although this is relatively uncommon. The reaction is freely reversible with an equilibrium constant of about 1.

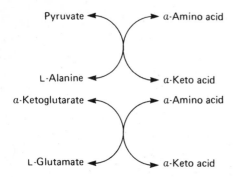

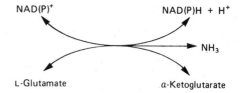

Figure 31–6. The L-glutamate dehydrogenase reaction. NAD(P)+ means that either NAD+ or NADP+ can serve as cosubstrate. The reaction is reversible, but the equilibrium constant favors glutamate formation.

Figure 31–5. Alanine transaminase *(top)* and glutamate transaminase *(bottom).*

levels of transaminases are elevated in some disease states.

Transaminases Are Specific for Only One Pair of α-Amino and α-Keto Acids

Each transaminase is specific for one pair of substrates but nonspecific for the other pair. Since alanine is also a substrate for glutamate transaminase, all the amino nitrogen from amino acids that can undergo transamination can be concentrated in glutamate. This is important, because L-glutamate is the only amino acid in mammalian tissues that undergoes oxidative deamination at an appreciable rate. The formation of ammonia from α-amino groups thus occurs mainly via conversion to the α-amino nitrogen of L-glutamate.

Non-α-Amino Groups May Transaminate

Transamination is not restricted to α-amino groups. The δ-amino group of ornithine (but not the ε-amino group of lysine) is readily transaminated, forming glutamate-γ-semialdehyde (Figure 32–3).

L-GLUTAMATE DEHYDROGENASE OCCUPIES A CENTRAL POSITION IN NITROGEN METABOLISM

The α-amino groups of most amino acids ultimately are transferred to α-ketoglutarate by transamination, forming L-glutamate (Figure 31–3). Release of this nitrogen as ammonia is then catalyzed by **L-glutamate dehydrogenase,** a ubiquitous enzyme of mammalian tissues that uses either NAD+ or NADP+ as oxidant (Figure 31–6). The net conversion of α-amino groups to ammonia thus requires the concerted action of glutamate transaminase and glutamate dehydrogenase. Liver glutamate dehydrogenase activity is regulated by the allosteric inhibitors ATP, GTP,

and NADH, and by the activator ADP. This freely reversible reaction functions both in amino acid catabolism and biosynthesis. Catabolically, it channels nitrogen from glutamate to urea. Anabolically, it catalyzes amination of α-ketoglutarate by free ammonia (see Chapter 30).

Amino Acid Oxidases Also Remove Ammonia From α-Amino Acids

Most of the ammonia released from L-α-amino acids reflects the coupled action of transaminases and L-glutamate dehydrogenase. However, L-amino acid oxidase is present in mammalian liver and kidney tissue. These autoxidizable flavoproteins oxidize amino acids to an α-imino acid that adds water and decomposes to the corresponding α-keto acid with release of ammonium ion (Figure 31–7). The reduced flavin is reoxidized directly by molecular oxygen, forming hydrogen peroxide (H_2O_2), which is split to O_2 and H_2O by the enzyme **catalase** present in many tissues, especially liver.

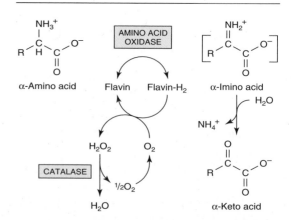

Figure 31–7. Oxidative deamination catalyzed by L-amino acid oxidase (L-α-amino acid:O_2 oxidoreductase). The α-imino acid, shown in brackets, is not a stable intermediate.

Ammonia Intoxication Is Life-Threatening

Ammonia generated by enteric bacteria is absorbed into the portal venous blood, which thus contains higher levels of ammonia than does systemic blood. Since a healthy liver promptly removes this ammonia from the portal blood, peripheral blood is virtually ammonia-free. This is essential, since even traces of ammonia are toxic to the central nervous system. Should portal blood bypass the liver, ammonia then may rise to toxic levels in the systemic blood. This follows severely impaired hepatic function or development of collateral communications between the portal and systemic veins, as may occur in cirrhosis. Symptoms of **ammonia intoxication** include tremor, slurred speech, blurred vision, and in severe cases, coma and death. These symptoms resemble those of hepatic coma, which occurs when blood and brain ammonia levels are elevated. Treatment stresses measures designed to reduce blood ammonia levels.

Glutamine Synthetase Fixes Ammonia as Glutamine

While ammonia is constantly produced in the tissues, it is rapidly removed from the circulation by the liver and converted to glutamate, glutamine, and ultimately to urea. Ammonia thus normally is present only in traces in peripheral blood (10–20 μg/dL). In addition to fixation of ammonia via the glutamate dehydrogenase reaction, formation of glutamine is catalyzed by **glutamine synthetase** (Figure 31–8), a mitochondrial enzyme present in high quantities in renal tissue. Synthesis of the amide bond of glutamine is accomplished at the expense of hydrolysis of one equivalent of ATP to ADP and P_i. The reaction is thus strongly favored in the direction of glutamine synthesis.

Although brain tissue can form urea, this does not appear to play a significant role in ammonia removal. In brain tissue, the major mechanism for detoxification of ammonia is glutamine formation. However, if blood ammonia levels are elevated, the supply of blood glutamate available to the brain is inadequate for formation of glutamine. The brain therefore also must synthesize glutamate from α-ketoglutarate. This would rapidly deplete citric acid cycle intermediates unless they were replaced by CO_2 fixation, converting pyruvate to oxaloacetate (see Chapter 18). Fixation of CO_2 into amino acids indeed occurs in brain tissue. After infusion of ammonia, citric acid cycle intermediates are diverted to the synthesis of α-ketoglutarate and subsequently of glutamine.

Glutaminase and Asparaginase Deamidate Glutamine and Asparagine

Hydrolytic release of the amide nitrogen of glutamine as ammonia, catalyzed by **glutaminase** (Figure 31–9), strongly favors glutamate formation. Glutamine synthetase and glutaminase thus catalyze interconversion of free ammonium ion and glutamine (Figure 31–10). An analogous reaction is catalyzed by L-asparaginase. Since certain tumors exhibit abnormally high requirements for glutamine and asparagine, asparaginase and glutaminase have been tested as antitumor agents.

Formation and Secretion of Ammonia Maintain Acid-Base Balance

Excretion into the urine of the ammonia produced by renal tubular cells facilitates cation conservation

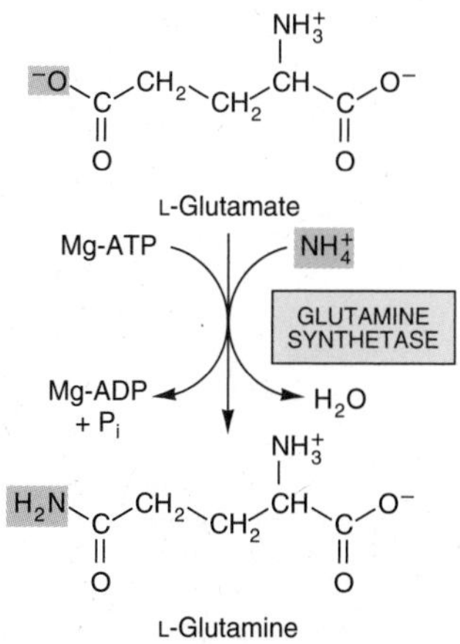

Figure 31–8. The glutamine synthetase reaction. The reaction strongly favors glutamine synthesis.

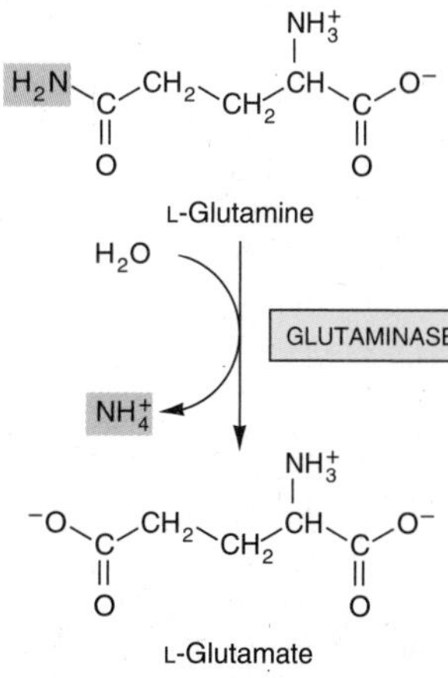

Figure 31–9. The glutaminase reaction proceeds essentially irreversibly in the direction of glutamate and NH_4^+ formation. Note that the *amide* nitrogen, not the α-amino nitrogen, is removed.

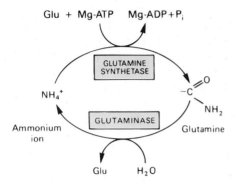

Figure 31–10. Interconversion of ammonia and of gluta-mine catalyzed by glutamine synthetase and glutami-nase. Both reactions are strongly favored in the direc-tions indicated by the arrows. Glutaminase thus serves solely for glutamine deamidation and glutamine syn-thetase solely for synthesis of glutamine from glutamate. (Glu, glutamate.)

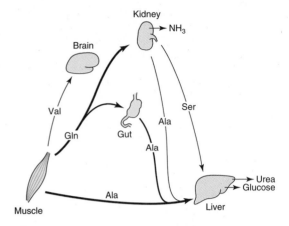

Figure 31–11. Interorgan amino acid exchange in nor-mal postabsorptive humans. The key role of alanine in amino acid output from muscle and gut and uptake by the liver is shown. (Reproduced, with permission, from Felig P: Amino acid metabolism in man. Annu Rev Biochem 1975;44:937. Copyright © 1975 by Annual Reviews, Inc.)

and regulation of acid-base balance. Derived from in-tracellular renal amino acids, particularly glutamine, from which it is released by renal glutaminase, am-monia production increases in **metabolic acidosis** and decreases in **metabolic alkalosis.**

Interorgan Exchanges Maintain Circulating Levels of Amino Acids

The maintenance of steady-state concentrations of circulating plasma amino acids between meals de-pends on the net balance between release from en-dogenous protein stores and utilization by various tis-sues. Muscle generates greater than 50% of the total body pool of free amino acids, while liver is the site of the urea cycle enzymes necessary for disposal of excess nitrogen. Muscle and liver thus play major roles in maintaining circulating amino acid levels.

Figure 31–11 summarizes the postabsorptive state. Free amino acids, particularly alanine and glutamine, are released from muscle into the circulation. Ala-nine, which appears to be the vehicle of nitrogen transport in the plasma, is extracted primarily by the liver. Glutamine is extracted by the gut and the kid-ney, both of which convert a significant portion to alanine. Glutamine also serves as a source of ammo-nia for excretion by the kidney. The kidney provides a major source of serine for uptake by peripheral tis-sues, including liver and muscle. Branched-chain amino acids, particularly valine, are released by mus-cle and taken up predominantly by the brain.

Alanine serves as a key **gluconeogenic amino acid** (Figure 31–12). In the liver, the rate of glucose synthesis from alanine is far higher than from all other amino acids. The capacity of the liver for glu-coneogenesis from alanine does not reach saturation

until the alanine concentration reaches 20–30 times its physiologic level.

Following a protein-rich meal, the splanchnic tis-sues release amino acids (Figure 31–13) while the peripheral muscles extract amino acids, in both in-

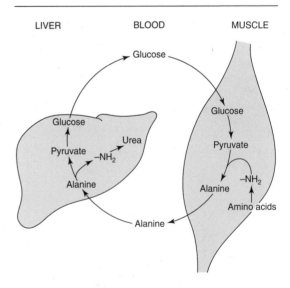

Figure 31–12. The glucose-alanine cycle. Alanine is syn-thesized in muscle by transamination of glucose-derived pyruvate, released into the bloodstream, and taken up by the liver. In the liver, the carbon skeleton of alanine is re-converted to glucose and released into the bloodstream, where it is available for uptake by muscle and resynthesis of alanine. (Reproduced, with permission, from Felig P: Amino acid metabolism in man. Annu Rev Biochem 1975;44:938. Copyright © 1975 by Annual Reviews, Inc.)

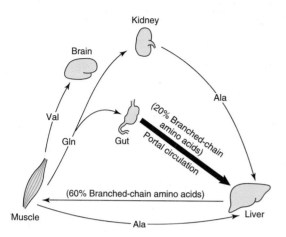

Figure 31–13. Summary of amino acid exchange between organs immediately after feeding.

stances predominantly branched-chain amino acids. Branched-chain amino acids thus serve a special role in nitrogen metabolism, both in the fasting state, when they provide the brain with an energy source, and after feeding, when they are extracted predominantly by muscles, having been spared by the liver.

UREA IS THE MAJOR END PRODUCT OF NITROGEN CATABOLISM IN HUMANS

A human subject who consumes 300 g of carbohydrate, 100 g of fat, and 100 g of protein daily excretes about 16.5 g of nitrogen each day, 95% in the urine and 5% in the feces. For subjects consuming occidental diets, urea synthesized in the liver, released into the blood, and cleared by the kidneys, constitutes 80–90% of the nitrogen excreted.

Urea Is Formed From Ammonia, Carbon Dioxide, and Aspartate

Synthesis of 1 mol of urea requires 3 mol of ATP and 1 mol each of ammonium ion and the α-amino nitrogen of aspartate. Five enzymes catalyze the numbered reactions of Figure 31–14. Of the six participating amino acids, *N*-acetylglutamate functions solely as an enzyme activator. The others serve as carriers of the atoms that ultimately become urea. In mammals, the major metabolic role of **ornithine, citrulline,** and **argininosuccinate** is urea synthesis. Urea biosynthesis is a cyclic process. Ornithine consumed in reaction 2 is regenerated in reaction 5, and there is no net loss or gain of ornithine, citrulline, argininosuccinate, or arginine. Ammonium ion, CO_2, ATP, and aspartate are, however, consumed. As shown in Figure 31–14, some reactions of urea syn-

thesis occur in the matrix of the mitochondrion while others occur in the cytosol.

Carbamoyl Phosphate Synthase I Initiates Urea Biosynthesis

The biosynthesis of urea begins with the condensation of carbon dioxide, ammonia, and ATP to form **carbamoyl phosphate,** a reaction catalyzed by **carbamoyl phosphate synthase I** (Figure 31–14). Human tissues contain two forms of carbamoyl phosphate synthase. Carbamoyl phosphate synthase I, the enzyme functional in urea synthesis, is a hepatic mitochondrial enzyme. Carbamoyl phosphate synthase II, a cytosolic enzyme that uses glutamine rather than ammonia as the nitrogen donor, functions in pyrimidine biosynthesis (see Chapter 36). Formation of carbamoyl phosphate requires 2 mol of ATP. One ATP serves as a source of phosphate. Conversion of the second ATP to AMP and pyrophosphate, together with the coupled hydrolysis of pyrophosphate to orthophosphate, provides the driving force for synthesis of the amide bond and the mixed acid anhydride bond of carbamoyl phosphate. The concerted action of glutamate dehydrogenase and carbamoyl phosphate synthase I thus shuttles nitrogen into carbamoyl phosphate, an intermediate with high group transfer potential.

This complex reaction proceeds stepwise, probably as follows. Reaction of bicarbonate and ATP forms carbonyl phosphate and ADP. Ammonia then displaces ADP, forming carbamate and orthophosphate. Finally, phosphorylation of carbamate by the second ATP forms carbamoyl phosphate. Carbamoyl phosphate synthase I is the rate-limiting, or pacemaker, enzyme of the urea cycle. This regulatory enzyme is active only in the presence of the allosteric activator **N-acetylglutamate,** whose binding induces a conformational change that enhances the affinity of the synthetase for ATP.

Carbamoyl Phosphate Plus Ornithine Forms Citrulline

L-Ornithine transcarbamoylase catalyzes transfer of the carbamoyl moiety of carbamoyl phosphate to ornithine, forming citrulline + orthophosphate (Figure 31–14). While this reaction takes place in the mitochondrial matrix, the compartment in which the substrate ornithine is formed and the product citrulline is further metabolized is the cytosol. Entry of ornithine into mitochondria and exodus of citrulline from mitochondria therefore involve mitochondrial inner membrane transport systems (Figure 31–14).

Citrulline Plus Aspartate Forms Argininosuccinate

The argininosuccinate synthase reaction links aspartate and citrulline via the amino group of aspartate (Figure 31–14) and provides the second nitrogen of urea. The reaction requires ATP and involves the in-

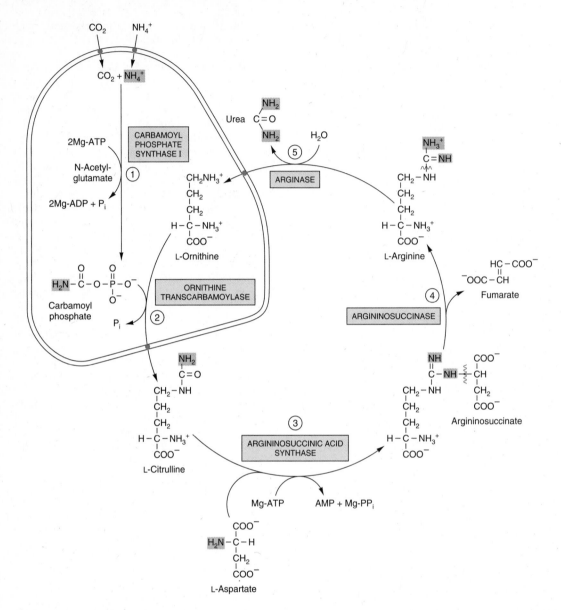

Figure 31–14. Reactions and intermediates of urea biosynthesis. The nitrogen-containing groups that contribute to the formation of urea are shaded. Reactions ① and ② occur in the matrix of liver mitochondria and reactions ③, ④, and ⑤ in liver cytosol. CO_2 (as bicarbonate), ammonium ion, ornithine, and citrulline traverse the mitochondrial matrix via specific carriers (●) present in the inner membrane of liver mitochondria.

termediate formation of citrullyl-AMP. Subsequent displacement of the AMP by aspartate then forms citrulline.

Cleavage of Argininosuccinate Forms Arginine and Fumarate

Cleavage of argininosuccinate, a reversible *trans* elimination reaction catalyzed by **argininosuccinase,** retains nitrogen in the product arginine and releases the aspartate skeleton as fumarate (Figure 31–14). Addition of water to fumarate forms L-malate, and subsequent NAD⁺-dependent oxidation of malate forms oxaloacetate. These two reactions, while analogous to reactions of the citric acid cycle, are catalyzed by cytosolic fumarase and malate dehydrogenase. Transamination of oxaloacetate by glutamate then re-forms aspartate. The carbon skeleton of aspartate/fumarate thus acts as a carrier for transport of the nitrogen of glutamate into a precursor of urea.

Cleavage of Arginine Releases Urea and Re-forms Ornithine

The final reaction of the urea cycle, the hydrolytic cleavage of the guanidino group of arginine catalyzed by liver **arginase,** releases urea. The other product, ornithine, reenters liver mitochondria for additional rounds of the urea cycle. Smaller quantities of arginase also occur in renal tissue, brain, mammary gland, testicular tissue, and skin. Ornithine and lysine are potent inhibitors of arginase competitive with arginine.

Carbamoyl Phosphate Synthase I Is the Pacemaker Enzyme of the Urea Cycle

The activity of carbamoyl phosphate synthase I is determined by the steady-state concentration of its allosteric activator *N*-acetylglutamate. This steady-state level is determined by the rates of its synthesis from acetyl-CoA and glutamate and hydrolysis to acetate and glutamate, reactions catalyzed by *N*-acetylglutamate synthase and *N*-acetylglutamate hydrolase, respectively. In addition, major changes in the diet of animal models, and presumably also of human subjects, can alter the concentrations of individual urea cycle enzymes 10- to 20-fold. Starvation, for example, elevates enzyme levels, presumably to cope with the increased production of ammonia that accompanies enhanced protein degradation.

METABOLIC DISORDERS ARE ASSOCIATED WITH EACH REACTION OF THE UREA CYCLE

The metabolic disorders of urea biosynthesis, while extremely rare, illustrate four medically relevant principles: (1) Defects in many enzymes of a metabolic pathway enzyme can result in essentially identical clinical signs and symptoms. (2) The accumulation of intermediates prior to a metabolic block, or the accumulation of ancillary products, provides insight into the reaction that is impaired. (3) Precise diagnosis requires quantitative assay of the reaction catalyzed by the enzyme thought to be defective. (4) Rational therapy must be based on a complete understanding of the underlying biochemical reactions in normal and impaired individuals.

Since urea synthesis converts toxic ammonia to nontoxic urea, all defects in urea synthesis result in ammonia intoxication. This intoxication is more severe when the metabolic block occurs at reactions 1 or 2, since some covalent linking of ammonia to carbon has already occurred if citrulline can be synthesized. Clinical symptoms common to all urea cycle disorders include vomiting, avoidance of high-protein foods, intermittent ataxia, irritability, lethargy, and mental retardation. The clinical features and treatment of all five disorders discussed below are similar. Significant improvement accompanies a low-protein diet, and much of the brain damage may thus be prevented. Food intake should be in frequent small meals to avoid sudden increases in blood ammonia levels.

(1) Hyperammonemia type 1: About 24 cases of **carbamoyl phosphate synthase I** deficiency (reaction 1, Figure 31–14) have been reported. This disorder probably is familial.

(2) Hyperammonemia type 2: A deficiency of **ornithine transcarbamoylase** (reaction 2, Figure 31–14) produces this X chromosome-linked deficiency. The mothers also exhibit hyperammonemia and an aversion to high-protein foods. The only consistent clinical finding is an elevation of glutamine in blood, cerebrospinal fluid, and urine, which probably reflects enhanced glutamine synthesis consequent to elevated tissue levels of ammonia.

(3) Citrullinemia: This rare disorder probably is recessively inherited. From 1 to 2 g of citrulline is excreted in the urine daily, and both plasma and cerebrospinal fluid citrulline levels are elevated. One patient lacked detectable **argininosuccinate synthase** activity (reaction 3, Figure 31–14). In another, the K_m for citrulline was 25 times normal, suggestive of a nonlethal modification of the catalytic site. Citrulline and argininosuccinate serve as alternative carriers of waste nitrogen, since they contain nitrogen intended for urea synthesis. Feeding arginine enhances excretion of citrulline in these patients. Similarly, feeding benzoate diverts ammonium nitrogen to hippurate via glycine (see Figure 33–1).

(4) Argininosuccinicaciduria: This rare, recessively inherited disease characterized by elevated levels of argininosuccinate in the blood, cerebrospinal fluid, and urine is associated with the occurrence of friable, tufted hair (trichorrhexis nodosa). While both early- and late-onset types are known, the disease is always manifest by age 2 and usually terminates fatally early in life. Argininosuccinicaciduria reflects the absence of **argininosuccinase** (reaction 4, Figure 31–14). While the diagnosis is readily made by two-dimensional chromatography of the urine, additional abnormal spots appear in urine on standing owing to the tendency of argininosuccinate to form cyclic anhydrides. Confirmatory diagnosis requires measurement of erythrocyte argininosuccinase activity. This test can be performed on umbilical cord blood or amniotic fluid cells. As for citrullinemia, feeding arginine and benzoate promotes nitrogen excretion in these patients.

(5) Hyperargininemia: This defect in urea synthesis is characterized by elevated blood and cerebrospinal fluid arginine levels, low erythrocyte levels of arginase (reaction 5, Figure 31–14), and a urinary amino acid pattern resembling that of lysine-cystinuria. Possibly this pattern reflects competition by arginine with lysine and cystine for reabsorption in the renal tubule. A low-protein diet lowers plasma ammonia levels and abolishes urinary lysine-cystin-uria.

SUMMARY

To characterize states of nitrogen nutrition, clinicians and nutritionists refer to positive nitrogen balance, negative nitrogen balance, and nitrogen equilibrium. Protein turnover, the continual synthesis and catabolism of proteins, occurs in all life forms. Each day, human subjects degrade 1–2% of their body protein, principally from skeletal muscle. Rates of protein degradation vary widely between proteins, and rates of degradation may vary in different physiologic states. The half-lives of proteins, the time required to degrade half of the existing protein, vary from 30 minutes to over 150 hours. Extremely short half-lives often characterize enzymes that perform key regulatory roles. Proteases and peptidases degrade proteins by both ATP-dependent and ATP-independent pathways. For circulating glycoproteins, liver cell surface asialoglycoprotein receptors bind and internalize asialoglycoproteins destined for degradation by lysosomal proteases. For many intracellular proteins, attachment of several molecules of ubiquitin targets them for degradation. Amino acids liberated by protein catabolism or ingested in excess of need are then degraded, not stored.

Ammonia is highly toxic to all animals. While fish excrete ammonia directly, the need to conserve water precludes this as a major route of nitrogen elimination in birds or land animals. Birds detoxify ammonia by converting it to uric acid. Humans and other higher vertebrates convert ammonia to urea. In either case, the initial reaction in amino acid catabolism is removal of the α-amino group by transamination, a reaction that requires pyridoxal phosphate. Transamination thus channels α-amino acid nitrogen into glutamate. L-Amino acid oxidase also deaminates α-amino acids, although this is of less certain physiologic significance. L-Glutamate dehydrogenase occupies a central position in nitrogen metabolism. Ammonia intoxication is life-threatening, and glutamine synthetase converts ammonia to nontoxic glutamine for transport to the liver. Liver glutaminase then releases ammonia from glutamine for use in urea synthesis. Exchange between organs maintains circulating levels of amino acids, and a complex set of interorgan exchanges characterizes the postabsorptive state.

Urea, the major end product of nitrogen catabolism in humans, is synthesized from ammonia, carbon dioxide, and the amide nitrogen of aspartate. The reactions take place in part in the mitochondrial matrix and in part in the cytosol. Synthesis of carbamoyl phosphate from ammonium ion and CO_2 takes place in liver mitochondria, as does condensation of carbamoyl phosphate with ornithine to form citrulline. The subsequent reactions are cytosolic. The final reaction, catalyzed by arginase, cleaves arginine to urea and ornithine and completes the cycle. Regulation of urea biosynthesis involves both changes in enzyme levels and allosteric regulation of carbamoyl phosphate synthase activity by *N*-acetylglutamate. Inborn errors of metabolism are associated with each reaction of the urea cycle. These include hyperammonemia types 1 and 2, citrullinemia, argininosuccinicaciduria, and hyperargininemia.

REFERENCES

Curthoys NP, Watford M: Regulation of glutaminase activity and glutamine metabolism. Annu Rev Nutr 1995;15:133.

Gebhardt R, Gaunitz F, Mecke D: Heterogeneous (positional) expression of hepatic glutamine synthetase: Features, regulation and implications for carcinogenesis. Adv Enzyme Regul 1994;34:27.

Hershko A, Ciehanover A: The ubiquitin system for protein degradation. Annu Rev Biochem 1992;61:761.

Morris SM Jr: Regulation of enzymes of urea and arginine biosynthesis. Annu Rev Nutr 1992;12:81.

Scriver CR et al (editors): *The Metabolic and Molecular Bases of Inherited Disease,* 7th ed. McGraw-Hill, 1995.

Torchinsky YM: Transamination: Its discovery, biological and clinical aspects (1937–1987). Trends Biochem Sci 1987;12:115.

Wilkinson KD: Role of ubiquitinylation in proteolysis and cellular regulation. Annu Rev Nutr 1995;15:161.

Catabolism of the Carbon Skeletons of Amino Acids

32

Victor W. Rodwell, PhD

INTRODUCTION

The previous chapter described the metabolic fate of the nitrogen atoms of amino acids. The present chapter considers conversion of the carbon skeletons of common L-amino acids to amphibolic intermediates and the metabolic diseases or "inborn errors of metabolism" associated with these catabolic pathways.

BIOMEDICAL IMPORTANCE

Certain disorders of amino acid metabolism have played major roles in elucidating the pathways by which amino acids are metabolized in normal human subjects. While most of these diseases are rare and unlikely to be encountered by most practicing physicians, they pose formidable challenges for psychiatrists, pediatricians, genetic counselors, and molecular biologists. Left untreated, many of these genetic disorders result in irreversible brain damage and early mortality. Prenatal or early postnatal detection and rapid initiation of appropriate treatment, if available, therefore are essential. Since several of the enzymes concerned are detectable in cultures of amniotic fluid cells, prenatal diagnosis by amniocentesis is possible. Treatment consists primarily of feeding diets low in the amino acids whose catabolism is impaired. However, recombinant DNA technology offers great promise for complementation of defective genes by "gene therapy."

Mutations in the exons or in the regulatory regions of a gene that encodes an enzyme of amino acid catabolism can result in a nonfunctional enzyme or in complete failure to synthesize that enzyme. While some changes in the primary structures of enzymes may have little effect, others modify the three-dimensional structure of catalytic or regulatory sites. The modified or mutant enzyme may possess altered catalytic efficiency (low V_{max} or high K_m) or altered ability to bind an allosteric regulator of its catalytic activity. A variety of mutations may cause the same clinical symptoms. For example, any mutation that significantly lowers the catalytic activity of argini-

nosuccinase will result in the metabolic disorder known as argininosuccinic acidemia. It is, however, most unlikely that all cases of argininosuccinic acidemia represent mutations at the same genetic locus. At a molecular level, these are therefore distinct molecular diseases. To supplement the disorders of amino acid metabolism discussed in this chapter, readers should consult major reference works such as Scriver et al, 1995.

AMINO ACIDS ARE CATABOLIZED TO SUBSTRATES FOR CARBOHYDRATE AND LIPID BIOSYNTHESIS

Nutritional studies in the period 1920–1940, reinforced by studies using isotopically labeled amino acids conducted from 1940 to 1950, established the interconvertibility of fat, carbohydrate, and protein carbons and revealed that each amino acid is convertible either to carbohydrate (13 amino acids), fat (one amino acid), or both (five amino acids) (Table 32–1). Figure 32–1 outlines overall aspects of these interconversions.

THE INITIAL REACTION OFTEN IS REMOVAL OF THE α-AMINO GROUP

Removal of the α-amino nitrogen via transamination generally is the first catabolic reaction. However, this is not the case for proline, hydroxyproline, threonine, or lysine. Depending on physiologic need, the released nitrogen may either be reutilized for anabolic processes such as protein synthesis, or converted to urea and excreted. The partially oxidized hydrocarbon skeleton that remains may then be degraded to amphibolic intermediates.

Asparagine and Aspartate Form Oxaloacetate

All four carbons of asparagine and of aspartate form oxaloacetate via successive reactions cat-

Table 32–1. Fates of the carbon skeletons of the common L-α-amino acids.

Converted to amphibolic intermediates forming:			
Glycogen ("Glycogenic")		Fat ("Ketogenic")	Glycogen and Fat ("Glycogenic" and "Ketogenic")
Ala	Hyp	Leu	Ile
Arg	Met		Lys
Asp	Pro		Phe
Cys	Ser		Trp
Glu	Thr		Tyr
Gly	Val		
His			

Glutamine and Glutamate Form α-Ketoglutarate

Catabolism of glutamine and of glutamate parallels that of asparagine and aspartate but forms **α-ketoglutarate** (Figure 32–2, bottom). While both glutamate and aspartate are substrates for the same transaminase, deamidation of glutamine is catalyzed by **glutaminase.** Possibly for the reason stated immediately above, there are no known metabolic defects of the glutamine-glutamate catabolic pathway.

Proline Forms α-Ketoglutarate

Rather than undergoing direct transamination, proline is oxidized to dehydroproline, which adds water, forming glutamate-γ-semialdehyde. This is then oxidized to glutamate and transaminated to α-ketoglutarate (Figure 32–3, left). Two autosomally recessive hyperprolinemias have been described. Mental retardation occurs in half of the known cases, but neither type is life-threatening.

alyzed by **asparaginase** and a **transaminase** (Figure 32–2, top). Probably because defects in transaminases, which fulfill central amphibolic functions, may be incompatible with life, no known metabolic defect is associated with this short catabolic pathway.

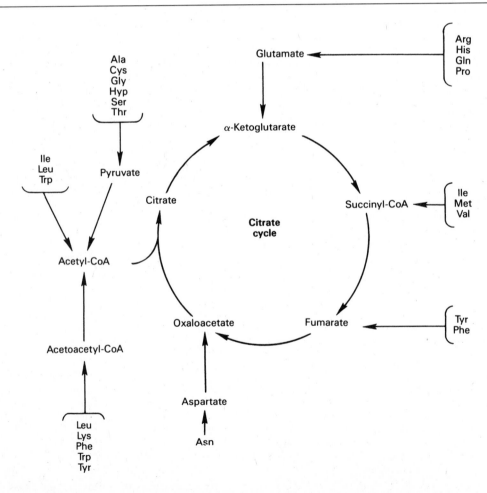

Figure 32–1. Amphibolic intermediates formed from the carbon skeleton of amino acids.

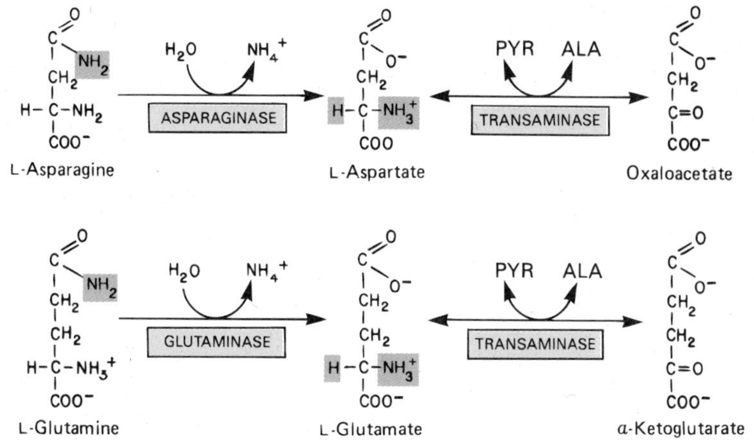

Figure 32–2. Catabolism of L-asparagine *(top)* and of L-glutamine *(bottom)* to amphibolic intermediates. (PYR, pyruvate; ALA, L-alanine.) In this and subsequent figures, shading on functional groups highlights portions of the molecules undergoing chemical change.

A. Hyperprolinemia Type I: The metabolic block in type I hyperprolinemia is at **proline dehydrogenase** (Figure 32–3). Type I heterozygotes exhibit only a mild hyperprolinemia. There is no associated impairment of hydroxyproline catabolism.

B. Hyperprolinemia Type II: The metabolic block occurs at **glutamate-γ-semialdehyde dehydrogenase** (Figure 32–3). Since the same dehydrogenase functions in hydroxyproline catabolism (Figure 32–12), both proline and hydroxyproline catabolism are affected. The urine contains the hydroxyproline catabolite Δ^1-pyrroline-3-hydroxy-5-carboxylate (Figure 32–12). Unlike type I heterozygotes, type II heterozygotes exhibit no hyperprolinemia.

Arginine and Ornithine Form α-Ketoglutarate

Catabolism of the six-carbon amino acid arginine forms α-ketoglutarate. A carbon atom and three nitrogens must first be removed. This requires merely the hydrolytic removal of the guanidino group catalyzed by **arginase,** forming ornithine. Ornithine then undergoes transamination of its 5-amino group, forming glutamate-γ-semialdehyde, which forms α-ketoglutarate as described above for the catabolism of proline (Figure 32–3). Two inherited disorders result in hyperornithinemia.

A. Gyrate Atrophy of the Retina: This inherited autosomal recessive trait involves chorioretinal degeneration with progressive loss of peripheral vision, tunnel vision, and ultimately blindness. Plasma ornithine levels are elevated and 1–10 mmol of ornithine are excreted daily. The defect is in **ornithine δ-transaminase.** The human transaminase has been sequenced, and over 50 mutations have been identified as responsible for gyrate atrophy. Therapy involves restricting dietary arginine.

B. The Hyperornithinemia-Hyperammonemia Syndrome: This recessive genetic disorder, characterized by elevated blood levels of ornithine and ammonia, appears to be due to impaired transport of ornithine into mitochondria. The result is impaired urea biosynthesis with consequent ammonemia (Chapter 30) and ornithinemia. The defect probably involves an **ornithine-citrulline antiporter** analogous to that purified from rat liver mitochondria. If so, this syndrome could also be considered a defect of the urea cycle.

Histidine Forms α-Ketoglutarate

Histidase-catalyzed deamination of histidine produces urocanate (Figure 32–4). Addition of H_2O plus an internal oxidation-reduction reaction, catalyzed by **urocanase,** then converts urocanate to 4-imidazolone-5-propionate. Hydrolysis of 4-imidazolone-5-propionate forms N-formiminoglutamate (Figlu). Subsequent transfer of the formimino group of Figlu to tetrahydrofolate forms glutamate. Transamination of glutamate then forms α-ketoglutarate. In folic acid deficiency, this reaction is partially or totally blocked, and Figlu is excreted in the urine. Excretion of Figlu following a test dose of histidine therefore provides a diagnostic test for folic acid deficiency. While a conspicuous increase in histidine excretion typifies normal pregnancy, this reflects temporary changes in renal function.

Two apparently benign disorders of histidine catabolism are known.

A. Histidinemia: Since the first description of histidinemia in 1961, screening of over 20 million newborn infants has revealed a worldwide incidence of 1:11,500. Characterized by elevated blood and urine histidine, the syndrome is benign in most individuals. The defective enzyme is **histidase,** resulting

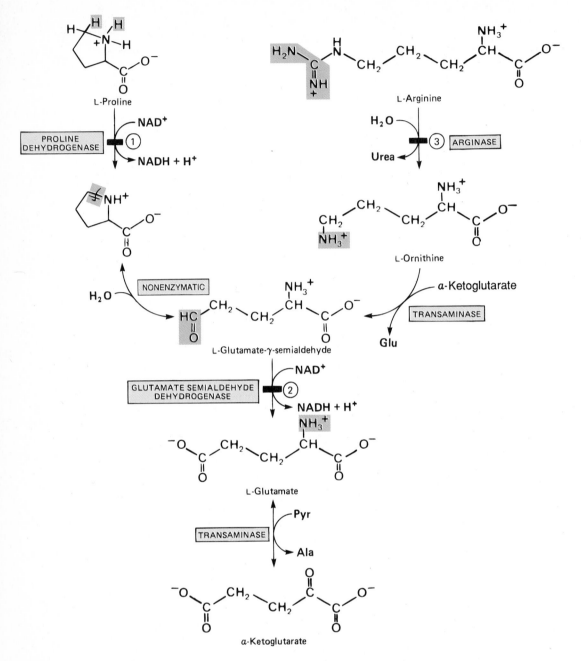

Figure 32–3. Catabolism of L-proline *(left)* and of L-arginine *(right)* to α-ketoglutarate. Numerals mark the sites of the metabolic defects in ① type I hyperprolinemia; ② type II hyperprolinemia; and ③ hyperargininemia (Chapter 31).

in impaired conversion of histidine to urocanate (Figure 32–4).

B. Urocanic Aciduria: In this apparently autosomal recessive disorder, elevated excretion of urocanate results from a defective **urocanase** (Figure 32–4). Elevated urocanic acid excretion is the only sign of this apparently benign disorder.

SIX AMINO ACIDS FORM PYRUVATE

All of the carbons of glycine, alanine, cysteine, and serine—but only two of the carbons of threonine—form pyruvate. Pyruvate may then be converted to acetyl-CoA.

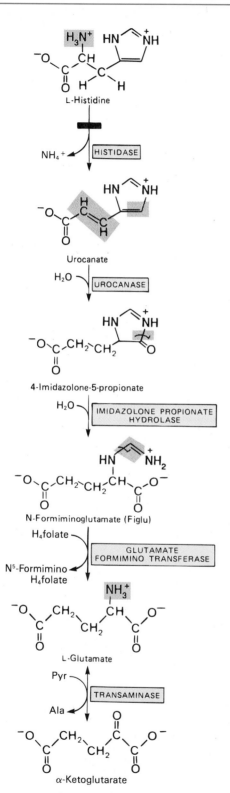

Figure 32–4. Catabolism of L-histidine to α-ketoglutarate. (H_4 folate, tetrahydrofolate.) The reaction catalyzed by histidase represents the site of the probable metabolic defect in histidinemia.

L-Threonine
↓
Glycine
↓
L-Serine   L-Cystine
↓   ↓
L-Alanine → Pyruvate ← L-Cysteine
↓
Acetyl-CoA

Catabolism of Glycine Proceeds Via Glycine Cleavage

While glycine can also form pyruvate via initial conversion to serine (Figure 32–5), glycine cleavage (Figure 32–6) probably constitutes the major route for glycine and for serine catabolism in humans and many other vertebrates. The **glycine synthase complex,** a macromolecular aggregate in liver mitochondria, splits glycine to CO_2 and NH_4^+ and forms N^5,N^{10}-methylene tetrahydrofolate in a reversible manner (Figure 32–6).

A. Glycinuria: Glycinuria is characterized by urinary excretion of 0.6–1 g of glycine per day and a tendency to form oxalate renal stones. Since plasma glycine levels are normal, glycinuria probably results from a defect in renal tubular reabsorption.

B. Primary Hyperoxaluria: In primary hyperoxaluria, urinary excretion of oxalate is unrelated to dietary intake of oxalate. Progressive bilateral calcium oxalate urolithiasis, nephrocalcinosis, and recurrent infection of the urinary tract are followed by early mortality from renal failure or hypertension. The oxalate apparently arises from deamination of glycine, forming glyoxylate (oxalate semialdehyde; see Figure 32–12). The metabolic defect involves a failure to catabolize glyoxylate, which therefore is oxidized to oxalate.

Alanine Forms Pyruvate

Transamination of α-alanine forms pyruvate (Figure 32–7, right), which may then be decarboxylated to acetyl-CoA. Possibly for the reasons advanced under glutamate and aspartate catabolism, there is no known metabolic defect of α-alanine catabolism.

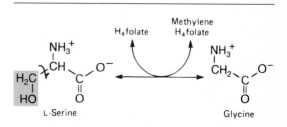

Figure 32–5. The freely reversible serine hydroxymethyltransferase reaction. (H_4 folate, tetrahydrofolate.)

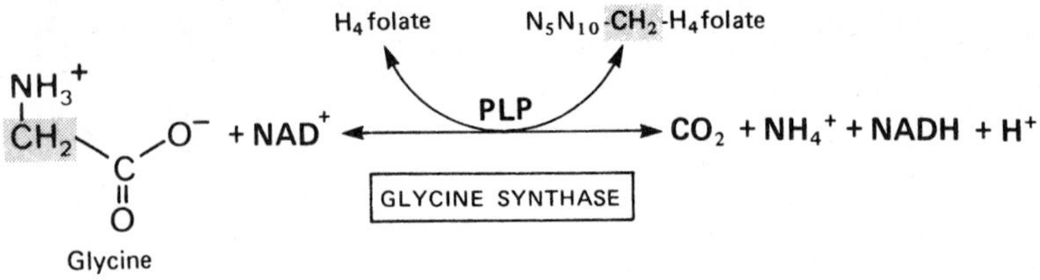

Figure 32–6. Reversible cleavage of glycine by the mitochondrial glycine synthase complex. (PLP, pyridoxal phosphate.)

Serine Is Catabolized Via Glycine

Humans and many other vertebrates degrade serine primarily to glycine and N^5,N^{10}-methylenetetrahydrofolate. Following the reaction catalyzed by **serine hydroxymethyltransferase** (Figure 32–5), serine catabolism merges with that of glycine (Figure 32–6). By contrast, rodent livers convert serine to pyruvate via serine dehydratase, a pyridoxal phosphate protein, via loss of water followed by hydrolytic loss of ammonia (Figure 32–7).

Cystine Reductase Reduces Cystine to Cysteine

Human subjects excrete 20–30 mmol of sulfur per day, mostly as inorganic sulfate. The major catabolic fate of cystine in mammals is conversion to cysteine, catalyzed by **cystine reductase** (Figure 32–8). Catabolism of cystine then merges with that of cysteine.

Two Pathways Convert Cysteine to Pyruvate

Cysteine is catabolized via 2 catabolic pathways: (1) the direct oxidative (cysteine sulfinate) pathway and (2) the transamination (3-mercaptopyruvate) pathway. Conversion of cystine to cysteine sulfinate (Figure 32–9, left) is catalyzed by **cysteine dioxygenase**, an enzyme that requires Fe^{2+} and NAD(P)H. Further catabolism of cysteine sulfinate probably involves its transamination to β-sulfinylpyruvate. Conversion of β-sulfinylpyruvate to pyruvate and sulfite, catalyzed by a desulfinase, occurs even in the absence of enzymatic catalysis.

Initial Transamination of Cysteine Forms 3-Mercaptopyruvate

Reversible transamination of cysteine to 3-mercaptopyruvate (thiolpyruvate) is catalyzed by specific **cysteine transaminases** or by glutamate or asparagine transaminases of mammalian liver and kidney (Figure 32–9, right). Reduction of 3-mercaptopyruvate by **L-lactate dehydrogenase** forms 3-mercaptolactate, present in normal human urine as its mixed disulfide with cysteine and excreted in quantity by patients with mercaptolactate-cysteine disulfiduria. Alternatively, 3-mercaptopyruvate undergoes desulfuration, forming pyruvate and H_2S (Figure 32–9, right). Table 32–2 summarizes known defects of sulfur-containing amino acid catabolism. These include the following:

A. Cystinuria (Cystine-Lysinuria): In this inherited metabolic disease, urinary excretion of cystine is up to 30 times normal. Excretion of lysine, arginine, and ornithine is also increased, suggesting a defect in the renal reabsorptive mechanisms for these 4 amino acids. Cystine-lysinuria therefore may be the preferred name. Since cystine is relatively insoluble, cystine calculi form in the renal tubules of cystinuric patients. Otherwise, cystinuria is a benign anomaly. The mixed disulfide of L-cysteine and L-homocysteine (Figure 32–10) present in the urine of cystinuric patients is more soluble than cystine and hence reduces formation of cystine crystals and calculi.

B. Cystinosis (Cystine Storage Disease): Cystinosis is a rare lysosomal disorder characterized by defective carrier-mediated transport of cystine. Cystine crystals are deposited in tissues and organs,

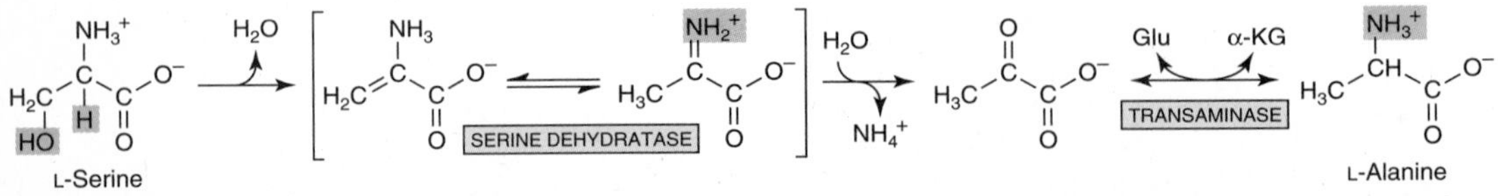

Figure 32–7. Conversion of alanine and serine to pyruvate. Both the alanine transaminase and serine dehydratase reactions require pyridoxal phosphate. The serine dehydratase reaction proceeds via elimination of H_2O from serine, forming an unsaturated amino acid. This rearranges to an α-imino acid that is spontaneously hydrolyzed to pyruvate plus ammonia. (Glu, glutamate; α-KG, α-ketoglutarate.)

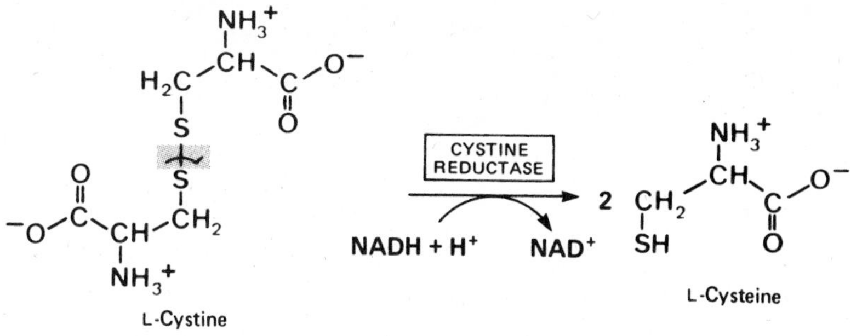

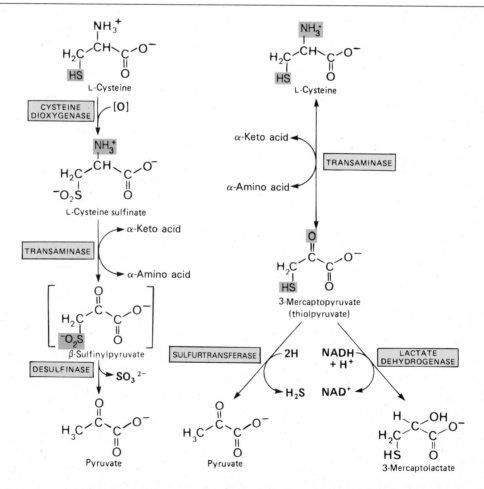

Figure 32–8. The cystine reductase reaction.

particularly the reticuloendothelial system. Cystinosis usually is accompanied by a generalized aminoaciduria. Other renal functions are also seriously impaired, and patients usually die young from acute renal failure.

C. Homocystinurias: In these heritable defects of methionine catabolism (incidence of about one per 160,000 births), up to 300 mg of homocystine, sometimes together with *S*-adenosylmethionine, is excreted daily in the urine. Plasma methionine levels

Figure 32–9. Catabolism of L-cysteine via the direct oxidative (cysteine sulfinate) pathway *(left)* and by the transamination (3-mercaptopyruvate) pathway *(right)*. β-Sulfinylpyruvate is a putative intermediate. Oxidation of the sulfite produced in the last reaction of the direct oxidative pathway is catalyzed by sulfite oxidase. (α-KA, α-keto acid; α-AA, α-amino acid.)

Table 32–2. Inborn errors of sulfur-containing amino acid metabolism.

Name	Defect	Reference
Homocystinuria I	Cystathionine β-synthase	
Homocystinuria II	N^5, N^{10}-methylenetetrahydrofolate reductase	
Homocystinuria III	Low N^5-methyltetrahydrofolate-homocysteine transmethylase owing to inability to synthesize methylcobalamin	
Homocystinuria IV	Low N^5-methyltetrahydrofolate-homocysteine transmethylase owing to defective intestinal absorption of cobalamin	
Hypermethioninemia	Liver methionine adenosyltransferase[1]	Figure 32–22
Cystathioninuria	Cystathionase	
Sulfituria (sulfocysteinuria)	Sulfite oxidase	Figure 32–9, legend
Cystinosis	Defect in lysosomal function	
3-Mercaptopyruvate-cysteine disulfiduria	3-Mercaptopyruvate sulfurtransferase	Figure 32–9
Methionine malabsorption syndrome	Inability to absorb methionine from gut	

[1]May also occur in cystathioninuria, tyrosinemia, and fructose intolerance.

also are elevated. At least four metabolic defects cause homocystinuria (Table 32–2). In type I homocystinuria, clinical findings include thromboses, osteoporosis, dislocated lenses in the eyes, and frequently mental retardation. Both vitamin B_6-responsive and vitamin B_6-unresponsive forms are known. A diet low in methionine and high in cystine prevents pathologic changes if initiated early in life. Other types of homocystinuria reflect defects in the remethylation cycle (Table 32–2).

Threonine Aldolase Initiates Threonine Catabolism

Threonine is cleaved to acetaldehyde and glycine by **threonine aldolase.** Acetaldehyde is then oxidized to acetate, which then is converted to acetyl-CoA (Figure 32–11). Catabolism of glycine was discussed above.

4-Hydroxyproline Forms Pyruvate and Glyoxylate

A mitochondrial dehydrogenase oxidizes 4-hydroxy-L-proline to L-Δ^1-pyrroline-3-hydroxy-5-carboxylate, which is in nonenzymatic equilibrium with γ-hydroxy-L-glutamate-γ-semialdehyde (Figure 32–12). The semialdehyde is oxidized to erythro-γ-hydroxy-L-glutamate, then transaminated to α-keto-γ-hydroxyglutarate. An aldol-type cleavage then forms glyoxylate plus pyruvate.

The site of the metabolic defect in **hyperhydroxyprolinemia,** an autosomal recessive trait, is **4-hydroxyproline dehydrogenase** (Figure 32–12). The disorder is characterized by high plasma levels of 4-hydroxyproline. There is no accompanying impairment of proline catabolism since the affected enzyme functions only in hydroxyproline catabolism. The condition has no effect on collagen metabolism and, like the hyperprolinemias, appears to be harmless.

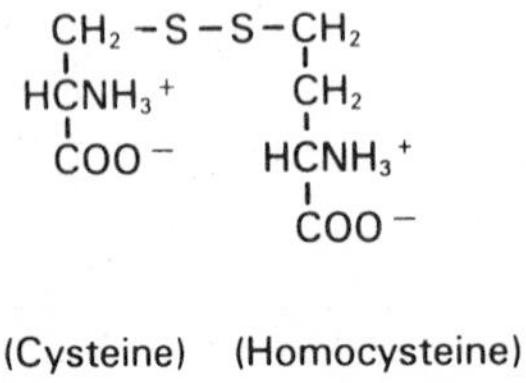

TWELVE AMINO ACIDS FORM ACETYL-CoA

All amino acids that form pyruvate (alanine, cysteine, cystine, glycine, hydroxyproline, serine, and threonine) also form acetyl-CoA via **pyruvate dehydrogenase** (Chapter 19). In addition, phenylalanine, tyrosine, tryptophan, lysine, and leucine form acetyl-CoA without first forming pyruvate.

(1) Transamination of Tyrosine Forms *p*-Hydroxyphenylpyruvate: Transamination of tyrosine to *p*-hydroxyphenylpyruvate is catalyzed by **tyrosine-α-ketoglutarate transaminase,** an inducible enzyme of mammalian liver.

(2) *p*-Hydroxyphenylpyruvate Forms Homogentisate: This unusual reaction (Figure 32–13) involves concerted ring hydroxylation and side-chain migration. Since the physiologic reductant is ascorbate, scorbutic patients excrete incompletely oxidized products of tyrosine metabolism. Several intermediates of tyrosine metabolism were

$$CH_2-S-S-CH_2$$
$$HCNH_3^+ \quad\quad CH_2$$
$$COO^- \quad\quad HCNH_3^+$$
$$COO^-$$

(Cysteine)   (Homocysteine)

Figure 32–10. Mixed disulfide of cysteine and homocysteine.

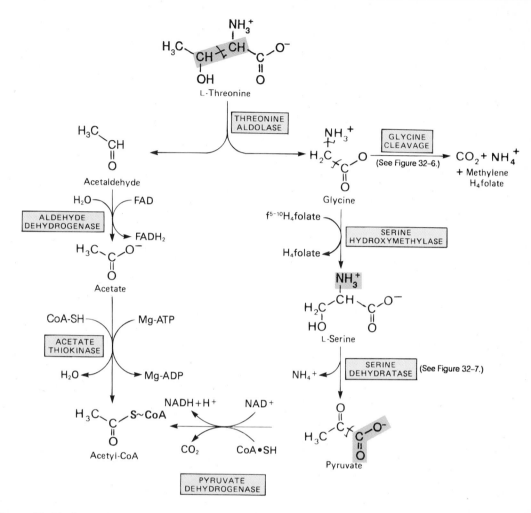

Figure 32–11. Conversion of threonine and glycine to serine, pyruvate, and acetyl-CoA. ($f^{5-10}H_4$ folate, formyl [5–10]tetrahydrofolic acid.)

initially identified by feeding suspected precursors of homogentisate to patients suffering from **alkaptonuria** (see below), who consequently excreted homogentisate.

(3) Homogentisate Oxidase Opens the Aromatic Ring: Oxidative rupture of the benzene ring of homogentisate, catalyzed by **homogentisate oxidase** of mammalian liver, forms maleylacetoacetate.

(4) *Cis,trans* Isomerization Forms Fumarylacetoacetate: Isomerization of maleylacetoacetate, catalyzed by **maleylacetoacetate *cis,trans* isomerase,** forms fumarylacetoacetate.

(5) Hydrolysis of Fumarylacetoacetate Forms Fumarate and Acetoacetate: Hydrolysis of fumarylacetoacetate by **fumarylacetoacetate hydrolase** forms fumarate and acetoacetate. The acetoacetate can then form acetyl-CoA plus acetate via **β-ketothiolase** (see Chapter 24).

TYROSINEMIA, TYROSINURIA, AND ALKAPTONURIA

(1) Tyrosinemia Type I (Tyrosinosis): The pathophysiology of tyrosinosis is complex, since accumulated metabolites affect the activities of several enzymes and transport systems. The probable metabolic defect is in **fumarylacetoacetate hydrolase** (Figure 32–13) and possibly also in **maleylacetoacetate hydrolase.** Plasma tyrosine levels are elevated (6–12 mg/dL), as are those of methionine. In **acute tyrosinosis,** infants exhibit diarrhea, vomiting, and a "cabbage-like" odor and fail to thrive. Without treatment, death from liver failure ensues in 6–8 months. In **chronic tyrosinemia,** symptoms are similar but milder, but death generally ensues by the age of 10 years. Therapy consists of a diet low in tyrosine and phenylalanine, and in some cases, low in methionine also.

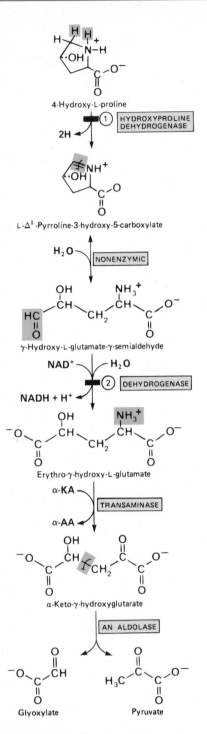

Figure 32–12. Intermediates in L-hydroxyproline catabolism. (α-KA, α-keto acid; α-AA, α-amino acid.) Numerals represent sites of metabolic defects in ① hyperhydroxyprolinemia and ② type II hyperprolinemia.

(2) Tyrosinemia Type II (Richner-Hanhart Syndrome): The probable site of the metabolic defect in type II tyrosinemia is hepatic **tyrosine transaminase** (Figure 32–13). Clinical findings include elevated plasma tyrosine levels (4–5 mg/dL), eye and skin lesions, and moderate mental retardation. Tyrosine is the only amino acid whose urinary concentration is elevated. However, renal clearance and reabsorption of tyrosine fall within normal limits. Urinary metabolites include *p*-hydroxyphenylpyruvate, *p*-hydroxyphenyllactate, *p*-hydroxyphenylacetate, *N*-acetyltyrosine, and tyramine (Figure 32–14).

(3) Neonatal Tyrosinemia: This disorder is thought to result from a relative deficiency of ***p*-hydroxyphenylpyruvate hydroxylase** (Figure 32–13). Blood levels of tyrosine and phenylalanine are elevated, as are urinary levels of tyrosine, *p*-hydroxyphenylacetate, *N*-acetyltyrosine, and tyramine (Figure 32–14). Therapy consists of a diet low in protein.

(4) Alkaptonuria: This inherited metabolic disorder, noted as early as the 16th century and characterized in 1859, formed the basis for Garrod's classic ideas concerning heritable metabolic disorders. Its most striking clinical manifestation is the darkening of urine that stands in air. Late in the disease, there is generalized pigmentation of connective tissues (ochronosis) and a form of arthritis. Ochronosis results from oxidation of homogentisate by polyphenol oxidase, forming benzoquinone acetate, which polymerizes and binds to connective tissue macromolecules. The metabolic defect is lack of **homogentisate oxidase** (Figure 32–13). Homogentisate in the urine is then oxidized by the O_2 in air to a brownish-black pigment. Over 600 cases of this autosomal recessive trait have been reported. Estimated incidence is 2–5 per million live births.

Benzoquinone acetate

Hydroxylation of Phenylalanine Forms Tyrosine

Phenylalanine is first converted to tyrosine by **phenylalanine hydroxylase** (Figure 30–10). The labeling pattern in the amphibolic products fumarate and acetoacetate (Figure 32–15) thus is identical to that for tyrosine (Figure 32–13).

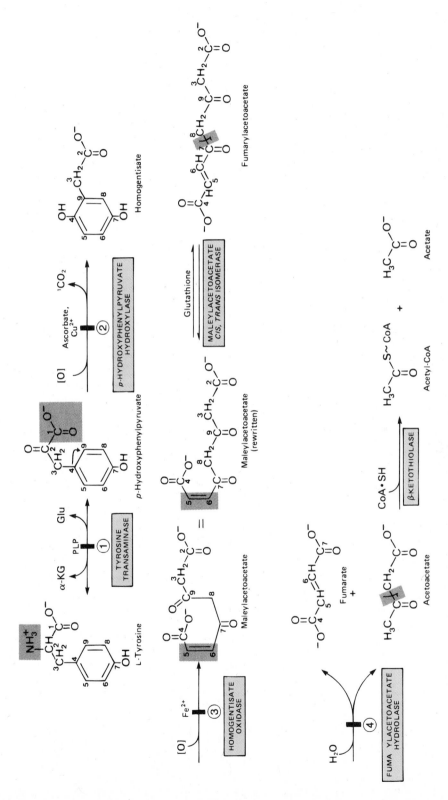

Figure 32–13. Intermediates in tyrosine catabolism. With the exception of β-ketothiolase, reactions are discussed in the text. Carbon atoms of intermediates are numbered to assist readers in determining the ultimate fate of each carbon (see also Figure 32–15). (α-KG, α-ketoglutarate; Glu, glutamate; PLP, pyridoxal phosphate.) The numerals represent the probable sites of the metabolic defects in ① type II tyrosinemia; ② neonatal tyrosinemia; ③ alkaptonuria; and ④ type I tyrosinemia, or tyrosinosis.

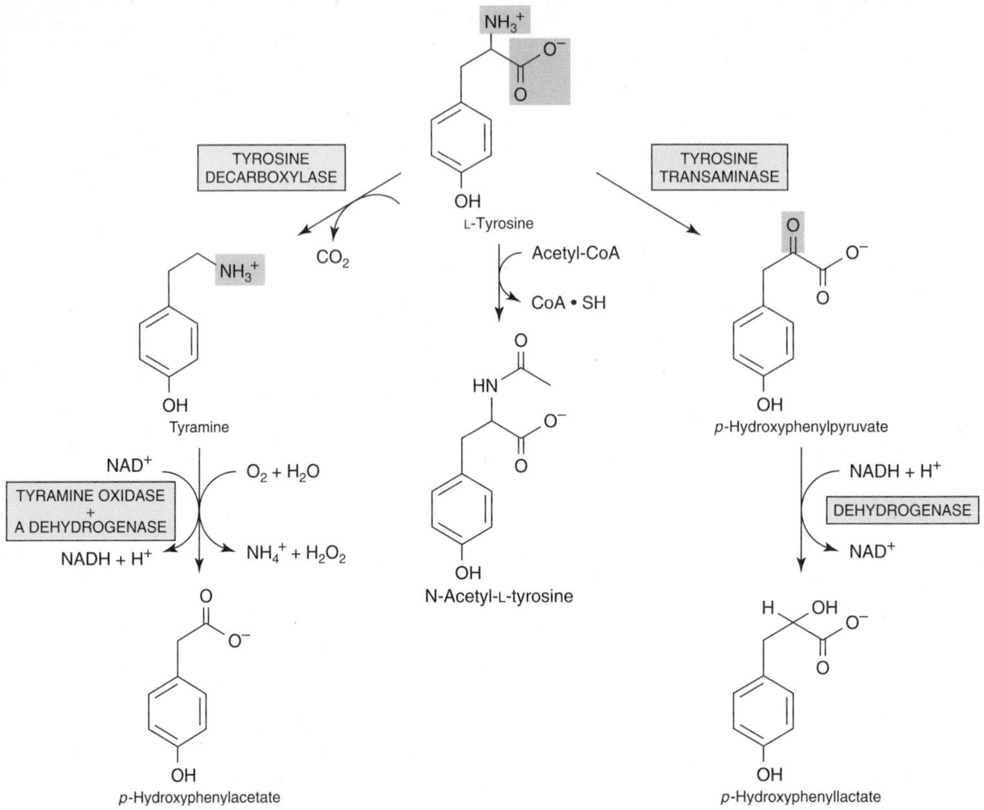

Figure 32–14. Alternative catabolites of tyrosine. *p*-Hydroxyphenylacetaldehyde is formed as an intermediate during oxidation of tyramine to *p*-hydroxyphenylacetate.

METABOLIC DISORDERS OF PHENYLALANINE CATABOLISM

While additional types have been identified (Table 32–3), the major metabolic disorders associated with

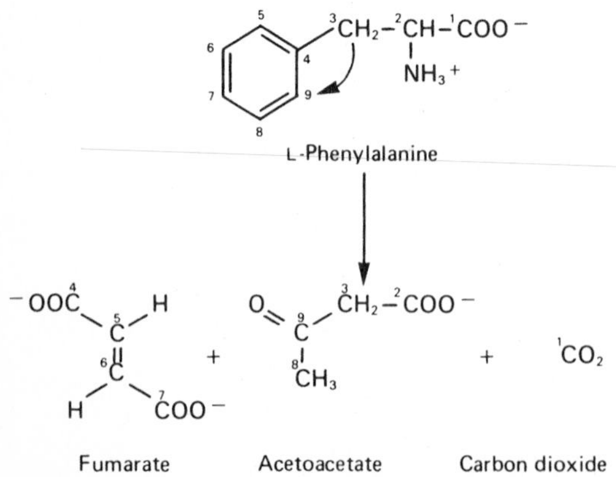

Figure 32–15. Ultimate catabolic fate of each carbon atom of phenylalanine. Numbers illustrate the pattern of isotopic labeling in the ultimate catabolites of phenylalanine (and tyrosine).

impaired ability to convert phenylalanine to tyrosine (see Figure 30–10) fall into three broad groups: (1) defects in **phenylalanine hydroxylase** (hyperphenylalaninemia type I, or classic phenylketonuria); (2) defects in **dihydrobiopterin reductase** (hyperphenylalaninemia types II and III); and (3) defects in **dihydrobiopterin biosynthesis** (hyperphenylalaninemia types IV and V).

The human genes that encode phenylalanine hydroxylase and dihydrobiopterin reductase have been cloned and sequenced; consequently, DNA probes are available for prenatal diagnosis of defects in either of these enzymes.

The major consequence of untreated **type I hyperphenylalaninemia (classic phenylketonuria, or PKU)** is mental retardation. Additional clinical signs include seizures, psychoses, eczema, and a "mousy" odor. Appropriate dietary intervention can ameliorate the mental retardation. In classic PKU, a heritable disorder with a frequency of about 1:10,000 live births, levels of component I of liver **phenylalanine hydroxylase** (see Figure 30–10) average approximately 25% of normal, and the hydroxylase is insensitive to regulation by phenylalanine. Since patients cannot convert phenylalanine to tyrosine, alternative catabolites are produced (Figure 32–16). These in-

Table 32–3. Hyperphenylalaninemias.[1]

Type	Condition	Defect	Treatment
I	Phenylketonuria	Phe hydroxylase absent	Low Phe diet
II	Persistent hyperphenylalaninemia	Decreased Phe hydroxylase	None, or temporary dietary therapy
III	Transient mild hyperphenylalaninemia	Maturational delay of hydroxylase	Same as type II
IV	Dihydrobiopteridine reductase deficiency	Deficient or absent dihydropteridine reductase	Dopa, 5-hydroxytryptophan, carbidopa
V	Abnormal dihydrobiopterin function	Dihydrobiopterin synthesis defect	Dopa, 5-hydroxytryptophan, carbidopa
VI	Persistent hyperphenylalaninemia and tyrosinemia	? Catabolism of tyrosine	Reduced Phe intake
VII	Transient neonatal tyrosinemia	p-Hydroxyphenylpyruvic oxidase inhibition	Vitamin C
VIII	Hereditary tyrosinemia	Deficiency: 1. p-Hydroxyphenylpyruvate oxidase 2. Cytoplasmic tyrosine aminotransferase 3. Fumarylacetoacetate hydrolase	Low Tyr diet plus glutathione injections

[1]Modified and reproduced, with permission, from Tourian A, Sidbury JB: Phenylketonuria and hyperphenylalaninemia. In: Scriver CR et al (editors): *The Metabolic Basis of Inherited Disease,* 6th ed. McGraw-Hill, 1989.

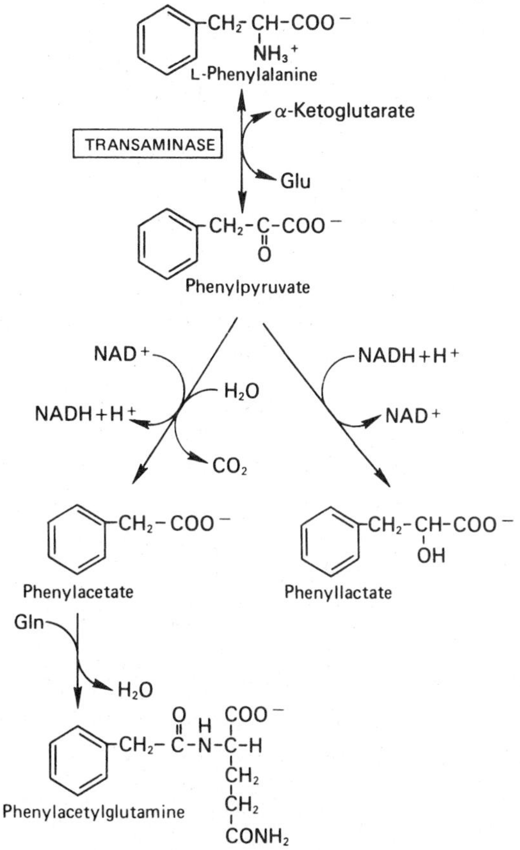

Figure 32–16. Alternative pathways of phenylalanine catabolism in phenylketonuria. The reactions also occur in the liver tissue of normal individuals but are of minor significance if a functional phenylalanine hydroxylase is present. (Glu, glutamate; Gln, glutamine.)

clude phenylpyruvic acid and phenyllactic acid. Much of the phenylacetate is excreted as phenacetylglutamine (Table 32–4). The mental deterioration of phenylketonuric children can be prevented by a diet containing low levels of phenylalanine. This diet may be terminated at 6 years of age, when high concentrations of phenylalanine and its derivatives no longer injure the brain.

Screening of newborn infants for PKU now is compulsory. Plasma phenylalanine can be measured in only 20 µL of blood. However, since intake of dietary protein is low, high blood phenylalanine levels may not occur in phenylketonuric infants until the third or fourth day of life. False-positive results may also occur in premature infants owing to delayed maturation of enzymes of phenylalanine catabolism. A useful, but less reliable, screening test depends on detecting elevated urinary levels of phenylpyruvate with ferric chloride.

Administration of phenylalanine to phenylketonuric subjects results in prolonged elevation of its level in the blood. Since phenylalanine tolerance and a high fasting level of phenylalanine also characterize the parents of phenylketonurics, the genetic defect responsible for phenylketonuria can be detected in phenotypically normal, heterozygous individuals.

Neither Nitrogen of Lysine Participates in Transamination

Mammals convert the intact carbon skeleton of L-lysine to α-aminoadipate and α-ketoadipate (Figure 32–17). The intermediates include saccharopine, which also is an intermediate in lysine biosynthesis by fungi. L-Lysine first forms a Schiff base with α-ketoglutarate that is reduced to saccharopine, then oxidized by a second dehydrogenase. Addition of

Table 32–4. Metabolites of phenylalanine that accumulate in the plasma and urine of phenylketonuric patients.

Metabolite	Plasma (mg/dL)		Urine (mg/dL)	
	Normal	Phenylketonuric	Normal	Phenylketonuric
Phenylalanine	1–2	15–63	30	300–1000
Phenylpyruvate		0.3–1.8		300–2000
Phenyllactate				290–550
Phenylacetate				Increased
Phenylacetylglutamine			200–300	2400

water forms L-glutamate and L-α-aminoadipate-γ-semialdehyde. These reactions thus mimic removal of the ε-nitrogen of lysine by transamination, a reaction that does not in fact occur in mammalian tissues.

Transamination of α-aminoadipate forms α-ketoadipate. This probably is followed by oxidative decarboxylation to glutaryl-CoA. While lysine is both glycogenic and ketogenic, the precise catabolites of glutaryl-CoA in mammalian systems are not known.

Two rare metabolic abnormalities result from impaired conversion of L-lysine and α-ketoglutarate to saccharopine (Figure 32–17).

A. Periodic Hyperlysinemia With Associated Hyperammonemia: Ingestion of normal levels of protein triggers hyperlysinemia. Elevated liver lysine levels then competitively inhibit liver **arginase,** causing hyperammonemia. Fluid therapy and restriction of lysine intake relieve both the hyperammonemia and its clinical manifestations. Conversely, administration of a lysine load precipitates severe crises and coma.

B. Persistent Hyperlysinemia Without Hyperammonemia: Some patients are mentally retarded. There is no associated hyperammonemia, even in response to a lysine load. Lysine catabolites may or may not accumulate in biologic fluids. Persistent hyperlysinemia is believed to be inherited as an autosomal recessive trait. In addition to impaired conversion of lysine and α-ketoglutarate to saccharopine, some patients cannot cleave saccharopine (Figure 32–17).

Tryptophan Oxygenase Initiates Tryptophan Catabolism

The carbon atoms both of the side chain and of the aromatic ring of tryptophan may be completely degraded to amphibolic intermediates via the kynurenine-anthranilate pathway (Figure 32–18), important both for tryptophan degradation and for conversion of tryptophan to **nicotinamide.**

Tryptophan oxygenase (tryptophan pyrrolase) catalyzes cleavage of the indole ring with incorporation of two atoms of molecular oxygen, forming N-formylkynurenine. The oxygenase, an iron porphyrin metalloprotein, is inducible in liver by adrenal corticosteroids and by tryptophan. A large fraction of newly synthesized enzyme is in a latent form that requires activation. Tryptophan also stabilizes the oxygenase toward proteolytic degradation. Tryptophan oxygenase is feedback-inhibited by nicotinic acid derivatives, including NADPH. Hydrolytic removal of the formyl group of N-formylkynurenine, catalyzed by **kynurenine formylase** of mammalian liver, produces kynurenine (Figure 32–18). Kynurenine may be deaminated by transamination. The resulting 2-amino-3-hydroxybenzoyl pyruvate loses water, and spontaneous ring closure forms the by-product kynurenic acid. Further metabolism of kynurenine involves conversion to 3-hydroxykynurenine, then 3-hydroxyanthranilate. Hydroxylation requires molecular oxygen in an NADPH-dependent reaction similar to that for hydroxylation of phenylalanine (Figure 30–10).

Xanthurenate Accumulates in Vitamin B_6 Deficiency

Kynurenine and hydroxykynurenine are converted to hydroxyanthranilate by **kynureninase,** a pyridoxal phosphate enzyme. A deficiency of vitamin B_6 results in partial failure to catabolize these kynurenine derivatives, forming xanthurenate (Figure 32–19), which occurs in the urine when dietary vitamin B_6 is inadequate. Feeding excess tryptophan induces excretion of xanthurenate in vitamin B_6 deficiency.

In many animals, conversion of tryptophan to nicotinic acid makes a dietary supply of the vitamin unnecessary. Tryptophan can completely replace vitamin B_6 in the diet of rodents, dogs, and pigs. For humans and other animals, tryptophan increases the urinary excretion of nicotinic acid derivatives (eg, N-methylnicotinamide). In vitamin B_6 deficiency, synthesis of NAD$^+$ and NADP$^+$ may be impaired by inadequate conversion of tryptophan to nicotinic acid for pyridine nucleotide synthesis. If adequate nicotinic acid is supplied, pyridine nucleotide synthesis proceeds normally even in the absence of vitamin B_6.

Hartnup disease, an autosomal recessive trait, results from defects in the intestinal and renal transport of neutral amino acids, including tryptophan. Signs include a general neutral aminoaciduria and generally also increased excretion of indole derivatives

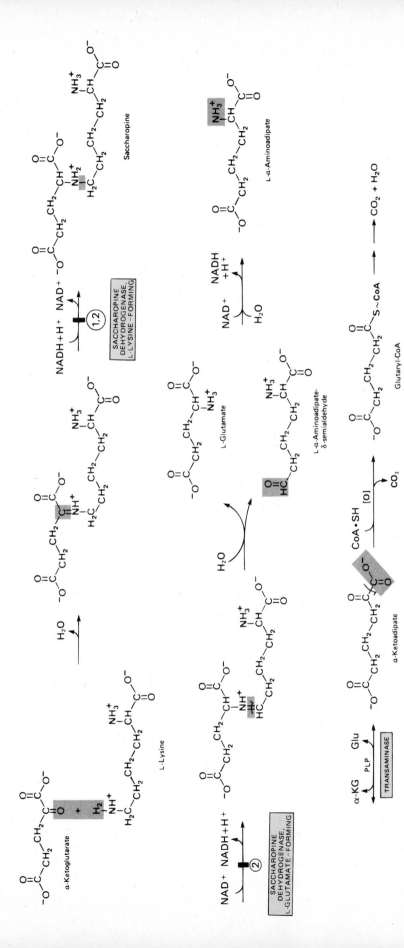

Figure 32–17. Catabolism of L-lysine. (α-KG, α-ketoglutarate; Glu, glutamate; PLP, pyridoxal phosphate.) Numerals indicate the probable sites of the metabolic defects in ① periodic hyperlysinemia with associated hyperammonemia; and ② persistent hyperlysinemia without associated hyperammonemia.

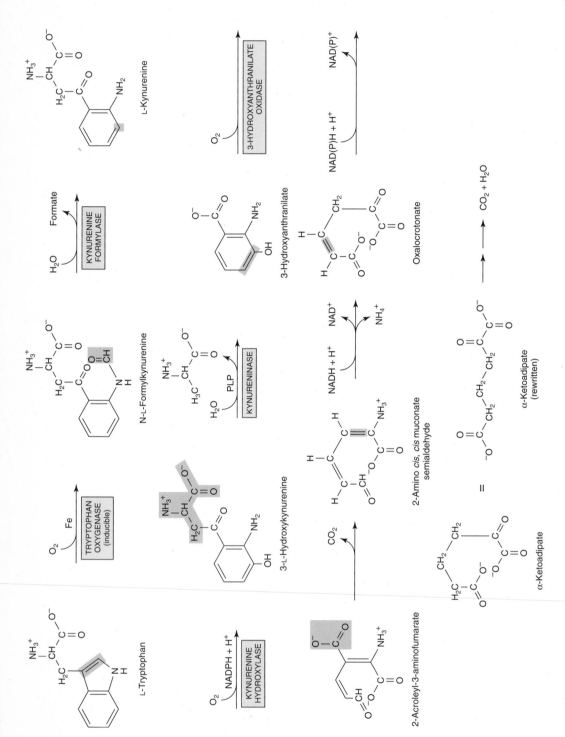

Figure 32–18. Catabolism of L-tryptophan. (PLP, pyridoxal phosphate.)

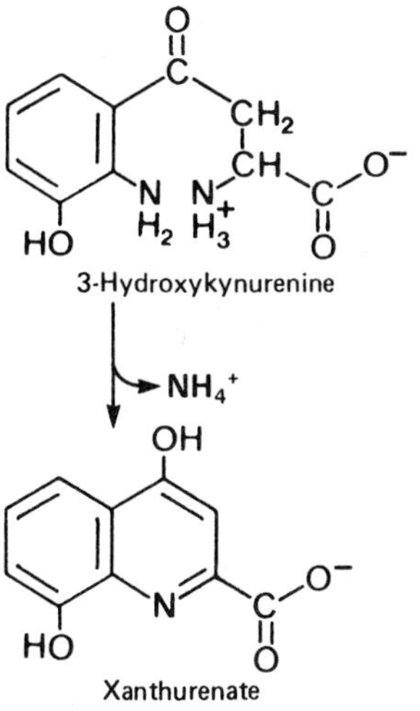

3-Hydroxykynurenine

Xanthurenate

Figure 32–19. Formation of xanthurenate in vitamin B_6 deficiency. Conversion of the tryptophan metabolite 3-hydroxykynurenine to 3-hydroxyanthranilate is impaired (see Figure 32–18). A large portion is therefore converted to xanthurenate.

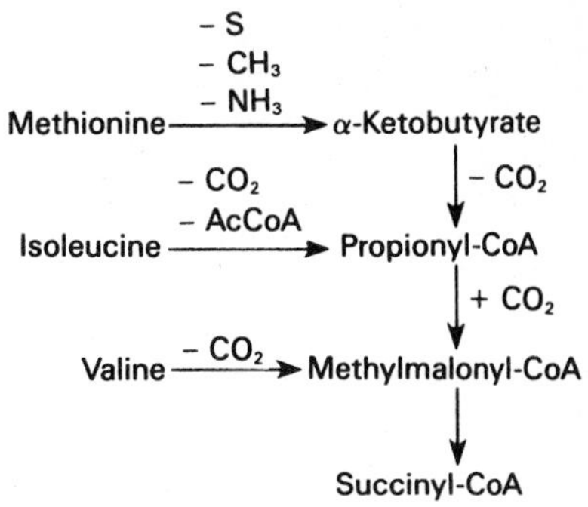

Figure 32–20. Overall catabolism of methionine, isoleucine, and valine.

that arise from intestinal bacterial degradation of unabsorbed tryptophan. The impaired intestinal absorption and renal reabsorption of tryptophan limits the tryptophan available for niacin biosynthesis and accounts for the accompanying pellagra-like signs and symptoms.

METHIONINE, ISOLEUCINE, AND VALINE ARE CATABOLIZED TO SUCCINYL-CoA

While succinyl-CoA is the amphibolic end product for catabolism of methionine, isoleucine, and valine, only portions of the skeletons are converted (Figure 32–20). Four-fifths of the carbons of valine, three-fifths of those of methionine, and half of those of isoleucine form succinyl-CoA. The carboxyl carbons of all three form CO_2. The terminal two carbons of isoleucine form acetyl-CoA, and the methyl group of methionine is removed as such.

What follows relates only to conversion of methionine and isoleucine to propionyl-CoA and of valine to methylmalonyl-CoA. The reactions leading from propionyl-CoA through methylmalonyl-CoA to succinyl-CoA are discussed in Chapter 24 in connection with catabolism of propionate and of fatty acids that contain an odd number of carbon atoms.

Methionyl Carbons Form Propionyl-CoA

L-Methionine first condenses with ATP, forming S-adenosylmethionine, "active methionine" (Figure 32–21). The activated S-methyl group may transfer to various acceptors. Removal of the methyl group forms S-adenosyl homocysteine. Hydrolysis of the S—C bond yields L-homocysteine plus adenosine. Homocysteine then condenses with serine, forming

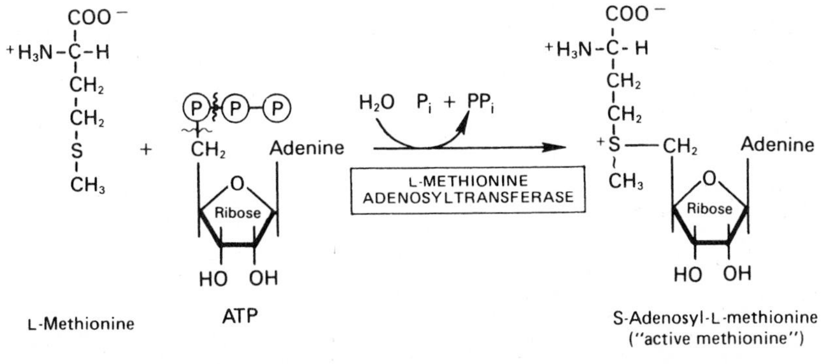

L-Methionine   ATP

S-Adenosyl-L-methionine ("active methionine")

Figure 32–21. Formation of S-adenosylmethionine. ~CH_3 represents the high transfer potential of "active methionine."

cystathionine (Figure 32–22). Hydrolytic cleavage of cystathionine forms L-homoserine plus cysteine, so that the net effect is conversion of homocysteine to homoserine and of serine to cysteine. These two reactions are therefore also involved in biosynthesis of cysteine from serine (see Chapter 30). Homoserine is converted to α-ketobutyrate by **homoserine deaminase** (Figure 32–23). Conversion of α-ketobutyrate to propionyl-CoA then occurs in the usual manner for oxidative carboxylation of α-keto acids (eg, pyruvate, α-ketoglutarate) to form acyl-CoA derivatives. Table 32–2 lists metabolic disorders of methionine catabolism.

THE TWO INITIAL CATABOLIC REACTIONS ARE COMMON TO ALL THREE BRANCHED-CHAIN AMINO ACIDS

Catabolism of leucine, valine, and isoleucine initially involves the same reactions. Subsequently, each amino acid skeleton follows a unique pathway to amphibolic intermediates (Figures 32–24 and 32–25) whose structures determine that valine is glycogenic, leucine is ketogenic, and isoleucine is both. Many of the reactions involved are analogous to reactions of straight- and branched-chain fatty acid catabolism. In what follows, reaction numbers refer to Figures 32–25 through 32–28.

The catabolism of branched-chain amino acids occurs in liver, kidney, muscle, heart, and adipose tissue and begins with their entry into cells via a cell membrane transporter. Following reversible transamination, the resulting α-keto acids enter mitochondria, where they are oxidatively decarboxylated by a single branched-chain α-keto acid decarboxylase, a multienzyme complex loosely associated with the mitochondrial inner membrane. The resulting branched-chain α-ketoacyl-CoA thioesters are then catabolized by distinct pathways.

One Enzyme Transaminates All Three Branched Amino Acids

Transamination of all three branched amino acids is catalyzed by a single transaminase (reaction 1, Figure 32–25). Since this reaction is reversible, the corresponding α-keto acids can replace their amino acids in the diet.

Oxidative Decarboxylation of Branched α-Keto Acids Is Analogous to Conversion of Pyruvate to Acetyl-CoA

A mitochondrial multienzyme complex, branched-chain α-keto acid dehydrogenase, catalyzes the oxidative decarboxylation of the α-keto acids derived from leucine, isoleucine, and valine (reaction 2, Figure 32–25). The structure and regulation of this dehy-

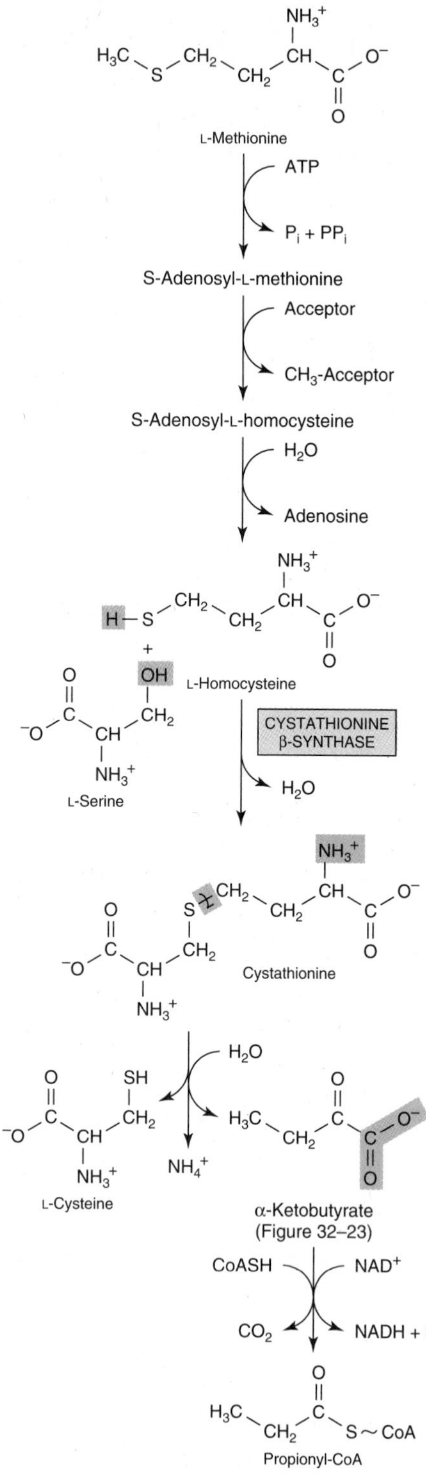

Figure 32–22. Conversion of methionine to propionyl-CoA.

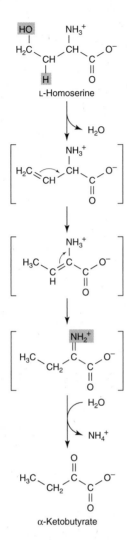

Figure 32–23. Conversion of L-homoserine to α-ketobutyrate, catalyzed by homoserine deaminase.

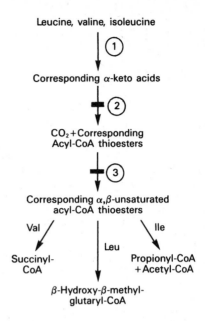

Figure 32–24. Catabolism of the branched-chain amino acids. Reactions ①, ②, and ③ are common to all three amino acids. Lines intersecting arrows mark sites of metabolic blocks in two rare human diseases; at ②, maple syrup urine disease, a defect in catabolism of all three amino acids; and at ③, isovaleric acidemia, a defect of leucine catabolism.

drogenase closely resemble pyruvate dehydrogenase. Its subunits are an α-keto acid decarboxylase, a transacylase, and a dihydrolipoyl dehydrogenase. The complex is inactivated when phosphorylated by ATP and a protein kinase and is reactivated by a Ca^{2+}-independent phosphoprotein phosphatase. The protein kinase is inhibited by ADP, branched-chain α-keto acids, the hypolipidemic agents clofibrate and dichloroacetate, and coenzyme A thioesters.

Dehydrogenation of Branched Acyl-CoA Thioesters Is Analogous to a Reaction of Fatty Acid Catabolism

Reaction 3 is analogous to dehydrogenation of acyl-CoA thioesters in fatty acid catabolism. Indirect

evidence implies at least two enzymes. In **isovaleric acidemia,** ingestion of protein-rich foods results in elevated blood levels of isovalerate, a deacylation product of isovaleryl-CoA.

Three Reactions Are Specific to Leucine Catabolism

A. Reaction 4L: The key to understanding the ketogenic action of leucine was the discovery that 1 mol of CO_2 was "fixed" per mole of isopropyl groups from leucine converted to acetoacetate. This CO_2 fixation (reaction 4L, Figure 32–26) requires biotinyl-CO_2 and forms β-methylglutaconyl-CoA.

B. Reaction 5L: Hydration of β-methylglutaconyl-CoA forms β-hydroxy-β-methylglutaryl-CoA (HMG-CoA), a precursor both of ketone bodies (reaction 6L, Figure 32–26) and of mevalonate and hence of cholesterol and all other polyisoprenoids (Chapter 28).

C. Reaction 6L: Cleavage of HMG-CoA to acetyl-CoA and acetoacetate by HMG-CoA lyase of liver, kidney, and heart mitochondria accounts for the strong ketogenic effect of leucine. Not only is 1 mol of acetoacetate formed per mole of leucine, but another half mole of ketone bodies may be formed from the remaining product, acetyl-CoA (Chapter 24).

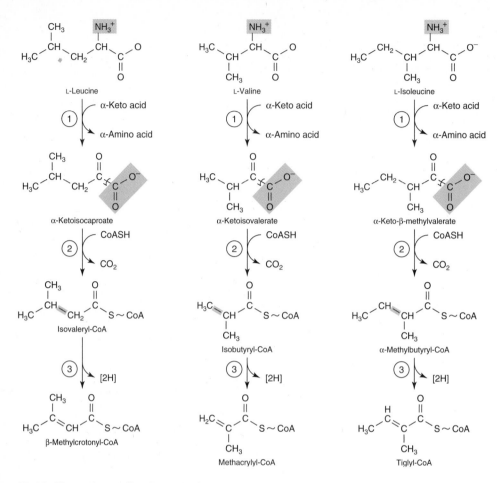

Figure 32–25. The analogous first three reactions in the catabolism of leucine, valine, and isoleucine. Note also the analogy in reactions ② and ③ to the catabolism of fatty acids (see Chapter 24). This analogy continues, as shown in subsequent figures. (α-KA, α-keto acid; α-AA, α-amino acid.)

Four Reactions Are Specific to Valine Catabolism

A. Reaction 4V: Hydration of methylacrylyl-CoA is catalyzed by crotonase, a hydrolase of broad specificity for L-β-hydroxyacyl-CoA thioesters with four to nine carbon atoms (Figure 32–27).

B. Reaction 5V: Since β-hydroxyisobutyryl-CoA is not a substrate for the subsequent reaction, it is deacylated to β-hydroxyisobutyrate.

C. Reaction 6V: The reversible, NAD⁺-dependent oxidation of the primary alcohol group of β-hydroxyisobutyrate to an aldehyde forms methylmalonate semialdehyde.

D. Reaction 7V: Transamination of methylmalonate semialdehyde forms α-aminoisobutyrate, a normal urinary amino acid.

E. Reaction 8V: Alternatively, methylmalonate semialdehyde is oxidatively acylated to methylmalonyl-CoA (reaction 8V).

F. Reaction 9V: Isomerization of methylmalonyl-CoA to succinyl-CoA is catalyzed by methylmalonyl-CoA mutase, an enzyme for which the vitamin B_{12} derivative adenosylcobalamin is the coenzyme. This isomerization reaction also functions in the conversion to succinyl-CoA of the propionyl-CoA derived from isoleucine (Figure 32–28). In vitamin B_{12} deficiency, mutase activity is impaired. This produces a "dietary metabolic defect" in ruminants that utilize propionate from fermentation in the rumen as an energy source.

Three Reactions Are Unique to Isoleucine Catabolism

A. Reaction 4L: Hydration of tiglyl-CoA, like the analogous reaction in valine catabolism (reaction 4V), is catalyzed by crotonase (Figure 32–28).

B. Reaction 5L: Dehydrogenation of α-methyl-β-hydroxybutyryl-CoA is analogous to reaction 5V of valine catabolism.

C. Reaction 6L: Thiolytic cleavage of the bond linking carbons 2 and 3 of α-methylacetoacetyl-CoA resembles thiolysis of acetoacetyl-CoA by β-keto-

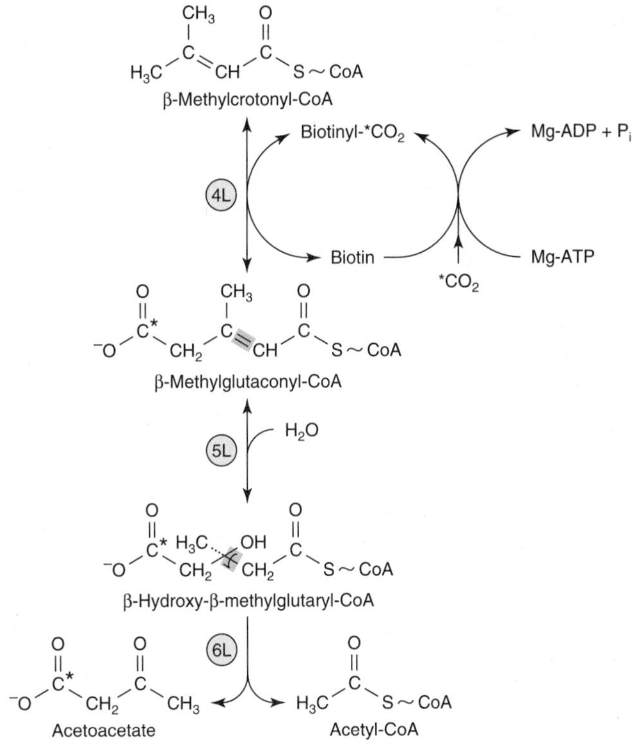

Figure 32–26. Catabolism of the β-methylcrotonyl-CoA formed from L-leucine. *Carbon atoms derived from CO_2.

thiolase. The products, acetyl-CoA (ketogenic) and propionyl-CoA (glycogenic), account for the ketogenic and glycogenic properties of isoleucine.

METABOLIC DISORDERS OF BRANCHED-CHAIN AMINO ACID CATABOLISM

(1) Maple Syrup Urine Disease (Branched-Chain Ketonuria): The most striking feature of this hereditary autosomal recessive disorder (incidence 1:185,000 worldwide) is the odor of the urine, which resembles that of maple syrup or burnt sugar. Plasma and urinary levels of leucine, isoleucine, valine, and their α-keto acids are elevated. Smaller quantities of branched-chain α-hydroxy acids, formed by reduction of the α-keto acids, also occur in the urine. The disease is evident by the end of the first week of extrauterine life. The infant is difficult to feed, may vomit, and may be lethargic. Diagnosis prior to 1 week of age is possible only by enzymatic analysis. Extensive brain damage occurs in surviving children. Without treatment, death usually occurs by the end of 1 year. The biochemical defect is the absence or greatly reduced activity of the **α-keto acid decarboxylase** (reaction 2, Figure 32–25). The

mechanism of toxicity is unknown. Treatment initiated in the first week of life largely averts dire consequences. Therapy involves replacing dietary protein by a mixture of amino acids that excludes leucine, isoleucine, and valine. When plasma levels of these amino acids fall to normal, they are restored in the form of milk and other foods in amounts that never exceed metabolic demand.

When the mutation impairs the dihydrolipoate reductase (E3), there is a combined inability to decarboxylate pyruvate, α-ketoglutarate, and branched-chain α-keto acids. Considering the complexity of the multienzyme complex that converts the branched-chain α-keto acids to their acyl-CoA thioesters, it is not surprising that numerous mutations, many of which have been identified, give rise to maple syrup urine disease. For example, the defect in the Old Order Mennonites of Lancaster and Lebanon counties, Pennsylvania, where the incidence of maple syrup urine disease is 1:175, is a homozygous Tyr 395 to Asn mutation in the E1 α-subunit of the decarboxylase that impairs its association with E1 β-subunits.

(2) Intermittent Branched-Chain Ketonuria: This variant of maple syrup urine disease probably reflects a less severe structural modification of the **α-keto acid decarboxylase.** Since affected individuals

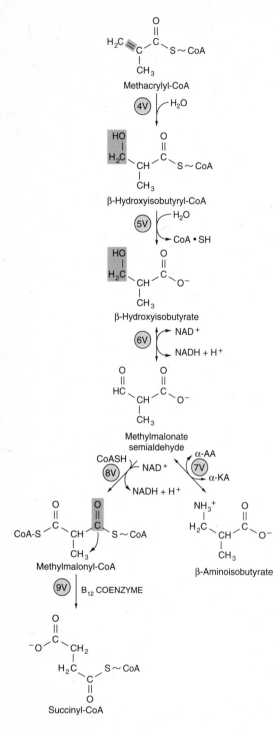

Figure 32–27. Subsequent catabolism of the methacrylyl-CoA formed from L-valine (see Figure 32–25). (α-KA, α-keto acid; α-AA, α-amino acid.)

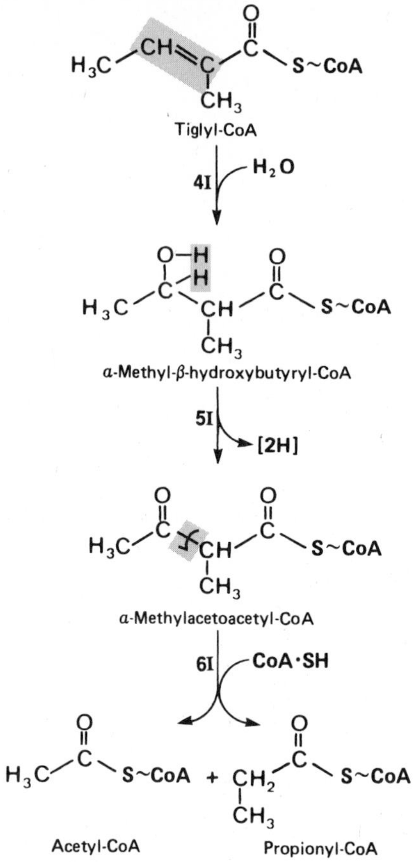

Figure 32–28. Subsequent catabolism of the tiglyl-CoA formed from L-isoleucine.

possess an impaired but distinct capability for catabolism of leucine, valine, and isoleucine, symptoms occur later in life and only intermittently, and the prognosis for dietary therapy is more favorable.

Maple syrup urine disease and intermittent branched-chain ketonuria illustrate mutations that result in different changes in the primary structure of the same enzyme. A spectrum of activities ranging from frank disease through intermittent manifestations to normal values thus can occur in individual subjects.

(3) Isovaleric Acidemia: The impaired enzyme is **isovaleryl-CoA dehydrogenase** (reaction 3, Figure 32–25). Isovaleryl-CoA accumulates, is hydrolyzed to isovalerate, and is excreted in the urine and sweat. Symptoms include a "cheesy" odor of the breath and body fluids, vomiting, acidosis, and coma precipitated by excessive ingestion of protein.

DISORDERS OF METHYLMALONYL-CoA CATABOLISM

(1) Methylmalonic Aciduria: A vitamin B_{12} coenzyme-dependent reaction isomerizes methylmalonyl-CoA to succinyl-CoA (reaction 9V). The methylmalonic aciduria of patients with acquired vitamin B_{12} deficiency disappears when sufficient vitamin B_{12} is administered. Two forms are known: one responds to pharmacologic levels of vitamin B_{12}; the other requires massive doses (1 g/d) of vitamin B_{12}.

(2) Propionic Acidemia: Propionyl-CoA carboxylase deficiency is characterized by high serum propionate levels. Treatment involves a low-protein diet and measures to counteract metabolic acidosis.

SUMMARY

When amino acids are present in excess of metabolic needs, their carbon skeletons are catabolized to amphibolic intermediates for use as sources of energy or as substrates for carbohydrate and lipid biosynthesis. Removal of the α-amino nitrogen by transamination is most often the initial reaction of amino acid catabolism. Subsequent reactions remove any additional nitrogens and restructure the remaining hydrocarbon skeleton for conversion to amphibolic intermediates such as oxaloacetate, α-ketoglutarate, pyruvate, and acetyl-CoA. For example, deamination of asparagine forms aspartate, which becomes oxaloacetate. Glutamine and glutamate form α-ketoglutarate by analogous reactions. No known metabolic disorders are associated with the catabolism of these four amino acids. Two hyperprolinemias are associated with the catabolic pathway that converts proline to α-ketoglutarate. The initial reaction of arginine catabolism is catalyzed by arginase, an enzyme of urea synthesis. Transamination of the δ-nitrogen of ornithine then forms glutamate-γ-semialdehyde, at which point the catabolism of arginine merges with that of histidine.

Six amino acids form pyruvate. Metabolic diseases associated with glycine catabolism include glycinuria and primary hyperoxaluria. Transamination of alanine forms pyruvate directly, while serine is catabolized via glycine. Following reduction of cystine to cysteine, two sets of catabolic reactions convert cysteine to pyruvate. Metabolic diseases associated with cysteine catabolism include cystine-lysinuria, cystine storage disease, and the homocystinurias. Threonine catabolism merges with that of glycine after threonine aldolase cleaves threonine to glycine and acetaldehyde. Oxidation of acetaldehyde forms acetate and ultimately acetyl-CoA. Hydroxyprolinemia is associated with defective ability to degrade 4-hydroxyproline to pyruvate and glyoxylate.

Twelve amino acids form acetyl-CoA. Following transamination, the carbon skeleton of tyrosine is degraded to fumarate and acetoacetate via a protracted reaction sequence. Metabolic diseases associated with enzymatic defects in tyrosine catabolism include tyrosinosis, Richner-Hanhart syndrome, neonatal tyrosinemia, and alkaptonuria. Hydroxylation of phenylalanine forms tyrosine. Metabolic disorders of phenylalanine catabolism include phenylketonuria (PKU) and several hyperphenylalaninemias.

Neither nitrogen of lysine undergoes transamination. The α-amino nitrogen is removed indirectly via saccharopine. Metabolic diseases of lysine catabolism include periodic and persistent forms of hyperlysinemia/ammonemia. The abnormal tryptophan metabolite xanthurenate accumulates in vitamin B_6 deficiency.

Methionine, isoleucine, and valine are catabolized to succinyl-CoA. The catabolism of leucine, valine, and isoleucine presents many analogies to fatty acid catabolism. Oxidative decarboxylation of branched α-keto acids is analogous to conversion of pyruvate to acetyl-CoA, and dehydrogenation of branched acyl-CoA thioesters is analogous to a reaction of fatty acid catabolism. Three reactions are specific to leucine catabolism, four to valine catabolism, and three to isoleucine catabolism. Metabolic disorders of branched-chain amino acid catabolism include hypervalinemia, maple syrup urine disease, intermittent branched-chain ketonuria, isovaleric acidemia, and methylmalonic aciduria.

REFERENCES

Cooper AJL: Biochemistry of the sulfur-containing amino acids. Annu Rev Biochem 1983;52:187.

Harris RA et al: Molecular cloning of the branched-chain α-ketoacid dehydrogenase kinase and the CoA-dependent methylmalonate semialdehyde dehydrogenase. Adv Enzyme Regul 1993;33:255.

Scriver CR et al (editors): *The Metabolic and Molecular Bases of Inherited Disease,* 7th ed. McGraw-Hill, 1995.

33 Conversion of Amino Acids to Specialized Products

Victor W. Rodwell, PhD

INTRODUCTION

This chapter presents examples of the biosynthesis of important nitrogenous compounds, other than proteins, derived from amino acids. In addition to topics discussed here, the reader's attention is directed to processes discussed in detail in other chapters of this book. For these processes, readers are referred to the appropriately cited chapters. The material therefore merges with metabolic pathways discussed elsewhere in this book.

BIOMEDICAL IMPORTANCE

Physiologically important products derived from amino acids include heme, purines, pyrimidines, hormones, neurotransmitters, and biologically active peptides. In addition, many proteins contain amino acids that have been modified for a specific function, eg, calcium binding, or as intermediates that serve to stabilize proteins, generally structural proteins, by subsequent covalent cross-linking. The amino acid residues in those proteins serve as precursors for these modified residues. Finally, there are small peptides or peptide-like molecules not synthesized on ribosomes that carry out specific functions in cells. Histamine, formed by decarboxylation of histidine, plays a central role in many allergic reactions. Specific neurotransmitters derived from amino acids include γ-aminobutyrate from glutamate; 5-hydroxytryptamine (serotonin) from tryptophan; and dopamine, norepinephrine, and epinephrine from tyrosine. Many drugs used to treat neurologic and psychiatric conditions affect the metabolism of the neurotransmitters mentioned above.

GLYCINE PARTICIPATES IN BIOSYNTHESIS OF GLYCINE CONJUGATES, CREATINE, HEME, AND PURINES

(1) Glycine Conjugates: Many metabolites and pharmaceuticals are excreted as water-soluble glycine conjugates. Examples include the conjugated bile acid **glycocholic acid** (Figure 28–7) and **hippuric acid** formed from the food additive benzoate (Figure 33–1). The capacity of the liver to convert benzoate to hippuric acid formerly was used to assess liver function. In addition to benzoate, many drugs and drug metabolites that contain carboxyl groups are excreted in the urine as glycine conjugates.

(2) Creatine: The sarcosine (*N*-methylglycine) component of creatine is derived from glycine and *S*-adenosylmethionine (Figure 33–10).

Topics discussed elsewhere in this book include the following:

(3) Heme: The nitrogen and α-carbon of glycine contribute the nitrogen and both an α-carbon of the pyrrole rings and the methylene bridge carbons of heme (Figure 34–5). Metabolic disorders of heme metabolism are discussed in Chapter 34.

(4) Purines: The entire glycine molecule becomes atoms 4, 5, and 7 of purines (Figure 36–1).

α-ALANINE IS A MAJOR PLASMA AMINO ACID

α-Alanine and glycine together make up a major fraction of the amino nitrogen of human plasma. In bacteria, both L-alanine and D-alanine are major components of cell walls.

MAMMALS CATABOLIZE β-ALANINE VIA MALONATE SEMIALDEHYDE

Most of the β-alanine of mammalian tissues is present as coenzyme A (Figure 52–6) and as β-alanyl dipeptides (see below). While microorganisms form β-alanine by α-decarboxylation of aspartate, in mammalian tissues β-alanine arises as a catabolite of cytosine (Figure 36–16), carnosine, and anserine (Figure 33–2). Mammalian tissues transaminate β-alanine, forming malonate semialdehyde, which is oxidized to acetate and thence to CO_2. In the rare metabolic disorder **hyper-β-alaninemia**, body fluid

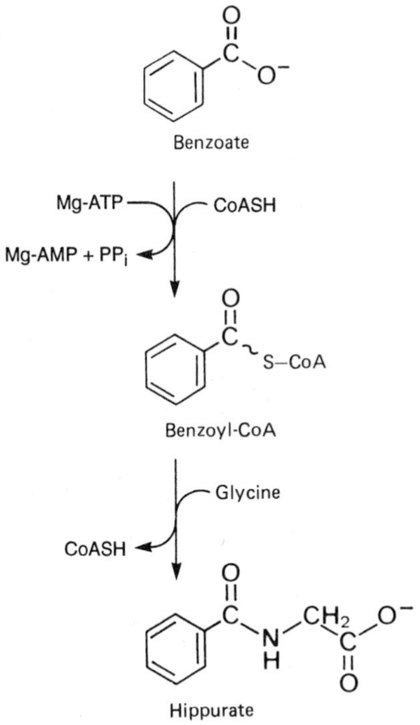

Figure 33–1. Hippurate biosynthesis.

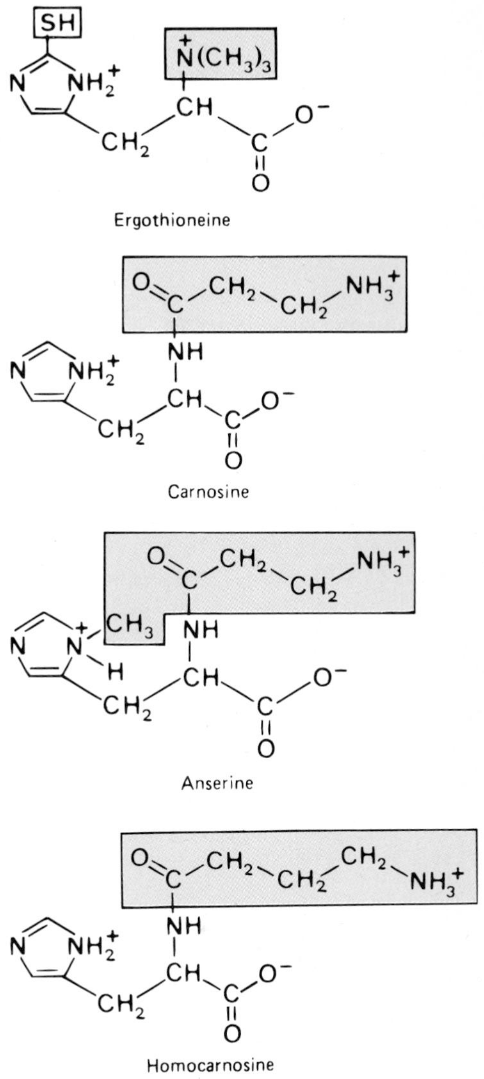

Figure 33–2. Compounds related to histidine. The boxes surround the components not derived from histidine.

and tissue levels of β-alanine, taurine, and β-aminoisobutyrate are elevated.

CARNOSINE IS A β-ALANYL DIPEPTIDE

The β-alanyl dipeptide carnosine (Figure 33–2) occurs in human skeletal muscle. When present in humans, the β-alanyl dipeptide anserine (*N*-methyl-carnosine, Figure 33–2) is derived from the diet, since it occurs in skeletal muscle characterized by rapid contractile activity (eg, rabbit limb and bird pectoral muscle). Anserine thus may fulfill physiologic functions distinct from those of carnosine. β-Alanyl-imidazole buffers the pH of anaerobically contracting skeletal muscle. Carnosine and anserine activate myosin ATPase activity. Both dipeptides also chelate copper and enhance copper uptake.

Short Pathways Synthesize and Degrade β-Alanyl Dipeptides

Biosynthesis of carnosine from β-alanine and L-histidine is catalyzed by carnosine synthetase:

ATP + L-Histidine + β-Alanine →

AMP + PP$_i$ + Carnosine

Some animals, but not humans, use *S*-adenosylmethionine to methylate carnosine, a reaction catalyzed by carnosine *N*-methyltransferase:

S-Adenosylmethionine + Carnosine →

S-Adenosylhomocysteine + Anserine

Carnosine is hydrolyzed to β-alanine and L-histidine by the serum zinc metalloenzyme **carnosinase** (carnosine hydrolase). The heritable disorder **carnosinase deficiency** is characterized by persistent carnosinuria that persists even if carnosine is excluded from the diet.

Homocarnosine Is a Central Nervous System Dipeptide

Present in human brain, where its concentration varies between regions, **homocarnosine** (Figure 33–2) is related structurally and metabolically to carnosine but is present in human brain at levels 100 times higher than carnosine levels. Synthesis of homocarnosine in brain tissue is catalyzed by carnosine synthetase. Serum carnosinase does not, however, hydrolyze homocarnosine.

Homocarnosinosis, an extremely rare genetic disorder presumably due to serum carnosinase deficiency, is associated with progressive spastic paraplegia and mental retardation.

PHOSPHOSERYL AND PHOSPHOTHREONYL RESIDUES OCCUR IN PROTEINS

Much of the serine in phosphoproteins is present as O-phosphoserine. Threonine also occurs in certain proteins as O-phosphothreonine.

Phosphorylation (by protein kinases) and subsequent dephosphorylation (by protein phosphatases) of specific seryl residues serve important regulatory functions. Phosphorylation achieves rapid changes in the activity of key metabolic enzymes, resulting in readily reversible and finely tuned control of metabolite flux through biosynthetic and degradative pathways such as those of carbohydrate and lipid metabolism.

Serine participates in biosynthesis of **sphingosine** (Chapter 26) and also of **purines** and **pyrimidines,** where its β-carbon provides the methyl groups of thymine (and of choline) and of carbons 2 and 8 of purines (Figure 36–2).

Since threonine does not transaminate, neither its α-keto acid nor D-threonine can be utilized by mammals.

S-ADENOSYLMETHIONINE PROVIDES METHYL GROUPS FOR BIOSYNTHESIS

S-Adenosylmethionine is the principal source of methyl groups in the body. In addition, it participates in the biosynthesis of the 3-diaminopropane portions of the polyamines **spermine** and **spermidine** (Figure 33–5).

URINARY SULFATE DERIVES FROM CYSTEINE

Urinary sulfate arises ultimately almost entirely from oxidation of L-cysteine. However, the sulfur of methionine, as homocysteine, is transferred to serine (Figure 30–9) and thus contributes to the urinary sulfate indirectly, ie, via cysteine. L-Cysteine serves as a precursor of the thioethanolamine portion of **coenzyme A** and of the **taurine** that conjugates with bile acids such as taurocholic acid.

DECARBOXYLATION OF HISTIDINE FORMS HISTAMINE

Decarboxylation of histidine forms **histamine,** a reaction catalyzed in mammalian tissues by a broad-specificity **aromatic L-amino acid decarboxylase** that also catalyzes decarboxylation of dopa, 5-hydroxytryptophan, phenylalanine, tyrosine, and tryptophan (see below). α-Methyl amino acids, which inhibit decarboxylase activity, find clinical application as antihypertensive agents. A different enzyme, **histidine decarboxylase,** present in most cells, also catalyzes decarboxylation of histidine.

Histidine compounds present in the human body include ergothioneine, carnosine, and dietary anserine (Figure 33–2). The 1-methylhistidine of normal human urine probably is derived from anserine. 3-Methylhistidine, present in normal human urine at a concentration of about 50 mg/dL, is unusually low in the urine of patients with **Wilson's disease** (Chapter 59).

ORNITHINE AND HENCE ARGININE FORM POLYAMINES

Arginine is the formamidine donor for creatine synthesis in primates (Figure 33–10) and for streptomycin synthesis in *Streptomyces*. Other fates include conversion, via ornithine, to putrescine, spermine, and spermidine (Figure 33–3) and in invertebrate muscle to **arginine phosphate,** a phosphagen whose function as a high-energy phosphate reserve is analogous to that of creatine phosphate in vertebrate muscle.

In addition to its role in urea biosynthesis (Chapter 31), ornithine (with methionine) serves as a precursor of the ubiquitous mammalian and bacterial polyamines **spermidine** and **spermine** (Figure 33–4). Humans synthesize about 0.5 mmol of spermine per day. Spermidine and spermine function in diverse physiologic processes that share as a common thread a close relationship to cell proliferation and growth. They are growth factors for cultured mammalian and bacterial cells and function in the stabilization of intact cells, subcellular organelles, and membranes. Pharmacologic doses of polyamines are hypothermic and hypotensive. Since they bear multiple positive charges, polyamines associate readily with polyanions such as DNA and RNA and may stimulate DNA and RNA biosynthesis, DNA stabilization, and packaging of DNA in bacteriophage.

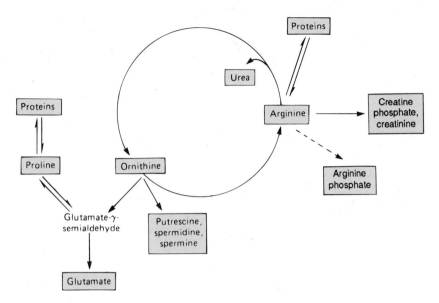

Figure 33–3. Arginine, ornithine, and proline metabolism. Reactions with solid arrows all occur in mammalian tissues. Putrescine and spermine synthesis occurs in both mammals and bacteria. Arginine phosphate of invertebrate muscle functions as a phosphagen analogous to creatine phosphate of mammalian muscle.

Polyamines also exert diverse effects on protein synthesis and inhibit certain enzymes, including protein kinases.

Certain experiments have suggested that polyamines are essential in mammalian metabolism. Addition to cultured mammalian cells of inhibitors of **ornithine decarboxylase** (eg, α-methylornithine or difluoromethylornithine), which catalyzes the initial reaction in polyamine biosynthesis (Figure 33–5), triggers overproduction of ornithine decarboxylase. This suggests an essential physiologic role for this enzyme, whose only known function is polyamine biosynthesis.

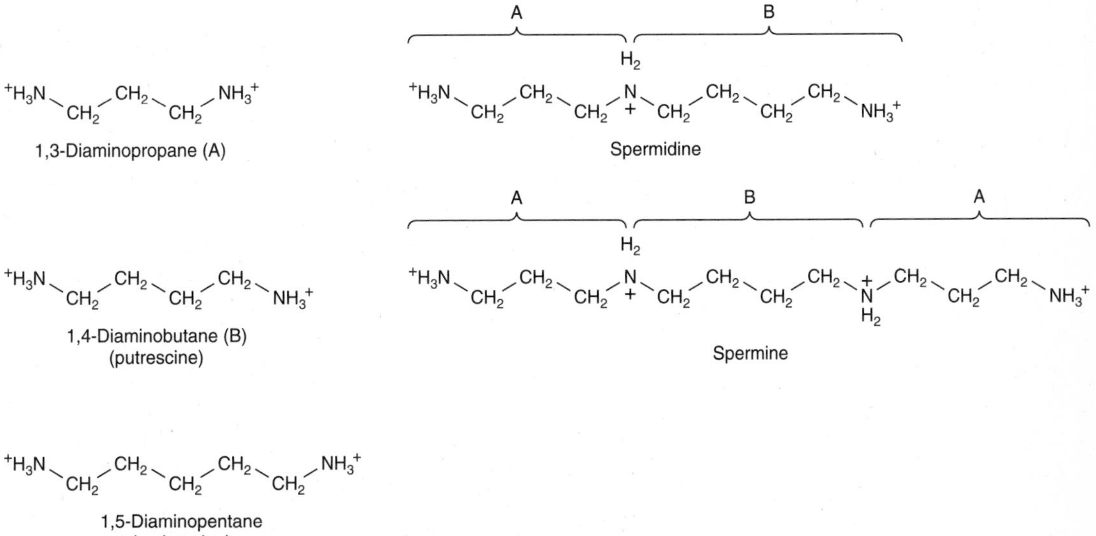

Figure 33–4. Structures of the natural polyamines. Note that spermidine and spermine are polymers of diaminopropane *(A)* and diaminobutane *(B).* Diaminopentane (cadaverine) also occurs in mammalian tissues.

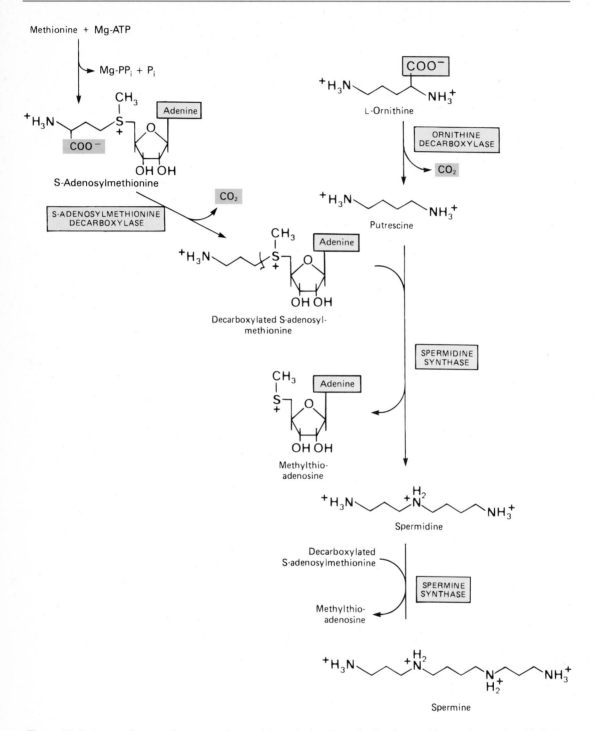

Figure 33–5. Intermediates and enzymes that participate in the biosynthesis of spermidine and spermine. Methylene groups are abbreviated to facilitate visualization of the overall process.

Polyamines Are Important in Cell and Tissue Growth

Figure 33–5 summarizes polyamine biosynthesis in mammalian tissues. The putrescine portion of spermidine and spermine derives from L-ornithine and the diaminopropane portion from L-methionine via *S*-adenosylmethionine. Ornithine decarboxylase and *S*-adenosylmethionine decarboxylase are inducible enzymes with short half-lives. Spermine and spermidine synthases are, by contrast, neither inducible nor unusually labile enzymes.

Of the enzymes of mammalian polyamine biosynthesis two, ornithine decarboxylase and *S*-adenosylmethionine decarboxylase, are of interest with respect to their regulation and their potential for enzyme-directed chemotherapy. The half-life of **ornithine decarboxylase,** about 10 minutes, is among the shortest of any mammalian enzyme, and its activity responds rapidly and dramatically to many stimuli. Ten- to 200-fold increases in ornithine decarboxylase activity follow administration to cultured mammalian cells of growth hormone, corticosteroids, testosterone, or epidermal growth factor. Polyamines added to cultured cells induce synthesis of a protein that inhibits ornithine decarboxylase activity.

S-Adenosylmethionine decarboxylase contains bound pyruvate rather than pyridoxal phosphate as its cofactor, has a half-life of 1–2 hours and responds to promoters of cell growth in a manner qualitatively similar to ornithine decarboxylase. *S*-Adenosylmethionine decarboxylase (Figure 33–5) is inhibited by decarboxylated *S*-adenosylmethionine and activated by putrescine.

Polyamine Catabolites Are Excreted in Urine

Figure 33–6 summarizes the catabolism of polyamines in mammalian tissues. **Polyamine oxidase** of liver peroxisomes oxidizes spermine to spermidine and subsequently oxidizes spermidine to putrescine. Both aminopropane moieties are converted to β-aminopropionaldehyde. Subsequently, putrescine is partially oxidized to NH_4^+ and CO_2 by obscure mechanisms. A major fraction of putrescine and spermidine are, however, excreted in urine as conjugates, principally as acetyl derivatives.

TRYPTOPHAN FORMS SEROTONIN

Hydroxylation of tryptophan to 5-hydroxytryptophan is catalyzed by liver tyrosine hydroxylase. Subsequent decarboxylation forms **serotonin** (5-hydroxytryptamine), a potent vasoconstrictor and stimulator of smooth muscle contraction (Figure 33–7). Catabolism of serotonin is initiated by **monoamine oxidase-**catalyzed oxidative deamination to 5-hydroxyindoleacetate (Figure 33–7), which humans excrete in their urine (2–8 mg/d). The psychic stimulation that

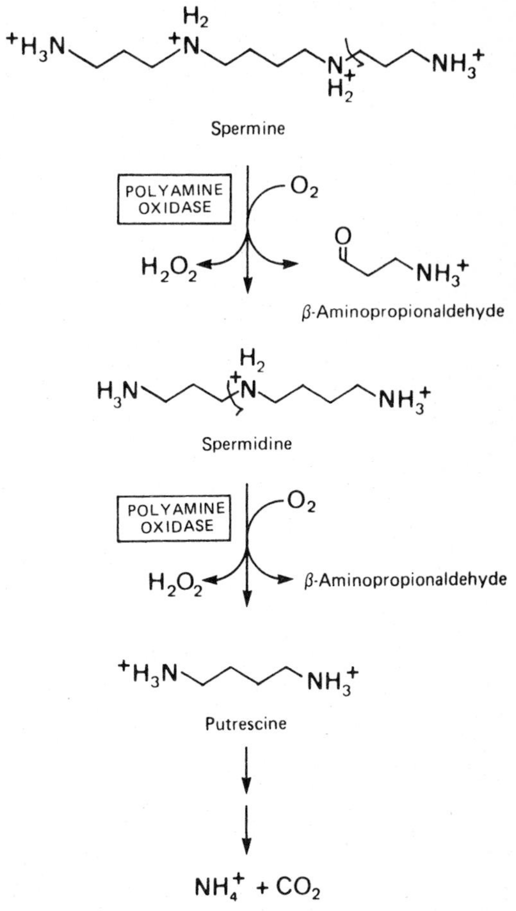

Figure 33–6. Catabolism of polyamines. Structures are abbreviated to facilitate presentation.

follows administration of iproniazid is attributed to its ability to prolong the action of serotonin by inhibiting monoamine oxidase.

Malignant Carcinoid Increases Serotonin Production

In carcinoid (argentaffinoma), serotonin-producing tumor cells in the argentaffin tissue of the abdominal cavity overproduce serotonin. Serotonin catabolites present in the urine of patients with carcinoid include *N*-acetylserotonin glucuronide and the glycine conjugate of 5-hydroxyindoleacetate, 5-hydroxyindoleaceturate. Since the enhanced conversion of tryptophan to serotonin reduces nicotinic acid synthesis, patients with carcinoid may exhibit symptoms of pellagra.

Serotonin Forms Melatonin

N-Acetylation of serotonin, followed by *O*-methylation in the pineal body, forms melatonin (Figure 33–7). Direct methylation of serotonin and of 5-hy-

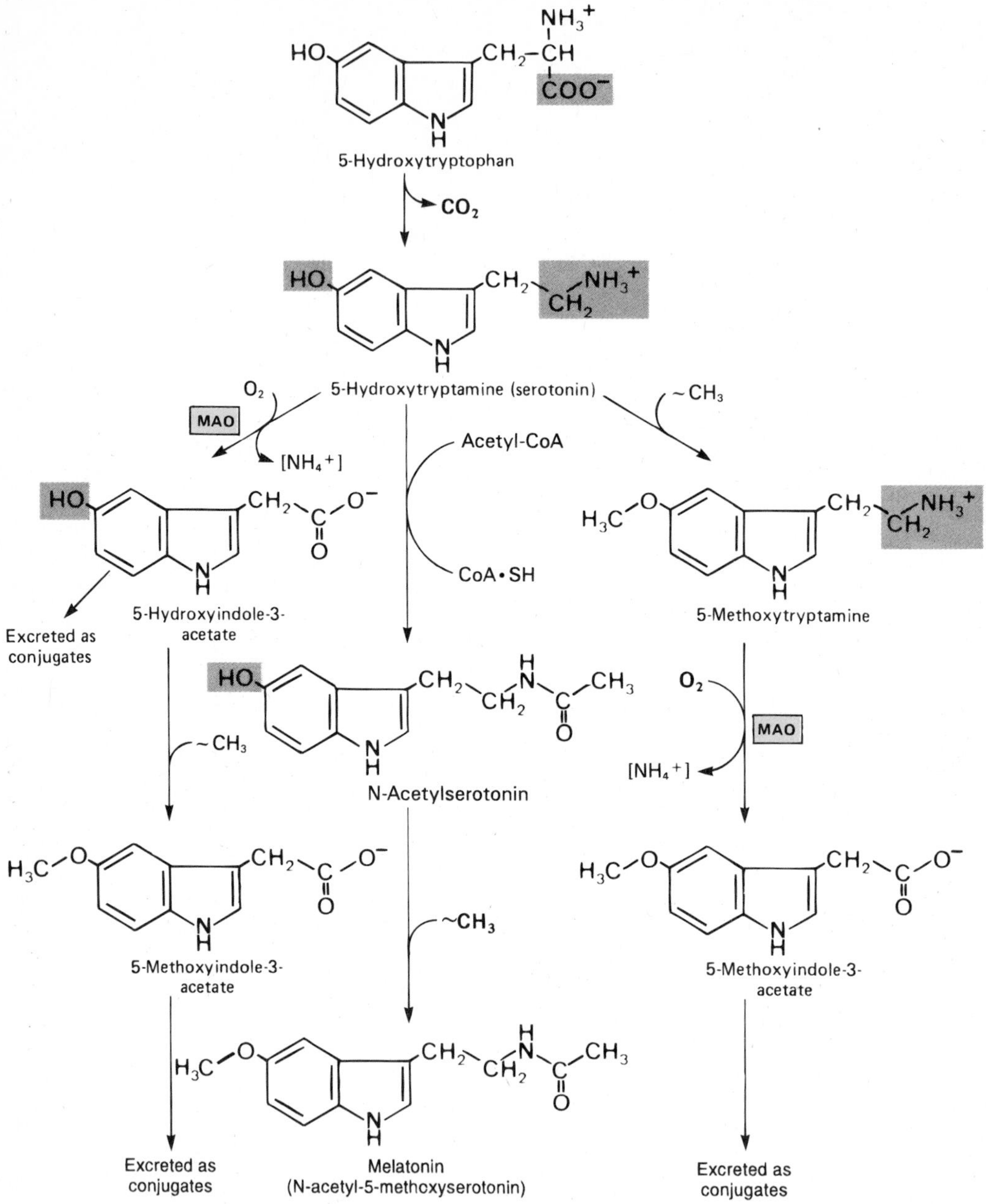

Figure 33–7. Biosynthesis and metabolism of melatonin. ([NH_4^+], by transamination; MAO, monoamine oxidase.)

droxyindoleacetate (Figure 33–7) also occurs. Serotonin and 5-methoxytryptamine are metabolized to the corresponding acids by **monoamine oxidase.** Circulating melatonin is taken up by all tissues, including brain, but is rapidly metabolized by hydroxylation at position 6, followed by conjugation with sulfate or with glucuronic acid.

Tryptophan Metabolites Are Excreted in Urine and Feces

Tryptophan forms additional indole derivatives (Figure 33–7). Mammalian kidney, liver, and human fecal bacteria decarboxylate tryptophan to tryptamine, whose oxidation forms indole-3-acetate. The principal normal urinary catabolites of trypto-

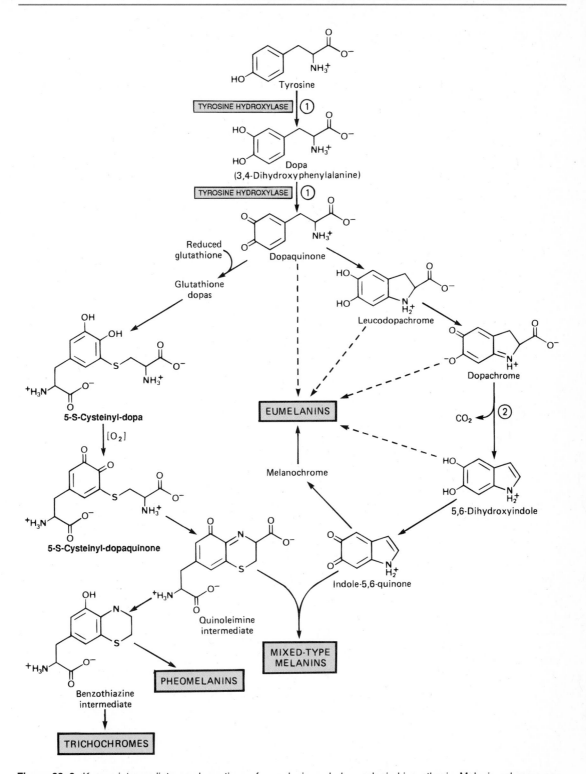

Figure 33–8. Known intermediates and reactions of eumelanin and pheomelanin biosynthesis. Melanin polymers contain both eumelanin and pheomelanin in varying proportions. Dotted arrows indicate that intermediates contribute toward synthesis of eumelanins in varying proportions. Circled numerals indicate probable regulated reactions of the biosynthetic pathway. Reaction ①, catalyzed by tyrosinase, is defective in tyrosinase-negative oculocutaneous albinism.

phan are 5-hydroxyindoleacetate and indole-3-acetate.

MELANINS ARE POLYMERS OF TYROSINE CATABOLITES

Figure 33–8 summarizes the known intermediates in eumelanin and pheomelanin biosynthesis in the membrane-associated melanosomes of pigment cells (melanocytes). The initial reaction is catalyzed by **tyrosine hydroxylase,** a copper-dependent enzyme. The developing eumelanin polymer is thought to entrap free radicals and to undergo partial degradation by H_2O_2 generated during the auto-oxidative process. Pheomelanins and eumelanins then form complexes with proteins of the melanosomal matrix, forming melanoproteins.

Albinism Accompanies Defective Melanin Biosynthesis

The term **albinism** includes a spectrum of clinical syndromes characterized by hypomelanosis due to heritable defects in eye and skin melanocytes. While all ten forms of human **oculocutaneous albinism** can be differentiated on the basis of their clinical, biochemical, ultrastructural, and genetic characteristics, all include decreased pigmentation of the skin and eye. Several are discussed below.

Tyrosine hydroxylase-negative albinos lack all visual pigment. Hair bulbs from these patients fail to convert added tyrosine to pigment, and the melanocytes contain unpigmented melanosomes.

Tyrosine hydroxylase-positive albinos have some visible pigment and white-yellow to light tan hair. Their hair bulb melanocytes may contain lightly pigmented melanosomes, which convert tyrosine to black eumelanin in vitro.

Ocular albinism occurs both as an autosomal recessive and as an X-linked trait. The melanocytes of X-linked and heterozygous (but not autosomal recessive) ocular albinos contain macromelanosomes. The retinas of females heterozygous for X-linked ocular albinism (Nettleship variety) exhibit a mosaic pattern of pigment distribution due to random X-chromosome inactivation. The precise metabolic defects that result in hypomelanosis in ocular albinism are unknown.

Tyrosine Forms Epinephrine and Norepinephrine

Cells of neural origin convert tyrosine to epinephrine and norepinephrine (Figure 33–9). While dopa is an intermediate in the formation of both melanin (Figure 33–8) and norepinephrine (Figure 33–9), different enzymes hydroxylate tyrosine in melanocytes and other cell types. Dopa decarboxylase, a pyridoxal phosphate-dependent enzyme, forms dopamine. Sub-

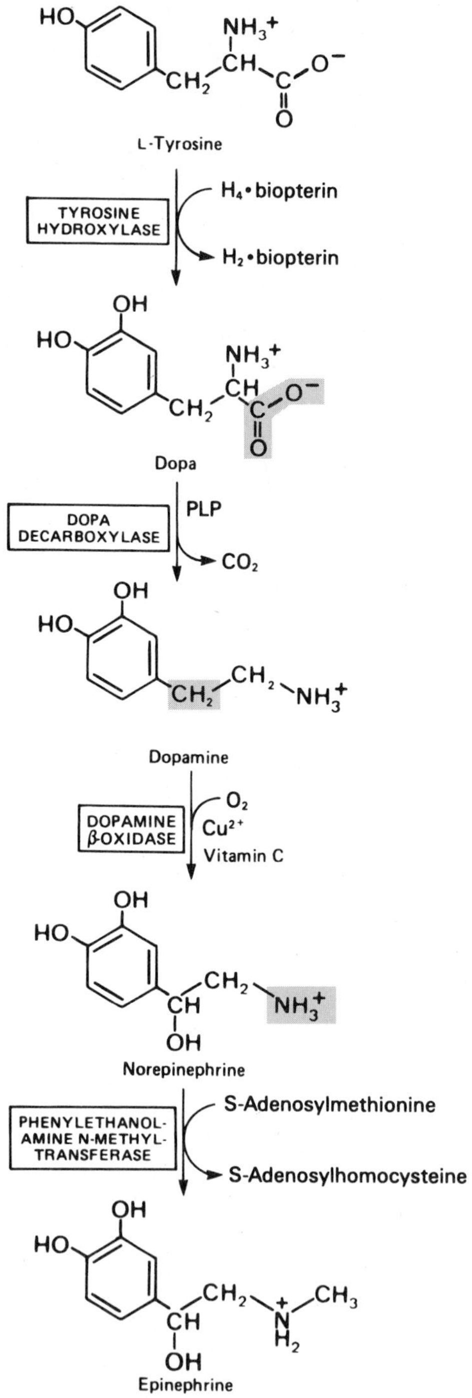

Figure 33–9. Conversion of tyrosine to epinephrine and norepinephrine in neuronal and adrenal cells. (PLP, pyridoxal phosphate.)

sequent hydroxylation by dopamine β-oxidase, a copper-dependent enzyme that requires vitamin C, then forms norepinephrine. In the adrenal medulla, phenylethanolamine-*N*-methyltransferase utilizes *S*-adenosylmethionine to methylate the primary amine of norepinephrine, forming epinephrine (Figure 33–9).

Tyrosine is also a precursor of the thyroid hormones triiodothyronine and thyroxine (see Chapter 46).

CREATININE EXCRETION IS A FUNCTION OF MUSCLE MASS

Both creatine and its energy-reserve form phosphocreatine are present in muscle, brain, and blood. Creatinine (creatine anhydride) is formed in muscle from creatine phosphate by irreversible, nonenzymatic dehydration and loss of phosphate (Figure 33–10). The 24-hour excretion of creatinine in the urine of a given subject is remarkably constant from day to day and proportionate to muscle mass. Traces of creatine also normally occur in urine.

Glycine, arginine, and methionine all participate in creatine biosynthesis. Transfer of a guanidino group from arginine to glycine, forming guanidoacetate (glycocyamine), occurs in the kidney but not in the liver or in heart muscle. Synthesis of creatine is completed by methylation of guanidoacetate by *S*-adenosylmethionine in the liver (Figure 33–10).

FORMATION AND CATABOLISM OF γ-AMINOBUTYRATE

While it occurs in kidney and pancreatic islet cells, γ-aminobutyric acid (GABA) occurs principally in brain tissue, where it functions as an inhibitory neurotransmitter by altering transmembrane potential differences.

Biosynthesis

γ-Aminobutyrate (GABA) is formed by decarboxylation of L-glutamate, a reaction catalyzed by the pyridoxal phosphate-dependent enzyme **L-glutamate decarboxylase** (Figure 33–11) present in tissues of the central nervous system, principally the gray matter. Two reaction sequences also convert putrescine (Figure 33–4) to γ-aminobutyrate, either by deamination by diamine oxidase or via *N*-acetylated intermediates. The relative importance of these three routes of γ-aminobutyrate biosynthesis varies among tissues and with developmental stage.

Catabolism

Transamination of γ-aminobutyrate, catalyzed by **γ-aminobutyrate transaminase,** forms succinate semialdehyde (Figure 33–11). Succinate semialdehyde may then undergo reduction to γ-hydroxybutyrate, a reaction catalyzed by **L-lactate dehydrogenase,** or oxidation to succinate and thence via the citric acid cycle to CO_2 and H_2O.

An extremely rare genetic disorder of GABA me-

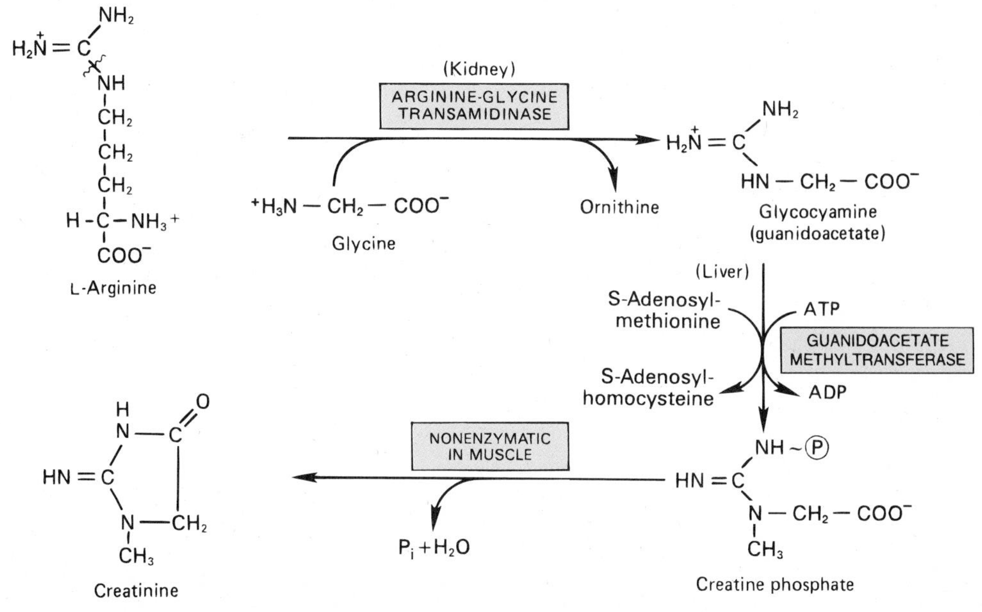

Figure 33–10. Biosynthesis of creatine and creatinine.

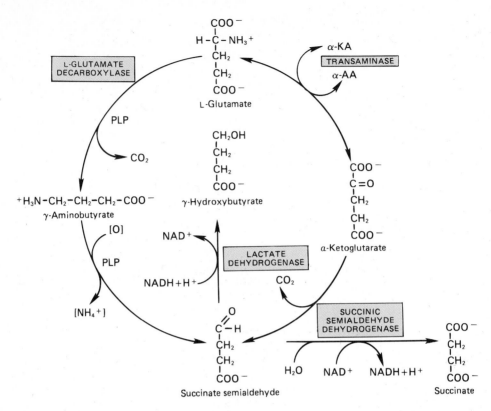

Figure 33–11. Metabolism of γ-aminobutyrate. (α-KA, α-keto acids; α-AA, α-amino acids; PLP, pyridoxal phosphate.)

tabolism involves a defective GABA transaminase, an enzyme that participates in the catabolism of GABA subsequent to its postsynaptic release in brain tissue.

SUMMARY

In addition to fulfilling specific structural and functional roles in proteins and polypeptides, amino acids participate in a wide variety of biosynthetic processes discussed throughout this book. For example, glycine participates in the biosynthesis of heme, purines, and creatine and is conjugated to bile acids and to the urinary metabolites of many drugs. In addition to its roles in phospholipid and sphingosine biosynthesis, serine provides carbons 2 and 8 of purines and the methyl group of thymine. The me-

thionine derivative *S*-adenosylmethionine, the methyl group donor for many biosynthetic processes, also participates directly in spermine and spermidine biosynthesis. Glutamate and ornithine form the neurotransmitter γ-aminobutyrate (GABA). Both the thioethanolamine of coenzyme A and the taurine of taurocholic and other taurine-linked bile acids arise from cysteine. Decarboxylation of histidine forms histamine, while several dipeptides are derived from histidine and the nonprotein amino acid and cytosine catabolite β-alanine. Arginine serves as the formamidine donor for creatine biosynthesis and, via ornithine, participates in polyamine biosynthesis. Important tryptophan metabolites include serotonin and melanin. The potent vasoconstrictor serotonin in turn serves as a precursor of melatonin. Tyrosine forms both epinephrine and norepinephrine, and its iodination forms thyroid hormone.

REFERENCES

Scriver CR et al (editors): *The Metabolic and Molecular Bases of Inherited Disease,* 7th ed. McGraw-Hill, 1995.

Tabor CW, Tabor H: Polyamines. Annu Rev Biochem 1984;53:749.

Porphyrins & Bile Pigments

34

Robert K. Murray, MD, PhD

INTRODUCTION

The biochemistry of the porphyrins and of the bile pigments is presented in this chapter. These topics are closely related, because heme is synthesized from porphyrins and iron, and the products of degradation of heme are the bile pigments and iron.

BIOMEDICAL IMPORTANCE

Knowledge of the biochemistry of the porphyrins and of heme is basic to understanding the varied functions of hemoproteins (see below) in the body. The **porphyrias** are a group of diseases caused by abnormalities in the pathway of biosynthesis of the various porphyrins. They are not very prevalent, but physicians must be aware of them, and in particular dermatologists, hepatologists, and psychiatrists will encounter patients with these conditions. A much more prevalent clinical condition is **jaundice,** due to elevation of bilirubin in the plasma. This elevation is due to overproduction of bilirubin or to failure of its excretion and is seen in numerous diseases, ranging from hemolytic anemias to viral hepatitis and to cancer of the pancreas.

METALLOPORPHYRINS AND HEMOPROTEINS ARE IMPORTANT IN NATURE

Porphyrins are cyclic compounds formed by the linkage of four pyrrole rings through methenyl bridges (—HC═; Figure 34–1). A characteristic property of the porphyrins is the formation of complexes with metal ions bound to the nitrogen atom of the pyrrole rings. Examples are the **iron porphyrins** such as **heme** of hemoglobin and the **magnesium**-containing porphyrin **chlorophyll,** the photosynthetic pigment of plants.

Proteins that contain heme (hemoproteins) are widely distributed in nature. Examples of importance in humans and animals are listed in Table 34–1.

Natural Porphyrins Have Substituent Side Chains on the Porphin Nucleus

The porphyrins found in nature are compounds in which various **side chains** are substituted for the eight hydrogen atoms numbered in the porphin nucleus shown in Figure 34–1. As a simple means of showing these substitutions, Fischer proposed a shorthand formula in which the methenyl bridges are omitted and each pyrrole ring is shown as a bracket with the eight substituent positions numbered as shown (Figure 34–2). Various porphyrins are represented in Figures 34–2, 34–3, and 34–4.

The arrangement of the acetate (A) and propionate (P) substituents in the uroporphyrin shown in Figure 34–2 is asymmetric (in ring IV, the expected order of the A and P substituents is reversed). A porphyrin with this type of **asymmetric substitution** is classified as a type III porphyrin. A porphyrin with a completely symmetric arrangement of the substituents is classified as a type I porphyrin. Only types I and III are found in nature, and the type III series is by far more abundant (Figure 34–3) and more important, because it includes heme.

Heme and its immediate precursor, protoporphyrin IX (Figure 34–4), are both type III porphyrins (ie, the methyl groups are asymmetrically distributed, as in type III coproporphyrin). However, they are sometimes identified as belonging to series IX, because they were designated ninth in a series of isomers postulated by Hans Fischer, the pioneer worker in the field of porphyrin chemistry.

HEME IS SYNTHESIZED FROM SUCCINYL-CoA AND GLYCINE

Heme is synthesized in living cells by a pathway that has been much studied. The two starting materials are **succinyl-CoA,** derived from the citric acid cycle in mitochondria, and the amino acid **glycine.** Pyridoxal phosphate is also necessary in this reaction to "activate" glycine. The product of the condensa-

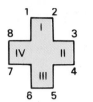

Pyrrole

Porphin
(C₂₀H₁₄N₄)

Figure 34–1. The porphin molecule. Rings are labeled I, II, III, IV. Substituent positions on rings are labeled 1, 2, 3, 4, 5, 6, 7, 8. Methenyl bridges (—HC=) are labeled α, β, γ, δ.

Figure 34–2. Uroporphyrin III. A (acetate) = — CH₂COOH; P (propionate) = — CH₂CH₂COOH.

tion reaction between succinyl-CoA and glycine is α-amino-β-ketoadipic acid, which is rapidly decarboxylated to form δ-aminolevulinate (ALA) (Figure 34–5). This reaction sequence is catalyzed by **ALA synthase,** which is the rate-controlling enzyme in porphyrin biosynthesis in mammalian liver. Synthesis of ALA occurs in **mitochondria.** In the cytosol, two molecules of ALA are condensed by the enzyme **ALA dehydratase** to form two molecules of water and one of **porphobilinogen** (PBG) (Figure 34–5). ALA dehydratase is a zinc-containing enzyme and is sensitive to inhibition by **lead,** as can occur in lead poisoning.

The formation of a tetrapyrrole, ie, a porphyrin, occurs by condensation of four molecules of PBG (Figure 34–6). These four molecules condense in a head-to-tail manner to form a linear tetrapyrrole, hy-

droxymethylbilane. The reaction is catalyzed by uroporphyrinogen I synthase, also known as PBG deaminase. Hydroxymethylbilane cyclizes spontaneously to form **uroporphyrinogen I** (left-hand side of Figure 34–6), or is converted to **uroporphyrinogen III** by the combined action of uroporphyrinogen I synthase and uroporphyrinogen III cosynthase (right-hand side of Figure 34–6). Under normal conditions, the uroporphyrinogen formed is almost exclusively the III isomer, but in certain of the porphyrias (discussed below), the type I isomers of porphyrinogens are formed in excess.

Note that both of these uroporphyrinogens have the pyrrole rings connected by methylene bridges (—CH₂—), which do not form a conjugated ring system. Thus, these compounds (as are all porphyrinogens) are colorless. However, the porphyrinogens are readily auto-oxidized to their respective colored porphyrins. These oxidations are catalyzed by light and by the porphyrins that are formed.

Uroporphyrinogen III is converted to coproporphyrinogen III by decarboxylation of all of the acetate (A) groups, which changes them to methyl (M) substituents. The reaction is catalyzed by **uroporphyrinogen decarboxylase,** which is also capable of converting uroporphyrinogen I to coproporphyrinogen I (Figure 34–7). Coproporphyrinogen III then enters the mitochondria, where it is converted to **protoporphyrinogen III** and then to **protoporphyrin III.** Several steps are involved in this conversion. The mitochondrial enzyme **coproporphyrinogen oxidase** catalyzes the decarboxylation and oxidation of two propionic side chains to form protoporphyrinogen. This enzyme is able to act only on type III coproporphyrinogen, which would explain why type I protoporphyrins do not generally occur in nature. The oxidation of protoporphyrinogen to protoporphyrin is catalyzed by another mitochondrial enzyme, **protoporphyrinogen oxidase.** In mammalian liver, the conversion of coproporphyrinogen to protoporphyrin requires molecular oxygen.

Formation of Heme Involves Incorporation of Iron Into Protoporphyrin

The final step in heme synthesis involves the incorporation of ferrous iron into protoporphyrin in a

Table 34–1. Examples of some important human and animal hemoproteins.[1]

Protein	Function
Hemoglobin	Transport of oxygen in blood
Myoglobin	Storage of oxygen in muscle
Cytochrome c	Involvement in electron transport chain
Cytochrome P450	Hydroxylation of xenobiotics
Catalase	Degradation of hydrogen peroxide
Tryptophan pyrrolase	Oxidation of tryptophan

[1]The functions of the above proteins are described in various chapters of this text.

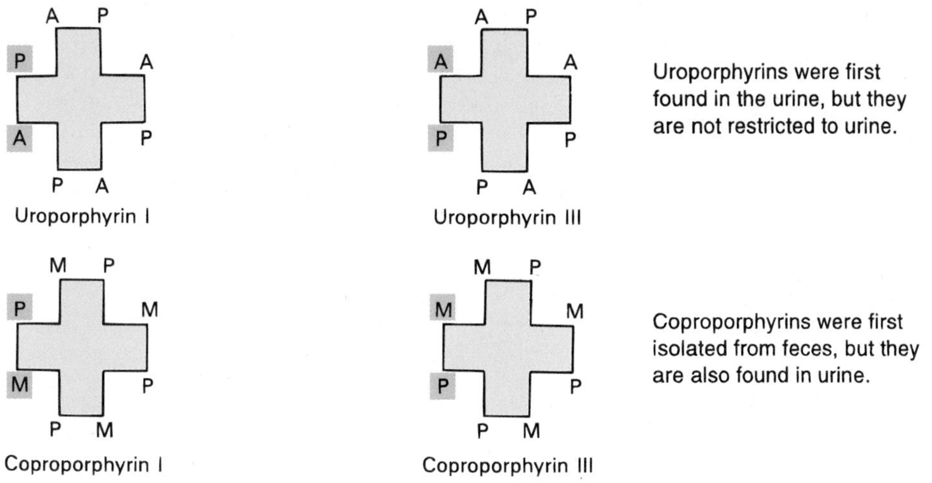

Uroporphyrin I

Uroporphyrin III

Uroporphyrins were first found in the urine, but they are not restricted to urine.

Coproporphyrin I

Coproporphyrin III

Coproporphyrins were first isolated from feces, but they are also found in urine.

Figure 34–3. Uroporphyrins and coproporphyrins. A (acetate); P (propionate); M (methyl) = —CH$_3$; V (vinyl) = —CH = CH$_2$.

reaction catalyzed by **heme synthase** or **ferrochelatase,** another mitochondrial enzyme (Figure 34–4).

A summary of the steps in the biosynthesis of the porphyrin derivatives from PBG is given in Figure 34–8. Heme biosynthesis occurs in most mammalian cells with the exception of mature erythrocytes, which do not contain mitochondria.

The porphyrinogens described above are colorless, containing six extra hydrogen atoms as compared to the corresponding colored porphyrins. These **reduced porphyrins** (the porphyrinogens) and not the corresponding porphyrins are the actual intermediates in the biosynthesis of protoporphyrin and of heme.

ALA Synthase Is the Key Regulatory Enzyme in Heme Biosynthesis

The rate-limiting reaction in the synthesis of heme is that catalyzed by ALA synthase (Figure 34–5), a regulatory enzyme. It appears that heme, probably acting through an aporepressor molecule, acts as a negative regulator of the synthesis of ALA synthase. This repression and derepression mechanism is depicted diagrammatically in Figure 34–9. It is possible that there is also significant feedback inhibition at this step, but the major regulatory effect of heme appears to be one in which the rate of synthesis of ALA synthase increases greatly in the absence of heme and is diminished in its presence. The turnover rate of ALA synthase is normally rapid (half-life is about 1 hour) in mammalian liver, a common feature of an enzyme catalyzing a rate-limiting reaction.

Many drugs (see Chapter 61), when administered to humans, can result in a marked increase in hepatic ALA synthase. Most of these drugs are metabolized by a system in the liver that utilizes a specific hemoprotein, **cytochrome P450**. During their metabolism, the utilization of heme by cytochrome P450 is greatly increased, which in turn diminishes the intracellular heme concentration. This latter event effects

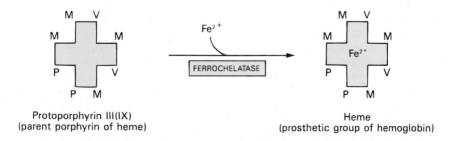

Protoporphyrin III(IX)
(parent porphyrin of heme)

Heme
(prosthetic group of hemoglobin)

Figure 34–4. Addition of iron to protoporphyrin to form heme.

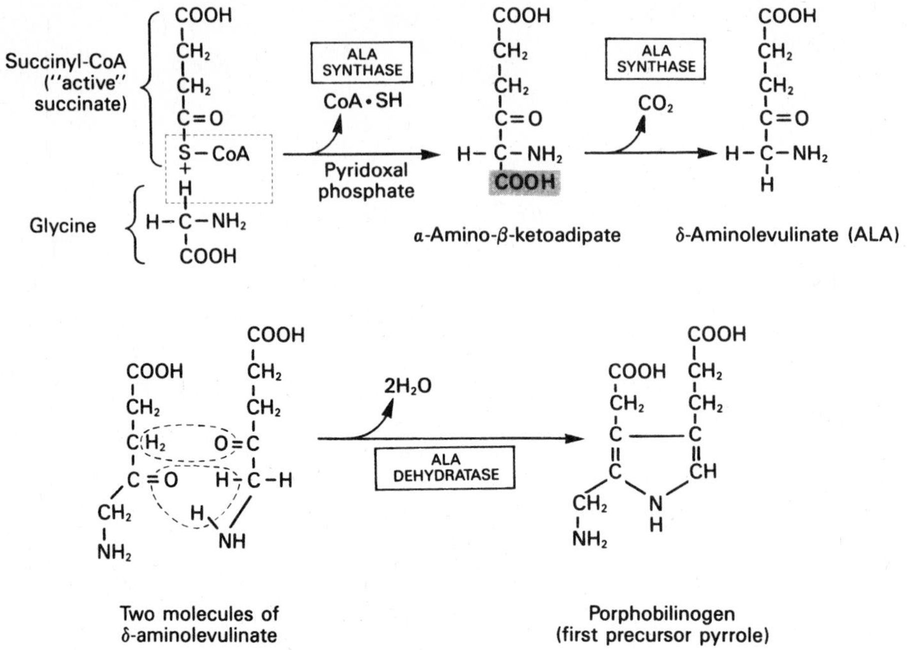

Figure 34–5. Biosynthesis of porphobilinogen. ALA synthase occurs in the mitochondria, whereas ALA dehydratase is present in the cytosol.

a derepression of ALA synthase with a corresponding increased rate of heme synthesis to meet the needs of the cells.

Several factors affect drug-mediated derepression of ALA synthase in the liver. In particular, the administration of glucose can prevent it, as can the administration of hematin (an oxidized form of heme).

The importance of some of these regulatory mechanisms is further discussed below when the porphyrias are described.

PORPHYRINS ARE COLORED AND FLUORESCE

The various porphyrinogens are colorless, whereas the various porphyrins are all colored. In the study of porphyrins or porphyrin derivatives, the characteristic absorption spectrum that each exhibits, in both the visible and the ultraviolet regions of the spectrum, is of great value. An example is the absorption curve for a solution of porphyrin in 5% hydrochloric acid (Figure 34–10). Note particularly the sharp absorption band near 400 nm. This is a distinguishing feature of the porphin ring and is characteristic of all porphyrins regardless of the side chains present. This band is termed the **Soret band,** after its discoverer.

When porphyrins dissolved in strong mineral acids or in organic solvents are illuminated by ultraviolet light, they emit a strong red fluorescence. This **fluo-**rescence is so characteristic that it is frequently used to detect small amounts of free porphyrins. The double bonds joining the pyrrole rings in the porphyrins are responsible for the characteristic absorption and fluorescence of these compounds; these double bonds are absent in the porphyrinogens.

An interesting application of the photodynamic properties of porphyrins is their possible use in the treatment of certain types of cancer, a procedure called **cancer phototherapy.** Tumors often take up more porphyrins than do normal tissues. Thus, hematoporphyrin or other related compounds are administered to a patient with an appropriate tumor. The tumor is then exposed to an argon laser, which excites the porphyrins, producing cytotoxic effects.

Spectrophotometry Is Used to Test for Porphyrins and Their Precursors

Coproporphyrins and uroporphyrins are of clinical interest because they are excreted in increased amounts in the porphyrias. These compounds, when present in urine or feces, can be separated from each other by extraction with appropriate solvent mixtures. They can then be identified and quantified using spectrophotometric methods.

ALA and PBG can also be measured in urine by appropriate colorimetric tests.

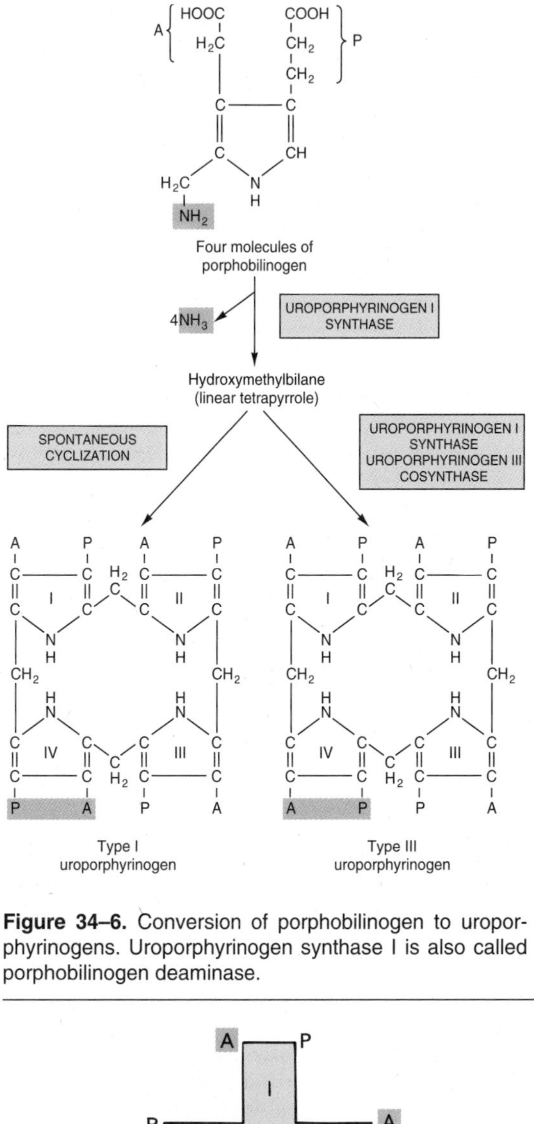

THE PORPHYRIAS ARE GENETIC DISORDERS OF HEME METABOLISM

The **porphyrias** are a group of inborn errors of metabolism due to mutations in the genes directing the synthesis of the enzymes involved in the biosynthesis of heme. They are not prevalent, but it is important to consider them in certain circumstances (eg, in the differential diagnosis of abdominal pain and of a variety of neuropsychiatric findings); otherwise patients will be subjected to inappropriate treatments. It has been speculated that King George III had variegate porphyria, which may account for his periodic confinements in Windsor Castle and possibly some of his views regarding American colonists. Also, the **photosensitivity** (favoring nocturnal activities) and severe **disfigurement** exhibited by some victims of congenital erythropoietic porphyria have led to the suggestion that these individuals may have been the prototypes of werewolves.

Biochemistry Underlies the Causes, Diagnoses, and Treatments of the Porphyrias

Six major types of porphyria have been described, resulting from depressions in the activities of enzymes 3 through 8 shown in Figure 34–9 (see also Table 34–1). Assay of the activity of one or more of these enzymes using an appropriate source (eg, red blood cells) is thus important in making a definitive diagnosis in a suspected case of porphyria. Individuals with low activities of enzyme 1 (ALA synthase) have not been reported. Subjects exhibiting depres-

Figure 34–6. Conversion of porphobilinogen to uroporphyrinogens. Uroporphyrinogen synthase I is also called porphobilinogen deaminase.

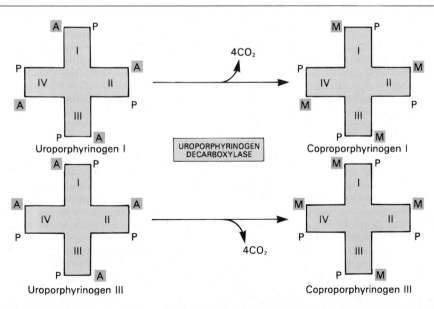

Figure 34–7. Decarboxylation of uroporphyrinogens to coproporphyrinogens in cytosol. (A, acetyl; M, methyl; P, propionyl.)

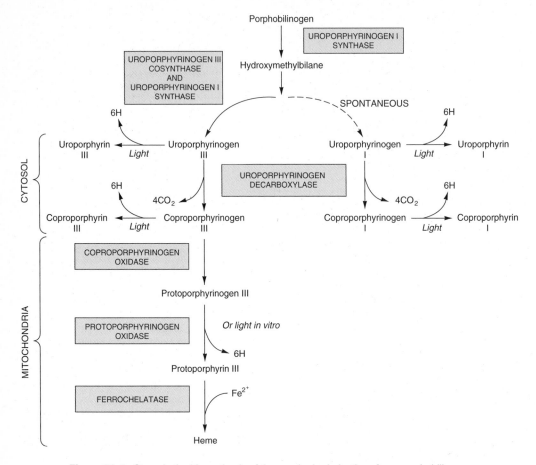

Figure 34–8. Steps in the biosynthesis of the porphyrin derivatives from porphobilinogen.

sions of the activity of enzyme 2 (ALA dehydratase) have been reported but are very rare.

In general, the porphyrias described are inherited in an autosomal dominant manner, with the exception of congenital erythropoietic porphyria, which is inherited in a recessive mode. The precise abnormalities in the genes directing synthesis of the enzymes involved in heme biosynthesis are being determined by the methods of recombinant DNA technology.

As is true of most inborn errors, the clinical signs and symptoms of porphyria result from either a deficiency of metabolic products beyond the enzymatic block or from an accumulation of metabolites behind the block.

Where the enzyme lesion occurs early in the pathway prior to the formation of porphyrinogens (eg, enzyme 3 of Figure 34–9, in intermittent acute porphyria), ALA and PBG will accumulate in body tissues and fluids (Figure 34–11). One or both of these compounds can cause toxic effects in abdominal nerves and in the central nervous system, resulting in the abdominal pain and neuropsychiatric

symptoms seen in this type of porphyria. Possible biochemical bases for these symptoms are that ALA may inhibit an ATPase in nervous tissue or that ALA may be taken up by brain and somehow cause a conduction paralysis.

On the other hand, enzyme blocks later in the pathway result in the accumulation of the porphyrinogens indicated in Figures 34–9 and 34–11. Their oxidation products, the corresponding porphyrin derivatives, cause photosensitivity, a reaction to visible light of about 400 nm. The porphyrins, when exposed to light of this wavelength, are thought to become "excited" and then react with molecular oxygen to form oxygen radicals. These latter species injure lysosomes and other organelles. Damaged lysosomes release their degradative enzymes, causing variable degrees of skin damage, including scarring.

The porphyrias can be classified on the basis of the organs or cells that are most affected. These are generally organs or cells in which synthesis of heme is particularly active. The bone marrow synthesizes

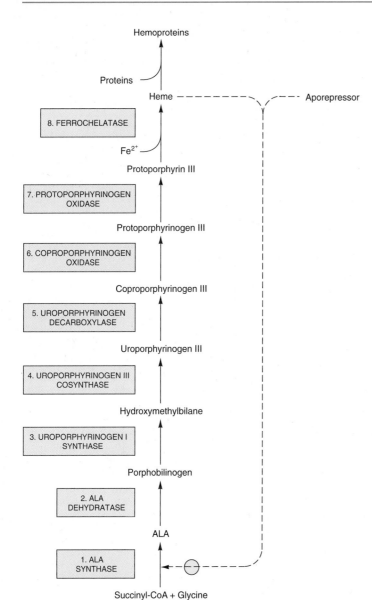

Figure 34–9. Intermediates, enzymes, and regulation of heme synthesis. The enzyme numbers are those referred to in Table 34–1. Enzymes 1, 6, 7, and 8 are located in mitochondria, the others in the cytosol. Mutations of enzymes 2–8 cause the porphyrias, though only a few cases due to deficiency of enzyme 2 have been reported. Regulation of heme synthesis occurs at ALA synthase by a repression-derepression mechanism mediated by heme and its hypothetical aporepressor. The dotted lines indicate the negative ($\ominus$) regulation by repression.

considerable hemoglobin and the liver is active in the synthesis of another hemoprotein, cytochrome P450. Thus, one classification of the porphyrias is to designate them as erythropoietic, hepatic, and erythrohepatic (mixed); the types of porphyria that fall into these classes are indicated in Table 34–1. Why do specific types of porphyria affect certain organs more markedly than others? A partial answer is that the levels of metabolites that cause damage (eg, ALA, PBG or specific porphyrins) can vary markedly in different organs or cells, depending upon the differing activities of their heme-forming enzymes.

As described above, **ALA synthase** is the key regulatory enzyme of the heme biosynthetic pathway.

Although this enzyme is not directly implicated as a cause of the porphyrias, it is important to understand its regulation in order to comprehend some features of these diseases. ALA synthase is subject to both induction and repression, and its activity can increase markedly (up to 50-fold) under certain conditions. A large number of drugs (eg, barbiturates, griseofulvin) induce the enzyme. Most of these drugs do so by inducing cytochrome P450 (see Chapter 61), which uses up heme and thus derepresses (induces) ALA synthase. In patients with porphyria, increased activities of ALA synthase result in increased levels of potentially harmful heme precursors prior to the metabolic block. Thus, taking drugs that cause induction

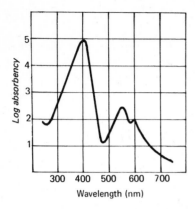

Figure 34–10. Absorption spectrum of hematoporphyrin (0.01% solution in 5% HCl).

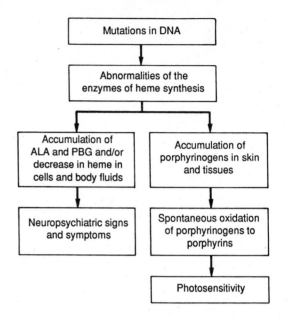

Figure 34–11. Biochemical causes of the major signs and symptoms of the porphyrias.

of cytochrome P450 (so-called microsomal inducers) can precipitate attacks of porphyria.

The diagnosis of a specific type of porphyria can generally be established by consideration of the clinical and family history, of the physical examination, and of appropriate laboratory tests. The major findings in the six principal types of porphyria are listed in Table 34–2.

High levels of lead can affect heme metabolism by combining with SH groups in enzymes such as ferrochelatase and ALA dehydratase. This affects porphyrin metabolism. Elevated levels of protoporphyrin are found in red blood cells, and elevated levels of ALA and of coproporphyrin are found in urine.

It is hoped that treatment of the porphyrias at the gene level will become possible. In the meantime, treatment is essentially symptomatic. It is important

for patients to avoid any anesthetics and drugs, including alcohol, that cause induction of cytochrome P450. Ingestion of large amounts of food rich in carbohydrates (glucose loading) or administration of hematin (a hydroxide of heme) may repress ALA synthase, resulting in diminished production of harmful heme precursors. Patients exhibiting photosensitivity may benefit from administration of β-carotene; this compound appears to lessen production of free radicals, thus diminishing photosensitivity. Sun-

Table 34–2. Summary of major findings in the porphyrias[1]

| Enzyme Involved[2] | Porphyria | | Results of Laboratory Tests[3] |
	Type and [Class]	Major Symptoms	
3. Uroporphyrinogen I synthetase	Acute intermittent porphyria [hepatic]	Abdominal pain Neuropsychiatric	Urinary PBG + Urinary uroporphyrin +
4. Uroporphyrinogen III cosynthase	Congenital erythropoietic [erythropoietic]	No photosensitivity	Urinary uroporphyrin + Urinary PBG –
5. Uroporphyrinogen decarboxylase	Porphyria cutanea tarda [hepatic]	Photosensitivity	Urinary uroporphyrin + Urinary PBG –
6. Coproporphyrinogen oxidase	Hereditary coproporphyria [hepatic]	Photosensitivity Abdominal pain Neuropsychiatric	Urinary PBG + Urinary uroporphyrin + Fecal coproporphyrin +
7. Protoporphyrinogen oxidase	Variegate porphyria [hepatic]	Photosensitivity Abdominal pain Neuropsychiatric	Urinary PBG + Urinary uroporphyrin + Fecal protoporphyrin +
8. Ferrochelatase	Protoporphyria (erythrohepatic)	Photosensitivity	Fecal protoporphyrin + Red cell protoporphyrin +

[1]Only the biochemical findings in the active stages of these diseases are indicated. Certain biochemical abnormalities are detectable in the latent stages of some of the above conditions.
[2]The numbering of the enzymes in this table corresponds to that used in Figure 34–9.
[3]PBG, porphobilinogen.

screens that filter out visible light can also be helpful to such patients.

CATABOLISM OF HEME PRODUCES BILIRUBIN

Under physiologic conditions in the human adult, $1–2 \times 10^8$ erythrocytes are destroyed per hour. Thus, in 1 day, a 70-kg human turns over approximately 6 g of hemoglobin. When hemoglobin is destroyed in the body, **globin** is degraded to its constituent amino acids, which are reused, and the **iron** of heme enters the iron pool, also for reuse. The iron-free **porphyrin**

portion of heme is also degraded, mainly in the reticuloendothelial cells of the liver, spleen, and bone marrow.

The catabolism of heme from all of the heme proteins appears to be carried out in the microsomal fractions of cells by a complex enzyme system called **heme oxygenase.** By the time the heme of heme proteins reaches the heme oxygenase system, the iron has usually been oxidized to the ferric form, constituting **hemin.** The heme oxygenase system is substrate-inducible. It is located in close proximity to the microsomal electron transport system. As depicted in Figure 34–12, the hemin is reduced to heme with NADPH, and, with the aid of more NADPH, oxygen

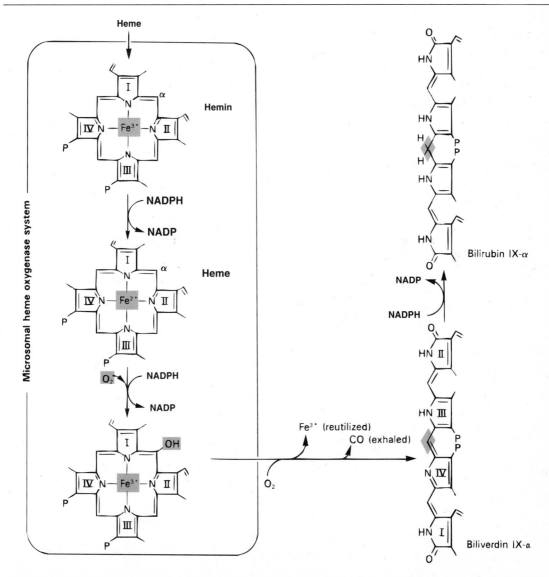

Figure 34–12. Schematic representation of the microsomal heme oxygenase system. (Modified from Schmid R, McDonough AF in: *The Porphyrins.* Dolphin D [editor]. Academic Press, 1978.)

is added to the α-methenyl bridge between pyrroles I and II of the porphyrin. The ferrous iron is again oxidized to the ferric form. With the further addition of oxygen, **ferric ion** is released, **carbon monoxide** is produced, and an equimolar quantity of **biliverdin IX-α** results from the splitting of the tetrapyrrole ring.

In birds and amphibia, the green biliverdin IX-α is excreted; in mammals, a soluble enzyme called **biliverdin reductase** reduces the methenyl bridge between pyrrole III and pyrrole IV to a methylene group to produce **bilirubin IX-a**, a yellow pigment (Figure 34–12).

It is estimated that 1 g of hemoglobin yields 35 mg of bilirubin. The daily bilirubin formation in human adults is approximately 250–350 mg, deriving mainly from hemoglobin but also from ineffective erythropoiesis and from various other heme proteins such as cytochrome P450.

The chemical conversion of heme to bilirubin by reticuloendothelial cells can be observed in vivo as the purple color of the heme in a hematoma is slowly converted to the yellow pigment of bilirubin.

Bilirubin formed in peripheral tissues is transported to the liver by plasma albumin. The further metabolism of bilirubin occurs primarily in the liver. It can be divided into three processes: (1) uptake of bilirubin by liver parenchymal cells, (2) conjugation of bilirubin in the smooth endoplasmic reticulum, and (3) secretion of conjugated bilirubin into the bile. Each of these processes will be considered separately.

THE LIVER TAKES UP BILIRUBIN

Bilirubin is only sparingly soluble in water, but its solubility in plasma is increased by noncovalent binding to albumin. Each molecule of albumin appears to have one high-affinity site and one low-affinity site for bilirubin. In 100 mL of plasma, approximately 25 mg of bilirubin can be tightly bound to albumin at its high-affinity site. Bilirubin in excess of this quantity can be bound only loosely and thus can easily be detached and diffuse into tissues. A number of compounds such as antibiotics and other drugs compete with bilirubin for the high-affinity binding site on albumin. Thus, these compounds can displace bilirubin from albumin and have significant clinical effects.

In the liver, the bilirubin is removed from albumin and taken up at the sinusoidal surface of the hepatocytes by a carrier-mediated saturable system. This **facilitated transport system** has a very large capacity, so that even under pathologic conditions the system does not appear to be rate-limiting in the metabolism of bilirubin.

Since this facilitated transport system allows the equilibrium of bilirubin across the sinusoidal membrane of the hepatocyte, the net uptake of bilirubin will be dependent upon the removal of bilirubin by subsequent metabolic pathways.

Conjugation of Bilirubin With Glucuronic Acid Occurs in the Liver

Bilirubin is nonpolar and would persist in cells (eg, bound to lipids) if not rendered water-soluble. Hepatocytes convert bilirubin to a polar form, which is readily excreted in the bile, by adding glucuronic acid molecules to it. This process is called conjugation and can employ polar molecules other than glucuronic acid (eg, sulfate). Many steroid hormones and drugs are also converted to water-soluble derivatives by conjugation in preparation for excretion (see Chapter 61).

Liver contains at least two isoforms of glucuronosyltransferase, both of which act on bilirubin. These enzymes are mainly located in the smooth endoplasmic reticulum and use UDP-glucuronic acid as the glucuronosyl donor. Bilirubin monoglucuronide is an intermediate and is subsequently converted to the diglucuronide (Figures 34–13 and 34–14). Most of the bilirubin excreted in the bile of mammals is in the

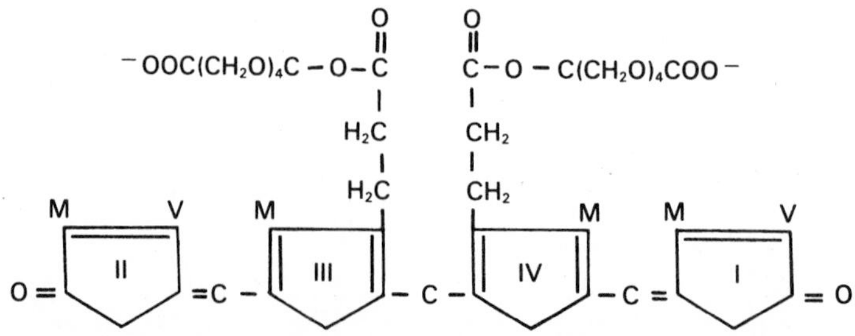

Figure 34–13. Structure of bilirubin diglucuronide (conjugated, "direct-reacting" bilirubin). Glucuronic acid is attached via ester linkage to the two propionic acid groups of bilirubin to form an acylglucuronide.

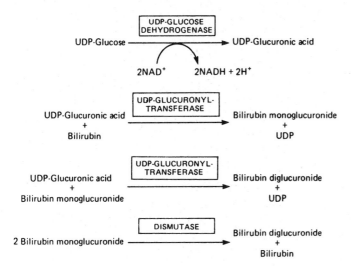

Figure 34–14. Conjugation of bilirubin with glucuronic acid. The glucuronate donor, UDP-glucuronic acid, is formed from UDP-glucose as depicted. At least two isoforms of glucuronosyltransferase that use bilirubin as a substrate are present in human liver.

form of bilirubin diglucuronide. However, when bilirubin conjugates exist abnormally in human plasma (eg, in obstructive jaundice), they are predominantly monoglucuronides. UDP-glucuronosyltransferase activity can be **induced** by a number of clinically useful drugs, including phenobarbital. More information about glucuronosylation is presented below in the discussion of the inherited disorders of bilirubin conjugation.

Bilirubin Is Secreted Into Bile

Secretion of conjugated bilirubin into the bile occurs by an active transport mechanism, which is probably rate-limiting for the entire process of hepatic bilirubin metabolism. The hepatic transport of conjugated bilirubin into the bile is inducible by those same drugs that are capable of inducing the conjugation of bilirubin. Thus, the conjugation and excretion systems for bilirubin behave as a coordinated functional unit.

Under physiologic conditions, essentially all of the bilirubin secreted into the bile is conjugated. Only after phototherapy can significant quantities of unconjugated bilirubin be found in bile.

In the liver, there are multiple systems for secreting naturally occurring and pharmaceutical compounds into the bile after their metabolism. Some of these secretory systems are shared by bilirubin diglucuronide, but others operate independently.

Figure 34–15 summarizes the three major processes involved in the transfer of bilirubin from blood to bile. Sites that are affected in a number of conditions causing jaundice (see below) are also indicated.

Conjugated Bilirubin Is Reduced to Urobilinogen by Intestinal Bacteria

As the conjugated bilirubin reaches the terminal ileum and the large intestine, the glucuronides are removed by specific bacterial enzymes (**β-glu-**

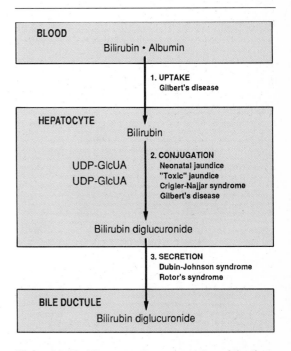

Figure 34–15. Diagrammatic representation of the three major processes (uptake, conjugation, and secretion) involved in the transfer of bilirubin from blood to bile. Certain intracellular proteins of hepatocytes, such as ligandin and Y protein, are involved in the uptake of bilirubin by these cells. The process affected in a number of conditions causing jaundice is also shown.

curonidases), and the pigment is subsequently reduced by the fecal flora to a group of colorless tetrapyrrolic compounds called **urobilinogens** (Figure 34–16). In the terminal ileum and large intestine, a small fraction of the urobilinogens is reabsorbed and reexcreted through the liver to constitute the **intrahepatic urobilinogen cycle.** Under abnormal conditions, particularly when excessive bile pigment is formed or liver disease interferes with this intrahepatic cycle, urobilinogen may also be excreted in the urine.

Normally, most of the colorless urobilinogens formed in the colon by the fecal flora are oxidized there to urobilins (colored compounds) and are excreted in the feces (Figure 34–16). Darkening of feces upon standing in air is due to the oxidation of residual urobilinogens to urobilins.

HYPERBILIRUBINEMIA CAUSES JAUNDICE

When bilirubin in the blood exceeds 1 mg/dL (17.1 μmol/L), hyperbilirubinemia exists. Hyperbilirubinemia may be due to the production of more bilirubin than the normal liver can excrete, or it may result from the failure of a damaged liver to excrete bilirubin produced in normal amounts. In the absence of hepatic damage, obstruction to the excretory ducts of the liver—by preventing the excretion of bilirubin—will also cause hyperbilirubinemia. In all these situations, bilirubin accumulates in the blood, and when it reaches a certain concentration (approximately 2–2.5 mg/dL), it diffuses into the tissues, which then become yellow. The condition is called **jaundice** or **icterus.**

In clinical studies of jaundice, measurement of bilirubin in the serum is of great value. A method for quantitatively assaying the bilirubin content of the serum was first devised by Van den Bergh by application of Erhlich's test for bilirubin in urine. The Ehrlich reaction is based on the coupling of diazotized sulfanilic acid (Ehrlich's diazo reagent) and bilirubin to produce a reddish-purple azo compound. In the original procedure as described by Ehrlich, methanol was used to provide a solution in which both bilirubin and the diazo reagent were soluble. Van den Bergh inadvertently omitted the methanol on an occasion when assay of bile pigment in human bile was being attempted. To his surprise, normal development of the color occurred "directly." This form of bilirubin that would react without the addition of methanol was thus termed **"direct-reacting."** It was then found that this same direct reaction would also occur in serum from cases of jaundice due to biliary obstruction. However, it was still necessary to add methanol to detect bilirubin in normal serum or that which was present in excess in serum from cases of hemolytic jaundice where no evidence of obstruction was to be found. To that form of bilirubin which could be measured only after the addition of methanol, the term **"indirect-reacting"** was applied.

It was subsequently discovered that the indirect bilirubin is "free" (unconjugated) bilirubin en route to the liver from the reticuloendothelial tissues, where the bilirubin was originally produced by the breakdown of heme porphyrins. Since this bilirubin is not water-soluble, it requires methanol to initiate coupling with the diazo reagent. In the liver, the free bilirubin becomes conjugated with glucuronic acid, and the conjugate, bilirubin glucuronide, can then be excreted into the bile. Furthermore, conjugated bilirubin, being water-soluble, can react directly with the diazo reagent, so that the "direct bilirubin" of Van den Bergh is actually a bilirubin conjugate (bilirubin glucuronide).

Depending on the type of bilirubin present in plasma, ie, unconjugated or conjugated, hyperbilirubinemia may be classified as **retention hyperbilirubinemia,** due to overproduction; or **regurgitation**

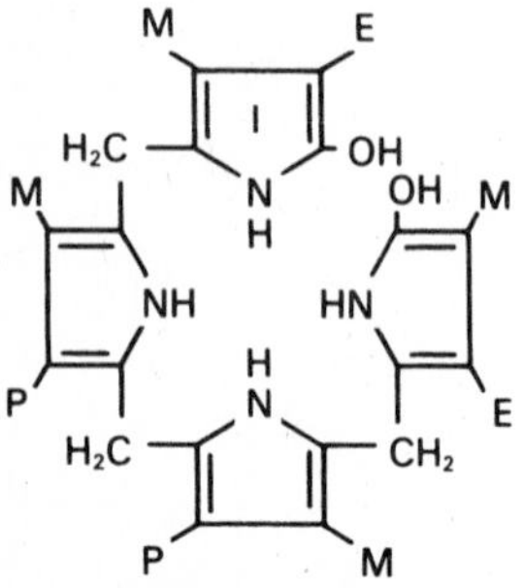

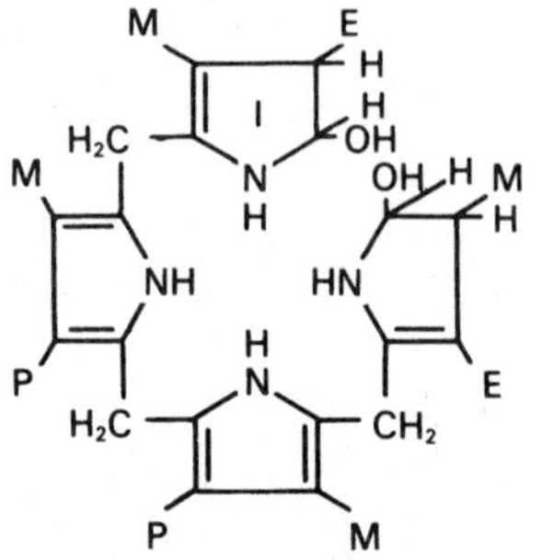

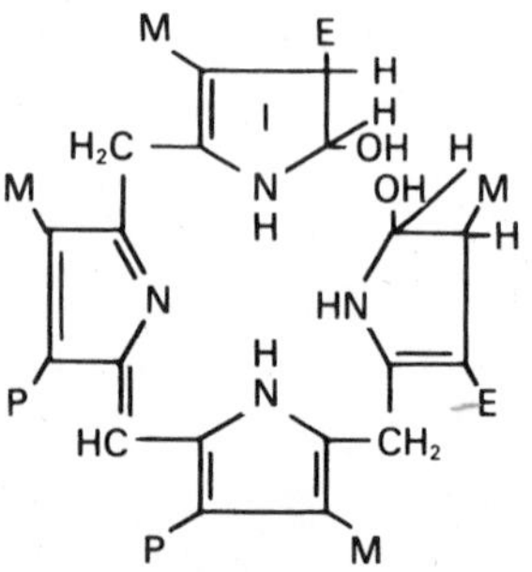

Figure 34–16. Structure of some bile pigments.

hyperbilirubinemia, due to reflux into the blood stream because of biliary obstruction.

Because of its hydrophobicity, only unconjugated bilirubin can cross the blood-brain barrier into the central nervous system; thus, encephalopathy due to hyperbilirubinemia (**kernicterus**) can occur only in connection with unconjugated bilirubin, as found in retention hyperbilirubinemia. On the other hand, because of its water-solubility, only conjugated bilirubin can appear in urine. Accordingly, **choluric jaundice** (choluria = presence of biliary derivatives in the urine) occurs only in regurgitation hyperbilirubinemia, and **acholuric jaundice** occurs only in the presence of an excess of unconjugated bilirubin.

Elevated Amounts of Unconjugated Bilirubin in Blood Occur in a Number of Conditions

A. Hemolytic Anemias: Hemolytic anemias are important causes of unconjugated hyperbilirubinemia although, even in the event of extensive hemolysis, unconjugated hyperbilirubinemia is usually only slight (< 4 mg/dL; < 68.4 μmol/L), because of the liver's large capacity for handling bilirubin. However, if the handling of bilirubin is defective owing to either an acquired defect or an inherited abnormality, unconjugated hyperbilirubinemia may occur. The following are the commonest causes of this condition.

B. Neonatal "Physiologic Jaundice": This transient condition is the most common cause of unconjugated hyperbilirubinemia. It results from an accelerated hemolysis and an immature hepatic system for the uptake, conjugation, and secretion of bilirubin. Not only is the UDP-glucuronosyltransferase activity reduced, but there probably is reduced synthesis of the substrate for that enzyme, UDP-glucuronic acid. Since the increased bilirubin is unconjugated, it is capable of penetrating the blood-brain barrier when its concentration in plasma exceeds that which can be tightly bound by albumin (20–25 mg/dL). This can result in a hyperbilirubinemic toxic encephalopathy, or kernicterus, which can cause mental retardation. Because of the recognized inducibility of this bilirubin metabolizing system, phenobarbital has been administered to jaundiced neonates and is effective in this disorder. In addition, exposure to visible light (phototherapy) can promote (by a mechanism that is not understood) the hepatic excretion of unconjugated bilirubin by converting some of the bilirubin to other derivatives such as maleimide fragments and geometric isomers that are excreted in the bile.

C. Crigler-Najjar Syndrome, Type I; Congenital Nonhemolytic Jaundice: Type I Crigler-Najjar syndrome, a rare autosomal recessive disorder of humans, is due to a primary metabolic defect in the conjugation of bilirubin. It is characterized by severe congenital jaundice due to the inherited absence of bilirubin UDP-glucuronosyltransferase activity in hepatic tissues. The disease is usually fatal within the first 15 months of life, but a few teenagers have been reported who did not develop difficulties until puberty. These children have been treated with phototherapy with some reduction in plasma bilirubin levels. Phenobarbital has no effect on the formation of bilirubin glucuronides in patients with type I Crigler-Najjar syndrome. Serum bilirubin usually exceeds 20 mg/dL when untreated.

D. Crigler-Najjar Syndrome, Type II: This rare inherited disorder seems to result from a **milder defect** in the bilirubin conjugating system and has a more benign course. The serum bilirubin concentrations usually do not exceed 20 mg/dL, but all of the bilirubin accumulated is of the unconjugated type. Surprisingly, the bile in these patients does contain bilirubin monoglucuronide, and it has been proposed that the genetic defect may involve the hepatic UDP-glucuronosyltransferase that adds the second glucuronyl group to bilirubin monoglucuronide. Patients with this syndrome can respond to treatment with large doses of phenobarbital.

E. Gilbert's Disease: Gilbert's disease is a heterogeneous group of disorders, many of which are now recognized to be due to a compensated hemolysis associated with unconjugated hyperbilirubinemia. There also appears to be a defect in the hepatic clearance of bilirubin, possibly due to a defect in the uptake of bilirubin by the liver parenchymal cells. However, bilirubin UDP-glucuronosyltransferase activities in the livers of those patients studied with this disease also were found to be reduced.

F. Toxic Hyperbilirubinemia: Unconjugated hyperbilirubinemia can result from toxin-induced liver dysfunction such as that caused by chloroform, arsphenamines, carbon tetrachloride, acetaminophen, hepatitis virus, cirrhosis, and *Amanita* mushroom poisoning. Although most of these acquired disorders are due to hepatic parenchymal cell damage, which impairs conjugation, there is frequently a component of obstruction of the biliary tree within the liver that results in the presence of some conjugated hyperbilirubinemia.

Obstruction in the Biliary Tree Is the Commonest Cause of Conjugated Hyperbilirubinemia

A. Obstruction of the Biliary Tree: Conjugated hyperbilirubinemia commonly results from blockage of the hepatic or common bile ducts. Because of the obstruction, bilirubin diglucuronide cannot be excreted. It thus regurgitates into the hepatic veins and lymphatics, and conjugated bilirubin appears in the blood and urine (choluric jaundice).

The term **cholestatic jaundice** is used to include all cases of extrahepatic obstructive jaundice. It also covers those cases of jaundice that exhibit conjugated hyperbilirubinemia due to micro-obstruction of intra-hepatic biliary ductules by swollen, damaged hepato-cytes (eg, as may occur in infectious hepatitis).

B. Chronic Idiopathic Jaundice (Dubin-Johnson Syndrome): This autosomal recessive disorder consists of conjugated hyperbilirubinemia in childhood or during adult life. The hyperbilirubin-emia is apparently caused by a defect in the hepatic secretion of conjugated bilirubin into the bile. This secretory defect of conjugated compounds is not re-stricted to bilirubin but also involves secretion of conjugated estrogens and test compounds such as the dye sulfobromophthalein. In patients with this syn-drome, the hepatocytes in the centrilobular area con-tain an abnormal black pigment that has not been identified.

C. Rotor's Syndrome: Rotor's syndrome is a rare condition characterized by chronic conjugated hyperbilirubinemia and normal liver histology. Its precise cause has not been identified, but it also may be due to a defect in transport by hepatocytes of or-ganic anions, including bilirubin.

Some Conjugated Bilirubin Can Bind Covalently to Albumin

When levels of conjugated bilirubin remain high in plasma, a fraction can bind covalently to albumin (delta bilirubin). Because it is bound covalently to al-bumin, this fraction has a longer half-life in plasma than does conventional conjugated bilirubin. Thus, it remains elevated during the recovery phase of ob-structive jaundice after the remainder of the conju-gated bilirubin has declined to normal levels; this ex-plains why some patients continue to appear jaundiced after conjugated bilirubin levels have re-turned to normal.

Urobilinogen and Bilirubin in Urine Are Clinical Indicators

Normally, there are mere traces of urobilinogen in the urine. In **complete obstruction of the bile duct,** no urobilinogen is found in the urine, since bilirubin has no access to the intestine, where it can be con-verted to urobilinogen. In this case, the presence of bilirubin (conjugated) in the urine without urobilino-gen suggests obstructive jaundice, either intrahepatic or posthepatic.

In **hemolytic jaundice,** the increased production of bilirubin leads to increased production of uro-bilinogen, which appears in the urine in large amounts. Bilirubin is not usually found in the urine in hemolytic jaundice (because unconjugated biliru-bin does not pass into the urine), so that the combina-tion of increased urobilinogen and absence of biliru-bin is suggestive of hemolytic jaundice. Increased blood destruction from any cause (eg, pernicious anemia) will, of course, also bring about an increase in urine urobilinogen.

Table 34–3 summarizes laboratory results ob-tained on patients with three different causes of jaun-dice—hemolytic anemia (a prehepatic cause), hepati-tis (a hepatic cause), and obstruction of the common bile duct (a posthepatic cause). Laboratory tests on blood (to evaluate the possibility of a hemolytic ane-mia) and on serum (eg, activities of the enzymes ALT and alkaline phosphatase) are also important in helping to distinguish among prehepatic, hepatic, and posthepatic causes of jaundice.

SUMMARY

Hemoproteins, such as hemoglobin and the cy-tochromes, contain heme. Heme is an iron-porphyrin (Fe^{2+}-protoporphyrin IX) in which four pyrrole rings are joined by methenyl bridges. The eight side groups (methyl, vinyl, and propionyl) on the four pyrrole rings of heme are arranged in a specific se-quence.

Biosynthesis of the heme ring occurs in mitochon-dria and cytosol via eight enzymatic steps. It com-mences with formation of δ-aminolevulinate (ALA) from succinyl-CoA and glycine in a reaction cat-alyzed by ALA synthase, the regulatory enzyme of the pathway.

Genetically determined abnormalities of seven of the eight enzymes involved in heme biosynthesis re-sult in the inherited porphyrias. Red blood cells and liver are the major sites of metabolic expression of the porphyrias. Photosensitivity and neurologic prob-

Table 34–3. Laboratory results in normal patients and patients with three different causes of jaundice.

Condition	Serum Bilirubin	Urine Urobilinogen	Urine Bilirubin	Fecal Urobilinogen
Normal	Direct: 0.1–0.4 mg/dL Indirect: 0.2–0.7 mg/dL	0–4 mg/24 h	Absent	40–280 mg/24 h
Hemolytic anemia	Elevation of indirect	Increased	Absent	Increased
Hepatitis	Elevations of direct and indirect	Decreased	Present	Decreased
Obstructive jaundice[1]	Elevation of direct	Absent	Present	Trace to absent

[1]The commonest causes of obstructive (posthepatic) jaundice are cancer of the head of the pancreas and a gallstone lodged in the common bile duct. The presence of bilirubin in the urine is sometimes referred to as choluria: hence, hepatitis and obstruc-tion of the common bile duct cause choluric jaundice, whereas the jaundice of hemolytic anemia is referred to as acholuric.

lems are common complaints. Intake of certain compounds (such as lead) can cause acquired porphyrias. Increased amounts of porphyrins or their precursors can be detected in blood and urine, facilitating diagnosis.

Catabolism of the heme ring is initiated by the enzyme heme oxygenase, producing a linear tetrapyrrole.

Biliverdin is an early product of catabolism and on reduction yields bilirubin. The latter is transported by albumin from peripheral tissues to the liver, where it is taken up by hepatocytes. The iron of heme and the amino acids of globin are conserved and reutilized.

In the liver, bilirubin is made water-soluble by conjugation with two molecules of glucuronic acid and is secreted into the bile. The action of bacterial enzymes in the gut produces urobilinogen and urobilin, which are excreted in the feces and urine.

Jaundice is due to elevation of the level of bilirubin in the blood. The causes of jaundice can be classified as prehepatic (eg, hemolytic anemias), hepatic (eg, hepatitis), and posthepatic (eg, obstruction of the common bile duct). Measurements of plasma total and nonconjugated bilirubin, of urinary urobilinogen and bilirubin, and of certain serum enzymes as well as inspection of stool samples help distinguish between these causes.

REFERENCES

Chowdhury JR et al: Hereditary jaundice and disorders of bilirubin metabolism. In: *The Metabolic and Molecular Bases of Inherited Disease,* 7th ed. Scriver CR et al (editors). McGraw-Hill, 1995.

Desnick RJ: The porphyrias. In: *Harrison's Principles of Internal Medicine,* 13th ed. Isselbacher KJ et al (editors). McGraw-Hill, 1994.

Elder GH: Haem synthesis and the porphyrias. In: *Scientific Foundations of Biochemistry in Clinical Practice,* 2nd ed. Williams DL, Marks V (editors). Butterworth-Heinemann, 1994.

Goldberg A et al: Porphyrin metabolism and the porphyrias. In: *Oxford Textbook of Medicine,* 2nd ed. Weatherall DJ et al (editors). Oxford Univ Press, 1987.

Isselbacher KJ: Bilirubin metabolism and hyperbilirubinemia. In: *Harrison's Principles of Internal Medicine,* 13th ed. Isselbacher KJ et al (editors). McGraw-Hill, 1994.

Kaplan LM, Isselbacher KJ: Jaundice. In: *Harrison's Principles of Internal Medicine,* 13th ed. Isselbacher KJ et al (editors). McGraw-Hill, 1994.

Kappas A et al: The porphyrias. In: *The Metabolic and Molecular Bases of Inherited Disease,* 7th ed. Scriver CR et al (editors). McGraw-Hill, 1995.

Nucleotides

35

Victor W. Rodwell, PhD

INTRODUCTION

This chapter introduces the aromatic heterocyclic bases purine and pyrimidine and their major derivatives, the nucleosides and nucleotides, which, in addition to supplying the monomer units or building blocks of nucleic acids, serve diverse functions essential to life and health.

Major biochemical functions of purine and pyrimidine nucleotides include the numerous phosphate transfer reactions of ATP and other nucleoside triphosphates that drive otherwise endergonic reactions. UDP-glucose and UDP-galactose function in carbohydrate biosynthesis, and CDP-acylglycerol in lipid biosynthesis, as "high-energy intermediates" for covalent bond synthesis. Nucleotides form a portion of coenzymes such as FAD, NAD^+, $NADP^+$, coenzyme A, and S-adenosylmethionine. Nucleotides also serve regulatory functions. ADP levels regulate the rate of mitochondrial oxidative phosphorylation, specific nucleotides act as allosteric regulators of enzyme activity, and cAMP and cGMP serve "second messenger" functions. Finally, nucleoside triphosphates serve as the monomer unit precursors of the nucleic acids RNA and DNA.

BIOMEDICAL IMPORTANCE

The ability of nucleotides to absorb ultraviolet light—a consequence of their chemical structure—makes ultraviolet light a potent mutagen. A battery of chemically synthesized purine and pyrimidine analogs is employed in the chemotherapy of cancer, AIDS, and other situations of medical interest. Examples include allopurinol used to treat hyperuricemia and gout and azathioprine used to suppress the immune response during organ transplantation.

While relatively soluble as a urate salt at alkaline pH, uric acid is insoluble in acidic urine. The purine catabolites xanthine and uric acid thus may occur in urinary tract stones. Synthetic analogs of naturally occurring bases, nucleosides, or nucleotides exploited to inhibit the growth of cancer cells or of certain viruses include 5-fluorouracil, 5'-iodo-2'-deoxyuridine, 6-thioguanine, 6-mercaptopurine, 6-azauridine, and arabinosyl cytosine—drugs that inhibit specific enzymes or replace natural purines or pyrimidines during synthesis of DNA or RNA. These analogs provide, in addition, valuable tools for biomedical research investigations.

CHEMISTRY OF PURINES, PYRIMIDINES, THEIR NUCLEOSIDES, AND NUCLEOTIDES

Purines and Pyrimidines Are Heterocyclic Compounds

Heterocyclic compounds are ring (cyclic) compounds that contain both carbon and noncarbon (hetero) atoms. While heterocycles that contain sulfur or oxygen are important both biologically and therapeutically, by far the most common hetero atom in biology is nitrogen.

Purines and pyrimidines constitute a class of nitrogen-containing heterocycles of major biologic importance. Their principal derivatives are nucleosides and nucleotides, both of which contain a cyclized sugar, often a pentose, linked to a nitrogen hetero atom by a β-N-glycosidic bond. Nucleotides contain, in addition, one or more phosphoryl groups esterified to hydroxy groups of the sugar.

Different Numbering Systems Are Used for Purines and Pyrimidines

Note the paradox that the smaller (six-atom) heterocycle pyrimidine has the longer name while the larger (nine-atom) heterocycle purine has the shorter name. Note also that the clockwise direction in which atoms of the pyrimidine ring are numbered is opposite to the counterclockwise direction of numbering of the atoms of purines (Figure 35–1).

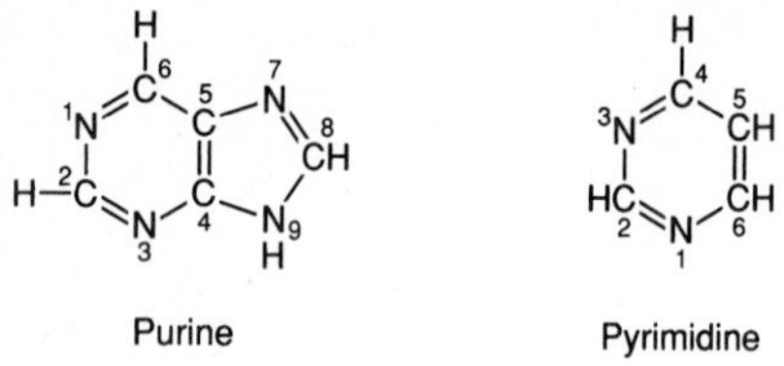

Figure 35–1. Purine and pyrimidine. The positions of the atoms are numbered according to the international system.

Purines and Pyrimidines Are Planar Molecules

Viewed edge on, purines and pyrimidines are essentially planar molecules. Were their rings puckered, purine and pyrimidine bases would stack relatively loosely in the interior of double-stranded helices of DNA or DNA-RNA hybrids, with a resulting loss of helix stability (Chapter 38).

Nucleic Acids Contain Five Major Heterocyclic Bases

The most abundant, or "major," heterocyclic bases of nucleic acids are the purines adenine and guanine and the pyrimidines cytosine, thymine, and uracil. All nucleic acids contain adenine, guanine, and cytosine. DNA (but not RNAs) also contains thymine, while RNAs (but not DNA) also contain uracil (Table 35–1).

Most Purines and Pyrimidines Exist in Cells as Nucleotides

Nucleosides consist of a purine or pyrimidine and a cyclized sugar, most often D-ribose or 2-deoxy-D-ribose, linked via a covalent, β-N-glycosidic bond to N-9 of a purine or to N-1 of a pyrimidine. Numbering of the sugar atoms employs a prime (eg, 3'- or 5'-) to distinguish sugar atoms from those of the heterocyclic base. Nucleotides are termed ribonucleotides or deoxyribonucleotides based on whether the sugar is ribose or 2-deoxyribose.

Base Formula	Base X = H	Nucleoside X = ribose or deoxyribose	Nucleotide, where X = ribose phosphate
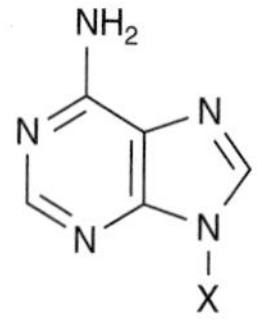	Adenine A	Adenosine A	Adenosine monophosphate AMP
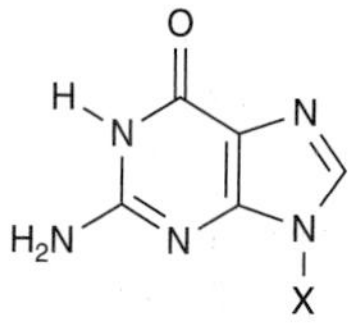	Guanine G	Guanosine G	Guanosine monophosphate GMP
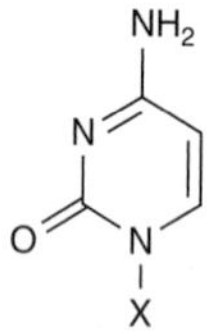	Cytosine C	Cytidine C	Cytidine monophosphate CMP
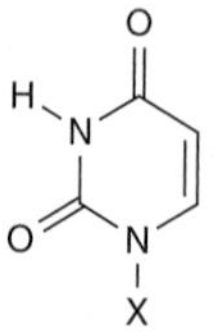	Uracil U	Uridine U	Uridine monophosphate UMP
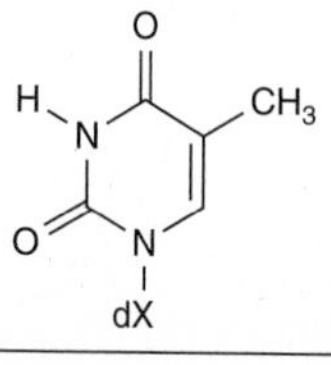	Thymine T	Thymidine T	Thymidine monophosphate TMP

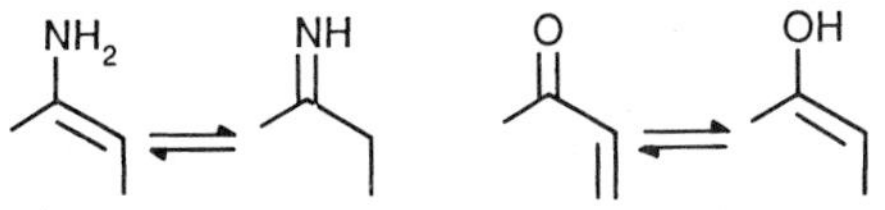

Figure 35–2. Tautomerism of the oxo- and amino- functional groups of purines and pyrimidines.

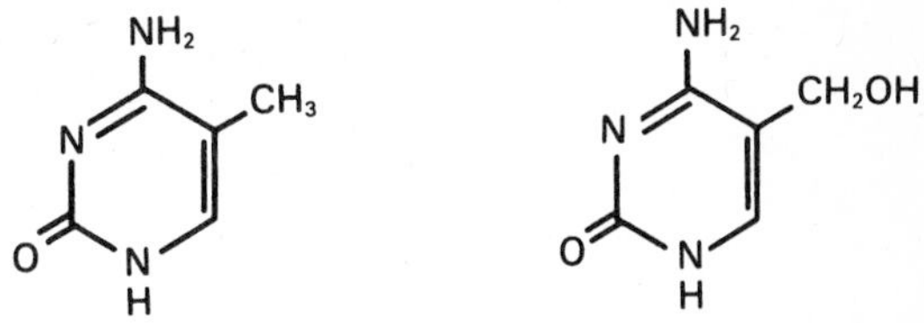

Figure 35–4. Two uncommon naturally occurring pyrimidines.

Heterocyclic Oxo- and Amino- Groups Exist as Tautomeric Mixtures

Chapter 15 described the keto-enol tautomerism of the oxo- and hydroxy- groups of aldoses and ketoses. Owing to the aromatic character of purines or pyrimidines, when oxo- or amino- substituents are present they participate in keto-enol and amine-imine tautomerism (Figure 35–2). These functional groups therefore exist as tautomeric mixtures of amino-imino and oxo-hydroxy pairs that differ with respect to the position of a hydrogen atom and certain electrons. Although physiologic conditions strongly favor the amino and lactam forms, the unfavored tautomers may participate in mutagenic events (Chapters 39 and 41).

Heterocyclic *N*-Glycosides Form *Syn* and *Anti* Conformers

Steric hindrance by the heterocyclic base dictates that once formed, there is no freedom of rotation about the β-*N*-glycosidic bond that links sugars to purines or pyrimidines. Nucleosides and nucleotides thus exist as stable, non-interconvertible *syn* and *anti* conformers (Figure 35–3) that can only be interconverted by rupture and re-formation of the glycosidic bond. While both conformers occur in nature, *anti* conformers predominate, and it is the *anti* conformers of nucleotides that participate in normal base pairing in double-stranded DNA (Chapter 38).

Nucleic Acids Also Contain "Minor" or "Unusual" Bases

In addition to the major bases, specific DNAs and RNAs of both prokaryotes and eukaryotes contain considerably smaller quantities of additional purines and pyrimidines termed "minor" or "unusual" bases. Neither term is particularly apt, since these bases are functionally important and hence not of minor physiologic importance, and they are widely distributed in nature. 5-Methylcytosine (Figure 35–4) is present in both bacterial and human DNA. Additional minor bases present in mammalian RNAs include N^6-methyladenine, N^6,N^6-dimethyladenine, and N^6,N^7-methylguanine of messenger RNAs (mRNAs) (Figure 35–5) and various derivatized bases in transfer RNAs (tRNAs). Other minor bases, eg, the 5-hydroxymethylcytosine of bacteriophage DNA, occur exclusively in the nucleic acids of bacteria and viruses.

Nucleotides present in the free state in cells include hypoxanthine and xanthine (Figure 35–6), intermediates in the metabolism of adenine and guanine, and uric acid, the oxidized end product of purine catabolism that human subjects excrete in their urine.

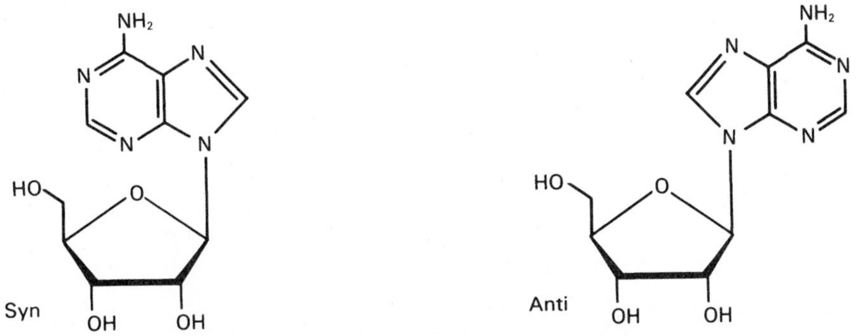

Figure 35–3. The *syn (left)* and *anti (right)* conformers of adenosine.

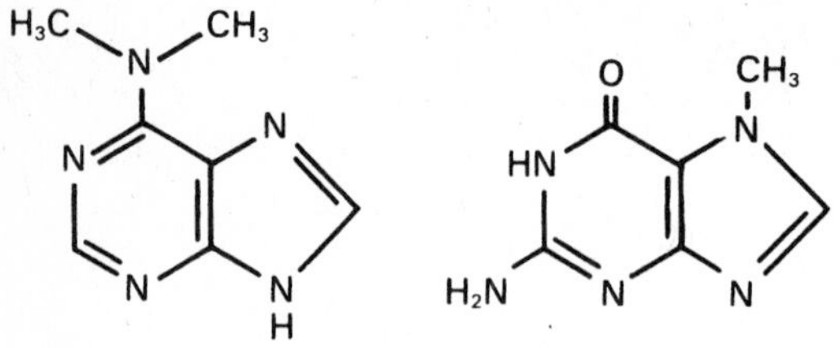

Figure 35–5. Two uncommon naturally occurring purines.

Methylated Purines of Plants Have Pharmacologic Properties

Plants contain additional heterocyclic bases, many of which are of pharmacologic interest. Examples include the methylated xanthine derivatives caffeine (1,3,7-trimethylxanthine) of coffee, theophylline (1,3-dimethylxanthine) of tea, and theobromine (3,7-dimethylxanthine) of cocoa (Figure 35–7).

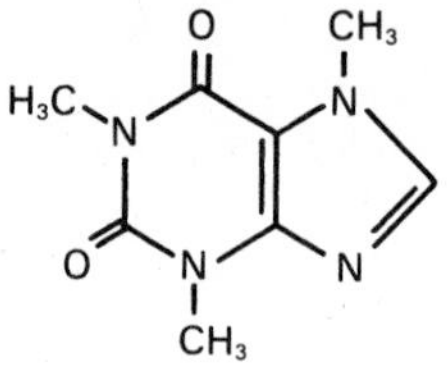

Caffeine (1,3,7-trimethyl-xanthine)

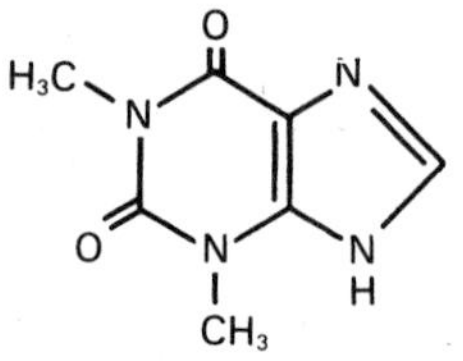

Theophylline· (1,3-dimethyl-xanthine)

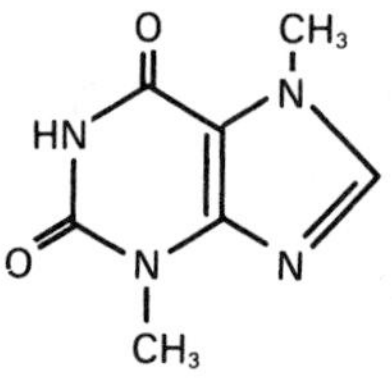

Theobromine (3,7-dimethyl-xanthine)

Figure 35–7. Methylxanthines present in foods.

PURINES AND PYRIMIDINES FORM NUCLEOSIDES AND NUCLEOTIDES

Nucleosides Contain Monosaccharides

The ribonucleosides adenosine, guanosine, cytidine, and uridine consist of D-ribose or 2-deoxy-D-ribose linked via a β-*N*-glycosidic bond to N-9 of a purine or to N-1 of a pyrimidine (Figure 35–8). The structures of the 2′-deoxyribonucleosides deoxyadenosine, deoxyguanosine, deoxycytidine, deoxyuridine, and (deoxy)thymidine are similar except that 2-deoxy-D-ribose rather than D-ribose forms the β-*N*-glycosidic bond to N-9 (purines) or N-1 (pyrimidines).

Nucleotides Are Phosphorylated Nucleosides

Mononucleotides are nucleosides singly phosphorylated on hydroxyl groups of the sugar (Figure 35–9). For example, AMP (adenosine monophosphate) is adenine + ribose + phosphate. Table 35–1 lists the major purines and pyrimidines and their nucleoside and nucleotide derivatives. The process of posttranslational modification of the bases present in preformed polynucleotides can generate additional bases. In pseudouridine, D-ribose is attached to carbon 5 of uracil by a carbon-to-carbon bond. Its mononucleotide, pseudouridylic acid (Ψ) arises by posttranslational rearrangement of uridylic acid in preformed tRNA. Similarly, TMP (thymidine monophosphate), which contains ribose rather than deoxyribose (Figure 35–10), arises when UMP of a preformed tRNA is methylated by *S*-adenosylmethionine.

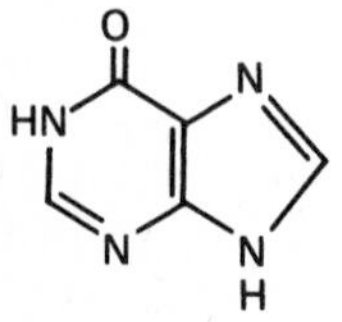

Hypoxanthine (6-oxypurine)

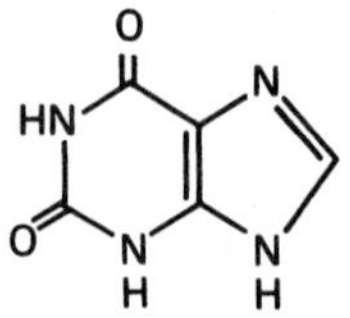

Xanthine (2,6-dioxypurine)

Figure 35–6. Structures of hypoxanthine and xanthine.

Nomenclature

The abbreviations A, G, C, T, and U refer to the bases, both free and present in nucleosides or nucleotides, of adenine, guanine, cytosine, thymine, and uracil, respectively. Where present, the prefix "d" (deoxy) indicates that the sugar is 2′-deoxy-D-ribose, eg, dGTP.

Nucleosides phosphorylated on the 3′- or 5′-carbon of ribose are termed nucleoside 3′-monophosphates and nucleoside 5′-monophosphates, respectively. An example is adenosine 3′-monophosphate, or 3′-AMP (Figure 35–11). However, since the 5′-hydroxyl is the one most commonly esterified, "5′-" is by convention omitted when naming 5′-nucleotides. Abbreviations such as "UMP" or "AMP"

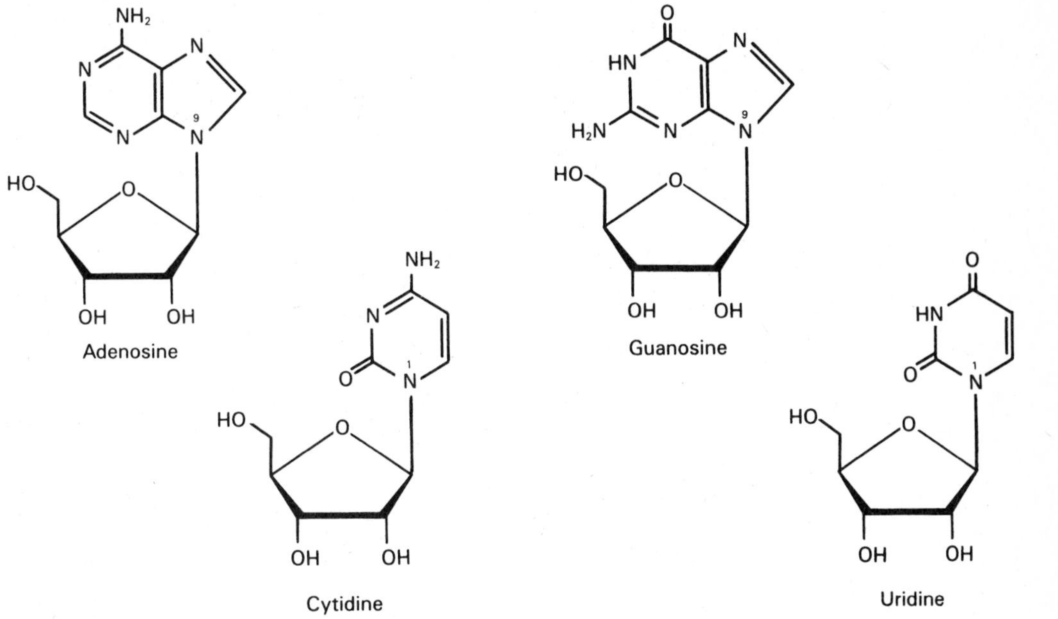

Figure 35–8. Structures of ribonucleosides. Adenosine and guanosine are shown as the most common *syn* conformers.

therefore denote nucleotides in which the phosphate is esterified to carbon 5 of the pentose.

Additional phosphates linked by acid anhydride bonds to the existing phosphate of a mononucleotide form nucleoside di- and triphosphates such as ADP (adenosine diphosphate) and ATP (adenosine triphosphate) (Figure 35–12).

Nucleotides Serve Diverse Physiologic Functions

Specific nucleotides participate in reactions that fulfill physiologic functions as diverse as protein and nucleic acid synthesis, regulatory cascades, and intra-

and intercellular signal transduction. Representative examples appear below.

Nucleoside Triphosphates Have High Group Transfer Potential

Acid anhydrides, unlike phosphate esters, have high group transfer potential. ΔG^0 for the hydrolysis of both terminal phosphates of all nucleoside triphosphates is about 7 kcal/mol. The high group transfer potential of purine and pyrimidine nucleoside triphosphates permits them to participate as group transfer reagents in numerous reactions that form covalent bonds. In these reactions, cleavage of an acid anhydride bond is coupled with a highly endergonic

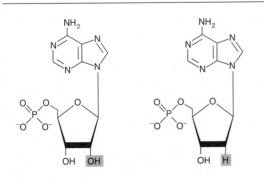

Figure 35–9. Adenylic acid (AMP) *(left)* and 2′-deoxyadenylic acid (dAMP) *(right).*

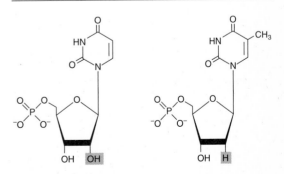

Figure 35–10. Uridylic acid (UMP) *(left)* and thymidylic acid (TMP) *(right).*

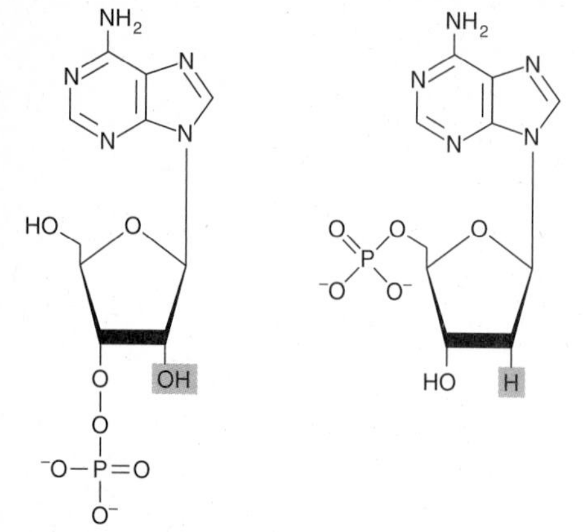

Figure 35–11. Adenosine 3'-monophosphate *(left)* and 2'-deoxyadenosine-5'-monophosphate *(right)*.

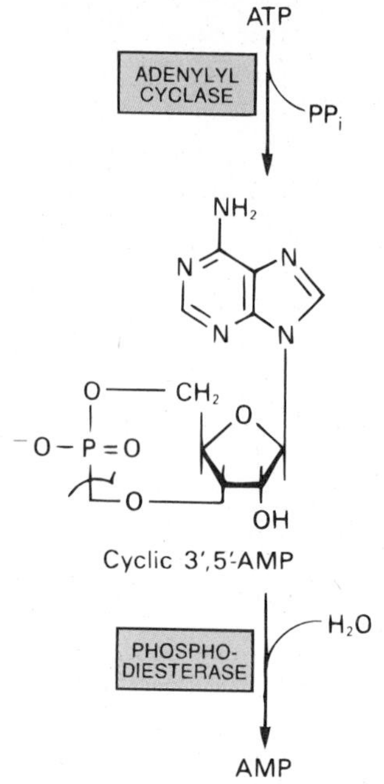

process such as covalent bond synthesis. A prominent example is polymerization of the major nucleoside triphosphates to form a nucleic acid.

A. Adenosine Derivatives: ADP and ATP are substrates and products, respectively, for oxidative phosphorylation, and ATP serves as the major biologic transducer of free energy. The mean intracellular concentration of ATP, the most abundant free nucleotide in mammalian cells, is about 1 mmol/L.

The cyclic phosphodiester **cAMP** (adenosine 3',5'-monophosphate) is formed from ATP in a reaction catalyzed by **adenylyl cyclase** (Figure 35–13). Adenylyl cyclase activity is regulated by complex interactions, many of which involve hormone receptors (Chapter 44). The "second messenger" cAMP participates in diverse regulatory functions in cells, eg, regulation of the activity of cAMP-dependent protein

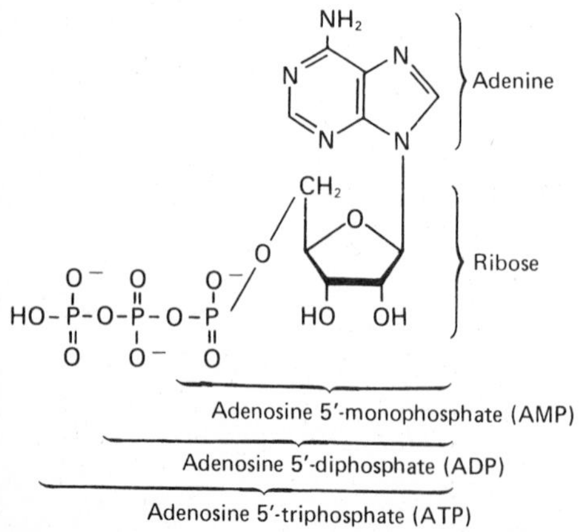

Figure 35–12. ATP, its diphosphate, and its monophosphate.

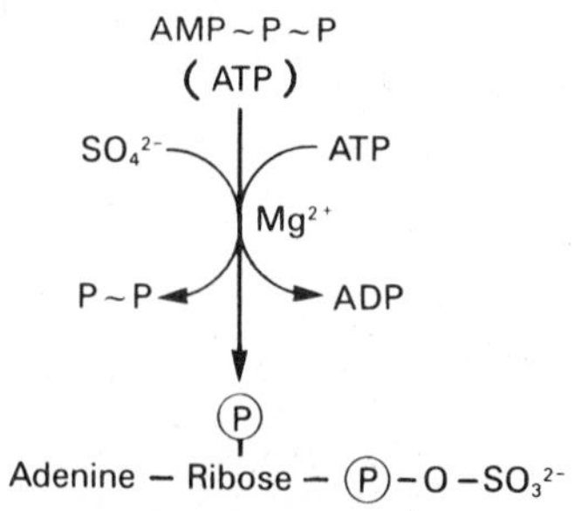

Cyclic 3',5'-AMP

Figure 35–13. Formation of cAMP from ATP by adenylyl cyclase and hydrolysis of cAMP by cAMP phosphodiesterase.

kinase activity. As a regulatory molecule, comparatively little cAMP is required. Consequently, the intracellular cAMP concentration (about 1 nmol/L) is three orders of magnitude below that of ATP. cAMP levels are maintained by the interaction of adenylyl cyclase and **cAMP phosphodiesterase,** which catalyzes hydrolysis of cAMP to 5'-AMP (Figure 35–13).

Adenosine 3'-phosphate-5'-phosphosulfate (phosphoadenosine phosphosulfate), or "active sulfate" (Figure 35–14), is the sulfate donor for the formation of sulfated proteoglycans (Chapter 57) or urinary metabolites of drugs excreted as sulfate conjugates.

Figure 35–14. Formation of adenosine 3'-phosphate-5'-phosphosulfate.

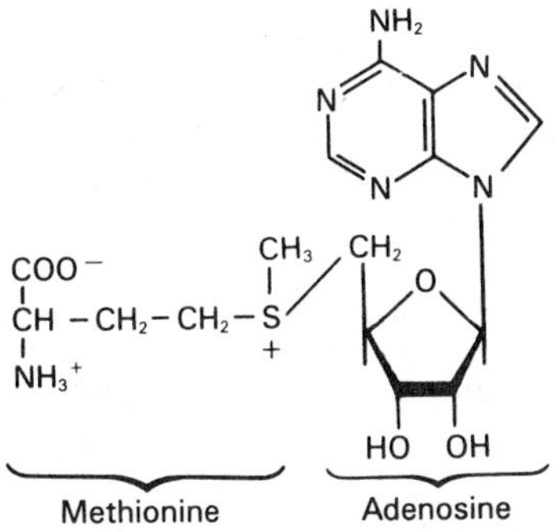

Figure 35–15. S-Adenosylmethionine.

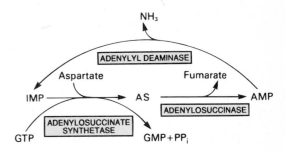

Figure 35–17. The purine nucleotide cycle.

of IMP forms the nucleoside inosine (hypoxanthine riboside), an intermediate in the purine salvage cycle (Chapter 36).

D. Uracil Derivatives: UDP-sugar derivatives participate in sugar epimerizations such as the interconversion of glucose 1-phosphate and galactose 1-phosphate. **UDP-glucose** is the glucosyl donor for biosynthesis of glycogen and glucosyl disaccharides, and other UDP-sugars act as sugar donors for biosynthesis of the oligosaccharides of glycoproteins and proteoglycans (Chapters 56 and 57). **UDP-glucuronic acid** is the glycosidic acid donor for conjugation reactions that form the urinary glucuronide conjugates of bilirubin (Chapter 34) or drugs such as aspirin.

E. Cytosine Derivatives: CTP is required for the biosynthesis of some phosphoglycerides in animal tissues, while reactions involving ceramide and CDP-choline form sphingomyelin and other substituted sphingosines (Chapter 26).

Many Coenzymes Are Nucleotide Derivatives

Many coenzymes incorporate nucleotides as well as structures similar to purine and pyrimidine nucleotides (see Table 35–2 and Chapter 8).

Nucleotides Are Polyfunctional Acids

The pKs of the primary phosphate groups (pK about 1.0) and secondary phosphate groups (pK about 6.2) of mononucleotides ensure that nucleotides bear a negative charge at physiologic pH. By contrast, nucleosides or free purine or pyrimidine bases are uncharged at physiologic pH. They can, however, act as proton donors or acceptors at pH values 2 or more units removed from neutrality.

Nucleotides Absorb Ultraviolet Light

The conjugated double bonds of the heterocyclic bases of purines and pyrimidines ensure that nucleosides, nucleotides, and polynucleotides absorb ultraviolet light. Their spectra are pH-dependent, since

S-Adenosylmethionine (Figure 35–15), a form of "active" methionine, serves as a methyl donor in methylation reactions and as a source of propylamine for the synthesis of polyamines (Chapter 33).

B. Guanosine Derivatives: Guanosine nucleotides participate in the conversion of succinyl-CoA to succinate, a reaction that is coupled to the substrate-level phosphorylation of GDP to GTP. GTP, which is required for activation of adenylyl cyclase by some hormones, serves as an allosteric regulator and as an energy source for protein synthesis.

Cyclic GMP (cGMP; guanosine 3′,5′-monophosphate; Figure 35–16) is an intracellular signal or second messenger that can act antagonistically to cAMP. cGMP is formed from GTP by **guanylyl cyclase** in a reaction analogous to that catalyzed by adenylyl cyclase. Both cyclases are regulated by effectors that include hormones. As for cAMP, a phosphodiesterase hydrolyzes cGMP to its 5′-monophosphate, GMP.

C. Hypoxanthine Derivatives: Hypoxanthine ribonucleotide (IMP), a precursor of purine ribonucleotides (Chapter 36), arises by deamination of AMP, a reaction that in muscle tissue forms part of the purine nucleotide cycle (Figure 35–17). Amination of IMP re-forms AMP, while dephosphorylation

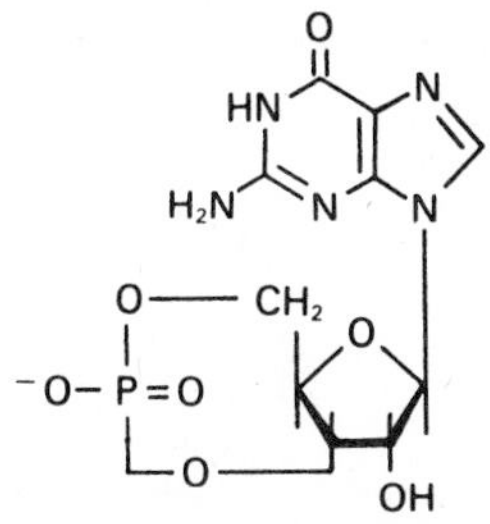

Figure 35–16. Cyclic 3′,5′-guanosine monophosphate (cyclic GMP; cGMP).

Table 35–2. Many coenzymes and related compounds are derivatives of adenosine monophosphate.

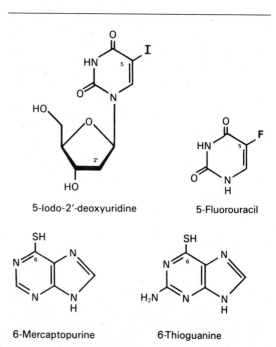

Coenzyme	R	R′	R″	n
Active methionine	Methionine*	H	H	0
Amino acid adenylates	Amino acid	H	H	1
Active sulfate	SO_3^{2-}	H	PO_3^{2-}	1
3′,5′-Cyclic AMP	H	H	PO_3^{2-}	
NAD*	†	H	H	2
NADP*	†	PO_3^{2-}	H	2
FAD	†	H	H	2
CoA-SH	†	H	PO_3^{2-}	2

*Replaces phosphate group.
†R is a B-vitamin derivative.

protonation and deprotonation affect charge distribution. However, at pH 7.0 all the common nucleotides absorb light at a wavelength close to 260 nm. Nucleotide and nucleic acid concentrations thus often are expressed in terms of "OD at 260 nm." Nucleotides exhibit different spectra as pH is varied. pH-dependent spectra thus assist in the identification of individual nucleotides. That ultraviolet light is a potent mutagen is also a consequence of the ability of the nucleotides present in DNA to absorb ultraviolet light.

SYNTHETIC NUCLEOTIDE ANALOGS ARE USED CHEMOTHERAPEUTICALLY

Chemically synthesized analogs of purines and pyrimidines, their nucleosides, and their nucleotides, find numerous applications in clinical medicine and medical science research. Administration of an analog in which either the heterocyclic ring or the sugar moiety has been altered induces toxic effects when the analog is incorporated into specific cellular constituents. Their effects reflect one of two processes: (1) inhibition by the drug of specific enzymes essential for nucleic acid synthesis, or (2) incorporation of metabolites of the drug into nucleic acids, where they affect the base pairing essential to accurate transfer of information. The oncologist's pharmacopeia includes a host of synthetic analogs of purines, pyrimidines, and their nucleosides. Most clinical applications exploit the role of nucleotides as precursors of nucleic acids and the fact that when a cell is about to divide, its DNA is replicated. Examples include the **5-fluoro** or **5-iodo** derivatives of uracil or deoxyuridine, which serve as thymine or thymidine analogs, respectively (Figure 35–18). Both **6-thioguanine** and **6-mercaptopurine,** in which thiol groups replace the hydroxyl groups at the 6-position, are widely used clinically. Analogs such as **5-** or **6-azauridine, 5-** or **6-azacytidine,** and **8-azaguanine** (Figure 35–19), compounds in which a nitrogen atom replaces a heterocyclic ring carbon atom, also are employed clinically.

The purine analog 4-hydroxypyrazolopyrimidine **(allopurinol),** used in treatment of hyperuricemia and gout, inhibits de novo purine biosynthesis and xanthine oxidase activity. The nucleoside **cytarabine** (arabinosyl cytosine; ara-C), in which arabinose replaces ribose, is used in the chemotherapy of cancer and viral infections (Figure 35–20).

Azathioprine, which is catabolized to 6-mercaptopurine, is used during organ transplantation to suppress events involved in immunologic rejection. Among several nucleoside analogs with antiviral activities, **5-iododeoxyuridine** (see above) is effective in the treatment of herpetic keratitis, an infection of the cornea by herpes virus.

Figure 35–18. Synthetic pyrimidine and purine analogs.

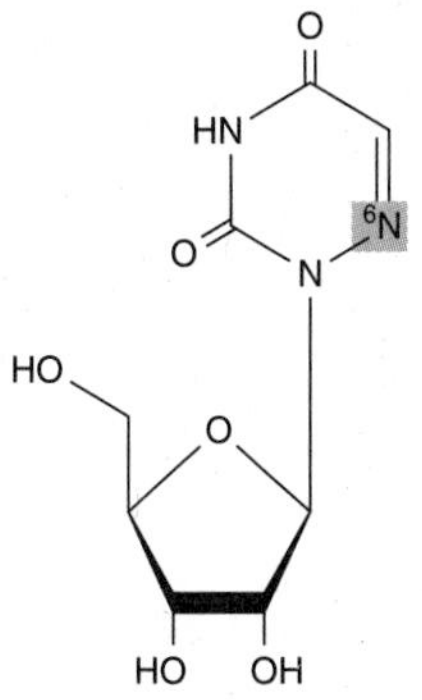

6-Azauridine

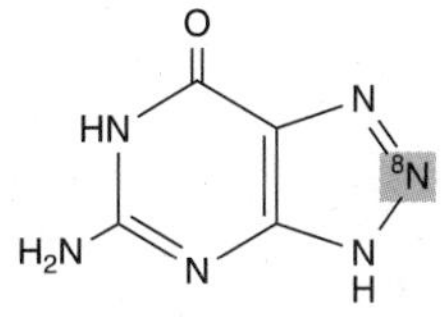

8-Azaguanine

Figure 35–19. 6-Azauridine ***(left)*** and 8-azaguanine ***(right).***

Nonhydrolyzable Nucleoside Triphosphate Analogs Provide Research Tools

Synthetic nonhydrolyzable analogs of nucleoside triphosphates provide valuable investigative tools for medical research. These nucleotide analogs (Figure 35–21) allow investigators to determine whether the effects of nucleoside di- or triphosphates require hydrolysis or whether their effects are mediated by occupying specific nucleotide-binding sites on enzymes or regulatory proteins.

POLYNUCLEOTIDES

The singly esterified 5′-phosphate of a nucleotide can esterify a second alcohol functional group (—OH), forming a **diester.** Most commonly, this second —OH group resides on the pentose of a polynucleotide. For example, in the cyclic nucleotide cAMP, the phosphate is doubly esterified to the 5′-OH and the 3′-OH of the same D-ribose moiety. Alternatively (and more commonly), the second —OH group is present on the pentose of a second polynucleotide. This results in a **dinucleotide** in which the pentose moieties are linked by a 3′ → 5′ phosphodiester bond. The 3′ → 5′-phosphodiester bond forms the "backbone" of polynucleotides such as RNA and DNA.

K_{eq} Favors Phosphodiester Hydrolysis

While we can represent the formation of a dinucleotide as the elimination of water between a pair of monomers, in an aqueous environment K_{eq} strongly favors phosphodiester hydrolysis. Although this might suggest that the phosphodiester bond is of insufficient stability to persist in cells for long periods of time, the contrary is true. Owing to the large energy barrier of the hydrolytic reaction, hydrolysis is,

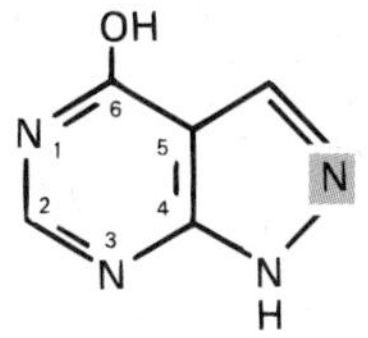

Allopurinol (lactim)

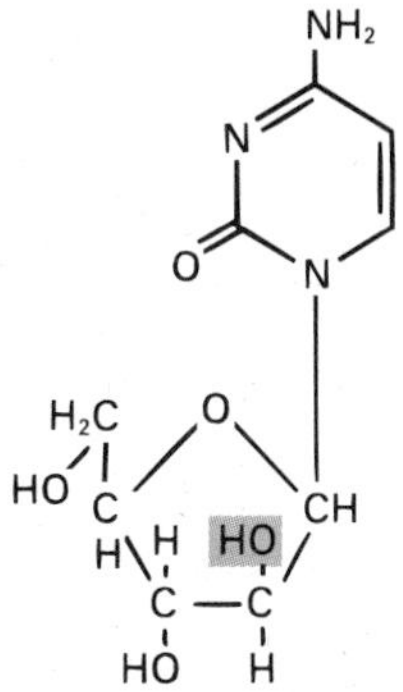

Arabinosyl cytosine

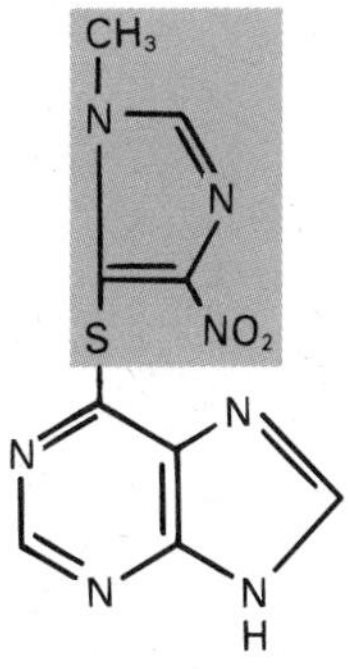

Azathioprine

Figure 35–20. 4-Hydroxypyrazolopyrimidine (allopurinol), arabinosyl cytosine (cytarabine), and azathioprine.

in the absence of catalysis by enzymes known as **phosphodiesterases,** an extremely slow process. Consequently, DNA persists for considerable periods and has even been detected in fossils. However, in the presence of catalysts, hydrolysis of phosphodiester bonds is rapid, ie, when nucleic acids are digested.

Polynucleotides Are Directional Macromolecules

Since the phosphodiester bond links 3′- and 5′-carbons of adjacent monomers, each end of a polymer is distinct. We therefore refer to the "5′- end" or the "3′-end" of polynucleotides. For example, the 5′- end is the end with a free or phosphorylated 5′-hydroxyl.

Polynucleotides Have Primary Structure

The individual character of polynucleotides derives from the sequence of their constituent bases, ie, their **primary structure.** There are several ways to

B—R—O—P(=O)(O⁻)—O—P(=O)(O⁻)—O—P(=O)(O⁻)—O⁻

Parent nucleoside triphosphate

B—R—O—P(=O)(O⁻)—O—P(=O)(O⁻)—CH₂—P(=O)(O⁻)—O⁻

β,γ-Methylene derivative

B—R—O—P(=O)(O⁻)—O—P(=O)(O⁻)—N(H)—P(=O)(O⁻)—O⁻

β,γ-Imino derivative

Figure 35–21. Synthetic derivatives of nucleoside triphosphates incapable of undergoing hydrolytic release of the terminal phosphate group. (B, a purine or pyrimidine base; R, ribose or deoxyribose.) Shown are the parent (hydrolyzable) nucleoside triphosphate *(top)* and the unhydrolyzable β,γ-methylene *(center)* and β,γ-imino derivatives *(bottom).*

represent the primary structures of polynucleotides. In the examples below, P or p represents the phosphodiester bond, bases are represented by single letter notations, and pentoses by a vertical line.

Where all the phosphodiester bonds are 5′ → 3′, a more compact notation is possible:

pGpGpApTpCpA

The above representation implies that the 5′-hydroxyl is phosphorylated and the 3′-hydroxyl is not derivatized.

In the most compact representation, which displays only the base sequence, the 5′- end is, by convention, shown on the left and the 3′- end on the right:

GGATCA

SUMMARY

The amino- and oxo- groups of purines, pyrimidines, and their derivatives exist as amino and imino (—NH₂/═NH) or keto and enol (—CH₂—CHO/—CH═CH—OH) tautomeric pairs. Under physiologic conditions, the amino- and oxo- tautomers predominate. As their nucleotide derivatives, purines and pyrimidines fulfill diverse metabolic functions. Nucleic acids contain, in addition to the purines adenine (A) and guanine (G) and the pyrimidines cytosine (C), thymine (T), and uracil (U), traces of derivatives such as 5-methylcytosine, 5-hydroxymethylcytosine, pseudouridine (Ψ), or variously N-methylated bases.

Nucleosides contain a monosaccharide, frequently D-ribose or 2-deoxy-D-ribose, linked to N-1 (pyrimidines) or N-9 (purines) by a β-glycosidic bond whose *syn* conformers predominate. Nucleotides are phosphate esters of nucleosides. In mononucleotides, a single phosphate is esterified to an —OH of the sugar. When naming mononucleotides, a primed numeral locates the phosphate (3′-AMP, 5′-GMP), although 5′ typically is omitted. Esterification by the same phosphate of a second —OH of the same sugar forms the cyclic phosphodiesters cAMP and cGMP, which function as intracellular "second messengers." Additional phosphates linked to the first by acid anhydride bonds form nucleoside di- and triphosphates. The second and third phosphates of nucleoside triphosphates have high group transfer potential and participate in covalent bond syntheses.

Mononucleotides linked by 3′ → 5′ phosphodiester bonds form polynucleotides, directional macromolecules with distinct 3′- and 5′- ends. In the absence of phosphodiesterases, which catalyze their hydrolysis, phosphodiester bonds are stable. The base sequence constitutes the primary structure of polynucleotides. For the notations pTpGpTp or TGCATCA, the 5′- end appears at the left, and all phosphodiester bonds are 3′ → 5′. Several synthetic analogs of purine and pyrimidine bases and their derivatives are used chemotherapeutically as anticancer drugs.

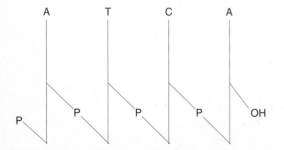

REFERENCES

Adams RLP, Knowler JT, Leader DP: *The Biochemistry of the Nucleic Acids,* 10th ed. Chapman & Hall, 1986.

Blackburn GM, Gait MJ: *Nucleic Acids in Chemistry & Biology.* IRL Press, 1990.

Bugg CE, Carson WM, Montgomery JA: Drugs by design. Sci Am 1992;269:92.

Saenger W: *Principles of Nucleic Acid Structure.* Springer, 1984.

36

Metabolism of Purine & Pyrimidine Nucleotides

Victor W. Rodwell, PhD

INTRODUCTION

This chapter concerns the digestion, biosynthesis, and catabolism of purine and pyrimidine nucleotides and selected diseases associated with genetic defects in these processes.

BIOMEDICAL IMPORTANCE

Even when humans consume a diet rich in nucleoproteins, dietary purine and pyrimidine bases are not incorporated into tissue nucleic acids. Humans biosynthesize the purines and pyrimidines of tissue nucleic acids, ATP, NAD^+, coenzyme A, etc, from amphibolic intermediates. However, *injected* purine or pyrimidine analogues, including potential anticancer drugs, may be incorporated into DNA. The biosyntheses of purine and pyrimidine oxy- and deoxyribonucleotides (NTPs and dNTPs) are precisely regulated events coordinated by feedback mechanisms that ensure production in appropriate quantities and at times appropriate to varying physiologic demand (eg, cell division). Human diseases that involve abnormalities in purine metabolism include gout, Lesch-Nyhan syndrome, adenosine deaminase deficiency, and purine nucleoside phosphorylase deficiency. Diseases of pyrimidine biosynthesis are more rare and include orotic acidurias. Since, unlike the urates, the products of pyrimidine catabolism are highly soluble (carbon dioxide, ammonia, and β-aminoisobutyrate), there are fewer clinically significant disorders of pyrimidine catabolism.

PURINES AND PYRIMIDINES ARE DIETARILY NONESSENTIAL

Although humans ingest dietary nucleic acids and nucleotides, survival does not require their absorption and utilization. Humans and most other vertebrates can synthesize ample amounts of purine and pyrimidine nucleotides de novo (ie, from amphibolic intermediates).

Ingested Nucleic Acids Are Degraded to Purines and Pyrimidines

Nucleic acids released from ingested nucleoproteins in the intestinal tract are degraded to mononucleotides by ribonucleases, deoxyribonucleases, and polynucleotidases. Nucleotidases and phosphatases hydrolyze the mononucleotides to nucleosides, which either are absorbed or are further degraded by intestinal phosphorylase to purine and pyrimidine bases. Purine bases are oxidized to uric acid, which may be absorbed and subsequently excreted in the urine.

While little or no dietary purine or pyrimidine is incorporated into tissue nucleic acids, parenterally administered compounds are incorporated. For example, injected [³H]thymidine is incorporated into newly synthesized DNA. This incorporation provides a technique for measuring rates of DNA synthesis in vivo and in vitro.

BIOSYNTHESIS OF PURINE NUCLEOTIDES

With the exception of parasitic protozoa, all forms of life synthesize purine and pyrimidine nucleotides. Synthesis from amphibolic intermediates proceeds at controlled rates appropriate for all cellular functions. Since demand for nucleotide triphosphates can vary—for example, during growth or when tissues are regenerating and cells are about to divide—rates of purine and pyrimidine biosynthesis are subject to intracellular mechanisms that sense and effectively regulate the pool sizes of these intermediates of nucleic acid synthesis.

Our understanding of the biosynthetic pathway for nucleotide biosynthesis and its regulation in human subjects draws on investigations of the same process in birds and in *Escherichia coli*. In uricotelic animals (birds, amphibians, and reptiles), nucleotides serve the additional function of being precursors of the purine uric acid, the end product of protein nitrogen catabolism (Chapter 31). The excretion by birds of large quantities of uric acid was exploited in early studies of purine biosynthesis. Feeding isotopic precursors to pigeons established the source of each atom of a purine base (Figure 36–1) and initiated study of the reactions and intermediates of purine biosynthesis. More recently, birds have again been exploited to clone the genes that encode enzymes of purine biosynthesis and the regulatory proteins that control the rate of purine biosynthesis.

The three processes that contribute to purine nucleotide biosynthesis, listed in order of decreasing importance, are (1) synthesis from amphibolic intermediates (synthesis de novo), (2) phosphoribosylation of purines, and (3) phosphorylation of purine nucleosides.

INOSINE MONOPHOSPHATE (IMP) ARISES FROM AMPHIBOLIC INTERMEDIATES

Inosine monophosphate (IMP) is the "parent" nucleotide from which both AMP and GMP are formed. Synthesis of IMP begins with the amphibolic intermediate α-D-ribose-5-phosphate and involves a linear sequence of 11 reactions (Figure 36–2). The pathway then branches, one path leading from IMP to AMP, the other from IMP to GMP (Figure 36–3).

In what follows, Arabic numerals designate correspondingly numbered *reactions* of Figures 36–2 and 36–3, and Roman numerals designate *structures* in Figure 36–2.

(1) In addition to being the first intermediate formed in the de novo pathway for purine biosynthesis, 5-phosphoribosyl-1-pyrophosphate (PRPP) (II) is an intermediate in the purine salvage pathway, in the biosynthesis of NAD^+ and $NADP^+$, and in the biosynthesis of pyrimidine nucleotides. Synthesis of PRPP involves transfer of pyrophosphate from ATP to carbon 1 of α-D-ribose-5-phosphate (I) and is catalyzed by PRPP synthetase.

(2) Synthesis of an *N*-glycosidic bond employs glutamine as the nitrogen donor and forms 5-phospho-β-D-ribosylamine (III). The 5′-phosphoribosylglycinamide synthetase reaction inverts configuration at the anomeric carbon of the sugar from α to β. The reaction is highly favored owing to the accompanying release of pyrophosphate and its subsequent hydrolysis to inorganic orthophosphate catalyzed by pyrophosphatase.

(3) Condensation of 5-phospho-β-D-ribosylamine (III) with glycine forms glycinamide ribosyl-5-phosphate (IV). In this reaction, glycine provides what will become carbons 4 and 5 and nitrogen 7 of IMP.

(4) Carbon 8 of IMP derives from the formyl group of N^5,N^{10}-methenyl-tetrahydrofolate, forming formylglycinamide ribosyl-5-phosphate (V), a reaction catalyzed by glycinamide ribosyl-5-phosphate formyltransferase.

(5) Transfer to (V) of the amide nitrogen of glutamine forms formylglycinamidine ribosyl-5-phosphate (VI). Catalyzed by formylglycinamidine ribosyl-5-phosphate synthetase, this reaction adds the atom that will become nitrogen 3 of IMP.

(6) In the reaction catalyzed by aminoimidazole ribosyl-5-phosphate synthetase, loss of water accompanied by ring closure forms aminoimidazole ribosyl-5-phosphate (VII). The initial event is phosphoryl transfer from ATP to the oxo function of (VI). Subsequent nucleophilic attack by the adjacent amino nitrogen results in ring closure and release of inorganic orthophosphate.

(7) Addition of CO_2 to (VII) adds the atom that will become carbon 6 of IMP. The reaction, catalyzed by aminoimidazole ribosyl-5-phosphate carboxylase, requires neither ATP nor biotin and forms aminoimidazole carboxylate ribosyl-5-phosphate (VIII).

(8) Condensation of aspartate with (VIII), catalyzed by succinyl carboxamide ribosyl-5-phosphate synthetase, forms aminoimidazole succinyl carboxamide ribosyl-5-phosphate (IX).

(9) Liberation of the succinyl group of (IX) as fumarate, catalyzed by adenylosuccinase, forms aminoimidazole carboxamide ribosyl-5-phosphate (X). Note that reactions 8 and 9, which add the atom that becomes nitrogen 1 of IMP, parallel the two reactions of the urea cycle that convert ornithine to arginine (Figure 31–14).

(10) Carbon 2 of IMP is added in a reaction that involves a second tetrahydrofolate derivative and a second formyltransferase. Transfer to (X) of the formyl group of N^{10}-formyl-tetrahydrofolate forms formimidoimidazole carboxamide ribosyl-5-phosphate (XI).

(11) Ring closure of (XI) catalyzed by IMP cyclo-

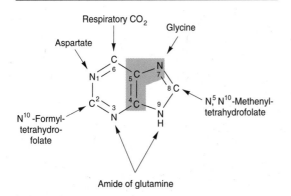

Figure 36–1. The sources of the nitrogen and carbon atoms of the purine ring. Atoms 4, 5, and 7 (shaded) derive from glycine.

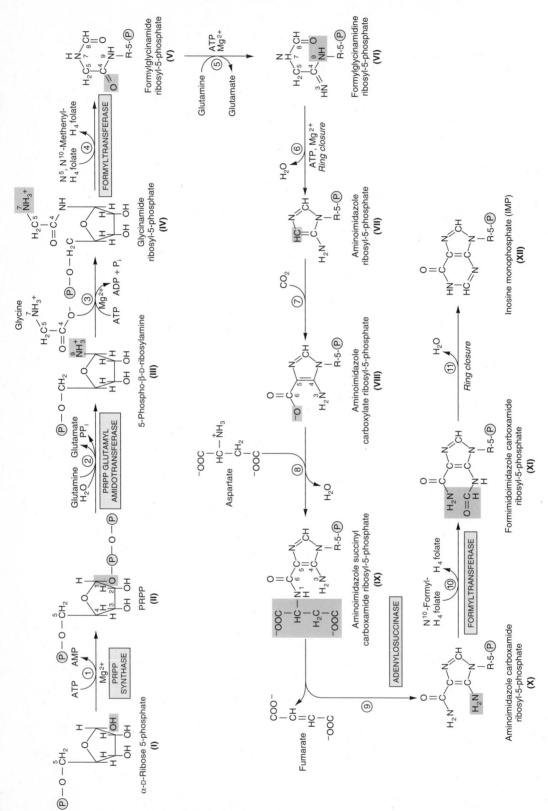

Figure 36–2. The pathway of de novo purine biosynthesis from ribose 5-phosphate and ATP. (See text for explanation.) Ⓟ, PO_3^{2-} or PO_2^-.)

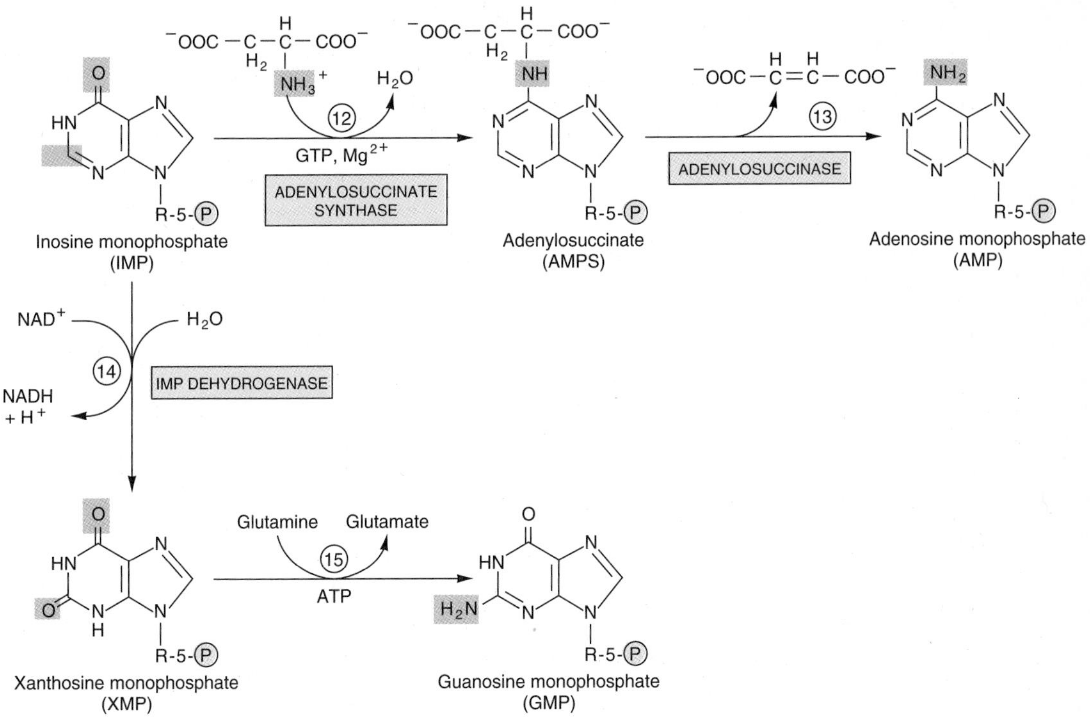

Figure 36–3. Conversion of IMP to AMP and GMP.

hydrolase forms the first purine nucleotide, inosine monophosphate, IMP (XII).

Conversion of IMP to AMP and GMP: Following synthesis of IMP, the pathway branches, and two short reaction sequences lead to the formation of AMP and GMP (Figure 36–3).

(12) Addition of aspartate to IMP forms adenylosuccinate. The adenylosuccinate synthase reaction, while superficially similar to reaction 8, requires GTP and thus provides a potential focus for regulation of adenine nucleotide biosynthesis.

(13) Release of fumarate forming adenosine 5'-monophosphate (AMP) is catalyzed by adenylosuccinase, the same enzyme that catalyzes reaction 9.

(14) Oxidation of IMP by NAD^+, catalyzed by IMP dehydrogenase, forms xanthosine monophosphate (XMP).

(15) Transamidation of the 6-oxo group of XMP by the amide nitrogen of glutamine proceeds by analogy to reaction 5.

PHOSPHORYL TRANSFER FROM ATP CONVERTS MONONUCLEOTIDES TO NUCLEOSIDE DI- AND TRIPHOSPHATES

The mononucleotides AMP and GMP are converted to their nucleoside diphosphates (ADP and GDP) via phosphoryl transfer from ATP catalyzed by

nucleoside monophosphate kinase. GDP is then converted to GTP by nucleoside diphosphate kinase at the expense of another ATP (Figure 36–4). Conversion of ADP to ATP is achieved primarily by oxidative phosphorylation and secondarily by reactions of glycolysis and the citric acid cycle.

Multifunctional Catalysts Participate in Purine Nucleotide Biosynthesis

In prokaryotes, each reaction of Figure 36–2 is catalyzed by a different polypeptide. By contrast, gene fusion has given rise in eukaryotes to single polypeptides with multiple catalytic functions. For purine biosynthesis, three multifunctional catalysts catalyze reactions 3, 4, and 6, reactions 7 and 8, and reactions 10 and 11, respectively. Multifunctionality confers several advantages. Adjacent catalytic sites facilitate rapid and complete transfer of intermedi-

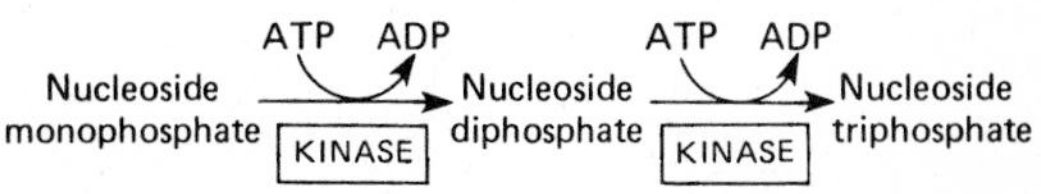

Figure 36–4. Conversion of nucleoside monophosphates to nucleoside diphosphates and nucleoside triphosphates.

ates, and gene fusion ensures production of equal quantities of different catalytic activities.

Antifolate Drugs or Glutamine Analogs Block Purine Nucleotide Biosynthesis

The two carbons inserted in reactions 4 and 10 (Figure 36–2) derive from N^5,N^{10}-methenyl- and N^{10}-formyl-tetrahydrofolate. Inhibiting formation of tetrahydrofolate compounds thus can block purine synthesis. Inhibitory compounds and the reactions they inhibit include azaserine (reaction 5), diazanorleucine (reaction 2), 6-mercaptopurine (reactions 13 and 14), and mycophenolic acid (reaction 14).

Purine Deficiency Is Rare in Humans

Purine deficiency states in humans are due primarily to deficiencies of folic acid and occasionally to a deficiency of vitamin B_{12} when this results in a secondary deficiency of folate derivatives.

"SALVAGE" REACTIONS CONVERT PURINES AND THEIR NUCLEOSIDES TO MONONUCLEOTIDES

Conversion of purines, purine ribonucleosides, and purine deoxyribonucleosides to mononucleotides involves so-called salvage reactions that require far less energy than does de novo synthesis. Quantitatively, the more important mechanism involves phosphoribosylation of a free purine (Pu) by PRPP, forming a purine 5′-mononucleotide (Pu-RP).

$$Pu + PP - RP \rightarrow Pu - RP + PP_i$$

PRPP-dependent phosphoribosylation of purines is catalyzed by adenine phosphoribosyltransferase (converts adenine to AMP; Figure 36–5) and hypo-

xanthine-guanine phosphoribosyltransferase (converts hypoxanthine or guanine to IMP or GMP; Figure 36–6).

A second salvage mechanism involves direct phosphorylation of a purine ribonucleoside (PuR) by ATP:

$$PuR + ATP \rightarrow PuR{-}P + ADP$$

Adenosine kinase catalyzes phosphorylation of adenosine to AMP or of deoxyadenosine to dAMP. Deoxycytidine kinase phosphorylates deoxycytidine, deoxyadenosine, and 2′-deoxyguanosine to dCMP, dAMP, and dGMP, respectively.

Mammalian liver, the major site of purine nucleotide biosynthesis, provides purines and their nucleosides for salvage and utilization by tissues incapable of their biosynthesis. For example, human brain has a low level of PRPP amidotransferase and hence depends, in part, on exogenous purines. Erythrocytes and polymorphonuclear leukocytes cannot synthesize 5-phosphoribosylamine and hence utilize exogenous purines to form nucleotides. Peripheral lymphocytes, however, possess some ability to synthesize purines de novo.

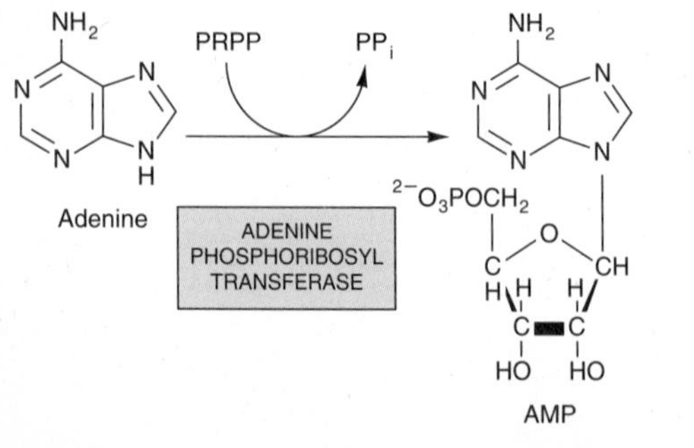

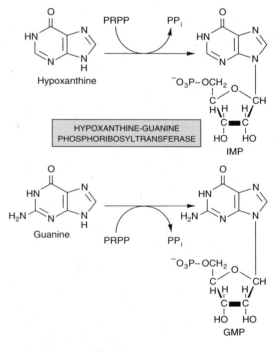

Figure 36–6. Phosphoribosylation of hypoxanthine and guanine to form IMP and GMP, respectively. Both reactions are catalyzed by hypoxanthine-guanine phosphoribosyl transferase.

Figure 36–5. Phosphoribosylation of adenine catalyzed by adenine phosphoribosyltransferase.

HEPATIC PURINE NUCLEOTIDE BIOSYNTHESIS IS STRINGENTLY REGULATED

PRPP Pool Size Regulates Purine Nucleotide Biosynthesis

Since the biosynthesis of IMP from amphibolic intermediates consumes glycine, glutamine, tetrahydrofolate derivatives, aspartate, and ATP, it is imperative that cells regulate purine biosynthesis. The major determinant of the overall rate of de novo purine nucleotide biosynthesis is the concentration of PRPP, a parameter that reflects the relative rates of PRPP synthesis, utilization, and degradation. The rate of PRPP synthesis depends both on the availability of ribose 5-phosphate and on the activity of PRPP synthetase, an enzyme sensitive both to phosphate concentration and to the purine ribonucleotides that act as its allosteric regulators (Figure 36–7).

AMP and GMP Feedback-Regulate PRPP Glutamyl Amidotransferase

PRPP glutamyl amidotransferase (reaction 2, Figure 36–2), the first enzyme uniquely committed to purine synthesis, is feedback-inhibited by purine nucleotides, particularly AMP and GMP, which inhibit competitively with PRPP (Figure 36–7). However, regulation of purine synthesis via the amidotransferase is probably of less physiologic importance than regulation of PRPP synthetase.

AMP and GMP Feedback-Regulate Their Formation From IMP

Two mechanisms regulate conversion of IMP to GMP and AMP (Figure 36–8). AMP feedback-regulates adenylosuccinate synthetase, and GMP feedback-inhibits IMP dehydrogenase. Furthermore, conversion of IMP to adenylosuccinate en route to AMP requires GTP, and conversion of xanthinylate (XMP) to GMP requires ATP. Cross-regulation between the pathways of IMP metabolism thus serves to decrease synthesis of one purine nucleotide when there is a deficiency of the other nucleotide. AMP and GMP also inhibit hypoxanthine-guanine phosphoribosyltransferase (converts hypoxanthine and guanine to IMP and GMP; Figure 36–6).

REDUCTION OF NDPs FORMS dNDPs

Reduction at the 2'-carbon of purine and pyrimidine ribonucleotides, catalyzed by the **ribonucleotide reductase complex** (Figure 36–9), forms the deoxyribonucleoside diphosphates (dNDPs). The

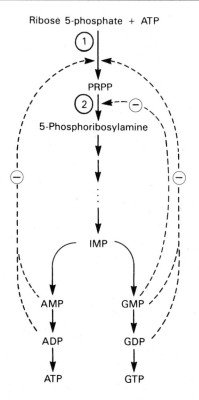

Figure 36–7. Control of the rate of de novo purine nucleotide synthesis. Solid lines represent chemical flow, and broken lines represent feedback inhibition (⊖) by end products of the pathway. Reactions ① and ② are catalyzed by PRPP synthetase and by PRPP glutamyl amidotransferase (Figure 36–2), respectively.

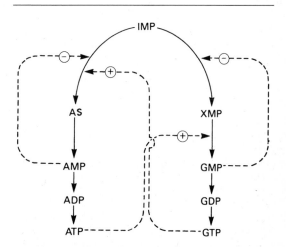

Figure 36–8. Regulation of the interconversion of IMP to adenosine nucleotides and guanosine nucleotides. Solid lines represent chemical flow, and broken lines represent both positive (⊕) and negative (⊖) feedback regulation.

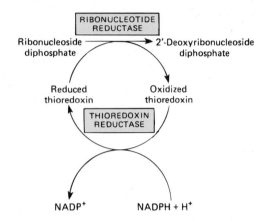

Figure 36–9. Reduction of ribonucleoside diphosphates to 2′-deoxyribonucleoside diphosphates.

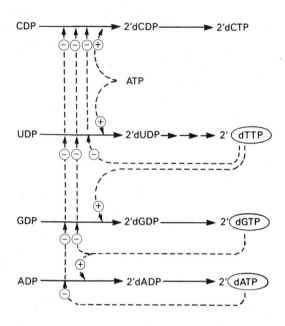

Figure 36–10. Regulation of the reduction of purine and pyrimidine ribonucleotides to their respective 2′-deoxyribonucleotides. Solid lines represent chemical flow, and broken lines represent negative ($\ominus$) or positive ($\oplus$) feedback regulation.

enzyme complex is active only when cells are actively synthesizing DNA preparatory to cell division. Reduction requires thioredoxin (a protein cofactor), thioredoxin reductase (a flavoprotein), and NADPH (and, in certain bacteria, though not in mammals, cobalamin or vitamin B_{12}). The immediate reductant of the NDP is reduced thioredoxin, produced in a reaction catalyzed by NADPH:thioredoxin reductase (Figure 36–9).

Reduction of ribonucleoside diphosphates (NDPs) to deoxyribonucleoside diphosphates (dNDPs) is subject to complex regulation (Figure 36–10), which achieves balanced production of deoxyribonucleotides for synthesis of DNA.

BIOSYNTHESIS OF PYRIMIDINE NUCLEOTIDES

Pyrimidine and purine nucleoside biosynthesis share several common precursors: PRPP, glutamine, CO_2, aspartate, and, for thymine nucleotides, tetrahydrofolate derivatives. However, while ribose phosphate is an early reactant in purine nucleotide synthesis (Figure 36–2), attachment of the ribose phosphate moiety to N-3 of a pyrimidine base occurs much later than in purine biosynthesis (Figure 36–11).

(1) Pyrimidine biosynthesis begins with the formation, from glutamine, ATP, and CO_2, of carbamoyl phosphate. This reaction is catalyzed by *cytosolic* carbamoyl phosphate synthase II, an enzyme distinct from the *mitochondrial* carbamoyl phosphate synthase I functional in urea synthesis (Figure 31–14). Compartmentation thus provides independent pools of carbamoyl phosphate for each process.

(2) Condensation of carbamoyl phosphate with aspartate forms carbamoyl aspartate in a reaction catalyzed by aspartate transcarbamoylase.

(3) Ring closure via loss of water, catalyzed by dihydroorotase, forms dihydroorotic acid.

(4) Abstraction of hydrogens from C-5 and C-6 by NAD^+ introduces a double bond, forming orotic acid, a reaction catalyzed by *mitochondrial* dihydroorotate dehydrogenase. All other enzymes of pyrimidine biosynthesis are *cytosolic*.

(5) Transfer of a ribose phosphate moiety from PRPP, forming orotidine monophosphate (OMP), is catalyzed by orotate phosphoribosyltransferase. Formation of the β-*N*-glycosidic bond thus is analogous to the transribosylation reactions of Figure 36–6. Only at the penultimate reaction of UMP synthesis is the pyrimidine ring phosphoribosylated.

(6) Decarboxylation of orotidylate forms uridine monophosphate (UMP), the first true pyrimidine ribonucleotide.

(7, 8) Phosphate transfer from ATP yields UDP and UTP in reactions analogous to those for phosphorylation of purine nucleoside monophosphates (Figure 36–4).

(9) UTP is aminated to CTP by glutamine and ATP.

(10) Reduction of ribonucleoside diphosphates (NDPs) to their corresponding dNDPs involves reac-

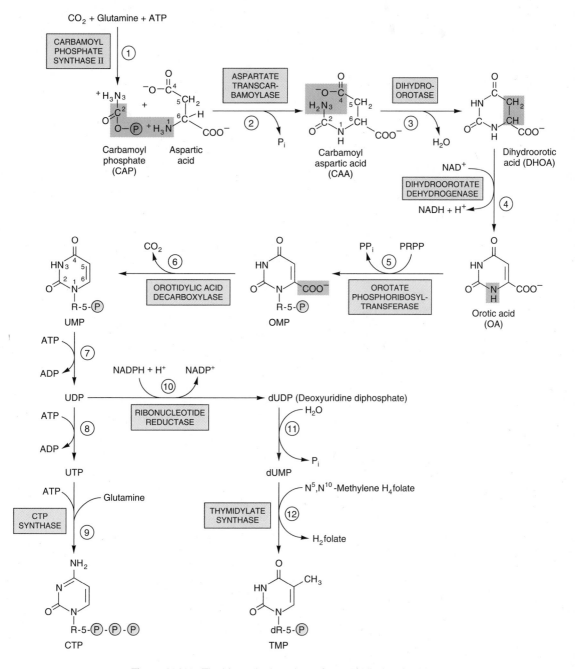

Figure 36–11. The biosynthetic pathway for pyrimidine nucleotides.

tions analogous to those for purine nucleosides (Figures 36–9 and 36–10).

(11) dUMP may accept a phosphate from ATP, forming dUTP (not shown). Alternatively, and since the substrate for thymidine monophosphate (TMP) synthesis is dUMP, dUDP is dephosphorylated to dUMP.

(12) Methylation of dUMP at C-5 by N^5,N^{10}-meth-

ylene-tetrahydrofolate, catalyzed by thymidylate synthase, forms thymidine monophosphate (TMP).

Multifunctional Proteins Catalyze the Early Reductions of de Novo Pyrimidine Biosynthesis

In man and other animals, five of the first six enzymes of de novo pyrimidine biosynthesis are organized as multifunctional polypeptides rather than as distinct enzymes. The sole exception is dihydroorotate dehydrogenase (reaction 4). A single gene encodes CAD, a 220 kDa polypeptide that contains the first three enzyme activities: carbamoyl phosphate synthase (CPS), aspartate transcarbamoylase (ATC), and dihydroorotase (DHO). Termed CAD (for *CPS*, *ATC*, *DHO*), this multifunctional enzyme consists of three distinct catalytic domains arranged in the order NH$_2$-DHO-CPS-ATC-COOH. The close association of these activities ensures that almost all of the carbamoyl phosphate produced by ATC, which is termed ATC-II to distinguish it from the ATC-I functional in urea biosynthesis (Chapter 31), is channeled to pyrimidine biosynthesis. An analogous bifunctional protein, UMP synthase, contains the activities for orotate phosphoribosyl transferase (reaction 5) and orotidine 5′-monophosphate decarboxylase (reaction 6).

URACIL AND CYTOSINE RIBO- AND DEOXYRIBONUCLEOSIDES ARE SALVAGED

While mammalian cells salvage few free pyrimidines, salvage reactions convert two pyrimidine ribonucleosides (uridine and cytidine) and two deoxyribonucleosides (thymidine and deoxycytidine) to their respective nucleotides (Figure 36–12). 2′-Deoxycytidine is phosphorylated by deoxycytidine kinase, an enzyme that also phosphorylates deoxyguanosine and deoxyadenosine. Orotate phosphoribosyltransferase (reaction 5, Figure 36–11), an enzyme of pyrimidine nucleotide synthesis, salvages orotic acid by converting it to OMP.

Methotrexate Blocks Reduction of Dihydrofolate

Reaction 12 of Figure 36–11 is the sole reaction of pyrimidine nucleotide biosynthesis that requires a tetrahydrofolate derivative. During the transfer process, the methylene group of N^5,N^{10}-methylenetetrahydrofolate is reduced to a methyl group, and the tetrahydrofolate carrier is oxidized to dihydrofolate. For further synthesis to occur, dihydrofolate must be reduced to tetrahydrofolate in a reaction catalyzed by dihydrofolate reductase. Consequently, dividing cells, which are by necessity generating TMP and dihydrofolate, are especially sensitive to inhibitors of dihydrofolate reductase. One such inhibitor is the widely used anticancer drug **methotrexate.**

Certain Pyrimidine Analogs Are Substrates for Enzymes of Pyrimidine Nucleotide Biosynthesis

Orotate phosphoribosyltransferase, while it cannot use normal pyrimidine bases as substrates, catalyzes conversion of the drug **allopurinol** (4-hydroxypyrazolopyramidine) to a nucleotide in which the ribosyl phosphate is attached to N-1 of the pyrimidine ring of allopurinol. The anticancer drug **5-fluorouracil** is also phosphoribosylated by orotate phosphoribosyl transferase.

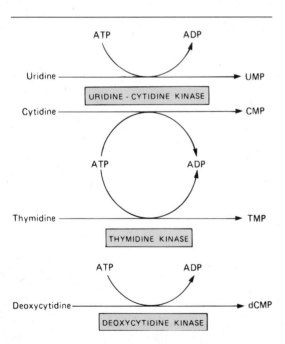

Figure 36–12. The pyrimidine nucleoside kinase reactions responsible for formation of pyrimidine nucleoside monophosphates.

REGULATION OF PYRIMIDINE NUCLEOTIDE BIOSYNTHESIS

Gene Expression and Enzyme Activity Both Are Regulated

The first two enzymes of pyrimidine nucleotide biosynthesis are sensitive to **allosteric regulation,** and the first three and last two enzymes of the pathway are regulated at the genetic level by apparently coordinate repression and derepression. Carbamoyl phosphate synthase II (reaction 1, Figure 36–11) is inhibited by UTP and purine nucleotides but activated by PRPP (Figure 36–13). Aspartate transcarbamoylase (reaction 2, Figure 36–11) is inhibited by CTP and activated by ATP, a classic example of allostery.

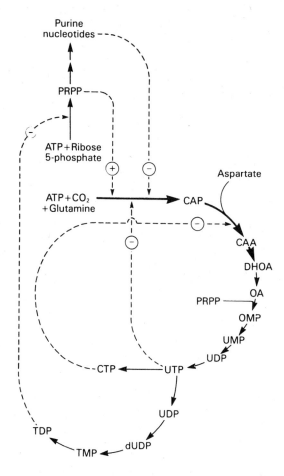

Figure 36–13. Control of pyrimidine nucleotide synthesis. Solid lines represent chemical flow. Broken lines represent positive ($\oplus$) and negative ($\ominus$) feedback regulation.

Purine and Pyrimidine Nucleotide Biosynthesis Are Coordinately Regulated Processes

Mole for mole, pyrimidine biosynthesis parallels purine biosynthesis, suggesting coordinate control. Several sites of cross-regulation characterize purine and pyrimidine nucleotide biosynthesis. The PRPP synthetase reaction (reaction 1, Figure 36–2), which forms a precursor essential for both processes, is subject to feedback inhibition by both purine and pyrimidine nucleotides, and is activated by PRPP.

HUMANS CATABOLIZE PURINES TO URIC ACID

Humans convert the major purine nucleosides adenosine and guanosine to the excreted end product uric acid via the intermediates and reactions shown in Figure 36–14. Adenosine is first deaminated to in-

osine by adenosine deaminase. Phosphorolysis of the *N*-glycosidic bonds of inosine and guanosine, catalyzed by purine nucleoside phosphorylase, releases ribose 1-phosphate and a purine base. Hypoxanthine and guanine next form xanthine in reactions catalyzed by xanthine oxidase and guanase, respectively. Xanthine is then oxidized to uric acid in a second reaction catalyzed by xanthine oxidase. Xanthine oxidase thus provides a potential locus for pharmacologic intervention in patients with **hyperuricemia** and **gout** (see below).

Net excretion of total uric acid in normal humans averages 400–600 mg/24 h. Many pharmacologic and naturally occurring compounds influence renal absorption and secretion of sodium urate. For example, high doses of aspirin competitively inhibit both urate excretion and reabsorption.

In mammals other than higher primates, the enzyme uricase cleaves uric acid, forming the highly water-soluble end product allantoin (Figure 36–15). However, since humans lack uricase, the end product of purine catabolism in man is uric acid. Amphibians, birds, and reptiles also lack uricase and excrete uric acid and guanine as end products of purine catabolism.

Urates Are More Soluble Than Uric Acid

As for any weak acid, the relative proportions of the undissociated weak acid uric acid and its conjugate base urate depend upon pH. Only the first proton dissociation ($pK_1 = 5.8$) need be considered here, since pK_2 for the second proton is 10.3, a value well above that of physiologic fluids. Thus, only uric acid and its monosodium urate salt are present in body fluids.

Urates are far more water-soluble than uric acid. Urine at pH 5 can dissolve only about one-tenth as much total urates (15 mg/dL) as urine at pH 7 (150–200 mg/dL), and the pH of normal urine typically is below 5.8. Urinary tract crystals thus are sodium urate anywhere proximal to the site of urine acidification (the distal tubule and collecting ducts) but uric acid at distal sites. Since most stones of the urinary collecting system are composed of uric acid, stone formation can be reduced by alkalinization of the urine.

GOUT IS A METABOLIC DISORDER OF PURINE CATABOLISM

In **hyperuricemia,** serum urate levels exceed the solubility limit. Resulting crystallization of sodium urate in soft tissues and joints forms deposits called tophi, causing an inflammatory reaction, **acute gouty arthritis,** which can progress to **chronic gouty arthritis.** Visualization under a polarizing light microscope of needle-shaped, intensively negatively

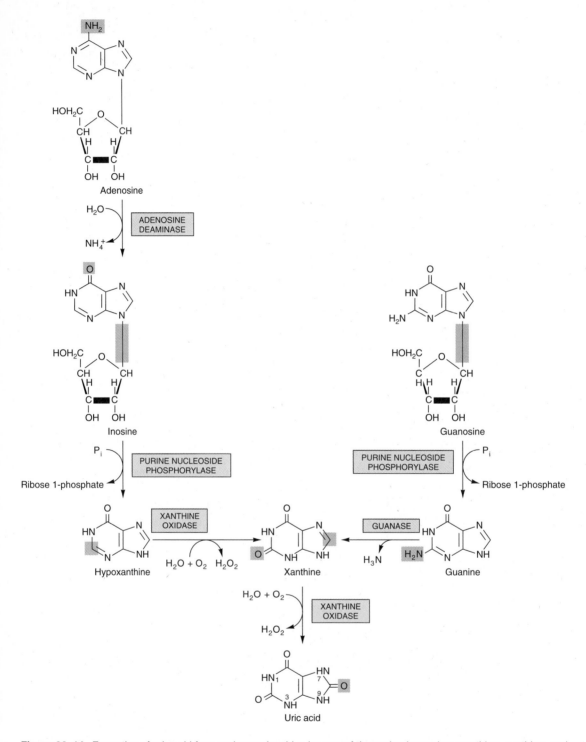

Figure 36–14. Formation of uric acid from purine nucleosides by way of the purine bases hypoxanthine, xanthine, and guanine. Purine deoxyribonucleosides are degraded by the same catabolic pathway and enzymes, all of which exist in the mucosa of the mammalian gastrointestinal tract.

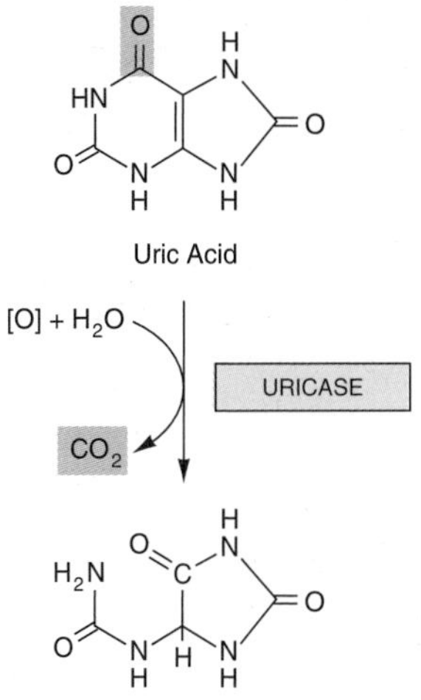

Figure 36–15. Conversion of uric acid to allantoin. The reaction does not occur in humans.

birefringent crystals of sodium urate in joint fluid is diagnostic of gout. The crystals appear yellow when their long axis is parallel to the plane of polarized light and blue when perpendicular to it.

Isotopic Methods Measure Miscible Urate Pools

The dilution of intravenously administered [^{15}N]uric acid may be used to calculate a subject's **miscible urate pool.** This pool averages 1200 mg in normal adult males and 600 mg in normal adult females but ranges from 2000 to 4000 mg in gouty patients without tophi and may be as high as 30,000 mg in patients with severe tophaceous gout.

OTHER DISORDERS OF PURINE CATABOLISM

While purine deficiency states are rare in human subjects, numerous genetic disorders of purine catabolism have been characterized. Table 36–1 summarizes the major symptomatic presentations and inheritance patterns of several diseases of purine catabolism. **Hyperuricemias** may be differentiated based on whether patients excrete normal or excessive quantities (over 600 mg/24 h) of total urates (Table 36–2). While some hyperuricemias reflect specific enzyme defects, several hyperuricemias are secondary to diseases such as cancer or psoriasis that enhance tissue turnover.

Lesch-Nyhan Syndrome

Lesch-Nyhan syndrome, an overproduction hyperuricemia with frequent uric acid lithiasis and a bizarre syndrome of self-mutilation, is due to a nonfunctional **hypoxanthine-guanine phosphoribosyl transferase,** an enzyme of purine salvage (Figure

Table 36–1. Inherited disorders of purine metabolism and their associated enzyme abnormalities.

Clinical Disorder	Defective Enzyme	Nature of the Defect	Characteristics of Clinical Disorder	Inheritance Pattern
Gout	PRPP synthetase	Superactive (increased V_{max})	Purine overproduction and overexcretion	X-linked recessive
Gout	PRPP synthetase	Resistance to feedback inhibition	Purine overproduction and overexcretion	X-linked recessive
Gout	PRPP synthetase	Low K_m for ribose 5-phosphate	Purine overproduction and overexcretion	Probably x-linked recessive
Gout	HGPRTase[1]	Partial deficiency	Purine overproduction and overexcretion	X-linked recessive
Lesch-Nyhan syndrome	HGPRTase[1]	Complete deficiency	Purine overproduction and overexcretion; self-mutilation	X-linked recessive
Immunodeficiency	Adenosine deaminase	Severe deficiency	Combined (T cell and B cell) immunodeficiency, deoxyadenosinuria	Autosomal recessive
Immunodeficiency	Purine nucleoside phosphorylase	Severe deficiency	T cell deficiency, inosinuria, deoxyinosinuria, guanosinuria, deoxyguanosinuria, hypouricemia	Autosomal recessive
Renal lithiasis	Adenine phosphoribosyltransferase	Complete deficiency	2,8-Dihydroxyadenine renal lithiasis	Autosomal recessive
Xanthinuria	Xanthine oxidase	Complete deficiency	Xanthine renal lithiasis, hypouricemia	Autosomal recessive

[1]HGPRTase = hypoxanthine-guanine phosphoribosyltransferase (Figure 36–6).

Table 36–2. Classification of patients with hyperuricemia.

I. Normal excretion of urate; renal disorder responsible for elevated serum urate.
II. Excessive excretion of urate because of overproduction.
   A. Secondary to other disease, eg, cancer, psoriasis.
   B. Known enzyme defects responsible for overproduction.
      1. PRPP synthetase abnormalities.
      2. Hypoxanthine-guanine phosphoribosyltransferase deficiencies.
      3. Glucose-6-phosphatase deficiencies.
   C. Unrecognized defects.

36–6). The accompanying rise in intracellular PRPP, which has been spared from purine salvage, results in purine overproduction. Less deleterious mutations that result in low enzyme activity are characterized by severe hyperuricemia but without neurologic symptoms.

Von Gierke's Disease

Purine overproduction and hyperuricemia in von Gierke's disease (**glucose-6-phosphatase deficiency**) occurs secondarily to enhanced generation of the PRPP precursor ribose 5-phosphate. In addition, associated lactic acidosis elevates the renal threshold for urate, elevating total body urates.

Hypouricemia

Hypouricemia and increased excretion of hypoxanthine and xanthine are associated with **xanthine oxidase deficiency,** due either to a genetic defect or to severe liver damage. In severe xanthine oxidase deficiency, patients may exhibit **xanthinuria** and **xanthine lithiasis.**

Adenosine Deaminase and Purine Nucleoside Phosphorylase Deficiency

Adenosine deaminase deficiency is associated with a severe combined immunodeficiency disease in which both thymus-derived lymphocytes (T cells) and bone marrow-derived lymphocytes (B cells) are sparse and dysfunctional. **Purine nucleoside phosphorylase deficiency** is associated with a severe thymus-derived lymphocyte deficiency with apparently normal B cell function. Both are autosomal recessive disorders. Immune dysfunctions appear to result from accumulation of dGTP and dATP, which allosterically inhibits ribonucleotide reductase, and thereby depletes cells of DNA precursors, particularly dCTP.

CATABOLISM OF PYRIMIDINES PRODUCES WATER-SOLUBLE METABOLITES

By contrast with the sparingly soluble products of purine catabolism, the end products of pyrimidine catabolism are highly water-soluble: CO_2, NH_3, β-alanine, and β-aminoisobutyrate (Figure 36–16).

Excretion of β-aminoisobutyrate increases in leukemia and severe x-ray radiation exposure due to increased destruction of DNA. Abnormally high excretion of β-aminoisobutyrate, traceable to a recessive gene, occurs in the heterozygous offspring of otherwise normal individuals. Approximately 25% of tested persons of Chinese or Japanese ancestry consistently excrete large amounts of β-aminoisobutyrate. By analogy to other animals, humans probably transaminate β-aminoisobutyrate to methylmalonate semialdehyde, which then forms succinyl-CoA (Figure 21–2).

Pseudouridine Is Excreted Unchanged

Since no human enzyme catalyzes hydrolysis or phosphorolysis of pseudouridine, this unusual nucleoside, which while it was first detected in human urine occurs only in tRNAs, is excreted unchanged in the urine of normal subjects.

OVERPRODUCTION OF PYRIMIDINE CATABOLITES IS RARELY ASSOCIATED WITH CLINICALLY SIGNIFICANT ABNORMALITIES

Since the end products of pyrimidine catabolism are highly water-soluble, pyrimidine overproduction results in few clinically detectable abnormalities (Table 36–3). In hyperuricemia associated with severe overproduction of PRPP, there is overproduction of pyrimidine nucleotides and increased excretion of β-alanine. Since N^5,N^{10}-methylene-tetrahydrofolate is required for thymidylate synthesis, disorders of folate and vitamin B_{12} metabolism result in deficiencies of TMP. The orotic aciduria that accompanies **Reye's syndrome** probably is secondary to the inability of severely damaged mitochondria to utilize carbamoyl phosphate, which then may become available for cytosolic overproduction of orotic acid.

Orotic Aciduria Responds to Dietary Pyrimidine Nucleosides

Type I orotic aciduria reflects a deficiency of both orotate phosphoribosyltransferase and orotidylate decarboxylase (reactions 5 and 6, Figure 36–11), and the rarer **type II orotic aciduria** reflects a deficiency only of orotidylate decarboxylase (reaction 6, Figure 36–11). Both type I and type II patients respond to oral uridine, following which the greatly increased activities of aspartate transcarbamoylase and dihydroorotase in patients with type I orotic aciduria return to normal.

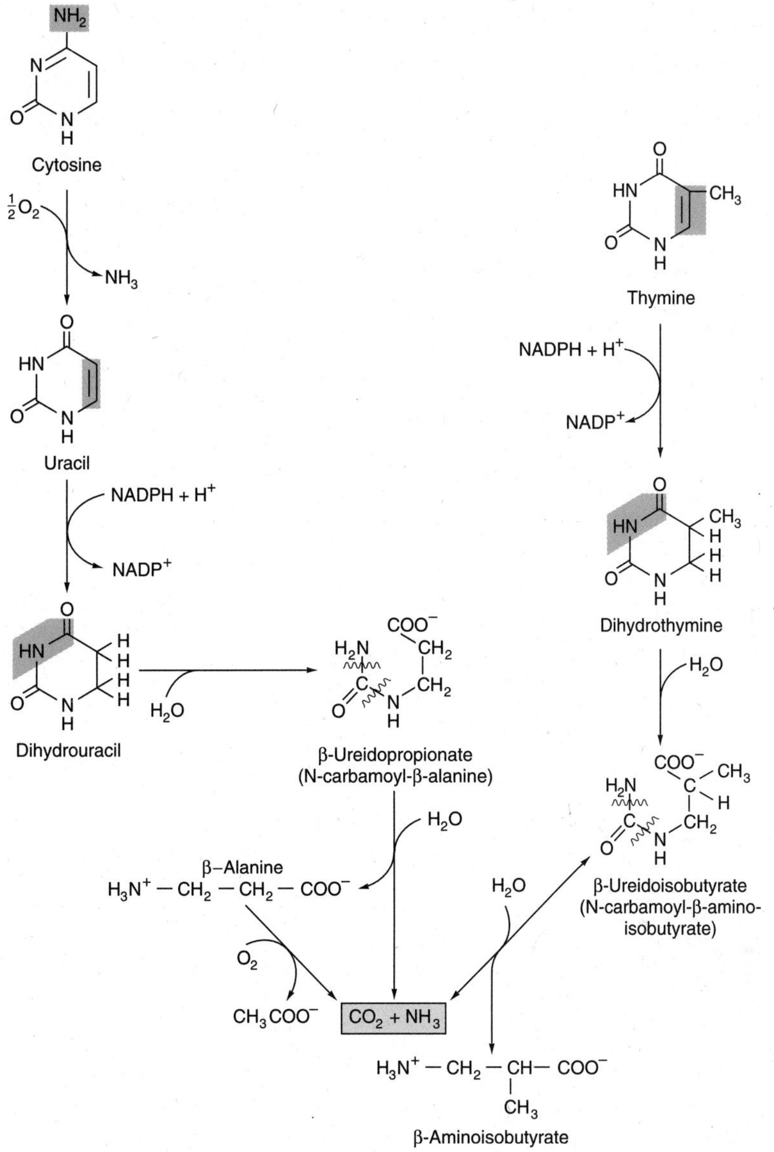

Figure 36–16. Catabolism of pyrimidines.

Deficiency of a Urea Cycle Enzyme Results in Excretion of Pyrimidine Precursors

Increased excretion of orotic acid, uracil, and uridine accompanies a deficiency in liver mitochondrial ornithine transcarbamoylase. The underutilized substrate, carbamoyl phosphate, exits to the cytosol, where it stimulates pyrimidine nucleotide biosynthesis. The resulting mild **orotic aciduria** is accentuated by high-nitrogen foods.

Drugs May Precipitate Orotic Aciduria

The purine analog **allopurinol** (Figure 35–20), a substrate for orotate phosphoribosyltransferase (reaction 5, Figure 36–11), competes for phosphoribosylation with the natural substrate, orotic acid. In addition, the resulting nucleotide product inhibits orotidylate decarboxylase (reaction 6, Figure 36–11), resulting in **orotic aciduria** and **orotidinuria.** Since the pathway for pyrimidine nucleotide biosynthesis

Table 36–3. Inherited disorders of pyrimidine metabolism and their associated enzyme abnormalities.

Clinical Disorder	Defective Enzyme	Characteristics of Clinical Disorder	Inheritance Pattern
β-Aminoisobutyric aciduria	Transaminase	No symptoms; frequent in Asians.	Autosomal recessive
Orotic aciduria, type I	Orotate phosphoribosyltransferase and orotidylate decarboxylase	Orotic acid crystalluria, failure to thrive, and megaloblastic anemia. Immunodeficiency. Remission with oral uridine.	Autosomal recessive
Orotic aciduria, type II	Orotidylate decarboxylase	Orotidinuria and orotic aciduria, megaloblastic anemia. Remission with oral uridine.	Autosomal recessive
Ornithine transcarbamoylase deficiency	Ornithine transcarbamoylase	Protein intolerance, hepatic encephalopathy, and mild orotic aciduria.	X-linked recessive

adjusts to this inhibition, human subjects are only transiently starved for pyrimidine nucleotides during early stages of treatment.

6-Azauridine, following its conversion to 6-azauridylate, competitively inhibits orotidylate decarboxylase (reaction 6, Figure 36–11), greatly enhancing excretion of orotic acid and orotidine.

SUMMARY

Purines and pyrimidines are not essential in the human diet, and purine deficiency is rare in humans. Dietary nucleic acids are degraded in the gastrointestinal tract to purines and pyrimidines. While "salvage" reactions convert purines and their ribo- and deoxyribonucleosides directly to the corresponding mononucleotides, most of the body's purines, pyrimidines, and their derivatives are formed by biosynthesis from amphibolic intermediates. Biosynthesis of the parent purine nucleotide inosine monophosphate (IMP) involves a protracted sequence of reactions, some of which are catalyzed by multifunctional catalysts. Since folate derivatives and glutamine participate in this reaction sequence, antifolate drugs or glutamine analogs inhibit purine biosynthesis. Oxidation and amination of IMP forms AMP and GMP, and subsequent phosphate transfer from ATP forms ADP and GDP. Further phosphoryl transfer from ATP to GDP forms GTP. ADP is converted to ATP by oxidative phosphorylation. Hepatic purine nucleotide biosynthesis is stringently regulated, principally by the pool size of phosphoribosyl pyrophosphate (PRPP) and by feedback-inhibition of PRPP-glutamyl amidotransferase by the end products AMP and GMP. Reduction of NDPs forms deoxyribonucleotide diphosphates (dNDPs).

While uridine and cytidine are salvaged, pyrimidine nucleotides arise primarily via biosynthesis from amphibolic intermediates. Biosynthesis involves a protracted set of reactions that differ from those of purine biosynthesis but involve analogous reactions once a mononucleotide has been formed. Certain

pyrimidine analogs are substrates for enzymes of pyrimidine nucleotide biosynthesis and hence inhibit this process. Regulation of pyrimidine nucleotide biosynthesis involves control both of gene expression and of enzymatic activity. Coordinated regulation of the biosynthesis of purine and pyrimidine nucleotides ensures their presence in proportions appropriate for nucleic acid biosynthesis and other metabolic needs.

While the reactions and intermediates of purine and pyrimidine biosynthesis are the same in bacteria and human subjects, the enzymes that catalyze these reactions are organized differently. In bacteria, each reaction is catalyzed by a distinct protein. By contrast, some enzymes of de novo purine and pyrimidine biosynthesis in humans are multifunctional polypeptides that catalyze contiguous reactions. Apparent advantages of multifunctional polypeptides include coordinate expression of several catalytic activities and channeling of the products of a reaction to the next reaction in sequence without dissociation from an enzyme.

Humans catabolize purines to the weak acid uric acid (pK 5.8), which, depending on urinary pH, exists as the relatively insoluble acid (at acidic pH) or its more soluble sodium urate salt (at pHs near neutrality). Urate crystals are diagnostic of gout, a metabolic disorder of purine catabolism. Other disorders of purine catabolism include Lesch-Nyhan syndrome, von Gierke's disease, and the hypouricemias.

Unlike uric acid and urate, the relatively insoluble products of purine catabolism, the end products of pyrimidine catabolism are highly water soluble: CO_2, NH_3, and β-aminoisobutyrate. Pseudouridine is, however, excreted unchanged. Overproduction of pyrimidine catabolites generally is not associated with clinically significant abnormalities. Patients with orotic aciduria respond to dietary pyrimidine nucleosides. Certain drugs also may precipitate orotic aciduria. Excretion of pyrimidine precursors can, however, result from a deficiency of the urea cycle enzyme ornithine transcarbamoylase, since the spared carbamoyl phosphate becomes available for pyrimidine biosynthesis.

REFERENCES

Benkovic SJ: The transformylase enzymes in de novo purine biosynthesis. Trends Biochem Sci 1984;9:320.

Holmgren A: Thioredoxin. Annu Rev Biochem 1985;54: 237.

Schinke RT: Methotrexate resistance and gene amplification: Mechanisms and implications. Cancer 1986;57: 1912.

Scriver CR et al (editors): *The Metabolic and Molecular Bases of Inherited Disease,* 7th ed. McGraw-Hill, 1995.

Seegmiller JE: Overview of the possible relation of defects in purine metabolism to immune deficiency. Ann N Y Acad Sci 1985;45:9.

Zalkin H, Dixon JE: De novo purine nucleotide synthesis. Prog Nucleic Acid Res Mol Biol 1992;42:259.

37

Nucleic Acid Structure & Function

Daryl K. Granner, MD

INTRODUCTION

The discovery that genetic information is coded along the length of a polymeric molecule composed of only four types of monomeric units is one of the major scientific achievements of this century. This polymeric molecule, **DNA,** is the chemical basis of heredity and is organized into genes, the fundamental units of genetic information. Genes control the synthesis of various types of RNA, most of which are involved in protein synthesis. Genes do not function autonomously; their replication and function are controlled by various gene products, often in collaboration with components of various signal transduction pathways. Knowledge of the structure and function of nucleic acids is essential in understanding genetics and many aspects of disease pathophysiology as well as the genetic basis of disease.

BIOMEDICAL IMPORTANCE

The chemical basis of heredity and of genetic diseases is found in the structure of DNA. The basic information pathway (ie, DNA directs the synthesis of RNA, which in turn directs protein synthesis) has been elucidated. This knowledge is being used to define normal cellular physiology and the pathophysiology of disease at the molecular level.

DNA CONTAINS THE GENETIC INFORMATION

The demonstration that DNA contained the genetic information was first made in 1944 in a series of experiments by Avery, MacLeod, and McCarty, who showed that the genetic determination of the character (type) of the capsule of a specific pneumococcus could be transmitted to another of a distinctly different capsular type by introducing purified DNA from the former coccus into the latter. These authors referred to the agent (later shown to be DNA) accomplishing the change as "transforming factor." Subsequently, this type of genetic manipulation has become commonplace. Similar experiments have recently been performed utilizing yeast, cultured mammalian cells, and insect and rodent embryos as recipients, and cloned DNA as the donor of genetic information.

DNA Contains Four Deoxynucleotides

The chemical nature of the monomeric deoxynucleotide units of DNA—**deoxyadenylate, deoxyguanylate, deoxycytidylate,** and **thymidylate**—is described in Chapter 35. These monomeric units of DNA are held in polymeric form by 3′,5′-phosphodiester bridges constituting a single strand, as depicted in Figure 37–1. The informational content of DNA (the genetic code) resides in the sequence in which these monomers—purine and pyrimidine deoxyribonucleotides—are ordered. The polymer as depicted possesses a polarity; one end has a 5′-hydroxyl or phosphate terminal while the other has a 3′-phosphate or hydroxyl moiety. The importance of this polarity will become evident. Since the genetic information resides in the order of the monomeric units within the polymers, there must exist a mechanism of reproducing or replicating this specific information with a high degree of fidelity. That requirement, together with x-ray diffraction data from the DNA molecule and the observation of Chargaff that in DNA molecules the concentration of deoxyadenosine (A) nucleotides equals that of thymidine (T) nucleotides (A = T), while the concentration of deoxyguanosine (G) nucleotides equals that of deoxycytidine (C) nucleotides (G = C), led Watson, Crick, and Wilkins to propose in the early 1950s a model of a double-stranded DNA molecule. The model they proposed is depicted in Figure 37–2. The two strands of this right-handed, double-stranded molecule are held in register by **hydrogen bonds** between the purine and pyrimidine bases of the respective linear molecules. The pairings between the purine and pyrimidine nucleotides on the opposite strands are very specific and are dependent upon hydrogen bonding of **A with T, and G with C** (Figure 37–3).

In the double-stranded molecule, restrictions imposed by the rotation about the phosphodiester bond, the favored anti configuration of the glycosidic bond (Figure 35–8), and the predominant tautomers (see

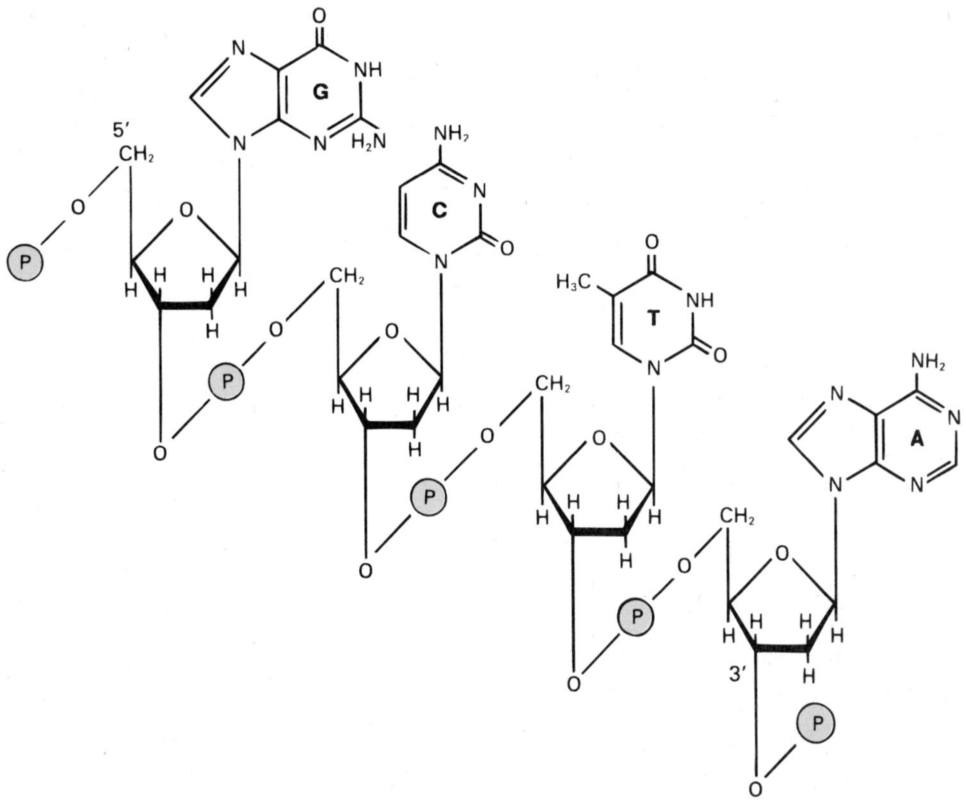

Figure 37–1. A segment of one strand of a DNA molecule in which the purine and pyrimidine bases adenine (A), thymine (T), cytosine (C), and guanine (G) are held together by a phosphodiester backbone between 2'-deoxyribosyl moieties attached to the nucleobases by an *N*-glycosidic bond. Note that the backbone has a polarity (ie, a direction). This is generally depicted in the 5' to 3' orientation.

Figure 35–3) of the four bases (A, G, T, and C) allow A to pair only with T, and G only with C, as depicted in Figure 37–3. This base-pairing restriction explains the earlier observation that in a double-stranded DNA molecule the content of A equals that of T and the content of G equals that of C. The two strands of the double-helical molecule, each of which possesses a polarity, are **antiparallel;** ie, one strand runs in the 5' to 3' direction and the other in the 3' to 5' direction. This is analogous to two parallel streets, each running one way but carrying traffic in opposite directions. In the double-stranded DNA molecules, the genetic information resides in the sequence of nucleotides on one strand, the **template strand;** the opposite strand is considered the **coding strand** because it matches the RNA transcript that encodes the protein.

The two strands, in which opposing bases are held together by hydrogen bonds, wind around a central axis in the form of a **double helix.** Double-stranded DNA exists in at least six forms (A—E and Z). The B form is usually found under physiologic conditions (low salt, high degree of hydration). A single turn of

B-form DNA about the axis of the molecule contains ten base pairs. The distance spanned by one turn of B-form DNA is 3.4 nm. The width (helical diameter) of the double helix in B-form DNA is 2 nm.

As depicted in Figure 37–3, three hydrogen bonds hold the deoxyguanosine nucleotide to the deoxycytidine nucleotide, whereas the other pair, the A—T pair, is held together by two hydrogen bonds. Thus, the G—C bond is stronger by approximately 50%. Because of this added strength and also because of stacking interactions, regions of DNA that are rich in G—C bonds are much more resistant to denaturation, or "melting," than A—T-rich regions.

The Denaturation (Melting) of DNA Is Used to Analyze Its Structure

The double-stranded structure of DNA can be melted in solution by increasing the temperature or decreasing the salt concentration. Not only do the two stacks of bases pull apart but the bases themselves unstack while still connected in the polymer by the phosphodiester backbone. Concomitant with this denaturation of the DNA molecule is an increase

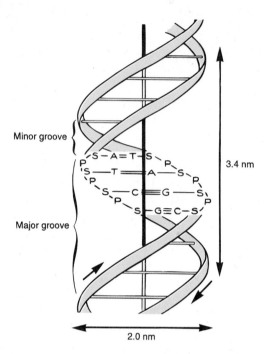

Figure 37–2. A diagrammatic representation of the Watson and Crick model of the double-helical structure of the B form of DNA. The horizontal arrow indicates the width of the double helix (2.0 nm), and the vertical arrow indicates the distance spanned by one complete turn of the double helix (3.4 nm). One turn of B DNA includes ten base pairs. The central axis of the double helix is indicated by the vertical rod. The short arrows designate the polarity of the antiparallel strands. (A, adenine; C, cytosine; G, guanine; T, thymine; P, phosphate; S, sugar [deoxyribose].)

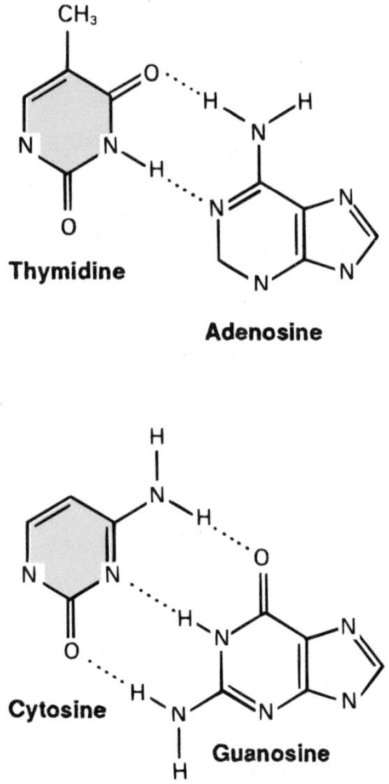

Figure 37–3. Base pairing between deoxyadenosine and thymidine involves the formation of two hydrogen bonds. Three such bonds form between deoxycytidine and deoxyguanosine. The broken lines represent hydrogen bonds.

in the optical absorbance of the purine and pyrimidine bases—a phenomenon referred to as hyperchromicity of denaturation. Because of the stacking of the bases and the hydrogen bonding between the stacks, the double-stranded DNA molecule exhibits properties of a rigid rod and in solution is a viscous material that loses its viscosity upon denaturation.

The strands of a given molecule of DNA separate over a temperature range. The midpoint is called the melting temperature, or T_m. The T_m is influenced by the base composition of the DNA and by the salt concentration of the solution. DNA rich in G—C pairs, which have 3 hydrogen bonds, melts at a higher temperature than that rich in A—T pairs, which have two hydrogen bonds. A tenfold increase of monovalent cation concentration increases the T_m by 16.6 °C. Formamide, which is commonly used in recombinant DNA experiments, destabilizes hydrogen bonding between bases, thereby lowering the T_m. This allows the strands of DNA or DNA-RNA hybrids to be separated at much lower temperatures and minimizes the strand breakage that occurs at high temperatures.

There Are Grooves in the DNA Molecule

Careful examination of the model depicted in Figure 37–2 reveals a **major groove** and a **minor groove** winding along the molecule parallel to the phosphodiester backbones. In these grooves, proteins can interact specifically with exposed atoms of the nucleotides (usually by H bonding) and thereby recognize and bind to specific nucleotide sequences without disrupting the base pairing of the double-helical DNA molecule. As discussed in Chapters 39 and 41, regulatory proteins can control the expression of specific genes via such interactions.

DNA Exists in Relaxed and Supercoiled Forms

In some organisms such as bacteria, bacteriophages, and many DNA-containing animal viruses, the ends of the DNA molecules are joined to create a closed circle with no terminal. This of course does not destroy the polarity of the molecules, but it eliminates all free 3′ and 5′ hydroxyl and phosphoryl groups. Closed circles exist in relaxed or supercoiled forms. Supercoils are introduced when a closed circle is twisted around its own axis or when a linear piece of duplex DNA, whose ends are fixed, is twisted. This energy-requiring process puts the molecule under stress, and the greater the number of supercoils, the greater the stress or torsion (test this with a rubber band). **Negative supercoils** are formed when the molecule is twisted in the direction opposite from the clockwise turns of the right-handed double helix found in B-DNA. Such DNA is said to be underwound. The energy required to achieve this state is, in a sense, stored in the supercoils. The transition to another form that requires energy is thereby facilitated by the underwinding. One such transition is strand separation, which is a prerequisite for DNA replication and transcription. Supercoiled DNA is therefore a preferred form in biologic systems. Enzymes that catalyze topologic changes of DNA are called **topoisomerases.** Topoisomerases can relax or insert supercoils. The best characterized is **bacterial gyrase,** which induces negative supercoiling in DNA using ATP as an energy source.

DNA PROVIDES A TEMPLATE FOR REPLICATION AND TRANSCRIPTION

The genetic information stored in the nucleotide sequence of DNA serves two purposes. It is the source of information for the synthesis of all protein molecules of the cell and organism, and it provides the information inherited by daughter cells or offspring. Both these functions require that the DNA molecule serve as a template—in the first case for the transcription of the information into RNA and in the second case for the replication of the information into daughter DNA molecules.

The complementarity of the Watson and Crick double-stranded model of DNA strongly suggests that replication of the DNA molecule occurs in a semiconservative manner. Thus, when each strand of the double-stranded parental DNA molecule separates from its complement during replication, each serves as a template on which a new complementary strand is synthesized (Figure 37–4). The two newly formed double-stranded daughter DNA molecules, each containing one strand (but complementary rather than identical) from the parent double-stranded DNA molecule, are then sorted between the two daughter cells (Figure 37–5). Each daughter cell contains DNA molecules with information identical to that which the parent possessed; yet in each daughter cell the DNA molecule of the parent cell has been only semiconserved.

THE CHEMICAL NATURE OF RNA DIFFERS FROM THAT OF DNA

Ribonucleic acid (RNA) is a polymer of purine and pyrimidine ribonucleotides linked together by

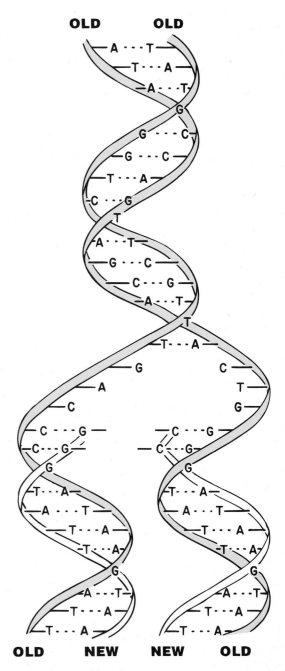

Figure 37–4. The double-stranded structure of DNA and the template function of each old strand (shaded) on which a new complementary strand is synthesized. (Modified from James D. Watson: *Molecular Biology of the Gene,* 3rd ed. Copyright © 1976, 1970, 1965, by W.A. Benjamin, Inc., Menlo Park, Calif.)

3′,5′-phosphodiester bridges analogous to those in DNA (Figure 37–6). Although sharing many features with DNA, RNA possesses several specific differences:

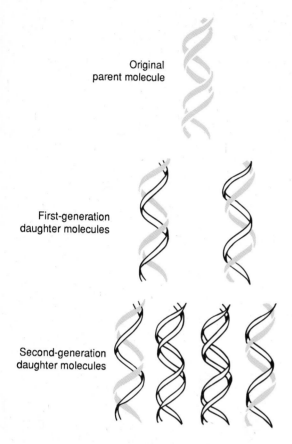

Original
parent molecule

First-generation
daughter molecules

Second-generation
daughter molecules

Figure 37–5. DNA replication is semiconservative. During a round of replication, each of the two strands of DNA is used as a template for synthesis of a new, complementary strand.

(1) In RNA, the sugar moiety to which the phosphates and purine and pyrimidine bases are attached is ribose rather than the 2′-deoxyribose of DNA.

(2) The pyrimidine components of RNA differ from those of DNA. Although RNA contains the ribonucleotides of adenine, guanine, and cytosine, it does not possess thymine except in the rare case mentioned below. Instead of thymine, RNA contains the ribonucleotide of uracil.

(3) RNA exists as a single strand, whereas DNA exists as a double-stranded helical molecule. However, given the proper complementary base sequence with opposite polarity, the single strand of RNA, as demonstrated in Figure 37–7, is capable of folding back on itself like a hairpin and thus acquiring double-stranded characteristics.

(4) Since the RNA molecule is a single strand complementary to only one of the two strands of a gene, its guanine content does not necessarily equal its cytosine content, nor does its adenine content necessarily equal its uracil content.

(5) RNA can be hydrolyzed by alkali to 2′,3′ cyclic diesters of the mononucleotides, compounds that cannot be formed from alkali-treated DNA because of the absence of a 2′-hydroxyl group. The alkali lability of RNA is useful both diagnostically and analytically.

Information within the single strand of RNA is contained in its sequence ("primary structure") of purine and pyrimidine nucleotides within the polymer. The sequence is complementary to the template strand of the gene from which it was transcribed. Because of this complementarity, an RNA molecule can bind specifically via the base-pairing rules to its template DNA strand; it will not bind ("hybridize") with the other (coding) strand of its gene. The sequence of the RNA molecule (except for U replacing T) is the same as that of the coding strand of the gene (Figure 37–8).

Nearly All of the Several Species of RNA Are Involved in Some Aspect of Protein Synthesis

Those cytoplasmic RNA molecules that serve as templates for protein synthesis (ie, that transfer genetic information from DNA to the protein-synthesizing machinery) are designated **messenger RNAs, or mRNAs.** Many other cytoplasmic RNA molecules (**ribosomal RNAs, or rRNAs**) have structural roles wherein they contribute to the formation of ribosomes (the organellar machinery for protein synthesis) or serve as adapter molecules (**transfer RNAs; tRNAs**) for the translation of RNA information into specific sequences of polymerized amino acids.

Some RNA molecules have intrinsic catalytic activity. An example is the role of RNA in catalyzing the processing of the primary transcript of a gene into mature messenger RNA.

Much of the RNA synthesized from DNA templates in eukaryotic cells, including mammalian cells, is degraded within the nucleus, and it never serves as either a structural or an informational entity within the cellular cytoplasm.

In human cells there are **small nuclear RNA (snRNA)** species that are not directly involved in protein synthesis but that may have roles in RNA processing and the cellular architecture. These relatively small molecules vary in size from 90 to about 300 nucleotides (Table 37–1).

The genetic material for some animal and plant viruses is RNA rather than DNA. Although some RNA viruses do not ever have their information transcribed into a DNA molecule, many animal RNA viruses—specifically the retroviruses (the HIV or AIDS virus, for example)—are transcribed by an RNA-dependent DNA polymerase, the so-called **reverse transcriptase,** to produce a double-stranded DNA copy of their RNA genome. In many cases, the resulting double-stranded DNA transcript is integrated into the host genome and subsequently serves

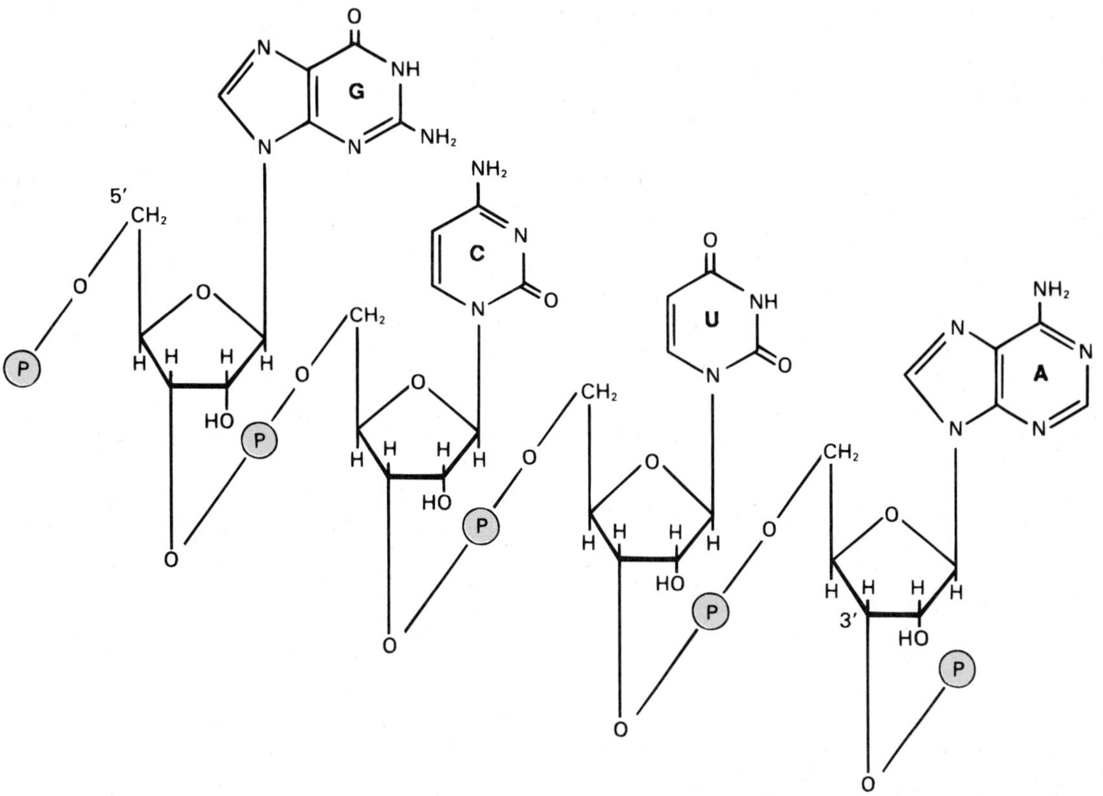

Figure 37–6. A segment of a ribonucleic acid (RNA) molecule in which the purine and pyrimidine bases—adenine (A), uracil (U), cytosine (C), and guanine (G)—are held together by phosphodiester bonds between ribosyl moieties attached to the nucleobases by *N*-glycosidic bonds. Note that the polymer has a polarity as indicated by the labeled 3′- and 5′-attached phosphates.

as a template for gene expression and from which new viral RNA genomes can be transcribed.

RNA Is Organized in Several Unique Structures

In all prokaryotic and eukaryotic organisms, three main classes of RNA molecules exist: messenger RNA (mRNA), transfer RNA (tRNA), and ribosomal RNA (rRNA). Each differs from the others by size, function, and general stability.

A. Messenger RNA (mRNA): This class is the most heterogeneous in size and stability. All members of the class function as messengers conveying the information in a gene to the protein-synthesizing machinery, where each serves as a template on which a specific sequence of amino acids is polymerized to form a specific protein molecule, the ultimate gene product (Figure 37–9).

Messenger RNAs, particularly in eukaryotes, have some unique chemical characteristics. The 5′ terminal of mRNA is "capped" by a 7-methylguanosine triphosphate that is linked to an adjacent 2′-*O*-methyl ribonucleoside at its 5′-hydroxyl through the three phosphates (Figure 37–10). The mRNA molecules frequently contain internal 6-methyladenylates and other 2′-*O*-ribose methylated nucleotides. The cap is probably involved in the recognition of mRNA by the translating machinery, and it helps stabilize the mRNA by preventing the attack of 5′-exonucleases. The protein-synthesizing machinery begins translating the mRNA into proteins at the 5′ or capped terminal. The other end of most mRNA molecules, the 3′-hydroxyl terminal, has attached a polymer of adenylate residues 20–250 nucleotides in length. The specific function of the **poly(A) "tail"** at the 3′-hydroxyl terminal of mRNAs is not fully understood, but it seems that it maintains the intracellular stability of the specific mRNA by preventing the attack of 3′-exonucleases. Some mRNAs, including those for some histones, do not contain poly(A). The poly(A) tail, because it will form a base pair with oligodeoxythymidine polymers attached to a solid substrate like cellulose, can be used to separate mRNA from other species of RNA, including mRNA molecules that lack this tail.

In mammalian cells, including cells of humans, the

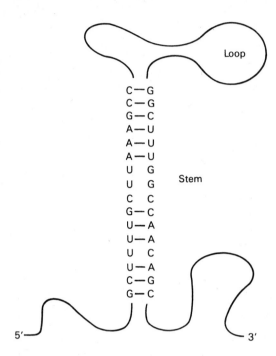

Figure 37–7. Diagrammatic representation of the secondary structure of a single-stranded RNA molecule in which a stem loop, or "hairpin," has been formed and is dependent upon the intramolecular base pairing.

Table 37–1. Some of the species of small stable RNAs found in mammalian cells.

Name	Length (nucleotides)	Molecules per Cell	Localization
U1	165	1×10^6	Nucleoplasm/hnRNA
U2	188	5×10^5	Nucleoplasm
U3	216	3×10^5	Nucleolus
U4	139	1×10^5	Nucleoplasm
U5	118	2×10^5	Nucleoplasm
U6	106	3×10^5	Perichromatin granules
4.5S	91–95	3×10^5	Nucleus and cytoplasm
7S	280	5×10^5	Nucleus and cytoplasm
7-2	290	1×10^5	Nucleus and cytoplasm
7-3	300	2×10^5	Nucleus

mRNA molecules present in the cytoplasm are not the RNA products immediately synthesized from the DNA template but must be formed by processing from a precursor molecule before entering the cytoplasm. Thus, in mammalian nuclei, the immediate products of gene transcription constitute a fourth class of RNA molecules. These nuclear RNA molecules are very heterogeneous in size and are quite large. The **heterogeneous nuclear RNA (hnRNA)** molecules may have a molecular weight in excess of 10^7, whereas the molecular weight of mRNA molecules is generally less than 2×10^6. As is discussed in Chapter 39, hnRNA molecules are processed to generate the mRNA molecules which then enter the cytoplasm to serve as templates for protein synthesis.

B. Transfer RNA (tRNA): tRNA molecules consist of approximately 75 nucleotides. They also are generated by nuclear processing of a precursor molecule (Chapter 39). The tRNA molecules serve as adapters for the translation of the information in the sequence of nucleotides of the mRNA into specific amino acids. There are at least 20 species of tRNA molecules in every cell, at least one (and often several) corresponding to each of the 20 amino acids required for protein synthesis. Although each specific tRNA differs from the others in its sequence of nucleotides, the tRNA molecules as a class have many features in common. The primary structure—ie, the nucleotide sequence—of all tRNA molecules allows extensive folding and intrastrand complementarity to generate a secondary structure that appears like a cloverleaf (Figure 37–11).

All tRNA molecules contain four main arms. The **acceptor arm** consists of a base-paired stem that terminates in the sequence CCA (5′ to 3′). It is through an ester bond to the 3′-hydroxyl group of the adenosyl moiety that the carboxyl groups of amino acids are attached. The other arms have base-paired stems and unpaired loops (Figure 37–7). The **anticodon arm** at the end of a base-paired stem recognizes the triplet nucleotide or codon (discussed in Chapter 40) of the template mRNA. It has a nucleotide sequence complementary to the codon and is responsible for the specificity of the tRNA. The **D arm** is named for the presence of the base dihydrouridine, and the **TΨC arm** for the sequence T, pseudouridine, and C.

Figure 37–8. The relationship between the sequences of an RNA transcript and its gene, in which the coding and template strands are shown with their polarities. The RNA transcript with a 5′ to 3′ polarity is complementary to the template strand with its 3′ to 5′ polarity. Note that the sequence in the RNA transcript and its polarity is the same as that in the coding strand, except that the U of the transcript replaces the T of the gene.

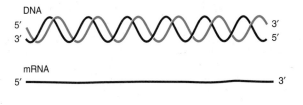

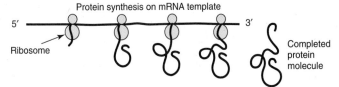

Figure 37–9. The expression of genetic information in DNA into the form of an mRNA transcript. This is subsequently translated by ribosomes into a specific protein molecule.

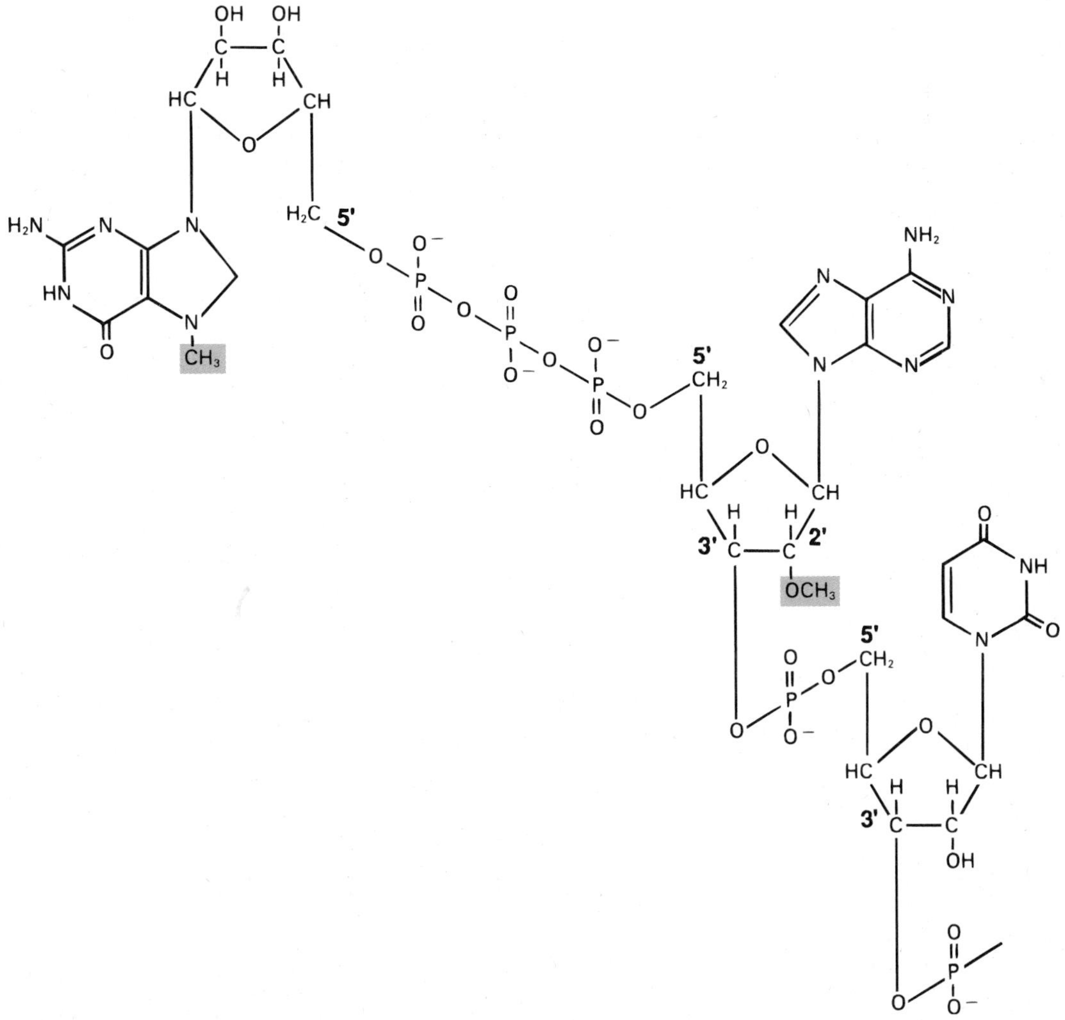

Figure 37–10. The cap structure attached to the 5' terminal of most eukaryotic messenger RNA molecules. A 7-methyl-guanosine triphosphate is attached at the 5' terminal of the mRNA, which usually contains a 2'-*O*-methylpurine nucleotide.

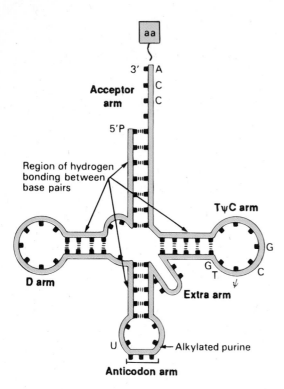

Figure 37–11. Typical aminoacyl tRNA in which the amino acid (aa) is attached to the 3′ CCA terminal. The anticodon, TΨC, and dihydrouracil (D) arms are indicated, as are the positions of the intramolecular hydrogen bonding between these base pairs. (From James D. Watson, *Molecular Biology of the Gene*, 3rd ed. Copyright © 1976, 1970, 1965, by W.A. Benjamin, Inc., Menlo Park, Calif.)

TΨC and anticodon arms 5 bp, and the D arm 3 (or 4) bp.

Although tRNAs are quite stable in prokaryotes, they are somewhat less stable in eukaryotes. The opposite is true for mRNAs, which are quite unstable in prokaryotes but generally stable in eukaryotic organisms.

C. Ribosomal RNA (rRNA): A ribosome is a cytoplasmic nucleoprotein structure that acts as the machinery for the synthesis of proteins from the mRNA templates. On the ribosomes, the mRNA and tRNA molecules interact to translate into a specific protein molecule information transcribed from the gene.

The components of the mammalian ribosome, which has a molecular weight of about 4.2×10^6 and a sedimentation velocity of 80S (Svedberg units), are shown in Table 37–2. The mammalian ribosome contains two major nucleoprotein subunits, a larger one with a molecular weight of 2.8×10^6 (60S) and a smaller subunit with a molecular weight of 1.4×10^6 (40S). The 60S subunit contains a 5S ribosomal RNA (rRNA), a 5.8S rRNA, and a 28S rRNA; there are also probably more than 50 specific polypeptides. The 40S subunit is smaller and contains a single 18S rRNA and approximately 30 polypeptide chains. All of the ribosomal RNA molecules except the 5S rRNA are processed from a single 45S precursor RNA molecule in the nucleolus (Chapter 39). The 5S rRNA apparently has its own precursor that is independently transcribed. The highly methylated ribosomal RNA molecules are packaged in the nucleolus with the specific ribosomal proteins. In the cytoplasm, the ribosomes remain quite stable and capable of many translation cycles. The functions of the ribosomal RNA molecules in the ribosomal particle are not fully understood, but they are necessary for ribosomal assembly and seem to play key roles in the binding of mRNA to ribosomes and its translation. Recent studies suggest that an rRNA component performs the peptidyl transferase activity and thus is an enzyme (a ribozyme).

D. Small Stable RNA: A large number of discrete, highly conserved, and small stable RNA species are found in eukaryotic cells. The majority of these molecules exist as ribonucleoproteins and are

The **extra arm** is the most variable feature of tRNA, and it provides a basis for classification. Class 1 tRNAs (about 75% of all tRNAs) have an extra arm that is 3–5 bp long. Class 2 tRNAs have an extra arm 13–21 bp long and often have a stem-loop structure.

The secondary structure of tRNA molecules is maintained by the base pairing in these arms, and this is a consistent feature. The acceptor arm has 7 bp, the

Table 37–2. Components of mammalian ribosomes.[1]

Component	Mass (mw)	Protein Number	Protein Mass	RNA Size	RNA Mass	Bases
40S subunit	1.4×10^6	~35	7×10^5	18S	7×10^5	1900
60S subunit	2.8×10^6	~50	1×10^6	5S	35,000	120
				5.8S	45,000	160
				28S	1.6×10^6	4700

[1]The ribosomal subunits are defined according to their sedimentation velocity in Svedberg units (40S or 60S). This table illustrates the total mass (MW) of each. The number of unique proteins and their total mass (MW) and the RNA components of each subunit in size (Svedberg units), mass, and base composition are listed.

distributed in the nucleus, in the cytoplasm, or in both. They range in size from 90 to 300 nucleotides and are present in 100,000–1,000,000 copies per cell.

Small nuclear RNAs (snRNAs) are significantly involved in mRNA processing and gene regulation. Of the several snRNAs, U1, U2, U4, U5, and U6 are involved in intron removal and the processing of hn-RNA into mRNA (Chapter 39). The U7 snRNA may be involved in production of the correct 3′ ends of histone mRNA (which lacks a poly[A] tail). The U4 and U6 snRNAs may also be required for poly(A) processing.

SUMMARY

The nucleic acids DNA and RNA are polymeric molecules. DNA consists of 4 bases, A, G, C, and T, which are held in linear array by phosphodiester bonds through the 3′ and 5′ positions of adjacent deoxyribose moieties. DNA is organized into two strands by the pairing of bases A to T and G to C on complementary strands. These strands form a double helix around a central axis. The 3×10^9 base pairs of DNA in humans are organized into the haploid complement of 23 chromosomes. The exact sequence of these 3 billion nucleotides defines the uniqueness of each individual. One function of DNA is to provide a template for replication and thus maintenance of the genotype. Another is to provide a template for transcription of the approximately 100,000 genes that encode a variety of RNA molecules.

In contrast to DNA, RNA exists in several different single-stranded structures, most of which are involved in protein synthesis. The linear array of nucleotides in RNA consists of A, G, C, and U, and the sugar moiety is ribose. The major forms of RNA include messenger RNA (mRNA), ribosomal RNA (rRNA), and transfer RNA (tRNA). These RNAs vary in size from tRNA, which consists of about 75 nucleotides, to mRNA, which can consist of several thousand nucleotides. The specific functions of these RNAs in protein synthesis are discussed in subsequent chapters.

REFERENCES

Guthrie C, Patterson B: Spliceosomal snRNAs. Ann Rev Genet 1988;22:387.
Hunt T: *DNA Makes RNA Makes Protein.* Elsevier, 1983.
Noller HF: Ribosomal RNA and translation. Annu Rev Biochem 1991;60:191.

Watson JD, Crick FHC: Molecular structure of nucleic acids. Nature 1953;171:737.
Watson JD: *The Double Helix.* Atheneum, 1968.

38

DNA Organization & Replication

Daryl K. Granner, MD

INTRODUCTION

The DNA in prokaryotic organisms is generally not combined with proteins other than those involved in DNA replication or transcription. Much of the DNA in eukaryotic organisms is covered with a variety of proteins. These proteins and DNA form a complex structure, chromatin, that allows for numerous configurations of the DNA molecule and types of control unique to the eukaryotic organism.

The genetic information in the DNA of a chromosome can be transmitted by exact replication, or it can be exchanged by a number of processes, including crossing over, recombination, transposition, and conversion. These provide a means of ensuring adaptability and diversity for the organism but can also result in disease.

DNA replication, a highly complex and ordered process, follows the 5′ to 3′ polarity typical of the synthesis of RNA and protein described in other chapters. In eukaryotic cells, replication of the DNA in a chromosome begins at multiple sites and proceeds in both directions simultaneously. A number of enzymes are required for the synthesis and repair of DNA, both of which follow the rules of Watson-Crick base pairing.

BIOMEDICAL IMPORTANCE

Mutations are due to a change in the base sequence of DNA. They may result from the faulty replication, movement, or repair of DNA and occur with a frequency of about one in every 10^6 cell divisions. An abnormal gene product can be the result of mutations that occur in coding or regulatory-region DNA. A mutation in a germ cell will be transmitted to offspring (so-called vertical transmission of hereditary disease). A number of factors, including viruses, chemicals, ultraviolet light, and ionizing radiation, increase the rate of mutation. Mutations often affect somatic cells and so are passed on to successive generations of cells within an organism. It is becoming apparent that a number of diseases, and perhaps most cancers, are due to this horizontal transmission of induced mutations.

CHROMATIN IS THE CHROMOSOMAL MATERIAL EXTRACTED FROM NUCLEI OF CELLS OF EUKARYOTIC ORGANISMS

Chromatin* consists of very long double-stranded **DNA molecules** and a nearly equal mass of rather small basic proteins termed **histones** as well as a smaller amount of **nonhistone proteins** (most of which are acidic and larger than histones) and a small quantity of **RNA.** The double-stranded DNA helix in each chromosome has a length that is thousands of times the diameter of the cell nucleus. One purpose of these molecules, particularly the histones, is to condense the DNA. Electron microscopic studies of chromatin have demonstrated dense spherical particles called **nucleosomes,** which are approximately 10 nm in diameter and connected by DNA filaments (Figure 38–1). Nucleosomes are composed of DNA wound around a collection of histone molecules.

Histones Are the Most Abundant Chromatin Proteins

The **histones** are a small family of closely related basic proteins. Among the histones, **H1** histones are the least tightly bound to chromatin and are, therefore, easily removed with a salt solution, after which chromatin becomes soluble. The isolated core nucleosomes contain four classes of histones: **H2A, H2B, H3,** and **H4.** The structures of slightly lysine-rich histones—H2A and H2B—appear to have been significantly conserved between species, while the structures of arginine-rich histones—H3 and H4— have been highly conserved between species. This severe conservation implies that the function of his-

---

*So far as is possible, the discussion in this chapter and in Chapters 39, 40, and 41 will pertain to mammalian organisms, which are, of course, among the higher eukaryotes. At times it will be necessary to refer to observations in prokaryotic organisms such as bacteria and viruses, but in such cases the information will be of a kind that can be extrapolated to mammalian organisms. The division of the material presented in Chapters 37–41 is somewhat arbitrary, and its separation in the text in this way should not be taken to mean that the processes described are not fully integrated and interdependent.

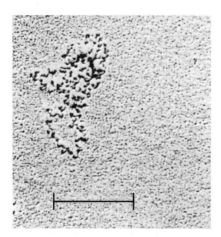

Figure 38–1. Electron micrograph of nucleosomes attached by strands of nucleic acid. (The bar represents 2.5 μm.) (Reproduced, with permission, from Oudet P, Gross-Bellard M, Chambon P: Electron microscopic and biochemical evidence that chromatin structure is a repeating unit. Cell 1975;4:281.)

tones is identical in all eukaryotes and that the entire molecule is involved quite specifically in carrying out this function. The C-terminal two-thirds of the molecules have a usual amino acid composition, while their amino terminal thirds are rich in basic amino acids. **These four core histones are subject to five types of covalent modification:** acetylation, methylation, phosphorylation, ADP-ribosylation, and covalent linkage (H2A only) to ubiquitin, the nuclear protein. These histone modifications likely play some role in chromatin structure and function, but little is currently understood about this.

The histones interact with each other in very specific ways. **H3 and H4 form a tetramer** containing two molecules of each $(H3/H4)_2$, while **H2A and H2B form dimers** (H2A-H2B) and higher oligomeric complexes ($[H2A-H2B]_n$). Under physiologic conditions, two of these histone tetramers associate to form the **histone octamer.**

The Nucleosome Contains Histone and DNA

When the histone octamer is mixed with purified, double-stranded DNA, the same x-ray diffraction pattern is formed as that observed in freshly isolated chromatin. Electron microscopic studies confirm the existence of reconstituted nucleosomes. Furthermore, the reconstitution of nucleosomes from DNA and histones H2A, H2B, H3, and H4 is independent of the organismal or cellular origin of the various components. The histone H1 and the nonhistone proteins are not necessary for the reconstitution of the nucleosome core.

In the nucleosome, the DNA is supercoiled in a left-handed helix over the surface of the disk-shaped histone octamer consisting of a central H3-H4 tetramer $(H3/H4)_2$ and two H2A-H2B dimers (Figure 38–2). The core histones interact with the DNA on the inside of the supercoil without protruding.

The $H3_2$-$H4_2$ tetramer itself can confer nucleosome-like properties on DNA and thus has a central role in the formation of the nucleosome. The addition of two H2A-H2B dimers stabilizes the primary particle and binds firmly two additional half-turns of DNA previously bound only loosely to the $(H3/H4)_2$. Thus, 1.75 superhelical turns of DNA are wrapped around the surface of the histone octamer, protecting 146 base pairs of DNA and forming the nucleosome core (Figure 38–2).

The assembly of nucleosomes is probably mediated by the anionic nuclear protein **nucleoplasmin.** Histones, which are strongly cationic, can bind nonspecifically to the strongly anionic DNA by forming salt bridges. Clearly, such a nonspecific interaction of histones and DNA would be detrimental to nucleosome formation and chromatin function. Nucleoplasmin is an anionic pentameric protein that binds neither to DNA nor to chromatin, but it can interact reversibly with a histone octamer in such a way that the histones no longer adhere nonspecifically to negatively charged surfaces such as DNA. It seems that nucleoplasmin thereby maintains in the nucleus an ionic environment conducive to the specific interaction of histones and DNA and the assembly of nucleosomes. As the nucleosome is assembled, nucleoplasmin must be released from the histones. Nucleosomes appear to exhibit preference for certain regions on specific DNA molecules, but the basis for this nonrandom distribution, termed **phasing,** is unknown. It is probably related to the relative physical flexibility of certain nucleotide sequences that are

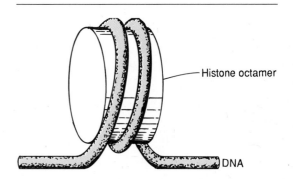

Histone octamer

DNA

Figure 38–2. Model for the structure of the nucleosome, in which DNA is wrapped around the surface of a flat protein cylinder consisting of two each of histones H2A, H2B, H3, and H4. The 146 base pairs of DNA, consisting of 1.75 superhelical turns, are in contact with the histone octamer. This protects the DNA from digestion by a nuclease.

able to accommodate the regions of kinking within the supercoil.

The super-packing of nucleosomes in nuclei is seemingly dependent upon the interaction of the H1 histones with adjacent nucleosomes.

HIGHER ORDER STRUCTURES PROVIDE FOR THE COMPACTION OF CHROMATIN

Electron microscopy of chromatin reveals two higher orders of structure—the 10-nm fibril and the 25- to 30-nm chromatin fiber—beyond that of the nucleosome itself. The disk-like nucleosome structure has a 10-nm diameter and a height of 5 nm. The **10-nm fibril** seems to consist of nucleosomes arranged with their edges touching and their flat faces parallel with the fibril axis (Figure 38–3). The 10-nm fibril is probably further supercoiled with 6–7 nucleosomes per turn to form the **30-nm chromatin fiber** (Figure 38–3). Each turn of the supercoil is relatively flat, and the faces of the nucleosomes of successive turns would be nearly parallel to each other. H1 histones appear to stabilize the 30-nm fiber, but their position and that of the variable length spacer DNA are not clear. It is probable that nucleosomes can form a variety of packed structures. In order to form a mitotic chromosome, the 30-nm fiber must be compacted in length another 100-fold (see below).

In **interphase chromosomes,** chromatin fibers appear to be organized into 30,000–100,000 base-pair **loops or domains** anchored in a scaffolding (or supporting matrix) within the nucleus. Within these domains, some DNA sequences may be located nonrandomly. It has been suggested that each looped domain of chromatin corresponds to a separate genetic function, containing both coding and noncoding regions of the gene.

SOME REGIONS OF CHROMATIN ARE "ACTIVE" AND OTHERS ARE "INACTIVE"

Generally, every cell of an individual metazoan organism contains the same genetic information in the form of the same DNA sequences. Thus, the differences between different cell types within an organism must be explained by differential expression of the common genetic information. Chromatin containing active genes (ie, transcriptionally active chromatin) has been shown to differ in several ways from that of nonactive regions. The nucleosome structure of active chromatin appears to be altered or even absent in highly active regions. DNA in active chromatin contains large regions (about 100,000 bases long) that are **sensitive to digestion by a nuclease** such as DNase I. The sensitivity to DNase I of chromatin regions being actively transcribed reflects only a potential for transcription rather than transcription itself and in several systems can be correlated with a relative lack of 5-methyldeoxycytidine in the DNA.

Within the large regions of active chromatin there exist shorter stretches of 100–300 nucleotides that exhibit an even greater (another 10-fold) sensitivity to DNase I. These **hypersensitive sites** probably result from a structural conformation that favors access of the nuclease to the DNA. These regions are often located immediately upstream from the active gene and are the location of interrupted nucleosomal structure caused by the binding of nonhistone proteins. (See Chapters 39 and 41.) In many cases, it seems that if a gene is capable of being transcribed, it must have a DNase hypersensitive site in the chromatin immediately upstream. The proteins involved in transcription, and those involved in maintaining access to the template strand, lead to the formation of hypersensitive sites. Hypersensitive sites often provide the first clue about the presence and location of a transcription control element.

Transcriptionally inactive chromatin is densely packed during interphase as observed by electron microscopic studies and is referred to as **heterochromatin;** transcriptionally active chromatin stains less densely and is referred to as **euchromatin.** Generally, euchromatin is replicated earlier in the mammalian cell cycle (see below) than is heterochromatin.

There are two types of heterochromatin: constitutive heterochromatin and facultative heterochromatin. **Constitutive heterochromatin** is always condensed and, thus, inactive. Constitutive heterochromatin is found in the regions near the chromosomal centromere and at chromosomal ends (telomeres). **Facultative heterochromatin** is at times

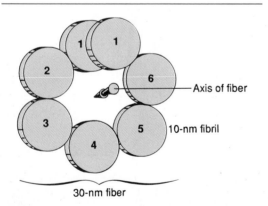

Figure 38–3. Proposed structure of the 30-nm chromatin fiber consisting of superhelices of six 10-nm fibrils of nucleosomes. The axis of the 30-nm fiber is perpendicular to the plane of the page.

condensed, but at other times it is actively transcribed and, thus, uncondensed and appears as euchromatin. Of the two members of the X chromosome pair in mammalian females, one X chromosome is almost completely inactive transcriptionally and is heterochromatic. However, the heterochromatic X chromosome decondenses during gametogenesis and becomes transcriptionally active during early embryogenesis; thus, it is facultative heterochromatin.

Certain cells of insects, eg, *Chironomus,* contain giant chromosomes that have been replicated for ten cycles without separation of daughter chromatids. These copies of DNA line up side by side in precise register and produce a banded chromosome containing regions of condensed chromatin and lighter bands of more extended chromatin. Transcriptionally active regions of these **polytene chromosomes** are especially decondensed into **"puffs,"** which can be shown to contain the enzymes responsible for transcription and to be the sites of RNA synthesis (Figure 38–4).

DNA IS ORGANIZED INTO CHROMOSOMES

At metaphase, mammalian **chromosomes** possess a twofold symmetry, with identical **sister chromatids** connected at a **centromere,** the relative position of which is characteristic for a given chromosome (Figure 38–5). Each sister chromatid contains one double-stranded DNA molecule. During interphase, the packing of the DNA molecule is less dense than it is in the condensed chromosome during the metaphase. Metaphase chromosomes are transcriptionally inactive.

The human haploid genome consists of about 3×10^9 base pairs or pairs of nucleotides and about 1.7×10^7 nucleosomes. Thus, each of the 23 chromatids in the human haploid genome would contain on the average 1.3×10^8 nucleotides in one double-stranded DNA molecule. The length of each DNA molecule must be **compressed about 8000-fold** to generate the structure of a condensed metaphase chromosome! In metaphase chromosomes, the 25- to 30-nm chromatin fibers are also folded into a series of **looped domains,** the proximal portions of which are anchored to a nonhistone proteinaceous scaffolding. The packing ratios of each of the orders of DNA structure are summarized in Table 38–1.

The packaging of nucleoproteins within chromatids is not random, as evidenced by the characteristic patterns observed when chromosomes are stained with specific dyes such as quinacrine or Giemsa stain (Figure 38–6).

From individual to individual within a single species, the pattern of staining (banding) of the entire chromosome complement is highly reproducible;

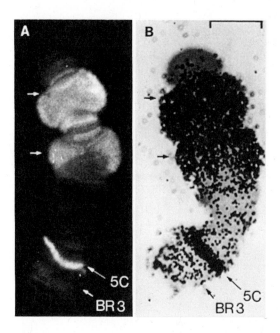

Figure 38–4. The correlation between RNA polymerase II activity and RNA synthesis. A number of genes are activated when *Chironomus tentans* larvae are subjected to heat shock (39 °C for 30 minutes). **A:** Distribution of RNA polymerase B (also called type II) in isolated chromosome IV from the salivary gland. The enzyme was detected by immunofluorescence using an antibody directed against the polymerase. The 5C and BR3 are specific bands of chromosome IV, and the arrows indicate puffs. **B:** Autoradiogram of a chromosome IV that was incubated in ^{3}H-uridine to label the RNA. Note the correspondence of the immunofluorescence and presence of the radioactive RNA (dots). Bar = 7 μm. (Reproduced, with permission, from Sass H: RNA polymerase B in polytene chromosomes. Cell 1982;28:274. Copyright © 1982 by the Massachusetts Institute of Technology.)

nonetheless, it differs significantly from other species, even those closely related. Thus, the packaging of the nucleoproteins in chromosomes of higher eukaryotes must in some way be dependent upon species-specific characteristics of the DNA molecules.

A combination of specialized staining techniques and high-resolution microscopy has allowed geneticists to quite precisely map thousands of genes to specific regions of mouse and human chromosomes.

Coding Regions Are Often Interrupted by Intervening Sequences

The **coding regions of DNA,** the transcripts of which ultimately appear in the cytoplasm as single mRNA molecules, are usually **interrupted in the eukaryotic genome by large intervening sequences of noncoding DNA.** Accordingly, the primary tran-

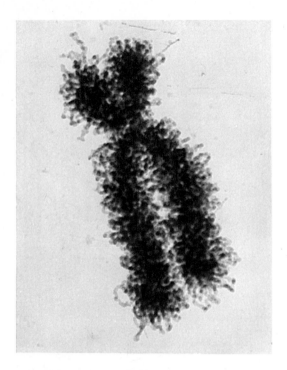

Figure 38–5. The two sister chromatids of human chromosome 12. (× 27,850.) (Reproduced, with permission, from DuPraw EJ: *DNA and Chromosomes.* Holt, Rinehart, & Winston, 1970.)

scripts of DNA—hnRNA—contain noncoding intervening sequences of RNA that must be removed in a process which also joins together the appropriate coding segments to form the mature mRNA. Most coding sequences for a single mRNA are interrupted in the genome (and thus in the primary transcript) by at least one—and as many as 50 in some cases—noncoding intervening sequences (**introns**). In most cases the introns are much longer than the continuous coding regions (**exons**). The processing of the primary transcript, which involves removal of introns and splicing of adjacent exons, is described in detail in Chapter 39.

The function of the intervening sequences, or introns, is not clear. They may serve to separate functional domains (exons) of coding information in a form that permits genetic rearrangement by recombination to occur more rapidly than if all coding regions for a given genetic function were contiguous. Such an enhanced rate of genetic rearrangement of functional domains might allow more rapid evolution of biologic function. The relationships among chromosomal DNA, gene clusters on the chromosome, the exon-intron structure of genes, and the final mRNA product are shown in Figure 38–7.

MUCH OF THE MAMMALIAN GENOME IS REDUNDANT AND MUCH IS NOT TRANSCRIBED

The haploid genome of each human cell consists of 3×10^9 base pairs of DNA, subdivided into 23 chromosomes. The entire haploid genome contains sufficient DNA to code for nearly 1.5 million pairs of genes. However, studies of mutation rates and of the complexities of the genomes of higher organisms strongly suggest that humans have only about 100,000 proteins. This implies that most of the DNA is noncoding; ie, its information is never translated into an amino acid sequence of a protein molecule. Certainly, some of the excess DNA sequences serve to regulate the expression of genes during development, differentiation, and adaptation to the environment. Some excess clearly makes up the intervening sequences that split the coding regions of genes, but much of the excess appears to be composed of many families of repeated sequences for which no functions have been clearly defined.

The DNA in a eukaryotic genome can be divided into different "sequence classes." These are **unique-sequence, or nonrepetitive, DNA** and **repetitive-sequence DNA.** In the haploid genome unique-sequence DNA generally includes the single copy genes that code for proteins. The repetitive DNA in the haploid genome includes sequences that vary in copy number from two to as many as 10^7 copies per cell.

More Than Half the DNA in Eukaryotic Organisms Is in Unique or Nonrepetitive Sequences

This estimation (and the distribution of repetitive-sequence DNA) is based on a variety of DNA-RNA hybridization techniques. Similar techniques are used to estimate the number of active genes in a population of unique-sequence DNA. In yeast, a lower eukaryote, about 4000 genes are expressed. In typical tissues in a higher eukaryote (eg, mammalian liver and kidney), between 10,000 and 15,000 genes are expressed. Different combinations of genes are expressed in each tissue, of course, and how this is accomplished is one of the major unanswered questions in biology.

Table 38–1. The packing ratios of each of the orders of DNA structure.

Chromatin Form	Packing Ratio
Bare double-helix DNA	~1.0
10-nm fibril of nucleosomes	7–10
25- to 30-nm chromatin fiber of superhelical nucleosomes	40–60
Condensed metaphase chromosome of loops	8000

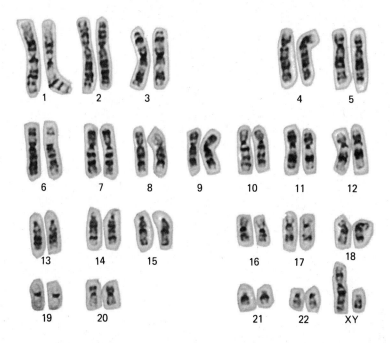

Figure 38–6. A human karyotype (of a man with a normal 46,XY constitution), in which the chromosomes have been stained by the Giemsa method and aligned according to the Paris Convention. (Courtesy of H Lawce and F Conte.)

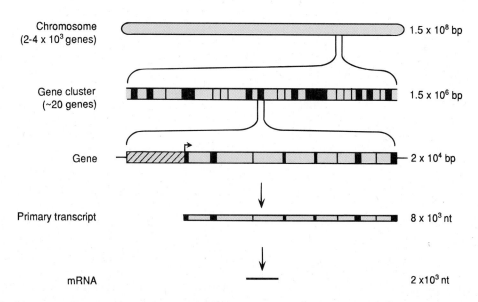

Figure 38–7. The relationship between chromosomal DNA and mRNA. The human haploid DNA complement of 3×10^9 base pairs (bp) is distributed between 23 chromosomes. Genes are clustered on these chromosomes. An average gene may be 20,000 bp in length, including the regulatory region (hatched area), which is usually located at the 5′ end of the gene. The regulatory region is shown here as being adjacent to the transcription initiation site (arrow). Most eukaryotic genes have alternating exons and introns. In this example, there are nine exons (solid areas) and eight introns (open areas). The introns are removed from the primary transcript by the processing reaction, and the exons are ligated together in sequence to form the mature mRNA. (nt, nucleotides.)

In Human DNA, at Least 20–30% of the Genome Consists of Repetitive Sequences

Repetitive-sequence DNA can be broadly classified as moderately repetitive or as highly repetitive. The highly repetitive sequences consist of 5–500 base pair lengths repeated many times in tandem. These sequences are usually clustered in centromeres and telomeres of the chromosome and are present in about 1–10 million copies per haploid genome. These sequences are transcriptionally inactive and may play a structural role in the chromosome.

The moderately repetitive sequences, which are defined as being present at less than 10^6 copies per haploid genome, are not clustered but are interspersed with unique sequences. In many cases, these long interspersed repeats are transcribed by RNA polymerase II and contain caps indistinguishable from those on mRNA.

Depending on their length, moderately repetitive sequences are classified as **long interspersed repeat sequences (LINEs)** or **short interspersed repeat sequences (SINEs).** Both types appear to be **retroposons,** that is, they arose from movement from one location to another **(transposition)** through an RNA intermediate by the action of reverse transcriptase that transcribes an RNA template into DNA. Mammalian genomes contain 20–50 thousand copies of the 6–7 kb LINEs. These represent species-specific families of repeat elements. SINEs are shorter (70–300 bp), and there may be more than 100,000 copies per genome. Of the SINEs in the human genome, one family, the **Alu family,** is present in about 500,000 copies per haploid genome and accounts for at least 5–6% of the human genome. Members of the human Alu family and their closely related analogs in other animals are transcribed as integral components of hnRNA or as discrete RNA molecules, including the well-studied 4.5S RNA and 7S RNA. These particular family members are highly conserved within a species as well as between mammalian species. Components of the short interspersed repeats, including the members of the Alu family, may be mobile elements, capable of jumping into and out of various sites within the genome (see below).

Microsatellite Repeat Sequences

One category of repeat sequences exists as both dispersed and grouped tandem arrays. The sequences consist of 2–5 bp repeated up to 50 times. These **microsatellite sequences** most commonly are found as dinucleotide repeats of AC on one strand and TG on the opposite strand, but several other forms occur. These AC repeat sequences are estimated to occur at 50,000–100,000 locations in the genome. At any locus, the number of these repeats may vary on the two chromosomes, thus providing heterozygosity of the number of copies of a particular microsatellite number in an individual. This is a heritable trait, and, because of their number and the ease of detecting them using the polymerase chain reaction (PCR) (Chapter 42), AC repeats are very useful in constructing genetic linkage maps. Most genes are associated with one or more microsatellite markers, so the relative position of genes on chromosomes can be assessed, as can the association of a gene with a disease. Using PCR, a large number of family members can be rapidly screened for a certain **microsatellite polymorphism.** The association of a specific polymorphism with a gene in affected family members—and the lack of this association in unaffected members—may be the first clue about the genetic basis of a disease.

Trinucleotide sequences that increase in number (microsatellite instability) can cause disease. The unstable p(CCG)n repeat sequence is associated with the fragile X syndrome. Other trinucleotide repeats that undergo dynamic mutation (usually an increase) are associated with Huntington's chorea, myotonic dystrophy, and spinobulbar muscular atrophy.

GENETIC MATERIAL CAN BE ALTERED AND REARRANGED

An alteration in the sequence of purine and pyrimidine bases in a gene due to a change, a removal, or an insertion of one or more bases may result in an altered gene product that in most instances ultimately is a protein. Such alteration in the genetic material results in a **mutation,** the consequences of which are discussed in detail in Chapter 40.

Chromosomal Recombination Is One Way of Rearranging Genetic Material

Genetic information can be exchanged between similar or homologous chromosomes. The exchange or **recombination** event occurs primarily during meiosis in mammalian cells and requires alignment of homologous chromosomes, an alignment that almost always occurs with great exactness. A process of crossing over occurs as shown in Figure 38–8. This usually results in an equal and reciprocal exchange of genetic information between homologous chromosomes. If the homologous chromosomes possess different alleles of the same genes, the crossover may produce noticeable and heritable genetic linkage differences. In the rare case where the alignment of homologous chromosomes is not exact, the crossing over or recombination event may result in an unequal exchange of information. One chromosome may receive less genetic material and thus a deletion, while the other partner of the chromosome pair receives more genetic material and thus an insertion or duplication (Figure 38–8). Unequal crossing over does occur in humans, as evidenced by the existence of he-

DNA ORGANIZATION & REPLICATION / **403**

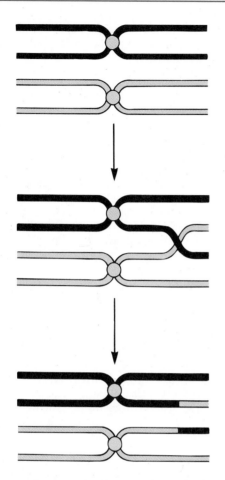

Figure 38–8. The process of crossing-over between homologous chromosomes to generate recombinant chromosomes.

pairing can result in expansion or contraction in the copy number of the repeat family and may contribute to the expansion and fixation of variant members throughout the array.

Chromosomal Integration Occurs With Some Viruses

Some bacterial viruses (bacteriophages) are capable of recombining with the DNA of a bacterial host in such a way that the genetic information of the bacteriophage is incorporated in a linear fashion into the genetic information of the host. This integration, which is a form of recombination, occurs by the mechanism illustrated in Figure 38–10. The backbone of the circular bacteriophage genome is broken, as is that of the DNA molecule of the host; the appropriate ends are resealed with the proper polarity. The bacteriophage DNA is figuratively straightened out ("linearized") as it is integrated into the bacterial DNA molecule—frequently a closed circle as well. The site at which the bacteriophage genome integrates or recombines with the bacterial genome is chosen by one of two mechanisms. If the bacteriophage contains a DNA sequence **homologous** to a sequence in the host DNA molecule, then a recombination event analogous to that occurring between homologous chromosomes can occur. However, some bacteriophages synthesize proteins that bind specific sites on bacterial chromosomes to a **nonhomologous** site characteristic of the bacteriophage DNA molecule. Integration occurs at the site and is said to be **"site specific."**

Many animal viruses, particularly the oncogenic viruses—either directly or, in the case of RNA viruses, their DNA transcripts—can be integrated into chromosomes of the mammalian cell. The integration of the animal virus DNA into the animal genome generally is not "site-specific."

Transposition Can Produce Processed Genes

In eukaryotic cells, small DNA elements that clearly are not viruses are capable of transposing

moglobins designated Lepore and anti-Lepore (Figure 38–9). **Unequal crossover** affects tandem arrays of repeated DNAs whether they are related globin genes, as in Figure 38–9, or more abundant repetitive DNA. The unequal crossover through slippage in the

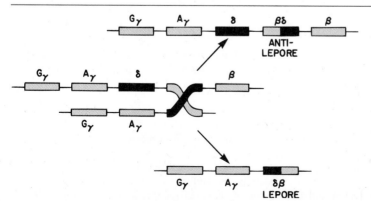

Figure 38–9. The process of unequal crossover in the region of the mammalian genome that harbors the structural genes for hemoglobin and the generation of the unequal recombinant products hemoglobin delta-beta Lepore and beta-delta anti-Lepore. The examples given show the locations of the crossover regions between amino acid residues. (Redrawn and reproduced, with permission, from Clegg JB, Weatherall DJ: β⁰ Thalassemia: Time for a reappraisal? Lancet 1974;2:133.)

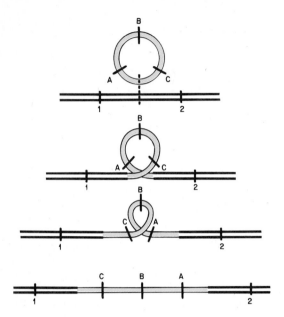

Figure 38–10. The integration of a circular genome from a bacteriophage (with genes A, B, and C) into the DNA molecule of a host (with genes 1 and 2) and the consequent ordering of the genes.

themselves in and out of the host genome in ways that affect the function of neighboring DNA sequences. These mobile elements, sometimes called "jumping DNA," can carry flanking regions of DNA and, therefore, profoundly affect evolution. As mentioned above, the Alu family of moderately repeated DNA sequences has structural characteristics similar to the termini of retroviruses, which would account for the ability of the latter to move into and out of the mammalian genome.

Direct evidence for the transposition of other small DNA elements into the human genome has been provided by the discovery of "processed genes" for immunoglobulin molecules, α-globin molecules, and several others. These **processed genes** consist of DNA sequences identical or nearly identical to those of the messenger RNA for the appropriate gene product. That is, the 5′ nontranscribed region, the coding region without intron representation, and the 3′ poly(A) tail are all present contiguously. This particular DNA sequence arrangement must have resulted from the reverse transcription of an appropriately processed messenger RNA molecule from which the intron regions had been removed and the poly(A) tail added. The only recognized mechanism that this reverse transcript could have used to integrate into the genome would have been a transposition event. In fact, these "processed genes" have short terminal repeats at each end, as do known transposed sequences in lower organisms. Some of the processed genes have been randomly altered through evolution so that

they now contain nonsense codons that preclude their expression (see Chapter 40). Thus, they are referred to as **"pseudogenes."**

Gene Conversion Produces Rearrangements

Besides unequal crossover and transposition, a third mechanism can effect rapid changes in the genetic material. Similar sequences on homologous or nonhomologous chromosomes may occasionally pair up and eliminate any mismatched sequences between them. This may lead to the accidental fixation of one variant or another throughout a family of repeated sequences and thereby homogenize the sequences of the members of repetitive DNA families. This latter process is referred to as **gene conversion.**

Sister Chromatids Exchange

In diploid eukaryotic organisms such as humans, after cells progress through the S phase they contain a tetraploid content of DNA. This is in the form of sister chromatids of chromosome pairs. Each of these sister chromatids contains identical genetic information, since each is a product of the semiconservative replication of the original parent DNA molecule of that chromosome. Crossing over occurs between these genetically identical sister chromatids. Of course, these **sister chromatid exchanges** (Figure 38–11) have no genetic consequence so long as the exchange is the result of an equal crossover.

Immunoglobulin Genes Rearrange

In mammalian cells, some interesting gene rearrangements occur normally during development and differentiation. For example, in mice the V_L and C_L genes for a single immunoglobulin molecule (see Chapter 41) are widely separated in the germ line DNA. In the DNA of a differentiated immunoglobulin-producing (plasma) cell, the same V_L and C_L genes have been moved physically closer together in the genome, and into the same transcription unit. However, even then, this rearrangement of DNA during differentiation does not bring the V_L and C_L genes into contiguity in the DNA. Instead, the DNA contains an interspersed or interruption sequence of about 1200 base pairs at or near the junction of the V and C regions. The interspersed sequence is transcribed into RNA along with the V_L and C_L genes, and the interspersed information is removed from the RNA during its nuclear processing (Chapters 39 and 41).

DNA SYNTHESIS AND REPLICATION ARE RIGIDLY CONTROLLED

The primary function of DNA replication is understood to be the provision of progeny with the genetic information possessed by the parent. Thus, the repli-

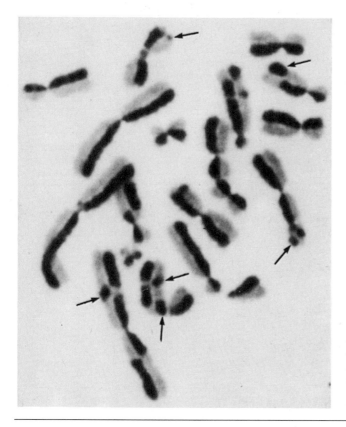

Figure 38–11. Sister chromatid exchanges between human chromosomes. These are detectable by Giemsa staining of the chromosomes of cells replicated for two cycles in the presence of bromodeoxyuridine. The small arrows indicate some regions of exchange. (Courtesy of S Wolff and J Bodycote.)

cation of DNA must be complete and carried out with high fidelity to maintain genetic stability within the organism and the species. The process of DNA replication is complex and involves many cellular functions and several verification procedures to ensure fidelity in replication. About 30 proteins are involved in the replication of the *E coli* chromosome, and this process is almost certainly more complex in eukaryotic organisms. The first enzymologic observations on DNA replication were made in *E coli* by Kornberg, who described in that organism the existence of an enzyme now called DNA polymerase I. This enzyme has multiple catalytic activities, a complex structure, and a requirement for the triphosphates of the four deoxyribonucleosides of adenine, guanine, cytosine, and thymine. The polymerization reaction catalyzed by DNA polymerase I of *E coli* has served as a prototype for all DNA polymerases of both prokaryotes and eukaryotes, even though it is now recognized that the major role of this polymerase is to ensure fidelity and to repair rather than to replicate DNA.

In all cells, replication can occur only from a single-stranded DNA (ssDNA) template. Mechanisms must exist to target the site of initiation of replication and to unwind the double-stranded DNA (dsDNA) in that region. The replication complex must then form. After replication is complete in an area, the parent

and daughter strands must re-form dsDNA. In eukaryotic cells, an additional step must occur. The dsDNA must precisely re-form the chromatin structure, including nucleosomes, that existed prior to the onset of replication. Although this entire process is not well understood in eukaryotic cells, it has been quite precisely described in prokaryotic cells, and the general principles are thought to be the same in both. The major steps are listed in Table 38–2, illustrated in Figure 38–12, and discussed, in sequence, below.

The Origin of Replication

At the **origin of replication (ori)**, there is an association of sequence-specific dsDNA-binding proteins with a series of direct repeat DNA sequences. In bacteriophage λ the oriλ binds the O protein to four adjacent sites. In *E coli*, the oriC binds the protein

Table 38–2. Steps involved in DNA replication in eukaryotes.

1. Identification of the origins of replication.
2. Unwinding (denaturation) of dsDNA to provide an ssDNA template.
3. Formation of the replication fork.
4. Initiation of DNA synthesis and elongation.
5. Formation of replication bubbles with ligation of the newly synthesized DNA segments.
6. Reconstitution of chromatin structure.

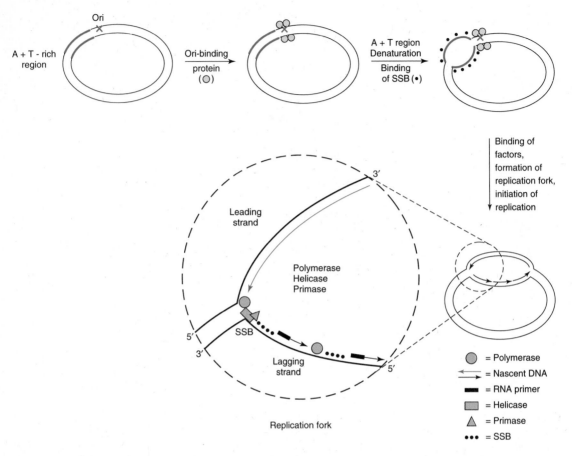

Figure 38–12. Steps involved in DNA replication. This figure describes DNA replication in an **E coli** cell, but the general steps are similar in eukaryotes. A specific interaction of a protein (the O protein) to the origin of replication (ori) results in local unwinding of DNA at an adjacent A+T-rich region. The DNA in this area is maintained in the single-strand confor-mation (ssDNA) by single-strand binding proteins (SSBs). This allows a variety of proteins, including helicase, primase, and DNA polymerase, to bind and to initiate DNA synthesis. The replication fork proceeds as DNA synthesis occurs con-tinuously *(solid arrow)* on the leading strand and discontinuously *(broken arrows)* on the lagging strand.

dnaA. In both cases, a complex is formed consisting of 150–250 bp of DNA and multimers of the DNA-binding protein. This leads to the local denaturation and unwinding of an adjacent A+T-rich region of DNA. Functionally similar **autonomously replicat-ing sequences (ARS)** have been identified in yeast cells. Each ARS is a somewhat degenerate 11 bp se-quence located adjacent to an approximately 80 bp sequence that is easily unwound. Consensus se-quences similar to ori or ARS in structure or function have not been precisely defined in mammalian cells, nor have the proteins that bind to such sequences been identified.

Unwinding of DNA

The interaction of proteins with ori defines the start site of replication and provides a short region of ssDNA essential for initiation of synthesis of the nascent DNA strand. This process requires the for-mation of a number of protein-protein and protein-DNA interactions. A critical step is provided by a DNA helicase that allows for processive unwinding of DNA. In uninfected *E coli*, this function is pro-vided by a complex of dnaB helicase and the dnaC protein. Single-stranded DNA-binding proteins (SSBs) stabilize this complex. In λ phage-infected *E coli*, the phage protein P binds to dnaB and the P•dnaB complex binds to oriλ by interacting with the O protein. dnaB is not an active helicase when in the P•dnaB•O complex. Three *E coli* heat shock pro-teins (dnaK, dnaJ, and GrpE) cooperate to remove the P protein and activate the dnaB helicase. In coop-eration with SSB, this leads to DNA unwinding and active replication. In this way, the replication of the λ phage is accomplished at the expense of replication of the host *E coli* cell.

Formation of the Replication Fork

A replication fork consists of four components that form in the following sequence: (1) the DNA helicase unwinds a short segment of the parental, duplex DNA; (2) a primase initiates synthesis of an RNA molecule that is essential for priming DNA synthesis; (3) the DNA polymerase initiates nascent, daughter strand synthesis; and (4) SSBs bind to ssDNA and prevent premature reannealing of ssDNA to dsDNA. These components are illustrated in Figure 38–12.

The polymerase III holoenzyme (the dnaE gene product in *E coli*) binds to template DNA as part of a multiprotein complex that consists of several polymerase accessory factors (β, γ, δ, δ′, and τ). DNA polymerases only synthesize DNA in the 5′ to 3′ direction, and only one of the several different types of polymerases is involved at the replication fork. Because the DNA strands are antiparallel (Chapter 37), the polymerase functions asymmetrically. On the **leading (forward) strand,** the DNA is synthesized continuously. On the **lagging (retrograde) strand,** the DNA is synthesized in short (1–5 kb) fragments, the so-called **Okazaki fragments.** Several Okazaki fragments (up to 250) must be synthesized, in sequence, for each replication fork. To ensure that this happens, the helicase acts on the lagging strand to unwind dsDNA in a 5′ to 3′ direction. The helicase associates with the primase to afford the latter proper access to the template. This allows the RNA primer to be made and, in turn, the polymerase to begin replicating the DNA. This is an important reaction sequence since DNA polymerases cannot initiate DNA synthesis de novo. The mobile complex between helicase and primase has been called a **primosome.** As the synthesis of an Okazaki fragment is completed and the polymerase is released, a new primer has been synthesized. The same polymerase molecule remains associated with the replication fork and proceeds to synthesize the next Okazaki fragment.

The DNA Polymerase Complex

The primary enzyme involved in DNA replication is DNA polymerase III, which is the 132 kDa product of the dnaE gene in *E coli*. This holoenzyme (also called polymerase α) is unable to copy long stretches of ssDNA. The functional complex consists of ten different proteins: α plus nine other subunits. These range in size from 12 to 71 kDa and the final complex is > 1 MDa. This complex forms a highly processive enzyme that is capable of polymerizing 0.5 mb of DNA following one primary event on the leading strand. The complex has several other activities that allow for proofreading and repair in the event a copying error is made.

In mammalian cells, the polymerase is capable of polymerizing about 100 nucleotides per second, a rate at least tenfold less than the rate of polymerization of deoxynucleotides by the bacterial DNA poly-

merase complex. This reduced rate may result from interference by nucleosomes. It is not known how the replication complex negotiates nucleosomes.

Initiation and Elongation of DNA Synthesis

The initiation of DNA synthesis (Figure 38–13) requires **priming by a short length of RNA,** about 10–200 nucleotides long. This priming process involves the nucleophilic attack by the 3′-hydroxyl group of the RNA primer on the α phosphate of the first entering deoxynucleoside triphosphate (N in Figure 38–13) with the splitting off of pyrophosphate. The 3′-hydroxyl group of the recently attached deoxyribonucleoside monophosphate is then free to carry out a **nucleophilic attack** on the next entering deoxyribonucleoside triphosphate (N + 1 in Figure 38–13), again at its α phosphate moiety, with the splitting off of pyrophosphate. Of course, the selection of the proper deoxyribonucleotide whose terminal 3′-hydroxyl group is to be attacked is dependent upon **proper base pairing with the other strand** of the DNA molecule according to the rules proposed originally by Watson and Crick (Figure 38–14). When an adenine deoxyribonucleoside monophosphoryl moiety is in the template position, a thymidine triphosphate will enter and its α phosphate will be attacked by the 3′-hydroxyl group of the deoxyribonucleoside monophosphoryl most recently added to the polymer. By this stepwise process, the template dictates which deoxyribonucleoside triphosphate is complementary and by hydrogen bonding holds it in place while the 3′-hydroxyl group of the growing strand attacks and incorporates the new nucleotide into the polymer. These fragments of DNA attached to an RNA initiator component, discovered by Okazaki, are referred to as **Okazaki fragments** (Figure 38–15). In mammals, after many Okazaki fragments are generated, the replication complex begins to remove the RNA primers, to fill in the gaps left by their removal with the proper base-paired deoxynucleotide, and then to seal the fragments of newly synthesized DNA by enzymes referred to as **DNA ligases.**

Replication Exhibits Polarity

As has already been noted, DNA molecules are double-stranded and the two strands are antiparallel, ie, running in opposite directions. The replication of DNA in prokaryotes and eukaryotes occurs on both strands simultaneously. However, an enzyme capable of polymerizing DNA in the 3′ to 5′ direction does not exist in any organism, so that both of the newly replicated DNA strands cannot grow in the same direction simultaneously. Nevertheless, the same enzyme does replicate both strands at the same time. The single enzyme replicates one strand ("leading strand") in a continuous manner in the 5′ to 3′ direction, with the same overall forward direction. It repli-

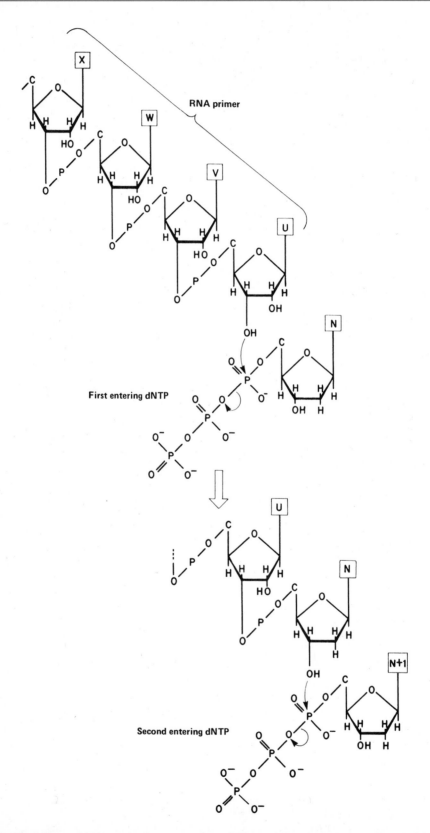

Figure 38–13. The initiation of DNA synthesis upon a primer of RNA and the subsequent attachment of the second deoxyribonucleoside triphosphate.

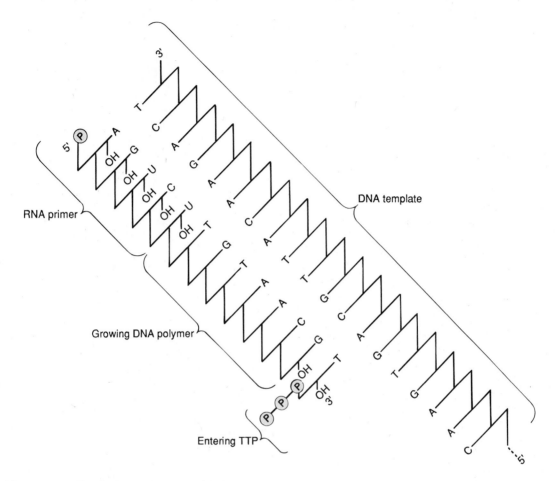

Figure 38–14. The RNA-primed synthesis of DNA demonstrating the template function of the complementary strand of parental DNA.

cates the other strand ("lagging strand") discontinuously while polymerizing the nucleotides in short spurts of 150–250 nucleotides again in the 5′ to 3′ direction, but at the same time it faces toward the back end of the preceding RNA primer rather than toward the unreplicated portion. This process of **semidiscontinuous DNA synthesis** is shown diagrammatically in Figure 38–12.

In the mammalian nuclear genome, most of the RNA primers are eventually removed as part of the replication process, whereas after replication of the mitochondrial genome the small piece of RNA remains as an integral part of the closed circular DNA structure.

Formation of Replication Bubbles

Replication proceeds from a single ori in the circular bacterial chromosome. The entire mammalian genome replicates in approximately 9 hours, the average period required for formation of a tetraploid

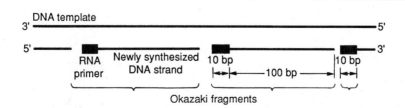

Figure 38–15. The discontinuous polymerization of deoxyribonucleotides and formation of Okazaki fragments.

genome from a diploid genome in a replicating cell. This requires the presence of **multiple origins** of DNA replication that occur in clusters of up to 100 of these replication units. Replication occurs in both directions along the chromosome, and both strands are replicated simultaneously. This replication process generates **"replication bubbles"** (Figure 38–16).

The multiple sites that serve as origins for DNA replication in eukaryotes are poorly defined except in a few animal viruses and in yeast. However, it is clear that initiation is regulated both spatially and temporally, since clusters of adjacent sites initiate synchronously. There are suggestions that functional domains of chromatin replicate as intact units, implying that the origins of replication are specifically located with respect to transcription units.

During the replication of DNA, there must be a separation of the two strands to allow each to serve as a template by hydrogen bonding its nucleotide bases to the incoming deoxynucleoside triphosphate. The separation of the DNA double helix is promoted by SSBs, specific protein molecules that stabilize the single-stranded structure as the replication fork progresses. These stabilizing proteins bind stoichiometrically to the single strands without interfering with the abilities of the nucleotides to serve as templates (Figure 38–12). In addition to separating the two strands of the double helix, there must be an unwinding of the molecule (once every 10 nucleotide pairs) to allow strand separation. This must happen in segments, given the time during which DNA replication occurs. There are multiple "swivels" interspersed in the DNA molecules of all organisms. The swivel function is provided by specific enzymes that introduce **"nicks" in one strand of the unwinding double helix,** thereby allowing the unwinding process to proceed. The nicks are quickly resealed without requiring energy input, because of the formation of a high-energy covalent bond between the nicked phosphodiester backbone and the nicking-sealing enzyme.

The nicking-resealing enzymes are called DNA **topoisomerases.** This process is depicted diagrammatically in Figure 38–17 and there compared to the ATP-dependent resealing carried out by the DNA ligases. Topoisomerases are also capable of unwinding supercoiled DNA. Supercoiled DNA is a higher ordered structure occurring in circular DNA molecules wrapped around a core, as depicted in Figure 38–18.

There exists in one species of animal viruses (retroviruses) a class of enzymes capable of synthesizing a single-stranded and then a double-stranded DNA molecule from a single-stranded RNA template. This polymerase, RNA-dependent DNA polymerase or **"reverse transcriptase,"** first synthesizes a DNA-RNA hybrid molecule utilizing the RNA genome as a template. A specific enzyme, RNase H, degrades the RNA strand, and the remaining DNA strand in turn serves as a template to form a double-stranded DNA molecule containing the information originally present in the RNA genome of the animal virus.

Reconstitution of Chromatin Structure

There is evidence that nuclear organization and chromatin structure are involved in determining the regulation and initiation of DNA synthesis. As noted above, the rate of polymerization in eukaryotic cells, which have chromatin and nucleosomes, is tenfold less than that in prokaryotic cells, which have naked DNA. It is also clear that chromatin structure must be re-formed after replication. Newly replicated DNA is rapidly assembled into nucleosomes, and the preexisting and newly assembled histone octamers are randomly distributed to each arm of the replication fork.

DNA Synthesis Occurs During the S Phase of the Cell Cycle

In animal cells, including human cells, the replication of the DNA genome occurs only at a specified

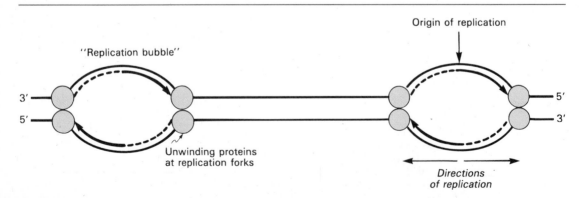

Figure 38–16. The generation of "replication bubbles" during the process of DNA synthesis. The bidirectional replication and the proposed positions of unwinding proteins at the replication forks are depicted.

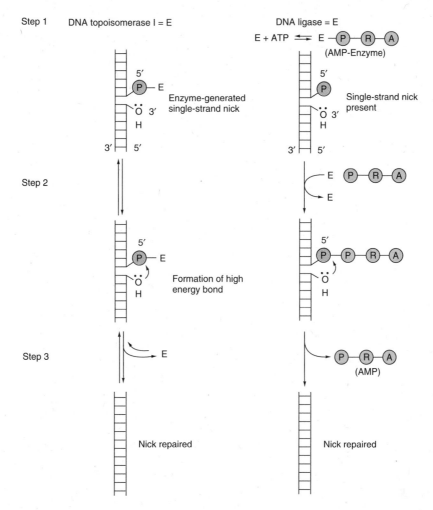

Figure 38–17. Comparison of two types of nick-sealing reactions on DNA. The series of reactions at left is catalyzed by DNA topoisomerase I, that at right by DNA ligase. (Slightly modified and reproduced, with permission, from Lehninger AL: *Biochemistry,* 2nd ed. Worth, 1975.)

time during the life span of the cell. This period is referred to as the synthetic or S phase. This is usually temporally separated from the mitotic phase by nonsynthetic periods referred to as gap 1 (G1) and gap 2 (G2), occurring before and after the S phase, respectively (Figure 38–19). The cell regulates its DNA synthesis grossly by allowing it to occur only at specific times and mostly in cells preparing to divide by a mitotic process.

It appears that all eukaryotic cells have gene products that govern the transition from one phase of the cell cycle to another. The **cyclins** are a family of proteins whose concentration increases and decreases throughout the cell cycle—thus their name. The cyclins turn on, at the appropriate time, different **cyclin-dependent protein kinases (CDKs)** that phosphorylate substrates essential for progression through the cell cycle. For example, cyclin D levels rise in

late G1 phase and allow progression beyond the **start (yeast)** or **restriction point (mammals),** the point beyond which cells irrevocably proceed into the S or DNA synthesis phase.

The D cyclins activate CDK4 and CDK6. These two kinases are also synthesized during G1 in cells undergoing active division. The D cyclins and CDK4 and CDK6 are nuclear proteins that assemble as a complex in late G1 phase. The complex is an active serine-threonine protein kinase. One substrate for this kinase is the retinoblastoma (Rb) protein. Rb is a cell cycle regulator because it binds to and inactivates a transcription factor (E2F) necessary for the transcription of genes and progression from G1 to S phase. The phosphorylation of Rb by CDK4 or CDK6 results in the release of E2F from Rb, gene activation (eg, the dihydrofolate reductase gene), and cell cycle progression.

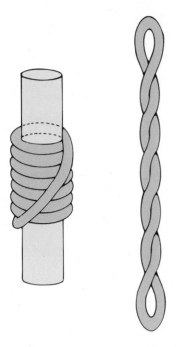

Figure 38–18. Supercoiling of DNA. A left-handed toroidal (solenoidal) supercoil, at left, will convert to a right-handed interwound supercoil, at right, when the cylindric core is removed. Such a transition is analogous to that which occurs when nucleosomes are disrupted by the high salt extraction of histones from chromatin.

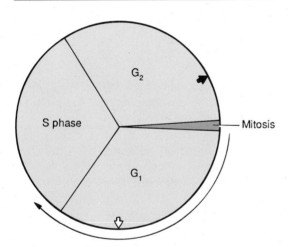

Figure 38–19. Mammalian cell cycle. The DNA synthetic phase (S phase) is separated from mitosis by gap 1 (G1) and gap 2 (G2). The arrow outside the circle indicates the direction of the progression of the cell cycle. The open arrowhead indicates the Start (yeast) or restriction point (mammals) site at which CDK4 or CDK6 acts, after activation by cyclin D. This protein kinase releases an inhibition, perhaps by the Rb gene product, and progression through the cell cycle commences. The closed arrowhead indicates when peak levels of the cyclin B/CDK1 complex are achieved. This allows the cell to undergo mitosis.

Other cyclins and CDKs are involved in different aspects of cell cycle progression (Table 38–3). Cyclin E and CDK2 form a complex in late G1. Cyclin E is rapidly degraded, and the released CDK2 then forms a complex with cyclin A. This sequence is necessary for the initiation of DNA synthesis in S phase. A complex between cyclin B and CDK1 is rate-limiting for the G2/M transition in eukaryotic cells.

Many of the cancer-causing viruses (oncoviruses) and oncogenes are capable of alleviating or disrupting the apparent restriction that normally controls the entry of mammalian cells from G1 into the S phase. From the foregoing, one might have surmised that excessive production of a cyclin, or production at an inappropriate time, might result in abnormal or unrestrained cell division. In this context it is noteworthy that the *bcl* oncogene associated with B cell lymphoma appears to be the cyclin D1 gene.

During the S phase, mammalian cells contain greater quantities of DNA polymerase than during the nonsynthetic phases of the cell cycle. Furthermore, those enzymes responsible for the formation of the substrates for DNA synthesis, ie, deoxyribonucleoside triphosphates, are also increased in activity, and their activity will diminish following the synthetic phase until the reappearance of the signal for renewed DNA synthesis. During the S phase, the nuclear DNA is **completely replicated once and only once.** It seems that once chromatin has been replicated, it is marked so as to prevent its further replication until it again passes through mitosis. It has been suggested that DNA methylation may serve as such a covalent marker.

In general, a given pair of chromosomes will replicate simultaneously and within a fixed portion of the S phase upon every replication. On a chromosome, clusters of replication units replicate coordinately. The nature of the signals that regulate DNA synthesis at these levels is unknown, but the regulation does appear to be an intrinsic property of each individual chromosome.

Enzymes Repair Damaged DNA

The maintenance of the integrity of the information in DNA molecules is of utmost importance to the survival of a particular organism as well as to survival of the species. Thus, it can be concluded that

Table 38–3. Cyclins and cyclin-dependent kinase involved in cell cycle progression.

Cyclin	Kinase	Function
D	CDK4, CDK6	Progression past restriction point at G1/S boundary
E, A	CDK2	Initiation of DNA synthesis in early S phase
B	CDK1	Transition from G2 to M

surviving species have evolved mechanisms for repairing DNA damage incurred as a result of either replication errors or environmental insults.

As described in Chapter 37, the major responsibility for the fidelity of replication resides in the specific pairing of nucleotide bases. Proper pairing is dependent upon the presence of the favored tautomers of the purine and pyrimidine nucleotides, but the equilibrium whereby one tautomer is more stable than another is only about 10^4 or 10^5 in favor of that with the greater stability. Although this is not sufficiently favorable to ensure the high fidelity that is necessary, the favoring of the preferred tautomers, and thus of the proper base pairing, could be ensured by monitoring the base pairing twice. Such double monitoring does appear to occur in both bacterial and mammalian systems: once at the time of insertion of the deoxyribonucleoside triphosphates, and later by a follow-up, energy-requiring mechanism which removes all improper bases that may occur in the newly formed strand. This double monitoring does not permit errors of mispairing due to the presence of the unfavored tautomers to occur more frequently than once every 10^8–10^{10} base pairs. The molecule responsible for this monitoring mechanism in *E coli* is the built-in $3' \rightarrow 5'$ exonuclease activity of DNA polymerase complex, but mammalian DNA polymerases do not seem to possess such a nuclease proofreading function. Other enzymes provide this repair function.

Replication errors, even with a very efficient repair system, lead to the accumulation of mutations. A human has 10^{14} nucleated cells with 3×10^9 base pairs per cell. If some 10^{16} cell divisions occur in a lifetime and 10^{-10} mutations per base pair per cell generation escape repair, there may eventually be as many as 1 mutation per 10^6 bp in the genome. Fortunately, most of these will probably occur in DNA that does not encode proteins or will not affect the function of encoded proteins and so are of no consequence. In addition, spontaneous and chemically induced damage to DNA must be repaired.

Damage to DNA by environmental, physical, and chemical agents may be classified into four types (Table 38–4). Abnormal regions of DNA, either from copying errors or DNA damage, are replaced by three mechanisms: (1) mismatch repair, (2) base excision repair, and (3) nucleotide excision repair (Table 38–5). These mechanisms exploit the redundancy of information inherent in the double helical DNA structure. The defective region in one strand can be returned to its original form by relying on the complementary information stored in the unaffected strand.

Mismatch Repair

Mismatch repair corrects errors made when DNA is copied. For example, a C could be inserted opposite an A, or the polymerase could slip or stutter and

Table 38–4. Types of damage to DNA.

I. Single-base alteration
   A. Depurination
   B. Deamination of cytosine to uracil
   C. Deamination of adenine to hypoxanthine
   D. Alkylation of base
   E. Insertion or deletion of nucleotide
   F. Base-analog incorporation
II. Two-base alteration
   A. UV light–induced thymine-thymine dimer
   B. Bifunctional alkylating agent cross-linkage
III. Chain breaks
   A. Ionizing radiation
   B. Radioactive disintegration of backbone element
IV. Cross-linkage
   A. Between bases in same or opposite strands
   B. Between DNA and protein molecules (eg, histones)

insert two to five extra unpaired bases. Specific proteins scan the newly synthesized DNA, using adenine methylation within a GATC sequence as the point of reference (Figure 38–20). If a mismatch or small loop is found, a GATC endonuclease cuts the strand bearing the mutation at a site corresponding to the GATC. An exonuclease then digests this strand from the GATC through the mutation, thus removing the faulty DNA. This can occur from either end if the defect is bracketed by two GATC sites. This defect is then filled in by normal cellular enzymes according to base pairing rules. In *E coli,* three proteins (Mut S, Mut C, and Mut H) are required for recognition of the mutation and nicking of the strand. Other cellular enzymes, including ligase, polymerase, and SSBs, remove and replace the strand. The process is somewhat more complicated in mammalian cells, as about six proteins are involved in the first steps.

Faulty mismatch repair has been linked to hereditary nonpolyposis colon cancer (HNPCC), one of the most common inherited cancers. Genetic studies linked HNPCC in some families to a region of chromosome 2. The gene located, designated hMSH2, was subsequently shown to be the human analog of the *E coli* MutS protein that is involved in mismatch repair (see above). hMSH2 accounts for 50–60% of HNPCC cases. Another gene, hMLH1, is associated with most other cases. hMLH1 is the human analog of the bacterial mismatch repair gene Mut L. How does faulty mismatch repair result in colon cancer? The human genes were localized because microsatellite instability was detected. That is, the cancer cells had a microsatellite of different length than that found in the normal cells of the individual. Apparently the affected cells, which harbor a mutated hMSH2 or hMLH1 mismatch repair enzyme, are unable to remove small loops of unpaired DNA, and the microsatellite thus increases in size. This must affect the function of a protein critical in surveillance of the cell cycle in these colon cells.

Table 38–5. Mechanism of DNA repair.

Mechanism	Problem	Solution
Mismatch repair	Copying errors (single base or two- to five-base unpaired loops)	Methyl-directed strand cutting, exonuclease digestion, and replacement
Base excision repair	Spontaneous, chemical, or radiation damage to a single base	Base removal by N-glycosylase, abasic sugar removal, replacement
Nucleotide excision repair	Spontaneous, chemical, or radiation damage to a DNA segment	Removal of an approximately 30-nucleotide oligomer and replacement

Base Excision-Repair

The **depuration of DNA,** which happens spontaneously owing to the thermal lability of the purine N-glycosidic bond, occurs at a rate of 5000–10,000/cell/d at 37 °C. Specific enzymes recognize a depurinated site and replace the appropriate purine

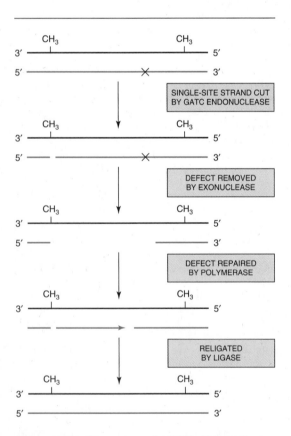

Figure 38–20. Mismatch repair of DNA. This mechanism corrects a single mismatch base pair (eg, C to A rather than T to A) or a short region of unpaired DNA. The defective region is recognized by an endonuclease that makes a single-strand cut at an adjacent methylated GATC sequence. The DNA strand is removed through the mutation, replaced, and religated.

directly, without interruption of the phosphodiester backbone.

Both cytosine and adenine bases in DNA spontaneously deaminate to form uracil and hypoxanthine, respectively. Since neither uracil nor hypoxanthine normally exists in DNA, it is not surprising that specific **N-glycosylases** can recognize these abnormal bases and remove the base itself from the DNA. This removal marks the site of the defect and allows an **apurinic or apyrimidinic endonuclease** to excise the abasic sugar. The proper base is then replaced by a repair DNA polymerase β, and a **ligase** returns the DNA to its original state (Figure 38–21). This series of events is called **base excision-repair.** By a similar series of steps involving initially the recognition of the defect, alkylated bases and base analogs can be removed from DNA and the DNA returned to its original informational content. This mechanism is suitable for replacement of a single base but is not effective at replacing regions of damaged DNA.

Nucleotide Excision-Repair

This mechanism is used to replace regions of damaged DNA up to 30 bases in length. Common examples of DNA damage include ultraviolet (UV) light, which induces the formation of cyclobutane pyrimidine-pyrimidine dimers, and smoking, which causes formation of benzo[a]pyrene-guanine adducts. Ionizing radiation, cancer chemotherapeutic agents, and a variety of chemicals found in the environment cause base modification, strand breaks, cross-linkage between bases on opposite strands or between DNA and protein, and numerous other defects. These are repaired by a process called nucleotide excision-repair (Figure 38–22). This complex process, which involves more gene products than the two other types of repair, essentially involves the hydrolysis of two phosphodiester bonds on the strand containing the defect. A special excision nuclease (exinuclease), consisting of at least three subunits in E $coli$ and 17 polypeptides in humans, accomplishes this task. After the strand is removed it is replaced, again by exact base pairing, through the action of yet another polymerase (δ/ϵ in humans), and the ends are joined to the existing strands by DNA ligase.

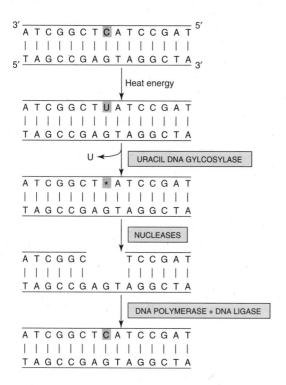

Figure 38–21. Base excision-repair of DNA. The enzyme uracil DNA glycosylase removes the uracil created by spontaneous deamination of cytosine in the DNA. An endonuclease cuts the backbone near the defect; then, after an endonuclease removes a few bases, the defect is filled in by the action of a repair polymerase and the strand is rejoined by a ligase. (Courtesy of B Alberts.)

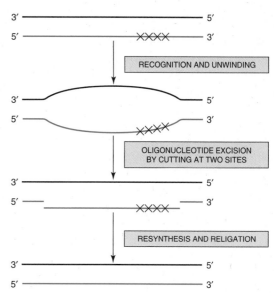

Figure 38–22. Nucleotide excision-repair. This mechanism is employed to correct larger defects in DNA and generally involves more proteins than either mismatch or base excision-repair. After defect recognition and unwinding of the DNA encompassing the defect, an excision nuclease (exinuclease) cuts the DNA above and below the defective region. This gap is then filled in by a special polymerase (δ/ε in humans) and religated.

Somewhat surprising is the recent observation that DNA repair proteins can serve other purposes. For example, some repair enzymes are also found as components of the large TFIIH complex that plays a central role in gene transcription (Chapter 39). Another component of TFIIH is involved in cell cycle regulation. Thus, three critical cellular processes may be linked through use of common proteins. There is also good evidence that some repair enzymes are involved in gene rearrangements that normally occur.

Xeroderma pigmentosum is an autosomal recessive genetic disease. (See also Case No. 1 in Chapter 65.) The clinical syndrome includes marked sensitivity to sunlight (ultraviolet) with subsequent formation of multiple skin cancers and premature death. The risk of developing skin cancer is increased 1000- to 2000-fold. The inherited defect seems to involve the repair of damaged DNA. Cells cultured from patients with xeroderma pigmentosum exhibit low activity for the nucleotide excision-repair process. Seven complementation groups have been identified using hybrid cell analyses, and six xeroderma pigmentosum genes have been identified.

In patients with **ataxia-telangiectasia,** an autosomal recessive disease in humans resulting in the development of cerebellar ataxia and lymphoreticular neoplasms, there appears to exist an increased sensitivity to damage by x-ray. Patients with **Fanconi's anemia,** an autosomal recessive anemia characterized also by an increased frequency of cancer and by chromosomal instability, probably have defective repair of cross-linking damage.

All three of these clinical syndromes are associated with increased frequency of cancer. It is likely that other human diseases resulting from disordered DNA repair capabilities will be found in the future.

SUMMARY

DNA in eukaryotic cells is associated with a variety of proteins, resulting in a structure called chromatin. The double-stranded DNA helix in the chromatin of each eukaryotic chromosome has a length that is thousands of times the diameter of the cell nucleus. Much of this DNA is associated with histone proteins to form a structure called the nucleosome. Nucleosomes serve to compact DNA. Other higher order structures, the 10-nm fibril and the 30-nm fiber, further compact the DNA. As much as 90% of

DNA may be transcriptionally inactive. This DNA is generally associated with nucleosomes and is not sensitive to digestion by nucleases. DNA in transcriptionally active regions is sensitive to nuclease attack, and some regions are exceptionally sensitive. These hypersensitive regions are often found to contain transcription control sites. Transcriptionally active DNA (the genes) is often clustered in regions of each chromosome. Within these regions, genes may be separated by inactive DNA in nucleosomal structures. A further subdivision occurs within genes. The transcription unit, that portion of a gene that is copied by RNA polymerase, consists of coding regions of DNA (exons) interrupted by intervening sequences of noncoding DNA (introns). During RNA processing, the introns are removed and the exons are ligated together to form the mature mRNA that appears in the cytoplasm.

Eukaryotic DNA in nondividing cells is located in chromosomes, which typically are associated as identical pairs. The DNA in each chromosome is exactly replicated according to the rules of base pairing during the S phase of the cell cycle. Each strand of the double helix is replicated simultaneously but by somewhat different mechanisms. A complex of proteins, including DNA polymerase, replicates the leading strand continuously in the 5' to 3' direction. The lagging strand is replicated discontinuously, in short pieces of 150–250 nucleotides, in the 3' to 5' direction. These are joined by DNA ligase. DNA replication occurs at several sites, called replication bubbles, in each chromosome. The entire process takes about 9 hours in a typical cell. A variety of mechanisms, employing different enzymes, repair damaged DNA, as after exposure to chemical mutagens or ultraviolet radiation.

REFERENCES

Coverly D, Laskey RA: Regulation of eukaryotic DNA replication. Annu Rev Biochem 1994;63:745.

DePamphilis ML: Origins of DNA replication in metazoan chromosomes. J Biol Chem 1993;268:1.

Gross DS, Garrard WT: Nuclease hypersensitive sites in chromatin. Annu Rev Biochem 1988;57:159.

Igo-Kemenes T, Horz W, Zachau HG: Chromatin. Annu Rev Biochem 1982;51:89.

Marians KJ: Prokaryotic DNA replication. Annu Rev Biochem 1992;61:673.

McGhee JD, Felsenfeld G: Nucleosome structure. Annu Rev Biochem 1980;49:1115.

Modrich P: Mismatch repair, genetic stability, and cancer. Science 1994;266:1959.

Sancar A: Mechanisms of DNA excision repair. Science 1994;266:1954.

Sherr CJ: Mammalian G1 cyclins and cell cycle progression. Proc Assoc Amer Phys 1995;107:181.

Stillman B: Initiation of chromosomal DNA replication in eukaryotes. J Biol Chem 1994;269:7047.

Wang TS: Eukaryotic DNA polymerases. Annu Rev Biochem 1991;60:513.

Weiner AM, Deininger PL, Efstratiadis A: Nonviral retrotransposons: Genes, pseudogenes and transposable elements generated by the reverse flow of genetic information. Annu Rev Biochem 1986;55:631.

RNA Synthesis, Processing, & Metabolism

39

Daryl K. Granner, MD

INTRODUCTION AND BIOMEDICAL IMPORTANCE

The synthesis of an RNA molecule from DNA is a very complex process involving one of the group of RNA polymerase enzymes and a number of associated proteins. The general steps required to synthesize the primary transcript are initiation, elongation, and termination. Most is known about initiation. A number of DNA regions (generally located upstream from the initiation site) and protein factors that bind to these sequences to regulate the initiation of transcription have been identified. This process is best understood in prokaryotes and viruses, but considerable progress has been made in deciphering mammalian cell transcription in recent years. Certain RNAs, mRNAs in particular, have very different life spans in a cell. It is important to understand the basic principles of RNA metabolism, for modulation of this process results in altered rates of protein synthesis and hence a variety of metabolic changes. This is how all organisms adapt to changes of environment. It is also how differentiated cell structures and functions are established and maintained in higher metazoans.

The RNA molecules synthesized in mammalian cells are often very different from those made in prokaryotic organisms, particularly the mRNA-encoding transcripts. Prokaryotic mRNA can be translated as it is being synthesized, whereas in mammalian cells most RNAs are made as precursor molecules that have to be processed into mature, active RNA. Erroneous processing and splicing of mRNA transcripts are a cause of disease, eg, certain types of thalassemia (Chapter 42).

RNA IS SYNTHESIZED FROM A DNA TEMPLATE BY AN RNA POLYMERASE

The process of synthesizing RNA from a DNA template has been characterized best in prokaryotes. Although in mammalian cells the regulation of RNA synthesis and the processing of the RNA transcripts are different from that in prokaryotes, the process of RNA synthesis per se is quite similar in these two classes of organisms. Therefore, the description of RNA synthesis in prokaryotes will be applicable to eukaryotes even though the enzymes involved and the regulatory signals are different.

The sequence of ribonucleotides in an RNA molecule is complementary to the sequence of deoxyribonucleotides in one strand of the double-stranded DNA molecule (Figure 37–8). The strand that is transcribed into an RNA molecule is referred to as the **template strand** of the DNA. The other DNA strand is frequently referred to as the **coding strand** of that gene. It is called this because, with the exception of T for U changes, it corresponds exactly to the sequence of the primary transcript, which encodes the protein product of the gene. In the case of a double-stranded DNA molecule containing many genes, the template strand for each gene will not necessarily be the same strand of the DNA double helix (Figure 39–1). Thus, a given strand of a double-stranded DNA molecule will serve as the template strand for some genes and the coding strand of other genes. Note that the nucleotide sequence of an RNA transcript will be the same (except for U replacing T) as that of the coding strand. The information in the template strand is read out in the 3′ to 5′ direction.

DNA-dependent RNA polymerase is the enzyme responsible for the polymerization of ribonucleotides into a sequence complementary to the template strand of the gene (see Figures 39–2 and 39–3). The enzyme attaches at a specific site, the promoter, on the template strand. This is followed by initiation of RNA synthesis at the starting point, and the process continues until a termination sequence is reached (Figure 39–3). A **transcription unit** is defined as that region of DNA that extends between the promoter and the terminator. The RNA product, which is synthesized in the 5′ to 3′ direction, is the **primary transcript.** In prokaryotes, this can represent the product of several contiguous genes; in mammalian cells, it usually represents the product of a single gene. The 5′ terminals of the primary RNA transcript and the mature cytoplasmic RNA are iden-

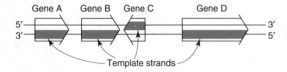

Figure 39–1. This figure illustrates that genes can be transcribed off both strands of DNA. The arrowheads indicate the direction of transcription (polarity). Note that the template strand is always read in the 3′ to 5′ direction. The opposite strand is called the coding strand because it is identical (except for T for U changes) to the mRNA transcript (the primary transcript in eukaryotic cells) that encodes the protein product of the gene.

tical. Thus, the **start point of transcription corresponds to the 5′ nucleotide of the mRNA.** This is designated position +1, as is the corresponding nucleotide in the DNA. The numbers increase as the sequence proceeds *downstream.* This convention makes it easy to locate particular regions, such as intron and exon boundaries. The nucleotide in the promoter adjacent to the transcription initiation site is designated −1, and these negative numbers increase as the sequence proceeds *upstream,* away from the initiation site. This provides a conventional way of defining the location of regulatory elements in the promoter.

The primary transcripts generated by RNA polymerase II are promptly capped by 7-methylguanosine triphosphate caps (Figure 37–10) that persist and eventually appear on the 5′ end of mature cytoplas-

mic mRNA. These caps are presumably necessary for the subsequent processing of the primary transcript to mRNA, for the translation of the mRNA, and for protection of the mRNA against 5′ → 3′exonucleolytic attack.

The DNA-dependent RNA polymerase (RNAP) of the bacterium *Escherichia coli* exists as a core molecule composed of four subunits; two of these are identical to each other (the α subunits), and two are similar in size to each other but not identical (the β subunit and β′ subunit) (Figure 39–2). RNAP also contains two zinc molecules. The core RNA polymerase utilizes a specific protein factor (the sigma [σ] factor) that helps the core enzyme attach more tightly to the specific deoxynucleotide sequence of the promoter region. Bacteria contain multiple σ factors, each of which acts as a regulatory protein that modifies the **promoter recognition specificity** of the RNA polymerase. The appearance of different σ factors can be correlated temporally with various programs of gene expression in prokaryotic systems, such as bacteriophage development, sporulation, and the response to heat shock.

RNA SYNTHESIS INVOLVES INITIATION, ELONGATION, AND TERMINATION

The process of RNA synthesis in bacteria, depicted in Figure 39–3, involves first the binding of the RNA holopolymerase molecule to the template at the promoter site. Binding is followed by a conformational change of the RNAP, and then the first nucleotide (almost always a purine) associates with the initiation site on the β subunit of the enzyme. In the presence of the four nucleotides, the RNAP moves to the second base in the template, a phosphodiester bond forms, and the nascent chain is now attached to the polymerization site on the β subunit of RNAP. (The analogy to the A and P sites on the ribosome should be noted; see Figures 40–7 and 40–8.)

Initiation of formation of the RNA molecule at its 5′ end then follows with the release of the σ factor, while the elongation of the RNA molecule from the 5′ to its 3′ end continues antiparallel to its template. The enzyme polymerizes the ribonucleotides in a specific sequence that is dictated by the template strand and interpreted by Watson-Crick base-pairing rules. Pyrophosphate is released in the polymerization reaction. In both prokaryotes and eukaryotes, a purine ribonucleotide is usually the first to be polymerized into the RNA molecule.

As the **elongation** complex containing the core RNA polymerase progresses along the DNA molecule, **DNA unwinding** must occur in order to provide access for the appropriate base pairing to the nucleotides of the coding strand. The extent of DNA unwinding is constant throughout transcription and

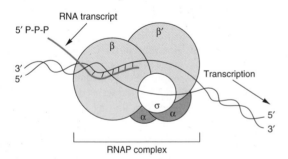

Figure 39–2. RNA polymerase (RNAP) catalyzes the polymerization of ribonucleotides into an RNA sequence that is complementary to the template strand of the gene. The RNA transcript has the same polarity (5′ to 3′) as the coding strand but contains U rather than T. *E coli* RNAP consists of a core complex of two α subunits and two β subunits (β and β′). The holoenzyme contains the σ subunit when near the transcription start site (within about 10 bp), but then transcription proceeds with just the core complex. The transcription "bubble" is a 17-bp area of melted DNA, and the entire complex covers 30–75 bp, depending on the conformation of RNAP.

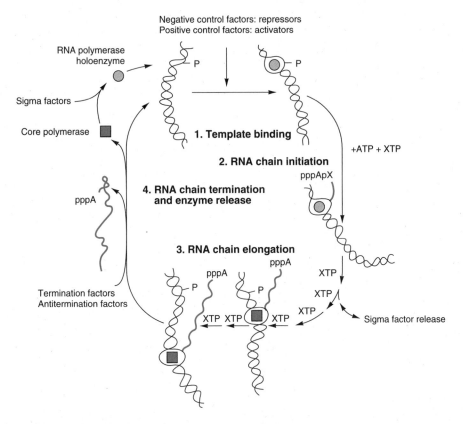

Figure 39–3. The transcription cycle in bacteria. Bacterial RNA transcription is described in four steps: **(1) Template binding:** RNA polymerase (RNAP) binds to DNA and locates a promoter. **(2) Chain initiation:** RNAP holoenzyme (core + sigma factors) catalyzes the coupling of the first base (usually ATP or GTP) to a second ribonucleoside triphosphate to form a dinucleotide. **(3) Chain elongation:** Successive residues are added to the 3′-OH terminus of the nascent RNA molecule; sigma factor dissociates from the holoenzyme after the chain length of about 10 is achieved. **(4) Chain termination and release:** The completed RNA chain and RNAP are released from the template. The RNAP holoenzyme re-forms, finds a promoter, and the cycle is repeated. (Slightly modified and reproduced, with permission, from Chamberlin M: Bacterial DNA-dependent RNA polymerases. Page 85 in: *The Enzymes,* vol 15. Academic Press, 1982.)

has been estimated to be about 17 base pairs per polymerase molecule. Thus, it appears that the size of the unwound DNA region is dictated by the polymerase and is independent of the DNA sequence in the complex. This suggests that RNA polymerase has associated with it an "unwindase" activity that opens the DNA helix. The fact that the DNA double helix must unwind and the strands part at least transiently for transcription implies some disruption of the nucleosome structure of eukaryotic cells.

Termination of the synthesis of the RNA molecule is signaled by a sequence in the template strand of the DNA molecule, a signal that is recognized by a termination protein, the rho (ρ) factor. After termination of synthesis of the RNA molecule, the core enzyme separates from the DNA template. With the assistance of another σ factor, the core enzyme then recognizes a promoter at which the synthesis of a new RNA molecule commences. More than one

RNA polymerase molecule may transcribe the same template strand of a gene simultaneously, but the process is phased and spaced in such a way that at any one moment each is transcribing a different portion of the DNA sequence. An electron micrograph of RNA synthesis is shown in Figure 39–4.

MAMMALIAN CELLS POSSESS SEVERAL DNA-DEPENDENT RNA POLYMERASES

The properties of mammalian polymerases are described in Table 39–1. Each of these DNA-dependent RNA polymerases seems to be responsible for the transcription of different sets of genes. The sizes of the RNA polymerases for the three major classes of eukaryotic RNA range from molecular weight 500,000 to 600,000. All these enzymes share the ba-

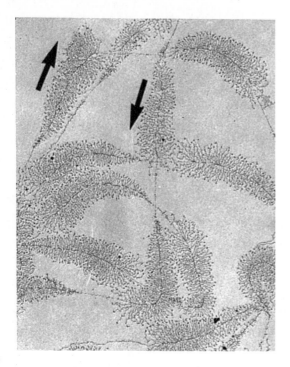

Figure 39–4. Electron photomicrograph of multiple copies of amphibian ribosomal RNA genes in the process of being transcribed. The magnification is about 6000×. Note that the length of the transcripts increases as the RNA polymerase molecules progress along the individual ribosomal RNA genes. Thus, the proximal end of the transcribed gene has short transcripts attached to it, while much longer transcripts are attached to the distal end of the gene. The arrows indicate the direction (5′ to 3′) of transcription. (Reproduced with permission, from Miller OL Jr, Beatty BR: Portrait of a gene. J Cell Physiol, 1969;74[Suppl 1]:225.)

sic subunit structural organization of bacterial RNA polymerase. They all have two large subunits and a number of smaller subunits. Recent DNA cloning and sequencing work indicates that eukaryotic RNA polymerases have extensive amino acid homologies with prokaryotic RNA polymerases. The functions of each of the subunits are not yet understood. Many could have regulatory functions, such as serving to

Table 39–1. Nomenclature of animal DNA-dependent RNA polymerases.

Class of Enzyme	Sensitivity to α-Amanitin	Major Products
I (A)	Insensitive	rRNA
II (B)	Sensitive to low concentrations (10^{-9} to 10^{-8} mol/L)	hnRNA (mRNA)
III (C)	Sensitive to high concentrations	tRNA and 5S RNA

assist the polymerase in the recognition of specific sequences like promoters and termination signals.

One peptide toxin from the mushroom *Amanita phalloides,* α-amanitin, is a specific inhibitor of the eukaryotic nucleoplasmic DNA-dependent RNA polymerase (RNA polymerase II) and as such has proved to be a powerful research tool (Table 39–1). α-Amanitin is thought to block the translocation of RNA polymerase II during transcription.

CERTAIN DNA SEQUENCES PROVIDE TRANSCRIPTION SIGNALS

The DNA sequence analysis of specific genes obtained by recombinant DNA technology has allowed the recognition of a number of sequences important in gene transcription. From the large number of bacterial genes studied it is possible to construct consensus models of transcription initiation and termination signals.

The question "How does RNAP find the correct site to initiate transcription?" is not trivial when the complexity of the genome is considered. E coli has 2×10^3 transcription initiation sites in 4×10^6 base pairs (bp) of DNA. The situation is even more complex in humans, where some 10^5 transcription initiation sites are scattered in 3×10^9 bp of DNA. RNAP can bind to many regions of DNA, but it scans the DNA sequence, at a rate of 10^3 bp/s, until it recognizes certain specific regions of DNA to which it binds with higher affinity. This region is called the **promoter,** and it is the association of RNAP with the promoter that ensures accurate initiation of transcription.

Bacterial promoters are approximately 40 nucleotides (40 bp or four turns of the DNA double helix) in length, a region sufficiently small to be covered by an E coli RNA holopolymerase molecule. In this consensus promoter region are two short, conserved sequence elements. Approximately 35 bp upstream of the transcription start site there is a consensus sequence of eight nucleotide pairs 5′-TGTTGACA-3′ (Figure 39–5). More proximal to the transcription start site, about 10 nucleotides upstream, is a 6-nucleotide-pair AT-rich sequence (5′-TATAAT-3′). The latter sequence has a low melting temperature because of its deficiency of GC nucleotide pairs. Thus, the **TATA box** is thought to ease the dissociation between the coding and noncoding strands so that RNA polymerase bound to the promoter region can have access to the nucleotide sequence of its immediately downstream coding strand. Other bacteria have different combinations (boxes), but all generally have two components to the promoter; these tend to be in the same position relative to the transcription start site, and in all cases the sequences between the boxes have no similarity.

Rho-dependent transcription **termination signals**

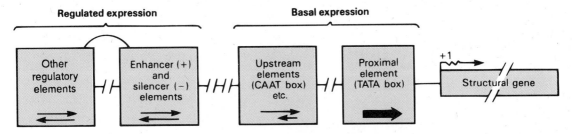

Figure 39–8. Schematic diagram showing the transcription control regions in a hypothetical class II (mRNA-producing) eukaryotic gene. Such a gene can be divided into its structural and regulatory regions, as defined by the transcription start site *(arrow)*. The structural gene contains the DNA sequence that is transcribed into mRNA, which is ultimately translated into protein. The regulatory region consists of two classes of elements. One class is responsible for ensuring basal expression. These elements generally have two components. The proximal component, generally the TATA box, directs RNA polymerase II to the correct site (fidelity). In TATA-less promoters, an initiator (Inr) element that spans the initiation site (+1) may direct the polymerase to this site. Another component, the upstream element, specifies the frequency of initiation. The best studied of these is the CAAT box, but several other elements (Sp1, NF1, AP1, etc) may be used in various genes. A second class is responsible for regulated expression. This class consists of elements that enhance or silence expression and of others that mediate the response to various signals, including hormones, heat shock, metals, and chemicals. Tissue-specific expression also involves specific sequences of this sort. It is possible that these two regulatory regions overlap in function (the connecting line). The orientation dependence of all the elements is indicated by the arrows within the boxes. For example, the proximal element must be in the 5′ to 3′ orientation. The upstream elements work best in the 5′ to 3′ orientation, but some of them can be reversed. The broken lines indicate that some elements are not fixed with respect to the transcription start site. Indeed, some elements responsible for regulated expression can be located either interspersed with the upstream elements, or they can be located downstream from the start site.

bacterial ρ factor are involved. However, it is known that the mRNA 3′ terminal is generated posttranscriptionally, and it appears to involve two steps. After RNA polymerase II has traversed the region of the transcription unit encoding the 3′ end of the transcript, an RNA endonuclease cleaves the primary transcript at a position about 15 bases 3′ to the consensus sequence **AAUAAA** that seems to serve in eukaryotic transcripts as a cleavage signal. Finally, this newly formed 3′ terminal is polyadenylated in the nucleoplasm, as described below.

THE EUKARYOTIC TRANSCRIPTION COMPLEX

A complex apparatus, consisting of as many as 50 unique proteins, provides accurate and regulatable transcription of eukaryotic genes. The RNA polymerase enzymes (pol I, pol II, and pol III for class I, II, and III genes, respectively) transcribe information contained in the template strand of DNA into RNA. These polymerases must recognize a specific site in the promoter in order to initiate transcription at the proper nucleotide. RNA polymerases, however, are not able to discriminate between promoter sequences and other regions of DNA; thus, other proteins facilitate promoter-specific binding of these enzymes. In bacteria this function is accomplished by a variety of so-called **sigma factors.** A sigma factor-polymerase complex selectively binds to DNA in the bacterial promoter.

Formation of the Basal Transcription Complex

The situation is more complex in eukaryotic genes. Class II genes—those transcribed by pol II to make mRNA—are described as an example. In class II genes, the function of sigma factors is assumed by a number of proteins. **Basal transcription requires, in addition to pol II, a number of factors called A, B, D, E, F, H and J,** some of which are composed of several different subunits. These factors are conventionally abbreviated as TFIIA, TFIIB, etc, for *tra*nscription *f*actor, class *II* gene, A. **TFIID, which binds to the TATA box, is the only one of these factors capable of binding to specific sequences of DNA.** As described below, TFIID consists of TATA binding protein (TBP) and several *T*BP associated *f*actors (TAFs). TBP binds to the TATA box in the minor groove of DNA (most transcription factors bind in the major groove). This event marks a specific promoter for transcription and, in sequence, TFIIA binds (upstream of TFIID), then TFIIB binds (downstream from TFIID), followed by the TFIIF pol II complex and then TFIIE (Figure 39–9). Each of these binding events extends the size of the complex so that finally about 75 bp (from −45 to +30 relative to +1, the nucleotide from which transcription commences) are covered. TFIIH and TFIIJ are added to the surface of the complex, which is now complete and capable of basal transcription initiated from the correct nucleotide. This collection of assembled components is termed the **preinitiation complex.** In genes that lack a TATA box, the same factors, in-

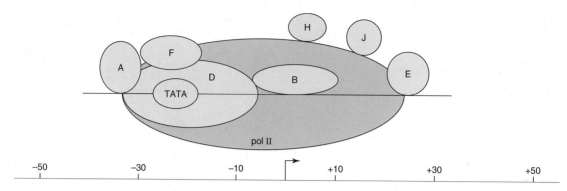

Figure 39–9. The eukaryotic basal transcription complex. Formation of the basal transcription complex begins when TFIID binds to the TATA box. It directs the assembly of several other components by protein-DNA and protein-protein interactions. The entire complex spans DNA from position –45 to +30 relative to the initiation site (+1, marked by bent arrow).

cluding TBP, are required. In such cases an **initiator (inr) sequence** positions the complex for accurate initiation of transcription.

The Role of Transcription Activators and Coactivators

TFIID was originally considered to be a single protein. However, several pieces of evidence led to the important discovery that TFIID is actually a complex consisting of TBP and several TAFs. The first evidence was that TBP binds to a 10 bp segment of DNA, immediately over the TATA box, whereas TFIID covers a 35 bp region (Figure 39–9). Second, TBP has a molecular mass of about 30 kDa, whereas the TFIID complex has a mass of about 700 kDa. Finally, and perhaps most importantly, TBP supports basal transcription but not the augmented transcription provided by certain activators, eg, Sp1 bound to the GC box. TFIID, on the other hand, supports both basal and enhanced transcription by Sp1, Oct1, AP1, CTF, ATF, etc (Table 39–2). The TAFs are essential for this activator-enhanced transcription. At present it is unclear whether there are one or several forms of TFIID that might differ in their complement of TAFs associated with TBP. It is conceivable that different combinations of TAFs with TBP may bind to different promoters, and this may account for selective activation noted in various promoters and for the different strength of certain promoters. **TAFs, since they are required for the action of activators, are often called coactivators.** There are thus three classes of transcription factors involved in the regulation of class II genes: basal factors, coactivators, and activators-repressors (Table 39–3).

A representation of how various activators might enhance gene transcription is shown in Figure 39–10. Hormones, by promoting the interaction of receptors and other factors with specific DNA elements, are also thought to influence transcription through the coactivator-activator mechanism. Repression of transcription, not as completely studied, could be accomplished through interference of the assembly of TAFs with TBP or through interruption of an activator TAF complex.

Table 39–2. Some of the transcription control elements, their consensus sequences, and the factors that bind to them which are found in mammalian genes transcribed by RNA polymerase II. The asterisks mean that there are several members of this family.

Element	Consensus Sequence	Factor
TATA box	TATAAA	TBP
CAAT box	CCAATC	C/EBP*, NF-Y*
GC box	GGGCGG	Sp 1
	CAACTGAC T/CGGA/CN$_5$GCCAA	Myo D NF1*
Ig octamer	ATGCAAAT	Oct-1,2,4,6*
AP 1	TGAG/CTC/AA	Jun, Fos, ATF*
Serum response	GATGCCCATA	SRF
Heat shock	(NGAAN)$_3$	HSF

THE LOCATION OF *CIS*-ACTING DNA ELEMENTS IS INTRAGENIC IN CLASS III GENES

DNA-dependent RNA polymerase III, which transcribes tRNA genes and low-molecular-weight

Table 39–3. Three classes of transcription factors in class II genes.

General Mechanisms	Specific Components
Basal components	TBP, TFIIA, B, E, G, F, H, and J
Coactivators	TAFs
Activators	SP1, ATF, CTF, AP1, etc.

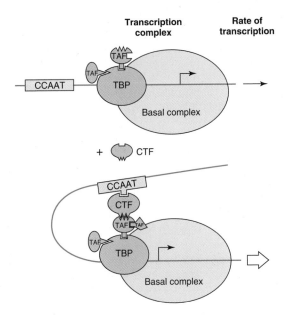

Transcription complex   **Rate of transcription**

Figure 39–10. A hypothetical view of how transcription activators and coactivators might enhance transcription. Shown here as a large oval are all the components of the basal transcription complex illustrated in Figure 39–9. The TBP component of TFIID is shown bound to the TATA box. A few TAFs are associated with TBP. A transcription activator, CTF bound to the CAAT box in this case, forms a loop complex by interacting with a TAF.

RNA genes (see Chapter 37), recognizes a promoter that is internal to the gene to be expressed rather than upstream of the transcription starting point. In the case of eukaryotic tRNA genes, two internal, separated blocks (A and B) of sequences exist that act as an intragenic promoter. The sequences within the A and B blocks exist in the mature tRNA molecule in regions that are highly conserved and participate in the formation of the D arm and the TΨC arm, respectively (Figure 37–11). By manipulating the structure of tRNA genes, it has been shown that for promoter function the optimal distance between the A and B blocks is 30–40 base pairs and that the transcription start point occurs between 10 and 16 base pairs upstream from the A block. For the 5S RNA gene, which is also transcribed by RNA polymerase III, there appears to be a specific transcription protein factor which, once bound to the intragenic promoter for that gene, probably interacts with an RNA polymerase III molecule to position its catalytic sites on the transcription start point of the DNA. A similar mechanism, employing a specific *trans*-acting transcription factor (or factors) may be involved in tRNA gene transcription.

RNA MOLECULES ARE OFTEN PROCESSED BEFORE THEY BECOME FUNCTIONAL

In prokaryotic organisms, the primary transcripts of mRNA-encoding genes begin to serve as translation templates even before their transcription has been completed. This is presumably because the site of transcription is not compartmentalized into a nucleus as it is in eukaryotic organisms. Thus, transcription and translation are coupled in prokaryotic cells. Consequently, prokaryotic mRNAs are subjected to little modification and processing prior to carrying out their intended function in protein synthesis. Prokaryotic rRNA and tRNA molecules are transcribed in units considerably longer than the ultimate molecule. In fact, many of the tRNA transcription units contain more than one molecule. Thus, in prokaryotes the processing of these rRNA and tRNA precursor molecules is required for the generation of the mature functional molecules.

Nearly all eukaryotic RNA primary transcripts undergo extensive processing between the time they are synthesized and the time at which they serve their ultimate function, whether it be as mRNA or as a structural molecule such as rRNA, 5S RNA, or tRNA. The processing occurs primarily within the nucleus. The processing includes **capping, nucleolytic and ligation reactions, terminal additions of nucleotides, and nucleoside modifications.** However, it is clear that, for mammalian cells, 50–75% of the nuclear RNA, including those RNAs with capped 5′ terminals, do not contribute to the cytoplasmic mRNA. This nuclear RNA loss is significantly greater than can be reasonably accounted for by the loss of intervening sequences alone (see below). Thus, the exact function of the seemingly excessive transcripts in the nucleus of a mammalian cell is not known.

THE CODING PORTIONS (EXONS) OF MOST EUKARYOTIC GENES ARE INTERRUPTED BY INTRONS

As a result of advances in techniques for DNA cloning and DNA sequencing, it is now apparent that interspersed within the amino acid–coding portions **(exons)** of many genes are long sequences of DNA that do not contribute to the genetic information ultimately translated into the amino acid sequence of a protein molecule (see Chapter 38). In fact these sequences actually interrupt the coding region of structural genes. These **intervening sequences (introns)** exist within most but not all genes of higher eukaryotes. The primary transcripts of the structural genes contain RNA complementary to the interspersed sequences. However, the intron RNA sequences are cleaved out of the transcript, and the exons of the transcript are appropriately spliced together in the nucleus before the resulting mRNA molecule appears in the cytoplasm for translation (Figures 39–11 and 39–12).

The mechanisms whereby the introns are removed from the primary transcript in the nucleus, the exons are ligated to form the mRNA molecule, and the mRNA molecule is transported to the cytoplasm are being elucidated. Although the sequences of nucleotides in the introns of the various eukaryotic transcripts, and even those within a single transcript, are very heterogeneous, there are reasonably conserved sequences at each of the two exon-intron (splice) junctions and at the branch site, which is located 20–40 nucleotides upstream from the 3' splice site (see consensus sequences in Figure 39–12). A special structure, the **spliceosome,** is involved in converting the primary transcript into mRNA. Spliceosomes consist of the primary transcript, five small nuclear RNAs (U1, U2, U4, U5, and U6), and an undetermined number of proteins. The splicing reaction starts with a cut at the junction of the 5' exon (donor or left) and intron (Figure 39–11). The free 5' terminal then forms a loop or lariat structure that is linked by a 5'–2' bond to the A in the PyNPyPyPuAPy branch site sequence (Figure 39–12). The branch site identifies the 3' splice site. A second cut is made at the junction of the intron with the 3' exon (donor or right), and the lariat structure containing the intron is released and hydrolyzed. The 5' and 3' exons are ligated to form a continuous sequence.

The snRNAs and associated proteins are required for formation of the various structures and intermediates. U1 binds first, by base pairing, to the 5' end of the intron. U2 then binds to the branch site and appears to direct U1 binding to this site also, perhaps through U1-associated proteins. The U2/U1 interaction at the branch site forms the loop and approximates the two splicing sites. A trimeric complex of U5 and the U4/U6 combination then binds to the complex. U4 is released and, after a series of movements and rearrangements, the two ends are cleaved, probably by the U2/U6 complex. It is important to note that RNA serves as the catalytic agent. This sequence is then repeated in genes containing multiple introns. In such cases, a definite pattern is followed for each gene, and the introns are not necessarily removed in sequence—1 then 2 then 3, etc.

Recently, it has been discovered that during the process of removing the intron sequence from pre-mRNA, there is formed an unusual RNA molecule resembling a lariat. It appears that the 5' end of the intervening sequence is joined via a 2'–5' phosphodiester linkage to an adenylyl residue 28–37 nucleotides upstream from the 3' end of the intervening sequence. This process and structure are diagrammed in Figure 39–11.

It seems that the mystery of the relationship be-

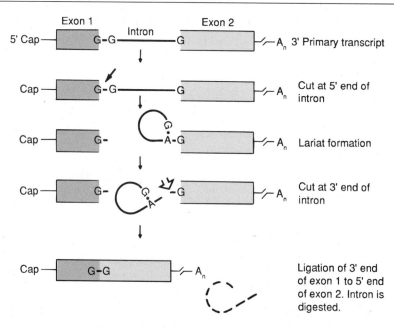

Figure 39–11. The processing of the primary transcript (hnRNA) to mRNA. In this hypothetical transcript, the 5' (left) end of the intron is cut (↓) and a lariat forms between the G at the 5' end of the intron and an A near the 3' end, in the consensus sequence UACUAAC. This sequence is called the branch site, and it is the 3' most A that forms the 5'–2' bond with the G. The 3' (right) end of the intron is then cut (⇓). This releases the lariat, which is digested, and exon 1 is joined to exon 2 at G residues.

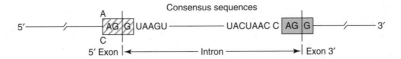

Figure 39–12. Consensus sequences at splice junctions. The 5′ (donor or left) and 3′ (acceptor or right) sequences are shown. Also shown is the yeast consensus sequences (UACUAAC) for the branch site. In mammalian cells, this consensus sequence is PyNPyPyPuAPy, where Py is a pyridine, Pu is a purine, and N is any nucleotide. The branch site is located 20–40 nucleotides upstream from the 3′ site.

tween hnRNA and the corresponding mature mRNA in eukaryotic cells is solved. The hnRNA molecules are the primary transcripts plus their early processed products, which, after the addition of caps and poly(A) tails and removal of the portion corresponding to the introns, are transported to the cytoplasm as mature mRNA molecules.

Alternative Splicing Provides a Form of Regulation

The processing of hnRNA molecules is a potential site for regulation of gene expression. In fact, it has been demonstrated that alternative patterns of RNA splicing are subject to developmental control. Three examples make the point. For example, the cytoplasmic mRNAs for α-amylase in rat salivary gland and in rat liver differ in their 5′ nucleotide sequence, while the remainder of the mRNA genes containing the coding region and poly(A) addition sites are identical. Further analysis reveals that although the primary transcripts are huge and overlapping, different splice sites are used to join two different cap and leader sequences to the same mRNA "body." In addition, alternative patterns of RNA splicing are used to generate two different immunoglobulin heavy chain mRNAs—one that codes for a membrane-bound heavy chain protein and another that codes for a secreted heavy chain protein (Chapter 41). RNA splicing is usually necessary for the generation of messenger RNA molecules, and this splicing requirement provides an additional mechanism of differential regulation of gene expression.

At least one form of β-thalassemia, a disease in which the β-globin gene of hemoglobin is severely underexpressed, appears to result from a nucleotide change at an exon-intron junction, precluding removal of the intron and therefore leading to diminished or absent synthesis of the β chain. This is a consequence of the fact that the normal translation reading frame of the mRNA is disrupted. Clearly a defect in this fundamental process (splicing) underscores the accuracy which the process of RNA-RNA splicing must attain.

RNA Editing Changes mRNA After Transcription

The central dogma states that, for a given gene and gene product, there is a linear relationship between the coding sequence in DNA, the mRNA sequence, and the protein sequence (Figure 38–7). Changes in the DNA sequence should be reflected in a change in the mRNA sequence, and, depending on codon usage, in protein sequence. A few exceptions to this dogma have been recently documented. Coding information can be changed at the mRNA level by **RNA editing.** In such cases, the coding sequence of the mRNA differs from that in the cognate DNA. An example is the apolipoprotein B (apo B) gene and mRNA. The single apo B gene encodes an mRNA of 4563 codons which is translated into a protein of 512 kDa in liver. The intestinal form of apo B is approximately 250 kDa. It is identical to the liver protein up to codon 2153, where it stops. This truncation arises from a C to T substitution in codon 2153, which converts the CAA = glutamine codon into a UAA stop codon. This C to U change occurs posttranscriptionally by mechanisms not yet defined. Other examples include a glutamine to arginine change in the glutamate receptor and several changes in trypanosome mitochondrial mRNAs, generally involving the addition or deletion of uridine.

RNA CAN ACT AS A CATALYST

In addition to the catalytic action served by the snRNAs in the formation of mRNA, several other enzymatic functions have been attributed to RNA. **Ribozymes** are RNA molecules with catalytic activity. These generally involve transesterification reactions, and most are concerned with RNA metabolism (splicing and endoribonuclease). Recently, a ribosomal RNA component was noted to hydrolyze an aminoacyl ester and thus to play a central role in peptide bond function. These observations, made in organelles from plants, yeast, viruses, and higher eukaryotic cells, show that **RNA can act as an enzyme.** This has revolutionized thinking about enzyme action and the origin of life itself.

MESSENGER RNA (mRNA) IS MODIFIED AT THE 5′ AND 3′ ENDS

As mentioned above, mammalian mRNA molecules contain a capped structure at their 5′ terminal, and most have a poly(A) tail at the 3' terminal. The cap structure is added to the 5' end of the newly transcribed mRNA precursor in the nucleus prior to transport of the mRNA molecule to the cytoplasm. Poly(A) tails (when present) appear to be added either in the nucleus or in the cytoplasm. The secondary methylations of mRNA molecules, those on the 2'-hydroxy groups and the N^6 of adenylyl residues, occur after the mRNA molecule has appeared in the cytoplasm. The **5' cap** of the RNA transcript appears to be required for the formation of the ribonucleoprotein complex necessary for the splicing reactions, may be involved in mRNA transport and translation initiation, and protects the 5' end of mRNA from attack by $5' \rightarrow 3'$ exonucleases.

The function of the **poly(A) tail** is unknown, but it appears to protect the 3' end of mRNA from $3' \rightarrow 5'$ exonuclease attack. In any event, the presence or absence of the poly(A) tail does not determine whether a precursor molecule in the nucleus appears in the cytoplasm, because all poly(A)-tailed hnRNA molecules do not contribute to cytoplasmic mRNA, nor do all cytoplasmic mRNA molecules contain poly(A) tails (the histones are most notable in this regard). Cytoplasmic processes in mammalian cells can both add and remove adenylate residues from the poly(A) tails; this has been associated with an alteration of mRNA stability.

The size of the cytoplasmic mRNA molecules even after the poly(A) tail is removed is still considerably greater than the size required to code for the specific protein for which it is a template, often by a factor of 2 or 3. **The extra nucleotides occur in untranslated (noncoding) regions** both 5' and 3' to the coding region; the longest untranslated sequences are usually at the 3' end. The exact function of these sequences is unknown, but they have been implicated in RNA processing, transport, degradation, and translation.

Transfer RNA (tRNA) Is Extensively Modified

The tRNA molecules, as described in Chapters 37 and 40, serve as adapter molecules for the translation of mRNA into protein sequences. The tRNAs contain many modifications of the standard bases A, U, G, and C. Some are simply methylated derivatives, and some possess rearranged glycosidic bonds. The tRNA molecules are transcribed in both prokaryotes and eukaryotes as large precursors, frequently containing the sequence for more than one tRNA, which are then subjected to **nucleolytic processing** and reduced in size by a specific class of ribonucleases. In addition, the genes of some tRNA molecules contain—very near the portion corresponding to the anticodon loop—a single intron 10–40 nucleotides long. These introns in the tRNA genes are transcribed; therefore, the processing of the precursor transcripts

of many tRNA molecules must include removal of the introns and proper splicing of the anticodon region to generate an active adapter molecule for protein synthesis. The nucleolytic processing enzymes of tRNA precursors apparently recognize three-dimensional structure and not just linear RNA sequences. This enzyme system thereby processes only molecules capable of folding into functionally competent products.

The further modification of the tRNA molecules includes nucleotide **alkylations and the attachment of the characteristic C • C • A terminal** at the 3' end of the molecule. The 3' OH of the A ribose is the point of attachment for the specific amino acid that is to enter into the polymerization reaction of protein synthesis. The methylation of mammalian tRNA precursors probably occurs in the nucleus, whereas the cleavage and attachment of **C • C • A** are cytoplasmic functions, since the terminals turn over more rapidly than do the tRNA molecules themselves. Enzymes within the cytoplasm of mammalian cells are required for the attachment of amino acids to the C • C • A residues.

Ribosomal RNA (rRNA) Is Synthesized as a Large Precursor

In mammalian cells, the three rRNA molecules are transcribed as part of a single large precursor molecule (Figure 39–13). The precursor is subsequently processed in the **nucleolus** to provide the RNA component for the ribosome subunits found in the cytoplasm. The rRNA genes are located in the nucleoli of mammalian cells. Hundreds of copies of these genes are present in every cell. This large number of genes is required to synthesize sufficient copies of each type of rRNA to form the 10^7 ribosomes required for each cell replication. Whereas a single mRNA molecule may be copied into 10^5 protein molecules, providing a large amplification, the rRNAs are end products. This lack of amplification requires a large number of genes.

The rRNA genes are transcribed as units, each of which encodes (5' to 3') an 18S, a 5.8S, and a 28S ribosomal RNA. The primary transcript is a 45S molecule that is highly methylated in the nucleolus. In the **45S precursor,** the eventual 28S segment contains 65 ribose-methyl groups and five base-methyl groups. Only those portions of the precursor that eventually become stable rRNA molecules are methylated. The 45S precursor is nucleolytically processed, but the processing signals are clearly distinct from those in hnRNA. Therefore, the processing of rRNA is mediated by a series of endonucleolytic and exonucleolytic reactions (see below) rather than by the RNA-catalyzed reaction involved in the formation of mRNA (Figure 39–11).

Nearly half the original primary transcript is degraded by these nucleolytic reactions, as shown in Figure 39–13. During the processing of rRNA, fur-

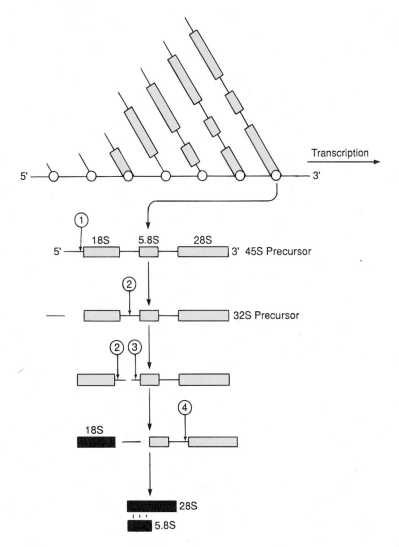

Figure 39–13. Diagrammatic representation of the processing of ribosomal RNA from large RNA precursor molecules. The primary transcript has a size of 45S and is about 13,000 nucleotides in length. It is synthesized from 5′ to 3′ by RNA polymerase I (open circles). The densely clustered genes are transcribed at a rapid rate, so the elongating RNA molecules, easily seen in chromatin preparations, resemble the "trees" in the electron micrograph in Figure 39–4. The final products are formed through a series of endo- and exonuclease reactions involving rRNA-specific ribonucleases. A possible sequence of these reactions is shown by the numbered circles and arrows. The sizes and nucleotide lengths of the 18S, 5.8S, and 28S products are shown. The 18S rRNA is incorporated into the small ribosomal subunit. The 28S and 5.8S rRNAs are incorporated into the large subunit.

ther methylation occurs, and eventually, in the nucleoli, the 28S chains self-assemble with ribosomal proteins to form the larger 60S subunit. The 5.8S rRNA molecule also formed from the 45S precursor RNA in the nucleolus becomes an integral part of the larger ribosomal subunit. The smaller ribosomal subunits (40S) are formed by the association of appropriate ribosome proteins with the 18S rRNA molecule.

SPECIFIC NUCLEASES DIGEST NUCLEIC ACIDS

Enzymes capable of degrading nucleic acids have been recognized for many years. These can be classified in several ways. Those which exhibit specificity for deoxyribonucleic acid are referred to as **deoxyribonucleases.** Those which specifically hydrolyze ribonucleic acids are **ribonucleases.** Within both of these classes are enzymes capable of cleaving internal phosphodiester bonds to produce either 3'-hydroxyl and 5'-phosphoryl terminals or 5'-hydroxyl and 3'-phosphoryl terminals. These are referred to as **endonucleases.** Some are capable of hydrolyzing both strands of a **double-stranded** molecule, whereas others can only cleave **single strands** of nucleic acids. Some nucleases can hydrolyze only unpaired single strands, while others are capable of hydrolyzing single strands participating in the formation of a double-stranded molecule. There exist classes of endonucleases that recognize specific sequences in DNA; the majority of these are the **restriction endonucleases,** which have in recent years become important tools in molecular genetics and medical sciences. A list of some currently recognized restriction endonucleases is presented in Table 42–2.

Some nucleases are capable of hydrolyzing a nucleotide only when it is present at a terminal of a molecule; these are referred to as **exonucleases.** Exonucleases act in one direction ($3' \rightarrow 5'$ or $5' \rightarrow 3'$) only. In bacteria, a $3' \rightarrow 5'$ exonuclease is an integral part of the DNA replication machinery and there serves to edit the most recently added deoxynucleotide for base-pairing errors.

REGULATION OF DEGRADATION PROVIDES ANOTHER MECHANISM FOR REGULATING THE AMOUNT OF MESSENGER RNA

Although most mRNAs in mammalian cells are very stable (half-lives measured in hours), some turn over very rapidly (half-lives of 10–30 minutes). In certain instances, mRNA stability is subject to regulation. This has important implications, since there is usually a direct relationship between mRNA amount and the translation of that mRNA into its cognate protein. Changes in the stability of a specific mRNA can therefore have major effects on biologic processes.

Messenger RNAs exist in the cytoplasm as ribonucleoprotein particles (RNPs). Some of these proteins protect the mRNA from digestion by nucleases, while others may, under certain conditions, promote nuclease attack. It is thought that mRNAs are stabilized, or destabilized, by the interaction of proteins with these various structures or sequences. Certain effectors, such as hormones, may regulate mRNA stability by increasing or decreasing the amount of these proteins.

It appears that the ends of mRNA molecules are involved in mRNA stability (Figure 39–14). The 5' cap structure in eukaryotic mRNA prevents attack by 5' exonucleases, and the poly(A) tail prohibits the action of 3' exonucleases. In mRNA molecules with those structures, it is presumed that a single endonucleolytic cut allows exonucleases to attack and digest the entire molecule. Other structures (sequences) in the 5' noncoding sequence (5' NCS), the coding region, and the 3' NCS are thought to promote or prevent this initial endonucleolytic action (Figure 39–14). A few illustrative examples will be cited.

Deletion of the 5' NCS results in a three- to fivefold prolongation of the half-life of c-*myc* mRNA. Shortening the coding region of histone mRNA results in a prolonged half-life. A form of autoregulation of mRNA stability indirectly involves the coding region. Free tubulin binds to the first four amino acids of a nascent chain of tubulin as it emerges from the ribosome. This appears to activate an RNase associated with the ribosome (RNP) which then digests the tubulin mRNA.

Structures at the 3' end, including the poly(A) tail, enhance or diminish the stability of specific mRNAs. The absence of a poly(A) tail is associated with rapid degradation of mRNA, and the removal of poly(A) from some RNAs results in their destabilization. Histone mRNAs lack a poly(A) tail but have a sequence near the 3' terminal that can form a stem-loop structure, and this appears to provide resistance to exonucleolytic attack. Histone H4 mRNA, for example, is degraded in the 3' to 5' direction, but only after a

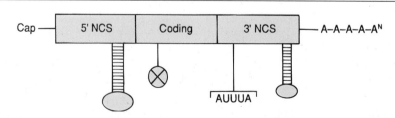

Figure 39–14. Structure of a typical eukaryotic mRNA showing elements that are involved in regulating mRNA stability. The typical eukaryotic mRNA has a 5' noncoding sequence (5' NCS), a coding region, and a 3' NCS. All are capped at the 5' end, and most have a polyadenylate sequence at the 3' end. The 5' cap and 3' poly(A) tail protect the mRNA against exonuclease attack. Stem-loop structures in the 5' and 3' NCS, features in the coding sequence, and the AU-rich region in the 3' NCS are thought to play roles in mRNA stability.

single endonucleolytic cut occurs about nine nucleotides from the 3′ end, in the region of the putative stem-loop structure. Stem-loop structures in the 3′ noncoding sequence are also critical for the regulation, by iron, of the mRNA encoding the transferrin receptor. Stem-loop structures are also associated with mRNA stability in bacteria, suggesting that this mechanism may be commonly employed.

Other sequences in the 3′ end of certain eukaryotic mRNAs appear to be involved in the destabilization of these molecules. Of particular interest are AU-rich regions, many of which contain the sequence AUUUA. This sequence appears in mRNAs that have a very short half-life, including some encoding oncogenes and cytokines. The importance of this region is underscored by an experiment in which a sequence corresponding to the 3′ noncoding region of the short half-life colony-stimulating factor (CSF) mRNA, which contains the AUUUA motif, was added to the 3′ end of the β-globin mRNA. Instead of becoming very stable, this hybrid β-globin mRNA now had the short half-life characteristic of CSF mRNA.

From the few examples cited, it is clear that a number of mechanisms are used to regulate mRNA stability, just as several mechanisms are used to regulate the synthesis of mRNA. Coordinate regulation of these two processes affords the cell remarkable adaptability.

SUMMARY

RNA is synthesized from a DNA template by the enzyme RNA polymerase. There are three RNA polymerases: type I transcribes ribosomal RNA; type II, messenger RNA; and type III, transfer RNA. The multisubunit polymerase enzymes attach to the promoter region of a gene and transcribe the sequence of the template strand of DNA (the transcription unit) into a complementary RNA strand (the primary transcript). The promoters of prokaryotic genes are simple. They generally consist of about 40 bp of DNA that contain two elements: a TATA box, located 10 bp upstream from the transcription initiation site, and a second element, located 35 bp in the 5′ direction upstream. Eukaryotic promoters are more complex and usually consist of several elements. One set of elements ensures that transcription initiates at the correct site. The elements in this set, which often include a TATA box, also provide for a basal rate of initiation and are usually found within 100–200 bp of the transcription start site. Another set of elements, usually located farther 5′ from the transcription initiation site, can enhance or silence transcription. These work with, or serve as, tissue-specific or signal-response elements.

The transcription process can be subdivided into initiation, elongation, and termination. Each step requires several enzymes or factors in addition to the RNA polymerase. In eukaryotic cells, the primary transcript usually consists of amino acid coding portions (exons) interrupted by intervening sequences (introns) that are removed in the processing of the primary transcript into the mature mRNA. Ribosomal RNA is also synthesized as a large 45S precursor molecule that is processed into the mature 28S, 18S, and 5.8S forms.

mRNA molecules turn over with half-lives ranging from a few minutes to days. Regulation of mRNA stability is therefore an important control step. RNA sequences in the 5′ and 3′ noncoding regions of the mRNA, and in the coding region, have been implicated in mRNA stability. These sequences, in association with binding proteins, may target nucleases to the mRNA or protect the mRNA against attack by nucleases.

REFERENCES

Faisst S, Meyer S: Compilation of vertebrate-encoded transcription factors. Nucleic Acids Res 1992;20:3.

Guthrie C: Messenger RNA splicing in yeast. Science 1991;253:157.

Javahery R et al: DNA sequence requirements for transcriptional initiator activity in mammalian cells. Mol Cell Biol 1994; 14:1116.

Pace NR: New horizons for RNA catalysis. Science 1992;256:1402.

Patrusky B: The intron story. Mosaic 1992;23:22.

Ross J: The turnover of messenger RNA. Sci Am 1989;260:48.

Shapiro DJ et al: Regulation of mRNA stability in eukaryotic cells. Bioessays 1987;6:221.

Tjian R: Molecular machines that control genes. Sci Am 1995;272:54.

Wobbe CR, Struhl K: Yeast and human TATA-binding proteins have nearly identical DNA sequence requirements for transcription in vitro. Mol Cell Biol 1990; 10:3859.

Young RA: RNA polymerase II. Annu Rev Biochem 1991;60:689.

40

Protein Synthesis & the Genetic Code

Daryl K. Granner, MD

INTRODUCTION

The letters A, G, T, and C correspond to the nucleotides found in DNA. They are organized into three-letter code words called **codons,** and the collection of these codons makes up the **genetic code.** A linear array of codons (a **gene**) specifies the synthesis of various RNA molecules, most of which are involved in some aspect of protein synthesis. Protein synthesis occurs in three major steps: initiation, elongation, and termination. This process resembles DNA replication and transcription in its general features and in the fact that it, too, follows a 5′ to 3′ polarity.

BIOMEDICAL IMPORTANCE

It was impossible to understand protein synthesis, or to explain mutations, before the genetic code was elucidated. The genetic code provides a foundation for explaining the way in which protein defects may cause genetic disease and for the diagnosis and perhaps treatment of these disorders. In addition, the pathophysiology of many viral infections is related to the ability of these agents to disrupt host cell protein synthesis. Many antibacterial agents are effective because they disrupt protein synthesis in the invading cell.

GENETIC INFORMATION FLOWS FROM DNA TO RNA TO PROTEIN

The genetic information within the nucleotide sequence of DNA is transcribed in the nucleus into the specific nucleotide sequence of an RNA molecule. The sequence of nucleotides in the RNA transcript is complementary to the nucleotide sequence of the template strand of its gene in accordance with the base-pairing rules. Several different classes of RNA combine to direct the synthesis of proteins.

In prokaryotes there is a linear correspondence between the gene, the **messenger RNA (mRNA)** tran-

scribed from the gene, and the polypeptide product. The situation is more complicated in higher eukaryotic cells, in which the primary transcript, **heterogeneous nuclear RNA (hnRNA),** is much larger than the mature mRNA. The large hnRNA contains coding regions **(exons)** that will form the mature mRNA and long intervening sequences **(introns)** that separate the exons. The hnRNA is processed within the nucleus, and the introns, which often make up much more of the hnRNA than the exons, are removed. Exons are spliced to form mature mRNA, which is transported to the cytoplasm, where it is translated into protein.

The cell must possess the machinery necessary to translate information accurately and efficiently from the nucleotide sequence of an mRNA into the sequence of amino acids of the corresponding specific protein. Clarification of our understanding of this process, which is termed **translation,** awaited deciphering of the genetic code. It was realized early that mRNA molecules in themselves have no affinity for amino acids and, therefore, that the translation of the information in the mRNA nucleotide sequence into the amino acid sequence of a protein requires an intermediate adapter molecule. This adapter molecule must recognize a specific nucleotide sequence on the one hand as well as a specific amino acid on the other. With such an adapter molecule, the cell can direct a specific amino acid into the proper sequential position of a protein as dictated by the nucleotide sequence of the specific mRNA. In fact, the functional groups of the amino acids do not themselves actually come into contact with the mRNA template.

THE NUCLEOTIDE SEQUENCE OF AN mRNA MOLECULE CONSISTS OF A SERIES OF CODONS THAT SPECIFY THE AMINO ACID SEQUENCE OF THE ENCODED PROTEIN

The adapter molecules that translate the codons into the amino acid sequence of a protein are the **transfer RNA (tRNA)** molecules. The **ribosome** is the cellular component on which these various functional entities interact to assemble the protein molecule. Many of these subcellular units (ribosomes) can assemble to translate simultaneously a single mRNA molecule and, in so doing, form a **polyribosome.** The **rough endoplasmic reticulum** is a compartment of membrane-attached polyribosomes that provides for the synthesis of integral membrane proteins and proteins to be exported. Polyribosomal structures also exist free in the cytoplasm, where they synthesize proteins that remain within the cell.

Twenty different amino acids are required for the synthesis of the cellular complement of proteins; thus, there must be at least 20 distinct codons that make up the genetic code. Since there are only four different nucleotides in mRNA, each codon must consist of more than a single purine or pyrimidine nucleotide. Codons consisting of two nucleotides each could provide for only 16 (4^2) specific codons, whereas codons of three nucleotides could provide 64 (4^3) specific codons.

It is now known that each codon consists of a sequence of three nucleotides; ie, **it is a triplet code.** The deciphering of the genetic code depended heavily on the chemical synthesis of nucleotide polymers, particularly triplets.

Table 40–1. The genetic code (codon assignments in messenger RNA).[1]

First Nucleotide	Second Nucleotide				Third Nucleotide
	U	C	A	G	
U	Phe	Ser	Tyr	Cys	U
	Phe	Ser	Tyr	Cys	C
	Leu	Ser	Term	Term[2]	A
	Leu	Ser	Term	Trp	G
C	Leu	Pro	His	Arg	U
	Leu	Pro	His	Arg	C
	Leu	Pro	Gln	Arg	A
	Leu	Pro	Gln	Arg	G
A	Ile	Thr	Asn	Ser	U
	Ile	Thr	Asn	Ser	C
	Ile[2]	Thr	Lys	Arg[2]	A
	Met	Thr	Lys	Arg[2]	G
G	Val	Ala	Asp	Gly	U
	Val	Ala	Asp	Gly	C
	Val	Ala	Glu	Gly	A
	Val	Ala	Glu	Gly	G

[1]The terms first, second, and third nucleotide refer to the individual nucleotides of a triplet codon. U, uridine nucleotide; C, cytosine nucleotide; A, adenine nucleotide; G, guanine nucleotide; Met, chain initiator codon; Term, chain terminator codon. AUG, which codes for Met, serves as the initiator codon in mammalian cells. (Abbreviations of amino acids are explained in Chapter 4.)

[2]In mammalian mitochondria, AUA codes for Met and UGA for Trp, and AGA and AGG serve as chain terminators.

THE GENETIC CODE IS DEGENERATE, UNAMBIGUOUS, NONOVERLAPPING, WITHOUT PUNCTUATION, AND UNIVERSAL

Three codons do not code for specific amino acids; these have been termed **nonsense codons.** At least two of these nonsense codons are utilized in the cell as **termination signals;** they specify where the polymerization of amino acids into a protein molecule is to stop. The remaining 61 codons code for 20 amino acids (Table 40–1). Thus, there must be **"degeneracy"** in the genetic code; ie, multiple codons must decode the same amino acid. Some amino acids are encoded by several codons; eg, six different codons specify serine. Other amino acids, such as methionine and tryptophan, have a single codon. In general, the third nucleotide in a codon is less important than the other two in determining the specific amino acid to be incorporated, and this accounts for most of the degeneracy of the code. However, for any specific codon, only a single amino acid is indicated; with rare exceptions, the genetic code is **unambiguous**—ie, given a specific codon, only a single amino acid is indicated. **The distinction between ambiguity and degeneracy is an important concept.**

The unambiguous but degenerate code can be explained in molecular terms. The recognition of specific codons in the mRNA by the tRNA adapter molecules is dependent upon their **anticodon region** and

specific base-pairing rules. Each tRNA molecule contains a specific sequence, complementary to a codon, which is termed its anticodon. For a given codon in the mRNA, only a single species of tRNA molecule possesses the proper anticodon. Since each tRNA molecule can be charged with only one specific amino acid, each codon therefore specifies only one amino acid. However, some tRNA molecules can utilize the anticodon to recognize more than one codon. **With few exceptions, given a specific codon, only a specific amino acid will be incorporated—although, given a specific amino acid, more than one codon may call for it.**

As discussed below, the reading of the genetic code during the process of protein synthesis does not involve any overlap of codons. **Thus, the genetic code is nonoverlapping.** Furthermore, once the reading is commenced at a specific codon, there is **no punctuation** between codons, and the message is read in a continuing sequence of nucleotide triplets until a nonsense codon is reached.

Until recently, the genetic code was thought to be

Table 40–2. Features of the genetic code.

- Degenerate
- Unambiguous
- Nonoverlapping
- Not punctuated
- Universal

universal. It has now been shown that the set of tRNA molecules in mitochondria (which contain their own separate and distinct set of translation machinery) from lower and higher eukaryotes, including humans, reads four codons differently from the tRNA molecules in the cytoplasm of even the same cells. As noted in Table 40–1, the codon AUA is read as Met, and UGA codes for Trp in mammalian mitochondria. In addition, in mitochondria, the codons AGA and AGG are read as stop or chain terminator codons rather than as Arg. As a result, mitochondria require only 22 tRNA molecules to read their genetic code, whereas the cytoplasmic translation system possesses a full complement of 31 tRNA species. These exceptions noted, **the genetic code is universal.** The frequency of use of each amino acid codon varies considerably between species and among different tissues within a species. Tables of **codon usage** are becoming more accurate as more genes are sequenced. This is of considerable importance because investigators often need to deduce mRNA structure from the primary sequence of a portion of protein in order to synthesize an oligonucleotide probe and initiate a recombinant DNA cloning project. The main features of the genetic code are listed in Table 40–2.

AT LEAST ONE SPECIES OF TRANSFER RNA (tRNA) EXISTS FOR EACH OF THE 20 AMINO ACIDS

tRNA molecules have extraordinarily similar functions and three-dimensional structures. The adapter function of the tRNA molecules requires the charging of each specific tRNA with its specific amino acid. Since there is no affinity of nucleic acids for specific functional groups of amino acids, this recognition must be carried out by a protein molecule capable of recognizing both a specific tRNA molecule and a specific amino acid. At least 20 specific enzymes are required for these specific recognition functions and for the proper attachment of the 20 amino acids to specific tRNA molecules. The process of recognition and attachment (charging) is carried out in two steps by one enzyme for each of the 20 amino acids. These enzymes are termed **aminoacyl-tRNA synthetases.** They form an activated intermediate of aminoacyl-AMP-enzyme complex (Figure 40–1). The specific aminoacyl-AMP-enzyme complex then recognizes a specific tRNA to which it attaches the aminoacyl moiety at the 3'-hydroxyl adenosine terminal. The amino acid remains attached to its specific tRNA in an ester linkage until it is polymerized at a specific position in the fabrication of a polypeptide precursor of a protein molecule.

The regions of the tRNA molecule referred to in Chapter 37 (and illustrated in Figure 37-11) now be-

come important. The thymidine-pseudouridine-cytidine (TΨC) arm is involved in binding of the aminoacyl-tRNA to the ribosomal surface at the site of protein synthesis. The D arm is one of the sites important for the proper recognition of a given tRNA species by its proper aminoacyl-tRNA synthetase. The acceptor arm, located at the 3'-hydroxyl adenosyl terminal, is the site of attachment of the specific amino acid.

The anticodon region consists of seven nucleotides, and it recognizes the three-letter codon in mRNA (Figure 40–2). The sequence read from the 3' to 5' direction in that anticodon loop consists of a variable base–modified purine–XYZ–pyrimidine–pyrimidine-5'. Note that this direction of reading the anticodon is 3' to 5', whereas the genetic code in Table 40–1 is read 5' to 3', since the codon and the anticodon loop of the mRNA and tRNA molecules, respectively, are **antiparallel** in their complementarity.

The degeneracy of the genetic code resides mostly in the last nucleotide of the codon triplet, suggesting that the base pairing between this last nucleotide and the corresponding nucleotide of the anticodon is not strict. This is called **wobble;** the pairing of the codon and anticodon can "wobble" at this specific nucleotide-to-nucleotide pairing site. For example, the two codons for arginine, AGA and AGG, can bind to the same anticodon having a uracil at its 5' end. Similarly, three codons for glycine, GGU, GGC, and GGA, can form a base pair from one anticodon, CCI. I is an inosine nucleotide, another of the peculiar bases appearing in tRNA molecules.

The codon recognition by a tRNA molecule does not depend upon the amino acid that is attached at its 3'-hydroxyl terminal. This has been ingeniously demonstrated by charging a tRNA specific for cysteine (tRNA$_{cys}$) with radioactively labeled cysteine. By chemical means, the cysteinyl residue was then altered to generate a tRNA molecule specific for cysteine but charged instead with alanine. The chemical transformation of the cysteinyl to the alanyl moiety did not alter the anticodon portion of the cysteine-specific tRNA molecule. When this alanyl-tRNA$_{cys}$ was used in the translation of a hemoglobin mRNA, a radioactive alanine was incorporated at what was normally a cysteine site in the hemoglobin protein molecule. The experiment demonstrated that the aminoacyl derivative of an aminoacyl-tRNA molecule does not play a role in the codon recognition. As already noted, the aminoacyl moiety never comes in contact with the template mRNA containing the codons.

MUTATIONS RESULT WHEN CHANGES OCCUR IN THE NUCLEOTIDE SEQUENCE

Although the initial change may not occur in the coding strand of the double-stranded DNA molecule

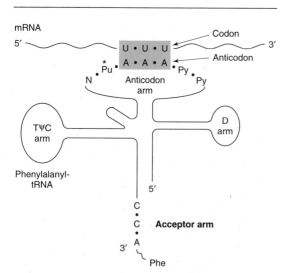

Figure 40–1. Formation of aminoacyl-tRNA. A two-step reaction, involving the enzyme aminoacyl-tRNA synthetase, results in the formation of aminoacyl-tRNA. The first reaction involves the formation of an AMP-amino acid-enzyme complex. This activated amino acid is next transferred to the corresponding tRNA molecule. The AMP and enzyme are released, and the latter can be reutilized.

for that gene, after replication, daughter DNA molecules with mutations in the coding strand will segregate and appear in the population of organisms.

Some Mutations Occur by Base Substitution

Single base changes (**point mutations**) may be **transitions** or **transversions.** In the former, a given pyrimidine is changed to the other pyrimidine or a given purine is changed to the other purine. Transversions are changes from a purine to either of the two pyrimidines or the change of a pyrimidine into either of the two purines, as shown in Figure 40–3.

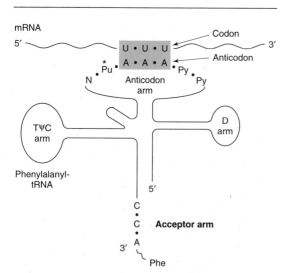

Figure 40–2. Recognition of the codon by the anticodon. One of the codons for phenylalanine is UUU. tRNA charged with phenylalanine (Phe) has the complementary sequence AAA; hence, it forms a base-pair complex with the codon. The anticodon region typically consists of a sequence of seven nucleotides: variable (N), modified purine (Pu*), X, Y, Z, and two pyrimidines (Py) in the 3′ to 5′ direction.

If the nucleotide sequence of the gene containing the mutation is transcribed into an RNA molecule, then the RNA molecule will possess a complementary base change at this corresponding locus.

Single base changes in the mRNA molecules may have one of several effects when translated into protein:

(1) There may be **no detectable effect** because of the degeneracy of the code. This would be more likely if the changed base in the mRNA molecule were to be at the third nucleotide of a codon. Because of wobble, the translation of a codon is least sensitive to a change at the third position.

(2) A **missense effect** will occur when a different amino acid is incorporated at the corresponding site in the protein molecule. This mistaken amino acid, or missense, depending upon its location in the specific protein, might be acceptable, partially acceptable, or unacceptable to the function of that protein molecule. From a careful examination of the genetic code, one can conclude that most single base changes would result in the replacement of one amino acid by another with rather similar functional groups. This is an effective mechanism to avoid drastic change in the physical properties of a protein molecule. If an acceptable missense effect occurs, the resulting protein molecule may not be distinguishable from the normal one. A partially acceptable missense will result in a protein molecule with partial but abnormal function. If an unacceptable missense effect occurs, then the

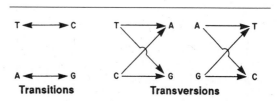

Figure 40–3. Diagrammatic representation of transition mutations and transversion mutations.

protein molecule will not be capable of functioning in its assigned role.

(3) A **nonsense** codon may appear that would then result in the **premature termination** of amino acid incorporation into a peptide chain and the production of only a fragment of the intended protein molecule. The probability is high that a prematurely terminated protein molecule or peptide fragment will not function in its assigned role.

Hemoglobin Illustrates the Effects of Single Base Changes in Structural Genes

Some mutations have no apparent effect. The lack of effect of a single base change would be demonstrable only by sequencing the nucleotides in the mRNA molecules or structural genes for hemoglobin from a large number of humans with normal hemoglobin molecules. However, it can be deduced that the codon for valine at position 67 of the β chain of hemoglobin is not identical in all persons possessing the normal β chain of hemoglobin. Hemoglobin Milwaukee has at position 67 a glutamic acid; hemoglobin Bristol contains aspartic acid at position 67. In order to account for the amino acid change by the change of a single nucleotide residue in the codon for amino acid 67, one must infer that the mRNA encoding hemoglobin Bristol possessed a GUU or GUC codon prior to a later change to GAU or GAC, both codons for aspartic acid (Figure 40–4). However, the mRNA encoding hemoglobin Milwaukee would have to possess at position 67 a codon GUA or GUG in order that a single nucleotide change could provide for

the appearance of the glutamic acid codons GAA or GAG. Hemoglobin Sydney, which contains an alanine at position 67, could have arisen by the change of a single nucleotide in any of the four codons for valine (GUU, GUC, GUA, or GUG) to the alanine codons (GCU, GCC, GCA, or GCG, respectively).

Substitution of Amino Acids Causes Missense Mutations

A. Acceptable Missense Mutations: An example of an acceptable missense mutation (Figure 40–5, top) in the structural gene for the β chain of hemoglobin could be detected by the presence of an electrophoretically altered hemoglobin in the red cells of an apparently healthy individual. Hemoglobin Hikari has been found in at least two families of Japanese people. This hemoglobin has asparagine substituted for lysine at the 61 position in the β chain. The corresponding transversion might be either AAA or AAG changed to either AAU or AAC. The replacement of the specific lysine with asparagine apparently does not alter the normal function of the β chain in these individuals.

B. Partially Acceptable Missense Mutations: A partially acceptable missense mutation (Figure 40–5, center) is best exemplified by **hemoglobin S,** sickle hemoglobin, in which the normal amino acid in position 6 of the β chain, glutamic acid, has been replaced by valine. The corresponding single nucleotide change within the codon would be GAA or GAG of glutamic acid to GUA or GUG of valine. Clearly, this missense mutation hinders normal function and results in sickle cell anemia

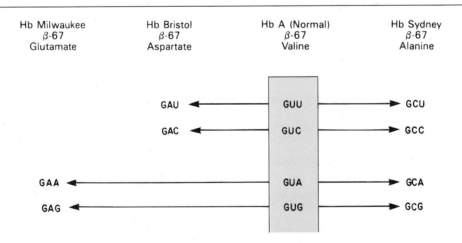

Figure 40–4. The normal valine at position 67 of the β chain of hemoglobin A can be coded for by one of the four codons shown in the box. In abnormal hemoglobin Milwaukee, the amino acid at position 67 of the β chain contains glutamate, coded for by GAA or GAG, either one of which could have resulted from a single-step transversion from the valine codons GUA or GUG. Similarly, the alanine present at position 67 of the β chain of hemoglobin Sydney could have resulted from a single-step transition from any one of the four valine codons. However, the aspartate residue at position 67 of hemoglobin Bristol could have resulted from a single-step transversion only from the GUU or GUC valine codons.

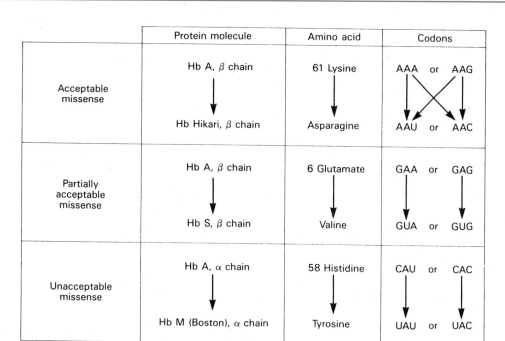

	Protein molecule	Amino acid	Codons
Acceptable missense	Hb A, β chain ↓ Hb Hikari, β chain	61 Lysine ↓ Asparagine	AAA or AAG ↓ AAU or AAC
Partially acceptable missense	Hb A, β chain ↓ Hb S, β chain	6 Glutamate ↓ Valine	GAA or GAG ↓ GUA or GUG
Unacceptable missense	Hb A, α chain ↓ Hb M (Boston), α chain	58 Histidine ↓ Tyrosine	CAU or CAC ↓ UAU or UAC

Figure 40–5. Examples of three types of missense mutations resulting in abnormal hemoglobin chains. The amino acid alterations and possible alterations in the respective codons are indicated. The hemoglobin Hikari β-chain mutation has apparently normal physiologic properties but is electrophoretically altered. Hemoglobin S has a β-chain mutation and partial function; hemoglobin S binds oxygen but precipitates when deoxygenated. Hemoglobin M Boston, an α-chain mutation, permits the oxidation of the heme ferrous iron to the ferric state and so will not bind oxygen at all.

when the mutant gene is present in the homozygous state. The glutamate-to-valine-change may be considered to be partially acceptable because hemoglobin S does bind and release oxygen, although abnormally.

C. Unacceptable Missense Mutations: An unacceptable missense mutation (Figure 40–5, bottom) in a hemoglobin gene generates a nonfunctioning hemoglobin molecule. For example, the hemoglobin M mutations generate molecules that allow the Fe^{2+} of the heme moiety to be oxidized to Fe^{3+}, producing methemoglobin. Methemoglobin cannot transport oxygen (see Chapter 7).

Frame Shift Mutations Result From Deletion or Insertion of Nucleotides in DNA That Generates Altered mRNAs

The deletion of a single nucleotide from the coding strand of a gene results in an altered reading frame in the mRNA. The machinery translating the mRNA does not recognize that a base was missing, since there is no punctuation in the reading of codons. Thus, a major alteration in the sequence of polymerized amino acids, as depicted in example 1, Figure 40–6, results. Altering the reading frame results in a garbled translation of the mRNA distal to the single nucleotide deletion. Not only is the se-

quence of amino acids distal to this deletion garbled, but reading of the message can also result in the appearance of a nonsense codon and thus the production of a polypeptide both garbled and prematurely terminated near its carboxyl terminal (example 3, Figure 40–6).

If three nucleotides or a multiple of three were deleted from a coding region, the corresponding mRNA when translated would provide a protein from which was missing the corresponding number of amino acids (example 2, Figure 40–6). Because the reading frame is a triplet, the reading phase would not be disturbed for those codons distal to the deletion. If, however, deletion of one or two nucleotides occurs just prior to or within the normal termination codon (nonsense codon), the reading of the normal termination signal is disturbed. Such a deletion might result in reading through a termination signal until another nonsense codon was encountered (example 1, Figure 40–6). Excellent examples of this phenomenon are described in discussions of hemoglobinopathies.

Insertions of one or two or nonmultiples of three nucleotides into a gene result in an mRNA in which the reading frame is distorted upon translation, and the same effects that occur with deletions are reflected in the mRNA translation. This may result in garbled amino acid sequences distal to the insertion,

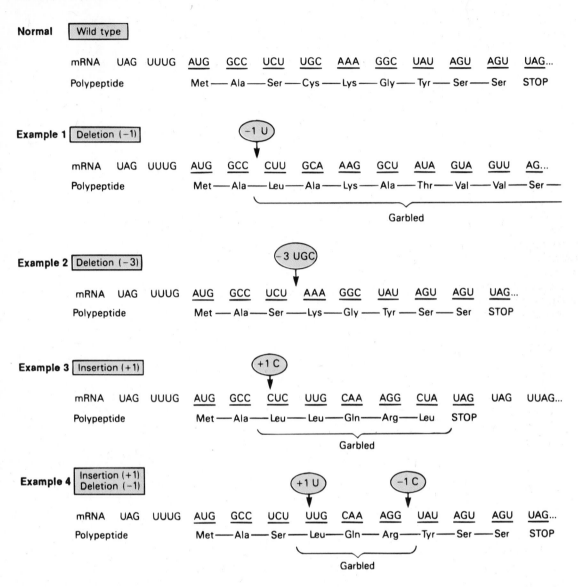

Figure 40–6. Demonstration of the effects of deletions and insertions in a gene on the sequence of the mRNA transcript and of the polypeptide chain translated therefrom. The arrows indicate the sites of deletions or insertions, and the numbers in the circles indicate the number of nucleotide residues deleted or inserted.

and the generation of a **nonsense codon** at or distal to the insertion, or perhaps reading through the normal termination codon. Following a deletion in a gene, an insertion (or vice versa) can reestablish the proper reading frame (example 4, Figure 40–6). The corresponding mRNA, when translated, would contain a garbled amino acid sequence between the insertion and deletion. Beyond the reestablishment of the reading frame, the amino acid sequence would be correct. One can imagine that different combinations of deletions, of insertions, or of deletions and insertions would result in formation of a protein wherein a portion is abnormal, but this portion is surrounded by

the normal amino acid sequences. Such phenomena have been demonstrated convincingly in the bacteriophage T4, a finding which contributed significantly to evidence that the reading frame was a triplet.

Suppressor Mutations Reduce the Effects of Missense, Nonsense, and Frame Shift Mutations

The above discussion of the altered protein products of gene mutations is based on the presence of normally functioning tRNA molecules. However, in prokaryotic and lower eukaryotic organisms abnormally functioning tRNA molecules have been discovered that are themselves the results of mutations. Some of these abnormal tRNA molecules are capable of suppressing the effects of mutations in distant structural genes. These **suppressor tRNA molecules,** usually formed as the result of alterations in their anticodon regions, are capable of suppressing missense mutations, nonsense mutations, and frame shift mutations. However, since the suppressor tRNA molecules are not capable of distinguishing between a normal codon and one resulting from a gene mutation, their presence in a cell usually results in decreased viability. For instance, the nonsense suppressor tRNA molecules can suppress the normal termination signals to allow a read-through when it is not desirable. Frame shift suppressor tRNA molecules may read a normal codon plus a component of a juxtaposed codon to provide a frame shift, also when it is not desirable. Suppressor tRNA molecules may exist in mammalian cells, since read-through transcription occurs.

PROTEIN SYNTHESIS CAN BE DESCRIBED IN THREE PHASES: INITIATION, ELONGATION, AND TERMINATION

The general structural characteristics of ribosomes and their self-assembly process are discussed in Chapter 39. These particulate entities serve as the machinery on which the mRNA nucleotide sequence is translated into the sequence of amino acids of the specified protein. The translation of the mRNA commences near its 5′ terminal with the formation of the corresponding amino terminal of the protein molecule. The message is read from 5′ to 3′, concluding with the formation of the carboxyl terminal of the protein. Again, the concept of **polarity** is apparent. As described in Chapter 39, the transcription of a gene into the corresponding mRNA or its precursor first forms the 5′ terminal of the RNA molecule. In prokaryotes, this allows for the beginning of mRNA translation before the transcription of the gene is completed. In eukaryotic organisms, the process of transcription is a nuclear one; mRNA translation occurs in the cytoplasm. This precludes simultaneous transcription and translation in eukaryotic organisms and makes possible the processing necessary to generate mature mRNA from the primary transcript—hnRNA.

Initiation Involves Several Protein-RNA Complexes (Figure 40–7)

Initiation of protein synthesis requires that an mRNA molecule be selected for translation by a ribosome. Once the mRNA binds to the ribosome, the latter finds the correct reading frame on the mRNA,

and translation begins. This process involves tRNA, rRNA, mRNA, and at least 10 eukaryotic initiation factors (eIFs), some of which have multiple (three to eight) subunits. Also involved are GTP, ATP, and amino acids. Initiation can be divided into four steps: (1) dissociation of the ribosome into its 40S and 60S subunits; (2) binding of a ternary complex consisting of met-tRNA S_i, GTP, and eIF-2 to the 40S ribosome to form a preinitiation complex; (3) binding of mRNA to the 40S preinitiation complex to form a 40S initiation complex; and (4) combination of the 40S initiation complex with the 60S ribosomal subunit to form the 80S initiation complex.

A. Ribosomal Dissociation: Two initiation factors, eIF-3 and eIF-1A, bind to the 40S subunit. This favors dissociation of the 80S ribosome into its 40S and 60S subunits and prevents reassociation. The binding of eIF-3A to the 60S subunit may also prevent reassociation.

B. Formation of the 40S Preinitiation Complex: The first step in this process involves the binding of GTP by eIF-2. This binary complex then binds to met-tRNAi, a tRNA specifically involved in binding to the initiation codon AUG. This ternary complex binds to the 40S ribosomal subunit to form the 40S preinitiation complex, which is stabilized by association with eIF-3 and eIF-1A.

eIF-2 is one of two control points for protein synthesis in eukaryotic cells. eIF-2 consists of α and β subunits. eIF-2α is phosphorylated (on serine 51) by a cAMP-independent protein kinase that is activated when a cell is under stress and when the energy expenditure required for protein synthesis would be deleterious. Such conditions include amino acid and glucose starvation, virus infection, serum deprivation, hyperosmolality, and heat shock. Phosphorylated eIF-2α binds tightly to, and inactivates, the GTP-GDP recycling protein eIF-2β. This prevents formation of the 40S preinitiation complex and blocks protein synthesis.

C. Formation of the 40S Initiation Complex: The 5′ terminals of most mRNA molecules in eukaryotic cells are "capped," as described in Chapter 39. This methyl-guanosyl triphosphate cap facilitates the binding of mRNA to the 40S preinitiation complex. A cap binding protein complex, eIF-4F, binds to the cap through one of its subunits. Then eIF-4A and eIF-4B bind and probably reduce the complex secondary structure of the 5′ end of the mRNA through their respective ATPase and helicase activities. Certainly the association of mRNA with the 40S preinitiation complex to form the 40S initiation complex requires ATP hydrolysis. eIF-3 is a key protein because it binds with high affinity to both mRNA and the 40S ribosomal subunit. Following association of the 40S preinitiation complex with the mRNA cap and reduction ("melting") of the secondary structure near the 5′ end of the mRNA, the complex scans the mRNA for a suitable initiation codon. Generally this

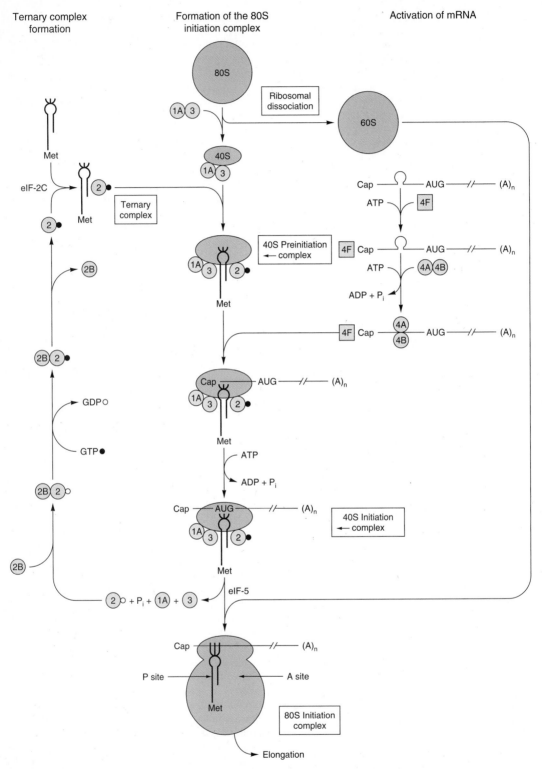

Figure 40–7. Diagrammatic representation of the initiation of protein synthesis on the mRNA template containing a 5′ cap (G^mTP-5′) and 3′ poly(A) terminal [3′ (A)_n]. This process proceeds in three steps: (1) activation of mRNA; (2) formation of the ternary complex consisting of tRNAmet_i, initiation factor eIF-2, and GTP; and (3) formation of the active 80S initiation complex. (See text for details.) GTP, ●; GDP, ○. The various initiation factors appear in abbreviated form as circles or squares, eg, eIF-3 (③), eIF-4F (④F).

is the 5′-most AUG, but the precise initiation codon is determined by so-called Kozak consensus sequences that surround the AUG:

$$\overset{-3}{} \quad \overset{+1}{} \quad \overset{+4}{}$$
$$\text{GCCA/GCCAUGG}$$

Most preferred is the presence of a purine at positions −3 and +4 relative to the AUG.

The eIF-4F (4F) complex is particularly important in controlling the rate of protein translation. 4F is a complex consisting of eIF-4E (4E), which binds to the m^7G cap structure at the 5′ end of the mRNA, eIF-4A, which is an RNA-dependent ATPase that helps unwind the RNA, and p220, whose exact function is unknown (Figure 40–8). 4E is important because cap binding is a rate-limiting step in translation. Overexpression of 4E causes malignant transformation of cells, and underexpression results in a prolongation of the cell cycle and disaggregation of polysomes.

This process is under regulation at two levels. Insulin and mitogenic growth factors result in phosphorylation of 4E (on serine 53) and activation of the protein. A recently discovered family of proteins binds to, and inactivates, 4E. These proteins include 4E-BP1 (and the closely related protein PHAS-1) and 4E-BP2. Phosphorylation of these proteins, again by insulin or growth factors via the stimulation of serine protein kinase, results in the dissociation of the binding protein from 4E, the assembly of 4E and other components of the 4F complex in the mRNA cap, and initiation of protein synthesis (Figure 40–8). This observation explains how insulin causes a marked posttranscriptional increase of protein synthesis in adipose tissue and muscle.

D. Formation of the 80S Initiation Complex: The binding of the 60S ribosomal subunit to the 40S initiation complex involves the hydrolysis of the GTP bound to eIF-2 by eIF-5. This reaction results in the release of the initiation factors bound to the 40S initiation complex (these factors then are recycled) and the rapid association of the 40S and 60S subunits to form the 80S ribosome. At this point, the met-tRNA is on the P site of the ribosome, ready for the elongation cycle to commence.

Elongation Also Is a Multistep Process (Figure 40–9)

Elongation, a cyclic process, involves several steps catalyzed by proteins called elongation factors (eEF). These steps are (1) binding of aminoacyl-tRNA to the A site, (2) peptide bond formation, and (3) translocation.

A. Binding of Aminoacyl-tRNA to the A Site: In the complete 80S ribosome formed during the process of initiation, the A site is free. The binding of

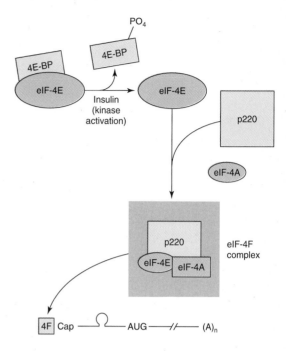

Figure 40–8. Activation of eIF-4E by insulin and formation of the cap binding eIF-4F complex. The 4F-cap mRNA complex is depicted as in Figure 40–7. The 4F complex consists of eIF-4E (4E), eIF-4A, and p220. 4E is inactive when bound by one of a family of binding proteins (4E-BPs). Insulin and mitogenic factors activate a serine protein kinase, and this results in the phosphorylation of 4E-BP. Phosphorylated 4E-BP dissociates from 4E, and the latter is then able to form the 4F complex and bind to the mRNA cap.

the proper aminoacyl-tRNA in the A site requires proper codon recognition. **Elongation factor eEF-1α** forms a complex with GTP and the entering aminoacyl-tRNA (Figure 40–9). This complex then allows the aminoacyl-tRNA to enter the A site with the release of eEF-1α–GDP and phosphate. As shown in Figure 40–9, eEF-1α–GDP then recycles to eEF-1α–GTP with the aid of other soluble protein factors and GTP.

B. Peptide Bond Formation: The α-amino group of the new aminoacyl-tRNA in the A site carries out a nucleophilic attack on the esterified carboxyl group of the peptidyl-tRNA occupying the P site. This reaction is catalyzed by a protein component, **peptidyltransferase,** of the 60S ribosomal subunit. This enzymatic activity may be performed by some RNA component of the ribosome—perhaps the 23S rRNA. This is another example of ribozyme activity and indicates an important, and previously unsuspected, direct role for RNA in protein synthesis (Table 40–3). Because the amino acid on the aminoacyl-tRNA is already "activated," no further energy source is required for this reaction. The reaction re-

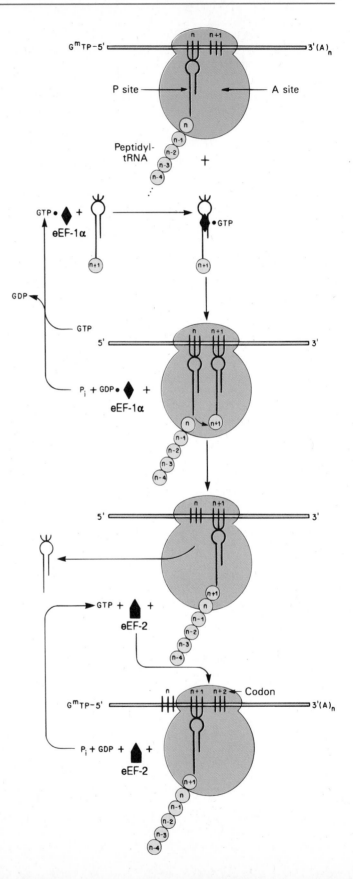

Figure 40–9. Diagrammatic representation of the peptide elongation process of protein synthesis. The small circles labeled n − 1, n, n + 1, etc, represent the amino acid residues of the newly formed protein molecule. eEF-1α and eEF-2 represent elongation factors 1 and 2, respectively. The peptidyl-tRNA and aminoacyl-tRNA sites on the ribosome are represented by P site and A site, respectively.

Table 40–3. Evidence that rRNA is peptidyltransferase.

- Ribosomes can make peptide bonds even when proteins are removed or inactivated.
- Certain parts of the rRNA sequence are highly conserved in all species.
- These conserved regions are on the surface of the RNA molecule.
- RNA can be catalytic.
- Mutations that result in antibiotic resistance at the level of protein synthesis are more often found in rRNA than in the protein components of the ribosome.

sults in attachment of the growing peptide chain to the tRNA in the A site.

C. Translocation: Upon removal of the peptidyl moiety from the tRNA in the P site, the discharged tRNA quickly dissociates from the P site. **Elongation factor 2 (eEF-2)** and GTP are responsible for the **translocation** of the newly formed peptidyl-tRNA at the A site into the empty P site. The GTP required for eEF-2 is hydrolyzed to GDP and phosphate during the translocation process. The translocation of the newly formed peptidyl-tRNA and its corresponding codon into the P site then frees the A site for another cycle of aminoacyl-tRNA codon recognition and elongation.

The charging of the tRNA molecule with the aminoacyl moiety requires the hydrolysis of an ATP to an AMP, equivalent to the hydrolysis of two ATPs to two ADPs and phosphates. The entry of the aminoacyl-tRNA into the A site results in the hydrolysis of one GTP to GDP. The translocation of the newly formed peptidyl-tRNA in the A site into the P site by eEF-2 similarly results in the hydrolysis of GTP to GDP and phosphate. Thus, the energy requirements for the formation of one peptide bond include the equivalent of the hydrolysis of two ATP molecules to ADP and two GTP molecules to GDP, or the hydrolysis of four high-energy phosphate bonds. This process occurs rapidly. A eukaryotic ribosome can incorporate as many as six amino acids per second; prokaryotic ribosomes incorporate as many as 18 per second. Thus, the process of peptide synthesis occurs with great speed and accuracy until a termination codon is reached.

Termination Occurs When a Nonsense Codon Is Recognized (Figure 40–10)

In comparison to initiation and elongation, termination is a relatively simple process. After multiple cycles of elongation culminating in polymerization of the specific amino acids into a protein molecule, the nonsense or terminating codon of mRNA (UUA, UAG, UGA) appears in the A site. Normally, there is no tRNA with an anticodon capable of recognizing such a termination signal. **Releasing factors (eRF)** are capable of recognizing that a termination signal

resides in the A site (Figure 40–10). The releasing factor, in conjunction with GTP and the peptidyl transferase, promotes the hydrolysis of the bond between the peptide and the tRNA occupying the P site. Thus, a water molecule, rather than an amino acid, is added. This hydrolysis releases the protein and the tRNA from the P site. Upon hydrolysis and release, the **80S ribosome dissociates** into its 40S and 60S subunits, which are then recycled. Therefore, the releasing factors are proteins that hydrolyze the peptidyl-tRNA bond when a nonsense codon occupies the A site. The mRNA is then released from the ribosome, which dissociates into its component 40S and 60S subunits, and another cycle can be repeated.

Polysomes Are Assemblies of Ribosomes

Many ribosomes can translate the same mRNA molecule simultaneously. Because of their relatively large size, the ribosome particles cannot attach to an mRNA any closer than 80 nucleotides apart. Multiple ribosomes on the same mRNA molecule form a **polyribosome,** or "polysome." In an unrestricted system, the number of ribosomes attached to an mRNA (and thus the size of polyribosomes) correlates positively with the length of the mRNA molecule. The mass of the mRNA molecule is, of course, quite small compared to the mass of even a single ribosome.

A single mammalian ribosome is capable of synthesizing about 100 peptide bonds each minute. Polyribosomes actively synthesizing proteins can exist as free particles in the cellular cytoplasm or may be attached to sheets of membranous cytoplasmic material referred to as **endoplasmic reticulum.** The attachment of the particulate polyribosomes to the endoplasmic reticulum is responsible for its "rough" appearance as seen by electron microscopy. The proteins synthesized by the attached polyribosomes are extruded into the cisternal space between the sheets of rough endoplasmic reticulum and are exported from there. Some of the protein products of the rough endoplasmic reticulum are packaged by the Golgi apparatus into zymogen particles for eventual export (see Chapter 43). The polyribosomal particles free in the cytosol are responsible for the synthesis of proteins required for intracellular functions.

The Machinery of Protein Synthesis Can Respond to Environmental Threats or Can Be Co-opted to Become a Part of a Disease Mechanism

Ferritin, an iron-binding protein, prevents ionized iron (Fe^{2+}) from reaching toxic levels within cells. Elemental iron stimulates ferritin synthesis by causing the release of a cytoplasmic protein that binds to a specific region in the $5'$ nontranslated region of ferritin mRNA. The disruption of this protein-mRNA

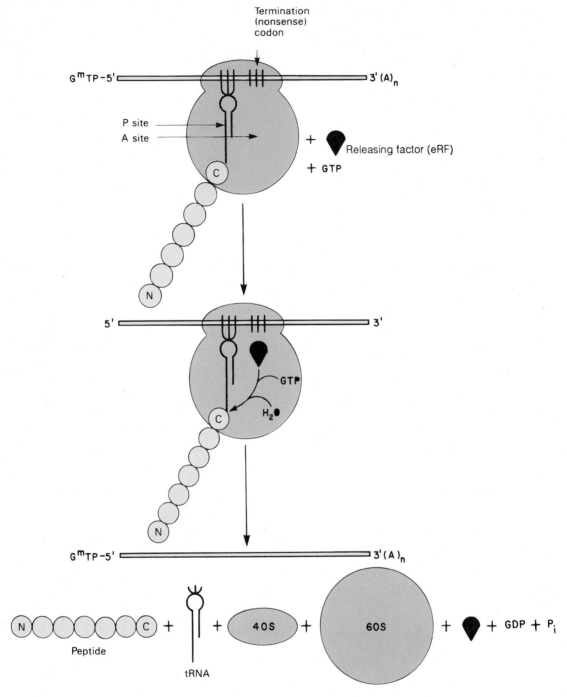

Figure 40–10. Diagrammatic representation of the termination process of protein synthesis. The peptidyl-tRNA and aminoacyl-tRNA sites are indicated as P site and A site, respectively. The hydrolysis of the peptidyl-tRNA complex is shown by the entry of H_2O. N and C indicate the amino and carboxyl terminal amino acids, respectively, and illustrate the polarity of protein synthesis.

interaction activates ferritin mRNA and results in its translation. This mechanism provides for rapid control of the synthesis of a protein that sequesters Fe^{2+}, a potentially toxic molecule.

The protein synthesis machinery can also be modified in deleterious ways. **Viruses replicate by using host cell processes,** including those involved in protein synthesis. Some viral mRNAs are translated much more efficiently than those of the host cell (eg, Mengo virus and encephalomyocarditis virus). Others, such as reovirus and vesicular stomatitis virus, replicate abundantly, and their mRNAs have a competitive advantage over host cell mRNAs for limited translation factors. Other viruses inhibit host cell protein synthesis by preventing the association of mRNA with the 40S ribosome. Poliovirus accomplishes this by activating a cellular protease that degrades the 220-kDa component of the eIF-4F cap binding complex described above.

POSTTRANSLATIONAL PROCESSING AFFECTS THE ACTIVITY OF MANY PROTEINS

Some animal viruses, notably poliovirus and hepatitis A virus, synthesize long polycistronic proteins from one long mRNA molecule. These protein molecules are subsequently cleaved at specific sites to provide the several specific proteins required for viral function. In animal cells, many proteins are synthesized from the mRNA template as a precursor molecule, which then must be modified to achieve the active protein. The prototype is insulin, which is a low-molecular-weight protein having two polypeptide chains with interchain and intrachain disulfide bridges. The molecule is synthesized as a single chain precursor, or **prohormone,** which folds to allow the disulfide bridges to form. A specific protease then clips out the segment that connects the two chains that form the functional insulin molecule (see Figure 51–3).

Many other peptides are synthesized as proproteins that require modifications before attaining biologic activity. Many of the posttranslational modifications involve the removal of amino terminal amino acid residues by specific aminopeptidases. Collagen, an abundant protein in the extracellular spaces of higher eukaryotes, is synthesized as procollagen. Three procollagen polypeptide molecules, frequently not identical in sequence, align themselves in a way dependent upon the existence of specific amino terminal peptides. Specific enzymes then carry out hydroxylations and oxidations of specific amino acid residues within the procollagen molecules to provide cross-links for greater stability. Amino terminal peptides are cleaved off the molecule to form the final product a strong, insoluble collagen molecule (Chapter 57). Many other posttranslational modifications of

proteins occur. Covalent modification by acetylation, phosphorylation, and glycosylation is common, for example.

MANY ANTIBIOTICS WORK BECAUSE THEY SELECTIVELY INHIBIT PROTEIN SYNTHESIS IN BACTERIA

Ribosomes in bacteria and in the mitochondria of higher eukaryotic cells differ from the mammalian ribosome described in Chapter 37. The bacterial ribosome is smaller (70S rather than 80S) and has a different, somewhat simpler, complement of RNA and protein molecules. This difference is exploited for clinical purposes because many effective antibiotics interact specifically with the proteins of prokaryotic ribosomes and thus inhibit protein synthesis. This results in growth arrest or death of the bacterium. The

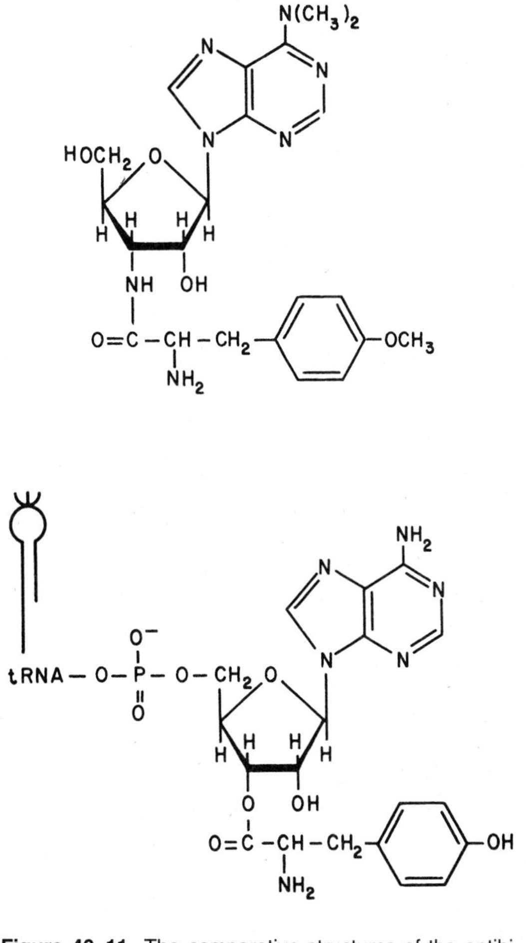

Figure 40–11. The comparative structures of the antibiotic puromycin *(top)* and the 3′ terminal portion of tyrosinyl-tRNA *(bottom).*

most useful of this class of antibiotics (eg, tetracyclines, lincomycin, erythromycin, and chloramphenicol) do not interact with components of eukaryotic ribosomal particles and thus are not toxic to eukaryotes. The macrolide class of antibiotics works by binding to 23S rRNA, which is interesting in view of the newly appreciated role of rRNA in peptide bond formation.

Other antibiotics inhibit protein synthesis on all ribosomes (**puromycin**) or only on those of eukaryotic cells (**cycloheximide**). Puromycin (Figure 40–11) is a structural analog of tyrosinyl-tRNA. Puromycin is incorporated via the A site on the ribosome into the carboxy-terminal position of a peptide but causes the premature release of the polypeptide. Puromycin, as a tyrosinyl-tRNA analog, effectively inhibits protein synthesis in both prokaryotes and eukaryotes. Cycloheximide inhibits peptidyltransferase in the 60S ribosomal subunit in eukaryotes, presumably by binding to an rRNA component.

Diphtheria toxin, an exotoxin of *Corynebacterium diphtheriae* infected with a specific lysogenic phage, catalyzes the ADP-ribosylation of eEF-2 in mammalian cells. This modification inactivates eEF-2 and thereby specifically inhibits mammalian protein synthesis. Many animals (eg, mice) are resistant to diphtheria toxin. This resistance is due to inability of diphtheria toxin to cross the cell membrane rather than to insensitivity of mouse eEF-2 to diphtheria toxin, catalyzed ADP-ribosylation by NAD.

Many of these compounds, puromycin and cycloheximide in particular, are not clinically useful but have been important in elucidating the role of protein synthesis in the regulation of metabolic processes, particularly enzyme induction by hormones.

SUMMARY

The flow of genetic information follows the sequence DNA → RNA → protein. The genetic information in the structural region of a gene is transcribed into an RNA molecule such that the sequence of the latter is complementary to that in the DNA.

Several different types of RNA, including ribosomal RNA (rRNA), transfer RNA (tRNA), and messenger RNA (mRNA), are involved in protein synthesis, perhaps in a direct catalytic role in the case of rRNA. The information in mRNA is in a tandem array of codons, each of which is three nucleotides long. These triplet codons are nonoverlapping, and there is no punctuation between codons; the mRNA is read continuously from a start codon (AUG) to a termination codon (UAA, UAG, UGA). These codons bracket the open reading frame of the mRNA, which is the series of codons, each specifying a certain amino acid, that determines the precise amino acid sequence of the protein to be synthesized. Some amino acids have a single codon (methionine, tryptophan), whereas others can be encoded by as many as six different codons (leucine, serine).

Protein synthesis, like DNA and RNA synthesis, follows a 5′ to 3′ polarity, and can be divided into three processes: initiation, elongation, and termination. Each process is complex. For example, the initiation of protein synthesis, in addition to requiring ribosomes, which in themselves consist of several types of RNA and dozens of proteins, also requires the specific mRNA, tRNAs corresponding to all 20 amino acids, at least 10 initiation factors (some of which are multisubunit complexes), ATP, and GTP. In spite of this complexity, a general understanding of the orderly process that results in initiation, and the separate factors involved in elongation and termination of protein synthesis, can be described.

The information summarized above makes it easier to understand how mutant proteins arise. Single base substitutions can result in codons that specify a different amino acid at a given position or in a stop codon that results in a truncated protein. Base additions or deletions alter the reading frame, so different codons are read. This can also result in an abnormal sequence of amino acids or of a truncated protein if a stop codon is created.

A variety of compounds, including several antibiotics, inhibit protein synthesis by affecting one or more of the steps described above.

REFERENCES

Crick F et al: The genetic code. Nature 1961;192:1227.

Forget BG: Molecular genetics of human hemoglobin synthesis. Ann Intern Med 1979;91:605.

Hershey JWB: Translational control in mammalian cells. Annu Rev Biochem 1991;60:717.

Hinnebusch A: Translational control of GCN4: An in vivo barometer of initial-factor activity. Trends Biochem Sci 1994;371:762.

Kozak M: Structural features in eukaryotic mRNAs that modulate the initiation of translation. J Biol Chem 1991;266:1986.

Noller HF: Ribosomal RNA and translation. Annu Rev Biochem 1991;60:191.

Pause A et al: Insulin-dependent stimulation of protein synthesis by phosphorylation of a regulator of 5′ cap function. Nature 1994;371:762.

Rhoads RE: Regulation of eukaryotic protein synthesis by initiation factors. J Biol Chem 1993;268:3017.

Schneider RJ, Shenk T: Impact of virus infection on host cell protein synthesis. Annu Rev Biochem 1987;56:317.

Shatkin AJ: mRNA cap binding proteins: Essential factors for initiating translation. Cell 1985;40:223.

Regulation of Gene Expression

41

Daryl K. Granner, MD

INTRODUCTION

Organisms adapt to environmental changes by altering gene expression. The process of alteration of gene expression has been studied in detail in bacteria and viruses, and it generally involves the interaction of specific binding proteins with various regions of DNA in the immediate vicinity of the transcription start site. This can have a positive or negative effect on transcription. Eukaryotic cells use this basic paradigm but employ other mechanisms as well to regulate transcription. Such processes as enhancement or silencing; tissue-specific expression; regulation by hormones, metals, and chemicals; gene amplification; gene rearrangement; and posttranscriptional modifications are also used to control gene expression.

BIOMEDICAL IMPORTANCE

Many of the mechanisms that control gene expression are used to respond to hormones and therapeutic agents. An understanding of these processes may lead to development of agents that alter pathophysiologic mechanisms or inhibit the function or arrest the growth of pathogenic organisms.

REGULATED EXPRESSION OF GENES IS REQUIRED FOR DEVELOPMENT, DIFFERENTIATION, AND ADAPTATION

The genetic information present in each somatic cell of a metazoan organism is practically identical. The exceptions are found in those few cells that have amplified or rearranged genes in order to carry out specialized cellular functions. The expression of the genetic information must be regulated during ontogeny and differentiation of the organism and its cellular components. Furthermore, in order for the organism to adapt to its environment and to conserve energy and nutrients, the expression of genetic information must be responsive to extrinsic signals. As organisms have evolved, more sophisticated regulatory mechanisms have appeared to provide the organism

and its cells with the responsiveness necessary for survival in its complex environment. Mammalian cells possess about 1000 times more genetic information than does the bacterium *Escherichia coli*. Much of this additional genetic information is probably involved in the regulation of gene expression during the differentiation of tissues and biologic processes in the multicellular organism and in ensuring that the organism can respond to complex environmental challenges.

In simple terms, there are only two types of gene regulation: **positive regulation** and **negative regulation** (Table 41–1). When the expression of genetic information is quantitatively **increased** by the presence of a specific regulatory element, regulation is said to be **positive;** whereas when the expression of genetic information is **diminished** by the presence of a specific regulatory element, regulation is said to be **negative.** The element or molecule mediating the negative regulation is said to be a negative regulator; that mediating positive regulation is a positive regulator. However, a **double negative** has the effect of acting as a positive. Thus, an effector that inhibits the function of a negative regulator will appear to bring about a positive regulation. Many regulated systems that appear to be induced are, in fact, derepressed at the molecular level. (See Chapter 11 for a description of these terms.)

BIOLOGIC SYSTEMS EXHIBIT THREE TYPES OF TEMPORAL RESPONSES TO A REGULATORY SIGNAL

These three responses are depicted diagrammatically in Figure 41–1 as rate of gene expression in temporal response to an inducing signal.

A **type A response** is characterized by an increased rate of gene expression that is **dependent** upon the continued presence of the inducing signal. When the inducing signal is removed, the rate of gene expression diminishes to its basal level, but the rate repeatedly increases in response to the reappearance of the specific signal. This type of response is commonly observed in many higher organisms after

Table 41–1. Effects of positive and negative regulation on gene expression.

	Rate of Gene Expression	
	Negative Regulation	Positive Regulation
Regulator present	Decreased	Increased
Regulator absent	Increased	Decreased

exposure to inducers such as steroid hormones (Chapter 44).

A **type B response** exhibits an increased rate of gene expression that is **transient** even in the continued presence of the regulatory signal. After the regulatory signal has terminated and the cell has been al-

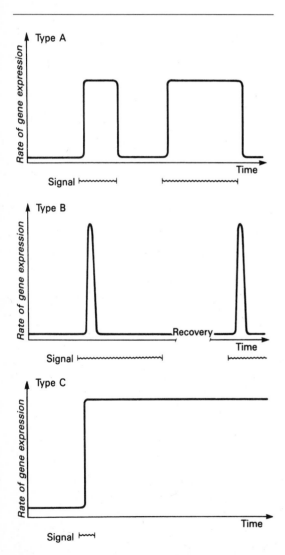

Figure 41–1. Diagrammatic representations of the responses of the rate of expression of a gene to specific regulatory signals such as a hormone.

lowed to recover, a second transient response to a subsequent regulatory signal may be observed. This phenomenon of response-desensitization-recovery characterizes the action of many pharmacologic agents, but it is also a feature of many naturally occurring processes. This type of response may commonly occur during development of an organism when only the transient appearance of a specific gene product is required although the signal persists.

The **type C response** pattern exhibits, in response to the regulatory signal, an increased rate of gene expression that persists **indefinitely** even after the termination of the signal. The signal acts as a trigger in this pattern. Once the gene expression is initiated in the cell, it cannot be terminated even in the daughter cells; it is therefore an irreversible and inherited alteration.

Prokaryotes Provide Models for the Study of Gene Expression in Mammalian Cells

In the last 20 years, with the understanding of how information flows from the gene through a messenger RNA to a specific protein molecule, there has developed sophisticated knowledge of the regulation of gene expression in prokaryotic cells. Most of the detailed knowledge about molecular mechanisms has been limited until recent years to prokaryotic and lower eukaryotic systems. This was due to the more advanced genetic analyses first available in these primitive organisms. Recent advances in recombinant DNA technology have allowed the sophisticated analysis of mammalian gene expression to begin. In this chapter, the initial discussion will center on prokaryotic systems. The impressive genetic studies will not be described, but rather what may be termed the physiology of gene expression will be discussed. However, nearly all of the conclusions about this physiology have been derived from genetic studies.

Before the physiology can be explained, a few specialized genetic terms must be defined for prokaryotic systems.

The **cistron** is the smallest unit of genetic expression. As described in Chapter 11, some enzymes and other protein molecules are composed of two or more nonidentical subunits. Thus, the "one gene, one enzyme" concept is now known to be not necessarily valid. The cistron is the genetic unit coding for the structure of the subunit of a protein molecule, acting as it does as the smallest unit of genetic expression. Thus, the one gene, one enzyme idea might more accurately be regarded as a **one cistron, one subunit concept.**

An **inducible gene** is a gene whose expression increases in response to an **inducer,** a specific regulatory signal.

The expression of some genes is **constitutive,** meaning that they are expressed at a reasonably constant rate and not known to be subject to regulation.

As the result of mutation, some inducible gene products become constitutively expressed. A mutation resulting in constitutive expression of what was formerly a regulated gene is called a constitutive mutation.

ANALYSIS OF LACTOSE METABOLISM IN *E COLI* LED TO THE OPERON HYPOTHESIS

Jacob and Monod in 1961 described their **operon** model in a classic paper. Their hypothesis was to a large extent based on observations on the regulation of lactose metabolism by the intestinal bacterium *E coli.* The molecular mechanisms responsible for the regulation of the genes involved in the metabolism of lactose are now among the best understood in any organism. β-Galactosidase hydrolyzes the β-galactoside lactose to galactose and glucose (Figure 41–2). The structural gene for β-galactosidase (the lac Z gene) is clustered with the genes responsible for the permeation of galactose into the cell (Y) and for galactoside acetylase (A), whose function is not understood. The structural genes for these three enzymes are physically associated to constitute the **lac operon** as depicted in Figure 41–3. This genetic arrangement of the structural genes and their regulatory genes allows for the **coordinate expression** of the three enzymes concerned with lactose metabolism. Each of these linked genes is transcribed into one large mRNA molecule that contains multiple, independent translation start (AUG) and stop (UAA) codons for each cistron. This type of mRNA molecule is called a **polycistronic mRNA.** Polycistronic mRNAs are predominantly found in prokaryotic organisms.

When *E coli* are presented with lactose or some specific lactose analogs, the expression of the activities of β-galactosidase, galactoside permease, and

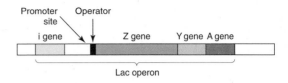

Figure 41–3. The positional relationships of the structural and regulatory genes of the lac operon. The Z gene encodes β-galactosidase, the Y gene encodes a permease, and the A gene encodes an acetylase. The i gene encodes the lac operon repressor protein.

galactoside acetylase is increased 10-fold to 100-fold. This is a type A response, as depicted in Figure 41–1. Upon removal of the signal, ie, the inducer, the rate of synthesis of these three enzymes declines. Since there is no significant degradation of these enzymes in bacteria, the level of β-galactosidase as well as that of the other two enzymes will remain the same unless they are diluted out by cell division.

When *E coli* are exposed to both lactose and glucose as sources of carbon, the organisms first metabolize the glucose and then temporarily cease growing until the genes of the lac operon become induced to provide the ability to metabolize lactose. Although lactose is present from the beginning of the bacterial growth phase, the cell does not induce those enzymes necessary for catabolism of lactose until the glucose has been exhausted. This phenomenon was first thought to be attributable to the repression of the lactose operon by some catabolite of glucose; hence, it was termed catabolite repression. It is now known that "catabolite repression" is in fact mediated by a **catabolite gene activator protein (CAP)** in conjunction with **cyclic AMP (cAMP)** (Figure 20–5). The expression of many inducible enzyme systems or operons in *E coli* and other prokaryotes is sensitive to catabolite repression, as discussed below.

The physiology of the induction of the lac operon

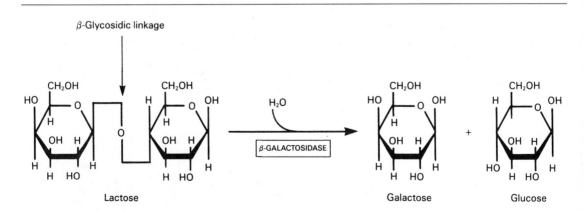

Figure 41–2. The hydrolysis of lactose to galactose and glucose by the enzyme β-galactosidase.

is well understood at the molecular level (Figure 41–4). The expression of the normal **i gene** of the lac operon is constitutive; it is expressed at a constant rate, resulting in the formation of the subunits of the **lac repressor.** Four identical subunits with molecular weights of 38,000 assemble into a lac repressor molecule. The repressor protein molecule, the product of the i gene, has a high affinity (K_d about 10^{-12} mol/L) for the operator locus. The **operator locus** is a region of double-stranded DNA 27 base pairs long with a twofold rotational symmetry (indicated by solid lines about the dotted axis) in a region that is 21 base pairs long, as shown below:

$$\begin{array}{c} \vdots \\ 5'\text{-}\underline{\text{AA}}\textbf{T}\ \overline{\textbf{TGTGAGC}}\ \textbf{G}\ \overline{\textbf{GATAACAA}}\underline{\text{TT}} \\ 3'\text{-}\underline{\text{TT}}\textbf{A}\ \overline{\textbf{ACACTCG}}\ \textbf{C}\ \overline{\textbf{CTATTGTT}}\underline{\text{AA}} \\ \vdots \end{array}$$

The minimum effective size of an operator for lac repressor binding is 17 base pairs (boldface letters in above sequence). At any one time, only two subunits of the repressors appear to bind to the operator, and within the 17-base-pair region at least one base of each base pair is involved in the lac repressor recognition and binding. The binding occurs mostly in the **major groove** without interrupting the base-paired, double helical nature of the operator DNA. The **operator locus** is between the **promoter site,** at which the DNA-dependent RNA polymerase attaches to commence transcription, and the transcription initiation site of the **Z gene,** the structural gene for β-galactosidase (Figure 41–3). When attached to the operator locus, the repressor molecule prevents the transcription of the operator locus as well as of the distal structural genes, Z, Y, and A. Thus, the repressor molecule is a **negative regulator;** in its pres-

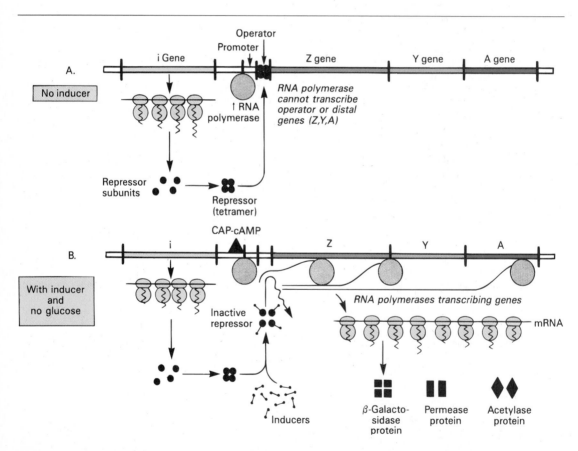

Figure 41–4. The mechanism of repression and derepression of the lactose operon. When no inducer is present **(A)**, the i gene products that are synthesized constitutively form a repressor molecule which binds at the operator locus to prevent the binding of RNA polymerase at the promoter locus and thus to prevent the subsequent transcription of the Z, Y, and A structural genes. When inducer is present **(B)**, the constitutively expressed i gene forms repressor molecules that are inactivated by the inducer and cannot bind to the operator locus. In the presence of cAMP and its binding protein (CAP), the RNA polymerase can transcribe the structural genes Z, Y, and A, and the polycistronic mRNA molecule formed can be translated into the corresponding protein molecules β-galactosidase, permease, and acetylase, allowing for the catabolism of lactose.

ence (and in the absence of inducer; see below), the expression of the Z, Y, and A genes is prevented. There are normally 20–40 repressor tetramer molecules and one operator locus per cell.

A lactose analog that is capable of inducing the lac operon while not itself serving as a substrate for β-galactosidase is an example of a **gratuitous inducer.** An example is isopropylthiogalactoside (IPTG). The addition of lactose or of a gratuitous inducer to bacteria growing on a poorly utilized carbon source (such as succinate) results in the prompt induction of the lac operon enzymes. Small amounts of the gratuitous inducer or of lactose are able to enter the cell even in the absence of permease. The repressor molecules, both those attached to the operator loci and those free in the cytosol, have a high affinity for the inducer. The binding of the inducer to a repressor molecule attached to the operator locus will induce a conformational change in the structure of the repressor and cause it to dissociate from the DNA. If DNA-dependent RNA polymerase has already attached to the coding strand at the promoter site, transcription will begin. The polymerase generates a polycistronic mRNA, the 5′ terminal of which is complementary to the template strand of the operator. In such a manner, an inducer derepresses the lac operon and allows the transcription of the structural genes for β-galactosidase, galactoside permease, and galactoside acetylase. The translation of the polycistronic mRNA can occur even before transcription is completed. Derepression of the lac operon allows the cell to synthesize the enzymes necessary to catabolize lactose as an energy source.

In order for the RNA polymerase to attach at the promoter site, there must also be present the catabolite gene activator protein (CAP) to which cAMP is bound. By an independent mechanism, the bacterium accumulates cAMP only when it is starved for a source of carbon. In the presence of glucose, or glycerol in concentrations sufficient for growth, the bacteria will lack sufficient cAMP to bind to CAP. Thus, in the presence of glucose or glycerol, cAMP-saturated CAP is lacking, so that the DNA-dependent RNA polymerase cannot initiate transcription of the lac operon. In the presence of the CAP-cAMP complex, which binds to DNA just upstream of the promoter site, transcription then occurs (Figure 41–4). Thus, the CAP-cAMP regulator is acting as a **positive regulator,** because its presence is required for gene expression. Hence, the lac operon is subject to both positive and negative regulation.

When the i gene has been mutated so that its product, the lac repressor, is not capable of binding to operator DNA, the organism will exhibit **constitutive expression** of the lac operon. In a contrary manner, an organism with an i gene mutation that prevents the binding of an inducer to the repressor will remain repressed even in the presence of the inducer molecule, because the inducer cannot bind to

the repressor on the operator locus in order to derepress the operon.

Bacteria harboring mutations in their operator locus such that the operator sequence will not bind a normal repressor molecule constitutively express the lac operon genes.

The Genetic Switch of Bacteriophage Lambda (λ) Provides a Paradigm for Protein-DNA Interactions in Eukaryotic Cells

Some bacteria harbor viruses that can reside in a dormant state within the bacterial chromosome or can replicate within the bacterium and eventually lead to lysis and killing of the bacterial host. Some *E coli* harbor such a "temperate" virus, bacteriophage lambda (λ). When a lambda infects a sensitive *E coli,* it injects its 45,000-base-pair, double-stranded, linear DNA genome into the cell (Figure 41–5). Depending upon the nutritional state of the cell, the lambda DNA will either **integrate** into the host genome **(lysogenic pathway)** and remain dormant until activated (see below), or it will commence **replicating** until it has made about 100 copies of complete, protein-packaged virus, at which point it effects lysis of its host **(lytic pathway).** The newly generated virus particles can then infect other sensitive hosts.

When integrated into the host genome in its dormant state, lambda will remain in such a state until activated by exposure of its lysogenic bacterial host to DNA-damaging agents. In response to such a noxious stimulus, the dormant bacteriophage becomes "induced" and begins to transcribe and subsequently translate those genes of its own genome which are necessary for its excision from the host chromosome, its DNA replication, and its protein coat and lysis enzymes. This event acts like a trigger or type C (Figure 41–1) response; ie, once lambda has committed itself to induction, there is no turning back until the cell is lysed and the replicated bacteriophage released. This **switch** from a dormant or **prophage state** to a **lytic infection** is well understood at the genetic and molecular levels and will be described in detail here.

The switching event in lambda is centered around an 80-base-pair region in its double-stranded DNA molecule referred to as the "right operator" (O_R) (Figure 41–6A). The **right operator** is flanked on its left side by the structural gene for the lambda repressor (cI) and on its right side by the structural gene for another regulatory protein called **cro.** When lambda is in its prophage state, ie, integrated into the host genome, the **repressor gene** is the *only* lambda gene that is expressed. When the bacteriophage is undergoing lytic growth, the repressor gene is not expressed, but the cro gene, as well as many other genes in lambda, is expressed. That is, **when the repressor gene is on, the cro gene is off, and when the cro gene is on, the repressor gene is off.** As we shall

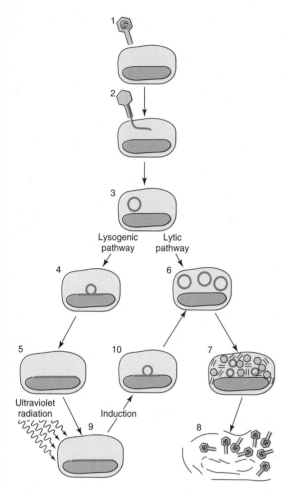

Figure 41–5. Infection of the bacterium *E coli* by phage lambda begins when a virus particle attaches itself to the bacterial cell (1) and injects its DNA (shaded line) into the cell (2, 3). Infection can take either of two courses depending on which of two sets of viral genes is turned on. In the lysogenic pathway, the viral DNA becomes integrated into the bacterial chromosome (4, 5), where it replicates passively as the bacterial cell divides. The dormant virus is called a prophage, and the cell that harbors it is called a lysogen. In the alternative lytic mode of infection, the viral DNA replicates itself (6) and directs the synthesis of viral proteins (7). About 100 new virus particles are formed. The proliferating viruses lyse, or burst, the cell (8). A prophage can be "induced" by an agent such as ultraviolet radiation (9). The inducing agent throws a switch, so that a different set of genes is turned on. Viral DNA loops out of the chromosome (10) and replicates; the virus proceeds along the lytic pathway. (Reproduced, with permission, from Ptashne M, Johnson AD, Pabo CO: A genetic switch in a bacterial virus. Sci Am [Nov] 1982;247:128.)

see, these two genes regulate each other's expression and thus, ultimately, the decision between lytic and lysogenic growth of lambda. **This decision between repressor gene transcription and cro gene transcription is an example of a molecular switch.**

The operator region can be subdivided into three discrete sites, each consisting of 17 base pairs of similar but not identical DNA sequence joined to one another (Figure 41–6B). Each of these three subregions, O_R1, O_R2, and O_R3, can bind either repressor or cro proteins predominantly through major groove contacts between repressor and the DNA double helix. The DNA region between the cro and repressor genes also contains two promoter sequences that direct the binding of RNA polymerase in a specified orientation, where it commences transcribing the adjacent genes. One promoter directs RNA polymerase to transcribe in the **rightward direction** and, thus, to transcribe cro and other distal genes, while the other promoter directs the transcription of the **repressor** gene in the **leftward direction** (Figure 41–6B).

The product of the repressor gene, the 236-amino-acid, 27 kDa **repressor protein,** exists as a **two-domain** molecule in which the **amino terminal domain binds to operator DNA** and the **carboxyl terminal domain promotes the association** of one repressor protein with another to form a dimer. A **dimer** of repressor molecules binds to **operator DNA** much more tightly than does the monomeric form (Figure 41–7A to 41–7C).

The product of the cro gene, the 66-amino-acid, 9 kDa **cro protein,** has a single domain but also binds the operator DNA more tightly as a **dimer** (Figure 41–7D). Obviously, the cro protein's single domain mediates both operator binding and dimerization.

In a lysogenic bacterium, ie, a bacterium containing a lambda prophage, the lambda repressor dimer binds **preferentially to O_R1** but in so doing, by a cooperative interaction, **enhances** the binding (by a factor of 10) of another repressor dimer to O_R2 (Figure 41–8). The affinity of repressor for O_R3 is the least of the three operator subregions. The binding of repressor to O_R1 has two major effects. The occupation of O_R1 by repressor **blocks the binding of RNA polymerase to the rightward promoter** and thereby prevents the expression of the cro gene. Second, as mentioned above, repressor dimer bound to O_R1 enhances the binding of repressor dimer to O_R2. The binding of repressor to O_R2 has the important added effect of **enhancing the binding of RNA polymerase to the leftward promoter** that overlaps O_R2 and thereby enhances the transcription and subsequent expression of the repressor gene. This enhancement of transcription is apparently mediated through direct protein-protein interactions between promoter-bound RNA polymerase and O_R2-bound repressor. Thus, the lambda repressor is both a **negative regulator,** by preventing transcription of the cro gene, and a **positive regulator,** by enhancing the

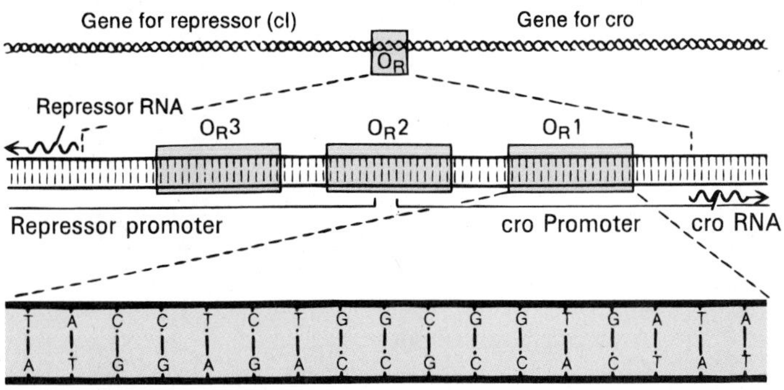

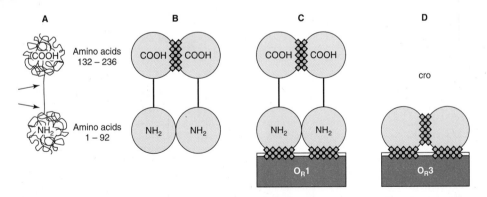

Figure 41–6. Right operator (O_R) is shown in increasing detail in this series of drawings. The operator is a region of the viral DNA some 80 base pairs long **(A).** To its left lies the gene encoding lambda repressor (cI), to its right the gene (cro) encoding the regulator protein cro. When the operator region is enlarged **(B)**, it is seen to include three subregions, O_R1, O_R2, and O_R3, each 17 base pairs long. They are recognition sites to which both repressor and cro can bind. The recognition sites overlap two promoters: sequences of bases to which the enzyme RNA polymerase binds in order to transcribe a gene into mRNA (wavy lines), which is translated into protein. Site O_R1 is enlarged **(C)** to show its base sequence. Note that in this region of the λ chromosome, both strands of DNA act as a template for transcription (Chapter 39). (Reproduced, with permission, from Ptashne M, Johnson AD, Pabo CO: A genetic switch in a bacterial virus. Sci Am [Nov] 1982;247:128.)

transcription of its own gene, the repressor gene. This dual effect of repressor is responsible for the stable state of the dormant lambda bacteriophage; not only does the repressor prevent the expression of the genes necessary for lysis, but it also promotes the expression of itself to stabilize this state of differentiation. In the event that the repressor protein concentration becomes very high, repressor can bind to O_R3 and by so doing diminish the transcription of the repressor gene from the leftward promoter, until the re-

pressor concentration drops and repressor dissociates itself from O_R3.

When a DNA-damaging signal, such as ultraviolet light, strikes the lysogenic host bacterium, fragments of single-stranded DNA are generated that activate a specific **protease** coded by a bacterial gene and referred to as **recA** (Figure 41–8). The activated recA protease hydrolyzes the portion of the repressor protein that connects the amino-terminal and carboxyterminal domains of that molecule. Such cleavage of

Figure 41–7. Schematic molecular structures of cI (lambda repressor, shown in **A, B,** and **C**) and cro. Lambda repressor protein is a polypeptide chain 236 amino acids long. The chain folds itself into a dumbbell shape with two substructures: an amino terminal (NH_2) domain and a carboxyl terminal (COOH) domain. The two domains are linked by a region of the chain that is susceptible to cleavage by proteases (indicated by the two arrows in **A**). Single repressor molecules (monomers) tend to associate to form dimers **(B)**; a dimer can dissociate to form monomers again. A dimer is held together mainly by contact between the carboxyl terminal domains (hatching). Repressor dimers bind to (and can fall off) the recognition sites in the operator region; their greatest affinity is for site O_R1 **(C)**. It is the amino terminal domain of the repressor molecule that makes contact with the DNA (hatching). Cro **(D)** has a single domain with sites that promote dimerization and other sites that promote binding of dimers to operator, preferentially to O_R3. (Reproduced, with permission, from Ptashne M, Johnson AD, Pabo CO: A genetic switch in a bacterial virus. Sci Am [Nov] 1982;247:128.)

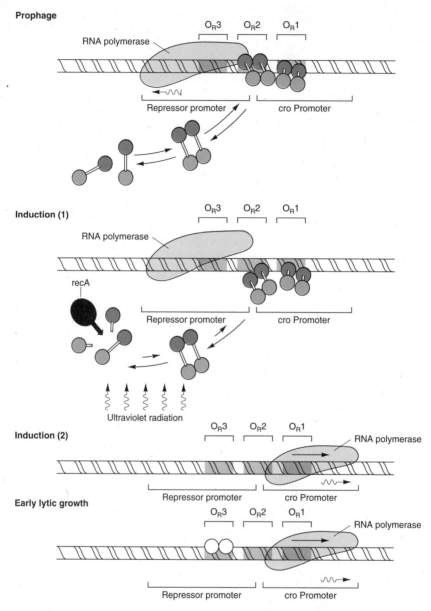

Figure 41–8. Configuration of the switch is shown at four stages of lambda's life cycle. The lysogenic pathway (in which the virus remains dormant as a prophage) is selected when a repressor dimer binds to O_R1, thereby making it likely that O_R2 will be filled immediately by another dimer. In the prophage (top), the repressor dimers bound at O_R1 and O_R2 prevent RNA polymerase from binding to the rightward promoter and so block the synthesis of cro (negative control). The repressors also enhance the binding of polymerase to the leftward promoter (positive control), with the result that the repressor gene is transcribed into RNA (wavy line) and more repressor is synthesized, maintaining the lysogenic state. The prophage is induced when ultraviolet radiation activates the protease recA, which cleaves repressor monomers. The equilibrium of free monomers, free dimers, and bound dimers is thereby shifted, and dimers leave the operator sites. Polymerase is no longer encouraged to bind to the leftward promoter, so that repressor is no longer synthesized. As induction proceeds, all the operator sites become vacant, and so polymerase can bind to the rightward promoter and cro is synthesized. During early lytic growth, a single cro dimer binds to O_R3, the site for which it has the highest affinity. Now polymerase cannot bind to the leftward promoter, but the rightward promoter remains accessible. Polymerase continues to bind there, transcribing cro and other early lytic genes. Lytic growth ensues. (Reproduced, with permission, from Ptashne M, Johnson AD, Pabo CO: A genetic switch in a bacterial virus. Sci Am [Nov] 1982;247:128.)

the repressor domains causes the **repressor dimers to dissociate,** which in turn causes a **dissociation of the repressor molecules from** O_R2 and eventually from O_R1. The effects of removal of repressor from O_R1 and O_R2 are predictable. RNA polymerase immediately has access to the rightward promoter and commences transcribing the **cro gene,** and the enhancement effect of the repressor at O_R2 on leftward transcription is lost (Figure 41–8).

The cro protein translated from the newly transcribed cro gene also binds to the operator region as dimers, but its order of preference is the opposite of that of repressor (Figure 41–8). That is, **cro binds most tightly to** O_R3, but there is no cooperative effect of cro at O_R3 on the binding of cro to O_R2. At increasingly higher concentrations of cro, the protein will bind to O_R2 and eventually to O_R1.

The occupancy of O_R3 by cro immediately turns off the transcription from the leftward promoter and, hence, **prevents any further expression of the repressor gene.** Thereby the switch is completely effected: the cro gene is now expressed, and the repressor gene is fully turned off. This event is irreversible, and the expression of other lambda genes begins as part of the lytic cycle. When cro repressor concentration becomes quite high, it will eventually occupy O_R1 and in so doing turn down the expression of its own gene, a process that is necessary in order to effect the final stages of the lytic cycle.

The three-dimensional structure of the cro protein and that of the lambda repressor protein have been determined by x-ray crystallography, and models for their binding and effecting the above-described molecular and genetic events have been proposed and tested. Both bind to DNA by the helix-turn-helix mechanism (see below). To date, this system provides the best understanding of the molecular events involved in gene regulation.

GENE REGULATION IN PROKARYOTES AND EUKARYOTES DIFFERS IN IMPORTANT RESPECTS

In addition to transcription, eukaryotic cells employ a variety of mechanisms to regulate gene expression (Table 41–2). The nuclear membrane of eukaryotic cells physically segregates gene transcrip-

Table 41–2. Gene expression is regulated by transcription and in numerous other ways in eukaryotic cells.

Other Methods of Gene Regulation
• Gene amplification
• Gene rearrangement
• RNA processing
• Alternate mRNA splicing
• Transport of mRNA from nucleus to cytoplasm
• Regulation of mRNA stability

tion from translation, since ribosomes exist only in the cytoplasm. Many more steps, especially in RNA processing, are involved in the expression of eukaryotic genes than of prokaryotic genes, and these steps provide additional sites for regulatory influences that cannot exist in prokaryotes. These RNA processing steps in eurkaryotes include capping of the $5'$ end of the primary transcript, addition of a polyadenylate tail to the $3'$ end of transcripts, and excision of intron regions to generate spliced exons in the mature mRNA molecule. To date, analyses of eukaryotic gene expression provide evidence that regulation occurs at the level of **transcription, nuclear RNA processing,** and **mRNA stability.** In addition, **gene amplification** and **rearrangement** have been shown to occur and to influence gene expression.

Owing to the advent of recombinant DNA technology, much progress has been made in recent years in the understanding of eukaryotic gene expression. However, because most eukaryotic organisms contain so much more genetic information than do prokaryotes and the manipulation of their genes is so much more limited, molecular aspects of eukaryotic gene regulation are less well understood than the examples discussed earlier in this chapter. This section briefly describes a few different types of eukaryotic gene regulation.

Eukaryotic Genes Can Be Amplified During Development or in Response to Drugs

During early development of metazoans, there is an abrupt increase in the need for specific molecules such as ribosomal RNA and messenger RNA molecules for proteins that make up such organs as the eggshell. One way to increase the rate at which such molecules can be formed is to increase the number of genes available for transcription of these specific molecules. Among the repetitive DNA sequences are hundreds of copies of ribosomal RNA genes and tRNA genes. These genes preexist repetitively in the genomic material of the gametes and, thus, are transmitted in high copy number from generation to generation. In some specific organisms such as the fruit fly *(Drosophila),* there occurs during oogenesis an amplification of a few preexisting genes, such as those for the chorion (eggshell) proteins. Subsequently, these amplified genes, presumably generated by a process of repeated initiations during DNA synthesis, provide multiple sites for gene transcription (Figures 38–15 and 41–9).

In recent years, it has been possible to promote the amplification of specific genetic regions in cultured mammalian cells. In some cases, a several thousand-fold increase in the copy number of specific genes can be achieved over a period of time involving increasing doses of selective drugs. In fact, it has been demonstrated in patients receiving methotrexate for treatment of cancer that malignant cells can develop

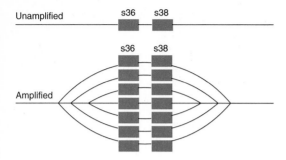

Figure 41–9. Schematic representation of the amplification of chorion protein genes s36 and s38. (Reproduced, with permission, from Chisholm R: Gene amplification during development. Trends Biochem Sci 1982;7:161.)

drug resistance by increasing the number of genes for dihydrofolate reductase, the target of methotrexate. Gene amplification events such as these occur spontaneously in vivo, ie, in the absence of exogenously supplied selective agents, and these unscheduled extra rounds of replication can become "frozen" in the genome under appropriate selective pressures.

The Formation of Active Immunoglobulin Genes Involves Selective DNA Rearrangement

Some of the most interesting and perplexing questions raised by biologists in recent decades concern the genetic and molecular basis of antibody diversity (Chapter 59). In addition, advances in immunology have made it apparent that as cells of the humoral immunity system differentiate, they produce antibodies with the same specificity but different effector functions. Within the last several years, many laboratories have contributed greatly to the understanding of the genetic basis of antibody diversity and regulation of the expression of immunoglobulin genes during development and differentiation.

As described in Chapter 39, the coding segments responsible for the generation of specific protein molecules are frequently not contiguous in the mammalian genome. The coding segments for the variable and the constant domains of the immunoglobulin (antibody) light chain were the first recognized to be separated in the genome. As described in more detail in Chapter 59, immunoglobulin molecules are composed of two types of polypeptide chains, light (L) and heavy (H) chains (Figure 59–8). The L and H chains are each divided into amino terminal variable (V) and carboxyl terminal constant (C) regions. The V regions are responsible for the recognition of antigens (foreign molecules) and the constant regions for effector functions that determine how the antibody molecule will dispense with the antigen.

There are three unlinked families of genes responsible for immunoglobulin molecule structure. Two families are responsible for the light chains (λ and κ chains) and one family for heavy chains.

Each **light chain** is encoded by three distinct segments: the variable (V_L), the joining (J_L), and the constant (C_L) segments. For λ and κ light chains, there are approximately 300 variable (V_L) segments each and five or six J_L segments. Lambda light chains are derived from fewer than ten C_L regions, whereas κ light chains come from a single C_L segment. During the differentiation of a lymphoid B cell, a V_L segment is brought from a distant site on the same chromosome to a position closer to the region of the genome containing the J_L and C_L segments. This **DNA rearrangement** then allows the V_L, J_L, and C_L segments to be transcribed as a single mRNA precursor and subsequently processed to generate the mRNA for a specific antibody light chain. By rearrangement of the various V_L, J_L, and C_L segments in the genome, the immune system can generate an immensely diverse library (millions) of antigen-specific immunoglobulin molecules. This DNA rearrangement is referred to as **V-J joining** of the light chain and is illustrated in Figure 41–10.

The **heavy chain** is encoded by four gene segments: the V_H, the D (diversity), the J_H, and the C_H DNA segments. In humans, there are about 1000 V_H segments, 12 or more D segments, and four J segments. The variable region of the heavy chain is generated by joining the V_H with a D and a J_H segment. The resulting V_H-D-J_H DNA region is in turn linked to a C_H gene, of which there are nine. These C_H genes (C_μ, C_δ, $C_\gamma3$, $C_\gamma1$, $C_{\alpha1}$, $C_\gamma2$, $C_\gamma4$, C_ε, and $C_{\alpha2}$) determine the immunoglobulin class or subclass—IgM, IgG, IgA, etc—of the immunoglobulin molecule (Chapter 59). Given the larger number of V, D, and J segments and the fact that there are nine constant regions rather than one, the recombination possibilities are far greater for heavy chains than for light chains.

Alternative RNA Processing Is Another Control Mechanism

In addition to affecting the efficiency of promoter utilization, eukaryotic cells employ alternative RNA processing to control gene expression. This can result when alternative promoters, intron-exon splice sites, or polyadenylation sites are used. Occasionally, heterogeneity within a cell results, but more commonly the same primary transcript is processed differently in different tissues. A few examples of each of these types of regulation are presented below.

The use of alternate **transcription start sites** results in a different 5′ exon on mRNAs corresponding to mouse amylase and myosin light chain, rat glucokinase, and *Drosophila* alcohol dehydrogenase and actin. **Alternative polyadenylation sites** in the μ immunoglobulin heavy-chain primary transcript result in mRNAs that are either 2700 bases long (μ_m) or 2400 bases long (μ_s). This results in a different carboxyl terminal region of the encoded proteins such

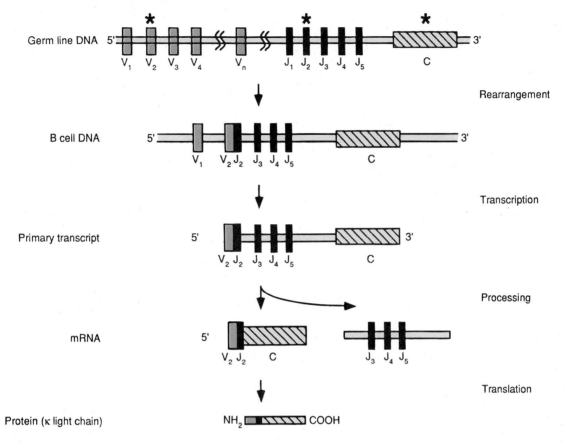

Figure 41–10. Recombination events leading to a V_2J_2 κ light chain. The 500 V_L (variable), 5–6 J_L (joining), and single C_L (constant) segments exist in a linear array in the chromosome. These components span thousands of kilobase pairs. Upon receipt of a signal to differentiate, the DNA in the B cell undergoes rearrangement so that, in this case, the 3' end of V_2 is ligated to the 5'end of J_2. Transcription initiates through a promoter at the 5' end of V_2 and proceeds through C. In order to produce a V_2J_2 C mRNA, the sequence from the 3' end of J_2 to the 5' end of C is excised, then V_2J_2 is spliced to C. This mRNA can then be translated into a κ light chain that contains the amino acid sequence encoded by V_2J_2 and C. The same general procedure can result in κ chains consisting of any combination of V and J segments with C.

that the μ_m protein remains attached to the membrane of the β lymphocyte and the μ_s immunoglobulin is secreted. **Alternative splicing and processing** results in the formation of seven unique α-tropomyosin mRNAs in seven different tissues. It is not clear how these processing-splicing decisions are made or whether these steps can be regulated.

Regulation of Messenger RNA Stability Provides Another Control Mechanism

The stability of messenger RNA molecules in the cytoplasm can clearly affect the level of gene expression in a positive or negative direction. Stabilization of mRNA, given a fixed rate of transcription, would lead to increased accumulation, and vice versa. The mechanisms involved in the regulation of mRNA stability were discussed in Chapter 39. Each of these is

a potential control site that can be influenced by hormones and other effectors. Some hormones influence the synthesis and degradation of specific mRNAs. For example, estradiol prolongs the half-life of vitellogenin mRNA from a few hours to more than 200 hours. This, coupled with the fact that estrogens enhance the rate of transcription of this gene by four- to sixfold, results in a tremendous increase of vitellogenin mRNA.

DIFFERENTIAL EFFECTS ON CHROMATIN STRUCTURE MAY BE INVOLVED IN DEVELOPMENT AND DIFFERENTIATION

Most of the DNA in prokaryotic cells is organized into genes, and the templates can always be transcribed. A very different situation exists in mammalian cells. Here relatively little of the total DNA is organized into genes and their associated regulatory regions. The function of the extra DNA is unknown (this is one of the reasons there is such interest in sequencing the entire human genome).

Chromatin structure provides an additional level of control. As discussed in Chapter 38, large regions of chromatin are transcriptionally inactive, while others are either active or potentially active. With few exceptions, each cell contains the same complement of genes (antibody-producing cells are a notable exception). The development of specialized organs, tissues, and cells and their function in the intact organism depend upon the differential expression of genes.

Some of this differential expression is accomplished by having different regions of chromatin available for transcription in cells from various tissues. For example, the DNA containing the β-globin gene cluster is in **"active" chromatin** in the reticulocyte but is in **"inactive" chromatin** in a muscle cell. The mechanisms that determine "active" versus "inactive" chromatin are not known, but protein-DNA interactions are probably involved.

Additionally, as described in Chapter 38, there is evidence that the **methylation** of deoxycytidine residues (in the sequence $5'$-^{m}CpG-$3'$) in DNA may effect gross changes in chromatin so as to preclude its active transcription. For example, in mouse liver only the unmethylated ribosomal genes can be expressed, and there is evidence that many animal viruses are not transcribed when their DNA is methylated. However, it is *not* possible to generalize that methylated DNA is transcriptionally inactive, that all inactive chromatin is methylated, or that active DNA is not methylated.

Eukaryotic DNA that is in an "active" region of chromatin can be transcribed. As in prokaryotic cells, a **promoter** dictates where the RNA polymerase will initiate transcription, but this promoter cannot be neatly defined as a −35 and −10 box, particularly in mammalian cells (Chapter 39). In addition, the *trans*-acting factors generally come from other chromosomes (so act in *trans*), whereas this consideration is moot in the case of the single chromosome-containing prokaryotic cells. Additional complexity is added by elements or factors that **enhance** or **silence** transcription, define tissue-specific expression, and modulate the actions of many effector molecules.

CERTAIN DNA ELEMENTS ENHANCE OR SILENCE TRANSCRIPTION OF EUKARYOTIC GENES

In addition to gross changes in chromatin affecting transcriptional activity, there is increasing evidence that certain DNA elements facilitate or enhance initiation at the promoter. For example, in simian virus 40 (SV40) there exists about 200 bp upstream from the promoter of the early genes a region of two identical, tandem 72-bp lengths that can greatly increase the expression of genes in vivo. Each of these 72-bp elements can be subdivided into a series of smaller elements; hence, some enhancers have a very complex structure. Enhancer elements differ from the promoter in two remarkable ways. They can exert their positive influence on transcription even when separated by thousands of base pairs from a promoter, they work when oriented in either direction, and they can work upstream ($5'$) or downstream ($3'$) from the promoter. Enhancers are promiscuous; they can stimulate any promoter in the vicinity, and may act on more than one promoter. The SV40 enhancer element can exert an influence on, for example, the transcription of β-globin by increasing its transcription 200-fold in cells containing both the enhancer and the β-globin gene on the same plasmid (see below and Figure 41–11). The enhancer element does not seem to be producing a product that in turn acts on the promoter, since it is active only when it exists within the same DNA molecule as (ie, *cis* to) the promoter. Enhancer binding proteins are now being isolated, and these should help elucidate how these elements work. Enhancer elements do appear to convey nuclease hypersensitivity to those regions where they reside (Chapter 38). A summary of the properties of enhancers is given in Table 41–3.

The *cis*-acting elements that decrease or **silence** the expression of specific genes have also been identified. Fewer of these elements have been studied, so it is not possible to state generalizations about their mechanism of action.

Tissue-Specific Expression May Result From the Action of Enhancers or Silencers

Many genes have now been recognized to harbor enhancer elements in various locations relative to their coding regions. In addition to being able to enhance gene transcription, some of these enhancer elements clearly possess the ability to do so in a tissue-specific manner. Thus, the enhancer element associated with the immunoglobulin genes between the J and C regions enhances the expression of those genes preferentially in lymphoid cells. Enhancer elements associated with the genes for pancreatic enzymes are capable of enhancing even unrelated but physically linked genes preferentially in the pancreatic cells of mice into which the specifically engineered gene constructions were introduced microsurgically at the single-cell embryo stage. This **transgenic animal** approach has proved useful in studying tissue-specific gene expression. For example, DNA containing a pancreatic β-cell tissue-specific enhancer (from the insulin gene), when ligated in a vector to polyoma large-T antigen, produced β-cell tumors in transgenic mice. Tumors did not de-

Response element   Promoter   Structural gene

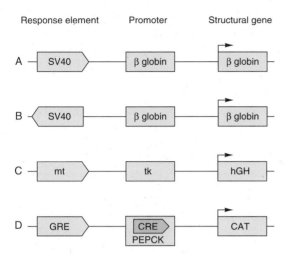

Table 41–3. Summary of the properties of enhancers.

Properties of Enhancers
• Work when located long distances from the promoter
• Work when upstream or downstream from the promoter
• Work when oriented in either direction
• Work through heterologous promoters
• Work by binding one or more proteins

on their expression. Pieces of DNA thought to harbor regulatory elements are ligated to a suitable reporter gene and introduced into a host cell (Figure 41–11). Basal expression of the reporter gene will be increased if the DNA contains an enhancer. Addition of a hormone or heavy metal to the culture medium will increase expression of the reporter gene if the DNA contains a hormone or metal response element

Figure 41–11. A schematic explanation of the action of enhancers and other *cis*-acting regulatory elements. This model chimeric gene consists of a reporter (structural) gene that encodes a protein which can be readily assayed, a promoter that ensures initiation of transcription, and the putative regulatory elements. Examples **A** and **B** illustrate the fact that enhancers (eg, SV40) work in either orientation, and upon a heterologous promoter. Example **C** illustrates that the metallothionein (mt) regulatory element (which under the influence of cadmium or zinc induces transcription of the endogenous mt gene and hence the metal-binding mt protein) will work through the thymidine kinase (tk) promoter to enhance transcription of the human growth hormone (hGH) gene. The engineered genetic constructions were introduced into the male pronuclei of single-cell mouse embryos and the embryos placed into the uterus of a surrogate mother to develop as transgenic animals. Offspring have been generated under these conditions, and in some the addition of zinc ions to their drinking water effects an increase in liver growth hormone. In this case, these transgenic animals have responded to the high levels of growth hormone by becoming twice as large as their normal litter mates. Example **D** illustrates that a glucocorticoid response element (GRE) will work through homologous (PEPCK gene) or heterologous promoters and that the PEPCK gene promoter also contains an element which functions as both a basal level enhancer and a cAMP response element (CRE).

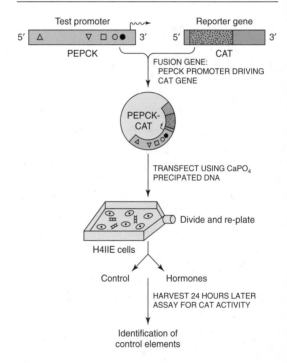

Figure 41–12. The use of fusion genes to define DNA regulatory elements. A DNA fragment from the phosphoenolpyruvate carboxykinase (PEPCK) gene promoter thought to contain one or more regulatory elements is ligated into a plasmid vector that contains a suitable reporter gene, the bacterial enzyme chloramphenicol transferase (CAT). CAT is not present in mammalian cells; hence, detection of this activity in a cell extract means that the cell was successfully transfected by the plasmid. An increase of CAT activity over the basal level, eg, after addition of a glucocorticoid hormone, means that the region of DNA inserted contains a functional glucocorticoid hormone response element (GRE). Progressively shorter pieces of DNA, regions with internal deletions, or regions with point mutations can be constructed and inserted to pinpoint the response element.

velop in any other tissue. Tissue-specific gene expression may therefore be mediated by enhancers or enhancer-like elements.

Fusion Genes Are Used to Define Enhancers and Other Regulatory Elements

By ligating regions of DNA suspected of harboring regulatory sequences to various reporter genes (the **fusion** or **chimeric gene approach**) (Figures 41–11 and 41–12), one can determine which regions in the vicinity of structural genes have an influence

(Figure 41–12). The location of the element can be pinpointed by using progressively shorter pieces of DNA, deletions, or point mutations (Figure 41–13).

This strategy, **using transfected cells in culture and transgenic animals,** has led to the identification of dozens of enhancers, silencers, tissue-specific elements; hormone-, metal-, drug-response elements; etc. The activity of a gene at any moment reflects the interaction of these numerous *cis*-acting DNA elements with their respective *trans*-acting factors. The challenge is to figure out how this occurs.

Transcription Domains Can Be Defined by Locus Control Regions and Insulators

The large number of genes in eukaryotic cells and the complex arrays of transcription regulatory factors presents an organizational problem. Why are some genes available for transcription in a given cell whereas others are not? If enhancers can regulate several genes and are not position- and orientation-dependent, how are they prevented from triggering transcription randomly? Part of the solution to these problems is arrived at by having the chromatin arranged in functional units that restrict patterns of gene expression. This is achieved by having the chromatin form a structure with the nuclear matrix or other physical entities. Alternatively, some regions are controlled by complex DNA elements called **locus control regions (LCRs).** An LCR—with associated bound proteins—controls the expression of a cluster of genes. The best-defined LCR regulates expression of the globin gene family over a large region of DNA. Another mechanism is provided by **insulators.** These DNA elements, also in association with one or more proteins, prevent an enhancer from acting on a promoter on the other side of an insulator, in another transcription domain.

SEVERAL MOTIFS MEDIATE THE BINDING OF REGULATORY PROTEINS TO DNA

The specificity involved in the control of transcription requires that regulatory proteins bind, with high affinity, to the correct region of DNA. Three unique motifs, the **helix-turn-helix,** the **zinc finger,** and the **leucine zipper,** account for many of these specific protein-DNA interactions. Examples of proteins containing these motifs are given in Table 41–4.

A comparison of the binding activities of the proteins that contain these motifs leads to several important generalizations.

(1) Binding must be of high affinity to the specific site and of low affinity to other DNA.

(2) Small regions of the protein make direct contact with DNA; the rest of the protein may ensure the proper information of the DNA recognition site or be involved in the dimerization of monomers of the binding protein.

(3) The protein-DNA interactions are maintained by hydrogen bonds and van der Waals forces.

(4) The motifs found in these proteins are unique; their presence in a protein of unknown function suggests that the protein may bind to DNA.

(5) Proteins with the helix-turn-helix or leucine zipper motifs form symmetric dimers, and their respective DNA binding sites are symmetric palindromes. In proteins with the zinc finger motif, the binding site is repeated two to nine times. These fea-

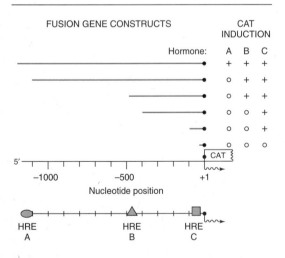

FUSION GENE CONSTRUCTS — CAT INDUCTION

Figure 41–13. Location of hormone response elements (HREs) **A, B,** and **C** using the fusion gene-transfection approach. A fusion gene, constructed as described in Figure 41–12, can be transfected into a recipient cell. By analyzing when certain hormone responses are lost in comparison to the 5′ deletion, specific hormone-responsive elements can be located.

Table 41–4. Examples of transcription regulatory proteins that contain the various binding motifs.

Binding Motif	Organism	Regulatory Protein
Helix-turn-helix	E coli	lac repressor CAP
	Phage	cro, λ, tryptophan, and 434 repressors
	Mammals	homeo box proteins Pit-1, Oct1, Oct2
Zinc finger	E coli	Gene 32 protein
	Yeast	Gal4
	Drosophila	Serendipity, Hunchback
	Xenopus	TFIIIA
	Mammals	steroid receptor family, Sn1
Leucine zipper	Yeast	GCN4
	Mammals	C/EBP, fos, Jun, Fra-1, CRE binding protein, c-*myc*, n-*myc*, l-*myc*

tures allow for cooperative interactions between binding sites and enhance the degree and affinity of binding.

The Helix-Turn-Helix Motif

The first motif described, and the one studied most extensively, is the helix-turn-helix. Analysis of the three-dimensional structure of cro has revealed that each monomer consists of three antiparallel β sheets and three α-helices (Figure 41–14). The dimer forms by the association of the antiparallel β$_3$ sheets. The α$_3$ helices form the DNA recognition surface, and the rest of the molecule appears to be involved in stabilizing these structures. The average diameter of an α-helix is 1.2 nm, which is the approximate width of the major groove in the B form of DNA. The DNA recognition domain of each cro monomer interacts with 5 bp and the dimer binding sites span 3.4 nm, allowing fit into successive half turns of the major groove on the same surface (Figure 41–14). X-ray analyses of λ repressor, CRP (the cAMP receptor protein of *E coli*), tryptophan repressor, and phage 434 repressor confirm this dimeric helix-turn-helix structure.

The Zinc Finger Motif

The zinc finger was the second DNA binding motif elucidated. It was known that the protein TFIIIA, which is a positive regulator of 5S RNA transcription, required zinc for activity. Structural and biophysical analyses revealed that each TFIIIA molecule contains nine zinc ions in a repeating coordination complex formed by closely spaced cysteine-cysteine residues followed 12–13 amino acids later by a histidine-histidine pair (Figure 41–15). In some instances, notably the steroid-thyroid receptor family, the His-His doublet is replaced by a second Cys-Cys pair. The protein containing zinc fingers appears to lie on one face of the DNA helix, with successive fingers alternatively positioned in one turn in the major groove. As in the case with the recognition domain in the helix-turn-helix protein, each TFIIIA zinc finger contacts about 5 bp of DNA. The importance of this motif in the action of steroid hormones is underscored by an "experiment of nature." A single amino acid mutation in either of the two zinc fingers of the calcitriol receptor protein results in resistance to the action of this hormone and the clinical syndrome of rickets (Chapter 47).

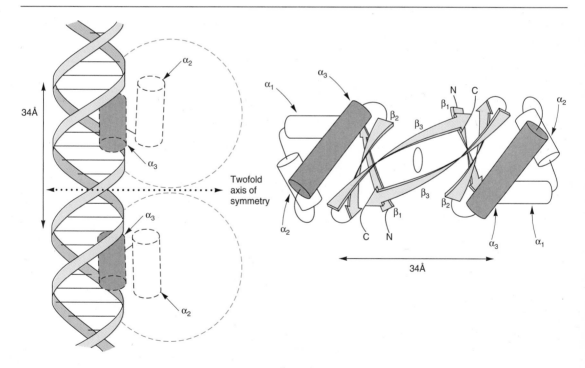

Figure 41–14. A schematic representation of the three-dimensional structure of cro protein and its binding to DNA by the helix-turn-helix motif. The cro monomer consists of three antiparallel β sheets (β$_1$–β$_3$) and three α-helices (α$_1$–α$_3$). The helix-turn-helix motif is formed because the α$_3$ and α$_2$ helices are held at 90 degrees to each other by a turn of four amino acids. The α$_3$ helix of cro is the DNA recognition surface (shaded). Two monomers associate through the antiparallel β$_3$ sheets to form a dimer that has a twofold axis of symmetry (right). A cro dimer binds to DNA through its α$_3$ helices, each of which contacts about 5 bp on the same surface of the major groove. The distance between comparable points on the two DNA α-helices is 34 Å, which is the distance required for one complete turn of the double helix. (Courtesy of B Mathews.)

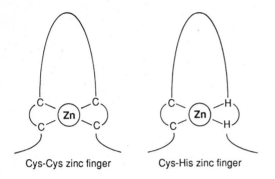

Figure 41–15. Zinc fingers are a series of repeated domains (two to nine) in which each is centered on a tetrahedral coordination with zinc. In the case of TFIIIA, the coordination is provided by a pair of cysteine residues (C) separated by 12–13 amino acids from a pair of histidine (H) residues. In other zinc finger proteins, the second pair also consists of C residues. Zinc fingers bind in the major groove, with adjacent fingers making contact with 5 bp along the same face of the helix.

The Leucine Zipper Motif

Careful analysis of a 30-amino acid sequence in the carboxyl terminal region of the enhancer binding protein C/EBP revealed a novel structure. As illustrated in Figure 41–16, this region of the protein forms an α-helix in which there is a periodic repeat of leucine residues at every seventh position. This occurs for eight helical turns and four leucine repeats. Similar structures have been found in a number of other proteins associated with the regulation of transcription in mammalian and yeast cells. It is thought that this structure allows two identical monomers or heterodimers (eg, Fos•Jun or Jun•Jun) to "zip together" in a coiled coil and form a tight dimeric complex (Figure 41–16). This protein-protein interaction may serve to enhance the association of the separate DNA binding domains with their target (Figure 41–16).

The DNA Binding and *Trans-Activation Domains of These Regulatory Proteins Are Separate and Noninteractive

DNA binding could result in a general conformational change that allows the bound protein to activate transcription, or these two functions could be

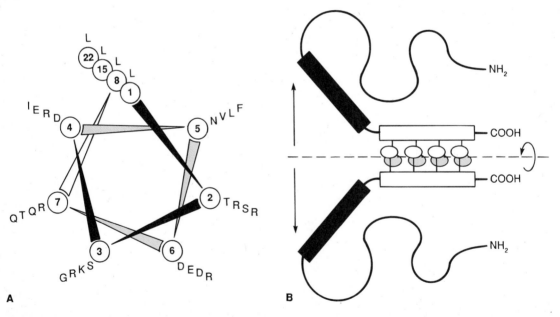

A

B

Figure 41–16. The leucine zipper motif. *A* shows a helical wheel analysis of a carboxyl terminal portion of the DNA binding protein C/EBP. The amino acid sequence is displayed end-to-end down the axis of a schematic α-helix. The helical wheel consists of seven spokes that correspond to the seven amino acids that comprise every two turns of the α-helix. Note that leucine residues (L) occur at every seventh position. Other proteins with "leucine zippers" have a similar helical wheel pattern. *B* is a schematic model of the DNA binding domain of C/EBP. Two identical C/EBP polypeptide chains are held in dimer formation by the leucine zipper domain of each polypeptide (denoted by the rectangles and attached ovals). This association is apparently required to hold the DNA binding domains of each polypeptide (the shaded rectangles) in the proper conformation for DNA binding. (Courtesy of S McKnight.)

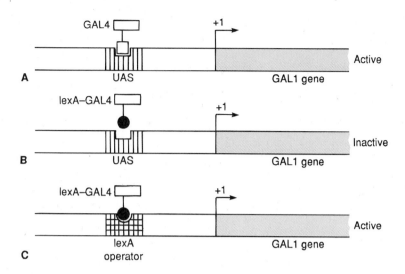

Figure 41–17. Domain-swap experiments show the separateness of DNA binding and transcription activation. The Gal1 gene promoter contains an upstream activating sequence (UAS) that binds the regulatory protein Gal4 *(A)*. This interaction results in a stimulation of Gal1 gene transcription. A fusion protein, in which the amino terminal DNA binding domain of Gal4 is removed and replaced with the DNA binding region of the *E coli* protein lexA, fails to stimulate Gal1 transcription because the lexA domain cannot bind to the UAS *(B)*. The lexA–Gal4 fusion protein does increase Gal1 transcription when the lexA operator (its natural target) is inserted into the Gal1 promoter region *(C)*.

served by separate and independent domains. Domain-swap experiments suggest the latter is the case.

The Gal1 gene product is involved in galactose metabolism in yeast. This gene is positively regulated by the Gal4 protein, which binds to an upstream activator sequence (UAS) through an amino-terminal domain. The 73-amino-acid DNA binding terminus of Gal4 was removed and replaced with the DNA binding domain of lexA in *E coli* protein. This resulted in a molecule that did not bind to the Gal1 UAS and, of course, did not activate the Gal1 gene (Figure 41–17). If, however, the lexA operator was inserted into the promoter region of the Gal gene, the hybrid protein bound to this promoter (at the lexA operator), and it activated transcription of Gal1. This experiment, which has been repeated a number of times (including lexA-glucocorticoid receptor fusion proteins that *trans*-activate glucocorticoid-responsive genes), afford solid evidence that the carboxyl terminal region of Gal4 causes transcriptional activation. The DNA binding and *trans*-activation domains appear to be independent and noninteractive. The carboxyl terminal regions appear to have dense concentrations of negatively charged amino acids. These domains, often referred to as "acid blobs" or "negative noodles," presumably interact with positively charged regions of some component of the transcription complex.

SUMMARY

The genetic information in each somatic cell is practically identical. The distinction between a brain, muscle, or liver cell depends on the pattern of genes expressed in these cells, so-called tissue-specific expression. The ability of an organism to respond to environmental challenges depends on being able to regulate, positively or negatively, those genes that are expressed. Both purposes are accomplished through the interaction of specific proteins with particular regions of DNA called promoters. Promoters are generally located 5′ from, and adjacent to, the coding segment of a gene. Much has been learned about how proteins contact DNA, and a number of classes of such interaction have been described. These include helix-turn-helix, zinc finger, and leucine zipper motifs. Protein-protein interactions, particularly between members of the leucine zipper family, are also important. Proteins that regulate gene transcription must have at least two domains: a recognition domain for binding to DNA or to another protein and a domain that affects some aspect of the transcription apparatus. These domains are distinct, as shown by experiments in which hybrid molecules are tested. Extensive information is available on how gene regulation is accomplished in *E coli* and in the lambda bacteriophage. These models are being applied to the much more complex process of gene regulation in eukaryotic cells, which remains an enigma.

REFERENCES

Breitbart RE, Andreadis A, Nadal-Ginard B: Alternative splicing: A ubiquitous mechanism for the generation of multiple protein isoforms from single genes. Annu Rev Biochem 1987;56:467.

Chung JH, Whiteley M, Felsenfeld G: A 5′ element of the chicken β-globin domain serves as an insulator in human erythroid cells and protects against position effect in *Drosophila.* Cell 1993;74:505.

Jacob F, Monod J: Genetic regulatory mechanisms in protein synthesis. J Mol Biol 1961;3:318.

Johnson PF, McKnight SL: Eukaryotic transcriptional regulatory proteins. Annu Rev Biochem 1989;58:799.

Landschulz WH, Johnson PF, McKnight SL: The leucine zipper: A hypothetical structure common to a new class of DNA binding proteins. Science 1989;240:1759.

Lieber MR: The mechanism of V(D)J recombination: A balance of diversity, specificity and stability. Cell 1992;70:873.

Ptashne M: *A Genetic Switch,* 2nd ed. Cell Press and Blackwell Scientific Publications, 1992.

Pugh BF, Tjian R: Diverse transcriptional functions of the multisubunit eukaryotic TFIID complex. J Biol Chem 1992;267:679.

Struhl K: Promoters, activator proteins and the mechanism of transcriptional initiation in yeast. Cell 1987;49:295.

Tjian R: Molecular machines that control genes. Sci Am (Feb) 1995;272:54.

Wu R, Bahl CP, Narang SA: Lactose operator-repressor interaction. Curr Top Cell Regul 1978;13:137.

Recombinant DNA Technology

42

Daryl K. Granner, MD

INTRODUCTION

Recombinant DNA technology has revolutionized biology and is having an ever-increasing impact on clinical medicine. Much has been learned about human genetic disease from pedigree analysis and study of affected proteins, but in many cases where the specific genetic defect is unknown, these approaches cannot be used. The new technology circumvents these limitations by going directly to the DNA molecule for information. Manipulation of a DNA sequence and the construction of chimeric molecules, so-called genetic engineering, provides a means of studying how a specific segment of DNA works.

This chapter is aimed at clarifying this rather complex topic. It presents the basic concepts of recombinant DNA technology, its applications to clinical medicine, and a glossary. In order for the chapter to be complete in itself, some repetition of subjects discussed in other chapters will be found.

BIOMEDICAL IMPORTANCE

Understanding recombinant DNA technology is important for several reasons. (1) The information explosion occurring in this area is truly staggering. To understand and keep up with this field, one must have an appreciation of the fundamental concepts involved. (2) It offers a rational approach to understanding the molecular basis of a number of diseases (eg, familial hypercholesterolemia, sickle cell disease, the thalassemias, cystic fibrosis, muscular dystrophy). (3) Using recombinant DNA technology, human proteins can be produced in abundance for therapy (eg, insulin, growth hormone, plasminogen activator). (4) Proteins for vaccines (eg, hepatitis B) and for diagnostic tests (eg, AIDS test) can be obtained. (5) Recombinant DNA technology is used to diagnose existing diseases and predict the risk of developing a given disease. (6) Special techniques have led to remarkable advances in forensic medicine. (7) Gene therapy for sickle cell disease, the thalassemias, adenosine deaminase deficiency, and other diseases may be devised.

ELUCIDATION OF THE BASIC FEATURES OF DNA LED TO RECOMBINANT DNA TECHNOLOGY

DNA Is a Complex Biopolymer Organized as a Double Helix

The fundamental organizational element is the sequence of purine (adenine [A] or guanine [G]) and pyrimidine (cytosine [C] or thymine [T]) bases. These bases are attached to the C-1′ position of the sugar deoxyribose, and the bases are linked together through joining of the sugar moieties at their 3′ and 5′ positions via a phosphodiester bond (Figure 37–1). The alternating deoxyribose and phosphate groups form the backbone of the double helix (Figure 37–2). These 3′–5′ linkages also define the orientation of a given strand of the DNA molecule, and since the two strands run in opposite directions, they are said to be antiparallel.

Base Pairing Is a Fundamental Concept of DNA Structure and Function

Adenine and thymine always pair, by hydrogen bonding, as do guanine and cytosine (Figure 37–3). These base pairs are said to be **complementary**, and the guanine content of a fragment of double-stranded DNA will always equal its cytosine content; likewise the thymine and adenine contents are equal. Base pairing and hydrophobic base-stacking interactions hold the two DNA strands together. These interactions can be reduced by heating the DNA to denature it. The laws of base pairing predict that two complementary DNA strands will reanneal exactly in register upon renaturation, as happens when the temperature of the solution is slowly reduced to normal. Indeed, the degree of base-pair matching (or mismatching) can be estimated from the temperature required for denaturation-renaturation. Segments of DNA with high degrees of base-pair matching require more energy input (heat) to accomplish denaturation, or, to put it another way, a closely matched segment will withstand more heat before the strands separate. This reaction is used to determine whether there are significant differences between two DNA sequences, and it underlies the concept of **hybridization,** which is fundamental to the processes described below.

There are about 3×10^9 base pairs (bp) in each human haploid genome. If an average gene length is 3×10^3 bp (3 kilobases [kb]), the genome could consist of 10^6 genes, assuming that there is no overlap and that transcription proceeds in only one direction. It is thought that there are only about 10^5 genes in the human and that only 10% of the DNA codes for proteins. The function of the remaining 90% of the human genome has not yet been defined.

The double-helical DNA is packaged into a more compact structure by a number of proteins, most notably the basic proteins called histones. This condensation may serve a regulatory role and certainly has a practical purpose. The DNA present within the nucleus of a cell, if simply extended, would be about a meter long. The chromosomal proteins compact this long length of DNA so that it can be packaged into a nucleus with a volume of a few cubic microns.

DNA Is Organized Into Genes

In general, prokaryotic genes consist of a small regulatory region (100–500 bp) and a large protein-coding segment (500–10,000 bp). Several genes are often controlled by a single regulatory unit. Most mammalian genes are more complicated, in that the coding regions are interrupted by noncoding regions that are eliminated when the primary RNA transcript is processed into mature **messenger RNA (mRNA).** The **coding regions** (those regions that appear in the mature RNA species) are called **exons,** and the **noncoding regions,** which interpose or intervene between the exons, are called **introns** (Figure 42–1). Introns are always removed from precursor RNA before transport into the cytoplasm occurs. The process by which introns are removed from precursor RNA and by which exons are ligated together is called **RNA splicing.** Incorrect processing of the primary transcript into the mature mRNA can result in disease in humans (see below); this underscores the importance of these posttranscriptional processing steps. The variation in size and complexity of some human genes is illustrated in Table 42–1. Although there is a 300-fold difference in the sizes of the genes illustrated, the mRNA sizes vary only about 20-fold. This is because most of the DNA in genes is present as introns, and introns tend to be much larger than exons. Regulatory regions for specific eukaryotic genes are usually located in the DNA that flanks the transcription initiation site at its 5′ end (**5′ flanking-sequence DNA).** Occasionally, such sequences are found within the gene itself or in the region that flanks the 3′ end of the gene. In mammalian cells, each gene has its own regulatory region. Many eukaryotic genes (and some viruses that replicate in mammalian cells) have special regions, called **enhancers,** that increase the rate of transcription. Some genes also have DNA sequences, known as **silencers,** that diminish transcription. Mammalian genes are obviously complicated, multicomponent structures.

Genes Are Transcribed Into RNA

Information generally flows from DNA to mRNA to protein, as illustrated in Figure 42–1 and discussed in more detail in Chapter 41. This is a rigidly controlled process involving a number of complex steps, each of which no doubt is regulated by one or more enzymes or factors; faulty function at any of these steps can cause disease.

RECOMBINANT DNA TECHNOLOGY INVOLVES ISOLATION AND MANIPULATION OF DNA TO MAKE CHIMERIC MOLECULES

Isolation and manipulation of DNA, including end-to-end joining of sequences from very different sources to make chimeric molecules (eg, molecules containing both human and bacterial DNA sequences in a sequence-independent fashion), is the essence of recombinant DNA research. This involves several unique techniques and reagents.

Restriction Enzymes Cut DNA Chains at Specific Locations

Certain endonucleases, enzymes that cut DNA at specific DNA sequences within the molecule (as opposed to exonucleases, which digest from the ends of DNA molecules), are a key tool in recombinant DNA research. These enzymes were originally called **restriction enzymes** because their presence in a given bacterium restricted the growth of certain bacterial viruses called bacteriophages. Restriction enzymes cut DNA of any source into short pieces in a sequence-specific manner, in contrast to most other enzymatic, chemical, or physical methods, which break DNA randomly. These defensive enzymes (hundreds have been discovered) protect the host bacterial DNA from DNA from foreign organisms (primarily infective phages). However, they are only present in cells that also have a companion enzyme that methylates the host DNA, rendering it an unsuitable substrate for digestion by the restriction enzyme. Thus, **site-specific DNA methylases** and restriction enzymes always exist in pairs in a bacterium.

Restriction enzymes are named after the bacterium from which they are isolated. For example, *EcoRI* is from *Escherichia coli,* and *BamHI* is from *Bacillus amyloliquefaciens* (Table 42–2). The first three letters in the restriction enzyme name consist of the first letter of the genus (E) and the first two letters of the species (co). These may be followed by a strain designation (R) and a roman numeral (I) to indicate the order of discovery (eg, *EcoRI, EcoRII*).

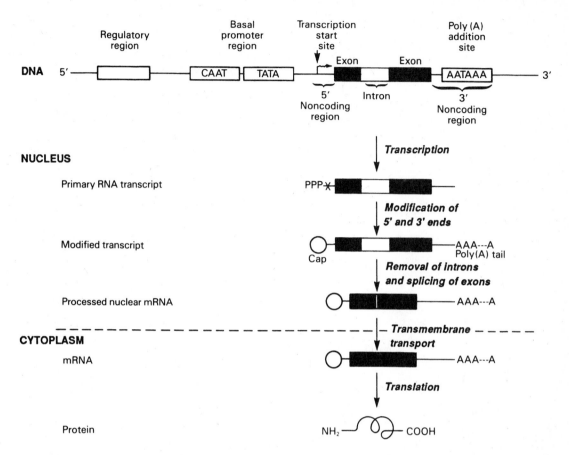

Figure 42–1. Organization of a eukaryotic transcription unit and the pathway of eukaryotic gene expression. Eukaryotic genes have structural and regulatory regions. The structural region consists of the coding DNA and 5′ and 3′ noncoding DNA sequences. The coding regions are divided into two parts: (1) exons, which eventually are ligated together to become mature RNA, and (2) introns, which are processed out of the primary transcript. The structural region is bounded at its 5′ end by the transcription initiation site and at its 3′ end by the polyadenylate addition or termination site. The promoter region, which contains specific DNA sequences that interact with various protein factors to regulate transcription, is discussed in detail in Chapters 39 and 41. The primary transcript has a special structure, a cap, at the 5′ end and a stretch of As at the 3′ end. This transcript is processed to remove the introns; and the mature mRNA is then transported to the cytoplasm, where it is translated into protein.

Each enzyme recognizes and cleaves a specific double-stranded DNA sequence that is 4–7 bp long. These DNA cuts result in **blunt ends** *(HpaI)* or overlapping **(sticky) ends** *(BamHI)* (Figure 42–2), depending on the mechanism used by the enzyme. Sticky ends are particularly useful in constructing hybrid or chimeric DNA molecules (see below). If the nucleotides are distributed randomly in a given DNA molecule, one can calculate how frequently a given enzyme would cut a length of DNA. For each position in the DNA molecule there are four possibilities (A, C, G, and T); therefore, a restriction enzyme that recognizes a 4-bp sequence will cut, on average, once every 256 bp (4^4), whereas another enzyme that recognizes a 6-bp sequence will cut once every 4096 bp (4^6). A given piece of DNA will have a characteristic linear array of sites for the various enzymes; hence, a **restriction map** can be constructed. When DNA is digested with a given enzyme, the ends of all the fragments will have the same DNA sequence. The fragments produced can be isolated by electrophoresis on agarose or polyacrylamide (see the discussion of blot transfer, below); this is an essential step in cloning and a major use of these enzymes.

A number of other enzymes that act on DNA and RNA are an important part of recombinant DNA technology. Many of these are referred to in this and subsequent chapters (Table 42–3).

Restriction Enzymes and DNA Ligase Are Used to Prepare Chimeric DNA Molecules

Sticky-end ligation is technically easy, but some special techniques are often required to overcome

Table 42–1. Variations in the size and complexity of some human genes and mRNAs.[1]

Gene	Gene Size (kb)	Number of Introns	mRNA Size (kb)
β-Globin	1.5	2	0.6
Insulin	1.7	2	0.4
β-Adrenergic receptor	3	0	2.2
Albumin	25	14	2.1
LDL receptor	45	17	5.5
Factor VIII	186	25	9.0
Thyroglobulin	300	36	8.7

[1]The sizes are given in kilobases (kb). The sizes of the genes include some proximal promoter and regulatory region sequences; these are generally about the same size for all genes. Genes vary in size from about 1500 base pairs (bp) to over 2×10^6 bp. There is also great variation in the number of introns and exons. The β-adrenergic receptor gene is intronless, and the thyroglobulin gene has 36 introns. As noted by the smaller difference in mRNA sizes, introns comprise most of the gene sequence.

Table 42–2. Selected restriction endonucleases and their sequence specificities.[1]

Endonuclease	Sequence Cleaved	Bacterial Source
BamHI	↓ G G A T C C C C T A G G ↑	Bacillus amyloliquefaciens H
BglII	↓ A G A T C T T C T A G A ↑	Bacillus globigii
EcoRI	↓ G A A T T C C T T A A G ↑	Escherichia coli RY13
EcoRII	↓ C C T G G G G A C C ↑	Escherichia coli R245
HindIII	↓ A A G C T T T T C G A A ↑	Haemophilus influenzae R_d
HhaI	↓ G C G C C G C G ↑	Haemophilus haemolyticus
HpaI	↓ G T T A A C C A A T T G ↑	Haemophilus parainfluenzae
MstII	↓ C C T N A G G G G A N T C C ↑	Microcoleus strain
PstI	↓ C T G C A G G A C G T C ↑	Providencia stuartii 164
TaqI	↓ T C G A A G C T ↑	Thermus aquaticus YTI

[1]A, adenine; C, cytosine; G, guanine; T, thymine. Arrows show the site of cleavage; depending on the site, sticky ends (BamHI) or blunt ends (HpaI) may result. The length of the recognition sequence can be 4 bp (TaqI), 5 bp (EcoRII), 6 bp (EcoRI), or 7 bp (MstII). By convention, these are written in the 5′ to 3′ direction for the upper strand of each recognition sequence, and the lower strand is shown with the opposite (ie, 3′ to 5′) polarity. Note that most recognition sequences are palindromes (ie, the sequence reads the same in opposite directions on the 2 strands). A residue designated N means that any nucleotide is permitted.

problems inherent in this approach. Sticky ends of a vector may reconnect with themselves, with no net gain of DNA. Sticky ends of fragments can also anneal, so that tandem heterogeneous inserts form. Also, sticky-end sites may not be available or in a convenient position. To circumvent these problems, an enzyme that generates blunt ends is used, and new ends are added using the enzyme terminal transferase. If poly d(G) is added to the 3′ ends of the vector and poly d(C) is added to the 3′ ends of the foreign DNA, the two molecules can only anneal to each other, thus circumventing the problems listed above. This procedure, called **homopolymer tailing,** also generates an *SmaI* restriction site, and so it is easy to retrieve the fragment. Sometimes, synthetic oligonucleotide linkers with a convenient restriction enzyme sequence are ligated to the blunt-ended DNA. Direct blunt-end ligation is accomplished using the enzyme bacteriophage T4 DNA ligase. This technique, though more difficult than sticky-end ligation, has the advantage of joining together any pairs of ends. The disadvantages are that there is no control over the orientation of insertion or the number of molecules annealed together, and there is no easy way of retrieving the insert.

Cloning Amplifies DNA

A clone is a large population of identical molecules, bacteria, or cells that arise from a common ancestor. Cloning allows for the production of a large number of identical DNA molecules, which can then be characterized or used for other purposes. This technique is based on the fact that chimeric or hybrid DNA molecules can be constructed in **cloning vectors,** typically bacterial plasmids, phages, or cosmids, which then continue to replicate in a host cell under their own control systems. In this way, the chimeric DNA is amplified. The general procedure is illustrated in Figure 42–3.

Bacterial **plasmids** are small, circular, duplex DNA molecules whose natural function is to confer antibiotic resistance to the host cell. Plasmids have several properties that make them extremely useful as cloning vectors. They exist as single or multiple copies within the bacterium and replicate independently from the bacterial DNA. The complete DNA

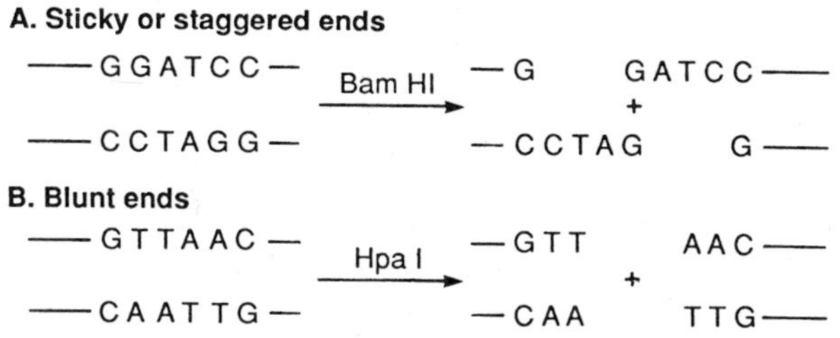

A. Sticky or staggered ends

```
——GGATCC—              — G          GATCC——
              Bam HI
                                       +
——CCTAGG—              — CCTAG       G——
```

B. Blunt ends

```
——GTTAAC—              —GTT         AAC——
              Hpa I
                                       +
——CAATTG—              —CAA         TTG——
```

Figure 42–2. Results of restriction endonuclease digestion. Digestion with a restriction endonuclease can result in the formation of DNA fragments with sticky, or cohesive, ends *(A)* or blunt ends *(B)*. This is an important consideration in devising cloning strategies.

sequence of many plasmids is known; hence, the precise location of restriction enzyme cleavage sites for inserting the foreign DNA is available. Plasmids are smaller than the host chromosome and are therefore easily separated from the latter, and the desired DNA is readily removed by cutting the plasmid with the enzyme specific for the restriction site into which the original piece of DNA was inserted.

Phages usually have linear DNA molecules into which foreign DNA can be inserted at several restriction enzyme sites. The chimeric DNA is collected after the phage proceeds through its lytic cycle and produces mature, infective phage particles. A major advantage of phage vectors is that while plasmids accept DNA pieces about 6–10 kb long, phages can accept DNA fragments 10–20 kb long, a limitation imposed by the amount of DNA that can be packed into the phage head.

Even larger fragments of DNA can be cloned in **cosmids,** which combine the best features of plasmids and phages. Cosmids are plasmids that contain the DNA sequences, so-called **cos sites,** required for packaging lambda DNA into the phage particle.

These vectors grow in the plasmid form in bacteria, but since much of the unnecessary lambda DNA has been removed, more chimeric DNA can be packaged into the particle head. It is not unusual for cosmids to carry inserts of chimeric DNA that are 35–50 kb long. A comparison of these vectors is shown in Table 42–4.

Insertion of DNA into a functional region of the vector will interfere with the action of this region, and so care must be taken not to interrupt an essential function of the vector. This concept can be exploited, however, to provide a selection technique. The common plasmid vector **pBR322** has both **tetracycline (tet)** and **ampicillin (amp) resistance genes.** A single *PstI* site within the amp resistance gene is commonly used as the insertion site for a piece of foreign DNA. In addition to having sticky ends (Table 42–2 and Figure 42–2), the DNA inserted at this site disrupts the amp resistance gene and makes the bacterium carrying this plasmid amp-sensitive (Figure 42–4). Thus, the parental plasmid, which provides resistance to both antibiotics, can be readily separated from the chimeric plasmid, which is resistant only to

Table 42–3. Enzymes used in recombinant DNA research.[1]

Enzyme	Reaction	Primary Use
Alkaline phosphatase	Dephosphorylates 5′ ends of RNA and DNA.	Removal of 5′-PO_4 groups prior to kinase labeling to prevent self-ligation.
BAL 31 nuclease	Degrades both the 3′ and 5′ ends of DNA.	Progressive shortening of DNA molecules.
DNA ligase	Catalyzes bonds between DNA molecules.	Joining of DNA molecules.
DNA polymerase I	Synthesizes double-stranded DNA from single-stranded DNA.	Synthesis of double-stranded cDNA; nick translation.
DNase I	Under appropriate conditions, produces single-stranded nicks in DNA.	Nick translation; mapping of hypersensitive sites.
Exonuclease III	Removes nucleotides from 3′ ends of DNA.	DNA sequencing; mapping of DNA-protein interactions.
λ Exonuclease	Removes nucleotides from 5′ ends of DNA.	DNA sequencing.
Polynucleotide kinase	Transfers terminal phosphate (γ position) from ATP to 5′-OH groups of DNA or RNA.	^{32}P labeling of DNA or RNA.
Reverse transcriptase	Synthesizes DNA from RNA template.	Synthesis of cDNA from mRNA; RNA (5′ end) mapping studies.
SI nuclease	Degrades single-stranded DNA.	Removal of "hairpin" in synthesis of cDNA; RNA mapping studies (both 5′ and 3′ ends).
Terminal transferase	Adds nucleotides to the 3′ ends of DNA.	Homopolymer tailing.

[1]Adapted and reproduced, with permission, from Emery AEH: Page 41 in: *An Introduction to Recombinant DNA.* Wiley, 1984.

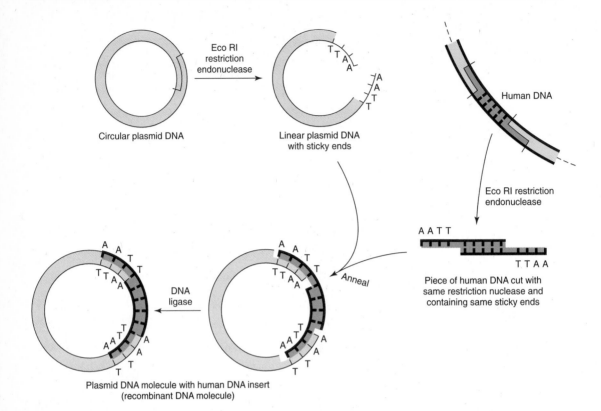

Figure 42–3. Use of restriction nucleases to make new recombinant or chimeric DNA molecules. When inserted back into a bacterial cell (by the process called transformation), the plasmid DNA replicates not only itself but also the physically linked new DNA insert. Since recombining the sticky ends, as indicated, regenerates the same DNA sequence recognized by the original restriction enzyme, the cloned DNA insert can be cleanly cut back out of the recombinant plasmid circle with this endonuclease. If a mixture of all of the DNA pieces created by treatment of total human DNA with a single restriction nuclease is used as the source of human DNA, a million or so different types of recombinant DNA molecules can be obtained, each pure in its own bacterial clone. (Modified and reproduced, with permission, from Cohen SN: The manipulation of genes. Sci Am [July] 1975;233:34.)

tetracycline. Additional confirmation that insertion has taken place comes from sizing the plasmid DNA obtained from the putative recombinant on an agarose gel, since the chimeric DNA molecule is now larger than the host vector DNA.

A Library Is a Collection of Recombinant Clones

The combination of restriction enzymes and various cloning vectors allows the entire genome of an organism to be packed into a vector. A collection of these different recombinant clones is called a library. A **genomic library** is prepared from the total DNA

of a cell line or tissue. A **cDNA library** represents the population of mRNAs in a tissue. Genomic libraries are prepared by performing partial digestion of total DNA with a restriction enzyme that cuts DNA frequently (eg, *SauIIIA*). The idea is to generate rather large fragments, so that most genes will be left intact. Phage vectors are preferred for these libraries because they accept large pieces of DNA (up to 20 kb). The goal is to achieve a complete library. The number of fragments required to attain this objective is inversely related to fragment size and directly related to genome size (Table 42–5). A human library that contains 10^6 recombinant fragments of large size has a 99% probability of being complete. Thus, the chances of finding any single-copy gene are excellent.

cDNA libraries are prepared by first isolating all the mRNAs in a tissue and then copying these molecules into double-stranded DNA, using (sequentially) the enzymes reverse transcriptase and DNA polymerase. For technical reasons, full-length cDNA

Table 42–4. Common cloning vectors.

Vector	DNA Insert Size
Plasmid pBR322	0.01–10 kb
Lambda charon 4A	10–20 kb
Cosmids	35–50 kb

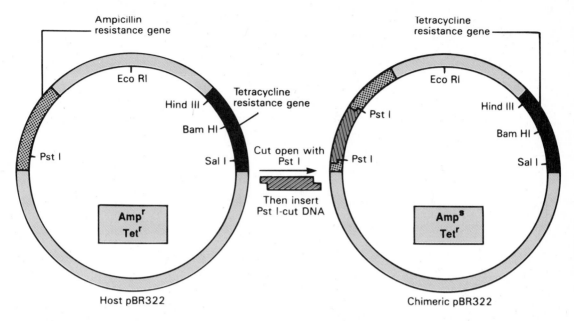

Figure 42–4. A method of screening recombinants for inserted DNA fragments. Using the plasmid pBR322, a piece of DNA is inserted into the unique *PstI* site. This insertion disrupts the gene coding for a protein that provides ampicillin resistance to the host bacterium. Hence, the chimeric plasmid will no longer survive when plated on a substrate medium that contains this antibiotic. The differential sensitivity to tetracycline and ampicillin can therefore be used to distinguish clones of plasmid that contain an insert.

copies are seldom obtained, and so smaller DNA fragments are cloned. Plasmids are often the favored vectors for cDNA libraries because they are much more convenient to work with than are phages or cosmids, although various lambda phage vectors have special advantages for cDNA cloning (see below).

A vector in which the protein coded by the gene introduced by recombinant DNA technology is actually synthesized is known as an **expression vector.** Such vectors are now commonly used to detect specific cDNA molecules in libraries and to produce proteins by genetic engineering techniques. These vectors are specially constructed to contain very active inducible promoters, proper in-phase translation initiation codons, both transcription and translation termination signals, and appropriate protein processing signals, if needed. Some expression vectors even contain genes that code for protease inhibitors, so that the final yield of product is enhanced. The vector λgt11 is popular for library construction because it accepts large cDNA molecules that are replicated and translated into proteins; thus, λgt11 recombinant libraries can be screened with either cDNA or antibody probes.

Probes Search Libraries for Specific Genes or cDNA Molecules

A variety of molecules can be used to "probe" libraries in search of a specific gene or cDNA mole-

cule or to define and quantitate DNA or RNA separated by electrophoresis through various gels. Probes are generally pieces of DNA or RNA labeled with a ^{32}P-containing nucleotide. The probe must recognize a complementary sequence to be effective. A cDNA synthesized from a specific mRNA can be used to screen either a cDNA library for a longer cDNA or a genomic library for a complementary sequence in the coding region of a gene. A popular technique for finding specific genes entails taking a short amino acid sequence and, using the codon usage for that species (see Chapter 40), making an oligonucleotide probe that will detect the corresponding DNA fragment in a genomic library. If the sequences match exactly, probes 15–20 nucleotides long will hybridize. cDNA probes are used to detect DNA fragments on Southern blot transfers and to detect and quantitate RNA on Northern blot transfers. Specific antibodies can also be used as probes, provided that the vector used synthesizes protein molecules that are recognized by them.

Blotting and Hybridization Techniques Allow Visualization of Specific Fragments

Visualization of a specific DNA or RNA fragment among the many thousands of "contaminating" molecules requires the convergence of a number of techniques, which are collectively termed **blot transfer.** Figure 42–5 illustrates the **Southern** (DNA), **North-**

Table 42–5. The composition of complete genomic libraries.[1]

Source	Complete Genomic Library
E coli	1500 fragments
Yeast	4500 fragments
Drosophila	50,000 fragments
Mammals	800,000 fragments

[1]The number of random fragments (unique clones) that a library should have to ensure that any single gene is represented is inversely related to the average fragment size used to construct the library and directly related to the number of genes in the organism. The numbers given above represent the number of fragments (independent clones) necessary to achieve a 99% probability of finding a given DNA sequence in a recombinant DNA library with an average insert size of 2×10^4 nucleotides. The differences represent the variations in genomic complexity between the creatures.

The number of clones necessary is calculated from the following formula:

$$N = \frac{\ln(1-P)}{\ln(1-f)}$$

where P is the probability desired and f is the fraction of the total genome in a single clone. In the case of the mammalian genomic library cited above, given the presence of 3×10^9 nucleotides in the haploid genome, the equation is as follows:

$$N = \frac{\ln(1 - 0.99)}{\ln\left(1 - \left[\dfrac{2 \times 10^4}{3 \times 10^9}\right]\right)}$$

The advantage of having a library composed of large DNA inserts is immediately apparent if this equation is solved using an average fragment size of 5×10^3 nucleotides rather than 2×10^4.

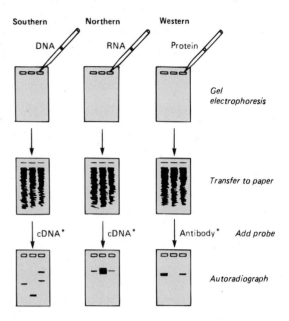

Gel electrophoresis

Transfer to paper

Add probe

Autoradiograph

Figure 42–5. The blot transfer procedure. In a Southern, or DNA, blot transfer, DNA isolated from a cell line or tissue is digested with one or more restriction enzymes. This mixture is pipetted into a well in an agarose or polyacrylamide gel and exposed to a direct electrical current. DNA, being negatively charged, migrates toward the cathode; the smaller fragments move the most rapidly. After a suitable time, the DNA is denatured by exposure to mild alkali and transferred to nitrocellulose paper, in an exact replica of the pattern on the gel, by the blotting technique devised by Southern. The DNA is annealed to the paper by exposure to heat, and the paper is then exposed to the labeled cDNA probe, which hybridizes to complementary fragments on the filter. After thorough washing, the paper is exposed to x-ray film, which is developed to reveal several specific bands corresponding to the DNA fragment that recognized the sequences in the cDNA probe. The RNA, or Northern, blot is conceptually similar. RNA is subjected to electrophoresis before blot transfer. This requires some different steps from those of DNA transfer, primarily to ensure that the RNA remains intact, and is generally somewhat more difficult. In the protein, or Western, blot, proteins are electrophoresed and transferred to nitrocellulose and then probed with a specific antibody or other probe molecule.

ern (RNA), and **Western** (protein) blot transfer procedures. (The first is named for the person who devised the technique, and the other names began as laboratory jargon but are now accepted terms.) These procedures are useful in determining how many copies of a gene are in a given tissue or whether there are any gross alterations in a gene (deletions, insertions, or rearrangements). Occasionally, if a specific base is changed and a restriction site is altered, these procedures can detect a point mutation. The Northern and Western blot transfer techniques are used to size and quantitate specific RNA and protein molecules, respectively.

Colony or **plaque hybridization** is the method by which specific clones are identified and purified. Bacteria are grown on colonies on an agar plate and overlaid with a nitrocellulose filter paper. Cells from each colony stick to the filter and are permanently fixed thereto by heat, which with NaOH treatment also lyses the cells and denatures the DNA so that it will hybridize with the probe. A radioactive probe is added to the filter, and after washing, the hybrid complex is localized by exposing the filter to x-ray film. By matching the spot on the autoradiograph to a colony, the latter can be picked from the plate. A

similar strategy is used to identify fragments in phage libraries. Successive rounds of this procedure result in a clonal isolate (bacterial colony) or individual phage plaque.

All of the hybridization procedures discussed in this section depend on the specific base-pairing properties of complementary nucleic acid strands described above. Perfect matches hybridize readily and withstand high temperatures in the hybridization and washing reactions. These complexes also form in the presence of low salt concentrations. Less than perfect matches do not tolerate these **stringent conditions** (ie, elevated temperatures and low salt concentrations); thus, hybridization either never occurs or is disrupted during the washing step. Gene families, in which there is some degree of homology, can be detected by varying the stringency of the hybridization and washing steps. Cross-species comparisons of a given gene can also be made using this approach.

Manual and Automatic Techniques Are Available to Determine the Sequence of DNA

The segments of specific DNA molecules obtained by recombinant DNA technology can be analyzed for their nucleotide sequence. This method depends upon having a large number of identical DNA molecules. This requirement can be satisfied by cloning the fragment of interest, using the techniques described above. The **manual, enzymatic method (Sanger's)** employs specific dideoxynucleotides that terminate DNA strand synthesis at specific nucleotides as the strand is synthesized on purified template nucleic acid. The reactions are adjusted so that a population of DNA fragments representing termination at every nucleotide is obtained. By having a radioactive label incorporated at the end opposite the termination site,

one can separate the fragments according to size using polyacrylamide gel electrophoresis. An autoradiograph is made, and each of the fragments produces an image (band) on an x-ray film. These are read in order to give the DNA sequence (Figure 42–6). Another manual method, that of **Maxam and Gilbert,** employs **chemical methods** to cleave the DNA molecules where they contain the specific nucleotides. Techniques that do not require the use of radioisotopes are commonly employed in automated DNA sequencing. Most commonly employed is a procedure in which four different fluorescent labels, one representing each nucleotide, are used. Each emits a specific signal upon excitation by a laser beam, and this can be recorded by a computer.

Oligonucleotide Synthesis Is Now Routine

The automated, chemical synthesis of moderately long oligonucleotides (about 100 nucleotides) of precise sequence is now a routine laboratory procedure. Each synthetic cycle takes a few minutes, so an entire molecule can be made by synthesizing relatively short segments that can then be ligated to one another. Oligonucleotides are now indispensable for DNA sequencing, library screening, DNA mobility

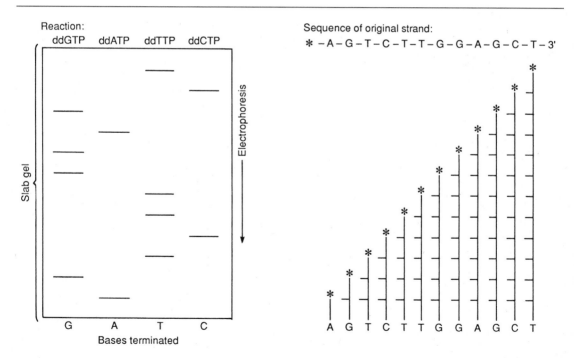

Figure 42–6. Sequencing of DNA by the method devised by Sanger. The ladder-like arrays represent from bottom to top all of the successively longer fragments of the original DNA strand. Knowing which specific dideoxynucleotide reaction was conducted to produce each mixture of fragments, one can determine the sequence of nucleotides from the labeled end toward the unlabeled end by reading up the gel. The base-pairing rules of Watson and Crick (A–T, G–C) dictate the sequence of the other (complementary) strand.

shift assays, the polymerase chain reaction (see below), and numerous other applications.

The Polymerase Chain Reaction (PCR) Amplifies DNA Sequences

The polymerase chain reaction (PCR) is a method of amplifying a target sequence of DNA. PCR provides a sensitive, selective, and extremely rapid means of amplifying a desired sequence of DNA. Specificity is based on the use of two oligonucleotide primers that hybridize to complementary sequences on opposite strands of DNA and flank the target sequence (Figure 42–7). The DNA sample is first heated to separate the two strands, the primers are allowed to bind to the DNA, and each strand is copied by a DNA polymerase, starting at the primer site. The two DNA strands each serve as a template for the synthesis of new DNA from the two primers. Repeated cycles of heat denaturation, annealing of the primers to their complementary sequences, and extension of the annealed primers with DNA polymerase result in the exponential amplification of DNA segments of defined length. Early PCR reactions used an *E coli* DNA polymerase that was destroyed by each heat denaturation cycle. Substitution of a heat-stable DNA polymerase from *Thermus aquaticus,* an organism that lives and replicates at 70–80 °C, obviates this problem and has allowed for automation of the reaction, since the polymerase reactions can be run at 70 °C. This has also improved the specificity and the yield of DNA.

DNA sequences as short as 50–100 bp and as long as 2.5 kbp can be amplified. Twenty cycles provide an amplification of 10^6 and 30 cycles of 10^9. The PCR allows the DNA in a single cell, hair follicle, or sperm to be amplified and analyzed. Thus, the applications of PCR to forensic medicine are obvious. The PCR is also used (1) to detect infectious agents, especially latent viruses; (2) to make prenatal genetic diagnoses; (3) to detect allelic polymorphisms; (4) to establish precise tissue types for transplants; and (5) to study evolution, using DNA from archeological samples. There are an equal number of applications of PCR to problems in basic science, and new uses are developed every year.

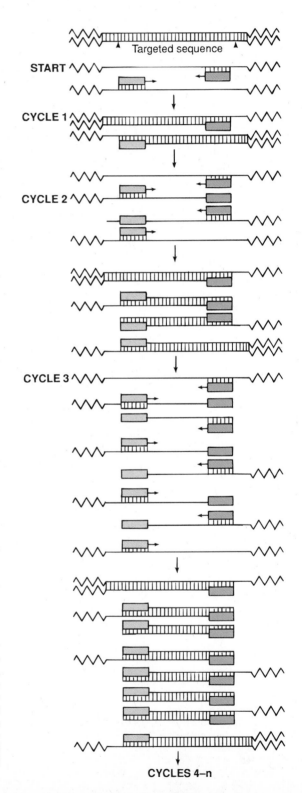

Figure 42–7. The polymerase chain reaction is used to amplify specific gene sequences. Double-stranded DNA is heated to separate it into individual strands. These bind two distinct primers that are directed at specific sequences on opposite strands and that define the segment to be amplified. DNA polymerase extends the primers in each direction and synthesizes two strands complementary to the original two. This cycle is repeated several times, giving an amplified product of defined length and sequence.

PRACTICAL APPLICATIONS OF RECOMBINANT DNA TECHNOLOGY ARE NUMEROUS

The isolation of a specific gene from an entire genome requires a technique that will discriminate one part in a million. The identification of a regulatory region that may be only 10 bp in length requires a sensitivity of one part in 3×10^8; a disease such as sickle cell anemia is caused by a single base change, or one part in 3×10^9. Recombinant DNA technology is powerful enough to accomplish all these things.

Gene Mapping Localizes Specific Genes to Distinct Chromosomes

Gene localizing thus can define a map of the human genome. This is already yielding useful information in the definition of human disease. Somatic cell hybridization and in situ hybridization are two techniques used to accomplish this. In **in situ hybridization,** the simpler and more direct procedure, a radioactive probe is added to a metaphase spread of chromosomes on a glass slide. The exact area of hybridization is localized by layering photographic emulsion over the slide and, after exposure, lining up the grains with some histologic identification of the chromosome. Fluorescence in situ hybridization (FISH) is a very sensitive technique that is also used for this purpose. This often places the gene at a location on a given band or region on the chromosome. Some of the human genes localized using these techniques are listed in Table 42–6.

This table represents only a sampling, since hundreds of genes have been mapped. This human map will become more complete in ensuing years, and the effort to sequence the entire human genome is well under way. The following conclusions can already be drawn: (1) Genes that code for proteins with similar functions can be located on separate chromosomes (α- and β-globin). (2) Genes that form part of a family can also be on separate chromosomes (growth hormone and prolactin). (3) The genes involved in many hereditary disorders known to be due to specific protein deficiencies, including X chromosome–linked conditions, are indeed located at specific sites. Of most interest, perhaps, is the fact that because of the availability of defined and cloned restriction fragments, the chromosomal location for many disorders is being defined, eg, Huntington's chorea, chromosome 4; cystic fibrosis, chromosome 7; adult polycystic kidney disease, chromosome 16; and Duchenne-type muscular dystrophy, chromosome X. Once the defect is localized to a region of DNA that has the characteristic structure of a gene (Figure 42–1), a synthetic gene can be constructed and expressed in an appropriate vector and its function can be assessed, or the putative peptide, deduced from the open reading frame in the coding region, can be synthesized. Antibodies directed against this peptide can be used to assess whether this peptide is expressed in normal persons and whether it is absent in those with the genetic syndrome.

Proteins Can Be Produced for Research and Diagnosis

A practical goal of recombinant DNA research is the production of materials for biomedical application. This technology has two distinct merits: (1) It can supply large amounts of material that could not be obtained by conventional purification methods (eg, interferon, plasminogen activating factor). (2) It can provide human material (eg, insulin, growth hormone). The advantages in both cases are obvious. Although the primary aim is to supply products, generally proteins, for treatment (insulin) and diagnosis (AIDS test) of human and other animal diseases and for disease prevention (hepatitis B vaccine), there are other real and potential commercial applications, especially in agriculture. An example of the latter is the attempt to engineer plants that are more resistant to

Table 42–6. Localization of human genes.[1]

Gene	Chromosome	Disease
Insulin	11p15	
Prolactin	6p23-q12	
Growth hormone	17q21-qter	Growth hormone deficiency
α-Globin	16p12-pter	α-Thalassemia
β-Globin	11p12	β-Thalassemia, sickle cell
Adenosine deaminase	20q13-qter	Adenosine deaminase deficiency
Phenylalanine hydroxylase	12q24	Phenylketonuria
Hypoxanthine-guanine phosphoribosyltransferase	Xq26-q27	Lesch-Nyhan syndrome
DNA segment G8	4p	Huntington's chorea

[1]This table indicates the chromosomal location of several genes and the diseases associated with deficient or abnormal production of the gene products. The chromosome involved is indicated by the first (underlined) number or letter. The other numbers and letters refer to precise localizations, as defined in McKusick VA: *Mendelian Inheritance in Man,* 6th ed. Johns Hopkins Univ Press, 1983.

drought or temperature extremes or more efficient at fixing nitrogen.

Recombinant DNA Technology Is Used in the Molecular Analysis of Disease

A. Normal Gene Variations: There is a normal variation of DNA sequence just as there is with more obvious aspects of human structure. Variations of DNA sequence, **polymorphisms,** occur approximately once in every 500 nucleotides, or about 10^7 times per genome. There are, no doubt, deletions and insertions of DNA as well as single-base substitutions. In healthy people, these alterations obviously occur in noncoding regions of DNA or at sites that cause no change in function of the encoded protein. This heritable polymorphism of DNA structure can be associated with certain diseases within a large kindred and can be used to search for the specific gene involved, as is illustrated below. It can also be used in a variety of applications in forensic medicine.

B. Gene Variations Causing Disease: Classical genetics taught that most genetic diseases were due to point mutations which resulted in an impaired protein. This may still be true, but if on reading the initial sections of this chapter one predicted that genetic disease could result from derangement of any of the steps illustrated in Figure 42–1, one would have made a proper assessment.

This point is nicely illustrated by an examination of the β-globin gene. This gene is located in a cluster on chromosome 11 (Figure 42–8), and an expanded version of the gene is illustrated in Figure 42–9. Defective production of β-globin results in a variety of diseases and is due to many different lesions in and around the β-globin gene (Table 42–7).

C. Point Mutations: The classic example is **sickle cell disease,** which is caused by mutation of a single base out of the 3×10^9 in the genome, a T-to-A DNA substitution, which in turn results in an A-to-U change in the mRNA corresponding to the sixth codon of the β-globin gene (see Figure 7–20). The altered codon specifies a different amino acid (valine rather than glutamic acid), and this causes a structural abnormality of the β-globin molecule. Other point mutations in and around the β-globin gene result in decreased or, in some instances, no production of β-globin; β-thalassemia is the result of these mutations. (The thalassemias are characterized by defects in the synthesis of hemoglobin subunits, and so β-thalassemia results when there is insufficient production of β-globin.) Figure 42–9 illustrates that point mutations affecting each of the many processes involved in generating a normal mRNA (and therefore a normal protein) have been implicated as a cause of β-thalassemia.

D. Deletions, Insertions, and Rearrangements of DNA: Studies of bacteria, viruses, yeasts, and fruit flies show that pieces of DNA can move from one place to another within a genome. The deletion of a critical piece of DNA, the rearrangement of DNA within a gene, or the insertion of a

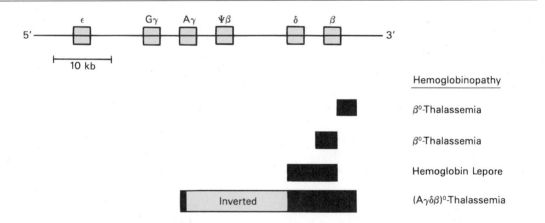

Figure 42–8. Schematic representation of the β-globin gene cluster and of the lesions in some genetic disorders. The β-globin gene is located on chromosome 11 in close association with the two γ-globin genes and the δ-globin gene. The β-gene family is arranged in the order 5′-ε-Gγ-Aγ-ψb-δ-β-3′. The ε locus is expressed in early embryonic life ($a_2\epsilon_2$). The γ genes are expressed in fetal life, making fetal hemoglobin (HbF, $\alpha_2\gamma_2$). Adult hemoglobin consists of HbA ($\alpha_2\beta_2$) or $HbA_2(\alpha_2\delta_2)$. The ψβ is a pseudogene that has sequence homology with β but contains mutations that prevent its expression. Deletions (solid bar) of the β locus cause β-thalassemia (deficiency or absence $[\beta^0]$ of β-globin). A deletion of δ and β causes hemoglobin Lepore (only hemoglobin α is present). An inversion $(A\gamma\delta\beta)^0$ in this region (open bar) disrupts gene function and also results in thalassemia (type III). Each type of thalassemia tends to be found in a certain group of people, eg, the $(A\gamma\delta\beta)^0$ deletion inversion occurs in persons from India. Many more deletions in this region have been mapped, and each causes some type of thalassemia.

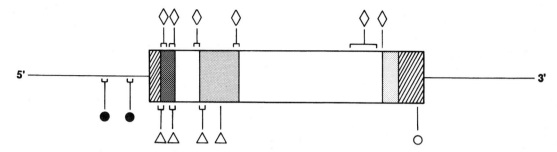

Figure 42–9. Mutations in the β-globin gene causing β-thalassemia. The β-globin gene is shown in the 5' to 3' orientation. The cross-hatched areas indicate the 5' and 3' nontranslated regions. Reading from the 5' to 3' direction, the shaded areas are exons 1–3 and the clear spaces are introns 1 and 2. Mutations that affect transcription control (●) are located in the 5' flanking-region DNA. Examples of nonsense mutations (△), mutations in RNA processing (◇), and RNA cleavage mutations (○) have been identified and are indicated. In some regions, many mutations have been found. These are indicated by the brackets.

piece of DNA within a coding or regulatory region can all cause changes in gene expression resulting in disease. Again, a molecular analysis of β-thalassemia produces numerous examples of these processes, particularly deletions, as a cause of disease (Figure 42–8). The globin gene clusters seem particularly prone to this lesion. Deletions in the α-globin cluster, located on chromosome 16, cause α-thalassemia. There is a strong ethnic association for many of these deletions, so that northern Europeans, Filipinos, blacks, and Mediterranean peoples have different lesions all resulting in the absence of hemoglobin A and α-thalassemia.

A similar analysis could be made for a number of other diseases. Point mutations are usually defined by sequencing the gene in question, though occasionally, if the mutation destroys or creates a restriction enzyme site, the technique of restriction fragment analysis can be used to pinpoint the lesion. Deletions or insertions of DNA larger than 50 bp can often be detected by the Southern blotting procedure.

E. Pedigree Analysis: Sickle cell disease again provides an excellent example of how recombinant DNA technology can be applied to the study of human disease. The substitution of T for A in the template strand of DNA in the β-globin gene changes

the sequence in the region that corresponds to the sixth codon from

$$\downarrow$$

CCTGAGG   coding strand
GGAC(T)CC   template strand
$$\uparrow$$

to

CCTGTGG   coding strand
GGAC(A)CC   template strand

and destroys a recognition site for the restriction enzyme *MstII* (CCTNAGG; denoted by the small vertical arrows; Table 42–2). Other *MstII* sites 5' and 3' from this site (Figure 42–10) are not affected and so will be cut. Therefore, incubation of DNA from normal (AA), heterozygous (AS), and homozygous (SS) individuals results in three different patterns on Southern blot transfer (Figure 42–10). This illustrates how a DNA pedigree can be established using the principles discussed in this chapter. Pedigree analysis has been applied to a number of genetic diseases and is most useful in those caused by deletions and insertions or the rarer instances in which a restriction endonuclease cleavage site is affected, as in the example cited in this paragraph. The analysis is facilitated by the PCR reaction, which can provide sufficient DNA for analysis from just a few nucleated red blood cells.

F. Prenatal Diagnosis: If the genetic lesion is understood and a specific probe is available, prenatal diagnosis is possible. DNA from cells collected from as little as 10 mL of amniotic fluid (or by chorionic villus biopsy) can be analyzed by Southern blot transfer. A fetus with the restriction pattern AA in Figure 42–10 does not have sickle cell disease, nor is

Table 42–7. Structural alterations of the β-globin gene.

Alteration	Function Affected	Disease
Point mutations	Protein folding	Sickle cell disease
	Transcriptional control	β-Thalassemia
	Frameshift and non-sense mutations	β-Thalassemia
	RNA processing	β-Thalassemia
Deletion	mRNA production	β°-Thalassemia Hemoglobin Lepore
Rearrangement	mRNA production	β-Thalassemia type III

A. Mst II restriction sites around and in the β-globin gene

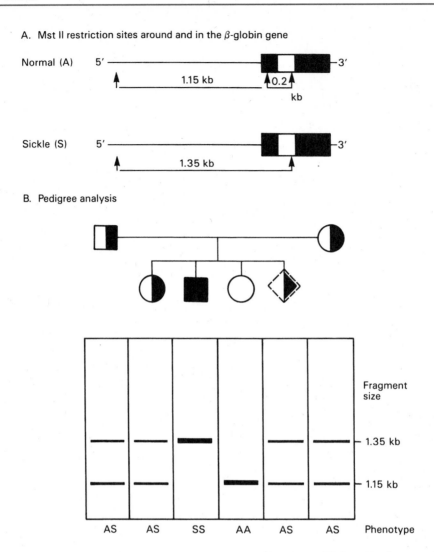

B. Pedigree analysis

Figure 42–10. Pedigree analysis of sickle cell disease. The top part of the figure *(A)* shows the first part of the β-globin gene and the *MstII* restriction enzyme sites (i) in the normal (A) and sickle cell (S) β-globin genes. Digestion with the restriction enzyme *MstII* results in DNA fragments 1.15 kb and 0.2 kb long in normal individuals. The T-to-A change in individuals with sickle cell disease abolishes one of the three *MstII* sites around the β-globin gene; hence, a single restriction fragment 1.35 kb in length is generated in response to *MstII*. This size difference is easily detected on a Southern blot *(B)*. (The 0.2-kb fragment would run off the gel in this illustration.) Pedigree analysis shows three possibilities: AA = normal (open circle); AS = heterozygous (half-solid circles, half-solid square); SS = homozygous (solid square). This approach allows for prenatal diagnosis of sickle cell disease (dash-sided square with solid triangle on right).

it a carrier. A fetus with the SS pattern will develop the disease. Probes are now available for this type of analysis of many genetic diseases.

G. Restriction Fragment Length Polymorphism (RFLP): The differences in DNA sequence cited above can result in variations of restriction sites and thus in the length of restriction fragments. An inherited difference in the pattern of restriction (eg, a DNA variation occurring in more than 1% of the general population) is known as a restriction fragment length polymorphism, or RFLP. A fairly extensive RFLP map of the human genome has been con-

structed. This is proving useful in the human genome sequencing project, and is an important component in the effort to understand various single and multigenic diseases. RFLPs result from single-base changes (eg, sickle cell disease) or from deletions or insertions of DNA into a restriction fragment (eg, the thalassemias) and are proving to be a useful diagnostic tool. They have been found at known gene loci and in sequences that have no known function; thus, RFLPs may disrupt the function of the gene or may have no biologic consequences.

RFLPs are inherited, and they segregate in a

Mendelian fashion. A major use of RFLPs (thousands are now known) is in the definition of inherited diseases in which the functional deficit is unknown. RFLPs can be used to establish linkage groups, which in turn, by the process of **chromosome walking,** will eventually define the disease locus. In chromosome walking (Figure 42–11), a fragment representing one end of a long piece of DNA is used to isolate another that overlaps but extends the first. The direction of extension is determined by restriction mapping, and the procedure is repeated sequentially until the desired sequence is obtained. The X chromosome-linked disorders are particularly amenable to this approach, since only a single allele is expressed. Hence, 20% of the defined RFLPs are on the X chromosome, and a reasonably complete linkage map of this chromosome exists. The gene for the X-linked disorder, Duchenne-type muscular dystrophy, has been found using RFLPs. Likewise, the defect in Huntington's chorea has been localized to the terminal region of the short arm of chromosome 4, and the defect that causes polycystic kidney disease is linked to the α-globin locus on chromosome 16.

H. RFLPs and VNTRs in Forensic Medicine: Variable numbers of tandemly repeated (VNTR) units are one common type of "insertion" that results in an RFLP. The VNTRs can be inherited, in which case they are useful in establishing genetic association with a disease in a family or kindred; or they can be unique to an individual and thus serve as a molecular fingerprint of that person.

I. Gene Therapy: Diseases caused by deficiency of a gene product (Table 42–6) are amenable to replacement therapy. The strategy is to clone a gene (eg, the gene that codes for adenosine deaminase) into a vector that will readily be taken up and incorporated into the genome of a host cell. Bone marrow precursor cells are being investigated for this

purpose because they presumably will resettle in the marrow and replicate there. The introduced gene would begin to direct the expression of its protein product, and this would correct the deficiency in the host cell.

J. Transgenic animals: The somatic cell gene replacement described above would obviously not be passed on to offspring. Other strategies to alter germ cell lines have been devised but have been tested only in experimental animals. A certain percentage of genes injected into a fertilized mouse ovum will be incorporated into the genome and found in both somatic and germ cells. Hundreds of transgenic animals have been established, and these are proving to be useful for analysis of tissue-specific effects on gene expression and effects of overproduction of gene products (eg, those from the growth hormone gene or oncogenes) and in discovering genes involved in development, a process that heretofore has been difficult to study. The transgenic approach has recently been used to correct a genetic deficiency in mice. Fertilized ova obtained from mice with genetic hypogonadism were injected with DNA containing the coding sequence for the gonadotropin-releasing hormone (GnRH) precursor protein. This gene was expressed and regulated normally in the hypothalamus of a certain number of the resultant mice, and these animals were in all respects normal. Their offspring also showed no evidence of GnRH deficiency. This is, therefore, evidence of somatic cell expression of the transgene and of its maintenance in germ cells.

Targeted Gene Disruption or Knockout

In transgenic animals, one is adding a gene to the genome, and there is no way to control where that gene eventually resides. A complementary—and

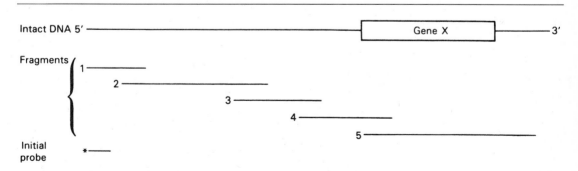

Figure 42–11. The technique of chromosome walking. Gene X is to be isolated from a large piece of DNA. The exact location of this gene is not known, but a probe (*——) directed against a fragment of DNA (shown at the 5′ end in this representation) is available, as is a library containing a series of overlapping DNA fragments. For the sake of simplicity, only five of these are shown. The initial probe will hybridize only with clones containing fragment 1, which can then be isolated and used as a probe to detect fragment 2. This procedure is repeated until fragment 4 hybridizes with fragment 5, which contains the entire sequence of gene X.

much more difficult—approach involves the selective removal of a gene from the genome. Gene knockout animals (usually mice) are made by creating a mutation that totally disrupts the function of a gene. This is then used to replace one of the two genes in an embryonic stem cell that can be used to create a heterozygous transgenic animal. The mating of two such animals will, by mendelian genetics, result in a homozygous mutation in 25% of offspring. Several hundred gene knockout mice have been developed.

SUMMARY

A variety of very sensitive techniques can now be applied to the isolation and characterization of genes and to the quantitation of gene products. A fundamental concept is central to these techniques, ie, complementary base pairs form hydrogen bonds with each other—A with T and G with C. In DNA cloning, a particular segment of DNA is removed from its normal environment using one of many restriction endonucleases. This is then ligated into one of several vectors in which the DNA segment can be amplified and produced in abundance. It is relatively easy to isolate the cloned DNA, which can then be sequenced and used as a probe in one of several types of hybridization reactions to detect other, related or adjacent pieces of DNA, or used to quantitate gene products such as mRNA. Manipulation of the DNA to change its structure, so-called genetic engineering, is a key element in cloning (eg, the construction of chimeric molecules) and can also be used to study the function of a certain fragment of DNA and to analyze how genes are regulated. Chimeric DNA molecules are introduced into cells to make transfected cells or into the fertilized oocyte to make transgenic animals. These approaches are used to study gene regulation in a normal cellular context and to alter cellular function. Techniques involving cloned DNA are used to locate genes to specific regions of chromosomes, to identify the genes responsible for diseases, to study how faulty gene regulation causes disease, to diagnose genetic diseases, and increasingly to treat genetic diseases.

GLOSSARY

ARS: Autonomously replicating sequence; the origin of replication in yeast.
Autoradiography: The detection of radioactive molecules (eg, DNA, RNA, protein) by visualization of their effects on photographic film.
Bacteriophage: A virus that infects a bacterium.
Blunt-ended DNA: Two strands of a DNA duplex having ends that are flush with each other.
cDNA: A single-stranded DNA molecule that is complementary to an mRNA molecule and is syn-

thesized from it by the action of reverse transcriptase.
Chimeric molecule: A molecule (eg, DNA, RNA, protein) containing sequences derived from two different species.
Clone: A large number of cells or molecules that are identical with a single parental cell or molecule.
Cosmid: A plasmid into which the DNA sequences from bacteriophage lambda that are necessary for the packaging of DNA (cos sites) have been inserted; this permits the plasmid DNA to be packaged in vitro.
Endonuclease: An enzyme that cleaves internal bonds in DNA or RNA.
Excinuclease: The excision nuclease involved in nucleotide exchange repair of DNA.
Exon: The sequence of a gene that is represented (expressed) as mRNA.
Exonuclease: An enzyme that cleaves nucleotides from either the 3′ or 5′ ends of DNA or RNA.
Fingerprinting: The use of RFLPs or repeat sequence DNA to establish a unique pattern of DNA fragments for an individual.
Footprinting: DNA with protein bound is resistant to digestion by DNase enzymes. When a sequencing reaction is performed using such DNA, a protected area, representing the "footprint" of the bound protein, will be detected.
Hairpin: A double-helical stretch formed by base pairing between neighboring complementary sequences of a single strand of DNA or RNA.
Hybridization: The specific reassociation of complementary strands of nucleic acids (DNA with DNA, DNA with RNA, or RNA with RNA).
Insert: An additional length of base pairs in DNA, generally introduced by the techniques of recombinant DNA technology.
Intron: The sequence of a gene that is transcribed but excised before translation.
Library: A collection of cloned fragments that represents the entire genome. Libraries may be either genomic DNA (in which both introns and exons are represented) or cDNA (in which only exons are represented).
Ligation: The enzyme-catalyzed joining in phosphodiester linkage of two stretches of DNA or RNA into one; the respective enzymes are DNA and RNA ligases.
Lines: Long interspersed repeat sequences.
Microsatellite polymorphism: Heterozygosity of a certain microsatellite repeat in an individual.
Microsatellite repeat sequences: Dispersed or group repeat sequences of 2–5 bp repeated up to 50 times. May occur at 50–100 thousand locations in the genome.
Nick translation: A technique for labeling DNA based on the ability of the DNA polymerase from *E coli* to degrade a strand of DNA that has been nicked and then to resynthesize the strand; if a radioactive nucleoside triphosphate is employed, the rebuilt strand becomes labeled and can be used as a radioactive probe.

Northern blot: A method for transferring RNA from an agarose gel to a nitrocellulose filter, on which the RNA can be detected by a suitable probe.

Oligonucleotide: A short, defined sequence of nucleotides joined together in the typical phosphodiester linkage.

Ori: The origin of replication in prokaryotes.

Palindrome: A sequence of duplex DNA that is the same when the two strands are read in opposite directions.

Plasmid: A small, extrachromosomal, circular molecule of DNA that replicates independently of the host DNA.

Polymerase chain reaction (PCR): An enzymatic method for the repeated copying (and thus amplification) of the two strands of DNA that make up a particular gene sequence.

Primosome: The mobile complex of helicase and primase that is involved in DNA replication.

Probe: A molecule used to detect the presence of a specific fragment of DNA or RNA in, for instance, a bacterial colony that is formed from a genetic library or during analysis by blot transfer techniques; common probes are cDNA molecules, synthetic oligodeoxynucleotides of defined sequence, or antibodies to specific proteins.

Pseudogene: An inactive segment of DNA arising by mutation of a parental active gene.

Recombinant DNA: The altered DNA that results from the insertion of a sequence of deoxynucleotides not previously present into an existing molecule of DNA by enzymatic or chemical means.

Restriction enzyme: An endodeoxynuclease that causes cleavage of both strands of DNA at highly specific sites dictated by the base sequence.

Reverse transcription: RNA-directed synthesis of DNA, catalyzed by reverse transcriptase.

Signal: The end product observed when a specific sequence of DNA or RNA is detected by autoradiography or some other method. Hybridization with a complementary radioactive polynucleotide (eg, by Southern or Northern blotting) is commonly used to generate the signal.

Sines: Short interspersed repeat sequences.

SnRNA: Small nuclear RNA. This family of RNAs is best known for its role in mRNA processing.

Southern blot: A method for transferring DNA from an agarose gel to nitrocellulose filter, on which the DNA can be detected by a suitable probe (eg, complementary DNA or RNA).

Splicing: The removal of introns from RNA accompanied by the joining of its exons.

Splicosome: The macromolecular complex responsible for precursor mRNA splicing. The splicosome consists of at least five small nuclear RNAs (snRNA; U1, U2, U4, U5, and U6) and many proteins.

Sticky-ended DNA: Complementary single strands of DNA that protrude from opposite ends of a DNA duplex or from the ends of different duplex molecules (see also Blunt-ended DNA, above).

Tandem: Used to describe multiple copies of the same sequence (eg, DNA) that lie adjacent to one another.

Terminal transferase: An enzyme that adds nucleotides of one type (eg, deoxyadenonucleotidyl residues) to the 3′ end of DNA strands.

Transcription: DNA-directed synthesis of RNA.

Transgenic: Describing the introduction of new DNA into germ cells by its injection into the nucleus of the ovum.

Translation: Synthesis of protein using mRNA as template.

Vector: A plasmid or bacteriophage into which foreign DNA can be introduced for the purposes of cloning.

Western blot: A method for transferring protein to a nitrocellulose filter, on which the protein can be detected by a suitable probe (eg, an antibody).

REFERENCES

Lewin B: *Genes V.* Oxford Univ Press, 1994.

Maniatis T, Fritsch EF, Sambrook J: *Molecular Cloning,* 2nd ed. Cold Spring Harbor Laboratory, 1989.

Martin JB, Gusella JF: Huntington's disease: Pathogenesis and management. N Engl J Med 1986:315:1267.

Watson JD et al: *Recombinant DNA,* 2nd ed. Scientific American Books. Freeman, 1992.

Weatherall DJ: *The New Genetics and Clinical Practice,* 3rd ed. Oxford Univ Press, 1991.

Section V.
Biochemistry of Extracellular and Intracellular Communication

Membranes: Structure, Assembly, and Function

43

Robert K. Murray, MD, PhD, Daryl K. Granner, MD

INTRODUCTION

Membranes are highly viscous yet plastic structures. Plasma membranes form closed compartments around cellular protoplasm to separate one cell from another and thus permit cellular individuality. The plasma membrane has selective permeabilities and acts as a barrier, thereby maintaining differences in composition between the inside and the outside of the cell. The selective permeabilities are provided by channels and pumps for ions and substrates and by specific receptors for signals, eg, hormones. Plasma membranes also exchange material with the extracellular environment by exocytosis and endocytosis, and there are special areas of membrane structures—the gap junctions—through which adjacent cells exchange material.

Membranes also form specialized compartments within the cell. Such intracellular membranes form many of the morphologically distinguishable structures (organelles), eg, mitochondria, endoplasmic reticulum, sarcoplasmic reticulum, Golgi complexes, secretory granules, lysosomes, and the nuclear membrane. Membranes localize enzymes, function as integral elements in excitation-response coupling, and provide sites of energy transduction, such as in photosynthesis and oxidative phosphorylation.

BIOMEDICAL IMPORTANCE

Gross alterations of membrane structure can affect water balance and ion flux and therefore every process within the cell. Specific deficiencies or alterations of certain membrane components lead to a variety of diseases. Examples include the lysosomal absence of α-glucosidase, causing type II glycogen storage disease; the lack of an iodide transporter, causing congenital goiter (Figure 46–2); and defec-

tive endocytosis of low-density lipoproteins, resulting in accelerated hypercholesterolemia and coronary artery disease. Normal cellular function obviously begins with normal membranes.

MAINTENANCE OF A NORMAL INTRA- AND EXTRACELLULAR ENVIRONMENT IS FUNDAMENTAL TO LIFE

Life originated in an aqueous environment; enzyme reactions, cellular and subcellular processes, and so forth have therefore evolved to work in this milieu. Since most mammals live in a gaseous environment, how is the aqueous state maintained? Membranes accomplish this by internalizing and compartmentalizing body water.

The Body's Internal Water Is Compartmentalized

Water makes up about 56% of the lean body mass of the human body (Chapters 2 and 3) and is distributed in two large compartments.

A. Intracellular Fluid (ICF): This compartment constitutes two-thirds of total water and provides the environment for the cell to (1) make, store, and utilize energy; (2) repair itself; (3) replicate; and (4) perform special functions.

B. Extracellular Fluid (ECF): This compartment contains about one-third of total water and is distributed between the plasma and interstitial compartments. The extracellular fluid is a delivery system. It brings to the cells nutrients (eg, glucose, fatty acids, amino acids), oxygen, various ions and trace minerals, and a variety of regulatory molecules (hormones) that coordinate the functions of widely separated cells. Extracellular fluid removes CO_2, waste products, and toxic or detoxified materials from the immediate cellular environment.

The Ionic Composition of Intracellular and Extracellular Fluids Differs Greatly

As illustrated in Table 43–1, the internal environment is rich in K^+ and Mg^{2+}, and phosphate is its major anion. Extracellular fluid is characterized by high Na^+ and Ca^{2+} content, and Cl^- is the major anion. Note also that glucose is higher in extracellular fluid than in the cell, whereas the opposite is true for proteins. Why is there such a difference? It is thought that the primordial sea in which life originated was rich in K^+ and Mg^{2+}. It therefore follows that enzyme reactions and other biologic processes evolved to function best in that environment, hence the high concentration of these ions within cells. Cells were faced with strong selection pressure as the sea gradually changed to a composition rich in Na^+ and Ca^{2+}. Vast changes would have been required for evolution of a completely new set of biochemical and physiologic machinery; instead, as it happened, cells developed barriers—membranes with associated "pumps"—to maintain the internal microenvironment.

MEMBRANES ARE COMPLEX STRUCTURES COMPOSED OF LIPIDS, PROTEINS, AND CARBOHYDRATES

Different membranes within the cell and between cells have different compositions, as reflected in the ratio of protein to lipid (Figure 43–1). The difference is not surprising given the very different functions of membranes. Membranes are asymmetric sheet-like enclosed structures with an inside and an outside surface. These sheet-like structures are noncovalent assemblies that are thermodynamically stable and metabolically active. Specific protein molecules are anchored in membranes, where they carry out specific functions of the organelle, the cell, or the organism.

Table 43–1. Comparison of the mean concentrations of various substances outside and inside a mammalian cell.

Substance	Extracellular Fluid		Intracellular Fluid	
Na^+	140	mmol/L	10	mmol/L
K^+	4	mmol/L	140	mmol/L
Ca^{2+} (free)	2.5	mmol/L	0.1	μmol/L
Mg^{2+}	1.5	mmol/L	30	mmol/L
Cl^-	100	mmol/L	4	mmol/L
HCO_3^-	27	mmol/L	10	mmol/L
PO_4^{3-}	2	mmol/L	60	mmol/L
Glucose	5.5	mmol/L	0–1	mmol/L
Protein	2	g/dL	16	g/dL

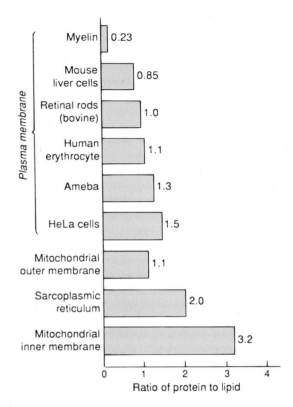

Figure 43–1. Ratio of protein to lipid in different membranes. Proteins equal or exceed the quantity of lipid in nearly all membranes. The outstanding exception is myelin, an electrical insulator found on many nerve fibers.

The Major Lipids in Mammalian Membranes Are Phospholipids, Glycosphingolipids, and Cholesterol

A. Phospholipids: Of the two major phospholipid classes present in membranes, **phosphoglycerides** are the more common and consist of a glycerol backbone to which are attached two fatty acids in ester linkage and a phosphorylated alcohol (Figure 43–2). The fatty acid constituents are usually even-numbered carbon molecules most commonly containing 16 or 18 carbons. They are unbranched and can be saturated or unsaturated. The simplest phosphoglyceride is phosphatidic acid, which is 1,2-diacylglycerol 3-phosphate, a key intermediate in the formation of all other phospholipids (Chapter 26). In other phospholipids, the 3-phosphate is esterified to an alcohol such as ethanolamine, choline, serine, glycerol, or inositol (Chapter 16).

The second class of phospholipids is composed of **sphingomyelins,** which contain a sphingosine backbone rather than glycerol. A fatty acid is attached by an amide linkage to the amino group of sphingosine. The primary hydroxyl group of sphingosine is esteri-

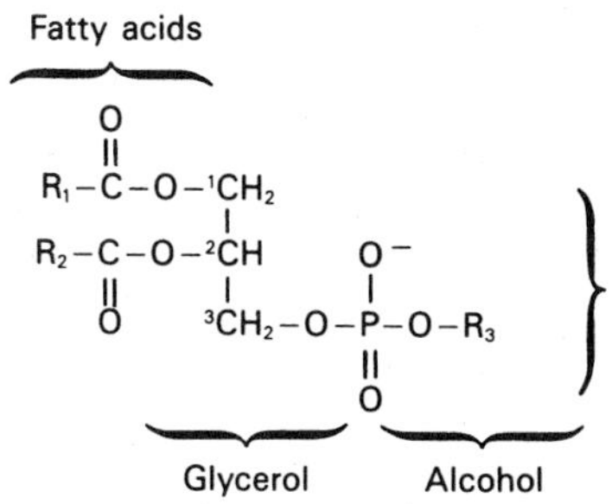

Fatty acids

$$R_1-\overset{\overset{O}{\|}}{C}-O-{}^1CH_2$$
$$R_2-\overset{}{C}-O-{}^2CH \qquad O^-$$
$$\overset{\|}{O} \qquad {}^3CH_2-O-\overset{}{P}-O-R_3$$
$$\overset{\|}{O}$$

Glycerol   Alcohol

Figure 43–2. A phosphoglyceride showing the fatty acids (R_1 and R_2), glycerol, and phosphorylated alcohol components. In phosphatidic acid, R_3 is hydrogen.

fied to phosphorylcholine. Sphingomyelins, as the name implies, are prominent in myelin sheaths.

B. Glycosphingolipids: The glycosphingolipids are sugar-containing lipids such as cerebrosides and gangliosides and are also derived from sphingosine. The cerebrosides and gangliosides differ from sphingomyelin in the moiety attached to the primary hydroxyl group of sphingosine. In sphingomyelin, a phosphorylcholine is attached to the alcohol group. A cerebroside contains a single hexose moiety, glucose or galactose, at that site. A ganglioside contains a chain of 3 or more sugars—at least one of which is a sialic acid—attached to the primary alcohol of sphingosine.

C. Sterols: The most common sterol in membranes is **cholesterol,** which exists almost exclusively in the plasma membranes of mammalian cells but can also be found in lesser quantity in mitochondria, Golgi complexes, and nuclear membranes. Cholesterol is generally more abundant toward the outside of the plasma membrane. Cholesterol intercalates among the phospholipids of the membrane, with its hydroxyl group at the aqueous interface and the remainder of the molecule within the leaflet. At temperatures above the transition temperature (see discussion of fluid mosaic model, below), its rigid sterol ring interacts with the acyl chains of the phospholipids, limits their movement, and thus decreases membrane fluidity. On the other hand, when the temperature approaches the transition temperature, the interaction of cholesterol with the acyl chains interferes with their alignment with each other; this phenomenon lowers the temperature at which the fluid → gel transition occurs, thus assisting in keeping the membrane fluid at lower temperatures.

Membranes Are Amphipathic

All major lipids in membranes contain both hydrophobic and hydrophilic regions and are therefore termed "amphipathic." Membranes themselves are thus amphipathic. If the hydrophobic regions were separated from the rest of the molecule, it would be insoluble in water but soluble in oil. Conversely, if the hydrophilic region were separated from the rest of the molecule, it would be insoluble in oil but soluble in water. The amphipathic membrane lipids have a polar head group and nonpolar tails, as represented in Figure 43–3. Saturated fatty acids have straight tails, whereas unsaturated fatty acids, which generally exist in the *cis* form in membranes, make kinked tails. As more kinks are inserted in the tails, the membrane becomes less tightly packed and therefore more fluid. Detergents are amphipathic molecules that are important in biochemistry and in the household. The molecular structure of a detergent is not unlike that of a phospholipid. Certain detergents are widely used to solubilize membrane proteins as a first step in their purification. The hydrophobic end of the detergent binds to hydrophobic regions of the proteins, displacing most of their bound lipids. The polar end of the detergent is free, bringing the proteins into solution as detergent-protein complexes, usually also containing some residual lipids.

Membrane Lipids Form Bilayers

The amphipathic character of phospholipids suggests that the two regions of the molecule have incompatible solubilities; however, in a solvent such as water, phospholipids organize themselves into a form that thermodynamically satisfies both regions. A micelle (Figure 43–4) is such a structure; the hydrophobic regions are shielded from water, while the hydrophilic polar groups are immersed in the aqueous environment.

As recognized about 60 years ago by Gorter and Grendel, a bimolecular layer, or bilayer, can also satisfy the thermodynamic requirements of amphipathic molecules in an aqueous environment. A bilayer exists as a sheet in which the hydrophobic regions of

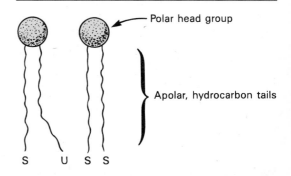

Figure 43–3. Diagrammatic representation of a phospholipid or other membrane lipid. The polar head group is hydrophilic, and the hydrocarbon tails are hydrophobic or lipophilic. The fatty acids in the tails are saturated (S) or unsaturated (U); the former are usually attached to carbon 1 of glycerol and the latter to carbon 2.

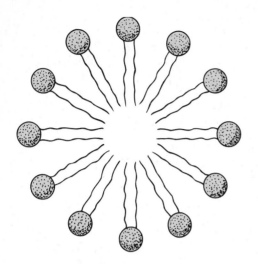

Figure 43–4. Diagrammatic cross-section of a micelle. The polar head groups are bathed in water, whereas the hydrophobic hydrocarbon tails are surrounded by other hydrocarbons and thereby protected from water. Micelles are spherical structures.

the phospholipids are protected from the aqueous environment, while the hydrophilic regions are immersed in water (Figure 43–5). Only the ends or edges of the bilayer sheet are exposed to an unfavorable environment, but even these exposed edges can be eliminated by folding the sheet back upon itself to form an enclosed vesicle with no edges. The closed bilayer provides one of the essential properties of membranes. It is impermeable to most water-soluble molecules, since they would be insoluble in the hydrophobic core of the bilayer.

Two questions immediately arise. First: How

many biologic materials are lipid-soluble and can therefore readily enter the cell? Gases such as oxygen, CO_2, and nitrogen—small molecules with little interaction with solvents—readily diffuse through the hydrophobic regions of the membrane. Lipid-derived molecules, eg, steroid hormones, readily traverse the bilayer. Organic nonelectrolyte molecules exhibit diffusion rates that are dependent upon their oil-water partition coefficients (Figure 43–6); the greater the lipid solubility of a molecule, the greater is its diffusion rate across the membrane.

The second question concerns molecules that are not lipid-soluble: How are the transmembrane concentration gradients for non-lipid-soluble molecules maintained? The answer is that membranes contain proteins, and proteins are also amphipathic molecules that can be inserted into the correspondingly amphipathic lipid bilayer. Proteins form channels for the movement of ions and small molecules and serve as transporters for larger molecules that otherwise could not pass the bilayer. These processes are described below.

Membrane Proteins Are Associated With the Lipid Bilayer

Membrane phospholipids act as a solvent for membrane proteins, creating an environment in which the latter can function. Of the 20 amino acids contributing to the primary structure of proteins, the functional groups attached to the α carbon are strongly hydrophobic in six, weakly hydrophobic in a few, and hydrophilic in the remainder. As described in Chapter 6, the α-helical structure of proteins minimizes the hydrophilic character of the peptide bonds themselves. Thus, proteins can be amphipathic and form an integral part of the membrane by having hydrophilic regions protruding at the inside and outside faces of the membrane but connected by a hydrophobic region traversing the hydrophobic core of the bi-

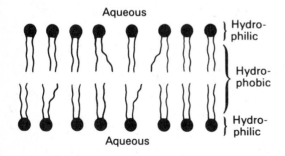

Figure 43–5. Diagram of a section of a bilayer membrane formed from phospholipid molecules. The unsaturated fatty acid tails are kinked and lead to more spacing between the polar head groups, hence to more room for movement. (Slightly modified and reproduced, with permission, from Stryer L: *Biochemistry,* 2nd ed. Freeman, 1981.)

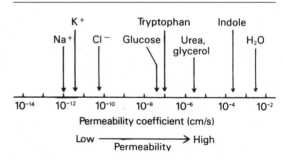

Figure 43–6. Permeability coefficients of water, some ions, and other small molecules in lipid bilayer membranes. Molecules that move rapidly through a given membrane are said to have a high permeability coefficient. (Slightly modified and reproduced, with permission, from Stryer L: *Biochemistry,* 2nd ed. Freeman, 1981.)

layer. In fact, those portions of membrane proteins that traverse membranes do contain substantial numbers of hydrophobic amino acids and a high α-helical or β-pleated sheet content.

The number of different proteins in a membrane varies from six to eight in the sarcoplasmic reticulum to over 100 in the plasma membrane. The proteins consist of enzymes, transport proteins, structural proteins, antigens (eg, for histocompatibility), and receptors for various molecules. Because every membrane possesses a different complement of proteins, there is no such thing as a typical membrane structure. The enzymatic properties of several different membranes are shown in Table 43–2.

Membranes and their components are dynamic structures. The lipids and proteins in membranes turn over, just as they do in other compartments of the cell. Different lipids have different turnover rates, and the turnover rates of individual species of membrane proteins may vary widely. The membrane itself can turn over even more rapidly than any of its constituents. This is discussed in more detail in the section on endocytosis.

Membranes Are Asymmetric Structures

This asymmetry can be partially attributed to the irregular distribution of proteins within the membranes. An **inside-outside asymmetry** is also provided by the external location of the carbohydrates attached to membrane proteins. In addition, specific enzymes are located exclusively on the outside or inside of membranes, as in the mitochondrial and plasma membranes.

There are **regional asymmetries** in membranes. Some, such as occur at the villous border of mucosal cells, are almost macroscopically visible. Others, such as those at gap junctions, tight junctions, and synapses, occupy much smaller regions of the membrane and generate correspondingly smaller local asymmetries.

There is also inside-outside (transverse) asymmetry of the phospholipids. The choline-containing phospholipids (phosphatidylcholine and sphingo-

myelin) are located mainly in the outer molecular layer; the aminophospholipids (phosphatidylserine and phosphatidylethanolamine) are preferentially located in the inner layer. Cholesterol is generally present in larger amounts on the outside than on the inside. Obviously, if this asymmetry is to exist at all, there must be limited transverse mobility (flip-flop) of the membrane phospholipids. In fact, phospholipids in synthetic bilayers exhibit an extraordinarily slow rate of flip-flop; the half-life of the asymmetry can be measured in several weeks. However, when certain membrane proteins such as the erythrocyte protein glycophorin are inserted artificially into synthetic bilayers, the frequency of phospholipid flip-flop may increase as much as 100-fold.

The mechanisms involved in the establishment of lipid asymmetry are not understood. The enzymes involved in the synthesis of phospholipids are located on the cytoplasmic side of microsomal membrane vesicles. It has therefore been postulated that translocases exist which transfer certain phospholipids from the inner leaflet to the outer. In addition, specific proteins that preferentially bind individual phospholipids may be present in the two leaflets, contributing to the asymmetric distribution of these lipid molecules.

Membranes Contain Integral and Peripheral Proteins (Figure 43–7)

Most membrane proteins are integral components of the membrane (they interact with the phospholipids and require the use of detergents for their solubilization), and in fact all those which have been adequately studied span the entire 5- to 10-nm transverse distance of the bilayer. These integral proteins are usually globular and are themselves amphipathic. They consist of two hydrophilic ends separated by an intervening hydrophobic region that traverses the hydrophobic core of the bilayer. As the structures of integral membrane proteins are being elucidated, it is apparent that some (notably transporter molecules) may span the bilayer many times (Figure 43–8).

Integral proteins are asymmetrically distributed across the membrane bilayer. These proteins are given their asymmetric orientation in the membrane at the time of their insertion in the lipid bilayer. The hydrophilic external region of an amphipathic protein, which is clearly synthesized inside the cell, must traverse the hydrophobic core of the membrane and eventually be found outside the membrane. The molecular mechanisms of membrane assembly are discussed below.

Peripheral proteins do not interact directly with the phospholipids in the bilayer and hence do not require use of detergents for their release. They are weakly bound to the hydrophilic regions of specific integral proteins and can be released from them by treatment with salt solutions of high ionic strength. For example, ankyrin, a peripheral protein, is bound to the in-

Table 43–2. Enzymatic markers of different membranes.[1]

Membrane	Enzyme
Plasma	5′-Nucleotidase Adenylyl cyclase Na⁺-K⁺ ATPase
Endoplasmic reticulum	Glucose-6-phosphatase
Golgi complex	Galactosyltransferase
Inner mitochondrial membrane	ATP synthase

[1]Membranes contain many proteins, some of which have enzymatic activity. Some of these enzymes are located only in certain membranes and can therefore be used as markers to follow the purification of these membranes.

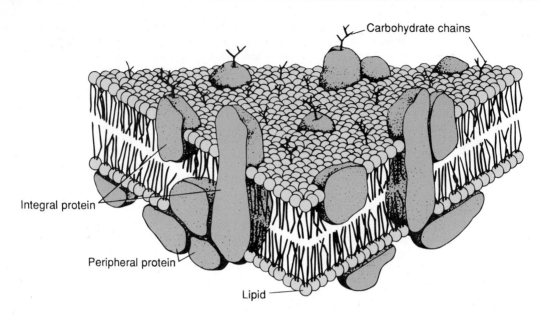

Figure 43–7. The fluid mosaic model of membrane structure. The membrane consists of a bimolecular lipid layer with proteins inserted in it or bound to either surface. Integral membrane proteins are firmly embedded in the lipid layers. Some of these proteins completely span the bilayer and are called transmembrane proteins, while others are embedded in either the outer or inner leaflet of the lipid bilayer. Loosely bound to the outer or inner surface of the membrane are the peripheral proteins. Many of the proteins and lipids have externally exposed oligosaccharide chains. (Reproduced, with permission, from Junqueira LC, Carneiro J, Kelly RO: *Basic Histology,* 7th ed. Appleton & Lange, 1992.)

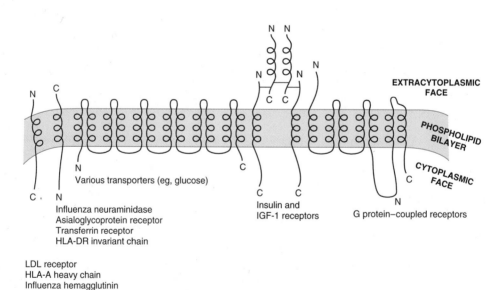

Figure 43–8. Variations in the way in which proteins are inserted into membranes. This schematic representation, which illustrates a number of possible orientations, shows the segments of the proteins within the membrane as α-helices and the other segments as lines. The LDL receptor, which crosses the membrane once and has its amino terminal on the exterior, is called a type I transmembrane protein. The asialoglycoprotein receptor, which also crosses the membrane once but has its carboxyl terminal on the exterior, is called a type II transmembrane protein. The various transporters indicated (eg, glucose) cross the membrane a number of times and are called type III transmembrane proteins; they are also referred to as polytopic membrane proteins. (N, amino terminal; C, carboxyl terminal.) (Adapted, with permission, from Wickner WT, Lodish HF: Multiple mechanisms of protein insertion into and across membranes. Science 1985;230:400. Copyright © 1985 by the American Association for the Advancement of Science.)

tegral protein "band 3" of erythrocyte membrane. Spectrin, a cytoskeletal structure within the erythrocyte, is in turn bound to ankyrin and thereby plays an important role in the maintenance of the biconcave shape of the erythrocyte. The immunoglobulin molecules on the plasma membranes of lymphocytes are integral membrane proteins and can be released by the shedding of small fragments of the membrane. Many hormone receptor molecules are integral proteins, and the specific polypeptide hormones that bind to these receptor molecules may therefore be considered peripheral proteins. Peripheral proteins, such as peptide hormones, may even organize the distribution of integral proteins, such as their receptors, within the plane of the bilayer (see below).

ARTIFICIAL MEMBRANES MODEL MEMBRANE FUNCTION

Artificial membrane systems can be prepared by appropriate techniques. These systems generally consist of mixtures of one or more phospholipids of natural or synthetic origin that can be treated (eg, by using mild sonication) to form spherical vesicles in which the lipids form a bilayer. Such vesicles, surrounded by a lipid bilayer, are termed liposomes.

Some of the advantages and uses of artificial membrane systems in the study of membrane function follow.

(1) The lipid content of the membranes can be varied, allowing systematic examination of the effects of varying lipid composition on certain functions. For instance, vesicles can be made that are composed solely of phosphatidylcholine or, alternatively, of known mixtures of different phospholipids, glycolipids, and cholesterol. The fatty acid moieties of the lipids used can also be varied by employing synthetic lipids of known composition to permit systematic examination of the effects of fatty acid composition on certain membrane functions (eg, transport).

(2) Purified membrane proteins or enzymes can be incorporated into these vesicles in order to assess what factors (eg, specific lipids or ancillary proteins) the proteins require to reconstitute their function. Investigations of purified proteins, eg, the Ca^{2+} ATPase of the sarcoplasmic reticulum, have in certain cases suggested that only a single protein and a single lipid are required to reconstitute an ion pump.

(3) The environment of these systems can be rigidly controlled and systematically varied (eg, ion concentrations). The systems can also be exposed to known ligands if, for example, the liposomes contain specific receptor proteins.

(4) When liposomes are formed, they can be made to entrap certain compounds inside themselves, eg, drugs and isolated genes. There is interest in using liposomes to distribute drugs to certain tissues, and if components (eg, antibodies to certain cell surface molecules) could be incorporated into liposomes so that they would be targeted to specific tissues or tumors, the therapeutic impact could be considerable. DNA entrapped inside liposomes appears to be less sensitive to attack by nucleases; this approach may prove useful in attempts at gene therapy.

FUNCTIONAL MEMBRANES CONTAIN GLOBULAR INTEGRAL PROTEINS DISPERSED IN A FLUID PHOSPHOLIPID MATRIX

The fluid mosaic model of membrane structure was proposed in 1972 by Singer and Nicolson (Figure 43–7). Early evidence for the model was the rapid and random redistribution of species-specific integral proteins in the plasma membrane of an interspecies hybrid cell formed by the artificially induced fusion of two different parent cells. It has subsequently been demonstrated that phospholipids also undergo rapid redistribution in the plane of the membrane. This diffusion within the plane of the membrane, termed translational diffusion, can be quite rapid for a phospholipid; in fact, within the plane of the membrane, one molecule of phospholipid can move several micrometers per second.

The phase changes, and thus the fluidity of membranes, are highly dependent upon the lipid composition of the membrane. In a lipid bilayer, the hydrophobic chains of the fatty acids can be highly aligned or ordered to provide a rather stiff structure. As the temperature increases, the hydrophobic side chains undergo a transition from the ordered state to a disordered one, taking on a more liquid-like or fluid arrangement. The temperature at which the structure undergoes the transition from ordered to disordered is called the "transition temperature." The longer and more saturated fatty acid chains exhibit higher transition temperatures, ie, higher temperatures are required to increase the fluidity of the structure. Unsaturated bonds that exist in the *cis* configuration tend to increase the fluidity of a bilayer by decreasing the compactness of the side chain packing without diminishing hydrophobicity (Figure 43–3). The phospholipids of cellular membranes generally contain at least one unsaturated fatty acid with at least one *cis* double bond.

Cholesterol acts as a modulator molecule in membranes, producing intermediate states of fluidity. If the acyl side chains exist in a disordered phase, cholesterol will have a condensing effect; if the acyl side chains are ordered or in a crystalline phase, cholesterol will induce disorder. At high cholesterol: phospholipid ratios, transition temperatures are abolished altogether.

The fluidity of a membrane significantly affects its functions. As membrane fluidity increases, so does its permeability to water and other small hydrophilic molecules. The lateral mobility of integral proteins increases as the fluidity of the membrane increases. If the active site of an integral protein involved in some given function resides exclusively in its hydrophilic regions, changing lipid fluidity will probably have little effect on the activity of the protein; however, if the protein is involved in a transport function in which transport components span the membrane, lipid phase effects may significantly alter the transport rate. The insulin receptor is an excellent example of altered function with changes in fluidity (see Chapter 51). As the concentration of unsaturated fatty acids in the membrane is increased (by growing cultured cells in a medium rich in such molecules), fluidity increases. This alters the receptor so that it binds more insulin.

A state of fluidity and thus of translational mobility in a membrane may be confined to certain regions of membranes under certain conditions. For example, protein-protein interactions may take place within the plane of the membrane, such that the integral proteins form a rigid matrix, in contrast to the more usual situation, where the lipid acts as the matrix. Such regions of rigid protein matrix can exist side by side in the same membrane with the usual lipid matrix. Gap junctions, tight junctions, and bacteriorhodopsin-containing regions of the purple membranes of halobacteria are clear examples of such side-by-side coexistence of different matrices.

Some of the protein-protein interactions taking place within the plane of the membrane may be mediated by interconnecting peripheral proteins, such as cross-linking antibodies or lectins that are known to patch or cap on membrane surfaces. Thus, peripheral proteins, by their specific attachments, may restrict the mobility of integral proteins within the membrane.

THE ASSEMBLY OF MEMBRANES IS COMPLEX

There are many cellular membranes, each with its own specific features. No satisfactory scheme describing the assembly of any one of these membranes is available. In this section, areas of major progress are described out of which it is hoped that coherent models of membrane assembly will eventually emerge. Both lipids and proteins must be considered in such models. Coverage of the former will be brief, since little is known about how lipids are assembled into membranes.

Asymmetry of Both Proteins and Lipids Is Maintained During Membrane Assembly

Vesicles formed from membranes of the endoplasmic reticulum and Golgi apparatus, either naturally or pinched off by homogenization, exhibit transverse asymmetries of both lipid and protein. These asymmetries are maintained during fusion of transport vesicles with the plasma membrane. The inside of the vesicles after fusion becomes the outside of the plasma membrane, and the cytoplasmic side of the vesicles remains the cytoplasmic side of the membrane (Figure 43–9). Since the transverse asymmetry of the membranes already exists in the vesicles of the endoplasmic reticulum well before they are fused to the plasma membrane, a major problem of membrane assembly becomes understanding how the integral proteins are inserted into the lipid bilayer of the endoplasmic reticulum. This problem is addressed below.

Phospholipids are the major class of lipid in membranes. The enzymes responsible for the synthesis of phospholipids reside in the cytoplasmic surface of the cisternae of the endoplasmic reticulum. As phospholipids are synthesized at that site, they probably self-assemble into thermodynamically stable bimolecular layers, thereby expanding the membrane and perhaps promoting the detachment of so-called lipid vesicles from it. It has been proposed that these vesicles travel to other sites, donating their lipids to other membranes; however, little is known about this matter. Cytosolic proteins that take up phospholipids from one membrane and release them to another (ie, phospholipid exchange proteins) have been demonstrated; they probably play a role in contributing to the specific lipid composition of various membranes.

Many Proteins Are Targeted by Signal Sequences to Their Correct Destinations

The protein biosynthetic pathways in cells can be considered to be one large sorting system. Many proteins carry signals (usually but not always specific sequences of amino acids) that target them to their destination, thus ensuring that they will end up in the correct membrane or cell compartment; these signals are a fundamental component of the sorting system.

A major sorting decision is made early in protein biosynthesis, when specific proteins are synthesized either on free or on membrane-bound polyribosomes. This results in two sorting branches, called the cytosolic and rough endoplasmic reticulum branch, respectively (Figure 43–10). This sorting occurs because proteins synthesized on membrane-bound polyribosomes contain a signal peptide that mediates their attachment to the membrane of the endoplasmic reticulum. Further details on the signal peptide are given below. Proteins synthesized on free polyribosomes lack this signal peptide and are delivered into the cytosol. There they are directed to mitochondria, nuclei, and peroxisomes by specific signals—or remain in the cytosol if they lack a signal.

Proteins synthesized and sorted in the rough endoplasmic reticulum branch (Figure 43–11) include many destined for various membranes (eg, of the en-

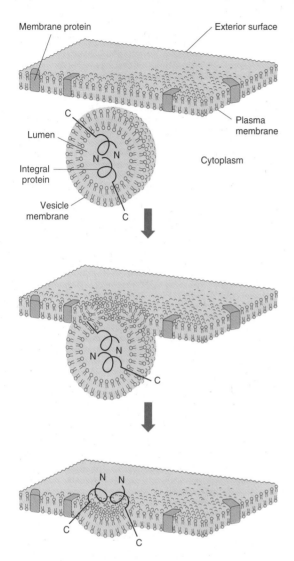

Figure 43–9. Fusion of a vesicle with the plasma membrane preserves the orientation of any integral proteins embedded in the vesicle bilayer. Initially, the amino terminal of the protein faces the lumen, or inner cavity, of such a vesicle. After fusion, the amino terminal is on the exterior surface of the plasma membrane. That the orientation of the protein has not been reversed can be perceived by noting that the other end of the molecule, the carboxyl terminal, is always immersed in the cytoplasm. The lumen of a vesicle and the outside of the cell are topologically equivalent. (Redrawn and modified, with permission, from Lodish HF, Rothman JE: The assembly of cell membranes. Sci Am [Jan] 1979;240:43.)

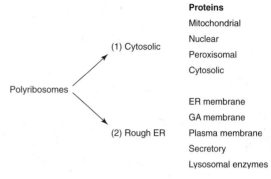

Figure 43–10. Diagrammatic representation of the two branches of protein sorting occurring by synthesis on (1) cytosolic and (2) membrane-bound polyribosomes. The mitochondrial proteins listed are encoded by nuclear genes. The signals used in further sorting of most of these proteins are listed in Table 43–9. (ER, endoplasmic reticulum.)

signal-mediated sorting of certain proteins occurs in the Golgi apparatus, resulting in delivery to lysosomes, membranes of the Golgi apparatus, and other sites. Proteins destined for the plasma membrane or for secretion pass through the Golgi apparatus but generally do not carry specific sorting signals and reach their destinations by default.

The entire pathway of endoplasmic reticulum → Golgi apparatus → plasma membrane is often called the secretory or exocytotic pathway. Events along this route will be given special attention. Most of the proteins reaching the Golgi apparatus or the plasma membrane are carried in transport vesicles; a brief description of the formation of these important particles will be given subsequently. Other proteins destined for secretion are carried in secretory vesicles (Figure 43–11). These are prominent in the pancreas and certain other glands. Their mobilization and discharge is regulated and is often referred to as "regulated secretion," whereas the secretory pathway involving transport vesicles is called "constitutive."

The sorting of proteins belonging to the cytosolic branch referred to above is described next.

The Mitochondrion Both Imports and Synthesizes Proteins

Mitochondria contain many proteins. Thirteen proteins are encoded by the mitochondrial genome (Table 64–3) and synthesized in that organelle using its own protein-synthesizing system. However, the majority are encoded by nuclear genes, are synthesized outside the mitochondria on cytosolic polyribosomes, and must be imported. Yeast cells have proved to be a particularly useful system for analyzing the mechanisms of import of such mitochondrial proteins, partly because of the ability to generate mu-

doplasmic reticulum, Golgi apparatus, lysosomes, and plasma membrane) and for secretion. Lysosomal enzymes are also included. Thus, such proteins may reside in the membranes or lumina of the endoplasmic reticulum or follow the major transport route of intracellular proteins to the Golgi apparatus. Further

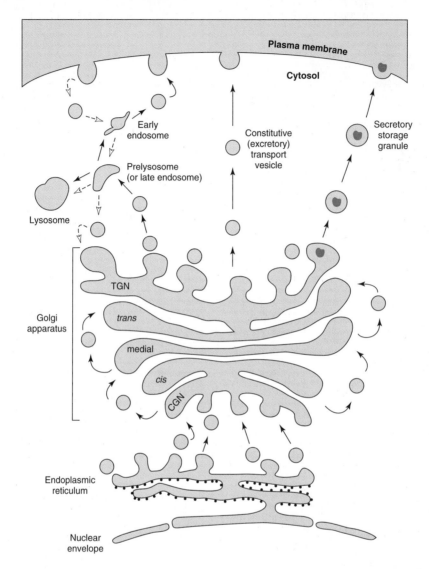

Figure 43–11. Diagrammatic representation of the rough endoplasmic reticulum branch of protein sorting. Newly synthesized proteins are inserted into the endoplasmic reticulum membrane or lumen from membrane-bound polyribosomes (small black circles studding the cytosolic face of the endoplasmic reticulum). Those proteins that are transported out of the endoplasmic reticulum (indicated by solid black arrows) do so from ribosome-free transitional elements. Such proteins may then pass through the various subcompartments of the Golgi until they reach the TGN, the exit side of the Golgi. In the TGN, proteins are segregated and sorted. Secretory proteins accumulate in secretory storage granules from which they may be expelled as shown in the upper right-hand side of the figure. Proteins destined for the plasma membrane or those that are secreted in a constitutive manner are carried out to the cell surface in transport vesicles, as indicated in the upper middle area of the figure. Some proteins may reach the cell surface via late and early endosomes. Other proteins enter prelysosomes (late endosomes) and are selectively transferred to lysosomes. The endocytic pathway illustrated in the upper left-hand area of the figure is considered elsewhere in this chapter. Retrieval from the Golgi apparatus to the endoplasmic reticulum is not considered in this scheme. (CGN, *cis*-Golgi network; TGN, *trans*-Golgi network.) (Courtesy of E Degen.)

tants that have illuminated the fundamental processes involved. Most progress has been made in the study of proteins present in the mitochondrial matrix, such as the F_1 ATPase subunits. These proteins must pass through the outer and inner mitochondrial membranes to reach their destination, and translocation occurs at sites where these membranes are in contact (translocation contact sites). They have an amino terminal leader sequence (presequence), up to 70 amino acids in length, which is not highly conserved but contains many positively charged amino acids (eg, Lys or Arg). This sequence is equivalent to a signal peptide, directing these proteins into the matrix; if the leader sequence is cleaved off, these proteins will not enter. There are receptors for these proteins on the surface of the outer mitochondrial membrane, and protein-lined channels in both membranes permit their import. Evidence indicates that they must be in an unfolded state to pass through these membranes, which is made possible by ATP-dependent binding to several chaperone proteins, one of which resembles one found in the lumen of the endoplasmic reticulum, BiP (see below). The roles of chaperone proteins in protein folding are discussed later in this chapter. A proton-motive force across the inner membrane is required for import; it is made up of the electric potential across the membrane (inside negative) and the pH gradient (Chapter 14). It is possible that the positively charged leader sequence may be helped through the membrane by the negative charge in the matrix. The presequence is split off by a metalloproteinase present in the matrix. Contact with two other chaperones present in the matrix is essential to complete the overall process of import. Interaction with hsp70 (hsp = "heat shock protein"), another BiP-like protein, ensures proper import into the matrix and prevents misfolding or aggregation, while interaction with hsp60 ensures proper folding. The latter is a very complex protein and resembles the bacterial chaperones called "GroEL." The interactions of imported proteins with both of these chaperones require hydrolysis of ATP to drive them.

Additional steps are required to target proteins to the inner membrane and intermembrane space. Certain other mitochondrial proteins do not contain presequences, so that a variety of mechanisms and routes are employed by proteins to enter mitochondria.

Nuclear and Peroxisomal Proteins Also Contain Signal Sequences

Nuclear pores allow free passage of proteins of small size, but larger proteins (approximately 60 kDa or more) require special signals called nuclear localization signals (NLSs). These sequences are diverse in structure, need not be present at the amino terminal, and are not removed. Nuclear proteins interact with cytosolic chaperones prior to interaction with receptors on the nuclear pores. Relatively little is known of the details of import except that it is energy-dependent.

Many peroxisomal proteins (Chapters 13 and 24) contain a peroxisomal targeting signal (PTS) at their carboxyl terminals. One frequent sequence is Ser-Lys-Leu. Receptor proteins are located on the peroxisomal membrane, and internal proteins involved in transport of imported proteins into the matrix are also present. Studies of Zellweger syndrome have illuminated this area. This condition is apparent at birth and is characterized by profound neurologic impairment, victims usually dying within 1 year after birth.

By electron microscopy, the peroxisomes are found to be empty, lacking their normal matrix content; their membranes are apparently normal. At the biochemical level, there is an accumulation of very long chain fatty acids, abnormalities of the synthesis of bile acids, and a marked reduction in plasmalogens, all reflecting impairments of peroxisomal functions. The basic defect is a failure to import peroxisomal proteins into the matrix.

Based on genetic studies, it has been suggested that about 13 genes may be involved in peroxisomal importation. One peroxisomal assembly factor has been cloned and found to be related to P-glycoprotein (Chapter 62), itself a transport protein. Various dietary regimens, in which the amounts of various lipids are carefully adjusted, are under investigation as treatment for Zellweger syndrome and closely related milder abnormalities of peroxisomal function (eg, neonatal adrenoleukodystrophy, infantile Refsum's disease, and hyperpipecolic acidemia).

The protein affected in X-linked adrenoleukodystrophy appears to be the transporter responsible for uptake of long chain fatty acyl-CoA synthase into peroxisomes. Adrenoleukodystrophy is characterized by adrenal atrophy and diffuse cerebral demyelination. It results in widespread neurologic damage and death at an early age.

The Signal Hypothesis Explains How Polyribosomes Bind to the Endoplasmic Reticulum

As indicated above, the rough endoplasmic reticulum branch is the second of the two branches involved in the synthesis and sorting of proteins. In this branch, proteins are synthesized on membrane-bound polyribosomes and translocated into the lumen of the rough endoplasmic reticulum prior to further sorting (Figure 43–11).

The signal hypothesis was proposed by Blobel and Sabatini, at least in part to explain the distinction between free and membrane-bound polyribosomes. They found that proteins synthesized on membrane-bound polyribosomes contained a peptide extension (signal peptide) at their amino terminals that mediated their attachment to the membranes of the endoplasmic reticulum. As noted above, proteins whose entire synthesis occurs on free polyribosomes lack this signal peptide. An important aspect of the signal hypothesis was that it suggested—as turns out to be the case—that all ribosomes have the same structure and that the distinction between membrane-bound and free ribosomes depends solely on the former carrying proteins that have signal peptides. Much evidence has confirmed the original hypothesis. Because many membrane proteins are synthesized on membrane-bound polyribosomes, the signal hypothesis plays an important role in concepts of membrane assembly. Some characteristics of signal peptides are summarized in Table 43–3.

Table 43–3. Some properties of signal peptides.

Usually, but not always, located at the N-terminus.
Contain approximately 12–35 amino acids.
Methionine is usually the N-terminal amino acid.
Contain a central cluster of hydrophobic amino acids.
Contain at least one positively charged amino acid near their N-terminus.
Usually cleaved off at the C-terminal end of an Ala residue by signal peptidase.

Figure 43–12 illustrates the principal features in relation to the passage of a secreted protein through the membrane of the endoplasmic reticulum. It incorporates features from the original signal hypothesis and from subsequent work. The mRNA for such a protein encodes an amino terminal signal peptide (also variously called a leader sequence, a transient insertion signal, a signal sequence, or a presequence). The signal hypothesis proposed that the protein is inserted into the endoplasmic reticulum membrane at the same time as its mRNA is being translated on polyribosomes, so-called cotranslational insertion. As the signal peptide emerges from the large subunit of the ribosome, it is recognized by a signal recognition particle (SRP) that blocks further translation af-

ter about 70 amino acids have been polymerized (40 buried in the large ribosomal subunit and 30 exposed). The block is referred to as elongation arrest. The SRP contains six proteins and has a 7S RNA associated with it that is closely related to the Alu family of highly repeated DNA sequences (Chapter 38). The SRP-imposed block is not released until the SRP-signal peptide-polyribosome complex has bound to the so-called docking protein (a receptor for the SRP) on the endoplasmic reticulum membrane; the SRP thus guides the signal peptide to its receptor in the endoplasmic reticulum membrane and prevents premature folding and expulsion of the protein being synthesized into the cytosol.

The receptor for the SRP is an integral membrane protein composed of α and β subunits. The α subunit binds GDP. When the SRP-signal peptide complex interacts with the receptor, the exchange of GDP for GTP is stimulated. This form of the receptor (with GTP bound) has a high affinity for the SRP and thus releases the signal peptide, which binds to the translocation machinery, also present in the endoplasmic reticulum membrane. The α subunit then hydrolyzes its bound GTP, restoring GDP and completing a GTP-GDP cycle. The unidirectionality of this cycle helps drive the interaction of the polyribosome

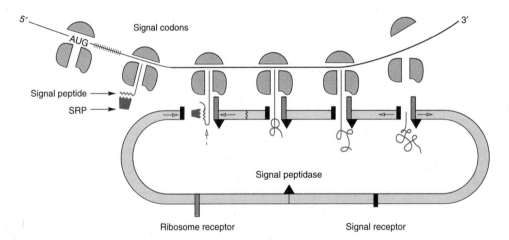

Figure 43–12. Diagram of the signal hypothesis for the transport of secreted proteins across the endoplasmic reticulum membrane. The ribosomes synthesizing a protein move along the messenger RNA specifying the amino acid sequence of the protein. (The messenger is represented by the line between 5′ and 3′.) The codon AUG marks the start of the message for the protein; the hatched lines that follow AUG represent the codons for the signal sequence. As the protein grows out from the larger ribosomal subunit, the signal sequence is exposed and bound by the signal recognition particle (SRP). Translation is blocked until the complex binds to the "docking protein" (represented by the solid bar) on the endoplasmic reticulum membrane. There is also a receptor (open bar) for the ribosome itself. The interaction of the ribosome and growing peptide chain with the endoplasmic reticulum membrane results in the opening of a pore through which the protein is transported to the interior space of the endoplasmic reticulum. During transport, the signal sequence of most proteins is removed by an enzyme called the "signal peptidase." The completed protein is eventually released by the ribosome, which then separates into its two components, the large and small ribosomal subunits. The protein ends up inside the endoplasmic reticulum. (Slightly modified and reproduced, with permission, from Marx JL: Newly made proteins zip through the cell. Science 1980;207:164. Copyright © 1980 by the American Association for the Advancement of Science.)

and its signal peptide with the endoplasmic reticulum membrane in the forward direction.

The translocation machinery consists of a number of membrane proteins that probably form a protein-conducting channel in the endoplasmic reticulum membrane through which the newly synthesized protein may pass. The channel appears to be open only when a signal peptide is present, preserving conductance across the ER membrane when it closes.

The insertion of the signal peptide into the conducting channel, while the other end of the parent protein is still attached to ribosomes, is termed "cotranslational insertion." The process of elongation of the remaining portion of the protein probably facilitates passage of the nascent protein across the lipid bilayer as the ribosomes remain attached to the membrane of the endoplasmic reticulum. Thus, the rough (or ribosome-studded) endoplasmic reticulum is formed. It is important that the protein be kept in an unfolded state while passing through the conducting channel—otherwise, it may not make it across.

Ribosomes remain attached to the endoplasmic reticulum during synthesis of the membrane protein but are released and dissociated into their two types of subunits when the process is completed. The signal peptide is hydrolyzed by signal peptidase (Figure 43–12); its precise fate has not been established.

Cytochrome P450 (Chapter 61), an integral endoplasmic reticulum membrane protein, does not completely cross the membrane. Instead, it resides in the membrane with its signal peptide intact. Its passage through the membrane is prevented by a sequence of amino acids called a halt- or stop-transfer signal.

Secretory proteins and proteins destined for membranes distal to the endoplasmic reticulum completely traverse the membrane bilayer and are discharged into the lumen of the endoplasmic reticulum. N-Glycan chains, if present, are added (Chapter 56) as these proteins traverse the inner part of the endoplasmic reticulum membrane—a process called "cotranslational glycosylation." Subsequently, the proteins are found in the lumen of the Golgi apparatus, where further changes in glycan chains occur (Figure 56–9) prior to intracellular distribution or secretion. There is strong evidence that the signal peptide is involved in the process of protein insertion into endoplasmic reticulum membranes. Mutant proteins, containing altered signal peptides in which a hydrophobic amino acid is replaced by a hydrophilic one, are not inserted into endoplasmic reticulum membranes. Nonmembrane proteins (eg, α-globin) to which signal peptides have been attached by genetic engineering can be inserted into the lumen of the endoplasmic reticulum or even secreted.

Proteins Follow Several Routes to Be Inserted Into or Attached to the Membranes of the Endoplasmic Reticulum

The routes that proteins follow to be inserted into the membranes of the endoplasmic reticulum include the following:

A. Cotranslational Insertion: Figure 43–8 shows a variety of ways in which proteins are disposed in the plasma membrane. In particular, the amino terminals of certain proteins (eg, the LDL receptor) can be seen to be on the extracytoplasmic face, whereas for other proteins (eg, the asialoglycoprotein receptor) the carboxyl terminals are on this face. To explain these dispositions, one must consider the initial biosynthetic events at the endoplasmic reticulum membrane. The LDL receptor enters the endoplasmic reticulum membrane in a manner analogous to a secretory protein (Figure 43–12); it partly traverses the endoplasmic reticulum membrane, its signal peptide is cleaved, and its amino terminal protrudes into the lumen. However, it is retained in the membrane because it contains a highly hydrophobic segment, the halt- or stop-transfer signal. This sequence forms the single transmembrane segment of the protein and is its membrane-anchoring domain. The small patch of endoplasmic reticulum membrane in which the newly synthesized LDL receptor is located subsequently buds off as a component of a transport vesicle, probably from the transitional elements of the endoplasmic reticulum (Figure 43–11). As described above in the discussion of asymmetry of proteins and lipids in membrane assembly, the disposition of the receptor in the endoplasmic reticulum membrane is preserved in the vesicle, which eventually fuses with the plasma membrane. In contrast, the asialoglycoprotein receptor possesses an internal insertion sequence, which inserts into the membrane but is not cleaved. This acts as an anchor, and its carboxyl terminal is extruded through the membrane. The more complex disposition of the transporters (eg, for glucose) can be explained by the fact that alternating transmembrane α-helices act as uncleaved insertion sequences and as halt-transfer signals, respectively. Each pair of helical segments is inserted as a hairpin. Sequences that determine the structure of a protein in a membrane are called topogenic sequences. As explained in the legend to Figure 43–8, the above three proteins are examples of type I, type II, and type III transmembrane proteins.

B. Synthesis on Free Polyribosomes and Subsequent Attachment to the Endoplasmic Reticulum Membrane: An example is cytochrome b_5.

C. Retention at the Luminal Aspect of the Endoplasmic Reticulum by Specific Amino Acid Sequences: A number of proteins possess the amino acid sequence KDEL (Lys-Asp-Glu-Leu) at their carboxyl terminal. This sequence specifies that such proteins will be attached to the inner aspect of the endoplasmic reticulum in a relatively loose manner. The chaperone BiP (see below) is one such protein.

D. Retrograde Transport From the Golgi Apparatus: Certain proteins destined for the membranes of the endoplasmic reticulum may pass to the Golgi and then return, by retrograde vesicular transport, to the endoplasmic reticulum to be inserted therein.

The foregoing paragraphs demonstrate that a variety of routes are involved in assembly of the proteins of the endoplasmic reticulum membranes; a similar situation probably holds for other membranes (eg, the mitochondrial membranes and the plasma membrane). Precise targeting sequences have been identified for only a few of the above mechanisms (eg, KDEL sequences). It has been shown that the half-lives of the lipids of the endoplasmic reticulum membranes of rat liver are generally shorter than those of its proteins, so that the turnover rates of lipids and proteins are independent. Indeed, different lipids have been found to exhibit different half-lives. Furthermore, the half-lives of the proteins of these membranes vary quite widely, some exhibiting short (hours) and others long (days) half-lives. Thus, individual lipids and proteins of the endoplasmic reticulum membranes appear to be inserted into it relatively independently.

Proteins Move Through Cellular Compartments to Specific Membranes

A scheme representing the possible flow of membrane proteins along the endoplasmic reticulum → Golgi apparatus → plasma membrane route is shown in Figure 43–13. The horizontal arrows denote transport steps that may be independent of targeting signals, whereas the vertical open arrows represent steps that depend on specific signals. Thus, flow of certain membrane proteins from the endoplasmic reticulum to the plasma membrane (designated "bulk flow," as it is nonselective) probably occurs without any targeting sequences being involved, ie, by default. On the other hand, insertion of resident proteins into the endoplasmic reticulum and Golgi membranes is dependent upon specific signals (eg, KDEL or halt-transfer sequences for the endoplasmic reticulum). Similarly, transport of many enzymes to lysosomes is dependent upon the Man 6-P signal (Chapter 56), and a signal may be involved for entry of proteins into secretory granules. Table 43–4 summarizes information on sequences that are known to be involved in targeting various proteins to their correct intracellular sites.

Chaperones Are Proteins That Prevent Faulty Folding and Unproductive Interactions of Other Proteins

Exit from the endoplasmic reticulum may be the rate-limiting step in the secretory pathway. In this context, it has been found that certain proteins play a role in the assembly or proper folding of other proteins without themselves being components of the latter. Such proteins are called molecular chaperones; a number of important properties of these proteins are listed in Table 43–5, and the names of some of particular importance in the endoplasmic reticulum are listed in Table 43–6. Basically, they stabilize unfolded or partially folded intermediates, allowing them time to fold properly, and prevent inappropriate interactions, thus combating the formation of nonfunctional structures. Most chaperones exhibit

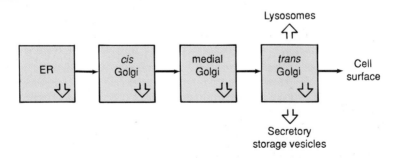

Figure 43–13. Flow of membrane proteins from the endoplasmic reticulum (ER) to the cell surface. Horizontal arrows denote steps that have been proposed to be signal independent and thus represent bulk flow. The open vertical arrows in the boxes denote retention of proteins that are resident in the membranes of the organelle indicated. The open vertical arrows outside the boxes indicate signal-mediated transport to lysosomes and secretory storage granules. (Reproduced, with permission, from Pfeffer SR, Rothman JE: Biosynthetic protein transport and sorting by the endoplasmic reticulum and Golgi. Annu Rev Biochem 1987;56:829.)

Table 43–4. Sequences or compounds that direct proteins to specific organelles.

Targeting Sequence or Compound	Organelle Targeted
Signal peptide sequence	Membrane of ER
C-terminal KDEL sequence (Lys·Asp·Glu·Leu)	Luminal surface of ER
N-terminal sequence (70-residue positive region)	Mitochondrion
Short, basic amino acid sequences	Nucleus
Mannose 6-phosphate	Lysosome

Table 43–6. Some chaperones and enzymes involved in folding that are located in the rough endoplasmic reticulum.

- BiP (immunoglobulin heavy chain binding protein)
- GRP94 (glucose-regulated protein)
- Calnexin
- PDI (protein disulfide isomerase)
- PPI (peptidyl prolyl *cis-trans* isomerase)

ATPase activity and bind ADP and ATP. This activity is important in their effect on folding. The ADP-chaperone complex often has a high affinity for the unfolded protein, which, when bound, stimulates release of ADP with replacement by ATP. The ATP-chaperone complex, in turn, releases segments of the protein that have folded properly, and the cycle involving ADP and ATP binding is repeated until the folded protein is released.

Several examples of chaperones were met when the sorting of mitochondrial proteins was discussed above. The immunoglobulin heavy chain binding protein (BiP) is located in the lumen of the endoplasmic reticulum. This protein will bind abnormally folded immunoglobulin heavy chains and certain other proteins and prevent them from leaving the endoplasmic reticulum, in which they are degraded. Another important chaperone is calnexin, located in the endoplasmic reticulum membrane. This protein binds a wide variety of proteins, including mixed histocompatibility (MHC) antigens and a variety of serum proteins. As mentioned in Chapter 56, calnexin binds the monoglucosylated species of glycoproteins that occur during processing of glycoproteins, retaining them in the endoplasmic reticulum until the glycoprotein has folded properly. Chaperones are not the only proteins in the endoplasmic reticulum lumen that are concerned with proper folding of proteins. Two enzymes are present that play an active role in folding. Protein disulfide isomerase (PDI) promotes rapid reshuffling of disulfide bonds until the correct set is achieved. Peptidyl prolyl isom-

Table 43–5. Some properties of chaperone proteins.

- Present in a wide range of species from bacteria to humans.
- Many are so-called heat shock proteins (hsp).
- Some are inducible by conditions that cause unfolding of newly synthesized proteins (eg, elevated temperature and various chemicals).
- They bind unfolded and aggregated proteins.
- Most chaperones show associated ATPase activity, with ATP or ADP being involved in the protein-chaperone interaction.
- Found in various cellular compartments such as cytosol, mitochondria, and the lumen of the endoplasmic reticulum.

erase (PPI) accelerates folding of proline-containing proteins by catalyzing the *cis-trans* isomerization of X-Pro bonds, where X is any amino acid residue.

Transport Vesicles Are Key Players in Intracellular Protein Traffic

Most proteins that are synthesized on membrane-bound polyribosomes and are destined for the Golgi apparatus or plasma membrane reach these sites inside transport vesicles. How proteins that are synthesized in the rough endoplasmic reticulum are inserted into these vesicles is not known. Those involved in transport to the Golgi apparatus and from the Golgi to the plasma membrane are mainly clathrin-free, unlike the coated vesicles involved in endocytosis (see discussion of the LDL receptor in Chapter 28). For the sake of clarity, the non-clathrin-coated vesicles will be referred to in this text as transport vesicles. There is evidence that proteins destined for the membranes of the Golgi apparatus have specific signal sequences. On the other hand, most proteins destined for the plasma membrane or for secretion do not appear to contain specific signals, reaching these destinations by default.

The Golgi Apparatus Is Involved in Glycosylation and Sorting of Proteins

The Golgi apparatus plays two important roles in membrane synthesis. First, it is involved in the processing of the oligosaccharide chains of membrane and other *N*-linked glycoproteins and also contains enzymes involved in *O*-glycosylation. Second, it is involved in the sorting of various proteins prior to their delivery to their appropriate intracellular destinations. All parts of the Golgi apparatus participate in the first role, whereas the *trans*-Golgi is particularly involved in the second and is very rich in vesicles. Because of their central role in protein transport, considerable research has been conducted in recent years concerning the formation and fate of transport vesicles.

A Model of Non Clathrin-Coated Vesicles Involves SNAREs and Other Factors

Vesicles—mainly but not exclusively of the non-clathrin-coated variety—lie at the heart of intracellular transport of many proteins. Recently, significant progress has been made in understanding the events involved in vesicle formation and transport. This has transpired because of the use of a number of approaches. These include establishment of cell-free systems with which to study vesicle formation. For instance, it is possible to observe, by electron microscopy, budding of vesicles from Golgi preparations incubated with cytosol and ATP. The development of genetic approaches for studying vesicles in yeast and the use of brefeldin A have also been crucial. The picture is complex, with its own nomenclature (Table 43–7) and involves a variety of cytosolic and membrane proteins, GTP, ATP, and accessory factors.

Based largely on a proposal by Rothman and Warren, anterograde vesicular transport can be considered to occur in eight steps (Figure 43–14). The basic concept is that each transport vesicle bears a unique address marker consisting of one or more v-SNAREs, while each target membrane bears one or more complementary t-SNAREs.

Step 1: Coat assembly is initiated when ARF (ADP-ribosylation factor) is activated by binding GTP, which is exchanged for GDP. This leads to the association of GTP-bound ARF with its putative receptor (hatched in Figure 43–14) in the donor membrane. To participate, ARF must first be modified by addition of myristic acid (C14:0), employing myristoyl-CoA as the acyl donor. Myristoylation is one of a number of enzyme-catalyzed posttranslational modifications, involving addition of certain lipids to specific residues of proteins, that facilitate the binding of proteins to the cytosolic surfaces of membranes or vesicles. Others are addition of palmitate, farnesyl, and geranylgeranyl; the two latter molecules are polyisoprenoids containing 15 and 20 carbon atoms, respectively.

Step 2: Membrane-associated ARF recruits the coat proteins that comprise the coatomer shell, forming a coated bud.

Step 3: The bud pinches off in a process involving acyl-CoA—and probably ATP—to complete the formation of the coated vesicle.

Step 4: Coat disassembly (involving dissociation of ARF and coatomer shell) follows hydrolysis of bound GTP; uncoating is necessary for fusion to occur.

Step 5: Vesicle targeting is achieved via members of a family of integral proteins, termed v-SNAREs, that tag the vesicle during its budding. v-SNAREs pair with cognate t-SNAREs in the target membrane to dock the vesicle.

It is presumed that steps 4 and 5 are closely coupled and that step 4 may follow step 5, with ARF and the coatomer shell rapidly dissociating after docking.

Step 6: The general fusion machinery then assembles on the paired SNARE complex; it includes an ATPase (NSF; NEM-sensitive factor) and the SNAP (soluble NSF attachment factor) proteins. SNAPs bind to the SNARE (SNAP receptor) complex, enabling NSF to bind.

Step 7: Hydrolysis of ATP by NSF is essential for fusion, a process that can be inhibited by NEM (N-ethylmaleimide). Certain other proteins and calcium are also required.

Step 8: Retrograde transport occurs to restart the cycle. This last step may retrieve certain proteins or recycle v-SNAREs. Nocodazole, a microtubule-disrupting agent, inhibits this step.

Brefeldin A Inhibits the Coating Process

The following points expand and clarify the above.

(1) The association between the ARF receptor and v-SNARE in step 1 is speculative, but some means is needed to pack v-SNAREs into buds. It is possible that other proteins are involved in this step, such as rab proteins (see below) or heterotrimeric G proteins.

(2) The fungal metabolite brefeldin A prevents GTP from binding to ARF in step 1 and thus inhibits the entire coating process. In its presence, the Golgi apparatus appears to disintegrate, and fragments are lost. It may do this by inhibiting the guanine nucleotide exchanger involved in step 1.

(3) GTP-γ-S (a nonhydrolyzable analog of GTP often used in investigations of the role of GTP in biochemical processes) blocks disassembly of the coat from coated vesicles, leading to a build-up of coated vesicles.

(4) A family of ras-like proteins (Chapter 62), called the rab protein family, are required in several steps of intracellular protein transport, regulated secretion, and endocytosis. They are small monomeric GTPases which attach to the cytosolic faces of membranes via geranylgeranyl chains. Their precise role in the cycle described above is not clear at present; one possibility is that they interact with SNAREs and catalyze or monitor their interactions.

Table 43–7. Factors involved in the formation of non–clathrin-coated vesicles and their transport.

- ARF: ADP-ribosylation factor, the name given to the original member of this family of proteins
- Coatomer: A family of at least seven coat proteins (α, β, γ, δ, ϵ, β', and ξ)
- SNAP: Soluble NSF attachment factor
- SNARE: SNAP receptor
- v-SNARE: Vesicle SNARE
- t-SNARE: Target SNARE
- GTP-γ-S: A nonhydrolyzable analog of GTP, used to test the involvement of GTP
- NEM: N-Ethylmaleimide, a chemical that alkylates sulfhydryl groups
- NSF: NEM-sensitive factor, an ATPase
- Rab proteins: A family of ras-related proteins first observed in rat brain

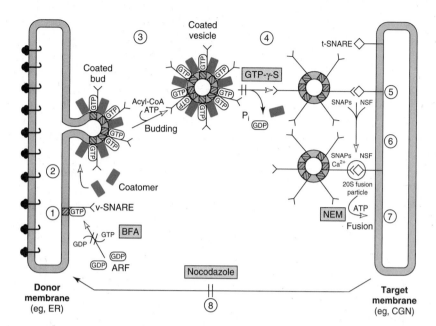

Figure 43–14. Model of the steps in a round of anterograde vesicular transport. The cycle starts in the bottom left-hand side of the figure, where two molecules of ARF are represented as small ovals containing GDP. The steps in the cycle are described in the text. The various abbreviations used are explained in Table 43–7. (Adapted from Rothman JE: Mechanisms of intracellular protein transport. Nature 1994;372:55.) (Courtesy of E Degen.)

(5) The fusion of synaptic vesicles with the plasma membrane of neurons (Chapter 64) involves a series of events similar to that described above. For example, one v-SNARE is designated synaptobrevin and two t-SNAREs are designated syntaxin and SNAP 25 (synaptosome-associated protein of 25 kDa). Botulinum B toxin is one of the most lethal toxins known and is the most serious cause of food poisoning. It has been found that one component of this toxin is a protease that appears to cleave only synaptobrevin, thus inhibiting release of acetylcholine at the neuromuscular junction and possibly proving fatal, depending on the dose taken.

(6) Although this model describes non-clathrin-coated vesicles, it appears likely that many of the events described above apply, at least in principle, to clathrin-coated vesicles.

The above discussion regarding the biogenesis of membranes has shown that this is a complex process about which much is still unknown. One indication of the complexity involved is to consider the number of posttranslational modifications that membrane proteins may be subjected to prior to attaining their mature state. These include proteolysis, glycosylation, addition of a glycophosphatidyl (GPI) anchor, sulfation on tyrosine or carbohydrate moieties, phosphorylation, acylation, and prenylation—a list that is undoubtedly not complete. However, significant progress has been made. Table 43–8 summarizes some of the major features of membrane assembly discussed above.

MEMBRANE SELECTIVITY ALLOWS SPECIALIZED FUNCTIONS

If the plasma membrane is relatively impermeable, how do most molecules enter a cell? How is selectivity of this movement established? Answers to such questions are important in understanding how cells adjust to a constantly changing extracellular environment. Metazoan organisms also must have means of communicating between adjacent and distant cells, so that complex biologic processes can be coordinated. These signals must arrive at and be transmitted by the membrane, or they must be generated as a consequence of some interaction with the membrane.

Table 43–8. Major features of membrane assembly.

- Lipids and proteins are inserted independently into membranes.
- Individual membrane lipids and proteins turn over independently and at different rates.
- Topogenic sequences (eg, signal [amino terminal or internal] and stop-transfer) are important in determining the insertion and disposition of proteins in membranes.
- Membrane proteins inside transport vesicles bud off the endoplasmic reticulum on their way to the Golgi; final sorting of many membrane proteins occurs in the *trans*-Golgi network.
- Specific sorting sequences guide proteins to particular organelles such as lysosomes, peroxisomes, and mitochondria.
- Transport of proteins across membranes into many organelles requires both the unfolded state and ATP.

Some of the major mechanisms used to accomplish these different objectives are listed in Table 43–9.

Passive Mechanisms Move Some Small Molecules Across Membranes

Molecules can passively traverse the bilayer down electrochemical gradients by simple or facilitated diffusion. This spontaneous movement toward equilibrium contrasts with active transport, which requires energy because it constitutes movement against an electrochemical gradient. Figure 43–15 provides a schematic representation of these mechanisms.

As described above, some solutes such as gases can enter the cell by diffusing down an electrochemical gradient across the membrane and do not require metabolic energy. The simple **passive diffusion** of a solute across the membrane is limited by the thermal agitation of that specific molecule, by the concentration gradient across the membrane, and by the solubility of that solute (the permeability coefficient, Figure 43–6) in the hydrophobic core of the membrane bilayer. Solubility is inversely proportionate to the number of hydrogen bonds that must be broken in order for a solute in the external aqueous phase to become incorporated in the hydrophobic bilayer. Electrolytes, poorly soluble in lipid, do not form hydrogen bonds with water, but they do acquire a shell of water from hydration by electrostatic interaction. The size of the shell is directly proportionate to the charge density of the electrolyte. Electrolytes with a large charge density have a larger shell of hydration and thus a slower diffusion rate. Na^+, for example, has a higher charge density than K^+. Hydrated Na^+ is therefore larger than hydrated K^+; hence, the latter tends to move more easily through the membrane.

In natural membranes, as opposed to synthetic membrane bilayers, there are transmembrane channels, pore-like structures composed of proteins that constitute selective ion-conductive pathways. Cation-conductive channels have an average diameter of about 5-8 nm and are negatively charged within the

channel. The permeability of a channel depends upon the size, extent of hydration, and extent of charge density on the ion. Specific channels for Na^+, K^+, and Ca^{2+} have been identified.

The membranes of nerve cells contain well-studied ion channels that are responsible for the action potentials generated across the membrane. The activity of some of these channels is controlled by neurotransmitters; hence, channel activity can be regulated. One ion can regulate the activity of the channel of another ion. For example, a decrease of Ca^{2+} concentration in the extracellular fluid increases membrane permeability and increases the diffusion of Na^+. This depolarizes the membrane and triggers nerve discharge. This may explain the numbness, tingling, and muscle cramps symptomatic of a low level of serum Ca^{2+}.

Channels are open transiently and thus are "gated." Gates can be controlled by opening or closing. In **ligand-gated channels,** a specific molecule binds to a receptor and opens the channel. **Voltage-gated channels** open (or close) in response to a change in membrane potential.

Some microbes synthesize small organic molecules, **ionophores,** that function as shuttles for the movement of ions across membranes. These ionophores contain hydrophilic centers that bind specific ions and are surrounded by peripheral hydrophobic regions; this arrangement allows the molecules to dissolve effectively in the membrane and diffuse transversely therein. Others, like the well-studied polypeptide gramicidin, form channels. Microbial toxins such as diphtheria toxin and activated serum complement components can produce large pores in cellular membranes and thereby provide macromolecules with direct access to the internal milieu.

In summary, net diffusion of a substance depends upon the following: (1) Its concentration gradient across the membrane. Solutes move from high to low concentration. (2) The electrical potential across the membrane. Solutes move toward the solution that has the opposite charge. The inside of the cell usually has a negative charge. (3) The permeability coefficient of the substance for the membrane. (4) The hydrostatic pressure gradient across the membrane. Increased pressure will increase the rate and force of the collision between the molecules and the membrane. (5) Temperature. Increased temperature will increase particle motion and thus increase the frequency of collisions between external particles and the membrane.

Table 43–9. Transfer of material and information across membranes.

Cross-membrane movement of small molecules
  Diffusion (passive and facilitated)
  Active transport
Cross-membrane movement of large molecules
  Endocytosis
  Exocytosis
Signal transmission across membranes
  Cell surface receptors
    1. Signal transduction (eg, glucagon → cAMP)
    2. Signal internalization (coupled with endocytosis, eg, the LDL receptor)
  Movement to intracellular receptors (steroid hormones; a form of diffusion)
Intercellular contact and communication

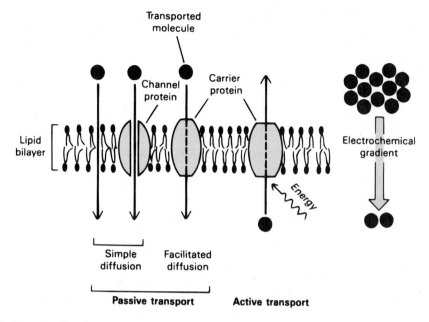

Figure 43–15. Many small uncharged molecules pass freely through the lipid bilayer. Charged molecules, larger uncharged molecules, and some small uncharged molecules are transferred through channels or pores or by specific carrier proteins. Passive transport is always down an electrochemical gradient, toward equilibrium. Active transport is against an electrochemical gradient and requires an input of energy, whereas passive transport does not. (Redrawn and reproduced, with permission, from Alberts B et al: *Molecular Biology of the Cell.* Garland, 1983.)

PLASMA MEMBRANES ARE INVOLVED IN FACILITATED DIFFUSION, ACTIVE TRANSPORT, AND OTHER PROCESSES

Transport systems can be described in a functional sense according to the number of molecules moved and the direction of movement (Figure 43–16) or according to whether movement is toward or away from equilibrium. A **uniport** system moves one type of molecule bidirectionally. In **cotransport** systems, the transfer of one solute depends upon the stoichiometric simultaneous or sequential transfer of another solute. A **symport** moves these solutes in the same direction. Examples are the proton-sugar transporter in bacteria and the Na^+-sugar transporters (glucose, mannose, galactose, xylose, and arabinose) and the Na^+-amino acid transporters in mammalian cells. **Antiport** systems move two molecules in opposite directions (eg, Na^+ in and Ca^{2+} out).

Molecules that cannot pass freely through the lipid bilayer membrane by themselves do so in association with carrier proteins. This involves two processes—facilitated diffusion and active transport—and highly specific transport systems.

Facilitated diffusion and active transport share

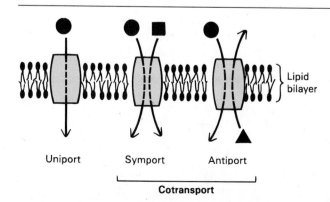

Figure 43–16. Schematic representation of types of transport systems. Transporters can be classified with regard to the direction of movement and whether one or more unique molecules are moved. (Redrawn and reproduced, with permission, from Alberts B et al: *Molecular Biology of the Cell.* Garland, 1983.)

many features. Both appear to involve carrier proteins, and both show specificity for ions, sugars, and amino acids. Mutations in bacteria and mammalian cells (including some that result in human disease) have supported these conclusions. Facilitated diffusion and active transport resemble a substrate-enzyme reaction except that no covalent interaction occurs. These points of resemblance are as follows: (1) There is a specific binding site for the solute. (2) The carrier is saturable, so it has maximum rate of transport (V_{max}, Figure 43–17). (3) There is a binding constant (K_m) for the solute, and so the whole system has a K_m (Figure 43–17). (4) Structurally similar competitive inhibitors block transport.

Major differences are the following: (1) Facilitated diffusion can operate bidirectionally, whereas active transport is usually unidirectional. (2) Active transport always occurs against an electrical or chemical gradient, and so it requires energy.

Facilitated Diffusion

Some specific solutes diffuse down electrochemical gradients across membranes more rapidly than might be expected from their size, charge, or partition coefficients. This facilitated diffusion exhibits properties distinct from those of simple diffusion. The rate of facilitated diffusion, a uniport system, can be saturated; ie, the number of sites involved in diffusion of the specific solutes appears finite. Many facilitated diffusion systems are stereospecific but, like simple diffusion, require no metabolic energy.

As described earlier, the inside-outside asymmetry of membrane proteins is stable, and mobility of proteins across (rather than in) the membrane is rare; therefore, transverse mobility of specific carrier proteins is not likely to account for facilitated diffusion processes except those for microbial ionophores (see above).

A **"Ping Pong" mechanism** (Figure 43–18) explains facilitated diffusion. In this model, the carrier protein exists in two principal conformations. In the "pong" state, it is exposed to high concentrations of solute, and molecules of the solute bind to specific sites on the carrier protein. Transport occurs when a conformational change exposes the carrier to a lower concentration of solute ("ping" state). This process is completely reversible, and net flux across the membrane depends upon the concentration gradient. The rate at which solutes enter a cell by facilitated diffusion is determined by the following factors: (1) The concentration gradient across the membrane. (2) The amount of carrier available (this is a key control step). (3) The rapidity of the solute-carrier interaction. (4) The rapidity of the conformational change for both the loaded and the unloaded carrier.

Hormones regulate facilitated diffusion by changing the number of transporters available. Insulin increases glucose transport in fat and muscle by recruiting transporters from an intracellular reservoir (Figure 51–9). Insulin also enhances amino acid transport in liver and other tissues. One of the coordinated actions of glucocorticoid hormones is to enhance transport of amino acids into liver, where the amino acids then serve as a substrate for gluconeogenesis. Growth hormone increases amino acid transport in all cells, and estrogens do this in the uterus. There are at least five different carrier systems for amino acids in animal cells. Each is specific for a group of closely related amino acids, and most operate as Na^+-symport systems (Figure 43–16).

Active Transport

The process of active transport differs from diffusion in that molecules are transported away from thermodynamic equilibrium; hence, energy is required. This energy can come from the hydrolysis of ATP, from electron movement, or from light. The maintenance of electrochemical gradients in biologic systems is so important that it consumes perhaps 30–40% of the total energy expenditure in a cell.

In general, cells maintain a low intracellular Na^+ concentration and a high intracellular K^+ concentration (Table 43–1), along with a net negative electrical potential inside. The pump that maintains these gradients is an ATPase that is activated by Na^+ and K^+ (Figure 43–19). The ATPase is an integral membrane protein and requires phospholipids for activity. The ATPase has catalytic centers for both ATP and Na^+ on the cytoplasmic side of the membrane, but the K^+ binding site is located on the extracellular side of the membrane. Ouabain (or digitalis) inhibits this ATPase by binding to the extracellular domain. Inhi-

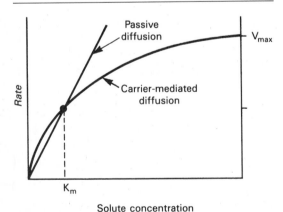

Figure 43–17. A comparison of the kinetics of carrier-mediated (facilitated) diffusion with passive diffusion. The rate of movement in the latter is directly proportionate to solute concentration, whereas the process is saturable when carriers are involved. The concentration at half-maximal velocity is equal to the binding constant (K_m) of the carrier for the solute. (V_{max}, maximal rate.)

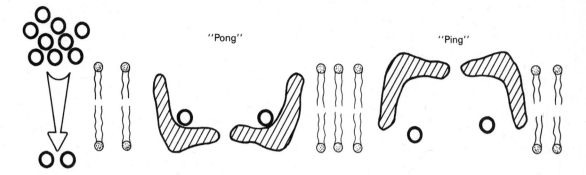

Figure 43–18. The "Ping-Pong" model of facilitated diffusion. A protein carrier (shaded structure) in the lipid bilayer associates with a solute in high concentration on one side of the membrane. A conformational change ensues ("pong" to "ping"), and the solute is discharged on the side favoring the new equilibrium. The empty carrier then reverts to the original conformation ("ping" to "pong") to complete the cycle.

bition of the ATPase by ouabain can be antagonized by extracellular K^+.

Nerve Impulses Are Transmitted Up and Down Membranes

The membrane forming the surface of neuronal cells maintains an asymmetry of inside-outside voltage (electrical potential) and is electrically excitable. When appropriately stimulated by a chemical signal mediated by a specific synaptic membrane receptor (see discussion of the transmission of biochemical signals, below), gates in the membrane are opened to allow the rapid influx of Na^+ or Ca^{2+} (with or without the efflux of K^+), so that the voltage difference rapidly collapses and that segment of the membrane

INSIDE   MEMBRANE   OUTSIDE

3 Na$^+$

ATP

3 Na$^+$

Mg^{2+}

2 K$^+$

ADP
+
P$_i$

2 K$^+$

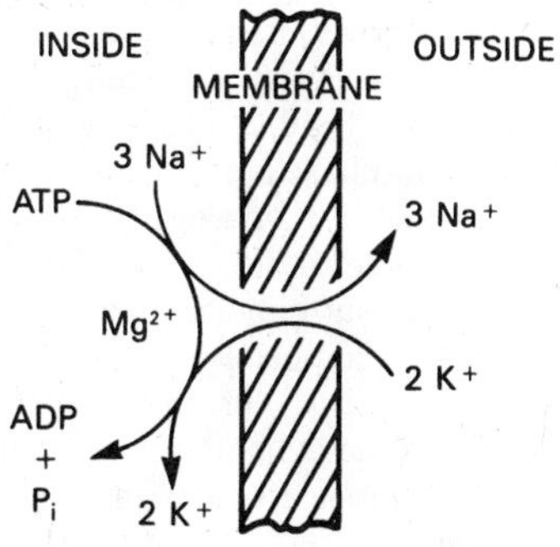

Figure 43–19. Stoichiometry of the Na^+-K^+ ATPase pump. This pump moves three Na^+ ions from inside the cell to the outside and brings two K^+ ions from the outside to the inside for every molecule of ATP hydrolyzed to ADP by the membrane-associated ATPase. Ouabain and other cardiac glycosides inhibit this pump by acting on the extracellular surface of the membrane. (Courtesy of R Post.)

is depolarized. However, as a result of the action of the ion pumps in the membrane, the gradient is quickly restored.

When large areas of the membrane are depolarized in this manner, the electrochemical disturbance propagates in wave-like form down the membrane, generating a nerve impulse. Myelin sheets, formed by Schwann cells, wrap around nerve fibers and provide an electrical insulator that surrounds most of the nerve and greatly speeds up the propagation of the wave (signal) by allowing ions to flow in and out of the membrane only where the membrane is free of the insulation. The myelin membrane is composed of phospholipids (including sphingomyelin), cholesterol, proteins, and glycosphingolipids. Relatively few integral and peripheral proteins are associated with the myelin membrane; those present appear to hold together multiple membrane bilayers to form the hydrophobic, insulating structure that is impermeable to ions and water. Certain diseases, eg, multiple sclerosis and the Guillain-Barré syndrome, are characterized by demyelination and impaired nerve conduction.

Glucose Transport Involves Several Mechanisms

A discussion of the transport of glucose summarizes many of the points made in this chapter. Glucose must enter cells as the first step in energy utilization. In adipocytes and muscle, glucose enters by a specific transport system that is enhanced by insulin. Changes in transport are primarily due to alterations of V_{max} (presumably from more or fewer active transporters) but changes in K_m may also be involved. Glucose transport involves different aspects of the principles of transport discussed above. Glucose and Na^+ bind to different sites on the glucose transporter. Na^+ moves into the cell down its electrochemical gradient and "drags" glucose with it

(Figure 43–20). Therefore, the greater the Na^+ gradient, the more glucose enters, and if Na^+ in extracellular fluid is low, glucose transport stops. To maintain a steep Na^+ gradient, this Na^+-glucose symport is dependent on gradients generated by an Na^+-K^+ pump that maintains a low intracellular Na^+ concentration. Similar mechanisms are used to transport other sugars as well as amino acids.

The transcellular movement of sugars involves one additional component, a uniport that allows the glucose accumulated within the cell to move across a different surface toward a new equilibrium; this occurs in intestinal and renal cells, for example.

Cells Transport Certain Macromolecules Across the Plasma Membrane

The process by which cells take up large molecules is called **"endocytosis."** Some of these molecules (eg, polysaccharides, proteins, and polynucleotides) can be sources of nutritional elements. Endocytosis provides a mechanism for regulating the content of certain membrane components, hormone receptors being a case in point. Endocytosis can be used to learn more about how cells function. DNA from one cell type can be used to transfect a different cell and alter the latter's function or phenotype. A specific gene is often employed in these experiments, and this provides a unique way to study and analyze the regulation of that gene. DNA transfection depends upon endocytosis; endocytosis is responsible for the entry of DNA into the cell. Such experiments commonly use calcium phosphate, since Ca^{2+} stimulates endocytosis and precipitates DNA, which makes the DNA a better object for endocytosis. Cells also release macromolecules by **exocytosis.** Endocytosis and exocytosis both involve vesicle formation with or from the plasma membrane.

A. Endocytosis: All eukaryotic cells are continuously ingesting parts of their plasma membranes. Endocytotic vesicles are generated when segments of the plasma membrane invaginate, enclosing a minute volume of extracellular fluid and its contents. The vesicle then pinches off as the fusion of plasma membranes seals the neck of the vesicle at the original site of invagination (Figure 43–21). This vesicle fuses with other membrane structures and thus achieves the transport of its contents to other cellular compartments or even back to the cell exterior. Most endocytotic vesicles fuse with primary lysosomes to form secondary lysosomes, which contain hydrolytic enzymes and are therefore specialized organelles for intracellular disposal. The macromolecular contents are digested to yield amino acids, simple sugars, and nucleotides, and they diffuse out of the vesicles to be reused in the cytoplasm. Endocytosis requires (1) energy, usually from the hydrolysis of ATP; (2) Ca^{2+} in extracellular fluid; and (3) contractile elements in the cell (probably the microfilament system) (Chapter 58).

There are two general types of endocytosis. **Phagocytosis** occurs only in specialized cells such as macrophages and granulocytes. Phagocytosis involves the ingestion of large particles such as viruses, bacteria, cells, or debris. Macrophages are extremely active in this regard and may ingest 25% of their volume per hour. In so doing, a macrophage may internalize 3% of its plasma membrane each minute or the entire membrane every 30 minutes.

Pinocytosis is a property of all cells and leads to the cellular uptake of fluid and fluid contents. There are two types. Fluid-phase pinocytosis is a nonselective process in which the uptake of a solute by formation of small vesicles is simply proportionate to its concentration in the surrounding extracellular fluid. The formation of these vesicles is an extremely active process. Fibroblasts, for example, internalize their plasma membrane at about one-third the rate of macrophages. This process occurs more rapidly than membranes are made. The surface area and volume of a cell do not change much, so membranes must be replaced by exocytosis or by being recycled as fast as they are removed by endocytosis.

The other type of pinocytosis, absorptive pinocytosis, is a receptor-mediated selective process primarily responsible for the uptake of macromolecules for which there are a finite number of binding sites on the plasma membrane. These high-affinity receptors per-

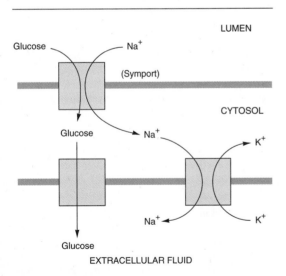

Figure 43–20. The transcellular movement of glucose in an intestinal cell. Glucose follows Na^+ across the luminal epithelial membrane. The Na^+ gradient that drives this symport is established by Na^+-K^+ exchange, which occurs at the basal membrane facing the extracellular fluid compartment. Glucose at high concentration within the cell moves "downhill" into the extracellular fluid by facilitated diffusion (a uniport mechanism).

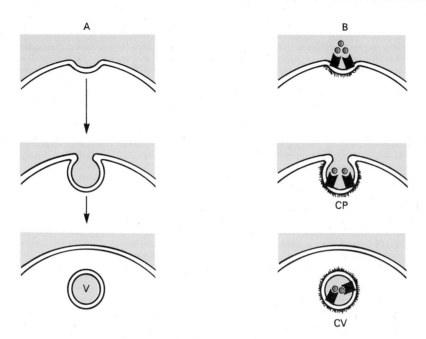

Figure 43–21. Two types of endocytosis. An endocytotic vesicle (V) forms as a result of invagination of a portion of the plasma membrane. Fluid-phase endocytosis **(A)** is random and nondirected. Receptor-mediated endocytosis **(B)** is selective and occurs in coated pits (CP) lined with the protein clathrin (the fuzzy material). Targeting is provided by receptors (solid symbols) specific for a variety of molecules. This results in the formation of a coated vesicle (CV).

mit the selective concentration of ligands from the medium, minimize the uptake of fluid or soluble unbound macromolecules, and markedly increase the rate at which specific molecules enter the cell. The vesicles formed during absorptive pinocytosis are derived from invaginations (pits) that are coated on the cytoplasmic side with a filamentous material. In many systems, clathrin is the filamentous material; it is probably a peripheral membrane protein. Coated pits may constitute as much as 2% of the surface of some cells.

For example, the low-density lipoprotein (LDL) molecule and its receptor (Chapter 27) are internalized by means of coated pits containing the LDL receptor. These endocytotic vesicles containing LDL and its receptor fuse to lysosomes in the cell. The receptor is released and recycled back to the cell surface membrane, but the apoprotein of LDL is degraded and the cholesteryl esters metabolized. Synthesis of the LDL receptor is regulated by secondary or tertiary consequences of pinocytosis, eg, by metabolic products, such as cholesterol, released during the degradation of LDL. Disorders of the LDL receptor and its internalization are medically important and are discussed in Chapter 27.

Other macromolecules, including several hormones, are subject to adsorptive pinocytosis and form **receptosomes,** vesicles that avoid lysosomes and deliver their contents to other intracellular sites, such as the Golgi system.

Adsorptive pinocytosis of extracellular glycopro-

teins requires that the glycoproteins carry specific carbohydrate recognition signals. These recognition signals are bound by membrane receptor molecules, which play a role analogous to that of the LDL receptor. A galactosyl receptor on the surface of hepatocytes is instrumental in the adsorptive pinocytosis of asialoglycoproteins from the circulation (Chapter 56). Acid hydrolases taken up by adsorptive pinocytosis in fibroblasts are recognized by their mannose 6-phosphate moieties. Interestingly, the mannose 6-phosphate moiety also seems to play an important role in the intracellular targeting of the acid hydrolases to the lysosomes of the cells in which they are synthesized (Chapter 56).

There is a dark side to receptor-mediated endocytosis, for viruses that cause such diseases as hepatitis (affecting liver cells), poliomyelitis (affecting motor neurons), and AIDS (affecting T cells) initiate their damage by this mechanism. Iron toxicity also begins with excessive uptake due to endocytosis.

B. Exocytosis: Most cells also release macromolecules to the exterior by exocytosis. This process is also involved in membrane remodeling when the components synthesized in the Golgi apparatus are carried in vesicles to the plasma membrane. The signal for exocytosis is often a hormone which, when it binds to a cell-surface receptor, induces a local and transient change in Ca^{2+} concentration. Ca^{2+} triggers exocytosis. Figure 43–22 provides a comparison of the mechanisms of exocytosis and endocytosis.

Figure 43–22. A comparison of the mechanisms of endocytosis and exocytosis. Exocytosis involves the contact of two inside surface (cytoplasmic side) monolayers, whereas endocytosis results from the contact of two outer surface monolayers.

Molecules released by exocytosis fall into three categories: (1) They can attach to the cell surface and become peripheral proteins, eg, antigens. (2) They can become part of the extracellular matrix, eg, collagen and glycosaminoglycans. (3) They can enter extracellular fluid and signal other cells. Insulin, parathyroid hormone, and the catecholamines are all packaged in granules and processed within cells, to be released upon appropriate stimulation (see Chapters 47, 49, and 51).

Some Signals Are Transmitted Across Membranes

Specific biochemical signals such as neurotransmitters, hormones, and immunoglobulins bind to specific **receptors** (integral proteins) exposed to the outside of cellular membranes and transmit information through these membranes to the cytoplasm. This mechanism involves the generation of a number of signals, including cyclic nucleotides, calcium, phosphoinositides, and diacylglycerol. It is discussed in detail in Chapter 44.

Information Can Be Communicated by Intercellular Contact

There are many areas of intercellular contact in a metazoan organism. This necessitates contact between the plasma membranes of the individual cells. Cells have developed specialized regions on their membranes for intracellular communication in close proximity. **Gap junctions** mediate and regulate the passage of ions and small molecules through a narrow hydrophilic core connecting the cytoplasm of adjacent cells. It is through this central opening that ions and small molecules can pass from one cell to another in a regulated fashion.

MUTATIONS AFFECTING MEMBRANE PROTEINS CAUSE DISEASES

In view of the fact that membranes are located in so many organelles and are involved in so many processes, it is not surprising that mutations affecting

Table 43–10. Some diseases or pathologic states resulting from or attributed to abnormalities of membranes.[1]

Disease	Abnormality
Achondroplasia	Mutations in the gene encoding the fibroblast growth factor receptor 3
Familial hypercholesterolemia	Mutations in the gene encoding the LDL receptor
Cystic fibrosis	Mutations in the gene encoding the CFTR protein, a Cl^- transporter
Congenital long QT syndrome	Mutations in genes encoding ion channels in the heart
Wilson's disease	Mutations in the gene encoding a copper-dependent ATPase
I-cell disease	Mutations in the gene encoding GlcNAc phosphotransferase, resulting in absence of the Man 6-P signal for lysosomal localization of certain hydrolases
Hereditary spherocytosis	Mutations in the gene encoding spectrin, a structural protein in the red cell membrane
Metastasis	Abnormalities in the oligosaccharide chains of membrane glycoproteins and glycolipids are thought to be of importance
Paroxysmal nocturnal hemoglobinuria	Mutation resulting in deficient attachment of the GPI anchor to certain proteins of the red cell membrane

[1]The disorders listed are discussed further in other chapters. The table lists examples of mutations affecting receptors, a transporter, ion channels, enzymes, and a structural protein. Examples of altered or defective glycosylation of glycoproteins are also presented. Most of the conditions listed affect the plasma membrane.

their protein constituents should result in many diseases or disorders. Proteins in membranes can be classified as receptors, transporters, ion channels, enzymes, and structural components. Members of all of these classes are often glycosylated, so that mutations affecting this process may alter their function. Examples of diseases or disorders due to abnormalities in membrane proteins are listed in Table 43–10; these mainly reflect mutations in proteins of the plasma membrane, with one affecting lysosomal function (I-cell disease). Almost 30 genetic diseases or disorders have been ascribed to mutations affecting various proteins involved in the transport of amino acids, sugars, lipids, urate, anions, cations, water, and vitamins across the plasma membrane. Mutations in genes encoding proteins in other membranes can also have harmful consequences. For example, mutations in genes encoding mitochondrial membrane proteins involved in oxidative phosphorylation can cause neurologic and other problems (Chapter 64). Membrane proteins can also be affected by conditions other than mutations. Formation of autoantibodies to the acetylcholine receptor in skeletal muscle causes myasthenia gravis (Chapter 64). Ischemia can quickly affect the integrity of various ion channels in membranes. Abnormalities of membrane constituents other than proteins can also be harmful. With regard to lipids, excess of cholesterol (eg, in familial hypercholesterolemia), of lysophospholipid (eg, after bites by certain snakes, whose venom contains phospholipases), or of glycosphingolipids (eg, in a sphingolipidosis) can all affect membrane function.

SUMMARY

Membranes are complex structures composed of lipids, carbohydrates, and proteins. The basic structure of all membranes is the lipid bilayer. This bilayer is formed by two sheets of phospholipids in which the hydrophilic polar head groups are directed away from each other and are exposed to the aqueous environment on the outer and inner surfaces of the membrane. The hydrophobic nonpolar tails of these molecules are oriented toward each other, in the direction of the center of the membrane. Proteins can be an integral component of the membrane and span the bilayer, or they can be attached, by electrostatic charge, to the outer or inner surface of the mem-

brane. The 20 or so different membranes in a mammalian cell have intrinsic functions (eg, enzymatic activity), and they define compartments, or specialized environments, within the cell that have specific functions (eg, lysosomes). Membrane assembly is discussed and shown to be complex. Asymmetry of both lipids and proteins is maintained during membrane assembly. Many proteins are targeted to their destinations by signal sequences. A major sorting decision is made when proteins are partitioned between cytosolic and membrane-bound polyribosomes by virtue of the absence or presence of a signal peptide. Many proteins synthesized on the latter polyribosomes proceed to the Golgi apparatus and the plasma membrane in transport vesicles. A number of glycosylation reactions occur in compartments of the Golgi, and proteins are further sorted in the *trans*-Golgi network. Most proteins destined for the plasma membrane and for secretion appear to lack specific signals, a default mechanism. The role of chaperone proteins in the folding of proteins is discussed, and a model describing budding and attachment of transport vesicles to a target membrane is summarized.

Certain molecules freely diffuse across membranes, but the movement of others is restricted because of size, charge, or solubility. Various passive and active mechanisms are employed to maintain gradients of such molecules across different membranes. Certain solutes, eg, glucose, enter cells by facilitated diffusion, along a downhill gradient from high to low concentration. Specific carrier molecules, or transporters, are involved in such processes. Ligand- or voltage-gated channels are often employed to move charged molecules (Na^+, K^+, Ca^{2+}, etc) across membranes. Large molecules can enter or leave cells through mechanisms such as endocytosis or exocytosis. These processes often require the binding of the molecule to a receptor, which affords specificity to the process. Finally, receptors may be integral components of membranes (particularly the plasma membrane). The interaction of a ligand with its receptor may not involve the movement of either into the cell, but the interaction results in the generation of a signal that influences intracellular processes. Mutations that affect the structure of membrane proteins (receptors, transporters, ion channels, enzymes, and structural proteins) may cause diseases; examples include cystic fibrosis and familial hypercholesterolemia.

REFERENCES

Balch WE, Farquhar MG: Beyond bulk flow. Trends Cell Biol 1995;5:16.

Dawidowicz EA: Dynamics of membrane lipid metabolism and turnover. Annu Rev Biochem 1987;56:43.

Ellis RJ, van der Vries SM: Molecular chaperones. Annu Rev Biochem 1991;60:321.

Gennis RB: *Biomembranes: Molecular Structure and Function.* Springer-Verlag, 1989.

Gilmore R: Protein translocation across the endoplasmic reticulum: A tunnel with toll booths at entry and exit. Cell 1993;75:589.

Jain MK: *Introduction to Biological Membranes,* 2nd ed. Wiley, 1988.

Jennings ML: Topography of membrane proteins. Annu Rev Biochem 1989;58:999.

Kirchhausen T: Coated pits and coated vesicles—sorting it all out. Curr Opin Struct Biol 1993;3:182.

Lazarow PB, Moser HW: Disorders of peroxisome biogenesis. In: *The Metabolic and Molecular Bases of Inherited Disease,* 7th ed. Scriver CR et al (editors). McGraw-Hill, 1995.

Pearse BMF: Clathrin, adaptors, and sorting. Annu Rev Cell Biol 1990;6:151.

Rosenberg LE, Short EM: Inherited defects of membrane transport. In: *Harrison's Principles of Internal Medicine,* 13th ed. Isselbacher KJ et al (editors). McGraw-Hill, 1994.

Rothman JE: Mechanisms of intracellular protein transport. Nature 1994;372:55.

Rothman JE, Warren G: Implications of the SNARE hypothesis for intracellular membrane topology. Curr Biol 1994;4:220.

Sabatini DD, Adesnik MB: The biogenesis of membranes and organelles. In: *The Metabolic and Molecular Bases of Inherited Disease,* 7th ed. Scriver CR et al (editors). McGraw-Hill, 1995.

Schatz G: The protein import machinery of mitochondria. Protein Sci 1993;2:141.

Schlesinger MJ (editor): *Lipid Modification of Proteins.* CRC Press, 1993.

Silverman M: Structure and function of hexose transporters. Annu Rev Biochem 1991;60:757.

Smythe E, Warren G: The mechanism of receptor-mediated endocytosis. Eur J Biochem 1992;202:689.

Stein WD: *Transport and Diffusion Across Cell Membranes.* Academic Press, 1986.

Vance DE, Vance JE (editors): *Biochemistry of Lipids, Lipoproteins and Membranes.* Elsevier, 1991.

Williams DB: Calnexin: A molecular chaperone with a taste for carbohydrate. Biochem Cell Biol 1995;73:123.

Hormone Action

44

Daryl K. Granner, MD

INTRODUCTION

Hormone action at the cellular level begins with the association of the hormone and its specific receptor. Hormones can be classified by the location of the receptor and by the nature of the signal or second messenger used to mediate hormone action within the cell. A number of these second messengers have been defined. Considerable progress has been made in elucidating how hormones work intracellularly, particularly in regard to the regulation of expression of specific genes.

BIOMEDICAL IMPORTANCE

The rational diagnosis and therapy of a disease depend upon understanding the pathophysiology involved and the ability to quantitate it. Diseases of the endocrine system, which are generally due to excessive or deficient production of hormones, are an excellent example of the application of basic principles to clinical medicine. Knowing the general aspects of hormone action and understanding the physiologic and biochemical effects of the individual hormones enable one to recognize endocrine disease syndromes that result from hormone imbalance and to apply effective therapy.

HORMONE RECEPTORS ARE OF CENTRAL IMPORTANCE

Receptors Discriminate Precisely

Hormones are present at very low concentrations in the extracellular fluid, generally in the range of 10^{-15} to 10^{-9} mol/L. This is a much lower concentration than that of the many structurally similar molecules (sterols, amino acids, peptides, proteins) and other molecules that circulate at concentrations in the 10^{-5} to 10^{-3} mol/L range. Target cells, therefore, must distinguish not only between different hormones present in small amounts but also between a given hormone and the 10^6- to 10^9-fold excess of other molecules. This high degree of discrimination is provided by cell-associated recognition molecules called receptors. Hormones initiate their biologic effects by binding to specific receptors, and since any effective control system also must provide a means of stopping a response, hormone-induced actions generally terminate when the effector dissociates from the receptor.

A target cell is defined by its ability to bind selectively a given hormone via such a receptor, an interaction that is often quantitated using radioactive ligands that mimic hormone binding. Several features of this interaction are important: (1) the radioactivity must not alter the biologic activity of the ligand; (2) the binding should be specific, ie, displaceable by unlabeled agonist or antagonist; (3) binding should be saturable; and (4) binding should occur within the concentration range of the expected biologic response.

Both Recognition and Coupling Domains Occur on Receptors

All receptors, whether for polypeptides or steroids, have at least two functional domains. A recognition domain binds the hormone, and a second region generates a signal that couples hormone recognition to some intracellular function. Coupling (signal transduction) occurs in two general ways. Polypeptide and protein hormones and the catecholamines bind to receptors located in the plasma membrane and thereby generate a signal that regulates various intracellular functions, often by changing the activity of an enzyme. Steroid and thyroid hormones interact with intracellular receptors, and this complex provides the signal (see below).

The amino acid sequences of these two domains in many polypeptide hormone receptors have been identified. Steroid hormone receptors have several functional domains: one site binds the hormone, another binds to specific DNA regions, a third activates (or represses) gene transcription, and a fourth may specify high-affinity binding to other proteins.

The dual functions of binding and coupling ultimately define a receptor, and it is the coupling of hormone binding to signal transduction, so-called **receptor-effector coupling,** that provides the first step in the amplification of the hormonal response. This dual purpose also distinguishes the target cell receptor from the plasma carrier proteins that bind hormone but do not generate a signal.

Receptors Are Proteins

The acetylcholine receptor, which was easy to purify, since it exists in relatively large amounts in the electric organ of the eel *Torpedo californica,* was one of the first receptors to be studied in detail. The acetylcholine receptor consists of four subunits in the configuration α_2, β, γ, δ. The two α subunits bind acetylcholine; the technique of site-directed mutagenesis has been used to show which regions of this subunit are involved in the formation of the transmembrane ion channel, which performs the major function of the acetylcholine receptor.

Other receptors are present in very small amounts; thus, purification and characterization by classical techniques were difficult. Recombinant DNA techniques provide the requisite amounts of material for such studies, and so this has become an active area of investigation. The insulin receptor is a heterotetramer $(\alpha_2\beta_2)$ linked by multiple disulfide bonds, in which the extramembrane α subunit binds insulin and the membrane-spanning β subunit transduces the signal, presumably through the tyrosine kinase component of the cytoplasmic portion of this polypeptide. The receptors for insulin-like growth factor I (IGF-I), epidermal growth factor (EGF), and low-density lipoprotein (LDL) are generally similar to the insulin receptor (Figure 51–12). Polypeptide hormone receptors that transduce signals by altering the rate of production of cAMP are characterized by the presence of seven domains that span the plasma membrane.

A comparison of several different steroid receptors with thyroid hormone receptors revealed a remarkable conservation of the amino acid sequence in regions, particularly in the DNA-binding domains. This led to the view that receptors of **the steroid-thyroid type constitute a large superfamily.** Many related members of this family have no known ligand, hence are called **orphan receptors.** Receptors of this class have several functional domains.

The glucocorticoid receptor is a good example (Figure 48–6). This molecule has several functional domains: (1) a hormone-binding region in the carboxyl terminal portion; (2) an adjacent DNA-binding region; (3) at least two regions that activate gene transcription; (4) at least two regions responsible for translocation of the receptor from the cytoplasm to the nucleus; and (5) a region that binds heat shock protein in the absence of ligand.

HORMONES CAN BE CLASSIFIED IN SEVERAL WAYS

Hormones can be classified according to chemical composition, solubility properties, location of receptors, and nature of the signal used to mediate hormone action within the cell. A classification based on the last two properties is illustrated in Table 44–1, and general features of each group are illustrated in Table 44–2.

The hormones in group I are lipophilic and, with the exception of T_3 and T_4, are derived from cholesterol. After secretion, these hormones associate with transport proteins, a process that circumvents the problem of solubility while prolonging the plasma half-life. The free hormone readily traverses the

Table 44–1. Classification of hormones by mechanism of action.

I. Hormones that bind to intracellular receptors
Androgens
Calcitriol (1,25[OH]$_2$-D$_3$)
Estrogens
Glucocorticoids
Mineralocorticoids
Progestins
Retinoic acid
Thyroid hormones (T_3 and T_4)

II. Hormones that bind to cell surface receptors
A. The second messenger is cAMP:
α_2-Adrenergic catecholamines
β-Adrenergic catecholamines
Adrenocorticotropic hormone (ACTH)
Angiotensin II
Antidiuretic hormone (ADH)
Calcitonin
Chorionic gonadotropin, human (hCG)
Corticotropin-releasing hormone (CRH)
Follicle-stimulating hormone (FSH)
Glucagon
Lipotropin (LPH)
Luteinizing hormone (LH)
Melanocyte-stimulating hormone (MSH)
Parathyroid hormone (PTH)
Somatostatin
Thyroid-stimulating hormone (TSH)
B. The second messenger is cGMP:
Atrial natriuretic factor (ANF)
Nitric oxide (NO)
C. The second messenger is calcium or phosphatidylinositols (or both):
Acetylcholine (muscarinic)
α_1-Adrenergic catecholamines
Angiotensin II
Antidiuretic hormone (ADH, vasopressin)
Cholecystokinin
Gastrin
Gonadotropin-releasing hormone (GnRH)
Oxytocin
Platelet-derived growth factor (PDGF)
Substance P
Thyrotropin-releasing hormone (TRH)
D. The second messenger is a kinase or phosphatase cascade:
Chorionic somatomammotropin (CS)
Epidermal growth factor (EGF)
Erythropoietin (EPO)
Fibroblast growth factor (FGF)
Growth hormone (GH)
Insulin
Insulin-like growth factors (IGF-I, IGF-II)
Nerve growth factor (NGF)
Platelet-derived growth factor (PDGF)
Prolactin (PRL)

Table 44–2. General features of hormone classes.

	Group I	Group II
Types	Steroids, iodothyro-nines, calcitriol, retinoids	Polypeptides, proteins, glycoproteins, catechol-amines
Solubility	Lipophilic	Hydrophilic
Transport proteins	Yes	No
Plasma half-life	Long (hours to days)	Short (minutes)
Receptor	Intracellular	Plasma membrane
Mediator	Receptor-hormone complex	cAMP, cGMP, Ca^{2+}, metabolites of complex phosphoinositols, kinase cascades

plasma membrane of all cells and encounters receptors in either the cytosol or nucleus of target cells.
The ligand-receptor complex is assumed to be the intracellular messenger in this group.

The second major group consists of water-soluble
hormones that bind to the plasma membrane of the
target cell. Hormones that bind to the surface of cells
communicate with intracellular metabolic processes
through intermediary molecules, so-called **second
messengers** (the hormone itself is the first messenger), which are generated as a consequence of the
ligand-receptor interaction. The second-messenger
concept arose from Sutherland's observation that epinephrine binds to the plasma membrane of pigeon
erythrocytes and increases intracellular cAMP. This
was followed by a series of experiments in which
cAMP was found to mediate the metabolic effects of
many hormones. Hormones that clearly employ this
mechanism are shown in group II.A of Table 44–1.
To date only one hormone, atrial natriuretic factor
(ANF), uses cGMP as its second messenger, but
other hormones will probably be added to group II.B.
Several hormones, many of which were previously
thought to affect cAMP, appear to use calcium or
metabolites of complex phosphatidylinositols (or
both) as the intracellular signal. These are shown in
group II.C. The intracellular messenger for group
II.D is a protein kinase-phosphatase cascade. Several
of these have been identified, and a given hormone
may use more than one kinase cascade. A few hormones fit into more than one category, and assignments change with new information.

GROUP I HORMONES HAVE
INTRACELLULAR RECEPTORS
AND AFFECT GENE EXPRESSION

The general features of the action of this group of
hormones are illustrated in Figure 44–1. These
lipophilic molecules diffuse through the plasma
membrane of all cells but only encounter their specific, high-affinity receptor in target cells. The hormone-receptor complex next undergoes a tempera-
ture- and salt-dependent **"activation" reaction** that
results in size, conformation, and surface charge
changes that render it able to bind to chromatin.
Whether this association and "activation" process occurs in the cytoplasm or nucleus is debatable but not
crucial to understanding the whole process. The hormone-receptor complex binds to a specific region of
DNA (called the "hormone response element") and
activates or inactivates specific genes. By selectively
affecting gene transcription and the production of the
respective mRNAs, the amounts of specific proteins
are changed and metabolic processes are influenced.
The effect of each of these hormones is quite specific; generally, the hormone affects less than 1% of
the proteins or mRNA in a target cell. This discussion has concentrated on nuclear actions of steroid,
thyroid, and retinoid hormones because these are
well defined. Direct actions in the cytoplasm and on
various organelles and membranes have also been
described. Most evidence suggests that steroid hormones exert their predominant effect on gene transcription, but these hormones, and many of those
found in the other classes discussed below, can act at
any step of the "information pathway" illustrated in
Figure 44–2. Although the biochemistry of gene transcription in mammalian cells is not well understood,
a general model of the structural requirements for
steroid and thyroid regulation of gene transcription
can be drawn (Figure 44–3). These genes must be in
regions of "open" or transcriptionally active chromatin (depicted as the bubble in Figure 44–1), as defined by their susceptibility to digestion by the enzyme DNase I. The genes studied to date have at
least two separate regulatory elements (control sites)
in the DNA sequence immediately 5′ of the transcription initiation site (Figure 44–3). The first of
these, the **promoter element (PE),** is generic, since
it is present in some form or other in all genes. This
element specifies the site of RNA polymerase II attachment to DNA and therefore the accuracy of transcript initiation (Chapter 41).

A second element, the **hormone response element (HRE),** has been identified in many genes regulated by steroid hormones. This is located slightly
farther 5′ than the promoter element and may consist
of several discrete elements. The hormone response
element (HRE) presumably modulates the frequency
of transcript initiation and is less dependent on position and orientation; in these respects, it resembles
the transcription **enhancer elements** found in other
genes (Chapter 41). Generally, it is found within a
few hundred nucleotides upstream of the transcription initiation site, but its precise location varies from
gene to gene. In some instances, it is located within
the gene. Genes controlled by several hormones have
a corresponding number of HREs. Although the initial reactions are different, peptide hormones also ex-

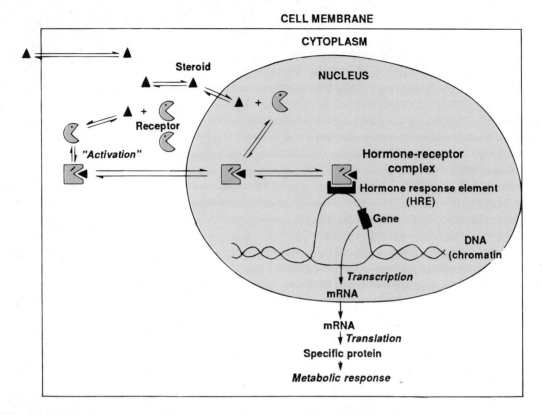

Figure 44–1. The steroid or thyroid hormone binds to an intracellular receptor and causes a conformational change of the latter. This complex then binds to a specific DNA region, the HRE, and this interaction results in the activation or repression of a restricted number of genes.

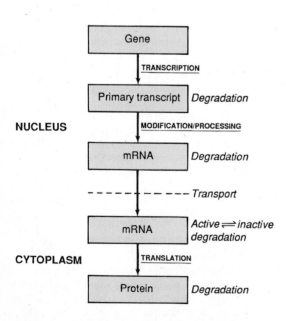

Figure 44–2. The "information pathway." Hormones can affect any of these steps.

ert their effects on transcription through HREs. For example, many of the hormones that use cAMP as a second messenger affect transcription. A special protein, the cAMP response element binding (CREB) protein, is the *trans*-acting factor (analogous to the steroid thyroid hormone receptor) in these instances. The consensus DNA sequences of several HREs have been defined (Table 44–3).

The identification of an HRE requires that it bind the hormone-receptor complex more avidly than does surrounding DNA or DNA from another source. In the cases just cited, such specific binding has been demonstrated. The HRE must also confer hormone responsiveness. Putative regulatory sequence DNA can be ligated to reporter genes to assess this point. Usually, these **"fusion genes"** contain reporter genes not ordinarily influenced by the hormone, and often these genes are not normally expressed in the tissue being tested. Commonly used reporter genes are globin, thymidine kinase, bacterial chloramphenicol acetyltransferase, and luciferase. The "fusion gene" is transfected into a target cell, and if the hormone now regulates the transcription of the reporter gene, one has functionally defined an HRE. Position, orientation, and base substitution effects can be precisely

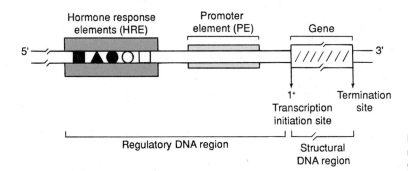

Figure 44–3. Structural requirements for hormonal regulation of gene transcription.

defined by this technique. Exactly how the hormone-receptor interaction with the HRE affects transcription is an area of active investigation. Transcript initiation is a probable control site, but effects on elongation and termination might also occur. Control sites farther 5' from the initiation site, or 3' downstream, either within or beyond the gene, have been proposed. Finally, *trans*-acting control mechanisms (eg, from another chromosome) may also be operative.

Recent evidence suggests that, although simple HREs do transmit a hormone response, matters may be much more complex in many genes. The HRE may have to exist in association with other elements (and associated binding proteins) to function optimally. Such assemblies of *cis*-acting DNA elements and *trans*-acting factors are called **hormone response units.**

GROUP II (PEPTIDE) HORMONES HAVE MEMBRANE RECEPTORS AND USE INTRACELLULAR MESSENGERS

The largest number of hormones are water-soluble, have no transport proteins (and therefore have a short plasma half-life), and initiate a response by binding to a receptor located in the plasma membrane (Tables 44–1 and 44–2). The mechanism of action of this group of hormones can best be discussed in terms of their intracellular messengers.

cAMP Is the Second Messenger for Many Hormones

cAMP (cyclic AMP, 3',5'-adenylic acid; see Figure 20–5), a ubiquitous nucleotide derived from ATP through the action of the enzyme adenylyl cyclase,

Table 44–3. The DNA sequences of several hormone response elements (HREs).[1]

Hormone or Effector	HRE	DNA Sequence
Glucocorticoids Progestins Mineralocorticoids Androgens	GRE PRE MRE ARE	←GGTACA NNN TGTTCT→
Estrogens	ERE	←AGGTCA – – – TGA/TCCT→
Thyroid hormone Retinoic acid Vitamin D	TRE RARE VDRE	AGGTCA→ N3,4,5 AGGTCA→
cAMP	CRE	TGACGTCA

[1]Letters indicate nucleotide; N means any one of the four can be used in that position. The arrows pointing in opposite directions illustrate the slightly imperfect inverted palindromes present in many HREs; in some cases these are called "half binding sites" because each binds one monomer of the receptor. The GRE, PRE, MRE, and ARE consist of the same DNA sequence. Specificity may be conferred by the intracellular concentration of the ligand or hormone receptor, by flanking DNA sequences not included in the consensus, or by other accessory elements. A second group of HREs includes those for thyroid hormones, estrogens, retinoic acid, and vitamin D. These HREs are similar except for the orientation and spacing between the half palindromes. Spacing determines the hormone specificity. VDRE (N=3), TRE (N=4) and RARE (N=5) bind to direct repeats rather than to inverted repeats. Another member of the steroid receptor superfamily, the retinoid X receptor (RXR), forms heterodimers with VDR, TR, and RARE, and these constitute the *trans*-acting factors. cAMP affects gene transcription through the CRE.

plays a crucial role in the action of a number of hormones. The intracellular level of cAMP is increased or decreased by various hormones (Table 44–4), and this effect varies from tissue to tissue. Epinephrine causes large increases of cAMP in muscle and relatively small changes in liver. The opposite is true of glucagon. Tissues that respond to several hormones of this group do so through unique receptors converging upon a single adenylyl cyclase molecule. The best example is the adipose cell, in which epinephrine, ACTH, TSH, glucagon, MSH, and vasopressin (ADH) stimulate adenylyl cyclase and increase cAMP. Combinations of maximally effective concentrations are not additive, and treatments that destroy one receptor have no effect on the cellular response to other hormones.

A. Adenylyl Cyclase System: The components of this system in mammalian cells are illustrated in Figure 44–4. The interaction of the hormone with its receptor results in the activation or inactivation of adenylyl cyclase or some other effector molecule. Receptors that couple to effectors through the GTP-binding protein intermediary described below typically have seven hydrophobic membrane-spanning domains. This is illustrated in Figure 44–4. The regulation of adenylyl cyclase is mediated by at least two GTP-dependent regulatory proteins, designated G_s (stimulatory) and G_i (inhibitory), each of which is composed of three subunits: α, β, and γ. Adenylyl cyclase, located on the inner surface of the plasma membrane, catalyzes the formation of cAMP from ATP in the presence of magnesium (Figures 44–4 and 35–13).

What was originally conceived of as a single protein with two functional domains is now viewed as a system of extraordinary complexity. Over the past 20 years, a number of studies have established the biochemical uniqueness of the hormone receptor, the GTP regulatory protein complex, and adenylyl cyclase (Figure 44–4).

Different peptide hormones can either stimulate (s) or inhibit (i) the production of cAMP (Table 44–4). Two parallel systems, a stimulatory (s) one and an inhibitory (i) one, converge upon a single catalytic molecule (C). Each consists of a receptor, R_s or R_i, and regulatory complex, G_s and G_i. G_s and G_i are each trimers composed of α, β, and γ subunits. Because the α subunit in G_s differs from that in G_i, the proteins are designated α_s (45 kDa) and α_i (41 kDa). The β and γ subunits are 37 kDa and 9 kDa proteins, respectively. The β and γ subunits are always associated ($\beta\gamma$) and appear to function as a heterodimer. The binding of a hormone to R_s or R_i results in a receptor-mediated activation of G, which entails Mg^{2+}-dependent binding of GTP by α and the concomitant dissociation of $\beta\gamma$ from α.

$$GTP$$
$$\alpha\beta\gamma \rightleftharpoons \alpha \cdot GTP + \beta\gamma$$
$$GTPase$$

The α_s has intrinsic GTPase activity, and the active form, $\alpha_s \cdot GTP$, is inactivated upon hydrolysis of the GTP to GDP, and the trimeric G_s complex is reformed. **Cholera toxin,** known to be an irreversible activator of cyclase, causes ADP-ribosylation of α_s and in so doing inactivates the GTPase; therefore, α_s is frozen in the active form. The α_i also has a GTPase activity; however, GDP does not freely dissociate from $\alpha_i \cdot GDP$. The α_i is reactivated by an exchange of GTP for GDP. **Pertussis toxin** irreversibly activates adenylyl cyclase by promoting the ADP-ribosylation of α_i subunit, preventing it from being activated. The exact role of the different α and $\beta\gamma$ subunits in the regulation of adenylyl cyclase has not been defined. The α_s can stimulate cyclase directly, and in certain instances $\beta\gamma$ augments this action. Inhibition of cyclase is more complex. Direct inhibitory effects of α_i on cyclase have been hard to detect. Although some forms of $\beta\gamma$ can inhibit cyclase, a more popular hypothesis is that the $\beta\gamma$ complex of G_i, which is much more abundant than G_s, binds to and inactivates α_s.

It is now apparent that there is a large family of G proteins and that these are part of the superfamily of GTPases. The G protein family can be classified according to sequence homology into four subfamilies, as illustrated in Table 44–5. There are more than 20 α subunits, at least four β subunits, and at least six γ subunits. Five unique adenylyl cyclase molecules have been identified. Various combinations of these provide a large number of possible $\alpha\beta\gamma$ complexes. The α subunits and the $\beta\gamma$ complex have actions independent of those on adenylyl cyclase. Some forms of α_i stimulate K^+ channels and inhibit Ca^{2+} channels and some α_s molecules have the opposite effects. Members of the G_q family activate the phospholipase C group of enzymes. $\beta\gamma$ complexes have been associated with K^+ channel stimulation and phospholipase

Table 44–4. Subclassification of group II.A hormones.

Hormones That Stimulate Adenylyl Cyclase (H_s)	Hormones That Inhibit Adenylyl Cyclase (H_i)
ACTH	Acetylcholine
ADH	α_2-Adrenergics
β-Adrenergics	Angiotensin II
Calcitonin	Somatostatin
CRH	
FSH	
Glucagon	
hCG	
LH	
LPH	
MSH	
PTH	
TSH	

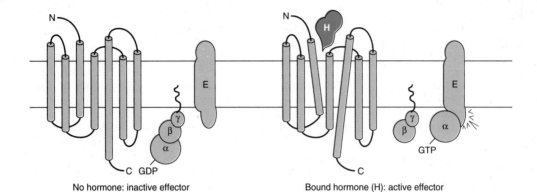

Figure 44–4. Components of the hormone receptor-G protein effector system. Receptors that couple to effectors through G proteins typically have seven membrane-spanning domains. In the absence of hormone (left), the heterotrimeric G-protein complex (α, β, γ) is in an inactive, guanosine diphosphate (GDP)-bound form and is probably not associated with the receptor. This complex is anchored to the plasma membrane through prenylated groups on the βγ subunits (wavy lines) and perhaps by myristoylated groups on α subunits. On binding of hormone to the receptor, there is a presumed conformational change of the receptor and activation of the G-protein complex. This results from the exchange of GDP and guanosine triphosphate (GTP) on the α subunit, after which α and βγ dissociate. The α subunit binds to and activates the effector (E). E can be adenylyl cyclase (α_s), a K$^+$ channel (α_i, α_o), phospholipase Cβ (α_q), or other molecule. The βγ subunit can also have direct actions on E. (Reproduced, with permission, from Granner DK in: *Principles and Practice of Endocrinology and Metabolism,* 2nd ed. Becker KL [editor]. Lippincott, 1995.)

C activation. G proteins are involved in many important biologic processes in addition to hormone action. Some of these are listed in Table 44–5.

The importance of these components is underscored by an "experiment of nature." Pseudohypoparathyroidism is a syndrome characterized by hypocalcemia and hyperphosphatemia, the biochemi-

cal hallmarks of hypoparathyroidism, and by a number of congenital defects. Affected individuals do not have defective parathyroid function; in fact, they secrete large amounts of biologically active PTH. Some have target organ resistance on the basis of a postreceptor defect. They are partially deficient in G protein (probably only the α_s subunit) and thus fail to

Table 44–5. Classes and medium functions of G proteins.[1,2]

Class or Type	Stimulus	Effector	Effect
G$_s$			
α_s	Glucagon, β-adrenergics	↑ Adenylyl cyclase	Gluconeogenesis Lipolysis, glycogenolysis
α_{olf}	Odorant	↑ Adenylyl cyclase	Olfaction
G$_i$			
α_{i1}	Acetylcholine	↓ Adenylyl cyclase	Slowed heart rate
α_{i2}	α_2-Adrenergics M$_2$ cholinergics	↑ Potassium channels	
α_o	Opioids, endorphins	↑ Potassium channels ↓ Calcium channels	Neuronal electrical activity
α_t	Light	↑ cGMP phosphodiesterase	Vision
G$_q$			
α_q	M$_1$ cholinergics α_1-Adrenergics	↑ Phospholipase C-β1	↑ Muscle contraction and
α_{11}	α_1-Adrenergics	↑ Phospholipase C-β2	↑ Blood pressure
G$_{12}$			
α_{12}	?	?	?

[1]Reproduced, with permission, from Granner DK in: *Principles and Practice of Endocrinology and Metabolism,* 2nd ed., Becker KL (editor). Lippincott, 1995.
[2]The four major classes or families of mammalian G proteins (G$_s$, G$_i$, G$_q$, and G$_{12}$) are based on protein sequence homology. Representative members of each are shown, along with known stimuli, effectors, and well-defined biologic effects. More than 20 different α subunits have been identified.

couple binding to adenylyl cyclase stimulation. Others appear to generate cAMP in response to PTH but fail to respond to the cAMP signal. The observation that patients with pseudohypoparathyroidism often show evidence of defective responses to other hormones, including TSH, glucagon, and β-adrenergic agents, is not surprising.

B. Protein Kinase: In prokaryotic cells, cAMP binds to a specific protein, called "catabolite regulatory protein (CRP)," which binds directly to DNA and influences gene expression. The analogy of this to steroid hormone action described above is apparent. In **eukaryotic cells,** cAMP binds to a protein kinase that is a heterotetrameric molecule consisting of two regulatory subunits (R) and two catalytic subunits (C). cAMP binding results in the following reaction:

$$4 \text{ cAMP} + R_2C_2 \rightleftharpoons R_2 \cdot (4 \text{ cAMP}) + 2 \text{ C}$$

The R_2C_2 complex has no enzymatic activity, but the binding of cAMP by R dissociates R from C, thereby activating the latter (Figure 44–5). The active C subunit catalyzes the transfer of the γ phosphate of ATP (Mg^{2+}) to a serine or threonine residue in a variety of proteins. The consensus phosphorylation sites are -Arg-Arg-X-Ser- and -Lys-Arg-X-X-Ser, where X can be any amino acid.

Protein kinase activities were originally described as being "cAMP-dependent" or "cAMP-independent." This too has become considerably more complex, as protein phosphorylation is now recognized as being an important regulatory mechanism. More than 100 protein kinases have been described, each a unique molecule with considerable variability with respect to subunit composition, molecular weight, autophosphorylation, K_m for ATP, and substrate specificity.

C. Phosphoproteins: The effects of cAMP in eukaryotic cells are all thought to be mediated by protein phosphorylation-dephosphorylation. The control of any of the effects of cAMP, including such diverse processes as steroidogenesis, secretion, ion transport, carbohydrate and fat metabolism, enzyme induction, gene regulation, and cell growth and replication, could be conferred by a specific protein kinase, a specific phosphatase, or by specific substrates for phosphorylation. In some instances, a phosphoprotein that is a known participant in a metabolic pathway has been identified; however, in most

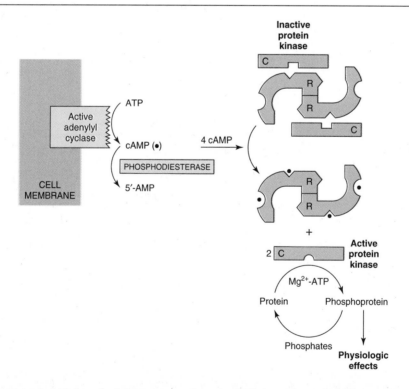

Figure 44–5. Hormonal regulation of cellular processes through cAMP-dependent protein kinases. The cAMP (•) generated by the action of adenylyl cyclase (activated as shown in Figure 44–4) binds to the regulatory (R) subunit of cAMP-dependent protein kinase. This results in the release and activation of the catalytic (C) subunit. (Courtesy of J Corbin.)

processes cited above, the phosphoproteins involved have not been identified. These substrates may help define a target tissue and certainly are involved in defining the extent of the response within a given cell. Many proteins can be phosphorylated, including casein, histones, and protamine; such phosphorylations may be epiphenomena, although they are useful for assaying protein kinase activity. Until recently, the only actions of cAMP that had been defined were actions that occurred outside the nucleus. Effects of cAMP on the transcription of several genes have now been described. These effects appear to be mediated by the CREB protein described above.

D. Phosphodiesterases: Actions caused by hormones that increase cAMP concentration can be terminated in a number of ways, including the hydrolysis of cAMP by phosphodiesterases. The presence of these hydrolytic enzymes ensures a rapid turnover of the signal (cAMP) and hence a rapid termination of the biologic process once the hormonal stimulus is removed. cAMP phosphodiesterases exist in low and high K_m forms and are themselves subject to regulation by hormones as well as by intracellular messengers such as calcium, probably acting through calmodulin. Inhibitors of phosphodiesterase, most notably methylated xanthine derivatives such as caffeine, increase intracellular cAMP and mimic or prolong the actions of hormones.

E. Phosphoprotein Phosphatases: Another means of controlling hormone action is the regulation of the protein dephosphorylation reaction. The phosphoprotein phosphatases are themselves subject to regulation by phosphorylation-dephosphorylation reactions and by a variety of other mechanisms, such as protein-protein interactions. In fact, the substrate specificity of the phosphoserine-phosphothreonine phosphatases may be dictated by distinct regulatory subunits, the binding of which is regulated hormonally. The best-studied role of regulation by the dephosphorylation of proteins is that of glycogen metabolism in muscle. In this tissue, two major types of phosphoserine-phosphothreonine phosphatases have been described. Type I preferentially dephosphorylates the β subunit of phosphorylase kinase, whereas type II dephosphorylates the α subunit. Type I phosphatase is implicated in the regulation of glycogen synthase, phosphorylase, and phosphorylase kinase. This phosphatase is itself regulated by phosphorylation of certain of its subunits, and these reactions are reversed by the action of one of the type II phosphatases. In addition, two heat-stable protein inhibitors regulate type I phosphatase activity. Inhibitor-1 is phosphorylated and activated by cAMP-dependent protein kinases, and inhibitor-2, which may be a subunit of the inactive phosphatase, is also phosphorylated, possibly by glycogen synthase kinase-3, although the role of this phosphorylation in vivo remains unclear.

F. Extracellular cAMP: Some cAMP leaves cells and can be readily detected in extracellular fluids. The action of glucagon on liver and vasopressin or PTH on kidney is reflected in elevated levels of cAMP in plasma and urine, respectively; this has led to diagnostic tests of target organ responsiveness. Extracellular cAMP has little if any biologic activity in mammals, but it is an extremely important intercellular messenger in lower eukaryotes and prokaryotes.

One Hormone Uses cGMP as the Second Messenger

Cyclic GMP is made from GTP by the enzyme guanylyl cyclase, which exists in soluble and membrane-bound forms. Each of these isozymes has unique kinetic, physiochemical, and antigenic properties. For some time, cGMP was thought to be the functional counterpart of cAMP. It now appears that cGMP has its unique place in hormone action. The atriopeptins, a family of peptides produced in cardiac atrial tissues, cause natriuresis, diuresis, vasodilation, and inhibition of aldosterone secretion. These peptides (eg, atrial natriuretic factor) bind to and activate the membrane-bound form of guanylyl cyclase. This results in an increase of cGMP of as much as 50-fold in some cases, which is thought to mediate these effects. Other evidence links cGMP to vasodilation. A series of compounds, including nitroprusside, nitroglycerin, nitric oxide, sodium nitrite, and sodium azide, all cause smooth muscle relaxation and are potent vasodilators. These agents increase cGMP by activating the soluble form of guanylyl cyclase, and inhibitors of cGMP phosphodiesterase enhance and prolong these responses. The increased cGMP activates cGMP-dependent protein kinase, which in turn phosphorylates a number of smooth muscle proteins, including the myosin light chain. Presumably, this is involved in relaxation of smooth muscle and vasodilation.

Several Hormones Act Through Calcium or Phosphatidylinositols

Ionized calcium is an important regulator of a variety of cellular processes including muscle contraction, stimulus-secretion coupling, the blood clotting cascade, enzyme activity, and membrane excitability. It is also an intracellular messenger of hormone action.

A. Calcium Metabolism: The extracellular calcium (Ca^{2+}) concentration is about 5 mmol/L and is very rigidly controlled (see Chapter 47). The intracellular concentration of this free ion is much lower, 0.1–10 μmol/L, and the concentration associated with intracellular organelles such as mitochondria and endoplasmic reticulum is in the range of 1–20 μmol/L. In spite of this 5000- to 10,000-fold concentration gradient and a favorable transmembrane electrical gradient, Ca^{2+} is restrained from entering the cell. There are three ways of changing cytosolic

Ca^{2+}. Certain hormones (class II.C) enhance membrane permeability to Ca^{2+} and thereby increase Ca^{2+} influx. This is probably accomplished by an Na^+-Ca^{2+} exchange mechanism that has a high capacity but a low affinity for Ca^{2+}. There also is a Ca^{2+}-$2H^+$ ATPase-dependent pump that extrudes Ca^{2+} in exchange for H^+. This has a high affinity for Ca^{2+} but a low capacity and is probably responsible for fine-tuning cytosolic Ca^{2+}. Finally, Ca^{2+} can be mobilized (or deposited) from (or into) the mitochondrial and endoplasmic reticulum pools.

Two observations led to the current understanding of how Ca^{2+} serves as an intracellular messenger of hormone action. First was the ability to quantitate the rapid changes of intracellular Ca^{2+} concentration that are implicit in a role for Ca^{2+} as an intracellular messenger. Such evidence was provided by a variety of techniques, including the use of Quin 2 or Fura 2, fluorescent Ca^{2+} chelators. Rapid changes of Ca^{2+} in the submicromolar range can be quantitated using these compounds. The second important observation linking Ca^{2+} to hormone action involved the definition of the intracellular targets of Ca^{2+} action. The discovery of a Ca^{2+}-dependent regulator of phosphodiesterase activity provided the basis for understanding how Ca^{2+} and cAMP interact within cells.

B. Calmodulin: The calcium-dependent regulatory protein is now referred to as calmodulin, a 17-kDa protein that is homologous to the muscle protein troponin C in structure and function. Calmodulin has four Ca^{2+} binding sites, and full occupancy of these sites leads to a marked conformational change, so that most of the molecule assumes an alpha-helical structure. This conformational change is presumably linked to calmodulin's ability to activate or inactivate enzymes. The interaction of Ca^{2+} with calmodulin (with the resultant change of activity of the latter) is conceptually similar to the binding of cAMP to protein kinase and the subsequent activation of this molecule. Calmodulin is often one of numerous subunits of complex proteins and is particularly involved in regulating various kinases and enzymes of cyclic nucleotide generation and degradation. A partial list of the enzymes regulated directly or indirectly by Ca^{2+}, probably through calmodulin, is given in Table 44–6.

In addition to its effects on enzymes and ion transport, Ca^{2+}/calmodulin regulates the activity of many structural elements in cells. These include the actin-myosin complex of smooth muscle, which is under β-adrenergic control, and various microfilament-mediated processes in noncontractile cells, including cell motility, conformation changes, mitosis, granule release, and endocytosis.

Calcium Is a Mediator of Hormone Action

A role for ionized calcium in hormone action is suggested by the observations that the effect of many

Table 44–6. Enzymes regulated by calcium or calmodulin.

- Adenylyl cyclase
- Ca^{2+}-dependent protein kinase
- Ca^{2+}-Mg^{2+} ATPase
- Ca^{2+}-phospholipid-dependent protein kinase
- Cyclic nucleotide phosphodiesterase
- Glycerol-3-phosphate dehydrogenase
- Glycogen synthase
- Guanylyl cyclase
- Myosin kinase
- NAD kinase
- Phospholipase A_2
- Phosphorylase kinase
- Phosphoprotein phosphatase 2B
- Pyruvate carboxylase
- Pyruvate dehydrogenase
- Pyruvate kinase

hormones (1) is blunted by Ca^{2+}-free media or when intracellular calcium is depleted; (2) can be mimicked by agents that increase cytosolic Ca^{2+}, such as the Ca^{2+} ionophore A23187; and (3) influences cellular calcium flux. These processes have been studied in some detail in pituitary, smooth muscle, platelets, and salivary gland, but most is probably known about how vasopressin and α-adrenergic catecholamines regulate glycogen metabolism in liver. This is shown schematically in Figures 20–6 and 20–7.

Addition of α_1 agonists or vasopressin to isolated hepatocytes results in a threefold increase of cytosolic Ca^{2+} (from 0.2 to 0.6 $\mu mol/L$) within a few seconds. This change precedes and equals the increase in phosphorylase a activity, and the hormone concentrations required for both processes are comparable. This effect on Ca^{2+} is inhibited by α_1 antagonists, and removal of the hormone results in a prompt decline of both cytosolic Ca^{2+} and phosphorylase a. The initial source of the Ca^{2+} appears to be the intracellular organelle reservoirs, which seem to be sufficient for the early effects of the hormones. More prolonged action appears to require enhanced influx or inhibition of Ca^{2+} efflux through the Ca^{2+} pump. The latter may depend upon concomitant increases of cAMP.

Phosphorylase activation results from the conversion of phosphorylase b to phosphorylase a through the action of the enzyme phosphorylase b kinase. This enzyme contains calmodulin as its δ subunit, and its activity is increased through a Ca^{2+} concentration range of 0.1–1 $\mu mol/L$, the range through which hormones increase Ca^{2+} in liver. The link between Ca^{2+} and phosphorylase activation is definite.

A number of critical metabolic enzymes are regulated by Ca^{2+}, phosphorylation, or both, including glycogen synthase, pyruvate kinase, pyruvate carboxylase, glycerol-3-phosphate dehydrogenase, and pyruvate dehydrogenase. It is uncertain whether calmodulin is directly involved or whether the newly discovered Ca^{2+}-calmodulin-dependent or Ca^{2+}-

phospholipid-dependent protein kinases are responsible.

Phosphatidylinositide Metabolism Affects Ca^{2+}-Dependent Hormone Action

Some signal must provide communication between the hormone receptor on the plasma membrane and the intracellular Ca^{2+} reservoirs. This is accomplished by products of phosphatidylinositol metabolism. Cell surface receptors such as those for acetylcholine, antidiuretic hormone, and α_1-type catecholamines are, when occupied by their respective ligands, potent activators of phospholipase C. Receptor binding and activation of phospholipase C are coupled by a unique G protein (Figure 44–6). Phospholipase C catalyzes the hydrolysis of phosphatidylinositol 4,5-bisphosphate to inositol trisphosphate and 1,2-diacylglycerol (Figure 44–7). The diacylglycerol is itself capable of activating protein kinase C, the activity of which also depends upon free ionic calcium. Inositol trisphosphate is an effective releaser of calcium from intracellular storage sites such as the sarcoplasmic reticulum and mitochondria. Thus, the hydrolysis of phosphatidylinositol 4,5-bisphosphate leads to activation of protein kinase C and promotes an increase of cytoplasmic calcium ion. As shown in Figure 44–6, the activated G protein complex can also have a direct action on Ca^{2+} channels.

Steroidogenic agents, including ACTH and cAMP in the adrenal cortex; angiotensin II, K^+, serotonin, ACTH, and cAMP in the zona glomerulosa of the adrenal; LH in the ovary; and LH and cAMP in the Leydig cells of the testes, have been associated with increased amounts of phosphatidic acid, phosphatidylinositol, and polyphosphoinositides in the respective target tissues. Several other examples could be cited.

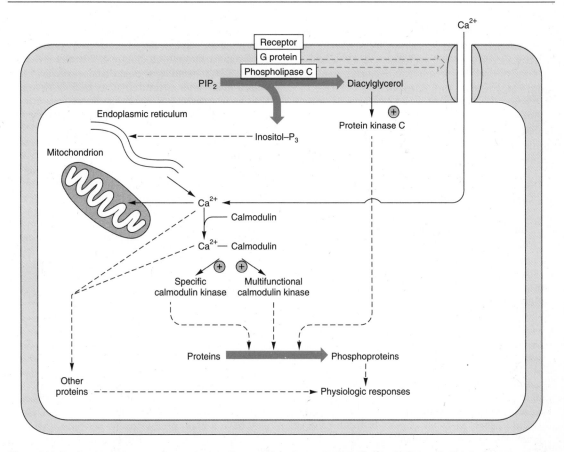

Figure 44–6. Certain hormone receptor interactions result in the activation of phospholipase C. This appears to involve a specific G protein, which also may activate a calcium channel. Phospholipase C results in the generation of inositol trisphosphate, which liberates stored intracellular Ca^{2+}, and diacylglycerol (DAG), which activates protein kinase C. In this scheme, the activated protein kinase C phosphorylates specific substrates, which then alter physiologic processes. Likewise, the Ca^{2+}-calmodulin complex can activate specific kinases. These actions result in the modification of substrates, and this leads to altered physiologic responses. (Courtesy of JH Exton.)

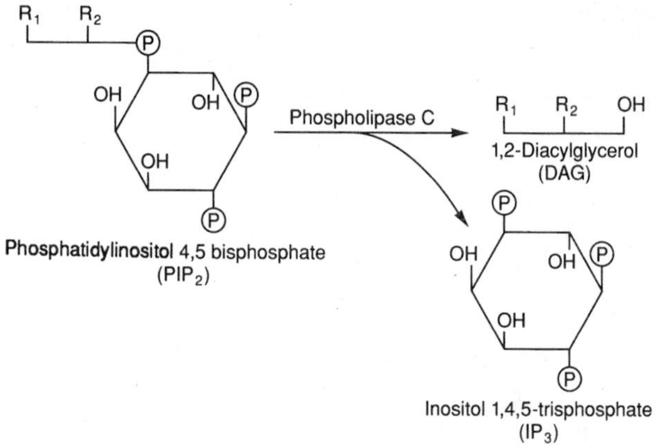

Figure 44–7. Phospholipase C cleaves PIP_2 into diacylglycerol and inositol trisphosphate. R_1 generally is stearate, and R_2 is usually arachidonate. IP_3 can be dephosphorylated (to the inactive I-$1,4$-P_2) or phosphorylated (to the potentially active I-$1,3,4,5$-P_4).

The roles that Ca^{2+} and polyphosphoinositide breakdown products might play in hormone action are presented in Figure 44–6. In this scheme the activated protein kinase C can phosphorylate specific substrates, which then alter physiologic processes. Likewise, the Ca^{2+}-calmodulin complex can activate specific kinases. These then modify substrates and thereby alter physiologic responses.

Some Hormones Act Through a Protein Kinase Cascade

In previous editions of this book, a number of hormones were listed under the category "intracellular mediator unknown." The discovery that the EGF receptor contains an intrinsic tyrosine kinase activity that is activated by the binding of the ligand, EGF, was an important breakthrough. The insulin and IGF-I receptors also contain intrinsic ligand-activated tyrosine kinase activity. Several receptors—generally those involved in binding ligands involved in growth control, differentiation, and the inflammatory response—either have intrinsic tyrosine kinase activity or are associated with proteins that are tyrosine kinases. Another distinguishing feature of this class of hormone action is that these kinases preferentially phosphorylate tyrosine residues, and tyrosine phosphorylation is infrequent (< 0.03% of total amino acid phosphorylation) in mammalian cells.

Some hormone receptors, such as those for insulin, EGF, and IGF-I, have intrinsic tyrosine kinase activity. Activation of this kinase results in the phosphorylation of protein substrates on tyrosine residues. This results in a cascade of events that is described in detail in Chapter 51 in the context of insulin action. A graphic depiction is shown in Figure 51–13.

Tyrosine kinase activation can also initiate a phosphorylation and dephosphorylation cascade that involves the action of several other protein kinases and the counterbalancing actions of phosphatases. Two mechanisms are employed to initiate this cascade. Some hormones, such as growth hormone, prolactin, erythropoietin, and the cytokines, initiate their action by activating tyrosine kinase, but this activity is not an integral part of the hormone receptor. The hormone-receptor interaction activates cytoplasmic protein tyrosine kinases, such as Tyk-2, JAK1, or JAK2. These kinases phosphorylate one or more cytoplasmic proteins, which then associate with other docking proteins through binding to Src homology 2 domains. These peptide segments, approximately 100 amino acids long, are referred to as SH2 domains. One such interaction results in the activation of a family of cytosolic proteins called signal transduction and activators of transcription (STATs). The phosphorylated STAT protein dimerizes and translocates into the nucleus, binds to a specific DNA element, such as the interferon response element or the serum response element, and activates transcription. This is illustrated in Figure 44–8. Other SH2 docking events may result in the activation of PI 3-kinase, the MAP kinase pathway (through SHC or GRB2), or G protein-mediated activation of phospholipase C (PLCγ) with the attendant production of diacylglycerol and activation of protein kinase C. It is apparent that there is a potential for crossover when different hormones activate these various signal transduction pathways.

SUMMARY

The cellular and subcellular actions of hormones require the binding of a hormone to its specific receptor. Receptors have the following characteristics: they have a high affinity for the hormone, the binding is readily reversible, it is saturable, and it is highly specific. Receptors are responsible for two basic functions: they bind the hormone, and they couple hormone binding to signal transduction.

Receptors can be a component of the plasma membrane, as in the case of the peptide hormones. Or the receptor can be located within the cell, as in

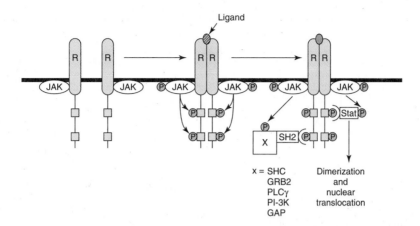

Figure 44–8. Initiation of signal transduction by receptors linked to JAK kinases. The receptors that bind prolactin, growth hormone, interferons, and cytokines lack endogenous tyrosine kinase. Upon the ligand binding, these receptors dimerize and an associated protein (JAK1, JAK2, or TYK) is phosphorylated. JAK•P, an active kinase, phosphorylates the receptor on tyrosine residues. The Stat proteins associate with the phosphorylated receptor and then are themselves phosphorylated by JAK•P. STAT•P dimerizes, translocates to the nucleus, binds to specific DNA elements, and regulates transcription. The phosphotyrosine residues of the receptor also bind to several SH2 domain-containing proteins. This results in activation of the MAP kinase pathway (through SHC or GRB2), PLCγ or PI-3 kinase.

the case of the steroid-thyroid-retinoid family of receptors. In the latter case, the hormone receptor complex is the intracellular signal. In general, the hormone-receptor complex binds to specific regions of DNA, called hormone response elements, or HREs. Each hormone effect, which involves the regulation of the transcription of specific genes, is mediated by a specific HRE. The interaction of peptide hormones with their receptors results in a variety of effects in addition to the regulation of the expression of specific genes. These effects include the regulation of ion and channel activity, the activity of intracellular proteins, and the secretion of various molecules. These effects are mediated by second messengers (the hormone being the first messenger) such as cAMP, cGMP, Ca^{2+}, various phosphatidylinositides, and protein kinase cascades.

REFERENCES

Argetsinger LS et al: Identification of JAK2 as a growth hormone receptor-associated tyrosine kinase. Cell 1993;74:237.

Berridge M: Inositol triphosphate and calcium signalling. Nature 1993;361:315.

Chinkers M, Garbers DL: Signal transduction by guanylyl cyclases. Annu Rev Biochem 1991;60:553.

Cobb MH, Robbins DJ, Boulton TG: ERKs, extracellular signal-regulated MAP-2 kinases. Curr Opin Cell Biol 1991;3:1025.

Darnell JE Jr, Kerr IM, Stark GR: Jak-STAT pathways and transcriptional activation in response to IFNs and other extracellular signaling proteins. Science 1994;264:1415.

Evans R: The steroid and thyroid hormone receptor superfamily. Science 1988;240:889.

Fantl WJ, Johnson DE, Williams LT: Signalling by receptor tyrosine kinases. Annu Rev Biochem 1993;62:453.

Gilman A: G proteins and dual control of adenylate cyclase. Cell 1984;36:577.

Hepler JR, Gilman AG: G proteins. Trends Biochem Sci 1992;17:383.

Hunter T: A thousand and one protein kinases. Cell 1987;50:823.

Lucas P, Granner D: Hormone response domains in gene transcription. Annu Rev Biochem 1992;61:1131.

Pawson T, Schlessinger J: SH2 and SH3 domains. Curr Biol 1993;3:434.

Rasmussen H: The calcium messenger system. (Two parts.) N Engl J Med 1986;314:1094, 1164.

Walton KM, Dixon JE: Protein tyrosine phosphatases. Annu Rev Biochem. 1993;62:101.

White MF, Kahn CR: The insulin signalling system. J Biol Chem 1994;269:1.

45

Pituitary & Hypothalamic Hormones

Daryl K. Granner, MD

INTRODUCTION

The anterior pituitary, under control of hypothalamic hormones, secretes a number of hormones (trophic hormones) that regulate the growth and function of other endocrine glands or influence metabolic reactions in other target tissues. The posterior pituitary produces hormones that regulate water balance and milk ejection from the lactating mammary gland.

BIOMEDICAL IMPORTANCE

The loss of anterior pituitary function (panhypopituitarism) results in atrophy of the thyroid, adrenal cortex, and gonads. Secondary effects due to the absence of the hormones secreted by these target glands affect most body organs and tissues and many general processes such as protein, fat, carbohydrate, and fluid and electrolyte metabolism. The loss of posterior pituitary function results in diabetes insipidus, the inability to concentrate the urine.

HYPOTHALAMIC HORMONES REGULATE THE ANTERIOR PITUITARY

The release (and in some cases production) of each of the pituitary hormones listed in Table 45–1 is under tonic control by at least one hypothalamic hormone. The hypothalamic hormones are released from the hypothalamic nerve fiber endings around the capillaries of the hypothalamic-hypophysial system in the pituitary stalk and reach the anterior lobe through the special portal system that connects the hypothalamus and the anterior lobe. The structures of several hypothalamic hormones are illustrated in Table 45–2.

The hypothalamic hormones are released in a pulsatile manner, and isolated anterior pituitary target cells respond better to pulsatile administration of these hormones than to continuous exposure. The release of LH and FSH is controlled by the concentration of one releasing hormone, GnRH; this in turn is primarily regulated by circulating levels of gonadal hormones that reach the hypothalamus. Similar feed-

ACRONYMS USED IN THIS CHAPTER	
ACTH	Adrenocorticotropic hormone
ADH	Antidiuretic hormone
CG	Chorionic gonadotropin
CLIP	Corticotropin-like intermediate lobe peptide
CRH	Corticotropin-releasing hormone
CS	Chorionic somatomammotropin; placental lactogen
FSH	Follicle-stimulating hormone
GAP	GnRH-associated peptide
GH	Growth hormone
GHRH or GRH	Growth hormone–releasing hormone
GHRIH	Growth hormone release–inhibiting hormone; somatostatin
GnRH	Gonadotropin-releasing hormone
IGF-I, -II	Insulin-like growth factors I and II
LH	Luteinizing hormone
LPH	Lipotropin
MSA	Multiplication-stimulating activity
MSH	Melanocyte-stimulating hormone
NGF	Nerve growth factor
POMC	Pro-opiomelanocortin peptide family
PRIH or PIH	Prolactin release–inhibiting hormone
PRL	Prolactin
SRIH	Somatotropin release–inhibiting hormone
T_3	Triiodothyronine
T_4	Thyroxine; tetraiodothyronine
TRH	Thyrotropin-releasing hormone
TSH	Thyroid-stimulating hormone
VIP	Vasoactive intestinal polypeptide

back loops exist for all of the hypothalamic-pituitary-target gland systems (Table 45–1).

The release of ACTH is primarily controlled by CRH, but a number of other hormones, including ADH, catecholamines, VIP, and angiotensin II, may be involved. CRH release is influenced by cortisol, a

Table 45–1. Hypothalamic-hypohysial-target gland hormones form integrated feedback loops.[1]

Hypothalamic Hormone	Acronym	Pituitary Hormone Affected[2]	Target Gland Hormone Affected
Corticotropin-releasing hormone	CRH	ACTH (LPH, MSH, endorphins)	Hydrocortisone
Thyrotropin-releasing hormone	TRH	TSH (PRL)	T_3 and T_4
Gonadotropin-releasing hormone	GnRH (LHRH, FSHRH)	LH, FSH	Androgens, estrogens, progestins
Growth hormone–releasing hormone	GHRH or GRH	GH	IGF-1; others (?)
Growth hormone release–inhibiting hormone; somatostatin; somatotropin release–inhibiting hormone	GHRIH or SRIH	GH (TSH, FSH, ACTH)	IGF-1; T_3 and T_4; others (?)
Prolactin release–inhibiting hormones; dopamine and GAP	PRIH or PIH	PRL	Neurohormones (?)

[1]The general features of each major feedback system can be deduced by substituting the corresponding hypothalamic, pituitary, or target gland hormone into the appropriate place in Figure 44–1.
[2]The hypothalamic hormone has a secondary or lesser effect on the hormones in parentheses.

glucocorticoid hormone secreted by the adrenal. TSH release is primarily affected by TRH, which in turn is regulated by the thyroid hormones T_3 and T_4; but TSH release is also inhibited by somatostatin. Growth hormone release and production are under tonic control by both stimulating and inhibiting hypothalamic hormones. In addition, a peripheral feedback loop is involved in GH regulation. IGF-I (somatomedin C), which mediates some of the effects of GH, stimulates the release of somatostatin (GHRIH) while inhibiting the release of GHRH. The regulation of PRL synthesis and secretion is primarily under tonic inhibition by hypothalamic agents. It is unique because of the combined neural (nipple stimulation) and neurotransmitter-neurohormone link. Dopamine (Table 45–1) inhibits PRL synthesis (by inhibiting transcription of the PRL gene) and release but does not account for overall PRL inhibition. Recently, a 56-amino-acid neuropeptide was discovered that has both GnRH and PRIH activities—thus the name GnRH-associated peptide (GAP). GAP is a potent inhibitor of PRL release and may be the elusive PRIH

peptide. GAP may explain the curious link between GnRH and PRL secretion that is particularly obvious in some species.

Many of the hypothalamic hormones, in particular TRH, CRH, and somatostatin, are found in other portions of the nervous system and in a variety of peripheral tissues.

Although cAMP was originally thought to mediate the action of releasing hormones on the adenohypophysis, recent studies with GnRH and TRH suggest that a calcium-phosphatidylinositol mechanism, similar to that described above, is involved. Whether the releasing hormones also affect the synthesis of the corresponding pituitary hormone has been argued, but recently GHRH has been shown to stimulate the rate of transcription of the GH gene, and TRH has a similar effect on the prolactin gene.

Table 45–2. Structures of hypothalamic releasing hormones.

Hormone	Structure
TRH	(pyro)Glu-His-Pro-NH$_2$
Somatostatin	Ala-Gly-Cys-Lys-Asn-Phe-Phe-Trp-Lys-Thr-Phe-Thr-Ser-Cys-NH$_2$ (with S—S bridge between the two Cys residues)
GnRH	(pyro)Glu-His-Trp-Ser-Tyr-Gly-Leu-Arg-Pro-Gly-NH$_2$
PRIH	HO—(dihydroxyphenyl)—CH$_2$CH$_2$NH$_2$; GnRH-associated peptide (GAP)
Ovine CRH	Ser-Gln-Glu-Pro-Pro-Ile-Ser-Leu-Asp-Leu-Thr-Phe-His-Leu-Leu-Arg-Glu-Val-Leu-Glu-Met-Thr-Lys-Ala-Asp-Gln-Leu-Ala-Gln-Gln-Ala-His-Ser-Asn-Arg-Lys-Leu-Leu-Asp-Ile-Ala-NH$_2$
Human GHRH	Try-Ala-Asp-Ala-Ile-Phe-Thr-Asn-Ser-Tyr-Arg-Lys-Val-Leu-Gly-Gln-Leu-Ser-Ala-Arg-Lys-Leu-Leu-Gln-Asp-Ile-Met-Ser-Arg-Gln-Gln-Gly-Glu-Ser-Asn-Gln-Glu-Arg-Gly-Ala-Arg-Ala-Arg-Leu-NH$_2$

THE ANTERIOR PITUITARY PRODUCES MANY HORMONES THAT STIMULATE VARIOUS PHYSIOLOGIC PROCESSES

The anterior pituitary hormones have traditionally been discussed individually, but recent studies dealing with the mechanism of synthesis and with the intracellular mediators of action (see Table 44–1) allow classification of these hormones into three categories: (1) the growth hormone-prolactin-chorionic somatomammotropin group, (2) the glycoprotein hormone group, and (3) the pro-opiomelanocortin peptide family.

Growth Hormone, Prolactin, and Chorionic Somatomammotropin Constitute One Hormone Group

Growth hormone (GH), prolactin (PRL), and chorionic somatomammotropin (CS; placental lactogen) are a family of protein hormones having considerable sequence homology. GH, CS, and PRL range in size from 190 to 199 amino acids in different species. Each has a single tryptophan residue (position 85 in GH and CS; position 91 in PRL), and each has two homologous disulfide bonds. The amino acid homology between hGH and hCS is 85%, whereas that between hGH and hPRL is 35%. In view of this homology, it is not surprising that these three hormones share common antigenic determinants and that all have growth-promoting and lactogenic activity. The hormones are produced in a tissue-specific manner, with GH and PRL produced in the anterior pituitary and CS in the syncytiotrophoblast cells of the placenta. Each appears to be under different regulation (Table 45–1).

On the basis of these striking similarities, it was postulated that these hormones may have arisen by duplication of an ancestral gene. Recombinant DNA technology has revealed that there are multiple genes for GH and CS in primates and humans; that the single PRL gene, while encoding a very similar protein, is five times as large as those for GH and CS; that hCS is a variant of hGH; and that the GH-CS group in humans is located on chromosome 17, while PRL in humans is found on chromosome 6. There is marked evolutionary divergence of these genes. Rat and bovine tissues have a single copy of GH and PRL per haploid genome, and humans have a single PRL gene. Humans have one functional GH gene (GH-N) and a variant (GH-V), two CS genes that are expressed (CS-A and CS-B), and one CS gene that is not expressed (CS-L). Several simian species have at least four of the genes from the GH-CS family. The coding sequence of all these genes is organized into five exons interrupted by four introns (Figure 45–1). The genes are highly homologous in the 5' flanking regions and the coding sequence areas (approximately 93% homology in the latter) and diverge in the 3' flanking regions. The splice junctions are highly conserved, even though the introns in the PRL gene are much longer.

The human GH-CS gene family is located on a linkage group in region q22–24 on the long arm of chromosome 17. Figure 45–2 indicates the relative positions of each of these genes in a 5' to 3' orientation. The genes are all transcribed in the 5' to 3' direction, and GH-N is separated from CS-B by about 45 kb.

The GH-N coding sequence matches the amino acid sequence for circulating GH, and the gene is DNase I-sensitive, signifying its location in a region of **"active chromatin."** The GH-V gene, if expressed, would encode for a protein with 13 amino acid differences. This gene is DNase I-resistant, and thus it may not be active. The GH-V gene is present in patients who lack the GH-N gene (inherited GH deficiency), but since these persons have complete GH deficiency, the GH-V gene either is silent or is producing an inactive GH molecule. The first possibility is most likely, because these individuals form antibodies in response to exogenous GH, an indication that this molecule has not previously been seen by the immune system.

The CS-A and CS-B genes are expressed in placenta; CS-L is silent.

Growth Hormone (GH)

A. Synthesis and Structure: Growth hormone is synthesized in **somatotropes,** a subclass of the pituitary acidophilic cells; somatotropes are the most abundant cells in the gland. The concentration

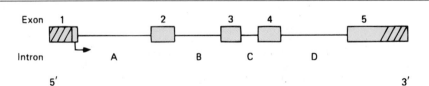

Figure 45–1. Schematic representation, drawn to scale, of the structure of the human growth hormone gene. The gene is about 45 kb in length and consists of five exons and four introns. Cross-hatching represents noncoding regions in exons 1 and 5. Arrow indicates direction of transcription.

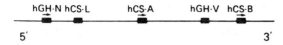

Figure 45–2. Location and orientation of the human GH-CS gene family on chromosome 17. Relative positions of hGH and hCS genes are shown in a 5′ to 3′ orientation. Arrows indicate direction of transcription.

of GH in the pituitary is 5–15 mg/g, which is much higher than the microgram per gram quantities of other pituitary hormones. GH is a single polypeptide with a molecular mass of about 22 kDa in all mammalian species. The general structure of the 191-amino-acid human growth hormone molecule is shown in Figure 45–3. Although there is a high degree of sequence homology between various mammalian growth hormones, only human growth hormone or that of other higher primates is active in humans. Human GH made by recombinant DNA techniques is now available for therapeutic use.

B. The GH Receptor: The GH receptor is a member of the cytokine-hematopoietin receptor superfamily. The GH receptor is a protein with a molecular mass of about 70 kDa with a single membrane spanning domain. The current view is that GH binding causes the dimerization of two GH receptors. This results in the activation of a GH receptor–associated JAK2 tyrosine kinase and phosphorylation of the receptor and JAK2 on tyrosyl residues. These events result in the activation of a number of signaling pathways (Figure 45–4), including (1) Stat protein phosphorylation and gene transcription; (2) the SHC/Grb2-associated activation of the MAP kinase pathway; (3) IRS phosphorylation with activation of PI3 kinase; and (4) PLC activation with production of diacylglycerol and activation of protein kinase C. The JAK kinase pathway is unique to this class of receptors, since the other pathways are activated by a number of different hormone receptors. There is thus the possibility of hormone cross-talk at the level of biologic responses.

C. Physiologic and Biochemical Actions: GH is essential for postnatal growth and for normal carbohydrate, lipid, nitrogen, and mineral metabolism. The growth-related effects are primarily mediated by **IGF-I,** a member of the insulin-like gene family. This was originally known as "sulfation factor" because of its ability to enhance the incorporation of sulfate into cartilage. It next was known as somatomedin C. Structurally, it is similar to proinsulin (see Chapter 51 and Figure 51–5). Another closely related peptide found in human plasma, **IGF-II,** has activity similar or identical to what is often referred to in the rat as multiplication-stimulating activity (MSA). IGF-I and IGF-II both bind to membrane receptors; however, they can be differentiated on the basis of specific radioimmunoassays. IGF-I has 70 amino acids, and IGF-II has 67. Plasma levels of IGF-II are twice those of IGF-I, but it is IGF-I that correlates most directly with GH effects. Individuals who lack sufficient IGF-I but have IGF-II (GH-deficient dwarfs and pygmies; see Table 45–3) fail to grow normally.

1. Protein synthesis–GH increases the transport of amino acids into muscle cells and also increases protein synthesis by a mechanism separate from the transport effect. Animals treated with GH show positive nitrogen balance, reflecting a generalized increase in protein synthesis and a decrease in plasma and urinary levels of amino acids and urea. This is accompanied by increased synthesis of RNA and DNA in some tissues. In these respects GH actions resemble some of the actions of insulin.

2. Carbohydrate metabolism–GH generally antagonizes the effects of insulin. Hyperglycemia after growth hormone administration is the combined result of decreased peripheral utilization of glucose and increased hepatic production via gluconeogene-

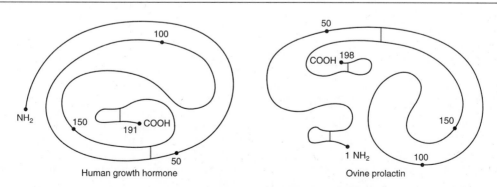

Figure 45–3. The structures of human growth hormone (left) and ovine prolactin (right) are compared. Growth hormone has disulfide bonds between residues 53–165 and 182–189. Prolactin has disulfide bonds between residues 4–11, 58–73, and 190–198.

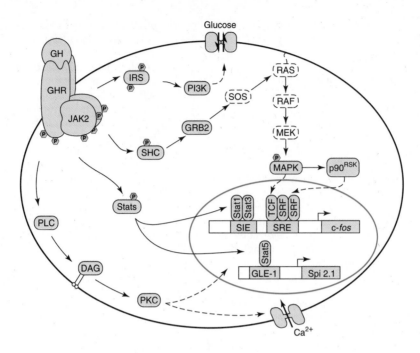

Figure 45–4. Signal transduction pathways activated by the interaction of growth hormone (GH) with its receptor (GHR). As described in the text, GH action may involve four different pathways, each of which is shown in this figure. The activation of JAK2, with subsequent activation of Stats 1 and 3, results in the binding of these transcription factors to specific genes, c-*fos* and Spi 2.1 in this case. The other pathways can be used by different hormones. (Courtesy of C Carter-Su.)

sis. In liver, GH increases liver glycogen, probably from activation of gluconeogenesis from amino acids. Impairment of glycolysis may occur at several steps, and the mobilization of fatty acids from tri-acylglycerol stores may also contribute to the inhibition of glycolysis in muscle. Prolonged administration of GH may result in diabetes mellitus.

3. Lipid metabolism–GH promotes the release of free fatty acids and glycerol from adipose tissue, increases circulating free fatty acids, and causes increased oxidation of free fatty acids in the liver. Under conditions of insulin deficiency (eg, diabetes), increased ketogenesis may occur. These effects and those on carbohydrate metabolism probably are not mediated by IGF-I.

4. Mineral metabolism–GH or, more likely, IFG-I promotes a positive calcium, magnesium, and phosphate balance and causes the retention of Na+, K+, and Cl−. The first effect probably relates to the action of GH in bone, where it promotes growth of long bones at the epiphysial plates in growing children and appositional or acral growth in adults. In children, GH also increases formation of cartilage.

5. Prolactin-like effects–GH binds to lactogenic receptors and thus has many of the properties of prolactin, such as stimulation of the mammary glands, lactogenesis.

D. Pathophysiology: Deficient amounts of GH, whether from panhypopituitarism or isolated GH deficiency, are most serious in infancy because affected infants fail to grow properly. The other metabolic effects are less troublesome. Several types of dwarfism help illustrate the importance of the various steps in GH action (Table 45–3). **GH-deficient dwarfs** respond normally to exogenous GH. Two types of target organ resistance have been described. **Laron type dwarfs** have excessive amounts of GH-N, but they lack functional hepatic GH receptors. **Pygmies** apparently have a post-GH receptor defect, and this may be limited to the action GH exerts through IGF-I.

GH excess, usually from an acidophilic tumor, causes **gigantism** if it occurs before the epiphysial plates close, since there is accelerated growth of the

Table 45–3. Relationship of GH, IGF-I, and IGF-II to dwarfism.

	Plasma Levels			Response to GH Stimulation
	GH	IGF-I	IGF-II	
GH-deficient dwarfs	Low	Low	Low to normal	Yes
Pygmies	Normal	Low	Normal	No
Laron type dwarfs	High	Low	Low	No

long bones. **Acromegaly** results from excessive release of GH that begins after epiphysial closure and the cessation of long bone growth. Acral bone growth causes the characteristic facial changes (protruding jaw, enlarged nose) and enlargement of the hands, feet, and skull. Other findings include enlarged viscera, thickening of the skin, and a variety of metabolic problems, including diabetes mellitus.

About 40% of persons with acromegaly have a G protein–linked disease. Such persons have one of two mutations in the α_s subunit that abolishes the intrinsic GTPase activity in the protein (Chapter 44). One mutation, at arginine position 201, affects the site that is ADP-ribosylated by cholera toxin. Since α_s with such a mutation is constitutively active, cAMP is overproduced. This results in excessive production and release of GH and in unrestrained growth and replication of the somatotroph cells. In this sense, the α_s is an oncogene.

A knowledge of GH regulation allows one to understand the clinical tests used to confirm these diagnoses. GH-deficient patients fail to increase GH levels in response to induced hypoglycemia or administration of arginine or levodopa. Patients with increased GH from a tumor (gigantism or acromegaly) fail to suppress GH levels in response to glucose administration.

Prolactin (PRL; Lactogenic Hormone, Mammotropin, Luteotropic Hormone)

A. Synthesis and Structure: PRL is a protein hormone with a molecular mass of about 23 kDa; its general structure is compared to that of GH in Figure 45–3. It is secreted by **lactotropes,** which are acidophilic cells in the anterior pituitary. The number of these cells and their size increase dramatically during pregnancy. The similarities between the structures and functions of PRL, GH, and CS are noted above.

B. The Prolactin Receptor: The prolactin receptor is similar in size to the GH receptor. It also has a single membrane spanning domain, and it signals through pathways similar to those illustrated in Figure 45–4.

C. Physiologic and Biochemical Actions: PRL is involved in the initiation and maintenance of lactation in mammals. Physiologic levels act only on breast tissue primed by female sex hormones, but excessive levels can trigger breast development in ovariectomized females or in males. In rodents, PRL is capable of maintaining the corpus luteum—hence the name **luteotropic hormone.** Related molecules appear to be responsible for the adaptation of salt-water fish to fresh water, for the molting of reptiles, and for crop-sac milk production in birds. The intracellular mediator of PRL action is unknown.

D. Pathophysiology: Tumors of prolactin-secreting cells cause **amenorrhea** (cessation of menses) and **galactorrhea** (milk discharge) in women. Excessive PRL has been associated with **gynecomastia** (breast enlargement) and **impotence** in men.

Chorionic Somatomammotropin (CS; Placental Lactogen)

The final member of the GH-PRL-CS family has no definite function in humans. In bioassays, CS has lactogenic and luteotropic activity and metabolic effects that are qualitatively similar to those of growth hormone, including inhibition of glucose uptake, stimulation of free fatty acid and glycerol release, enhancement of nitrogen and calcium retention (despite increased urinary calcium excretion), and reduction in the urinary excretion of phosphorus and potassium.

The Glycoprotein Hormones Are Another Group

The most complex protein hormones yet discovered are the pituitary and placental glycoproteins: **thyroid-stimulating hormone (TSH), luteinizing hormone (LH), follicle-stimulating hormone (FSH), and chorionic gonadotropin (CG).** These hormones affect diverse biologic processes and yet have remarkable structural similarities. This class of hormones is found in all mammals. These molecules, like other peptide and protein hormones, interact with cell surface receptors and activate adenylyl cyclase; thus, they employ cAMP as their intracellular messenger.

Each of these hormones consists of two subunits, α and β, joined by noncovalent bonding. The α subunits are identical for all of these hormones within a species, and there is considerable interspecies homology. The specific biologic activity is determined by the β subunit, which also is highly conserved between hormones but to a lesser extent than that noted for the α subunit. The β subunit is not active by itself, and receptor recognition involves the interaction of regions of both subunits. Interhormone and interspecies hybrid molecules are fully active; eg, $TSH_\alpha LH_\beta$ = LH activity, and $hTSH_\alpha mTSH_\beta$ = mouse TSH activity. Thus, interspecies differences between α and β do not affect subunit association or the biologic function domain on β. Each subunit is synthesized from unique mRNAs from separate genes. It is thought that all hormones in this class evolved from a common ancestral gene that resulted in two molecules, α and β, and that the latter evolved further to provide the separate hormones.

A great deal is known about the structure of these molecules. For example, the carboxyl terminal pentapeptide of α is essential for receptor binding but not for α/β association. The feature that distinguishes hormones in the glycoprotein group from hormones in other groups is their glycosylation. In each glycoprotein hormone, the α subunit contains two complex asparagine-linked oligosaccharides, and the β subunit has either one or two. The glycosylation may be necessary for α/β interaction. The α subunit has five S—S bridges, and the β moiety has six.

Free α subunits are found in the pituitary and placenta. This finding and the observation that α and β are translated from separate mRNAs support the concept that the syntheses of α and β are under separate control and that β is limiting for the production of the complete hormone. All are synthesized as preprohormones and are subject to posttranslational processing within the cell to yield the glycosylated proteins.

A. The Gonadotropins (FSH, LH, and hCG): These hormones are responsible for gametogenesis and steroidogenesis in the gonads. Each is a glycoprotein with a molecular mass of about 25 kDa.

1. Follicle-stimulating hormone (FSH)–FSH binds to specific receptors on the plasma membranes of its target cells, the **follicular cells** in the ovary and the **Sertoli cells** in the testis. This results in activation of adenylyl cyclase and increased cAMP production. The actions of FSH are described in more detail in Chapter 50.

2. Luteinizing hormone (LH)–LH binds to specific plasma membrane receptors and stimulates the production of progesterone by **corpus luteum** cells and of testosterone by the **Leydig cells.** The intracellular signal of LH action is cAMP. This nucleotide mimics the actions of LH, which include enhanced conversion of acetate to squalene (the precursor for cholesterol synthesis) and enhanced conversion of cholesterol to pregnenolone, a necessary step in the formation of progesterone and testosterone. The actions of LH are described in more detail in Chapter 50. There is tight coupling between the binding of LH and the production of cAMP, but steroidogenesis occurs when very small increases of cAMP have occurred. Prolonged exposure to LH results in desensitization, perhaps owing to down-regulation of LH receptors. This phenomenon may be exploited as an effective means of birth control.

3. Human chorionic gonadotropin (hCG)–hCG is a glycoprotein synthesized in the **syncytiotrophoblast cells** of the placenta. It has the αβ dimer structure characteristic of this class of hormones and most closely resembles LH. It increases in blood and urine shortly after implantation (see above); hence, its detection is the basis of many pregnancy tests.

B. Thyroid-Stimulating Hormone (TSH): TSH is a glycoprotein of αβ dimer structure with a molecular mass of about 30 kDa. Like other hormones of this class, TSH binds to plasma membrane receptors and activates adenylyl cyclase. The consequent increase of cAMP is responsible for the action of TSH in thyroid hormone biosynthesis. Its relationship to the trophic effects of TSH on the thyroid is less certain.

TSH has several acute effects on thyroid function. These occur in minutes and involve increases of all phases of T_3 and T_4 biosynthesis, including iodide concentration, organification, coupling, and thyroglobulin hydrolysis. TSH also has several chronic effects on the thyroid. These require several days and include increases in the synthesis of proteins, phospholipids, and nucleic acids and in the size and number of thyroid cells. Long-term metabolic effects of TSH are due to the production and action of the thyroid hormones.

Complex Processing Generates the Pro-opiomelanocortin (POMC) Peptide Family

The POMC family consists of peptides that act as hormones (ACTH, LPH, MSH) and others that may serve as neurotransmitters or neuromodulators (endorphins). POMC is synthesized as a precursor molecule of 285 amino acids and is processed differently in various regions of the pituitary.

A. Distribution, Processing, and Functions of the POMC Gene Products: The POMC gene is expressed in the anterior and intermediate lobes of the pituitary. The most conserved sequences between species are within the amino terminal fragment, the ACTH region, and the β-endorphin region. POMC or related products are found in several other vertebrate tissues, including the brain, placenta, gastrointestinal tract, reproductive tract, lung, and lymphocytes. This is presumably due to gene expression in these tissues (rather than to absorption from plasma). Related peptides have also been found in many invertebrate species.

The POMC protein is processed differently in the anterior lobe than in the intermediate lobe. The intermediate lobe is rudimentary in adult humans, but it is active in human fetuses and in pregnant women during late gestation and is also active in many animal species. Processing of the POMC protein in the peripheral tissues (gut, placenta, male reproductive tract) resembles that in the intermediate lobe. There are three basic peptide groups: (1) ACTH, which can give rise to α-MSH and corticotropin-like intermediate lobe peptide (CLIP); (2) β-lipotropin (β-LPH), which can yield γ-LPH, β-MSH, and β-endorphin (and thus α- and γ-endorphins); and (3) a large amino terminal peptide, which generates γ-MSH. The diversity of these products is due to the many dibasic amino acid clusters that are potential cleavage sites for trypsin-like enzymes. Each of the peptides mentioned is preceded by Lys-Arg, Arg-Lys, Arg-Arg, or Lys-Lys residues. The prehormone segment is cleaved, and modification by glycosylation, acetylation, and phosphorylation occurs after translation. The next cleavage, in both anterior and intermediate lobes, is between ACTH and β-LPH, resulting in an amino terminal peptide with ACTH and a β-LPH segment (Figure 45–5). $ACTH_{1-39}$ is subsequently cleaved from the amino terminal peptide, and in the

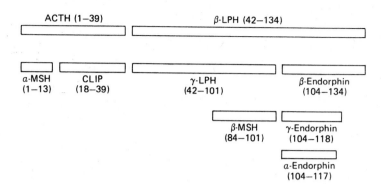

Figure 45–5. Products of pro-opiomelanocortin (POMC) cleavage. (MSH, melanocyte-stimulating hormone. CLIP, corticotropin-like intermediate lobe peptide; LPH, lipotropin.)

anterior lobe essentially no further cleavages occur. In the intermediate lobe, $ACTH_{1-39}$ is cleaved into α-MSH (residues 1–13) and CLIP (18–39); β-LPH (42–134) is converted to γ-LPH (42–101) and β-endorphin (104–134). β-MSH (84–101) is derived from γ-LPH.

There are extensive additional modifications of these peptides. Much of the amino terminal peptide and $ACTH_{1-39}$ in the anterior pituitary is glycosylated. α-MSH is found predominantly in an *N*-acetylated and carboxyl terminal amidated form; deacetylated α-MSH is much less active. β-Endorphin is rapidly acetylated in the intermediate lobe; acetylated β-endorphin, in contrast to α-MSH, is 1000 times less active than the unmodified form. β-Endorphin may therefore be inactive in the pituitary. In the hypothalamus, these molecules are not acetylated and presumably are active. β-Endorphin is also trimmed at the carboxyl terminal end to form α- and γ-endorphin (Figure 45–5). These form the three major endorphins in the rodent intermediate lobe. The large amino terminal fragment is probably also extensively cleaved, but while γ-MSH has been found in rat and bovine pituitaries, less is known about this fragment. This structural information has come largely from studies of the rodent pituitary, but the general scheme is thought to apply to other species. Precise functions for most of the POMC peptides have not been established.

B. Action and Regulation of Specific Peptides:

1. Adrenocorticotropic hormone (ACTH) structure and mechanism of action–ACTH, a single-chain polypeptide consisting of 39 amino acids (Figure 45–6), regulates the growth and function of the adrenal cortex. The 24 amino terminal amino acids are required for full biologic activity and are invariant between species, whereas the 15 carboxyl terminal amino acids are quite variable. A synthetic $ACTH_{1-24}$ analog is widely used in diagnostic testing.

ACTH increases the synthesis and release of adrenal steroids by enhancing the conversion of cholesterol to pregnenolone. This step entails the conversion from a C_{27} to a C_{21} steroid by removal of a six-carbon side chain. Since pregnenolone is the precursor of all adrenal steroids (Figure 48–3), prolonged ACTH stimulation results in excessive production of glucocorticoids, mineralocorticoids, and dehydroepiandrosterone (an androgen precursor). However, the contribution of ACTH to the last two classes of steroids is minimal under physiologic conditions. ACTH increases adrenal cortical growth (the trophic effect) by enhancing protein and RNA synthesis.

ACTH, like other peptide hormones, binds to a plasma membrane receptor. Within a few seconds of this interaction, intracellular cAMP levels increase markedly. cAMP analogs mimic the action of ACTH, but calcium also is involved.

2. ACTH pathophysiology–Excessive production of ACTH by the pituitary or by ectopic production from a tumor results in **Cushing's syndrome.** The weak MSH-like activity of ACTH or associated release of β- or α-MSH results in hyperpigmentation. The metabolic manifestations are due to excessive production of adrenal steroids and include (1) negative nitrogen, potassium, and phosphorus balance; (2) sodium retention, which can result in hypertension, edema, or both; (3) glucose intolerance or overt

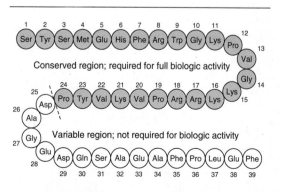

Figure 45–6. Structure of human ACTH.

diabetes mellitus; (4) increased plasma fatty acids; and (5) decreased circulating eosinophils and lymphocytes, with increased polymorphonuclear leukocytes. Patients with Cushing's syndrome may have muscle atrophy and a peculiar redistribution of fat, ie, truncal obesity. Loss of ACTH owing to tumor, infection, or infarction of the pituitary results in an opposite constellation of findings.

C. β-Lipotropin (β-LPH): This peptide consists of the carboxyl terminal 91 amino acids of POMC (Figure 45–5). β-LPH contains the sequences of β-MSH, γ-LPH, metenkephalin, and β-endorphin. Of these, β-LPH, γ-LPH, and β-endorphin have been found in human pituitary, but β-MSH has not been detected. β-LPH is found only in the pituitary, since it is rapidly converted to γ-LPH and β-endorphin in other tissues. β-LPH contains a seven-amino-acid sequence (β-LPH$_{47-53}$) that is identical to ACTH$_{4-10}$. β-LPH causes lipolysis and fatty acid mobilization, but its physiologic role is minimal. It probably serves only as the precursor for β-endorphin.

D. Endorphins: β-Endorphin consists of the carboxyl terminal 31 amino acids of β-LPH (Figure 45–5). The α- and γ-endorphins are modifications of β-endorphin from which 15 and 14 amino acids, respectively, are removed from the carboxyl terminal end. These peptides are found in the pituitary, but they are acetylated there (see above) and probably are inactive. In other sites (eg, central nervous system neurons), they are modified and hence probably serve as neurotransmitters or neuromodulators. Endorphins bind to the same central nervous system receptors as do the opiates and may play a role in endogenous control of pain perception. They have higher analgesic potencies (18–30 times on a molar basis) than morphine. The sequence for enkephalin is present in POMC, but it is not preceded by dibasic amino acids and presumably is not cleaved or expressed.

E. Melanocyte-Stimulating Hormone (MSH): MSH stimulates **melanogenesis** in some species by causing the dispersion of intracellular melanin granules, resulting in darkening of the skin. Three different MSH molecules, α, β, and γ, are contained within the POMC molecule, and two of these, α and β, are secreted in some nonhuman species. In humans, the actual circulating MSH activity is contained within the larger molecules γ- or β-LPH. α-MSH contains an amino acid sequence that is identical to ACTH$_{1-13}$, but it has an acetylated amino terminal. α-MSH (and CLIP) are generally found in animals that have a well-developed intermediate lobe. These peptides are not found in postnatal humans.

Patients with insufficient production of glucocorticoids (**Addison's disease**) have hyperpigmentation associated with increased plasma MSH activity. This could be due to ACTH but is more likely the result of concomitant secretion of β- and γ-LPH, with their associated MSH activity.

THE POSTERIOR PITUITARY CONTAINS TWO ACTIVE HORMONES, VASOPRESSIN AND OXYTOCIN

Vasopressin, originally so named because of its ability to increase blood pressure when administered in pharmacologic amounts, is more appropriately called **antidiuretic hormone (ADH)** because its most important physiologic action is to promote reabsorption of water from the distal renal tubules. **Oxytocin** is also named for an effect of questionable physiologic significance, the acceleration of birth by stimulation of uterine smooth muscle contraction. Its probable physiologic role is to promote milk ejection from the mammary gland.

Each hormone is produced in a specific cell type in the hypothalamus and transported by axoplasmic flow to nerve endings in the posterior pituitary where, upon appropriate stimulation, the hormones are released into the circulation. The probable reason for this arrangement is to escape the blood-brain barrier. ADH is primarily synthesized in the **supraoptic nucleus** and oxytocin in the **paraventricular nucleus.** Each is transported through axons in association with specific carrier proteins called **neurophysins.** Neurophysins I and II are synthesized with oxytocin and ADH, respectively, each as a part of a single protein (sometimes referred to as propressophysin) from a single gene. Neurophysins I and II are unique proteins with molecular masses of 19 kDa and 21 kDa, respectively. ADH and oxytocin are secreted separately into the blood stream along with their appropriate neurophysins. They circulate unbound to proteins and have very short plasma half-lives, on the order of 2–4 minutes. The structures for ADH and oxytocin are shown below.

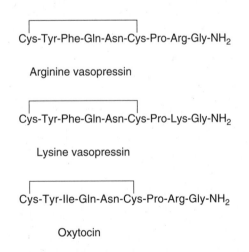

Cys-Tyr-Phe-Gln-Asn-Cys-Pro-Arg-Gly-NH$_2$

Arginine vasopressin

Cys-Tyr-Phe-Gln-Asn-Cys-Pro-Lys-Gly-NH$_2$

Lysine vasopressin

Cys-Tyr-Ile-Gln-Asn-Cys-Pro-Arg-Gly-NH$_2$

Oxytocin

Each is a nonapeptide containing cysteine molecules at positions 1 and 6 linked by an S—S bridge. Most animals have arginine vasopressin; however, the hormone in pigs and related species has a lysine substi-

tuted in position 8. Because of the close structural similarity, it is not surprising that ADH and oxytocin each exhibit some of the effects of the other molecule. These peptides are primarily metabolized in the liver, though renal excretion of ADH accounts for a significant part of its loss from blood.

Oxytocin

A. Regulation of Secretion: The neural impulses that result from stimulation of the nipples are the primary stimulus for oxytocin release. Vaginal and uterine distention are secondary stimuli. PRL is released by many of the stimuli that release oxytocin, and a fragment of oxytocin has been proposed as prolactin-releasing factor. Estrogen stimulates the production of oxytocin and of neurophysin I, and progesterone inhibits the production of these compounds.

A. Mechanism of Action: The mechanism of action of oxytocin is unknown. It causes contraction of uterine smooth muscle and thus is used in pharmacologic amounts to induce labor in humans. Interestingly, pregnant animals in which the hypothalamic-hypophysial tract has been destroyed do not necessarily have trouble delivering their young. The most likely physiologic function of oxytocin is to stimulate contraction of myoepithelial cells surrounding the mammary alveoli. This promotes the movement of milk into the alveolar duct system and allows for milk ejection. Membrane receptors for oxytocin are found in both uterine and mammary tissues. These receptors are increased in number by estrogens and decreased by progesterone. The concomitant rise in estrogens and fall in progesterone occurring immediately before parturition probably explains the onset of lactation prior to delivery. Progesterone derivatives are commonly used to inhibit postpartum lactation in humans. Oxytocin and neurophysin I appear to be produced in the ovary, wherein oxytocin may inhibit steroidogenesis.

The chemical groups important for oxytocin action include the primary amino group of the amino terminal cysteine; the phenolic group of tyrosine; the three carboxyamide groups of asparagine, glutamine, and glycinamide; and the disulfide (S—S) linkage. By deleting or substituting these groups, numerous analogs of oxytocin have been produced. For example, deletion of the free primary amino group of the terminal half cysteine residue (position 1) results in desamino oxytocin, which has four to five times the antidiuretic activity of oxytocin.

Antidiuretic Hormone (ADH; Vasopressin)

A. Regulation of Secretion: The neural impulses that trigger ADH release are activated by a number of different stimuli. Increased osmolality of plasma is the primary physiologic stimulus. This is mediated by **osmoreceptors** located in the hypothal-

amus and by **baroreceptors** located in the heart and other regions of the vascular system. Hemodilution (decreased osmolality) has the opposite effect. Other stimuli include emotional and physical stress and pharmacologic agents including acetylcholine, nicotine, and morphine. Most of these effects involve increased synthesis of ADH and neurophysin II, since the depletion of stored hormone is not associated with this action. Epinephrine and agents that expand plasma volume inhibit ADH secretion, as does ethanol.

B. Mechanism of Action: The most important physiologic target cells of ADH in mammals are those of the distal convoluted tubules and collecting structures of the kidney. These ducts pass through the renal medulla, in which the extracellular solute pool has an osmolality gradient up to four times that of plasma. These cells are relatively impermeable to water, so that in the absence of ADH, the urine is not concentrated and may be excreted in amounts exceeding 2 L/d, occasionally up to 15 L/d. ADH increases the permeability of the cells to water and permits osmotic equilibration of the collecting tubule urine with the hypertonic interstitium, resulting in urine volumes in the range of 0.5–1 L/d.

There are two types of ADH or vasopressin receptors: V1 and V2. V2 receptors are found only on the surface of renal epithelial cells. This receptor is linked to adenylyl cyclase, and cAMP is thought to mediate the effects of ADH in the renal tubule. This physiologic action is the basis of the name "antidiuretic hormone." cAMP and inhibitors of phosphodiesterase activity (caffeine, for example) mimic the actions of ADH. In vivo, an elevated level of calcium in the medium bathing the mucosal surface of the tubular cells inhibits the action of ADH on water movement, apparently by inhibiting the action of adenylyl cyclase, since it does not diminish the action of cAMP per se. This may account, in part, for the excessive volumes of urine that are characteristic of patients with hypercalcemia.

All extrarenal ADH receptors are of the V1 type. Binding of ADH to the V1 receptor causes activation of phospholipase C, which results in the generation of IP_3 and diacylglycerol. This results in an increase of intracellular Ca^{2+} and activation of protein kinase C. A major effect of V1 receptors is vasoconstriction and increased peripheral vascular resistance—hence the name vasopressin that is also used to denote this hormone.

C. Pathophysiology: Abnormalities of ADH secretion or action lead to **diabetes insipidus,** which is characterized by the excretion of large volumes of dilute urine. Primary diabetes insipidus, an insufficient amount of the hormone, is usually due to destruction of the hypothalamic-hypophysial tract from a basal skull fracture, tumor, or infection, but it can be hereditary. In **hereditary nephrogenic diabetes insipidus,** ADH is secreted normally but the target cell is incapable of responding, presumably because of a receptor defect. This hereditary lesion is distinguished from **acquired nephrogenic diabetes insipidus,** which most often is due to the pharmacologic administration of lithium for manic-depressive illness. The **inappropriate secretion of ADH** occurs in association with ectopic production by a variety of tumors (usually tumors of the lung) but can also occur in conjunction with diseases of the brain, pulmonary infections, or hypothyroidism. It is called inappropriate secretion because ADH is produced at a normal or increased rate in the presence of hypoosmolality, thus causing a persistent and progressive dilutional hyponatremia with excretion of hypertonic urine.

SUMMARY

Specific interactions between the hypothalamus, pituitary, and target endocrine glands form a series of closed-loop regulatory units that are the core of the endocrine system. The purpose of this arrangement is to provide different levels of target gland hormones in response to various metabolic and environmental challenges and to ensure a normal reproductive cycle. The hypothalamic hormones involved in this type of regulation are small, labile peptides that are delivered to the anterior pituitary through a special vascular portal system. These hormones stimulate the synthesis and release of anterior pituitary hormones, which are delivered to the target glands by the systemic circulation. Examples of these coupled systems include CRH-ACTH-cortisol, TRH-TSH-T_3/T_4, and GnRH-LH/FSH-testosterone/estradiol/progesterone. The hormones involved in these regulatory systems include small and large monomeric peptides and proteins, heterodimeric glycoproteins, steroids, and amino acid-derived hormones. All the basic biochemical mechanisms of action of hormones are represented in the actions of these hormones, eg, those involving cAMP, Ca^{2+}, diacylglycerol, IP_3, kinase cascades, and direct ligand-receptor interactions. In addition, ACTH is synthesized as part of a very large precursor molecule that is processed into a number of hormones.

Very different functional units are formed between specific regions of the hypothalamus and the posterior pituitary. Cells in the regions of the supraoptic and paraventricular nuclei synthesize large molecules that contain ADH and oxytocin, respectively. These large precursor molecules also contain neurophysins I and II, respectively, which are involved in transporting ADH and oxytocin through the axons of the producing neurons to storage sites in the posterior pituitary. The hormones are released from these depots by changes of serum osmolality (ADH) and nipple stimulation (oxytocin).

REFERENCES

Amselem S et al: Laron dwarfism and mutations of the growth hormone-receptor gene. N Engl J Med 1989; 321:989.

Argetsinger LS et al: Identification of JAK2 as a growth hormone receptor-associated tyrosine kinase. Cell 1993; 74:237.

Barbieri RL: Clinical application of GnRH and its analogs. Trends Endocrinol Metab 1992;3:30.

Chord IT: The posterior pituitary gland. Clin Endocrinol 1975;4:89.

Crowley WF et al: The physiology of gonadotropin-releasing hormone secretion in men and women. Recent Prog Horm Res 1985;41:473.

Douglass J, Civelli O, Herbert E: Polyprotein gene expression: Generation of diversity of neuroendocrine peptides. Annu Rev Biochem 1984;53:665.

Imura H et al: Effect of CNS peptides on hypothalamic regulation of pituitary secretion. Adv Biochem Psychopharmacol 1981;28:557.

Kelly PA et al: The growth hormone/prolactin receptor family. Rec Prog Horm Res 1993;48:123.

Labrie F et al: Mechanism of action of hypothalamic hormones in the adenohypophysis. Annu Rev Physiol 1979; 41:555.

Reichlin S: Systems for the study of regulation of neuropeptide secretion. In: *Neurosecretion and Brain Peptides: Implications for Brain Function and Neurological Disease.* Martin JB, Reichlin S, Bick KL (editors). Raven Press, 1981.

Robertson GL: Regulation of vasopressin function in health and disease. Recent Prog Horm Res 1977;33:333.

Thyroid Hormones

46

Daryl K. Granner, MD

INTRODUCTION

Thyroid hormones regulate gene expression, tissue differentiation, and general development. The thyroid gland produces two iodoamino acid hormones, **3,5,3′-triiodothyronine (T_3)** and **3,5,3′,5′-tetraiodothyronine (T_4, thyroxine),** which have long been recognized for their importance in regulating general metabolism, development, and tissue differentiation. These hormones, whose structures are illustrated in Figure 46–1, regulate gene expression using mechanisms similar to those employed by steroid hormones.

BIOMEDICAL IMPORTANCE

Diseases of the thyroid are among the most common afflictions involving the endocrine system. Diagnosis and therapy are firmly based on the principles of thyroid hormone physiology and biochemistry. The availability of radioisotopes of iodine has greatly aided in the elucidation of these principles. Radioactive iodine, because it localizes in the gland, is widely used in the diagnosis and treatment of thyroid disorders. Radioiodine has a dangerous aspect as well, since excessive exposure, such as from nuclear fallout, is a major risk factor for thyroid cancer. This is especially true in infants and adolescents, whose thyroid cells are still actively dividing.

THYROID HORMONE BIOSYNTHESIS INVOLVES THYROGLOBULIN AND IODIDE METABOLISM

Thyroid hormones are unique in that they require the trace element **iodine** for biologic activity. In most parts of the world, iodine is a scarce component of soil, and hence there is little in food. A complex mechanism has evolved to acquire and retain this crucial element and to convert it into a form suitable for incorporation into organic compounds. At the same time, the thyroid must synthesize thyronine, and this synthesis takes place in thyroglobulin. These processes will be discussed separately, although they occur concurrently.

ACRONYMS USED IN THIS CHAPTER	
DIT	Diiodotyrosine
MIT	Monoiodotyrosine
T_3	Triiodothyronine
T_4	Thyroxine; tetraiodothyronine
TBG	Thyroxine-binding globulin
TBPA	Thyroxine-binding prealbumin
TSH	Thyroid-stimulating hormone
TSI	Thyroid-stimulating IgG

Thyroglobulin Is a Complex Protein

A. Biosynthesis: Thyroglobulin is the precursor of T_4 and T_3. It is a large, iodinated, glycosylated protein with a molecular mass of 660 kDa. Carbohydrate accounts for 8–10% of the weight of thyroglobulin and iodide for about 0.2–1%, depending upon the iodine content in the diet. Thyroglobulin is composed of two subunits. It contains 115 tyrosine residues, each of which is a potential site of iodination. About 70% of the iodide in thyroglobulin exists in the inactive precursors, **monoiodotyrosine (MIT)** and **diiodotyrosine (DIT),** while 30% is in the **iodothyronyl residues,** T_4 and T_3. When iodine supplies are sufficient, the T_4:T_3 ratio is about 7:1. In **iodine deficiency,** this ratio decreases, as does the DIT:MIT ratio. This large molecule of 5000 amino acids provides the proper conformation required for tyrosyl coupling and iodide organification necessary in the formation of the diamino acid thyroid hormones. Thyroglobulin is a prohormone. Thyroglobulin is synthesized in the basal portion of the cell, moves to the lumen, where it is stored in the extracellular colloid, and reenters the cell and moves in an apical to basal direction during its hydrolysis into the active T_3 and T_4 hormones. All of these steps are enhanced by TSH, and this hormone (or cAMP) also enhances transcription of the thyroglobulin gene.

B. Hydrolysis: Thyroglobulin is a storage form of T_3 and T_4 in the colloid: a several weeks' supply of these hormones exists in the normal thyroid.

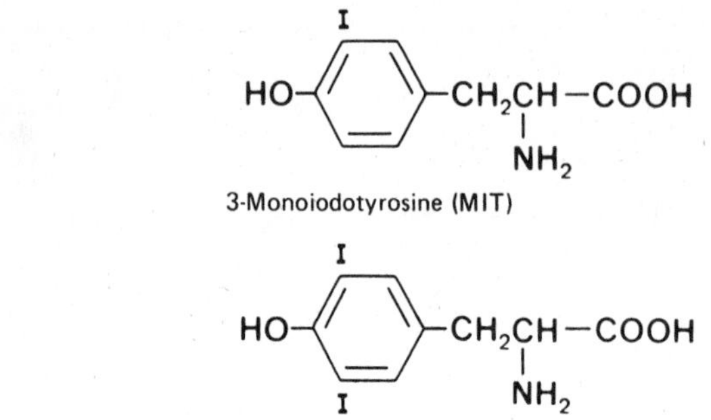

Figure 46–1. Structure of thyroid hormones and related compounds.

Within minutes after the stimulation of the thyroid by TSH (or cAMP), there is a marked increase of microvilli on the apical membrane. This microtubule-dependent process entraps thyroglobulin, and subsequent pinocytosis brings it back into the follicular cell. These phagosomes fuse with lysosomes to form **phagolysosomes** in which various acid proteases and peptidases hydrolyze the thyroglobulin into amino acids, including the iodothyronines. T_4 and T_3 are discharged from the basal portion of the cell, perhaps by a facilitated process, into the blood. The $T_4:T_3$ ratio in this blood is lower than that in thyroglobulin, so that some selective deiodination of T_4 must occur in the thyroid. About 50 μg of thyroid hormone iodide is secreted each day. With an average uptake of iodide (25–30% of the iodide ingested), the daily iodide requirement is between 150 and 200 μg.

As mentioned above, most of the iodide in thyroglobulin is not in iodothyronine; about 70% is in the inactive compounds MIT and DIT. These amino acids are released when thyroglobulin is hydrolyzed and the iodide is scavenged by a **deiodinase.** Other forms of this NADPH-dependent enzyme are present in the pituitary, kidney, and liver. The iodide removed from MIT and DIT constitutes an important pool within the thyroid, as distinguished from that I^- which enters from blood. Under steady-state conditions, the amount of iodide that enters the thyroid matches the amount that leaves. If one-third of the iodide in thyroglobulin leaves (as T_4 and T_3), it follows that two-thirds of the iodide available for biosynthesis comes from the deiodination of MIT and DIT within the thyroid.

Iodide Metabolism Involves Several Discrete Steps (Figure 46–2)

A. Concentration of Iodide (I⁻): The thyroid, along with several other epithelial tissues including mammary gland, chorion, salivary gland, and stomach, is able to concentrate I^- against a strong electrochemical gradient. This is an energy-dependent process and is linked to the ATPase-dependent Na^+-K^+ pump. The activity of the **thyroidal I^- pump** can be isolated from subsequent steps in hormone biosynthesis by inhibiting organification of I^- with drugs of the thiourea class (Figure 46–3). The ratio of iodide in thyroid to iodide in serum (T:S ratio) is a reflection of the activity of this pump or concentrating mechanism. This activity is primarily controlled by TSH and ranges from 500 in animals chronically stimulated with TSH to 5 or less in hypophysectomized animals. The T:S ratio in humans on a normal iodine diet is about 25:1.

A very small amount of iodide also enters the thyroid by diffusion. Any intracellular I^- that is not incorporated into MIT or DIT (generally < 10%) is free to leave by this mechanism.

The transport mechanism is inhibited by two classes of molecules. The first group consists of perchlorate (ClO_4^-), perrhenate (ReO_4^-), and pertechnetate (TcO_4^-), all anions with a similar partial specific volume to I^-. These anions compete with I^- for its carrier and are concentrated by the thyroid. A radioisotope of TcO_4^- is commonly used to study iodide transport in humans. The linear anion thiocyanate (SCN^-), an example of the second class, is a competitive inhibitor of I^- transport but is not concentrated by the thyroid.

B. Oxidation of I⁻: The thyroid is the only tissue that can oxidize I^- to a higher valence state, an obligatory step in I^- organification and thyroid hormone biosynthesis. This step involves a heme-containing peroxidase and occurs at the luminal surface of the follicular cell.

Thyroperoxidase, a tetrameric protein with a molecular mass of 60 kDa, requires hydrogen peroxide as an oxidizing agent. The H_2O_2 is produced by an NADPH-dependent enzyme resembling cytochrome c reductase. A number of compounds in-

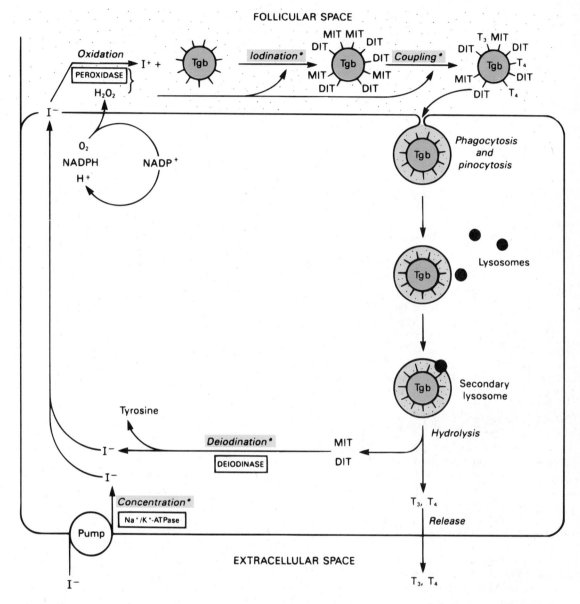

FOLLICULAR SPACE

Figure 46–2. Model of iodide metabolism in the thyroid follicle. A follicular cell is shown facing the follicular lumen (stippled area) and the extracellular space (at bottom). Iodide enters the thyroid by a pump and by passive diffusion. Thyroid hormone synthesis occurs in the follicular space through a series of reactions, many of which are peroxidase-mediated. Thyroid hormones are released from thyroglobulin by hydrolysis. (Tgb, thyroglobulin; MIT, monoiodotyrosine; DIT, diiodotyrosine; T_3, triiodothyronine; T_4, tetraiodothyronine.) Asterisks indicate steps or processes that are inherited enzyme deficiencies which cause congenital goiter and often result in hypothyroidism.

hibit I^- oxidation and therefore its subsequent incorporation into MIT and DIT. The most important of these clinically are the thiourea drugs, some of which are shown in Figure 46–3. They are known as **antithyroid drugs** because of their ability to inhibit thyroid hormone biosynthesis at this step.

C. Iodination of Tyrosine: Oxidized iodide re-

acts with the tyrosyl residues in thyroglobulin in a reaction that probably also involves thyroperoxidase. The 3 position of the aromatic ring is iodinated first and then the 5 position to form MIT and DIT, respectively. This reaction, sometimes called **organification,** occurs within seconds in luminal thyroglobulin. Once iodination occurs, the iodine does not readily

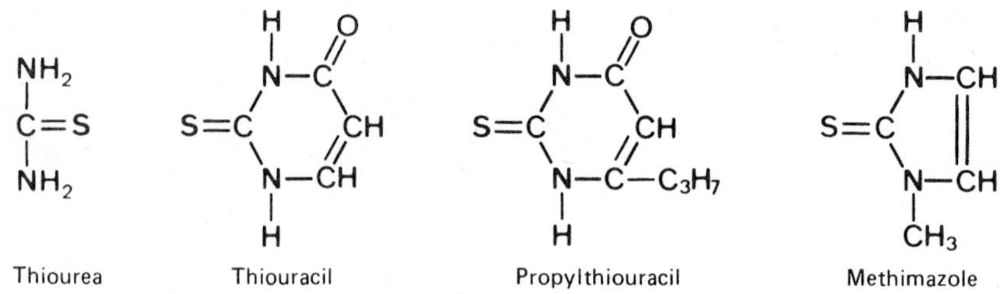

Figure 46–3. Thiourea class of antithyroid drugs.

leave the thyroid. Free tyrosine can be iodinated, but it is not incorporated into proteins, since no tRNA recognizes iodinated tyrosine.

D. Coupling of Iodotyrosyls: The coupling of two DIT molecules to form T_4 or of an MIT and DIT to form T_3 occurs within the thyroglobulin molecule, although the addition of a free MIT or DIT to a bound DIT has not been conclusively excluded. A separate coupling enzyme has not been found, and since this is an oxidative process, it is assumed that the same thyroperoxidase catalyzes this reaction by stimulating free radical formation of iodotyrosine. This hypothesis is supported by the observation that the same drugs which inhibit I^- oxidation also inhibit coupling. The formed thyroid hormones remain as integral parts of thyroglobulin until the latter is degraded, as described above. Thyroglobulin hydrolysis is stimulated by TSH but is inhibited by I^-; this latter effect is occasionally exploited by using potassium iodide to treat hyperthyroidism.

THYROID HORMONES ARE TRANSPORTED BY THYROID-BINDING GLOBULIN

One-half to two-thirds of T_4 and T_3 in the body is extrathyroidal, and most of this circulates in bound form, ie, bound to two specific binding proteins, **thyroxine-binding globulin** (**TBG**) and **thyroxine-binding prealbumin** (**TBPA**). TBG, a glycoprotein with a molecular mass of 50 kDa, is quantitatively the more important. It binds T_4 and T_3 with 100 times the affinity of TBPA and has the capacity to bind 20 μg/dL of plasma. Under normal circumstances, TBG binds, noncovalently, nearly all of the T_4 and T_3 in plasma (Table 46–1). The small, unbound (free) fraction is responsible for the biologic activity. In spite of the great difference in total amount, the free fraction of T_3 approximates that of T_4, but the plasma half-life of T_4 is four to five times that of T_3.

TBG is also subject to regulation, an important consideration in diagnostic testing of thyroid function, since most assays of T_4 or T_3 measure the total amount in plasma rather than the free hormone. TBG is produced in liver, and its synthesis is increased by estrogens (pregnancy and birth control pills). Decreased production of TBG occurs following androgen or glucocorticoid therapy and in certain liver diseases. Inherited increases or decreases of TBG also occur. All of these conditions result in changes of total T_4 and T_3 without a change of the free level. Phenytoin and salicylates compete with T_3 and T_4 for binding to TBG. This decreases the total level of hormone without changing the free fraction and must be considered when interpreting diagnostic tests.

Extrathyroidal deiodination converts T_4 to T_3. Since T_3 binds to the thyroid receptor in target cells with 10 times the affinity of T_4, T_3 is thought to be the preponderant metabolically active form of the molecule. About 80% of circulating T_4 is converted to T_3 or reverse T_3 (rT_3) in the periphery, and this conversion accounts for most of the production of T_3. Reverse T_3 is a very weak agonist that is made in relatively larger amounts in chronic disease, in carbohydrate starvation, and in the fetus. Propylthiouracil and propranolol decrease the conversion of T_4 to T_3.

Other forms of thyroid hormone metabolism include total deiodination and inactivation by deamination or decarboxylation. Hepatic glucuronidation and sulfation result in a more hydrophilic molecule that is excreted into bile, reabsorbed in the gut, deiodinated in the kidney, and excreted as the glucuronide conjugate in the urine.

THYROID HORMONES ACT VIA A NUCLEAR MECHANISM

Thyroid hormones bind to specific high-affinity receptors in the target cell nucleus. T_3 binds with ap-

Table 46–1. Comparison of T_4 and T_3 in plasma.

	Total Hormone (μg/dL)	Free Hormone			$t_{1/2}$ in Blood (days)
		Percent of Total	ng/dL	Molarity	
T_4	8	0.03	~2.24	3.0×10^{-11}	6.5
T_3	0.15	0.3	~0.4	$~0.6 \times 10^{-11}$	1.5

proximately ten times the affinity of T_4 and has proportionately greater biologic activity. Thyroid hormones bind to low-affinity sites in cytoplasm, but this is apparently not the same protein as the nuclear receptor. The cytoplasmic binding may serve to keep thyroid hormones "in the neighborhood."

A major effect of T_3 and T_4 is to enhance general protein synthesis and cause positive nitrogen balance. Thyroid hormones, like steroids, induce or repress proteins by increasing or decreasing gene transcription (Figure 44–1). In the case of T_3 and T_4 the *trans*-acting factor is the hormone-receptor complex, which always seems to reside in the nucleus. The *cis*-acting DNA hormone response element that binds this complex consists of the core sequence shown in Table 44–3 AGGTCANNNNAGGTCA.

There is a curious association between the two classes of hormones related to growth: the thyroid hormones and growth hormone itself. T_3 and glucocorticoids enhance transcription of the GH gene, so that more GH is produced. This explains a classic observation in which the pituitaries of T_3-deficient animals were found to lack GH, and it may account for some of the general anabolic effects of T_3. Very high concentrations of T_3 inhibit protein synthesis and cause negative nitrogen balance.

Thyroid hormones are known to be important modulators of developmental processes. This is most apparent in amphibian metamorphosis. Thyroid hormones are required for the conversion of a tadpole into a frog, a process that involves resorption of the tail, limb bud proliferation, conversion from fetal to adult hemoglobin, stimulation of urea cycle enzymes (carbamoyl phosphate synthase) so that urea is excreted rather than ammonia, and epidermal changes. These effects probably result from the regulation of specific gene expression. Thyroid hormones are required for normal development in humans. Intrauterine or neonatal hypothyroidism results in **cretinism,** a condition characterized by multiple congenital defects and severe, irreversible mental retardation.

THE PATHOPHYSIOLOGY OF MANY THYROID DISEASES RELATES TO TSH, T_3, AND T_4

A Goiter Is an Enlarged Thyroid

Any enlargement of the thyroid is referred to as a goiter. Simple goiter represents an attempt to compensate for decreased thyroid hormone production; thus, in all of these situations, elevated TSH is the common denominator. Causes include iodide deficiency; iodide excess, when an autoregulatory mechanism fails; and a variety of rare inherited metabolic defects that illustrate the importance of various steps in thyroid hormone biosynthesis. These defects include (1) I^- transport defect; (2) iodination defect; (3) coupling defect; (4) deiodinase deficiency; and

(5) production of abnormal iodinated proteins. Partial deficiencies of these functions may cause simple goiter in adults. Any of these causes of simple goiter can, when severe, cause hypothyroidism. Simple goiter is treated with exogenous thyroid hormone. Supplementation or restriction of iodide intake is appropriate for specific types of goiter.

Insufficient Free T_3 or T_4 Results in Hypothyroidism

This is usually due to thyroid failure but can be due to disease of the pituitary or hypothalamus. In hypothyroidism, the basal metabolic rate is decreased, as are other processes dependent upon thyroid hormones. Prominent features include slow heart rate, diastolic hypertension, sluggish behavior, sleepiness, constipation, sensitivity to cold, dry skin and hair, and a sallow complexion. Other features depend upon the age at onset. Hypothyroidism later in childhood results in short stature but no mental retardation. The various kinds of hypothyroidism are treated with exogenous thyroid hormone replacement.

Hyperthyroidism, or Thyrotoxicosis, Is Due to Excessive Production of Thyroid Hormone

There are many causes, but most cases in the USA are due to **Graves' disease,** which results from the production of **thyroid-stimulating IgG (TSI)** that activates the TSH receptor. This causes a diffuse enlargement of the thyroid and excessive, uncontrolled production of T_3 and T_4, since the production of TSI is not under feedback control. Findings are multisystemic and include rapid heart rate, widened pulse pressure, nervousness, inability to sleep, weight loss in spite of increased appetite, weakness, excessive sweating, sensitivity to heat, and red, moist skin. The hyperthyroidism of Graves' disease is treated by blocking hormone production with an antithyroid drug, by ablating the gland with a radioactive isotope of iodide (such as ^{131}I), or by a combination of these two methods. Occasionally, the gland is removed surgically.

SUMMARY

The thyroid hormones triiodothyronine (T_3) and tetraiodothyronine (T_4) require the rare element iodine (in the form of iodide) for biologic activity. An extensive series of physiologic and biochemical reactions has evolved to ensure that sufficient quantities of iodide are available for T_3 and T_4 biosynthesis. This process, which involves thyroglobulin, one of the largest proteins known, includes (1) the active transport of iodide into the thyroid cell; (2) the oxidation of iodide to a higher valence state by a peroxidase enzyme; (3) the iodination of tyrosine, perhaps by the same enzyme; and (4) the coupling of two iodinated tyrosyl moieties to form iodothyronines. The latter steps occur in thyroglobulin, which must be proteolyzed to component amino acids within the cell in order to release T_3 and T_4. TSH stimulates all these steps.

Thyroid hormones cause their many effects by binding to specific intranuclear receptors of the steroid and thyroid superfamily of receptors. This ligand-receptor complex binds to thyroid hormone response elements in target genes to regulate the rate of synthesis of specific mRNAs. This results in a change in the amount and activity of the cognate protein, which in turn alters the rate of a metabolic process.

REFERENCES

Chopra IJ et al: Pathways of metabolism of thyroid hormones. Recent Prog Horm Res 1978;34:531.

Larsen PR: Thyroid-pituitary interaction: Feedback regulation of thyrotropin secretion by thyroid hormones. N Engl J Med 1982;396:23.

Oppenheimer JH: Thyroid hormone action at the nuclear level. Ann Intern Med 1985;102:374.

Samuels H: Regulation of gene expression by thyroid hormone. J Clin Invest 1988;81:957.

Hormones That Regulate Calcium Metabolism

47

Daryl K. Granner, MD

INTRODUCTION

Calcium ions regulate a number of important physiologic and biochemical processes. These include neuromuscular excitability, blood coagulation, secretory processes, membrane integrity and plasma membrane transport, enzyme reactions, the release of hormones and neurotransmitters, and the intracellular action of a number of hormones. In addition, the proper extracellular fluid and periosteal concentrations of Ca^{2+} and PO_4^{3-} are required for bone mineralization. To ensure that these processes operate normally, the plasma Ca^{2+} concentration is maintained within very narrow limits. The purpose of this chapter is to explain how this is accomplished.

BIOMEDICAL IMPORTANCE

Deviations of the ionized calcium from the normal range cause many disorders and can be life-threatening. As many as 3% of hospitalized patients may have disorder of calcium homeostasis.

CALCIUM OCCURS IN BONE AND EXTRACELLULAR FLUID

There is approximately 1 kg of calcium in the human body. Ninety-nine percent of this amount is located in bone where, with phosphate, it forms the **hydroxyapatite crystals** that provide the inorganic and structural component of the skeleton. Bone is a dynamic tissue, and it undergoes constant remodeling as stresses change; in the steady-state condition, there is a balance between new bone formation and bone resorption. Most of the calcium in bone is not freely exchangeable with **extracellular fluid calcium.** Thus, in addition to its mechanical role, bone serves as a large reservoir of calcium. About 1% of skeletal Ca^{2+} is in a freely exchangeable pool and this, with another 1% of the total found in the periosteal space, constitutes the **miscible pool** of Ca^{2+}. The hormones discussed in this chapter regulate the

ACRONYMS USED IN THIS CHAPTER

CBP	Calcium-binding protein
CT	Calcitonin
ECF	Extracellular fluid
250H-D$_3$	Vitamin D$_3$
1,25(OH)$_2$D$_3$	Calcitriol
PTH	Parathyroid hormone

amount of calcium in the extracellular fluid by influencing the transport of calcium across the membrane that separates the extracellular fluid space from the periosteal fluid space. This transport is primarily stimulated by parathyroid hormone (PTH), but calcitriol is also involved.

Plasma calcium exists in three forms: (1) complexed with organic acids, (2) protein-bound, and (3) ionized. About 6% of total calcium is complexed with citrate, phosphate, and other anions. The remainder is divided nearly equally between a protein-bound form (bound primarily to albumin) and an ionized (unbound) form. The ionized calcium, which is maintained at concentrations between 1.1 and 1.3 mmol/L in most mammals, birds, and freshwater fish, is the biologically active fraction. The organism has very little tolerance for significant deviation from this normal range. If the ionized calcium level falls, the animal becomes increasingly hyperexcitable and may develop tetanic convulsions. A marked elevation of plasma calcium may result in death owing to muscle paralysis and coma.

Calcium ion and the counter-ion, phosphate, exist at or near their solubility product in plasma; hence, protein binding may protect against precipitation and **ectopic calcification.** An alteration of the plasma protein concentration (primarily albumin, but globulins also bind calcium) results in parallel changes in total plasma calcium. For example, hypoalbuminemia results in a decrease of total plasma calcium of approximately 0.8 mg/dL for each gram per deciliter of albumin decrease. The converse is noted when the plasma albumin is increased. The association of calcium with plasma proteins is pH-dependent; acidosis favors the ionized form, whereas alkalosis enhances binding and causes a concomitant decrease in Ca^{2+}. The latter probably accounts for the numbness and tingling associated with the **hyperventilation syndrome,** which causes acute respiratory alkalosis.

TWO MAIN HORMONES ARE INVOLVED IN CALCIUM HOMEOSTASIS

Parathyroid Hormone (PTH) Is an 84-Amino-Acid Peptide

This hormone (molecular mass 9.5 kDa) contains no carbohydrate or other covalently bound molecules (Figure 47–1). Full biologic activity resides in the amino terminal third of the molecule; PTH_{1-34} has full biologic activity. The region 25–34 is primarily responsible for receptor binding.

PTH is synthesized as a 115-amino-acid precursor molecule (Figure 47–1). The immediate precursor of PTH is **proPTH,** which differs from the native hormone by having an amino terminal highly basic hexapeptide extension whose function is obscure. The primary gene product and the immediate precursor for proPTH is **preproPTH.** This differs from proPTH by having an additional 25-amino-acid amino terminal extension that, in common with the other leader or signal sequences characteristic of secreted proteins, is hydrophobic. The complete structure of preproPTH and the sequences of proPTH and PTH are illustrated in Figure 47–1. The sequence of events involved in the conversion of PTH preprohormone to PTH is shown in schematic form in Figure 47–2. PreproPTH is transferred to the cisternal space of the endoplasmic reticulum while the molecule is still being translated from PTH mRNA by the ribosomes. During this transfer, the 25-amino-acid prepeptide (signal or leader peptide) is removed to yield proPTH. ProPTH is then transported to the Golgi apparatus, where an enzyme removes the pro-

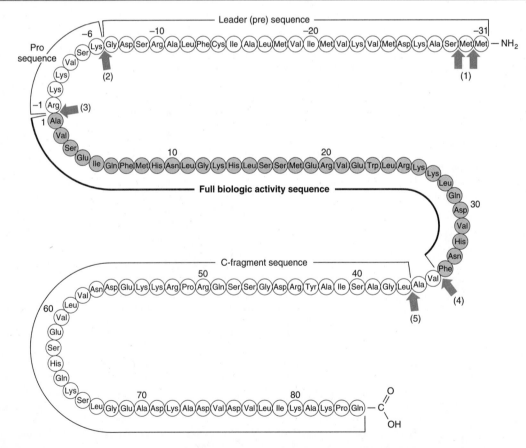

Figure 47–1. Structure of bovine preproparathyroid hormone. Arrows indicate sites cleaved by processing enzymes in the parathyroid gland (1–5) and in the liver after secretion of the hormone (4–5). The biologically active region of the molecule is flanked by sequence not required for activity on target receptors. (Slightly modified and reproduced, with permission, from Habener JF: Recent advances in parathyroid hormone research. Clin Biochem 1981;14:223.)

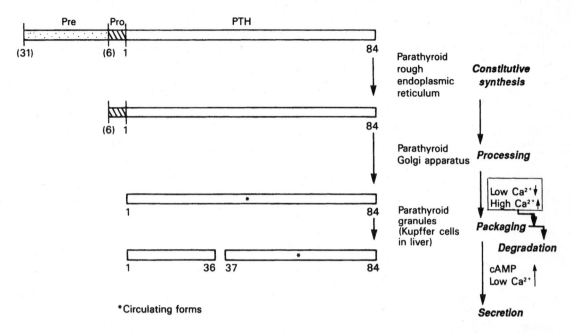

Figure 47–2. The precursors and cleavage products of PTH and the location of these steps in the parathyroid gland and liver. The numbers in parentheses indicate the number of amino acids in the pre (31) and pro (6) fragments.

hormone extension to yield the mature PTH molecule. The PTH released from the Golgi apparatus in secretory vesicles has three possible fates: (1) transport into a storage pool; (2) degradation; or (3) immediate secretion.

The Synthesis, Secretion and Metabolism of PTH Are All Regulated

A. Regulation of Synthesis: The rate of synthesis and degradation of proPTH is unaffected by the ambient Ca^{2+} concentration, even though the rate of formation and secretion of PTH is always markedly enhanced at low Ca^{2+} concentrations. Indeed, 80–90% of the proPTH synthesized cannot be accounted for as intact PTH in cells or in the incubation medium of experimental systems. This finding led to the conclusion that most of the proPTH synthesized is quickly degraded. It was later discovered that this rate of degradation decreases when Ca^{2+} concentrations are low and increases when Ca^{2+} concentrations are high, indicating that calcium affects PTH production through control of degradation and not synthesis. The constitutive synthesis of proPTH is reflected in PTH mRNA levels, which also do not change in spite of wide fluctuations of extracellular Ca^{2+}. It appears that the only way for the organism to enhance PTH synthesis is to increase the size and number of PTH-producing chief cells in the parathyroid glands.

B. Regulation of Metabolism: The degrada-

tion of PTH begins about 20 minutes after proPTH is synthesized, is initially unaffected by Ca^{2+} concentration, and occurs after the hormone is in secretory vesicles. Newly formed PTH can either be secreted immediately or be placed in storage vesicles for subsequent secretion. Degradation occurs as soon as the secretory vesicle begins to enter the storage compartment.

Very specific fragments of PTH are generated during its proteolytic digestion (Figures 47–1 and 47–2), and large amounts of carboxyl terminal fragments of PTH are found in the circulation. These molecules, with molecular masses of about 7 kDa, consist of PTH_{37-84} and lesser amounts of PTH_{34-84}. Most of the newly synthesized PTH is degraded. About 2 mol of the carboxyl terminal fragments is secreted for each mole of intact PTH; hence, the bulk of circulating PTH consists of the carboxyl terminal molecules. No biologic function for the carboxyl terminal fragment of PTH has been defined, but it may prolong the half-life of the hormone in the circulation. A number of proteolytic enzymes, including **cathepsins B and D,** have been identified in parathyroid tissue. Cathepsin B cleaves PTH into two fragments: PTH_{1-36} and PTH_{37-84}. PTH_{37-84} is not further degraded; however, PTH_{1-36} is rapidly and progressively cleaved into di- and tripeptides. ProPTH has never been found in circulation, and little (if any) PTH_{1-34} escapes from the gland. PreproPTH was identified by deciphering the coding sequence of the PTH gene.

Most of the proteolysis of PTH occurs within the gland; however, a number of studies confirm that PTH, once secreted, is proteolytically degraded in other tissues. The exact contribution of extraglandular proteolysis has not been defined, nor is it clear whether the proteolytic enzymes in the two sites are similar or whether the patterns and products of cleavage are identical.

The liver and kidneys are involved in peripheral metabolism of secreted PTH. After hepatectomy, no 34–84 or 37–84 fragments are detected, indicating that the liver is the principal organ involved in the generation of these fragments. The role of the kidneys may be to remove and excrete these fragments. The principal site of peripheral proteolysis appears to be the **Kupffer cells** lining the intrasinusoidal passages of the liver. The endopeptidase responsible for the initial cleavage into the amino and carboxyl terminal fragments is located on the surface of these macrophage-like cells, which are in intimate contact with plasma. This enzyme, also a cathepsin B, cleaves PTH between residues 36 and 37; as in the parathyroid, the resulting carboxyl terminal fragment continues to circulate, whereas the amino terminal fragment is rapidly degraded.

C. Regulation of Secretion: PTH secretion is inversely related to the ambient concentration of ionized calcium and magnesium, as is the circulating level of immunoreactive PTH. Serum PTH declines in a rectilinear fashion in relation to serum calcium levels between 4 mg/dL and 10.5 mg/dL. Sensing is accomplished by a special G protein-associated Ca^{2+} receptor located on parathyroid cells. G protein activation stimulates phospholipase C_β, and this causes the generation of IP_3. IP_3 results in increased intracellular Ca^{2+}; this leads to increased cAMP, and PTH secretion ensues (see below). The presence of biologically active PTH when the serum calcium level is 10.5 mg/dL or greater is an indication of **hyperparathyroidism.**

There is a linear relationship between PTH release and the parathyroid intracellular level of cAMP. The intracellular Ca^{2+} level may be involved in this process, since there is an inverse relationship between the intracellular concentrations of calcium and cAMP. Calcium may exert this effect through its known action on phosphodiesterase (via Ca^{2+}-calmodulin-dependent protein kinase) or through a similar mechanism by inhibiting adenylyl cyclase. Phosphate has no effect on PTH secretion.

Parathyroid glands have relatively few storage granules and contain enough hormone to maintain maximal secretion for only 1.5 hours. This is in contrast to the pancreatic islets, which contain insulin stores sufficient for several days, and to the thyroid, which contains hormone stores adequate for several weeks. PTH must therefore be continually synthesized and secreted.

PTH Acts via a Membrane Receptor

PTH binds to a single membrane receptor protein with a molecular mass of approximately 70 kDa. This receptor appears to be identical in bone and kidney, and it is not found in nontarget cells. The hormone-receptor interaction initiates a typical cascade: activation of adenylyl cyclase $\to$ increased intracellular cAMP $\to$ increased intracellular calcium $\to$ phosphorylation of specific intracellular proteins by kinases $\to$ activation of specific genes and the intracellular enzymes that finally mediate the biologic actions of the hormone. The PTH response system, like that for many other peptide and protein hormones, is subject to **down-regulation** of receptor number and to **"desensitization,"** which may involve a post-cAMP mechanism.

PTH Affects Calcium Homeostasis

The central role of PTH in calcium metabolism is underscored by the observation that the first evolutionary appearance of this hormone was in animals attempting to adapt to a terrestrial existence. The physiologic maintenance of calcium balance depends on the long-term effects of PTH acting on intestinal absorption through the formation of calcitriol. If in the face of prolonged dietary Ca^{2+} deficiency intestinal calcium absorption is inadequate, a complex regulatory system involving PTH is brought into play. PTH restores normal extracellular fluid calcium concentration by acting directly on bone and kidney and by acting indirectly on the intestinal mucosa (through stimulation of synthesis of calcitriol). PTH (1) increases the rate of dissolution of bone, including both organic and inorganic phases, which moves Ca^{2+} into extracellular fluid; (2) reduces the renal clearance or excretion of calcium, hence increasing the extracellular fluid concentration of this action; and (3) increases the efficiency of calcium absorption from the intestine by promoting the synthesis of calcitriol. The most rapid changes occur through the action on the kidney, but the largest effect is from bone. Therefore, although PTH prevents hypocalcemia in the face of dietary calcium deficiency, it does so at the expense of bone substance.

PTH Affects Phosphate Homeostasis

The usual counter-ion for Ca^{2+} is phosphate, and the hydroxyapatite crystal in bone consists of calcium phosphate. Phosphate is released with calcium from bone whenever PTH increases dissolution of the mineral matrix. PTH increases renal phosphate clearance; thus, the net effect of PTH on bone and kidney is to increase the ECF calcium concentration and decrease the extracellular fluid phosphate concentration. Importantly, this prevents the development of a supersaturated concentration of calcium and phosphate in plasma.

Pathophysiology

Insufficient amounts of PTH result in **hypoparathyroidism.** The biochemical hallmarks of this condition are decreased serum ionized calcium and elevated serum phosphate levels. Symptoms include neuromuscular irritability which, when mild, causes muscle cramps and **tetany.** Severe, acute hypocalcemia results in tetanic paralysis of the respiratory muscles, laryngospasm, severe convulsions, and death. Long-standing hypocalcemia results in cutaneous changes, cataracts, and calcification of the basal ganglia of the brain. The usual cause of hypoparathyroidism is accidental removal or damage of the glands during neck surgery (secondary hypoparathyroidism), but the disorder occasionally results from **autoimmune destruction** of the glands (primary hypoparathyroidism).

In **pseudohypoparathyroidism,** an inherited disorder, biologically active PTH is produced, but there is end-organ resistance to its effects. The biochemical consequences are the same, however. There are usually associated developmental anomalies including short stature, short metacarpal or metatarsal bones, and mental retardation. There are several types of pseudohypoparathyroidism, and they have been attributed to (1) partial deficiency of the G_s adenylyl cyclase regulatory protein and (2) a defective step beyond the formation of cAMP.

Hyperparathyroidism, the excessive production of PTH, is usually due to the presence of a functioning **parathyroid adenoma** but can be due to **parathyroid hyperplasia** or to **ectopic production** of PTH or of PTH-related peptide (PTHRP). PTHRP is a 141-amino-acid protein that is structurally and functionally similar to PTH, especially in its amino terminal region. PTHRP has been found in various carcinomas. It is associated with the hypercalcemia of malignancy. The biochemical hallmarks of hyperparathyroidism are elevated serum ionized calcium and PTH and depressed serum phosphate levels. In long-standing hyperparathyroidism, findings include extensive resorption of bone and a variety of renal effects, including kidney stones, nephrocalcinosis, frequent urinary tract infections, and (in severe cases) decreased renal function. **Secondary hyperparathyroidism,** characterized by hyperplasia of the glands and hypersecretion of PTH, may be seen in patients with progressive renal failure. Hyperparathyroidism in these patients is presumably due to the decreased conversion of 25OH-D_3 to 1,25(OH)$_2$-D_3 in the diseased renal parenchyma, which results in inefficient calcium absorption in the gut and the secondary release of PTH in a compensatory attempt to maintain normal extracellular fluid calcium levels.

CALCITRIOL AFFECTS SEVERAL ASPECTS OF CALCIUM HOMEOSTASIS

Historical Perspective

Rickets, a childhood disorder characterized by deficient mineralization of the skeleton and severe, crippling bone deformities, was epidemic in North America and Western Europe early in this century. Results of a series of studies suggested that rickets was due to a dietary deficiency. After the discovery that rickets could be prevented by ingestion of cod liver oil and that the active ingredient in this agent was not vitamin A, the preventive factor was termed fat-soluble **vitamin D.** About the same time, it was found that ultraviolet light, either artificial or from sunlight, would also prevent the disorder. It was subsequently determined that there was an adult equivalent to rickets. **Osteomalacia,** in which there is a failure to mineralize bone, also responded to vitamin D. Clues to further developments resulted from the observation that patients with liver or kidney disease did not respond normally to vitamin D. Efforts to elucidate the structure of vitamin D and to define its mechanism of action have gone forward at a greatly accelerated rate during the last several years.

Calcitriol Stimulates Intestinal Absorption of Calcium and Phosphate

Calcitriol is the only hormone that can promote this translocation of calcium against the concentration gradient which exists across the intestinal cell membrane. Since the production of calcitriol is tightly regulated (Figure 47–3), a fine mechanism exists for controlling extracellular fluid Ca^{2+} in spite of marked fluctuations of the calcium content of food. This ensures a proper concentration of calcium and phosphate for deposition, as hydroxyapatite crystals, onto the collagen fibrils in bone. In vitamin D deficiency (calcitriol deficiency), new bone formation slows and bone remodeling is also impaired. These processes are primarily regulated by PTH acting on bone cells, but small concentrations of calcitriol are also required. Calcitriol may also augment the actions of PTH on renal calcium reabsorption.

The Synthesis and Metabolism of Calcitriol Involve Several Tissues and Are Highly Regulated Processes

A. Biosynthesis: Calcitriol is a hormone in every respect. It is produced by a complex series of enzymatic reactions that involve the plasma transport of precursor molecules to a number of different tissues (Figure 47–3). The active molecule, calcitriol, is transported to other organs where it activates biologic processes in a manner similar to that employed by the steroid hormones.

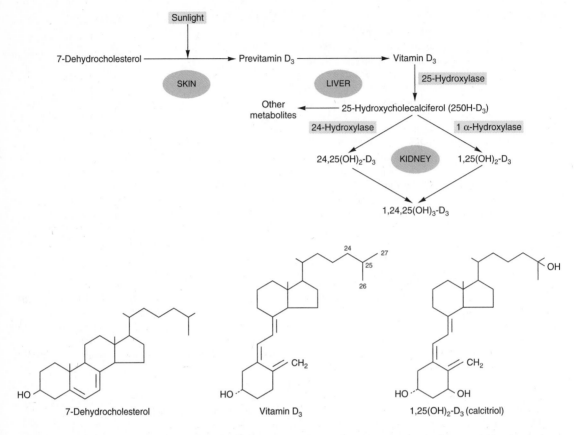

Figure 47–3. Formation and hydroxylation of vitamin D_3. 25-Hydroxylation takes place in the liver, and the other hydroxylations occur in the kidneys. $25,26(OH)_2$-D_3 and $1,25,26(OH)_3$-D_3 are probably formed as well. The formulas of 7-dehydrocholesterol, vitamin D_3, and $1,25(OH)_2$-D_3 (calcitriol) are also shown. (Modified and reproduced, with permission, from Ganong WF: *Review of Medical Physiology,* 17th ed. Appleton & Lange, 1995.)

1. Skin–Small amounts of vitamin D occur in food (fish liver oil, egg yolk), but most of the vitamin D available for calcitriol synthesis is produced in the malpighian layer of the epidermis from 7-dehydrocholesterol in an ultraviolet light-mediated, nonenzymatic **photolysis reaction.** The extent of this conversion is directly related to the intensity of the exposure and inversely related to the extent of pigmentation in the skin. There is an age-related loss of 7-dehydrocholesterol in the epidermis that may be related to the negative calcium balance associated with old age.

2. Liver–A specific transport protein called the **vitamin D-binding protein** binds vitamin D_3 and its metabolites and moves vitamin D_3 from the skin or intestine to the liver, where it undergoes 25-hydroxylation, the first obligatory reaction in the production of calcitriol. 25-Hydroxylation occurs in the endoplasmic reticulum in a reaction that requires magnesium, NADPH, molecular oxygen, and an uncharacterized cytoplasmic factor. Two enzymes, an NADPH-dependent cytochrome P450 reductase and

a cytochrome P450, are involved. This reaction is not regulated, and it also occurs with low efficiency in kidney and intestine. The $25OH$-D_3 enters the circulation, where it is the major form of vitamin D found in plasma, and is transported to the kidney by the vitamin D-binding protein.

3. Kidney–$25OH$-D_3 is a weak agonist and must be modified by hydroxylation at position C_1 for full biologic activity. This is accomplished in mitochondria of the renal proximal convoluted tubule in a complex, three-component monooxygenase reaction that requires NADPH, Mg^{2+}, molecular oxygen, and at least three enzymes: (1) a flavoprotein, renal ferredoxin reductase; (2) an iron sulfur protein, renal ferredoxin; and (3) cytochrome P450. This system produces $1,25(OH)_2$-D_3, which is the most potent naturally occurring metabolite of vitamin D.

4. Other tissues–The placenta has a 1α-hydroxylase that appears to be an important extrarenal source of calcitriol. Enzyme activity is found in a variety of other tissues, including bone; however, the physiologic significance appears to be minimal, since very little calcitriol is found in nonpregnant, nephrectomized animals.

B. Regulation of Metabolism and Synthesis: Like other steroid hormones, calcitriol is subject to tight feedback regulation (Figure 47–3 and Table 47–1). Low-calcium diets and hypocalcemia result in marked increases of 1α-hydroxylase activity in intact animals. This effect requires PTH, which is also released in response to hypocalcemia. The action of PTH is as yet unexplained, but it stimulates 1α-hydroxylase activity in both vitamin D-deficient and vitamin D-treated animals. Low-phosphorus diets and hypophosphatemia also induce 1α-hydroxylase activity, but this appears to be a weaker stimulus than that provided by hypocalcemia.

Calcitriol is an important regulator of its own production. High levels of calcitriol inhibit renal 1α-hydroxylase and stimulate the formation of a 24-hydroxylase that leads to the formation of $24,25(OH)_2$-D_3, an apparently inactive by-product. Estrogens, progestins, and androgens cause marked increases of 1α-hydroxylase in ovulating birds. The role that these hormones, along with insulin, growth hormone, and prolactin, play in mammals is uncertain.

The basic sterol molecule can be modified by alternative metabolic pathways, ie, by hydroxylation at positions 1, 23, 24, 25, and 26 and by the formation of a number of lactones. Over 20 metabolites have been found; none have unequivocally been shown to have biologic activity.

Calcitriol Acts at the Cellular Level in a Manner Similar to Steroid Hormones

Studies using radioactive calcitriol revealed localization in the nuclei of intestinal villus and crypt cells, osteoblasts, and distal renal tubular cells. There also is nuclear accumulation of this hormone in cells not previously suspected of being targets, including cells in the malpighian layer of the skin; pancreatic islet cells; some brain cells; some cells in the pituitary, ovary, testis, placenta, uterus, mammary gland, and thymus; and myeloid precursors. Calcitriol binding has also been noted in parathyroid cells, which leads to the intriguing possibility that it might be involved in PTH metabolism.

A. The Calcitriol Receptor: The calcitriol receptor is a member of the steroid receptor family (Figure 48–7). The ligand-binding domain of this receptor binds calcitriol with high affinity and low capacity. This binding is saturable, specific, and reversible. The receptor has a DNA-binding domain that appears to contain the zinc finger motif characteristic of other steroid receptors. The calcitriol-receptor complex binds to a vitamin D response element of the sequence AGGTCANNNAGGTCA (Table 44–3).

B. Calcitriol-Dependent Gene Products: It has been known for several years that the response of intestinal transport to calcitriol requires RNA and protein synthesis. The observation of binding of the calcitriol receptor to chromatin in the nucleus suggests that calcitriol stimulates gene transcription and the formation of specific mRNAs. One such example, the induction of an mRNA that codes for a calcium-binding protein (CBP), has been reported.

Several cytosolic proteins bind Ca^{2+} with high affinity. One group, composed of several proteins of different molecular mass, antigenicity, and tissue location (intestine, skin, and bone), is calcitriol-dependent. Of these, intestinal CBP has been studied most intensively. No CBP is found in the intestine of vitamin D-deficient rats, and the concentration of CBP is highly correlated with the extent of nuclear localization of calcitriol.

C. Effects of Calcitriol on Intestinal Mucosa: The transfer of Ca^{2+} or PO_4^{3-} across the intestinal mucosa requires (1) uptake across the brush border and microvillar membrane; (2) transport across the mucosal cell membrane; and (3) efflux across the basal lateral membrane into the extracellular fluid. It is clear that calcitriol enhances one or more of these steps, but the precise mechanism has not been established. CBP was thought to be actively involved until it was observed that Ca^{2+} translocation occurs within 1–2 hours after administration of calcitriol, well before CBP increases in response to calcitriol. CBP may bind Ca^{2+} and protect the mucosa cell against the large fluxes of Ca^{2+} coincident with the transport process. Several investigators are searching for other proteins that may be involved in Ca^{2+} transport, whereas others suggest that the process, particularly the early increase of Ca^{2+} flux, may be mediated by a membrane change. Metabolites of polyphosphoinositides have been implicated.

D. Pathophysiology: Rickets is a childhood disorder characterized by low plasma calcium and phosphorus levels and by poorly mineralized bone with associated skeletal deformities. Rickets is most commonly due to **vitamin D-deficiency.** There are two types of **vitamin D-dependent rickets.** Type I is an inherited autosomal recessive trait characterized by a defect in the conversion of 25OH-D_3 to calcitriol. Type II is an autosomal recessive disorder in which there is a single amino acid change in one of the zinc fingers of the DNA-binding domain. This results in a nonfunctional receptor.

Table 47–1. Regulation of renal 1α-hydroxylase.

Primary Regulators	Secondary Regulators
Hypocalcemia ($\uparrow$) PTH ($\uparrow$) Hypophosphatemia ($\uparrow$) Calcitriol ($\downarrow$)	Estrogens Androgens Progesterone Insulin Growth hormone Prolactin Thyroid hormone

Vitamin D deficiency in the adult results in **osteomalacia.** Calcium and phosphorus absorption are decreased, as are the extracellular fluid levels of these ions. Consequently, mineralization of osteoid to form bone is impaired, and such undermineralized bone is structurally weak.

When substantial renal parenchyma is lost or diseased, the formation of calcitriol is reduced and calcium absorption decreases. When hypocalcemia ensues, there is a compensatory increase of PTH, which acts on bone in an attempt to increase extracellular fluid Ca^{2+}. The associated extensive bone turnover, structural changes, and symptoms are known as **renal osteodystrophy.** Early treatment with vitamin D will blunt this process.

THE ROLE OF CALCITONIN IN HUMAN CALCIUM HOMEOSTASIS IS UNCLEAR

Calcitonin (CT) is a 32-amino-acid peptide secreted by the parafollicular C cells of the human thyroid (less commonly, the parathyroid or thymus) or by similar cells located in the ultimobranchial gland of other species. These cells originate in the neural crest and are biochemically related to cells in a variety of other endocrine glands.

The entire CT molecule, including the seven-member amino terminal loop, formed by a Cys-Cys bridge, is required for biologic activity. There is tremendous interspecies variation of the amino acid sequence of CT (human and porcine CT share only 14 of 32 amino acids), but in spite of these differences there is cross-species bioactivity. The most potent naturally occurring CT is isolated from salmon.

CT has a history unmatched by any other hormone. Within a 7-year span (1962–1968), CT was discovered, isolated, sequenced, and synthesized, yet its role in human physiology is still uncertain.

SUMMARY

Numerous physiologic and biochemical processes require calcium. Calcium mainly resides in bone in mammals, but a very small percentage of the total is found in the extracellular fluid. Extracellular fluid calcium is about equally partitioned between protein-bound and a free or ionized form (Ca^{2+}). It is the latter that is biologically active. There is little tolerance for deviation from the normal range of Ca^{2+}, which is 1.1–1.3 mmol/L in most species. This rigid control is maintained by a multiorgan (liver, skin, kidney, bone, gut, and parathyroid), multihormone (parathyroid hormone [PTH], calcitriol, and ?calcitonin) system.

PTH is synthesized as a 115-amino-acid preprohormone. It is successively cleaved at two sites to form the mature 84-amino-acid hormone. PTH is inactivated by cleavage between residues 36 and 37. This process, which occurs in the parathyroid and peripheral tissues, is regulated by Ca^{2+}, as is the secretion of the hormone from the parathyroids. A newly discovered cell surface Ca^{2+} receptor initiates these events. PTH is synthesized constitutively, ie, there is no regulation of its rate of synthesis. PTH, a peptide hormone, binds to cell surface receptors on bone and kidney cells. This interaction results in the generation of cAMP, which is the intracellular messenger of PTH action.

Calcitriol is synthesized through a series of reactions that occur in skin, liver, and kidney. The most critical, a 1α-hydroxylation, occurs in kidney and is regulated by Ca^{2+}, phosphate, and calcitriol itself. Calcitriol is a hormone of the steroid-thyroid class. Thus, it acts by binding to an intracellular receptor, and the ligand-receptor complex binds to a specific vitamin D response element to affect gene transcription. The primary targets of this action are genes involved in calcium transport into and out of the intestinal villus cell.

REFERENCES

Brown EM et al: Calcium-ion-sensing cell-surface receptors. N Engl J Med 1995;333:234.

Burtis WF et al: Immunochemical characterization of circulating parathyroid hormone-related protein in patients with humoral hypercalcemia of cancer. N Engl J Med 1990;332:1106.

Donahue HJ et al: Differential effects of parathyroid and its analogues on cytosolic calcium ion and cAMP levels in cultured rat osteoblast-like cells. J Biol Chem 1988; 263:13522.

Habener JF et al: Parathyroid hormone: Biochemical aspects of biosynthesis, secretion, action, and metabolism. Physiol Rev 1983;64:985.

Norman AW: The vitamin D endocrine system. Physiologist 1985;28:219.

Hormones of the Adrenal Cortex

48

Daryl K. Granner, MD

INTRODUCTION

The adrenal cortex synthesizes dozens of different steroid molecules, but only a few of these have biologic activity. These sort into three classes of hormones: glucocorticoids, mineralocorticoids, and androgens. These hormones initiate their actions by combining with specific intracellular receptors, and this complex binds to specific regions of DNA to regulate gene expression. This results in altered rates of synthesis of a small number of proteins, which in turn affect a variety of metabolic processes, eg, gluconeogenesis and Na^+ and K^+ balance.

BIOMEDICAL IMPORTANCE

The hormones of the adrenal cortex, particularly the glucocorticoids, are an essential component of adaptation to severe stress. The mineralocorticoids are required for normal Na^+ and K^+ balance. Synthetic analogs of both classes are used therapeutically. In particular, many glucocorticoid analogs are potent anti-inflammatory agents. Excessive or deficient plasma levels of any of these three classes of hormones, whether due to disease or therapeutic use, result in serious, sometimes life-threatening, complications. A series of inherited enzyme deficiencies helps define the steps involved in steroidogenesis and illustrates the capacity of the adrenal cortex to alter the relative rates of production of these different hormones.

THE ADRENAL CORTEX MAKES THREE KINDS OF HORMONES

The adult cortex has three distinct layers or zones. The subcapsular area is called the **zona glomerulosa** and is associated with the production of mineralocorticoids. Next is the **zona fasciculata,** which, with the **zona reticularis,** produces glucocorticoids and androgens.

Some 50 steroids have been isolated and crystallized from adrenal tissue. Most of these are intermediates; only a small number are secreted in signifi-

ACRONYMS USED IN THIS CHAPTER	
ACE	Angiotensin converting enzyme
ACTH	Adrenocorticotropic hormone
ADH	Antidiuretic hormone
CBG	Corticosteroid-binding globulin
CRH	Corticotropin-releasing hormone
DHEA	Dehydroepiandrosterone
DOC	Deoxycorticosterone
GH	Growth hormone
GRE	Glucocorticoid response element
3β-OHSD	3β-Hydroxysteroid dehydrogenase
PEPCK	Phosphoenolpyruvate carboxykinase
POMC	Pro-opiomelanocortin
PTH	Parathyroid hormone
TBG	Thyroid-binding globulin

cant amounts; and few have significant hormonal activity. The adrenal cortex makes three general classes of steroid hormones, which are grouped according to their dominant action. There is an overlap of biologic activity, since all natural glucocorticoids have mineralocorticoid activity and vice versa. This can now be understood on the basis of the commonality of the hormone response elements that mediate the effects of these hormones (and progestins) at the gene level (see Table 44–3).

The **glucocorticoids** are 21-carbon steroids with many actions, the most important of which is to promote gluconeogenesis. **Cortisol** is the predominant glucocorticoid in humans, and it is made in the zona fasciculata. **Corticosterone,** made in the zonae fasciculata and glomerulosa, is less abundant in humans, but it is the dominant glucocorticoid in rodents. **Mineralocorticoids** are also 21-carbon steroids. The primary action of these hormones is to promote retention of Na^+ and excretion of K^+ and H^+, particularly in the kidney. **Aldosterone** is the most potent hormone in this class, and it is made exclusively in the zona glomerulosa. The zonae fasciculata and reticularis of the adrenal cortex also produce significant amounts of the androgen precursor **dehydroepi-androsterone** and of the weak androgen **andro-stenedione.** These steroids are converted into more potent androgens in extraadrenal tissues and become pathologic sources of androgens when specific steroidogenic enzymes are deficient. Estrogens are not made in the normal adrenal in significant amounts, but in certain cancers of the adrenal they may be produced, and androgens of adrenal origin are important precursors of estrogen (converted by peripheral aromatization) in postmenopausal women.

A SPECIAL NOMENCLATURE DESCRIBES THE CHEMISTRY OF STEROIDS

All steroid hormones have in common the 17-carbon **cyclopentanoperhydrophenanthrene** structure with the four rings labeled A–D (Figure 48–1). Additional carbons can be added at positions 10 and 13 or as a side chain attached to C_{17}. Steroid hormones and their precursors and metabolites differ in number and type of substituted groups, number and location of double bonds, and stereochemical configuration. A precise nomenclature for designating these chemical formulations has been devised. The asymmetric carbon atoms (shaded on the C_{21} molecule in Figure 48–1) allow for **stereoisomerism.** The angular methyl groups (C_{19} and C_{18}) at positions 10 and 13 project in front of the ring system and serve as the point of reference. Nuclear substitutions in the same plane as these groups are designated *cis* or "β" and are represented in drawings by solid lines. Substitutions that project behind the plane of the ring system are designated *trans* or "α" and are represented as a dashed line. Double bonds are referred to by the number of the preceding carbon (eg, Δ^3, Δ^4). The steroid hormones are named according to whether they have one angular methyl group (estrane, 18 carbons), two angular methyl groups (androstane, 19 carbons), or two angular groups plus a 2-carbon side

chain at C_{17} (pregnane, 21 carbons). This information (Figure 48–2), together with the glossary provided in Table 48–1, should allow one to understand the chemical names of the natural and synthetic hormones listed in Table 48–2.

SEVERAL ENZYMES ARE INVOLVED IN THE BIOSYNTHESIS OF ADRENAL STEROID HORMONES

The adrenal steroid hormones are synthesized from cholesterol that is mostly derived from the plasma, but a small portion is synthesized in situ from acetyl-CoA via mevalonate and squalene. Much of the cholesterol in the adrenal is esterified and stored in cytoplasmic lipid droplets. Upon stimulation of the adrenal by ACTH (or cAMP), an esterase is activated, and the free cholesterol formed is transported into the mitochondrion, where a **cytochrome P450 side chain cleavage enzyme (P450$_{scc}$)** converts cholesterol to pregnenolone. Cleavage of the side chain involves sequential hydroxylations, first at C_{22} and then at C_{20}, followed by side chain cleavage (removal of the six-carbon fragment isocaproaldehyde) to give the 21-carbon steroid (Figure 48–2). An ACTH-dependent protein may bind and activate cholesterol or P450$_{scc}$. Amino-glutethimide is a very efficient inhibitor of P450$_{scc}$ and of steroid biosynthesis.

All mammalian steroid hormones are formed from cholesterol via pregnenolone through a series of reactions that occur in either the mitochondria or endoplasmic reticulum of the adrenal cell. **Hydroxylases** that require molecular oxygen and NADPH are essential, and **dehydrogenases,** an **isomerase,** and a **lyase** are also necessary for certain steps. There is some cellular specificity in steroidogenesis. For instance, 18-hydroxylase and 18-hydroxysteroid dehydrogenase, which are required for aldosterone synthesis, are found only in glomerulosa cells, so that the biosynthesis of this mineralocorticoid is confined

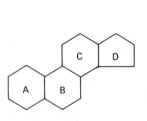

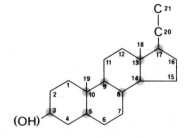

Figure 48–1. Structural features of steroid molecules.

Cyclopentaoper-hydrophenanthrene nucleus

Numbering of the carbon atoms. Asymmetric carbons are shaded.

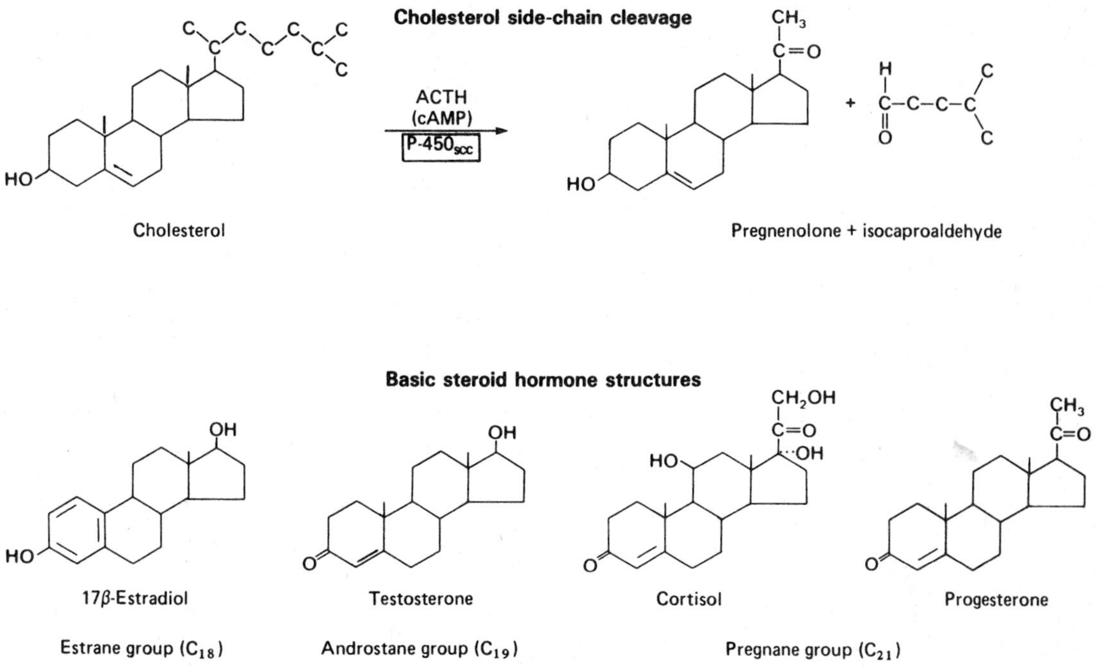

Figure 48–2. Cholesterol side-chain cleavage and basic steroid hormone structures.

to this region. A schematic representation of the pathways involved in the synthesis of the three major classes of adrenal steroids is presented in Figure 48–3. The enzymes are shown in the rectangular boxes, and the modifications at each step are shaded.

Table 48–1. Nomenclature of steroids.

Prefix	Suffix	Chemical Nature
Hydroxy-	-ol	Alcohols
Dihydroxy-	-diol	
Oxo-	-one	Ketones (eg, -dione = 2 keto groups)
Cis-		Arrangement of 2 groups in same plane as C_{19}
Trans-		Arrangement of 2 groups in opposing plane to C_{19}
α-		A group trans to the 19-methyl
β-		A group cis to the 19-methyl
Deoxy-		Lacking a hydroxy group
Iso- or epi-		Isomerism at a C—C, C—OH, or C—H bond, eg, androsterone (5α) versus isoandrosterone (5β)
Dehydro-		Removal of 2 hydrogen atoms to form a double bond
Dihydro-		Addition of 2 hydrogen atoms to a double bond
Allo-		Trans configuration of the A and B rings

Mineralocorticoid Synthesis Occurs in the Zona Glomerulosa

Synthesis of aldosterone follows the mineralocorticoid pathway and occurs in the zona glomerulosa. Pregnenolone is converted to progesterone by the action of two smooth endoplasmic reticulum enzymes, **3β-hydroxysteroid dehydrogenase (3β-OHSD)** and **$\Delta^{5,4}$ isomerase.** Progesterone is hydroxylated at the C_{21} position to form 11-deoxycorticosterone (DOC), which is an active (Na^+-retaining) mineralocorticoid. The next hydroxylation, at C_{11}, produces corticosterone, which has glucocorticoid activity and is a weak mineralocorticoid (it has less than 5% of the potency of aldosterone). In some species (eg, rodents), it is the most potent glucocorticoid. C_{21} hydroxylation is necessary for both mineralocorticoid and glucocorticoid activity, but most steroids with a C_{17} hydroxyl group have more glucocorticoid and less mineralocorticoid action. In the zona glomerulosa, which does not have the smooth endoplasmic reticulum enzyme 17α-hydroxylase, a mitochondrial 18-hydroxylase is present. The **18-hydroxylase** acts on corticosterone to form 18-hydroxycorticosterone, which is changed to aldosterone by the conversion of the 18-alcohol to an aldehyde. This unique distribution of enzymes and the special regulation of the zona glomerulosa (see below) have led some investigators to suggest that, in addition to the adrenal being two glands, the adrenal cortex is actually two separate organs.

Table 48–2. Trivial and chemical names of some steroids.

Trivial Name	Chemical Name
Aldosterone	11β,21-Dihydroxy-3,20-dioxo-4-pregnen-18-al
Androstenedione	4-Androstene-3,17-dione
Cholesterol	5-Cholesten-3β-ol
Corticosterone (compound B)	11β,21-Dihydroxy-4-pregnene-3,20-dione
Cortisol (compound F)	11β,17α,21-Trihydroxy-4-pregnene-3,20-dione
Cortisone (compound E)	17α,21-Dihydroxy-4-pregnene-3,11,20-trione
Dehydroepiandrosterone (DHEA)	3β-Hydroxy-5-androsten-17-one
11-Deoxycorticosterone (DOC)	21-Hydroxy-4-pregnene-3,20-dione
11-Deoxycortisol (compound S)	17,21-Dihydroxy-4-pregnene-3,20-dione
Dexamethasone	9α-Fluoro-16α-methyl-11β,17α,21-trihydroxypregna-1,4-diene-3,20-dione
Estradiol	1,3,5(10)-Estratriene-3,17β-diol
Estriol	1,3,5(10)-Estratriene-3,16α,17β-triol
Estrone	3-Hydroxy-1,3,5(10)-estratriene-3-ol-17-one
Etiocholanolone	3α-Hydroxy-5β-androstan-17-one
9α-Fluorocortisol	9α-Fluoro-11β,17α,21-trihydroxypregn-4-ene,3,20-dione
Prednisolone	11β,17α,21-Trihydroxypregna-1,4-diene-3,20-dione
Prednisone	17α,21-Dihydroxypregna-1,4-diene-3,11,20-trione
Pregnanediol	5β-Pregnane-3α,20α-diol
Pregnanetriol	5β-Pregnane-3α,17α,20α-triol
Pregnenolone	3β-Hydroxy-5-pregnen-20-one
Progesterone	4-Pregnene-3,20-dione
Testosterone	17β-Hydroxy-4-androsten-3-one
Triamcinolone	9α-Fluoro-11β,16α,17α,21-tetrahydroxypregna-1,4-diene-3,20-dione

A. Glucocorticoid Synthesis: Cortisol synthesis requires three hydroxylases that act sequentially on the C_{17}, C_{21}, and C_{11} positions. The first two reactions are rapid, while C_{11} hydroxylation is relatively slow. If the C_{21} position is hydroxylated first, the action of 17α-hydroxylase is impeded and the mineralocorticoid pathway is followed (forming corticosterone or aldosterone, depending on the cell type). 17α-Hydroxylase is a smooth endoplasmic reticulum enzyme that acts upon either progesterone or, more commonly, pregnenolone. 17α-Hydroxyprogesterone is hydroxylated at C_{21} to form 11-deoxycortisol, which is then hydroxylated at C_{11} to form cortisol, the most potent natural glucocorticoid hormone in humans. The **21-hydroxylase** is a smooth endoplasmic reticulum enzyme, whereas the **11β-hydroxylase** is a mitochondrial enzyme. Steroidogenesis thus involves the repeated shuttling of substrates into and out of the mitochondria of the fasciculata and reticularis cells (Figure 48–4).

B. Androgen Synthesis: The major androgen or androgen precursor produced by the adrenal cortex is **dehydroepiandrosterone (DHEA).** Most 17-hydroxypregnenolone follows the glucocorticoid pathway, but a small fraction is subjected to oxidative fission and removal of the two-carbon side chain through the action of 17,20-lyase. This enzyme is found in the adrenals and gonads and acts exclusively on 17α-hydroxy-containing molecules. Adrenal androgen production increases markedly if glucocorticoid biosynthesis is impeded by the lack of one of the hydroxylases. Most DHEA is rapidly modified by the addition of sulfate, about half of which occurs in the adrenal and the rest in the liver. DHEA sulfate is inactive, but removal of the sulfate results in reactivation. DHEA is really a prohormone, since the actions of 3β-OHSD and $\Delta^{5,4}$ isomerase convert the weak androgen DHEA into the more potent **androstenedione.** Small amounts of androstenedione are also formed in the adrenal by the action of the lyase on 17α-hydroxyprogesterone. Reduction of androstenedione at the C_{17} position results in the formation of **testosterone,** the most potent adrenal androgen. Small amounts of testosterone are produced in the adrenal by this mechanism, but most of this conversion occurs in other tissues.

Small amounts of other steroids can be isolated from adrenal venous blood, including 11-deoxycorticosterone, progesterone, pregnenolone, 17α-hydroxyprogesterone, and a very small amount of estradiol (from the aromatization of testosterone). None of these amounts are important in relation to production from other glands, however.

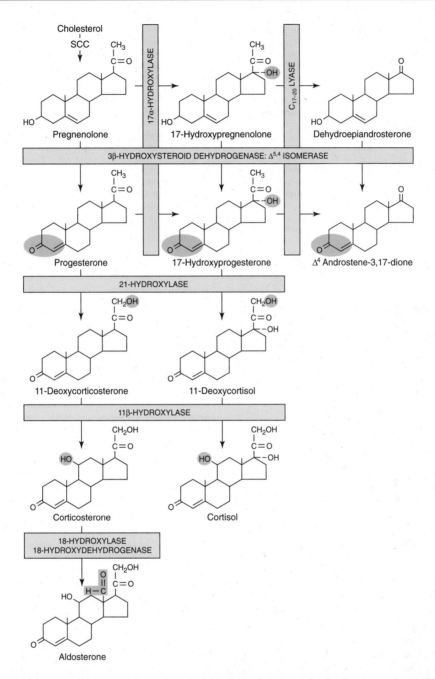

Figure 48–3. Pathways involved in the synthesis of the three major classes of adrenal steroids. Enzymes are shown in the rectangular boxes, and the modifications at each step are shaded. (Slightly modified and reproduced, with permission, from Harding BW: In: *Endocrinology,* vol 2. DeGroot LJ [editor]. Grune & Stratton, 1979.)

THE SECRETION, TRANSPORT, AND METABOLISM OF ADRENAL STEROID HORMONES AFFECT BIOAVAILABILITY

Secretion of Steroid Hormones

There is little, if any, storage of steroid hormones within the adrenal (or gonad) cell, since these

hormones are released into the plasma when they are made. Cortisol release occurs with a periodicity that is regulated by the **diurnal rhythm** of ACTH release. Consequently, cortisol levels are highest in the AM, shortly after awakening, and are lowest in the late afternoon and early evening.

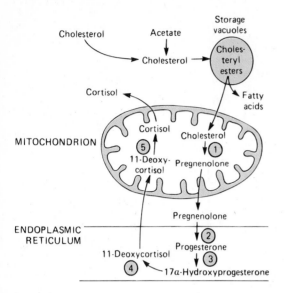

Figure 48–4. Subcellular compartmentalization of glucocorticoid biosynthesis. Adrenal steroidogenesis involves the shuttling of precursors between mitochondria and the endoplasmic reticulum. The enzymes involved are (1) C_{20-22} lyase, (2) 3β-hydroxysteroid dehydrogenase and $\Delta^{5,4}$ isomerase, (3) 17α-hydroxylase, (4) 21-hydroxylase, and (5) 11β-hydroxylase. (Slightly modified and reproduced, with permission, from Harding BW: In: *Endocrinology,* vol 2. DeGroot LJ [editor]. Grune & Stratton, 1979.)

Plasma Transport

A. Glucocorticoids: Cortisol circulates in plasma in protein-bound and free forms. The main plasma-binding protein is an α-globulin called **transcortin** or **corticosteroid-binding globulin (CBG).** CBG is produced in the liver, and its synthesis, like that of thyroid-binding globulin (TBG), is increased by estrogens. CBG binds most of the hormone when plasma cortisol levels are within the normal range; much smaller amounts of cortisol are bound to albumin. The avidity of binding helps determine the biologic half-lives of various glucocorticoids. Cortisol binds tightly to CBG and has a $t_{1/2}$ of 1.5–2 hours, while corticosterone, which binds less tightly, has a $t_{1/2}$ of less than 1 hour. Binding to CBG is not restricted to glucocorticoids. Deoxycorticosterone and progesterone interact with CBG with sufficient affinity to compete for cortisol binding. The unbound, or free, fraction constitutes about 8% of the total plasma cortisol and represents the biologically active fraction of cortisol.

B. Mineralocorticoids: Aldosterone, the most potent natural mineralocorticoid, does not have a specific plasma transport protein, but it forms a very weak association with albumin. Corticosterone and 11-deoxycorticosterone, other steroids with mineralocorticoid effects, bind to CBG. These observations are important in understanding the mechanism of action of aldosterone (see below).

Metabolism and Excretion Rates Depend on the Presence or Absence of Carrier Proteins

A. Glucocorticoids: Cortisol and its metabolites constitute about 80% of the 17-hydroxycorticoids in plasma; the other 20% consist of cortisone and 11-deoxycortisol. About half of the cortisol (as well as cortisone and 11-deoxycortisol) circulates in the form of the reduced dihydro- and tetrahydrometabolites that are produced from reduction of the A ring double bond by NADPH-requiring dehydrogenases and from reduction of the 3-ketone group by a reversible dehydrogenase reaction. Substantial amounts of all these compounds are also modified by conjugation at the C_3 position with glucuronide or, to a lesser extent, with sulfate. These modifications occur primarily in the liver and make the lipophilic steroid molecule water soluble and excretable. In humans, most of the conjugated steroids that enter the intestine by biliary excretion are reabsorbed by the enterohepatic circulation. About 70% of the conjugated steroids are excreted in the urine, 20% leave in feces, and the rest exit through the skin.

B. Mineralocorticoids: Aldosterone is very rapidly cleared from the plasma by the liver, no doubt because it lacks a plasma carrier protein. The liver forms tetrahydroaldosterone 3-glucuronide, which is excreted in the urine.

C. Androgens: Androgens are excreted as 17-keto compounds including DHEA (sulfate) as well as androstenedione and its metabolites. Testosterone, secreted in small amounts by the adrenal, is not a 17-keto compound, but the liver converts about 50% of testosterone to androsterone and etiocholanolone, which are 17-keto compounds.

THE SYNTHESIS OF ADRENAL STEROID HORMONES IS REGULATED BY DIFFERENT MECHANISMS

Glucocorticoid Hormones

The secretion of cortisol is dependent on ACTH, which in turn is regulated by corticotropin-releasing hormone (CRH). These hormones are linked by a classic negative feedback loop (Table 45–1).

Mineralocorticoid Hormones

The production of aldosterone by the glomerulosa cells is regulated in a completely different manner. The primary regulators are the renin-angiotensin system and potassium. Sodium, ACTH, and neural mechanisms are also involved.

A. The Renin-Angiotensin System: This system is involved in the regulation of blood pressure and electrolyte metabolism. The primary hormone in these processes is **angiotensin II,** an octapeptide made from **angiotensinogen** (Figure 48–5). Angiotensinogen, an α_2-globulin made in liver, is the substrate for renin, an enzyme produced in the **juxtaglomerular cells** of the renal afferent arteriole. The position of these cells makes them particularly sensitive to blood pressure changes, and many of the physiologic regulators of renin release act through renal **baroreceptors** (Table 48–3). The juxtaglomerular cells are also sensitive to changes of Na⁺ and Cl⁻ concentration in the renal tubular fluid; therefore, any combination of factors that decreases fluid volume (dehydration, decreased blood pressure, fluid or blood loss) or decreases NaCl concentration stimulates renin release. Renal sympathetic nerves that terminate in the juxtaglomerular cells mediate the central nervous system and postural effects on renin release independent of the baroreceptor and salt effects, a mechanism that involves the β-adrenergic receptor.

Renin acts upon the substrate angiotensinogen to produce the decapeptide **angiotensin I.** The synthesis of angiotensinogen in liver is enhanced by glucocorticoids and estrogen. Hypertension associated with these hormones may be due in part to increased plasma levels of angiotensinogen. Since this protein circulates at about the K_m for its interaction with renin, small changes could markedly affect the generation of angiotensin II.

Angiotensin-converting enzyme, a glycoprotein found in lung, endothelial cells, and plasma, removes

Table 48–3. Factors that influence renin release.

Stimulators	Inhibitors
Decreased blood pressure	Increased blood pressure
Change from supine to erect posture	Change from erect to supine posture
Salt depletion	Salt loading
β-Adrenergic agents	β-Adrenergic antagonists
Prostaglandins	Prostaglandin inhibitors
	Potassium
	Vasopressin
	Angiotensin II

two carboxyl terminal amino acids from the decapeptide angiotensin I to form angiotensin II in a step that is not thought to be rate-limiting. Various nonapeptide analogs of angiotensin I and other compounds act as competitive inhibitors of converting enzyme and are used to treat **renin-dependent hypertension.** These are referred to as angiotensin-converting enzyme inhibitors (ACEs). Converting enzyme also degrades bradykinin, a potent vasodilator; thus, this enzyme increases blood pressure in two distinct ways.

Angiotensin II increases blood pressure by causing vasoconstriction of the arteriole and is a very potent vasoactive substance. It inhibits renin release from the juxtaglomerular cells and is a potent stimulator of aldosterone production. Although angiotensin II stimulates the adrenal directly, it has no effect on cortisol production.

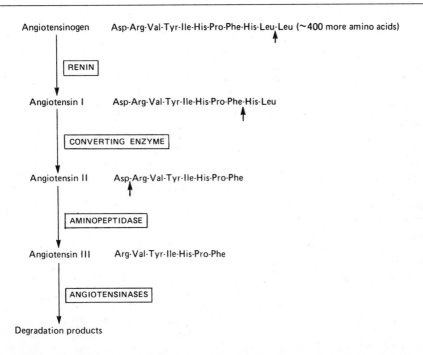

Figure 48–5. Formation and metabolism of angiotensins. Small arrows indicate cleavage sites.

In some species, angiotensin II is converted to the des-Asp^1 heptapeptide **angiotensin III** (Figure 48–5), an equally potent stimulator of aldosterone production. In humans, the plasma level of angiotensin II is four times greater than that of angiotensin III, so most effects are exerted by the octapeptide. Angiotensins II and III are rapidly inactivated by **angiotensinases.**

Angiotensin II binds to specific glomerulosa cell receptors. The hormone-receptor interaction does not activate adenylyl cyclase, and cAMP does not appear to mediate the action of this hormone. The actions of angiotensin II, which are to stimulate the conversion of cholesterol to pregnenolone and of corticosterone to 18-hydroxycorticosterone and aldosterone, may involve changes in the concentration of intracellular calcium and of phospholipid metabolites by mechanisms similar to those described in Chapter 44.

B. Potassium: Aldosterone secretion is sensitive to changes in plasma potassium level; an increase as small as 0.1 meq/L stimulates production, whereas a similar decrease reduces aldosterone production and secretion. K^+ affects the same enzymatic steps as does angiotensin II, although the mechanism involved is obscure. Like angiotensin II, K^+ does not affect the biosynthesis of cortisol.

C. Other Effectors: In special circumstances, ACTH and sodium may be involved in aldosterone production in humans.

ADRENAL STEROID HORMONES HAVE NUMEROUS AND DIVERSE METABOLIC EFFECTS

Loss of adrenal cortical function results in death unless replacement therapy is instituted. In humans, treatment of adrenal insufficiency with mineralocorticoids is generally not sufficient; glucocorticoids seem to be more critical in this regard. Rats, in contrast, do well with mineralocorticoid replacement. Excessive or deficient plasma levels of either class of hormone, whether due to disease or therapeutic use, cause a number of serious complications directly related to their metabolic actions.

Glucocorticoid Hormones Affect Basal Metabolism, Host Defense Mechanisms, Blood Pressure, and Response to Stress

A detailed discussion of the various metabolic effects of the glucocorticoid hormones is found in standard physiology texts. A brief description of the effects is presented in Table 48–4.

Mineralocorticoid Hormones Affect Electrolyte Balance and Ion Transport

Mineralocorticoid hormones act in the kidney to stimulate active Na^+ transport by the distal convo-

Table 48–4. The diverse effects of glucocorticoids.

Effects on intermediary metabolism
1. Increase glucose production (1) by increasing the delivery of amino acids (the gluconeogenic substrate) from peripheral tissues; (2) by increasing the rate of gluconeogenesis by increasing the amount (and activity) of several key enzymes; and (3) by "permitting" other metabolic reactions to operate at maximal rates.
2. Increase hepatic glycogen deposition by promoting the activation of glycogen synthetase.
3. Promote lipolysis (in extremities), but can cause lipogenesis in other sites (face and trunk) especially at higher than physiologic levels.
4. Promote protein and RNA metabolism. This is an anabolic effect at physiologic levels but can be catabolic in certain conditions and at higher than physiologic levels.

Effects on host mechanisms
1. Suppress the immune response. These hormones cause a species- and cell type-specific lysis of lymphocytes.
2. Suppress the inflammatory response (1) by decreasing the number of circulating leukocytes and the migration of tissue leukocytes; (2) by inhibiting fibroblast proliferation; and (3) by inducing lipocortins, which by inhibiting phospholipase A2, blunt the production of the potent anti-inflammatory molecules, the prostaglandins and leukotrienes.

Other effects
1. Necessary for maintenance of normal blood pressure and cardiac output.
2. Required for maintenance of normal water and electrolyte balance, perhaps by restraining ADH release (H_2O) and by increasing angiotensinogen (Na^+). These effects contribute to the effect on blood pressure.
3. Necessary, with the hormones of the adrenal medulla, in allowing the organism to respond to stress.

luted tubules and collecting tubules, the net result being Na^+ retention. These hormones also promote the secretion of K^+, H^+, and NH_4^+ by the kidney and affect ion transport in other epithelial tissues including sweat glands, intestinal mucosa, and salivary glands. Aldosterone is 30–50 times more potent than 11-deoxycorticosterone (DOC) and 1000 times more potent than cortisol or corticosterone. As the most potent naturally occurring mineralocorticoid, aldosterone accounts for most of this action in humans. Cortisol, though far less potent, has a much higher production rate and thus has a significant effect on Na^+ retention and K^+ excretion. Since the amount of DOC produced is very small, it is much less important in this regard.

RNA and protein synthesis are required for the action of aldosterone, which appears to involve the production of specific gene products (see below).

ADRENAL STEROID HORMONES BIND TO INTRACELLULAR RECEPTORS

Glucocorticoid hormones initiate their action in a target cell by interacting with a specific receptor. This step is necessary for entry into the nucleus and DNA binding. There is generally a high correlation between the association of a steroid with receptor and the elicitation of a given biologic response. This correlation holds true for a wide range of activities, so that a steroid with one-tenth the binding affinity evokes a correspondingly decreased biologic effect at a given steroid concentration.

The biologic effect of a steroid depends upon both its ability to bind to the receptor and the concentration of free hormone in the plasma. Cortisol, corticosterone, and aldosterone all bind with high affinity to the glucocorticoid receptor, but in physiologic circumstances cortisol is the dominant glucocorticoid because of its much greater plasma concentration. Corticosterone is an important glucocorticoid in certain pathologic conditions (17α-hydroxylase deficiency), but aldosterone never reaches a concentration in plasma sufficient to exert glucocorticoid effects.

The Functional Domains of the Glucocorticoid Receptor Have Been Defined

A number of biochemical, immunologic, and genetic studies have led to the formulation of the glucocorticoid receptor illustrated in Figure 48–6. The amino terminal half contains most of the antigenic sites and has a region that modulates promoter function (*trans*-activation). The carboxyl terminal half contains the DNA- and hormone-binding domains. The DNA-binding domain is closer to the center of the molecule, while the hormone-binding domain is near the carboxyl terminal. Both of these domains are required for *trans*-activation of gene transcription. A sequence of amino acids in the carboxyl terminal region is required for dimerization of two receptor molecules, a reaction thought to be required for binding to each of the two "half" binding sites in the glucocorticoid response element (GRE) (see the arrows in Table 44–3). Two separate regions appear to be necessary for entry of the receptor into the nucleus (nuclear localization).

The amino acid sequence of the receptor, deduced by analysis of appropriate cDNA molecules, reveals two regions with an abundance of Cys-Lys-Arg residues in the DNA-binding domain. By comparing these regions with other known DNA-binding proteins (especially transcription factor IIIA [TFIIIA]), it is possible to hypothesize a protein structure having two of the "fingers" (each with a coordinated zinc in its center) that bind to a turn of DNA. This has now been directly confirmed, and the **zinc finger motif** is one of the forms of protein-DNA interaction discussed in Chapter 41.

The Steroid-Thyroid Hormone Receptor Superfamily

Steroid and thyroid hormones regulate a variety of processes involved in development, differentiation, growth, reproduction, and adaptation to environmental changes. In recent years, it has become obvious that a general mechanism could explain how these hormones work at the molecular level (Chapter 44). An essential component in this mechanism is the hormone receptor. These molecules are not abundant, so structural analysis awaited the isolation of cDNA clones for each. The first structures deduced were those for the glucocorticoid, estrogen, and progesterone receptors. The homology in the DNA-binding domains of these and the close similarity of each to v-*erb*-A, a DNA-binding oncogene protein, led to the hypothesis that these receptors might belong to a **supergene family.** If so, a corollary hypothesis was that other receptors should be isolated from cDNA libraries using probes directed against the common region (DNA domain) under conditions of reduced hybridization stringency (Chapter 42). This hypothesis proved to be true. As illustrated in Figure 48–7, structures of all of the steroid receptors have been deduced, along with those of several potential DNA binding proteins whose ligand has not yet been identified. These are called **orphan receptors.** The homology between the DNA-binding domains of these

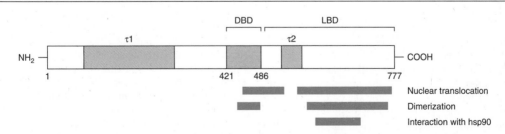

Figure 48–6. Schematic representation of the 777-amino-acid human glucocorticoid receptor. The NH₂-terminal amino acid is shown as number 1, the DNA-binding domain (DBD) is located between amino acids 421 and 486, and the carboxyl terminal amino acid is number 777. The receptor consists of several domains, the approximate limits of which are shown on the figure. Members of this family of receptors all have ligand-binding, transactivation (tau 1 and tau 2 in GR), and DNA-binding domains, and many have dimerization domains. GR dissociates from hsp 90 when ligand binds, and then it moves from the cytosol to the nucleus. The regions of the receptor necessary for these functions are also shown.

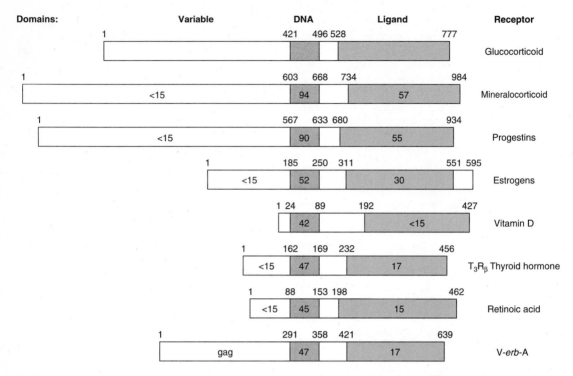

Figure 48–7. Schematic comparison of the steroid-thyroid hormone receptor superfamily. The sequences of the human receptors (v-*erb*-A is avian) are aligned by their DNA-binding domains, which show the highest amino acid similarity (numbers inside the rectangles are percentage similarities to the corresponding region of the glucocorticoid receptor). The numbers above the vertical lines that separate the domains show the amino acid positions. The amino terminal position is designated as 1. Many other similar molecules have been isolated, but the ligands and functions for these have not been determined. (Modified and reproduced, with permission, from Evans R: The steroid and thyroid hormone receptor superfamily. Science 1988;240:889.)

receptors is striking, and the general organization of each is the same. There is considerable variation in the total length of the receptors, most of which is due to the amino terminal half of the molecule. This observation has greatly accelerated understanding of how this class of hormones regulates gene transcription.

Glucocorticoid Hormones Regulate Gene Expression

The general features of glucocorticoid hormone action are described in Chapter 44 and illustrated in Figure 44–1. Numerous examples support the concept that this class of hormones affects specific cellular processes by influencing the amount of critical proteins, usually enzymes, within the cell. Glucocorticoids usually accomplish this by regulating the rate of transcription of specific genes in the target cell, but they also affect other steps in the "information flow" (see Figure 44–2). Regulation of transcription requires that the steroid-receptor complex bind to specific regions of DNA in the vicinity of the transcription initiation site and that such regions confer specificity to the response. How this interaction actu-

ally enhances or inhibits transcription, how tissue specificity is accomplished, and how a given gene can be stimulated in one tissue and inhibited in another are a few of the important questions that remain unanswered.

A brief description of how glucocorticoid hormones affect transcription of mouse mammary tumor virus DNA provides a good illustration of what is known about steroid hormone action. The mammary tumor virus system has been useful because the steroid effect is rapid and large and the molecular biology of the virus has been extensively studied. The glucocorticoid hormone receptor complex binds with high selectivity and specificity to a region, the glucocorticoid response element (GRE), a few hundred base pairs upstream from the transcription initiation site. Within the GRE are sequences closely related to the consensus sequence GGTACANNNTGTTCT that is found in the regulatory elements of most glucocorticoid-regulated genes. The receptor-charged glucocorticoid response element enhances transcription initiation of the mouse mammary tumor virus genome and will also activate heterologous promoters. This *cis*-acting element also works when moved to different regions upstream and downstream, and it works in either the forward or the backward orientation. In these respects, the glucocorticoid response element qualifies as a transcription enhancer. These features have also been demonstrated in several other glucocorticoid-regulated genes and are summarized in Figure 48–8. Most genes regulated by glucocorticoids and other members of the steroid-thyroid-retinoid family of hormones require additional DNA-binding proteins and cognate DNA elements to fully activate transcription. These complexes are referred to as **hormone response units.**

Control of the rate of gene transcription appears to be the major action of the glucocorticoid hormones, but it is not the sole mechanism employed. The ability to measure specific processes has revealed that these hormones also regulate the rate of degradation of specific mRNAs (eg, growth hormone and phospho-enolpyruvate-carboxykinase), and posttranslational processing (various mammary tumor virus proteins). These and other classes of steroid hormones appear able to act at any level of the "information flow" from DNA to protein (Figure 44–2) and the relative importance of each varies from system to system.

Broad Features of Mineralocorticoid Hormone (Aldosterone) Action Resemble Those of Other Steroid Hormones (Figures 44–1, 44–2, and 44–3; Table 44–3)

Although specific gene products have not been isolated, protein and RNA synthesis are known to be required for aldosterone action, and it is presumed that specific proteins are involved in mediating the effects of aldosterone on ion transport.

Receptors That Bind Aldosterone With High Affinity (K_d About 1 nmol/L) Occur in the Cytoplasm and Nuclei of Target Cells

These cells are found in the kidney, parotid, and colon and other organs not thought to be targets of aldosterone action (hippocampus and heart). These receptors have equal affinity for aldosterone, cortisol, and corticosterone and are called type I receptors to distinguish them from the classic glucocorticoid receptor (type II).

Given the fact that the plasma level of aldosterone is much lower than that of either of the other two steroids, one might suppose that these would preferentially occupy the type I sites and that aldosterone would exert little effect. Recall that DOC and corticosterone are avidly bound to corticosteroid-binding globulin, the plasma glucocorticoid transport protein,

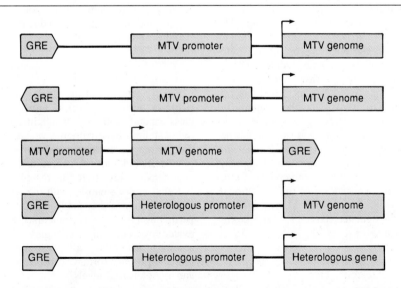

Figure 48–8. The glucocorticoid response element (GRE) is a transcription enhancer. The wild-type mammary tumor virus (MTV) contains a region of DNA that is copied into the transcription unit (MTV genome), the promoter, and a GRE. The GRE is normally situated within 200 bp 5′ of the transcription start site, but it works in either orientation and can be located downstream from the transcription start site. It will also work on heterologous promoters and coding units. These observations mean that this GRE, and others isolated from different genes, function as transcription enhancers. The GRE often works in conjunction with other DNA elements and associated proteins to modulate (increase or decrease) transcription.

while aldosterone has no specific carrier protein. Consequently, the effective "free" concentration of aldosterone in plasma is greater than that of either corticosterone or DOC. Aldosterone therefore is readily able to enter cells, and this ensures a competitive advantage for aldosterone with respect to occupying the type I receptor in vivo. The important action of aldosterone is assured by an additional "fail-safe" mechanism. The receptor in mineralocorticoid target tissues has absolute selectivity for aldosterone because of the presence of the enzyme 11β-hydroxysteroid dehydrogenase. This enzyme converts cortisol and corticosterone to their 11β metabolites but it is not active on aldosterone. These metabolites cannot bind to the type I receptor, so aldosterone has unimpeded access.

The Major Actions of Aldosterone Are on Ion Transport

The molecular mechanisms of aldosterone action on Na^+ transport have not been elucidated, but several studies point to the following hypothesis.

Na^+ from the luminal fluid bathing the apical surface of the renal cell enters passively through Na^+ channels. Na^+ is then transported into the interstitial fluid through the serosal side of the cell by the Na^+-K^+-dependent ATPase pump. ATP provides the energy required for this active process.

Aldosterone increases the number of apical membrane Na^+ channels, and this presumably increases intracellular Na^+. Aldosterone also increases the activity of several mitochondrial enzymes, and this could result in the generation of the ATP required to drive the serosal membrane Na^+-K^+ pump. The NADH:NAD ratio increases as a result of aldosterone action, as do the activities of several mitochondrial enzymes including citrate synthase. The increased activity of citrate synthase involves a true induction (perhaps mediated by the gene transcription effects alluded to above), and the temporal increase of this protein correlates highly with the effect of aldosterone on Na^+ transport. Aldosterone has not been shown to have an effect on the Na^+ pump itself; therefore, it appears that the hormone increases the intracellular concentration of Na^+ and creates the energy source required for removal of this ion through the serosal pump. Other mechanisms, involving different aldosterone-regulated proteins, may be involved in the handling of K^+ and H^+.

PATHOPHYSIOLOGY OF THE ADRENAL CORTEX

Disorders of Glucocorticoid Hormone Insufficiency and Excess

Primary adrenal insufficiency (**Addison's disease**) results in hypoglycemia, extreme sensitivity to in-sulin, intolerance to stress, anorexia, weight loss, nausea, and severe weakness. Patients with Addison's disease have low blood pressure, decreased glomerular filtration rate, and decreased ability to excrete a water load. They often have a history of salt craving. Plasma Na^+ levels are low, K^+ levels are high, and blood lymphocyte and eosinophil counts are increased. Such patients often show increased pigmentation of skin and mucous membranes because of the exaggerated compensatory secretion of ACTH and associated products of the POMC gene. **Secondary adrenal insufficiency** is due to a deficiency of ACTH resulting from tumor, infarction, or infection. This results in a similar metabolic syndrome without hyperpigmentation.

Glucocorticoid excess, commonly called **Cushing's syndrome,** is usually due to the pharmacologic use of steroids, but it may result from an ACTH-secreting pituitary adenoma, from adrenal adenomas or carcinomas, or from the ectopic production of ACTH by a neoplasm. Patients with Cushing's syndrome typically lose the diurnal pattern of ACTH and cortisol secretion. They have hyperglycemia or glucose intolerance (or both) because of accelerated gluconeogenesis. Related to this are severe protein catabolic effects, which result in thinning of the skin, muscle wasting, osteoporosis, extensive lymphoid tissue involution, and generally a negative nitrogen balance. There is a peculiar redistribution of fat, with truncal obesity and the typical "buffalo hump." Resistance to infections and inflammatory responses is impaired, as is wound healing. Several findings, including hypernatremia, hypokalemia, alkalosis, edema, and hypertension, are due to the mineralocorticoid actions of cortisol.

Disorders of Mineralocorticoid Excess

Small adenomas of the glomerulosa cells result in **primary aldosteronism (Conn's syndrome),** the classic manifestations of which include hypertension, hypokalemia, hypernatremia, and alkalosis. Patients with primary aldosteronism do not have evidence of glucocorticoid hormone excess, and plasma renin and angiotensin II levels are suppressed.

Renal artery stenosis, with the attendant decrease in perfusion pressure, can lead to hyperplasia and hyperfunction of the juxtaglomerular cells and cause elevated levels of renin and angiotensin II. This action results in **secondary aldosteronism,** which resembles the primary form, except for the elevated renin and angiotensin II levels.

Congenital Adrenal Hyperplasia Is Due to Enzyme Deficiency

Insufficient amounts of steroidogenic enzymes result in the deficiency of end products, the accumulation of intermediates, and the exaggerated production of steroids from alternative pathways. A common feature of most of these syndromes, which develop in utero, is deficient cortisol production with ACTH overproduction and adrenal hyperplasia—hence the term **congenital adrenal hyperplasia.** The overproduction of adrenal androgens is another common feature. This hormone excess results in increased body growth, virilization, and ambiguous external genitalia—hence the alternative designation **adrenogenital syndrome.**

Two types of **21-hydroxylase deficiency** (partial, or simple virilizing, and complete, or salt wasting) account for more than 90% of cases of congenital adrenal hyperplasia, and most of the rest are due to **11β-hydroxylase deficiency.** Only a few cases of other deficiencies (3β-hydroxysteroid dehydrogenase, 17α-hydroxylase, cholesterol desmolase, 18-hydroxylase, and 18-dehydrogenase) have been described. The **18-hydroxylase** and **18-dehydrogenase deficiencies** affect only aldosterone biosynthesis and so do not cause adrenal hyperplasia. The **cholesterol desmolase deficiency** prevents any steroid biosynthesis and so is usually incompatible with extrauterine life.

SUMMARY

The adrenal cortex has enzymes that convert cholesterol into dozens of different steroid molecules. Of these, three classes have hormone activity: (1) the glucocorticoids, (2) the mineralocorticoids, and (3) the androgens. All these hormones share the basic 17-carbon cyclopentanoperhydrophenanthrene structure, which consists of four rings labeled A–D. Additional carbons make the C_{19} androgens and the C_{21} glucocorticoids and mineralocorticoids.

The major glucocorticoid made in the human adrenal cortex is cortisol. The production of cortisol is regulated by a negative feedback loop consisting of CRH (hypothalamus) and ACTH (anterior pituitary). ACTH catalyzes cleavage of the side chain from cholesterol, which is the rate-limiting step in adrenal steroidogenesis. The numerous effects of cortisol include those in intermediary metabolism and the suppression of host defense mechanisms, and it is critically involved in the response to stress. The effects of cortisol are mediated by its interaction with a specific receptor located in target cells. The ligand-receptor complex binds to specific regions of DNA, called glucocorticoid response elements, to affect the rate of transcription of specific genes. The resultant changes in the rate of synthesis of select proteins mediate most of the effects of the hormone.

Aldosterone, the most potent mineralocorticoid, is synthesized in the zona glomerulosa of the adrenal cortex. This hormone is produced in response to changes in the plasma levels of K^+ and angiotensin II. Aldosterone, which also acts by regulating gene expression through a receptor-mediated mechanism, is the primary hormone responsible for Na^+ retention by the kidney.

The adrenal hormones play a central role in glucose homeostasis, sodium retention and blood pressure regulation, host defense mechanisms, stress response, and general protein anabolism. The absence of adrenal gland function is a life-threatening condition in humans.

REFERENCES

Carson-Junica MA, Schrader WT, O'Malley BW: Steroid receptor family: Structure and function. Endocr Rev 1990;11:201.

Dallman MF: Stress update: Adaptation of the hypothalamic-pituitary-adrenal axis to chronic stress. Trends Endocrinol Metab 1993;4:62.

Evans R: The steroid and thyroid hormone receptor superfamily. Science 1988;240:889.

Freedman LP: Anatomy of the steroid receptor zinc finger region. Endocr Rev 1992;13:129.

Funder JW: Target tissue specificity of mineralocorticoids. Trends Endocr Metab 1990;1:145.

Granner DK, Stromstedt P-E: Glucocorticoid hormone action. In: *Therapeutic Immunology.* Austen KF et al (editors). Blackwell, 1995.

Gustafsson JA et al: Biochemistry, molecular biology, and physiology of the glucocorticoid receptor. Endocr Rev 1987;8:185.

Lucas PC, Granner DK: Hormone response domains in gene transcription. Annu Rev Biochem 1992;61:1131.

Pearce D, Yamamoto KR: Mineralocorticoid and glucocorticoid receptor activities distinguished by nonreceptor factors at a composite response element. Science 1993;259:1161.

Truss M, Beato M: Steroid hormone receptors: Interaction with deoxyribonucleic acid and transcription factors. Endocr Rev 1993;14:459.

49 Hormones of the Adrenal Medulla

Daryl K. Granner, MD

INTRODUCTION

The sympathoadrenal system consists of the parasympathetic nervous system with cholinergic pre- and postganglionic nerves, the sympathetic nervous system with cholinergic preganglionic and adrenergic postganglionic nerves, and the adrenal medulla. The last is actually an extension of the sympathetic nervous system, since preganglionic fibers from the splanchnic nerve terminate in the adrenal medulla, where they innervate the chromaffin cells that produce the catecholamine hormones dopamine, norepinephrine, and epinephrine. The adrenal medulla is thus a specialized ganglion without axonal extensions. Its chromaffin cells synthesize, store, and release products that act on distant sites, so that it also functions as an endocrine organ-a perfect illustration of the enmeshing of the nervous and endocrine systems alluded to in Chapter 44.

BIOMEDICAL IMPORTANCE

The hormones of the sympathoadrenal system, while not necessary to life, are required for adaptation to acute and chronic stress. Epinephrine, norepinephrine, and dopamine are the major elements in the response to severe stress. This response involves an acute, integrated adjustment of many complex processes in the organs vital to the response (brain, muscles, cardiopulmonary system, and liver) at the expense of other organs that are less immediately involved (skin, gastrointestinal system, and lymphoid tissue). Catecholamines do not facilitate the stress response alone but are aided by the glucocorticoids, growth hormone, vasopressin, angiotension II, and glucagon.

CATECHOLAMINE HORMONES ARE 3,4-DIHYDROXY DERIVATIVES OF PHENYLETHYLAMINE

These amines—dopamine, norepinephrine, and epinephrine—are synthesized in the chromaffin cells of the adrenal medulla, so named because they con-

ACRONYMS USED IN THIS CHAPTER

COMT	Catechol-*O*-methyltransferase
DBH	Dopamine β-hydroxylase
MAO	Monoamine oxidase
PNMT	Phenylethanolamine-N-methyltransferase
VMA	Vanillylmandelic acid

tain granules that develop a red-brown color when exposed to potassium dichromate. Collections of these cells are also found in the heart, liver, kidney, gonads, adrenergic neurons of the postganglionic sympathetic system, and central nervous system.

The major product of the adrenal medulla is epinephrine. This compound constitutes about 80% of the catecholamines in the medulla, and it is not made in extramedullary tissue. In contrast, most of the norepinephrine present in organs innervated by sympathetic nerves is made in situ (about 80% of the total), and most of the rest is made in other nerve endings and reaches the target sites via the circulation. Epinephrine and norepinephrine may be produced and stored in different cells in the adrenal medulla and other chromaffin tissues.

The conversion of tyrosine to epinephrine requires 4 sequential steps: (1) ring hydroxylation; (2) decarboxylation; (3) side-chain hydroxylation; and (4) *N*-methylation. The biosynthetic pathway and the enzymes involved are illustrated in Figure 49–1, and a schematic representation is illustrated in Figure 49–2.

Tyrosine Hydroxylase Is Rate-Limiting for Catecholamine Biosynthesis

Tyrosine is the immediate precursor of catecholamines, and tyrosine hydroxylase is the rate-limiting enzyme in catecholamine biosynthesis. Tyrosine hydroxylase is found in both soluble and particle-bound forms only in tissues that synthesize catecholamines; it functions as an oxidoreductase, with tetrahydropteridine as a cofactor, to convert L-tyrosine to L-dihydroxyphenylalanine (L-dopa). As the rate-limiting enzyme, tyrosine hydroxylase is reg-

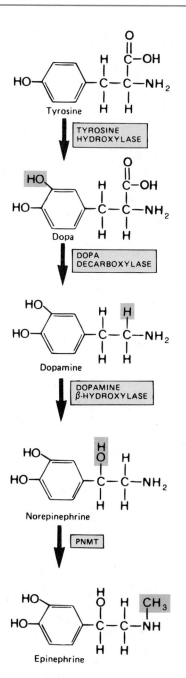

Figure 49–1. Biosynthesis of catecholamines. (PNMT, phenylethanolamine-*N*-methyltransferase.) (Modified and reproduced, with permission, from Goldfien A: The adrenal medulla. In: *Basic & Clinical Endocrinology,* 3rd ed. Greenspan FS [editor]. Appleton & Lange, 1991.)

α-methyltyrosine. This compound is occasionally used to treat catecholamine excess in pheochromocytoma, but other agents are more effective and have fewer side effects. A third group of compounds inhibit tyrosine hydroxylase by chelating iron and thus removing available cofactor. An example is α,α′-dipyridyl.

Catecholamines cannot cross the blood-brain barrier; hence, in the brain they must be synthesized locally. In certain central nervous system diseases, eg, Parkinson's disease, there is a local deficiency of dopamine synthesis. L-Dopa, the precursor of dopamine, readily crosses the blood-brain barrier and so is an important agent in the treatment of Parkinson's disease (Chapter 64).

Dopa Decarboxylase Is Present in All Tissues

This soluble enzyme requires pyridoxal phosphate for the conversion of L-dopa to 3,4-dihydroxyphenylethylamine (dopamine). Compounds that resemble L-dopa, such as α-methyldopa, are competitive inhibitors of this reaction. Halogenated compounds form a Schiff base with L-dopa and also inhibit the decarboxylase reaction.

α-Methyldopa and other related compounds, such as 3-hydroxytyramine (from tyramine), α-methyltyrosine, and metaraminol, are effective in treating some kinds of hypertension.

Dopamine β-Hydroxylase (DBH) Catalyzes the Conversion of Dopamine to Norepinephrine

DBH is a mixed-function oxidase and uses ascorbate as an electron donor, copper at the active site, and fumarate as modulator. DBH is in the particulate fraction of the medullary cells, probably in the secretion granule; thus, the conversion of dopamine to norepinephrine occurs in this organelle. DBH is released from the adrenal medulla or nerve endings with norepinephrine, but (unlike norepinephrine) it cannot reenter nerve terminals via the reuptake mechanism.

Phenylethanolamine-*N*-Methyltransferase Catalyzes the Production of Epinephrine

The soluble enzyme phenylethanolamine-*N*-methyltransferase (PNMT) catalyzes the *N*-methylation of norepinephrine to form epinephrine in the epinephrine-forming cells of the adrenal medulla. Since PNMT is soluble, it is assumed that norepinephrine-to-epinephrine conversion occurs in the cytoplasm. The synthesis of PNMT is induced by glucocorticoid hormones that reach the medulla via the intra-adrenal portal system. This system provides for a 100-fold steroid concentration gradient over systemic arterial blood, and this high intra-adrenal concentration appears to be necessary for the induction of PNMT.

ulated in a variety of ways. The most important mechanism involves feedback inhibition by the catecholamines, which compete with the enzyme for the pteridine cofactor by forming a Schiff base with the latter. Tyrosine hydroxylase is also competitively inhibited by a series of tyrosine derivatives, including

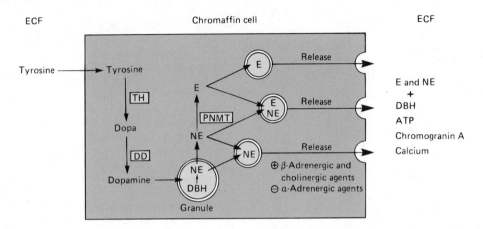

Figure 49–2. Schematic representation of catecholamine biosynthesis. (TH, tyrosine hydroxylase; DD, dopa decarboxylase; PNMT, phenylethanolamine-*N*-methyltransferase; DBH, dopamine β-hydroxylase; ATP, adenosine triphosphate.) The biosynthesis of catecholamines occurs within the cytoplasm and in various granules of the adrenal medullary cell. Some granules contain epinephrine (E), others have norepinephrine (NE), while still others have both hormones. Upon stimulation, all contents of the granules are released into the extracellular fluid (ECF).

CATECHOLAMINES ARE STORED AND RELEASED

Storage Is in Chromaffin Granules

The adrenal medulla contains the **chromaffin granules**—organelles capable of the biosynthesis, uptake, storage, and secretion of catecholamines. These granules contain a number of substances in addition to the catecholamines, including ATP-Mg^{2+}, Ca^{2+}, DBH, and the protein chromagranin A. Catecholamines enter the granule via an ATP-dependent transport mechanism and bind this nucleotide in a 4:1 ratio (hormone:ATP). Norepinephrine is stored in these granules but can exit to be *N*-methylated; the epinephrine formed then enters a new population of granules.

Release Is Calcium-Dependent

Neural stimulation of the adrenal medulla results in the fusion of the membranes of the storage granules with the plasma membrane, and this leads to the exocytotic release of norepinephrine and epinephrine. This process is calcium-dependent and, like most exocytotic events, is stimulated by cholinergic and β-adrenergic agents and inhibited by α-adrenergic agents (Figure 49–2). Catecholamines and ATP are released in proportion to their intragranular ratio, as are the other contents including DBH, calcium, and chromagranin A.

Neuronal reuptake of catecholamines is an important mechanism for conserving these hormones and for quickly terminating hormonal or neurotransmitter activity. The adrenal medulla, unlike the sympathetic nerves, does not have a mechanism for the reuptake and storage of discharged catecholamines. The epinephrine discharged from the adrenal goes to the liver and skeletal muscle but then is rapidly metabolized. Very little adrenal norepinephrine reaches distal tissues. Catecholamines circulate in plasma in a loose association with albumin. They have an extremely short biologic half-life (10–30 seconds).

CATECHOLAMINES ARE RAPIDLY METABOLIZED

Very little epinephrine (< 5%) is excreted in the urine. Catecholamines are rapidly metabolized by catechol-*O*-methyltransferase and monoamine oxidase to form the inactive *O*-methylated and deaminated metabolites (Figure 49–3). Most catecholamines are substrates for both of these enzymes, and these reactions can occur in any sequence.

Catechol-*O*-methyltransferase (COMT) is a cytosolic enzyme found in many tissues. It catalyzes the addition of a methyl group, usually at the 3 position (meta) on the benzene ring, to a variety of catecholamines. The reaction requires a divalent cation, and *S*-adenosylmethionine is the methyl donor. The result of this reaction, depending on the substrate, is the production of homovanillic acid, normetanephrine, and metanephrine.

Monoamine oxidase (MAO) is an oxidoreductase that deaminates monoamines. It is located in many tissues, but it occurs in highest concentrations in the liver, stomach, kidney, and intestine. At least two isozymes of MAO have been described. MAO-A is found in neural tissue and deaminates serotonin, epinephrine, and norepinephrine, while MAO-B is found in extraneural tissues and is most active against 2-phenylethylamine and benzylamine. Dopamine and tyramine are metabolized by both

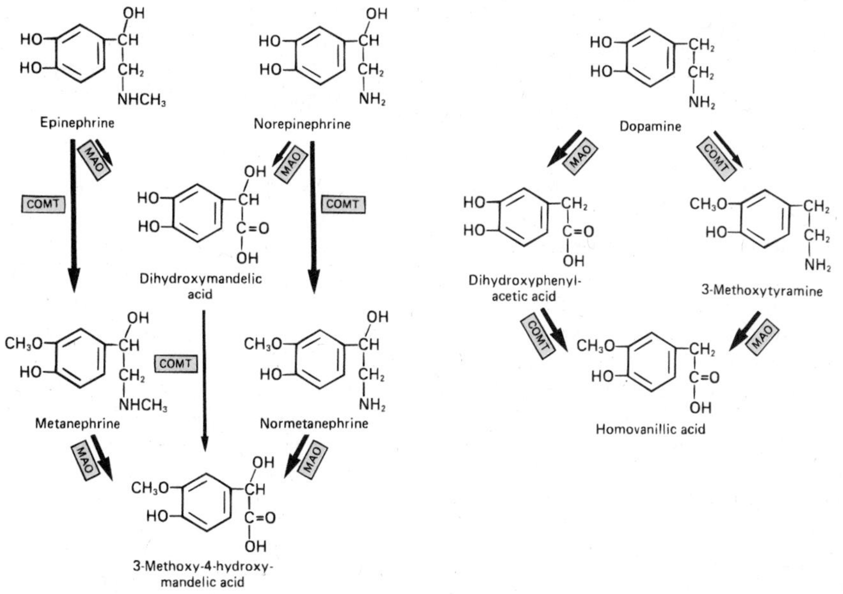

Figure 49–3. Metabolism of catecholamines by catechol-*O*-methyltransferase (COMT) and monoamine oxidase (MAO). (Reproduced, with permission, from Goldfien A: The adrenal medulla. In: *Basic & Clinical Endocrinology,* 3rd ed. Greenspan FS [editor]. Appleton & Lange, 1991.)

forms. Much research effort is directed at correlating affective disorders with increases or decreases of the activity of these isozymes. MAO inhibitors have been used to treat hypertension and depression, but serious reactions with foods or drugs that contain sympathomimetic amines limit their usefulness.

O-Methoxylated derivatives are further modified by conjugation with glucuronic or sulfuric acid.

A bewildering number of metabolites of catecholamines are formed. Two classes of these have diagnostic significance, since they are found in readily measurable amounts in urine. **Metanephrines** represent the methoxy derivatives of epinephrine and norepinephrine, while the *O*-methylated deaminated product of epinephrine and norepinephrine is **3-methoxy-4-hydroxymandelic acid** (also called **vanillylmandelic acid [VMA]**) (Figure 49–3). The concentration of metanephrines or VMA in urine is elevated in more than 95% of patients with pheochromocytoma. These tests have excellent diagnostic precision, particularly when coupled with a measurement of plasma or urine catecholamines.

NERVE IMPULSES REGULATE CATECHOLAMINE SYNTHESIS

Stimulation of the splanchnic nerve, which supplies the preganglionic fibers to the adrenal medulla, results in the exocytotic release of catecholamines, the granule carrier protein, and DBH. Such stimulation is controlled by the hypothalamus and brain stem, but the exact feedback loop has not been described.

Nerve stimulation also results in increased synthesis of catecholamines. Norepinephrine synthesis increases after acute stress, but the amount of tyrosine hydroxylase is unchanged even though tyrosine hydroxylase activity increases. Tyrosine hydroxylase is a substrate for cAMP-dependent protein kinase, and so this activation may involve phosphorylation. Prolonged stress accompanied by chronic sympathetic nerve activity results in an induction (increased amount) of tyrosine hydroxylase. A similar induction of DBH has also been reported. The induction of these enzymes of the catecholamine biosynthetic pathway is a means of adapting to physiologic stress and depends on neural (tyrosine hydroxylase and DBH induction) and endocrine (PNMT induction) factors.

CATECHOLAMINES CAN BE CLASSIFIED BY THEIR MECHANISM OF ACTION

The mechanism of action of the catecholamines has attracted the attention of investigators for nearly a century. Indeed, many of the general concepts of receptor biology and hormone action can be traced to these early studies.

The catecholamines act through two major classes of receptors. These are designated α-adrenergic and β-adrenergic, and each consists of two subclasses, ie,

α_1, α_2, β_1, and β_2. This classification is based on the relative order of binding of various agonists and antagonists. Epinephrine binds to and activates both α and β receptors, so that its action in a tissue having both depends on the relative affinity of these receptors for the hormone. Norepinephrine at physiologic concentrations primarily binds to α receptors.

The Structure of the β-Adrenergic Receptor Is Known

Molecular cloning of the gene and cDNA for the mammalian β-adrenergic receptor revealed some surprising features. First, the gene has no introns and thus joins the histone and interferon genes as mammalian genes that lack these structures. Second, the β-adrenergic receptor is closely homologous to rhodopsin (in three peptide regions at least), the protein that initiates the process which converts light into the visual response.

The catecholamine receptors are members of the G protein-linked class of receptor. The most striking feature of these receptors is a series of domains that span the plasma membrane seven times (Figure 44–4). These membrane-spanning domains assume the configuration of α-helices.

Three Adrenergic Receptor Subgroups Are Coupled to the Adenylyl Cyclase System

Hormones that bind to the β_1 and β_2 receptors activate adenylyl cyclase, whereas hormones that bind to α_2 receptors inhibit this enzyme (Table 44–4). Catecholamine binding induces the coupling of the receptor to a G protein that then binds GTP. This either stimulates (G_s) or inhibits (G_i) adenylyl cyclase, thus stimulating or inhibiting the synthesis of cAMP. The response terminates when the α subunit-associated GTPase hydrolyzes the GTP (Chapter 44). α_1 Receptors are coupled to processes that alter intracellular calcium concentrations or modify phosphatidylinositide metabolism (or both). A separate G protein complex is involved in this response.

THERE IS FUNCTIONAL SIMILARITY BETWEEN THE CATECHOLAMINE RECEPTOR AND THE VISUAL RESPONSE SYSTEM

The stimulation of rhodopsin by light couples it to **transducin**, a G protein complex whose α subunit also binds GTP. The activated G protein in turn activates a phosphodiesterase that hydrolyzes cGMP. This results in the closure of ion channels on the rod cell membrane and produces the visual response. The response terminates when the α subunit associated GTPase hydrolyzes the bound GTP. A partial list of the biochemical and physiologic effects mediated by each of these receptors is provided in Table 49–1. Activation of phosphoproteins by cAMP-dependent protein kinase (Figure 44–5) accounts for many of the biochemical effects of epinephrine.

PHEOCHROMOCYTOMAS ARE TUMORS OF THE ADRENAL MEDULLA

These tumors are usually not detected unless they produce and secrete enough epinephrine or norepinephrine to cause a severe hypertension syndrome. The ratio of norepinephrine to epinephrine is often increased in pheochromocytoma. This may account for differences in clinical presentation, since norepinephrine is thought to be primarily responsible for hypertension and epinephrine for hypermetabolism.

SUMMARY

The adrenal medulla contains enzymes capable of converting tyrosine to epinephrine. The rate-limiting enzyme in this process is tyrosine hydroxylase, but the enzyme phenylethanolamine-N-methyltransferase (PNMT) also plays a critical role. PNMT is induced by glucocorticoids, which reach the adrenal medulla

Table 49–1. Actions mediated through various adrenergic receptors.

Alpha$_1$	Alpha$_2$	Beta$_1$	Beta$_2$
Increased glycogenolysis	Smooth muscle relaxation	Stimulation of lipolysis	Increased hepatic gluconeogenesis
Smooth muscle contraction	Gastrointestinal tract	Myocardial contraction	Increased hepatic glycogenolysis
Blood vessels	Smooth muscle contraction	Increased rate	Increased muscle glycogenolysis
Genitourinary tract	Some vascular beds	Increased force	Increased release of:
	Inhibition of:		Insulin
	Lipolysis		Glucagon
	Renin release		Renin
	Platelet aggregation		Smooth muscle relaxation
	Insulin secretion		Bronchi
			Blood vessels
			Genitourinary tract
			Gastrointestinal tract

from the adrenal cortex through a specialized portal system, thus ensuring a high local concentration of cortisol. PNMT catalyzes the conversion of norepinephrine to epinephrine. This is important because the former, also made and released from neurons elsewhere in the body, primarily activates the α-adrenergic system, which mediates effects through inhibition of adenylyl cyclase and release of intracellular Ca^{2+}. Epinephrine primarily activates the β-adrenergic system, whose effects are mediated by increases of cAMP. These mechanisms of action are in distinct contrast to those of the thyroid hormones, which are also derivatives of the amino acid tyrosine.

The catecholamines, which include norepinephrine and epinephrine, are involved in a number of neuropsychiatric disorders and in hypertension. Many therapeutic agents have been devised on the basis of their ability to alter the synthesis, metabolism, or action of the catecholamines at their respective receptors.

REFERENCES

Dixon RAF et al: Cloning of the gene and cDNA for mammalian β-adrenergic receptor and homology with rhodopsin. Nature 1986;321:75.

Dohlman HG et al: Model systems for the study of seven-transmembrane segment receptors. Annu Rev Biochem 1991;60:653.

Kaupp UB: Mechanism of photoreception in vertebrate vision. Trends Biochem Sci 1986;11:43.

Sibley DR, Lefkowitz RJ: Molecular mechanisms of receptor desensitization using the β-adrenergic receptor-coupled adenylate cyclase system as a model. Nature 1985;317:124.

Stiles GL, Caron MG, Lefkowitz RK: The β-adrenergic receptor: Biochemical mechanisms of physiological regulation. Physiol Rev 1984;64:661.

50

Hormones of the Gonads

Daryl K. Granner, MD

INTRODUCTION

The gonads are bifunctional organs that produce germ cells and the sex hormones. These two functions are closely approximated, for high local concentrations of the sex hormones are required for germ cell development. The ovaries produce ova and the steroid hormones estrogen and progesterone; the testes produce spermatozoa and testosterone. As in the adrenal, a number of steroids are produced, but only a few are active as hormones. The production of these hormones is tightly regulated through a feedback loop that involves the pituitary and the hypothalamus. The gonadal hormones act by a nuclear mechanism similar to that employed by the adrenal steroid hormones.

BIOMEDICAL IMPORTANCE

Proper functioning of the gonads is crucial for reproduction and, hence, survival of the species. Conversely, an understanding of the endocrine physiology and biochemistry of the reproductive process is the basis of many approaches to contraception. The gonadal hormones have other important actions; for example, they are anabolic and thus are required for maintenance of metabolism in skin, bone, and muscle.

THE TESTES PRODUCE TESTOSTERONE AND SPERMATOZOA

These functions are carried out by three specialized cell types: (1) the **spermatogonia** and more differentiated germ cells, which are located in the seminiferous tubules; (2) the **Leydig cells** (also called interstitial cells), which are scattered in the connective tissue between the coiled seminiferous tubules and which produce testosterone in response to LH; and (3) the **Sertoli cells,** which form the basement membrane of the seminiferous tubules and provide the environment necessary for germ cell differentiation and maturation. Spermatogenesis is stimulated

ACRONYMS USED IN THIS CHAPTER	
ACTH	Adrenocorticotropic hormone
ABP	Androgen-binding protein
CBG	Corticosteroid-binding globulin
DHEA	Dehydroepiandrosterone
DHT	Dihydrotestosterone
E₂	Estradiol
FSH	Follicle-stimulating hormone
GnRH	Gonadotropin-releasing hormone
hCG	Human chorionic gonadotropin
hCS	Human chorionic somatomammotropin
LH	Luteinizing hormone
MIF	Müllerian inhibiting factor
3β-OHSD	3β-Hydroxysteroid dehydrogenase
17β-OHSD	17β-Hydroxysteroid dehydrogenase
PL	Placental lactogen
SHBG	Sex hormone–binding globulin
TBG	Thyroid-binding globulin
TEBG	Testosterone-estrogen–binding globulin

by FSH and LH from the pituitary. It requires an environment conducive to germ cell differentiation and a concentration of testosterone in excess of that found in the systemic circulation—a requirement that can be met because the Leydig cells and seminiferous tubules are in close approximation.

Cholesterol Side-Chain Cleavage Enzyme and 3β-Hydroxysteroid Dehydrogenase Catalyze the Critical Steps in the Synthesis of Gonadal Steroids

Testicular androgens are synthesized in the interstitial tissue by the Leydig cells. The immediate precursor of the gonadal steroids, as for the adrenal steroids, is cholesterol. The rate-limiting step, as in the adrenal, is cholesterol side chain cleavage. The conversion of cholesterol to pregnenolone is identical in adrenal, ovary, and testis. In the latter two tissues, however, the reaction is promoted by LH rather than ACTH.

The conversion of pregnenolone to testosterone requires the action of four enzymes: (1) 3β-hydroxysteroid dehydrogenase (3β-OHSD) and $\Delta^{5,4}$ isomerase, (2) 17α-hydroxylase, (3) C_{17-20} lyase, and (4) 17β-hydroxysteroid dehydrogenase (17β-OHSD). This sequence, referred to as the progesterone (or Δ^4) pathway, is shown on the right side of Figure 50–1. Pregnenolone can also be converted to testosterone

by the dehydroepiandrosterone (or Δ^5) pathway, which is illustrated on the left side of Figure 50–1. The Δ^4 route appears to be preferred in human testes; however, since these are seldom available for study, most information about these pathways comes from studies in other animals. Significant species differences may exist.

The four enzymes are localized in the microsomal fraction in rat testes, and there is a close functional association between the activities of 3β-OHSD and

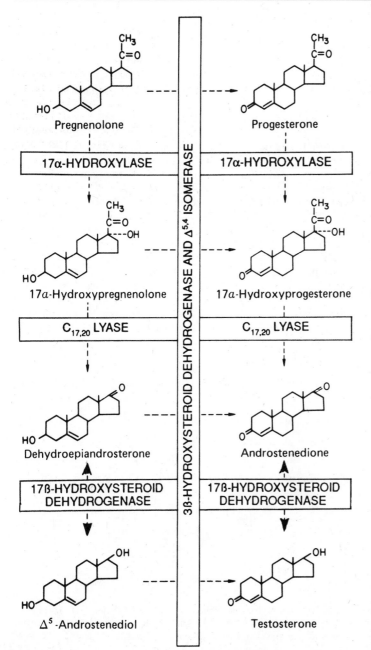

Figure 50–1. Pathways of testosterone biosynthesis. The pathway on the left side of the figure is called the Δ^5 or dehydroepiandrosterone pathway; the pathway on the right side is called the Δ^4 or progesterone pathway.

$\Delta^{5,4}$ isomerase and between those of a 17α-hydroxylase and C_{17-20} lyase. These enzyme pairs are shown in the general reaction sequence in Figure 50–1 and in the schematic representation of androgen biosynthesis in the testicular microsomal membrane in Figure 50–2. The latter shows how the various substrates for testosterone biosynthesis can enter the microsomal compartment and how they might proceed via the Δ^4 pathway from one reaction to the next. Since there are four potential substrates for what appears to be a single 3β-OHSD, multiple alternative pathways exist; thus, the route taken probably depends on the substrate concentration in the vicinity of the various enzymes. Partitioning in the microsomal membrane may provide these gradients.

A. Dihydrotestosterone (DHT) Is Formed From Testosterone by Reduction of the A Ring: Human testes secrete about 50–100 μg of DHT per day, but most DHT is derived from peripheral conversion (see below). The testes also make small but significant amounts of 17β-estradiol (E_2), the female sex hormone, but most of the E_2 produced by the male is derived from peripheral aromatization of testosterone and androstenedione. The Leydig cells, the Sertoli cells, and the seminiferous tubules are thought to be involved in E_2 production. The role of E_2 in the male has not been determined, but it may contribute to FSH regulation. Abnormally high plasma levels of E_2 and changes in the free E_2:testosterone ratio have been associated with pubertal or postpubertal gynecomastia (male breast enlargement), particularly in older individuals and in patients with chronic liver disease or hyperthyroidism.

B. Testicular Hormone Production Undergoes Remarkable Age-Related Changes: Testosterone is the dominant hormone in the fetal and neonatal rat, but the testes make only androsterone soon after birth. The ability to produce testosterone is restored at puberty and continues throughout life. Similar observations have been made in other species, and these age-related changes may also occur in humans.

Testosterone Binds to a Specific Plasma Protein

Most mammals, humans included, have a plasma β-globulin that binds testosterone with specificity, relatively high affinity, and limited capacity (Table 50–1). This protein, usually called **sex hormone-binding globulin (SHBG)** or **testosterone-estrogen-binding globulin (TEBG),** is produced in the liver. Its production is increased by estrogens (women have twice the serum concentration of SHBG as men), certain types of liver disease, and hyperthyroidism; it is decreased by androgens, advancing age, and hypothyroidism. Many of these conditions also affect the production of CBG (Chapter 48) and TBG (Chapter 46). Since SHBG and albumin bind 97–99% of circulating testosterone, only a small fraction of the hormone in circulation is in the free (biologically active) form. The primary function of SHBG may be to restrict the free concentration of testosterone in the serum. Testosterone binds to SHBG with higher affinity than does estradiol (Table 50–2). Therefore, a change in the level of SHBG causes a greater change in the free testosterone level than in the free estradiol level. An increase of SHBG may contribute to the increased free E_2:testosterone ratio noted in aging, cirrhosis, and hyperthyroidism and hence contribute to the attendant signs and symptoms of "estrogenization" alluded to above.

A number of steroids are present in testicular venous blood, but testosterone is the major steroid secreted by the adult testes.

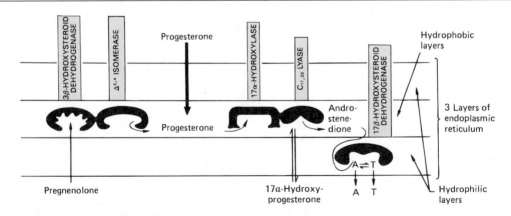

Figure 50–2. Schematic representation of androgen biosynthesis in the testicular microsomal membrane. The membrane is shown as horizontal, which may be the case in the cell; in microsomal preparations, however, it forms vesicles. (A, androstenedione; T, testosterone.) (Reproduced, with permission, from DeGroot LJ: *Endocrinology,* vol 3. Grune & Stratton, 1979.)

Table 50–1. Hormone binding to sex hormone–binding globulin (SHBG).

Steroids Bound	Steroids Not Bound
Testosterone	Conjugated androgens
17β-Estradiol	17α-Testosterone
Dihydrotestosterone	Dehydroisoandrosterone
Other 17β-hydroxysteroids	Cortisol
Estrone	Progesterone

The secretion rate of testosterone is about 5 mg/d in normal adult men. Like other steroid hormones, testosterone seems to be released as it is produced.

Many Testosterone Metabolites Are Inactive; Others Have Increased or Different Activities

A. Metabolic Pathways: Testosterone is metabolized by two pathways. One involves oxidation at the 17-position, and the other involves reduction of the A ring double bond and the 3-ketone. Metabolism via the first pathway occurs in many tissues, including liver, and produces 17-ketosteroids that are generally inactive or less active than the parent compound. Metabolism via the second pathway, which is less efficient, occurs primarily in target tissues and produces the potent metabolite DHT.

B. Metabolites of Testosterone: The most significant metabolic product of testosterone is DHT, since in many tissues, including seminal vesicles, prostate, external genitalia, and some areas of the skin, this is the active form of the hormone. The plasma content of DHT in the adult male is about one-tenth that of testosterone, and approximately 400 μg of DHT is produced daily, as compared to about 5 mg of testosterone. The reaction is catalyzed by a 5α-reductase, an NADPH-dependent enzyme (see below).

Testosterone
1–5%   ~2%   4%
Estradiol Androstanediol Dihydrotestosterone (DHT)

Table 50–2. Approximate affinities of steroids for serum-binding proteins.[1]

	SHBG[2]	CBG[2]
Estradiol	5	>10
Estrone	>10	>100
Androstenedione	. . .	. . .
Testosterone	2	>100
Dihydrotestosterone	1	>100
Progesterone	>100	2
Cortisol	>100	3

[1]Adapted from Siiteri PK, Febres F: Ovarian hormone synthesis, circulation and mechanisms of action. Page 1401 in: *Endocrinology*, Vol 3. DeGroot LJ (editor). Grune & Stratton, 1979.
[2]Affinity expressed as K_d of molar quantity $\times 10^9$.

Testosterone can thus be considered a prohormone, since it is converted into a much more potent compound (dihydrotestosterone) and since most of this conversion occurs outside the testes. A small percentage of testosterone is also converted into estradiol by aromatization, a reaction that is especially important in the brain, where these hormones help determine the sexual behavior of the animal.

The major 17-ketosteroid metabolites, **androsterone** and **etiocholanolone,** are conjugated with glucuronide and sulfate in the liver to make water-soluble, excretable compounds.

REGULATION OF TESTICULAR FUNCTION IS MULTIHORMONAL

Testicular Steroidogenesis Is Stimulated by LH

LH stimulates steroidogenesis and testosterone production by binding to receptors on the plasma membrane of the Leydig cells (an analogous LH receptor is found in the ovary on cells of the corpus luteum) and activating adenylyl cyclase, thus increasing intracellular cAMP. This action enhances the rate of cholesterol side-chain cleavage. The similarity between this action of LH and that of ACTH on the

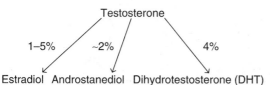

5α-REDUCTASE
NADPH

Testosterone   Dihydrotestosterone (DHT)

adrenal is apparent. Testosterone provides for feedback control at the hypothalamus through inhibition of GnRH release, GnRH production, or both (Figure 50–3).

Spermatogenesis Is Regulated by FSH and Testosterone

FSH binds to the Sertoli cells and promotes the synthesis of androgen-binding protein (ABP). ABP is a glycoprotein that binds testosterone. It is distinct from the intracellular androgen receptor but is identical to SHBG. ABP is secreted into the lumen of the seminiferous tubule, and in this process testosterone produced by the Leydig cells is transported in very high concentration to the site of spermatogenesis. This appears to be a critical step, since normal systemic levels of testosterone, such as might be achieved by replacement therapy, do not support spermatogenesis.

Androgens Affect Several Complex Physiologic Processes

The androgens, principally testosterone and DHT, are involved in (1) sexual differentiation, (2) spermatogenesis, (3) development of secondary sexual organs and ornamental structures, (4) anabolic metabolism and gene regulation, and (5) male-pattern behavior (Figure 50–3). The numerous target tissues involved in these complex processes are defined according to whether they are affected by testosterone or DHT. The classic target cells for DHT (and those which coincidentally have the highest 5α-reductase activity) are the prostate, seminal vesicles, external genitalia, and genital skin. Targets for testosterone include the embryonic Wolffian structures, spermatogonia, muscles, bone, kidney, and brain. The specific

androgen involved in regulating the many other processes mentioned above has not been determined.

Androgens Act by a Nuclear Mechanism Similar to That Employed by Adrenal Steroids

The current concept of androgen action is shown in Figure 50–3. Free testosterone enters cells through the plasma membrane by either passive or facilitated diffusion. Target cells retain testosterone, presumably because the hormone associates with a specific intracellular receptor. Although there is considerable tissue-to-tissue variability, most of the retained hormone is found in the cell nucleus. The cytoplasm of many (but not all) target cells contains the enzyme 5α-reductase, which converts testosterone to DHT. The consensus is that while there is but a single class of receptors, the affinity of the receptor for DHT exceeds that for testosterone. Single gene mutations in humans and mice result in loss of binding of both testosterone and DHT to the receptor in various tissues, suggesting that a single protein is involved. The affinity difference, coupled with the ability of a target tissue to form DHT from testosterone, may determine whether the testosterone-receptor complex or the DHT-receptor complex is active.

Nuclear localization of the testosterone-DHT-receptor complex is a prerequisite for androgen action. Binding of the receptor-steroid complex to chromatin may involve a prior activation step, and specificity is afforded by the androgen response element (Table 44–3).

It has recently become apparent that the DHT-receptor complex binds with higher affinity to the androgen response element than does the testosterone-receptor complex. This may also explain why DHT is the more potent androgen in some tissues.

In keeping with other steroid (and some peptide)

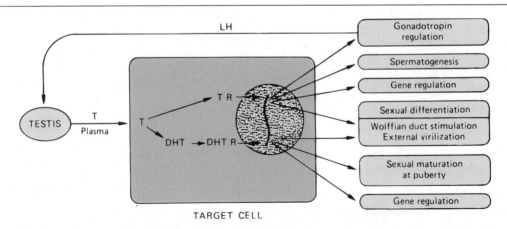

Figure 50–3. Mechanism of androgen action. (LH, luteinizing hormone; T, testosterone; DHT, dihydrotestosterone; R, androgen receptor.) (Modified and reproduced, with permission, from Wilson JD et al: The endocrine control of male phenotypic development. Aust J Biol Sci 1983;36:101.)

hormones, the testosterone-DHT receptor activates specific genes. The protein products of these genes mediate many (if not all) of the effects of the hormone. Testosterone stimulates protein synthesis in male accessory organs, an effect that is usually associated with increased accumulation of total cellular RNA, including mRNA, tRNA, and rRNA. A more specific example involves the effect of testosterone on the synthesis of ABP. The hormone increases the rate of transcription of the ABP gene, which results in an increased amount of the mRNA that codes for this protein. Another well-studied example is α_{2u} globulin, the major protein excreted in the urine of male mice. The rate of synthesis of α_{2u} globulin is directly related to the amount of the cognate mRNA, which in turn is related to the rate of transcription of the α_{2u} globulin gene. All are stimulated by androgens.

The kidney is a major target tissue for androgens. These hormones cause a general enlargement of the kidney and induce the synthesis of a number of enzymes in various species.

Androgens also stimulate the replication of cells in some target tissues, an effect that is poorly understood. Testosterone or DHT, in combination with E_2, appears to be implicated in the extensive and uncontrolled division of prostate cells that results in **benign prostatic hypertrophy,** a condition that afflicts as many as 75% of men over the age of 60 years. Inhibitors of 5α-reductase have recently been introduced in the treatment of this condition.

THE PATHOPHYSIOLOGY OF THE MALE REPRODUCTIVE SYSTEM RELATES TO HORMONAL DEFECTS

The lack of testosterone synthesis is called **hypogonadism.** If this occurs before puberty, secondary sex characteristics fail to develop, and if it occurs in adults, many of these features regress. **Primary hypogonadism** is due to processes that affect the testes differently and cause testicular failure, whereas secondary hypogonadism is due to defective secretion of the gonadotropins. Isolated genetic deficiencies help establish the importance of specific steps in the biosynthesis and action of androgens. Figure 50–4 represents the pathway involved in androgen action from testosterone biosynthesis through postreceptor actions of testosterone and DHT. At least five distinct genetic defects in testosterone biosynthesis have been described. In addition, a 5α-reductase deficiency is known.

There are a number of instances in which either no testosterone/DHT receptor is detected or the receptor is abnormal in some manner; and there are a number of cases in which all measurable entities, including the receptor, are normal, but the patients (always genetic males) have variable degrees of feminization. Persons who completely lack receptor activity appear to be phenotypic females but have an XY (male) genotype, while the mildest cases may have only an abnormally located penile urethra. Genetic males who completely lack functioning receptors have testes and produce testosterone but have complete feminization of the external genitalia (the so-called **testicular feminization syndrome**).

THE OVARIES PRODUCE THE FEMALE SEX HORMONES AND THE FEMALE GERM CELLS

Biosynthesis and Metabolism of Ovarian Hormones Are Similar to Those of Male Hormones

The estrogens are a family of hormones synthesized in a variety of tissues. 17β-Estradiol is the pri-

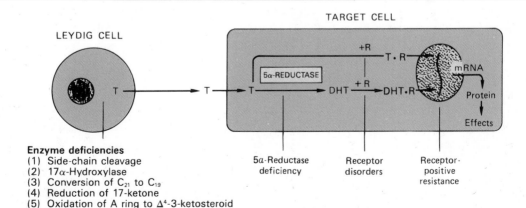

Enzyme deficiencies
(1) Side-chain cleavage
(2) 17α-Hydroxylase
(3) Conversion of C_{21} to C_{19}
(4) Reduction of 17-ketone
(5) Oxidation of A ring to Δ^4-3-ketosteroid

Figure 50–4. Steps involved in androgen resistance. Four stages at which mutations have been identified are shown. (T, testosterone; DHT, dihydrotestosterone; R, androgen receptor.) (Modified and reproduced, with permission, from Wilson JD et al: The endocrine control of male phenotypic development. Aust J Biol Sci 1983;36:101.)

mary estrogen of ovarian origin. In some species, estrone, synthesized in numerous tissues, is more abundant. In pregnancy, relatively more estriol is produced and this comes from the placenta. The general pathway and the subcellular localization of the enzymes involved in the early steps of estradiol synthesis are the same as those involved in androgen biosynthesis. Features unique to the ovary are illustrated in Figure 50–5.

Estrogens are formed by the aromatization of androgens in a complex process that involves three hydroxylation steps, each of which requires O_2 and NADPH. The aromatase enzyme complex is thought to include a P450 mixed-function oxidase. Estradiol is formed if the substrate of this enzyme complex is testosterone, whereas estrone results from the aromatization of androstenedione.

The cellular source of the various ovarian steroids has been difficult to unravel, but it now appears that the transfer of substrates between two cell types is involved. Theca cells are the source of androstenedione and testosterone. These are converted by the aromatase enzyme in granulosa cells to estrone and

estradiol, respectively. Progesterone is produced and secreted by the corpus luteum, which also makes some estradiol.

Significant amounts of estrogens are produced by the peripheral aromatization of androgens. In human males, the peripheral aromatization of testosterone to estradiol (E_2) accounts for 80% of the production rate of the latter. In females, adrenal androgens are important substrates, since as much as 50% of the E_2 produced during pregnancy comes from the aromatization of androgens. Finally, the conversion of androstenedione to estrone is the major source of estrogens in postmenopausal women. Aromatase activity is present in adipose cells and also in liver, skin, and other tissues. Increased activity of this enzyme may contribute to the "estrogenization" that characterizes such diseases as cirrhosis of the liver, hyperthyroidism, aging, and obesity.

Estrogens and Progestins Bind to Plasma Transport Proteins

Estrogens are bound to SHBG and progestins to CBG. SHBG binds estradiol about five times less

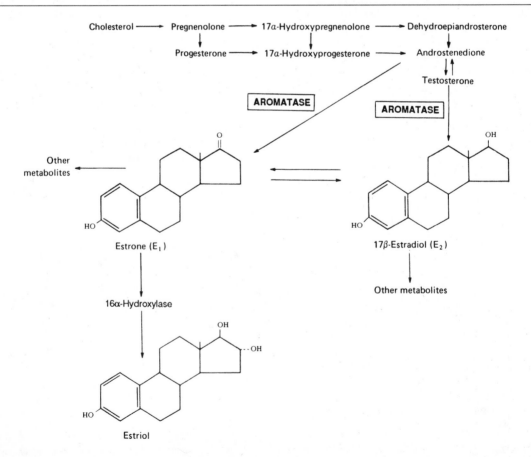

Figure 50–5. Biosynthesis of estrogens. (Slightly modified and reproduced, with permission, from Ganong WF: *Review of Medical Physiology,* 17th ed. Appleton & Lange, 1991.)

17β-Estradiol

Progesterone

avidly than it binds testosterone or DHT, while progesterone and cortisol have little affinity for this protein (Table 50–2). In contrast, progesterone and cortisol bind with nearly equal affinity to CBG, which in turn has little avidity for estradiol and even less for testosterone, DHT, or estrone.

The binding proteins provide a circulating reservoir of hormone, and because of the relatively large binding capacity, they probably buffer against sudden changes in the plasma level. The metabolic clearance rates of these steroids are inversely related to the affinity of their binding to SHBG; hence, estrone is cleared more rapidly than estradiol, which in turn is cleared more rapidly than testosterone or DHT. In this regard, the conjugated derivatives of these hormones (see above) are not bound by either SHBG or CBG. The factors that regulate the production of SHBG are discussed above. There is some evidence for the presence of specific cell surface receptors for SHBG (and CBG), but their function has not been established. There is no doubt that free hormone is biologically active.

The rate of secretion of ovarian steroids varies considerably during the menstrual (or estrous) cycle and is directly related to rate of production in the ovary. There is no storage of these compounds; they are secreted when they are produced.

Estrogens and Progestins Are Actively Metabolized by the Liver

A. Estrogens: The liver converts estradiol and estrone to estriol by the pathways shown in Figure 50–5. Estradiol, estrone, and estriol are substrates for hepatic enzymes that add glucuronide or sulfate moieties. Activity of these conjugating enzymes varies among species. Rodents have such active metabolizing enzyme systems that estrogens are almost completely metabolized by the liver and thus are essentially without activity when given orally. These enzyme systems are less active in primates, so that oral estrogens are more effective. The conjugated steroids are water-soluble and do not bind to transport proteins; thus, they are excreted readily in the bile, feces, and urine.

B. Progestins: Because the liver actively metabolizes progesterone to several compounds, progesterone is ineffective when given orally. Sodium preg-

nanediol-20-glucuronide is the major progestin metabolite found in human urine (Figure 50–6). Certain synthetic steroids, eg, derivatives of 17α-hydroxyprogesterone and 17α-alkyl-substituted 19-nortestosterone compounds, have progestational activity and avoid hepatic metabolism. Thus, they are widely used in oral contraceptives.

THE MATURATION AND MAINTENANCE OF THE FEMALE REPRODUCTIVE SYSTEM IS THE MAJOR FUNCTION OF THE OVARIAN HORMONES

These hormones prepare the structural components of the female reproductive system (see below) for reproduction by (1) maturing the primordial germ cells, (2) developing the tissues that will allow for implantation of the blastocyst, (3) providing the "hormonal timing" for ovulation, (4) establishing the milieu required for the maintenance of pregnancy, and (5) providing the hormonal influences for parturition and lactation.

Estrogens stimulate the development of tissues involved in reproduction. In general, these hormones stimulate the size and number of cells by increasing the rate of synthesis of protein, rRNA, tRNA, mRNA, and DNA. Under estrogen stimulation, the vaginal epithelium proliferates and differentiates; the uterine endometrium proliferates and the glands hypertrophy and elongate; the myometrium develops an intrinsic, rhythmic motility; and breast ducts proliferate. Estradiol also has anabolic effects on bone and cartilage, and so it is growth promoting. By affecting peripheral blood vessels, estrogens typically cause vasodilation and heat dissipation.

Progestins reduce the proliferative activity of the estrogens on the vaginal epithelium and convert the uterine epithelium from proliferative to secretory (increased size and function of secretory glands and increased glycogen content), thus preparing the uterine epithelium for implantation of the fertilized ovum. Progestins enhance development of the acinar portions of breast glands after estrogens have stimulated ductal development. Progestins decrease peripheral blood flow, thereby decreasing heat loss, so that

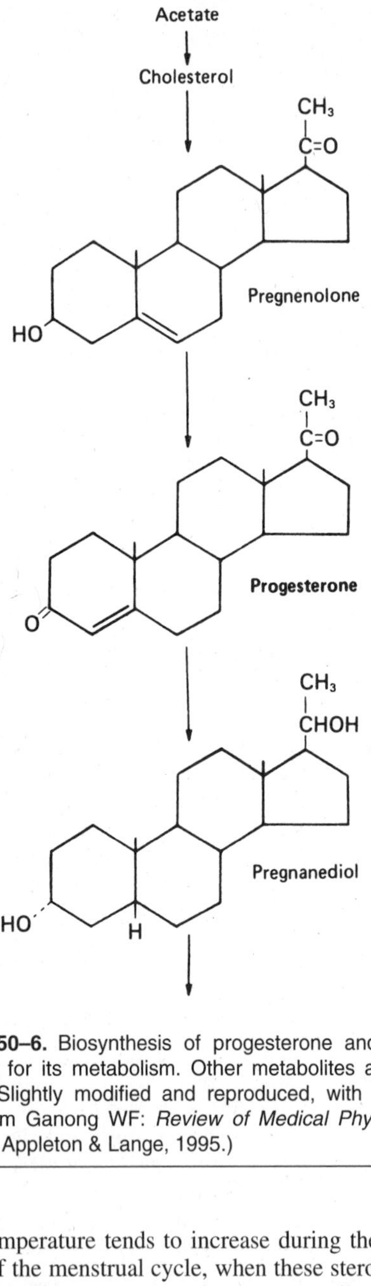

Acetate

↓

Cholesterol

↓

Pregnenolone

HO

↓

Progesterone

O

↓

Pregnanediol

HO · · · H

↓

Figure 50–6. Biosynthesis of progesterone and major pathway for its metabolism. Other metabolites are also found. (Slightly modified and reproduced, with permission, from Ganong WF: *Review of Medical Physiology*, 15th ed. Appleton & Lange, 1995.)

body temperature tends to increase during the luteal phase of the menstrual cycle, when these steroids are produced. This temperature increase, usually 0.5 °C, is used as an indicator of ovulation.

Progestins generally require the previous or concurrent presence of estrogens, perhaps because estrogens stimulate production of the progesterone receptor. The two classes of hormones often act synergistically, although they can be antagonists.

The number of oogonia in the human female ovary reaches a maximum of 6–7 million at about the fifth month of gestation. This decreases to about 2 million by birth and is further diminished to 100,000–200,000 by the onset of menarche. Some 400–500 of these develop into mature oocytes; the rest are gradu-

ally lost through a process that is not understood, although ovarian androgens have been implicated. Follicular maturation begins in infancy, and the ovaries gradually enlarge in prepubertal years owing to increased volume of the follicles because of the growth of granulosa cells, to the accumulation of tissue from atretic follicles, and to the increased mass of medullary stromal tissue with the interstitial and theca cells that will produce the steroids.

The concentration of sex hormones is low in childhood, though exogenous gonadotropins increase production; therefore, the immature ovary has the capacity to synthesize estrogen. It is thought that these low levels of sex steroids inhibit gonadotropin production in prepubertal girls and that at puberty the hypothalamic-pituitary system becomes less sensitive to suppression. At puberty, the pulsatile release of GnRH begins stimulating LH and this causes a dramatic increase of ovarian hormone production. FSH, the main stimulus for estrogen secretion, stimulates a follicle to ripen, and ovulation ensues.

The Menstrual Cycle Depends on a Complex Interaction Among Three Endocrine Glands

Hormones determine the frequency of ovulation and receptivity to mating. Monestrous species ovulate and mate once a year, whereas polyestrous species repeat this cycle several times a year. Primates have menstrual cycles, with shedding of the endometrium at the end of each cycle, and mating behavior is not tightly coupled to ovulation. The human menstrual cycle results from complex interactions between the hypothalamus, pituitary, and ovary. The cycle normally varies between 25 and 35 days in length (average, 28 days). It can be divided into a **follicular phase**, a **luteal phase**, and **menstruation** (Figure 50–7).

A. Follicular Phase: For reasons that are not clear, a particular follicle begins to enlarge under the general influence of FSH. E_2 levels are low during the first week of the follicular phase, but they begin to rise progressively as the follicle enlarges. E_2 reaches its maximal level 24 hours before the LH (FSH) peak and sensitizes the pituitary to GnRH. LH is released either in response to this high level of E_2 in a "positive feedback" manner or in response to a sudden decline of E_2 from this high level. Continual administration of high doses of estrogen (as in oral contraceptives) suppresses LH and FSH release and inhibits the action of GnRH on the pituitary. Progesterone levels are very low during the follicular phase. The LH peak heralds the end of the follicular phase and precedes ovulation by 16–18 hours.

B. Luteal Phase: After ovulation, the granulosa cells of the ruptured follicle luteinize and form the corpus luteum, a structure that soon begins to produce progesterone and some estradiol. Estradiol peaks about midway through the luteal phase and

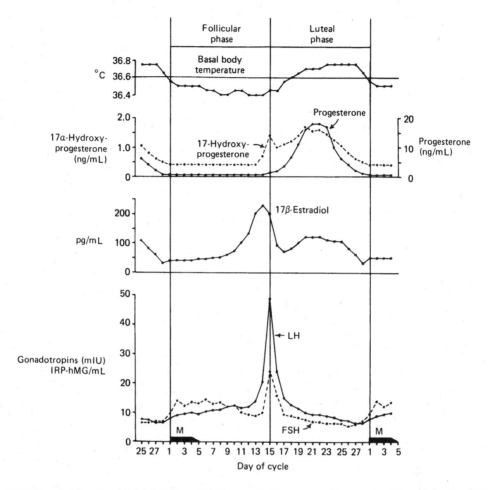

Figure 50–7. Hormonal and physiologic changes during a typical human menstrual cycle. (M, menstruation; IRP-hMG, international reference standard for gonadotropins.) (Reproduced, with permission, from Midgley AR in: *Human Reproduction.* Hafez ESE, Evans TN [editors]. Harper & Row, 1973.)

then declines to a very low level. The major hormone of the luteal portion of the cycle is **progesterone,** which (as noted above) is required for preparation and maintenance of the secretory endometrium that provides early nourishment for the implanted blastocyst. LH is required for the early maintenance of the corpus luteum, and the pituitary supplies it for about 10 days. If implantation occurs (day 22–24 of the average cycle), this LH function is assumed by chorionic gonadotropin (hCG), a placental hormone that is very similar to LH and is made by the cytotrophoblastic cells of the implanted early embryo. hCG stimulates progesterone synthesis by the corpus luteum until the placenta begins making large amounts of this steroid. In the absence of implantation (and hCG), the corpus luteum regresses and menstruation ensues; after the endometrium is shed, a new cycle commences. The luteal phase is always 14 ± 2 days in length. Variations in cycle length are almost always due to an altered follicular phase.

Placental Hormones Maintain Pregnancy

The implanted blastocyst forms the trophoblast, which is subsequently organized into the placenta. The placenta provides the nutritional connection between the embryo and the maternal circulation and produces a number of hormones.

A. Human Chorionic Gonadotropin (hCG): The primary function of the glycoprotein hormone hCG (the structural similarity of hCG to LH is discussed in Chapter 45) is to support the corpus luteum until the placenta produces amounts of progesterone sufficient to support the pregnancy. hCG can be detected within a few days of implantation, and this provides the basis of early diagnostic tests for pregnancy. Peak hCG levels are reached in the middle of the first trimester, after which there is a gradual decline throughout the remainder of pregnancy. Changes in hCG and other hormone levels in pregnancy are illustrated in Figure 50–8.

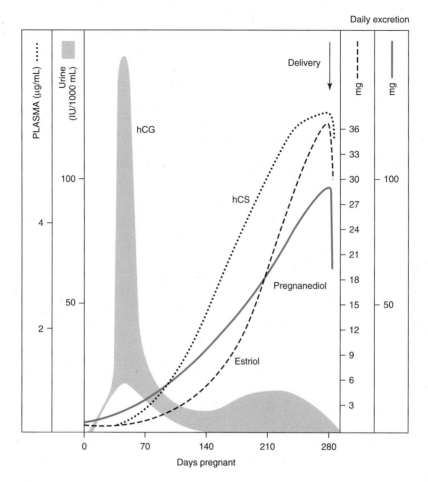

Figure 50–8. Hormone levels during normal pregnancy. (hCG, human chorionic gonadotropin; hCS, human chorionic somatomammotropin.) (Data from various authors.) (Reproduced, with permission, from Ganong WF: *Review of Medical Physiology,* 13th ed. Appleton & Lange, 1987.)

B. Progestins: The corpus luteum is the major source of progesterone for the first 6–8 weeks of the pregnancy, and then the placenta assumes this function. The corpus luteum continues to function, but late in pregnancy the placenta makes 30–40 times more progesterone than does the corpus luteum. The placenta does not synthesize cholesterol and so depends upon a maternal supply.

C. Estrogens: Plasma concentrations of estradiol, estrone, and estriol gradually increase throughout pregnancy. Estriol is produced in the largest amount, and its formation reflects a number of fetoplacental functions. The fetal adrenal produces DHEA and DHEA sulfate, which are converted to 16α-hydroxy derivatives by the fetal liver. These are converted to estriol by the placenta; travel via the placental circulation to the maternal liver, where they are conjugated to glucuronides; and then are excreted in the urine (Figure 50–9). The measurement of urinary estriol levels is used to document the function of a number of maternal-fetal processes.

Another interesting exchange of substrates is required for fetal cortisol production. The fetal adrenal expresses 3β-hydroxysteroid dehydrogenase Δ5,4 isomerase complex at very low levels, and this enzyme is repressed by estradiol, which effectively keeps this enzyme turned off. The adrenal thus depends upon the placenta for the progesterone required for cortisol synthesis (Figure 50–9).

D. Placental Lactogens: The placenta makes a hormone called placental lactogen (PL). PL is also called chorionic somatomammotropin or placental growth hormone because it has biologic properties of prolactin and growth hormone. The genetic relationship of these hormones is discussed in Chapter 45. The physiologic function of PL is uncertain, since women who lack this hormone appear to have normal pregnancies and deliver normal babies.

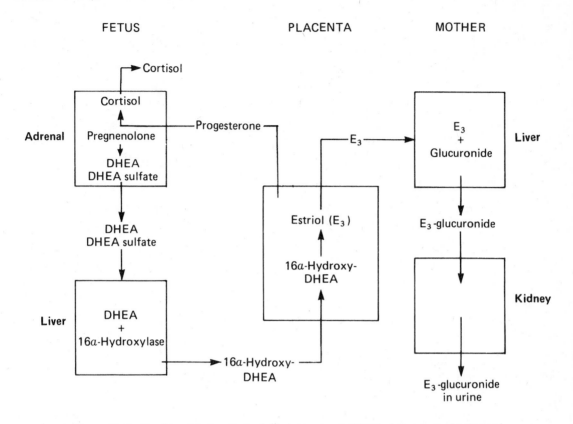

FETUS   PLACENTA   MOTHER

Figure 50–9. Steroid metabolism by the fetal-maternal unit. (DHEA, dehydroepiandrosterone.)

The Trigger for Parturition Is Unknown

Pregnancy lasts a predetermined number of days for each species, but the factors responsible for its termination are unknown. Hormonal influences are suspected but unproved. Estrogens and progestins are candidates, since they affect uterine contractility, and there is evidence that catecholamines are involved in induction of labor. Since oxytocin stimulates uterine contractility, it is used to facilitate delivery, but it will not initiate labor unless the pregnancy is at term. There are 100 times more oxytocin receptors in the uterus at term than there are at the onset of pregnancy.

The increased amount of estrogen at term may increase the number of oxytocin receptors (Chapter 45). Once labor begins, the cervix dilates, initiating a neural reflex that stimulates oxytocin release and hence further uterine contraction. Mechanical factors, such as the amount of stretch or force applied to the muscle, may be important. A sudden and dramatic change in the hormonal milieu of both the mother and newborn occurs with parturition, and plasma levels of progesterone (measured as pregnanediol) and estriol decline rapidly after the placenta is delivered (Figure 50–8).

Mammary Gland Development Is Stimulated by Estradiol and Progesterone and Lactation by Prolactin

The differentiation and function of the mammary gland are regulated by the concerted action of several hormones. The female sex hormones initiate this process, since estrogens are responsible for ductal growth and progestins stimulate alveolar proliferation. Some growth of glandular tissue occurs during puberty, along with deposition of adipose tissue, but extensive development occurs during pregnancy when glandular tissue is exposed to high concentrations of estradiol and progesterone. Complete differentiation, studied mostly in rat mammary gland explants, requires the additional action of prolactin, glucocorticoids, insulin or a growth peptide, and an unidentified serum factor. Of these hormones, only the concentration of prolactin changes dramatically in pregnancy; it increases from less than 2 ng/dL to over 200 ng/dL in late pregnancy. The effects of these hormones on the synthesis of various milk proteins, including lactalbumin, lactoglobulin, and casein, have been studied in detail. These hormones increase the rate of synthesis of these proteins by increasing amounts of the specific mRNAs, and in the case of casein at least, this is due to an increase in gene transcription and to stabilization of the mRNA. Progesterone, required for alveolar differentiation, inhibits milk production and secretion in late pregnancy. Lactation commences when levels of this hormone decrease abruptly after delivery. Prolactin levels also fall rapidly postpartum but are stimulated with each episode of suckling (see Chapter 45), thereby ensuring continual lactation. Lactation gradually decreases if suckling is not allowed and can be rapidly terminated by administration of a large parenteral dose of an androgen before suckling is allowed.

Suckling also results in the release of oxytocin from the posterior pituitary. Oxytocin stimulates contraction of the myoepithelial cells that surround the alveolar ducts, thus expelling milk from the gland. The regulation of oxytocin synthesis and secretion is discussed in Chapter 45.

Menopause Is Complete With the Loss of Ovarian Estrogen Production

Women in the Western hemisphere cease having regular menstrual cycles at about age 53, coincident with loss of all follicles and ovarian estrogen production. There is no alternative source of progesterone, but substantial amounts of a weak estrogen, estrone, are produced by the aromatization of androstenedione (Figure 50–5). The levels of estrone are not sufficient to suppress pituitary gonadotropin levels; thus, marked increases of LH and FSH are characteristic of the postmenopausal years. GnRH may also be involved in the onset of menopause. This may be due to changes in the absolute level of the hormone or to alterations in the periodicity of its secretion. For example, ovaries from young rats, which have numerous follicles, do not function well in older animals. Similarly, ovaries taken from old animals will resume some production of E_2 when placed in young rats.

Postmenopausal women are particularly prone to two problems associated with tissue catabolism. Estrone is not always able to prevent the atrophy of secondary sex tissues, particularly the epithelium of the lower urinary tract and vagina. **Osteoporosis** is a major health problem in older individuals, and women with the most severe decrease in bone mass have lower than normal estrone levels.

Synthetic Agonists and Antagonists Promote and Prevent Conception and Inhibit Tumor Growth

A. Estrogens: Several synthetic compounds have estrogenic activity and one or more favorable pharmacologic features. Most modifications are designed to retard hepatic metabolism, so that the compounds can be given orally. One of the first devel-

oped was diethylstilbestrol. Other examples of modified steroids include 17α-ethinyl estradiol and mestranol, which are used in oral contraceptives.

Diethylstilbestrol

Numerous compounds with antiestrogenic activity have been synthesized, and several of these have clinical applications. Most of these antagonists act by competing with estradiol for its intracellular receptor (see below).

17α-Ethinyl estradiol

Mestranol

Clomiphene citrate (Clomid) has a particular affinity for the estrogen receptor in the hypothalamus. Clomiphene was originally designed as an antifertility drug but, interestingly, it is now used for the opposite effect. Clomiphene competes with estradiol for hypothalamic receptor sites; thus, GnRH release is not restrained and increased amounts of LH and FSH are released. Multiple follicles often mature simultaneously in response to clomiphene, and multiple pregnancies can ensue. Nafoxidine, a nonsteroidal compound, and tamoxifen combine with the estrogen receptor to form very stable complexes with chromatin; hence, the receptor cannot recycle and these agents inhibit the action of estradiol for prolonged periods. These antagonists are used in the treatment of estrogen receptor-dependent breast cancer.

Clomiphene citrate

B. Progestins: It has been difficult to synthesize compounds that have progestin activity but no estrogenic or androgenic action. The 17 α-alkyl-substituted 19-nortestosterone derivatives (eg, norethindrone) have minimal androgenic activity in most women and are used in oral contraceptives. Another potent progestin is medroxyprogesterone acetate (Provera). Medroxyprogesterone inhibits ovulation for several months when given as an intramuscular depot injection. However, since progestins inhibit cell growth, this compound is more frequently used for treating well-differentiated endometrial carcinoma.

Medroxyprogesterone acetate

Norethindrone

Estrogens and Progestins Regulate Gene Expression

These hormones act through their ability to combine with intracellular receptors that then bind to specific regions of chromatin or DNA (or both) to effect changes in the rate of transcription of specific genes. Much information has been learned from the analysis of how estradiol and progesterone stimulate transcription of the avian egg white protein genes, especially ovalbumin and conalbumin. The determination of exactly how these hormones activate gene transcription is under intense investigation.

The Estrogen and Progesterone Receptors Are Part of a Gene Family

The sequence of the receptors for estrogens (ER) and progestins (PR) has been deduced by analysis of the corresponding cDNA sequences. These receptors are part of the steroid and thyroid hormone receptor gene family discussed in Chapter 48. As illustrated in Figure 44–3, each receptor has several functional domains. The steroids bind to the ligand site in the carboxy-terminal portion of the receptor molecule. This causes a conformational change that allows the receptor to bind to DNA. The DNA-binding domain of the ER recognizes the sequence AGGTCANNN-TGACCT (the estrogen response element, or ERE), whereas the cognate domain in the PR recognizes the sequence GGTACANNNTGTTCT, the PRE. The receptor-DNA interaction allows various *trans*-activating domains in each receptor to influence the activity of genes adjacent to the hormone response elements. The increased (or decreased) activity of specific genes results in altered rates of synthesis of specific proteins. This eventuates in altered metabolic responses.

Some points, which may have application to the mechanism of action of other hormones, bear noting:

(1) There is considerable cross-talk between the sex hormone receptors. Progesterone binds to the androgen receptor and thus is a weak androgen; some androgens bind to the estrogen receptor and mimic the action of the latter in the uterus. Inspection of Figure 44–3 reveals close similarity in the core sequences of the hormone response elements. This could explain why some hormones exert "crossover" actions in certain instances.

(2) Estrogens increase the concentration of both the estrogen and the progesterone receptor.

(3) Progesterone appears to enhance the rate of turnover of its receptor.

(4) So-called weak estrogens, such as estriol, act as potent estrogens when given frequently.

SOME PATHOPHYSIOLOGY OF THE FEMALE REPRODUCTIVE SYSTEM HAS A HORMONAL CONNECTION

A discussion of all disorders that affect the female reproductive system is beyond the scope of this chapter, but a few illustrative disorders follow. **Primary hypogonadism** is due to processes that directly involve the ovaries and thus cause ovarian deficiency (decreased ovulation, decreased hormone production, or both), whereas **secondary hypogonadism** is due to the loss of pituitary gonadotropin function. **Gonadal dysgenesis (Turner's syndrome)** is a relatively frequent genetic disorder in which individuals have an XO karyotype, female internal and external genitalia, several developmental abnormalities, and delayed puberty.

Several syndromes are related to abnormal amounts of hormones. The most frequent is **polycystic ovary syndrome** (Stein-Leventhal syndrome), in which overproduction of androgens causes hirsutism, obesity, irregular menses, and impaired fertility. The rare **Leydig cell tumors** and **arrhenoblastoma** produce testosterone; **granulosa–theca cell tumors** produce estrogens; and **intraovarian adrenal rests** produce cortisol. Persistent trophoblastic tissue results in the benign **hydatidiform mole** or its malignant transformation into **choriocarcinoma;** both produce enormous quantities of hCG. The radioimmunoassay of hCG is a diagnostic test for these dangerous conditions and can also be used to monitor efficacy of therapy.

SUMMARY

The gonads produce germ cells and sex hormones. These two functions are necessarily approximated geographically (spermatogenesis and testosterone production in the testes and oogenesis and estradiol production in the ovary) because maturation of germ cells requires a high local concentration of the respective sex hormone.

LH binds to its cell-surface receptor on Leydig cells (testes) and granulosa and theca cells (ovary). The resultant increase of cAMP results in removal of the side chain from cholesterol, again, as in the case of the adrenal cortex, the rate-limiting step in steroidogenesis. A series of enzyme reactions, most notably that catalyzed by 3β-hydroxysteroid dehydrogenase, results in production of testosterone in the testes. A small amount of testosterone is converted by aromatase to estradiol. This reaction is much more robust in the ovary, wherein estradiol is the main hormone produced during the follicular phase of the menstrual cycle. After ovulation occurs, progesterone is made by follicular cells, which now constitute the corpus luteum. Progesterone is the major hormone involved in sustaining pregnancy. It initially is made in the corpus luteum under the influence of pituitary LH and later by hCG from the placenta. After the second month of pregnancy, the placenta makes progesterone directly. Interruption of the production or action of progesterone results in cessation of the pregnancy.

The sex hormones, as members of the steroid/thyroid/retinoid family, act by binding to their cognate receptors in target cells. The ligand-receptor complexes bind to the androgen, estrogen, or progesterone response element in the promoter of target genes. This results in increased (or decreased) activity of these genes. Numerous metabolic, growth, and reproductive processes are affected in this manner.

REFERENCES

Adler AJ, Danielsen M, Robins DM: Androgen-specific gene activation via a consensus glucocorticoid response element is determined by interaction with nonreceptor factors. Proc Natl Acad Sci USA 1992;89:11660.

Chang C et al: Structural analysis of complementary DNA and amino acid sequences of human and rat androgen receptors. Proc Natl Acad Sci USA 1988;85:7211.

Green S et al: Human estrogen receptor DNA: Sequence, expression and homology to V-*erb*-A. Nature 1986;320: 134.

Hall PF: Testicular hormones: Synthesis and control. In: *Endocrinology,* vol 3. DeGroot LJ (editor). Grune & Stratton, 1979.

Huggins C: Two principles in endocrine therapy of cancer: Hormone deprival and hormone interference. Cancer Res 1965;25:1163.

McEwen BS: Steroid hormones are multifunctional messengers to the brain. Trends Endocrinol Metab 1991;2:62.

O'Malley BW: Steroid hormone action in eucaryotic cells. J Clin Invest 1984;74:307.

Siiteri PK, Febres F: Ovarian hormone synthesis, circulation and mechanisms of action. In: *Endocrinology,* vol 3. DeGroot LJ (editor). Grune & Stratton, 1979.

Toft D, Gorski J: A receptor molecule for estrogens. Proc Natl Acad Sci USA 1966;55:1574.

Wilson J: Metabolism of testicular androgens. In: *Handbook of Endocrinology.* Section 7: *Endocrinology,* vol 5: *Male Reproductive System.* Hamilton DW, Greep RP (editors). American Physiological Society, Washington DC, 1975.

Wilson JD et al: The endocrine control of male phenotypic development. Aust J Biol Sci 1983;36:101.

Hormones of the Pancreas & Gastrointestinal Tract

51

Daryl K. Granner, MD

INTRODUCTION

The pancreas consists of very different organs contained within one structure. The acinar portion of the pancreas has an **exocrine** function, secreting into the duodenal lumen the enzymes and ions used for the digestive process. The **endocrine** portion consists of the islets of Langerhans. The 1–2 million islets of the human pancreas make up 1–2% of its weight and are collections of the several different cell types listed in Table 51–1.

The pancreatic islets secrete at least four hormones: insulin, glucagon, somatostatin, and pancreatic polypeptide. The hormones are released into the pancreatic vein, which empties into the portal vein—a convenient arrangement, since the liver is a primary site of action of insulin and glucagon. These two hormones are chiefly involved in regulating carbohydrate metabolism but affect many other processes. Somatostatin, first identified in the hypothalamus as the hormone that inhibits growth hormone secretion, is present in higher concentration in the pancreatic islets than in the hypothalamus and is involved in the local regulation of insulin and glucagon secretion. Pancreatic polypeptide affects gastrointestinal secretion.

The gastrointestinal tract secretes many hormones, perhaps more than any other single organ. The purpose of the gastrointestinal tract is to propel foodstuffs to sites of digestion, to provide the proper milieu (enzymes, pH, salt, etc) for the digestive process, to move the digested products across the intestinal mucosa through the mucosal cells and into the extracellular space, to move those products to distant cells via the circulation, and to expel waste products. The gastrointestinal hormones assist in all these functions.

BIOMEDICAL IMPORTANCE

Insulin has been the model peptide hormone in many ways, being the first purified, crystallized, and synthesized by chemical and molecular biologic

ACRONYMS USED IN THIS CHAPTER

ACTH	Adrenocorticotropic hormone
CCK	Cholecystokinin
EGF	Epidermal growth factor
FGF	Fibroblast growth factor
GIP	Gastric inhibitory polypeptide
IGF	Insulin-like growth factor
LDL	Low-density lipoproteins
IDDM	Insulin-dependent diabetes mellitus
NIDDM	Non-insulin-dependent diabetes mellitus
PDGF	Platelet-derived growth factor
PEPCK	Phosphoenolpyruvate carboxykinase
PGF$_{2\alpha}$	Prostaglandin F$_{2\alpha}$
PP	Pancreatic polypeptide
VIP	Vasoactive intestinal polypeptide
VLDL	Very low density lipoproteins

techniques. Studies of its biosynthesis led to the important concept of the propeptide. Insulin has important medical implications. In developed countries the incidence of diabetes is 5%, and an equal number are liable to develop the disease. Diabetes mellitus is due to insufficient action of insulin, owing either to its absence or to resistance to its action. Glucagon, acting unopposed, aggravates this condition.

Disease syndromes due to excessive production of several of the gastrointestinal hormones have been described. Signs and symptoms often involve many organ systems, and accurate diagnosis can be difficult unless the physician is aware of these syndromes. The gastrointestinal hormones are of interest also because of their close link to neuropeptides.

Table 51–1. Cell types in the islets of Langerhans.

Cell Type	Relative Abundance	Hormone Produced
A (or α)	~25%	Glucagon
B (or β)	~70%	Insulin
D (or δ)	<5%	Somatostatin
F	Trace	Pancreatic polypeptide

THE PANCREATIC HORMONES ARE INSULIN, GLUCAGON, SOMATOSTATIN, AND PANCREATIC POLYPEPTIDE

Insulin Was Linked to Diabetes in 1921

Langerhans identified the islets in the 1860s but did not understand their function—nor did von Mering and Minkowski, who demonstrated in 1889 that pancreatectomy produced diabetes. The link between the islets and diabetes was suggested by de Mayer in 1909 and by Sharpey-Schaffer in 1917, but it was Banting and Best who proved this association in 1921. These investigators used acid-ethanol to extract from the tissue an islet cell factor that had potent hypoglycemic activity. The factor was named "insulin," and it was quickly learned that bovine and porcine islets contained insulin that was active in humans. Within a year, insulin was in widespread use for the treatment of diabetes and proved to be lifesaving.

Having large quantities of bovine or porcine insulin to study had an equally dramatic effect on biomedical research. Insulin was the first protein proved to have hormonal action, the first protein crystallized (Abel, 1926), the first protein sequenced (Sanger et al, 1955), the first protein synthesized by chemical techniques (Du et al, Zahn, Katsoyanis, ca 1964), the first protein shown to be synthesized as a large precursor molecule (Steiner et al, 1967), and the first protein prepared for commercial use by recombinant DNA technology. In spite of this impressive list of "firsts," less is known about how insulin works at the molecular level than about how most other hormones work at that level.

Insulin Is a Heterodimeric Polypeptide

Insulin is a polypeptide consisting of two chains, A and B, linked by two interchain disulfide bridges that connect A7 to B7 and A20 to B19. A third intra-chain disulfide bridge connects residues 6 and 11 of the A chain. The location of these three disulfide bridges is invariant, and the A and B chains have 21 and 30 amino acids, respectively, in most species. The covalent structure of human insulin (molecular mass 5.734 kDa) is illustrated in Figure 51–1, and a comparison of the amino acid substitutions found in a variety of species is presented in Table 51–2. Substitutions occur at many positions within either chain without affecting bioactivity and are particularly common at positions 8, 9, and 10 of the A chain. Thus, this region is not crucial for bioactivity.

Several positions and regions are highly conserved, however, including (1) the positions of the three disulfide bonds, (2) the hydrophobic residues in the carboxyl terminal region of the B chain, and (3) the amino and carboxyl terminal regions of the A chain. Chemical modification or substitution of specific amino acids in these regions has allowed investigators to formulate a composite active region (Figure 51–2). The carboxyl terminal hydrophobic region of the B chain is also involved in the dimerization of insulin.

A Close Similarity Exists Among Human, Porcine, & Bovine Insulins

Porcine insulin differs by a single amino acid, an alanine for threonine substitution at B30, while bovine insulin has this modification plus the substitutions of alanine for threonine at A8 and valine for isoleucine at A10 (Table 51–2). These modifications result in no appreciable change in biologic activity and very little antigenic difference. Although all pa-

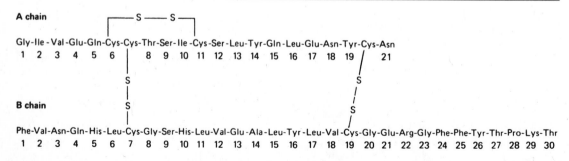

Figure 51–1. Covalent structure of human insulin. (Reproduced, with permission, from Ganong WF: *Review of Medical Physiology,* 15th ed. Appleton & Lange, 1991.)

Table 51–2. Variations in the structure of insulin in mammalian species.[1]

Species	Variations From Human Amino Acid Sequence	
	A-Chain Position 8 9 10	B-Chain Position 30
Human	Thr-Ser-Ile	Thr
Pig, dog, sperm whale	Thr-Ser-Ile	Ala
Rabbit	Thr-Ser-Ile	Ser
Cattle, goat	Ala-Ser-Val	Ala
Sheep	Ala-Gly-Val	Ala
Horse	Thr-Gly-Ile	Ala
Sei whale	Ala-Ser-Thr	Ala

[1]Modified and reproduced, with permission, from Ganong WF: *Review of Medical Physiology,* 15th ed. Appleton & Lange, 1991.

tients given heterologous insulin develop low titers of circulating antibodies against the molecule, few develop clinically significant titers. Porcine and bovine insulins were standard therapy for diabetes mellitus until human insulin was produced by recombinant DNA technology. Despite a wide variation in primary structure, biologic activity is about 25–30 IU/mg dry weight for all insulins.

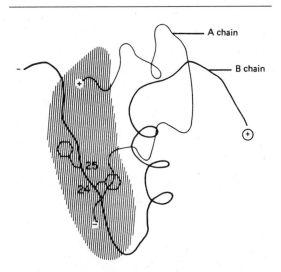

Figure 51–2. Region of the insulin molecule required for biologic activity. Diagrammatic structure of insulin as determined by x-ray crystallography. The shaded area illustrates the portion of insulin that is thought to be most important in conferring biologic activity to the hormone. The Phe residues B24 and B25 are the sites of mutations that affect insulin bioactivity. The amino terminals of the insulin A and B chains are indicated by ⊕, whereas the carboxyl terminals are indicated by ⊖. (Redrawn and reproduced, with permission, from Tager HS: Abnormal products of the human insulin gene. Diabetes 1984;33: 693.)

Insulin forms very interesting complex structures. Zinc is present in high concentration in the B cell and forms complexes with insulin and proinsulin. Insulins from all vertebrate species form isologous dimers through hydrogen bonding between the peptide groups of the B24 and B26 residues of two monomers, and at high concentrations these are organized as hexamers, each with two atoms of zinc. This higher-order structure made studies of the crystalline structure of insulin feasible. Insulin is probably in the monomeric form at physiologic concentrations.

Insulin Is Synthesized as a Preprohormone

Insulin is synthesized as a **preprohormone** (molecular weight approximately 11,500) and is the prototype for peptides that are processed from larger precursor molecules. The hydrophobic 23-amino-acid pre-, or leader, sequence directs the molecule into the cisternae of the endoplasmic reticulum and then is removed. This results in the 9000-molecular-weight proinsulin molecule that provides the conformation necessary for the proper disulfide bridges. As shown in Figure 51–3, the arrangement of proinsulin, starting from the amino terminal, is B chain–connecting (C) peptide–A chain. The proinsulin molecule undergoes a series of site-specific peptide cleavages that result in the formation of equimolar amounts of mature insulin and C-peptide. These enzymatic cleavages are summarized in Figure 51–3.

Other Islet Cell Hormones Are Also Synthesized as Precursor Molecules

The synthesis of other islet cell hormones also requires posttranslational enzymatic processing of higher molecular weight precursor molecules. Diagrammatic structures of pancreatic polypeptide, glucagon, and somatostatin are compared with that of insulin in Figure 51–4. Several combinations of endoproteolytic (trypsin-like) and exoproteolytic (carboxypeptidase B-like) cleavages are involved, since the hormone sequence may occur at the carboxyl terminal of the precursor (somatostatin), at the amino terminal (pancreatic polypeptide), at both ends (insulin), or in the middle (glucagon).

Insulin Synthesis and Granule Formation Occur in Subcellular Organelles

Proinsulin is synthesized by ribosomes on the rough endoplasmic reticulum, and the enzymatic removal of the leader peptide (pre) segment disulfide bond formation, and folding (Figure 51–3), occur in the cisternae of this organelle. The proinsulin molecule is transported to the Golgi apparatus wherein proteolysis and packaging into the secretory granules begin. Granules continue to mature as they traverse the cytoplasm toward the plasma membrane. Proin-

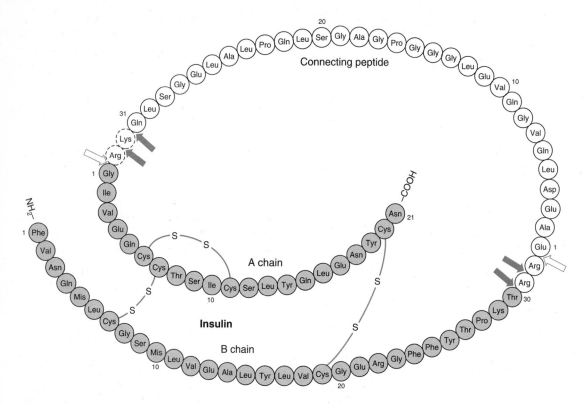

Figure 51–3. Structure of human proinsulin. Insulin and C-peptide molecules are connected at two sites by dipeptide links. An initial cleavage by a trypsin-like enzyme (open arrows) followed by several cleavages by a carboxypeptidase-like enzyme (solid arrows) results in the production of the heterodimeric (AB) insulin molecule and the C-peptide.

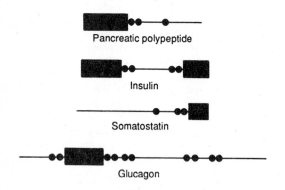

Figure 51–4. Diagrammatic structures of the precursors for the four major endocrine cell products of the pancreatic islet. Portions of the precursors that correspond to the named hormones are indicated by heavy bars, whereas portions that correspond to peptide extensions are indicated by lines; dibasic amino acid residues (arginine or lysine) corresponding to precursor conversion sites are shown as filled circles. Note that proinsulin has been drawn in an extended form that does not show disulfide bonds; the structure of proinsulin illustrated has the sequence B chain–C-peptide–A chain. (Redrawn and reproduced, with permission, from Tager HS: Abnormal products of the human insulin gene. Diabetes 1984;33: 693.)

sulin and insulin both combine with zinc to form hexamers, but since about 95% of the proinsulin is converted to insulin, it is the crystals of the latter that confer morphologic distinctness to the granules. Equimolar amounts of C-peptide are present within these granules, but these molecules do not form a crystalline structure. Upon appropriate stimulation (see below), the mature granules fuse with the plasma membrane and discharge their contents into the extracellular fluid by **emeiocytosis.**

The Properties of Proinsulin and C-Peptide Differ From Those of Insulin

Proinsulins vary in length from 78 to 86 amino acids, with the variation occurring in the length of the C-peptide region. Proinsulin has the same solubility and isoelectric point as insulin; it also forms hexamers with zinc crystals, and it reacts strongly with insulin antisera. Proinsulin has less than 5% of the bioactivity of insulin, indicating that most of the active site of the latter is occluded in the precursor molecule. Some proinsulin is released with insulin and in certain conditions (islet cell tumors) in larger than usual amounts. Since the plasma half-life of proinsulin is significantly longer than that of insulin and since proinsulin is strongly cross-reactive with insulin antisera, a radioimmunoassay for "insulin" may occasionally overestimate the bioactivity of "insulin" in plasma.

The C-peptide has no known biologic activity. It is a distinct molecule from an antigenic standpoint. Thus, C-peptide immunoassays can distinguish insulin secreted endogenously from insulin administered exogenously and can quantitate the former when anti-insulin antibodies preclude the direct measurement of insulin. The C-peptides of different species have a high rate of amino acid substitution, an observation which underscores the statement that this fragment probably has no biologic activity.

Insulin-Related Peptides Have Precursors

The structural arrangement of the precursor molecule is not unique to insulin; very closely related peptide hormones (relaxin and the insulin-like growth factors) show the same arrangement (Figure 51–5). All these hormones have highly homologous B- and A-chain regions at the amino and carboxyl terminals of a precursor molecule, and these are joined by a connecting segment. In the relaxin and insulin precursor peptides, this connecting segment is bound on both ends by two basic amino acids. After the B and A chains are joined by disulfide bonds, this piece is removed by endoproteolytic action, and these molecules are converted to two-chain peptide hormones.

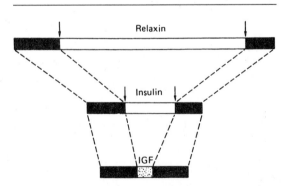

Figure 51–5. Diagrammatic structures of precursors for insulin-related peptides. Homologous regions of relaxin, insulin, and insulin-like growth factor (IGF) are shown as solid bars. Amino acid sequences connecting B chain and A chain sequences in the precursors for relaxin and insulin are shown as open bars; these sequences are removed during processing of the precursors to their corresponding two-chain products (vertical arrows). The amino acid sequence of IGF that corresponds to these connecting peptides, but is not removed by proteolytic processing events, is shown as a stippled bar; IGF is a single-chain peptide hormone. (Redrawn and reproduced, with permission, from Tager HS: Abnormal products of the human insulin gene. Diabetes 1984;33:693.)

The insulin-like growth factors, though highly homologous to insulin and relaxin in primary structure, lack these dibasic cleavage sites and thus remain single-chain peptide hormones.

The Human Insulin Gene Has Been Isolated

The human insulin gene (Figure 51–6) is located on the short arm of chromosome 11. Most mammals express a single insulin gene that is organized like the human gene, but rats and mice have two nonallelic genes. Each codes for a unique proinsulin that is processed into two distinct, active insulin molecules. The synthesis of human insulin in bacterial expression systems, using recombinant DNA technology, affords an excellent source of this hormone for diabetic patients.

Insulin Secretion Is Precisely Regulated

The human pancreas secretes 40–50 units of insulin daily, which represents about 15–20% of the hormone stored in the gland. Insulin secretion is an energy-requiring process that involves the microtubule-microfilament system in the B cells of the islets. A number of mediators have been implicated in insulin release.

A. Glucose: An increase in plasma glucose concentration is the most important physiologic regulator of insulin secretion. The threshold concentration for secretion is the fasting plasma glucose level (80–100 mg/dL), and the maximal response is obtained at glucose levels between 300 and 500 mg/dL. Two different mechanisms have been proposed to explain how glucose regulates insulin secretion. One hypothesis suggests that glucose combines with a receptor, possibly located on the B cell membrane, that activates the release mechanism. The second hypothesis suggests that intracellular metabolites or the rate of metabolite flux through a pathway such as the pentose phosphate shunt, the citric acid cycle, or the

Figure 51–6. Diagrammatic structure of the human insulin gene. Areas with diagonal stripes correspond to untranslated regions of the corresponding mRNA, open regions correspond to intervening sequences, and stippled regions correspond to coding sequences. L, B, C, and A identify coding sequences for the leader (or signal) peptide, the insulin B chain, the C-peptide, and the insulin A chain, respectively. Note that the coding sequence for the C-peptide is split by an intervening sequence. The diagrammatic structure is drawn to scale. (Redrawn and reproduced, with permission, from Tager HS: Abnormal products of the human insulin gene. Diabetes 1984;33: 693.)

glycolytic pathway, is involved. There is experimental evidence to support both positions.

B. Hormonal Factors: Numerous hormones affect insulin release. α-Adrenergic agonists, principally epinephrine, inhibit insulin release even when this process has been stimulated by glucose. β-Adrenergic agonists stimulate insulin release, probably by increasing intracellular cAMP (see below).

Chronic exposure to excessive levels of growth hormone, cortisol, placental lactogen, estrogens, and progestins also increases insulin secretion. It is therefore not surprising that insulin secretion increases markedly during the later stages of pregnancy.

C. Pharmacologic Agents: Many drugs stimulate insulin secretion, but the **sulfonylurea compounds** are used most frequently for therapy in humans. Drugs such as tolbutamide stimulate insulin release by a mechanism different from that employed by glucose and have achieved widespread use in the treatment of type II (non-insulin-dependent) diabetes mellitus. A receptor that binds this class of drugs has recently been cloned from the pancreatic B cells.

$$H_3C - \overset{}{\underset{}{\bigcirc}} - SO_2 - NH - \overset{O}{\underset{\|}{C}} - NH - (CH_2)_3 - CH_3$$

Tolbutamide

Insulin Is Rapidly Metabolized

Unlike the insulin-like growth factors, insulin has no plasma carrier protein; thus, its plasma half-life is less than 3–5 minutes under normal conditions. The major organs involved in insulin metabolism are the liver, kidneys, and placenta; about 50% of insulin is removed in a single pass through the liver.

Mechanisms involving two enzyme systems are responsible for the metabolism of insulin. The first involves an insulin-specific **protease** found in many tissues but in highest concentration in those listed above. This protease has been purified from skeletal muscle and is known to be sulfhydryl-dependent and active at physiologic pH. The second mechanism involves hepatic **glutathione-insulin transhydrogenase.** This enzyme reduces the disulfide bonds, and then the individual A and B chains are rapidly degraded. It is not clear which of these mechanisms is most active under physiologic conditions, nor is it clear whether either process is regulated.

Insulin Deficiency

The central role of insulin in carbohydrate, lipid, and protein metabolism can be best appreciated by examining the consequences of insulin deficiency in humans. The cardinal manifestation of diabetes mellitus is **hyperglycemia,** which results from (1) decreased entry of glucose into cells, (2) decreased utilization of glucose by various tissues, and (3) increased production of glucose (gluconeogenesis) by the liver (Figure 51–7). Each of these is discussed in more detail below.

Polyuria, polydipsia, and weight loss in spite of adequate caloric intake are the major symptoms of insulin deficiency. How is this explained? The plasma glucose level rarely exceeds 120 mg/dL in normal humans, but much higher levels are routinely found in patients with deficient insulin action. After a certain plasma glucose level is attained (generally > 80 mg/dL in humans), the maximum level of renal tubular reabsorption of glucose is exceeded, and sugar is excreted in the urine (glycosuria). The urine volume is increased owing to osmotic diuresis and coincident obligatory water loss (polyuria) and this in turn leads to dehydration (hyperosmolarity), increased thirst, and excessive drinking (polydipsia). Glycosuria causes a substantial loss of calories (4.1 kcal for every gram of glucose excreted); this loss, when coupled with the loss of muscle and adipose tissue, results in severe weight loss in spite of increased appetite (polyphagia) and normal or increased caloric intake.

Protein synthesis decreases in the absence of insulin, partly because the transport of amino acids into muscle is diminished (the amino acids serve as gluconeogenic substrates). Thus, insulin-deficient persons are in negative nitrogen balance. The antilipolytic action of insulin is lost, as is its lipogenic effect; hence, plasma fatty acid levels rise. When the capacity of the liver to oxidize fatty acids to CO_2 is exceeded, **β-hydroxybutyric acid and acetoacetic acid accumulate (ketosis).** The organism initially compensates for the accumulation of these organic acids by increasing respiratory losses of CO_2, but if unchecked by the administration of insulin, severe **metabolic acidosis** supervenes and the patient dies in **diabetic coma.** The pathophysiology of insulin deficiency is summarized in Figure 51–7.

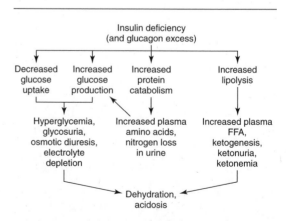

Figure 51–7. Pathophysiology of insulin deficiency.

A. Effects on Membrane Transport: The intracellular free glucose concentration is very low compared with the extracellular concentration. The rate of glucose transport across the plasma membrane of muscle and adipose cells determines the rate of phosphorylation of glucose and its further metabolism when glucose and insulin levels are normal. When glucose or insulin is elevated, such as after a meal, phosphorylation becomes rate-limiting. D-Glucose and other sugars with a similar configuration at the C_1–C_3 positions (galactose, D-xylose, and L-arabinose) enter cells by **carrier-mediated facilitated diffusion,** a process enhanced in many cells by insulin (Figure 51–8). This involves a V_{max} effect (increased number of transporters) rather than a K_m effect (increased affinity of binding). Data suggest that in adipose cells this is accomplished by recruiting **glucose transporters** from an inactive pool in the Golgi fraction and then moving them to an inactive site in the plasma membrane. This transporter translocation is temperature- and energy-dependent and is protein synthesis-independent (Figure 51–9).

The hepatic cell represents a notable exception to this scheme. Insulin does not promote the facilitated diffusion of glucose into hepatocytes, but it indirectly enhances net inward flux by converting intracellular glucose to glucose 6-phosphate through the action of glucokinase, an enzyme induced by insulin. This rapid phosphorylation keeps the free glucose concentration very low in the hepatocyte, thus favoring entry by simple diffusion down a concentration gradient.

Insulin also promotes the entry of amino acids into cells, particularly in muscle, and enhances the movement of K^+, Ca^{2+}, nucleosides, and inorganic phosphate. These effects are independent of the action of insulin on glucose entry.

B. Effects on Glucose Utilization: Insulin influences the intracellular utilization of glucose in a number of ways, as illustrated below.

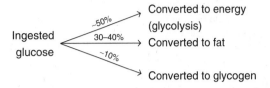

In a normal person, about half the glucose ingested is converted to energy through the glycolytic pathway and about half is stored as fat or glycogen. Glycolysis decreases in the absence of insulin, and the anabolic processes of glycogenesis and lipogenesis are impeded. Indeed, only 5% of an ingested glucose load is converted to fat in an insulin-deficient diabetic.

Insulin increases hepatic glycolysis by increasing the activity and amount of several key enzymes, including glucokinase, phosphofructokinase, and pyruvate kinase. Enhanced glycolysis increases glucose utilization and thus indirectly decreases glucose release into plasma. Insulin also decreases the activity of glucose-6-phosphatase, an enzyme found in liver

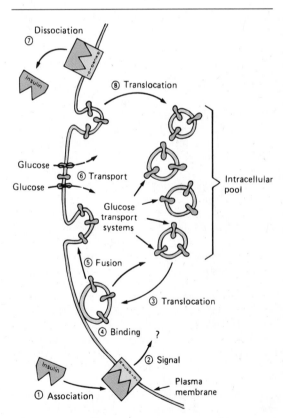

Figure 51–9. Translocation of glucose transporters by insulin. (Reproduced, with permission, from Karnieli E et al: Insulin-stimulated translocation of glucose transport systems in the isolated rat adipose cell. J Biol Chem 1981;256:4772. Courtesy of S Cushman.)

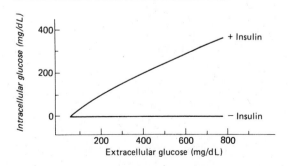

Figure 51–8. Entry of glucose into muscle cells.

but not in muscle. Since glucose 6-phosphate cannot exit from the plasma membrane, this action of insulin results in the retention of glucose within the liver cell.

In skeletal muscle, insulin promotes glucose entry through the transporter and also increases hexokinase II, which phosphorylates glucose and initiates glucose metabolism. Insulin stimulates lipogenesis in adipose tissue (1) by providing the acetyl-CoA and NADPH required for fatty acid synthesis; (2) by maintaining a normal level of the enzyme acetyl-CoA carboxylase, which catalyzes the conversion of acetyl-CoA to malonyl-CoA; and (3) by providing the glycerol involved in triacylglycerol synthesis. In insulin deficiency, all of these are decreased; thus, lipogenesis decreases. Another reason for the decreased lipogenesis in insulin deficiency is that fatty acids, released in large amounts by several hormones when unopposed by insulin, feedback-inhibit their own synthesis by inhibiting acetyl-CoA carboxylase. The net effect of insulin on fat is therefore anabolic.

The final action of insulin on glucose utilization involves another anabolic process. In liver and muscle, insulin stimulates the conversion of glucose to glucose 6-phosphate (by the actions of glucokinase and hexokinase II, respectively), which then undergoes isomerization to glucose 1-phosphate and is incorporated into glycogen by the enzyme glycogen synthase, the activity of which is stimulated by insulin. This action is indirect and dual in nature. Insulin decreases intracellular cAMP levels by activating a phosphodiesterase. Since cAMP-dependent phosphorylation inactivates glycogen synthase, low levels of this nucleotide allow the enzyme to stay in the active form. Insulin also activates a phosphatase that dephosphorylates glycogen synthase, thereby resulting in the activation of this enzyme. Finally, insulin inhibits phosphorylase by a mechanism involving cAMP and phosphatase as described above, and this decreases glucose liberation from glycogen. The net effect of insulin on glycogen metabolism is also anabolic.

C. Effects on Glucose Production (Gluconeogenesis): The actions of insulin on glucose transport, glycolysis, and glycogenesis occur within seconds or minutes, since they primarily involve the activation or inactivation of enzymes by phosphorylation or dephosphorylation. A more long-term effect on plasma glucose involves the inhibition of gluconeogenesis by insulin. The formation of glucose from noncarbohydrate precursors involves a series of enzymatic steps, many of which are stimulated by glucagon (acting through cAMP), by glucocorticoid hormones, and to a lesser extent by α- and β-adrenergic agents, angiotensin II, and vasopressin. Insulin inhibits these same steps. The key gluconeogenic enzyme in the liver is phosphoenolpyruvate carboxykinase (PEPCK), which converts oxaloacetate to phosphoenolpyruvate. Insulin decreases the amount of this enzyme by selectively inhibiting transcription of the gene that codes for the mRNA for PEPCK (see below).

D. Effects on Glucose Metabolism: The net action of all of the above effects of insulin is to decrease the blood glucose level. In this action, insulin stands alone against an array of hormones that attempt to counteract this effect. This no doubt represents one of the organism's most important defense mechanisms, since prolonged hypoglycemia poses a potentially lethal threat to the brain and must be avoided.

E. Effects on Lipid Metabolism: The lipogenic actions of insulin were discussed in the context of glucose utilization. Insulin also is a potent inhibitor of lipolysis in liver and adipose tissue and thus has an indirect anabolic effect. This is partly due to the ability of insulin to decrease tissue cAMP levels (which are increased in these tissues by the lipolytic hormones glucagon and epinephrine) but also to the fact that insulin inhibits hormone-sensitive lipase activity. This inhibition is presumably due to the activation of a phosphatase that dephosphorylates and thereby inactivates the lipase or cAMP-dependent protein kinase. Insulin therefore decreases circulating free fatty acids. This contributes to the action of insulin on carbohydrate metabolism, since fatty acids inhibit glycolysis at several steps and stimulate gluconeogenesis. Thus, metabolic regulation cannot be discussed in the context of a single hormone or metabolite. Regulation is a complex process in which the flux through a given pathway is the result of the interplay of a number of hormones and metabolites.

In patients with insulin deficiency, lipase activity increases, resulting in enhanced lipolysis and increased concentration of free fatty acids in plasma and liver. Glucagon levels also increase in these patients, and this enhances the release of free fatty acids. (Glucagon opposes most actions of insulin, and the metabolic state in the diabetic is a reflection of the relative levels of glucagon and insulin.) A portion of the free fatty acids is metabolized to acetyl-CoA (the reverse of lipogenesis) and then to CO_2 and H_2O via the citric acid cycle. In patients with insulin deficiency, the capacity of this process is rapidly exceeded and the acetyl-CoA is converted to acetoacetyl-CoA and then to acetoacetic and β-hydroxybutyric acids. Insulin reverses this pathway.

Insulin apparently affects the formation or clearance of VLDL and LDL, since levels of these particles, and consequently the level of cholesterol, are often elevated in poorly controlled diabetics. Accelerated atherosclerosis, a serious problem in many diabetics, is attributed to this metabolic defect.

The actions of insulin can be inferred by inspecting Figure 51–10, which depicts the flux through several critical pathways in the absence of the hormone.

F. Effects on Protein Metabolism: Insulin

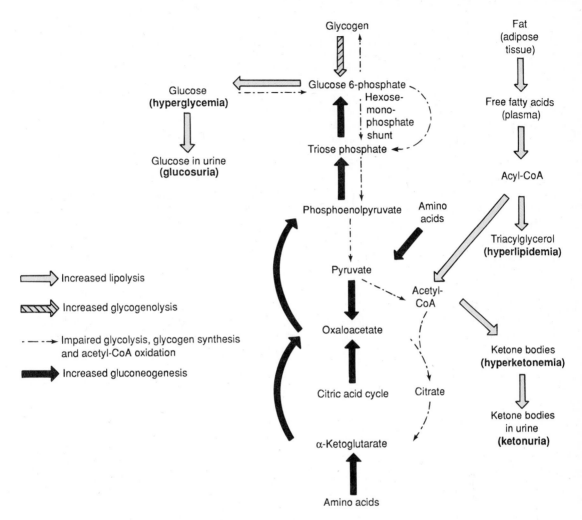

Figure 51–10. In severe insulin deficiency, there is accelerated lipolysis. This results in elevated plasma triacylglycerol levels (hyperlipidemia). Little of the acetyl-CoA can be metabolized by the citric acid cycle, so the remainder is converted to ketone bodies (ketonemia), and some is excreted (ketonuria). Since glycolysis is inhibited, the glucose 6-phosphate formed from accelerated glycogenolysis is converted to glucose. This, combined with accelerated gluconeogenesis, results in hyperglycemia (from increased availability of amino acids and increased amount of PEPCK). Insulin essentially reverses all these processes.

generally has an anabolic effect on protein metabolism in that it stimulates protein synthesis and retards protein degradation. Insulin stimulates the uptake of neutral amino acids into muscle, an effect that is not linked to glucose uptake or to subsequent incorporation of the amino acids into protein. The effects of insulin on general protein synthesis in skeletal and cardiac muscle and in liver are thought to be exerted at the level of mRNA translation.

In recent years, insulin has been shown to influence the synthesis of specific proteins by effecting changes in the corresponding mRNAs. This action of insulin, which may ultimately explain many of the effects the hormone has on the activity or amount of specific proteins, is discussed in more detail below.

G. Effects on Cell Replication: Insulin stimulates the proliferation of a number of cells in culture, and it may also be involved in the regulation of growth in vivo. Cultured fibroblasts are the most frequently used cells in studies of growth control. In such cells, insulin potentiates the ability of fibroblast growth factor (FGF), platelet-derived growth factor (PDGF), epidermal growth factor (EGF), tumor-promoting phorbol esters, prostaglandin $F_{2\alpha}$ ($PGF_{2\alpha}$), vasopressin, and cAMP analogs to stimulate cell cycle progression of cells arrested in the G_1 phase of the cycle by serum deprivation.

An exciting new area of research involves the investigation of tyrosine kinase activity. The insulin receptor, along with receptors for many other growth-promoting peptides including those of PDGF and EGF, has tyrosine kinase activity. Interestingly, many oncogene products, some of which are suspected to be involved in stimulating malignant cell replication, are also tyrosine kinases. Mammalian cells contain analogs of these oncogenes (**proto-oncogenes**), which may be involved in the replication of normal cells. Support for the theory that they are involved comes from recent observations that the expression of at least two proto-oncogene products, c-*fos* and c-*myc*, increases following addition of serum PDGF or insulin to growth-arrested cells.

The Mechanism of Action of Insulin Is Being Elucidated

Insulin action begins when the hormone binds to a specific glycoprotein receptor on the surface of the target cell. The diverse actions of the hormone (Figure 51–11) can occur within seconds or minutes (transport, protein phosphorylation, enzyme activation and inhibition, RNA synthesis) or after a few hours (protein and DNA synthesis and cell growth).

The insulin receptor has been studied in great detail using biochemical and recombinant DNA tech-

niques. It is a heterodimer consisting of two subunits, designated α and β, in the configuration $\alpha_2\beta_2$, linked by disulfide bonds (Figure 51–12). Both subunits are extensively glycosylated, and removal of sialic acid and galactose decreases insulin binding and insulin action. Each of these glycoprotein subunits has a unique structure and function. The α subunit (135 kDa) is entirely extracellular, and it binds insulin, probably via a cysteine-rich domain. The β subunit (95 kDa) is a transmembrane protein that performs the second major function of a receptor (Chapter 44), ie, signal transduction. **The cytoplasmic portion of the β subunit has tyrosine kinase activity and an autophosphorylation site.** Both of these are thought to be involved in signal transduction and insulin action (see below). The striking similarity between three receptors with very different functions is illustrated in Figure 51–12. Indeed, several regions of the β subunit have sequence homology with the EGF receptor.

The insulin receptor is constantly being synthesized and degraded, and its half-life is 7–12 hours. The receptor is synthesized as a single-chain peptide in the rough endoplasmic reticulum and is rapidly glycosylated in the Golgi region. The precursor of

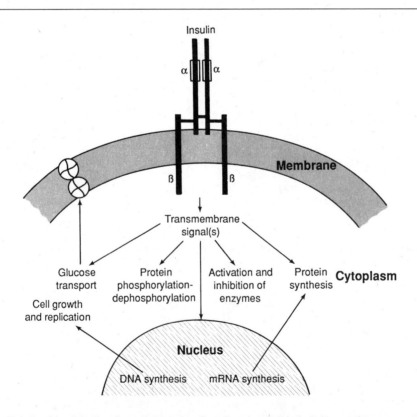

Figure 51–11. Relationship of the insulin receptor to insulin action. Insulin binds to its membrane receptor, and this interaction generates one or more transmembrane signals. This signal (or signals) modulates a wide variety of intracellular events.

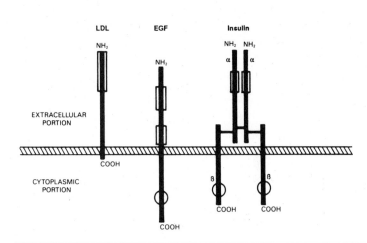

Figure 51–12. Schematic representation of the structure of the LDL, EGF, and insulin receptors. The amino terminal of each is in the extracellular portion of the molecule. The boxes represent cysteine-rich regions, which are thought to be involved in ligand binding. Each receptor has a short domain (about 25 amino acids) that traverses the plasma membrane (the hatched line), and there is an intracellular domain of variable length. The EGF and insulin receptors have tyrosine kinase activity associated with the cytoplasmic domain (circles) and also have autophosphorylation sites in this region. The insulin receptor is a heterotetramer connected by disulfide bridges (vertical bars).

the human insulin receptor has 1382 amino acids, has a molecular weight of 190,000, and is cleaved to form the mature α and β subunits. The human insulin receptor gene is located on chromosome 19.

Insulin receptors are found on most mammalian cells, in concentrations of up to 20,000 per cell, and often on cells not typically thought of as being insulin targets. Insulin has a well-known set of effects on metabolic processes but also is involved in growth and replication of cells (see above) as well as in fetal organogenesis and differentiation and in tissue repair and regeneration. The structure of the insulin receptor and the ability of different insulins to bind to receptors and elicit biologic responses are virtually identical in all cells and all species. Thus, porcine insulin is always 10–20 times more effective than porcine proinsulin, which in turn is 10–20 times more effective than guinea pig insulin, even in the guinea pig. The insulin receptor has apparently been highly conserved, more so than even insulin itself.

When insulin binds to the receptor, several events occur. (1) There is a conformational change of the receptor; (2) the receptors cross-link and form microaggregates; (3) the receptor is internalized; and (4) one or more signals are generated. The significance of the conformational change is unknown, and internalization probably represents a means of controlling receptor concentration and turnover. In conditions in which plasma insulin levels are high, eg, obesity or acromegaly, the number of insulin receptors is decreased and target tissues become less sensitive to insulin. This down-regulation results from the loss of receptors by internalization, the process whereby insulin-receptor complexes enter the cell through endocytosis in clathrin-coated vesicles (Chapter 43). Down-regulation explains part of the insulin resistance in obesity and type II diabetes mellitus.

Insulin Transmits Signals by One or More Kinase Cascades

The insulin and IGF-I receptors have intrinsic protein tyrosine kinase activities located in their cytoplasmic domains. These activities are put in motion when the receptor binds ligand. The receptors are then autophosphorylated, and this initiates a complex set of events (summarized in Figure 51–13). The phosphorylated insulin receptor next phosphorylates insulin receptor substrate 1 (IRS-1; other similar proteins may also be so affected) on tyrosine residues.

Phosphorylated IRS-1 binds to the SH2 domains of a variety of proteins that are directly involved in mediating different effects of insulin. One of these proteins, PI 3-kinase, may link insulin receptor activation to insulin action through activation of a number of molecules, including the p70 S6 kinase. Another SH2 domain-containing protein is GRB2, which links tyrosine phosphorylation to several proteins, the result of which is activation of a cascade of threonine or serine kinases. A pathway showing how the insulin-receptor interaction activates the mitogen-activated protein (MAP) kinase pathway is illustrated in Figure 51–13. The exact role of many of these docking proteins, kinases, and phosphatases remains to be established. It is particularly important to link these various pathways to the well-established physiologic and biochemical actions of this hormone. Insulin can be distinguished from other hormones that activate tyrosine kinases, such as EGF, as it does not activate phospholipase C and thus exert its effects through Ca^{2+} and IP_3 or diacylglycerol.

Protein Phosphorylation-Dephosphorylation Is Involved in Some Actions of Insulin

Many of the metabolic effects of insulin, particu-

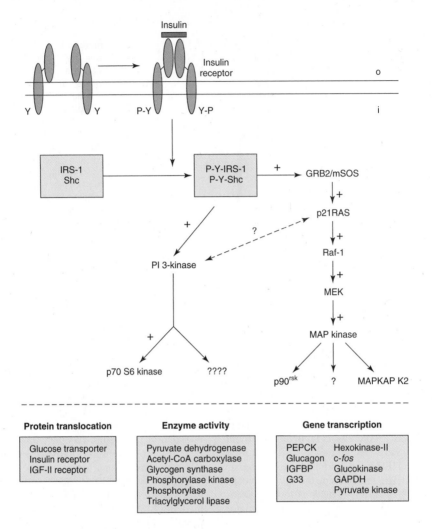

Figure 51–13. Insulin signaling pathways. Binding of insulin to its specific cell membrane receptor results in a cascade of intracellular events. The stimulation of the intrinsic tyrosine kinase activity of the insulin receptor marks the initial event, resulting in increased tyrosine phosphorylation (Y → Y-P) of both the receptor and specific signaling molecules. This increase in phosphotyrosine stimulates the activity of many intracellular molecules such as GTPases, protein kinases, and lipid kinases, all of which have a role to play in certain metabolic actions of insulin. The two best-described pathways are shown. First, activation of the lipid kinase, PI 3-kinase, generates novel inositol lipids that may act as "second messenger" molecules, which in turn activate a variety of poorly described signaling pathways (eg, p70 S6 kinase). Second, activation of the small GTPase, p21RAS, stimulates a protein kinase cascade that activates the p42/p44 MAP kinase isoforms, protein kinases that are important in the regulation of proliferation and differentiation of several cell types. (IRS-1, insulin receptor substrate-1; GRB2, growth factor receptor-binding protein 2; mSOS, mammalian son of sevenless; MEK, MAP kinase and ERK kinase; MAP kinase, mitogen-activated protein kinase; p90rsk, p90 ribosomal protein S6 kinase; MAPKAPK2, MAP kinase-activated protein kinase-2; PI 3-kinase, phosphatidylinositol 3-kinase; p70 S6 kinase, p70 ribosomal protein S6 kinase.)

larly those which occur rapidly, are mediated by influencing protein phosphorylation and dephosphorylation reactions that in turn alter the enzymatic activity of the protein. A list of enzymes affected in this way is presented in Table 51–3. In some instances, insulin decreases intracellular cAMP levels (by activating a cAMP-phosphodiesterase), thereby decreasing the activity state of cAMP-dependent protein ki-

nase; examples of this action include glycogen synthase and phosphorylase. In other instances, this action is independent of cAMP and is exerted by activating other protein kinases (as is the case with the insulin receptor, tyrosine kinase); by inhibiting other protein kinases (Table 44–5); or, more commonly, by stimulating the activity of phosphoprotein phosphatases. Dephosphorylation increases the activity of

Table 51–3. Enzymes whose degree of phosphorylation and activity are altered by insulin.[1]

Enzyme	Change in Activity	Possible Mechanism
cAMP metabolism		
Phosphodiesterase (low K_m)	Increase	Phosphorylation
Protein kinase (cAMP-dependent)	Decrease	Association of R and C subunits
Glycogen metabolism		
Glycogen synthase	Increase	Dephosphorylation
Phosphorylase kinase	Decrease	Dephosphorylation
Phosphorylase	Decrease	Dephosphorylation
Glycolysis and gluconeogenesis		
Pyruvate dehydrogenase	Increase	Dephosphorylation
Pyruvate kinase	Increase	Dephosphorylation
6-Phosphofructo-2-kinase	Increase	Dephosphorylation
Fructose-2,6-bisphosphatase	Decrease	Dephosphorylation
Lipid metabolism		
Acetyl-CoA carboxylase	Increase	Dephosphorylation
HMG-CoA reductase	Increase	Dephosphorylation
Triacylglycerol lipase	Decrease	Dephosphorylation
Signaling molecules		
p42/44MAP kinase	Increase	Dephosphorylation
p90RSK	Increase	Dephosphorylation
GSK3	Decrease	Dephosphorylation
p70 S6 kinase	Increase	Dephosphorylation
Phosphoprotein phosphatase 1G	Increase	Dephosphorylation

[1]Modified and reproduced, with permission, from Denton RM et al: A partial view of the mechanism of insulin action. Diabetologia 1981;21:347.

a number of key enzymes (Table 51–3). These covalent modifications allow for almost immediate changes in the activity of enzymes.

Insulin Affects mRNA Translation

Insulin is known to affect the activity or amount of at least 50 proteins in a variety of tissues, and many of these effects involve covalent modification. A role for insulin in the translation of mRNA has been proposed, largely based on studies of ribosomal protein S6, a component of the 40S ribosomal subunit. Such a mechanism could account for the general effect insulin has on protein synthesis in liver, skeletal muscle, and cardiac muscle.

Insulin Affects Gene Expression

The actions of insulin discussed heretofore all occur at the plasma membrane level or in the cytoplasm. In addition, insulin affects specific nuclear processes, presumably through its intracellular mediator. The enzyme **phosphoenolpyruvate carboxykinase (PEPCK)** catalyzes a rate-limiting step in gluconeogenesis. The synthesis of PEPCK is decreased by insulin; hence, gluconeogenesis decreases. Recent studies show that the rate of transcription of the PEPCK gene is selectively decreased within minutes after the addition of insulin to cultured hepatoma cells (Figure 51–14). The decrease in transcription accounts for the decreased amount of the primary

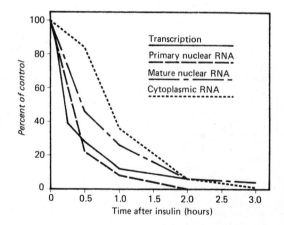

Figure 51–14. Effect of insulin on specific gene transcription. The addition of insulin to H4IIE hepatoma cells results in a rapid decrease in the rate of transcription of the PEPCK gene. This is followed by decreases in the amounts of primary transcript in the nucleus and mature mRNA^{PEPCK}. The rate of synthesis of PEPCK protein declines after the amount of cytoplasmic mRNA^{PEPCK} decreases. (Reproduced, with permission, from Sasaki K et al: Multihormonal regulation of phosphoenolpyruvate carboxykinase gene transcription. J Biol Chem 1984;259: 15242.)

594 / CHAPTER 51

transcript and of mature mRNA^{PEPCK}, which in turn is directly related to the decreased rate of PEPCK synthesis. This effect occurs at physiologic levels of insulin (10^{-12} to 10^{-9} mol/L), is mediated through the insulin receptor, and appears to be due to a decreased rate of mRNA^{PEPCK} transcript initiation.

Although studies of PEPCK regulation provided the first example of an effect of insulin on gene transcription, this case is no longer unique. Indeed, it appears that regulation of mRNA synthesis is a major action of insulin. More than 100 specific mRNAs are affected by insulin, and a number of mRNAs in liver, adipose tissue, skeletal muscle, and cardiac muscle, as yet unidentified, are also affected by the hormone. In several instances, the hormone is known to affect gene transcription. Some examples are shown in Table 51–4.

This effect of insulin involves enzymes retained in the cells, secreted enzymes and proteins, proteins involved in the reproductive process, and structural proteins (Table 51–4). A number of organs or tissues are involved, and the effect occurs in many species. The regulation of specific mRNA transcription by insulin is now well established, and as a means of modulating enzyme activity, it equals phosphorylation-dephosphorylation in importance. The effect of insulin on gene transcription may also explain its effect on embryogenesis, differentiation, and growth and replication of cells.

Table 51–4. Messenger RNAs regulated by insulin.[1]

Intracellular enzymes
  Tyrosine aminotransferase
  Phosphoenolpyruvate carboxykinase
  Fatty acid synthase
  Pyruvate kinase
  Glycerol-3-phosphate dehydrogenase
  Glyceraldehyde-1-dehydrogenase
  Glucokinase
Secreted proteins and enzymes
  Albumin
  Adipsin
  Amylase
  α_{2u} Globulin
  Growth hormone
Proteins involved in reproduction
  Ovalbumin
  Casein
Structural proteins
  δ-Crystallin
Other proteins
  Liver (p33, etc)
  Adipose tissue
  Cardiac muscle
  Skeletal muscle

[1]This list is selective. Insulin has been shown to regulate at least 60 different mRNAs.

Pathophysiology Involving Insulin Is Mostly Expressed as Diabetes Mellitus

Insulin deficiency or resistance to the action of insulin results in **diabetes mellitus.** About 90% of persons with diabetes have **non-insulin-dependent (type II) diabetes mellitus (NIDDM).** Such patients are usually obese, have elevated plasma insulin levels, and have down-regulated insulin receptors. The other 10% have **insulin-dependent (type I) diabetes mellitus (IDDM).**

Certain rare conditions illustrate essential features about insulin action. A few individuals produce antibodies directed against their insulin receptors. These antibodies prevent insulin from binding to the receptor, so that such persons develop a syndrome of severe insulin resistance. Tumors of B cell origin cause hyperinsulinism and a syndrome characterized by severe hypoglycemia. The role of insulin (or perhaps of IGF-I or IGF-II) in organogenesis and development is illustrated by the rare cases of **leprechaunism.** This syndrome is characterized by low birth weight, decreased muscle mass, decreased subcutaneous fat, elfin facies, insulin resistance with markedly elevated plasma levels of biologically active insulin, and early death. Several individuals with leprechaunism have been shown to lack insulin receptors or to have defective receptors.

IGF-I AND IGF-II ARE RELATED TO INSULIN IN STRUCTURE AND FUNCTION

It is difficult to separate the effects of insulin on cell growth and replication from similar actions exerted by IGF-I and IGF-II. Indeed, insulin and the IGFs may interact in this process. The structural similarity of these proteins was alluded to above and in Figure 51–5. A more detailed comparison is presented in Table 51–5. IGF-I and IGF-II are single-chain peptides of 70 and 67 amino acids, respectively. There is 62% homology between IGF-I and IGF-II, and these two hormones are identical with insulin in 50% of their residues. These molecules have unique antigenic sites and are regulated in different ways (Table 51–5). Insulin is the more potent metabolic hormone, whereas the IGFs are more potent in stimulating growth. Each hormone has a unique receptor. The IGF-I receptor, like the insulin receptor, is a heterodimer of $\alpha_2\beta_2$ structure and is a tyrosine kinase. IGF-I and insulin appear to use the same—or a very similar—signal transduction cascade. The IGF-II receptor, in contrast, is a single-chain polypeptide with a molecular weight of 260,000 and is not a tyrosine kinase. It is closely related to, if not identical with, the mannose 6-phosphate receptor.

Table 51–5. Comparison of insulin and the insulin-like growth factors. (Courtesy of CR Kahn.)

	Insulin	IGF-I	IGF-II
Other names	. . .	Somatomedin C	Multiplication-stimulating activity (MSA)
Number of amino acids	51	70	67
Source	Pancreatic B cells	Liver and other tissues	Diverse tissues
Level regulated by	Glucose	Growth hormone, nutritional status	Unknown
Plasma levels	0.3–2 ng/mL	ng/mL range	ng/mL range
Plasma binding protein	No	Yes	Yes
Major physiologic role	Control of metabolism	Skeletal and cartilage growth	Unknown; perhaps a role in embryonic development

GLUCAGON IS AN INSULIN ANTAGONIST

The early commercial preparations of insulin increased the plasma glucose level before lowering it, owing to the presence of a contaminating peptide, glucagon, which was the second pancreatic islet cell hormone discovered.

Glucagon Is Also Synthesized as a Precursor Molecule

Glucagon, synthesized mainly in the A cells of the pancreatic islets, is a single-chain polypeptide (molecular 3.485 kDa) consisting of 29 amino acids (Figure 51–15). Glucagon is synthesized as much larger (about 9 kDa) proglucagon precursor. Molecules larger than this have been detected, but whether they represent glucagon precursors or closely related peptides is unclear. Only 30–40% of the immunoreactive "glucagon" in plasma is pancreatic glucagon; the rest consists of biologically inactive larger molecules.

Glucagon shares some immunologic and physiologic properties with enteroglucagon, a peptide extracted from the duodenal mucosa, and 14 of the 27 amino acid residues of secretin are identical to those of glucagon.

Glucagon circulates in plasma in the free form. Since it does not associate with a transport protein, its plasma half-life is short (about 5 minutes). Glucagon is inactivated by the liver, which has an enzyme that removes the first two amino acids from the aminoterminal end by cleaving between Ser 2 and Gln 3. Since the liver is the first stop for glucagon after it is secreted and since the liver rapidly inactivates the hormone, the level of glucagon in the portal vein is much higher than that in the peripheral circulation.

Secretion of Glucagon Is Inhibited by Glucose, an Action That Emphasizes the Opposing Metabolic Roles of Glucagon and Insulin

It is not clear whether glucose directly inhibits glucagon secretion or whether this is mediated through the actions of insulin or IGF-I, since both of these islet cell hormones directly inhibit glucagon release. Many other substances, including amino acids, fatty acids and ketones, gastrointestinal tract hormones, and neurotransmitters, affect glucagon secretion.

In General, the Actions of Glucagon Oppose Those of Insulin

Whereas insulin promotes energy storage by stimulating glycogenesis, lipogenesis, and protein synthesis, glucagon causes the rapid mobilization of potential energy sources into glucose by stimulating glycogenolysis and into fatty acids by stimulating

Glucagon

$$NH_2 - \overset{1}{His} - Ser - Gln - Gly - Thr - Phe - Thr - Ser - Asp - \overset{10}{Tyr} -$$

$$\overset{11}{Ser} - Lys - Tyr - Leu - Asp - Ser - Arg - Arg - Ala - \overset{20}{Gln} -$$

$$\overset{21}{Asp} - Phe - Val - Gln - Trp - Leu - Met - Asn - \overset{29}{Thr} - COOH$$

Somatostatin

$$NH_2 - \overset{1}{Ala} - Gly - Cys - Lys - Asn - Phe - Phe - Trp - Lys - \overset{10}{Thr} -$$

$$\overset{11}{Phe} - Thr - Ser - \overset{14}{Cys} - COOH$$

Figure 51–15. The amino acid sequences of glucagon and somatostatin.

lipolysis. Glucagon is also the most potent gluconeogenic hormone, and it is ketogenic.

The liver is the primary target of glucagon action. Glucagon binds to specific receptors in the hepatic cell plasma membrane, and this activates adenylyl cyclase through a G protein-linked mechanism. The cAMP generated activates phosphorylase, which enhances the rate of glycogen degradation while inhibiting glycogen synthase and thus glycogen formation (Chapter 44). There is hormone and tissue specificity in this effect, since glucagon has no effect on glycogenolysis in muscle, whereas epinephrine is active in both muscle and liver.

The elevated cAMP level stimulates the conversion of amino acids to glucose by inducing a number of enzymes involved in the gluconeogenic pathway. Principal among these is PEPCK. Glucagon, through cAMP, increases the rate of transcription of mRNA from the PEPCK gene, and this stimulates the synthesis of more PEPCK. This is the opposite of the effect of insulin, which decreases PEPCK gene transcription. Other examples are illustrated in Table 51–6. The net action of glucagon in the liver is increased glucose production; since much of this glucose exits the liver, the plasma glucose concentration increases in response to glucagon.

Glucagon is a potent lipolytic agent. It increases adipose cell cAMP levels, and this activates the hormone-sensitive lipase. The increased fatty acids can be metabolized for energy or converted to the ketone bodies acetoacetate and β-hydroxybutyrate. This is an important aspect of metabolism in the diabetic, since glucagon levels are always increased in insulin deficiency.

Table 51–6. Enzymes induced or repressed by insulin or glucagon.[1]

Enzymes induced by a high insulin:glucagon ratio and repressed by a low insulin:glucagon ratio
  Glucokinase
  Citrate cleavage enzyme
  Acetyl-CoA carboxylase
  HMG-CoA reductase
  Pyruvate kinase
  6-Phosphofructo-1-kinase
  6-Phosphofructo-2-kinase/fructose-2,6-bisphosphatase
Enzymes induced by a low insulin:glucagon ratio and repressed by a high insulin:glucagon ratio
  Glucose-6-phosphatase
  Phosphoenolpyruvate carboxykinase (PEPCK)
  Fructose-1,6-bisphosphatase

[1]Slightly modified and reproduced, with permission, from Karam JH, Salber PR, Forsham PH: Pancreatic hormones and diabetes mellitus. In: *Basic & Clinical Endocrinology*, 3rd ed. Greenspan FS (editor). Appleton & Lange, 1991.

SOMATOSTATIN INHIBITS GROWTH HORMONE SECRETION

Somatostatin, so named because it was first isolated from the hypothalamus as the factor that inhibited growth hormone secretion, is a cyclic peptide synthesized as a large somatostatin prohormone (about 11.5 kDa) in the D cells of the pancreatic islets. The rate of transcription of the prosomatostatin gene is markedly enhanced by cAMP. The prohormone is first processed into a 28-amino-acid peptide and finally into a molecule that has a molecular weight of 1640 and contains 14 amino acids (Figure 51–15). All forms have biologic activity.

In addition to its presence in the hypothalamus and pancreatic islets, somatostatin is found in many gastrointestinal tissues, where it is thought to regulate a variety of functions, and in multiple sites in the central nervous system, where it may be a neurotransmitter.

Somatostatin inhibits the release of the other islet cell hormones through a paracrine action. In pharmacologic amounts, somatostatin significantly blunts the ketosis associated with acute insulin deficiency. This is apparently due to its ability to inhibit the glucagon release that accompanies insulinopenia. It also decreases the delivery of nutrients from the gastrointestinal tract into the circulation because it (1) prolongs gastric emptying, (2) decreases gastrin secretion and therefore gastric acid production, (3) decreases pancreatic exocrine (digestive enzyme) secretion, (4) decreases splanchnic blood flow, and (5) slows sugar absorption. Little is known about the biochemical and molecular actions of this hormone.

THE FUNCTION OF PANCREATIC POLYPEPTIDE IS UNKNOWN

Pancreatic polypeptide (PP), a 36-amino-acid peptide (about 4.2 kDa), is a product of the pancreatic F cells. Its secretion in humans is increased by a protein meal, fasting, exercise, and acute hypoglycemia and is decreased by somatostatin and intravenous glucose. The function of pancreatic polypeptide is unknown, but effects on hepatic glycogen levels and gastrointestinal secretion have been suggested.

THERE ARE MANY GASTROINTESTINAL HORMONES

The Discipline of Endocrinology Began With the Discovery of a Gastrointestinal Hormone

In 1902, Bayliss and Starling instilled hydrochloric acid into a denervated loop of dog jejunum and showed that this increased the secretion of fluid from the pancreas. Intravenous injection of hydrochloric acid did not mimic this effect, but the intravenous injection of an extract of the jejunal mucosa did. These investigators postulated that "secretin," released from the mucosa of the upper intestine in response to a stimulus, moved to the pancreas through the circulation, where it exerted its effect. Bayliss and Starling were the first to use the word "hormone," and secretin was the first hormone whose function was identified.

Although the activity of secretin was identified in 1902, it took 60 years before its chemical identity was proved. The reasons for the 60-year time span are now apparent; families of closely related gastrointestinal peptides have overlapping chemical structures and biologic functions, and most of these peptides exist in multiple forms.

Of the major gastrointestinal hormones, only secretin exists in a single form. The presence of multiple forms of gastrointestinal peptides in gastrointestinal tissues and in the circulation impeded the definition of the number and nature of these molecules. The concept of precursor molecules helped clarify this issue; much of tissue heterogeneity is due to this feature. Isolation techniques developed only recently also have helped to differentiate among them.

Gastrointestinal Hormones Have Some Special Features

More than a dozen peptides with unique actions have been isolated from gastrointestinal tissues (Table 51–7). A unique feature of this group of hormones is that many fit the classic definition of a hormone, some have paracrine actions, and others act in a neurocrine fashion (as local neurotransmitters or neuromodulators).

Another unique aspect of the gastrointestinal endocrine system is that the cells are scattered throughout the gastrointestinal tract rather than collected in discrete organs as in more typical endocrine glands. The distribution of the gastrointestinal hormones is outlined in Table 51–7.

Since many gastrointestinal peptides are found in the nerves in gastrointestinal tissues, it is not surprising that most are also present in the central nervous system.

There Are Families of Gastrointestinal Hormones

Many of these hormones can be placed in one of two families based on amino acid sequence and functional similarity. These are the **gastrin family,** which consists of gastrin and cholecystokinin (CCK), and the **secretin family,** which includes secretin, glucagon, gastric inhibitory polypeptide (GIP), vasoactive intestinal polypeptide (VIP), and glicentin (which has glucagon-like immunoreactivity but is a distinct peptide). The neurocrine peptides neurotensin, bombesin-like peptides, substance P, and somatostatin bear no structural similarity to any other gastrointestinal peptide. A final general characteristic of this last group of molecules is that they have very short plasma half-lives and may play no physiologic role in plasma.

Table 51–7. Gastrointestinal hormones.

Hormone	Location	Major Action
Gastrin	Gastric antrum, duodenum	Gastric acid and pepsin secretion
Cholecystokinin (CCK)	Duodenum, jejunum	Pancreatic amylase secretion
Secretin	Duodenum, jejunum	Pancreatic bicarbonate secretion
Gastric inhibitory polypeptide (GIP)	Small bowel	Enhances glucose-mediated insulin release; inhibits gastric acid secretion
Vasoactive intestinal polypeptide (VIP)	Pancreas	Smooth muscle relaxation; stimulates pancreatic bicarbonate secretion
Motilin	Small bowel	Initiates interdigestive intestinal motility
Somatostatin	Stomach, duodenum, pancreas	Numerous inhibitory effects
Pancreatic polypeptide (PP)	Pancreas	Inhibits pancreatic bicarbonate and protein secretion
Enkephalins	Stomach, duodenum, gallbladder	Opiate-like actions
Substance P	Entire gastrointestinal tract	Physiologic actions uncertain
Bombesin-like immunoreactivity (BLI)	Stomach, duodenum	Stimulates release of gastrin and CCK
Neurotensin	Ileum	Physiologic actions unknown
Enteroglucagon	Pancreas, small intestine	Physiologic actions unknown

Relatively Little Is Known About the Mechanism of Action of the Gastrointestinal Hormones

Studies of the mechanism of action of the gastrointestinal peptide hormones have lagged behind those of other hormones, no doubt because most attention to date has been directed toward cataloging the various molecules and establishing their physiologic action. A notable exception to this statement involves the regulation of secretion of enzymes by the pancreatic acinar cell.

Six different classes of receptors on pancreatic acinar cells have been identified. These are for (1) muscarinic cholinergic agents, (2) the gastrin-CCK family, (3) bombesin and related peptides, (4) the physalaemin-substance P family, (5) secretin and VIP, and (6) cholera toxin. Classes 1–4 appear to act through the calcium-phosphoinositide mechanism, whereas groups 5 and 6 act through cAMP.

SUMMARY

The pancreatic islet cells make and secrete at least four hormones: insulin, glucagon, somatostatin, and pancreatic polypeptide. Of these, insulin and glucagon have the best-defined metabolic roles.

Insulin is synthesized as a single-chain prohormone (proinsulin) with little biologic activity and processed into an active heterodimer of $\alpha\beta$ structure linked together by two interchain disulfide bonds. Proteins closely related to insulin, including relaxin and the insulin-like growth factors, share this precursor molecule feature. It was this example that led to the more general concept of precursor protein molecules and their processing into smaller, active molecules.

The insulin receptor contains an intrinsic tyrosine kinase activity, and current evidence suggests that receptor autophosphorylation initiates a series of protein-protein interactions that result in the activation of PI 3-kinase and the MAP kinase cascade. The physiologic effects of insulin are well known. Insulin affects ion and substrate transport, cellular trafficking, enzyme activity, protein synthesis, and gene expression. In the larger context, these effects concern most aspects of energy accumulation and utilization, and they may be involved in cell replication.

Glucagon is a small peptide hormone, but it too is first synthesized as a larger prohormone. Glucagon binds to a specific cell receptor that is linked to the G-protein system. Cyclic AMP is the intracellular mediator of glucagon action. In general, the metabolic actions of glucagon oppose those of insulin.

The gastrointestinal tract is not thought of as a traditional endocrine organ, yet it produces many different hormones. Many of these are also produced in brain and other neural tissue, where they act as neurotransmitters or neuromodulators. The gastrointestinal hormones act locally, on various organs and processes within the gastrointestinal system. Relatively little is known about the mechanisms of action of many of these hormones.

REFERENCES

Lawrence JC Jr: Signal transduction and protein phosphorylation in the regulation of cellular metabolism by insulin. Annu Rev Physiol 1992;54:177.

LeRoith D et al: Insulin-like growth factors and their receptors as growth regulators in normal physiology and pathologic states. Trends Endocrinol Metab 1992;2:134.

Makino H, Manganiello VC, Kono T: Role of ATP in insulin actions. Ann Rev Physiol 1994;56:273.

O'Brien R, Granner DK: Regulation of gene expression by insulin. Biochem J 1991;278:609.

Said SI: Vasoactive intestinal polypeptide: Biologic role in health and disease. Trends Endocrinol Metab 1992; 2:107.

Steiner DF et al: The new enzymology of precursor processing endoproteases. J Biol Chem 1992;267:23435.

White MF, Kahn CR: The insulin signaling system. J Biol Chem 1994;269:1.

Section VI.
Special Topics

Structure & Function
of the Water-Soluble Vitamins

52

Peter A. Mayes, PhD, DSc

INTRODUCTION

Vitamins are organic nutrients that are required in small quantities for a variety of biochemical functions and which, generally, cannot be synthesized by the body and must therefore be supplied by the diet. The first discovered vitamins, A and B, were found to be fat and water soluble, respectively. As more vitamins were discovered, they were also shown to be either fat or water soluble, and this property was used as a basis for their classification. The water-soluble vitamins were all designated members of the B complex (apart from vitamin C), and the newly discovered fat-soluble vitamins were given alphabetic designations (eg, vitamins D, E, K). Apart from their solubility characteristics, the water-soluble vitamins have little in common from the chemical point of view.

BIOMEDICAL IMPORTANCE

Absence or relative deficiency of vitamins in the diet leads to characteristic deficiency states and diseases. Deficiency of a single vitamin of the B complex is rare, since poor diets are most often associated with **multiple deficiency states.** Nevertheless, definite syndromes are characteristic of deficiencies of specific vitamins. Among the water-soluble vitamins, the following deficiency states are recognized: beriberi (thiamin deficiency); cheilosis, glossitis, seborrhea, and photophobia (riboflavin deficiency); pellagra (niacin deficiency); peripheral neuritis (pyridoxine deficiency); megaloblastic anemia, methylmalonic aciduria, and pernicious anemia (cobalamin deficiency); megaloblastic anemia (folic acid deficiency); and scurvy (ascorbic acid deficiency). Vitamin deficiencies are avoided by consumption of food of a wide variety in adequate amounts.

VITAMINS OF THE B COMPLEX ARE COFACTORS IN ENZYMATIC REACTIONS

The B vitamins essential for human nutrition are (1) thiamin (vitamin B_1), (2) riboflavin (vitamin B_2), (3) niacin (nicotinic acid, nicotinamide) (vitamin B_3), (4) pantothenic acid (vitamin B_5), (5) vitamin B_6 (pyridoxine, pyridoxal, pyridoxamine), (6) biotin, (7) vitamin B_{12} (cobalamin), and (8) folic acid (pteroylglutamic acid).

Because of their water solubility, excesses of these vitamins are excreted in urine and so rarely accumulate in toxic concentrations. For the same reason, their storage is limited (apart from cobalamin), and as a consequence they **must be provided regularly.**

THIAMIN

Thiamin consists of a substituted pyrimidine joined by a methylene bridge to a substituted thiazole (Figure 52–1).

Active Thiamin Is Thiamin Diphosphate

An ATP-dependent thiamin diphosphotransferase present in brain and liver is responsible for the conversion of thiamin to its active form, thiamin diphosphate (pyrophosphate) (Figure 52–1).

Thiamin Diphosphate Is a Coenzyme in Enzymatic Reactions in Which an Activated Aldehyde Unit Is Transferred

There are two types of such reactions: (1) an **oxidative decarboxylation** of α-keto acids (eg, α-ketoglutarate, pyruvate, and the α-keto analogs of leucine, isoleucine, and valine); and (2) **transketolase reactions** (eg, in the pentose phosphate pathway). All of these reactions are inhibited in thiamin

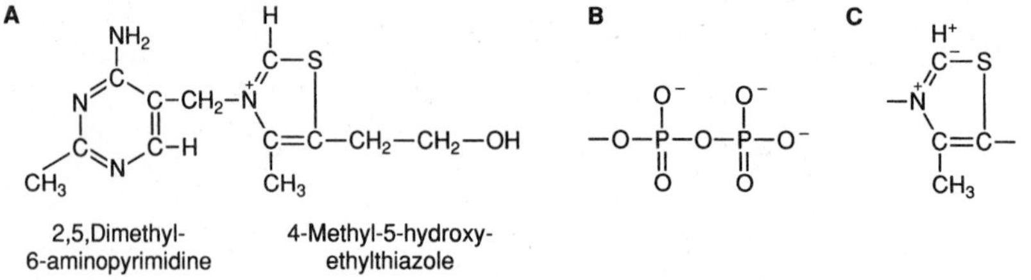

Figure 52–1. Thiamin. **A:** The free vitamin. **B:** In thiamin diphosphate, the —OH group is replaced by pyrophosphate. **C:** Carbanion form.

deficiency. In each case, the thiamin diphosphate provides a reactive carbon on the thiazole that forms a carbanion (Figure 52–1), which is then free to add to the carbonyl group of, for instance, pyruvate (Figure 19–5). The addition compound then decarboxylates, eliminating CO_2. This reaction occurs in a multienzyme complex known as the pyruvate dehydrogenase complex. (See Chapter 19 for further details.)

The oxidative decarboxylation of α-ketoglutarate to succinyl-CoA and CO_2 (Chapter 18) is catalyzed by an enzyme complex structurally similar to the pyruvate dehydrogenase complex. Again, the thiamin diphosphate provides a stable carbanion to react with the α carbon of α-ketoglutarate. A similar oxidative decarboxylation of the α-ketocarboxylic acid derivatives of the branched-chain amino acids (Chapter 32) utilizes thiamin diphosphate. The role of thiamin diphosphate as a coenzyme in the transketolase reactions (Chapter 22) is similar to that described above for the oxidative decarboxylations.

Lack of Thiamin Causes Beriberi and Related Deficiency Syndromes

In the thiamin-deficient human, thiamin diphosphate-dependent reactions are prevented or severely limited, leading to accumulation of the substrates of the reactions, eg, pyruvate, pentose sugars, and the α-ketocarboxylate derivatives of the branched-chain amino acids leucine, isoleucine, and valine.

Thiamin is present in almost all plant and animal tissues commonly used as food, but the content is usually small. Unrefined cereal grains and meat are good sources of the vitamin. **Beriberi** is caused by carbohydrate-rich low-thiamin diets, eg, polished rice or other highly refined foods such as sugar and white flour acting as the staple food sources. Early symptoms include peripheral neuropathy, exhaustion, and anorexia, which progress to edema and cardiovascular, neurologic, and muscular degeneration. **Wernicke's encephalopathy** is associated with thiamin deficiency. It is found frequently in chronic alcoholics consuming little other food. Certain raw fish contain a heat-labile enzyme (thiaminase) that de-

stroys thiamin, but this is not considered to be critical in human nutrition.

Erythrocyte transketolase activity is used as a measure of thiamin deficiency, as are thiamin excretion in urine and blood thiamin concentration.

RIBOFLAVIN

Riboflavin consists of a heterocyclic isoalloxazine ring attached to the sugar alcohol, ribitol (Figure 52–2). It is a colored, fluorescent pigment that is relatively heat-stable but decomposes in the presence of visible light.

Active Riboflavin Is Flavin Mononucleotide (FMN) or Flavin Adenine Dinucleotide (FAD)

FMN is formed by ATP-dependent phosphorylation of riboflavin (Figure 52–2), whereas FAD is synthesized by a further reaction with ATP in which

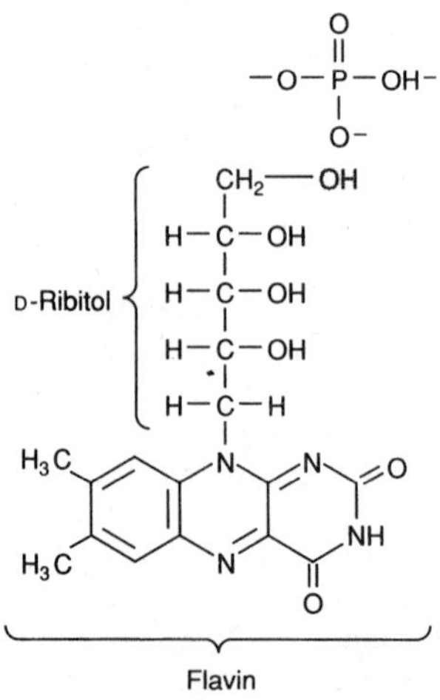

Figure 52–2. Riboflavin. In riboflavin phosphate (flavin mononucleotide, FMN), the —OH is replaced by phosphate.

the AMP moiety of ATP is transferred to FMN (Figure 52–3).

FMN and FAD Serve as Prosthetic Groups of Oxidoreductase Enzymes

These enzymes are known as **flavoproteins.** The prosthetic groups are usually tightly but not covalently bound to their apoproteins. Many flavoprotein enzymes contain one or more metals, eg, molybdenum and iron, as essential cofactors and are known as **metalloflavoproteins.**

Flavoprotein enzymes are widespread and are represented by several important oxidoreductases in mammalian metabolism, eg, α-amino acid oxidase in amino acid deamination (Figure 31–7), xanthine oxidase in purine degradation (Chapter 36), aldehyde dehydrogenase in the degradation of aldehydes, mitochondrial glycerol-3-phosphate dehydrogenase in transporting reducing equivalents from the cytosol into mitochondria (Figure 14–14), succinate dehydrogenase in the citric acid cycle (Table 18–1), acyl-CoA dehydrogenase and the electron-transferring flavoprotein in fatty acid oxidation (Chapter 24), and dihydrolipoyl dehydrogenase in the oxidative decarboxylation of pyruvate and α-ketoglutarate (Chapter 19); NADH dehydrogenase is a major component of the respiratory chain in mitochondria (Chapter 14). All of these enzyme systems are impaired in riboflavin deficiency.

In their role as coenzymes, flavoproteins undergo reversible reduction of the isoalloxazine ring to yield the reduced forms $FMNH_2$ and $FADH_2$ (Figure 13–2).

Lack of Riboflavin Causes a General Nonfatal Deficiency Syndrome

In view of its widespread metabolic functions, it is surprising that riboflavin deficiency does not lead to major life-threatening conditions. However, when there is deficiency, various symptoms are seen, including angular stomatitis, cheilosis, glossitis, seborrhea, and photophobia.

Riboflavin is synthesized by plants and microorganisms but not by mammals. Yeast, liver, and kidney are good sources of the vitamin, which is absorbed in the intestine by a phosphorylation-dephosphorylation sequence in the mucosa. Hormones (eg, thyroid hormone and ACTH), drugs (eg, chlorpromazine, a competitive inhibitor), and nutritional factors affect the conversion of riboflavin to its cofactor forms. Because of its light sensitivity, riboflavin deficiency may occur in newborn infants with hyperbilirubinemia who are treated by phototherapy.

Erythrocyte glutathione reductase activity is used as an assay of riboflavin status (Figure 22–3).

NIACIN

Niacin is the generic name for nicotinic acid and nicotinamide, either of which may act as a source of the vitamin in the diet. Nicotinic acid is a monocarboxylic acid derivative of pyridine (Figure 52–4).

Active Niacin Is Nicotinamide Adenine Dinucleotide (NAD⁺) and Nicotinamide Adenine Dinucleotide Phosphate (NADP⁺)

Nicotinate is the form of niacin required for the synthesis of NAD^+ and $NADP^+$ by enzymes present

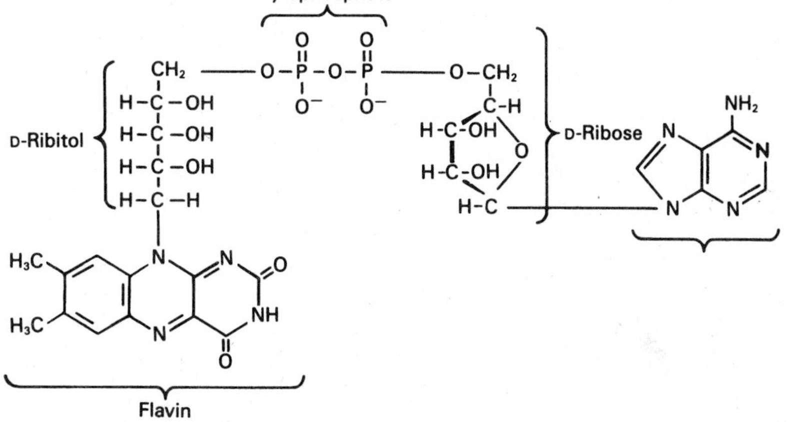

Figure 52–3. Flavin adenine dinucleotide (FAD).

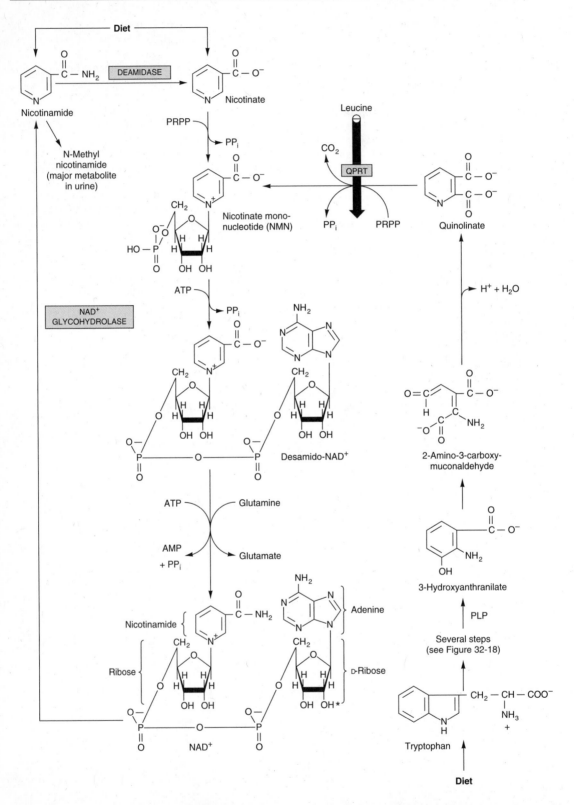

Figure 52–4. The biosynthesis and breakdown of nicotinamide adenine dinucleotide (NAD$^+$). The 2′-hydroxyl group (*) of the ribose of the adenosine moiety is phosphorylated in nicotinamide dinucleotide phosphate (NADP$^+$). Humans, but not cats, can provide all of their niacin requirement from tryptophan if there is a sufficient amount in the diet. Normally, about two-thirds comes from this source. (PRPP, 5-phosphoribosyl-1-pyrophosphate; QPRT, quinolinate phosphoribosyl transferase; PLP, pyridoxal phosphate.)

in the cytosol of most cells. Therefore, any dietary nicotinamide must first undergo deamidation to nicotinate (Figure 52–4). In the cytosol, nicotinate is converted to desamido-NAD$^+$ by reaction first with 5-phosphoribosyl 1-pyrophosphate (PRPP) and then by adenylylation with ATP. The amido group of glutamine then contributes to form the coenzyme NAD$^+$. This may be phosphorylated further to form NADP$^+$.

NAD$^+$ and NADP$^+$ Are Coenzymes for Many Oxidoreductase Enzymes

The nicotinamide nucleotides play a widespread role as coenzymes to many dehydrogenase enzymes occurring both in the cytosol (eg, lactate dehydrogenase) and within the mitochondria (eg, malate dehydrogenase). They are therefore key components of many metabolic pathways affecting carbohydrate, lipid, and amino acid metabolism. Generally, NAD-linked dehydrogenases catalyze oxidoreduction reactions in oxidative pathways (eg, the citric acid cycle), whereas NADP-linked dehydrogenases or reductases are often found in pathways concerned with reductive syntheses (eg, the pentose phosphate pathway).

The mechanism of oxidoreduction involves a reversible addition of a hydride ion (H$^-$) to the pyridine ring plus the generation of a free hydrogen ion (H$^+$) (see Figure 13–5), eg,

$$NAD^+ + AH_2 \leftrightarrow NADH + H^+ + A$$

Lack of Niacin Causes the Deficiency Syndrome Pellagra

Symptoms include weight loss, digestive disorders, dermatitis, depression, and dementia.

Niacin is found widely in most animal and plant foods. However, assessment of niacin value of a food must take account of the fact that the essential amino acid **tryptophan** can be converted to NAD$^+$ (Figure 52–4). For every 60 mg of tryptophan, 1 mg equivalent of niacin can be generated. Thus, in order to produce niacin deficiency, a diet must be poor in both available niacin and tryptophan. Such criteria occur in populations dependent on maize (corn) as the staple food, resulting in **pellagra.** In maize, niacin is in fact present, but it is in a bound unavailable form, niacytin, from which niacin can be released by pretreatment with alkali. Dependence on sorghum is also pellagragenic not because of low tryptophan but because of sorghum's high leucine content. Apparently, excess dietary leucine can bring about niacin deficiency by inhibiting **quinolinate phosphoribosyl transferase,** a key enzyme in the conversion of tryptophan to NAD$^+$ (Figure 52–4). It is also of note that **pyridoxal phosphate,** the active form of vitamin B$_6$, is involved as a cofactor in the pathway of synthesis of NAD$^+$ from tryptophan (Figure 52–4), and vitamin B$_6$ deficiency can therefore potentiate a deficiency in niacin.

Other conditions leading to symptoms of pellagra include administration of some drugs such as isoniazid, malignant carcinoid syndrome, in which tryptophan metabolism is diverted to serotonin, and Hartnup disease, in which tryptophan absorption is impaired.

Nicotinic acid (but not nicotinamide) has been used therapeutically for lowering plasma cholesterol. This is due to inhibition of the flux of free fatty acids from adipose tissue, which leads to less formation of the cholesterol-bearing lipoproteins, VLDL, IDL, and LDL (Chapter 28).

PANTOTHENIC ACID

Pantothenic acid is formed by combination of pantoic acid and β-alanine (Figure 52–5.)

Active Pantothenic Acid Is Coenzyme A (CoA) and the Acyl Carrier Protein (ACP)

Pantothenic acid is absorbed readily in the intestines and subsequently phosphorylated by ATP to form 4′-phosphopantothenate (Figure 52–6). Addition of cysteine and removal of its carboxyl group results in the net addition of thioethanolamine, generating **4′-phosphopantetheine,** the prosthetic group of both CoA and ACP. Like the active coenzymes of so many other water-soluble vitamins, CoA contains an adenine nucleotide. Thus, 4′-phosphopantetheine is adenylylated by ATP to form dephospho-CoA. The final phosphorylation occurs with ATP adding phosphate to the 3′-hydroxyl group of the ribose moiety to generate CoA (Figure 52–6).

The Thiol Group Acts as a Carrier of Acyl Radicals in Both CoA and ACP

This occurs with CoA in reactions of the citric acid cycle (Chapter 18), fatty acid synthesis (Chapter 23) and oxidation (Chapter 24), acetylation reactions (eg, of drugs), and cholesterol synthesis (Chapter 28). ACP participates in reactions concerned with fatty acid synthesis. It is customary to abbreviate the structure of the free (ie, reduced) CoA as CoA-SH, in which the reactive SH group of the coenzyme is designated.

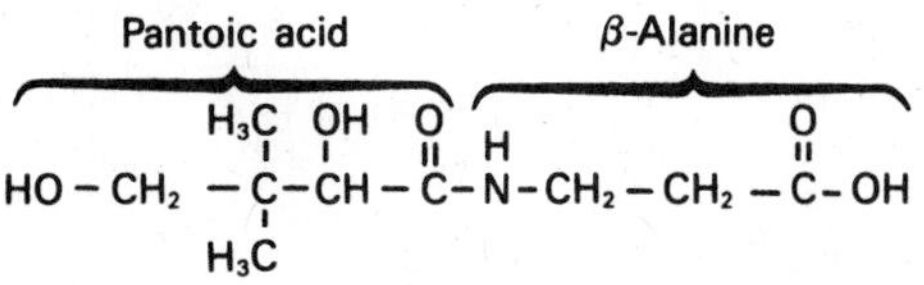

Figure 52–5. Pantothenic acid.

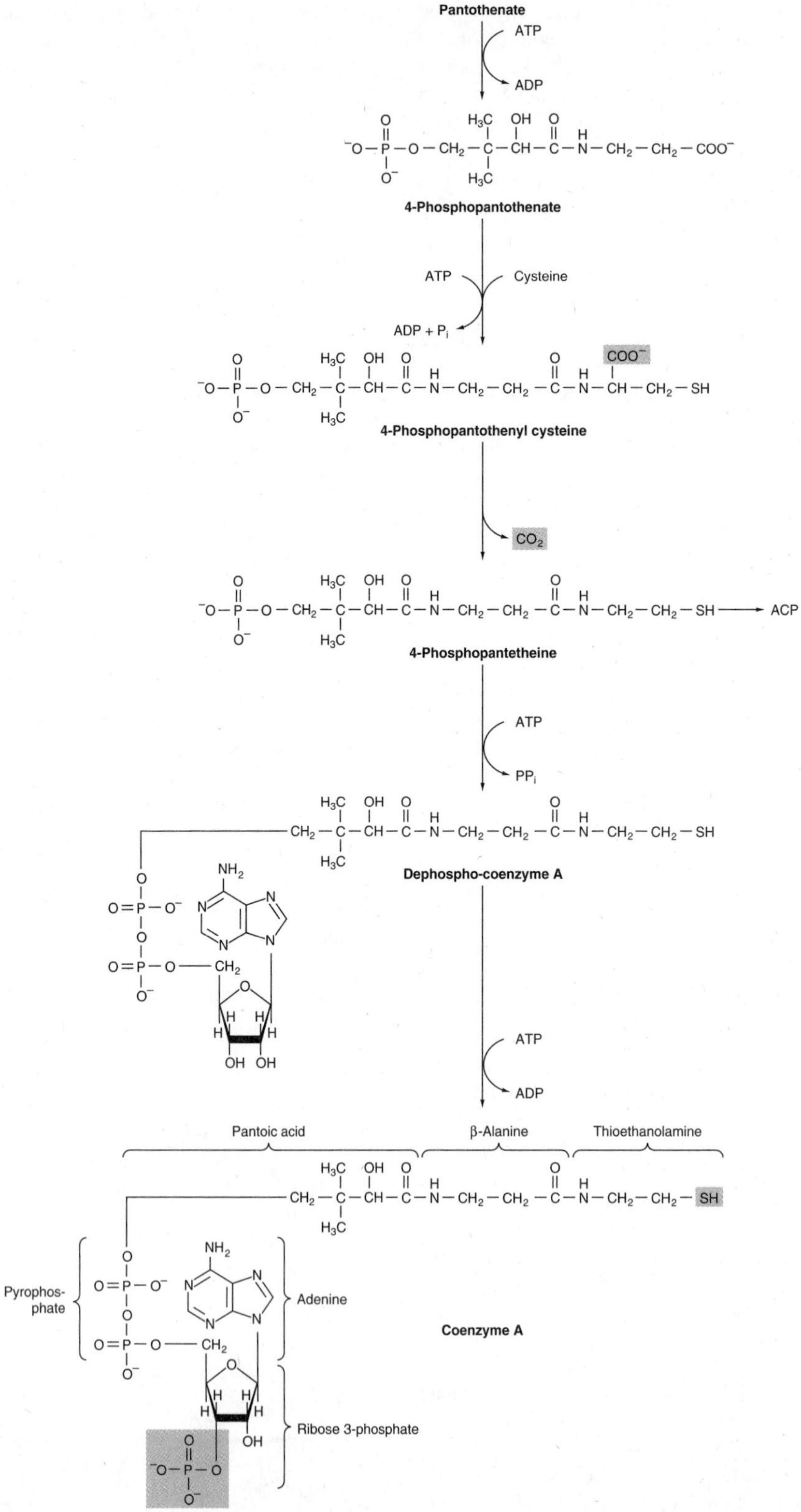

Figure 52–6. The biosynthesis of coenzyme A from pantothenic acid. (ACP, acyl carrier protein.)

Deficiency of Pantothenic Acid Is Rare

This is because the substance is widely distributed in foods, being particularly abundant in animal tissues, whole grain cereals, and legumes. However, the burning foot syndrome has been ascribed to pantothenate deficiency in prisoners of war and is associated with reduced capacity for acetylation.

VITAMIN B$_6$

Vitamin B$_6$ consists of three closely related pyridine derivatives: **pyridoxine, pyridoxal,** and **pyridoxamine** (Figure 52–7) and their corresponding phosphates. Of these, pyridoxine, pyridoxal phosphate, and pyridoxamine phosphate are the main representatives of the vitamin in the diet. All three have equal vitamin activity, as they can be interconverted in the body.

Active Vitamin B$_6$ Is Pyridoxal Phosphate

All forms of vitamin B$_6$ are absorbed from the intestine, but some hydrolysis of the phosphate esters occurs during digestion. Pyridoxal phosphate is the major form transported in plasma. Most tissues contain the enzyme **pyridoxal kinase,** which is able to catalyze the phosphorylation by ATP of the unphosphorylated forms of the vitamin to their respective phosphate esters (Figure 52–8). While pyridoxal phosphate is the major coenzyme expressing vitamin B$_6$ activity, pyridoxamine phosphate may also act as an active coenzyme.

Pyridoxal Phosphate Is the Coenzyme of Several Enzymes of Amino Acid Metabolism

By entering into a **Schiff base** combination between its aldehyde group and the amino group of an

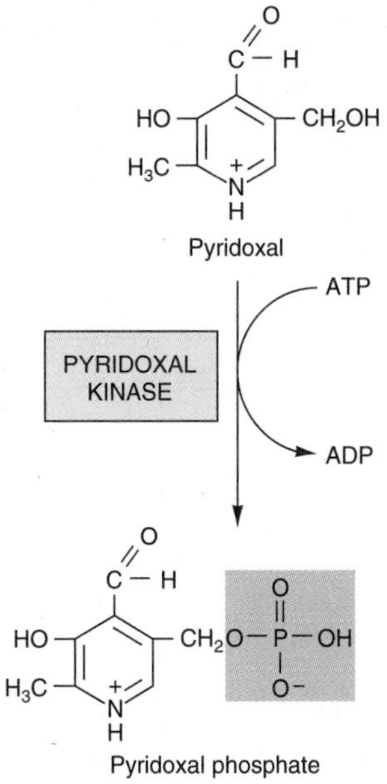

Figure 52–8. The phosphorylation of pyridoxal by pyridoxal kinase to form pyridoxal phosphate.

α-amino acid (Figure 52–9), pyridoxal phosphate can facilitate changes in the three remaining bonds of the α-amino carbon to allow either transamination (Figure 31–4), decarboxylation (Figure 33–9), or threonine aldolase activity (Figure 32–11), respectively. The role of pyridoxal phosphate in transamination is illustrated in Figure 52–10.

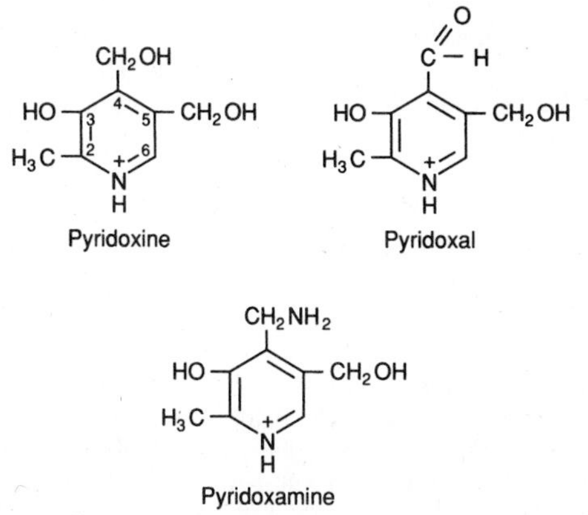

Pyridoxal Phosphate Also Functions in Glycogenolysis

The coenzyme is an integral part of the mechanisms of action of phosphorylase, the enzyme mediating the breakdown of glycogen (Chapter 20). In

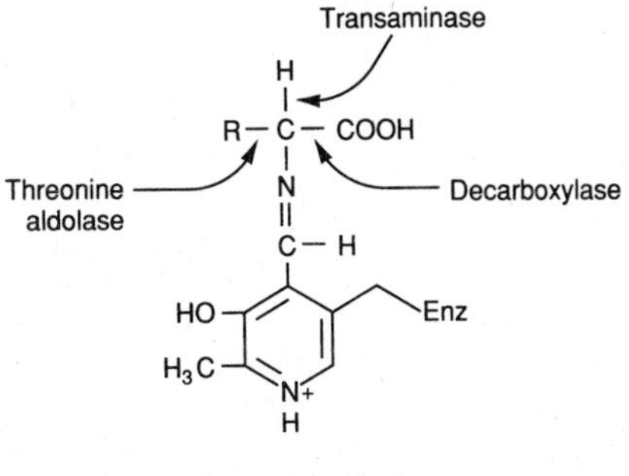

Figure 52–7. Naturally occurring forms of vitamin B$_6$.

Figure 52–9. The covalent bonds of an α-amino acid that can be made reactive by its binding to various pyridoxal phosphate-specific enzymes.

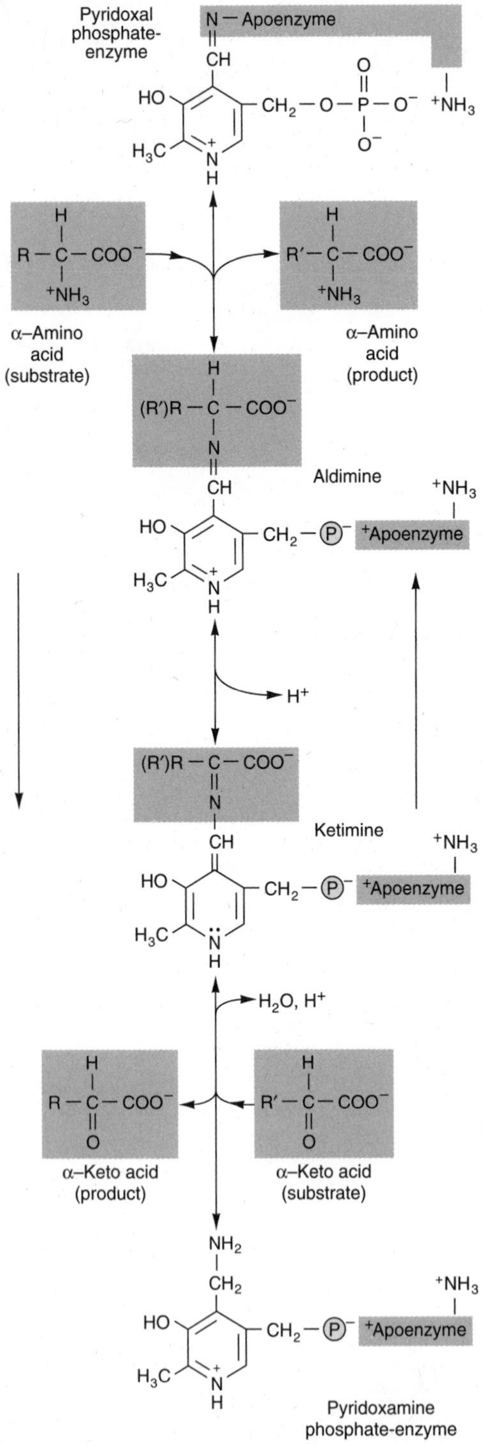

Pyridoxal phosphate-enzyme

N — Apoenzyme
‖
CH

α–Amino acid (substrate)

$R - \overset{\overset{\displaystyle H}{|}}{\underset{\underset{\displaystyle +NH_3}{|}}{C}} - COO^-$

α–Amino acid (product)

$R' - \overset{\overset{\displaystyle H}{|}}{\underset{\underset{\displaystyle +NH_3}{|}}{C}} - COO^-$

$(R')R - \overset{\overset{\displaystyle H}{|}}{\underset{\underset{\displaystyle N}{\|}}{C}} - COO^-$

Aldimine

$CH_2 - \text{\textcircled{P}}^- \quad {}^+Apoenzyme$

$+ H^+$

$(R')R - \overset{}{\underset{\underset{\displaystyle N}{\|}}{C}} - COO^-$

Ketimine

$CH_2 - \text{\textcircled{P}}^- \quad {}^+Apoenzyme$

$+ H_2O, H^+$

α–Keto acid (product)

$R - \overset{\overset{\displaystyle H}{|}}{\underset{\underset{\displaystyle O}{\|}}{C}} - COO^-$

α–Keto acid (substrate)

$R' - \overset{\overset{\displaystyle H}{|}}{\underset{\underset{\displaystyle O}{\|}}{C}} - COO^-$

$CH_2 - \text{\textcircled{P}}^- \quad {}^+Apoenzyme$

Pyridoxamine phosphate-enzyme

Figure 52–10. The role of pyridoxal phosphate coenzyme in the transamination of an α-amino acid. The first phase involves the production of the corresponding α-keto acid and pyridoxamine phosphate enzyme, followed by reversal of the process using a new α-keto acid as substrate. It will be noted that initially pyridoxal phosphate binds to its apoenzyme by a Schiff base link between its aldehyde group and an amino group of the enzyme (an ε-amino group of a lysine residue) and via an ionic bond between its phosphate and the enzyme. The α-amino group of a substrate amino acid displaces the ε-amino group, forming a new Schiff base.

this action, it also forms an initial Schiff base with an ε-amino group of a lysine residue of the enzyme, which, however, remains intact throughout the phosphorolysis of the $1 \rightarrow 4$ glycosidic bond to form glucose 1-phosphate. Muscle phosphorylase may account for as much as 70–80% of total body vitamin B_6.

Deficiency of Vitamin B_6 May Occur During Lactation, in Alcoholics, and During Isoniazid Therapy

A deficiency due to lack of vitamin B_6 alone is rare, and any deficiency is usually part of a general deficiency of B complex vitamins. Liver, mackerel, avocados, bananas, meat, vegetables, and eggs are good sources of the vitamin. A possibility of deficiency is recognized in nursing infants whose mothers are depleted of the vitamin owing to long-term use of oral contraceptives. Alcoholics may also be deficient owing to metabolism of ethanol to acetaldehyde, which stimulates hydrolysis of the phosphate of the coenzyme. A widely used antituberculosis drug, **isoniazid,** can induce vitamin B_6 deficiency by forming a hydrazone with pyridoxal (Figure 52–11).

BIOTIN

Biotin is an imidazole derivative widely distributed in natural foods (Figure 52–12). As a large portion of the human requirement for biotin is met by **synthesis from intestinal bacteria,** biotin deficiency is caused not by simple dietary deficiency but by defects in utilization.

Biotin Is a Coenzyme of Carboxylase Enzymes

Biotin functions as a component of specific multisubunit enzymes (Table 52–1) that catalyze carboxylase reactions. Each unit is a multienzyme complex containing three components on one polypeptide chain, comprising a biotin carrier protein, biotin carboxylase, and a transcarboxylase. A carboxylate ion is attached to the N^1 of the biotin, generating an acti-

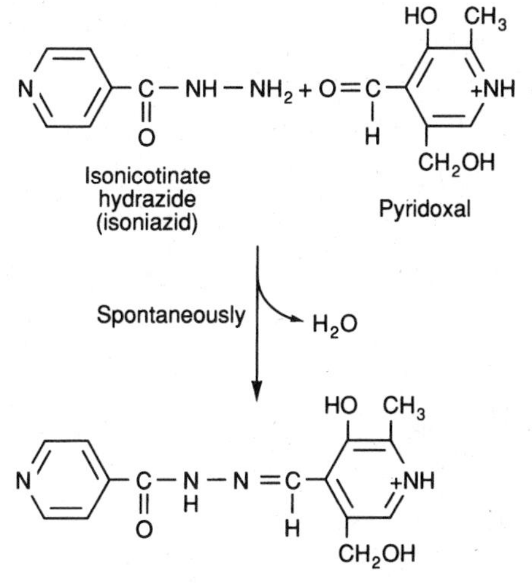

Figure 52–11. The formation of the rapidly excreted pyridoxal-hydrazone from pyridoxal and isonicotinate hydrazine (isoniazid).

Table 52–1. Biotin-dependent enzymes in animals.

Enzyme	Role
Pyruvate carboxylase	Is first reaction in pathway that converts 3-carbon precursors to glucose (gluconeogenesis) Replenishes oxaloacetate for citric acid cycle
Acetyl-CoA carboxylase	Commits acetyl units to fatty acid synthesis by forming malonyl-CoA
Propionyl-CoA carboxylase	Converts propionyl-CoA to D-methylmalonyl-CoA in the pathway of conversion of propionate to succinate, which can then enter citric acid cycle
β-Methylcrotonyl-CoA carboxylase	Catabolizes leucine and certain isoprenoid compounds

vated intermediate, **carboxybiotin,** attached to the biotin carrier protein (Figure 52–13). This step requires HCO_3^-, ATP, Mg^{2+}, and acetyl-CoA (as an allosteric effector). The activated carboxyl group is then transferred to the substrate of the reaction, eg, pyruvate.

Consumption of Raw Eggs Can Cause Biotin Deficiency

Egg white contains a heat-labile protein, **avidin,** which combines very tightly with biotin, preventing its absorption and inducing biotin deficiency. The symptoms include depression, hallucinations, muscle pain, and dermatitis. Absence of the enzyme **holocarboxylase synthase,** which attaches biotin to the lysine residue of the biotin carrier protein, is the cause of **multiple carboxylase deficiency** and also causes biotin deficiency symptoms, including accumulation of substrates of the biotin-dependent enzymes, which can be detected in urine. These metabolites include lactate, β-methylcrotonate, β-hydroxyisovalerate, and β-hydroxypropionate. Children with this deficiency sometimes have immunodeficiency diseases. There are also inherited disorders involving only a single carboxylase deficiency.

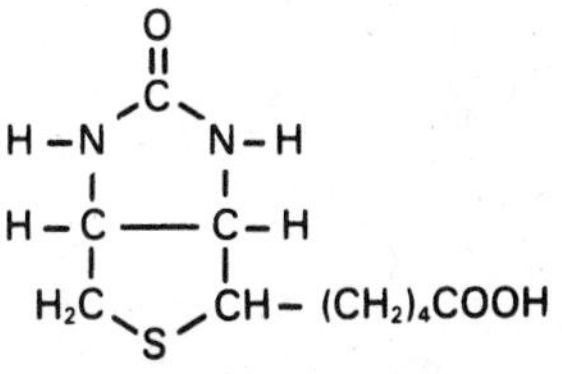

VITAMIN B₁₂

Vitamin B_{12} (cobalamin) has a complex ring structure (corrin ring), similar to a porphyrin ring, to which is added a **cobalt ion** at its center (Figure 52–14). The vitamin is synthesized exclusively by microorganisms. Thus, it is absent from plants—unless they are contaminated by microorganisms—but is conserved in animals in the liver, where it is found as **methylcobalamin, adenosylcobalamin,** and **hydroxocobalamin.** Liver is therefore a good source of the vitamin, as is yeast. The commercial preparation is cyanocobalamin.

Intrinsic Factor Is Necessary for Absorption of Vitamin B₁₂

The intestinal absorption of vitamin B_{12} is mediated by receptor sites in the ileum that require it to be bound by a highly specific glycoprotein, **intrinsic factor,** secreted by parietal cells of the gastric mucosa. After absorption, the vitamin is bound by a plasma protein known as **transcobalamin.** Transcobalamin II is needed for transport to the tissues. It is stored in the liver (unique for a water-soluble vitamin) bound to transcobalamin I.

The Active B₁₂ Coenzymes Are Methylcobalamin and Deoxyadenosylcobalamin

After transport in the blood, free cobalamin is released into the cytosol of cells as hydroxocobalamin. It is either converted in the cytosol to methylcobalamin or it enters mitochondria for conversion to 5'-deoxyadenosylcobalamin.

Figure 52–12. Biotin.

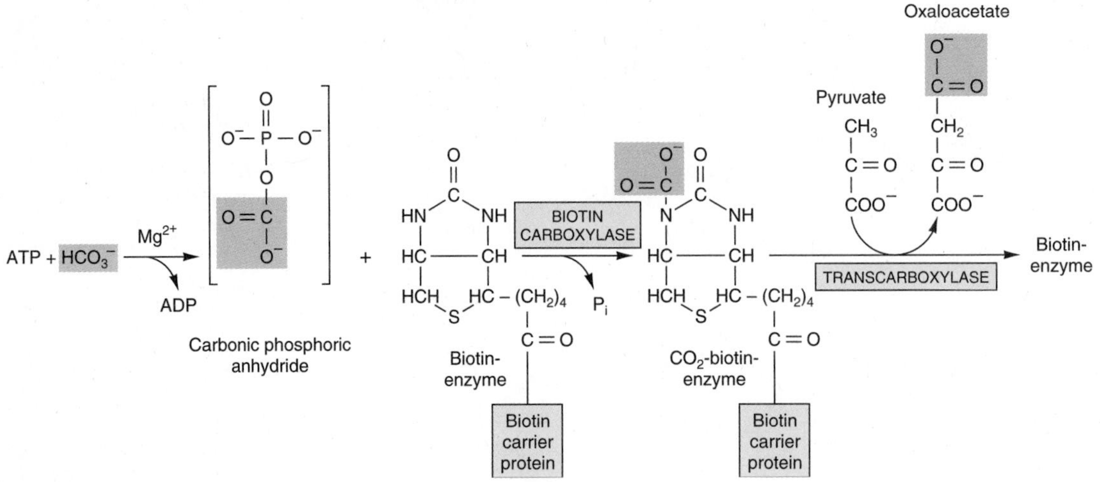

Figure 52–13. Formation of the CO_2-biotin complex and its participation in pyruvate carboxylation. In this example, the enzyme is pyruvate carboxylase.

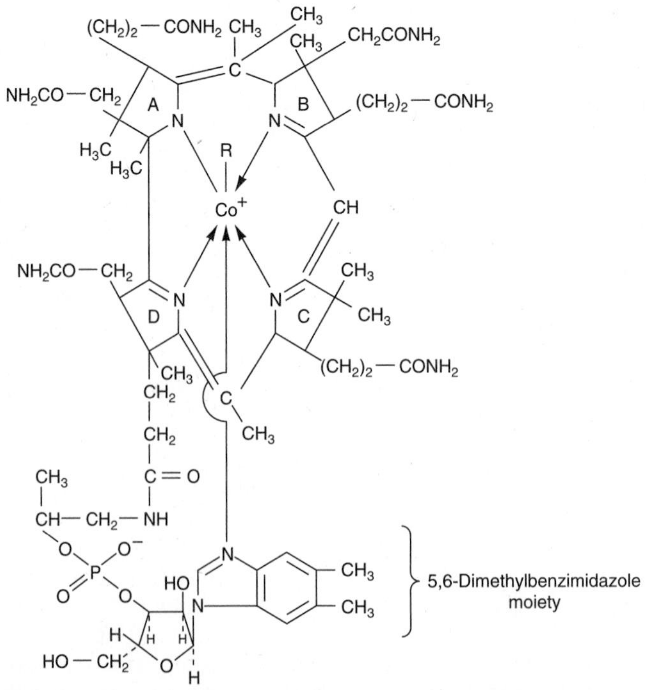

Figure 52–14. Vitamin B_{12} (cobalamin). R may be varied to give the various forms of the vitamin, eg, R = CN in cyanocobalamin; R = OH in hydroxocobalamin; R = 5′-deoxyadenosyl in 5′-deoxyadenosylcobalamin; and R = CH_3 in methylcobalamin.

Deoxyadenosylcobalamin Is the Coenzyme for Conversion of Methylmalonyl-CoA to Succinyl-CoA (Figure 52–15)

This is a key reaction in the pathway of conversion of propionate to a member of the citric acid cycle and is therefore of significance in the process of **gluconeogenesis** (Figure 21–2). It is of particular importance in ruminants, since propionate is a major product of microbial fermentation in the rumen.

Methylcobalamin Is Coenzyme in the Combined Conversion (1) of Homocysteine to Methionine and (2) of Methyltetrahydrofolate to Tetrahydrofolate (Figure 52–15)

In this reaction, the methyl group bound to cobalamin is transferred to homocysteine to form methionine, and the cobalamin then removes the methyl group from N^5-**methyltetrahydrofolate** to form tetrahydrofolate. The metabolic benefits of this reaction are that stores of methionine are maintained and tetrahydrofolate is made available to participate in purine, pyrimidine, and nucleic acid syntheses.

Deficiency of Vitamin B_{12} Leads to Megaloblastic Anemia

When absorption is prevented by lack of intrinsic factor (or by gastrectomy), the condition is called **pernicious anemia.** Vegans are at risk of actual dietary deficiency, since the vitamin is found only in foods of animal origin or in microorganisms—so that foods contaminated with microorganisms are of benefit in that way. The deficiency leads to impairment of the methionine synthase reaction. Anemia results from impaired DNA synthesis, preventing cell division and formation of the nucleus of new erythrocytes with consequent accumulation in the bone marrow of megaloblasts. The impaired purine and pyrimidine synthesis resulting from tetrahydrofolate deficiency is a consequence of folate being trapped as methyltetrahydrofolate (known as the "folate trap") (Figure 52–15). Homocystinuria and methylmalonic aciduria also occur. The neurologic disorder associated with vitamin B_{12} deficiency may be secondary to a relative deficiency of methionine.

Four inherited disorders of cobalamin metabolism have been described. Two affect synthesis of deoxyadenosylcobalamin only; in the other two, patients are unable to synthesize either deoxyadenosylcobalamin or methylcobalamin.

FOLIC ACID

Folacin is the generic term for folic acid and related substances having the biochemical activity of folic acid.

Folic acid, or folate, consists of the base **pteridine** attached to one molecule each of *p*-**aminobenzoic acid (PABA)** and **glutamic acid** (Figure 52–16). Animals are not capable of synthesizing PABA or of attaching glutamate to pteroic acid and, therefore, require folate in their diet; yeast, liver, and leafy

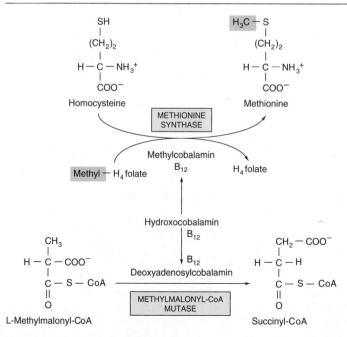

Figure 52–15. The two important reactions catalyzed by vitamin B_{12} coenzyme-dependent enzymes. Vitamin B_{12} deficiency leads to inhibition of both methylmalonyl-CoA mutase and methionine synthase activity, leading to methylmalonic aciduria, homocysteinuria, and the trapping of folate as methyl-H_4folate (the folate trap).

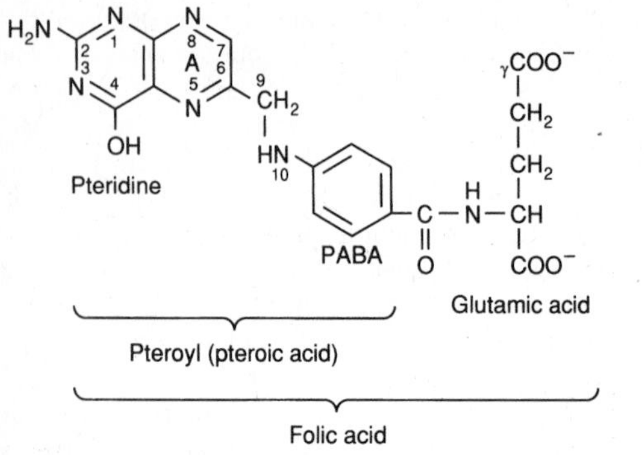

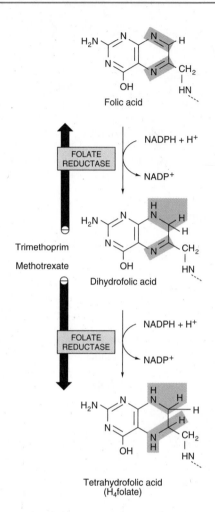

Figure 52–16. The structure and numbering of atoms of folic acid.

vegetables are major sources. In plants, folic acid is present as a polyglutamate conjugate consisting of a γ-linked polypeptide chain of seven glutamate residues. In the liver, the major folate is a pentaglutamyl conjugate.

Active Folate Is Tetrahydrofolate (H$_4$folate)

Folate derivatives in the diet are cleaved by specific intestinal enzymes to monoglutamyl folate for absorption. Most of this is reduced to **tetrahydrofolate** in the intestinal cell (Figure 52–17) by the enzyme **folate reductase,** which uses NADPH as donor of reducing equivalents. Tetrahydrofolate polyglutamates are probably the functional coenzymes in tissues.

H$_4$folate Is the Carrier of Activated One-Carbon Units

The one-carbon units carried by H$_4$folate represent a series in various states of oxidation, namely, **methyl, methylene, methenyl, formyl,** and **formimino.** All are metabolically interconvertible (Figure 52–18).

Serine is the major source of a one-carbon unit in the form of a methylene group, which it transfers reversibly to H$_4$folate to form glycine and N^5,N^{10}-**methylene-H$_4$folate,** which plays a central role in one-carbon unit metabolism. It can be reduced to N^5-**methyl-H$_4$folate,** which has an important role in methylation of homocysteine to methionine involving methylcobalamin as a cofactor (Figure 52–15). Alternatively, it can be oxidized to N^5,N^{10}-**methenyl-H$_4$folate,** which can then be hydrated to either N^{10}-**formyl-H$_4$folate** or to N^5-**formyl-H$_4$folate.** The latter is also known as **folinic acid,** a stable form that can be used for administration of reduced folate.

Formiminoglutamate (Figlu), a catabolite of histidine, transfers its formimino group to H$_4$folate to form N^5-**formimino-H$_4$folate.** In folate deficiency,

Figure 52–17. The reduction of folic acid to dihydrofolic acid and dihydrofolic acid to tetrahydrofolic acid by the enzyme folate reductase. Trimethoprim is a selective inhibitor of folate reductase in gram-negative bacteria and has little effect on the mammalian enzyme. It is therefore used as an antibiotic. Methotrexate binds more strongly and is used as an anticancer drug.

Figlu will accumulate after oral challenge with histidine.

Folate Deficiency Causes Megaloblastic Anemia

The explanation for this is similar to that advanced above to account for the effects of deficiency of vitamin B$_{12}$. N^5,N^{10}-Methylene-H$_4$folate provides the methyl group in the formation of thymidylate, a necessary precursor of DNA synthesis and erythrocyte formation (Figure 52–19). Concomitantly with the reduction of the methylene to the methyl group, there is oxidation of H$_4$folate to dihydrofolate, which must be reconverted to H$_4$folate for further use. Therefore,

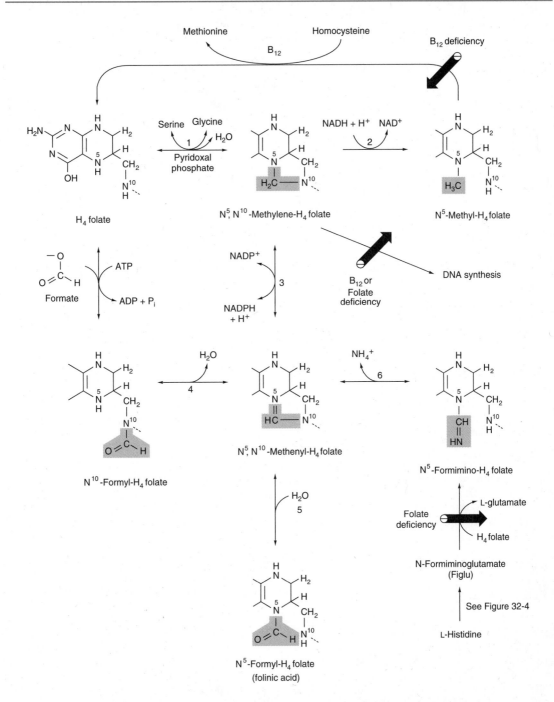

Figure 52–18. The interconversions of one-carbon units attached to tetrahydrofolate.

cells that synthesize thymidylate (for DNA) are particularly vulnerable to inhibitors of folate reductase such as **methotrexate** (Figures 52–17 and 52–19).

The complexities of the interaction of vitamin B_{12} and folate are a consequence of their common participation in the methionine synthase reaction. Thus, the megaloblastic anemia caused by B_{12} deficiency may

be alleviated by extra folate in the diet, but this treatment will not cure either homocystinuria, methylmalonic aciduria, or the neurologic disorders of B_{12} deficiency.

Supplementation with 400 μg of folic acid per day during the periconceptual period can markedly reduce the incidence of neural tube defects such as

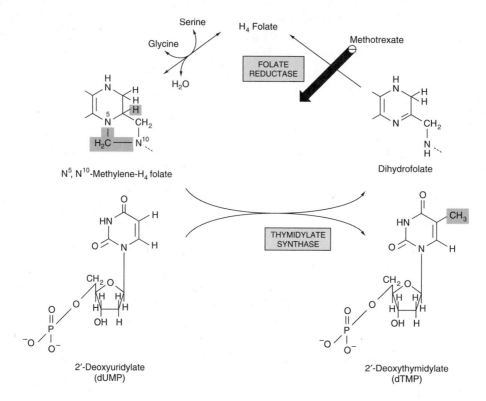

Figure 52–19. The transfer of a methyl group from N^5, N^{10}-methylene-H_4folate to deoxyuridylate to generate deoxythymidylate and dihydrofolate (H_2folate).

spina bifida. It is also important to maintain adequate folic acid supplementation in the later stages of pregnancy and beyond, when many women on a low plane of nutrition suffer from megaloblastic changes.

ASCORBIC ACID
(Vitamin C)

The structure of ascorbic acid (Figure 52–20) is reminiscent of glucose, from which it is derived in the majority of mammals (see Figure 22–4). However, in primates, including humans, and a number of other animals—eg, guinea pigs, some bats, birds, fishes, and invertebrates—the absence of the enzyme L-gulonolactone oxidase prevents this synthesis.

Active Vitamin C Is Ascorbic
Acid Itself, a Donor of
Reducing Equivalents

When ascorbic acid acts as a donor of reducing equivalents, it is oxidized to dehydroascorbic acid, which itself can act as a source of the vitamin. Ascorbic acid is a reducing agent with a hydrogen potential of +0.08 V, making it capable of reducing such compounds as molecular oxygen, nitrate, and cytochromes a and c. The mechanism of action of ascorbic acid in many of its activities is far from clear, but the following are some of the better-documented processes requiring ascorbic acid. In many of these processes, ascorbic acid does not participate directly but is required to maintain a metal cofactor in the reduced state. This includes Cu^+ in monooxygenases and Fe^{2+} in dioxygenases.

(1) In **collagen synthesis,** it is required for hydroxylation of proline (Chapter 57; Figure 30–11).

(2) In the **degradation of tyrosine,** the oxidation of p-hydroxyphenylpyruvate to homogentisate requires vitamin C, which may maintain the reduced state of copper necessary for maximal activity (Figure 32–13). The subsequent step is catalyzed by homogentisate dioxygenase, which is a ferrous iron-containing enzyme that also requires ascorbic acid.

(3) In the **synthesis of epinephrine from tyrosine,** it is required at the dopamine β-hydroxylase step (Figure 49–1).

(4) In **bile acid formation,** it is required at the initial 7α-hydroxylase step (Chapter 28).

(5) The **adrenal cortex** contains large amounts of vitamin C, which are rapidly depleted when the gland is stimulated by adrenocorticotropic hormone. The reason for this is obscure, but steroidogenesis involves several reductive syntheses.

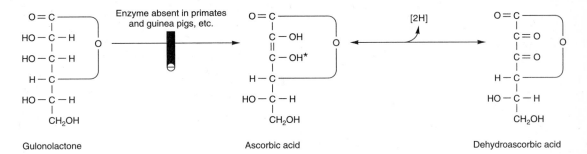

Figure 52–20. Ascorbic acid, its source in nonprimates, and its oxidation to dehydroascorbic acid. (★, ionizes in ascorbate.)

(6) The **absorption of iron** is significantly enhanced by the presence of vitamin C.

(7) Ascorbic acid may act as a general water-soluble **antioxidant** and may inhibit the formation of nitrosamines during digestion.

Ascorbic Acid Deficiency Causes Scurvy

Scurvy is the classic syndrome of vitamin C deficiency. It is related to defective collagen synthesis, which is indicated by subcutaneous and other hemorrhages, muscle weakness, soft swollen gums, and loose teeth. It is cured by consumption of fruits and fresh vegetables. Normal stores of vitamin C are sufficient to last 3–4 months before signs of scurvy appear.

SUMMARY

1. Vitamins are all organic nutrients with various essential metabolic functions, required in small amounts in the diet because they cannot be synthesized by the body.

2. Apart from vitamin C, the water-soluble vitamins are all members of the B complex and act as enzyme cofactors.

3. Thiamin is a cofactor in oxidative decarboxylation of α-keto acids and of an important enzyme of the pentose phosphate pathway, transketolase.

4. Riboflavin and niacin are each important cofactors in oxidoreduction reactions. Riboflavin is present as prosthetic groups in flavoprotein enzymes as flavin mononucleotide and flavin adenine dinucleotide, whereas niacin is present in the NAD and NADP cofactors of many dehydrogenase enzymes.

5. Pantothenic acid is present in coenzyme A and acyl carrier protein, which act as carriers for acyl groups in many important reactions, whereas pyridoxal phosphate is the coenzyme for several enzymes of amino acid metabolism including the transaminases.

6. Biotin is the coenzyme for several carboxylase enzymes, including acetyl-CoA carboxylase, the rate-controlling enzyme in lipogenesis, and pyruvate carboxylase, important in gluconeogenesis.

7. As well as having separate functions, vitamin B_{12} and folic acid take part in providing one-carbon residues for nucleic acid synthesis.

8. Ascorbic acid is a water-soluble antioxidant that maintains many metal cofactors in the reduced state.

9. Absence of the water-soluble vitamins from the diet provokes multiple deficiency states. Absence of a single vitamin leads to a characteristic deficiency syndrome.

REFERENCES

Benkovic SJ: On the mechanism of action of folate and biopterin-requiring enzymes. Annu Rev Biochem 1980; 49:227.

Dakshinamurti K, Chauhan J: Biotin. Vitam Horm 1989; 45:337.

Hayashi H et al: Recent topics in pyridoxal 5′-phosphate enzyme studies. Annu Rev Biochem 1990;59:87.

Knowles JR: The mechanism of biotin-dependent enzymes. Annu Rev Biochem 1989;58:195.

Olson RE et al (editors): *Present Knowledge in Nutrition,* 5th ed. The Nutrition Foundation, Inc, 1984.

Padh H: Vitamin C: Newer insights into its biochemical functions. Nutr Rev 1991;49:65.

Passmore R, Eastwood, MA: *Human Nutrition and Dietetics,* Churchill Livingstone, 1986.

Rubin RH, Swartz MN: Trimethoprim-sulfamethoxazole. N Engl J Med 1980;303:426.

Sebrell WH Jr: History of pellagra. Fed Proc 1981;40: 1520.

Seetharam B, Alpers DH: Absorption and transport of cobalamin (vitamin B_{12}). Annu Rev Nutr 1982;2:343.

Shane B: Folylpolyglutamate synthesis and role in the regulation of one-carbon metabolism. Vitam Horm 1989;45:263.

Structure & Function of the Lipid-Soluble Vitamins

Peter A. Mayes, PhD, DSc

INTRODUCTION

The lipid-soluble (fat-soluble) vitamins are **apolar hydrophobic** molecules which are all **isoprene derivatives** (Figure 53–1). They cannot be synthesized by the body in adequate amounts and must, therefore, be supplied by the diet. They can only be absorbed efficiently when normal fat absorption is taking place. Once absorbed, they must be transported in the blood, like any other apolar lipid, in **lipoproteins** or attached to **specific binding proteins.** The lipid-soluble vitamins have diverse functions, eg, vitamin A, vision; vitamin D, calcium and phosphate metabolism; vitamin E, antioxidant; vitamin K, blood clotting. Although once thought of solely as a vitamin, vitamin D is in reality a **prohormone.**

BIOMEDICAL IMPORTANCE

Conditions affecting the digestion and absorption of the lipid-soluble vitamins such as steatorrhea and disorders of the biliary system can all lead to deficiencies. Dietary inadequacy or deficiencies due to malabsorption cause syndromes consequent on the vitamins not carrying out their physiologic functions; viz, vitamin A deficiency causes night blindness and xerophthalmia; vitamin D deficiency leads to rickets in young children and osteomalacia in adults; in vitamin E deficiency, which is rare, neurologic disorders and anemia of the newborn may arise; vitamin K deficiency, again very rare in adults, leads to hemorrhage of the newborn. Because of the body's ability to store surplus lipid-soluble vitamins, toxicity can result from excessive intake of vitamins A and D. Vitamin A and β-carotene, provitamin A, and vitamin E are antioxidants. Their roles in atherosclerosis and cancer prevention have been ascribed to their antioxidant properties.

VITAMIN A

Vitamin A, or retinol, is a polyisoprenoid compound containing a cyclohexenyl ring (Figure 53–2).

Vitamin A is a generic term referring to all compounds from animal sources that exhibit the biologic activity of vitamin A. It is stored mainly as retinol esters in the liver. In the body, the main functions of vitamin A are carried out by **retinol** and its two derivatives **retinal** and **retinoic acid.** The term **retinoids** has been used to describe both the natural forms and the synthetic analogs of retinol.

Vitamin A Has a Provitamin, β-Carotene

In vegetables, vitamin A exists as a **provitamin** in the form of the yellow pigment β-carotene, which consists of two molecules of retinal joined at the aldehyde end of their carbon chains (Figure 53–3). However, because β-carotene is not efficiently metabolized to vitamin A, β-carotene is only about one-sixth as effective a source of vitamin A as retinol, weight for weight. β-Carotene-like compounds are known as **carotenoids.**

Digestion of Vitamin A Accompanies That of Lipids, Followed by Transformations in the Intestinal Mucosa

Retinol esters dissolved in the fat of the diet are dispersed in bile droplets and hydrolyzed in the intestinal lumen, followed by absorption directly into the intestinal epithelium. Ingested β-carotenes may be oxidatively cleaved by **β-carotene dioxygenase** (Figure 53–3). This cleavage utilizes molecular oxygen, is enhanced by the presence of bile salts, and generates two molecules of retinaldehyde (retinal). This reaction cannot occur in the cat, which must have preformed vitamin A in its diet. Also, in the intestinal mucosa, retinal is reduced to retinol by a specific **retinaldehyde reductase** utilizing NADPH. A small fraction of the retinal is oxidized to **retinoic acid.** Most of the retinol is esterified with saturated fatty acids and incorporated into lymph chylomicrons (Chapter 27), which enter the bloodstream. These are converted to chylomicron remnants, which are taken up by the liver together with their content of retinol. Carotenoids may escape some of these processes and pass directly into the chylomicrons.

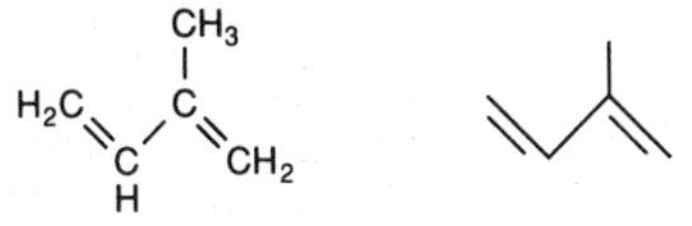

Figure 53–1. Two representations of isoprene.

Vitamin A Is Stored in the Liver and Released Into Blood Attached to Binding Proteins

In the liver, vitamin A is stored as an ester in the **lipocytes** (perisinusoidal stellate cells), probably as a lipoglycoprotein complex. For transport to the tissues, it is hydrolyzed and the retinol bound to **apo-retinol-binding protein** (RBP). The resulting **holo-RBP** is processed in the Golgi apparatus and secreted into the plasma. It is taken up into tissues via cell surface receptors. Retinoic acid is transported in plasma bound to albumin. Once inside extrahepatic cells, retinol is bound by a **cellular retinol-binding protein (CRBP).**

Vitamin A toxicity (hypervitaminosis A) occurs after the capacity of RBP has been exceeded and the cells are exposed to **unbound retinol.** This can occur with excessive use of vitamin A supplements and has been noted in Arctic explorers consuming polar bear liver, polar bears being at the end of the vitamin A food chain.

Retinol, Retinal, and Retinoic Acid All Have Unique Biologic Functions

Retinol and retinal are interconverted in the presence of NAD- or NADP-requiring dehydrogenases or reductases, present in many tissues. However, once formed from retinal, retinoic acid cannot be converted back to retinal, or to retinol. Thus, retinoic acid can support growth and differentiation but cannot replace retinal in its role in vision or retinol in its support of the reproductive system.

Retinol and Retinoic Acid Act Like Steroid Hormones

When retinol is taken up into CRBP it is transported around the cell and binds to **nuclear proteins,** where it is probably involved in the control of the expression of certain genes. Thus, in this respect vita-

min A behaves similarly to steroid hormones. Nuclear receptors for (all-*trans*) retinoic acid and 9-*cis* retinoic acid have been described. These are members of the steroid, thyroid, and retinoic acid receptor superfamily of proteins. Retinoic acid has been implicated in promoting amphibian limb regeneration and in controlling lung surfactant phospholipid synthesis. The requirement of vitamin A for normal reproduction may be ascribed to this function.

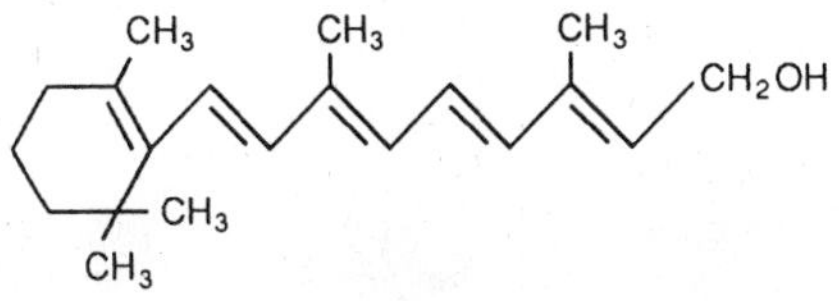

Retinal Is a Component of the Visual Pigment Rhodopsin

Rhodopsin occurs in the rod cells of the retina, which are responsible for vision in poor light. **11-*cis*-Retinal,** an isomer of **all-*trans*-retinal,** is specifically bound to the visual protein **opsin** to form **rhodopsin** (Figure 53–4). When rhodopsin is exposed to light, it dissociates as it bleaches and forms all-*trans*-retinal and opsin. This reaction is accompanied by a conformational change that induces a **calcium ion channel** in the membrane of the rod cell. The rapid influx of calcium ions triggers a nerve impulse, allowing light to be perceived by the brain.

Retinoic Acid Participates in Glycoprotein Synthesis

This may account, in part, for the action of retinoic acid in promoting growth and differentiation of tissues. It has been proposed that retinoyl phosphate functions as a carrier of oligosaccharides across the lipid bilayer of the cell, by way of an enzymatic **trans-cis isomerization** analogous to that described above in the *trans-cis* isomerization of rhodopsin. The evidence that retinoic acid is involved in glycoprotein synthesis is compelling, since a deficiency of vitamin A results in the accumulation of abnormally low molecular weight oligosaccharide-lipid intermediates of glycoprotein synthesis (Chapter 56).

Lack of Vitamin A Causes Characteristic Deficiency Symptoms

Vitamin A deficiency causes symptoms due to malfunction of the various cellular mechanisms in which retinoids participate. One of the first indications is **defective night vision,** which occurs when liver stores are nearly exhausted. Further depletion leads to keratinization of epithelial tissues of the eye, lungs, gastrointestinal, and genitourinary tracts, coupled with reduction in mucous secretion. Deterioration in the tissues of the eye, **xerophthalmia,** leads to blindness. Vitamin A deficiency occurs mainly in populations subsisting on poor basic diets coupled with a lack of vegetables that would otherwise provide the provitamin β-carotene. Alcoholics are particularly susceptible to vitamin A deficiency but also are more prone to hypervitaminosis when supplementation is attempted.

Figure 53–2. Retinol (vitamin A).

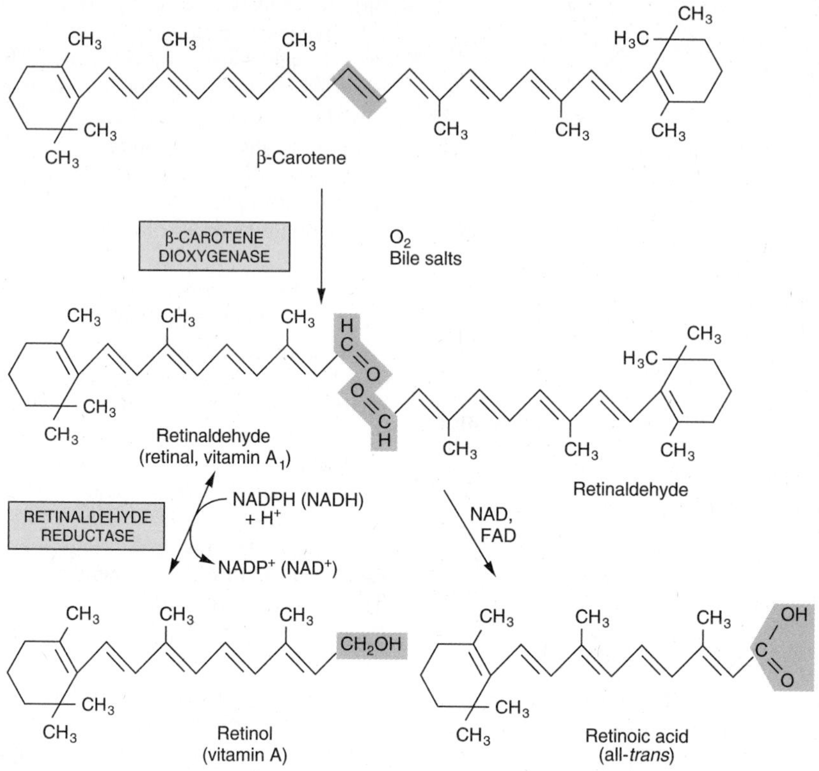

Figure 53–3. β-Carotene and its cleavage to retinaldehyde. The reduction of retinaldehyde to retinol and the oxidation of retinaldehyde to retinoic acid are also shown. 9-*cis* and 13-*cis*-retinoic acid are also formed.

Both Retinoids and Carotenoids Have Anticancer Activity

Many human cancers arise in epithelial tissues that depend on retinoids for normal cellular differentiation. Some epidemiologic studies have shown an inverse relationship between the vitamin A content of the diet and the risk of cancer, and experiments have shown that retinoid administration diminishes the effect of some carcinogens.

β-Carotene is an antioxidant and may play a role in trapping peroxy free radicals in tissues at low partial pressures of oxygen. The ability of β-carotene to

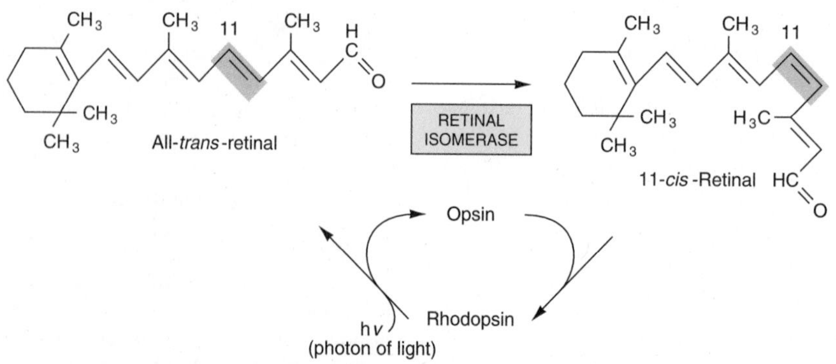

Figure 53–4. 11-*cis*-Retinal, formed from all-*trans*-retinal, combines with opsin to form rhodopsin in the rod cell of the eye. The absorption of a photon of light by rhodopsin causes it to bleach, generating opsin and all-*trans*-retinal. Retinal is required to maintain this cycle of reactions. The term "retinal isomerase" denotes a series of reactions in which retinyl ester is an intermediate.

act as an antioxidant is due to the stabilization of organic peroxide free radicals within its conjugated alkyl structure (Figure 53–5). Since β-carotene is effective at low oxygen concentrations, it complements the antioxidant properties of vitamin E, which is effective at higher oxygen concentrations (Chapter 16). The antioxidant properties of these two lipid-soluble vitamins may well account for their possible anticancer activity. Low serum concentrations of β-carotene and α-tocopherol are associated with development of senile cataract. LDL is the major carrier of β-carotene.

VITAMIN D

Vitamin D is a steroid prohormone. It is represented by steroids that occur in animals, plants, and yeast. By various metabolic changes in the body, they give rise to a hormone known as **calcitriol,** which plays a central role in **calcium and phosphate metabolism** (Chapter 47).

Vitamin D Is Generated From the Provitamin Dehydrocholesterol by the Action of Sunlight

Ergosterol occurs in plants and 7-dehydrocholesterol in animals. Ergosterol differs from 7-dehydrocholesterol only in its side chain, which is unsaturated and contains an extra methyl group (Figure 53–6). Ultraviolet irradiation cleaves the B ring of both compounds. **Ergocalciferol** (vitamin D_2) may be made commercially from plants in this way, whereas in animals, **cholecalciferol** (vitamin D_3) is formed from 7-dehydrocholesterol in exposed skin. Both vitamins D_2 and D_3 are of equal potency.

Both the Liver and the Kidney Are Involved in Calcitriol Synthesis

Vitamin D_3, formed from 7-dehydrocholesterol by the action of sunlight, and dietary vitamin D_3 (or D_2), after absorption from micelles in the intestine followed by transport in the lymphatics, circulate in the blood bound to a specific globulin, **vitamin D-binding protein.** Vitamin D_3 is taken up by the liver, where it is hydroxylated on the 25 position by **vitamin D_3-25-hydroxylase,** an enzyme of the endoplasmic reticulum considered to be rate-limiting in the pathway (Figure 53–7). 25-Hydroxyvitamin D_3 is the major form of vitamin D in the circulation and the major storage form in the liver, though adipose tissue and skeletal muscle have been reported as major sites of storage. A significant fraction of 25-hydroxyvitamin D_3 undergoes enterohepatic circulation, and disturbances of this process can lead to deficiency of vitamin D.

In the renal tubules, bone, and placenta, the 25-hydroxyvitamin D_3 is further hydroxylated in position 1 by **25-hydroxyvitamin D_3-1-hydroxylase,** a mitochondrial enzyme. The product is 1α,25-dihydroxyvitamin D_3 (calcitriol), the most potent vitamin D metabolite. Its production is regulated by its own concentration, parathyroid hormone, and serum phosphate.

25-Hydroxyvitamin D_3 can also be hydroxylated at the 24 position by a mitochondrial enzyme present in renal tubules, cartilage, intestine, and placenta. The level of the product 24,25-dihydroxyvitamin D_3 is reciprocally related to the level of 1,25-dihydroxyvitamin D_3 in serum and is biologically inactive.

For further details of the regulation and role of calcitriol in calcium and phosphate metabolism, see Chapter 47. Evidence has accumulated that vitamin

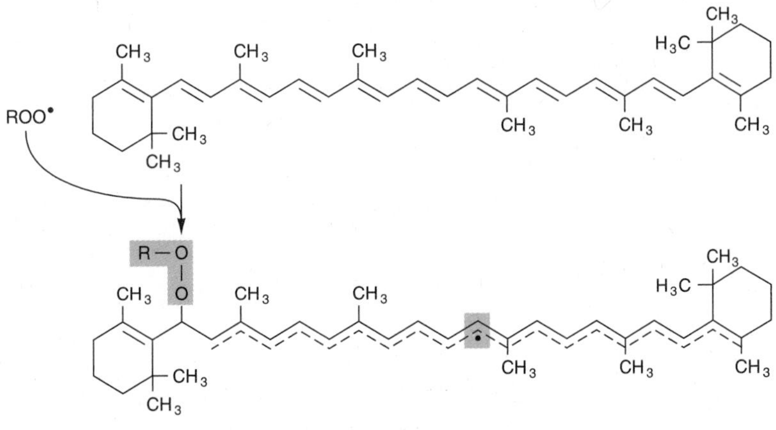

Figure 53–5. The formation of a resonance-stabilized carbon-centered radical from a peroxyl radical (ROO•) and β-carotene. (Slightly modified and reproduced, with permission, from Burton GW, Ingold KU: β-Carotene: An unusual type of lipid antioxidant. Science 1984;224:569. Copyright © 1984 by The American Association for the Advancement of Science.)

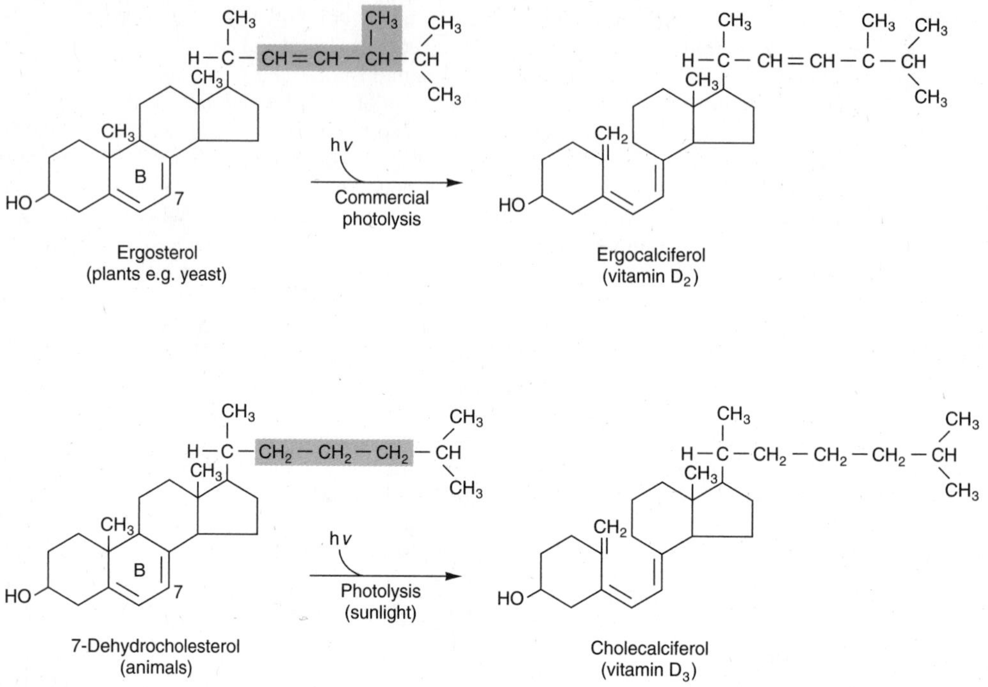

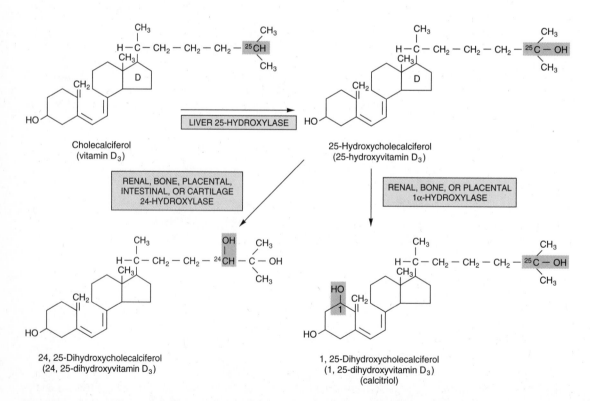

Figure 53–6. Ergosterol and 7-dehydrocholesterol and their conversion by photolysis to ergocalciferol and cholecalciferol, respectively.

Figure 53–7. Cholecalciferol can be hydroxylated at the C_{25} position by a liver enzyme. The 25-hydroxycholecalciferol is further metabolized to $1\alpha,25$-dihydroxycholecalciferol or to 24,25-dihydroxycholecalciferol.

D, as well as vitamin A, carries out its functions by activating specific genes. Vitamin D is also involved in cell differentiation and immune function.

Deficiency of Vitamin D Causes Rickets and Osteomalacia

Rickets occurs in young children and osteomalacia in adults who are not exposed to sunlight or who do not receive adequate amounts of vitamin D in the diet. It is due to softening of bones resulting from lack of calcium and phosphate. Oily fish, egg yolk, and liver are good sources of the vitamin. An individual's exposure to sunlight, which is governed by latitude, season, and other factors, influences the relative dependence on dietary sources to meet vitamin D requirements.

VITAMIN E
(Tocopherol)

There are several naturally occurring tocopherols. All are isoprenoid-substituted 6-hydroxychromanes (tocols) (Figure 53–8).

D-α-Tocopherol has the widest natural distribution and the greatest biologic activity. Other tocopherols of dietary significance are indicated in Table 53–1.

Active Fat Absorption Promotes the Absorption of Vitamin E

Impaired fat absorption leads to vitamin E deficiency because tocopherol is found dissolved in the fat of the diet and is liberated and absorbed during fat digestion. Furthermore, it is transported in the blood by lipoproteins—first, by incorporation into chylomicrons, which distribute the vitamin to the tissues containing lipoprotein lipase and then to the liver in chylomicron remnants; and second, by export from the liver in very low density lipoproteins. It is stored in adipose tissue. Thus, vitamin E deficiency may be found in situations associated with dysfunction of the above processes, eg, in chronic steatorrhea, abetalipoproteinemia, cholestatic liver disease, cystic fibrosis, and in patients who have undergone intestinal resection.

Table 53–1. Naturally occurring tocopherols of dietary significance.

Tocopherol	Substituents
Alpha	5,7,8-Trimethyl tocol
Beta	5,8-Dimethyl tocol
Gamma	7,8-Dimethyl tocol
Delta	8-Methyl tocol

Vitamin E Is a Most Important Natural Antioxidant

Vitamin E appears to be the first line of defense against **peroxidation of polyunsaturated fatty acids** contained in cellular and subcellular membrane phospholipids (Chapter 16). The phospholipids of mitochondria, endoplasmic reticulum, and plasma membranes possess affinities for α-tocopherol, and the vitamin appears to concentrate at these sites. The tocopherols act as antioxidants, breaking free-radical chain reactions as a result of their ability to transfer a phenolic hydrogen to a peroxyl free radical of a peroxidized polyunsaturated fatty acid (Figure 53–9; Chapter 16). The phenoxy free radical formed may react with vitamin C to regenerate tocopherol (Figure 53–12), or it reacts with a further peroxyl free radical so that the chromane ring and the side chain are oxidized to the non-free-radical product shown in Figure 53–10. This oxidation product is conjugated with glucuronic acid via the 2-hydroxyl group and excreted in the bile. If it reacts in this manner, tocopherol is not recycled after carrying out its function but must be replaced totally to continue its biologic role in the cell. The antioxidant action of tocopherol is effective at high oxygen concentrations, and thus it is not surprising that it tends to be concentrated in those lipid structures that are exposed to the highest O_2 partial pressures, eg, the erythrocyte membrane, the membranes of the respiratory tree, and the retina.

Vitamin E and Selenium Act Synergistically

Glutathione peroxidase, of which selenium is an integral component (Chapter 22), provides a second line of defense against hydroperoxides before they can damage membranes and other cell components (Figure 53–12). Thus, tocopherol and selenium rein-

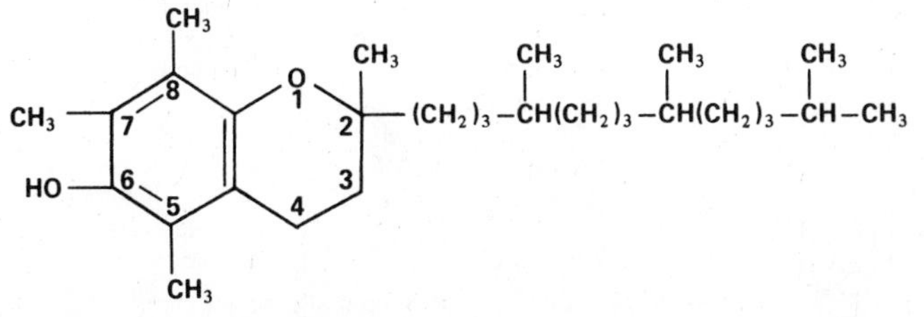

Figure 53–8. α-Tocopherol.

ROO• + TocOH ⟶ ROOH + TocO•

ROO• + TocO• ⟶ ROOH +
                    Non-free radical product

Figure 53–9. The chain-breaking antioxidant activity of tocopherols (TocOH) toward peroxyl radicals (ROO•).

force each other in their actions against lipid peroxides. In addition, selenium is required for normal pancreatic function, which is necessary for the digestion and absorption of lipids, including vitamin E. Conversely, vitamin E reduces selenium requirements by preventing loss of selenium from the body or maintaining it in an active form.

Deficiency of Vitamin E May Give Rise to Anemia of the Newborn

There is a possible need for supplementation of tocopherol in the diets of pregnant and lactating women and for newborn infants where anemia can arise as a result of insufficiency of vitamin E. Anemia may be due to decreased production of hemoglobin and a shortened erythrocyte life span.

The requirement for vitamin E is increased with greater intake of polyunsaturated fat. Intake of mineral oils, exposure to oxygen (as in oxygen tents), or diseases leading to inefficient lipid absorption may cause deficiencies of the vitamin leading to neurologic disorder.

Vitamin E is destroyed by commercial cooking and food processing, including deep-freezing. Wheat germ, sunflower seed and safflower seed oils, and corn and soya bean oil are all good sources of the vitamin. Although fish liver oils are rich sources of vitamins A and D, they have insignificant amounts of vitamin E.

Reactive Oxygen Species May Initiate Disease

A free radical is an atom or molecule that has one or more unpaired electrons. Its consequent tendency to acquire an electron from other substances makes it highly reactive. However, not all reactive oxygen species are free radicals, eg, singlet oxygen and H_2O_2. When oxygen is reduced to water by cytochrome oxidase, four electrons are acquired (Figure 53–11). Electrons, however, can be gained one at a time by univalent reduction, which may account for 1–5% of total oxygen consumption. The individual molecules in univalent reduction are highly reactive and potentially damaging to tissues. They are the superoxide free radical, hydrogen peroxide, and the hydroxyl free radical. The last is extremely toxic but short-lived. Other sources of reactive species are xanthine oxidase, which generates superoxide (eg, during reperfusion injury of ischemic organs), and cyclooxygenase and lipoxygenase (Chapter 25), which produce hydroxyl and peroxyl radicals. Stimulated neutrophils produce superoxide, which is one mechanism for the destruction of bacteria (Chapter 61). Superoxide may also be produced during the metabolism of xenobiotics by cytochrome P450. Because these molecules are so reactive, they act in situ very close to where they are generated. Therefore, most cell structures are vulnerable, including membranes, structural proteins, enzymes, and nucleic acids, which can lead to mutation and cell death.

Antioxidant Nutrients May Prevent Disease

The individual mechanisms for defense of the tissues by preventing the initiation of free-radical chain reactions and for their termination have been discussed, ie, superoxide dismutase (Chapter 13), lipid antioxidants (Chapter 16), glutathione peroxidase and selenium (see above), vitamin C (Chapter 52), vitamin A and β-carotene (see above), and vitamin E (see above). A summary of the principal reactive species and the mechanisms available for their destruction is given in Table 53–2. The interaction of some of the antioxidant mechanisms is depicted in Figure 53–12. There is growing evidence of the involvement of free radicals and other reactive molecules in disease processes. The main evidence comes from epidemiologic studies showing statistical correlations between the incidence of disease and low levels of antioxidant

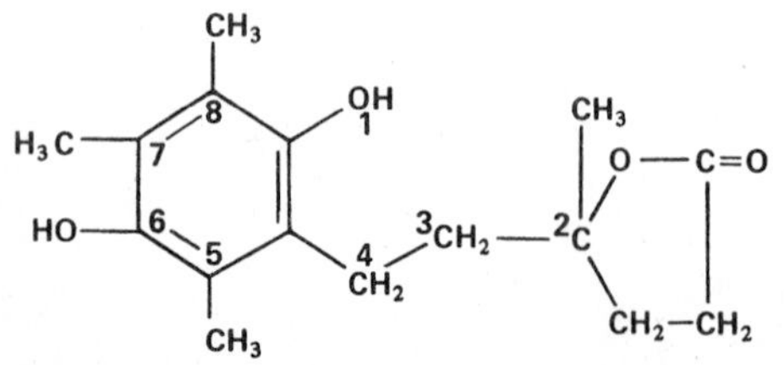

Figure 53–10. The oxidation product of α-tocopherol. The numbers allow one to relate the atoms to those in the parent compound.

Table 53–2. Reactive oxygen species and their antioxidants.

	Reactive Species	Antioxidant
$^1O_2$	Singlet oxygen*	Vitamin A, β-carotene, vitamin E
O_2^{-}	Superoxide free radical	Superoxide dismutase, vitamin E, β-carotene
OH•	Hydroxyl free radical	
RO•	Alkoxyl free radical	
ROO•	Peroxyl free radical	Vitamin E, vitamin C
H_2O_2	Hydrogen peroxide	Catalase, glutathione peroxidase
LOOH	Lipid peroxides	Glutathione peroxidase

*Electrons in the O_2 molecule are at a higher energy level.

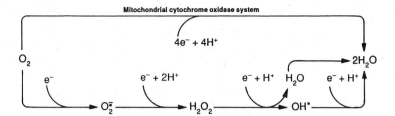

Figure 53–11. Production of reactive oxygen species during reduction of oxygen to water. Reduction of 1 mol of O_2 via the cytochrome oxidase system of the respiratory chain requires $4e^-$. However, certain reactions permit this reduction to take place by a series of univalent reductions each of which needs a single e^-. Reactive oxygen species are generated in this pathway. (•, free radical.)

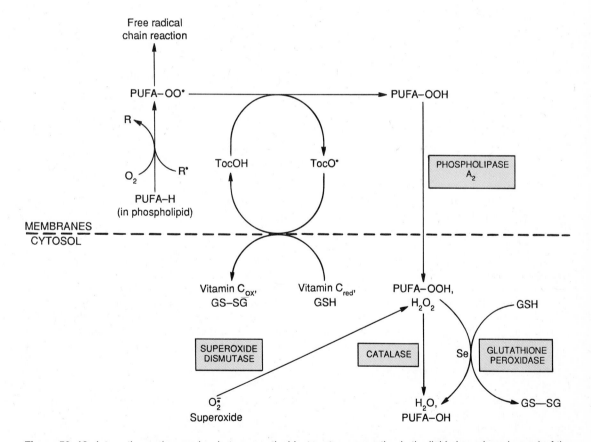

Figure 53–12. Interaction and synergism between antioxidant systems operating in the lipid phase (membranes) of the cell and the aqueous phase (cytosol). (R•}, free radical; PUFA-OO•}, peroxyl free radical of polyunsaturated fatty acid in membrane phospholipid; PUFA-OOH, hydroperoxy polyunsaturated fatty acid in membrane phospholipid released as hydroperoxy free fatty acid into cytosol by the action of phospholipase A_2; PUFA-OH, hydroxy polyunsaturated fatty acid; TocOH, vitamin E (α-tocopherol); TocO•, free radical of α-tocopherol; Se, selenium; GSH, reduced glutathione; GS—SG, oxidized glutathione, which is returned to the reduced state after reaction with NADPH catalyzed by glutathione reductase; PUFA-H, polyunsaturated fatty acid.)

nutrients in the blood or diet. This is the case with respect to cancer and selenium, vitamin A, β-carotene, vitamin C, and vitamin E. There is also an inverse relationship between the incidence of cardiovascular disease and the status of vitamins E and C. This finding agrees with other studies showing that oxidized LDL is taken up by macrophages and foam cells more readily than is normal LDL and that the antioxidant drug probucol has a beneficial effect on these processes. There is some evidence that topical vitamin E can protect skin against damaging effects of ultraviolet rays. At present, the case for general supplementation with any or all of the above antioxidants has not been made, and decisions must await the results of long-term intervention trials now in progress. However, it is recommended that the consumption of cereals, nuts, fruits, and vegetables—all good sources of antioxidants—should be increased.

VITAMIN K

Vitamins belonging to the K group are polyisoprenoid-substituted napthoquinones (Figure 53–13). Menadione (K_3), the parent compound of the vitamin K series, is not found naturally, but if administered it is alkylated in vivo to one of the menaquinones (K_2). Phylloquinone (K_1) is the major form of vitamin K found in plants. Menaquinone-7 is one of the series of polyprenoid unsaturated forms of vitamin K found in animal tissues and synthesized by bacteria in the intestine.

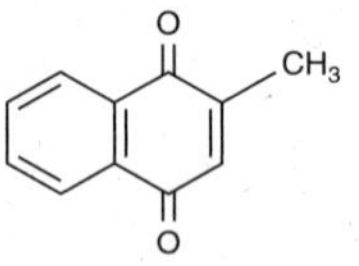

Menadione (vitamin K_3)

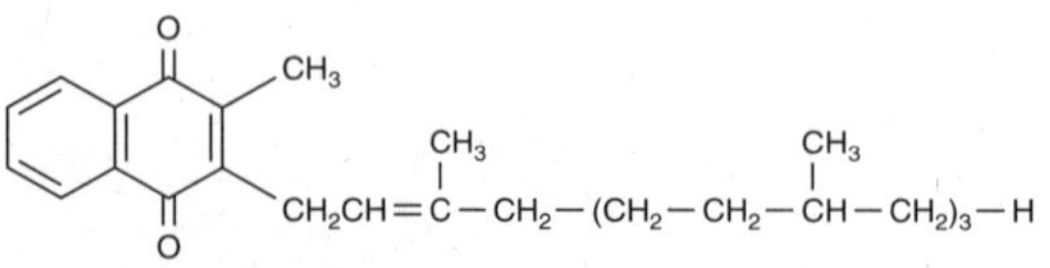

Phylloquinone (vitamin K_1, phytonadione, Mephyton)

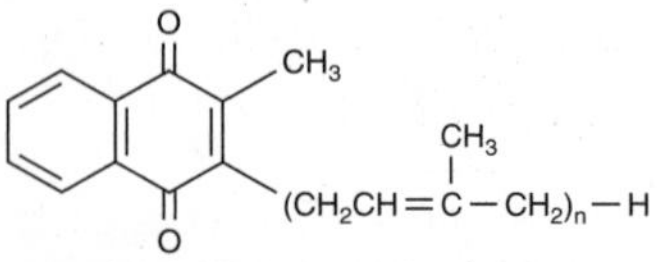

Menaquinone-n (vitamin K_2; n = 6, 7, or 9)

Figure 53–13. The vitamins K. Menadione is 2-methyl-1,4-napththoquinone. The other K vitamins are polyisoprenoid-substituted.

Absorption of Vitamin K Requires Normal Fat Absorption

Fat malabsorption is the commonest cause of vitamin K deficiency. The naturally occurring K derivatives are absorbed only in the presence of bile salts, like other lipids, and are distributed in the bloodstream via the lymphatics, in chylomicrons. Menadione, being water soluble, is absorbed **even in the absence of bile salts,** passing directly into the hepatic portal vein. Although vitamin K accumulates initially in the liver, its hepatic concentration declines rapidly and storage is limited.

Vitamin K Is Required for Biosynthesis of Blood Clotting Factors

Vitamin K has been shown to be involved in the maintenance of normal levels of **blood clotting factors II, VII, IX, and X,** all of which are synthesized in the liver initially as **inactive precursor proteins** (Chapter 59).

Vitamin K Acts as Cofactor of the Carboxylase That Forms γ-Carboxyglutamate in Precursor Proteins

Generation of the biologically active clotting factors involves the **posttranslational modification** of glutamate (Glu) residues of the precursor proteins to γ-carboxyglutamate (Gla) residues by a specific vitamin K dependent carboxylase (Figure 53–14). Prothrombin (factor II) contains ten of these residues, which allow **chelation of calcium** in a specific protein-calcium-phospholipid interaction that is essential to their biologic role (Figure 53–15). Other proteins containing K-dependent Gla residues have now been identified in several tissues.

The Vitamin K Cycle Allows Reduced Vitamin K to Be Regenerated

The vitamin K-dependent carboxylase reaction occurs in the endoplasmic reticulum of many tissues and requires molecular oxygen, carbon dioxide (not HCO_3^-), and the **hydroquinone** (reduced) form of vitamin K. In the endoplasmic reticulum of liver

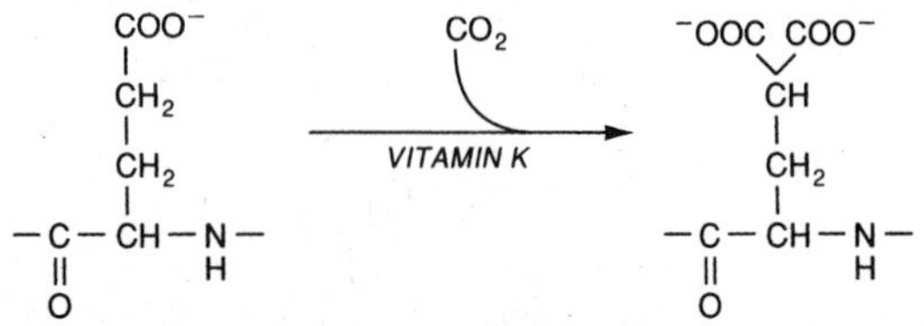

Figure 53–14. Carboxylation of a glutamate residue catalyzed by vitamin K-dependent carboxylase.

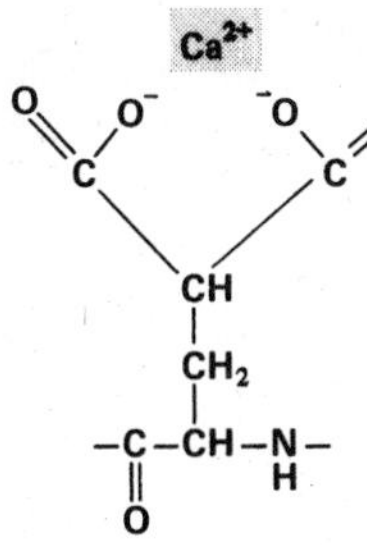

Figure 53–15. The chelation of calcium ion by the γ-carboxyglutamyl residue in clotting-factor proteins.

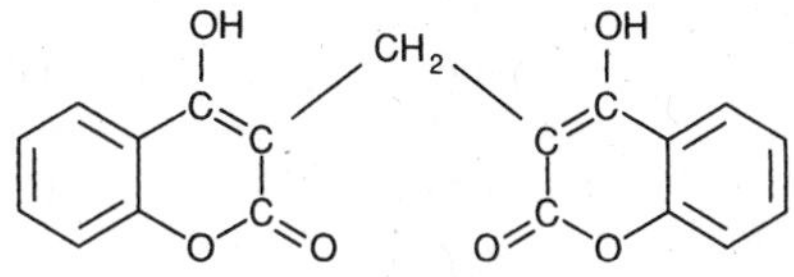

Figure 53–17. Dicumarol (bishydroxycoumarin).

epoxide reductase and provide a potential source of the active hydroquinone form of vitamin K.

Recent evidence has pointed to a role for vitamin K in the synthesis of bone proteins, eg, osteocalcin.

there exists a **vitamin K cycle** (Figure 53–16) in which the **2,3-epoxide** product of the carboxylation reaction is converted by 2,3-epoxide reductase to the quinone form of vitamin K, using an as yet unidentified dithiol reductant. This reaction is sensitive to inhibition by the 4-hydroxydicoumarin (dicumarol) type of anticoagulant, such as **warfarin** (Figure 53–17). Subsequent reduction of the quinone form to the hydroquinone by NADH completes the vitamin K cycle for regenerating the active form of the vitamin.

An important therapeutic use of vitamin K is as an antidote to poisoning by dicumarol-type drugs. The quinone forms of vitamin K will bypass the inhibited

Hemorrhagic Disease of the Newborn Is Caused by Deficiency of Vitamin K

Vitamin K is widely distributed in plant and animal tissues used as food, and production of the vitamin by the microflora of the intestine virtually ensures that dietary deficiency does not occur in adults. However, newborn infants are vulnerable to the deficiency, because the placenta does not pass the vitamin to the fetus efficiently and the gut is sterile immediately after birth. In normal infants, the plasma concentration decreases immediately after birth but recovers after food is absorbed. If the prothrombin level drops too low, the hemorrhagic syndrome may appear.

Vitamin K deficiency can be caused by fat malabsorption, which may be associated with pancreatic dysfunction, biliary disease, atrophy of the intestinal mucosa, or any cause of steatorrhea. In addition, sterilization of the large intestine by antibiotics can result in deficiency when dietary intake is limited.

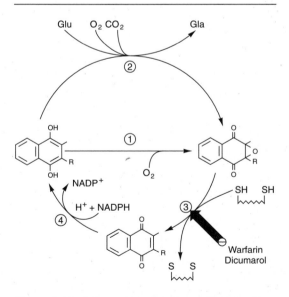

Figure 53–16. Vitamin K cycle in liver. The locus of action of the dicumarol-type anticoagulants is indicated. The details of some of the reactions are still uncertain. (① monoxygenase; ② carboxylase; ③ 2,3-epoxide reductase; ④ reductase.) (Modified and reproduced, with permission, from Suttie JW: The metabolic role of vitamin K. Fed Proc 1980;39:2730.)

SUMMARY

1. The lipid-soluble vitamins have the common features of being apolar, hydrophobic molecules and also of being isoprene derivatives. They all require normal fat absorption to be occurring for efficient absorption, and if this mechanism is defective, deficiency symptoms are likely to occur.

2. Vitamin A (retinol) is represented not only as such in the diet but also by the provitamin (β-carotene) in plants. Retinol and retinoic acid are considered to act by controlling gene expression, whereas retinal is utilized in vision and has a role in glycoprotein synthesis.

3. Vitamin D is a steroid prohormone whose activity is carried out by its hormone derivative, calcitriol. It is utilized in the regulation of calcium and phosphate metabolism, and its omission from the diet leads to rickets and osteomalacia.

4. Vitamin E (tocopherol) is the most important antioxidant in the body, acting in the lipid phase of

membranes throughout the cell. It protects against the effects of toxic radicals such as the peroxyl free radical, mainly as a breaker of free-radical chain reactions. The requirement for tocopherol increases with increased intake of polyunsaturated fat.

5. Vitamin K is needed for the synthesis of several blood clotting factors (eg, II, VII, IX, and X). It functions as cofactor to a carboxylase that acts on glutamate residues of clotting factor precursor proteins to enable them to chelate calcium. Interruption of the vitamin K regeneration cycle by dicumarol type compounds is the basis for their anticoagulant properties.

REFERENCES

Adams JS et al: Vitamin D synthesis and metabolism after ultraviolet irradiation of normal and vitamin D-deficient subjects. N Engl J Med 1982;306:722.

Duthie GG et al: Oxidants, antioxidants and cardiovascular disease. Nutr Res Rev 1989;2:51.

Farber JL et al: Biology of disease: Mechanisms of cell injury by activated oxygen species. Lab Invest 1990; 62:670.

Halliwell B, Chirico S: Lipid peroxidation: Its mechanism, measurement, and significance. Am J Clin Nutr 1993;57:715S.

Inhibition of free radical chain oxidation by α-tocopherol and other plasma antioxidants. Nutr Rev 1988;46:206.

Jamieson D: Oxygen toxicity and reactive oxygen metabolites in mammals. Free Radical Biol Med 1989;7:87.

Krinsky NI: Actions of carotenoids in biological systems. Annu Rev Nutr 1993;13:561.

Leo MA, Lieber CS: Hypervitaminosis A: A liver lover's lament. Hepatology 1988;8:412.

Olson RE et al (editors): *Present Knowledge in Nutrition,* 5th ed, The Nutrition Foundation, 1984.

Passmore R, Eastwood MA: *Human Nutrition and Dietetics.* Churchill Livingstone, 1986.

Reilly PM et al: Pharmacologic approach to tissue injury mediated by free radicals and other reactive oxygen metabolites. Am J Surg 1991;161:488.

Season, latitude, and ability of sunlight to promote synthesis of vitamin D_3 in skin. Nutr Rev 1989;47:252.

Slater TF, Block G (editors): Antioxidant vitamins and β-carotene in disease prevention. Am J Clin Nutr 1991;53: 1985.

Sokol RJ: Vitamin E deficiency and neurologic disease. Annu Rev Nutr 1988;8:351.

Sporn MB, Roberts AB: Role of retinoids in differentiation and carcinogenesis. Cancer Res 1983;43:3034.

Suttie JW: The metabolic role of vitamin K. Fed Proc 1980;39:2730.

Nutrition

54

Peter A. Mayes, PhD, DSc

INTRODUCTION

The science of nutrition examines the qualitative and quantitative requirements of the diet necessary to maintain good health. All components of the diet needed to maintain life appear to be known, since it is possible to sustain humans or other animals on chemically defined diets. However, considerable discussion and controversy still surround the quantitative requirements of each component of the diet, particularly as it varies with the age, sex, and lifestyle of the individual. Metabolic biochemistry provides much of the understanding of modern concepts of nutrition and has been discussed in earlier parts of this book. In particular, the biochemical roles played by the water-soluble and lipid-soluble vitamins have been described in the preceding two chapters.

BIOMEDICAL IMPORTANCE

Overt nutritional deficiency is rare in more affluent populations, although some degree of nutritional deficiency may be present among the poor or the elderly and among groups with specialized nutritional requirements, eg, growing children, pregnant or lactating women, ill and convalescing patients, alcoholics, or individuals on restricted diets from necessity, eg, patients on intravenous feeding, or from choice, eg, vegans (total vegetarians). In more deprived populations, overt deficiencies are more widespread, eg, deficiency of protein **(kwashiorkor),** of vitamins (vitamin A in **xerophthalmia**), of minerals (iron, giving rise to **anemia**), and of energy **(starvation).** Malabsorption may lead to deficiency of nutrients and cause pathologic conditions; eg, malabsorption of vitamin B_{12} causes **anemia.** Although **obesity** has always been associated with dietary excess, the concept of excess intake of particular nutrients and their association with the prevalence of certain diseases in developed societies is gaining recognition, eg, with atherosclerosis and coronary heart disease, diabetes, cancer of the breast and colon, cerebrovascular disease and strokes, and cirrhosis of the liver.

NUTRITIONAL REQUIREMENTS CAN NOW BE DEFINED

Table 54–1 summarizes nutritional requirements.

ENERGY IS REQUIRED TO POWER ALL BODY FUNCTIONS

The mammalian body requires nutrients sufficient to provide free energy to manufacture the daily requirement of high-energy phosphate (mainly ATP) and reducing equivalents (2H) needed to power all body functions (see Figure 17–1).

Carbohydrate and Fat Are the Principal Energy Sources in the Diet

Energy-yielding nutrients are provided by dietary carbohydrate, fat, and, to a lesser extent, protein in widely varying proportions among different human and animal populations. Consumption of alcohol can also provide a significant proportion of energy intake.

A constant body weight under conditions of unaltered energy requirements indicates that there is enough energy in the diet for immediate needs.

The amount of energy available in the major food sources is indicated in Table 54–2. The large energy content per gram of fat compared to that of protein or carbohydrate and the relatively high energy content of alcohol are two notable facts. The recommended energy intake per head for selected groups of people is given in Table 54–3.

Several Factors Affect Expenditure of Energy

Under conditions of **energy equilibrium** (calorie balance), energy intake must equal energy expenditure. Energy expenditure varies widely in different conditions and may be measured by placing an animal inside an insulated chamber and measuring the energy output represented by heat loss and excretory products. It is usually more convenient to measure **oxygen consumption,** since under most conditions

Table 54–1. Essential nutritional requirements.

	Humans	Selected Differences in Other Species
Amino acids	Histidine,[1] isoleucine, leucine, lysine, methionine (cysteine[3]), phenylalanine (tyrosine[3]), threonine, tryptophan, valine	Arginine[2] required by growing rats and adult and growing cats. Glycine required in chicks and taurine in cats. Most amino acids not essential in ruminants; requirement spared in other herbivores with substantial population of microorganisms in the gut.
Fatty acids	Linoleic acid (arachidonic acid[3]), α-linolenic acid[4]	Arachidonic acid is a specific requirement in cats.
Vitamins Water-soluble	Ascorbic acid (C), biotin,[5] cobalamin (B_{12}), folic acid, niacin, pantothenic acid, pyridoxine (B_6), riboflavin (B_2), thiamin (B_1)	Most mammals can synthesize ascorbic acid, but it is essential in the diet of primates, guinea pigs, and Indian fruit bats. Water-soluble vitamins are not essential in ruminants; requirements are spared in other herbivores with substantial populations of microorganisms in the gut.
Fat-soluble	Vitamin A, D,[6] E, K[5]	Most species can utilize β-carotene as a source of vitamin A (retinol); must be supplied as retinol in cats.
Minerals Macrominerals	Calcium, chloride, magnesium, phosphorus, potassium, sodium	
Microminerals (trace elements)	Chromium, copper, iodine, iron, manganese, molybdenum, selenium, zinc	Silicon, vanadium, nickel, arsenic, fluoride, and tin have been shown to be essential in various species and may be required in humans. Cobalt is required for synthesis of cobalamin by ruminal microorganisms.
Fiber	Required for optimal health	
Water	The most critical component of the diet	
Energy	Utilization of carbohydrates, fats, and protein in variable proportions	

[1]Required in infants and probably in children and adults.
[2]May be partly essential in infants.
[3]Cysteine, tyrosine, and arachidonic acid spare the requirement for methionine, phenylalanine, and linoleic acid, respectively.
[4]Workers disagree whether α-linolenic acid is essential in the human diet.
[5]Synthesized by intestinal microorganisms; therefore, dietary requirement uncertain.
[6]Exposure of the skin to sunlight reduces dietary requirement.

Table 54–2. Heats of combustion and energy available from the major food sources.[1]

	Energy kcal/g (kJ/g)		
	Heat of Combustion (Bomb Calorimeter)	Human Oxidation	Standard Conversion Factors[2]
Protein	5.4 (22.6)	4.1 (17.2)[3]	4 (17)
Fat	9.3 (38.9)	9.3 (38.9)	9 (38)
Carbohydrate	4.1 (17.2)	4.1 (17.2)	4 (17)
Ethanol	7.1 (29.7)	7.1 (29.7)	7 (29)

[1]Adapted from Davidson S et al: *Human Nutrition and Dietetics,* 7th ed. Churchill Livingstone, 1979.
[2]Conversion factors are obtained by rounding off heats of combustion and correcting for estimates of absorption efficiency.
[3]Protein oxidation corrected for loss of amino groups excreted in urine as urea.

Table 54–3. Recommended energy intake for men and women.[1]

		Weight		Energy Needs		
		(kg)	(lb)	(kcal)		(MJ)
Category	Age (years)			Mean	Range	
Men	23–50	70	154	2900	2300–3100	12.1
Women	23–50	55	120	2200	1600–2400	9.2
Pregnant				+300		
Lactating				+500		

[1]Data from *Recommended Dietary Allowances,* 10th ed. Food and Nutrition Board. National Research Council—National Academy of Sciences, 1989.

1 L of O_2 consumed accounts for approximately 20 kJ (4.83 kcal) of energy expended.

The energy expended by an individual depends on four main factors:

(1) The **basal metabolic rate** is the energy expenditure necessary to maintain basic physiologic functions under standardized conditions; the subject should be at rest, awake, and in a warm environment, and measurements should be taken at least 12 hours after the last meal. The basal metabolic rate is proportionate to lean body weight and to surface area. It is higher in males than females, in young children, and in people with **fever** and **hyperthyroidism.** It is lower in **hypothyroidism** and in **starvation.**

(2) The **thermogenic effect** (specific dynamic action) of food is equivalent to about 5–10% of total energy expenditure and is attributed to the energy expenditure due to digestion and to any stimulation of metabolism caused by the influx of new substrate.

(3) **Physical activity** is the largest variable affecting energy expenditure; the range is over tenfold between resting and maximum athletic activity.

(4) When **environmental temperature** is low, it causes increased energy expenditure owing to shivering, and to nonshivering thermogenesis in animals having brown fat (Figure 27–10). At temperatures above blood heat, extra energy is expended in cooling.

PROTEINS SUPPLY SPECIFIC AMINO ACIDS AND AMINO NITROGEN FOR THE SYNTHESIS OF KEY NITROGENOUS COMPOUNDS

Protein normally provides the body's requirement for amino acid nitrogen and for specific amino acids. Dietary protein is digested and enters the circulation as individual amino acids. The tissues require 20 amino acids to synthesize specific proteins and other nitrogen-containing compounds such as purines, pyrimidines, and heme.

Essential Amino Acids Are Required Amino Acids That Cannot Be Synthesized by the Body and Must Therefore Be Supplied by the Diet

There are nine essential amino acids in humans: histidine, isoleucine, leucine, lysine, methionine, phenylalanine, threonine, tryptophan, and valine (Table 54–1). Two other amino acids, cysteine and tyrosine, may be formed from the essential amino acids methionine and phenylalanine, respectively. If sufficient cysteine and tyrosine are present in the diet, they spare the requirement for methionine and phenylalanine.

As long as sufficient amounts of essential amino acids are present in the diet, the remaining amino acids required for protein synthesis and other purposes can be formed through transamination and other reactions (Figure 31–4).

Nitrogen Balance Is Maintained by Dietary Intake
(See also Chapter 31)

An adult animal in a state of metabolic equilibrium requires dietary protein to replace the essential amino acids and amino acid nitrogen lost during metabolic turnover. Nitrogen is lost in the urine, feces, saliva, desquamated skin, hair, and nails. The daily requirements for total protein and essential amino acids in humans are set forth in Table 54–4. When these requirements are calculated on the basis of body weight, the extra growth needs of infants and children are clearly evident. Pregnancy, lactation, tis-

Table 54–4. Estimated protein and amino acid requirements and intakes in humans.[1]

	Requirements (mg/kg body weight/d)			Intake (g/d)	
	Infant (4–6 months)	Child (10–12 years)	Adult	Adult (70 kg) Allowance[1]	Estimated Adult Intake in USA[2]
Protein	1100	1000	800	56	101
Animal	...	...	...	...	71
Vegetable	...	...	...	...	30
Essential amino acids	(3–4 months)				
Histidine	28	?	10	0.70	?
Isoleucine	70	28	10	0.70	5.3
Leucine	161	42	14	0.98	8.2
Lysine	103	44	12	0.84	6.7
Methionine (and cysteine)	58	22	13	0.91	2.1
Phenylalanine (and tyrosine)	125	22	14	0.98	4.7
Threonine	87	28	7	0.49	4.1
Tryptophan	17	3.3	3.5	0.25	1.2
Valine	93	25	10	0.70	5.7

[1]Data from *Recommended Dietary Allowances,* 10th ed. Food and Nutrition Board, National Research Council—National Academy of Sciences, 1989.
[2]Data from Munro HN, Crim M: The proteins and amino acids. In: Goodhart RS, Shils ME: *Modern Nutrition in Health and Disease,* 6th ed. Lea & Febiger, 1980.

sue repair after injury, recovery from illness, and increased physical activity all require more dietary protein. For most situations, a diet in which 12% of the energy is supplied as protein is adequate in humans.

The Efficiency With Which Dietary Protein Is Used Determines the Total Quantity of Protein Required

The quantity of protein required is affected by three major factors: protein quality, energy intake, and physical activity.

A. Protein Quality: The quality of protein is measured by comparing the proportions of essential amino acids in a food with the proportions required for good nutrition. The closer the proportions are, the higher the protein quality. Egg and milk proteins are **high-quality proteins** that are efficiently utilized by the body and are used as reference standards against which other proteins can be compared. Meat protein is of high protein quality, whereas several proteins from plants used as major food sources are relatively deficient in certain essential amino acids, eg, tryptophan and lysine in maize (corn), lysine in wheat, and methionine in some beans. In a mixed diet, a deficiency of an amino acid in one protein is made up by its abundance in another; such proteins are described as **complementary;** eg, the protein of wheat and beans combined provides a satisfactory amino acid intake. Under such circumstances, a greater *total* amount of protein must be consumed to satisfy requirements. Amino acids that are not incorporated into new protein and are unnecessary for immediate requirements cannot be stored and are rapidly degraded, and the nitrogen is excreted as urea and other products.

B. Energy Intake: The energy derived from carbohydrate and fat affects protein requirements because it spares the use of protein as an energy source. To use expensive (high-quality) dietary protein efficiently and to reduce requirements for it to a minimum, it is necessary to ensure adequate provision of energy from nonprotein sources, some of which should be carbohydrate in order to spare protein from gluconeogenesis.

C. Physical Activity: Physical activity increases nitrogen retention from dietary protein.

Protein-Energy Malnutrition Causes Marasmus and Kwashiorkor

Protein-energy malnutrition encompasses a range of disorders of starvation and malnutrition that involve deficiencies of other nutrients such as vitamins and minerals in addition to protein. In severe form, it occurs in growing children, usually under 5 years of age, in developing areas of Asia, Africa, and South America. Two extreme forms are recognized: marasmus and kwashiorkor (Chapter 65). In **marasmus,** there is generalized wasting due to deficiency of both energy and protein. In **kwashiorkor,** which is char-

acterized by edema, while energy intake may be adequate, there is a deficiency in both the **quantity and the quality of protein.** Conditions are often encountered that are intermediate between typical marasmus and typical kwashiorkor. Clearly, the two conditions are aggravated by general deficiency of other essential nutrients such as vitamins and minerals.

GLUCOSE REQUIREMENTS CAN BE MET BY MANY CARBOHYDRATES

Glucose is specifically required by many tissues but does not have to be provided as such in the diet since other dietary carbohydrates are readily converted to glucose, either during digestion (eg, starch) or subsequently in the liver (eg, fructose, galactose; Chapter 22). Glucose is also formed from the glycerol moiety of fats and from glucogenic amino acids by gluconeogenesis (Chapter 21). Although a minimum daily intake of carbohydrate (50–100 g) is recommended in humans to prevent **ketosis** (Chapter 29) and loss of muscle protein, a balanced diet should contain more carbohydrate than this in the form of polysaccharide in order to reduce the amount of fat that would otherwise be required for energy. The major foodstuffs containing carbohydrates are described in Chapter 15.

FIBER IS REQUIRED FOR OPTIMAL HEALTH

Dietary fiber consists of all plant cell wall components that cannot be digested by an animal's own enzymes, eg, cellulose, hemicellulose, lignin, gums, pectins, and pentosans. In herbivores such as ruminants, fiber (mainly as cellulose) is a major source of energy after it has been digested by rumen microorganisms to acetate, propionate, and butyrate, which are absorbed into the portal vein. Colonic fermentation may also contribute to human energy requirements (2–7% on low-fiber intakes). Gases such as CO_2, H_2, and sometimes CH_4 are also produced.

In humans, a high-fiber diet aids water retention during passage of food along the gut, producing larger, softer feces. A high-fiber diet is associated with reduced incidence of **diverticulosis, cancer of the colon, cardiovascular disease,** and **diabetes mellitus.** The more insoluble fibers such as cellulose and lignin found in wheat bran are beneficial with regard to colonic function, whereas the more soluble fibers found in legumes and fruit, eg, gums and pectins, lower blood cholesterol, possibly by binding bile acids and dietary cholesterol. The soluble fibers also slow stomach emptying, and they delay and attenuate the postprandial rise in blood glucose, with consequent reduction in insulin secretion. This effect is beneficial to diabetics and to dieters because it reduces the rebound fall in blood glucose that stimulates appetite.

LIPIDS ARE REQUIRED AS A VEHICLE FOR LIPID-SOLUBLE VITAMINS AND TO SUPPLY ESSENTIAL FATTY ACIDS

Although lipid frequently provides a significant proportion of the dietary requirement for energy, this is not an essential function. Apart from increasing the palatability of food and producing a feeling of satiety, dietary lipid has two essential functions in mammalian nutrition. It acts as the dietary vehicle for the **lipid-soluble vitamins,** and it supplies **essential polyunsaturated fatty acids** that the body is unable to synthesize. Three polyunsaturated fatty acids have been recognized as essential in the diet of at least some animals: **linoleic acid** (ω6, 18:2), **α-linolenic acid** (ω3, 18:3), and **arachidonic acid** (ω6, 20:4). These acids are found in the lipids of plant and animal food (see Table 16–2). See also Chapter 25 for a discussion of their metabolism.

In humans, arachidonic acid may be formed from linoleic acid and is not essential if sufficient linoleic acid is present in the diet. Debate continues about whether α-linolenic acid is truly essential in humans. **Deficiency of linoleic acid** is rare but may occur in infants on skim milk diets and in patients fed intravenously on lipid-free diets.

A principal function of essential fatty acids is to serve as precursors of **leukotrienes, lipoxins, prostaglandins,** and **thromboxanes** (see Figure 25–6), which function as "local hormones." A dietary intake in which 1–2% of the total energy requirement is supplied as essential fatty acid prevents clinical deficiency.

There Is an Association Between Fat Consumption and Disease

Numerous studies have shown a correlation between **coronary heart disease,** blood cholesterol, and the consumption of fat, particularly of saturated fat (Chapter 28). High fat consumption is also associated with cancer of the breast and colon. The main source of saturated fat in the human diet is the meat of ruminants, dairy products, and hard margarine. Cholesterol is found only in foods of animal origin and particularly in egg yolk.

VITAMINS PERFORM A VARIETY OF BIOCHEMICAL FUNCTIONS

Vitamins are organic nutrients that are required in small quantities for many different biochemical functions, generally cannot be synthesized by the body, and must therefore be supplied by the diet. Humans

require either milligram or microgram quantities of each vitamin per day. Vitamins are classified into two main groups: the **water-soluble vitamins,** more fully described in Chapter 52; and the **lipid-soluble vitamins,** more fully described in Chapter 53.

Water-soluble vitamins include the vitamin B complex (thiamin, riboflavin, niacin, pantothenic acid, vitamin B_6, biotin, vitamin B_{12}, and folic acid) and ascorbic acid (vitamin C). Water-soluble vitamins are absorbed into the hepatic portal vein, and any surplus of most of them is excreted in the urine. There is thus little storage of the free vitamin, which in most instances needs to be continually supplied in the diet. Some storage of folic acid occurs in the liver. Depletion may take several months for ascorbic acid and several years for B_{12} (also stored in the liver). Excess intake is generally well tolerated except for side effects occurring with large doses of niacin (in the form of nicotinic acid), ascorbic acid, or pyridoxine (vitamin B_6).

The **fat-soluble vitamins** (vitamins A, D, E, and K), present in food lipids of both plant and animal origin, are digested with fat and absorbed by the intestine and incorporated into chylomicrons. They are consequently transported mainly in chylomicron remnants, initially to the liver, which serves as a major store of vitamins A, D, and K. Adipose tissue is the major storage source of vitamin E. Fat-soluble vitamins are not excreted in the urine and, if taken in excess, are toxic (particularly vitamins A and D).

Some diseases concerned with cofactor metabolism that respond to treatment with specific vitamins are listed in Table 54–5.

Table 54–5. Vitamin-responsive syndromes. Examples of specific defects in vitamin cofactor metabolism that can be corrected by vitamin therapy, usually requiring very large doses.[1]

Vitamin	Disease	Biochemical Defect
Biotin	Propionic acidemia	Propionyl-CoA carboxylase
Vitamin B_{12}	Methylmalonic aciduria	Formation of cobamide co-enzyme
Folic acid	Folate malabsorption	Folic acid transport
Niacin	Hartnup disease	Tryptophan transport
Pyridoxine (vitamin B_6)	Infantile convulsions	Glutamic acid decarboxylase (?)
	Cystathioninuria	Cystathioninase
	Homocystinuria	Cystathionine synthase
Thiamin	Hyperalaninemia Thiamin-responsive lactic acidosis	Pyruvate dehydrogenase

[1]From: Herman RH, Stifel FB, Greene HL: Vitamin-deficient state and other related diseases. In: *Disorders of the Gastrointestinal Tract; Disorders of the Liver; Nutritional Disorders.* Dietschy JM (editor). Grune & Stratton, 1976.

Nonavailability of vitamins, whether due to dietary or other reasons (eg, defects in absorption), results in characteristic **deficiency syndromes.** Details of these disorders are given in Chapters 52 and 53.

MINERALS ARE REQUIRED FOR BOTH PHYSIOLOGIC AND BIOCHEMICAL FUNCTIONS

Minerals may be divided arbitrarily into two groups: (1) **macrominerals,** which are required in amounts greater than 100 mg/d, and (2) **microminerals (trace elements),** which are required in amounts less than 100 mg/d. Table 54–6 summarizes the properties of the macrominerals, and Table 54–7 does the same for the microminerals.

RECOMMENDED DIETARY ALLOWANCES (RDAs)

A review of the daily needs for essential nutrients has been published by the Food and Nutrition Board of the National Research Council as **recommended dietary allowances** (Table 54–8). The allowances are to provide for individual variations among most normal persons as they live under their usual environmental conditions. They do not allow for extra requirements in illness or pathologic disorders. **Diets should be based on a variety of foods,** both to cover known requirements and to provide other nutrients for which human requirements have been less well defined. Table 54–8 covers protein, ten vitamins, and six minerals. Too few data are available to ascribe RDAs to the remaining vitamins and minerals. However, ranges of intake of these nutrients that appear safe and adequate are given in Table 54–9. All nutri-

Table 54–6. Essential macrominerals: Summary of major characteristics.

Elements	Functions	Metabolism[1]	Deficiency Disease or Symptoms	Toxicity Disease or Symptoms[2]	Sources[3]
Calcium	Constituent of bones, teeth; regulation of nerve, muscle function.	Absorption requires calcium-binding protein. Regulated by vitamin D, parathyroid hormone, calcitonin, etc.	Children: rickets. Adults: osteomalacia. May contribute to osteoporosis.	Occurs with excess absorption due to hypervitaminosis D or hypercalcemia due to hyperparathyroidism, or idiopathic hypercalcemia.	Dairy products, beans, leafy vegetables.
Phosphorus	Constituent of bones, teeth, ATP, phosphorylated metabolic intermediates. Nucleic acids.	Control of absorption unknown (vitamin D?). Serum levels regulated by kidney reabsorption.	Children: rickets. Adults: osteomalacia.	Low serum Ca^{2+}:P_i ratio stimulates secondary hyperparathyroidism; may lead to bone loss.	Phosphate food additives.
Sodium	Principal cation in extracellular fluid. Regulates plasma volume, acid-base balance, nerve and muscle function, Na^+/K^+-ATPase.	Regulated by aldosterone.	Unknown on normal diet; secondary to injury or illness.	Hypertension (in susceptible individuals).	Table salt; salt added to prepared food.
Potassium	Principal cation in intracellular fluid; nerve and muscle function, Na^+/K^+-ATPase.	Also regulated by aldosterone.	Occurs secondary to illness, injury, or diuretic therapy; muscular weakness, paralysis, mental confusion.	Cardiac arrest, small bowel ulcers.	Vegetables, fruit, nuts.
Chloride	Fluid and electrolyte balance; gastric fluid; chloride shift in HCO_3^- transport in erythrocytes.		Infants fed salt-free formula. Secondary to vomiting, diuretic therapy, renal disease.		Table salt.
Magnesium	Constituent of bones, teeth; enzyme cofactor (kinases, etc).		Secondary to malabsorption or diarrhea, alcoholism.	Depressed deep tendon reflexes and respiration.	Leafy green vegetables (containing chlorophyll).

[1]In general, minerals require carrier proteins for absorption. Absorption is rarely complete; it is affected by other nutrients and compounds in the diet (eg, oxalates and phytates that chelate divalent cations). Transport and storage also require special proteins. Excretion occurs in feces (unabsorbed minerals and from bile) and in urine and sweat.
[2]Excess mineral intake produces toxic symptoms. Unless otherwise specified, symptoms include nonspecific nausea, diarrhea, and irritability.
[3]Mineral requirements are met by a varied intake of adequate amounts of whole-grain cereals, legumes, leafy green vegetables, meat, and dairy products.

Table 54–7. Essential microminerals (trace elements): Summary of major characteristics.

Elements	Functions	Metabolism[1]	Deficiency Disease or Symptoms	Toxicity Disease or Symptoms[1]	Good Sources[2]
Chromium	Trivalent chromium, a constituent of "glucose tolerance factor," which binds to and potentiates insulin.		Impaired glucose tolerance; secondary to parenteral nutrition.		Meat, liver, brewer's yeast, whole grains, nuts, cheese.
Cobalt	Required only as a constituent of vitamin B_{12}.	As for vitamin B_{12}.	Vitamin B_{12} deficiency.		Foods of animal origin.
Copper	Constituent of oxidase enzymes: cytochrome c oxidase, etc. Cystosolic superoxide dismutase. Role in iron absorption.	Transported by albumin; bound to ceruloplasmin.	Anemia (hypochromic, microcytic); secordary to malnutrition. Menke's syndrome.	Rare; secondary to Wilson's disease.	Liver.
Iodine	Constituent of thyroxine, triiodothyronine.	Stored in thyroid as thyroglobulin.	Children: cretinism. Adults: goiter and hypothyroidism, myxedema.	Thyrotoxicocis, goiter.	Iodized salt, seafood.
Iron	Constituent of heme enzymes (hemoglobin, cytochromes, etc).	Transported as transferrin; stored as ferritin or hemosiderin; lost in sloughed cells and by bleeding.	Anemia (hypochromic, microcytic).	Siderosis; hereditary hemochromatosis.	Red meat, liver, eggs. Iron cookware.
Manganese	Cofactor of hydrolase, decarboxylase, and transferase enzymes. Glycoprotein and proteoglycan synthesis. Mitochondrial superoxide dismutase.		Unknown in humans.	Inhalation poisoning produces psychotic symptoms and parkinsonism.	
Molybdenum	Constituent of oxidase enzymes (xanthine oxidase).		Secondary to parenteral nutrition.		
Selenium	Constituent of glutathione peroxidase.	Synergistic antioxidant with vitamin E.	Marginal deficiency when soil content is low; secondary to parenteral nutrition, protein-energy malnutrition.	Megadose supplementation induces hair loss, dermatitis, and irritability.	Plants, but varies with soil content. Meat.
Silicon[3]	Role in calcification of bone and in glycosaminoglycan metabolism in cartilage and connective tissue.		Impairment of normal growth.	Silicosis due to long-term inhalation of silica dust.	Plant foods.
Zinc	Cofactor of many enzymes: lactate dehydrogenase, alkaline phosphatase, carbonic anhydrase, etc.		Hypogonadism, growth failure, impaired wound healing, decreased taste and smell acuity; secondary to acrodermatitis enteropathica, parenteral nutrition.	Gastrointestinal irritation, vomiting.	
Fluoride[4]	Increases hardness of bones and teeth.		Dental caries; osteoporosis (?).	Dental fluorosis.	Drinking water.

[1]Excess mineral intake produces toxic symptoms. Unless otherwise specified, symptoms include nonspecific nausea, diarrhea, and irritability.

[2]Mineral requirements are met by a varied intake of adequate amounts of whole-grain cereals, legumes, leafy green vegetables, meat, and dairy products.

[3]Not yet demonstrated to be essential to humans but necessary in several animals.

[4]Fluoride is essential for rat growth. While not proved to be strictly essential for human nutrition, fluorides have a well-defined role in prevention and treatment of dental caries.

Table 54–8. Recommended daily dietary allowances. Designed for the maintenance of good nutrition of the majority of healthy people in the USA.[1]

	Age (years)	Weight (kg)	Weight (lb)	Height (cm)	Height (in)	Protein (g)	Vitamin A (µg RE)[2]	Vitamin D (µg)[3]	Vitamin E (mg α-TE)[4]	Vitamin K (µg)	Vitamin C (mg)	Thiamin (mg)	Riboflavin (mg)	Niacin (mg NE)[5]	Vitamin B6 (mg)	Folate (µg)	Vitamin B12 (µg)	Calcium (mg)	Phosphorus (mg)	Magnesium (mg)	Iron (mg)	Zinc (mg)	Iodine (µg)	Selenium (µg)
Infants	0.0–0.5	6	13	60	24	13	375	7.5	3	5	30	0.3	0.4	5	0.3	25	0.3	400	300	40	6	5	40	10
	0.5–1.0	9	20	71	28	14	375	10	4	10	35	0.4	0.5	6	0.6	35	0.5	600	500	60	10	5	50	15
Children	1–3	13	29	90	35	16	400	10	6	15	40	0.7	0.8	9	1.0	50	0.7	800	800	80	10	10	70	20
	4–6	20	44	112	44	24	500	10	7	20	45	0.9	1.1	12	1.1	75	1.0	800	800	120	10	10	90	20
	7–10	28	62	132	52	28	700	10	7	30	45	1.0	1.2	13	1.4	100	1.4	800	800	170	10	10	120	30
Males	11–14	45	99	157	62	45	1000	10	10	45	50	1.3	1.5	17	1.7	150	2.0	1200	1200	270	12	15	150	40
	15–18	66	145	176	69	59	1000	10	10	65	60	1.5	1.8	20	2.0	200	2.0	1200	1200	400	12	15	150	50
	19–24	72	160	177	70	58	1000	10	10	70	60	1.5	1.7	19	2.0	200	2.0	1200	1200	350	10	15	150	70
	25–50	79	174	176	70	63	1000	5	10	80	60	1.5	1.7	19	2.0	200	2.0	800	800	350	10	15	150	70
	51+	77	170	173	68	63	1000	5	10	80	60	1.2	1.4	15	2.0	200	2.0	800	800	350	10	15	150	70
Females	11–14	46	101	157	62	46	800	10	8	45	50	1.1	1.3	15	1.4	150	2.0	1200	1200	280	15	12	150	45
	15–18	55	120	163	64	44	800	10	8	55	60	1.1	1.3	15	1.5	180	2.0	1200	1200	300	15	12	150	50
	19–24	58	128	164	65	46	800	10	8	60	60	1.1	1.3	15	1.6	180	2.0	1200	1200	280	15	12	150	55
	25–50	63	138	163	64	50	800	5	8	65	60	1.1	1.3	15	1.6	180	2.0	800	800	280	15	12	150	55
	51+	65	143	160	66	50	800	5	8	65	60	1.0	1.2	13	1.6	180	2.0	800	800	280	10	12	150	55
Pregnant						60	800	10	10	65	70	1.5	1.6	17	2.2	400	2.2	1200	1200	320	30	15	175	65
Lactating						65	1300	10	12	65	95	1.6	1.8	20	2.1	280	2.6	1200	1200	355	15	19	200	75

[1] Data from Recommended Dietary Allowances, 10th ed. Food and Nutrition Board, National Research Council—National Academy of Sciences, 1989.
[2] Retinol equivalents. 1 retinol equivalent = 1 µg retinol or 6 µg β-carotene.
[3] As cholecalciferol. 10 µg cholecalciferol = 400 IU of vitamin D.
[4] α-Tocopherol equivalents. 1 mg α-tocopherol = 1 α-TE.
[5] 1 NE (niacin equivalent) = 1 mg of niacin or 60 mg of dietary tryptophan.

Table 54–9. Estimated safe and adequate daily dietary intakes of selected vitamins and minerals.[1]

	Age (years)	Vitamins		Trace Elements					Electrolytes		
		Biotin (µg)	Pantothenic Acid (mg)	Copper (mg)	Manganese (mg)	Fluoride (mg)	Chromium (mg)	Molybdenum (µg)	Sodium (mg)	Potassium (mg)	Chloride (mg)
Infants	0–0.5	10	2	0.4–0.6	0.3–0.6	0.1–0.5	0.01–0.04	15–30	115–350	350–925	275–700
	0.5–1	15	3	0.6–0.7	0.6–1.0	0.2–1.0	0.02–0.06	20–40	250–750	425–1275	400–1200
Children and adolescents	1–3	20	3	0.7–1.0	1.0–1.5	0.5–1.5	0.02–0.08	25–50	325–975	550–1650	500–1500
	4–6	25	3–4	1.0–1.5	1.5–2.0	1.0–2.5	0.03–0.12	30–75	450–1350	775–2325	700–2100
	7–10	30	4–5	1.0–2.0	2.0–3.0	1.5–2.5	0.05–0.2	50–150	600–1800	1000–3000	925–2775
	11+	30–100	4–7	1.5–2.5	2.0–5.0	1.5–2.5	0.05–0.2	75–250	900–2700	1525–4575	1400–4200
Adults		30–100	4–7	1.5–3.0	2.0–5.0	1.5–4.0	0.05–0.2	75–250	1100–3300	1875–5625	1700–5100

[1] Data from *Recommended Dietary Allowances*, 10th ed. Food and Nutrition Board, National Research Council—National Academy of Sciences, 1989.

tional requirements must be met to prevent deficiency diseases and ill health. Ignorance or poor economic conditions are almost always the underlying cause of failure to satisfy nutritional requirements. On the other hand, certain common diseases are associated with excess intake of nutrients. Obesity generally reflects excess intake of energy and is often associated with the development of **non-insulin-dependent diabetes mellitus. Atherosclerosis** and **coronary heart disease** are associated with diets high in total fat and saturated fat. **Cancer** of the breast, colon, and prostate correlates with high fat intake. A high incidence of **cerebrovascular disease** and of hypertension is associated with a high salt intake.

Many committees throughout the world have investigated the composition of human diets and made recommendations for improvement. Their recommendations may be summarized as follows: (1) If overweight, total energy intake should be reduced to achieve optimum weight. (2) There should be a general shift away from fat consumption to consumption of more carbohydrate. (3) A greater proportion of carbohydrate should be consumed as complex carbohydrates and less as sugars. (4) A greater proportion of dietary fat should be in the form of polyunsaturated and monounsaturated fat and less as saturated fat. (5) Consumption of cholesterol and salt should be reduced. (6) Dietary fiber should be increased. (7) Consumption of fruits and vegetables, particularly for their antioxidant nutrients, should be increased.

SUMMARY

1. Nutrition is concerned with both the qualitative and quantitative requirements of the diet. Virtually all qualitative requirements are now known, but there is still considerable debate about the quantities needed of each nutrient for optimal health.

2. The diet must supply enough energy to power all body functions. The requirement for energy varies with age, sex, physical activity, and temperature of the environment.

3. Twenty different amino acids are required for protein synthesis, of which nine, the nutritionally essential amino acids, must be supplied in the human diet. The quantity of protein required is affected by protein quality, energy intake, and physical activity.

4. Glucose requirements can be supplied by other carbohydrates, particularly starch, but lipid requirements are more specific, since certain polyunsaturated fatty acids of the n-3 and n-6 families cannot be synthesized and must be supplied in the diet.

5. Vitamins are also essential dietary requirements. Those of the water-soluble B group are mainly enzyme cofactors, whereas ascorbic acid (vitamin C), also water-soluble and an antioxidant, is needed to keep metal cofactors in the reduced state. The lipid-soluble vitamins have a variety of functions from vision (vitamin A), calcium and phosphate metabolism (vitamin D), and clotting (vitamin K) to the antioxidant properties of vitamins E and A.

6. Modern nutrition is also concerned with excesses in the diet, since they have been associated with diseases such as obesity, non-insulin-dependent diabetes mellitus, atherosclerosis, cancer, and hypertension.

REFERENCES

Burk RF, Hill KE: Regulation of selenoproteins. Annu Rev Nutr 1993;13:65.

Cummings JH: Dietary fibre. Br Med Bull 1981;37:65.

Forbes JM: Metabolic aspects of the regulation of voluntary food intake and appetite. Nutr Res Rev 1988;1:145.

Fuller MF, Garlick PJ: Human amino acid requirements. Annu Rev Nutr 1994;14:217.

Kritchevsky D: Dietary fiber. Annu Rev Nutr 1988;8:301.

National Academy of Sciences report on diet and health. Nutr Rev 1989;47:142.

Nestle M: *Nutrition in Clinical Practice.* Jones Medical Publications, 1985.

Nielsen FH: Nutritional significance of the ultratrace elements. Nutr Rev 1988;46:337.

Olson RE et al (editors): *Present Knowledge in Nutrition,* 5th ed. Nutrition Foundation, 1984.

Passmore R, Eastwood MA: *Human Nutrition and Dietetics,* 8th ed. Churchill Livingstone, 1986.

Stadtman TC: Selenium biochemistry. Annu Rev Biochem 1990;59:111.

Taylor A: Associations between nutrition and cataract. Nutr Rev 1989;47:225.

Woo R, Daniels-Kush R, Horton ES: Regulation of energy balance. Annu Rev Nutr 1985;5:411.

Digestion & Absorption

55

Peter A. Mayes, PhD, DSc

INTRODUCTION

Most foodstuffs are ingested in forms that are unavailable to the organism, since they cannot be absorbed from the digestive tract until they have been broken down into smaller molecules. This disintegration of the naturally occurring foodstuffs into assimilable forms constitutes the process of digestion.

The chemical changes incident to digestion are accomplished with the aid of hydrolase enzymes of the digestive tract that catalyze the hydrolysis of native proteins to amino acids, of starches to monosaccharides, and of triacylglycerols to monoacylglycerols, glycerol, and fatty acids. In the course of these digestive reactions, the minerals and vitamins of the foodstuffs are also made more assimilable.

A systematic account of the nature and functions of the gastrointestinal hormones is given in Chapter 51.

BIOMEDICAL IMPORTANCE

Some clinical conditions arise from defects in the digestive processes, such as **ulceration** by gastric HCl or diminished secretion of HCl causing **achlorhydria.** Defects of bile secretion lead to **gallstones** or possibly defective lipid digestion. Exocrine pancreatic insufficiency in **cystic fibrosis** causes steatorrhea. Malabsorption of nutrients is due to a wide variety of defects and often leads to **nutritional deficiency;** eg, malabsorption of vitamin B_{12} and folate causes **anemia;** defects in calcium and magnesium lead to tetany; and vitamin D malabsorption leads to **rickets** and **osteomalacia;** and a general **malabsorption syndrome** includes these and other defects. Lactase deficiency gives rise to **milk intolerance,** and defects in absorption of neutral amino acids are involved in **Hartnup disease.**

DIGESTION BEGINS IN THE ORAL CAVITY

Saliva, secreted by the salivary glands, consists of about 99.5% water. It acts as a lubricant for mastication and for swallowing. Adding water to dry food provides a medium in which food molecules can dissolve and in which hydrolases can initiate digestion. Mastication subdivides the food, increasing its solubility and surface area for enzyme attack. The saliva is also a vehicle for the excretion of certain drugs (eg, ethanol and morphine), of inorganic ions such as K^+, Ca^{2+}, HCO_3^-, thiocyanate (SCN^-), and iodine, and of immunoglobulins (IgA).

The pH of saliva is usually about 6.8, though it may vary on either side of neutrality.

Saliva Contains an Amylase

Salivary amylase is capable of bringing about the hydrolysis of starch and glycogen to maltose; however, this is of little significance in the body because of the short time during which it can act on the food. Salivary amylase is readily inactivated at pH 4.0 or less, so that digestive action on food in the mouth will soon cease in the acid environment of the stomach. In many animals, a salivary amylase is entirely absent. A **lingual lipase** is secreted by the dorsal surface of the tongue (Ebner's glands), but investigations indicate that this enzyme is not of such significance in the human as compared with the rat or mouse, in which it is the only preduodenal lipase.

PROTEIN DIGESTION BEGINS IN THE STOMACH

The gastric secretion is known as **gastric juice.** It is a clear, pale yellow fluid of 0.2–0.5% HCl, with a pH of about 1.0. The gastric juice is 97–99% water. The remainder consists of mucin and inorganic salts, the digestive enzymes (pepsin and rennin), and a lipase.

Hydrochloric Acid Denatures Protein and Kills Bacteria

The **parietal (oxyntic) cells** are the source of gastric HCl, which originates according to the reactions shown in Figure 55–1. The process is similar to that of the "chloride shift" described for the red blood cell. There is also a resemblance to the renal tubular mechanisms for secretion of H^+, wherein the source of H^+ is also the **carbonic anhydrase** catalyzed for-

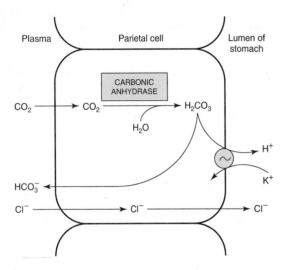

Figure 55–1. Production of gastric hydrochloric acid. (⊖, H^+-K^+ ATPase.)

mation of H_2CO_3 from H_2O and CO_2. An alkaline urine often follows the ingestion of a meal ("alkaline tide"), as a result of the formation of bicarbonate in the process of hydrochloric acid secretion. Secretion of H^+ into the lumen is an active process driven by a membrane-located **H^+-K^+ ATPase,** which, unlike the Na^+-K^+ ATPase, is ouabain-insensitive. Parietal cells contain numerous mitochondria needed to generate the ATP used for powering the H^+-K^+ ATPase. HCO_3^- passes into the plasma in exchange for Cl^-, which is coupled to the secretion of H^+ into the lumen.

As a result of contact with gastric HCl, proteins are denatured; ie, the tertiary protein structure is lost as a result of the destruction of hydrogen bonds. This allows the polypeptide chain to unfold, making it more accessible to the actions of proteolytic enzymes (proteases). The low pH also has the effect of destroying most microorganisms entering the gastrointestinal tract.

Pepsin Initiates Protein Digestion

This is the major digestive function of the stomach. Pepsin is produced in the chief cells as the inactive **zymogen, pepsinogen.** This is activated to pepsin by H^+, which splits off a protective polypeptide to expose active pepsin; and by pepsin, which rapidly activates further molecules of pepsinogen (**autocatalysis**). Pepsin splits denatured protein into large polypeptide derivatives. Pepsin is an **endopeptidase,** since it hydrolyzes peptide bonds within the main polypeptide structure rather than adjacent to amino or carboxyl terminal residues, which is characteristic of **exopeptidases.** It is specific for peptide bonds formed by aromatic amino acids (eg, tyrosine) or dicarboxylic amino acids (eg, glutamate).

Rennin (Chymosin, Rennet) Coagulates Milk

Rennin is important in the digestive processes of infants because it prevents the rapid passage of milk from the stomach. In the presence of calcium, rennin changes the casein of milk irreversibly to a **paracasein,** which is then acted on by pepsin. Rennin is reported to be absent from the stomach of adults. It is used in the making of cheese (rennet).

Lipases Continue the Digestion of Triacylglycerols

The heat of the stomach is important in liquidizing the bulk of dietary lipids; emulsification takes place aided by peristaltic contractions. The stomach secretes a **gastric lipase** that in humans is the main preduodenal lipase. Lingual and gastric lipases initiate lipid digestion by hydrolyzing triacylglycerols containing short- medium-, and generally unsaturated long-chain fatty acids, to form mainly free fatty acids and 1,2-diacylglycerols, the *sn*-3 ester bond being the primary site of hydrolysis. The enzymes are destroyed at low pH but are active after feeding because of the buffering action of dietary proteins in the stomach. The pH optimum is broad, from approximately 3.0 to 6.0. Preduodenal lipases appear to be of particular importance during the neonatal period when pancreatic lipase may be low in activity and milk fat needs to be digested. Owing to a retention time of 2-4 hours in the stomach, some 30% of dietary triacylglycerol can be digested in this time, most within the first hour. Milk fat contains short- and medium-chain fatty acids, which tend to be esterified in the *sn*-3 position. Therefore, milk fat seems to be a particularly good substrate for this enzyme. The released hydrophilic short- and medium-chain fatty acids are absorbed via the stomach wall and enter the portal vein, whereas longer-chain fatty acids dissolve in the fat droplets and pass on to the duodenum. Analysis of cDNA of rat lingual lipase and of human gastric lipase indicates a 78% homology in amino acid sequence between the two enzymes.

DIGESTION CONTINUES IN THE INTESTINE

The stomach contents, or **chyme,** are intermittently introduced during digestion into the duodenum through the pyloric valve. The **alkaline** content of pancreatic and biliary secretions neutralizes the acid of the chyme and changes the pH of this material to the alkaline side; this shift of pH is necessary for the activity of the enzymes contained in pancreatic and intestinal juice, but **it inhibits further action of pepsin.**

Bile Emulsifies, Neutralizes, and Excretes Cholesterol and Bile Pigments

In addition to many functions in intermediary metabolism, the **liver,** by producing bile, plays an important role in digestion. The gallbladder stores bile produced by the liver between meals. During digestion, the gallbladder contracts and supplies bile rapidly to the duodenum by way of the common bile duct. The pancreatic secretions mix with the bile, since they empty into the common duct shortly before its entry into the duodenum.

A. Composition of Bile: The composition of hepatic bile differs from that of gallbladder bile. As shown in Table 55–1, the latter is more concentrated.

B. Properties of Bile:

1. Emulsification–The bile salts have considerable ability to lower surface tension. This enables them to emulsify fats in the intestine and to dissolve fatty acids and water-insoluble soaps. The presence of bile in the intestine is an important adjunct to accomplish the digestion and absorption of fats as well as the absorption of the fat-soluble vitamins A, D, E, and K. When fat digestion is impaired, other foodstuffs are also poorly digested, since the fat covers the food particles and prevents enzymes from attacking them. Under these conditions, the activity of the intestinal bacteria causes considerable putrefaction and production of gas.

2. Neutralization of acid–In addition to its function in emulsification, the bile, having a pH slightly above 7, neutralizes the acid chyme from the stomach and prepares it for digestion in the intestine.

3. Excretion–Bile is an important vehicle for bile acid and cholesterol excretion, but it also removes many drugs, toxins, bile pigments, and various inorganic substances such as copper, zinc, and mercury. The origin of the bile pigments from hemoglobin is discussed in Chapter 34.

The Pancreatic Secretion Contains Enzymes for Attacking All the Major Foodstuffs

Pancreatic secretion is a nonviscid watery fluid that is similar to saliva in its content of water and contains some protein and other organic and inorganic compounds—mainly Na^+, K^+, HCO_3^-, and Cl^- —but Ca^{2+}, Zn^{2+}, HPO_4^{2-}, and SO_4^{2-} are also present in small amounts. The pH of pancreatic secretion is distinctly alkaline, 7.5–8.0 or higher.

Many enzymes are found in pancreatic secretion; some are secreted as zymogens.

Trypsin, chymotrypsin, and elastase are endopeptidases. The proteolytic action of pancreatic secretion is due to the three endopeptidases trypsin, chymotrypsin, and elastase, which attack proteins and polypeptides released from the stomach to produce polypeptides, peptides, or both. **Trypsin** is specific for peptide bonds of basic amino acids, and **chymotrypsin** is specific for peptide bonds containing uncharged amino acid residues, such as aromatic amino acids. **Elastase,** in spite of its name, has rather broad specificity in attacking bonds next to small amino acid residues such as glycine, alanine, and serine. All three enzymes are secreted as zymogens. Activation of **trypsinogen** is due to another proteolytic enzyme, **enterokinase,** secreted by the intestinal mucosa. This hydrolyzes a lysine peptide bond in the zymogen, releasing a small polypeptide that allows the molecule to unfold as active trypsin. Once trypsin is formed, it will attack not only additional molecules of trypsinogen but also the other zymogens in the pancreatic secretion, **chymotrypsinogen, proelastase,** and **procarboxypeptidase,** liberating chymotrypsin, elastase, and **carboxypeptidase,** respectively.

Carboxypeptidase is an exopeptidase. The further attack on the polypeptides produced by the action of endopeptidases is carried on by the exopeptidase carboxypeptidase, which attacks the carboxyl terminal peptide bond, liberating single amino acids.

Amylase attacks starch and glycogen. The starch-splitting action of the pancreatic secretion is due to a **pancreatic α-amylase.** It is similar in action to salivary amylase, hydrolyzing starch and glycogen to maltose, maltotriose (three α-glucose residues linked by $\alpha 1 \rightarrow 4$ bonds) and a mixture of branched $(1 \rightarrow 6)$ oligosaccharides (α-limit dextrins), nonbranched oligosaccharides, and some glucose.

Lipase attacks the primary ester link of triacylglycerols. The **pancreatic lipase** acts at the oil-water interface of the finely emulsified lipid droplets formed by mechanical agitation in the gut in the presence of the products of lingual and gastric lipase activity, **bile salts, colipase** (a protein present in pancreatic secretion), **phospholipids,** and **phospholipase A_2** (also present in the pancreatic secretion). The presence of free fatty acids from the actions of lingual and gastric lipases facilitates hydrolysis by pancreatic lipase, particularly of milk triacylglycerol. Phospholipase A_2 and colipase are secreted in proforms and require activation by tryptic hydrolysis of specific peptide bonds. Activation of prolipase involves the removal of a pentapeptide from the amino terminal end. This pentapeptide acts as a satiety signal for lipids and has been named **enterostatin.**

Table 55–1. The composition of hepatic and of gallbladder bile.

	Hepatic Bile (as secreted)		Bladder Bile
	Percent of Total Bile	Percent of Total Solids	Percent of Total Bile
Water	97.00	. . .	85.92
Solids	2.52	. . .	14.08
Bile acids	1.93	36.9	9.14
Mucin and pigments	0.53	21.3	2.98
Cholesterol	0.06	2.4	0.26
Esterified and non-esterified fatty acids	0.14	5.6	0.32
Inorganic salts	0.84	33.3	0.65
Specific gravity	1.01	. . .	1.04
pH	7.1–7.3	. . .	6.9–7.7

However, its physiologic status remains to be ascertained. Ca^{2+} is necessary for phospholipase A_2 activity. A limited hydrolysis of the ester bond in the 2 position of the phospholipid by phospholipase A_2 (Figure 26–6) results in the binding of lipase to the substrate interface and a rapid rate of hydrolysis of triacylglycerol. Apparently, pancreatic lipase is actually inhibited by bile salts. The function of colipase is to overcome this inhibition by binding in a 1:1 molar ratio with the lipase and also binding to the bile salt-covered triacylglycerol interface. Thus, it anchors the lipase to its triacylglycerol substrate. The complete hydrolysis of triacylglycerols produces glycerol and fatty acids. However, the second and third fatty acids are hydrolyzed from the triacylglycerols with increasing difficulty. Pancreatic lipase is virtually specific for the hydrolysis of primary ester linkages, ie, at positions 1 and 3 of triacylglycerols. During fat digestion, the aqueous or "micellar phase" contains mixed **disk-like micelles** and **liposomes** of bile salts saturated with lipolytic products (Figure 16–29).

The presence in human milk of a **bile salt-activated lipase** is an added factor ensuring complete digestion of milk fat when it comes into contact with bile salts in the duodenum. It is of particular importance in premature infants whose pancreatic secretions are not fully operational, and it has a wide specificity in triacylglycerol substrates. It has recently been demonstrated as identical to a pancreatic bile salt-activated lipase, also of wide specificity.

Because of the difficulty of hydrolysis of the secondary ester linkage in the triacylglycerol by pancreatic lipase, the digestion of triacylglycerol proceeds by removal of the terminal fatty acids to produce 2-monoacylglycerol. Since this last fatty acid is linked by a secondary ester bond, its removal requires isomerization to a primary ester linkage to achieve complete hydrolysis. This is a relatively slow process; as a result, **2-monoacylglycerols** are major end products of triacylglycerol digestion, and less than one-fourth of the ingested triacylglycerol is completely broken down to glycerol and fatty acids (Figure 55–2).

Cholesteryl esters are broken down by a specific hydrolase. Under the conditions within the lumen of the intestine, **cholesteryl ester hydrolase** (cholesterol esterase) catalyzes the hydrolysis of cholesteryl esters, which are thus absorbed from the intestine in the nonesterified free form.

Ribonuclease (RNase) and deoxyribonuclease (DNase) are responsible for the digestion of dietary nucleic acids (Chapters 38 and 39).

Phospholipase A_2 hydrolyzes the ester bond in the 2 position of glycerophospholipids of both biliary and dietary origins to form lysophospholipids, which, being detergents, aid emulsification and digestion of lipids.

Intestinal Secretions Complete the Digestive Process

The intestinal juice secreted by the glands of Brunner and Lieberkühn contains digestive enzymes, including the following:

Aminopeptidase, which is an exopeptidase attacking peptide bonds next to amino terminal amino acids of polypeptides and oligopeptides; and **dipeptidases** of various specificity, some of which may be within the intestinal epithelium. The latter complete digestion of dipeptides to free amino acids.

Specific **disaccharidases** and **oligosaccharidases,** ie, **α-glucosidase (maltase),** which removes single glucose residues from $\alpha(1 \rightarrow 4)$-linked oligosaccharides and disaccharides, starting from the nonreducing ends; **sucrase-isomaltase complex,** which is found as the proenzyme on one polypeptide chain but as active enzymes on separate polypeptides and hydrolyzes sucrose and $1 \rightarrow 6$ bonds in α-limit dextrins; β-glycosidase (lactase), for removing galactose from lactose, but which also attacks cellobiose and other β-glycosides and in addition has a second catalytic site that splits glycosylceramides; and **trehalase** for hydrolyzing trehalose. Many of these hydrolases remain attached to the brush border of the enterocyte while the catalytic domains are free in the lumen to react with the substrate.

A **phosphatase,** which removes phosphate from certain organic phosphates such as hexose phosphates, glycerophosphate, and the nucleotides derived from the diet and the digestion of nucleic acids by nucleases.

Polynucleotidases, which split nucleic acids into nucleotides.

Nucleosidases (nucleoside phosphorylases) that catalyze the phosphorolysis of nucleosides to give the free nitrogen base plus a pentose phosphate.

A **phospholipase** that attacks phospholipids to produce glycerol, fatty acids, phosphoric acid, and bases such as choline.

The Major Products of Digestion Are Assimilated

The final result of the action of the digestive enzymes is to reduce the foodstuffs of the diet to forms that can be absorbed and assimilated. These end products of digestion are, for carbohydrates, the monosaccharides (principally glucose); for proteins, the amino acids; for triacylglycerol, the fatty acids, glycerol, and monoacylglycerols; and for nucleic acids, the nucleobases, nucleosides, and pentoses.

The plant cell wall polysaccharides and lignin of the diet that cannot be digested by mammalian enzymes constitute **dietary fiber** and make up the bulk of the residues from digestion. Fiber performs an important function in adding bulk to the diet and is discussed in the preceding chapter. Table 55–2 summarizes the major digestive processes.

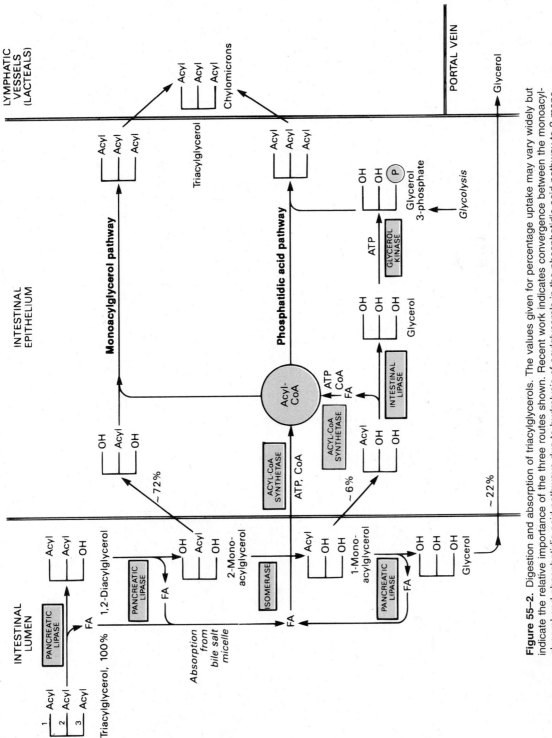

Figure 55–2. Digestion and absorption of triacylglycerols. The values given for percentage uptake may vary widely but indicate the relative importance of the three routes shown. Recent work indicates convergence between the monoacyl-glycerol and phosphatidic acid pathways due to hydrolysis of acylglycerols in the phosphatidic acid pathway to 2-mono-acylglycerol. (FA, long-chain fatty acid.) (Modified from Mattson FH, Volpenheim RA: The digestion and absorption of triglycerides. J Biol Chem 1964;239:2772.)

Table 55–2. Summary of digestive processes.

Source of Secretion and Stimulus for Secretion	Enzyme	Method of Activation and Optimal Conditions for Activity	Substrate	End Products or Action
Salivary glands: Secrete saliva in reflex response to presence of food in oral cavity.	Salivary amylase	Chloride ion necessary. pH 6.6–6.8.	Starch Glycogen	Maltose plus 1:6 glucosides (oligosaccharides) plus maltotriose.
Lingual glands	Lingual lipase	pH range 2.0–7.5; optimal, 3.0–6.0.	Primary ester link at sn-3 of triacylglycerols	Fatty acids plus 1,2-diacylglycerols.
Stomach glands: Chief cells and parietal cells secrete gastric juice in response to reflex stimulation and action of gastrin.	Pepsin A (fundus) Pepsin B (pylorus)	Pepsinogen converted to active pepsin by HCl. pH 1.0–2.0.	Protein	Peptides.
	Gastric lipase	As for lingual lipase	As for lingual lipase	As for lingual lipase.
	Rennin	Calcium necessary for activity. pH 4.0.	Casein of milk	Coagulates milk.
Pancreas: Presence of acid chyme from the stomach activates duodenum to produce (1) secretin, which hormonally stimulates flow of pancreatic juice; (2) cholecystokinin, which stimulates the production of enzymes.	Trypsin	Trypsinogen converted to active trypsin by enterokinase of intestine at pH 5.2–6.0. Autocatalytic at pH 7.9.	Protein Peptides	Polypeptides. Dipeptides.
	Chymotrypsin	Secreted as chymotrypsinogen and converted to active form by trypsin. pH 8.0.	Protein Peptides	Same as trypsin. More coagulating power for milk.
	Elastase	Secreted as proelastase and converted to active form by trypsin.	Protein Peptides	Polypeptides. Dipeptides.
	Carboxypeptidase	Secreted as procarboxypeptidase, activated by trypsin.	Polypeptides at the free carboxyl end of the chain.	Small peptides. Free amino acids.
	Pancreatic amylase	pH 7.1.	Starch Glycogen	Maltose plus 1:6 glucosides (oligosaccharides) plus maltotriose.
	Lipase	Combined activation by bile salts, phospholipids, colipase. pH 8.0.	Primary ester linkages of triacylglycerol	Fatty acids, 2-monoacylglycerols, glycerol.
	Bile salt-activated lipase[1] (the same enzyme is present in milk)	Activated by bile salts.	Triacylglycerol; esters, eg, cholesteryl ester, vitamin esters; lysophospholipids	Free fatty acids, vitamins, cholesterol.
	Ribonuclease		Ribonucleic acid	Nucleotides.
	Deoxyribonuclease		Deoxyribonucleic acids	Nucleotides.
	Cholesteryl ester hydrolase[1]	Activated by bile salts.	Cholesteryl esters	Free cholesterol plus fatty acids.
	Phospholipase A_2	Secreted as proenzyme, activated by trypsin and Ca^{2+}.	Phospholipids	Fatty acids, lysophospholipids.
Liver and gallbladder: Cholecystokinin, a hormone from the intestinal mucosa—and possibly also gastrin and secretin—stimulate the gallbladder and secretion of bile by the liver.	(Bile salts and alkali)		Fats—also neutralize acid chyme	Fatty acid-bile salt complexes and finely emulsified neutral fat-bile salt micelles and liposomes.

[1]Perhaps the same enzyme.

Table 55–2. Summary of digestive processes (*continued*).

Source of Secretion and Stimulus for Secretion	Enzyme	Method of Activation and Optimal Conditions for Activity	Substrate	End Products or Action
Small intestine: Secretions of Brunner's glands of the duodenum and glands of Lieberkühn.	Aminopeptidase		Polypeptides at the free amino end of the chain	Small peptides. Free amino acids.
	Dipeptidases		Dipeptides	Amino acids.
	Sucrase	pH 5.0–7.0.	Sucrose	Fructose, glucose.
	Maltase	pH 5.8–6.2.	Maltose	Glucose.
	Lactase	pH 5.4–6.0.	Lactose	Glucose, galactose.
	Trehalase		Trehalose	Glucose.
	Phosphatase	pH 8.6.	Organic phosphates	Free phosphate.
	Isomaltase or 1:6 glucosidase		1:6 glucosides	Glucose.
	Polynucleotidase		Nucleic acid	Nucleotides.
	Nucleosidases (nucleoside phosphorylases)		Purine or pyrimidine nucleosides	Purine or pyrimidine bases, pentose phosphate.

ABSORPTION FROM THE GASTRO-INTESTINAL TRACT RESULTS IN PASSAGE OF NUTRIENTS INTO THE HEPATIC PORTAL VEIN OR THE LYMPHATICS

There is little absorption from the stomach apart from short- and medium-chain fatty acids and ethanol.

The small intestine is the main absorptive organ. About 90% of the ingested foodstuffs are absorbed in the course of passage through the small intestine, and water is absorbed at the same time. Considerably more water is absorbed after the foodstuff residues pass into the large intestine, so that the contents, which were fluid in the small intestine, gradually become more solid in the colon.

There are two pathways for the transport of materials absorbed by the intestine: the **hepatic portal system,** which leads directly to the liver, transporting water-soluble nutrients, and the **lymphatic vessels,** which lead to the blood by way of the thoracic duct and transport lipid-soluble nutrients.

Carbohydrates Are Absorbed as Monosaccharides

The products of carbohydrate digestion are absorbed from the jejunum into the blood of the portal venous system in the form of monosaccharides, chiefly as hexose (glucose, fructose, mannose, and galactose) and as pentose sugars (ribose). The oligosaccharides (compounds derived from starches that yield three to ten monosaccharide units upon hydrolysis) and the disaccharides are hydrolyzed by appropriate enzymes derived from the mucosal surfaces of the small intestine, which may include pancreatic amylase adsorbed onto the mucosa. There is little free disaccharidase activity in the intestinal lumen. Most of the activity is associated with small "knobs" on the brush border of the intestinal epithelial cell.

Two mechanisms are responsible for the absorption of monosaccharides: **active transport** against a concentration gradient and **simple diffusion.** The molecular configurations that seem necessary for active transport, which are present in glucose and galactose, are the following: The OH on carbon 2 should have the same configuration as in glucose, a pyranose ring should be present, and a methyl or substituted methyl group should be present on carbon 5. Fructose is absorbed more slowly than glucose and galactose. Its absorption appears to proceed by diffusion with the concentration gradient by means of a sodium-independent facilitative transporter (GLUT 5). Glucose facilitates fructose absorption. Normally, there is little fructose in blood apart from that derived from the diet.

Active Absorption of Glucose Is Powered by the Sodium Pump

The brush border of the enterocyte contains several transporter systems, some very similar to those of the renal brush border membranes, which specialize in the uptake of the different amino acids and sugars. A sodium-dependent glucose transporter (SLGT 1) binds both glucose and Na^+ at separate sites and transports them both through the plasma membrane of the intestinal cell. It is envisaged that both glucose and Na^+ are released into the cytosol, allowing the transporter to take up more "cargo." The Na^+ is transported down its concentration gradient and at the same time causes the transporter to carry glucose against its concentration gradient. The free energy required for this active transport is obtained from the hydrolysis of ATP linked to a sodium pump that expels Na^+ from the cell in exchange for K^+ (Figure 55–3). The active transport of glucose is inhibited by **ouabain** (a cardiac glycoside), an inhibitor of the sodium pump, and by **phlorhizin,** a known inhibitor of glucose reabsorption in the kidney tubule. There is also a sodium-independent transporter of glucose.

Hydrolysis of polysaccharides, oligosaccharides, and disaccharides is rapid; therefore, the absorptive mechanisms for glucose and fructose are quickly saturated. A conspicuous exception is the hydrolysis of lactose, which proceeds at only half the rate for sucrose, accounting for the fact that digestion of lactose does not lead to saturation of the transport mechanisms for glucose and galactose. Adaptation to increase in carbohydrate in the diet occurs by an increase in the number of transporters in the brush border.

The Products of Lipid Digestion Are Absorbed From Bile Salt Micelles

The 2-monoacylglycerols, fatty acids, and small amounts of 1-monoacylglycerols leave the oil phase of the lipid emulsion and diffuse into the mixed mi-

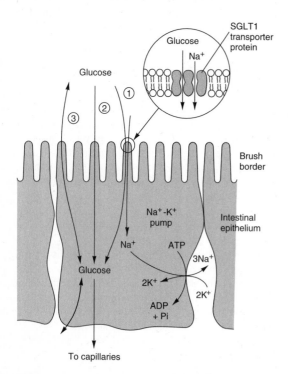

Figure 55–3. Transport of glucose across the intestinal epithelium. Active glucose transport by the SGLT 1 transporter is coupled to the Na^+-K^+ pump, ①, or to an Na^+-independent system (GLUT 5), ②. Diffusion is represented by ③.

celles and liposomes consisting of bile salts, phosphatidylcholine, and cholesterol, furnished by the bile (Figure 55–4). Because the micelles are soluble, they allow the products of digestion to be transported through the aqueous environment of the intestinal lumen to the brush border of the mucosal cells, where they are absorbed into the intestinal epithelium. The bile salts pass on to the ileum, where most are absorbed into the **enterohepatic circulation** by an active transport process (Chapter 27). Phospholipids of dietary and biliary origin (eg, phosphatidylcholine) are hydrolyzed by phospholipase A_2 of the pancreatic secretion to fatty acids and lysophospholipids, which are also absorbed from the micelles. Cholesteryl esters are hydrolyzed by cholesteryl ester hydrolase of the pancreatic juice, and the free cholesterol, together with most of the biliary cholesterol, is absorbed through the brush border after transportation in the micelles. Over 98% of dietary lipid is normally absorbed.

Within the intestinal wall, 1-monoacylglycerols are further hydrolyzed to produce free glycerol and fatty acids by an intestinal lipase (glycerol esterhydrolase), which is distinct from pancreatic lipase. 2-Monoacylglycerols are reconverted to triacylglycerols via the **monoacylglycerol pathway** (Figure 55–2). The utilization of fatty acids for resynthesis of triacylglycerols first requires their conversion to acyl-CoA by acyl-CoA synthetase. Short- and medium-chain triacylglycerols can be absorbed as such and are then hydrolyzed by the glycerol ester hydrolase. This enzyme assumes greater importance in patients suffering from pancreatic lipase deficiency who are fed on medium-chain triacylglycerols.

It is likely that the synthesis of triacylglycerols proceeds in the intestinal mucosa in a manner similar to that which takes place in other tissues, as described in Chapter 26. The absorbed lysophospholipids, together with much of the absorbed cholesterol, are also reacylated with acyl-CoA to regenerate phospholipids and cholesteryl esters.

The free glycerol released in the intestinal lumen is not reutilized but passes directly to the portal vein. However, the glycerol released within the intestinal cells can be reutilized for triacylglycerol synthesis after activation to glycerol 3-phosphate by ATP. Thus, all long-chain fatty acids absorbed by intestinal wall mucosal cells are utilized in the re-formation of acylglycerols, particularly triacylglycerols.

Triacylglycerols, having been synthesized in the intestinal mucosa, are not transported to any extent in the portal venous blood. Instead, the great majority of absorbed lipids, including phospholipids, cholesteryl esters, cholesterol, and fat-soluble vitamins, generate chylomicrons that form a milky fluid, the **chyle,** that is collected by the lymphatic vessels of the abdominal region and passed to the systemic blood via the thoracic duct.

The majority of absorbed fatty acids of more than

ten carbon atoms in length, irrespective of the form in which they are absorbed, are found as esterified fatty acids in the lymph of the thoracic duct. Fatty acids with carbon chains **shorter than 10–12 carbons** are transported in the portal venous blood as unesterified (free) fatty acids. One reason for this is that acyl-CoA synthetase is specific for fatty acids having 12 or more carbon atoms. Some short- or medium-chain fatty acids present in mixed triacylglycerols may be absorbed as 2-monoacylglycerols and enter the thoracic duct via the monoacylglycerol pathway.

Of the plant sterols (phytosterols), none are absorbed from the intestine except activated ergosterol (provitamin D).

The Products of Protein Digestion Are Absorbed as Individual Amino Acids

Under normal circumstances, the dietary proteins are almost completely digested to their constituent amino acids, and these end products of protein digestion are then rapidly absorbed from the intestine into the portal blood. It is possible that some hydrolysis, eg, of dipeptides, is completed in the intestinal wall.

The natural (L) isomer—but not the D isomer—of an amino acid is actively transported across the intestine from the mucosa to the serosa; vitamin B_6 (pyridoxal phosphate) may be involved in this transfer. This active transport of the L-amino acids is energy-dependent, as evidenced by the fact that 2,4-dinitrophenol, the uncoupler of oxidative phosphorylation (Figure 14–9), inhibits transport. Amino acids are transported through the brush border by a multiplicity of carriers (transporters), many having Na^+-dependent mechanisms similar to the glucose transporter system (Figure 55–3). Of the Na^+-dependent carriers, there is a neutral amino acid carrier, a phenylalanine and methionine carrier, and a carrier specific for imino acids such as proline and hydroxyproline. Na^+-independent carriers specializing in the transport of neutral and lipophilic amino acids (eg, phenylalanine and leucine) or of cationic amino acids (eg, lysine) have been characterized. Table 55–3 summarizes the sites of intestinal absorption of some common nutrients.

BACTERIA IN THE LARGE INTESTINE CAUSE PUTREFACTION AND FERMENTATION

Most ingested food is absorbed from the small intestine. The residue passes into the large intestine. Here, considerable absorption of water takes place, and the semiliquid intestinal contents gradually become more solid. During this period, considerable bacterial activity occurs. By fermentation and putrefaction, the bacteria produce various gases, such as CO_2, methane, hydrogen, nitrogen, and hydrogen sulfide, as well as acetic, lactic, propionic, and butyric acids. The bacterial decomposition of phosphatidylcholine may produce choline and related toxic amines such as neurine.

Choline   Neurine

Many amino acids undergo decarboxylation as a result of the action of intestinal bacteria to produce toxic amines (ptomaines).

Such decarboxylation reactions produce cadaverine from lysine; agmatine from arginine; tyramine from tyrosine; putrescine from ornithine; and histamine from histidine. Many of these amines are powerful vasopressor substances.

The amino acid tryptophan undergoes a series of reactions to form indole and methylindole (skatole), the substances particularly responsible for the odor of feces.

Indole   Skatole

The sulfur-containing amino acid cysteine undergoes a series of transformations to form mercaptans such as ethyl and methyl mercaptan as well as H_2S.

Table 55–3. Site of absorption of nutrients.

Site	Nutrient
Jejunum	Glucose and other monosaccharides, some disaccharides
	Monoacylglycerols, fatty acids, glycerol, cholesterol
	Amino acids, peptides
	Vitamins, folate
	Electrolytes, iron, calcium, water
Ileum	Bile acids
	Vitamin B_{12}
	Electrolytes
	Water

$$CH_3$$
$$|$$
$$CH_2SH$$

$$CH_3SH$$

Ethyl mercaptan   Methyl mercaptan

[2H]

$$CH_3SH \xrightarrow{\quad} CH_4 + H_2S$$

Methyl   Methane and
mercaptan   hydrogen sulfide

The large intestine is a source of considerable quantities of **ammonia,** a product of bacterial activity on nitrogenous substrates. This is absorbed into the portal circulation, but under normal conditions it is rapidly removed from the blood by the liver. In **liver disease,** this function of the liver may be impaired, in which case the concentration of ammonia in the peripheral blood will rise to toxic levels. It is believed that ammonia intoxication may play a role in the genesis of hepatic coma in some patients. The oral administration of **neomycin** has been shown to reduce the quantity of ammonia delivered from the intestine to the blood, owing to the antibacterial action of the drug. The feeding of **high-protein diets** to patients suffering from advanced liver disease, or the occurrence of gastrointestinal hemorrhage in such patients, may contribute to the development of ammonia intoxication. Neomycin is also beneficial under these circumstances.

Intestinal Bacteria Are Also Beneficial

The intestinal flora may make up as much as 25% of the dry weight of the feces. In herbivora, whose diet consists largely of cellulose, the intestinal or ruminal bacteria are essential to digestion, since they decompose the polysaccharide and make it available for absorption. In addition, these symbiotic bacteria accomplish the synthesis of essential amino acids and vitamins. In humans, although the intestinal flora is not as important as in the herbivora, nevertheless some nutritional benefit is derived from bacterial activity in the synthesis of certain vitamins, particularly vitamins K and some members of the B complex (eg, biotin), which are made available to the body.

CLINICAL ASPECTS

The Limited Solubility of Cholesterol in Bile Causes Gallstones

Free cholesterol is, for practical purposes, insoluble in water; consequently, it is incorporated into a phospholipid-bile salt micelle (Figure 16–29). Indeed, phosphatidylcholine, the predominant phos-

pholipid in bile, is itself insoluble in aqueous systems but can be dissolved by bile salts in micelles. The large quantities of cholesterol present in the bile of humans are solubilized in these water-soluble mixed micelles, allowing cholesterol to be transported in bile via the biliary tract to the intestine. However, the actual solubility of cholesterol in bile depends on the relative proportions of bile salt, phosphatidylcholine, and cholesterol. The solubility also depends on the water content of bile. This is especially important in dilute hepatic bile.

Using triangular coordinates (Figure 55–4), Redinger and Small were able to determine the maximum solubility of cholesterol in human gallbladder bile. Reference to the figure indicates that any triangular point falling above the line ABC would represent a bile whose composition is such that cholesterol is either supersaturated or precipitated.

It is believed that at some time during the life of a patient with gallstones there is formed an abnormal bile that has become supersaturated with cholesterol.

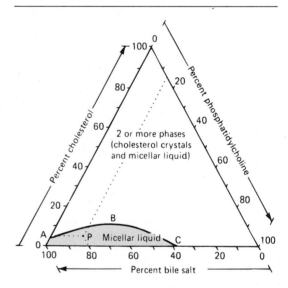

Figure 55–4. Method for presenting three major components of bile (bile salts, phosphatidylcholine, and cholesterol) on triangular coordinates. Each component is expressed as a percentage mole of total bile salt, phosphatidylcholine, and cholesterol. Line ABC represents maximum solubility of cholesterol in varying mixtures of bile salt and phosphatidylcholine. Point P represents normal bile composition, containing 5% cholesterol, 15% phosphatidylcholine, and 80% bile salt, and falls within the zone of a single phase of micellar liquid. Bile having a composition falling above the line would contain excess cholesterol in either supersaturated or precipitated form (crystals or liquid crystals). (Reproduced, with permission, from Redinger RN, Small DM: Bile composition, bile salt metabolism, and gallstones. Arch Intern Med 1972;130:620. Copyright © 1972 by the American Medical Association.)

With time, various factors such as infection, for example, serve as seeding agents to cause the supersaturated bile to precipitate the excess cholesterol as crystals. Unless the newly formed crystals are promptly excreted into the intestine with the bile, the crystals will grow to form stones. When the activities of key enzymes in bile acid formation were measured in the livers of patients with gallstones, cholesterol synthesis was elevated but bile acid synthesis was reduced, causing liver cholesterol concentrations to increase. It seems that decreased 7α-hydroxylase activity leads to a diminished enterohepatic bile acid pool that signals the liver to produce more cholesterol. The bile then becomes overloaded with cholesterol, which is unable to dissolve completely in the mixed micelles.

The above information concerning cholesterol solubility has been used in attempts to dissolve gallstones or to prevent their further formation. **Chenodeoxycholic acid** appears to offer specific medical treatment for asymptomatic radiolucent gallstones in functioning gallbladders because of its specific inhibition of HMG-CoA reductase in the liver, with consequent reduction in cholesterol synthesis.

Defects in Enzymes of Carbohydrate Digestion Cause Specific Disorders

A. Lactose Intolerance: Intolerance to lactose, the sugar of milk, may be attributable to a deficiency of lactase. Because lactase activity is rate-limiting for lactose absorption, any deficiency in the enzyme is directly reflected in a diminished rate of absorption of the sugar. The syndrome should not be confused with intolerance to milk resulting from a sensitivity to milk proteins, usually to the β-lactoglobulin. The signs and symptoms of lactose intolerance are the same regardless of the cause. These include abdominal cramps, diarrhea, and flatulence. They are attributed to accumulation of undigested lactose, which is osmotically active, so that it holds water, and to the fermentative action on the sugar of the intestinal bacteria which produce gases and other products that serve as intestinal irritants.

There are three types of lactase deficiency (hypolactasia):

1. Inherited lactase deficiency–In this syndrome, which is relatively rare, symptoms of intolerance develop very soon after birth. The feeding of a lactose-free diet results in disappearance of the symptoms. Consumption of live-culture yogurt provides not only active β-galactosidase to attack any small amounts of lactose in the diet but also calcium and energy to replace milk. Commercial preparations of β-galactosidase are also available.

2. Secondary low-lactase activity–Because digestion of lactose is limited even in normal humans, intolerance to milk is not uncommon as a consequence of intestinal diseases. Examples are tropical and nontropical (celiac) sprue, kwashiorkor, colitis, and gastroenteritis. The disorder may be noted also after surgery for peptic ulcer.

3. Primary low-lactase activity–This is a relatively common syndrome, particularly among nonwhite populations. Since intolerance to lactose was not a feature of the early life of adults with this disorder, it is presumed to represent a gradual decline in activity of lactase in susceptible individuals owing to reduction in expression of the enzyme. However, this is not due to lack of lactase mRNA, and a failure in translation appears to be a likely cause of the deficiency.

B. Sucrase Deficiency: There is an inherited deficiency of the disaccharidases sucrase and isomaltase. These two deficiencies coexist, because sucrase and isomaltase occur together as a complex enzyme. Symptoms occur in early childhood and are the same as those described in lactase deficiency.

C. Disacchariduria: An increase in the excretion of disaccharides may be observed in some patients with disaccharidase deficiencies. As much as 300 mg or more of disaccharide may be excreted in the urine of these people and in patients with intestinal damage (eg, sprue).

D. Monosaccharide Malabsorption: There is a congenital condition due to a single mutation in which glucose and galactose are absorbed only slowly, owing to a defect in the Na^+-glucose cotransporter carrier mechanism. Because fructose is not absorbed via this transporter, its absorption is normal. In a few children and adults, **fructose-sorbitol malabsorption** causes watery feces, abdominal discomfort, and gas production owing to bacterial fermentation.

In Chyluria, Chylomicrons Are Present in Urine

Chyluria is an abnormality in which the patient excretes milky urine because of the presence of an abnormal connection between the urinary tract and the lymphatic drainage system of the intestine, a so-called chylous fistula. In a similar abnormality, **chylothorax,** there is an abnormal connection between the pleural space and the lymphatic drainage of the small intestine that results in the accumulation of milky pleural fluid. Feeding triacylglycerols in which the fatty acids are of medium-chain length (fewer than 12 carbons) in place of dietary fat results in disappearance of chyluria. In chylothorax, the use of triacylglycerol with short-chain fatty acids results in the appearance of clear pleural fluid. Both these effects are due to the fact that medium-chain fatty acids are absorbed into the hepatic portal vein rather than as chylomicrons in the thoracic duct.

In **colipase deficiency,** patients suffer from steatorrhoea as a result of defective pancreatic lipase activity.

Table 55–4. Summary of disturbances due to malabsorption.

Sign or Symptom	Substance Malabsorbed
Anemia	Iron, vitamin B_{12}, folate
Edema	Products of protein digestion
Tetany	Calcium, magnesium, vitamin D
Osteoporosis	Calcium, products of protein digestion, vitamin D
Milk intolerance	Lactose
Bleeding, bruising	Vitamin K
Steatorrhea (fatty stools)	Lipids and fat-soluble vitamins
Hartnup disease (defect in intestinal neutral amino acid carrier)	Neutral amino acids

Absorption of Undigested Polypeptides May Cause Antigenic Reactions

Individuals in whom an immunologic response to ingested protein occurs must be able to absorb some unhydrolyzed protein, since digested protein is nonantigenic. This is not entirely undocumented, since the antibodies of the colostrum are known to be available to the infant.

There is increasing support for the hypothesis that the basic defect in **nontropical sprue** is located within the mucosal cells of the intestine and permits the polypeptides resulting from the peptic and tryptic digestion of gluten, the principal protein of wheat, not only to exert a local harmful effect within the intestine but also to be absorbed into the circulation and thus to elicit the production of antibodies. It has been established that circulating antibodies to wheat gluten or its fractions are frequently present in patients with nontropical sprue. The harmful entity is a peptide composed of six or seven amino acids.

These observations on a disease entity that is undoubtedly the adult analog of **celiac disease** in children advance the possibility that protein fragments of larger molecular size than amino acids are absorbed from the intestine under certain conditions.

Table 55–4 summarizes some disorders resulting from malabsorption.

SUMMARY

1. Digestion involves splitting of food molecules by hydrolysis into smaller molecules that can be absorbed through the epithelium of the gastrointestinal tract.

2. Carbohydrates such as starch and glycogen are attacked initially by salivary amylase followed by pancreatic amylase in the intestine. The resulting maltose plus oligosaccharides together with other disaccharides from the diet are attacked by specific intestinal hydrolases to form monosaccharides, chiefly glucose, fructose, and galactose.

3. Protein digestion is initiated by the endopeptidase pepsin in the acid medium of the stomach and continued by other endopeptidases in the pancreatic secretion (eg, trypsin, chymotrypsin, and elastase) together with the exopeptidase carboxypeptidase, which splits off successive amino acids from the carboxyl terminal. Protein digestion is completed by intestinal secretions including the exopeptidase, aminopeptidase, plus dipeptidases with various specificities, which result in formation of free amino acids.

4. Lingual and gastric lipases initiate triacylglycerol hydrolysis in the stomach, after which the contents are emulsified with bile salts in the duodenum and finally attacked by pancreatic lipase.

5. The products of digestion are absorbed in the intestine, ie, monosaccharides, amino acids, some dipeptides, together with glycerol, free fatty acids, and 2-monoacylglycerols from triacylglycerol digestion. The water-soluble monosaccharides, amino acids, glycerol, and short- and medium-chain fatty acids are transported to the liver in the hepatic portal vein. Long-chain fatty acids and monoacylglycerols are resynthesized to triacylglycerol in the intestinal epithelium to form chylomicrons. These are secreted into the circulation via the thoracic duct.

6. Digestive disorders include those due to deficiency of enzymes such as lactase and sucrase. Others are due to malabsorption, eg, defects in the Na^+-glucose cotransporter (SGLT 1) leading to glucose and galactose malabsorption. Still others may be due to immunologic responses to the absorption of unhydrolyzed polypeptides, such as in nontropical sprue. Gallstones are caused because cholesterol, owing to its low solubility, is precipitated in bile, leading to crystallization and stone formation.

REFERENCES

Börgstrom B: The micellar hypothesis of fat absorption: Must it be revisited? Scand J Gastroenterol 1985;20:389.

Büller HA, Grand RJ: Lactose intolerance. Annu Rev Med 1990;41:141.

Eggermont E: Problems of transfer of carbohydrates at the level of the intestinal mucosa. In: *Inborn Errors of Metabolism.* Schaub J et al (editors). Raven Press, 1991.

Erlanson-Albertsson C: Pancreatic colipase. Structure and physiological aspects. Biochim Biophys Acta 1993;1125:1.

Foltmann B: Gastric proteinases. Essays Biochem 1981;17:52.

Hamosh M: *Lingual & Gastric Lipases: Their Role in Fat Digestion.* CRC Press, 1990.

Holdsworth CD: In vitro studies of sugar absorption in humans. In: *Sugars in Nutrition.* Gracey M et al (editors). Raven Press, 1991.

Kuksis A (editor): *Fat Absorption.* CRC Press, 1987.

Tso P: Gastrointestinal digestion and absorption of lipid. Adv Lipid Res 1985;21:143.

Wang CS, Hartsuck JA: Bile salt-activated lipase: A multiple function lipolytic enzyme. Biochim Biophys Acta 1993;1166:1.

Wolfe MM, Soll AH: The physiology of gastric acid secretion. N Engl J Med 1988;319:1707.

Würsch P: Dietary fiber and unabsorbed carbohydrates. In: *Sugars in Nutrition.* Gracey M et al (editors). Raven Press, 1991.

Yang LY, Kuksis A: Apparent convergence (at 2-monoacylglycerol level) of phosphatidic acid and 2-monoacylglycerol pathways of synthesis of chylomicron triacylglycerols. J Lipid Res 1991;32:1173.

56

Glycoproteins

Robert K. Murray, MD, PhD

INTRODUCTION

Glycoproteins are proteins that have oligosaccharide (glycan) chains covalently attached to their polypeptide backbones. Glycoproteins are one class of **glycoconjugate** or **complex carbohydrates.** These are equivalent terms used to denote molecules containing one or more carbohydrate chains covalently linked to protein (to form glycoproteins or proteoglycans) or lipid (to form glycolipids). **Proteoglycans** are discussed in Chapter 57 and **glycolipids** in Chapter 16.

BIOMEDICAL IMPORTANCE

Almost all the **plasma proteins** of humans, except albumin, are glycoproteins. Many **proteins of cellular membranes** (Chapter 43) contain substantial amounts of carbohydrate. A number of the **blood group substances** are glycoproteins, whereas others are glycosphingolipids. Certain **hormones** (eg, chorionic gonadotropin) are glycoproteins. **Cancer** is increasingly recognized as a disorder resulting from abnormal genetic regulation (Chapter 62). The major problem in cancer is **metastasis,** the phenomenon whereby cancer cells leave their tissue of origin (eg, the breast), migrate through the bloodstream to some distant site in the body (eg, the brain), and grow there in a completely unregulated manner, with catastrophic results for the affected individual. Many cancer researchers think that alterations in the structures of glycoproteins and other glycoconjugates on the surfaces of cancer cells are important in the phenomenon of metastasis.

GLYCOPROTEINS OCCUR WIDELY AND PERFORM NUMEROUS FUNCTIONS

Glycoproteins occur in most organisms, from bacteria to humans. Many animal viruses also contain glycoproteins, some of which have been much investigated, in part because they are very suitable for biosynthetic studies. Numerous proteins with diverse functions are glycoproteins (Table 56–1); their car-

Table 56–1. Some functions served by glycoproteins.

Function	Glycoproteins
Structural molecule	Collagens
Lubricant and protective agent	Mucins
Transport molecule	Transferrin, ceruloplasmin
Immunologic molecule	Immunoglobulins, histocompatibility antigens
Hormone	Chorionic gonadotropin, thyroid-stimulating hormone (TSH)
Enzyme	Various, eg, alkaline phosphatase
Cell attachment-recognition site	Various proteins involved in cell-cell (eg, sperm-oocyte), virus-cell, bacterium-cell, and hormone-cell interactions
Antifreeze	Certain plasma proteins of cold water fish
Interact with specific carbohydrates	Some lectins

bohydrate content ranges from 1% to over 85% by weight.

Many studies have been performed to define the precise roles oligosaccharide chains play in the functions of glycoproteins. Table 56–2 summarizes results from such studies; some of the functions listed are firmly established, and others are still under investigation.

OLIGOSACCHARIDE CHAINS ENCODE BIOLOGIC INFORMATION

An enormous number of glycosidic linkages can be generated between sugars. For example, three different hexoses may be linked to each other to form over one thousand different trisaccharides. The conformations of the sugars in oligosaccharide chains vary depending on their linkages and proximity to other molecules with which the oligosaccharides may interact. A widely held belief is that oligosaccharide chains encode considerable **biologic information** and that this depends upon the constituent sugars, their sequences, and their conformations. For instance, mannose 6-

Table 56–2. Some functions of the oligosaccharide chains of glycoproteins.[1]

- Modulate physicochemical properties, eg, solubility, viscosity, charge, and denaturation
- Protect against proteolysis, from inside and outside of cell
- Affect proteolytic processing of precursor proteins to smaller products
- Are involved in biologic activity, eg, of human chorionic gonadotropin (hCG)
- Affect insertion into membranes, intracellular migration, sorting and secretion
- Affect embryonic development and differentiation
- May affect sites of metastases selected by cancer cells

[1]Adapted from Schachter H. Biosynthetic controls that determine the branching and heterogeneity of protein-bound oligosaccharides. Biochem Cell Biol 1986;64:163.

phosphate residues **target** newly synthesized lysosomal enzymes to that organelle (see below).

TECHNIQUES ARE AVAILABLE FOR DETECTION, PURIFICATION, AND STRUCTURAL ANALYSIS OF GLYCOPROTEINS

A variety of methods that are used in the detection, purification and structural analysis of glycoproteins are listed in Table 56–3. The conventional methods used to purify proteins and enzymes (Chapter 8) are also applicable to the purification of glycoproteins. Once a glycoprotein has been purified, the use of **mass spectrometry** and **high-resolution NMR spectroscopy** can often identify the structures of the glycan chains present in a glycoprotein. Analysis of glycoproteins can be complicated by the fact that they often exist as **glycoforms;** these are proteins with identical amino acid sequences, but different oligosaccharide compositions. Although linkage details are not stressed in this particular chapter, it is critical to appreciate that the precise natures of the **linkages** between the sugars of glycoproteins are of fundamental importance in determining the structures and functions of these molecules.

EIGHT SUGARS PREDOMINATE IN HUMAN GLYCOPROTEINS

About 200 monosaccharides are found in nature; however, only eight are commonly found in the oligosaccharide chains of glycoproteins (Table 56–4). Most of these sugars were described in Chapter 15. N-Acetylneuraminic acid (NeuAc) is found at the terminals of oligosaccharide chains, usually attached to subterminal galactose (Gal) or N-acetylgalactosamine (GalNAc) residues. The other sugars listed are generally found in more internal positions. **Sulfate** is often found in glycoproteins, usually attached to Gal, GalNAc, or GlcNAc.

NUCLEOTIDE SUGARS ACT AS SUGAR DONORS IN MANY BIOSYNTHETIC REACTIONS

The first nucleotide sugar to be reported was uridine diphosphate glucose (UDPGlc); its structure is shown in Figure 20–2. The common nucleotide sugars involved in the biosynthesis of glycoproteins are listed in Table 56–4; the reasons some contain UDP

Table 56–3. Some important methods used to study glycoproteins.

Method	Use
Periodic acid-Schiff reagent	Detects glycoproteins as pink bands after electrophoretic separation.
Incubation of cultured cells with a radioactive sugar	Leads to detection of glycoproteins as radioactive bands after electrophoretic separation.
Treatment with appropriate glycosidase or phospholipase	Resultant shifts in electrophoretic migration help distinguish among proteins with N-glycan, O-glycan, or GPI linkages and also between high mannose and complex N-glycans.
Sepharose-lectin column chromatography	To purify glycoproteins or glycopeptides that bind the particular lectin used.
Compositional analysis following acid hydrolysis	Identifies sugars that the glycoprotein contains and their stoichiometry.
Mass spectrometry	Provides information on molecular mass, composition, sequence and sometimes branching of a glycan chain.
NMR spectroscopy	To identify specific sugars, their sequence, linkages, and the anomeric nature of glycosidic linkages.
Methylation (linkage) analysis	To determine linkages between sugars.
Amino acid or cDNA sequencing	Determination of amino acid sequence.

Table 56–4. The principal sugars found in human glycoproteins. Their structures are illustrated in Chapter 15.

Sugar	Type	Abbreviation	Nucleotide Sugar	Comments
Galactose	Hexose	Gal	UDPGal	Often found subterminal to NeuAc in *N*-linked glycoproteins. Also found in core trisaccharide of proteoglycans.
Glucose	Hexose	Glc	UDPGlc	Present during the biosynthesis of *N*-linked glycoproteins but not usually present in mature glycoproteins.
Mannose	Hexose	Man	GDP-Man	Common sugar in *N*-linked glycoproteins.
N-Acetylneur-aminic acid	Sialic acid (nine C atoms)	NeuAc	CMP-NeuAc	Often the terminal sugar in both *N*- and *O*-linked glycoproteins. Other types of sialic acid are also found, but NeuAc is the major species found in humans.
Fucose	Deoxyhexose	Fuc	GDP-Fuc	May be external in both *N*- and *O*-linked glyco-proteins or linked to the GlcNAc residue attached to Asn in *N*-linked species.
N-Acetylgalactos-amine	Aminohexose	GalNAc	UDP-GalNAc	Present in both *N*- and *O*-linked glycoproteins.
N-Acetylglucos-amine	Aminohexose	GlcNAc	UDP-GlcNAc	The sugar attached to the polypeptide chain via Asn in *N*-linked glycoproteins; also found at other sites in the oligosaccharides of these proteins.
Xylose	Pentose	Xyl	UDP-Xyl	Xyl is attached to the OH of Ser in many pro-teoglycans. Xyl in turn is attached to two Gal residues forming a link trisaccharide.

and others guanosine diphosphate (GDP) or cytidine monophosphate (CMP) are obscure. Many, but not all, of the glycosylation reactions involved in the biosynthesis of glycoproteins utilize these compounds (see below). The anhydro nature of the linkage between the phosphate group and the sugars is of the high-energy, high-group-transfer-potential type (Chapter 12). The sugars of these compounds are thus "activated" and can be transferred to suitable acceptors provided appropriate transferases are available.

The nucleotide sugars are formed in the cytosol, generally from reactions involving the corresponding nucleoside triphosphate. Formation of uridine diphosphate galactose (UDPGal) requires the following two reactions:

$$\boxed{\begin{array}{c}\text{UDPGlc}\\\text{PYROPHOSPHORYLASE}\end{array}}$$

UTP + Glucose 1-phosphate $\longleftrightarrow$
UDPGlc + Pyrophosphate

$$\boxed{\begin{array}{c}\text{UDPGlc}\\\text{EPIMERASE}\end{array}}$$

UDPGlc $\leftrightarrow$ UDPGal

Because many glycosylation reactions occur within the lumen of the Golgi apparatus, **carrier systems** (permeases, transporters) are necessary to transport nucleotide sugars across the Golgi mem-

brane. Systems transporting UDPGal, GDP-Man, and CMP-NeuAc into the cisternae of the Golgi apparatus have been described. They are **antiport** systems; ie, the influx of one molecule of nucleotide sugar is balanced by the efflux of one molecule of the corresponding nucleotide (eg, UMP, GMP, or CMP) formed from the nucleotide sugars. This mechanism ensures an adequate concentration of each nucleotide sugar inside the Golgi. UMP is formed from UDPGal in the above process as follows:

$$\boxed{\begin{array}{c}\text{GALACTOSYL-}\\\text{TRANSFERASE}\end{array}}$$

UDPGal + Protein $\rightarrow$ Protein—Gal + UDP

$$\boxed{\begin{array}{c}\text{NUCLEOSIDE}\\\text{DIPHOSPHATE}\\\text{PHOSPHATASE}\end{array}}$$

UDP $\rightarrow$ UMP + P_i

EXO- AND ENDOGLYCOSIDASES FACILITATE STUDY OF GLYCOPROTEINS

A number of **glycosidases** of defined specificity have proved useful in examining structural and functional aspects of glycoproteins (Table 56–5). These

Table 56–5. Some glycosidases used to study the structure and function of glycoproteins. The enzymes are available from a variety of sources and are often specific for certain types of glycosidic linkages and also for their anomeric natures. The sites of action of endoglycosidases F and H are shown in Figure 56–5. F acts on both high-mannose and complex oligosaccharides, whereas H acts on the former.

Enzymes	Type
Neuraminidases	Exoglycosidase
Galactosidases	Exoglycosidase
Endoglycosidase F	Endoglycosidase
Endoglycosidase H	Endoglycosidase

enzymes act at either external (exoglycosidases) or internal (endoglycosidases) positions of oligosaccharide chains. Examples of exoglycosidases are **neuraminidases** and **galactosidases;** their sequential use removes terminal NeuAc and subterminal Gal residues from most glycoproteins. **Endoglycosidases F** and **H** are examples of the latter class; these enzymes cleave the oligosaccharide chains at specific GlcNAc residues close to the polypeptide backbone (ie, at internal sites; Figure 56–5) and are thus useful in releasing large oligosaccharide chains for structural analyses. A glycoprotein can be treated with one or more of the above glycosidases to analyze the effects on its biologic behavior of removal of specific sugars.

THE MAMMALIAN ASIALOGLYCOPROTEIN RECEPTOR IS INVOLVED IN CLEARANCE OF CERTAIN GLYCOPROTEINS FROM PLASMA BY HEPATOCYTES

Experiments performed by Ashwell and his colleagues in the early 1970s played an important role in focusing attention on the functional significance of the oligosaccharide chains of glycoproteins. Initially, these workers studied the effect on the clearance from the circulation of rabbits of removing NeuAc residues (ie, **desialylation**) from the plasma protein **ceruloplasmin** (Chapter 59) by treatment in vitro with neuraminidase. This procedure exposed subterminal Gal residues that were normally masked by terminal NeuAc residues. Neuraminidase-treated radioactive ceruloplasmin was found to disappear rapidly from the circulation, in contrast to the slow clearance of the untreated protein. Further studies demonstrated that liver cells contain a **mammalian asialoglycoprotein receptor** that recognizes the galactosyl moiety of many desialylated plasma proteins and leads to their endocytosis. This work indicated that an individual sugar, such as galactose, could play an important role in governing at least one of the biologic properties (ie, time of residence in the circulation) of certain glycoproteins.

LECTINS CAN BE USED TO PURIFY GLYCOPROTEINS AND TO PROBE THEIR FUNCTIONS

Lectins are sugar-binding proteins that agglutinate cells or precipitate glycoconjugates; a number of lectins are themselves glycoproteins. Immunoglobulins that react with sugars are not considered lectins. Lectins contain at least two sugar-binding sites; proteins with a single sugar-binding site will not agglutinate cells or precipitate glycoconjugates. The specificity of a lectin is usually defined by the sugars that are best at inhibiting its ability to cause agglutination or precipitation. Enzymes, toxins and transport proteins can be classed as lectins if they have multiple binding sites. Lectins were first discovered in plants and microbes, but many lectins of animal origin are now known. The mammalian asialoglycoprotein receptor described above is an important example of an animal lectin. Some important lectins are listed in Table 56–6. Much current research is centered on the roles of various animal lectins (eg, the selectins) in cell-cell interactions that occur in pathologic conditions such as inflammation and cancer metastasis (see below).

Numerous lectins have been purified and are commercially available; three plant lectins that have been widely used experimentally are listed in Table 56–7. The following are three of the biochemical uses to which such purified lectins have been put:

(1) In purification and analysis of glycoproteins: Lectins such as concanavalin A (ConA) can be attached covalently to inert supporting media such as Sepharose. The resulting Sepharose-ConA may be

Table 56–6. Some important lectins.

Lectins	Examples or Comments
Legume lectins	Concanavalin A, pea lectin
Wheat germ agglutinin	Widely used in studies of surfaces of normal cells and cancer cells
Ricin	Cytotoxic glycoprotein derived from seeds of the castor plant
Bacterial toxins	Heat-labile enterotoxin of *E coli* and cholera toxin
Influenza virus hemagglutinin	Responsible for host-cell attachment and membrane fusion
C-type lectins	Characterized by a Ca^{2+}-dependent carbohydrate recognition domain (CRD); includes the mammalian asialoglycoprotein receptor, the selectins, and the mannose-binding protein
S-type lectins	β-Galactoside-binding animal lectins with roles in cell-cell and cell-matrix interactions
P-type lectins	Mannose 6-P receptor
I-type lectins	Members of the immunoglobulin superfamily, eg, sialoadhesin mediating adhesion of macrophages to various cells

Table 56–7. Three plant lectins and the sugars with which they interact. In most cases, lectins show specificity for the anomeric nature of the glycosidic linkage (α or β); this is not indicated in the table.

Lectin	Abbreviation	Sugars
Concanavalin A	ConA	Man and Glc
Soybean lectin		Gal and GalNAc
Wheat germ agglutinin	WGA	Glc and NeuAc

used for the purification of glycoproteins that contain oligosaccharide chains which interact with Con A.

(2) As tools for probing the surfaces of cells: Since lectins recognize specific sugars, they can be used to probe, at a general level, the sugar residues exposed on the surface membranes of cells. Numerous studies have been performed using lectins to compare the surfaces of normal and cancer cells (Chapter 62). A general finding has been that smaller amounts of certain lectins are required to cause agglutination of tumor cells than of normal cells. This suggests that the organization or structure of a number of glycoproteins on the surfaces of tumor cells may be different from that found in normal cells.

(3) To generate mutant cells lacking certain enzymes of oligosaccharide synthesis: When mammalian cells in tissue culture are exposed to appropriate concentrations of certain lectins (eg, ConA), most are killed, but a few resistant cells survive. Such cells are often found to lack certain enzymes involved in oligosaccharide synthesis. The cells are resistant because they do not produce oligosaccharide chains that interact with the lectin used. The use of such mutant cells has been important in elucidating a number of aspects of the biosynthesis of glycoproteins.

THERE ARE THREE MAJOR CLASSES OF GLYCOPROTEINS

Based on the nature of the linkage between their polypeptide chains and their oligosaccharide chains, glycoproteins can be divided into three major classes (Figure 56–1): (1) those containing an *O*-glycosidic

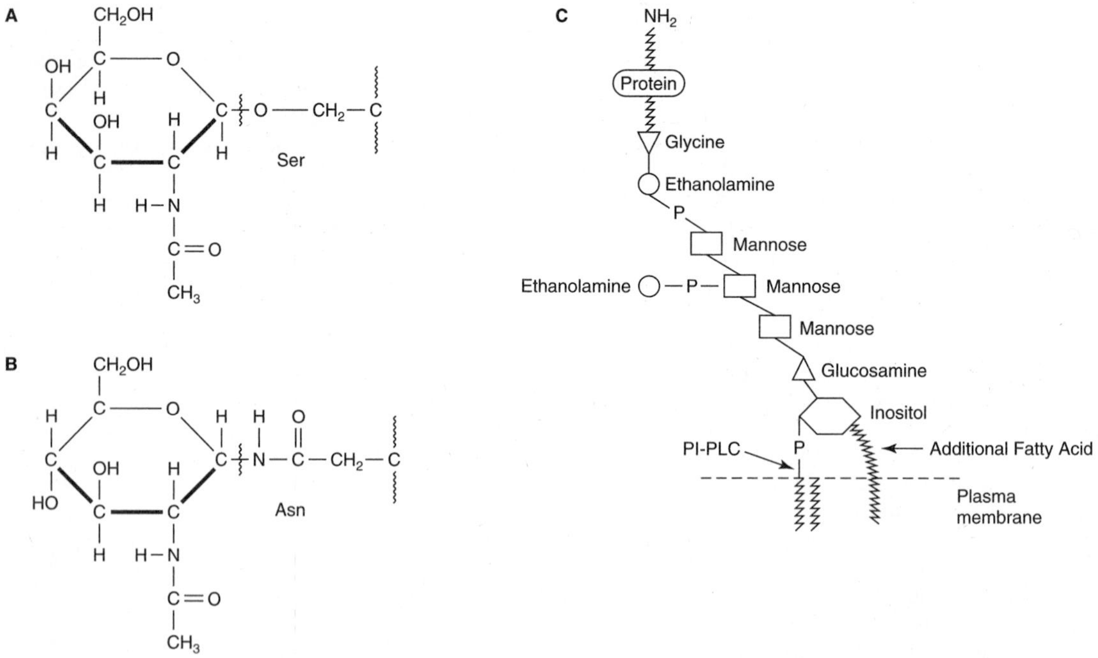

Figure 56–1. Depictions of *A* an *O*-linkage (*N*-acetylgalactosamine to serine); *B* an *N*-linkage (*N*-acetylglucosamine to asparagine) and *C* a glycosylphosphatidylinositol (GPI) linkage. The GPI structure shown is that linking acetylcholinesterase to the plasma membrane of the human red blood cell. The carboxyl terminal amino acid is glycine joined in amide linkage via its COOH group to the NH₂ group of phosphorylethanolamine, which in turn is joined to a mannose residue. The core glycan contains three mannose and one glucosamine residues. The glucosamine is linked to inositol, which is attached to phosphatidic acid. The site of action of PI-phospholipase C (PI-PLC) is indicated. The structure of the core glycan is shown in the text. This particular GPI contains an extra fatty acid attached to inositol and also an extra phosphorylethanolamine moiety attached to the middle of the three mannose residues. Variations found among different GPI structures include the identity of the carboxyl terminal amino acid, the molecules attached to the mannose residues, and the precise nature of the lipid moiety.

linkage (ie, *O*-linked), involving the hydroxyl side chain of serine or threonine and a sugar such as *N*-acetylgalactosamine (GalNAc-Ser[Thr]); (2) those containing an *N*-glycosidic linkage (ie, *N*-linked), involving the amide nitrogen of asparagine and *N*-acetylglucosamine (GlcNAc-Asn); and (3) those linked to the carboxyl terminal amino acid of a protein via a phosphorylethanolamine moiety joined to an oligosaccharide (glycan), which in turn is linked via glucosamine to phosphatidylinositol (PI). This latter class is referred to as **glycosylphosphatidyl-inositol-anchored** (**GPI-anchored,** or **GPI-linked**) glycoproteins. Other minor classes of glycoproteins exist.

The number of oligosaccharide chains attached to one protein can vary from one to 30 or more, with the sugar chains ranging from one or two residues in length to much larger structures. Some proteins can contain more than one type of linkage; for instance, **glycophorin,** an important red cell membrane glycoprotein (Chapter 60), contains both *O*- and *N*-linked oligosaccharides.

GLYCOPROTEINS CONTAIN SEVERAL TYPES OF *O*-GLYCOSIDIC LINKAGES

At least four subclasses of *O*-glycosidic linkages are found in human glycoproteins: (1) The **GalNAc-Ser(Thr)** linkage shown in Figure 56–1 is the predominant linkage. Two typical oligosaccharide chains found in members of this sub-class are shown in Figure 56–2. Usually a Gal or a NeuAc residue is attached to the GalNAc, but many variations in the sugar composition and length of such oligosaccharide chains are found. This type of linkage is found in **mucins** (see below). (2) **Proteoglycans** contain a **Gal-Gal-Xyl-Ser** trisaccharide (the so-called link trisaccharide). (3) **Collagens** contain a **Gal-hydroxy-lysine (Hyl)** linkage. (Subclasses [2] and [3] are discussed further in Chapter 57.) (4) Many **nuclear proteins** (eg, certain transcription factors) and **cytosolic proteins** contain side chains consisting of a single GlcNAc attached to a serine or threonine residue **(GlcNAc-Ser[Thr]).**

Mucins Have a High Content of *O*-Linked Oligosaccharides and Exhibit Repeating Amino Acid Sequences

Mucins are glycoproteins with two major characteristics: (1) a high content of *O*-linked oligosaccharides (the carbohydrate content of mucins is generally more than 50%), and (2) the presence of **repeating amino acid sequences** (tandem repeats) in the center of their polypeptide backbones, to which the *O*-glycan chains are attached in clusters (Figure 56–3). These sequences are rich in serine, threonine, and proline. Although *O*-glycans predominate, mucins often contain a number of *N*-glycan chains.

Both **secretory** and **membrane-bound** mucins occur. The former are found in the mucus present in the secretions of the gastrointestinal, respiratory and reproductive tracts. **Mucus** consists of about 94% water and 5% mucins, with the remainder being a mixture of various cell molecules, electrolytes, and remnants of cells. Secretory mucins generally have an oligomeric structure and thus often have a very high molecular mass. The oligomers are composed of monomers linked by disulfide bonds. Mucus exhibits a high **viscosity** and often forms a **gel.** These qualities are functions of its content of mucins. The high content of *O*-glycans confers an extended structure on mucins. This is in part explained by steric interactions between their GalNAc moieties and adjacent amino acids, resulting in a chain-stiffening effect so

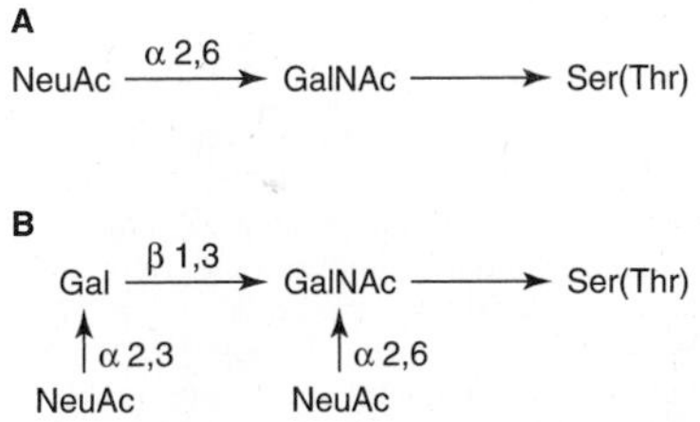

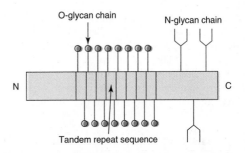

Figure 56–3. Schematic diagram of a mucin. *O*-glycan chains are shown attached to the central region of the extended polypeptide chain and *N*-glycan chains to the carboxyl terminal region. The narrow rectangles represent a series of tandem repeat amino acid sequences. Many mucins contain cysteine residues whose SH groups form interchain linkages; these are not shown in the figure. (Adapted from Strous GJ, Dekker J: Mucin-type glycoproteins. Crit Rev Biochem Mol Biol 1992;27:57.)

A
NeuAc $\xrightarrow{\alpha 2,6}$ GalNAc $\longrightarrow$ Ser(Thr)

B
Gal $\xrightarrow{\beta 1,3}$ GalNAc $\longrightarrow$ Ser(Thr)
  $\uparrow \alpha 2,3$ $\uparrow \alpha 2,6$
  NeuAc   NeuAc

Figure 56–2. Structures of two *O*-linked oligosaccharides found in **(A)** submaxillary mucins and **(B)** fetuin and in the sialoglycoprotein of the membrane of human red blood cells. (Modified and reproduced, with permission, from Lennarz WJ: *The Biochemistry of Glycoproteins and Proteoglycans.* Plenum Press, 1980.)

that the conformations of mucins often become those of rigid rods. Intermolecular noncovalent interactions between various sugars on neighboring glycan chains contribute to gel formation. The high content of **NeuAc** and **sulfate** residues found in many mucins confers a negative charge on them. With regards to function, mucins help **lubricate** and form a **protective physical barrier** on epithelial surfaces. Membrane-bound mucins participate in various **cell-cell interactions** (eg, involving selectins, see below). The density of oligosaccharide chains makes it difficult for **proteases** to approach their polypeptide backbones, so that mucins are often resistant to their action. Mucins also tend to "mask" certain surface antigens. Many cancer cells form excessive amounts of mucins; perhaps the mucins may mask certain surface antigens on such cells and thus protect the cells from immune surveillance.

The **genes** encoding the polypeptide backbones of a number of mucins have been studied. At least seven human genes (MUC1–7) for mucins have been cloned and sequenced, including these encoding pancreatic, small intestinal, tracheobronchial, gastric, and salivary mucins. These studies have revealed new information about the polypeptide backbones of mucins (size of tandem repeats, potential sites of N-glycosylation, etc) and ultimately should reveal aspects of their genetic control. Some important properties of mucins are summarized in Table 56–8.

The Biosynthesis of O-Linked Glycoproteins Uses Nucleotide Sugars

The polypeptide chains of O-linked and other glycoproteins are encoded by mRNA species; because most glycoproteins are membrane-bound or secreted, they are generally translated on membrane-bound polyribosomes (Chapter 40). The sugars of the oligosaccharide chains of the O-glycosidic type of glycoprotein are built up by the **stepwise donation of sugars from nucleotide sugars,** such as UDP-GalNAc, UDPGal, and CMP-NeuAc. The enzymes catalyzing this type of reaction are membrane-bound **glycoprotein glycosyltransferases.** The synthesis of each such enzyme is controlled by one specific gene. Generally, synthesis of one specific type of linkage requires the activity of a correspondingly specific

transferase (ie, the **one linkage, one glycosyl transferase** hypothesis). The enzymes catalyzing addition of the inner sugar residues are located in the endoplasmic reticulum, and addition of the first sugars occurs during translation (ie, **cotranslational** modification of the protein occurs). The factors that determine which specific serine and threonine residues are glycosylated have not been identified. The enzymes adding the terminal sugars (such as NeuAc) are located in the Golgi apparatus.

The major features of the biosynthesis of O-linked glycoproteins are summarized in Table 56–9.

N-LINKED GLYCOPROTEINS CONTAIN AN Asn-GlcNAc LINKAGE

N-Linked glycoproteins are distinguished by the presence of the Asn-GlcNAc linkage (Figure 56–1). It is the major class of glycoproteins and has been much studied, since the most readily accessible glycoproteins (eg, plasma proteins) mainly belong to this group. It includes both **membrane-bound** and **circulating** glycoproteins. The principal difference between this and the previous class, apart from the nature of the amino acid to which the oligosaccharide chain is attached (Asn versus Ser or Thr), concerns their biosynthesis.

Complex, Hybrid, and High-Mannose Are the Three Major Classes of N-Linked Glycoproteins

There are three major classes of N-linked glycoproteins: complex, hybrid, and high-mannose (Figure 56–4). Each type shares a common pentasaccharide, $Man_3GlcNAc_2$—shown within the boxed area in Figure 56–4 and depicted also in Figure 56–5—but they differ in their outer branches. The presence of the common pentasaccharide is explained by the fact that all three classes share an initial common mechanism of biosynthesis. Glycoproteins of the complex type generally contain terminal NeuAc residues and underlying Gal and GlcNAc residues, the latter often constituting the disaccharide lactosamine. (The presence of repeating **lactosamine units,** [Galβ1–

Table 56–9. Summary of main features of O-glycosylation.

- Involves a battery of membrane-bound glycoprotein glycosyltransferases acting in a stepwise manner; each transferase is generally specific for a particular type of linkage ("one linkage, one glycosyltransferase" hypothesis).
- Most of the enzymes involved are located in various subcompartments of the Golgi apparatus.
- Each glycosylation reaction involves the appropriate nucleotide-sugar.
- Oligosaccharide-P-P-dolichol is not involved, nor are glycosidases; and the reactions are not inhibited by tunicamycin.
- O-Glycosylation occurs posttranslationally at certain Ser and Thr residues.

Table 56–8. Some properties of mucins.

- Found in secretions of the gastrointestinal, respiratory, and reproductive tracts and also in membranes of various cells.
- Exhibit high content of O-glycan chains, and usually of NeuAc and sulfate.
- Contain repeating amino acid sequences rich in serine, threonine, and proline.
- Extended structure contributes to their high viscoelasticity.
- Form protective physical barrier on epithelial surfaces, are involved in cell-cell interactions, and may mask certain surface antigens.

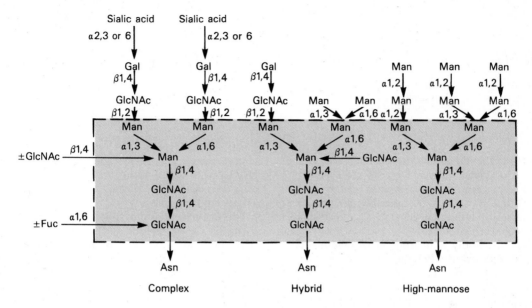

Figure 56–4. Structures of the major types of asparagine-linked oligosaccharides. The boxed area encloses the pentasaccharide core common to all *N*-linked glycoproteins. (Reproduced, with permission, from Kornfeld R, Kornfeld S: Assembly of asparagine-linked oligosaccharides. Annu Rev Biochem 1985;54:631.)

$4GlcNAc\beta1-3]_n$, characterizes a fourth class of *N*-linked glycoprotein, the **polylactosamine** class; the I/i blood group substances belong to it.) The majority of complex-type oligosaccharides contain two (Figure 56–4), three, or four outer branches, but structures containing five branches have also been described. The oligosaccharide branches are often referred to as **antennae,** so that bi-, tri-, tetra-, and penta-antennary structures may all be found. A bewildering number of chains of the complex type exist, and that indicated in Figure 56–4 is only one of many. Other complex chains may terminate in Gal or Fuc. High-mannose oligosaccharides typically have two to six additional Man residues linked to the pentasaccharide core. Hybrid molecules contain features of both of the two other classes.

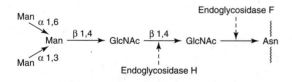

Figure 56–5. Schematic diagram of the pentasaccharide core common to all *N*-linked glycoproteins and to which various outer chains of oligosaccharides may be attached. The sites of action of endoglycosidases F and H are also indicated.

The Biosynthesis of *N*-Linked Glycoproteins Involves Oligosaccharide-P-P-Dolichol

Leloir and his colleagues described the occurrence of an **oligosaccharide-pyrophosphoryldolichol (oligosaccharide-P-P-Dol),** which subsequent research showed to play a key role in the biosynthesis of *N*-linked glycoproteins. The oligosaccharide chain of this compound generally has the structure Glc_3Man_9 $GlcNAc_2$—R (R = P-P-Dol). The sugars of this compound are first assembled on the pyrophosphoryldolichol backbone, and the oligosaccharide chain is then transferred en bloc to suitable Asn residues of acceptor apoglycoproteins during their synthesis on membrane-bound polyribosomes.

To form an oligosaccharide chain of the **complex** type, the Glc residues and six of the Man residues are removed by enzymes, forming the core pentasaccharide $Man_3GlcNAc_2$ (Figure 56–5). The sugars characteristic of complex chains (GlcNAc, Gal, NeuAc) are then added by the action of individual glycosyltransferases, mostly located in the Golgi apparatus.

To form **high-mannose** chains, only the Glc residues plus or minus certain of the peripheral Man residues are removed. The phenomenon whereby the glycan chains of *N*-linked glycoproteins are first partially degraded and then in some cases rebuilt is referred to as **oligosaccharide processing.** Thus, the initial steps involved in the biosynthesis of the *N*-linked glycoproteins differ markedly from those involved in the biosynthesis of the *O*-linked glyco-

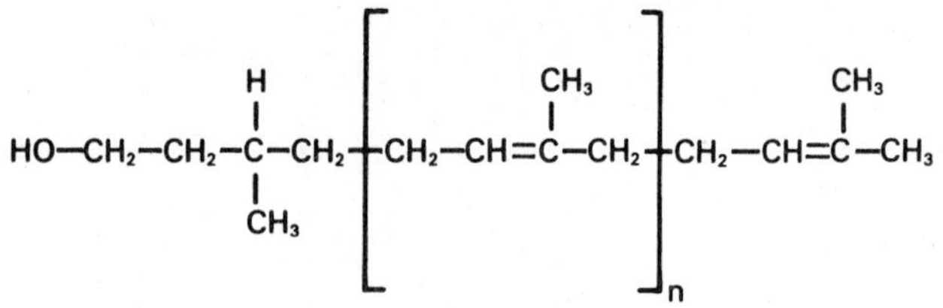

Figure 56–6. The structure of dolichol. The phosphate in dolichol phosphate is attached to the primary alcohol group at the left-hand end of the molecule. The group within the brackets is an isoprene unit (n = 17–20 isoprenoid units).

$$HO-CH_2-CH_2-\underset{\underset{CH_3}{|}}{\overset{\overset{H}{|}}{C}}-CH_2\left[CH_2-CH=\underset{}{\overset{\overset{CH_3}{|}}{C}}-CH_2\right]_nCH_2-CH=\overset{\overset{CH_3}{|}}{C}-CH_3$$

proteins. The former involves the oligosaccharide-P-P-Dol; the latter, as described earlier, does not.

The process of *N*-glycosylation can be broken down into two stages: (1) assembly and transfer of oligosaccharide-P-P-dolichol and (2) processing of the oligosaccharide chain.

A. Assembly and Transfer of Oligosaccharide-P-P-Dolichol: Polyisoprenol compounds exist in both bacteria and eukaryotic tissues. They participate in the synthesis of bacterial cell walls and in the biosynthesis of *N*-linked glycoproteins. The polyisoprenol used in eukaryotic tissues is **dolichol,** which is, next to rubber, the longest naturally occurring hydrocarbon made up of a single repeating unit. Dolichol is composed of 17–20 repeating isoprenoid units (Figure 56–6).

Before it participates in the biosynthesis of oligosaccharide-P-P-Dol, dolichol must first be phosphorylated to form dolichol phosphate (Dol-P) in a reaction catalyzed by **dolichol kinase** and using ATP as the phosphate donor.

GlcNAc-pyrophosphoryl-dolichol (GlcNAc-P-P-Dol) is the key lipid that acts as an acceptor for other sugars in the assembly of oligosaccharide-P-P-Dol. It is synthesized in the membranes of the endoplasmic reticulum from Dol-P and UDPGlcNAc in the following reaction:

Dol-P + UDP-GlcNAc → GlcNAc-P-P-Dol + UMP

The above reaction—which is the first step in the assembly of oligosaccharide-P-P-Dol—and the other later reactions are summarized in Figure 56–7. The essential features of the subsequent steps in the as-

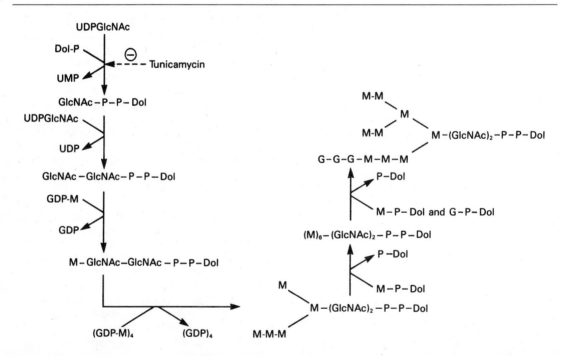

Figure 56–7. Pathway of biosynthesis of oligosaccharide-P-P-dolichol. The specific linkages formed are indicated in Figure 56–8. Note that the internal mannose residues are donated by GDP-mannose, whereas the more external mannose residues and the glucose residues are donated by dolichol-P-mannose and dolichol-P-glucose. (UDP, uridine diphosphate; Dol, dolichol; P, phosphate; UMP, uridine monophosphate; GDP, guanosine diphosphate; M, mannose; G, glucose.)

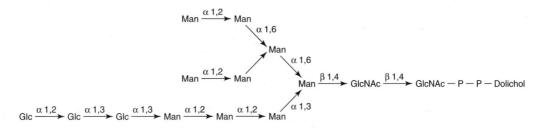

Figure 56–8. Structure of oligosaccharide-P-P-dolichol. (Reproduced, with permission, from Lennarz WJ: *The Biochemistry of Glycoproteins and Proteoglycans.* Plenum Press, 1980.)

sembly of oligosaccharide-P-P-dolichol are as follows:

(1) A second GlcNAc residue is added to the first, again using UDPGlcNAc as the donor.

(2) Five Man residues are added, using GDP-mannose as the donor.

(3) Four additional Man residues are next added, using Dol-P-Man as the donor. Dol-P-Man is formed by the following reaction:

$$Dol\text{-}P + GDP\text{-}Man \rightarrow Dol\text{-}P\text{-}Man + GDP$$

(4) Finally, the three peripheral glucose residues are donated by Dol-P-Glc, which is formed in a reaction analogous to that just presented, except that Dol-P and UDPGlc are the substrates.

It should be noted that the first seven sugars (two GlcNAc and five Man residues) are donated by nucleotide sugars, whereas the last seven sugars (four Man and three Glc residues) added are donated by dolichol-P-sugars. The net result is assembly of the compound illustrated in Figure 56–8 and referred to in shorthand as $Glc_3Man_9GlcNAc_2$-P-P-Dol.

The oligosaccharide linked to dolichol-P-P is transferred en bloc to form an *N*-glycosidic bond with one or more specific Asn residues of an acceptor protein emerging from the luminal surface of the membrane of the endoplasmic reticulum. The reaction is catalyzed by **oligosaccharide:protein transferase,** a membrane-associated enzyme. The transferase will recognize and utilize any glycolipid with the general structure R—(GlcNAc)$_2$-P-P-Dol. Glycosylation occurs at the Asn residue of an Asn-X-Ser/Thr tripeptide sequence, where X is any amino acid except proline, aspartic acid, or glutamic acid. A tripeptide site contained within a β turn is favored. Only about one-third of the Asn residues that are potential acceptor sites are actually glycosylated. The acceptor proteins are of both the secretory and integral membrane class. Cytosolic proteins are rarely glycosylated. The transfer reaction and subsequent processes in the glycosylation of *N*-linked glycoproteins, along with their subcellular locations, are depicted in Figure 56–9. The other product of the oligosaccharide:protein trans-

ferase reaction is dolichol-P-P, which is subsequently converted to dolichol-P by a phosphatase. The dolichol-P can serve again as an acceptor for the synthesis of another molecule of oligosaccharide-P-P-dolichol.

B. Processing of the Oligosaccharide Chain:

1. Early phase–The various reactions involved are indicated in Figure 56–9. The oligosaccharide:protein transferase catalyzes reaction 1 (see above). Reactions 2 and 3 involve the removal of the terminal Glc residue by glucosidase I and of the next two Glc residues by glucosidase II, respectively. In the case of **high-mannose** glycoproteins, the process may stop here, or up to four Man residues may also be removed. However, to form **complex** chains, additional steps are necessary, as follows. Four external Man residues are removed in reactions 4 and 5 by two different mannosidases. In reaction 6, a GlcNAc residue is added to one of the Man residues by GlcNAc transferase I. The action of this latter enzyme permits the occurrence of reaction 7, a reaction catalyzed by yet another mannosidase (Golgi α-mannosidase II) and which results in a reduction of the Man residues to the core number of three (Figure 56–5).

An important additional pathway is indicated in reactions I and II of Figure 56–9. This involves enzymes destined for **lysosomes.** Such enzymes are targeted to the lysosomes by a specific chemical marker. In reaction I, a residue of GlcNAc-1-P is added to carbon 6 of one or more specific Man residues of these enzymes. The reaction is catalyzed by a GlcNAc phosphotransferase, which uses UDPGlcNAc as the donor and generates UMP as the other product:

GlcNAc PHOSPHOTRANSFERASE

UDP-GlcNAc + Man—Protein →
GlcNAc-1-P-6-Man—Protein + UMP

In reaction II, the GlcNAc is removed by the action of a phosphodiesterase, leaving the Man residues phosphorylated in the 6 position:

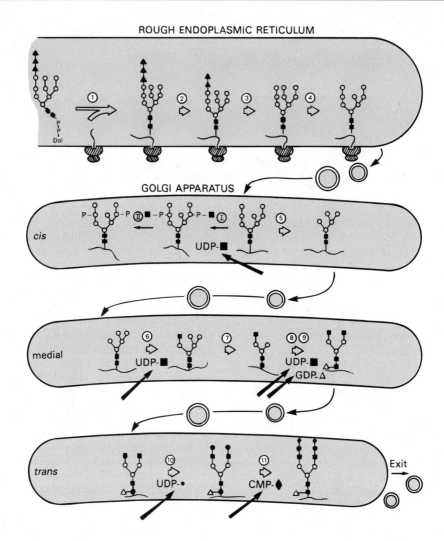

Figure 56–9. Schematic pathway of oligosaccharide processing. The reactions are catalyzed by the following enzymes: ①, oligosaccharide:protein transferase; ②, α-glucosidase I; ③, α-glucosidase II; ④, endoplasmic reticulum α1,2 mannosidase; Ⓘ, *N*-acetylglucosaminylphosphotransferase; Ⓘⓘ, *N*-acetylglucosamine-1-phosphodiester α-*N*-acetylglucosaminidase; ⑤, Golgi apparatus α-mannosidase I; ⑥, *N*-acetylglucosaminyltransferase I; ⑦, Golgi apparatus α-mannosidase II; ⑧, *N*-acetyl-glucosaminyltransferase II; ⑨, fucosyltransferase; ⑩, galactosyltransferase; ⑪, sialyltransferase. (Solid square, *N*-acetylglucosamine; open circle, mannose; solid triangle, glucose; open triangle, fucose; solid circle, galactose; solid diamond, sialic acid.) (Reproduced, with permission, from Kornfeld R, Kornfeld S: Assembly of asparagine-linked oligosaccharides. Annu Rev Biochem 1985;54:631.)

PHOSPHODIESTERASE

GlcNAc-1-P-6-Man—Protein →

P-6-Man—Protein + GlcNAc

Man 6-P receptors, located in the Golgi apparatus, bind the Man 6-P residue of these enzymes and direct them to the lysosomes. Fibroblasts from patients with **I-cell disease** (see below) are severely deficient in the activity of the GlcNAc phosphotransferase.

    2. Late phase–To assemble a typical complex oligosaccharide chain, additional sugars must be added to the structure formed in reaction 7. Hence, in reaction 8, a second GlcNAc is added to the peripheral Man residue of the other arm of the bi-antennary structure shown in Figure 56–9; the enzyme catalyzing this step is GlcNAc transferase II. Reactions 9, 10, and 11 involve the addition of Fuc, Gal, and NeuAc residues at the sites indicated, in reactions catalyzed by fucosyl, galactosyl, and sialyl transferases, respectively.

The Endoplasmic Reticulum and Golgi Apparatus Are the Major Sites of Glycosylation

As indicated in Figure 56–9, the endoplasmic reticulum and the Golgi apparatus are the major sites involved in glycosylation processes. Addition of the oligosaccharide occurs in the rough endoplasmic reticulum during or after translation. Removal of the Glc and some of the peripheral Man residues also occurs in the endoplasmic reticulum. The Golgi apparatus is composed of *cis,* medial, and *trans* cisternae; these can be separated by appropriate centrifugation procedures. Vesicles containing glycoproteins appear to bud off in the endoplasmic reticulum and are transported to the *cis* Golgi. Various studies have shown that the enzymes involved in glycoprotein processing show differential locations in the cisternae of the Golgi. As indicated in Figure 56–9, Golgi α-mannosidase I (catalyzing reaction 5) is located mainly in the *cis* Golgi, whereas GlcNAc transferase I (catalyzing reaction 6) appears to be located in the medial Golgi, and the fucosyl, galactosyl, and sialyl transferases (catalyzing reactions 9, 10, and 11) are located in the *trans* Golgi. The major features of the biosynthesis of *N*-linked glycoproteins are summarized in Table 56–10 and should be contrasted with those previously listed (Table 56–9) for *O*-linked glycoproteins.

Some Glycan Intermediates Formed During *N*-Glycosylation Have Specific Functions

The following are a number of specific functions of *N*-glycan chains that have been established or are under investigation. (1) The involvement of the **mannose 6-P signal** in targeting of certain lysosomal enzymes is clear (see above and discussion of I-cell disease, below). It is possible that other specific sequences of sugars in *N*-glycan chains may play roles in targeting specific glycoproteins to other or-

Table 56–10. Summary of main features of *N*-glycosylation.

- The oligosaccharide $Glc_3Man_9(GlcNAc)_2$ is transferred from oligosaccharide-P-P-dolichol in a reaction catalyzed by oligosaccharide:protein transferase, which is inhibited by tunicamycin.
- Transfer occurs to specific Asn residues in the sequence Asn-X-Ser/Thr, where X is any residue except Pro, Asp, or Glu.
- Transfer occurs cotranslationally in the endoplasmic reticulum.
- The protein-bound oligosaccharide is then partially processed by glucosidases and mannosidases; if no additional sugars are added, this results in a high-mannose chain.
- If processing occurs down to the core pentasaccharide $(Man_3[GlcNAc]_2)$, complex chains are synthesized by the stepwise addition of individual sugars in reactions catalyzed by specific transferases (eg, GlcNAc, Gal, NeuAc transferases) that employ appropriate nucleotide sugars.

ganelles. (2) It is likely that the large *N*-glycan chains present on newly synthesized glycoproteins may assist in keeping these proteins in a soluble state inside the lumen of the endoplasmic reticulum. (3) One species of *N*-glycan chains has been shown to play a role in the folding and retention of certain glycoproteins in the lumen of the endoplasmic reticulum. **Calnexin** is a protein present in the endoplasmic reticulum membrane that acts as a "chaperone" (Chapter 43). A chaperone is a protein that binds to a newly synthesized protein and assists in its proper folding. It has been found that calnexin will bind specifically to a number of glycoproteins (eg, the influenza virus hemagglutinin [HA]) that possess the **monoglycosylated core structure.** This species is the product of reaction 2 shown in Figure 56–9 but from which the terminal glucose residue has been removed, leaving only the innermost glucose attached. The release of fully folded HA from calnexin requires the enzymatic removal of this last glucosyl residue by α-glucosidase II. In this way, calnexin retains certain partly folded (or misfolded) glycoproteins and releases them when proper folding has occurred.

Several Factors Regulate the Glycosylation of Glycoproteins

It is evident that glycosylation of glycoproteins is a complex process, involving a large number of enzymes. One index of its complexity is that seven distinct GlcNAc transferases involved in glycoprotein biosynthesis have been reported, and a number of others are theoretically possible. Multiple species of the other glycosyltransferases (eg, sialyltransferases) also exist. Controlling factors of the first stage of *N*-linked glycoprotein biosynthesis (ie, oligosaccharide **assembly and transfer**) include (1) the presence of suitable acceptor sites in proteins, (2) the tissue level of Dol-P, and (3) the activity of the oligosaccharide transferase.

Some factors known to be involved in the regulation of **oligosaccharide processing** are listed in Table 56–11. Two of the points listed merit further comment. Firstly, **species variations** among processing enzymes have assumed importance in relation to production of glycoproteins of therapeutic use by means of recombinant DNA technology. For instance, recombinant erythropoietin (EPO) is sometimes administered to patients with certain types of chronic anemia in order to stimulate erythropoiesis. The half-life of EPO in plasma is influenced by the nature of its glycosylation pattern, with certain patterns being associated with a short half-life, thus appreciably limiting its period of therapeutic effectiveness. It is thus important to harvest EPO from host cells that confer a pattern of glycosylation consistent with a normal half-life in plasma. Secondly, there is great interest in analysis of the activities of glycoprotein-processing enzymes in various types of cancer cells. These cells have often been found to synthesize

Table 56–11. Some factors affecting the activities of glycoprotein processing enzymes.

Factor	Comment
Cell type	Different cell types contain different profiles of processing enzymes.
Previous enzyme	Certain glycosyltransferases will only act on an oligosaccharide chain if it has already been acted upon by another processing enzyme.[1]
Development	The cellular profile of processing enzymes may change during development if their genes are turned on or off.
Intracellular location	For instance, if an enzyme is destined for insertion into the membrane of the ER (eg, HMG-CoA reductase), it may never encounter Golgi-located processing enzymes.
Protein conformation	Differences in conformation of different proteins may facilitate or hinder access of processing enzymes to identical oligosaccharide chains.
Species	Same cells (eg, fibroblasts) from different species may exhibit different patterns of processing enzymes.
Cancer	Cancer cells may exhibit processing enzymes different from those of corresponding normal cells.

[1]For example, prior action of GlcNAc transferase I is necessary for the action of Golgi α-mannosidase II.

different oligosaccharide chains (eg, they often exhibit greater branching) from those made in control cells. This could be due to cancer cells containing different patterns of glycosyltransferases from those exhibited by corresponding normal cells, due to specific gene activation or repression. The differences in oligosaccharide chains could affect adhesive interactions between cancer cells and their normal parent tissue cells, contributing to metastasis. If a correlation could be found between the activity of particular processing enzymes and the **metastatic properties** of cancer cells (Chapter 62), this could be important as it might permit synthesis of drugs to inhibit these enzymes and, secondarily, metastasis. At the gene level, many glycosyltransferases have already been cloned, and others are under study. Cloning has already revealed new information on both protein and gene structures. The latter should also cast light on the mechanisms involved in their transcriptional control, and gene knockout studies are being used to evaluate the biologic importance of various glycosyltransferases.

Tunicamycin Inhibits *N*- but Not *O*-Glycosylation

A number of compounds are known to inhibit various reactions involved in glycoprotein processing. **Tunicamycin, deoxynojirimycin,** and **swainsonine** are three such agents. The reactions they inhibit are indicated in Table 56–12. These agents can be used experimentally to inhibit various stages of glycoprotein biosynthesis and to study the effects of specific alterations upon the process. For instance, if cells are grown in the presence of tunicamycin, no glycosylation of their normally *N*-linked glycoproteins will occur. In certain cases, lack of glycosylation has been

shown to increase the susceptibility of these proteins to proteolysis. Inhibition of glycosylation does not appear to have a consistent effect upon the secretion of glycoproteins that are normally secreted. Inhibitors of glycoprotein processing, including these listed in Table 56–12, do not affect the biosynthesis of *O*-linked glycoproteins.

SOME PROTEINS ARE ANCHORED TO THE PLASMA MEMBRANE BY GLYCOSYLPHOSPHATIDYLINOSITOL STRUCTURES

Glycosylphosphatidylinositol (GPI)-linked glycoproteins comprise the third major class of glycoprotein. The GPI structure (sometimes called a "sticky foot") involved in linkage of the enzyme acetylcholinesterase (ACh esterase) to the plasma membrane of the red blood cell is shown in Figure 56–1. GPI-linked proteins are anchored to the outer leaflet of the plasma membrane by the fatty acids of phosphatidylinositol (PI). The PI is linked via a GlcNH$_2$

Table 56–12. Three inhibitors of enzymes involved in the glycosylation of glycoproteins and their sites of action.

Inhibitor	Site of Action
Tunicamycin	Inhibits the enzyme catalyzing addition of GlcNac to dolichol-P, the first step in the biosynthesis of oligosaccharide-P-P-dolichol
Deoxynojirimycin	Inhibitor of glucosidases I and II
Swainsonine	Inhibitor of mannosidase II

moiety to a glycan chain that contains various sugars (eg, Man, GlcNH$_2$). In turn, the oligosaccharide chain is linked via phosphorylethanolamine in an amide linkage to the carboxyl terminal amino acid of the attached protein. The core of most GPI structures contains one molecule of phosphorylethanolamine, three Man residues, one molecule of GlcNH$_2$, and one molecule of phosphatidylinositol, as follows:

Ethanolamine-phospho $\rightarrow$ 6Manα1 $\rightarrow$ 2Manα1 $\rightarrow$ 6Manα1 $\rightarrow$ GlcNα1 $\rightarrow$ 6—*myo*-inositol-1-phospholipid

Additional constituents are found in many GPI structures; for example, that shown in Figure 56–1 contains an extra phosphorylethanolamine attached to the middle of the three Man moieties of the glycan and an extra fatty acid attached to GlcNH$_2$. The functional significance of these variations among structures is not understood. This type of linkage was first detected by the use of bacterial PI-specific phospholipase C (PI-PLC), which was found to release certain proteins from the plasma membrane of cells by splitting the bond indicated in Figure 56–1. Examples of some proteins that are anchored by this type of linkage are given in Table 56–13. At least three possible functions of this type of linkage have been suggested: (1) The GPI anchor may allow greatly enhanced mobility of a protein in the plasma membrane compared with that observed for a protein that contains transmembrane sequences. This is perhaps not surprising, as the GPI anchor is attached only to the outer leaflet of the lipid bilayer, so that it is freer to diffuse than a protein anchored via both leaflets of the bilayer. Increased mobility may be important in facilitating rapid responses to appropriate stimuli. (2) Some GPI anchors may connect with **signal transduction** pathways. (3) It has been shown that GPI structures can target certain proteins to apical domains of the plasma membrane of certain epithelial cells. The biosynthesis of GPI anchors is complex and begins in the endoplasmic reticulum. The GPI anchor is assembled independently by a series of enzyme-catalyzed reactions and then transferred to the carboxyl terminal end of its acceptor protein, accompanied by cleavage of the preexisting carboxyl terminal hydrophobic peptide from that protein. This process is sometimes called **glypiation.** An acquired defect in an early stage of the biosynthesis of the GPI

Table 56–13. Some GPI-linked proteins.

- Acetylcholinesterase (red cell membrane)
- Alkaline phosphatase (intestinal, placental)
- Decay-accelerating factor (red cell membrane)
- 5′-Nucleotidase (T lymphocytes, other cells)
- Thy-1 antigen (brain, T lymphocytes)
- Variable surface glycoprotein (*Trypanosoma brucei*)

structure has been implicated in the causation of **paroxysmal nocturnal hemoglobinuria** (see below).

GLYCOPROTEINS ARE INVOLVED IN MANY BIOLOGIC PROCESSES AND IN MANY DISEASES

As listed in Table 56–1, glycoproteins have many different functions, some of which have already been addressed in this chapter and others of which are described elsewhere in this text (eg, transport molecules, immunologic molecules, and hormones). Here, their involvement in two specific processes—fertilization and inflammation—will be briefly described. In addition, the bases of a number of diseases that are due to abnormalities in the synthesis and degradation of glycoproteins will be summarized.

Glycoproteins Are Important in Fertilization

To reach the plasma membrane of an oocyte, a sperm has to traverse the **zona pellucida (ZP),** a thick, transparent, noncellular envelope that surrounds the oocyte. The zona pellucida contains three glycoproteins, ZP1–3. Of particular interest is ZP3, an *O*-linked glycoprotein that functions as a receptor for the sperm. A protein on the sperm surface, possibly galactosyl transferase, interacts specifically with oligosaccharide chains of ZP3; in at least certain species (eg, the mouse), this interaction, by transmembrane signaling, induces the **acrosomal reaction,** in which enzymes, such as proteases and hyaluronidase, and other contents of the acrosome of the sperm are released. Liberation of these enzymes helps the sperm to pass through the zona pellucida and reach the plasma membrane of the oocyte. In hamsters, it has been shown that another glycoprotein, PH-30, is important in both the binding of the PM of the sperm to the PM of the oocyte and also in the subsequent fusion of the two membranes. These interactions enable the sperm to enter and thus fertilize the oocyte. It may be possible to inhibit fertilization by developing drugs or antibodies that interfere with the normal functions of ZP3 and PH-30 and which would thus act as contraceptive agents.

Selectins Play Key Roles in Inflammation and in Lymphocyte Homing

Leukocytes play important roles in many inflammatory and immunologic phenomena. The first steps in many of these phenomena are interactions between circulating leukocytes and endothelial cells prior to passage of the former out of the circulation. Work done to identify specific molecules on the surfaces of the cells involved in such interactions has revealed that leukocytes and endothelial cells contain on their

Table 56–14. Some molecules involved in leukocyte-endothelial cell interactions.[1]

Molecule	Cell	Ligands
Selectins		
L-selectin	PMN, lymphs	CD34, Gly-CAM-1[2]
P-selectin	EC, platelets	P-selectin glycoprotein ligand-1 (PSGL-1) Sialyl-Lewis[x] and others
E-selectin	EC	Sialyl-Lewis[x] and others
Integrins		
LFA-1	PMN, lymphs	ICAM-1, ICAM-2 (CD11a/CD18)
Mac-1	PMN	ICAM-1 and others (CD11b/CD18)
Immunoglobulin superfamily		
ICAM-1	Lymphs, EC	LFA-1, Mac-1
ICAM-2	Lymphs, EC	LFA-1
PECAM-1	EC, PMN, lymphs	Various platelets

[1]Modified from Albelda SM, Smith CW, Ward PA: Adhesion molecules and inflammatory injury. FASEB J 1994;8:504.
[2]These are ligands for lymphocyte L-selectin; the ligands for neutrophil L-selectin have not been identified.
Key: PMN, polymorphonuclear leukocytes; EC, endothelial cell; lymphs, lymphocytes; CD, cluster of differentiation; ICAM, intercellular adhesion molecule; LFA-1, lymphocyte function-associated antigen-1; PECAM-1, platelet endothelial cell adhesion cell molecule-1.

surfaces specific lectins, called **selectins,** that participate in their intercellular adhesion. Features of the three major classes of selectins are summarized in Table 56–14. Selectins are single-chain Ca^{2+}-binding transmembrane proteins that contain a number of domains (Figure 56–10). Their amino terminal ends contain the lectin domain, which is involved in binding to specific carbohydrate ligands.

The adhesion of neutrophils to endothelial cells of postcapillary venules can be considered to occur in four stages, as shown in Figure 56–11. The initial baseline stage is succeeded by slowing or rolling of the neutrophils, mediated by selectins. Interactions between L-selectin on the neutrophil surface and CD34 and GlyCAM-1 on the endothelial surface are

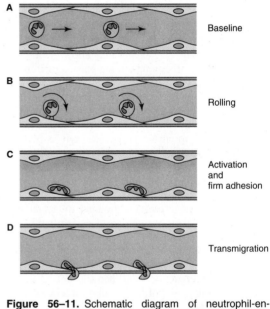

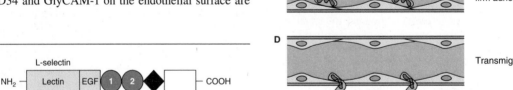

Figure 56–10. Schematic diagram of the structure of human L-selectin. The extracellular portion contains an amino terminal domain homologous to C-type lectins and an adjacent epidermal growth factor-like domain. These are followed by a variable number of complement regulatory-like modules (numbered circles) and a transmembrane sequence (black diamond). A short cytoplasmic sequence (open rectangle) is at the carboxyl terminal. The structures of P- and E-selectin are similar to that shown except that they contain more complement-regulatory modules. The numbers of amino acids in L-, P- and E-selectins, as deduced from the cDNA sequences, are 385, 789, and 589, respectively. (Reproduced, with permission, from Bevilacqua MP, Nelson RM: Selectins. J Clin Invest 1993;91:370.)

Figure 56–11. Schematic diagram of neutrophil-endothelial cell interactions. **A:** Baseline conditions: Neutrophils do not adhere to the vessel wall. **B:** The first event is the slowing or rolling of the neutrophils within the vessel (venule) mediated by selectins. **C:** Activation occurs resulting in neutrophils firmly adhering to the surfaces of endothelial cells and also assuming a flattened shape. This requires interaction of activated CD18 integrins on neutrophils with ICAM-1 on the endothelium. **D:** The neutrophils then migrate through the junctions of endothelial cells into the interstitial tissue; this requires involvement of PECAM-1. Chemotaxis is also involved in this latter stage. (Reproduced, with permission, from Albelda SM, Smith CW, Ward PA: Adhesion molecules and inflammatory injury. FASEB J 1994;8:504.)

involved. These particular interactions are initially short-lived, and the overall binding is of relatively low affinity, permitting rolling. However, during this stage, activation of the neutrophils by various chemical mediators (discussed below) occurs, resulting in a change of shape of the neutrophils and firm adhesion of these cells to the endothelium. An additional set of adhesion molecules is involved in firm adhesion, namely, LFA-1 and Mac-1 on the neutrophils and ICAM-1 and ICAM-2 on endothelial cells. LFA-1 and Mac-1 are CD11/CD18 integrins (see Chapter 60 for a discussion of integrins), whereas ICAM-1 and ICAM-2 are members of the immunoglobulin superfamily. The fourth stage is transmigration of the neutrophils across the endothelial wall. For this to occur, the neutrophils insert pseudopods into the junctions between endothelial cells, squeeze through these junctions, cross the basement membrane, and then are free to migrate in the extravascular space. Platelet-endothelial cell adhesion molecule-1 (PECAM-1) has been found to be localized at the junctions of endothelial cells and thus may have a role in transmigration. A variety of biomolecules have been found to be involved in activation of neutrophil and endothelial cells, including tumor necrosis factor α, various interleukins, platelet activating factor (PAF), leukotriene B$_4$, and certain complement fragments. These compounds stimulate various signaling pathways, resulting in changes in cell shape and function, and some are also chemotactic. One important functional change is recruitment of selectins to the cell surface, as in some cases selectins are stored in granules (eg, in endothelial cells and platelets).

The precise chemical nature of some of the ligands involved in selectin-ligand interactions has been determined. All three selectins bind **sialylated and fucosylated oligosaccharides,** and in particular all three bind **sialyl Lewis^X** (Figure 56–12), a structure present on both glycoproteins and glycolipids. Whether this compound is the actual ligand involved in vivo is not established. Sulfated molecules, such as the sulfatides (Chapter 16), may be ligands in certain instances. This basic knowledge is being used in attempts to synthesize compounds that block selectin-ligand interactions and thus may inhibit the inflammatory response. Approaches include administration of specific monoclonal antibodies or of chemically synthesized analogs of sialyl Lewis^X, both of which bind selectins.

Abnormalities in the Synthesis of Glycoproteins Underlie Certain Diseases

Table 56–15 lists a number of conditions in which abnormalities in the synthesis of glycoproteins are of importance. As mentioned above, many cancer cells exhibit different profiles of oligosaccharide chains on their surfaces, some of which may contribute to metastasis. In both **carbohydrate-deficient glycoprotein syndrome (CDGS)** and **hereditary erythroblastic multinuclearity with a positive acidified lysis test (HEMPAS)**—congenital dyserythropoietic anemia type II—abnormalities in specific processing enzymes have been discovered. I-cell disease is discussed further below. **Paroxysmal nocturnal hemoglobinuria** is an acquired mild anemia characterized by the presence of hemoglobin in urine due to hemolysis of red cells, particularly during sleep. This latter phenomenon may reflect a slight drop in plasma pH during sleep, which increases susceptibility to lysis by the complement system (Chapter 59). The basic defect in paroxysmal nocturnal hemoglobinuria is the acquisition of somatic mutations in the *PIG*-A (for phosphatidylinositol glycan class A) gene of certain hematopoietic cells. The product of this gene appears to be the enzyme that links the GlcNAc moiety to phosphatidylinositol in the GPI structure (Figure

Table 56–15. Some diseases due to or involving abnormalities in the biosynthesis of glycoproteins.

Disease	Abnormality
Cancer	Increased branching of cell surface glycans may be important in metastasis.
Carbohydrate-deficient glycoprotein syndrome (CDGS) (eg, due to deficiency of GlcNAc transferase II)[1]	Defects in the biosynthesis of N-glycan chains (eg, due to deficiency of GlcNAc transferase II) particularly affecting the central nervous system.
HEMPAS[2]	Abnormality in the biosynthesis of N-glycans (eg, possibly of mannosidase II), particularly affecting the red blood cell membrane.
Paroxysmal nocturnal hemoglobinuria	Acquired defect in biosynthesis of the GPI[3] structures of decay accelerating factor (DAF) and CD59.
I-cell disease	Deficiency of GlcNAc phosphotransferase, resulting in abnormal targeting of certain lysosomal enzymes.

[1]Three sub-types (I–III) of this syndrome have been recognized. The central nervous system is affected in all three, pointing to the importance of various glycoproteins in the normal development of this system.
[2]Hereditary erythroblastic multinuclearity with a positive acidified serum lysis test (congenital dyserythropoietic anemia type II). This is a relatively mild form of anemia. It reflects at least in part the presence in the red cell membranes of various glycoproteins with abnormal N-glycan chains, which contribute to the susceptibility to lysis.
[3]Glycosylphosphatidylinositol.

NeuAcα2 ⟶ 3Galβ1 ⟶ 4GlcNAc - - -

α 1–3

Fuc

Figure 56–12. Schematic representation of the structure of sialyl-Lewis^X.

56–1). Thus, proteins that are anchored by a GPI linkage are deficient in the red cell membrane. Two proteins are of particular interest: **decay accelerating factor (DAF)** and another protein designated CD59. They normally interact with certain components of the complement system (Chapter 58) to prevent the hemolytic actions of the latter. However, when they are deficient, the complement system can act on the red cell membrane to cause hemolysis. PNH can be diagnosed relatively simply, as the red cells are much more sensitive to hemolysis in normal serum acidified to pH 6.2 (Ham's test); the complement system is activated under these conditions, but normal cells are not affected. A variety of treatment modalities (transfusion, steroids, possible bone marrow transplant) are available. Figure 56–13 summarizes the etiology of paroxysmal nocturnal hemoglobinuria.

I-Cell Disease Results From Faulty Targeting of Lysosomal Enzymes

As indicated above, Man 6-P serves as a chemical marker to target certain lysosomal enzymes to that organelle. Analysis of cultured fibroblasts derived from patients with I-cell (inclusion cell) disease played a large part in revealing the above role of Man 6-P. I-cell disease is an uncommon condition characterized by severe progressive psychomotor retardation and a variety of physical signs, with death often occurring in the first decade. Cultured cells from patients with I-cell disease were found to lack almost all of the normal lysosomal enzymes; the lysosomes thus accumulate many different types of undegraded molecules, forming inclusion bodies. Samples of plasma from patients with the disease were observed to contain very high activities of lysosomal enzymes; this suggested that the enzymes were being synthesized but were failing to reach their proper intracellular destination and were instead being secreted. Cultured cells from patients with the disease were noted to take up exogenously added lysosomal enzymes obtained from normal subjects, indicating that the cells contained a nor-

mal receptor on their surfaces for endocytic uptake of lysosomal enzymes. In addition, this finding suggested that lysosomal enzymes from patients with I-cell disease might lack a recognition marker. Further studies revealed that lysosomal enzymes from normal individuals carried the Man 6-P recognition marker described above, which interacted with a specific intracellular protein, the Man 6-P receptor. Cultured cells from patients with I-cell disease were then found to be deficient in the activity of the *cis* Golgi-located GlcNAc phosphotransferase, explaining how their lysosomal enzymes failed to acquire the Man 6-P marker. It is now known that there are two Man 6-P receptor proteins, one of high (275 kDa) and one of low (46 kDa) molecular mass. These proteins are lectins, recognizing Man 6-P. The former is cation-independent and also binds IGF-II (hence it is named the Man 6-P–IGF–II receptor), whereas the latter is cation-dependent in some species and does not bind IGF-II. It appears that both receptors function in the intracellular sorting of lysosomal enzymes into clathrin-coated vesicles, which occurs in the *trans* Golgi subsequent to addition of Man 6-P in the *cis* Golgi. These vesicles then leave the Golgi and fuse with a prelysosomal compartment. The low pH in this compartment causes the lysosomal enzymes to dissociate from their receptors and subsequently enter into lysosomes. The receptors are recycled and reused. Only the smaller receptor functions in the endocytosis of extracellular lysosomal enzymes, which is a minor pathway for lysosomal location. Not all cells employ the Man 6-P receptor to target their lysosomal enzymes (eg, hepatocytes use a different but undefined pathway); furthermore, not all lysosomal enzymes are targeted by this mechanism. Thus, biochemical investigations of I-cell disease not only led to elucidation of its basis but also contributed significantly to knowledge of how newly synthesized proteins are targeted to specific organelles, in this case the lysosome. Figure 56–14 summarizes the causation of I-cell disease.

Pseudo-Hurler polydystrophy is another genetic disease closely related to I-cell disease. It is a milder condition, and patients may survive to adulthood. Studies have suggested that the GlcNAc phosphotransferase involved in I-cell disease has several domains, including a catalytic domain and a domain that specifically recognizes and interacts with lysosomal enzymes. It has been proposed that the defect in pseudo-Hurler polydystrophy lies in the latter domain, and the retention of some catalytic activity results in a milder condition.

Genetic Deficiencies of Glycoprotein Lysosomal Hydrolases Cause Diseases Such as α-Mannosidosis

Glycoproteins, like most other biomolecules, undergo both synthesis and degradation (ie, turnover). Degradation of the oligosaccharide chains of glyco-

Acquired mutations in the PIG-A gene of certain hematopoietic cells
↓
Defective synthesis of the GlcNH$_2$-PI linkage of GPI anchors
↓
Decreased amounts in the red blood membrane of GPI-anchored proteins, with decay accelerating factor (DAF) and CD59 being of especial importance
↓
Certain components of the complement system are not opposed by DAF and CD59, resulting in complement-mediated lysis of red cells

Figure 56–13. Scheme of causation of paroxysmal nocturnal hemoglobinuria.

Mutations in DNA

↓

Mutant GlcNAc phosphotransferase

↓

Lack of normal transfer of GlcNAc 1-P to specific mannose residues of certain enzymes destined for lysosomes

↓

These enzymes consequently lack Man 6-P and are secreted from cells (eg, into the plasma) rather than targeted to lysosomes

↓

Lysosomes are thus deficient in certain hydrolases, do not function properly, and accumulate partly digested cellular material, manifesting as inclusion bodies

Figure 56–14. Summary of the causation of I-cell disease.

Table 56–16. Major features of some diseases (eg, α-mannosidosis, β-mannosidosis, fucosidosis, sialidosis, aspartylglycosaminuria, and Schindler's disease) due to deficiencies of glycoprotein hydrolases.

- Usually exhibit mental retardation or other neurologic abnormalities, and in some disorders coarse features or visceromegaly (or both).
- Variations in severity from mild to rapidly progressive.
- Autosomal recessive inheritance.
- May show ethnic distribution (eg, aspartylglycosaminuria is common in Finland).
- Vacuolization of cells observed by microscopy in some disorders.
- Presence of abnormal degradation products (eg, oligosaccharides that accumulate because of the enzyme deficiency) in urine, detectable by TLC and characterizable by GLC-MS.
- Definitive diagnosis made by assay of appropriate enzyme, often using leukocytes.
- Possibility of prenatal diagnosis by appropriate enzyme assays.
- No definitive treatment at present.

proteins involves a battery of lysosomal hydrolases, including α-neuraminidase, β-galactosidase, β-hexosaminidase, α- and β-mannosidases, α-*N*-acetylgalactosaminidase, α-fucosidase, endo-β-*N*-acetylglucosaminidase, and aspartylglucosaminidase. The sites of action of the last two enzymes are indicated in the legend to Figure 56–5. Genetically determined defects of the activities of these enzymes can occur, resulting in abnormal degradation of glycoproteins. The accumulation in tissues of such abnormally degraded glycoproteins can lead to various diseases. Among the best-recognized of these diseases are **mannosidosis, fucosidosis, sialidosis, aspartylglycosaminuria,** and Schindler's disease, due respectively to deficiencies of α-mannosidase, α-fucosidase, α-neuraminidase, aspartylglucosaminidase, and a-*N*-acetylgalactosaminidase. These diseases, which are relatively uncommon, have a variety of manifestations; some of their major features are listed in Table 56–16. The fact that patients affected by these disorders all show signs referable to the central nervous system indicates the importance of glycoproteins in the development and normal function of that system.

From the above, it should be apparent that glycoproteins are involved in a wide variety of biologic processes and diseases. Glycoproteins play direct or indirect roles in a number of other diseases, as shown in the following examples. (1) The **influenza virus** possesses a neuraminidase that plays a key role in elution of newly synthesized progeny from infected cells. (2) **HIV-1,** thought by many to be the causative agent of AIDS, attaches to cells via one of its surface glycoproteins, gp120. (3) **Rheumatoid arthritis** is associated with an alteration in the glycosylation of circulating immunoglobulin-γ (IgG) molecules (Chapter 59), such that they lack galactose in their Fc regions and terminate in GlcNAc. **Mannose-binding protein** (not to be confused with the mannose-6-P re-

ceptor), a C-lectin synthesized by liver cells and secreted into the circulation, binds mannose, GlcNAc, and certain other sugars. It can thus bind the agalactosyl IgG molecules, which subsequently activate the complement system, contributing to chronic inflammation in the synovial membranes of joints. This protein can also bind the above sugars when they are present on the surfaces of certain bacteria, fungi, and viruses, preparing these pathogens for opsonization or for destruction by the complement system. This is an example of innate immunity, not involving immunoglobulins. Deficiency of this protein in young infants, due to mutation, renders them very susceptible to **recurrent infections.** It is hoped that basic studies of glycoproteins and other glycoconjugates (ie, the field of glycobiology) will lead to effective treatments for diseases in which these molecules are involved.

SUMMARY

Glycoproteins are widely distributed proteins, with diverse functions, that contain one or more covalently linked carbohydrate chains. The carbohydrate components of a glycoprotein range from 1% to more than 85% of its weight and may be simple or very complicated in structure. The oligosaccharide chains of glycoproteins encode biologic information. These chains are important to glycoproteins in modulating their solubility and viscosity, in protecting them against proteolysis, in their biologic actions, and in their participation in normal and abnormal cell-cell interactions (eg, sperm-egg interaction, development, and cancer, respectively). The structures of many oligosaccharide chains can be elucidated by gas-liquid chromatography, mass spectrometry, and high-resolution NMR spectrometry. Endo- and exoglycosidases hydrolyze specific linkages in oligosaccharides and are used to explore both the structures and functions of glycoproteins. Lectins are proteins that bind one or more specific sugars of oligosaccharide chains. The major classes of glycoproteins are *O*-linked (involving an OH of serine or threonine), *N*-linked (involving the N of the amide group of asparagine), and glycosylphosphatidylinositol (GPI)-linked. Mucins are a class of *O*-linked glycoproteins that are distributed on the surfaces of epithelial cells of the respiratory, gastrointestinal, and reproductive tracts.

The Golgi apparatus plays a major role in glycosylation reactions involved in the biosynthesis of glycoproteins. The oligosaccharides of *O*-linked glycoproteins are synthesized by the stepwise addition of sugars donated by nucleotide sugars in reactions catalyzed by individual specific glycoprotein glycosyltransferases. The biosynthesis of *N*-linked glycoproteins is more intricate. An oligosaccharide chain is donated by dolichol-pyrophosphate-oligosaccharide to the protein chain as it traverses the endoplasmic reticulum membrane. This chain is then trimmed down and rebuilt in the Golgi, but different external sugars are added. Depending on the enzymes and precursor proteins synthesized by a tissue, it can synthesize complex, hybrid, or high-mannose types of *N*-linked oligosaccharides.

Glycoproteins are involved in many biologic processes. For instance, the glycoprotein ZP3 found in the zona pellucida is involved in the interaction of spermatozoa with the ovum. Also, a group of lectins called selectins has been found to play important roles in the interactions of neutrophils and lymphocytes with endothelial cells, such as occur in inflammation. The carbohydrate ligand sialyl Lewis^X is involved in some of these interactions. New approaches to contraception and new therapies for inflammation may emerge from these findings.

A number of diseases involving abnormalities in the synthesis and degradation of glycoproteins have been recognized. The cell surface glycoproteins of metastatic cancer cells often exhibit differences in the structures of their oligosaccharides as compared with those of normal cells. Defects in the synthesis of glycoproteins appear to be implicated in carbohydrate-deficient glycoprotein syndrome, HEMPAS disease, and paroxysmal nocturnal hemoglobinuria. In I-cell disease, lysosomal membrane proteins are not properly targeted to the lysosome because of an absence of their Man 6-P recognition signal owing to a genetically determined deficiency of GlcNAc phosphotransferase. Some other relatively rare diseases are due to genetic deficiencies in the activities of specific glycoprotein lysosomal hydrolases. These include α- and β-mannosidosis, fucosidosis, sialidosis, aspartylglycosaminuria, and Schindler's disease. Glycoproteins are also involved in many other diseases, including influenza, AIDS, and rheumatoid arthritis.

REFERENCES

Beaudet AL, Thomas GH: Disorders of glycoprotein degradation and structure: α-mannosidosis, β-mannosidosis, fucosidosis, sialidosis, aspartylglycosaminuria, and carbohydrate-deficient glycoprotein syndrome. In: *The Metabolic and Molecular Bases of Inherited Disease,* 7th ed. Scriver CR et al (editors). McGraw-Hill, 1995.

Brockhausen I: Clinical aspects of glycoprotein biosynthesis. Crit Rev Clin Lab Sci 1993;30:65.

Charuk JHM et al: Carbohydrate-deficient glycoprotein syndrome type II. An autosomal recessive *N*-acetylglucosaminyl-transferase II deficiency different from typical hereditary erythroblastic multinuclearity, with a positive acidified-serum lysis test (HEMPAS). Eur J Biochem 1995. 1995;230:797.

Drickamer K, Taylor ME: Biology of animal lectins. Annu Rev Cell Biol 1993;9:237.

Elbein AD: Glycosidase inhibitors: Inhibitors of *N*-linked oligosaccharide processing. FASEB J 1991;5:3055.

Englund PT: The structure and biosynthesis of glycosyl phosphatidyl-inositol protein anchors. Annu Rev Biochem 1993;62:291.

Fiedler K, Simons K: The role of *N*-glycans in the secretory pathway. Cell 1995;81:309.

Jentoft N: Why are proteins O-glycosylated? Trends Biochem Sci 1990;15:291.

Kornfeld R, Kornfeld S: Assembly of asparagine-linked oligosaccharides. Annu Rev Biochem 1985;54:631.

Kornfeld S, Sly WS: I-cell disease and pseudo-Hurler polydystrophy: Disorders of lysosomal enzyme phosphorylation. In: *The Metabolic and Molecular Bases of Inherited Disease,* 7th ed. Scriver CR et al (editors). McGraw-Hill, 1995.

Lasky LA: Selectin-carbohydrate interactions and the initiation of the inflammatory response. Annu Rev Biochem 1995;64:113.

Rini JM: Lectin structure. Annu Rev Biophys Biomol Struct 1995;24:551.

Schachter H: Enzymes associated with glycosylation. Curr Opin Struct Biol 1991;1:755.

Strous GJ, Dekker J: Mucin-type glycoproteins. CRC Crit Rev Biochem Mol Biol 1992;27:57.

Takeda J et al: Deficiency of the GPI anchor caused by a somatic mutation of the *PIG*-A gene in paroxysmal nocturnal hemoglobinuria. Cell 1993;73:703.

Ware FE et al: The molecular chaperone calnexin binds GlcMan$_9$GlcNAc$_2$ as an initial step in recognizing unfolded proteins. J Biol Chem 1995;270:4697.

The Extracellular Matrix

57

Robert K. Murray, MD, PhD, & Frederick W. Keeley, PhD*

INTRODUCTION

Cells are the basic units of life. Most mammalian cells are located in tissues, where they are surrounded by a complex extracellular matrix, often referred to as "connective tissue." This matrix has a variety of important functions—to be described subsequently—apart from acting as a supporting scaffolding for the cells it surrounds. Extracellular matrix contains three major classes of biomolecules: (1) the **structural proteins,** collagen, elastin, and fibrillin; (2) certain **specialized proteins,** such as fibrillin, fibronectin, and laminin, which have specific functions in the extracellular matrix; and (3) **proteoglycans,** which consist of long chains of repeating disaccharides (glycosaminoglycans [GAGs], formerly called mucopolysaccharides) attached to specific core proteins. This chapter describes the basic biochemistry of these three classes of biomolecules and illustrates their biomedical significance. Major biochemical features of two specialized forms of extracellular matrix, bone and cartilage, and of a number of diseases involving them, are also briefly considered.

BIOMEDICAL IMPORTANCE

The extracellular matrix is receiving increasing attention as appreciation increases of its important roles in many normal and pathologic processes. During development, certain embryonic cells must migrate considerable distances through the extracellular matrix to find their ultimate destinations. Inflammatory states, both acute and chronic, involve many alterations in the biochemistry of the extracellular matrix. For cancer cells to metastasize, they must first detach from their organ or tissue of origin, migrate through the extracellular matrix, and then enter small blood vessels or lymphatics. There is considerable evidence of involvement of molecules of the extracellular matrix in both rheumatoid arthritis and os-

teoarthritis, two of the major diseases afflicting society. Several diseases (eg, osteogenesis imperfecta and a number of types of the Ehlers-Danlos syndrome) are due to genetic disturbances of the synthesis of collagen. Genetic deficiencies of the lysosomal hydrolases involved in the degradation of GAGs result in a number of diseases (designated **mucopolysaccharidoses** on the basis of the former name for GAGs), which are due to impaired degradation of GAGs and their consequent abnormal accumulation in various tissues. Finally, it is apparent that many changes occur in the extracellular matrix during the aging process. Claims by the cosmetics industry that certain products have anti-aging effects on collagen and other constituents of the extracellular matrix are unproved.

COLLAGEN IS THE MOST ABUNDANT PROTEIN IN THE ANIMAL WORLD

Collagen, the major component of most connective tissues, constitutes approximately 25% of the protein of mammals. It provides an extracellular framework for all metazoan animals and exists in virtually every animal tissue. About 19 distinct types of collagen made up of about 30 distinct polypeptide chains (encoded by the same number of collagen genes) have been identified in human tissues. Although several of these are present only in small proportions, they may play important roles in determining the physical properties of the tissues. In addition, a number of proteins (eg, the C1q component of the complement system, pulmonary surfactant proteins SP-A and SP-D) that are not classified as collagens have collagen-like domains in their structures; these proteins are sometimes referred to as "noncollagen" collagens.

Table 57–1 summarizes the types of collagens found in human tissues; the nomenclature used to designate types of collagen and their genes is indicated in the legend.

The 19 types of collagen mentioned above can be subdivided into a number of classes based primarily on the structures that they form (Table 57–2). In this chapter, we shall be primarily concerned with the

*Research Institute, Hospital for Sick Children, Toronto, and Department of Biochemistry, University of Toronto.

Table 57–1. Types of collagen and their genes.[1,2]

Type	Genes	Tissue
I	COL1A1, COL1A2,	Most connective tissues, including bone
II	COL2A1	Cartilage, vitreous humor
III	COL3A1	Extensible connective tissue such as skin, lung, and the vascular system
IV	COL4A1–COL4A6	Basement membranes
V	COL5A1–COL5A3	Minor component in tissues containing collagen I
VI	COL6A1–COL6A3	Most connective tissues
VII	COL7A1	Anchoring fibrils
VIII	COL8A1–COL8A2	Endothelium, other tissues
IX	COL9A1–COL9A3	Tissues containing collagen II
X	COL10A1	Hypertrophic cartilage
XI	COL11A1, COL11A2, COL2A1	Tissues containing collagen II
XII	COL12A1	Tissues containing collagen I
XIII	COL13A1	Many tissues
XIV	COL14A1	Tissues containing collagen I
XV	COL15A1	Many tissues
XVI	COL16A1	Many tissues
XVII	COL17A1	Skin hemidesmosomes
XVIII	COL18A1	Many tissues (eg, liver, kidney)
XIX	COL19A1	Rhabdomyosarcoma cells

[1]Adapted slightly from Prockop DJ, Kivirrikko KI: Collagens: Molecular biology, diseases, and potentials for therapy. Annu Rev Biochem 1995;64:403.

[2]The types of collagen are designated by Roman numerals. Constituent procollagen chains, called proα chains, are numbered using Arabic numerals, followed by the collagen type in parentheses. For instance, type I procollagen is assembled from two proα1(I) and one proα2(I) chain. It is thus a heterotrimer, whereas type 2 procollagen is assembled from three proα1(II) chains and is thus a homotrimer. The collagen genes are named according to the collagen type, written in Arabic numerals for the gene symbol, followed by an A and the number of the proα chain that they encode. Thus, the COL1A1 and COL1A2 genes encode the α1 and α2 chains of type I collagen, respectively.

fibril-forming collagens I and II, the major collagens of skin and bone and of cartilage, respectively. However, mention will be made of some of the other collagens when appropriate.

COLLAGEN TYPE I IS COMPOSED OF A TRIPLE HELIX STRUCTURE AND FORMS FIBRILS

All collagen types have a triple helical structure. In some collagens, the entire molecule is triple helical, whereas in others the triple helix may involve only a fraction of the structure. Mature collagen type I, containing approximately 1000 amino acids, be-

Table 57–2. Classification of collagens, based primarily on the structures that they form.[1]

Class	Type
Fibril-forming	I, II, III, V, and XI
Network-like	IV, VIII, X
FACITs[2]	IX, XII, XIV, XVI, XIX
Beaded filaments	VI
Anchoring fibrils	VII
Transmembrane domain	XIII, XVII
Others	XV, XVIII

[1]Based on Prockop DJ, Kivirrikko KI: Collagens: Molecular biology, diseases, and potentials for therapy. Annu Rev Biochem 1995;64:403.
[2]FACITs = fibril-associated collagens with interrupted triple helices.

longs to the former type; in it, each polypeptide subunit or alpha chain is twisted into a left-handed helix of three residues per turn (Figure 57–1). Three of these alpha chains are then wound into a right-handed superhelix, forming a rod-like molecule 1.4 nm in diameter and about 300 nm long. A striking characteristic of collagen is the occurrence of glycine residues at every third position of the triple helical portion of the alpha chain. This is necessary because glycine is the only amino acid small enough to be ac-

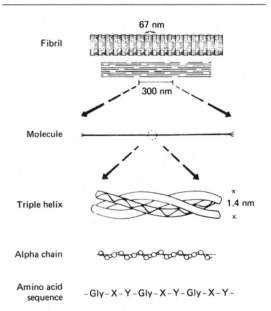

Figure 57–1. Molecular features of collagen structure from primary sequence up to the fibril. (Slightly modified and reproduced, with permission, from Eyre DR: Collagen: Molecular diversity in the body's protein scaffold. Science 1980;207:1315. Copyright © 1980 by the American Association for the Advancement of Science.)

commodated in the limited space available down the central core of the triple helix. This repeating structure, represented as $(Gly-X-Y)_n$, is an absolute requirement for the formation of the triple helix. While X and Y can be any other amino acids, about 100 of the X positions are proline and about 100 of the Y positions are hydroxyproline. Proline and hydroxyproline confer rigidity on the collagen molecule. Hydroxyproline is formed by the posttranslational hydroxylation of peptide-bound proline residues catalyzed by the enzyme prolyl hydroxylase, whose cofactors are ascorbic acid (vitamin C) and α-ketoglutarate. Lysines in the Y position may also be posttranslationally modified to hydroxylysine through the action of lysyl hydroxylase, an enzyme with similar cofactors. Some of these hydroxylysines may be further modified by the addition of galactose or galactosyl-glucose through an O-glycosidic linkage, a glycosylation site that is unique to collagen.

Collagen types that form long rod-like fibers in tissues are assembled by lateral association of these triple helical units into a **"quarter staggered"** alignment such that each is displaced longitudinally from its neighbor by slightly less than one-quarter of its length (Figure 57–1, upper part). This arrangement is responsible for the banded appearance of these fibers in connective tissues. Collagen fibers are further stabilized by the formation of covalent cross-links, both within and between the triple helical units. These cross-links form through the action of lysyl oxidase, a copper-dependent enzyme that oxidatively deaminates the ε-amino groups of certain lysine and hydroxylysine residues, yielding reactive aldehydes. Such aldehydes can form aldol condensation products with other lysine- or hydroxylysine-derived aldehydes or form Schiff bases with the ε-amino groups of unoxidized lysines or hydroxylysines. These reactions, after further chemical rearrangements, result in the stable, covalent cross-links that are important for the tensile strength of the fibers.

Several collagen types do not form fibrils in tissues (Table 57–2). They are characterized by interruptions of the triple helix with stretches of protein lacking Gly-X-Y repeat sequences. These non-Gly-X-Y sequences result in areas of globular structure interspersed in the triple helical structure.

Type IV collagen, the best-characterized example of collagens with discontinuous triple helices, is an important component of basement membranes where it forms a mesh-like network.

Collagen Undergoes Extensive Posttranslational Modifications

Newly synthesized collagen undergoes extensive posttranslational modification before becoming part of a mature, extracellular collagen fiber (Table 57–3). Like most secreted proteins, collagen is synthesized on ribosomes in a precursor form, **preprocollagen,** which contains a leader or signal sequence

Table 57–3. Order and location of processing of the fibrillar collagen precursor.

Intracellular
1. Cleavage of signal peptide
2. Hydroxylation of Y-prolyl residues and some Y-lysyl residues; glycosylation of some hydroxylysyl residues
3. Formation of intrachain and interchain S-S bonds in extension peptides
4. Formation of triple helix

Extracellular
1. Cleavage of amino and carboxyl terminal propeptides
2. Assembly of collagen fibers in quarter-staggered alignment
3. Oxidative deamination of ε-amino groups of lysyl and hydroxylysyl residues to aldehydes
4. Formation of intra- and interchain cross-links via Schiff bases and aldol condensation products

that directs the polypeptide chain into the vesicular space of the endoplasmic reticulum. As it enters the endoplasmic reticulum, this leader sequence is enzymatically removed. Hydroxylation of proline and lysine residues and glycosylation of hydroxylysines in this **procollagen** molecule also take place at this site. The procollagen molecule contains polypeptide extensions (extension peptides) of 20–35 kDa at both its amino and carboxyl terminal ends, neither of which is present in mature collagen. Both extension peptides contain cysteine residues. While the amino terminal propeptide forms only intrachain disulfide bonds, the carboxyl terminal propeptides form both intrachain and interchain disulfide bonds. Formation of these disulfide bonds assists in the registration of the three collagen molecules to form the triple helix, winding from the carboxyl terminal end. After formation of the triple helix, no further hydroxylation of proline or lysine or glycosylation of hydroxylysines can take place. Self-assembly is a cardinal principle in the biosynthesis of collagen.

Following secretion from the cell by way of the Golgi apparatus, extracellular enzymes called **procollagen aminoproteinase** and **procollagen carboxyproteinase** remove the extension peptides at the amino and carboxyl terminal ends, respectively. Cleavage of these propeptides may occur within crypts or folds in the cell membrane. Once the propeptides are removed, the triple helical collagen molecules, containing approximately 1000 amino acids per chain, spontaneously assemble into collagen fibers. These are further stabilized by the formation of inter- and intrachain cross-links through the action of lysyl oxidase, as described previously.

The same cells that secrete collagen also secrete **fibronectin,** a large glycoprotein present on cell surfaces, in the extracellular matrix, and in blood (see below). Fibronectin binds to aggregating precollagen fibers and alters the kinetics of fiber formation in the pericellular matrix. Associated with fibronectin and procollagen in this matrix are the **proteoglycans** heparan sulfate and chondroitin sulfate (see below). In fact, type IX collagen, a minor collagen type from cartilage, contains attached proteoglycan chains. Such interactions may serve to regulate the formation of collagen fibers and to determine their orientation in tissues.

Once formed, collagen is relatively metabolically stable. However, its breakdown is increased during starvation and various inflammatory states. Excessive production of collagen occurs in a number of conditions, eg, hepatic cirrhosis.

A Number of Genetic Diseases Result From Abnormalities in the Synthesis of Collagen

About 30 genes encode collagen, and its pathway of biosynthesis is complex, involving at least eight enzyme-catalyzed posttranslational steps. Thus, it is not surprising that a number of diseases (Table 57–4) are due to mutations in collagen genes or in genes encoding some of the enzymes involved in these posttranslational modifications. The diseases affecting bone and cartilage will be discussed later in this chapter.

Ehlers-Danlos syndrome comprises a group of inherited disorders whose principal clinical features are hyperextensibility of the skin, abnormal tissue fragility, and increased joint mobility. The clinical picture is variable, reflecting underlying extensive genetic heterogeneity. At least 11 types have been recognized, most of which reflect a variety of lesions in the synthesis of collagen. Type IV is the most serious because of its tendency for spontaneous rupture

Table 57–4. Diseases caused by mutations in collagen genes or by deficiencies in the activities of posttranslational enzymes involved in the biosynthesis of collagen.[1]

Gene or Enzyme	Disease[2]
COL1A1, COL1A2	Osteogenesis imperfecta Osteoporosis[3] Ehlers-Danlos syndrome type VIIA, VIIB
COL2A1	Severe chondrodysplasias Osteoarthritis[3]
COL3A1	Ehlers-Danlos syndrome type IV
COL4A3–COL4A6	Alport's syndrome (including both autosomal and X-linked forms)
COL7A1	Epidermolysis bullosa, dystrophic forms
COL10A1	Schmid metaphysial chondrodysplasia
Lysyl hydroxylase	Ehlers-Danlos syndrome type VI
Procollagen N-proteinase	Ehlers-Danlos syndrome type VIIC
Lysyl hydroxylase	Menkes' syndrome[4]

[1]Adapted from Prockop DJ, Kivirrikko KI: Collagens: Molecular biology, diseases, and potentials for therapy. Annu Rev Biochem 1995;64:403.
[2]Genetic linkage to collagen genes has been shown for a few other conditions not listed here.
[3]At present applies to only a relatively small number of such patients.
[4]Secondary to a deficiency of copper (Chapter 59).

of arteries or the bowel, reflecting abnormalities in type III collagen. Patients with type VI, due to a deficiency of lysyl hydroxylase, exhibit marked joint hypermobility and a tendency to ocular rupture. A deficiency of procollagen N-proteinase, causing formation of abnormal thin, irregular collagen fibrils, results in type VIIC Ehlers-Danlos syndrome, manifested by marked joint hypermobility and soft skin.

Alport's syndrome refers to a number of genetic disorders (both X-linked and autosomal) affecting the structure of type IV collagen fibers, the major collagen found in the basement membranes of the renal glomeruli (see discussion of laminin, below). Mutations in several genes encoding type IV collagen fibers have been demonstrated. The presenting sign is hematuria, and patients may eventually develop end-stage renal disease. Electron microscopy reveals characteristic abnormalities of the structure of the basement membrane and lamina densa.

In epidermolysis bullosa, the skin breaks and blisters as a result of minor trauma. The dystrophic form is due to mutations in COL7A1, affecting the structure of type VII collagen. This collagen forms delicate fibrils that anchor the basal lamina to collagen fibrils in the dermis. These anchoring fibrils have been shown to be markedly reduced in this form of the disease, probably resulting in the blistering. Epidermolysis bullosa simplex, another variant, is due to mutations in keratin 5 (Chapter 58).

Scurvy affects the structure of collagen. However, it is due to a deficiency of ascorbic acid (Chapter 52) and is not a genetic disease. Its major signs are bleeding gums, subcutaneous hemorrhages, and poor wound healing. These signs reflect impaired synthesis of collagen due to deficiencies of prolyl and lysyl hydroxylases, both of which require ascorbic acid as a cofactor.

ELASTIN CONFERS EXTENSIBILITY AND RECOIL ON LUNG, BLOOD VESSELS, AND LIGAMENTS

Elastin is a connective tissue protein that is responsible for properties of extensibility and elastic recoil in tissues. Although not as widespread as collagen, elastin is present in large amounts, particularly in tissues that require these physical properties, eg, lung, large arterial blood vessels, and some elastic ligaments. Smaller quantities of elastin are also found in skin, ear cartilage, and several other tissues. In contrast to collagen, there appears to be only one genetic type of elastin, although variants arise by differential processing of the hnRNA for elastin (Chapter 39). Elastin is synthesized as a soluble monomer of 70 kDa called "tropoelastin." Some of the prolines of tropoelastin are hydroxylated to **hydroxyproline** by prolyl hydroxylase, though hydroxylysine and glycosylated hydroxylysine are not present. Unlike collagen, tropoelastin is not synthesized in a pro-form with extension peptides. Furthermore, elastin does *not* contain repeat Gly-X-Y sequences, triple helical structure, or carbohydrate moieties.

After secretion from the cell, certain lysyl residues of tropoelastin are oxidatively deaminated to aldehydes by lysyl oxidase, the same enzyme involved in this process in collagen. However, the major cross-links formed in elastin are the **desmosines,** which result from the condensation of three of these lysine-derived aldehydes with an unmodified lysine to form a tetrafunctional cross-link unique to elastin. Once cross-linked in its mature, extracellular form, elastin is highly insoluble and extremely stable and has a very low turnover rate. Elastin exhibits a variety of random coil conformations that permit the protein to stretch and subsequently recoil during the performance of its physiologic functions.

Table 57–5 summarizes the main differences between collagen and elastin.

Deletions in the elastin gene (located at 7q11.23) have been found in approximately 90% of subjects with Williams syndrome, a developmental disorder affecting connective tissue and the central nervous system. The mutations, by affecting synthesis of elastin, probably play a causative role in the supravalvular aortic stenosis often found in this condition. A number of skin diseases (eg, scleroderma) are associated with accumulation of elastin. Fragmentation or, alternatively, a decrease of elastin is found in conditions such as pulmonary emphysema, cutis laxa, and aging of the skin.

MARFAN'S SYNDROME IS DUE TO MUTATIONS IN THE GENE FOR FIBRILLIN, A PROTEIN PRESENT IN MICROFIBRILS

Marfan's syndrome is a relatively frequent inherited disease affecting connective tissue; it is inherited as an autosomal dominant trait. It affects the eyes (eg, causing dislocation of the lens, known as ectopia lentis), the skeletal system (most patients are tall and exhibit long digits [arachnodactyly] and hyperextensibility of the joints), and the cardiovascular system (eg, causing weakness of the aortic media, leading to dilatation of the ascending aorta). Abraham Lincoln may have had this condition. Most cases are caused by mutations in the gene (on chromosome 15) for fibrillin; missense mutations have been detected in several patients with Marfan's syndrome.

Fibrillin is a large glycoprotein (about 350 kDa) that is a structural component of microfibrils, 10- to 12-nm fibers found in many tissues. Fibrillin is secreted (subsequent to a proteolytic cleavage) into the extracellular matrix by fibroblasts and becomes incorporated into the insoluble microfibrils, which appear to provide a scaffold for deposition of elastin. The distribution of fibrillin in tissues can be studied by immunohistochemical methods. Of special relevance to Marfan's syndrome, fibrillin is found in the zonular fibers of the lens, in the periosteum, and associated with elastin fibers in the aorta (and elsewhere); these locations respectively explain the ectopia lentis, arachnodactyly, and cardiovascular problems found in the syndrome. Other proteins are also present in these microfibrils, and it appears likely that abnormalities of them may cause other connective tissue disorders. Another gene for fibrillin exists on chromosome 5; mutations in this gene are linked to causation of congenital contractural arachnodactyly but not to Marfan's syndrome. The probable sequence of events leading to Marfan's syndrome is summarized in Figure 57–2.

FIBRONECTIN IS AN IMPORTANT GLYCOPROTEIN INVOLVED IN CELL ADHESION AND MIGRATION

Fibronectin is a major glycoprotein of the extracellular matrix, also found in a soluble form in plasma. It consists of two identical subunits, each of

Table 57–5. Major differences between collagen and elastin.

Collagen	Elastin
1, Many different genetic types	One genetic type
2. Triple helix	No triple helix; random coil conformations permitting stretching
3. (Gly-X-Y)$_n$ repeatiing structure	No (Gly-X-Y)$_n$ repeating structure
4. Presence of hydroxy-lysine	No hydroxylysine
5. Carbohydrate-containing	No carbohydrate
6. Intramolecular aldol cross-links	Intramolecular desmosine cross-links
7. Presence of extension peptides during biosynthesis	No extension peptides present during biosynthesis

Mutations in gene (on chromosome 15) for fibrillin, a large glycoprotein present in elastin-associated microfibrils

↓

Abnormalities of the structure of fibrillin

↓

Structures of the suspensory ligament of the eye, the periosteum, and the media of the aorta affected

↓

Ectopia lentis, arachnodactyly, and dilation of the ascending aorta

Figure 57–2. Probable sequence of events in the causation of the major signs exhibited by patients with Marfan's syndrome.

RGD

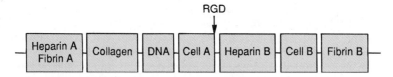

Figure 57–3. Schematic representation of fibronectin. Seven functional domains of fibronectin are represented; two different types of domain for heparin, cell-binding, and fibrin are shown. The domains are composed of various combinations of three structural motifs (I, II, and III), not depicted in the figure. Also not shown is the fact that fibronectin is a dimer joined by disulfide bridges near the carboxyl terminals of the monomers. The approximate location of the RGD sequence of fibronectin, which interacts with a variety of fibronectin integrin receptors on cell surfaces, is indicated by the arrow. (Redrawn after Yamada KM: Adhesive recognition sequences. J Biol Chem 1991;266:12809.)

about 230 kDa, joined by two disulfide bridges near their carboxyl terminals. The gene encoding fibronectin is very large, containing some 50 exons; the RNA produced by its transcription is subject to considerable alternative splicing, and as many as 20 different mRNAs have been detected in various tissues. Fibronectin contains three types of repeating motifs (I, II, and III), which are organized into functional domains (at least seven); functions of these domains include binding **heparin** (see below) and fibrin, collagen, DNA, and cell surfaces (Figure 57–3). The amino acid sequence of the fibronectin receptor of fibroblasts has been derived, and the protein is a member of the transmembrane integrin class of proteins (Chapter 60). The integrins are heterodimers, containing various types of α and β polypeptide chains. Fibronectin contains an Arg-Gly-Asp (RGD) sequence that binds to the receptor. The RGD sequence is shared by a number of other proteins present in the extracellular matrix that bind to integrins present in cell surfaces. Synthetic peptides containing the RGD sequence inhibit the binding of fibronectin to cell surfaces. Figure 57–4 illustrates the interaction of collagen, fibronectin, and laminin, all major proteins of the extracellular ma-

trix, with a typical cell (eg, fibroblast) present in the extracellular matrix.

The fibronectin receptor interacts indirectly with **actin** microfilaments (Chapter 58) present in the cytosol (Figure 57–5). A number of proteins, collectively known as **attachment proteins,** are involved; these include talin, vinculin, an actin-filament capping protein, and α-actinin. Talin interacts with the receptor and vinculin, whereas the latter two interact with actin. The interaction of fibronectin with its receptor provides one route whereby the exterior of the

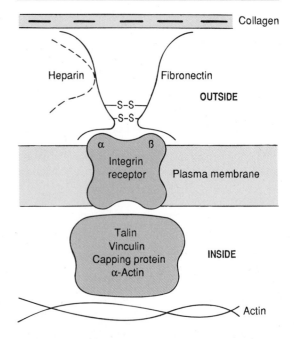

Figure 57–5. Schematic representation of fibronectin interacting with an integrin fibronectin receptor situated in the exterior of the plasma membrane of a cell of the extracellular matrix and of various attachment proteins interacting indirectly or directly with an actin microfilament in the cytosol. For simplicity, the attachment proteins are represented as a complex.

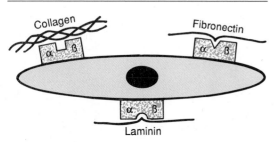

Figure 57–4. Schematic representation of a cell interacting through various integrin receptors with collagen, fibronectin, and laminin present in the extracellular matrix. (Specific subunits are not indicated.) (Redrawn after Yamada KM: Adhesive recognition sequences. J Biol Chem 1991;266:12809.)

cell can communicate with the interior and thus affect cell behavior. Via the interaction with its cell receptor, fibronectin plays an important role in the adhesion of cells to the extracellular matrix. It is also involved in cell migration, by providing a binding site for cells and thus helping them to steer their way through the extracellular matrix. The amount of fibronectin around many transformed cells is sharply reduced, partly explaining their faulty interaction with the extracellular matrix (Chapter 62).

LAMININ IS A MAJOR PROTEIN COMPONENT OF RENAL GLOMERULAR AND OTHER BASAL LAMINAE

Basal laminae are specialized areas of the extracellular matrix that surround epithelial and some other cells (eg, muscle cells); here we discuss only the laminae found in the **renal glomerulus.** In that structure, the basal lamina is contributed by two separate sheets of cells (one endothelial and one epithelial), each disposed on opposite sides of the lamina; these three layers make up the **glomerular membrane.** The primary components of the basal lamina are three proteins—laminin, entactin, and type IV collagen—and the GAG **heparin** or **heparan sulfate.** These components are synthesized by the underlying cells.

Laminin (about 850 kDa, 70 nm long) consists of three distinct elongated polypeptide chains (A, B_1, and B_2) linked together to form an elongated cruciform shape. It has binding sites for type IV collagen, heparin, and integrins on cell surfaces. The collagen interacts with laminin (rather than directly with the cell surface), which in turn interacts with integrins or other laminin receptor proteins, thus anchoring the lamina to the cells. **Entactin,** also known as "nidogen," is a glycoprotein containing an RGD sequence; it binds to laminin and is a major cell attachment factor. The relatively thick basal lamina of the renal glomerulus has an important role in **glomerular filtration,** regulating the passage of large molecules (most plasma proteins) across the glomerulus into the renal tubule. The glomerular membrane allows small molecules, such as inulin (5.2 kDa), to pass through as easily as water. On the other hand, only a small amount of the protein albumin (69 kDa), the major plasma protein, passes through the normal glomerulus. This is explained by two sets of facts: (1) The pores in the glomerular membrane are large enough to allow molecules up to about 8 nm to pass through. (2) Albumin is smaller than this pore size, but it is prevented from passing through easily by the negative charges of heparan sulfate and of certain sialic acid-containing glycoproteins present in the lamina. These negative charges repel albumin and most plasma proteins, which are negatively charged at the pH of blood. The normal structure of the glomerulus

may be severely damaged in certain types of **glomerulonephritis** (eg, caused by antibodies directed against various components of the glomerular membrane). This alters the pores and the amounts and dispositions of the negatively charged macromolecules referred to above, and relatively massive amounts of albumin (and of certain other plasma proteins) can pass through into the urine, resulting in severe **albuminuria.**

PROTEOGLYCANS AND GLYCOSAMINOGLYCANS

The Glycosaminoglycans Found in Proteoglycans Are Built Up of Repeating Disaccharides

Proteoglycans are proteins that contain covalently linked glycosaminoglycans. A number of them have been characterized and given names such as syndecan, betaglycan, serglycin, perlecan, aggrecan, versican, decorin, biglycan, and fibromodulin. They vary in tissue distribution, nature of the core protein, attached glycosaminoglycans, and function. The proteins bound covalently to glycosaminoglycans are called "core proteins"; they have proved difficult to isolate and characterize, but the use of recombinant DNA technology is beginning to yield important information about their structures. The amount of carbohydrate in a proteoglycan is usually much greater than is found in a glycoprotein and may comprise up to 95% of its weight. Figures 57–6 and 57–7 show the general structure of one particular proteoglycan, aggrecan, the major type found in cartilage. It is very large (about 2×10^3 kDa), with its overall structure resembling that of a bottle brush. It contains a long strand of hyaluronic acid (one type of GAG) to which link proteins are attached noncovalently. In turn, these latter interact noncovalently with core protein molecules from which chains of other GAGs (keratan sulfate and chondroitin sulfate in this case) project. More details on this macromolecule are given when cartilage is discussed below.

There are at least seven **glycosaminoglycans (GAGs):** hyaluronic acid, chondroitin sulfate, keratan sulfates I and II, heparin, heparan sulfate, and dermatan sulfate. A GAG is an unbranched polysaccharide made up of repeating disaccharides, one component of which is always an amino sugar (hence the name GAG), either D-glucosamine or D-galactosamine. The other component of the repeating disaccharide (except in the case of keratan sulfate) is a uronic acid, either L-glucuronic acid (GlcUA) or its 5′-epimer, L-iduronic acid (IdUA). With the exception of hyaluronic acid, all the GAGs contain sulfate groups, either as O-esters or as N-sulfate (in heparin and heparan sulfate). Hyaluronic acid affords another exception because there is no clear evidence that it is attached covalently to protein, as the definition of a

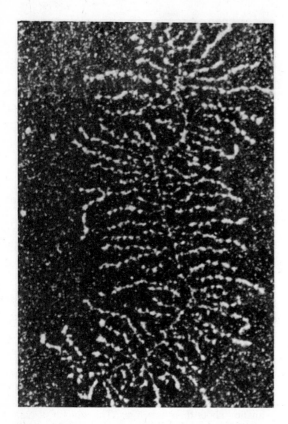

Figure 57–6. Darkfield electron micrograph of a proteoglycan aggregate in which the proteoglycan subunits and filamentous backbone are particularly well extended. (Reproduced, with permission, from Rosenberg L, Hellman W, Kleinschmidt AK: Electron microscopic studies of proteoglycan aggregates from bovine articular cartilage. J Biol Chem 1975;250:1877.)

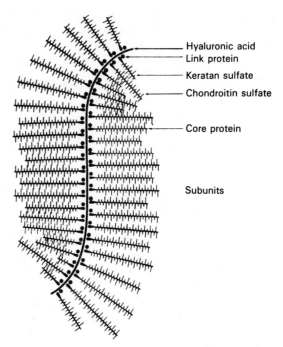

Figure 57–7. Schematic representation of the proteoglycan aggrecan. (Reproduced, with permission, from Lennarz WJ: *The Biochemistry of Glycoproteins and Proteoglycans.* Plenum Press, 1980.)

proteoglycan given above specifies. Both GAGs and proteoglycans have proved difficult to work with, partly because of their complexity. However, they are major components of the ground substance, they have a number of important biologic roles, and they are involved in a number of disease processes, so that interest in them is steadily increasing.

Biosynthesis of Glycosaminoglycans Involves Attachment to Core Proteins, Chain Elongation, and Chain Termination

A. Attachment to Core Proteins: The linkage between GAGs and their core proteins is generally one of three types.

1. An *O*-glycosidic bond between xylose (Xyl) and Ser, a bond that is unique to proteoglycans. This linkage is formed by transfer of a Xyl residue to Ser from UDP-xylose. Two residues of Gal are then added to the Xyl residue, forming a link trisaccharide, Gal-Gal-Xyl-Ser. Further chain growth of the GAG occurs on the terminal Gal.

2. An *O*-glycosidic bond forms between GalNAc (*N*-acetylgalactosamine) and Ser (Thr) (Figure 56–1[a]), present in keratan sulfate II. This bond is formed by donation to Ser (or Thr) of a GalNAc residue, employing UDP-GalNAc as its donor.

3. An *N*-glycosylamine bond between GlcNAc (*N*-acetylglucosamine) and the amide nitrogen of Asn, as found in *N*-linked glycoproteins (Figure 56–1[b]). Its synthesis is believed to involve dolichol-P-P-oligosaccharide.

The synthesis of the core proteins occurs in the endoplasmic reticulum, and formation of at least some of the above linkages also occurs there. Most of the later steps in the biosynthesis of GAG chains and their subsequent modifications occur in the Golgi apparatus.

B. Chain Elongation: Appropriate nucleotide sugars and highly specific Golgi-located glycosyltransferases are employed to synthesize the oligosaccharide chains of GAGs. The "one enzyme, one linkage" relationship appears to hold here, as in the case of certain types of linkages found in glycoproteins. The enzyme systems involved in chain elongation are capable of high fidelity reproduction of complex GAGs.

C. Chain Termination: This appears to result

from (1) sulfation, particularly at certain positions of the sugars, and (2) the progression of the growing GAG chain away from the membrane site where catalysis occurs.

D. Further Modifications: After formation of the GAG chain, numerous chemical modifications occur, such as the introduction of sulfate groups onto GalNAc and other moieties and the epimerization of GlcUA to IdUA residues. The enzymes catalyzing sulfation are designated sulfotransferases and use 3′-phosphoadenosine-5′-phosphosulfate (PAPS, active sulfate) as the sulfate donor. These Golgi-located enzymes are highly specific, and distinct enzymes catalyze sulfation at different positions (eg, carbons 2, 3, 4, and 6) on the acceptor sugars. An **epimerase** catalyzes conversions of glucuronyl to iduronyl residues.

Glycosaminoglycans Exhibit Subtle Differences in Structure and Have Characteristic Distributions

The seven GAGs named above differ from each other in a number of the following properties: amino sugar composition, uronic acid composition, linkages between these components, chain length of the disac-

charides, the presence or absence of sulfate groups and their positions of attachment to the constituent sugars, the nature of the core proteins to which they are attached, the nature of the linkage to core protein, their tissue and subcellular distribution, and their biologic functions.

The structures (Figure 57–8) and the distributions of each of the GAGs will now be briefly discussed. The major features of the seven GAGs are summarized in Table 57–6.

A. Hyaluronic Acid: Hyaluronic acid consists of an unbranched chain of repeating disaccharide units containing GlcUA and GlcNAc. Hyaluronic acid is present in bacteria and is widely distributed among various animals and tissues, including synovial fluid, the vitreous body of the eye, cartilage, and loose connective tissues.

B. Chondroitin Sulfates (Chondroitin 4-Sulfate and Chondroitin 6-Sulfate): Proteoglycans linked to chondroitin sulfate by the Xyl-Ser O-glycosidic bond are prominent components of cartilage (see below). The repeating disaccharide is similar to that found in hyaluronic acid, containing GlcUA but with GalNAc replacing GlcNAc. The GalNAc is substituted with sulfate at either its 4′ or its 6′ position,

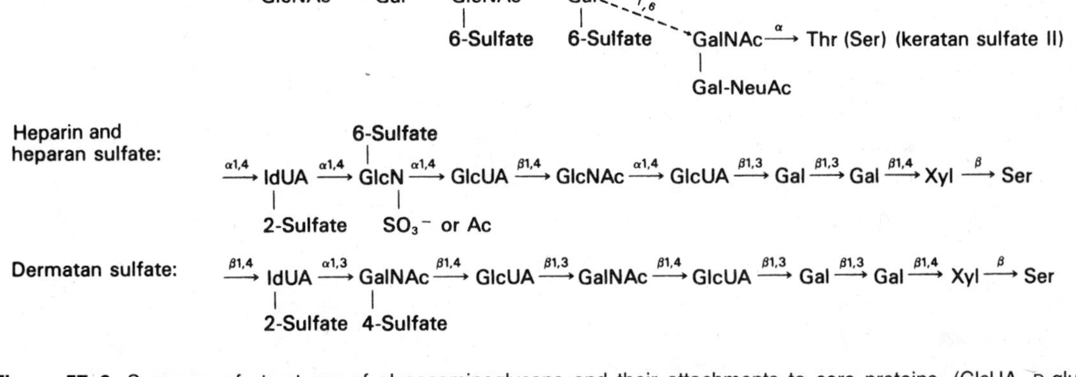

Figure 57–8. Summary of structures of glycosaminoglycans and their attachments to core proteins. (GlcUA, D-glucuronic acid; IdUA, L-iduronic acid; GlcN, D-glucosamine; GalN, D-galactosamine; Ac, acetyl; Gal, D-galactose; Xyl, D-xylose; Ser, L-serine; Thr, L-threonine; Asn, L-asparagine; Man, D-mannose; NeuAc, N-acetylneuraminic acid.) The summary structures are qualitative representations only and do not reflect, for example, the uronic acid composition of hybrid glycosaminoglycans such as heparin and dermatan sulfate, which contain both L-iduronic and D-glucuronic acid. Neither should it be assumed that the indicated substituents are always present, eg, whereas most iduronic acid residues in heparin carry a 2′-sulfate group, a much smaller proportion of these residues are sulfated in dermatan sulfate. The presence of link trisaccharides (Gal-Gal-Xyl) in the chondroitin sulfates, heparin, and heparan and dermatan sulfates is shown. (Slightly modified and reproduced, with permission, from Lennarz WJ: *The Biochemistry of Glycoproteins and Proteoglycans.* Plenum Press, 1980.)

Table 57–6. Major properties of the glycosaminoglycans.

GAG	Sugars	Sulfate[1]	Linkage to Protein	Location
HA	GlcNAc, GlcUA	Nil	No firm evidence	Synovial fluid, vitreous humor, loose connective tissue
CS	GalNAc, GlcUA	GalNAc	Xyl-Ser; associated with HA via link proteins	Cartilage, bone, cornea
KS I	GlcNAc, Gal	GlcNAc	GlcNAc-Asn	Cornea
KS II	GlcNAc, Gal	Same as KS I	GalNAc-Thr	Loose connective tissue
Heparin	GlcN, IdUA	GlcN GlcN IdUA	Ser	Mast cells
Heparan sulfate	GlcN, GlcUA	GlcN	Xyl-Ser	Skin fibroblasts, aortic wall
Dermatan sulfate	GalNAc, IdUA, (GlcUA)	GalNAc IdUA	Xyl-Ser	Wide distribution

[1]The sulfate is attached to various positions of the sugars indicated (see Figure 57–7).

with approximately one sulfate being present per disaccharide unit.

C. Keratan Sulfates I and II: As shown in Figure 57–8, the keratan sulfates consist of repeating Gal-GlcNAc disaccharide units containing sulfate attached to the 6' position of GlcNAc or occasionally of Gal. Type I is abundant in cornea, and type II is found along with chondroitin sulfate attached to hyaluronic acid in loose connective tissue. Types I and II have different attachments to protein (Figure 57–8).

D. Heparin: The repeating disaccharide contains glucosamine (GlcN) and either of the two uronic acids (Figure 57–9). Most of the amino groups of the GlcN residues are N-sulfated, but a few are acetylated. The GlcN also carries a C_6 sulfate ester.

Approximately 90% of the uronic acid residues are IdUA. Initially, all of the uronic acids are GlcUA, but a 5'-epimerase converts approximately 90% of the GlcUA residues to IdUA after the polysaccharide chain is formed. The protein molecule of the heparin proteoglycan is unique, consisting exclusively of serine and glycine residues. Approximately two-thirds of the serine residues contain GAG chains, usually of 5–15 kDa but occasionally much larger. Heparin is found in the granules of mast cells and also in liver, lung, and skin.

E. Heparan Sulfate: This molecule is present on many cell surfaces as a proteoglycan and is extracellular. It contains GlcN with fewer N-sulfates than

heparin, and unlike heparin, its predominant uronic acid is GlcUA.

F. Dermatan Sulfate: This substance is widely distributed in animal tissues. Its structure is similar to that of chondroitin sulfate, except that in place of a GlcUA in β-1,3 linkage to GalNAc, it contains an IdUA in an α-1,3 linkage to GalNAc. Formation of the IdUA occurs, as in heparin and heparan sulfate, by 5'-epimerization of GlcUA. Because this is regulated by the degree of sulfation, and sulfation is incomplete, dermatan sulfate contains both IdUA-GalNAc and GlcUA-GalNAc disaccharides.

Deficiencies of Enzymes That Degrade Glycosaminoglycans Result in Mucopolysaccharidoses

Both exo- and endoglycosidases degrade GAGs. Like most other biomolecules, GAGs are subject to turnover, being both synthesized and degraded. In adult tissues, GAGs generally exhibit relatively slow turnover, their half-lives being in the order of days to weeks.

Understanding of the degradative pathways for GAGs, as in the case of glycoproteins (Chapter 56) and glycosphingolipids (Chapter 26), has been greatly aided by elucidation of the specific enzyme deficiencies that occur in certain inborn errors of metabolism. When GAGs are involved, these inborn errors are called mucopolysaccharidoses (Table 57–7).

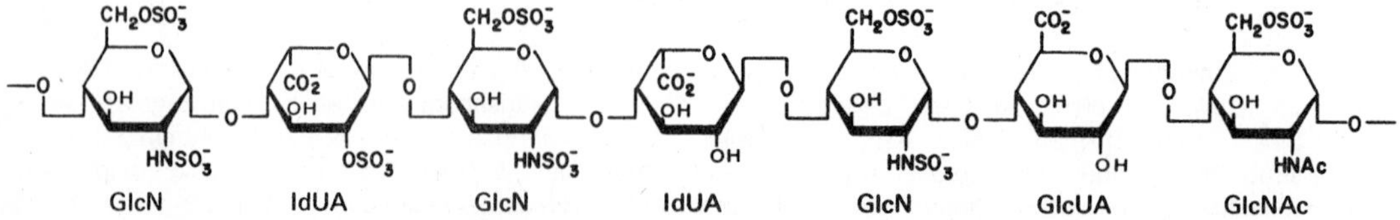

Figure 57–9. Structure of heparin. The polymer section illustrates structural features typical of heparin; however, the sequence of variously substituted repeating disaccharide units has been arbitrarily selected. In addition, non-O-sulfated or 3-O-sulfated glucosamine residues may also occur. (Modified, redrawn, and reproduced, with permission, from Lindahl U et al: Structure and biosynthesis of heparin-like polysaccharides. Fed Proc 1977;36:19.)

Table 57–7. Biochemical defects and diagnostic tests in mucopolysaccharidoses and mucolipidoses.[1]

Name	Alternative Designation[2]	Enzymatic Defect[3]	Urinary Metabolites
Mucopolysaccharidoses			
Hurler, Scheie, Hurler-Scheie	MPS I	α-L-Iduronidase	Dermatan sulfate, heparan sulfate
Hunter	MPS II	Iduronate sulfatase	Dermatan sulfate, heparan sulfate
Sanfilippo A	MPS III A	Heparan sulfate N-sulfatase (sulfamidase)	Heparan sulfate
Sanfilippo B	MPS III B	α-N-Acetylglucosaminidase	Heparan sulfate
Sanfilippo C	MPS III C	Acetyltransferase	Heparan sulfate
Morquio A	MPS IV A	Galactose 6-sulfatase	Keratan sulfate, C6-SO_4
Morquio B	MPS IF B	β-Galactosidase	Keratan sulfate
Maroteaux-Lamy	MPS VI	N-Acetylgalactosamine 4-sulfatase (arylsulfatase B)	Dermatan sulfate
Sly	MPS VII	β-Glucuronidase	Dermatan sulfate, heparan sulfate, C4- and C6-SO_4
Mucolipidoses			
Sialidosis	ML I	Sialidase (neuraminidase)	Glycoprotein fragments
I-cell disease	ML II	UDP N-acetylglucosamine: glycoprotein N-acetylglu-cosamininylphosphotrans-ferase (acid hydrolases thus lack phosphoman-nosyl residue)	Glycoprotein fragments
Pseudo-Hurler polydystrophy	ML H1	As for ML II but deficiency is incomplete	Glycoprotein fragments

[1]Modified and reproduced, with permission, from DiNatale P, Neufeld EF: The biochemical diagnosis of mucopolysaccharidoses, mucolipidosis and related disorders. In: *Perspectives in Inherited Metabolic Diseases,* vol 2. Barr B et al (editors). Editiones Ermes (Milan), 1979.
[2]MPS, mucopolysaccharidosis; ML, mucolipidosis; C6-SO_4, chondroitin 6-sulfate; C4-SO_4, chondroitin 4-sulfate. **Note:** The term MPS V is no longer used.
[3]Fibroblasts, leukocytes, tissues, amniotic fluid cells, or serum can be used for the assay of many of the above enzymes. Patients with these disorders exhibit a variety of clinical findings that may include cloudy corneas, mental retardation, stiff joints, cardiac abnormalities, hepatosplenomegaly, and short stature, depending on the specific disease and its severity.

Degradation of GAGs is carried out by a battery of lysosomal hydrolases. These include certain endoglycosidases, various exoglycosidases, and sulfatases, generally acting in sequence to degrade the various GAGs. A number of them are indicated in Table 57–7.

The mucopolysaccharidoses share a common mechanism of causation, as illustrated in Figure 57–10. They are inherited in an autosomal recessive manner, with Hurler's and Hunter's syndromes being perhaps the most carefully studied. None are common. In some cases, a family history of a mucopolysaccharidosis is obtained. Specific laboratory investigations of help in their diagnosis are urine testing for the presence of increased amount of GAGs and assay of suspected enzymes in white cells, fibroblasts, or sometimes in serum. In certain cases, a tissue biopsy is performed and the GAG that has accumulated can be determined by electrophoresis. DNA tests are increasingly available. Prenatal diagnosis can be made using amniotic cells or chorionic villus biopsy.

The term "mucolipidosis" was introduced to denote diseases that combined features common to both mucopolysaccharidoses and sphingolipidoses (Chapter 26). Three mucolipidoses are listed in Table 57–7. In sialidosis (mucolipidosis I, ML-I), various oligosaccharides derived from glycoproteins and certain gangliosides can accumulate in tissues. I-cell disease (ML-II) and pseudo-Hurler polydystrophy (ML-III) are described in Chapter 56. The term "mucolipidosis" is retained because it is still in relatively widespread clinical usage, but it is not appropriate for these two latter diseases since the mechanism of their causation involves mislocation of certain lysosomal enzymes. Genetic defects of the catabolism of the oligosaccharide chains of glycoproteins (eg, mannosidosis, fucosidosis) are also described in Chapter 56. Most of these defects are characterized by increased excretion of various fragments of glycoproteins in the urine, which accumulate because of the metabolic block, as in the case of the mucolipidoses.

Hyaluronidase is one important enzyme involved in the catabolism of certain GAGs that has not been

Mutation(s) in a gene encoding a lysosomal hydrolase
involved in the degradation of one or more GAGs
↓
Defective lysosomal hydrolase
↓
Accumulation of substrate in various tissues, including
liver, spleen, bone, skin, and central nervous system

Figure 57–10. Simplified scheme of causation of a mucopolysaccharidosis, such as Hurler's syndrome, in which the affected enzyme is α-L-iduronidase. Marked accumulation of the GAGs in the tissues mentioned in the figure could cause hepatomegaly, splenomegaly, disturbances of growth, coarse facies, and mental retardation, respectively.

implicated in any mucopolysaccharidosis. It is a widely distributed endoglycosidase that cleaves hexosaminidic linkages. From hyaluronic acid, the enzyme will generate a tetrasaccharide with the structure $(GlcUA-\beta-1,3-GlcNAc-\beta-1,4)_2$. Hyaluronidase acts on both hyaluronic acid and chrondroitin sulfate. The tetrasaccharide described above can be degraded further by a β-glucuronidase and β-N-acetylhexosaminidase.

Proteoglycans Have Numerous Functions

As indicated above, proteoglycans are remarkably complex molecules and are found in every tissue of the body, mainly in the **extracellular matrix** or "ground substance." There they are associated with each other and also with the other major structural components of the matrix, collagen and elastin, in quite specific manners. Some proteoglycans bind to collagen and others to elastin. These interactions are important in determining the structural organization of the matrix. In addition, some of them interact with certain adhesive proteins, such as fibronectin and laminin (see above), also found in the matrix. The GAGs present in the proteoglycans are polyanions and hence bind polycations and cations such as Na^+ and K^+. This latter ability attracts water by osmotic pressure into the extracellular matrix and contributes to its turgor. GAGs also gel at relatively low concentrations. Because of the long extended nature of the polysaccharide chains of GAGs and their ability to gel, the proteoglycans can act as sieves, restricting the passage of large macromolecules into the extracellular matrix but allowing relatively free diffusion of small molecules. Again, because of their extended structures and the huge macromolecular aggregates they often form, they occupy a large volume of the matrix relative to proteins.

A. Some Functions of Specific GAGs and Proteoglycans: Hyaluronic acid is especially high in concentration in embryonic tissues and is thought to play an important role in permitting cell migration during morphogenesis and wound repair. Its ability to attract water into the extracellular matrix and thereby "loosen it up" may be important in this regard. The high concentrations of hyaluronic acid and chondroitin sulfates present in cartilage contribute to its compressibility (see below).

Chondroitin sulfates are located at sites of calcification in endochondral bone and are also found in cartilage. They are also located inside certain neurons and may provide an endoskeletal structure, helping to maintain their shape.

Both **keratan sulfate I** and **dermatan sulfate** are present in the cornea. They lie between collagen fibrils and play a critical role in corneal transparency. Changes in proteoglycan composition found in corneal scars disappear when the cornea heals. The presence of dermatan sulfate in the sclera may also play a role in maintaining the overall shape of the eye. Keratan sulfate I is also present in cartilage.

Heparin is an important anticoagulant. It binds with factors IX and XI but its most important interaction is with plasma antithrombin III (Chapter 59). The 1:1 binding of heparin to this plasma protein greatly accelerates the ability of the latter to inactivate serine proteases, particularly thrombin. The binding of heparin to lysine residues in antithrombin III appears to induce a conformational change in the latter that favors its binding to the serine proteases. Heparin can also bind specifically to lipoprotein lipase present in capillary walls, causing a release of this enzyme into the circulation.

Certain proteoglycans (eg, **heparan sulfate**) are associated with the plasma membrane of cells, with their core proteins actually spanning that membrane. In it they may act as receptors and may also participate in the mediation of cell growth and cell-cell communication. The attachment of cells to their substratum in culture is mediated, at least in part, by heparan sulfate. This proteoglycan is also found in the basement membrane of the kidney, along with type IV collagen and laminin (see above), where it plays a major role in determining the charge selectiveness of glomerular filtration.

Proteoglycans are also found in intracellular locations such as the nucleus; their function in this organelle has not been elucidated. They are present in some storage or secretory granules, such as the chromaffin granules of the adrenal medulla. It has been postulated that they play a role in the release of the contents of such granules. The various functions of GAGs are summarized in Table 57–8.

B. Associations With Major Diseases and With Aging: Hyaluronic acid may be important in permitting tumor cells to migrate through the extracellular matrix. Tumor cells can induce fibroblasts to synthesize greatly increased amounts of this GAG, thereby perhaps facilitating their own spread. Some tumor cells have less heparan sulfate at their surfaces, and this may play a role in the lack of adhesiveness that these cells display.

Table 57–8. Some functions of glycosaminoglycans and/or proteoglycans.[1]

- Act as structural components of the extracellular (EC) matrix
- Have specific interactions with collagen, elastin, fibronectin, laminin, and other proteins of the matrix
- As polyanions, bind polycations and cations
- Contribute to the characteristic turgor of various tissues
- Act as sieves in the EC matrix
- Facilitate cell migration (HA)
- Have role in compressibility of cartilage in weight-bearing (HA, CS)
- Play role in corneal transparency (KS I and DS)
- Have structural role in sclera (DS)
- Act as anticoagulant (heparin)
- Are components of plasma membranes, where they may act as receptors and participate in cell adhesion and cell-cell interactions (eg, HS)
- Determine charge-selectiveness of renal glomerulus (HS)
- Are components of synaptic and other vesicles (eg, HS)

[1]HA, hyaluronic acid; CS, chondroitin sulfate; KS I, keratan sulfate I; DS, dermatan sulfate; HS, heparan sulfate.

The intima of the arterial wall contains hyaluronic acid and chondroitin sulfate, dermatan sulfate, and heparan sulfate proteoglycans. Of these proteoglycans, dermatan sulfate binds plasma low-density lipoproteins. In addition, dermatan sulfate appears to be the major GAG synthesized by arterial smooth muscle cells. Because it is these cells that proliferate in atherosclerotic lesions in arteries, dermatan sulfate may play an important role in the development of the atherosclerotic plaque.

In various types of **arthritis,** proteoglycans may act as autoantigens, thus contributing to the pathologic features of these conditions. The amount of chondroitin sulfate in cartilage diminishes with age, whereas the amounts of keratan sulfate and hyaluronic acid increase. These changes may contribute to the development of **osteoarthritis.** Changes in the amounts of certain GAGs in the skin are also observed with age and help to account for the characteristic changes noted in this organ in the elderly.

BONE IS A MINERALIZED CONNECTIVE TISSUE

Bone contains both organic and inorganic material. The organic matter is mainly protein. The principal proteins of bone are listed in Table 57–9; type I collagen is the major protein, comprising 90–95% of the organic material. Type V collagen is also present in small amounts, as are a number of noncollagen proteins, some of which are relatively specific to bone. The inorganic or mineral component is mainly crystalline hydroxyapatite ($Ca_{10}[PO_4]_6[OH]_2$), along with sodium, magnesium, carbonate, and fluoride; approximately 99% of the body's calcium is contained in bone (Chapter 47). Hydroxyapatite confers on bone the strength and resilience required by its physiologic roles.

Bone is a dynamic structure that undergoes continuing cycles of remodeling, consisting of resorption followed by deposition of new bone tissue. This re-

Table 57–9. The principal proteins found in bone.[1]

Proteins	Comments
Collagens	
Collagen type I	Approximately 90% of total bone protein. Composed of two α1(I) and one α2(I) chains.
Collagen type V	Minor component.
Noncollagen proteins	
Plasma proteins	Mixture of various plasma proteins.
Proteoglycans[2]	
CS-PG I (biglycan)	Contains two GAG chains; found in other tissues.
CS-PG II (decorin)	Contains one GAG chain; found in other tissues.
CS-PG III	Bone-specific.
Bone SPARC[3] protein (osteonectin)	Not bone-specific.
Osteocalcin (bone Gla protein)	Contains γ-carboxyglutamate residues that bind to hydroxyapatite. Bone-specific.
Osteopontin	Not bone-specific. Glycosylated and phosphorylated.
Bone sialoprotein	Bone-specific. Heavily glycosylated, and sulfated on tyrosine.

[1]Various functions have been ascribed to the noncollagen proteins, including roles in mineralization; however, most of them are still speculative. it is considered unlikely that the noncollagen proteins that are not bone-specific play a key role in mineralization. A number of other proteins are also present in bone, including a tyrosine-rich acidic matrix protein (TRAMP), some growth factors (eg, TGFβ), and enzymes involved in collagen synthesis (eg, lysyl oxidase).
[2]CS-PG, chondroitin sulfate–proteoglycan; these are similar to the dermatan sulfate PGs (DS-PGs) of cartilage (Table 57–12).
[3]SPARC, secreted protein acidic and rich in cysteine.

modeling permits bone to adapt to both physical (eg, increases in weight-bearing) and hormonal signals.

The major cell types involved in bone resorption and deposition are osteoclasts and osteoblasts (Figure 57–11). The former are associated with resorption and the latter with deposition of bone. Osteocytes are descended from osteoblasts; they also appear to be involved in maintenance of bone matrix but will not be discussed further here.

Osteoclasts are multinucleated cells derived from pluripotent hematopoietic stem cells. Osteoclasts possess an apical membrane domain, exhibiting a ruffled border that plays a key role in bone resorption (Figure 57–12). A proton-translocating ATPase expels protons across the ruffled border into the resorption area, which is the microenvironment of low pH shown in the figure. This lowers the local pH to 4.0 or less, thus increasing the solubility of hydroxyapatite and allowing demineralization to occur. Lysosomal acid proteases are released that digest the now accessible matrix proteins. Osteoblasts— mononuclear cells derived from pluripotent mesenchymal precursors—synthesize most of the proteins found in bone (Table 57–9) as well as various growth factors and cytokines. They are responsible for the deposition of new bone matrix (osteoid) and its subsequent mineralization. Osteoblasts control mineralization by regulating the passage of calcium and phosphate ions across their surface membranes. The latter contain alkaline phosphatase, which is used to generate phosphate ions from organic phosphates. The mechanisms involved in mineralization are not fully understood, but several factors have been implicated. Alkaline phosphatase contributes to mineralization but in itself is not sufficient. Small vesicles (matrix vesicles) containing calcium and phosphate have been described at sites of mineralization, but their role is not clear. Type I collagen appears to be necessary, with mineralization being first evident in the gaps between successive molecules. Recent interest has focused on acidic phosphoproteins, such as bone sialoprotein, acting as sites of nucleation. These proteins contain motifs (eg, poly-Asp and poly-Glu stretches) that bind calcium and may provide an initial scaffold for mineralization. Some macromolecules, such as certain proteoglycans and glycoproteins, can also act as inhibitors of nucleation.

It is estimated that approximately 4% of compact bone is renewed annually in the typical healthy adult, whereas approximately 20% of trabecular bone is replaced.

Many factors are involved in the regulation of bone metabolism (Table 57–10), some stimulating or inhibiting osteoblasts and others stimulating or inhibiting osteoclasts. The roles of growth hormone and insulin-like growth factors in bone metabolism are discussed in Chapters 45 and 51 and those of parathormone and calcitriol (1,25[OH]$_2$ vitamin D) in Chapter 47.

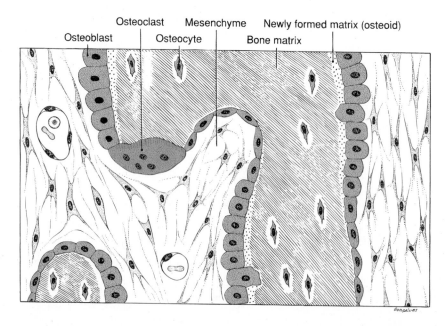

Figure 57–11. Schematic illustration of the major cells present in membranous bone. Osteoblasts (lighter color) are synthesizing type I collagen, which forms a matrix that traps cells. As this occurs, osteoblasts gradually differentiate to become osteocytes. (Reproduced, with permission, from Junqueira LC, Carneiro J, Kelley RO: *Basic Histology,* 8th ed. Appleton & Lange, 1995.)

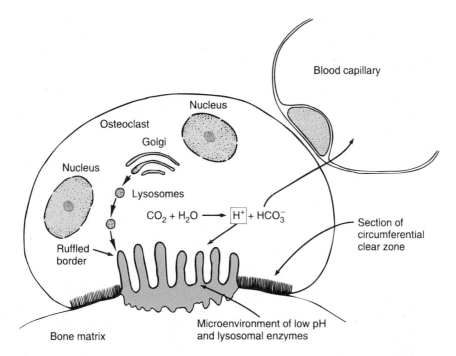

Figure 57–12. Schematic illustration of some aspects of the role of the osteoclast in bone resorption. Lysosomal enzymes and hydrogen ions are released into the confined microenvironment created by the attachment between bone matrix and the peripheral clear zone of the osteoclast. The acidification of this confined space facilitates the dissolution of calcium phosphate from bone and is the optimal pH for the activity of lysosomal hydrolases. Bone matrix is thus removed, and the products of bone resorption are taken up into the cytoplasm of the osteoclast, probably digested further, and transferred into capillaries. The chemical equation shown in the figure refers to the action of carbonic anhydrase II, described in the text. (Reproduced, with permission, from Junqueira LC, Carneiro J, Kelley RO: *Basic Histology,* 8th ed. Appleton & Lange, 1995.)

Table 57–10. Factors affecting osteoblasts and osteoclasts.[1]

Stimulate osteoblasts
  PTH
  1,25-Dihydroxycholecalciferol
  T_3, T_4
  hGH, IGF-I
  PGE_2
  $TGF\beta$
  Estrogens?
Inhibit osteoblasts
  Corticosteroids
Stimulate osteoclasts
  PTH
  1,25-Dihydroxycholecalciferol
  IL-1, IL-6
  TNF
  $TGF\alpha$
Inhibit osteoclasts
  Calcitonin
  Estrogens (by inhibiting IL-6 production)
  $TGF\beta$
  $IFN\alpha$
  PGE_2

[1]Reproduced, with permission, from Ganong WF: *Review of Medical Physiology,* 17th ed. Appleton & Lange, 1995.

MANY METABOLIC AND GENETIC DISORDERS INVOLVE BONE

A number of the more important examples of metabolic and genetic disorders that affect bone are listed in Table 57–11. Dwarfism is discussed in Chapter 45 and rickets, osteomalacia, and hyperparathyroidism in Chapter 47.

Osteogenesis imperfecta (brittle bones) is characterized by abnormal fragility of bones. The sclerae are often abnormally thin and translucent and may appear blue owing to a deficiency of connective tissue. Four types of this condition (mild, extensive, severe, and variable) have been recognized, of which the extensive type occurring in the newborn is the most ominous. Affected infants may be born with multiple fractures and not survive. Over 90% of patients with osteogenesis imperfecta have mutations in the COL1A1 and COL1A2 genes, encoding $pro\alpha1(I)$ and $pro\alpha2(I)$ chains, respectively. Over 100 mutations in these two genes have been documented and include partial gene deletions and duplications. Other mutations affect RNA splicing, and the most frequent type results in the replacement of glycine by another bulkier amino acid. In general, these mutations result in decreased expression of collagen or in structurally

Table 57–11. Some metabolic and genetic diseases affecting bone and cartilage.

Disease	Comments
Dwarfism	Often due to deficiency of growth hormone, but has many other causes.
Rickets	Due to a deficiency of vitamin D during childhood.
Osteomalacia	Due to a deficiency of vitamin D during adulthood.
Hyperparathyroidism	Excess parathormone causes bone resorption.
Osteogenesis imperfecta	Due to a variety of mutations in the COL1A1 and COL1A2 genes affecting the synthesis and structure of type I collagen.
Osteoporosis	Commonly postmenopausal or in other cases is more gradual and related to age; a small number of cases are due to mutations in the COL1A1 and COL1A2 genes.
Osteoarthritis	A small number of cases are due to mutations in the COL1A genes.
Several chondro-dysplasias	Due to mutations in COL2A1 genes.
Pfeiffer's syndrome[1]	Mutations in the gene encoding fibroblast growth receptor 1 (FGFR1).
Jackson-Weiss and Crouzon's syndromes[1]	Mutations in the gene encoding FGFR2.
Achondroplasia and thanatophoric dysplasia[2]	Mutations in the gene encoding FGFR3.

[1]The Pfeiffer, Jackson-Weiss, and Crouzon syndromes are craniosynostosis syndromes; craniosynostosis is a term signifying premature fusion of sutures in the skull.
[2]Thanatophoric (Gk *thanatos* "death"; *pherein* "to bear") dysplasia is the most common neonatal lethal skeletal dysplasia, displaying features similar to those of homozygous achondroplasia.

abnormal proα chains that assemble into abnormal fibrils, weakening the overall structure of bone. When one abnormal chain is present, it may interact with two normal chains, but folding may be prevented, resulting in enzymatic degradation of all of the chains. This is called "procollagen suicide" and is an example of a dominant negative mutation, a result often seen when a protein consists of multiple different subunits.

Osteopetrosis (marble bone disease), characterized by increased bone density, is due to an inability to resorb bone. One form occurs along with renal tubular acidosis and cerebral calcification. It is due to mutations in the gene (located on chromosome 8q22) encoding carbonic anhydrase II (CA II), one of four isozymes of carbonic anhydrase present in human tissues. The reaction catalyzed by carbonic anhydrase is shown below:

$$\overset{I}{} \qquad \overset{II}{}$$
$$CO_2 + H_2O \leftrightarrow H_2CO_3 \leftrightarrow H^+HCO_3^-$$

Table 57–12. The principal proteins found in cartilage.

Proteins	Comments
Collagen proteins	
Collagen type II	90–98% of total articular cartilage collagen. Composed of three α1(II) chains.
Collagens V, VI, IX, X, XI	Type IX cross-links to type II collagen. Type XI may help control diameter of type II fibrils.
Noncollagen proteins	
Proteoglycans	
Aggrecan	The major proteoglycan of cartilage.
Large non-aggregating proteoglycan	Found in some types of cartilage.
DS-PG I (biglycan)[1]	Similar to CS-PG I of bone.
DS-PG II (decorin)	Similar to CS-PG II of bone.
Chondronectin	May play role in binding type II collagen to surface of cartilage.
Anchorin C II	May bind type II collagen to surface of chondrocyte.

[1]The core proteins of DS-PG I and DS-PG II are homologous to those of CS-PG I and CS-PG II found in bone (Table 57–9). A possible explanation is that osteoblasts lack the epimerase required to convert glucuronic acid to iduronic acid, the latter of which is found in dermatan sulfate.

Reaction II is spontaneous. In osteoclasts involved in bone resorption, CA II apparently provides protons to neutralize the OH⁻ ions left inside the cell when H⁺ ions are pumped across their ruffled borders (see above). Thus, if CA II is deficient in activity in osteoclasts, normal bone resorption does not occur, and osteopetrosis results. The mechanism of the cerebral calcification is not clear, whereas the renal tubular acidosis reflects deficient activity of CA II in the renal tubules.

Osteoporosis is a generalized progressive reduction in bone tissue mass per unit volume causing skeletal weakness. The ratio of mineral to organic elements is unchanged in the remaining normal bone. Fractures of various bones, such as the head of the femur, occur very easily and represent a huge burden to both the affected patients and to the health care budget of society. Some of the factors listed in Table 57–10, such as estrogens and interleukins-1 and -6, appear to be intimately involved in the causation of osteoporosis.

THE MAJOR COMPONENTS OF CARTILAGE ARE TYPE II COLLAGEN AND CERTAIN PROTEOGLYCANS

The principal proteins of hyaline cartilage (the major type of cartilage) are listed in Table 57–12. Type II collagen is the principal protein (Figure 57–13), and a number of other minor types of collagen are also present. In addition to these components, elastic cartilage contains elastin and fibroelastic cartilage contains type I collagen. Cartilage contains a number

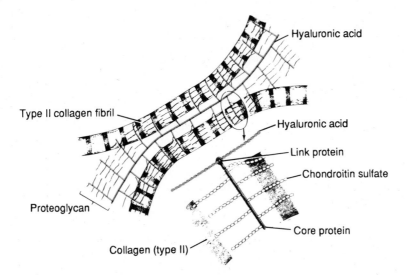

Figure 57–13. Schematic representation of the molecular organization in cartilage matrix. Link proteins noncovalently bind the core protein (lighter color) of proteoglycans to the linear hyaluronic acid molecules (darker color). The chondroitin sulfate side chains of the proteoglycan electrostatically bind to the collagen fibrils, forming a cross-linked matrix. The oval outlines the area enlarged in the lower part of the figure. (Reproduced, with permission, from Junqueira LC, Carneiro J, Kelley RO: *Basic Histology,* 8th ed. Appleton & Lange, 1995.)

of proteoglycans, which play an important role in its compressibility. Aggrecan (about 2×10^3 kDa) is the major proteoglycan. As shown in Figure 57–14, it has a very complex structure containing several GAGs (hyaluronic acid, chondroitin sulfate, and ker-

atan sulfate) and both link and core proteins. The core protein contains three domains: A, B, and C. The hyaluronic acid binds noncovalently to domain A of the core protein as well as to the link protein, which stabilizes the hyaluronate core protein interac-

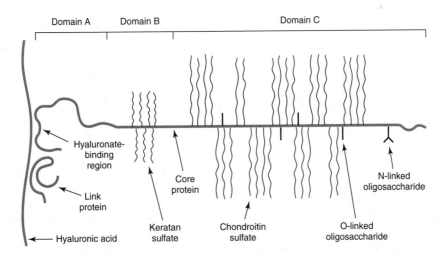

Figure 57–14. Schematic diagram of the aggrecan from bovine nasal cartilage. A strand of hyaluronic acid is shown on the left. The core protein (about 210 kDa) has three major domains. Domain A, at its amino terminal end, interacts with approximately five repeating disaccharides in hyaluronate. The link protein interacts with both hyaluronate and domain A, stabilizing their interactions. Approximately 30 keratan sulfate chains are attached, via GalNAc-Ser linkages, to domain B. Domain C contains about 100 chondroitin sulfate chains attached via Gal-Gal-Xyl-Ser linkages and about 40 *O*-linked oligosaccharide chains. One or more *N*-linked glycan chains are also found near the carboxyl terminal of the core protein. (Reproduced, with permission, from Moran LA et al: *Biochemistry,* 2nd ed. Neil Patterson Publishers, 1994.)

tions. The keratan sulfate chains are located in domain B, whereas the chondroitin sulfate chains are located in domain C; both of these types of GAGs are bound covalently to the core protein. The core protein also contains both *O*- and *N*-linked oligosaccharide chains.

The other proteoglycans found in cartilage have simpler structures than aggrecan.

Chondronectin is involved in the attachment of type II collagen to chondrocytes.

Cartilage is an avascular tissue and obtains most of its nutrients from synovial fluid. It exhibits slow but continuous turnover. Various proteases (eg, collagenases and stromalysin) synthesized by chondrocytes can degrade collagen and the other proteins found in cartilage. IL-1 and TNFα appear to stimulate the production of such proteases, whereas TGFβ and IGF-1 generally exert an anabolic influence on cartilage.

THE MOLECULAR BASES OF THE CHONDRODYSPLASIAS INCLUDE MUTATIONS IN GENES ENCODING TYPE II COLLAGEN AND FIBROBLAST GROWTH FACTOR RECEPTORS

Chondrodysplasias are a mixed group of hereditary disorders affecting cartilage. They are manifested by short-limbed dwarfism and numerous skeletal deformities. A number of them are due to a variety of mutations in the COL2A1 gene, leading to abnormal forms of type II collagen. One example is the Stickler syndrome, manifested by degeneration of joint cartilage and of the vitreous body of the eye.

The best-known of the chondrodysplasias is **achondroplasia,** the commonest cause of dwarfism. Individuals affected by this condition have short limbs, normal trunk size, macrocephaly, and a variety of other skeletal abnormalities. The condition is often inherited as an autosomal dominant trait, but many cases are due to new mutations. The recently elucidated molecular basis of achondroplasia is outlined in Figure 57–15. Achondroplasia is not a collagen disorder but is due to very specific mutations in the gene encoding fibroblast growth factor receptor 3 (FGFR3). Fibroblast growth factors are a family of at least seven proteins that affect the growth and differentiation of cells of mesenchymal and neuroectodermal origin. Their receptors are transmembrane proteins and form a subgroup of the family of receptor tyrosine kinases. FGFR3 is one member of this subgroup and mediates the actions of FGF3 on cartilage. In all cases of achondroplasia that have been investigated, the mutations were found to involve nucleotide 1138 and resulted in substitution of arginine for glycine (residue number 380) in the transmembrane domain of the protein, rendering it inactive. No such mutation was found in unaffected individuals.

Mutations of nucleotide 1138 in the gene encoding FGFR3 on chromosome 4
↓
Replacement in FGFR3 of Gly (residue number 380) by Arg
↓
Defective function of FGFR3
↓
Abnormal development and growth of cartilage leading to short-limbed dwarfism and other features

Figure 57–15. Simplified scheme of the causation of achrondroplasia. In most cases studied so far, the mutation has been a G to A transition at nucleotide 1138. In a few cases, the mutation was a G to C transversion at the same nucleotide. This particular nucleotide is a real "hot spot" for mutation. Both mutations result in replacement of a Gly residue by an Arg residue in the transmembrane segment of the receptor.

As indicated in Table 57–11, other skeletal dysplasias (including certain craniosynostosis syndromes) are also due to mutations in genes encoding FGF receptors. Another type of skeletal dysplasia (diastrophic dysplasia) has been found to be due to mutation in a sulfate transporter. Thus, thanks to recombinant DNA technology, a new era in understanding of skeletal dysplasias has begun.

SUMMARY

The major components of the extracellular matrix are the structural proteins collagen, elastin, and fibrillin; a number of specialized proteins (eg, fibronectin and laminin); and various proteoglycans.

Collagen is the most abundant protein in the animal kingdom. Approximately 19 types have been isolated, and 30 genes encode procollagen polypeptides. All collagens contain greater or lesser stretches of triple helix. The repeating structure $(Gly-X-Y)_n$ is present in all collagens, with X and Y frequently being proline and hydroxyproline. Hydroxylysine is also found and is a potential site of glycosylation. The biosynthesis of collagen is complex, featuring many posttranslational events, including hydroxylation of proline and lysine. Diseases associated with impaired synthesis of collagen include scurvy, osteogenesis imperfecta, Ehlers-Danlos syndrome (many types), and Menkes' syndrome. Elastin confers extensibility and elastic recoil on tissues. Only one gene for elastin has been detected in humans. Elastin contains hydroxyproline but lacks hydroxylysine, Gly-X-Y sequences, triple helical structure, and sugars. It also contains desmosine and isodesmosine cross-links not found in collagen.

Fibrillin is located in microfibrils associated with elastin. Mutations in the gene for fibrillin cause Marfan's syndrome. Fibronectin is a glycoprotein that is important in cell adhesion and migration. Laminin is an important component of basal laminae, such as are present in the renal glomerulus.

The glycosaminoglycans (GAGs) are made up of repeating disaccharides containing a uronic acid (glucuronic or iduronic) or hexose (galactose) and a hexosamine (galactosamine or glucosamine). Sulfate is also frequently present. The major GAGs are hyaluronic acid, chondroitin 4- and 6-sulfates, keratan sulfates I and II, heparin, heparan sulfate, and dermatan sulfate. The GAGs are synthesized by the sequential actions of a battery of specific enzymes (glycosyltransferases, epimerases, sulfotransferases, etc) and are degraded by the sequential action of lysosomal hydrolases. Genetic deficiencies of the latter result in mucopolysaccharidoses (eg, Hurler's syndrome). GAGs occur in tissues bound to various proteins (linker proteins and core proteins), constitut-

ing proteoglycans. These structures are often of very high molecular weight and serve many functions in tissues. Many components of the extracellular matrix (eg, certain GAGs and proteins such as fibronectin and laminin) bind to proteins of the cell surface named integrins; this constitutes one pathway by which the exteriors of cells can communicate with their interiors.

Bone and cartilage are specialized forms of the extracellular matrix. Collagen I and hydroxyapatite are the major constituents of bone. Bone also contains a number of minor proteins that may serve important functions. Collagen II and certain proteoglycans are major constituents of cartilage. The molecular causes of a number of heritable diseases of bone (eg, osteogenesis imperfecta) and of cartilage (eg, the chondrodystrophies) are being revealed by the application of recombinant DNA technology.

REFERENCES

Anwar R: Elastin: A brief review. Biochem Educ 1990;18: 162.

Byers PH: Disorders of collagen biosynthesis and structure. In: *The Metabolic and Molecular Bases of Inherited Disease,* 7th ed. Scriver CR et al (editors). McGraw-Hill, 1995.

Caplan A: Cartilage. Sci Am (Oct) 1984;250:84.

Clark EA, Brugge JS: Integrins and signal transduction pathways: The road taken. Science 1995;268:233.

Francomano CA: The genetic basis of dwarfism. N Engl J Med 1995;332:58.

Hardingham TE, Fosang AJ: Proteoglycans: Many forms and many functions. FASEB J 1992;6:861.

Hay ED (editor): *Cell Biology of the Extracellular Matrix,* 2nd ed. Plenum Press, 1993.

Manolagas SC, Jilka RL: Bone marrow, cytokines, and bone remodeling: Emerging insights into the pathophysiology of osteoporosis. N Engl J Med 1995;332:305.

Neufeld EF, Muenzer J: The mucopolysaccharidoses. In: *The Metabolic and Molecular Bases of Inherited Disease,* 7th ed. Scriver CR et al (editors). McGraw-Hill, 1995.

Prockop DJ, Kivirrikko KI: Collagens: Molecular biology, diseases, and potential therapy. Annu Rev Biochem 1995;64:403.

Raghow R: The role of extracellular matrix in postinflammatory wound healing and fibrosis. FASEB J 1994;8: 823.

Ramirez F et al: Marfan syndrome and related disorders. In: *The Metabolic and Molecular Bases of Inherited Disease,* 7th ed. Scriver CR et al (editors). McGraw-Hill, 1995.

Rosenbloom J, Abrams WR, Mecham R: Extracellular matrix 4: The elastic fiber. FASEB J 1993;7:1208.

Royce PM, Steinmann B (editors): *Connective Tissue and Its Heritable Disorders,* Wiley-Liss, 1993.

Sakai LY: The extracellular matrix. Sci Am Sci Med May–June 1995;58.

Schumacher HR, Klippel JH, Koopman WJ (editors): *Primer on the Rheumatic Diseases,* 10th ed. Arthritis Foundation, 1993. (Contains sections on articular cartilage, bone, collagen and elastin, proteoglycans, heritable disorders of connective tissue, and metabolic bone diseases.)

Tilstra DJ, Byers PH: Molecular basis of hereditary disorders of connective tissue. Annu Rev Med 1994;45: 149.

Yamada KM: Adhesive recognition sequences. J Biol Chem 1991;266:12809.

58

Muscle

Robert K. Murray, MD, PhD

INTRODUCTION

Proteins serve functions other than catalysis. For instance, regulatory, signaling, recognition, transport, and structural functions of proteins are described in earlier chapters. Proteins play an important role in movement, at both the organ (eg, skeletal muscle, heart, and gut) and cellular levels. In this chapter, the roles of specific proteins and certain other key molecules (eg, Ca^{2+}) in muscular contraction are described.

BIOMEDICAL IMPORTANCE

Understanding of the molecular basis of major genetic diseases received significant impetus in 1986 with the successful cloning of the gene for **Duchenne-type muscular dystrophy,** an achievement that holds high promise for accurate diagnosis and possible therapy for this disease (see Case No. 6 in Chapter 65). Significant progress has also been made in understanding the molecular basis of **malignant hyperthermia,** a serious complication for some patients undergoing certain types of anesthesia. This condition has been shown to be due to abnormal accumulation of Ca^{2+} in the cytoplasm of skeletal muscle cells, causing rigor and the generation of excessive heat. The elevation of Ca^{2+} is due, at least in certain cases, to mutation in the structural gene for the Ca^{2+} release channel of the sarcoplasmic reticulum. This condition also has serious financial implications for the swine industry, since a similar disorder occurs in pigs subjected to stress. **Heart failure** is a very common medical condition, with a variety of causes; its rational therapy requires understanding of the biochemistry of heart muscle. One group of conditions that cause heart failure are the **cardiomyopathies,** certain of which are genetically determined. Recent work has shown that mutations in the cardiac β-myosin heavy-chain gene are one cause of familial hypertrophic cardiomyopathy. A surprising recent development is the finding that the **endothelium-derived relaxing factor,** a compound synthesized by endothelial cells that regulates smooth muscle tone, is the gas **nitric oxide (NO).** Many widely used **vasodilators,** such as nitroglycerin, used in the treatment of angina pectoris, act by increasing the formation of this compound. Equally surprising is the finding that NO is apparently a neurotransmitter. **Myocardial infarction** is extremely common, and certain biochemical aspects of this condition are described in Case No. 4 (Chapter 65).

MUSCLE TRANSDUCES CHEMICAL ENERGY INTO MECHANICAL ENERGY

Muscle is the major biochemical transducer (machine) that converts potential (chemical) energy into kinetic (mechanical) energy. Muscle, the largest single tissue in the human body, makes up somewhat less than 25% of body mass at birth, more than 40% in the young adult, and somewhat less than 30% in the aged adult.

An effective chemical-mechanical transducer must meet several requirements: (1) There must exist a constant supply of chemical energy. In vertebrate muscle, ATP and creatine phosphate supply chemical energy. (2) There must be a means of regulating the mechanical activity—ie, the speed, duration, and force of contraction in the case of muscle. (3) The machine must be connected to an operator, a requirement met in biologic systems by the nervous system. (4) There must be a way of returning the machine to its original state.

Muscle is a pulling, not a pushing, machine. Therefore, a given muscle must be antagonized by another group of muscles or another force such as gravity or elastic recoil.

In vertebrates, the above requirements and the specific needs of the organisms are met by three types of muscles: skeletal muscle, cardiac muscle, and smooth muscle. Both skeletal and cardiac muscle appear striated upon microscopic observation; smooth muscle is nonstriated. Although skeletal muscle is under voluntary nervous control, the control of both cardiac and smooth muscle is involuntary.

The Sarcoplasm of Muscle Cells Contains ATP, Phosphocreatine, and Glycolytic Enzymes

Striated muscle is composed of multinucleated muscle fiber cells surrounded by an electrically excitable plasma membrane, the **sarcolemma.** An individual muscle fiber cell, which may extend the entire length of the muscle, contains a bundle of many **myofibrils** arranged in parallel, embedded in intracellular fluid termed **sarcoplasm.** Within this fluid is contained glycogen, the high-energy compounds ATP and phosphocreatine, and the enzymes of glycolysis.

The Sarcomere Is the Functional Unit of Muscle

An overall view of voluntary muscle at several levels of organization is shown in Figure 58–1.

When the myofibril is examined by electron microscopy, alternating dark and light bands (anisotropic bands, meaning birefringent in polarized light; and isotropic bands, meaning not altered by polarized light) can be observed. These bands are thus referred to as A and I bands, respectively. The central region of the A band (the H band, or H zone) appears less dense than the rest of the band. The I band is bisected by a very dense and narrow Z line (Figure 58–2).

The sarcomere is defined as the region between two Z lines (Figures 58–1 and 58–2) and is repeated along the axis of a fibril at distances of 1500–2300 nm, depending upon the state of contraction.

The striated appearance of voluntary and cardiac muscle in light microscopic studies results from their high degree of organization, in which most muscle fiber cells are aligned so that their sarcomeres are in parallel register (Figure 58–1).

Thick Filaments Contain Myosin, and Thin Filaments Contain Actin, Tropomyosin, and Troponin

When myofibrils are examined by electron microscopy, it appears that each one is constructed of two types of longitudinal filaments. One type, the thick filament, confined to the A band, contains chiefly the protein myosin. These filaments are about 16 nm in diameter and arranged in cross-section as a hexagonal array (Figure 58–2, center; right-hand cross-section).

The other filament, the thin filament, lies in the I band and extends into the A band but not into its H zone (Figure 58–2). The thin filaments contain the proteins actin, tropomyosin, and troponin (Figure 58–3). In the A band, the thin filaments are arranged around the thick (myosin) filament as a secondary hexagonal array. Each thin filament lies symmetrically between three thick filaments (Figure 58–2,

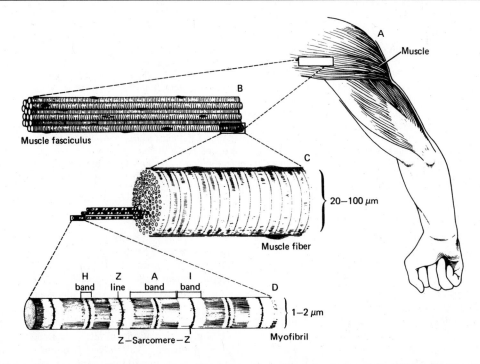

Figure 58–1. The structure of voluntary muscle. The sarcomere is the region between the Z lines. (Drawing by Sylvia Colard Keene. Reproduced, with permission, from Bloom W, Fawcett DW: *A Textbook of Histology,* 10th ed. Saunders, 1975.)

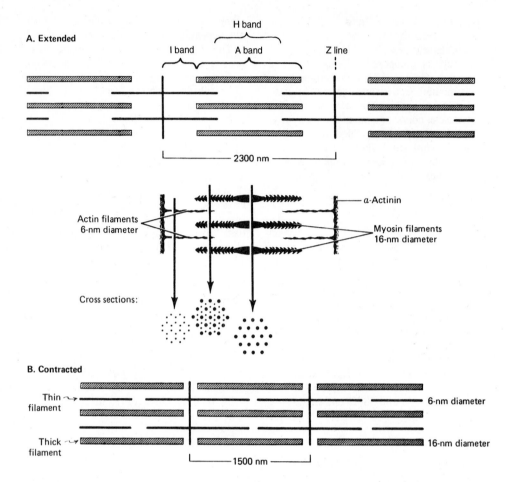

Figure 58–2. Arrangement of filaments in striated muscle. **A:** Extended. The positions of the I, A, and H bands in the extended state are shown. The thin filaments partly overlap the ends of the thick filaments, and the thin filaments are shown anchored in the Z lines (often called Z disks). In the lower part of the figure, arrowheads, pointing in opposite directions, are shown emanating from the myosin (thick) filaments. Four actin (thin) filaments are shown attached to two Z lines via α-actinin. The central region of the three myosin filaments, free of arrowheads, is called the M band (not labeled). Cross-sections through the M bands, through an area where myosin and actin filaments overlap and through an area in which solely actin filaments are present, are shown. **B:** Contracted. The actin filaments are seen to have slipped along the sides of the myosin fibers toward each other. The lengths of the thick filaments (indicated by the A bands) and the thin filaments (distance between Z lines and the adjacent edges of the H bands) have not changed. However, the lengths of the sarcomeres have been reduced (from 2300 nm to 1500 nm), and the lengths of the H and I bands are also reduced because of the overlap between the thick and thin filaments. These morphologic observations provided part of the basis for the sliding filament model of muscle contraction.

center; mid cross-section), and each thick filament is surrounded symmetrically by six thin filaments.

The thick and thin filaments interact via cross-bridges that emerge at intervals of 14 nm along the thick filaments. As depicted in Figure 58–2, the cross-bridges (drawn as arrowheads at each end of the myosin filaments but not shown extending fully across to the thin filaments) have opposite polarities at the two ends of the thick filaments. The two poles of the thick filaments are separated by a 150-nm segment (the M band, not labeled in the figure) that is free of projections.

The Sliding Filament Cross-Bridge Model Is the Foundation on Which Current Thinking About Muscle Contraction Is Built

This model was proposed independently in the 1950s by Henry Huxley and Andrew Huxley and their colleagues. It was largely based on careful morphologic observations on resting, extended, and contracting muscle. Basically, when muscle contracts, there is no change in the lengths of the thick and thin filaments, but the H zones and the I bands shorten (see legend to Figure 58–2). Thus, the arrays of inter-

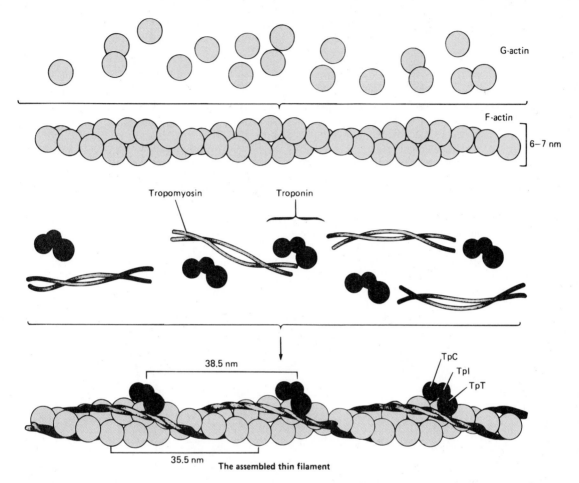

Figure 58–3. Schematic representation of the thin filament, showing the spatial configuration of its three major protein components: actin, myosin, and tropomyosin. The upper panel shows individual molecules of G-actin. The middle panel shows actin monomers assembled into F-actin. Individual molecules of tropomyosin (two strands wound around one another) and of troponin (made up of its three subunits) are also shown. The lower panel shows the assembled thin filament, consisting of F-actin, tropomyosin, and the three subunits of troponin (TpC, TpI, and TpT).

digitating filaments must slide past one another during contraction. Cross-bridges that link thick and thin filaments at certain stages in the contraction cycle generate and sustain the tension. The tension developed during muscle contraction is proportionate to the filament overlap and the number of cross-bridges. Each cross-bridge head is connected to the thick filament via a flexible fibrous segment that can bend outward from the thick filament. This flexible segment facilitates contact of the head with the thin filament when necessary but is also sufficiently pliant to be accommodated in the interfilament spacing.

ACTIN AND MYOSIN ARE THE MAJOR PROTEINS OF MUSCLE

The mass of a fresh muscle is made up of 75% water and more than 20% protein. The two major proteins are actin and myosin.

Monomeric G-actin (43 kDa; G, globular) makes up 25% of muscle protein by weight. At physiologic ionic strength and in the presence of Mg^{2+}, G-actin polymerizes noncovalently to form an insoluble double helical filament called F-actin (Figure 58–3). The F-actin fiber is 6–7 nm thick and has a pitch or repeating structure every 35.5 nm.

Myosins constitute a family of proteins, with about ten members having been identified. The myosin discussed in this chapter is myosin-II, and when myosin is referred to in this text, it is this species that is meant unless otherwise indicated.

Myosin-I is a monomeric species that binds to cell membranes. It may serve as a linkage between microfilaments and the cell membrane in certain locations.

Myosin contributes 55% of muscle protein by weight and forms the thick filaments. It is an asymmetric hexamer with a molecular mass of approximately 460 kDa. Myosin has a fibrous tail consisting of two intertwined helices. Each helix has a globular head portion attached at one end (Figure 58–4). The hexamer consists of one pair of heavy (H) chains each of approximately 200 kDA molecular mass, and two pairs of light (L) chains each with a molecular mass of approximately 20 kDa. The L chains differ, one being called the essential light chain and the other the regulatory light chain. Skeletal muscle myosin binds actin to form actomyosin (actin-myosin), and its intrinsic ATPase activity is markedly enhanced in this complex. Isoforms of myosin exist whose amounts can vary in different anatomic, physiologic, and pathologic situations.

The crystallographic structures of actin and of the head of myosin have been determined; these studies have verified a number of earlier findings concerning their structures and have also given rise to much new information.

Limited Digestion of Myosin With Proteases Has Helped to Elucidate Its Structure and Function

When myosin is digested with trypsin, two myosin fragments (meromyosins) are generated. Light meromyosin (LMM) consists of aggregated, insoluble α-helical fibers from the tail of myosin (Figure 58–4). LMM exhibits no ATPase activity and does not bind to F-actin.

Heavy meromyosin (HMM; molecular mass about 340 kDa) is a soluble protein that has both a fibrous portion and a globular portion (Figure 58–4). It exhibits ATPase activity and binds to F-actin. Digestion of HMM with papain generates two subfragments, S-1 and S-2. The S-2 fragment is fibrous in character, has no ATPase activity, and does not bind to F-actin.

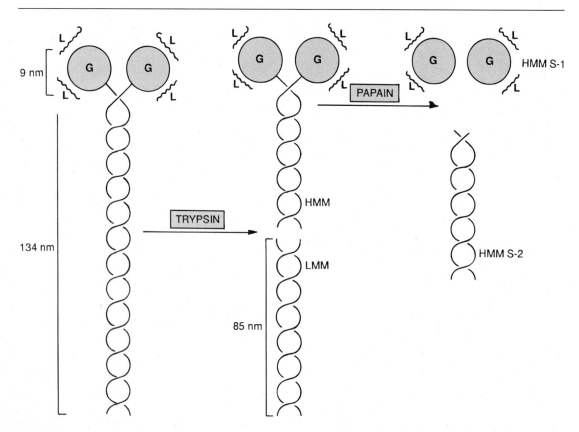

Figure 58–4. Diagram of a myosin molecule showing the two intertwined α-helices (fibrous portion), the globular region or head (G), the light chains (L), and the effects of proteolytic cleavage by trypsin and papain. The globular region (myosin head) contains an actin-binding site and an L chain-binding site and also attaches to the remainder of the myosin molecule.

S-1 (molecular mass approximately 115 kDa) does exhibit ATPase activity, binds L chains, and in the absence of ATP will bind to and decorate actin with "arrowheads" (Figure 58–5). Both S-1 and HMM exhibit ATPase activity, which is accelerated 100- to 200-fold by complexing with F-actin. As discussed below, F-actin greatly enhances the rate at which myosin ATPase releases its products, ADP and P_i. Thus, although F-actin does not affect the hydrolysis step per se, its ability to promote release of the products produced by the ATPase activity greatly accelerates the overall rate of catalysis.

CHANGES IN THE CONFORMATION OF THE HEAD OF MYOSIN DRIVE MUSCLE CONTRACTION

How can hydrolysis of ATP produce macroscopic movement? Muscle contraction essentially consists of the cyclic attachment and detachment of the S-1 head of myosin to the F-actin filaments. This process can also be referred to as the making and breaking of cross-bridges. The attachment of actin to myosin is followed by conformational changes which are of particular importance in the S-1 head and are dependent upon which nucleotide is present (ADP or ATP). These changes result in the power stroke, which drives movement of actin filaments past myosin filaments. The energy for the power stroke is ultimately supplied by ATP, which is hydrolyzed to ADP and P_i. However, the power stroke itself occurs as a result of conformational changes in the myosin head when ADP leaves it.

The major biochemical events occurring during one cycle of muscle contraction and relaxation can be represented in the five steps shown in Figure 58–6:

Figure 58–5. The decoration of actin filaments with the S-1 fragments of myosin to form "arrowheads." (Courtesy of JA Spudich.)

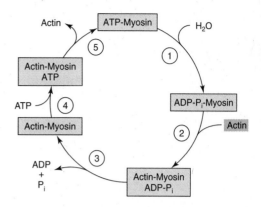

Figure 58–6. The hydrolysis of ATP drives the cyclic association and dissociation of actin and myosin in five reactions described in the text. (Modified from Stryer L: *Biochemistry,* 2nd ed. Freeman, 1981.)

(1) In the relaxation phase of muscle contraction, the S-1 head of myosin hydrolyzes ATP to ADP and P_i, but these products remain bound. The resultant ADP-P_i-myosin complex has been energized and is in a so-called high-energy conformation.

(2) When contraction of muscle is stimulated (via events involving Ca^{2+}, troponin, tropomyosin, and actin, which are described below), actin becomes accessible and the S-1 head of myosin finds it, binds it, and forms the actin-myosin-ADP-P_i complex indicated.

(3) Formation of this complex promotes the release of P_i, which initiates the power stroke. This is followed by release of ADP and is accompanied by a large conformational change in the head of myosin in relation to its tail (Figure 58–7), pulling actin about 10 nm toward the center of the sarcomere. This is the power stroke. The myosin is now in a so-called low-energy state, indicated as actin-myosin.

(4) Another molecule of ATP binds to the S-1 head, forming an actin-myosin-ATP complex.

(5) Myosin-ATP has a low affinity for actin, and actin is thus released. This last step is a key component of relaxation and is dependent upon the binding of ATP to the actin-myosin complex.

Another cycle then commences with the hydrolysis of ATP (step [1] of Figure 58–6), re-forming the high-energy conformation.

Thus, hydrolysis of ATP is used to drive the cycle, with the actual power stroke being the conformational change in the S-1 head that occurs upon the release of ADP. The hinge regions of myosin (referred to as flexible points at each end of S-2 in the legend to Figure 58–7) permit the large range of movement of S-1 and also allow S-1 to find actin filaments.

If intracellular levels of ATP drop (eg, after death), ATP is not available to bind the S-1 head (step 4 above), actin does not dissociate and relax-

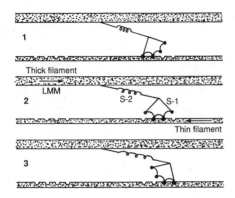

Figure 58–7. Representation of the active cross-bridges between thick and thin filaments. This diagram was adapted by A.F. Huxley from H.E. Huxley: The mechanism of muscular contraction. Science 1969;164:1356. The latter proposed that the force involved in muscular contraction originates in a tendency for the myosin head (S-1) to rotate relative to the thin filament and is transmitted to the thick filament by the S-2 portion of the myosin molecule acting as an inextensible link. Flexible points at each end of S-2 permit S-1 to rotate and allow for variations in the separation between filaments. The present figure is based on H.E. Huxley's proposal but also incorporates elastic (the coils in the S-2 portion) and stepwise-shortening elements (depicted here as four sites of interaction between the S-1 portion and the thin filament). (See Huxley AF, Simmons RM: Proposed mechanism of force generation in striated muscle. Nature [Lond] 1971;233:533.) The strengths of binding of the attached sites are higher in position 2 than in position 1 and higher in position 3 than position 2. The myosin head can be detached from position 3 with the utilization of a molecule of ATP; this is the predominant process during shortening. The myosin head is seen to vary in its position from about 90° to about 45°, as indicated in the text. (S-1, myosin head; S-2, portion of the myosin molecule; LMM, light meromyosin.) (See legend to Figure 58–4.) (Reproduced from Huxley AF: Muscular contraction. J Physiol 1974;243:1. By kind permission of the author and the *Journal of Physiology*.)

ation (step 5) does not occur. This is the explanation for rigor mortis, the stiffening of the body that occurs after death.

Calculations have indicated that the efficiency of contraction is about 50%; that of the internal combustion engine is less than 20%.

Tropomyosin and the Troponin Complex Present in Thin Filaments Perform Key Functions in Striated Muscle

In striated muscle, there are two other proteins that are minor in terms of their mass but important in terms of their function. Tropomyosin is a fibrous molecule that consists of two chains, alpha and beta, that attach to F-actin in the groove between its fila-

ments (Figure 58–3). Tropomyosin is present in all muscular and muscle-like structures. The troponin complex is unique to striated muscle and consists of three polypeptides. Troponin T (TpT) binds to tropomyosin as well as to the other two troponin components. Troponin I (TpI) inhibits the F-actin-myosin interaction and also binds to the other components of troponin. Troponin C (TpC) is a calcium-binding polypeptide that is structurally and functionally analogous to **calmodulin,** an important calcium-binding protein widely distributed in nature. Four molecules of calcium ion are bound per molecule of troponin C or calmodulin, and both molecules have a molecular mass of 17 kDa.

Ca²⁺ Plays a Central Role in Regulation of Muscle Contraction

The contraction of muscles from all sources occurs by the general mechanism described above. Muscles from different organisms and from different cells and tissues within the same organism may have different molecular mechanisms responsible for the regulation of their contraction and relaxation. In all systems, Ca^{2+} plays a key regulatory role. There are two general mechanisms of regulation of muscle contraction: actin-based and myosin-based.

Actin-Based Regulation Occurs in Striated Muscle

Actin-based regulation of muscle occurs in vertebrate skeletal and cardiac muscles, both striated. In the general mechanism described above (Figure 58–6), the only potentially limiting factor in the cycle of muscle contraction might be ATP. The skeletal muscle system is inhibited at rest; this inhibition is relieved to activate contraction. The inhibitor of striated muscle is the troponin system, which is bound to tropomyosin and F-actin in the thin filament (Figure 58–3). In striated muscle, there is no control of contraction unless the tropomyosin-troponin systems are present along with the actin and myosin filaments. As described above, tropomyosin lies along the groove of F-actin, and the three components of troponin—TpT, TpI, and TpC—are bound to the F-actin-tropomyosin complex. TpI prevents binding of the myosin head to its F-actin attachment site either by altering the conformation of F-actin via the tropomyosin molecules or by simply rolling tropomyosin into a position that directly blocks the sites on F-actin to which the myosin heads attach. Either way prevents activation of the myosin ATPase that is mediated by binding of the myosin head to F-actin. Hence, the TpI system blocks the contraction cycle at step 2 of Figure 58–6. This accounts for the inhibited state of relaxed striated muscle.

The Sarcoplasmic Reticulum Regulates Intracellular Levels of Ca^{2+} in Skeletal Muscle

In resting muscle sarcoplasm, the concentration of Ca^{2+} is 10^{-7} to 10^{-8} mol/L. The resting state is achieved because Ca^{2+} is pumped into the sarcoplasmic reticulum through the action of an active transport system, called the Ca^{2+} ATPase (Figure 58–8), initiating relaxation. The sarcoplasmic reticulum is a network of fine membranous sacs. Inside the sarcoplasmic reticulum, Ca^2 is bound to a specific Ca^{2+}-binding protein designated calsequestrin. The sarcomere is surrounded by an excitable membrane (the T tubule system) composed of transverse (T) channels closely associated with the sarcoplasmic reticulum.

When the sarcolemma is excited by a nerve impulse, the signal is transmitted into the T tubule system and a **Ca^{2+} release channel** in the nearby sarcoplasmic reticulum opens, releasing Ca^{2+} from the sarcoplasmic reticulum into the sarcoplasm. The concentration of Ca^{2+} in the sarcoplasm rises rapidly to 10^{-5} mol/L. The Ca^{2+}-binding sites on TpC in the thin filament are quickly occupied by Ca^{2+}. The $TpC·4Ca^{2+}$ interacts with TpI and TpT to alter their interaction with tropomyosin. Accordingly, tropomyosin moves out of the way or alters the conformation of F-actin so that the myosin head-$ADP-P_i$ (Figure 58–6) can interact with F-actin to start the contraction cycle.

The Ca^{2+} release channel is also known as the ryanodine receptor (RYR). There are two isoforms of this receptor, RYR1 and RYR2, the former being present in skeletal muscle and the latter in heart muscle and brain. Ryanodine is a plant alkaloid that binds to RYR1 and RYR2 specifically and modulates their activities. The Ca^{2+} release channel is a homotetramer made up of four subunits of 565 kDa. It has transmembrane sequences at its carboxyl terminal, and these probably form the Ca^{2+} channel. The remainder of the protein protrudes into the cytosol, bridging the gap between the sarcoplasmic reticulum and the transverse tubular membrane. The channel is ligand-gated, Ca^{2+} and ATP working synergistically in vitro, although how it operates in vivo is not clear. A possible sequence of events leading to opening of the channel is shown in Figure 58–9. The channel lies very close to the dihydropyridine receptor (slow Ca^{2+} channels) of the transverse tubule system (Figure 58–8); the two proteins may be coupled.

Relaxation occurs when (1) sarcoplasmic Ca^{2+} falls below 10^{-7} mol/L owing to its resequestration into the sarcoplasmic reticulum by the Ca^{2+} ATPase; (2) $TpC·4Ca^{2+}$ loses its Ca^{2+}; (3) troponin, via its interaction with tropomyosin, inhibits further myosin head-F-actin interaction; and (4) in the presence of ATP, the myosin head detaches from the F-actin. Thus, Ca^{2+} controls muscle contraction by an allosteric mechanism mediated in muscle by TpC, TpI, TpT, tropomyosin, and F-actin.

A decrease in the concentration of ATP in the sarcoplasm (eg, by excessive usage during the cycle of contraction-relaxation or by diminished formation, such as might occur in ischemia) has two major effects: (1) The Ca^{2+} ATPase (Ca^{2+} pump) in the sarcoplasmic reticulum ceases to maintain the low con-

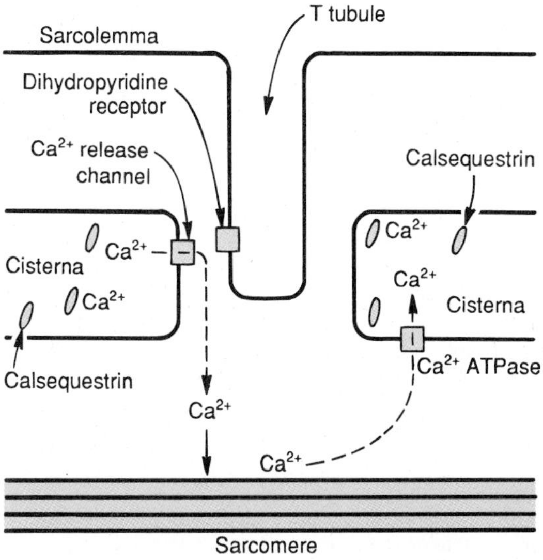

Figure 58–8. Diagram of the relationships among the sarcolemma (plasma membrane), a T tubule, and two cisternae of the sarcoplasmic reticulum of skeletal muscle (not to scale). The T tubule extends inward from the sarcolemma. A wave of depolarization, initiated by a nerve impulse, is transmitted from the sarcolemma down the T tubule. It is then conveyed to the Ca^{2+} release channel (ryanodine receptor), perhaps by interaction between it and the dihydropyridine receptor (slow Ca^{2+} voltage channel), which are shown in close proximity. Release of Ca^{2+} from the Ca^{2+} release channel into the cytosol initiates contraction. Subsequently, Ca^{2+} is pumped back into the cisternae of the sarcoplasmic reticulum by the Ca^{2+} ATPase (Ca^{2+} pump) and stored there, in part bound to calsequestrin.

Depolarization of nerve
↓
Depolarization of skeletal muscle
↓
Depolarization of the transverse tubular membrane
↓
Charge movement of the slow Ca^{2+} voltage channel (the dihydropyridine receptor) of the transverse tubular membrane
↓
Opening of the Ca^{2+} release channel

Figure 58–9. Possible chain of events leading to opening of the Ca^{2+} release channel.

centration of Ca^{2+} in the sarcoplasm. Thus, the interaction of the myosin heads with F-actin is promoted. (2) The ATP-dependent detachment of myosin heads from F-actin cannot occur, and rigidity (contracture) sets in. The condition of **rigor mortis,** following death, is an extension of these events.

Muscle contraction is a delicate dynamic balance of the attachment and detachment of myosin heads to F-actin, subject to fine regulation via the nervous system.

Mutations in the Gene Encoding the Ca^{2+} Release Channel Are One Cause of Human Malignant Hyperthermia

Some genetically predisposed patients experience a severe reaction, designated malignant hyperthermia, on exposure to certain anesthetics (eg, halothane) and depolarizing skeletal muscle relaxants (eg, succinylcholine). The reaction consists primarily of rigidity of skeletal muscles, hypermetabolism, and high fever. A high cytosolic concentration of Ca^{2+} in skeletal muscle is a major factor in its causation. Unless malignant hyperthermia is recognized and treated immediately, patients may die acutely of ventricular fibrillation or survive to succumb subsequently from other serious complications. Appropriate treatment is to stop the anesthetic and administer the drug dantrolene intravenously. **Dantrolene** (a hydantoin derivative) is a skeletal muscle relaxant that acts to inhibit release of Ca^{2+} from the sarcoplasmic reticulum into the cytosol, thus preventing the increase of cytosolic Ca^{2+} found in malignant hyperthermia.

Malignant hyperthermia also occurs in swine. Susceptible animals homozygous for malignant hyperthermia respond to stress with a fatal reaction **(porcine stress syndrome)** similar to that exhibited by humans. If the reaction occurs prior to slaughter, it affects the quality of the pork adversely, resulting in an inferior product. Both events can result in considerable financial loss to the swine industry.

The finding of a high level of cytosolic Ca^{2+} in muscle in malignant hyperthermia suggested that the condition might be caused by abnormalities of the Ca^{2+} ATPase or of the Ca^{2+} release channel. No abnormalities have been detected in the former, but sequencing of cDNAs for the latter protein has proved insightful, particularly in swine. All cDNAs from swine with malignant hyperthermia so far examined have shown a **substitution of T for C1843,** resulting in the substitution of Cys for Arg^{615} in the Ca^{2+} release channel. The mutation affects the function of the channel in that it opens more easily and remains open longer; the net result is massive release of Ca^{2+} into the cytosol, ultimately causing sustained muscle contraction.

The picture is more complex in humans, since malignant hyperthermia exhibits genetic heterogeneity. Members of a number of families who suffer from malignant hyperthermia have not shown genetic linkage to the RYR1 gene. Some humans susceptible to malignant hyperthermia already have been found to exhibit the same mutation found in swine. However, others may have mutations at different loci in the RYR1 gene, and still others may have mutations in genes for other proteins expressed in muscle and presumably involved in other aspects of Ca^{2+} metabolism or closely related metabolic processes. Figure 58–10 summarizes the probable chain of events in malignant hyperthermia. The major promise of these findings is that, once additional mutations are detected, it will be possible to **screen,** using suitable DNA probes, for individuals at risk of developing malignant hyperthermia during anesthesia. Current screening tests (eg, the in vitro caffeine-halothane test) are relatively unreliable. Affected individuals could then be given alternative anesthetics, which would not endanger their lives. It should also be possible, if desired, to eliminate malignant hyperthermia from swine populations using suitable breeding practices.

TITIN IS THE LARGEST PROTEIN KNOWN, AND IT AND OTHER ACCESSORY PROTEINS ARE IMPORTANT IN MUSCLE STRUCTURE AND FUNCTION

A number of accessory proteins play various roles in the structure and function of muscle (Table 58–1). **Titin** forms a third filament system, perhaps allowing muscle to spring back into shape after being stretched. It reaches from the Z line to the M line, with a single filament measuring over 1 mm in length. Part of the molecule overlaps the A band (where actin and myosin overlap) and part overlaps the I band (mainly actin). Titin is the largest protein known, the human heart isoform comprising 26,926 amino acids with a molecular mass of 2993 kDa. Over 90% of its mass is made up of a repetitive structure composed of 244

Mutations in the RYR1 gene (or probably in genes for other proteins involved in certain aspects of muscle metabolism)

↓

Altered Ca^{2+} release channel protein (RyR1) (eg, substitution of Cys for Arg^{615})

↓

Mutated channel opens more easily and stays open longer, thus flooding the cytosol with Ca^{2+}

↓

High intracellular levels of Ca^{2+} stimulate sustained muscle contraction (rigidity); high Ca^{2+} also stimulates breakdown of glycogen, glycolysis, and aerobic metabolism (resulting in excess production of heat)

Figure 58–10. Simplified scheme of the causation of malignant hyperthermia.

Table 58–1. Some accessory proteins of muscle.

Protein	Location	Comment or Function
Titin	Reaches from the Z line to the M line	Largest protein in body. Role in relaxation of muscle.
Nebulin	From Z line along length of actin filaments	May regulate assembly and length of actin filaments.
α-Actinin	Anchors actin to Z lines	Stabilizes actin filaments.
Desmin	Lies alongside actin filaments	Attaches to plasma membrane (plasma-lemma).
Dystrophin	Attached to plasma-lemma	Deficient in Duchenne muscular dystrophy.

copies of 100-residue repeats. These repeats consist of 112 immunoglobulin-like and 132 fibronectin-3-like (FN-3-like) domains. Titin appears to be involved in the assembly of muscle, acting as a template for insertion of other accessory A-band proteins. Phosphorylation of specific Ser residues located at opposite ends of titin may help to control its insertion into its correct position during myogenesis. Titin is also involved in regulating resting tension. A central region contains PEVK (Pro,Glu,Val,Lys) repeats and the tandem immunoglobulin domains. These may act in parallel, like two springs in series. Supporting the role of the PEVK region in determining relaxation, in stiff cardiac muscle it was found to measure 163 amino acids in length, whereas in more elastic skeletal muscle it measured about 2000 amino acids in length. Isoforms of titin can be generated by differential splicing; these isoforms, with variable lengths of stretch zones, may explain the diversity of the length of sarcomeres in vertebrate striated muscles and also variations in resting tension.

Nebulin is another giant protein that extends from the Z line along the length of an actin filament, perhaps controlling the length of thin filaments. It is largely made up of repeating units of 35 amino acids that constitute actin-binding domains. α-Actinin plays an important role in muscle by anchoring actin to the Z lines (Z disks). Another important accessory protein is **desmin,** an intermediate filament (see below). These filaments bind myofibrils side by side and are involved in their attachment to the cell surface. Another protein involved in attachment to the plasmalemma is **dystrophin,** mutations in the gene for which are the fundamental causes of Duchenne muscular dystrophy (Case No. 6, Chapter 65).

CARDIAC MUSCLE RESEMBLES SKELETAL MUSCLE IN MANY RESPECTS

The general picture of muscle contraction in the heart resembles that of skeletal muscle. Cardiac muscle, like skeletal muscle, is striated and uses the actin-myosin-tropomyosin-troponin system described above. Unlike skeletal muscle, cardiac muscle exhibits intrinsic rhythmicity, and individual myocytes communicate with each other because of its syncytial nature. The T tubular system is more developed in cardiac muscle, whereas the **sarcoplasmic reticulum** is less extensive and consequently the intracellular supply of Ca^{2+} for contraction is less. Cardiac muscle thus relies on extracellular Ca^{2+} for contraction; if isolated cardiac muscle is deprived of Ca^{2+}, it ceases to beat within approximately 1 minute, whereas skeletal muscle can continue to contract without an extracellular source of Ca^{2+}. Cyclic AMP plays a more prominent role in cardiac than in skeletal muscle. It modulates intracellular levels of Ca^{2+} through the activation of protein kinases; these enzymes phosphorylate various transport proteins in the sarcolemma and sarcoplasmic reticulum and also in the troponin-tropomyosin regulatory complex, affecting intracellular levels of Ca^{2+} or responses to it. There is a rough correlation between the phosphorylation of TpI and the increased contraction of cardiac muscle induced by catecholamines. This may account for the inotropic effects (increased contractility) of β-adrenergic compounds on the heart. Some differences among skeletal, cardiac, and smooth muscle are summarized in Table 58–2.

Ca²⁺ Enters Myocytes via Ca²⁺ Channels and Leaves via the Na⁺-Ca²⁺ Exchanger and the Ca²⁺ ATPase

As stated above, extracellular Ca^{2+} plays an important role in contraction of cardiac muscle but not in skeletal muscle. This means that Ca^{2+} both enters and leaves myocytes in a regulated manner. We shall briefly consider three transmembrane proteins that play roles in this process.

A. Ca²⁺ Channels: Ca^{2+} enters myocytes via these channels, which allow entry only of Ca^{2+} ions. The major portal of entry is the L-type (long-duration current, large conductance) or slow Ca^{2+} channel, which is voltage-gated, opening during depolarization induced by spread of the cardiac action potential and closing when the action potential declines. These channels are equivalent to the dihydropyridine receptors of skeletal muscle (Figure 58–8). Slow Ca^{2+} channels are regulated by cAMP-dependent protein kinases (stimulatory) and cGMP-protein kinases (inhibitory) and are blocked by so-called calcium channel blockers (eg, verapamil). Fast (or T, transient) Ca^{2+} channels are also present in the plasmalemma, though in much lower numbers; they probably contribute to the early phase of increase of myoplasmic Ca^{2+}.

The resultant increase of Ca^{2+} in the myoplasm acts on the Ca^{2+} release channel of the sarcoplasmic reticulum to open it. This is called Ca^{2+}-induced Ca^{2+} release (CICR). It is estimated that approxi-

Table 58–2. Some differences between skeletal, cardiac, and smooth muscle.

Skeletal Muscle	Cardiac Muscle	Smooth Muscle
Striated.	Striated.	Nonstriated.
No syncytium.	Syncytial.	Syncytial.
Small T tubules.	Large T tubules.	Generally rudimentary T tubules.
SR well developed and Ca^{2+} pump acts rapidly.	SR present and Ca^{2+} pump acts relatively rapidly.	SR often rudimentary and Ca^{2+} pump acts slowly.
Plasmalemma lacks many hormone receptors.	Plasmalemma contains a variety of receptors (eg, α- and β-adrenergic).	Plasmalemma contains a variety of receptors (eg, α- and β-adrenergic).
Nerve impulse initiates contraction.	Has intrinsic rhythmicity.	Contraction initiated by nerve impulses, hormones, etc.
ECF Ca^{2+} not important for contraction.	ECF Ca^{2+} important for contraction.	ECF Ca^{2+} important for contraction.
Troponin system present.	Troponin system present.	Lacks troponin system; uses regulatory head of myosin.
Caldesmon not involved.	Caldesmon not involved.	Caldesmon is important regulatory protein.
Very rapid cycling of the cross-bridges.	Relatively rapid cycling of the cross-bridges.	Slow cycling of the cross-bridges permits slow prolonged contraction and less utilization of ATP.

SR, sarcoplasmic reticulum; ECF, extracellular fluid.

mately 10% of the Ca^{2+} involved in contraction enters the cytosol from the extracellular fluid and 90% from the sarcoplasmic reticulum. However, the former 10% is important, as the rate of increase of Ca^{2+} in the myoplasm is important, and entry via the Ca^{2+} channels contributes appreciably to this.

B. Ca^{2+}-Na^+ Exchanger: This is the principal route of exit of Ca^{2+} from myocytes. In resting myocytes, it helps to maintain a low level of free intracellular Ca^{2+} by exchanging one Ca^{2+} for three Na^+. The energy for the uphill movement of Ca^{2+} out of the cell comes from the downhill movement of Na^+ into the cell from the plasma. This exchange contributes to relaxation but may run in the reverse direction during excitation. Because of the Ca^{2+}-Na^+ exchanger, anything that causes intracellular Na^+ (Na^+_i) to rise will secondarily cause Ca^{2+}_i to rise, causing more forceful contraction. This is referred to as a positive inotropic effect. One example is when the drug digitalis is used to treat heart failure. Digitalis inhibits the sarcolemmal Na^+-K^+ ATPase, diminishing exit of Na^+ and thus increasing Na^+_i. This in turn causes Ca^{2+} to increase, via the Ca^{2+}-Na^+ exchanger. The increased Ca^{2+}_i results in increased force of cardiac contraction, of benefit in heart failure.

C. Ca^{2+} ATPase: This Ca^{2+} pump, situated in the sarcolemma, also contributes to Ca^{2+} exit but is believed to play a relatively minor role as compared with the Ca^{2+}-Na^+ exchanger.

It should be noted that there are a variety of **ion channels** (Chapter 43) in most cells, for Na^+, K^+, Ca^{2+}, etc. Many of them have been cloned in recent years and their dispositions in their respective membranes worked out (number of times each one crosses its membrane, location of the actual ion transport site in the protein, etc). They can be classified as indicated in Table 58–3. Cardiac muscle is rich in ion channels, and they are also important in skeletal mus-

cle. Mutations in genes encoding ion channels have been shown to be responsible for a number of relatively rare conditions affecting muscle, such as hyperkalemic periodic paralysis (sodium channel), myotonia congenita (a chloride channel, the ClC-1 channel), and hypokalemic periodic paralysis (the dihydropyridine receptor).

Inherited Cardiomyopathies Are Due to Disorders of Cardiac Energy Metabolism or to Abnormal Myocardial Proteins

An inherited cardiomyopathy is any structural or functional abnormality of the ventricular myocardium due to an inherited cause. There are nonheritable types of cardiomyopathy, but these will not be described here. As shown in Table 58–4, the causes of inherited cardiomyopathies fall into two broad classes: (1) disorders of cardiac energy metabolism, mainly reflecting mutations in genes encoding enzymes or proteins involved in fatty acid oxidation (a major source of energy for the myocardium) and oxidative phosphorylation; and (2) mutations in genes encoding proteins involved in or affecting myocardial contraction, such as myosin, tropomyosin,

Table 58–3. Major types of ion channels found in cells.

Type	Comment
External ligand-gated	Open in response to a specific extracellular molecule, eg, acetylcholine.
Internal ligand-gated	Open or close in response to a specific intracellular molecule, eg, a cyclic nucleotide.
Voltage-gated	Open in response to a change in membrane potential, eg, Na^+, K^+, and Ca^{2+} channels in heart.
Mechanically gated	Open in response to change in mechanical pressure.

Table 58–4. Biochemical causes of inherited cardiomyopathies.[1,2]

Inborn errors of fatty acid oxidation
  Affecting carnitine entry into cells and mitochondria
  Affecting certain enzymes of fatty acid oxidation
Disorders of mitochondrial oxidative phosphorylation
  Proteins encoded by mitochondrial genes
  Proteins encoded by nuclear genes
Abnormalities of myocardial contractile and structural proteins
  β-Myosin heavy chains, troponin, tropomyosin, dystrophin

[1]Based on Kelly DP, Strauss AW: Inherited cardiomyopathies. N Engl J Med 1994;330:913.
[2]Mutations (eg, point mutations, or in some cases deletions) in the genes (nuclear or mitochondrial) encoding various proteins, enzymes, or tRNA molecules are the fundamental causes of the inherited cardiomyopathies. Some conditions are mild, whereas others are severe and may be part of a syndrome affecting other tissues (eg, MELAS; see Chapter 64).

troponin, and dystrophin. Mutations in the genes encoding these latter proteins cause familial hypertrophic cardiomyopathy, which will now be discussed.

Mutations in the Cardiac β-Myosin Heavy-Chain Gene Are One Cause of Familial Hypertrophic Cardiomyopathy

Familial hypertrophic cardiomyopathy is one of the most frequent hereditary cardiac diseases. Patients exhibit hypertrophy—often massive—of one or both ventricles, starting early in life, and not related to any extrinsic cause such as hypertension. Most cases are transmitted in an autosomal dominant manner; the rest are sporadic. Until recently, its cause was obscure.

The beginning of a molecular understanding of familial hypertrophic cardiomyopathy dawned in 1989 when genetic linkage of the disorder to a polymorphism on the long arm of chromosome 14 was found in an affected family by use of a battery of RFLP markers. Previous work had shown that two genes encoding the α and β isoforms of cardiac myosin heavy chains were situated in tandem near the site of the polymorphism, thus indicating that they were candidate genes. The β isoform predominates in human ventricular muscle. Subsequent work revealed that a missense mutation (ie, substitution of one amino acid by another) in the β-myosin heavy chain gene was probably responsible for familial hypertrophic cardiomyopathy in the family under investigation. Additional studies have shown a number of missense mutations in the gene, all coding for highly conserved residues. Some individuals have shown other mutations, such as formation of an α/β-myosin heavy chain hybrid gene. Patients with familial hypertrophic cardiomyopathy can show great variation in clinical picture. This in part reflects genetic het-

erogeneity; ie, mutation in a number of other genes may also cause familial hypertrophic cardiomyopathy. In addition, mutations at different sites in the gene for β-myosin heavy chain may affect the function of the protein to a greater or lesser extent. The missense mutations are clustered in the head and head-rod regions of myosin heavy chain. One hypothesis is that the mutant polypeptides ("poison polypeptides") cause formation of abnormal myofibrils, eventually resulting in compensatory hypertrophy. Interestingly, none of the mutations affect the three major functional domains of the protein (ATPase activity, actin binding, and myosin light chain binding), perhaps because these would be lethal. Some mutations alter the charge of the amino acid (eg, substitution of arginine for glutamine), presumably affecting the conformation of the protein more markedly and thus affecting its function. Patients with these mutations have a significantly shorter life expectancy than patients in whom the mutation produced no alteration in charge. Thus, definition of the precise mutations involved in the genesis of familial hypertrophic cardiomyopathy may prove to be of important prognostic value; it can be accomplished by appropriate use of the polymerase chain reaction on genomic DNA obtained from one sample of blood lymphocytes. Figure 58–11 is a simplified scheme of the events causing familial hypertrophic cardiomyopathy.

Ca²⁺ Also Regulates Contraction of Smooth Muscle

While all muscles contain actin, myosin, and tropomyosin, only vertebrate striated muscles contain the troponin system. Thus, the mechanisms that regulate contraction must differ in various contractile systems.

Smooth muscles have molecular structures similar to those in striated muscle, but the sarcomeres are not

Predominantly missense mutations in the β-myosin heavy-chain gene on chromosome 14
↓
Mutant polypeptide chains ("poison polypeptides") that lead to formation of defective myofibrils
↓
Compensatory hypertrophy of one or both cardiac ventricles
↓
Cardiomegaly and various cardiac signs and symptoms, including sudden death

Figure 58–11. Simplified scheme of the causation of familial hypertrophic cardiomyopathy due to mutations in the gene encoding β-myosin heavy chain. Mutations in genes encoding other proteins, such as troponin, tropomyosin, and dystrophin, can also cause this condition.

aligned so as to generate the striated appearance. Smooth muscles contain α-actinin and tropomyosin molecules, as do skeletal muscles. They do **not** have the troponin system, and the light chains of smooth muscle myosin molecules differ from those of striated muscle myosin. However, like striated muscle, smooth muscle contraction is regulated by Ca^{2+}.

Phosphorylation of Myosin p-Light Chains Initiates Contraction of Smooth Muscle

When smooth muscle myosin is bound to F-actin in the absence of other muscle proteins such as tropomyosin, there is no detectable ATPase activity. This absence of ATPase is quite unlike the situation described for striated muscle myosin and F-actin, which has abundant ATPase activity. Smooth muscle myosin contains a light chain (p-light chain) that prevents the binding of the myosin head to F-actin. The p-light chain must be phosphorylated before it allows F-actin to activate myosin ATPase. The ATPase activity then attained hydrolyzes ATP about tenfold more slowly than the corresponding activity in skeletal muscle. The phosphate on the myosin light chains may form a chelate with the Ca^{2+} bound to the tropomyosin-TpC-actin complex, leading to an increased rate of formation of cross-bridges between the myosin heads and actin. The phosphorylation of p-light chains commences the attachment-detachment contraction cycle of smooth muscle.

Myosin Light-Chain Kinase Is Activated by Calmodulin·4Ca^{2+} and Then Phosphorylates the p-Light Chain

Smooth muscle sarcoplasm contains a myosin light chain kinase that is calcium-dependent. The Ca^{2+} activation of myosin light chain kinase requires binding of **calmodulin·4Ca^{2+}** to its kinase subunit (Figure 58–12). The calmodulin·4Ca^{2+}-activated light chain kinase phosphorylates the p-light chain, which then ceases to inhibit the myosin–F-actin interaction. The contraction cycle then begins.

Smooth Muscle Relaxes When the Concentration of Ca^{2+} Falls Below 10^{-7} Molar

Relaxation of smooth muscle occurs when (1) sarcoplasmic Ca^{2+} falls below 10^{-7} mol/L. The Ca^{2+} dissociates from calmodulin, which in turn dissociates from the myosin light chain kinase, (2) inactivating the kinase. (3) No new phosphates are attached to the p-light chain, and light chain protein phosphatase, which is continually active and calcium-independent, removes the existing phosphates from the p-light chain. (4) Dephosphorylated myosin p-light chain then inhibits the binding of myosin heads to F-actin and the ATPase activity. (5) The myosin head detaches from the F-actin in the presence of ATP, but it

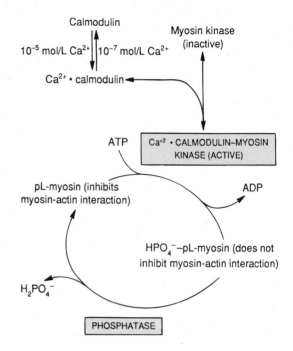

Figure 58–12. Regulation of smooth muscle contraction by Ca^{2+}. (Adapted from Adelstein RS, Eisenberg R: Regulation and kinetics of actin-myosin ATP interaction. Annu Rev Biochem 1980;49:921.)

cannot reattach because of the presence of dephosphorylated p-light chain; hence, relaxation occurs.

Table 58–5 summarizes and compares the regulation of actin-myosin interactions (activation of myosin ATPase) in striated and smooth muscles.

The myosin light-chain kinase is not directly affected or activated by cAMP. However, the usual cAMP-activated protein kinase (Chapter 44) can phosphorylate the myosin light-chain kinase (not the p-light chain itself). The phosphorylated myosin light-chain kinase exhibits a significantly lower affinity for calmodulin·Ca^{2+} and thus is less sensitive to activation. Accordingly, an increase in cAMP dampens the contraction response of smooth muscle to a given elevation of sarcoplasmic Ca^{2+}. This molecular mechanism can explain the relaxing effect of β-adrenergic stimulation on smooth muscle.

Another protein that appears to play a Ca^{2+}-dependent role in the regulation of smooth muscle contraction is caldesmon (87 kDa). This protein is ubiquitous in smooth muscle and is also found in nonmuscle tissue. At low concentrations of Ca^{2+}, it binds to tropomyosin and actin. This prevents interaction of actin with myosin, keeping muscle in a relaxed state. At higher concentrations of Ca^{2+}, Ca^{2+}-calmodulin binds caldesmon, releasing it from actin. The latter is then free to bind to myosin, and contraction can occur. Caldesmon is also subject to phosphorylation-dephosphorylation; when phosphory-

Table 58–5. Actin-myosin interactions in striated and smooth muscle.

	Striated Muscle	Smooth Muscle (and Nonmuscle Cells)
Proteins of muscle filaments	Actin Myosin Tropomyosin Troponin (TpI, TpT, TpC)	Actin Myosin[1] Tropomyosin
Spontaneous interaction of F-actin and myosin alone (spontaneous activation of myosin ATPase by F-actin)	Yes	No
Inhibitor of F actin–myosin interaction (inhibitor of F actin–dependent activation of ATPase)	Troponin system (TpI)	Unphosphorylated myosin p-light chain
Contraction activated by	Ca^{2+}	Ca^{2+}
Direct effect of Ca^{2+}	$4Ca^{2+}$ bind to TpC	$4Ca^{2+}$ bind to calmodulin
Effect of protein-bound Ca^{2+}	TpC · $4Ca^{2+}$ antagonizes TpI inhibition of F actin–myosin interaction (allows F-actin activation of ATPase)	Calmodulin · $4Ca^{2+}$ activates myosin light chain kinase that phosphorylates myosin p-light chain. The phosphorylated p-light chain no longer inhibits F actin–myosin interaction (allows F-actin activation of ATPase).

[1]Light chains of myosin are different in striated and smooth muscles.

lated, it cannot bind actin, again freeing the latter to interact with myosin. Caldesmon may also participate in organizing the structure of the contractile apparatus in smooth muscle. Many of its effects have been demonstrated in vitro, and its physiologic significance is still under investigation.

Nitric Oxide Relaxes the Smooth Muscle of Blood Vessels and Also Has Many Other Important Biologic Functions

Acetylcholine is a vasodilator that acts by causing relaxation of the smooth muscle of blood vessels. However, it does not act directly on smooth muscle. A key observation was that if endothelial cells were stripped away from underlying smooth muscle cells, acetylcholine no longer exerted its vasodilator effect. This finding indicated that vasodilators such as acetylcholine initially interact with the endothelial cells of small blood vessels via receptors. The receptors are coupled to the phosphoinositide cycle (Chapter 44), leading to the intracellular release of Ca^{2+} through the action of inositol triphosphate. In turn, the elevation of Ca^{2+} leads to the liberation of endothelium-derived relaxing factor (EDRF), which diffuses into the adjacent smooth muscle. There, it reacts with the heme moiety of a soluble guanylyl cyclase, resulting in activation of the latter, with a consequent elevation of intracellular levels of cGMP (Figure 58–13). This in turn stimulates the activities of certain cGMP-dependent protein kinases, which probably phosphorylate specific muscle proteins, causing relaxation; however, the details have not been clarified. The important coronary artery vasodilator nitroglycerin, widely used to relieve angina

pectoris, acts to increase intracellular release of EDRF and thus of cGMP.

Unexpectedly, EDRF was found to be the gas nitric oxide (NO). NO is formed by the action of the enzyme NO synthase, which is cytosolic. The endothelial and neuronal forms of NO synthase are activated by Ca^{2+} (Table 58–6). The substrate is arginine, and the products are citrulline and NO:

$$\text{Arginine} \xrightarrow{\boxed{\text{NO synthase}}} \text{Citrulline} + \text{NO}$$

NO synthase catalyzes a five-electron oxidation of an amidine nitrogen of arginine. L-Hydroxyarginine is an intermediate that remains tightly bound to the enzyme. NO synthase is a very complex enzyme, employing five redox cofactors: NADPH, FAD, FMN, heme, and tetrahydrobiopterin. NO can also be formed from nitrite, derived from vasodilators such as glyceryl trinitrate during their metabolism. NO has a very short half-life (approximately 3–4 seconds) in tissues because it reacts with oxygen and superoxide. The product of the reaction with superoxide is peroxynitrite ($ONOO^-$), which decomposes to form the highly reactive $OH^{\cdot}$ radical. NO is inhibited by hemoglobin and other heme proteins, which bind it tightly. Chemical inhibitors of NO synthase are now available that can markedly decrease formation of NO. Administration of such inhibitors to animals and humans leads to vasoconstriction and a marked elevation of blood pressure, indicating that NO is of major importance in the maintenance of blood pressure in vivo. Another important cardiovascular effect is that by increasing synthesis of cGMP, it acts as an inhibitor of platelet aggregation (Chapter 59).

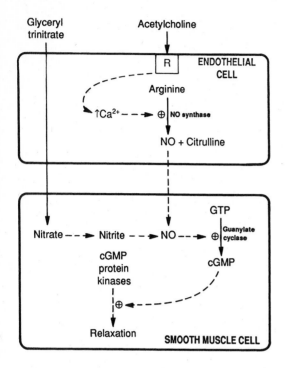

Figure 58–13. Diagram showing formation in an endothelial cell of nitric oxide (NO) from arginine in a reaction catalyzed by NO synthase. Interaction of an agonist (eg, acetylcholine) with a receptor (R) probably leads to intracellular release of Ca^{2+} via inositol triphosphate generated by the phosphoinositide pathway, resulting in activation of NO synthase. The NO subsequently diffuses into adjacent smooth muscle, where it leads to activation of guanylyl cyclase, formation of cGMP, stimulation of cGMP-protein kinases, and subsequent relaxation. The vasodilator nitroglycerin is shown entering the smooth muscle cell, where its metabolism also leads to formation of NO.

Since the discovery of the role of NO as a vasodilator, there has been intense experimental interest in this substance. It has turned out to have a variety of physiologic roles, involving virtually every tissue of the body (Table 58–7). Three major isoforms of NO synthase have been identified, each of which has been cloned, and the chromosomal locations of their genes in humans have been determined. Gene knockout experiments, using homologous recombination (Chapter 42), have been performed on each of the three isoforms and have helped establish some of the postulated functions of NO.

To summarize, research in the past decade has shown that NO occupies an important role in many physiologic and pathologic processes.

SEVERAL MECHANISMS REPLENISH STORES OF ATP IN MUSCLE

The ATP required as the constant energy source for the contraction-relaxation cycle of muscle can be generated (1) by glycolysis, using blood glucose or muscle glycogen, (2) by oxidative phosphorylation, (3) from creatine phosphate, and (4) from two molecules of ADP in a reaction catalyzed by adenylyl kinase (Figure 58–14). The amount of ATP in skeletal muscle is only sufficient to provide energy for contraction for 1–2 seconds, so that ATP must be constantly renewed from one or more of the above sources, depending upon metabolic conditions. As discussed below, there are at least two distinct types of fibers in skeletal muscle, one predominantly active in aerobic conditions and the other in anaerobic conditions; not unexpectedly, they use each of the above sources of energy to different extents.

Table 58–6. Summary of the nomenclature of the NO synthases and of the effects of knockout of their genes in mice.[1]

Subtype	Name[2]	Comments	Result of Gene Knockout in Mice[3]
1	nNOS	Activity depends on elevated Ca^{2+}. First identified in neurons. Calmodulin-activated.	Pyloric stenosis, resistant to vascular stroke, aggressive sexual behavior (males).
2	iNOS[4]	Independent of elevated Ca^{2+}. Prominent in macrophages.	More susceptible to certain types of infection.
3	eNOS	Activity depends on elevated Ca^{2+}. First identified in endothelial cells.	Elevated mean blood pressure.

[1]Adapted from Snyder SH: No endothelial NO. Nature 1995;377:196.
[2]n, neuronal; i, inducible; e, endothelial.
[3]Gene knockouts were performed by homologous recombination in mice (Chapter 42). The enzymes are characterized as neuronal, inducible (macrophage), and endothelial because these were the sites in which they were first identified. However, all three enzymes have been found in other sites, and the neuronal enzyme is also inducible. Each enzyme has been cloned, and its chromosomal location in humans has been determined.
[4]iNOS is Ca^{2+}-independent but binds calmodulin very tightly.

Table 58–7. Some physiologic functions and pathologic involvements of nitric oxide (NO).

- Vasodilator, important in regulation of blood pressure
- Involved in penile erection
- Neurotransmitter in the brain and peripheral autonomic nervous system
- Role in long-term potentiation
- Role in neurotoxicity
- Low level of NO involved in causation of pylorospasm in infantile hypertrophic pyloric stenosis
- May have role in relaxation of skeletal muscle
- May constitute part of a primitive immune system
- Inhibits adhesion, activation, and aggregation of platelets

Skeletal Muscle Contains Large Supplies of Glycogen

The sarcoplasm of skeletal muscle contains large stores of glycogen, located in granules close to the I bands. The release of glucose from glycogen is dependent on a specific muscle glycogen phosphorylase (Chapter 20), which can be activated by Ca^{2+}, epinephrine, and AMP. To generate glucose 6-phosphate for glycolysis in skeletal muscle, glycogen phosphorylase b must be activated to phosphorylase a via phosphorylation by phosphorylase b kinase (Chapter 20). Ca^{2+} promotes the activation of phosphorylase b kinase, also by phosphorylation. Thus, Ca^{2+} both initiates muscle contraction and activates a pathway to provide necessary energy. The hormone epinephrine also activates glycogenolysis in muscle. AMP, produced by breakdown of ADP during muscular exercise, can also activate phosphorylase b without causing phosphorylation. Muscle glycogen phosphorylase b is inactive in **McArdle's disease,** one of the glycogen storage diseases (Chapter 20).

Under Aerobic Conditions, Muscle Generates ATP Mainly by Oxidative Phosphorylation

Synthesis of ATP via oxidative phosphorylation requires a supply of oxygen. Muscles that have a high demand for oxygen as a result of sustained contraction (eg, to maintain posture) store it attached to the heme moiety of **myoglobin** (Chapter 7). Because of the heme moiety, muscles containing myoglobin are red, whereas muscles with little or no myoglobin are white. Glucose, derived from the blood glucose or from endogenous glycogen, and fatty acids derived from the triacylglycerols of adipose tissue, are the principal substrates used for aerobic metabolism in muscle.

Creatine Phosphate Constitutes a Major Energy Reserve in Muscle

Creatine phosphate prevents the rapid depletion of ATP by providing a readily available high-energy phosphate, which can be used to regenerate ATP from ADP. Creatine phosphate is formed from ATP and creatine (Figure 58–14) at times when the muscle is relaxed and demands for ATP are not so great. The enzyme catalyzing the phosphorylation of creatine is creatine kinase (CK), a muscle-specific enzyme with clinical utility in the detection of acute or chronic diseases of muscle.

Adenylyl Kinase Interconverts Adenosine Mono-, Di-, and Triphosphates

Adenylyl kinase catalyzes formation of one molecule of ATP and one of AMP from two molecules of ATP. This reaction (Figure 58–14) is coupled with

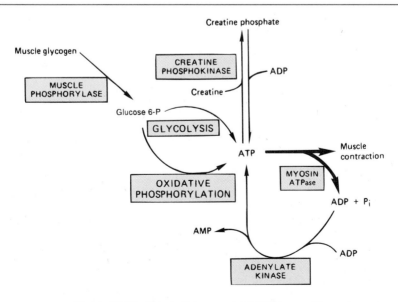

Figure 58–14. The multiple sources of ATP in muscle.

the hydrolysis of ATP by myosin ATPase during muscle contraction.

The AMP produced above can be deaminated by AMP deaminase, forming IMP and ammonia:

$$AMP + H_2O \rightarrow IMP + NH_3$$

Thus, muscle is a source of ammonia (also produced by the reaction catalyzed by adenosine deaminase, described below) to be disposed of by the urea cycle in the liver (Chapter 31). AMP can also be acted upon by 5′-nucleotidase, hydrolyzing the phosphate and producing adenosine:

$$AMP + H_2O \rightarrow Adenosine + P_i$$

Adenosine acts as a vasodilator, increasing the blood flow and supply of nutrients to muscle. Adenosine, in turn, is a substrate for adenosine deaminase, producing inosine and ammonia:

$$Adenosine + H_2O \rightarrow Inosine + NH_3$$

AMP, P_i, and NH_3 formed during various of the above reactions activate phosphofructokinase-1 (PFK-1), thus increasing the rate of glycolysis in rapidly exercising muscle, such as during a sprint.

SKELETAL MUSCLE CONTAINS SLOW (RED) AND FAST (WHITE) TWITCH FIBERS

Different types of fibers have been detected in skeletal muscle. One classification subdivides them into type I (slow twitch), type IIA (fast twitch-oxidative), and type IIB (fast twitch-glycolytic). For the sake of simplicity, we shall consider only two types: type I (slow twitch, oxidative) and type II (fast twitch, glycolytic) (Table 58–8). The type I fibers are red because they contain myoglobin and mitochondria; their metabolism is aerobic, and they maintain relatively sustained contractions. The type II fibers, lacking myoglobin and containing few mitochondria, are white: they derive their energy from anaerobic glycolysis and exhibit relatively short durations of contraction. The proportion of these two types of

fibers varies among the muscles of the body, depending on function (eg, whether or not a muscle is involved in sustained contraction, such as maintaining posture). The proportion also varies with training; for example, the number of type I fibers in certain leg muscles increases in athletes training for marathons, whereas the number of type II fibers increases in sprinters.

A Sprinter Uses Creatine Phosphate and Anaerobic Glycolysis to Make ATP, Whereas a Marathon Runner Uses Oxidative Phosphorylation

In view of the two types of fibers in skeletal muscle and of the various energy sources described above, it is of interest to compare their involvement in a sprint (eg, 100 meters) and in the marathon (42.2 km; just over 26 miles) (Table 58–9).

The major sources of energy in the 100-m sprint are creatine phosphate (first 4–5 seconds) and then anaerobic glycolysis, using muscle glycogen as the source of glucose. The two main sites of metabolic control are at glycogen phosphorylase and at PFK-1. As discussed above, the former is activated by Ca^{2+} (released from the sarcoplasmic reticulum during contraction), epinephrine, and AMP. PFK-1 is activated by AMP, P_i, and NH_3. Attesting to the efficiency of these processes, the flux through glycolysis can increase as much as 1000-fold during a sprint.

In contrast, in the marathon, aerobic metabolism is the principal source of ATP. The major fuel sources are blood glucose and free fatty acids, largely derived from the breakdown of triacylglycerols in adipose tissue, stimulated by epinephrine. Hepatic glycogen is degraded to maintain the level of blood glucose. Muscle glycogen is also a fuel source, but it is degraded much more gradually than in a sprint. It has been calculated that the amounts of glucose in the blood, of glycogen in the liver, of glycogen in muscle, and of triacylglycerol in adipose tissue are sufficient to supply muscle with energy during a marathon for 4 minutes, 18 minutes, 70 minutes, and approximately 4000 minutes, respectively. However,

Table 58–8. Characteristics of type I and type II fibers of skeletal muscle.

	Type I Slow Twitch	Type II Fast Twitch
Myosin ATPase	Low	High
Energy utilization	Low	High
Mitochondria	Many	Few
Color	Red	White
Myoglobin	Yes	No
Contraction rate	Slow	Fast
Duration	Prolonged	Short

Table 58–9. Types of muscle fibers and major fuel sources used by a sprinter and by a marathon runner.

Sprinter (100 m)	Marathon Runner
Type II (glycolytic) fibers are used predominantly. Creatine phosphate is the major energy source during the first 4–5 seconds. Glucose derived from muscle glycogen and metabolized by anaerobic glycolysis is the major fuel source. Muscle glycogen is rapidly depleted.	Type I (oxidative) fibers are used predominantly. ATP is the major energy source throughout. Blood glucose and free fatty acids are the major fuel sources. Muscle glycogen is slowly depleted.

the rate of oxidation of fatty acids by muscle is slower than that of glucose, so that oxidation of glucose and of fatty acids are both major sources of energy in the marathon.

Muscle Fatigue, Carbohydrate Loading, Soda Loading, and Blood Doping Have Biochemical Explanations

Fatigue of muscles during exercise is a phenomenon that almost everyone has experienced. What is its cause? The primary cause is accumulation in muscle tissue not of lactate (due to anaerobic glycolysis) but rather of protons. This fact has been demonstrated by infusing lactate and observing that fatigue does not necessarily follow. Increase of protons (decreased pH) can affect the function of muscle in a number of ways, including the following: (1) lowering the V_{max} of PFK-1, (2) lessening the release of Ca^{2+} from the sarcoplasmic reticulum, (3) lessening the activity of the actomyosin ATPase, (4) possibly by affecting the conformation of some of the muscle proteins involved in contraction.

Carbohydrate loading (also known as "glycogen stripping" and "unloading-loading") has been popular among long-distance runners. The goal is to fill the muscles with the maximum amount of glycogen possible before a race. One possible way of achieving this is by taking very little carbohydrate in the diet for 3 days, further depleting the glycogen stores by a run to exhaustion, and then eating mostly carbohydrate during the last 3 days prior to the race.

Soda loading consists of ingesting sodium bicarbonate in an attempt to buffer the production of protons during exercise. It is unlikely to have any effect on the sprint, since most protons produced in that effort remain in the muscle. It may be of some benefit in the 800-m run, since protons can move out of muscle into the bloodstream during its duration. It would appear to be of little benefit in long-distance runs, which involve predominantly aerobic metabolism, with lesser production of lactate and protons.

Blood doping consists of administering samples of their own red blood cells to athletes just before a race; the samples (approximately 1 L) are withdrawn about 5 weeks prior to a race and preserved at low temperature. There is some evidence that this may be beneficial in long-distance runs, suggesting that performance is limited by the availability of oxygen.

SKELETAL MUSCLE CONSTITUTES THE MAJOR RESERVE OF PROTEIN IN THE BODY

In humans, skeletal muscle protein is the major nonfat source of stored energy. This explains the very large losses of muscle mass, particularly in adults, resulting from prolonged caloric undernutrition.

The study of tissue protein breakdown in vivo is difficult, because amino acids released during intracellular breakdown of proteins can be extensively reutilized for protein synthesis within the cell, or the amino acids may be transported to other organs where they enter anabolic pathways. However, actin and myosin are methylated by a posttranslational reaction, forming 3-methylhistidine. During intracellular breakdown of actin and myosin, 3-methylhistidine is released and excreted into the urine. The urinary output of the methylated amino acid provides a reliable index of the rate of myofibrillar protein breakdown in the musculature of human subjects.

Skeletal muscle is active in the degradation of certain amino acids as well as in the synthesis of others. In mammals, muscle appears to be the primary site of catabolism of the branched-chain amino acids. Muscle oxidizes leucine to CO_2 and converts the carbon skeletons of isoleucine and valine, as well as aspartate, asparagine, and glutamate, into amphibolic intermediates of the tricarboxylic acid cycle. The capacity of muscles to degrade branched-chain amino acids increases three- to fivefold during fasting and in diabetes.

Muscle also synthesizes and releases large amounts of alanine and glutamine. These compounds are synthesized utilizing amino groups that are generated in the breakdown of branched-chain amino acids, and the amino nitrogen is then transferred to α-ketoglutarate and to pyruvate by transamination. Glycolysis from exogenous glucose provides almost all of the pyruvate for synthesis of alanine. These reactions constitute the so-called glucose-alanine cycle, wherein alanine from muscle is utilized in hepatic gluconeogenesis while simultaneously bringing amino groups to the liver for removal as urea.

Features of muscle metabolism are summarized in Table 58–10.

THE CYTOSKELETON PERFORMS MULTIPLE CELLULAR FUNCTIONS

Nonmuscle cells perform mechanical work, including self-propulsion, morphogenesis, cleavage, endocytosis, exocytosis, intracellular transport, and changing cell shape. These cellular functions are carried out by an extensive intracellular network of filamentous structures constituting the cytoskeleton. The cell cytoplasm is not a sac of fluid, as once thought. Essentially all eukaryotic cells contain three types of filamentous structures: actin filaments (7–9.5 nm in diameter; also known as microfilaments), microtubules (25 nm), and intermediate filaments (10–12 nm). Each type of filament can be distinguished biochemically and by the electron microscope.

Table 58–10. Summary of major features of the biochemistry of skeletal muscle related to its metabolism.[1]

- Skeletal muscle functions under both aerobic (resting) and anaerobic (eg, sprinting) conditions, so both aerobic and anaerobic glycolysis operate, depending on conditions.
- Skeletal muscle contains myoglobin as a reservoir of oxygen.
- Skeletal muscle contains different types of fibers primarily suited to anaerobic (fast-twitch fibers) or aerobic (slow-twitch fibers) conditions.
- Actin, myosin, tropomyosin, troponin complex (TpT, TpI, and TpC), ATP, and Ca^{2+} are key constituents in relation to contraction.
- The Ca^{2+} ATPase, the Ca^{2+} release channel, and calsequestrin are proteins involved in various aspects of Ca^{2+} metabolism in muscle.
- Insulin acts on skeletal muscle to increase uptake of glucose.
- In the fed state, most glucose is used to synthesize glycogen, which acts as a store of glucose for use in exercise; "preloading" with glucose is used by some long-distance athletes to build up stores of glycogen.
- Epinephrine stimulates glycogenolysis in skeletal muscle, whereas glucagon does not because of absence of its receptors.
- Skeletal muscle cannot contribute directly to blood glucose because it does not contain glucose-6-phosphatase.
- Lactate produced by anaerobic metabolism in skeletal muscle passes to liver, which uses it to synthesize glucose, which can then return to muscle (the Cori cycle).
- Skeletal muscle contains phosphocreatine, which acts as an energy store for short-term (seconds) demands.
- Free fatty acids in plasma are a major source of energy, particularly under marathon conditions and in prolonged starvation.
- Skeletal muscle can utilize ketone bodies during starvation.
- Skeletal muscle is the principal site of metabolism of branched-chain amino acids, which are used as an energy source.
- Proteolysis of muscle during starvation supplies amino acids for gluconeogenesis.
- Major amino acids emanating from muscle are alanine (destined mainly for gluconeogenesis in liver and forming part of the glucose-alanine cycle) and glutamine (destined mainly for the gut and kidneys).

[1]This table brings together material from various chapters in this book.

Nonmuscle Cells Contain Actin That Forms Microfilaments

G-actin is present in most if not all cells of the body. With appropriate concentrations of magnesium and potassium chloride, it spontaneously polymerizes to form double helical F-actin filaments like those seen in muscle. There are at least two types of actin in nonmuscle cells: β-actin and γ-actin. Both types can coexist in the same cell and probably even copolymerize in the same filament. In the cytoplasm, F-actin forms microfilaments of 7–9.5 nm that frequently exist as bundles of a tangled-appearing meshwork. These bundles are prominent just underlying the plasma membrane of many cells and are there referred to as stress fibers. The stress fibers disappear as cell motility increases or upon malignant transformation of cells by chemicals or oncogenic viruses.

Although not organized as in muscle, actin filaments in nonmuscle cells interact with myosin to cause cellular movements.

Microtubules Contain α- and β-Tubulins

Microtubules, an integral component of the cellular cytoskeleton, consist of cytoplasmic tubes 25 nm in diameter and often of extreme length. Microtubules are necessary for the formation and function of the mitotic spindle and thus are present in all eukaryotic cells. They are also involved in the intracellular movement of endocytic and exocytic vesicles and form the major structural components of cilia and flagella. Microtubules are a major component of axons and dendrites, in which they maintain structure and participate in the axoplasmic flow of material along these neuronal processes.

Microtubules are cylinders of 13 longitudinally arranged protofilaments, each consisting of dimers of α-tubulin and β-tubulin, closely related proteins, each of approximately 50 kDa molecular mass. The tubulin dimers assemble into protofilaments and subsequently into sheets and then cylinders. A microtubule-organizing center, located around a pair of centrioles, nucleates the growth of new microtubules. A third species of tubulin, γ-tubulin, appears to play an important role in this assembly. GTP is required for assembly. A variety of proteins are associated with microtubules (microtubule-associated proteins [MAPs], one of which is tau) and play important roles in microtubule assembly and stabilization. Microtubules are in a state of dynamic instability, constantly assembling and disassembling. They exhibit polarity (plus and minus ends); this is important in their growth from centrioles and in their ability to direct intracellular movement. For instance, in axonal transport, the protein **kinesin,** with a myosin-like ATPase activity, uses hydrolysis of ATP to move vesicles down the axon toward the positive end of the microtubular formation. Flow of materials in the opposite direction, toward the negative end, is powered by **cytosolic dynein,** another protein with ATPase activity. Similarly, **axonemal dyneins** power ciliary and flagellar movement. Another protein, **dynamin,** uses GTP and is involved in endocytosis. Kinesins, dyneins, dynamin, and myosins are referred to as **molecular motors.**

An absence of dynein in cilia and flagella results in immotile cilia and flagella, leading to male sterility and chronic respiratory infection, a condition known as Kartagener's syndrome.

Certain drugs bind to microtubules and thus interfere with their assembly or disassembly. These include colchicine (used for treatment of acute gouty arthritis), vinblastine (a *Vinca* alkaloid used for treating certain types of cancer), paclitaxel (Taxol) (effective against ovarian cancer), and griseofulvin (an antifungal agent).

Intermediate Filaments Differ From Microfilaments and Microtubules

An intracellular fibrous system exists of filaments with an axial periodicity of 21 nm and a diameter of 8–10 nm that is intermediate between that of microfilaments (6 nm) and microtubules (23 nm). Four classes of intermediate filaments are found, as indicated in Table 58–11. They are all elongated, fibrous molecules, with a central rod domain, an amino terminal head and a carboxyl terminal tail. They form a structure like a rope, and the mature filaments are composed of tetramers packed together in a helical manner. They are important structural components of cells, and most are relatively stable components of the cytoskeleton, not undergoing rapid assembly and disassembly and not disappearing during mitosis, as do actin and many microtubular filaments. An important exception to this is provided by the lamins, which, subsequent to phosphorylation, disassemble at mitosis and reappear when it terminates.

Keratins form a large family, with about 30 members being distinguished. As indicated in Table 58–11, two major types of keratins are found; all individual keratins are heterodimers made up of one member of each class.

Vimentins are widely distributed in mesodermal cells, and desmin, glial fibrillary acidic protein, and peripherin are related to them. All members of the vimentin-like family can copolymerize with each other. Intermediate filaments are very prominent in nerve cells; neurofilaments are classified as low, medium, and high on the basis of their molecular masses. Lamins form a meshwork in apposition to the inner nuclear membrane. The distribution of intermediate filaments in normal and abnormal (eg, cancer) cells can be studied by the use of immunofluorescent techniques, using antibodies of appropriate specificities. These antibodies to specific intermediate filaments can also be of use to pathologists in helping to decide the origin of certain dedifferentiated malignant tumors. These tumors may still retain the type of intermediate filaments found in their cell of origin.

Three skin diseases (epidermolysis bullosa simplex, epidermolytic hyperkeratosis, and epidermolytic palmoplantar keratoderma), all characterized by blistering, have been found to be due to mutations in genes encoding various keratins. The blistering probably reflects a diminished capacity of various layers of the skin to resist mechanical stresses due to abnormalities in microfilament structure.

SUMMARY

The myofibrils of skeletal muscle contain thick and thin filaments. The thick filaments contain myosin. The thin filaments contain actin, tropomyosin, and the troponin complex (troponins T, I, and C). Actin binds to myosin, tropomyosin binds to actin, troponin T binds to tropomyosin and the other troponins, troponin I inhibits the actin-myosin interaction, and troponin C binds Ca^{2+}, a key player in the overall process of contraction. The sliding filament cross-bridge model is the foundation of current thinking about muscle contraction. The basis of this model is that the interdigitating filaments slide past one another during contraction and cross-bridges between myosin and actin generate and sustain the tension. The hydrolysis of ATP is used to drive movement of the filaments. ATP binds to myosin heads and is hydrolyzed to ADP and P_i by the ATPase activity of the actomyosin complex. Ca^{2+} plays a key role in the initiation of muscle contraction by binding to troponin C. In skeletal muscle, the sarcoplasmic reticulum regulates distribution of Ca^{2+} to the sarcomeres, whereas inflow of Ca^{2+} via Ca^{2+} channels in the sarcolemma is of major importance in cardiac and smooth muscle. Proteins of the sarcoplasmic reticulum with specialized roles to play in Ca^{2+} metabo-

Table 58–11. Classes of intermediate filaments of eukaryotic cells and their distributions.

Proteins	Molecular Mass	Distributions

Keratins
  Type I (acidic)
  Type II (basic)
	40–60 kDa
	50–70 kDa
		Epithelial cells, hair, nails

Vimentin-like
  Vimentin
  Desmin
  Glial fibrillary acid protein
  Peripherin
	54 kDa
	53 kDa
	50 kDa
	66 kDa
		Various mesenchymal cells
		Muscle
		Glial cells
		Neurons

Neurofilaments
  Low (L), medium (M), and
  high (H)[1]
	60–130 kDa	Neurons

Lamins A, B, and C	65–75 kDa	Nuclear laminae

[1]Refers to their molecular masses.

lism include Ca^{2+} ATPase, calsequestrin, and the Ca^{2+} release channel. Certain cases of malignant hyperthermia in humans are due to mutations in the gene encoding the Ca^{2+} release channel. A number of differences exist between skeletal and cardiac muscle; in particular, the latter contains a variety of receptors on its surface. Some cases of familial hypertrophic cardiomyopathy are due to missense mutations in the gene coding for β-myosin heavy chain. Smooth muscle, unlike skeletal and cardiac muscle, does not contain the troponin system; instead, phosphorylation of myosin p-light chains initiates contraction. A newly discovered regulator of vascular smooth muscle is nitric oxide; blockage of its formation from arginine causes an acute elevation of blood pressure, indicating that regulation of blood pressure is one of its many functions. Duchenne-type muscu-

lar dystrophy is due to mutations in the gene, located on the X chromosome, encoding the protein dystrophin. Two major types of muscle fibers are found in humans: white (anaerobic) and red (aerobic). The former are particularly used in sprints and the latter in prolonged aerobic exercise. During a sprint, muscle uses creatine phosphate and glycolysis as energy sources; in the marathon, oxidation of fatty acids is of major importance during the later phases.

Nonmuscle cells perform various types of mechanical work carried out by the structures constituting the cytoskeleton. These structures include actin filaments (microfilaments), microtubules (composed primarily of α-tubulin and β-tubulin), and intermediate filaments. The latter include keratins, vimentin-like proteins, neurofilaments, and lamins.

REFERENCES

Bredt DS, Snyder SH: Nitric oxide: A physiologic messenger molecule. Annu Rev Biochem 1994;63:175.

Cooke R: The actomyosin engine. FASEB J 1995;9:636.

Fuchs E, Weber K: Intermediate filaments: Structure, dynamics, function, and disease. Annu Rev Biochem 1994;63:345.

Hoffman EP, Lehmann-Horn F, Rudel R: Overexcited or inactive: Ion channels in muscle disease. Cell 1995;80: 681.

Huxley AF: Crossbridge tilting confirmed. Nature 1995; 375:631

Huxley HE: Sliding filaments and molecular motile systems. J Biol Chem 1990;265:8347.

Labeit S, Kolmerer B: Titins: Giant proteins in charge of muscle ultrastructure and elasticity. Science 1995;270: 293.

Langer GA: Calcium and the heart: Exchange at the tissue, cell and organelle levels. FASEB J 1992;6:893.

Loscalzo J: Nitric oxide and vascular disease. N Engl J Med 1995;333:251.

MacLennan DH, Phillips MS: Malignant hyperthermia. Science 1992;256:789.

McPherson PS, Campbell KP: The ryanodine receptor/Ca^{2+} release channel. J Biol Chem 1993;329:2002.

Mongada S, Higgs A: The L-arginine nitric oxide pathway. N Engl J Med 1993;329:2002.

Murphy RA: What is special about smooth muscle ? The significance of covalent crossbridge regulation. FASEB J 1994;8:311.

Rayment I et al: Structure of the actin-myosin complex and its implications for muscle contraction. Science 1993; 261:58.

Snyder SH: No endothelial NO. Nature 1995;377:196.

Sobue K, Sellers JR: Caldesmon: A novel regulatory protein in smooth muscle and nonmuscle actomyosin systems. J Biol Chem 1991;266:12115.

Somlyo AP, Somlyo AV: Signal transduction and regulation in smooth muscle. Nature 1994;372:231.

Spudich JA: How molecular motors work. Nature 1994; 372:515.

Walker RA, Sheetz MP: Cytoplasmic microtubule-associated motors. Annu Rev Biochem 1993;62:429.

Worton RG, Brooke MH: The X-linked muscular dystrophies. In: *The Metabolic and Molecular Bases of Inherited Disease*, 7th ed. Scriver CR et al (editors). McGraw-Hill, 1995.

Plasma Proteins, Immunoglobulins, & Blood Coagulation

59

Margaret L. Rand, PhD,* Elizabeth J. Harfenist, PhD,* & Robert K. Murray, MD, PhD

INTRODUCTION

The blood circulates in what is virtually a closed system of blood vessels. Blood consists of solid elements, the red and white blood cells and the platelets, suspended in a liquid medium, the plasma. As indicated below, blood—and plasma in particular—performs many functions that are absolutely critical for the maintenance of health.

Once the blood has clotted (coagulated), the remaining liquid phase (serum) lacks the clotting factors (including fibrinogen) that are normally present in plasma but have been consumed during the process of coagulation. Serum does contain some degradation products of clotting factors—products that have been generated during the coagulation process and thus are not normally present in plasma.

BIOMEDICAL IMPORTANCE

The fundamental role of blood in the maintenance of **homeostasis** and the ease with which blood can be obtained have meant that the study of its constituents has been of central importance in the development of biochemistry and clinical biochemistry. Hemoglobin, albumin, the immunoglobulins, and the various clotting factors are among the most assiduously studied of all proteins. Changes in the amounts of various plasma proteins occur in many diseases and can be monitored by electrophoresis. Alterations of the activities of certain enzymes found in plasma are of diagnostic use in a number of pathologic conditions. Hemorrhagic and thrombotic states can pose serious medical emergencies, and thromboses in the coronary and cerebral arteries are major causes of death in many parts of the world. Rational management of these conditions requires a clear understanding of the bases of blood clotting and fibrinolysis.

THE BLOOD HAS MANY FUNCTIONS

The functions of blood—except for specific cellular ones such as oxygen transport and cell-mediated immunologic defense—are carried out by plasma and its constituents (Table 59–1).

Plasma consists of water, electrolytes, metabolites, nutrients, proteins, and hormones. The water and electrolyte composition of plasma is practically the same as that of all extracellular fluids. Laboratory determinations of serum levels of Na^+, K^+, Ca^{2+}, Cl^-, and carbon dioxide and of blood pH are important in the management of many patients.

PLASMA CONTAINS A COMPLEX MIXTURE OF PROTEINS

The concentration of total protein in human plasma is approximately 7.0–7.5 g/dL and comprises the major part of the solids of the plasma. The proteins of the plasma are actually a complex mixture that includes not only simple proteins but also conjugated proteins such as **glycoproteins** and various

Table 59–1. Major functions of blood.

(1) Respiration—transport of oxygen from the lungs to the tissues and of CO_2 from the tissues to the lungs
(2) Nutrition—transport of absorbed food materials
(3) Excretion—transport of metabolic waste to the kidneys, lungs, skin, and intestines for removal
(4) Maintenance of the normal acid-base balance in the body
(5) Regulation of water balance through the effects of blood on the exchange of water between the circulating fluid and the tissue fluid
(6) Regulation of body temperature by the distribution of body heat
(7) Defense against infection by the white blood cells and circulating antibodies
(8) Transport of hormones and regulation of metabolism
(9) Transport of metabolites
(10) Coagulation

---

*Department of Biochemistry, University of Toronto.

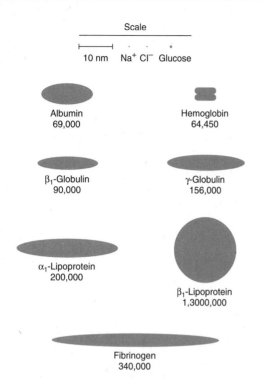

Scale

10 nm   Na⁺ Cl⁻ Glucose

Albumin
69,000

Hemoglobin
64,450

β₁-Globulin
90,000

γ-Globulin
156,000

α₁-Lipoprotein
200,000

β₁-Lipoprotein
1,3000,000

Fibrinogen
340,000

Figure 59–1. Relative dimensions and approximate molecular masses of protein molecules in the blood (Oncley).

types of **lipoproteins.** Thousands of **antibodies** are present in human plasma, though the amount of any one antibody is usually quite low under normal circumstances. The relative dimensions and molecular masses of some of the most important plasma proteins are shown in Figure 59–1.

The separation of individual proteins from a complex mixture is frequently accomplished by the use of solvents or electrolytes (or both) to remove different protein fractions in accordance with their solubility characteristics. This is the basis of the so-called salting-out methods, which find some usage in the determination of protein fractions in the clinical laboratory. Thus, one can separate the proteins of the plasma into three major groups—**fibrinogen, albumin,** and **globulins**—by the use of varying concentrations of sodium or ammonium sulfate.

The most common method of analyzing plasma proteins is by electrophoresis. There are many types of electrophoresis, each using a different supporting medium. In clinical laboratories, cellulose acetate is widely used as a supporting medium. Its use permits resolution, after staining, of plasma proteins into five bands, designated albumin, α_1, α_2, β, and γ fractions, respectively (Figure 59–2). The stained strip of cellulose acetate (or other supporting medium) is called an electrophoretogram. The amounts of these five bands can be conveniently quantified by use of den-

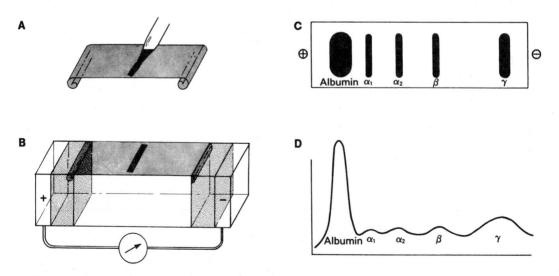

Figure 59–2. Technique of cellulose acetate zone electrophoresis. **A:** A small amount of serum or other fluid is applied to a cellulose acetate strip. **B:** Electrophoresis of sample in electrolyte buffer is performed. **C:** Separated protein bands are visualized in characteristic positions after being stained. **D:** Densitometer scanning from cellulose acetate strip converts bands to characteristic peaks of albumin, α_1-globulin, α_2-globulin, β-globulin, and γ-globulin. (Reproduced, with permission, from Stites DP, Terr AI, Parslow TG [editors]: *Basic & Clinical Immunology,* 8th ed. Appleton & Lange, 1994.)

sitometric scanning machines. Characteristic changes in the amounts of one or more of these five bands are found in many diseases.

The Concentration of Protein in Plasma Is Important in Determining the Distribution of Fluid Between Blood and Tissues

In arterioles, the hydrostatic pressure is about 37 mm Hg, with an interstitial (tissue) pressure of 1 mm Hg opposing it. The osmotic pressure (oncotic pressure) exerted by the plasma proteins is approximately 25 mm Hg. Thus, a net outward force of about 11 mm Hg drives fluid out into the interstitial spaces. In venules, the hydrostatic pressure is about 17 mm Hg, with the oncotic and interstitial pressures as described above; thus, a net force of about 9 mm Hg attracts water back into the circulation. The above pressures are often referred to as the **Starling forces.** If the concentration of plasma proteins is markedly diminished (eg, due to severe protein malnutrition), fluid is not attracted back into the intravascular compartment and accumulates in the extravascular tissue spaces, a condition known as **edema.** Edema has many causes; protein deficiency is one of them.

Plasma Proteins Have Been Studied Extensively

Because of the relative ease with which they can be obtained, plasma proteins have been studied extensively in both humans and animals. Considerable information is available about the biosynthesis, turnover, structure, and functions of the major plasma proteins. Alterations of their amounts and of their metabolism in many disease states have also been investigated. In recent years, many of the genes for plasma proteins have been cloned and their structures determined.

Many of the preparations listed in Table 2–7 have been used in the study of the plasma proteins. The preparation of antibodies specific for the individual plasma proteins has greatly facilitated their study, allowing the precipitation and isolation of pure proteins from the complex mixture present in tissues or plasma. In addition, the use of isotopes has made possible the determination of their pathways of biosynthesis and of their turnover rates in plasma.

The following generalizations have emerged from studies of plasma proteins.

A. Most Plasma Proteins Are Synthesized in the Liver: This has been established by experiments at the whole-animal level (eg, hepatectomy) and by use of the isolated perfused liver preparation, of liver slices, of liver homogenates, and of in vitro translation systems using preparations of mRNA extracted from liver. However, the γ-globulins are synthesized in plasma cells and certain plasma proteins are synthesized in other sites, such as endothelial cells.

B. Plasma Proteins Are Generally Synthesized on Membrane-Bound Polyribosomes: They then traverse the major secretory route in the cell (rough endoplasmic membrane → smooth endoplasmic membrane → Golgi apparatus → secretory vesicles) prior to entering the plasma. Thus, most plasma proteins are synthesized as **preproteins** and initially contain amino terminal signal peptides (Chapter 43). They are usually subjected to various posttranslational modifications (proteolysis, glycosylation, phosphorylation, etc) as they travel through the cell. Transit times through the hepatocyte from the site of synthesis to the plasma vary from 30 minutes to several hours or more for individual proteins.

C. Almost All Plasma Proteins Are Glycoproteins: Accordingly, they contain either *N*- or *O*-linked oligosaccharide chains, or both (Chapter 56). Albumin is the major exception; it does not contain sugar residues. The oligosaccharide chains have various functions (Table 56–2). An important finding is that removal of terminal sialic acid residues from certain plasma proteins (eg, ceruloplasmin) by exposure to neuraminidase can markedly shorten their half-lives in plasma (Chapter 56).

D. Many Plasma Proteins Exhibit Polymorphism: A polymorphism is a mendelian or monogenic trait that exists in the population in at least two phenotypes, neither of which is rare (ie, neither of which occurs with frequency of less than 1–2%). The ABO blood group substances (Chapter 60) are the best-known examples of human polymorphisms. Human plasma proteins that exhibit polymorphism include α_1-antitrypsin, haptoglobin, transferrin, ceruloplasmin, and immunoglobulins. The polymorphic forms of these proteins were often first recognized by use of **starch gel electrophoresis,** in which each polymorphic form may show a characteristic migration. Analyses of these human polymorphisms have proved to be of genetic and anthropologic interest and, in certain cases (eg, α_1-antitrypsin; see below), of clinical interest.

E. Each Plasma Protein Has a Characteristic Half-Life in the Circulation: The half-life of a plasma protein can be determined by labeling the isolated pure protein with ^{131}I under mild, nondenaturing conditions. This isotope unites covalently with tyrosine residues in the protein. The labeled protein is freed of unbound ^{131}I and its specific activity (dpm per mg protein) determined. A known amount of the radioactive protein is then injected into a normal adult subject, and samples of blood are taken at various time intervals for determinations of radioactivity. The values for radioactivity are plotted against time, and the half-life of the protein (the time for the radioactivity to decline from its peak value to one-half of its peak value) can be calculated from the resulting graph, discounting the times for the injected protein to equilibrate (mix) in the blood and in the extravascular spaces. The half-lives obtained for albumin and haptoglobin in normal healthy adults are approximately 20 and 5 days, respectively. In certain diseases, the half-life of a protein may be markedly altered. For instance, in some gastrointestinal diseases such as regional ileitis (Crohn's disease), considerable amounts of plasma proteins, including albumin, may be lost into the bowel through the inflamed intestinal mucosa. Patients with this condition have a **protein-losing gastroenteropathy,** and the half-life of injected iodinated albumin in these subjects may be reduced to as little as 1 day.

F. The Levels of Certain Proteins in Plasma Increase During Acute Inflammatory States or Secondary to Certain Types of Tissue Damage: These proteins are called "acute phase proteins" (or reactants) and include C-reactive protein (CRP, so-named because it reacts with the C polysaccharide of pneumococci), α_1-antitrypsin, haptoglobin, α_1-acid glycoprotein, and fibrinogen. The elevations of the levels of these proteins vary from as little as 50% to as much as 1000-fold in the case of CRP. Their levels are also usually elevated during chronic inflammatory states and in patients with cancer. These proteins are believed to play a role in the body's response to inflammation. For example, C-reactive protein can stimulate the classical complement pathway, and α_1-antitrypsin can neutralize certain proteases released during the acute inflammatory state. Interleukin 1 (IL-1), a polypeptide released from mononuclear phagocytic cells, is the principal—but not the sole—stimulator of the synthesis of the majority of acute phase reactants by hepatocytes. Additional molecules such as IL-6 are involved, and they as well as IL-1 appear to work at the level of gene transcription.

Table 59–2 summarizes the functions of many of the plasma proteins. The following section presents basic information regarding albumin, haptoglobin, transferrin, ceruloplasmin, α_1-antitrypsin, α_2-macroglobulin, the immunoglobulins, and the complement system. The lipoproteins are discussed in Chapter 27.

Albumin Is the Major Protein in Human Plasma

Albumin (69 kDa) is the major protein of human plasma (3.4–4.7 g/dL) and makes up approximately 60% of the total plasma protein. Some 40% of albumin is present in the plasma, and the other 60% is present in the extracellular space. The liver produces about 12 g of albumin per day, representing about 25% of total hepatic protein synthesis and half its secreted protein. Albumin is initially synthesized as a preproprotein. Its signal peptide is removed as it passes into the cisternae of the rough endoplasmic reticulum, and a hexapeptide at the resulting amino terminal is subsequently cleaved off farther along the secretory pathway. The synthesis of albumin is depressed in a variety of diseases, particularly those of the liver. The plasma of patients with liver disease

Table 59–2. Some functions of plasma proteins.

Function	Plasma Proteins
Antiproteases	Antichymotrypsin α_1-Antitrypsin (α_1-antiproteinase) α_2-Macroglobulin Antithrombin
Blood clotting	Various coagulation factors, fibrinogen
Enzymes	Function in blood, eg, coagulation factors, cholinesterase Leakage from cells or tissues, eg, aminotransferases
Hormones	Erythropoietin[1]
Immune defense	Immunoglobulins, complement proteins, β_2-microglobulin
Involvement in inflammatory responses	Acute phase response proteins (eg, C-reactive protein, α_1-acid glycoprotein [orosomucoid])
Oncofetal	α_1-Fetoprotein (AFP)
Transport or binding proteins	Albumin (various ligands, including bilirubin, free fatty acids, ions [Ca^{2+}], metals [eg, Cu^{2+}, Zn^{2+}], metheme, steroids, other hormones and a variety of drugs) Ceruloplasmin (contains Cu^{2+}; albumin probably more important in physiologic transport of Cu^{2+}) Corticosteroid-binding globulin (transcortin) (binds cortisol) Haptoglobin (binds extracorpuscular hemoglobin) Lipoproteins (chylomicrons, VLDL, LDL, HDL) Hemopexin (binds heme) Retinol-binding protein (binds retinol) Sex hormone-binding globulin (binds testosterone, estradiol) Thyroid-binding globulin (binds T_4, T_3) Transferrin (transports iron) Transthyretin (formerly prealbumin; binds T_4 and forms a complex with retinol-binding protein)

[1]Various other protein hormones circulate in the blood but are not usually designated as plasma proteins. Similarly, ferritin is also found in plasma in small amounts, but it too is not usually characterized as a plasma protein.

often shows a decrease in the ratio of albumin to globulins (decreased albumin/globulin ratio). The synthesis of albumin decreases relatively early in conditions of protein malnutrition, such as kwashiorkor (see Case No. 2 in Chapter 65).

Mature human albumin consists of one polypeptide chain of 585 amino acids and contains 17 disulfide bonds. By the use of proteases, albumin can be subdivided into three domains, which have different functions. Albumin has an ellipsoidal shape, which means that it does not increase the viscosity of the plasma as much as an elongated molecule such as fibrinogen does. Because of its relatively low molecular mass (about 69 kDa) and high concentration, albumin is thought to be responsible for 75–80% of the **osmotic pressure** of human plasma. Electrophoretic studies have shown that the plasma of certain humans lacks albumin. These subjects are said to exhibit analbuminemia. One cause of this condition is a mutation that affects splicing. Subjects with analbuminemia show only moderate edema, despite the fact that albumin is the major determinant of plasma osmotic pressure. It is thought that the amounts of the other plasma proteins increase and compensate for the lack of albumin.

Another important function of albumin is its ability to bind various ligands. These include free fatty acids (FFA), calcium, certain steroid hormones, bilirubin, and some of the plasma tryptophan. In addition, albumin appears to play an important role in transport of copper in the human body (see below). A variety of drugs, including sulfonamides, penicillin G, dicumarol, and aspirin, are bound to albumin; this finding has important pharmacologic implications.

Preparations of human albumin are of use in the treatment of hemorrhagic shock and of burns.

Haptoglobin Binds Extracorpuscular Hemoglobin, Preventing Free Hemoglobin From Entering the Kidney

Haptoglobin (Hp) is a plasma glycoprotein that binds extracorpuscular hemoglobin (Hb) in a tight noncovalent complex (Hb-Hp). The amount of haptoglobin in human plasma ranges from 40 to 180 mg of hemoglobin-binding capacity per deciliter. Approximately 10% of the hemoglobin that is degraded each day is released into the circulation and is thus extracorpuscular. The other 90% is present in old, damaged red blood cells, which are degraded by cells of the histiocytic system. The molecular mass of hemoglobin is approximately 65 kDa, whereas the molecular mass of the simplest polymorphic form of haptoglobin (Hp 1-1) found in humans is approximately 90 kDa. Thus, the Hb-Hp complex has a molecular mass of approximately 155 kDa. Free hemoglobin passes through the glomerulus of the kidney, enters the tubules, and tends to precipitate therein (as can happen after a massive incompatible blood transfusion, when the capacity of haptoglobin to bind hemoglobin is grossly exceeded) (Figure 59–3). However, the Hb-Hp complex is too large to pass through the glomerulus. The function of Hp thus appears to be to prevent loss of free hemoglobin into the kidney. This conserves the valuable iron present in hemoglobin, which would otherwise be lost to the body.

Human haptoglobin exists in three polymorphic forms, known as Hp 1-1, Hp 2-1, and Hp 2-2. Hp 1-1 migrates in starch gel electrophoresis as a single band, whereas Hp 2-1 and Hp 2-2 exhibit much more complex band patterns. Two genes, designated Hp^1 and Hp^2, direct these three phenotypes, with Hp 2-1 being the heterozygous phenotype. No significant functional differences among the polymorphic forms of Hp have been established.

The levels of haptoglobin in human plasma vary and are of some diagnostic use. Low levels of haptoglobin are found in patients with **hemolytic anemias.** This is explained by the fact that whereas the half-life of haptoglobin is approximately 5 days, the half-life of the Hb-Hp complex is about 90 minutes, the complex being rapidly removed from plasma by hepatocytes. Thus, when haptoglobin is bound to hemoglobin, it is cleared from the plasma about 80 times faster than normally. Accordingly, the level of haptoglobin falls rapidly in situations where hemoglobin is constantly being released from red blood cells, such as occurs in hemolytic anemias. Haptoglobin is an acute phase protein, and its plasma level is elevated in a variety of inflammatory states.

Certain other plasma proteins bind heme but not hemoglobin. Hemopexin is a β_1-globulin that binds free heme. Albumin will bind some metheme (ferric heme) to form methemalbumin, which then transfers the metheme to hemopexin.

Transferrin Shuttles Iron to Sites Where It Is Needed

A. Transferrin: Transferrin is a β_1-globulin with a molecular mass of approximately 80 kDa. It is a glycoprotein and is synthesized in the liver. More than 20 polymorphic forms of transferrin have been found. Transferrin plays a central role in the body's metabolism of iron because it transports iron (2 mol of Fe^{3+} per mole of transferrin) in the circulation to sites where iron is required, eg, from the gut to the bone marrow and other organs. Iron is important in the human body because of its occurrence in many hemoproteins, such as hemoglobin, myoglobin, and the cytochromes. Iron is ingested in the diet, and its absorption as Fe^{2+} is tightly regulated at the level of the intestinal mucosa. Under normal circumstances,

A. Hb → Kidney → Excreted in urine or precipitates in tubules;
(MW 65,000)   iron is lost to body

B. Hb + Hp → Hb : Hp complex ⇻ Kidney
(MW 65,000) (MW 90,000) |(MW 155,000)
↓
Catabolized by liver cells;
iron is conserved and reused

Figure 59–3. Different fates of free hemoglobin and of the hemoglobin-haptoglobin complex.

the body guards its content of iron zealously, so that a healthy adult male loses about 1 mg/d, which is replaced by absorption. Adult females are more prone to states of iron deficiency because some may lose excessive blood during menstruation. The amounts of iron in transferrin and in various other body compartments are shown in Table 59–3.

Approximately 200 billion red blood cells (about 20 mL) are catabolized per day, releasing about 25 mg of iron into the body. Free iron is toxic, but association with transferrin diminishes its potential toxicity and also directs iron to where it is required in the body. There are receptors on the surfaces of many cells for transferrin. The protein binds to these receptors and is internalized by receptor-mediated endocytosis (compare the fate of LDL, Chapter 27). The acid pH inside the lysosome causes the iron to dissociate from the protein. However, unlike the protein component of LDL, apotransferrin is not degraded within the lysosome. Instead, it remains associated with its receptor, returns to the plasma membrane, dissociates from its receptor, reenters the plasma, picks up more iron, and again delivers the iron to needy cells.

B. Problems of Iron Metabolism: Attention to iron metabolism is **particularly important in women** for the reason mentioned above. Additionally, in **pregnancy** allowances must be made for the growing fetus. Older people with poor dietary habits ("tea and toasters") may develop iron deficiency. Iron deficiency anemia due to inadequate intake, inadequate utilization, or excessive loss of iron is one of the commonest conditions seen in medical practice.

The concentration of transferrin in plasma is approximately 300 mg/dL. This amount of transferrin can bind 300 μg of iron per deciliter, so that this represents the **total iron-binding capacity** of plasma. However, the protein is normally only one-third saturated with iron. In **iron deficiency anemia,** the protein is even less saturated with iron, whereas in conditions of storage of excess iron in the body (eg, hemochromatosis) the saturation with iron is much greater than one-third.

C. Ferritin: Ferritin is another protein that is important in the metabolism of iron. Under normal conditions, it stores iron that can be called upon for use as conditions require. In conditions of excess iron (eg, hemochromatosis), body stores of iron are greatly increased and much more ferritin is present in the tissues, such as the liver and spleen. Ferritin contains approximately 23% iron, and apoferritin (the protein moiety free of iron) has a molecular mass of approximately 440 kDa. Ferritin is composed of 24 subunits of 18.5 kDa, which surround in a micellar form some 3000–4500 ferric atoms. Normally, there is a little ferritin in human plasma. However, in patients with excess iron, the amount of ferritin in plasma is markedly elevated. The amount of ferritin in plasma can be conveniently measured by a sensitive and specific radioimmunoassay and serves as an index of body iron stores.

Synthesis of the transferrin receptor (TfR) and that of ferritin are reciprocally linked to cellular iron content. Specific untranslated sequences of the mRNAs for both proteins (named **iron response elements**) interact with a cytosolic protein sensitive to variations in levels of cellular iron. When iron levels are high, cells use stored ferritin mRNA to synthesize ferritin, and the TfR mRNA is degraded. In contrast, when iron levels are low, the TfR mRNA is stabilized and increased synthesis of receptors occurs, while ferritin mRNA is apparently stored in an inactive form. This is an important example of control of expression of proteins at the **translational** level.

D. Hemosiderin: Hemosiderin is a somewhat ill-defined molecule; it appears to be a partly degraded form of ferritin but still containing iron. It can be detected by histologic stains (eg, Prussian blue) for iron, and its presence is determined histologically when excessive storage of iron occurs.

E. Primary Hemochromatosis: Primary hemochromatosis is a genetic disorder characterized by excessive storage of iron in tissues, leading to tissue damage. The cause appears to be excessive absorption of iron from the intestinal mucosa, though little is known about the molecular nature of the abnormality. The organs most affected are liver, skin, and pancreas; patients usually develop cirrhosis of the liver. In addition, they may acquire pigmentation of the skin and diabetes mellitus (bronze diabetes). The disorder can be kept under control by periodic withdrawals of blood (phlebotomies).

Table 59–4 summarizes laboratory tests that are useful in the assessment of patients with abnormalities of iron metabolism.

Table 59–3. Distribution of iron in a 70-kg adult male.[1]

Transferrin	3–4 mg
Hemoglobin in red blood cells	2500 mg
In myoglobin and various enzymes	300 mg
In stores (ferritin and hemosiderin)	1000 mg
Absorption	1 mg/d
Losses	1 mg/d

[1]In an adult female of similar weight, the amount in stores would generally be less (100–400 mg) and the losses would be greater (1.5–2 mg/d).

Table 59–4. Laboratory tests for assessing patients with disorders of iron metabolism.

- Red blood cell count and estimation of hemoglobin
- Determinations of plasma iron, total iron-binding capacity (TIBC), and % transferrin saturation
- Determination of ferritin in plasma by radioimmunoassay
- Prussian blue stain of tissue sections
- Determination of amount of iron (μg/g) in a tissue biopsy

Ceruloplasmin Binds Copper, and Low Levels of this Plasma Protein Are Associated With Wilson's Disease

Ceruloplasmin (about 160 kDa) is an α_2-globulin. It has a blue color because of its high copper content and carries 90% of the copper present in plasma. Each molecule of ceruloplasmin binds six atoms of copper very tightly, so that the copper is not readily exchangeable. Albumin carries the other 10% of the plasma copper but binds the metal less tightly than does ceruloplasmin. Albumin thus donates its copper to tissues more readily than ceruloplasmin and appears to be more important than ceruloplasmin in copper transport in the human body. Ceruloplasmin exhibits a copper-dependent **oxidase** activity, but its physiologic significance has not been clarified. The amount of ceruloplasmin in plasma is decreased in liver disease. In particular, low levels of ceruloplasmin are found in Wilson's disease (hepatolenticular degeneration), a disease due to abnormal metabolism of copper. In order to clarify the description of Wilson's disease, we shall first consider the metabolism of copper in the human body and then Menkes' disease, another condition involving abnormal copper metabolism.

Copper Is a Cofactor for Certain Enzymes

Copper is an essential trace element. It is required in the diet because it is the metal cofactor for a variety of enzymes (see Table 59–5). Copper accepts and donates electrons and is involved in reactions involving dismutation, hydroxylation, and oxygenation. However, excess copper can cause problems because it can oxidize proteins and lipids, bind to nucleic acids, and enhance the production of free radicals. It is thus important to have mechanisms that will maintain the amount of copper in the body within normal limits. The body of the normal adult contains about 100 mg of copper, located mostly in bone, liver, kidney, and muscle. The daily intake of copper is about 2–4 mg, with about 50% being absorbed in the stomach and upper small intestine and the remainder excreted in the feces. Copper is carried to the liver bound to albumin, taken up by liver cells, and part of it is excreted in the bile. Copper also leaves the liver attached to ceruloplasmin, which is synthesized in that organ.

Table 59–5. Some important enzymes that contain copper.

- Amine oxidase
- Copper-dependent superoxide dismutase
- Cytochrome oxidase
- Tyrosinase

The Tissue Levels of Copper and of Certain Other Metals Are Regulated in Part by Metallothioneins

Metallothioneins are a group of small proteins (about 6.5 kDa), found in the cytosol of cells, particularly of liver, kidney, and intestine. They have a high content of cysteine and can bind copper, zinc, cadmium, and mercury. The SH groups of cysteine are involved in binding the metals. Acute intake (eg, by injection) of copper and of certain other metals increases the amount (induction) of these proteins in tissues, as does administration of certain hormones or cytokines. These proteins may function to store the above metals in a nontoxic form and are involved in their overall metabolism in the body. Sequestration of copper also diminishes the amount of this metal available to generate free radicals.

Menkes' Disease Is Due to Mutations in the Gene for a Copper-Binding P-Type ATPase

Menkes' disease ("kinky" or "steely" hair disease) is a disorder of copper metabolism. It is X-linked, affects only male infants, involves the nervous system, connective tissue, and vasculature, and is usually fatal in infancy (see also Chapter 57). A similar condition was observed in Australia in sheep raised on a diet deficient in copper, and the hair of patients with Menkes' disease was observed to resemble the abnormal wool of these animals. This suggested that Menkes' disease was due to an abnormality in copper metabolism. In 1993, it was reported that the basis of Menkes' disease was mutations in the gene for a copper-binding P-type ATPase. Interestingly, the enzyme showed structural similarity to certain metal-binding proteins in microorganisms. This ATPase is thought to be responsible for directing the efflux of copper from cells. When altered by mutation, copper is not mobilized normally from the intestine, in which it accumulates, as it does in a variety of other cells and tissues, from which it cannot exit. Despite the accumulation of copper, the activities of many copper-dependent enzymes are decreased, perhaps because of a defect of its incorporation into the apoenzymes. Normal liver expresses very little of the ATPase, which explains the absence of hepatic involvement in Menkes' disease. This work led to the suggestion that liver might contain a different copper-binding ATPase, which could be involved in the causation of Wilson's disease. As described below, this turned out to be the case. Table 59–6 compares important features of Menkes' disease and Wilson's disease.

Wilson's Disease Is Also Due to Mutations in a Gene Encoding a Copper-Binding P-Type ATPase

Wilson's disease is a genetic disease in which copper fails to be excreted in the bile and accumu-

Table 59–6. Comparison of Menkes' disease and Wilson's disease.[1]

	Menkes' Disease	Wilson's Disease
Location of gene	Xq13.3	13q14.3
Inheritance	X-linked recessive	Autosomal recessive
Gene product	Cu^{2+}-binding P-type ATPase	Cu^{2+}-binding P-type ATPase
Expression	All tissues except liver	Liver, kidney, placenta
Mutations	Variety	Variety
Onset	At birth	Late childhood
Clinical findings	Cerebral degeneration, abnormal hair, other signs, early death	Liver disease, neurologic signs, can survive to late adulthood
Laboratory findings	↓ Serum Cu^{2+} ↓ Serum ceruloplasmin ↑ Intestinal and kidney Cu^{2+} ↓ Liver Cu^{2+}	↓ Serum Cu^{2+} ↓ Serum ceruloplasmin ↑ Urinary Cu^{2+} ↑ Liver Cu^{2+}
Cultured cells	↑ Cu^{2+} accumulation ↓ Cu^{2+} release	Normal in most patients
Defect	Intestinal Cu^{2+} absorption, deficiency of Cu^{2+}-dependent enzymes	Biliary Cu^{2+} excretion, Cu^{2+} incorporation into ceruloplasmin
Treatment	No effective treatment	Penicillamine

[1]Adapted from Chelly J, Monaco AP: Cloning the Wilson disease gene. Nat Genet 1993;5:317.

lates in liver, brain, kidney, and red blood cells. It can be regarded as an inability to maintain a near-zero copper balance, resulting in copper toxicosis. The increase of copper in liver cells appears to inhibit the coupling of copper to apoceruloplasmin and leads to low levels of ceruloplasmin in plasma. As the amount of copper accumulates, patients may develop a hemolytic anemia, chronic liver disease (cirrhosis, hepatitis), and a neurologic syndrome owing to accumulation of copper in the basal ganglia and other centers. A frequent clinical finding is the Kayser-Fleischer ring. This is a green or golden pigment ring around the cornea due to deposition of copper in Descemet's membrane. The major laboratory tests of copper metabolism are listed in Table 59–7. If Wilson's disease is suspected, a liver biopsy should be performed; a value for liver copper of over 250 μg/g dry weight along with a plasma level of ceruloplasmin of under 20 mg/dL is diagnostic.

The cause of Wilson's disease was also revealed in 1993. Following the discovery of the gene causing

Table 59–7. Major laboratory tests used in the investigation of diseases of copper metabolism.[1]

Test	Normal Adult Range
Serum copper	10–22 μmol/L
Ceruloplasmin	200–600 mg/L
Urinary copper	<1 μmol/24 h
Liver copper	20–50 μg/g dry weight

[1]Based on Gaw A et al: *Clinical Biochemistry.* Churchill Livingstone, 1995.

Menkes' disease, it was found that a variety of mutations in a gene encoding a similar copper-binding P-type ATPase were responsible for this condition. The gene is estimated to encode a protein of 1411 amino acids, highly homologous to the product of the Menkes' gene. In a manner not yet fully explained, a nonfunctional ATPase causes defective excretion of copper into the bile, a reduction of incorporation of copper into apoceruloplasmin, and the accumulation of copper in liver and subsequently in other organs such as brain.

Treatment for Wilson's disease consists of a diet low in copper along with lifelong administration of penicillamine, which chelates copper, is excreted in the urine, and thus depletes the body of the excess of this mineral.

Deficiency of α₁-Antiproteinase (α₁-Antitrypsin) Is Associated With Emphysema and One Type of Liver Disease

α_1-Antiproteinase (about 52 kDa) was formerly called α_1-antitrypsin, and this name is retained here. It is a single-chain protein of 394 amino acids, contains three oligosaccharide chains, and is the major component (> 90%) of the α_1 fraction of human plasma. It is synthesized by hepatocytes and macrophages and is the principal serine protease inhibitor (serpin or Pi) of human plasma. It inhibits trypsin, elastase, and certain other proteases by forming complexes with them. The protein is highly polymorphic; the multiple forms can be separated by electrophoresis. The major genotype is MM, and its phenotypic product is PiM. There are two areas of

A. Active elastase + α_1–AT → Inactive elastase: α_1–AT complex → No proteolysis of lung → No tissue damage

B. Active elastase + ↓ or no α_1–AT → Active elastase → Proteolysis of lung → Tissue damage

Figure 59–4. Scheme illustrating *(A)* normal inactivation of elastase by α_1-antitrypsin and *(B)* situation in which the amount of α_1-antitrypsin is substantially reduced, resulting in proteolysis by elastase and leading to tissue damage.

clinical interest concerning α_1-antitrypsin. A deficiency of this protein has a role in certain cases (approximately 5%) of **emphysema.** This occurs mainly in subjects with the ZZ genotype, who synthesize PiZ. Considerably less of this protein is secreted as compared with PiM. When the amount of α_1-antitrypsin is deficient and polymorphonuclear white blood cells increase in the lung (eg, during pneumonia), the affected individual lacks a countercheck to proteolytic damage of the lung by proteases such as elastase (Figure 59–4). It is of considerable interest that a particular methionine (residue 358) of α_1-antitrypsin is involved in its binding to proteases. Smoking oxidizes this methionine to methionine sulfoxide and thus inactivates it. As a result, affected molecules of α_1-antitrypsin no longer neutralize proteases. This is particularly devastating in patients (PiZZ phenotype) who already have low levels of α_1-antitrypsin. The further diminution in α_1-antitrypsin brought about by smoking results in increased proteolytic destruction of lung tissue, accelerating the development of emphysema. Intravenous administration of α_1-antitrypsin (augmentation therapy) has been used as an adjunct in the treatment of patients with emphysema due to α_1-antitrypsin deficiency. Attempts are being made, using the techniques of protein engineering (Chapter 42), to replace methionine 358 by another residue that would not be subject to oxidation. The resulting "mutant" α_1-antitrypsin would thus afford protection against proteases for a much longer period of time than would native α_1-antitrypsin. Attempts are also being made to develop **gene therapy** for this condition. One approach is to use a modified adenovirus (a pathogen of the respiratory tract) into which the gene for α_1-antitrypsin has been inserted. The virus would then be introduced into the respiratory tract (eg, by an aerosol). The hope is that pulmonary epithelial cells would express the gene and secrete α_1-antitrypsin locally. Experiments in animals have indicated the feasibility of this approach.

Deficiency of α_1-antitrypsin is also implicated in one type of liver disease (α_1-antitrypsin deficiency liver disease). In this condition, molecules of the ZZ phenotype accumulate and aggregate in the cisternae of the endoplasmic reticulum of hepatocytes. Recent work has shown that aggregation is due to formation of polymers of mutant α_1-antitrypsin, the polymers forming via a strong interaction between a specific loop in one molecule and a prominent β-pleated sheet in another (loop-sheet polymerization). By

mechanisms that are not understood, **hepatitis** results and consequent **cirrhosis** (accumulation of massive amounts of collagen, resulting in fibrosis) of the liver. It is possible that administration of a synthetic peptide resembling the loop sequence could inhibit loop-sheet polymerization. At present, severe α_1-antitrypsin deficiency liver disease can be successfully treated by liver transplantation. In the future, introduction of the gene for normal α_1-antitrypsin into hepatocytes may become possible. Figure 59–5 is a scheme of the causation of this disease.

α_2-Macroglobulin Neutralizes Many Proteases and Targets Certain Cytokines to Tissues

α_2-Macroglobulin is a large plasma glycoprotein (720 kDa) made up of four identical subunits of 180 kDa. Its gene is located on human chromosome 12p12–13. α_2-Macroglobulin comprises 8–10% of the total plasma protein in humans. Approximately 10% of the zinc in plasma is transported by α_2-macroglobulin, the remainder being transported by albumin. The protein is synthesized by a variety of cell types, including monocytes, hepatocytes, and astrocytes. It is the major member of a group of plasma proteins that include complement proteins C3 and C4. These proteins contain a unique internal cyclic thiol ester bond (formed between a cysteine and a glutamine residue) and for this reason have been designated as the **thiol ester plasma protein family.**

α_2-Macroglobulin binds many proteinases and is thus an important **panproteinase inhibitor.** The α_2-macroglobulin-proteinase complexes are rapidly cleared from the plasma by a specific receptor, lo-

GAG to AAG mutation in exon 5 of gene for α_1-AAT on chromosome 14

↓

Results in Glu^{342} to Lys^{342} substitution in α_1-AAT

↓

α_1-AAT accumulates in cisternae of endoplasmic reticulum and aggregates via loop-sheet polymerization

↓

Leads to hepatitis (mechanism unknown) and often cirrhosis in 10% of ZZ homozygotes

Figure 59–5. Scheme of causation of α_1-antitrypsin deficiency liver disease. (AAT, α_1-antitrypsin.)

cated on many cell types, which is identical to the low-density lipoprotein receptor-related protein (LRP). LRP may be the ancestral gene from which the LDL receptor gene evolved. In addition, α_2-macroglobulin binds many cytokines (eg, platelet-derived growth factor, transforming growth factor-β, etc) and appears to be involved in targeting them toward particular tissues or cells. Once taken up by cells, the cytokines can dissociate from α_2-macroglobulin and subsequently exert a variety of effects on cell growth and function. The binding of proteinases and cytokines by α_2-macroglobulin involves different mechanisms, which will not be considered here.

Amyloidosis Occurs by the Deposition of Fragments of Various Plasma Proteins in Tissues

Amyloidosis is the accumulation of various insoluble fibrillar proteins between the cells of tissues to an extent that affects function. The fibrils generally represent proteolytic fragments of various plasma proteins and are nonbranching and possess a β-pleated sheet structure. The term "amyloidosis" is a misnomer, as it was originally thought that the fibrils were starch-like in nature. Among the most common precursor proteins are immunoglobulin light chains (see below), amyloid-associated protein derived from serum amyloid-associated protein (a plasma glycoprotein), and transthyretin (Table 59–2). The precursor proteins in plasma are generally either increased in amount (eg immunoglobulin light chains in multiple myeloma or β_2-microglobulin in patients being maintained on chronic dialysis) or mutant forms (eg, of transthyretin in familial amyloidotic neuropathies). The precise factors that determine the deposition of proteolytic fragments in tissues await elucidation. Other proteins have been found in amyloid fibrils, such as calcitonin and amyloid β protein (not derived from a plasma protein) in Alzheimer's disease (Chapter 64); a total of about 15 different proteins have been found. All fibrils have a P component associated with them, which is derived from serum amyloid P component, a plasma protein closely related to C-reactive protein. Tissue sections containing amyloid fibrils interact with Congo red stain and display striking green birefringence when viewed by polarizing microscopy. Deposition of amyloid occurs mainly (1) in patients with monoclonal B cell proliferations (primary amyloidosis), (2) in patients with chronic inflammatory processes such as rheumatoid arthritis (secondary amyloidosis), (3) in hereditary circumstances (eg, the neuropathies mentioned above), (4) in patients being maintained on chronic dialysis, and (5) in patients with Alzheimer's disease. Treatment of the underlying disorder should be provided if possible.

PLASMA IMMUNOGLOBULINS PLAY A MAJOR ROLE IN THE BODY'S DEFENSE MECHANISMS

The immune system of the body consists of two major components: **B lymphocytes** and **T lymphocytes.** The B lymphocytes are mainly derived from bone marrow cells in higher animals and from the bursa of Fabricius in birds. The T lymphocytes are of thymic origin. The **B cells** are responsible for the synthesis of circulating, humoral antibodies, also known as **immunoglobulins.** The **T cells** are involved in a variety of important **cell-mediated immunologic processes** such as graft rejection, hypersensitivity reactions, and defense against malignant cells and many viruses. This section considers only the plasma immunoglobulins, which are synthesized mainly in **plasma cells,** specialized cells of B cell lineage that synthesize and secrete immunoglobulins into the plasma.

All Immunoglobulins Contain a Minimum of Two Light and Two Heavy Chains

All immunoglobulin molecules consist of two identical light (L) chains (23 kDa) and two identical heavy (H) chains (53–75 kDa), held together as a tetramer (L_2H_2) by disulfide bonds (Figure 59–6). Each chain can be divided conceptually into specific domains, or regions, that have structural and functional significance. The half of the light (L) chain toward the carboxyl terminal is referred to as the constant region (C_L), while the amino terminal half is the variable region of the light chain (V_L). Approximately one-quarter of the heavy (H) chain at the amino terminal end is referred to as its variable region (V_H), and the other three-quarters of the heavy chain are referred to as the constant regions (C_H1, C_H2, C_H3) of that H chain. The portion of the immunoglobulin molecule that binds the specific antigen is formed by the amino terminal portions (variable regions) of both the H and L chains—ie, the V_H and V_L domains. The domains of the protein chains do not simply exist as linear sequences of amino acids but form globular regions with secondary and tertiary structures in order to effect binding of specific antigens.

As depicted in Figure 59–6, digestion of an immunoglobulin by the enzyme papain produces two antigen-binding fragments (Fab) and one crystallizable fragment (Fc). The area in which papain cleaves the immunoglobulin molecule—ie, the region between the C_H1 and C_H2 domains—is referred to as the "hinge region."

All Light Chains Are Either Kappa or Lambda in Type

There are two general types of light chains, kappa (κ) and lambda (λ) (Tables 59–4 and 59–5), which can be distinguished on the basis of structural differ-

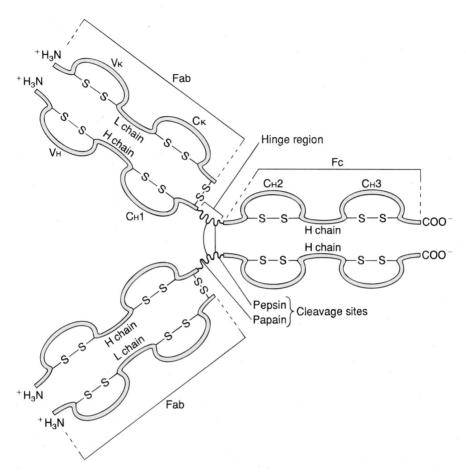

Figure 59–6. Simplified model for an IgG1 (κ) human antibody molecule showing the basic four-chain structure and domains (V_H, C_H1, etc). V indicates the variable region. C indicates the constant region. Sites of enzyme cleavage by pepsin and papain are shown. Note positions of inter- and intrachain disulfide bonds. (Reproduced, with permission, from Stites DP, Terr AI, Parslow TG [editors]: *Basic & Clinical Immunology,* 8th ed. Appleton & Lange, 1994.)

ences in their C_L regions. A given immunoglobulin molecule always contains two κ or two λ light chains—never a mixture of κ and λ. In humans, the κ chains are more frequent than λ chains in immunoglobulin molecules.

The Five Types of Heavy Chain Determine Immunoglobulin Class

Five classes of H chain have been found in humans (Tables 59–4 and 59–5), and they can be distinguished by differences in their C_H regions. They are designated γ, α, μ δ, and ε and vary in molecular mass from 50 kDa to 70 kDa (Table 59–4). The μ and ε chains each have four C_H domains rather than the usual three. The type of H chain determines the class of immunoglobulin and thus its effector function. There are five immunoglobulin classes: **IgG, IgA, IgM, IgD,** and **IgE.** As shown in Table 59–4, many of the H chain classes can be further divided

into subclasses on the basis of subtle structural differences in the C_H regions.

No Two Variable Regions Are Identical

The variable regions of immunoglobulin molecules consist of the V_L and V_H domains and are quite heterogeneous. In fact, no two variable regions from different humans have been found to have identical amino acid sequences. However, there are discernible patterns between the regions from different individuals, and these shared patterns have been divided into three main groups based on the degree of amino acid sequence homology. There is a $V_κ$ group for kappa L chains, a $V_λ$ group for lambda L chains, and a V_H group for the H chains. At higher resolution, there are subgroups within each of these three groups (Table 59–8).

Thus, within the variable regions there are some

Table 59–8. Properties of human immunoglobulin chains.[1]

Designation	H Chains					L Chains		Secretory Component	J Chain
	γ IgG	α IgA	μ IgM	δ IgD	ε IgE	κ All classes	λ All classes	SC IgA	J IgA, IgM
Classes in which chains occur	IgG	IgA	IgM	IgD	IgE	All classes	All classes	IgA	IgA, IgM
Subclasses or subtypes	1,2,3,4	1,2	1,2	...	...	...	1,2,3,4	...	...
Allotypic variants	Gm(1)–(25)	A2m(1),(2)	...	...	...	Km(1)–(3)[2]	...	...	...
M_r (approximate)	50,000[3]	55,000	70,000	62,000	70,000	23,000	23,000	70,000	15,000
V region subgroups	$V_HI–V_HIV$					$V_\kappa I–V_\kappa IV$	$V_\lambda I–V_\lambda VI$		
Carbohydrate (average percentage)	4	10	15	18	18	0	0	16	8
Number of oligosaccharides	1	2 or 3	5	?	5	0	0	?	1

[1]Modified and reproduced, with permission, from Stites DP, Terr AI, Parslow TG: (editors): *Basic & Clinical Immunology*, 8th ed. Appleton & Lange, 1994.
[2]Formerly Inv(1)–(3).
[3]60,000 for γ3.

positions that are relatively invariable to account for the groups and subgroups. Upon comparing variable regions from different light chains of the same group or subgroup or different heavy chains from the same group or subgroup, it is apparent that there are hypervariable regions interspersed between the relatively invariable (subgroup-determining) positions (Figure 59–7). L chains have three hypervariable regions (in V_L), and H chains have four (in V_H).

The Constant Regions Determine Class-Specific Effector Functions

The constant regions of the immunoglobulin molecules, particularly the C_H2 and C_H3 (and C_H4 of IgM

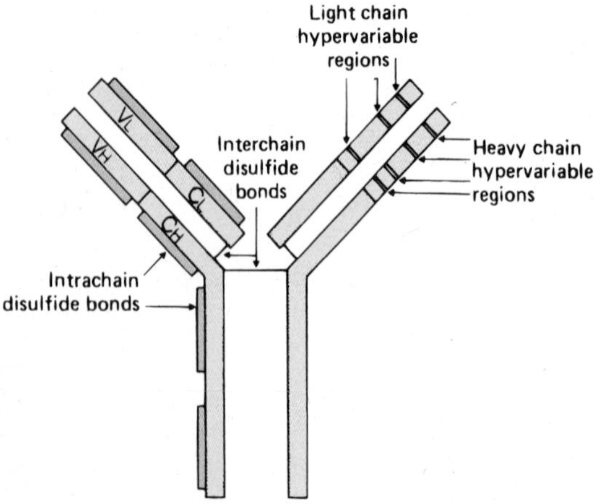

Figure 59–7. Schematic model of an IgG molecule showing approximate positions of the hypervariable regions in heavy and light chains. (Modified and reproduced, with permission, from Stites DP, Terr AI, Parslow TG [editors]: *Basic & Clinical Immunology*, 8th ed. Appleton & Lange, 1994.)

and IgE), which constitute the Fc fragment, are responsible for the class-specific effector functions of the different immunoglobulin molecules (Table 59–9, bottom part), eg, complement fixation or placental transfer.

Some immunoglobulins such as immune IgG exist only in the basic tetrameric structure, while others such as IgA and IgM can exist as higher order polymers of two, three (IgA), or five (IgM) tetrameric units (Figure 59–8).

The L chains and H chains are synthesized as separate molecules and are subsequently assembled within the B cell or plasma cell into mature immunoglobulin molecules, all of which are **glycoproteins** (Table 59–8).

Both Light and Heavy Chains Are Products of Multiple Genes

Each immunoglobulin light chain is the product of at least three separate structural genes: a variable region (V_L) gene, a joining region (J) gene (bearing no relationship to the J chain of IgA or IgM), and a constant region (C_L) gene. Each heavy chain is the product of at least four different genes: a variable region (V_H) gene, a diversity region (D) gene, a joining region (J) gene, and a constant region (C_H) gene. Thus, the "one gene, one protein" concept is not valid. The molecular mechanisms responsible for the generation of the single immunoglobulin chains from multiple structural genes are discussed in Chapters 38 and 41.

Antibody Diversity Depends on Gene Rearrangements

Each person is capable of generating antibodies directed against perhaps 1 million different antigens. The generation of such immense antibody diversity appears to depend upon the combinations of the vari-

Table 59–9. Properties of human immunoglobulins.[1]

	IgG	IgA	IgM	IgD	IgE
H chain class	γ	α	μ	δ	ε
H chain subclass	$\gamma1, \gamma2, \gamma3, \gamma4$,	$\alpha1, \alpha2$	$\mu1, \mu2$		
L chain type	κ and λ	κ and λ	κ and λ	κ and λ	κ and λ
Molecular formula	γ_2L_2	α_2L^2 or $(\alpha_2L_2)_2SC^4J^3$	$(\alpha_2L_2)_5J^3$	δ_2L_2	ε_2L_2
Sedimentation coefficient (S)	6–7	7	19	7–8	8
M_r (approximate)	150,000	160,000[2] 400,000[5]	900,000	180,000	190,000
Electrophoretic mobility (average)	γ	Fast γ to β	Fast γ to β	Fast γ	
Complement fixation (classic)	+	0	++++	0	0
Serum concentration (approximate; mg/dL)	1000	200	120	3	0.05
Placental transfer	+	0	0	0	0
Reaginic activity	?	0	0	0	++++
Antibacterial lysis	+	+	+++	?	?
Antiviral activity	+	+++	+	?	?

[1]Modified and reproduced, with permission, from Stites DP, Terr AI, Parslow TG (editors): *Basic & Clinical Immunology,* 8th ed. Appleton & Lange, 1994.
[2]For monomeric serum IgA.
[3]J chain.
[4]Secretory component.
[5]For secretory IgA.

ous structural genes contributing to the formation of each immunoglobulin chain and upon a high frequency of somatic mutational events in the rearranged V_H and V_L genes.

Class Switching Occurs During Immune Responses

In most humoral immune responses, antibodies with identical specificity but of different classes are generated in a specific chronologic order in response to the immunogen (immunizing antigen). For instance, antibodies of the IgM class normally precede molecules of the IgG class. The switch from one class to another is designated "class switching," and its molecular basis has been investigated extensively. A single type of immunoglobulin light chain can combine with an antigen-specific μ chain to generate a specific IgM molecule. Subsequently, the same antigen-specific light chain combines with a γ chain with an identical V_H region to generate an IgG molecule with antigen specificity identical to that of the original IgM molecule. The same light chain can also combine with an α heavy chain, again containing the identical V_H region, to form an IgA molecule with identical antigen specificity. These three classes (IgM, IgG, and IgA) of immunoglobulin molecules against the same antigen have identical variable domains of both their light (V_L) chains and heavy (V_H) chains and are said to share an **idiotype.** The different classes of these three immunoglobulins (called **isotypes**) are thus determined by the C_H regions that are combined with the same antigen-specific V_H regions. One aspect of the genetic regulatory mechanisms responsible for the switching of the C_H region gene is discussed in Chapter 41.

Both Over- and Underproduction of Immunoglobulins May Result in Disease States

Disorders of immunoglobulins include increased production of specific classes of immunoglobulins or even specific immunoglobulin molecules, the latter by clonal tumors of plasma cells called myelomas. **Hypogammaglobulinemia** may be restricted to a single class of immunoglobulin molecules (eg, IgA or IgG) or may involve underproduction of all classes of immunoglobulins (IgA, IgD, IgE, IgG, and IgM). The disorders of immunoglobulin levels are almost without exception due to disordered rates of immunoglobulin production or secretion, for which there can be many causes.

Hybridomas Provide Long-Term Sources of Highly Useful Monoclonal Antibodies

When an antigen is injected into an animal, the resulting antibodies are polyclonal, being synthesized by a mixture of B cells. Polyclonal antibodies are directed against a number of different sites (epitopes or determinants) on the antigen and thus are not monospecific. However, by means of a method developed by Kohler and Milstein, large amounts of a single monoclonal antibody specific for one epitope can be obtained.

The method involves **cell fusion,** and the resulting permanent cell line is called a **hybridoma.** Typically, B cells are obtained from the spleen of a mouse (or other suitable animal) previously injected with an antigen or mixture of antigens (eg, foreign cells). The B cells are mixed with mouse **myeloma cells** and exposed to polyethylene glycol, which causes cell fusion. A summary of the principles involved in gener-

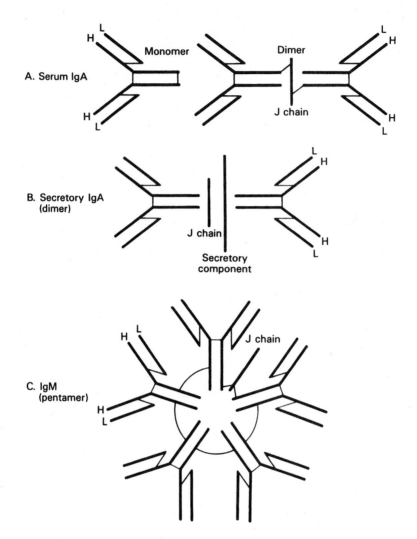

Figure 59–8. Highly schematic illustration of polymeric human immunoglobulins. Polypeptide chains are represented by thick lines; disulfide bonds linking different polypeptide chains are represented by thin lines. (Reproduced, with permission, from Stites DP, Terr AI, Parslow TG [editors]: *Basic & Clinical Immunology,* 8th ed. Appleton & Lange, 1994.)

ating hybridoma cells is given in Figure 59–9. Under the conditions used, only the hybridoma cells multiply in cell culture. This involves plating the hybrid cells into hypoxanthine-aminopterin-thymidine (HAT)-containing medium at a concentration such that each dish contains approximately one cell. Thus, a **clone** of hybridoma cells multiplies in each dish. The culture medium is harvested and screened for antibodies that react with the original antigen or antigens. If the immunogen is a mixture of many antigens (eg, a cell membrane preparation), an individual culture dish will contain a clone of hybridoma cells synthesizing a monoclonal antibody to one specific antigenic determinant of the mixture. By harvesting the media from many culture dishes, a battery of monoclonal antibodies can be obtained, many of which are specific for individual components of the immunogenic mixture. The hybridoma cells can be frozen and stored and subsequently thawed when more of the antibody is required; this ensures its long-term supply. The hybridoma cells can also be grown in the abdomen of mice, providing relatively large supplies of antibodies.

Because of their specificity, monoclonal antibodies have become useful reagents in many areas of biology and medicine. For example, they can be used to measure the amounts of many individual proteins (eg, plasma proteins), to determine the nature of infectious agents (eg, types of bacteria), and to subclassify both normal (eg, lymphocytes) and tumor cells (eg, leukemic cells). In addition, they are being used to direct therapeutic agents to tumor cells and also to

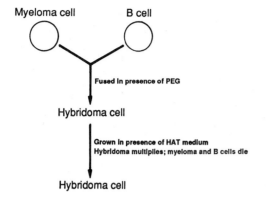

Figure 59–9. Scheme of production of a hybridoma cell. The myeloma cells are immortalized, do not produce antibody, and are HGPRT$^-$ (rendering the salvage pathway of purine synthesis [Chapter 36] inactive). The B cells are not immortalized, each produces a specific antibody, and they are HGPRT$^+$. Polyethylene glycol (PEG) stimulates cell fusion. The resulting hybridoma cells are immortalized (via the parental myeloma cells), produce antibody, and are HGPRT$^+$ (both latter properties gained from the parental B cells). The B cells will die in the medium because they are not immortalized. In the presence of HAT, the myeloma cells will also die, since the aminopterin in HAT suppresses purine synthesis by the de novo pathway by inhibiting reutilization of tetrahydrofolate (Chapter 36). However, the hybridoma cells will survive, grow (because they are HGPRT$^+$), and—if cloned—produce monoclonal antibody. (HAT, hypoxanthine, aminopterin, and thymidine; HGPRT, hypoxanthine-guanine phosphoribosyl transferase.)

accelerate removal of drugs from the circulation when they reach toxic levels (eg, digoxin).

The Complement System Comprises Some 20 Plasma Proteins and Is Involved in Cell Lysis, Inflammation, and Other Processes

Plasma contains approximately 20 proteins that are members of the complement system. This system was discovered when it was observed that addition of fresh serum containing antibodies directed to a bacterium caused its lysis. Unlike antibodies, the factor was labile when heated at 56°C. Subsequent work has resolved the proteins of the system and how they function; most have been cloned and sequenced. The major protein components are designated C1–9, with C9 associated with the C5–8 complex (together constituting the membrane attack complex) being involved in generating a lipid-soluble pore in the cell membrane that causes osmotic lysis.

The details of this system are relatively complex, and a textbook of immunology should be consulted. The basic concept is that the normally inactive proteins of the system, when triggered by a stimulus, be-

come activated by proteolysis and interact in a specific sequence with one or more of the other proteins of the system. This results in cell lysis and generation of fragments that are involved in various aspects of inflammation (chemotaxis, phagocytosis, etc). The system has other functions, such as clearance of antigen-antibody complexes from the circulation. Activation of the complement system is triggered by one of two routes, called the **classic** and the **alternative pathways.** The first involves interaction of C1 with antigen-antibody complexes, and the second (not involving antibody) involves direct interaction of bacterial cell surfaces or polysaccharides with a component designated C3b.

The complement system resembles blood coagulation (see below) in that it involves both conversion of inactive precursors to active products by proteases and a cascade with amplification.

HEMOSTASIS AND THROMBOSIS HAVE FOUR PHASES

Hemostasis is the cessation of bleeding from a cut or severed vessel, whereas thrombosis occurs when the endothelium lining blood vessels is damaged or removed. These processes encompass blood clotting (coagulation) and involve blood vessels, platelet aggregation, and plasma proteins that cause both formation and dissolution of platelet aggregates.

There are four phases to hemostasis and thrombosis:

(1) Constriction of the injured vessel to diminish blood flow distal to the injury.

(2) Formation of a loose and temporary platelet aggregate at the site of injury. Platelets bind to collagen at the site of vessel wall injury and are activated by thrombin (the mechanism of activation of platelets is described below), formed in the coagulation cascade at the same site, or by ADP released from other activated platelets. Upon activation, platelets change shape and, in the presence of fibrinogen, aggregate to form the hemostatic plug (in hemostasis) or thrombus (in thrombosis).

(3) Formation of a fibrin mesh that binds to the platelet aggregate, forming a more stable hemostatic plug or thrombus.

(4) Partial or complete dissolution of the hemostatic plug or thrombus by plasmin.

There Are Three Types of Thrombi

Three types of thrombi or clots are distinguished. All three contain **fibrin** in variable proportions.

(1) The **white** thrombus is composed of platelets and fibrin and is relatively poor in erythrocytes. It forms at the site of an injury or abnormal vessel wall, particularly in areas where blood flow is rapid (arteries).

(2) The **red** thrombus consists primarily of red

cells and fibrin. It morphologically resembles the clot formed in a test tube and may form in vivo in areas of retarded blood flow or stasis (eg, veins) with or without vascular injury, or it may form at a site of injury or in an abnormal vessel in conjunction with an initiating platelet plug.

(3) A third type is a disseminated **fibrin deposit** in very small blood vessels or capillaries.

We shall first describe the coagulation pathway leading to the formation of fibrin. Later, we shall briefly describe some aspects of the involvement of platelets and blood vessel walls in the overall process of coagulation. This separation of clotting factors and platelets is artificial, since both play intimate and often mutually interdependent roles in coagulation, but it facilitates description of the overall processes involved.

Both Intrinsic and Extrinsic Pathways Result in the Formation of Fibrin

Two pathways lead to fibrin clot formation: the intrinsic and the extrinsic pathways. These pathways are not independent, as previously thought. However, this artificial distinction is retained in the following text to facilitate their description.

The initiation of the red thrombus in an area of restricted blood flow or in response to an abnormal vessel wall without tissue injury is carried out by the intrinsic pathway. Initiation of the fibrin clot in response to tissue injury is carried out by the extrinsic pathway. These pathways converge in a **final common pathway** involving the activation of prothrombin to thrombin and the thrombin-catalyzed cleavage of fibrinogen to form the fibrin clot. The intrinsic, extrinsic, and final common pathways are complex and involve many different proteins (Figure 59–10 and Table 59–10). In general, as shown in Table 59–11, these proteins can be classified into five types: (1) zymogens of serine-dependent proteases, which become activated during the process of coagulation; (2) cofactors; (3) fibrinogen; (4) a transglutaminase, which stabilizes the fibrin clot; and (5) regulatory and other proteins.

The Intrinsic Pathway Leads to Activation of Factor X

The intrinsic pathway (Figure 59–10) involves factors XII, XI, IX, VIII, and X as well as prekallikrein, high-molecular-weight (HMW) kininogen, Ca^{2+}, and platelet phospholipids. It results in the production of factor Xa (by convention, activated clotting factors are referred to by use of the suffix a).

This pathway commences with the "contact phase" in which prekallikrein, HMW kininogen, factor XII, and factor XI are exposed to a negatively charged activating surface. Collagen on the exposed surface of a blood vessel probably provides this site in vivo, whereas glass or kaolin can be used for in vitro tests of the intrinsic pathway. When the components of the contact phase assemble on the activating surface, factor XII is activated to factor XIIa upon proteolysis by kallikrein. This factor XIIa, generated by kallikrein, attacks prekallikrein to generate more kallikrein, setting up a reciprocal activation. Factor XIIa, once formed, activates factor XI to XIa and also releases bradykinin (a nonapeptide with potent vasodilator action) from HMW kininogen.

Factor XIa in the presence of Ca^{2+} activates factor IX (55 kDa, a zymogen containing vitamin K-dependent γ-carboxyglutamate [Gla] residues), to the serine protease, factor IXa. This in turn cleaves an Arg-Ile bond in factor X (56 kDa) to produce the two-chain serine protease, factor Xa. This latter reaction requires the assembly of components, called the tenase complex, on the surface of activated platelets: Ca^{2+} and factor VIIIa, as well as factors IXa and X. It should be noted that in all reactions involving the Gla-containing zymogens (factors II, VII, IX, and X), the Gla residues in the amino terminal regions of the molecules serve as high-affinity binding sites for Ca^{2+}. For assembly of the tenase complex, the platelets must first be activated to expose the acidic (anionic) phospholipids, phosphatidylserine and phosphatidylinositol, that are normally on the internal side of the plasma membrane of resting, nonactivated platelets. Factor VIII (330 kDa), a glycoprotein, is not a protease precursor but a cofactor that serves as a receptor for factors IXa and X on the platelet surface. Factor VIII is activated by minute quantities of thrombin to form factor VIIIa, which is in turn inactivated upon further cleavage by thrombin.

The Extrinsic Pathway Also Leads to Activation of Factor X But by a Different Mechanism

Factor Xa occurs at the site where the intrinsic and extrinsic pathways converge (Figure 59–10) and lead into the final common pathway of blood coagulation. The extrinsic pathway involves tissue factor, factors VII and X, and Ca^{2+} and results in the production of factor Xa. It is initiated at the site of tissue injury with the expression of tissue factor (Figure 59–10), which acts as a cofactor in the factor VIIa catalyzed activation of factor X. Factor VIIa cleaves the same Arg-Ile bond in factor X that is cleaved by the tenase complex of the intrinsic pathway. Factor VII (53 kDa), a circulating Gla-containing glycoprotein synthesized in the liver, is a zymogen, but it has rather a high endogenous activity; this activity is increased by conversion to the active serine protease, factor VIIa, by **thrombin,** or **factor Xa.** Activation of factor X provides an important link between the intrinsic and extrinsic pathways.

An important interaction between the extrinsic and intrinsic pathways is that complexes of tissue factor and factor VIIa also activate factor IX in the intrinsic

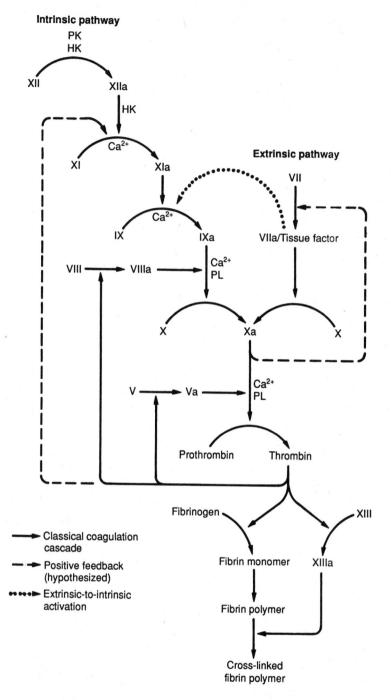

Figure 59–10. The pathways of blood coagulation. The intrinsic and extrinsic pathways are indicated. The events depicted below factor Xa are designated the final common pathway, culminating in the formation of cross-linked fibrin. New observations (dotted arrow) include the finding that complexes of tissue factor and factor VIIa activate not only factor X (in the classic extrinsic pathway) but also factor IX in the intrinsic pathway. In addition, thrombin and factor Xa feedback-activate at the two sites indicated (dashed arrows). (PK, prekallikrein; HK, HMW kininogen; PL, phospholipids.) (Reproduced, with permission, from Roberts HR, Lozier JN: New perspectives on the coagulation cascade. Hosp Pract [Off Ed] Jan 1992;27:97.)

Table 59–10. Numerical system for nomenclature of blood clotting factors. The numbers indicate the order in which the factors have been discovered and bear no relationship to the order in which they act.

Factor	Common Name
I	Fibrinogen ⎫ These factors are usually referred
II	Prothrombin ⎭ to by their common names.
III	Tissue factor ⎫ These factors are usually not re-
IV	Ca^{2+} ⎭ ferred to as coagulation factors.
V	Proaccelerin, labile factor, accelerator (Ac-) globulin
VII[1]	Proconvertin, serum prothrombin conversion accelerator (SPCA), cothromboplastin
VIII	Antihemophilic factor A, antihemophilic globulin (AHG)
IX	Antihemophilic factor B, Christmas factor, plasma thromboplastin component (PTC)
X	Stuart-Prower factor
XI	Plasma thromboplastin antecedent (PTA)
XII	Hageman factor
XIII	Fibrin stabilizing factor (FSF), fibrinoligase

[1]There is no factor VI.

pathway. Indeed, the formation of complexes between tissue factor and factor VIIa may be the key process involved in initiation of blood coagulation in vivo. The physiologic significance of the initial steps of the intrinsic pathway, in which factor XII, prekallikrein, and HMW kininogen are involved, has been called into question because patients with a hereditary deficiency of these components do not exhibit bleeding problems. Similarly, patients with a deficiency of factor XI may not have bleeding problems.

The Final Common Pathway of Blood Clotting Involves Activation of Prothrombin to Thrombin

In the final common pathway, factor Xa, produced by either the intrinsic or the extrinsic pathway, activates **prothrombin** (factor II) to **thrombin** (factor IIa), which then converts fibrinogen to fibrin (Figure 59–10).

The activation of prothrombin, like that of factor X, occurs on the surface of activated platelets and requires the assembly of a prothrombinase complex, consisting of platelet anionic phospholipids, Ca^{2+}, factor Va, factor Xa, and prothrombin.

Factor V (330 kDa), a glycoprotein with homology to factor VIII and ceruloplasmin, is synthesized in the liver, spleen, and kidney and is found in platelets as well as in plasma. It functions as a cofactor in a manner similar to that of factor VIII in the tenase complex. When activated to factor Va by traces of thrombin, it binds to specific receptors on the platelet membrane (Figure 59–11) and forms a complex with factor Xa and prothrombin. It is subsequently inactivated by further action of thrombin, thereby providing a means of limiting the activation of prothrombin to thrombin. **Prothrombin** (72 kDa;

Table 59–11. The functions of the proteins involved in blood coagulation.

Zymogens of serine proteases	
Factor XII	Binds to exposed collagen at site of vessel wall injury; activated by high-MW kininogen and kallikrein.
Factor XI	Activated by factor XIIa.
Factor IX	Activated by factor XIa in presence of Ca^{2+}.
Factor VII	Activated by thrombin in presence of Ca^{2+}.
Factor X	Activated on surface of activated platelets by prothrombinase complex (Ca^{2+}, factors VIIIa and IXa) and by factor VIIa in presence of tissue factor and Ca^{2+}.
Factor II	Activated on surface of activated platelets by prothrombinase complex (Ca^{2+}, factors Va and Xa). [Factors II, VII, IX, and X are Gla-containing zymogens.] (Gla = γ-carboxyglutamate.)
Cofactors	
Factor VIII	Activated by thrombin; factor VIIIa is a cofactor in the activation of factor X by factor IXa.
Factor V	Activated by thrombin; factor Va is a cofactor in the activation of prothrombin by factor Xa.
Tissue factor (factor III)	A protein exposed on the surface of stimulated endothelial cells that requires phospholipid to act as a cofactor for factor VII.
Fibrinogen	
Factor I	Cleaved by thrombin to form fibrin clot.
Thiol-dependent transglutaminase	
Factor XIII	Activated by thrombin in presence of Ca^{2+}; stabilizes fibrin clot by covalent cross-linking.
Regulatory and other proteins	
Protein C	Activated to protein Ca by thrombin bound to thrombomodulin; then degrades factors VIIIa and Va.
Protein S	Acts as a cofactor of protein C; both proteins contain Gla (γ-carboxyglutamate) residues.
Thrombomodulin	Protein on the surface of endothelial cells; binds thrombin, which then activates protein C.

Figure 59–12) is a single-chain glycoprotein synthesized in the liver. The amino terminal region of prothrombin (1 in Figure 59–12) contains ten Gla residues, and the serine-dependent active protease site (indicated by the arrowhead) is in the carboxyl terminal region of the molecule. Upon binding to the complex of factors Va and Xa on the platelet membrane, prothrombin is cleaved by factor Xa at two sites to generate the active, two-chain thrombin molecule, which is then released from the platelet surface. The A and B chains of thrombin are held together by a disulfide bond.

Conversion of Fibrinogen to Fibrin Is Catalyzed by Thrombin

Fibrinogen (factor I, 340 kDa; see Figures 59–11 and 59–12 and Tables 59–10 and 59–11) is a soluble plasma glycoprotein, 47.5 nm in length, that consists

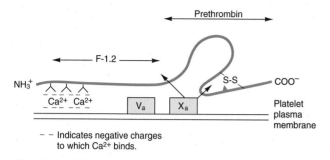

Figure 59–11. Diagrammatic representation (not to scale) of the binding of factors V_a, Ca^{2+}, and prothrombin to the plasma membrane of the activated platelet. The sites of cleavage of prothrombin by factor X_a are indicated by two arrows. The part of prothrombin destined to form thrombin is labeled prethrombin. The Ca^{2+} is bound to anionic phospholipids of the plasma membrane of the activated platelet.

of three nonidentical pairs of polypeptide chains $(A\alpha,B\beta\gamma)_2$ covalently linked by disulfide bonds. The $B\beta$ and γ chains contain asparagine-linked complex oligosaccharides. All three chains are synthesized in the liver; the three structural genes involved are on the same chromosome, and their expression is coordinately regulated in humans. The amino terminal regions of the six chains are held in close proximity by a number of disulfide bonds, while the carboxyl terminal regions are spread apart, giving rise to a highly asymmetric, elongated molecule (Figure 59–13). The A and B portions of the $A\alpha$ and $B\beta$ chains, designated fibrinopeptides A (FPA) and B (FPB), respectively, at the amino terminal ends of the chains, bear excess negative charges as a result of the presence of aspartate and glutamate residues, as well as an unusual tyrosine *O*-sulfate in FPB. These negative charges contribute to the solubility of fibrinogen in plasma and also serve to prevent aggregation by causing electrostatic repulsion between fibrinogen molecules.

Thrombin (34 kDa) is a serine protease present in plasma that hydrolyzes the four Arg-Gly bonds between the fibrinopeptides and the α and β portions of the $A\alpha$ and $B\beta$ chains of fibrinogen (Figure 59–14A). The release of the fibrinopeptides by

thrombin generates fibrin monomer, which has the subunit structure $(\alpha, \beta, \gamma)_2$. Since FPA and FPB contain only 16 and 14 residues, respectively, the fibrin molecule retains 98% of the residues present in fibrinogen. The removal of the fibrinopeptides exposes binding sites that allow the molecules of fibrin monomers to aggregate spontaneously in a regularly staggered array, forming an insoluble fibrin clot. It is the formation of this insoluble fibrin polymer that traps platelets, red cells, and other components to form the white or red thrombi. This initial fibrin clot is rather weak, held together only by the noncovalent association of fibrin monomers.

In addition to converting fibrinogen to fibrin, thrombin also converts factor XIII to factor XIIIa. This factor is a highly specific transglutaminase that covalently cross-links fibrin molecules by forming peptide bonds between the amide groups of glutamine and the ϵ-amino groups of lysine residues (Figure 59–14B), yielding a more stable fibrin clot with increased resistance to proteolysis.

Levels of Circulating Thrombin Must Be Carefully Controlled or Clots May Form

Once active thrombin is formed in the course of hemostasis or thrombosis, its concentration must be carefully controlled to prevent further fibrin formation or platelet activation. This is achieved in two ways. Thrombin circulates as its inactive precursor, prothrombin, which is activated as the result of a cascade of enzymatic reactions, each converting an inactive zymogen to an active enzyme and leading finally to the conversion of prothrombin to thrombin (Figure 59–10). At each point in the cascade, feedback mechanisms produce a delicate balance of activation and inhibition. The concentration of factor XII in plasma is approximately 30 µg/mL, while that of fibrinogen is 3 mg/mL, with intermediate clotting factors increasing in concentration as one proceeds down the cascade, showing that the clotting cascade provides amplification. The second means of controlling thrombin activity is the inactivation of any thrombin formed by circulating inhibitors, the most important of which is antithrombin III (see below).

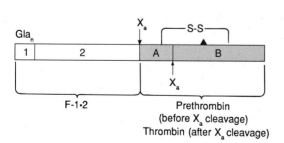

Figure 59–12. Diagrammatic representation (not to scale) of prothrombin. The amino terminus is to the left; region 1 contains all ten Gla residues. The sites of cleavage by factor X_a are shown and the products named. The site of the catalytically active serine residue is indicated by the solid triangle. The A and B chains of active thrombin (shaded) are held together by the disulfide bridge.

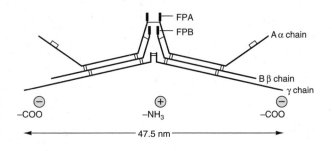

Figure 59–13. Diagrammatic representation (not to scale) of fibrinogen showing pairs of Aα, Bβ, and γ chains linked by disulfide bonds. (FPA, fibrinopeptide A; FPB, fibrinopeptide B.)

The Activity of Antithrombin III, an Inhibitor of Thrombin, Is Increased by Heparin

Four naturally occurring thrombin inhibitors exist in normal plasma. The most important is antithrombin III, which contributes approximately 75% of the antithrombin activity. Antithrombin III can also inhibit the activities of factors IXa, Xa, XIa, and XIIa. α_2-Macroglobulin contributes most of the remainder of the antithrombin activity, with heparin cofactor II and α_1-antitrypsin acting as minor inhibitors under physiologic conditions.

The endogenous activity of antithrombin III is greatly potentiated by the presence of acidic proteoglycans such as **heparin** (Chapter 57). These bind to a specific cationic site of antithrombin III, inducing a conformational change and promoting its binding to thrombin as well as to its other substrates. This is the basis for the use of heparin in clinical medicine to inhibit clotting. In addition, heparin in low doses appears to coat the endothelial lining of blood vessels, perhaps thereby reducing activation of the intrinsic pathway. The anticoagulant effects of heparin can be antagonized by strongly cationic polypeptides such as protamine, which bind strongly to heparin, thus inhibiting its binding to antithrombin III. Individuals with inherited deficiencies of antithrombin III are prone to develop frequent and widespread clots, providing evidence that antithrombin III has a physiologic function and that the clotting system in humans is normally in a dynamic state.

Thrombin is involved in an additional regulatory mechanism that operates in coagulation. It combines with thrombomodulin, a glycoprotein present on the surfaces of endothelial cells. The complex converts protein C to protein Ca. In combination with protein S, a cofactor, protein Ca degrades factors Va and VIIIa, limiting their actions in coagulation. A genetic

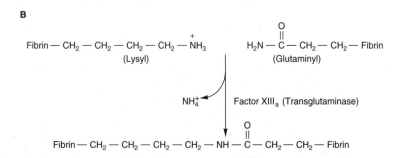

Figure 59–14. Formation of a fibrin clot. **A:** Thrombin-induced cleavage of Arg-Gly bonds of the Aα and Bβ chains of fibrinogen to produce fibrinopeptides (left-hand side) and the α and β chains of fibrin monomer (right-hand side). **B:** Cross-linking of fibrin molecules by activated factor XIII (factor XIIIa).

deficiency of either protein C or protein S can cause serious thrombotic episodes.

Coumarin Anticoagulants Inhibit the Vitamin K-Dependent Carboxylation of Factors II, VII, IX, and X

The coumarin drugs (eg, warfarin), which are used as anticoagulants, inhibit the vitamin K-dependent carboxylation of Glu to Gla residues (see Chapter 53) in the amino terminal regions of factors II, VII, IX, and X and also proteins C and S. These proteins, all of which are synthesized in the liver, are dependent on the Ca^{2+}-binding properties of the Gla residues for their normal function in the coagulation pathways. The coumarins act by inhibiting the reduction of the quinone derivatives of vitamin K to the active hydroquinone forms (Chapter 53). Thus, the administration of vitamin K will bypass the coumarin-induced inhibition and allow maturation of the Gla-containing factors. Reversal of coumarin inhibition by vitamin K requires 12–24 hours, whereas reversal of the anticoagulant effects of heparin by protamine is almost instantaneous.

Heparin and warfarin are widely used in the treatment of thrombotic and thromboembolic conditions, such as deep vein thrombosis and pulmonary embolus. Heparin is administered first, because of its prompt onset of action, whereas warfarin takes several days to reach full effect. Their effects are closely monitored by use of appropriate tests of coagulation (see below) because of the risk of producing hemorrhage.

Hemophilia A Is Due to a Genetically Determined Deficiency of Factor VIII

Inherited deficiencies of the clotting system are found in humans. The most common is deficiency of factor VIII, causing hemophilia A, an X chromosome-linked disease that has played a major role in the history of the royal families of Europe. Hemophilia B is due to a deficiency of factor IX; its clinical features are almost identical to those of hemophilia A, but the conditions can be separated on the basis of specific assays that distinguish between the two factors.

The gene for human factor VIII has been cloned and is one of the largest so far studied, measuring 186 kb in length and containing 26 exons. A variety of mutations have been detected leading to diminished activity of factor VIII; these include partial gene deletions and point mutations resulting in premature chain termination. Prenatal diagnosis by DNA analysis after chorionic villus sampling is now possible.

In recent years, treatment for patients with hemophilia A has consisted of administration of cryoprecipitates (enriched in factor VIII) prepared from individual donors or lyophilized factor VIII concentrates prepared from plasma pools of up to 5000 donors. It is now possible to prepare factor VIII by **recombinant DNA technology.** Such preparations are free of contaminating viruses (eg, HIV-1) found in human plasma but are at present expensive; their use may increase if cost of production decreases.

Fibrin Clots Are Dissolved by Plasmin

As stated above, the coagulation system is normally in a state of dynamic equilibrium in which fibrin clots are constantly being laid down and dissolved. Plasmin, the serine protease mainly responsible for degrading fibrin and fibrinogen, circulates in the form of its inactive zymogen, plasminogen (90 kDa), and any small amounts of plasmin that are formed in the fluid phase under physiologic conditions are rapidly inactivated by the fast-acting plasmin inhibitor, α_2-antiplasmin. Plasminogen binds to both fibrinogen and fibrin and thus becomes incorporated in clots as they are produced; since plasmin that is formed when bound to fibrin is protected from α_2-antiplasmin, it remains active. Activators of plasminogen of various types are found in most body tissues, and all cleave the same Arg-Val bond in plasminogen to produce the two-chain serine protease, plasmin (Figure 59–15).

Tissue plasminogen activator (alteplase, t-PA) is a serine protease that is released into the circulation from vascular endothelium under conditions of injury or stress and is catalytically inactive unless bound to fibrin. Upon binding to fibrin, t-PA cleaves plasminogen within the clot to generate plasmin, which in turn digests the fibrin to form soluble degradation products and thus dissolves the clot. Neither plasmin nor the plasminogen activator can remain bound to these degradation products, and so they are released into the fluid phase where they are inacti-

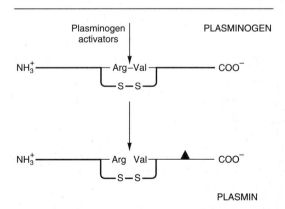

Figure 59–15. Activation of plasminogen. The same Arg-Val bond is cleaved by all plasminogen activators to give the two-chain plasmin molecule. The solid triangle indicates the serine residue of the active site. The two chains of plasmin are held together by a disulfide bridge.

vated by their natural inhibitors. Prourokinase is the precursor of a second activator of plasminogen, urokinase, which does not display the same high degree of selectivity for fibrin. Urokinase, which is secreted by certain epithelial cells lining excretory ducts (eg, renal tubules), is probably involved in lysing any fibrin that is deposited in such ducts. Figure 59–16 indicates the sites of action of five proteins that influence the formation and action of plasmin.

Recombinant t-PA and Streptokinase Are Used as Clot Busters

Alteplase (t-PA), produced by recombinant DNA technology, is used therapeutically as a fibrinolytic agent, as is streptokinase. However, the latter is less selective than t-PA, activating plasminogen in the fluid phase (where it can degrade circulating fibrinogen) as well as plasminogen that is bound to a fibrin clot. The amount of plasmin produced by therapeutic doses of streptokinase may exceed the capacity of the circulating α_2-antiplasmin, causing fibrinogen as well as fibrin to be degraded and resulting in the bleeding often encountered during fibrinolytic therapy. Because of its selectivity for degrading fibrin, there is considerable therapeutic interest in the use of recombinant t-PA to restore the patency of coronary arteries following thrombosis. If administered early enough, before irreversible damage of heart muscle occurs (about 6 hours after onset of thrombosis), t-PA can significantly reduce the mortality rate from myocardial damage following coronary thrombosis. It remains to be determined whether recombinant t-PA will prove superior to the much less expensive streptokinase in the treatment of acute coronary thrombosis. Table 59–12 compares some thrombolytic features of streptokinase and t-PA.

Table 59–12. Comparison of some properties of streptokinase (SK) and tissue plasminogen activator (t-PA) with regard to their use as thrombolytic agents.[1]

	SK	t-PA
Selective for fibrin clot	–	+
Produces plasminemia	+	–
Reduces mortality	+	+
Causes allergic reaction	+	–
Causes hypotension	+	–
Cost per treatment (approximate)	$400	$2900

[1]Data from Webb J, Thompson C: Thrombolysis for acute myocardial infarction. Can Fam Physician 1992;38:1415. Cost of treatment is in Canadian dollars.

There are a number of disorders, including cancer and shock, in which the concentrations of plasminogen activators increase. In addition, the antiplasmin activities contributed by α_1-antitrypsin and α_2-antiplasmin may be impaired in diseases such as cirrhosis. Since certain bacterial products, such as streptokinase, are capable of activating plasminogen, they may be responsible for the diffuse hemorrhage sometimes observed in patients with disseminated bacterial infections.

Activation of Platelets Involves Stimulation of the Polyphosphoinositide Pathway

In the formation of a hemostatic plug or thrombus, platelets must undergo three processes for hemostasis to occur: (1) adhesion to exposed collagen in blood vessels, (2) release of the contents of their granules, and (3) aggregation.

Platelets adhere to collagen via specific receptors on the platelet surface, including the glycoprotein complex GPIa–IIa ($\alpha_2\beta_1$ integrin; Chapter 60), in a reaction that involves von Willebrand factor. This is a glycoprotein, secreted by endothelial cells into the

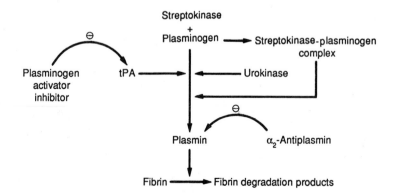

Figure 59–16. Scheme of sites of action of streptokinase, tissue plasminogen activator (t-PA), urokinase, plasminogen activator inhibitor, and α_2-antiplasmin (the last two proteins exert inhibitory actions). Streptokinase forms a complex with plasminogen, which exhibits proteolytic activity; this cleaves some plasminogen to plasmin, initiating fibrinolysis.

plasma, which stabilizes factor VIII and binds to collagen and the subendothelium. Platelets bind to von Willebrand factor via a glycoprotein complex (GPIb–V–IX) on the platelet surface; this interaction is especially important in platelet adherence to the subendothelium under conditions of high shear stress that occur in small vessels and stenosed arteries.

Platelets normally circulate in an unstimulated state. During hemostasis and thrombosis, however, involved platelets become activated. Platelet activation is a complex phenomenon that includes changes in platelet shape, secretion of contents of their granules, and aggregation.

Thrombin, the most potent activator of platelets, formed from the coagulation cascade, initiates platelet activation in vivo by interacting with its receptor on the plasma membrane (Figure 59–17). The further events leading to platelet activation are an example of transmembrane signaling, in which a chemical messenger outside the cell generates effector molecules inside the cell. In this instance, thrombin acts as the external chemical messenger (stimulus or agonist). The interaction of thrombin with its receptor stimulates the activity of a phospholipase C in the plasma membrane. This enzyme hydrolyzes phosphatidylinositol 4,5-bisphosphate (PIP_2, a polyphosphoinositide) to form the two internal effector molecules, 1,2-diacylglycerol and 1,4,5-inositol triphosphate.

Hydrolysis of PIP_2 is also involved in the action of many hormones (Chapter 44) and drugs. Diacylglycerol stimulates protein kinase C, which phosphorylates **pleckstrin**, a platelet protein (47 kDa). Phosphorylation of this protein results in aggregation and release of the contents of the various types of platelet granules (dense granules and alpha granules; Table 59–13). ADP released from dense granules can also activate platelets, resulting in activation of additional platelets. IP_3 causes release of Ca^{2+} into the cytosol mainly from the dense tubular system (or residual smooth endoplasmic reticulum from the megakaryocyte), which then interacts with calmodulin and myosin light chain kinase, leading to phosphorylation of the light chains of myosin. These chains then interact with actin, causing changes of platelet shape.

Collagen-induced activation of a platelet phospholipase A_2 by increased levels of cytosolic Ca^{2+} results in liberation of arachidonic acid from platelet

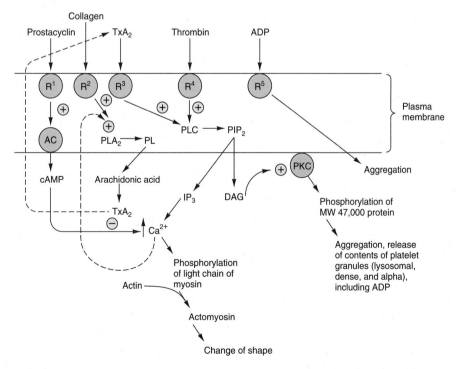

Figure 59–17. Diagrammatic representation of platelet activation. The external environment, the plasma membrane, and the inside of a platelet are depicted from top to bottom. Thrombin and collagen are the two most important platelet activators. ADP is considered a weak agonist; it causes aggregation but not granule release. (GP, glycoprotein; R^1–R^5, various receptors; AC, adenylyl cyclase; PLA_2, phospholipase A_2; PL, phospholipids; PLC, phospholipase C; PIP_2, phosphatidylinositol 4,5-bisphosphate; cAMP, cyclic AMP; PKC, protein kinase C; TxA_2, thromboxane A_2; IP_3, inositol 1,4,5-triphosphate; DAG, 1,2-diacylglycerol. The G proteins that are involved are not shown.)

Table 59–13. Some contents of platelet granules.

Granule	Content
Dense	ADP, ATP, Ca^{2+}, GTP, serotonin.
Alpha	Platelet factor 4 (PF-4),[1] β-thromboglobulin (β-TG), and platelet-derived growth factor (PDGF) (*not* plasma proteins). Albumin, fibrinogen, fibronectin, and von Willebrand factor (plasma proteins).
Lysosomal[2]	Various hydrolytic enzymes.

[1]PF-4 and β-TG are platelet proteins that bind heparin. PDGF is a growth factor.
[2]Contents released only partially by high concentrations of strong agonists.

phospholipids, leading to the formation of thromboxane A$_2$ (Chapter 25), which in turn can further activate phospholipase C, promoting platelet aggregation. Activated platelets, besides forming a platelet aggregate, are required, via their newly expressed anionic phospholipids, for acceleration of the activation of factors X and II in the coagulation cascade (Figure 59–10). All of the aggregating agents, including thrombin, collagen, ADP, and others such as platelet-activating factor will modify the platelet surface so that fibrinogen will attach to a glycoprotein complex, GPIIb–IIIa (α_{IIb}–β_3 integrin) (Chapter 60), on the activated platelet surface. Molecules of fibrinogen then link adjacent activated platelets to each other, forming a platelet aggregate. Some agents, including epinephrine, serotonin and vasopressin, exert synergistic effects with other aggregating agents.

Collagen-induced activation of a platelet phospholipase A$_2$ by increased levels of Ca^{2+} results in liberation of arachidonic acid from platelet phospholipids, leading to the formation of thromboxane A$_2$ (Chapter 25), which in turn can further activate phospholipase C, promoting platelet aggregation. Activated platelets, besides forming a platelet aggregate, are required, via their newly expressed phospholipids.

Endothelial Cells Synthesize Prostacyclin and Other Compounds That Affect Clotting and Thrombosis

The endothelial cells in the walls of blood vessels make important contributions to the overall regulation of clotting and thrombosis. As described in Chapter 25, these cells synthesize prostacyclin (PGI$_2$), a potent inhibitor of platelet aggregation, opposing the action of thromboxane A$_2$. Prostacyclin probably acts by stimulating the activity of adenylyl cyclase in the surface membranes of platelets. The resulting increase of intraplatelet cAMP opposes the increase in the level of intracellular Ca^{2+} produced by IP$_3$ and thus inhibits platelet activation (Figure 59–17). Endothelial cells play other roles in the regulation of thrombosis. For instance, these cells possess an ADPase, which hydrolyzes ADP, and thus op-

poses its aggregating effect on platelets. In addition, these cells appear to synthesize heparan sulfate, an anticoagulant, and they also synthesize plasminogen activators, which may help dissolve thrombi. Table 59–14 lists some compounds produced by endothelial cells that affect thrombosis and fibrinolysis. Endothelium-derived relaxing factor (nitric oxide) is discussed in Chapter 58.

Analysis of the mechanisms of uptake of atherogenic lipoproteins, such as LDL, by endothelial, smooth muscle, and monocytic cells of arteries, along with detailed studies of how these lipoproteins damage such cells is a key area of study in elucidating the mechanisms of atherosclerosis (Chapter 28).

Aspirin Is an Effective Antiplatelet Drug

Certain drugs (antiplatelet drugs) modify the behavior of platelets. The most important is aspirin (acetylsalicylic acid), which irreversibly acetylates and thus inhibits the platelet cyclooxygenase system involved in formation of thromboxane A$_2$ (Chapter 16), a potent aggregator of platelets and also a vasoconstrictor. Platelets are very sensitive to aspirin; as little as 30 mg/d (one aspirin tablet usually contains 325 mg) effectively eliminates the synthesis of thromboxane A$_2$. Aspirin also inhibits production of prostacyclin (PGI$_2$, which opposes platelet aggregation and is a vasodilator) by endothelial cells, but unlike platelets, these cells regenerate cyclooxygenase within a few hours. Thus, the overall balance between thromboxane A$_2$ and prostacyclin can be

Table 59–14. Molecules synthesized by endothelial cells that play a role in the regulation of thrombosis and fibrinolysis.[1]

Molecule	Action
ADPase (an ectoenzyme)	Degrades ADP (an activator of platelets) to AMP + P$_i$.
Endothelium-derived relaxing factor (nitric oxide)	Inhibits platelet adhesion and activation by elevating levels of cyclic GMP.
Heparan sulfate (a glycosaminoglycan)	Anticoagulant; combines with antithrombin III to inhibit thrombin.
Prostacyclin (PGI$_2$, a prostaglandin)	Inhibits platelet aggregation by elevating levels of cAMP.
Thrombomodulin (a glycoprotein)	Binds protein C, which is then cleaved by thrombin to yield activated protein C; this in combination with protein S degrades factors Va and VIIIa, limiting their actions.
Tissue plasminogen activator (t-PA, a protease)	Activates plasminogen to plasmin, which digests fibrin; the action of t-PA is opposed by plasminogen activator inhibitor–1 (PAI-1).

[1]Adapted from Wu KK: Endothelial cells in hemostasis, thrombosis and inflammation. Hosp Pract (Apr) 1992;27:145.

shifted in favor of the latter, opposing platelet aggregation. Indications for treatment with aspirin thus include management of angina and evolving myocardial infarction and also prevention of stroke and death in patients with transient cerebral ischemic attacks.

Laboratory Tests Measure Coagulation and Thrombolysis

A number of laboratory tests are available to measure the four phases of hemostasis described above. The tests include platelet count, bleeding time, partial thromboplastin time (PTT), activated partial thromboplastin time (APTT), prothrombin time (PT), thrombin time, concentration of fibrinogen, fibrin clot stability, and measurement of fibrin degradation products. The platelet count quantitates the number of platelets, and the bleeding time is an overall test of platelet function. PTT is a measure of the intrinsic pathway and PT of the extrinsic pathway. PT is used to measure the effectiveness of oral anticoagulants such as warfarin, and APTT is used to monitor heparin therapy. The reader is referred to a textbook of hematology for a discussion of these tests.

SUMMARY

Plasma contains many proteins with a variety of functions. Most are synthesized in the liver and are glycosylated. Albumin, which is not glycosylated, is the major protein and is the principal determinant of intravascular osmotic pressure; it also binds many ligands, such as drugs and bilirubin. Haptoglobin binds extracorpuscular hemoglobin, prevents its loss into the kidney and urine, and hence preserves its iron for reutilization. Transferrin binds iron, transporting it to sites where it is required. Ferritin is an intracellular store of iron. Iron deficiency anemia is a frequently occurring disorder. Ceruloplasmin contains substantial amounts of copper, but albumin appears to be more important with regard to its transport. Both Wilson's disease and Menkes' disease, which reflect abnormalities of copper metabolism,

have been found to be due to mutations in genes encoding copper-binding P-type ATPase. α_1-Antitrypsin is the major serine protease inhibitor of plasma, in particular inhibiting the elastase of neutrophils. Genetic deficiency of this protein is a cause of emphysema and can also lead to liver disease. α_2-Macroglobulin is a major plasma protein that neutralizes many proteases and targets certain cytokines to specific organs. Immunoglobulins play a key role in the defense mechanisms of the body, as do proteins of the complement system; some of the principal features of these proteins were described.

Blood coagulation is a complex process involving coagulation factors, platelets, and blood vessels. Many coagulation factors are zymogens of serine proteases, becoming activated during the overall process. Both intrinsic and extrinsic pathways of coagulation exist, the former initiated by exposure of subendothelial collagen and the latter by tissue factor. The pathways converge at factor Xa, starting the common final pathway resulting in thrombin-catalyzed conversion of fibrinogen to fibrin, which is strengthened by cross-linking, catalyzed by factor XIII. Genetic disorders of coagulation factors occur, and the two most common involve factors VIII (hemophilia A) and IX (hemophilia B). Among natural inhibitors of coagulation, the most important is antithrombin III; genetic deficiency of this protein can result in thromboses. For activity, factors II, VII, IX, and X and proteins C and S require vitamin K-dependent γ-carboxylation of certain glutamate residues, a process that is inhibited by the anticoagulant warfarin. Fibrin is dissolved by plasmin. Plasmin exists as an inactive precursor, plasminogen, which can be activated by tissue plasminogen activator (t-PA). Both t-PA and streptokinase are widely used to treat early thromboses in the coronary arteries. Thrombin and other agents cause platelet activation, which involves a variety of biochemical and morphologic events. Stimulation of phospholipase C and the polyphosphoinositide pathway is a key event in platelet activation, but other processes are also involved. Aspirin is an important antiplatelet drug that acts by inhibiting production of thromboxane A_2.

REFERENCES

Austen DEG: Clinical chemistry of blood coagulation. In: *Scientific Foundations of Biochemistry in Clinical Practice,* 2nd ed. Williams DL, Marks V (editors). Butterworth Heinemann, 1994.

Bennett JS: Mechanisms of platelet adhesion and aggregation: An update. Hosp Pract (Off Ed) Apr 1992;27: 124.

Broze GJ: Tissue factor pathway inhibitor and the revised theory of coagulation. Ann Rev Med 1995;46:103.

Collen D, Lijnen HR: Basic and clinical aspects of fibrinolysis and thrombolysis. Blood 1991;78:3114.

Crystal RG: α_1-Antitrypsin deficiency: Pathogenesis and treatment. Hosp Pract (Off Ed) Feb 1991;26:81.

Furie B, Furie BC: Molecular and cellular biology of blood coagulation. N Engl J Med 1992;326:800.

Handlin RI: Anticoagulant, fibrinolytic, and antiplatelet therapy. In: *Harrison's Principles of Internal Medicine,* 13th ed. Isselbacher KJ et al (editors). McGraw-Hill, 1994.

Handlin RI: Disorders of coagulation and thrombosis. In: *Harrison's Principles of Internal Medicine,* 13th ed. Isselbacher KJ et al (editors). McGraw-Hill, 1994.

Handlin RI: Disorders of the platelet and vessel wall. In:

Harrison's Principles of Internal Medicine, 13th ed. Isselbacher KJ et al (editors). McGraw-Hill, 1994.

Kroll MH, Schafer AI: Biochemical mechanisms of platelet activation. Blood 1989;74:1181.

Lomas DA et al: The mechanism of Z α_1-antitrypsin accumulation in the liver. Nature 1992;357:605.

Reiner AP, Davie EW: Introduction to hemostasis and the vitamin K-dependent coagulation factors. In: *Harrison's Principles of Internal Medicine,* 13th ed. Isselbacher KJ et al (editors). McGraw-Hill, 1994.

Roberts HR, Lozier JN: New perspectives on the coagulation cascade. Hosp Pract (Off Ed) Jan 1992;27:97.

Roth GJ, Calverley DC: Aspirin, platelets, and thrombosis: Theory and practice. Blood 1994;83:885.

Sifers RN: Z and the insoluble answer. Nature 1992;357:541.

Stites DP, Terr AI, Parslow TG (editors): *Basic & Clinical Immunology,* 8th ed. Appleton & Lange, 1994.

Venable ME et al: Platelet-activating factor: A phospholipid autacoid with diverse actions. J. Lipid Res 1993;34:691.

Whicher JT. Abnormalities of plasma proteins. In: *Scientific Foundations of Biochemistry in Clinical Practice,* 2nd ed. Williams DL, Marks V (editors). Butterworth Heinemann, 1994.

Worwood M: Disorders of iron metabolism. In: *Scientific Foundations of Biochemistry in Clinical Practice,* 2nd ed. Williams DL, Marks V (editors). Butterworth Heinemann, 1994.

Wu KK: Endothelial cells in hemostasis, thrombosis and inflammation. Hosp Pract (Off Ed) Apr 1992;27:145.

Red & White Blood Cells

60

Robert K. Murray, MD, PhD

INTRODUCTION

Blood is composed of plasma and various cells. Some plasma proteins are described in Chapter 59. Here, we discuss major biochemical features of red blood cells and of one class of white blood cells (leukocytes), namely, neutrophils. White blood cells are considered under three groups: granulocytes, monocytes, and lymphocytes. The granulocytes (also known as polymorphonuclear leukocytes [PMNs] because of their multilobed nuclei) contain numerous lysosomes and granules (secretory vesicles) and are divided into three classes. These classes (neutrophils, basophils, and eosinophils) are distinguished by their morphology and by the staining properties of their granules. Neutrophils phagocytose bacteria and play a major role in acute inflammation. Basophils, which resemble mast cells, contain histamine and heparin and play a role in some types of immunologic hypersensitivity reactions. Eosinophils are involved in certain allergic reactions and parasitic infections. Monocytes are precursors of macrophages, which, like neutrophils, are heavily involved in phagocytosis. Lymphocytes are classified as B or T lymphocytes. B lymphocytes synthesize antibodies (Chapter 59). T lymphocytes play major roles in various cellular immune mechanisms, such as killing virally infected cells and some cancer cells; the reader is referred to a textbook of immunology for detailed discussions. Neutrophils and lymphocytes are quantitatively by far the two major classes of white blood cells in normal blood. The role of platelets in coagulation is dealt with in Chapter 59 and will not be further discussed here. The differentiation of blood cells from stem cells is an important subject that is only briefly referred to in this chapter; readers are referred to textbooks of histology, cell biology, or hematology for more complete coverage.

BIOMEDICAL IMPORTANCE

Blood cells have been studied intensively because they are obtained easily, because of their functional importance, and because of their involvement in many disease processes. The structure and function of hemoglobin, the porphyrias, jaundice, and aspects of iron metabolism are discussed in previous chapters. Reduction of the number of red blood cells and of their content of hemoglobin is the cause of the anemias, a diverse and important group of conditions, some of which are seen very commonly in clinical practice. Certain of the blood group systems, present on the membranes of erythrocytes and other blood cells, are of extreme importance in relation to blood transfusion and tissue transplantation. Table 60–1 summarizes the causes of a number of important diseases affecting red blood cells; some are discussed in this chapter, and the remainder are discussed elsewhere in this text. Every organ in the body can be affected by inflammation; neutrophils play a central role in acute inflammation, and other white blood cells, such as lymphocytes, play important roles in chronic inflammation. Leukemias, defined as malignant neoplasms of blood-forming tissues, can affect precursor cells of any of the major classes of white blood cells; common types are acute and chronic myelocytic leukemia, affecting precursors of the neutrophils; and acute and chronic lymphocytic leukemias. Combination chemotherapy, using combinations of various chemotherapeutic agents, all of which act at one or more biochemical loci, has been remarkably effective in the treatment of certain of these types of leukemias. Understanding the role of red and white cells in health and disease requires a knowledge of certain fundamental aspects of their biochemistry.

THE RED BLOOD CELL IS SIMPLE IN TERMS OF ITS STRUCTURE AND FUNCTION

The major functions of the red blood cell are relatively simple, consisting of delivering oxygen to the tissues and of helping in the disposal of carbon dioxide and protons formed by tissue metabolism. Thus, it has a much simpler structure than most human cells, being essentially composed of a membrane surrounding a solution of hemoglobin (this protein forms about 95% of the intracellular protein of the red cell). There are no intracellular organelles, such

Table 60–1. Summary of the causes of some important disorders affecting RBCs.

Disorder	Sole or Major Cause
Iron deficiency anemia	Inadequate intake or excessive loss of iron
Methemoglobinemia	Intake of excess oxidants (various chemicals and drugs)
	Genetic deficiency in the NADH-dependent methemoglobin reductase system
	Inheritance of HbM
Sickle cell anemia	Sequence for codon 6 of the β chain changed from GAG in the normal gene to GTG in the sickle cell gene, resulting in substitution of valine for glutamic acid
α-Thalassemias	Mutations in the α-globin genes, mainly unequal crossing-over and large deletions and less commonly nonsense and frameshift mutations
β-Thalassemias	A very wide variety of mutations in the β-globin gene, including deletions, nonsense and frameshift mutations, and others affecting every aspect of its structure (eg, splice sites, promoter mutants)
Megaloblastic anemias	
Deficiency of vitamin B_{12}	Decreased absorption of B_{12}, often due to a deficiency of intrinsic factor, normally secreted by gastric parietal cells
Deficiency of folic acid	Decreased intake, defective absorption, or increased demand (eg, in pregnancy) for folate
Hereditary spherocytosis	Deficiencies in the amount or in the structure of spectrin
Glucose-6-phosphate dehydrogenase (G6PD) deficiency	A variety of mutations in the gene (X-linked) for G6PD, mostly single point mutations
Pyruvate kinase (PK) deficiency	Presumably a variety of mutations in the gene for the R (red cell) isozyme of PK

The last 3 disorders cause hemolytic anemias, as do some of the other disorders listed.

as mitochondria, lysosomes, or Golgi apparatus. Human red blood cells, like most red cells of animals, are nonnucleated. However, the red cell is not metabolically inert. ATP is synthesized from glycolysis and is important in processes that help the red blood cell maintain its biconcave shape and also in the regulation of the transport of ions (eg, by the Na^+-K^+ ATPase and the anion exchange protein [see below]) and of water in and out of the cell. The biconcave shape increases the surface-to-volume ratio of the red blood cell, thus facilitating gas exchange. The red cell contains cytoskeletal components (see below) that play an important role in determining its shape.

About Two Million Red Blood Cells Enter the Circulation per Second

The normal red blood cell count in men is 4.6–6.2 million/μL; in women, 4.2–5.4 million/μL. The total number of red blood cells in the circulation is approximately 2.5×10^{13}. The normal level of hemoglobin is 14–18 g/dL for men and 12–16 g/dL for women. The hematocrit values (the volume of packed red blood cells) for men and women are 42–52% and 37–47%, respectively. The life span of the normal red blood cell is 120 days; this means that slightly less than 1% of the population of red cells (200 billion cells, or 2 million per second) is replaced daily. The new red cells that appear in the circulation still contain ribosomes and elements of the endoplasmic reticulum. The RNA of the ribosomes can be detected by suitable stains (such as cresyl blue), and cells containing it are termed reticulocytes; they normally number about 1% of the total red blood cell count. The life span of the red blood cell can be dramatically shortened in a variety of **hemolytic anemias.** The number of reticulocytes is markedly increased in these conditions, as the bone marrow attempts to compensate for rapid breakdown of red blood cells by increasing the amount of new, young red cells in the circulation.

Erythropoietin Regulates Production of Red Blood Cells

Human erythropoietin is a glycoprotein of 166 amino acids (molecular mass about 34 kDa). Its amount in plasma can be measured by radioimmunoassay. It is the major regulator of human erythropoiesis. A cDNA encoding for erythropoietin has been isolated and sequenced. Erythropoietin is synthesized mainly by the kidney and is released in response to hypoxia into the bloodstream, in which it travels to the bone marrow. There it interacts with progenitors of red blood cells via a specific receptor. The receptor is a transmembrane protein consisting of two different subunits and a number of domains. It may be a tyrosine kinase, but this has not been established definitely. The identity of the second messenger and of the specific genes in the target cells that are affected are also not known. Erythropoietin interacts with a red cell progenitor, known as the burst-forming unit-erythroid (BFU-E), causing it to proliferate and differentiate. In addition, it interacts with a later progenitor of the red blood cell, called the colony-forming unit-erythroid (CFU-E), also causing it to proliferate and further differentiate. For these effects, erythropoietin requires the cooperation of other factors (eg, inter-

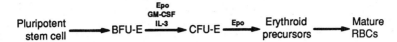

Figure 60–1. Greatly simplified scheme of differentiation of stem cells to red blood cells. (BFU-E, burst-forming unit-erythroid; CFU-E, colony-forming unit-erythroid; Epo, erythropoietin; GM-CSF, granulocyte-macrophage colony stimulating factor; IL-3, interleukin-3.)

leukin-3 and insulin-like growth factor; Figure 60–1).

The availability of a cDNA for erythropoietin has made it possible to produce substantial amounts of this hormone for analysis and for therapeutic purposes; previously the isolation of erythropoietin from human urine yielded very small amounts of the protein. The major use of recombinant erythropoietin has been in the treatment of a small number of **anemic states,** such as that due to renal failure.

MANY GROWTH FACTORS REGULATE PRODUCTION OF WHITE BLOOD CELLS

A large number of hematopoietic growth factors have been identified in recent years in addition to erythropoietin. This area of study adds to knowledge about the differentiation of blood cells, provides factors that may be useful in treatment, and also has implications for understanding of the abnormal growth of blood cells (eg, the leukemias). Like erythropoietin, most of the growth factors isolated have been glycoproteins, are very active in vivo and in vitro, interact with their target cells via specific cell surface receptors, and ultimately (via intracellular signals) affect gene expression, thereby promoting differentiation. Many have been cloned, permitting their production in relatively large amounts. Two of particular interest are **granulocyte-** and **granulocyte-macrophage colony-stimulating factors** (G-CSF and GM-CSF, respectively). G-CSF is relatively specific, inducing mainly granulocytes. GM-CSF affects a variety of progenitor cells and induces granulocytes, macrophages, and eosinophils. When the production of neutrophils is severely depressed, this condition is referred to as **neutropenia.** It is particularly likely to occur in patients treated with certain chemotherapeutic regimens and after bone marrow transplantation. Such patients are liable to develop overwhelming infections. G-CSF has been administered to such patients to boost production of neutrophils, and some studies have indicated that it is useful. One concern has been that if G-CSF is administered to patients who have developed severe neutropenia during treatment for leukemia, the growth factor may stimulate the growth of the leukemic cells. Recent work indicates that this problem may not occur if the dosage of G-CSF is kept low, but caution must be exercised.

THE RED BLOOD CELL HAS A UNIQUE AND RELATIVELY SIMPLE METABOLISM

Various aspects of the metabolism of the red cell, many of which are discussed in other chapters of this text, are summarized in Table 60–2.

The Red Blood Cell Has a Glucose Transporter in Its Membrane

The entry rate of glucose into red blood cells is far greater than would be calculated for simple diffusion. Rather, it is an example of facilitated diffusion (Chapter 43). The specific protein involved in this

Table 60–2. Summary of important aspects of the metabolism of the red blood cell.

The RBC is highly dependent upon glucose as its energy source; its membrane contains high affinity glucose transporters.

Glycolysis, producing lactate, is the site of production of ATP.

Because there are no mitochondria in RBCs, there is no production of ATP by oxidative phosphorylation.

The RBC has a variety of transporters that maintain ionic and water balance.

Production of 2,3-bisphosphoglycerate, by reactions closely associated with glycolysis, is important in regulating the ability of Hb to transport oxygen.

The pentose phosphate pathway is operative in the RBC (it metabolizes about 5–10% of the total flux of glucose) and produces NADPH; hemolytic anemia due to a deficiency of the activity of glucose-6-phosphate dehydrogenase is common.

Reduced glutathione (GSH) is important in the metabolism of the RBC, in part to counteract the action of potentially toxic peroxides; the RBC can synthesize GSH and requires NADPH to return oxidized glutathione (G-S-S-G) to the reduced state.

The iron of Hb must be maintained in the ferrous state; ferric iron is reduced to the ferrous state by the action of an NADH-dependent methemoglobin reductase system involving cytochrome b_5 reductase and cytochrome b_5.

Synthesis of glycogen, fatty acids, protein, and nucleic acids does not occur in the RBC; however, some lipids (eg, cholesterol) in the red cell membrane can exchange with corresponding plasma lipids.

The RBC contains certain enzymes of nucleotide metabolism (eg, adenosine deaminase, pyrimidine nucleotidase, and adenylate kinase); deficiencies of these enzymes are involved in some cases of hemolytic anemia.

When RBCs reach the end of their life span, the globin is degraded to amino acids (which are reutilized in the body), the iron is released from heme and also reutilized, and the tetrapyrrole component of heme is converted to bilirubin, which is mainly excreted into the bowel via the bile.

process is called the **glucose transporter** or glucose permease. Some of its properties are summarized in Table 60–3. The process of entry of glucose into red blood cells is of major importance because it is the major fuel supply for these cells. Five different but related glucose transporters have been isolated from various tissues; unlike the red cell transporter, some of these are insulin-dependent (eg, in muscle and adipose tissue). There is considerable interest in the latter types of transporter because defects in their recruitment from intracellular sites to the surface of skeletal muscle cells may help explain the **insulin resistance** displayed by patients with non-insulin-dependent diabetes mellitus (Chapter 51).

Reticulocytes Are Active in Protein Synthesis

The mature red blood cell cannot synthesize protein. Reticulocytes are active in protein synthesis. Once reticulocytes enter the circulation, they lose their intracellular organelles (ribosomes, mitochondria, etc) within about 24 hours, becoming young red blood cells and concomitantly losing their ability to synthesize protein. Extracts of rabbit reticulocytes (obtained by injecting rabbits with a chemical—phenylhydrazine—that causes a severe hemolytic anemia, so that the red cells are almost completely replaced by reticulocytes) are widely used as an in vitro system for synthesizing proteins. Endogenous mRNAs present in these reticulocytes are destroyed by use of a nuclease, whose activity can be inhibited by addition of Ca^{2+}. The system is then programmed by adding purified mRNAs or whole-cell extracts of mRNAs, and radioactive proteins are synthesized in the presence of ^{35}S-labeled L-methionine or other radiolabeled amino acids. The radioactive proteins synthesized are separated by SDS-PAGE and detected by radioautography.

Superoxide Dismutase, Catalase, and Glutathione Protect Blood Cells From Oxidative Stress and Damage

Several powerful oxidants are produced during the course of metabolism, in both blood cells and most other cells of the body. These include superoxide (O_2^-), hydrogen peroxide (H_2O_2), peroxyl radicals ($ROO^\bullet$), and hydroxyl radical ($OH^\bullet$). The last is a particularly reactive molecule and can react with proteins, nucleic acids, lipids, and other molecules to alter their structure and produce tissue damage. The reactions listed in Table 60–4 play an important role in forming these oxidants and in disposing of them; each of these reactions will now be considered in turn.

Superoxide is formed (reaction 1) in the red blood cell by the auto-oxidation of hemoglobin to methemoglobin (approximately 3% of hemoglobin in human red blood cells has been calculated to auto-oxidize per day); in other tissues, it is formed by the action of enzymes such as cytochrome P450 reductase and xanthine oxidase. When stimulated by contact with bacteria, neutrophils exhibit a respiratory burst (see below) and produce superoxide in a reaction catalyzed by NADPH-oxidase (reaction 2). Superoxide spontaneously dismutates to form H_2O_2 and O_2; however, the rate of this same reaction is speeded up tremendously by the action of the enzyme superoxide dismutase (reaction 3). Hydrogen peroxide is subject to a number of fates. The enzyme **catalase,** present in many types of cells, converts it to H_2O and O_2 (reaction 4). Neutrophils possess a unique enzyme, myeloperoxidase, that uses H_2O_2 and halides to produce hypohalous acids (reaction 5); this subject is discussed further below. The selenium-containing enzyme glutathione peroxidase (Chapter 22) will also act on reduced glutathione (GSH) and H_2O_2 to produce oxidized glutathione (GSSG) and H_2O (reaction 6); this enzyme can also use other peroxides as substrates. $OH^\bullet$ and OH^- can be formed from H_2O_2 in a nonenzymatic reaction catalyzed by Fe^{2+} (the **Fenton reaction,** reaction 7). O_2^- and H_2O_2 are the substrates in the iron-catalyzed **Haber-Weiss reaction** (reaction 8), which also produces $OH^\bullet$ and OH^-. Superoxide can release iron ions from ferritin. Thus, production of $OH^\bullet$ may be one of the mechanisms involved in tissue injury due to iron overload (eg, hemochromatosis; Chapter 59).

Chemical compounds and reactions capable of generating potential toxic oxygen species can be referred to as **pro-oxidants.** On the other hand, compounds and reactions disposing of these species, scavenging them, suppressing their formation, or opposing their actions are **antioxidants** and include compounds such as NADPH, GSH, ascorbic acid, and vitamin E.

Table 60–3. Some properties of the glucose transporter of the membrane of the RBC.

It accounts for about 2% of the protein of the membrane of the RBC.

It exhibits specificity for glucose and related D-hexoses (L-hexoses are not transported).

The transporter functions at approximately 75% of its V_{max} at the physiologic concentration of blood glucose, is saturable, and can be inhibited by certain analogs of glucose.

Five similar but distinct glucose transporters have been detected to date in mammalian tissues, of which the red cell transporter is one.

It is not dependent upon insulin, unlike the corresponding carrier in muscle and adipose tissue.

Its complete amino acid sequence (492 amino acids) has been determined.

It transports glucose when inserted into artificial liposomes.

It is estimated to contain 12 transmembrane helical segments.

It functions by generating a gated pore in the membrane to permit passage of glucose; the pore is conformationally dependent on the presence of glucose and can oscillate rapidly (about 900 times/s).

Table 60–4. Reactions of importance in relation to oxidative stress in blood cells and various tissues.

(1)	Production of superoxide (by-product of various reactions)	$O_2 + e^- \rightarrow O_2^-$
(2)	NADPH-oxidase	$2\ O_2 + NADPH \rightarrow 2\ O_2^- + NADP + H^+$
(3)	Superoxide dismutase	$O_2^- + O_2^- + 2H^+ \rightarrow H_2O_2 + O_2$
(4)	Catalase	$H_2O_2 \rightarrow 2\ H_2O + O_2$
(5)	Myeloperoxidase	$H_2O_2 + X^- + H^+ \rightarrow HOX + H_2O\ (X^- = Cl^-,\ Br^-,\ SCN^-)$
(6)	Glutathione peroxidase (Se-dependent)	$2\ GSH + R\text{-}O\text{-}OH \rightarrow GSSG + H_2O + ROH$
(7)	Fenton reaction	$Fe^{2+} + H_2O_2 \rightarrow Fe^{3+} + OH^{\cdot} + OH^-$
(8)	Iron-catalyzed Haber-Weiss reaction	$O_2^- + H_2O_2 \rightarrow O_2 + OH^{\cdot} + OH^-$
(9)	Glucose-6-phosphate dehydrogenase (G6PD)	$G6P + NADP \rightarrow 6\ Phosphogluconate + NADPH + H^+$
(10)	Glutathione reductase	$G\text{-}S\text{-}S\text{-}G + NADPH + H^+ \rightarrow 2\ GSH + NADP$

In a normal cell, there is an appropriate pro-oxidant:antioxidant balance. However, this balance can be shifted toward the pro-oxidants when production of oxygen species is increased greatly (eg, following ingestion of certain chemicals or drugs) or when levels of antioxidants are diminished (eg, by inactivation of enzymes involved in disposal of oxygen species and by conditions that cause low levels of the antioxidants mentioned above). This state is called "oxidative stress" and can result in serious cell damage if the stress is massive or prolonged.

Oxygen species are now thought to play an important role in many types of **cellular injury** (eg, resulting from administration of various toxic chemicals or from ischemia), some of which can result in cell death. Indirect evidence supporting a role for these species in generating cell injury is provided if administration of an enzyme such as superoxide dismutase or catalase is found to protect against cell injury in the situation under study.

Deficiency of Glucose-6-Phosphate Dehydrogenase Is Frequent in Certain Areas and Is an Important Cause of Hemolytic Anemia

NADPH, produced in the reaction catalyzed by the X-linked glucose-6-phosphate dehydrogenase (Table 60–4, reaction 9) in the pentose phosphate pathway (Chapter 22), plays a key role in supplying reducing equivalents in the red cell and in other cells such as the hepatocyte. Because the pentose phosphate pathway is virtually its sole means of producing NADPH, the red blood cell is very sensitive to oxidative damage if the function of this pathway is impaired (eg, by enzyme deficiency). One function of NADPH is to reduce GSSG to GSH, a reaction catalyzed by glutathione reductase (reaction 10).

Deficiency of the activity of glucose-6-phosphate dehydrogenase, owing to mutation, is extremely frequent in some regions of the world (eg, tropical Africa, the Mediterranean, certain parts of Asia, and in North America among blacks). It is the most common of all enzymopathies (diseases caused by abnormalities of enzymes), and over 300 genetic variants of the enzyme have been distinguished; at least 100 million people are deficient in this enzyme owing to these variants. The disorder resulting from deficiency of glucose-6-phosphate dehydrogenase is **hemolytic anemia.** Consumption of broad beans (*Vicia faba*) by individuals deficient in activity of the enzyme can precipitate an attack of hemolytic anemia (most likely because the beans contain potential oxidants). In addition, a number of drugs (eg, the antimalarial drug **primaquine** [the condition caused by intake of primaquine is called **primaquine-sensitive hemolytic anemia**] and **sulfonamides**) and chemicals (eg, naphthalene) precipitate an attack, because their intake leads to generation of H_2O_2 or O_2^-. Normally, H_2O_2 is disposed of by catalase and glutathione peroxidase (Table 60–4, reactions 4 and 6), the latter causing increased production of GSSG. GSH is regenerated from GSSG by the action of the enzyme glutathione reductase, which depends on the availability of NADPH (reaction 10). The red blood cells of individuals who are deficient in the activity of glucose-6-phosphate dehydrogenase cannot generate sufficient NADPH to regenerate GSH from GSSG, which in turn impairs their ability to dispose of H_2O_2 and of oxygen radicals. These compounds can cause oxidation of critical SH groups in proteins and possibly peroxidation of lipids in the membrane of the red cell, causing lysis of the red cell membrane. Some of the SH groups of hemoglobin become oxidized, and the protein precipitates inside the red blood cell, forming **Heinz bodies,** which stain purple with cresyl violet. The presence of Heinz bodies indicates that red blood cells have been subjected to oxidative stress. Figure 60–2 summarizes the possible chain of events in hemolytic anemia due to deficiency of glucose-6-phosphate dehydrogenase.

Methemoglobin Is Useless in Transporting Oxygen

The ferrous iron of hemoglobin is susceptible to oxidation by superoxide and other oxidizing agents, forming methemoglobin, which cannot transport oxygen. Only a very small amount of methemoglobin is present in normal blood, as the red blood cell possesses an effective system (the NADH-cytochrome b_5 methemoglobin reductase system) for reducing heme Fe^{3+} back to the Fe^{2+} state. This system consists of NADH (generated by glycolysis), a flavopro-

Mutations in the gene for G6PD
↓
Decreased activity of G6PD
↓
Decreased levels of NADPH
↓
Decreased regeneration of GSH from GSSG by glutathione reductase (which uses NADPH)
↓
Oxidation, due to decreased levels of GSH and increased levels of intracellular oxidants (eg, O_2^-), of SH groups of Hb (forming Heinz bodies) and of membrane proteins, altering membrane structure and increasing susceptibility to ingestion by macrophages (peroxidative damage to lipids in the membrane also possible)
↓
Hemolysis

Figure 60–2. Summary of probable events causing hemolytic anemia due to deficiency of the activity of glucose-6-phosphate dehydrogenase (G6PD).

tein named cytochrome b_5 reductase (also known as methemoglobin reductase), and cytochrome b_5. The Fe^{3+} of methemoglobin is reduced back to the Fe^{2+} state by the action of reduced cytochrome b_5:

$$Hb\text{-}Fe^{3+} + Cyt\ b_{5\ red} \rightarrow Hb\text{-}Fe^{2+} + Cyt\ b_{5\ ox}$$

Reduced cytochrome b_5 is then regenerated by the action of cytochrome b_5 reductase:

$$Cyt\ b_{5\ ox} + NADH \rightarrow Cyt\ b_{5\ red} + NAD$$

Methemoglobinemia Is Inherited or Acquired

Methemoglobinemia can be classified as either inherited or acquired by ingestion of certain drugs and chemicals. Neither type occurs frequently, but physicians must be aware of them. The inherited form is usually due to deficient activity of methemoglobin reductase, transmitted in an autosomal recessive manner. Certain abnormal hemoglobins (Hb M) are also rare causes of methemoglobinemia. In Hb M, mutation changes the amino acid residue to which heme is attached, thus altering its affinity for oxygen and favoring its oxidation. Ingestion of certain drugs (eg, sulfonamides) or chemicals (eg, aniline) can cause acquired methemoglobinemia. Cyanosis (bluish discoloration of the skin and mucous membranes due to increased amounts of deoxygenated hemoglobin in arterial blood, or in this case due to increased amounts of methemoglobin) is usually the presenting sign in both types and is evident when over 10% of hemoglobin is in the "met" form. Diagnosis is made by spectroscopic analysis of blood, which reveals the characteristic absorption spectrum of methemoglobin. Additionally, a sample of blood containing methemo-

globin cannot be fully reoxygenated by flushing oxygen through it, whereas normal deoxygenated blood can. Electrophoresis can be used to confirm the presence of an abnormal hemoglobin. Ingestion of methylene blue or ascorbic acid (reducing agents) is used to treat mild methemoglobinemia due to enzyme deficiency. Acute massive methemoglobinemia (due to ingestion of chemicals) should be treated by intravenous injection of methylene blue.

MORE IS KNOWN ABOUT THE MEMBRANE OF THE HUMAN RED BLOOD CELL THAN ABOUT THE SURFACE MEMBRANE OF ANY OTHER HUMAN CELL

A variety of biochemical approaches have been used to study the membrane of the red blood cell (Chapter 43). These include analysis of membrane proteins by SDS-PAGE, the use of specific enzymes (proteinases, glycosidases, and others) to determine the location of proteins and glycoproteins in the membrane, and various techniques to study both the lipid composition and disposition of individual lipids. Morphologic (eg, electron microscopy, freeze-fracture electron microscopy) and other techniques (eg, use of antibodies to specific components) have also been widely used. When red blood cells are lysed under specific conditions, their membranes will reseal in their original orientation to form **ghosts** (right-side-out ghosts). By altering the conditions, ghosts can also be made to reseal with their cytosolic aspect exposed on the exterior (inside-out ghosts). Both types of ghosts have been useful in analyzing the disposition of specific proteins and lipids in the membrane. In recent years, cDNAs for many proteins of this membrane have become available, permitting the deduction of their amino sequences and domains. All in all, more is known about the membrane of the red blood cell than about any other membrane of human cells (Table 60–5).

Analysis by SDS-PAGE Resolves the Proteins of the Membrane of the Red Blood Cell

When the membranes of red blood cells are analyzed by SDS-PAGE, about ten major proteins are resolved (Figure 60–3), several of which have been shown to be **glycoproteins** by use of the periodic acid-Schiff (PAS) reagent (Chapter 56). Their migration on SDS-PAGE was used to name these proteins, with the slowest migrating (and hence highest molecular mass) being designated band 1 or **spectrin**. All these major proteins have been isolated, most of them have been identified, and considerable insight has been obtained about their functions (Table 60–6). Many of their amino acid sequences also have been established. In addition, it has been determined which are integral or

Table 60–5. Summary of biochemical information about the membrane of the human red blood cell.

The membrane is a bilayer composed of about 50% lipid and 50% protein.

The major lipid classes are phospholipids and cholesterol; the major phospholipids are phosphatidylcholine (PC), phosphatidylethanolamine (PE), and phosphatidylserine (PS) along with sphingomyelin (Sph).

The choline-containing phospholipids, PC and Sph, predominate in the outer leaflet and the amino-containing phospholipids (PE and PS) in the inner leaflet.

Glycosphingolipids (GSLs) (neutral GSLs, gangliosides, and complex species, including the ABO blood group substances) constitute about 5–10% of the total lipid.

Analysis by SDS-PAGE shows that the membrane contains about 10 major proteins and more than 100 minor species.

The major proteins (which include spectrin, ankyrin, the anion exchange protein, actin, and band 4.1) have been studied intensively, and the principal features of their disposition (eg, integral or peripheral), structure, and function have been established.

Many of the proteins are glycoproteins (eg, the glycophorins) containing O- and/or N-linked oligosaccharide chains located on the external surface of the membrane.

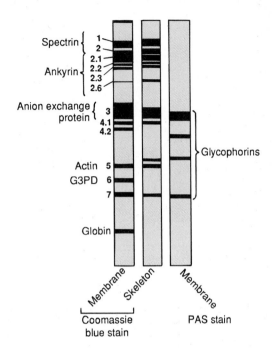

Figure 60–3. Diagrammatic representation of the major proteins of the membrane of the human red blood cell separated by SDS-PAGE. The bands detected by staining with Coomassie blue are shown in the two left-hand channels, and the glycoproteins detected by staining with periodic acid-Schiff reagent are shown in the right-hand channel. (Reproduced, with permission, from Beck WS, Tepper RI: Hemolytic anemias III: Membrane disorders. In: *Hematology,* 5th ed. Beck WS [editor]. The MIT Press, 1991.)

peripheral membrane proteins, which are situated on the external surface, which are on the cytosolic surface, and which span the membrane (Figure 60–4). Many minor components can also be detected in the red cell membrane by use of sensitive staining methods or two-dimensional gel electrophoresis. One of these is the glucose transporter described above.

The Major Integral Proteins of the Red Blood Cell Membrane Are the Anion Exchange Protein and the Glycophorins

The anion exchange protein (band 3) is a transmembrane glycoprotein, with its carboxyl terminal

Table 60–6. Principal proteins of the red cell membrane.[1]

Band Number[2]	Protein	Integral (I) or Peripheral (P)	Approximate Molecular Mass (kDa)
1	Spectrin (α)	P	240
2	Spectrin (β)	P	220
2.1	Ankyrin	P	210
2.2	"	P	195
2.3	"	P	175
2.6	"	P	145
3	Anion exchange protein	I	100
4.1	Unnamed	P	80
5	Actin	P	43
6	Glyceraldehyde-3-phosphate dehydrogenase	P	35
7	Tropomyosin	P	29
8	Unnamed	P	23
	Glycophorins A, B, and C	I	31, 23, and 28

[1]Adapted from Lux SE, Becker PS: Disorders of the red cell membrane skeleton: Hereditary spherocytosis and hereditary elliptocytosis. Chapter 95 in: *The Metabolic Basis of Inherited Disease,* 6th ed. Scriver CR et al (editors). McGraw-Hill, 1989.

[2]The band number refers to the position of migration on SDS-PAGE (see Figure 60–3). The glycophorins are detected by staining with the periodic acid–Schiff reagent. A number of other components (eg, 4.2 and 4.9) are not listed. Native spectrin is $\alpha_2\beta_2$.

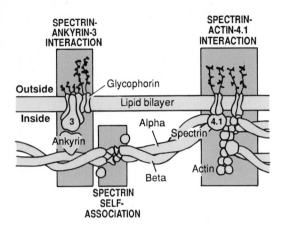

Figure 60–4. Diagrammatic representation of the interaction of cytoskeletal proteins with each other and with certain integral proteins of the membrane of the red blood cell. (Reproduced, with permission, from Beck WS, Tepper RI: Hemolytic anemias III: Membrane disorders. In: *Hematology,* 5th ed. Beck WS [editor]. The MIT Press, 1991.)

end on the external surface of the membrane and its amino terminal end on the cytoplasmic surface. It is an example of a **multipass** membrane protein, extending across the bilayer at least ten times. It probably exists as a dimer in the membrane, in which it forms a tunnel, permitting the exchange of chloride for bicarbonate. Carbon dioxide, formed in the tissues, enters the red cell as bicarbonate, which is exchanged for chloride in the lungs, where carbon dioxide is exhaled. The amino terminal end binds many proteins, including hemoglobin, proteins 4.1 and 4.2, ankyrin, and several glycolytic enzymes. Purified band 3 has been added to lipid vesicles in vitro and has been shown to perform its transport functions in this reconstituted system.

Glycophorins A, B, and **C** are also transmembrane glycoproteins but of the **single-pass** type, extending across the membrane only once. A is the major glycophorin, is made up of 131 amino acids, and is heavily glycosylated (about 60% of its mass). Its amino terminal end, which contains 16 oligosaccharide chains (15 of which are *O*-glycans), extrudes out from the surface of the red blood cell. Approximately 90% of the sialic acid of the red cell membrane is located in this protein. Its transmembrane segment (23 amino acids) is α-helical. The carboxyl terminal end extends into the cytosol and binds to protein 4.1, which in turn binds to spectrin. Polymorphism of this protein is the basis of the MN blood group system (see below). Glycophorin A contains binding sites for influenza virus and for *Plasmodium falciparum,* one cause of malaria. Intriguingly, the function of red blood cells of individuals who lack glycophorin A does not appear to be affected.

Spectrin, Ankyrin, and Other Peripheral Membrane Proteins Help Determine the Shape and Flexibility of the Red Blood Cell

The red blood cell must be able to squeeze through some tight spots in the microcirculation during its numerous passages around the body; the sinusoids of the spleen are of special importance in this regard. For the red cell to be easily and reversibly deformable, its membrane must be both fluid and flexible; it should also preserve its biconcave shape, since this facilitates gas exchange. Membrane lipids help determine membrane fluidity. Attached to the inner aspect of the membrane of the red blood cell are a number of peripheral cytoskeletal proteins (Table 60–6) that play important roles in respect to preserving shape and flexibility; these will now be described.

Spectrin is the major protein of the cytoskeleton. It is composed of two polypeptides: spectrin 1 (α chain) and spectrin 2 (β chain). These chains, measuring approximately 100 nm in length, are aligned in an antiparallel manner and are loosely intertwined, forming a dimer. Both chains are made up of segments of 106 amino acids that appear to fold into triple-stranded α-helical coils joined by nonhelical segments. One dimer interacts with another, forming a head-to-head tetramer. The overall shape confers flexibility on the protein and in turn on the membrane of the red blood cell. At least four binding sites can be defined in spectrin: (1) for self-association, (2) for ankyrin (bands 2.1, etc), (3) for actin (band 5), and (4) for protein 4.1.

Ankyrin is a pyramid-shaped protein that binds spectrin. In turn, ankyrin binds tightly to band 3, securing attachment of spectrin to the membrane. Ankyrin is sensitive to proteolysis, accounting for the appearance of bands 2.2, 2.3, and 2.6, all of which are derived from band 2.1.

Actin (band 5) exists in red blood cells as short, double-helical filaments of F-actin. The tail end of spectrin dimers binds to actin. Actin also binds to protein 4.1.

Protein 4.1, a globular protein, binds tightly to the tail end of spectrin, near the actin-binding site of the latter, and thus is part of a protein 4.1-spectrin-actin ternary complex. Protein 4.1 also binds to the integral proteins, glycophorins A and C, thereby attaching the ternary complex to the membrane. In addition, protein 4.1 may interact with certain membrane phospholipids, thus connecting the lipid bilayer to the cytoskeleton.

Certain other proteins (4.9, adducin, and tropomyosin) also participate in cytoskeletal assembly.

Abnormalities in the Amount or Structure of Spectrin Cause Hereditary Spherocytosis and Elliptocytosis

Hereditary spherocytosis is a genetic disease, transmitted as an autosomal dominant, that affects about 1:5000 North Americans. It is characterized by the presence of spherocytes (spherical red blood cells, with a low surface-to-volume ratio) in the peripheral blood, by a hemolytic anemia, and by splenomegaly. The spherocytes are not as deformable as are normal red blood cells, and they are subject to destruction in the spleen, thus greatly shortening their life in the circulation. Hereditary spherocytosis is curable by splenectomy because the spherocytes can persist in the circulation if the spleen is absent.

The spherocytes are much more susceptible to osmotic lysis than are normal red blood cells. This is assessed in the **osmotic fragility test,** in which red blood cells are exposed in vitro to decreasing concentrations of NaCl. The physiologic concentration of NaCl is 0.85 g/dL. When exposed to a concentration of NaCl of 0.50 g/dL, very few normal red blood cells are hemolyzed, whereas approximately 50% of spherocytes would lyse under these conditions. The explanation is that the spherocyte, being almost circular, has little potential extra volume to accommodate additional water and thus lyses readily when exposed to a slightly lower osmotic pressure than is normal.

The cause of hereditary spherocytosis (Figure 60–5) is a deficiency in the amount of spectrin or abnormalities of its structure, so that it no longer tightly binds the other proteins with which it normally interacts. This weakens the membrane and leads to the spherocytic shape. Abnormalities in certain other proteins (eg, ankyrin) are also involved in some cases.

Hereditary elliptocytosis is a genetic disorder

Mutations in DNA affecting the amount or structure of spectrin or of certain other cytoskeletal proteins (eg, ankyrin)
↓
Weakens interactions among the peripheral and integral proteins of the red cell membrane
↓
Weakens the structure of the red cell membrane
↓
Adopts spherocytic shape and is subject to destruction in the spleen
↓
Hemolytic anemia

Figure 60–5. Summary of the causation of hereditary spherocytosis.

that is similar to hereditary spherocytosis except that affected red blood cells assume an elliptic, disk-like shape, recognizable by microscopy. It is also due to abnormalities in spectrin; some cases reflect abnormalities of band 4.1 or of glycophorin C.

THERE ARE MANY BLOOD GROUP SYSTEMS, INCLUDING THE ABO, Rh, and MN SYSTEMS

At least 21 human blood group systems are recognized, the best known of which are the ABO, Rh (Rhesus), and MN systems. The term "blood group" applies to a defined system of red blood cell antigens (blood group substances) controlled by a genetic locus having a variable number of alleles (eg, A, B, and O in the ABO system). The term "blood type" refers to the antigenic phenotype, usually recognized by the use of appropriate antibodies. For purposes of blood transfusion, it is important to know the basics of the ABO and Rh systems. However, knowledge of blood group systems is also of biochemical, genetic, immunologic, anthropologic, obstetric, pathologic, and forensic interest. Here, we briefly discuss some key features of the ABO system and mention the biochemical nature of the Rh and MN systems. From a biochemical viewpoint, the major interests in the ABO substances have been in isolating and determining their structures, elucidating their pathways of biosynthesis, and determining the natures of the products of the A, B, and O genes.

The ABO System Is of Crucial Importance in Blood Transfusion

This system was first discovered by Landsteiner in 1900 when investigating the basis of compatible and incompatible transfusions in humans. The membranes of the red blood cells of most individuals contain one blood group substance of type A, type B, type AB, or type O. Individuals of type A have anti-B antibodies in their plasma and will thus agglutinate type B or type AB blood. Individuals of type B have anti-A antibodies and will agglutinate type A or type AB blood. Type AB blood has neither anti-A nor anti-B antibodies and has been designated the **universal recipient.** Type O blood has neither A nor B substances and has been designated the **universal donor.** The explanation of these findings is related to the fact that the body does not usually produce antibodies to its own constituents. Thus, individuals of type A do not produce antibodies to their own blood group substance, A, but do possess antibodies to the foreign blood group substance, B, possibly because similar structures are present in microorganisms to which the body is exposed early in life. Since individuals of type O have neither A nor B substances, they possess antibodies to both these foreign substances. The above description has been simplified considerably; eg, there are two subgroups of type A–A$_1$ and A$_2$.

The genes responsible for production of the ABO substances are present on the long arm of chromosome 9. There are three alleles, two of which are codominant (A and B) and the third (O) recessive; these ultimately determine the four phenotypic products: the A, B, AB, and O substances.

The ABO Substances Are Glycosphingolipids and Glycoproteins Sharing Common Oligosaccharide Chains

The ABO substances are complex oligosaccharides present in most cells of the body and in certain secretions. On membranes of red blood cells, the oligosaccharides that determine the specific natures of the ABO substances appear to be mostly present in **glycosphingolipids,** whereas in secretions the same oligosaccharides are present in **glycoproteins.** Their presence in secretions is determined by a gene designated **Se (for secretor),** which codes for a specific fucosyl (Fuc) transferase in secretory organs, such as the exocrine glands but which is not active in red blood cells. Individuals of *SeSe* or *Sese* genotypes secrete A or B antigens (or both), whereas individuals of the sese genotype do not secrete A or B substances, but their red blood cells can express the A and B antigens.

H Substance Is the Biosynthetic Precursor of Both the A and B Substances

The ABO substances have been isolated and their structures determined; simplified versions, showing only their nonreducing ends, are presented in Figure

60–6. It is important to first appreciate the structure of the H substance, since it is the precursor of both the A and B substances and is the blood group substance found in persons of type O. H substance itself is formed by the action of a **fucosyltransferase,** which catalyzes the addition of the terminal fucose in $\alpha1\rightarrow2$ linkage onto the terminal Gal residue of its precursor:

$$GDP\text{-}Fuc + Gal\text{-}\beta\text{-}R \rightarrow Fuc\text{-}\alpha1,2\text{-}Gal\text{-}\beta\text{-}R + GDP$$
$$\text{Precursor} \qquad \text{H substance}$$

The H locus codes for this fucosyltransferase. The h allele of the H locus codes for an inactive fucosyltransferase; therefore, individuals of the hh genotype cannot generate H substance, the precursor of the A and B antigens. Thus, individuals of the hh genotype will have red blood cells of type O, even though they may possess the enzymes necessary to make the A or B substances (see below).

THE *A* GENE ENCODES A GalNAc TRANSFERASE, THE *B* GENE A Gal TRANSFERASE, AND THE *O* GENE AN INACTIVE PRODUCT

In comparison with H substance (Figure 60–6), **A substance** contains an additional GalNAc and **B substance** an additional Gal, linked as indicated. Anti-A antibodies are directed to the additional GalNAc residue found in the A substance, and anti-B antibodies are directed toward the additional Gal residue found in the B substance. Thus, GalNAc is the immunodominant sugar (ie, the one determining the specificity of the antibody formed) of blood group A substance, whereas Gal is the immunodominant

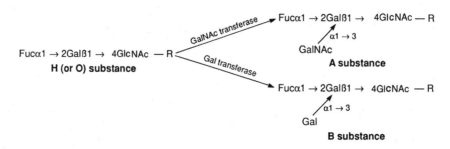

Figure 60–6. Diagrammatic representation of the structures of the H, A, and B blood group substances. R represents a long complex oligosaccharide chain, joined either to ceramide where the substances are glycosphingolipids, or to the polypeptide backbone of a protein via a serine or threonine residue where the substances are glycoproteins. Note that the blood groups substances are biantennary; ie, they have two arms, formed at a branch point (not indicated) between the GlcNAc—R, and only one arm of the branch is shown. Thus, the H, A, and B substances each contain two of their respective short oligosaccharide chains shown above. The AB substance contains one type A chain and one type B chain.

sugar of the B substance. In view of the structural findings, it is not surprising that A substance can be synthesized in vitro from O substance in a reaction catalyzed by a GalNAc transferase, employing UDP-GalNAc as the sugar donor. Similarly, blood group B can be synthesized from O substance by the action of a Gal transferase, employing UDP-Gal. It is crucial to appreciate that the product of the *A* gene is the GalNAc transferase that adds the terminal GalNAc to the O substance. Similarly, the product of the *B* gene is the Gal transferase adding the Gal residue to the O substance. Individuals of type AB possess both enzymes and thus have two oligosaccharide chains (Figure 60–6), one terminated by a GalNAc and the other by a Gal. Individuals of type O apparently synthesize an inactive protein, detectable by immunologic means; thus, H substance is their ABO blood group substance.

In 1990, a study using cloning and sequencing technology described the nature of the differences between the glycosyltransferase products of the *A, B,* and *O* genes. A difference of four nucleotides is apparently responsible for the distinct specificities of the A and B glycosyltransferases. On the other hand, the O allele has a single base-pair mutation, causing a **frameshift mutation** resulting in a protein lacking transferase activity.

The Rh Factor (D Antigen) Is an Integral Protein of the Red Blood Cell Membrane and Can Cause Serious Transfusion Problems

The Rh system is complex, and neither its genetic basis nor the biochemical nature of the gene products has yet been clearly elucidated. Three closely linked genes appear to be involved, located on chromosome 1. In one system of nomenclature, the products of these alleles are designated C or c, E or e, and D or d; no product corresponding to d has yet been identified. The **D(Rh$_0$) antigen** is the major antigen of interest and is also called the **Rh factor.** It appears to be an integral membrane protein of the red blood cell, with a molecular mass of about 30 kDa; interaction with membrane phospholipids may be important in its antigenicity. Approximately 15% of Caucasians lack this antigen and are designated **Rh-negative.** If these individuals are transfused even once with Rh-positive blood, they will probably form antibodies to the D(Rh$_0$) antigen. Thus, it is important for these individuals to receive Rh-negative blood, and it is crucial if the individual is a premenopausal female who may become pregnant. If such an individual has been carelessly transfused with Rh-positive blood, her blood will probably contain anti-D(Rh$_0$) antibodies. In turn, if her infant is Rh-positive, these antibodies can cause lysis of the infant's red cells, a condition called **hemolytic disease of the newborn.** After a delivery or an abortion, Rh-negative women should routinely receive **Rh$_0$(D) immune globulin,** which is effective in preventing active formation of antibodies by the mother to any Rh$_0$(D) antigens to which she has been exposed.

The MNSs Blood Group System Involves Polymorphic Forms of Glycophorins A and B

The MN blood groups are recognized by antisera that distinguish between polymorphic forms of glycophorin A. At the gene level are two codominant alleles, M and N, with three possible genotypes, M/M, M/N, and N/N. In turn, there are three phenotypes, M, N, and N. Although glycophorin A is heavily glycosylated, the polymorphism does not involve oligosaccharide heterogeneity but instead involves differences in amino acid sequence at the amino terminal end of the protein, as follows:

	Residue 1	Residue 5
M	Ser	Gly
N	Leu	Glu

The Ss system is a subclass of the MN group and comprises polymorphic forms of glycophorin B. The S and s forms differ by one amino acid.

The MNSs system is not frequently involved in transfusion reactions, but its members are useful genetic markers, their genes being closely linked on chromosome 4.

HEMOLYTIC ANEMIAS ARE CAUSED BY ABNORMALITIES OUTSIDE, WITHIN, OR INSIDE THE RED BLOOD CELL MEMBRANE

Causes outside the membrane include hypersplenism, a condition in which the spleen is enlarged from a variety of causes and red blood cells become sequestered in it. Immunologic abnormalities (eg, transfusion reactions, the presence in plasma of warm and cold antibodies that lyse red blood cells, and unusual sensitivity to complement) also fall in this class, as do toxins released by various infectious agents, such as certain bacteria (eg, *Clostridium*). Some snakes release venoms that act to lyse the red cell membrane (eg, via the action of phospholipases or proteinases).

Causes within the membrane include abnormalities of proteins. The most important conditions are hereditary spherocytosis and hereditary elliptocytosis, principally caused by abnormalities in the amount or structure of spectrin (see above).

Causes inside the red blood cell include **hemoglobinopathies** and **enzymopathies.** Sickle cell anemia is the most important hemoglobinopathy. Abnormalities of enzymes in the pentose phosphate pathway and in glycolysis are the most frequent enzymopathies involved, particularly the former. Defi-

Table 60–7. Laboratory investigations that assist in the diagnosis of hemolytic anemia.

General Tests	Specific Tests
Increased nonconjugated (indirect) bilirubin; shortened red cell survival time, as measured by injection of autologous ^{51}Cr-labeled RBCs; reticulocytosis; hemoglobinemia; low level of plasma haptoglobin.	Hb electrophoresis (eg, HbS), RBC enzymes (eg, G-6-P dehydrogenase or pyruvate kinase deficiency), osmotic fragility (eg, hereditary spherocytosis), Coombs' test, cold agglutinins. The direct Coombs test detects the presence of antibodies on RBCs, whereas the indirect test detects the presence of circulating antibodies to antigens present on RBCs.

Table 60–8. Summary of major biochemical features of neutrophils.

Active glycolysis
Active pentose phosphate pathway
Moderate oxidative phosphorylation
Rich in lysosomes and their degradative enzymes
Contain certain unique enzymes (eg, myeloperoxidase and NADPH-oxidase) and proteins
Contain CD11/CD18 integrins in plasma membrane

ciency of glucose-6-phosphate dehydrogenase is prevalent in certain parts of the world and is a frequent cause of hemolytic anemia (see above). Deficiency of pyruvate kinase is not frequent, but it is the second commonest enzyme deficiency resulting in hemolytic anemia; the mechanism appears to be due to impairment of glycolysis, resulting in decreased formation of ATP, affecting various aspects of membrane integrity.

Laboratory investigations that aid in the diagnosis of hemolytic anemia are listed in Table 60–7.

NEUTROPHILS HAVE AN ACTIVE METABOLISM AND CONTAIN SEVERAL UNIQUE ENZYMES AND PROTEINS

The major biochemical features of neutrophils are summarized in Table 60–8. Prominent features are active aerobic glycolysis, active pentose phosphate pathway, moderately active oxidative phosphorylation (because mitochondria are relatively sparse), and a high content of lysosomal enzymes. Many of the enzymes listed in Table 60–4 are also of importance in the oxidative metabolism of neutrophils (see below). Table 60–9 summarizes the functions of some proteins that are relatively unique to neutrophils.

Neutrophils Are Key Players in the Body's Defense Against Bacterial Infection

Neutrophils are motile phagocytic cells that play a key role in acute inflammation. When bacteria enter

Table 60–9. Some important enzymes and proteins of neutrophils. The expression of many of these molecules has been studied during the various stages of differentiation of normal neutrophils and also of corresponding leukemic cells employing state-of-the-art molecular biology techniques (eg, measurements of their specific mRNAs). For the majority, cDNAs have been isolated and sequenced, amino acid sequences deduced, their genes localized to specific chromosomal locations, and exons and intron sequences defined. Some important proteinases of neutrophils are listed in Table 60–12.

Enzyme or Protein	Reaction Catalyzed or Function	Comment
Myeloperoxidase (MPO)	$H_2O_2 + X^-$ (halide) $+ H^+ \rightarrow HOX + H_2O$ where $X^- = Cl^-$, HOX = hypochlorous acid)	Responsible for the green color of pus Genetic deficiency can cause recurrent infections
NADPH-oxidase	$2 O_2 + NADPH \rightarrow 2 O_2^{\bar{}} + NADP + H^+$	Key component of the respiratory burst Deficient in chronic granulomatous disease
Lysozyme	Hydrolyzes link between N-acetylmuramic acid and N-acetyl-D-glucosamine found in certain bacterial cell walls	Abundant in macrophages
Defensins	Basic antibiotic peptides of 29–33 amino acids	Apparently kill bacteria by causing membrane damage
Lactoferrin	Iron-binding protein	May inhibit growth of certain bacteria by binding iron and may be involved in regulation of proliferation of myeloid cells
CD 11a/CD 18, CD11b/CD 18[1]	Adhesion molecules (members of the integrin family)	Deficient in leukocyte adhesion deficiency
Receptors for Fc fragments of IgGs	Bind Fc fragments of IgG molecules	Target antigen-antibody complexes to myeloid and lymphoid cells, eliciting phagocytosis and other responses

[1]CD = cluster of differentiation. This refers to a uniform system of nomenclature that has been adopted to name surface markers of leukocytes. A specific surface protein (marker) that identifies a particular lineage or differentiation stage of leukocytes and that is recognized by a group of monoclonal antibodies is called a member of a cluster of differentiation. The system is particularly helpful in classifying subclasses of lymphocytes. Many CD antigens are involved in cell-cell interactions, adhesion, and transmembrane signaling.

Table 60–10. Sources of biomolecules with vasoactive properties involved in acute inflammation.

Mast cells and Basophils	Platelets	Neutrophils	Plasma Proteins
Histamine	Serotonin	Platelet-activating factor (PAF) Eicosanoids (Various prostaglandins and leukotrienes)	C3a, C4a, and C5a from the complement system Bradykinin and fibrin split products from the coagulation system

tissues, a number of phenomena result that are collectively known as the "acute inflammatory response." They include (1) increase of vascular permeability, (2) entry of activated neutrophils into the tissues, (3) activation of platelets, and (4) spontaneous subsidence (resolution) if the invading microorganisms have been dealt with successfully.

A variety of agents are released from cells and plasma proteins during acute inflammation whose net overall effect is to increase vascular permeability, resulting in tissue edema (Table 60–10).

In acute inflammation, neutrophils are recruited from the bloodstream into the tissues to help eliminate the foreign invaders. The neutrophils are attracted into the tissues by **chemotactic factors,** including complement fragment C5a, small peptides derived from bacteria (eg, *N*-formyl-methionyl-leucyl-phenylalanine), and a number of leukotrienes. To reach the tissues, circulating neutrophils must pass through the capillaries. To achieve this, they marginate along the vessel walls and then adhere to endothelial (lining) cells of the capillaries.

Integrins Mediate Adhesion of Neutrophils to Endothelial Cells

Adhesion of neutrophils to endothelial cells employs specific adhesive proteins (integrins) located on their surface and also specific receptor proteins in the endothelial cells. (See also the discussion of selectins in Chapter 56.)

The integrins are a superfamily of surface proteins present on a wide variety of cells. They are involved in the adhesion of cells to other cells or to specific components of the extracellular matrix. They are het-

erodimers, containing an α and a β subunit linked noncovalently. The subunits contain extracellular, transmembrane, and intracellular segments. The extracellular segments bind to a variety of ligands such as specific proteins of the extracellular matrix and of the surfaces of other cells. These ligands often contain Arg-Gly-Asp (R-G-D) sequences. The intracellular domains bind to various proteins of the cytoskeleton, such as actin and vinculin. The integrins are proteins that link the outsides of cells to their insides, thereby helping to integrate responses of cells (eg, movement, phagocytosis) to changes in the environment.

Three subfamilies of integrins were recognized initially. Members of each subfamily were distinguished by containing a common β subunit, but they differed in their α subunits. However, more than three β subunits have now been identified, and the classification of integrins has become rather complex. Some integrins of specific interest with regard to neutrophils are listed in Table 60–11.

A deficiency of the β2 subunit (also designated CD18) of LFA-1 and of two related integrins found in neutrophils and macrophages, Mac-1 (CD11bCD18) and p150,95 (CD11cCD18), causes leukocyte adhesion deficiency, a disease characterized by recurrent bacterial and fungal infections. Among various results of this deficiency, the adhesion of affected white blood cells to endothelial cells is diminished, and thus lesser numbers of neutrophils enter the tissues to combat infection.

Once having passed through the walls of small blood vessels, the neutrophils migrate toward the highest concentrations of the chemotactic factors, en-

Table 60–11. Examples of integrins that are important in the function of neutrophils, of other WBCs, and of platelets.[1]

Integrin	Cell	Subunit	Ligand	Function
VLA-1	WBCs, others	α1β1	Collagen, laminin	Cell-ECM adhesion
VLA-5	WBCs, others	α5β1	Fibronectin	Cell-ECM adhesion
VLA-6	WBCs, others	α6β1	Laminin	Cell-ECM adhesion
LFA-1 (CD11a/CD18)	WBCs	αLβ2	ICAM-1, ICAM-2	Adhesion of WBCs
Glycoprotein IIb/IIIa	Platelets	αIIbβ3	Fibrinogen, fibronectin, von Willebrand factor	Platelet adhesion and aggregation

[1]LFA-1, lymphocyte function–associated antigen 1; VLA, very late antigen; CD, cluster of differentiation; ICAM, intercellular adhesion molecule; ECM, extracellular matrix. A deficiency of LFA-1 and related integrins is found in leukocyte adhesion deficiency. A deficiency of platelet glycoprotein IIb/IIIa complex is found in Glanzmann's thrombasthenia, a condition characterized by a history of bleeding, normal platelet count, but abnormal clot retraction. These findings illustrate how fundamental knowledge of cell surface adhesion proteins is shedding light on the causation of a number of diseases.

counter the invading bacteria, and attempt to attack and destroy them. The neutrophils must be activated in order to turn on many of the metabolic processes involved in phagocytosis and killing of bacteria.

Activation of Neutrophils Is Similar to Activation of Platelets and Involves Hydrolysis of Phosphatidylinositol Bisphosphate

The mechanisms involved in platelet activation are discussed in Chapter 59 (Figure 59–17). The process involves interaction of the stimulus (eg, thrombin) with a receptor, activation of G proteins, stimulation of phospholipase C, and liberation from phosphatidylinositol bisphosphate of inositol triphosphate and diacylglycerol. These two second messengers result in an elevation of intracellular Ca^{2+} and activation of protein kinase C. In addition, activation of phospholipase A_2 produces arachidonic acid that can be converted to a variety of biologically active eicosanoids.

The process of activation of neutrophils is essentially similar. They are activated, via specific receptors, by interaction with bacteria, binding of chemotactic factors, or antibody-antigen complexes. The resultant rise in intracellular Ca^{2+} affects many processes in neutrophils, such as assembly of microtubules and the actin-myosin system. These processes are respectively involved in secretion of contents of granules and in motility, which enables neutrophils to seek out the invaders. The activated neutrophils are now ready to destroy the invaders by mechanisms that include production of active derivatives of oxygen.

The Respiratory Burst of Phagocytic Cells Involves NADPH-Oxidase and Helps Kill Bacteria

When neutrophils and other phagocytic cells engulf bacteria, they exhibit a rapid increase in oxygen consumption known as the respiratory burst. This phenomenon reflects the rapid utilization of oxygen (following a lag of 15–60 seconds) and production from it of large amounts of reactive derivatives, such as O_2^-, H_2O_2, $OH^·$, and OCl^- (hypochlorite ion). Some of these products are potent microbicidal agents.

The electron transport chain system responsible for the respiratory burst contains several components, including a flavoprotein, NADPH:O_2-oxidoreductase (often called NADPH-oxidase) and a b-type cytochrome (named cytochrome b_{558} because of a characteristic spectral peak at this wavelength or, alternatively, cytochrome b_{245} because of its midpoint oxidoreduction potential of 245 mV, the lowest of any b cytochrome, providing it with the capability of reducing oxygen to superoxide). This system catalyzes the one-electron reduction of oxygen to superoxide anion (Table 60–4, reaction 2). The oxidoreductase is reduced by NADPH, and the cytochrome

accomplishes the one-electron reduction of oxygen to form superoxide. The system is situated in the plasma membrane of neutrophils and other phagocytic cells. The NADPH is generated by the pentose phosphate cycle, the activity of which increases markedly during phagocytosis.

The above reaction is followed by the spontaneous production (by spontaneous dismutation) of hydrogen peroxide from two molecules of superoxide:

$$O_2^- + O_2^- + 2\,H^+ \rightarrow H_2O_2 + O_2$$

The superoxide ion is discharged to the outside of the cell or into phagolysosomes, where it encounters ingested bacteria. Killing of bacteria within phagolysosomes appears to depend on the combined action of elevated pH, superoxide ion, or further oxygen derivatives (H_2O_2, $OH^·$, and HOCl [hypochlorous acid; see below]) and on the action of certain bactericidal peptides (defensins) and other proteins (eg, cathepsin G and certain cationic proteins) present in phagocytic cells. Any superoxide that enters the cytosol of the phagocytic cell is converted to H_2O_2 by the action of **superoxide dismutase,** which catalyzes the same reaction as the spontaneous dismutation shown above. In turn, H_2O_2 is used by myeloperoxidase (see below) or disposed of by the action of glutathione peroxidase or catalase.

The NADPH-oxidase is inactive in resting phagocytic cells and is activated upon contact with various ligands (complement fragment C5a, chemotactic peptides, etc) with receptors in the plasma membrane. The events resulting in activation of the oxidase system have been much studied—though not yet completely resolved—and are similar to those described above for the process of activation of neutrophils; this is expected, since the respiratory burst is an integral component of activation. They involve **G proteins,** activation of **phospholipase C,** and generation of **inositol 1,4,5-triphosphate** (IP_3). The last mediates a transient increase in the level of cytosolic Ca^{2+}, which is essential for induction of the respiratory burst. **Diacylglycerol** is also generated and induces the translocation of protein kinase C into the plasma membrane from the cytosol, where it catalyzes the phosphorylation of various proteins, some of which are probably components of the oxidase system. The picture is further complicated by data indicating that activation of the oxidase system depends on a dual pathway, also involving a Ca^{2+}-independent transduction sequence. In addition, two cytosolic polypeptides have been shown to be important components of the total NADPH-oxidase system, being recruited, during activation, into the plasma membrane to form the active system.

Mutations in the Genes for Components of the NADPH-Oxidase System Cause Chronic Granulomatous Disease

The importance of the NADPH-oxidase system was clearly shown when it was observed that the respiratory burst was defective in chronic granulomatous disease, a relatively uncommon condition characterized by recurrent infections and widespread granulomas (chronic inflammatory lesions) in the skin, lungs, and lymph nodes. The granulomas form as attempts to wall off bacteria that have not been killed, owing to genetic deficiencies in the NADPH-oxidase system. Most patients are males, and the condition is X-linked. The fundamental causes of disease in these patients are mutations in the gene for the b-type cytochrome. In other patients, the condition is autosomal recessive, and mutations in one or another of the structural genes for the two cytosolic polypeptides are responsible. Some patients have responded to treatment with gamma interferon, which appears to increase amounts of the b-type cytochrome by affecting transcription of its gene. The probable sequence of events involved in the causation of chronic granulomatous disease is shown in Figure 60–7.

Neutrophils Contain Myeloperoxidase, Which Catalyzes the Production of Chlorinated Oxidants

The enzyme myeloperoxidase, present in large amounts in neutrophil granules and responsible for the green color of pus, can act on H_2O_2 to produce hypohalous acids:

$$H_2O_2 + X^- + H^+ \xrightarrow{\text{Myeloperoxidase}} HOX + H_2O$$

$(X^- = Cl^-, Br^-, I^-$ or SCN^-; $HOCl$ = hypochlorous acid)

The H_2O_2 used as substrate is generated by the NADPH-oxidase system. Cl^- is the halide usually

---

Mutations in genes for the protein components (ie, cytochrome b_{558} [also called b_{245}] and 2 cytosolic polypeptides) of the NADPH-oxidase system
$\downarrow$
Diminished production of superoxide ion and other active derivatives of oxygen
$\downarrow$
Diminished killing of certain bacteria
$\downarrow$
Recurrent infections and formation of tissue granulomas in order to wall off surviving bacteria

Figure 60–7. Simplified scheme of the sequence of events involved in the causation of chronic granulomatous disease.

---

employed, since it is present in relatively high concentration in plasma and body fluids. HOCl, the active ingredient of household liquid bleach, is a powerful oxidant and is highly microbicidal. When applied to normal tissues, its potential for causing damage is diminished because it reacts with primary or secondary amines present in neutrophils and tissues to produce various nitrogen-chlorine derivatives; these chloramines are also oxidants, though less powerful than HOCl, and act as microbicidal agents (eg, in sterilizing wounds) without causing tissue damage.

The Proteinases of Neutrophils Can Cause Serious Tissue Damage If Their Actions Are Not Checked

Neutrophils contain a number of proteinases (Table 60–12) that can hydrolyze elastin, various types of collagens, and other proteins present in the extracellular matrix. Such enzymatic action, if allowed to proceed unopposed, can result in serious damage to tissues. Most of these proteinases are lysosomal enzymes and exist mainly as inactive precursors in normal neutrophils. Small amounts of these enzymes are released into normal tissues, with the amounts increasing markedly during inflammation. The activities of elastase and other proteinases are normally kept in check by a number of **antiproteinases** (also listed in Table 60–11) present in plasma and the extracellular fluid. Each of them can combine—usually forming a noncovalent complex—with one or more specific proteinases and thus cause inhibition. In Chapter 59 it was shown that a genetic deficiency of **α_1-antiproteinase inhibitor** permits

Table 60–12. Proteinases of neutrophils and antiproteinases of plasma and tissues.[1]

Proteinases	Antiproteinases
Elastase	α_1-Antiproteinase
Collagenase	α_2-Macroglobulin
Gelatinase	Secretory leukoproteinase
Cathepsin G	inhibitor
Plasminogen activator	α_1-Antichymotrypsin
	Plasminogen activator
	inhibitor–1
	Tissue inhibitor of
	metalloproteinase

[1]The table lists some of the important proteinases of neutrophils and some of the proteins that can inhibit their actions. Most of the proteinases listed exist inside neutrophils as precursors. Plasminogen activator is not a proteinase, but it is included because it influences the activity of plasmin, which is a proteinase. The proteinases listed can digest many proteins of the extracellular matrix, causing tissue damage. The overall balance of proteinase:antiproteinase action can be altered by activating the precursors of the proteinases, or by inactivating the antiproteinases. The latter can be caused by proteolytic degradation or chemical modification, eg, Met-358 of α_1-antiproteinase inhibitor is oxidized by cigarette smoke.

elastase to act unopposed and digest pulmonary tissue, thereby participating in the causation of emphysema. α_2-Macroglobulin is a plasma protein that plays an important role in the body's defense against excessive action of proteases; it combines with and thus neutralizes the activities of a number of important proteases (Chapter 59).

When increased amounts of chlorinated oxidants are formed during inflammation, they affect the proteinase:antiproteinase equilibrium, tilting it in favor of the former. For instance, certain of the proteinases listed in Table 60–11 are activated by HOCl, whereas certain of the antiproteinases are inactivated by this compound. In addition, the tissue inhibitor of metalloproteinases and α_1-antichymotrypsin can be hydrolyzed by activated elastase, and α_1-antiproteinase inhibitor can be hydrolyzed by activated collagenase and gelatinase. In most circumstances, an appropriate balance of proteinases and antiproteinases is achieved. However, in certain instances, such as in the lung when α_1-antiproteinase inhibitor is deficient or when large amounts of neutrophils accumulate in tissues because of inadequate drainage, considerable tissue damage can result from the unopposed action of proteinases.

RECOMBINANT DNA TECHNOLOGY HAS HAD A PROFOUND IMPACT ON HEMATOLOGY

Recombinant DNA technology has had a major impact on many aspects of hematology. The bases of the thalassemias (Chapter 42) and of many disorders of coagulation (Chapter 59) have been greatly clarified by investigations using cloning and sequencing. The study of oncogenes and chromosomal translocations has advanced understanding of the leukemias (Chapter 62). As discussed above, cloning techniques have made available therapeutic amounts of erythropoietin and other growth factors. Deficiency of adenosine deaminase, which affects lymphocytes particularly, is the first disease to be treated by gene therapy. (See Case No. 8 in Chapter 65.) Like many other areas of biology and medicine, hematology has been and will continue to be revolutionized by this technology.

SUMMARY

The red blood cell is simple in terms of its structure and function, consisting principally of a concentrated solution of hemoglobin surrounded by a membrane. The production of red cells is regulated by erythropoietin, whereas other growth factors (eg, granulocyte- and granulocyte-macrophage colony-stimulating factors) regulate the production of white blood cells. The red cell contains a battery of cytosolic enzymes, such as superoxide dismutase, catalase, and glutathione peroxidase, to dispose of powerful oxidants generated during its metabolism. Genetically determined deficiency of the activity of glucose-6-phosphate dehydrogenase, which produces NADPH, is an important cause of hemolytic anemia. Methemoglobin is unable to transport oxygen; both genetic and acquired causes of methemoglobinemia are recognized. Considerable information has been acquired concerning the proteins and lipids of the red cell membrane. A number of cytoskeletal proteins, such as spectrin, ankyrin, and actin interact with specific integral membrane proteins to help regulate the shape and flexibility of the membrane. Deficiency of spectrin results in hereditary spherocytosis, another important cause of hemolytic anemia. The ABO blood group substances present in the red cell membrane are complex glycosphingolipids; the immunodominant sugar of A substance is N-acetylgalactosamine, whereas that of the B substance is galactose. The Rh antigens appear to be integral membrane proteins, and the MN substances are polymorphic forms of glycophorin A. Neutrophils play a major role in the body's defense mechanisms. Integrins present on surface membranes determine specific interactions with various cell and tissue components. Leukocytes are activated on exposure to bacteria and other stimuli; NADPH oxidase plays a key role in the process of activation (the respiratory burst). Mutations in this enzyme and associated proteins cause chronic granulomatous disease. The proteinases of neutrophils can digest many tissue proteins; normally, this is kept in check by a battery of antiproteinases. However, this defense mechanism can be overcome in certain circumstances, resulting in extensive tissue damage. The application of recombinant DNA technology is revolutionizing the study of hematology.

REFERENCES

Anderson DC, Kishimoto TK, Smith CW: Leukocyte adhesion deficiency and motility. In: *The Metabolic and Molecular Bases of Inherited Disease,* 7th ed. Scriver CR et al (editors). McGraw-Hill, 1995.

Beck WS (editor). *Hematology,* 5th ed. MIT Press, 1991.

Becker PS, Lux SE: Hereditary spherocytosis and hereditary elliptocytosis. In: *The Metabolic and Molecular*

Bases of Inherited Disease, 7th ed. Scriver CR et al (editors). McGraw-Hill, 1995.

Chanock SFJ et al: The respiratory burst oxidase. J Biol Chem 1994;269:24519.

Forehand JR et al: Inherited disorders of phagocyte killing. In: *The Metabolic and Molecular Bases of Inherited*

Disease, 7th ed. Scriver CR et al (editors). McGraw-Hill, 1995.

Jaffe ER, Hultquist DE: Cytochrome b_5 reductase deficiency and enzymopenic hereditary methemoglobinemia. In: *The Metabolic and Molecular Bases of Inherited Disease,* 7th ed. Scriver CR et al (editors). McGraw-Hill, 1995.

Krantz SB: Erythropoietin. Blood 1991;77:419.

Lienhard GE et al: How cells absorb glucose. Sci Am (Jan) 1991;266:86.

Lubbert M, Herrmann F, Koeffler HP: Expression and regulation of myeloid-specific genes in normal and leukemic myeloid cells. Blood 1991;77:909.

Luzzato L, Mehta A: Glucose-6-phosphate dehydrogenase deficiency. In: *The Metabolic and Molecular Bases of Inherited Disease,* 7th ed. Scriver CR et al (editors). McGraw-Hill, 1995.

Rosen GM et al: Free radicals and phagocytic cells. FASEB J 1995;9:20.

Spitznagel JK: Antibiotic proteins of human neutrophils. J Clin Invest 1990;86:1381.

Tanaka KR, Paglia DE: Pyruvate kinase and other enzymopathies of the erythrocyte. In: *The Metabolic and Molecular Bases of Inherited Disease,* 7th ed. Scriver CR et al (editors). McGraw-Hill, 1995.

Weatherall DJ et al: The hemoglobinopathies. In: *The Metabolic and Molecular Bases of Inherited Disease,* 7th ed. Scriver CR et al (editors). McGraw-Hill, 1995.

Yamamoto F et al: Molecular genetic basis of the histo-blood group ABO system. Nature 1990;345:229.

61

Metabolism of Xenobiotics

Robert K. Murray, MD, PhD

INTRODUCTION

Increasingly, humans are subjected to exposure to various foreign chemicals (xenobiotics)—drugs, food additives, pollutants, etc. The situation is well summarized in the following quotation from Rachel Carson: "As crude a weapon as the cave man's club, the chemical barrage has been hurled against the fabric of life." Understanding how xenobiotics are handled at the cellular level is important in learning how to cope with the chemical onslaught.

BIOMEDICAL IMPORTANCE

Knowledge of the metabolism of xenobiotics is basic to a rational understanding of pharmacology and therapeutics, pharmacy, toxicology, cancer research, and drug addiction. All these areas involve administration of, or exposure to, xenobiotics.

HUMANS ENCOUNTER THOUSANDS OF XENOBIOTICS THAT MUST BE METABOLIZED BEFORE BEING EXCRETED

A xenobiotic (Gk *xenos* "stranger") is a compound that is foreign to the body. The principal classes of xenobiotics of medical relevance are drugs, chemical carcinogens, and various compounds that have found their way into our environment by one route or another, such as polychlorinated biphenyls (PCBs) and certain insecticides. More than 200,000 manufactured environmental chemicals exist. Most of these compounds are subject to metabolism (chemical alteration) in the human body, with the liver being the main organ involved; occasionally, a xenobiotic may be excreted unchanged. Approximately 30 different enzymes catalyze reactions involved in xenobiotic metabolism; however, this chapter will only cover a selected group of them. It is convenient to consider the metabolism of xenobiotics in two phases.

In phase 1, the major reaction involved is **hydroxylation,** catalyzed by members of a class of enzymes referred to as monooxygenases or cytochrome P450s.

Hydroxylation may terminate the action of a drug, though this is not always the case. In addition to hydroxylation, these enzymes catalyze an astonishingly wide range of reactions, including those involving deamination, dehalogenation, desulfuration, epoxidation, peroxygenation, and reduction. Reactions involving hydrolysis (eg, catalyzed by esterases) and certain other non-P450-catalyzed reactions also occur in phase 1.

In phase 2, the hydroxylated or other compounds produced in phase 1 are converted by specific enzymes to various polar metabolites by **conjugation** with glucuronic acid, sulfate, acetate, glutathione, or certain amino acids, or by **methylation.**

The overall purpose of the two phases of metabolism of xenobiotics is to increase their water solubility (polarity) and thus facilitate their excretion from the body. Very hydrophobic xenobiotics would persist in adipose tissue almost indefinitely if they were not converted to more polar forms. In certain cases, phase 1 metabolic reactions convert xenobiotics from inactive to biologically active compounds. In these instances, the original xenobiotics are referred to as "prodrugs" or "procarcinogens." In other cases, additional phase 1 reactions (eg, further hydroxylation reactions) convert the active compounds to less active or inactive forms prior to conjugation. In yet other cases, it is the conjugation reactions themselves that convert the active products of phase 1 reactions to less active or inactive species, which are subsequently excreted in the urine or bile. In a very few cases, conjugation may actually increase the biologic activity of a xenobiotic.

The term "detoxification" is sometimes used for many of the reactions involved in the metabolism of xenobiotics. However, the term is not always appropriate because, as mentioned above, in some cases the reactions to which xenobiotics are subject actually increase their biologic activity and toxicity.

ISOFORMS OF CYTOCHROME P450 HYDROXYLATE A MYRIAD OF XENOBIOTICS IN PHASE 1 OF THEIR METABOLISM

Hydroxylation is the chief reaction involved in phase 1. The responsible enzymes are called monooxygenases or cytochrome P450; the human genome contains at least 11 families of these enzymes. The reaction catalyzed by a monooxygenase (cytochrome P450) is as follows:

$$RH + O_2 + NADPH + H^+ \rightarrow R{-}OH + H_2O + NADP$$

RH above can represent a very wide variety of xenobiotics, including drugs, carcinogens, pesticides, petroleum products, and pollutants (such as a mixture of PCBs). In addition, endogenous compounds, such as certain steroids, eicosanoids, fatty acids, and retinoids, are also substrates. The substrates are generally lipophilic and are rendered more hydrophilic by hydroxylation.

Cytochrome P450 is considered the most versatile biocatalyst known. The actual reaction mechanism is complex and has been briefly described previously (Figure 13–8). It has been shown by the use of $^{18}O_2$ that one atom of oxygen enters R–OH and one atom enters water. This dual fate of the oxygen accounts for the former naming of monooxygenases as "mixed-function oxidases." The reaction catalyzed by cytochrome P450 can also be represented as follows:

$$\overset{\text{Reduced cytochrome P-450}\quad\text{Oxidized cytochrome P-450}}{RH + O_2 \rightarrow R{-}OH + H_2O}$$

The major monooxygenases in the endoplasmic reticulum are **cytochrome P450s**—so named because the enzyme was discovered when it was noted that preparations of microsomes that had been chemically reduced and then exposed to carbon monoxide exhibited a distinct peak at 450 nm. Among reasons that this enzyme is important is the fact that approximately 50% of the drugs that humans ingest are metabolized by isoforms of cytochrome P450; these enzymes also act on various carcinogens and pollutants.

Isoforms of Cytochrome P450 Make Up a Superfamily of Heme-Containing Enzymes

The following are important points concerning cytochrome P450s.

(1) Because of the large number of isoforms (about 150) that have been discovered, it became important to have a systematic nomenclature for isoforms of P450 and for their genes. This is now available and in wide use and is based on structural homology. The abbreviated root symbol CYP denotes a cytochrome P450. This is followed by an Arabic number designating the family; cytochrome P450s are included in the same family if they exhibit 40% or more sequence identity. The Arabic number is followed by a capital letter indicating the subfamily, if two or more members exist; P450s are in the same subfamily if they exhibit greater than 55% sequence identity. The individual P450s are then arbitrarily assigned Arabic numerals. Thus, CYPA1 denotes a cytochrome P450 that is a member of family 1 and subfamily A and is the first individual member of that subfamily. The nomenclature for the genes encoding cytochrome P450s is identical to that described above except that italics are used; thus the gene encoding CYP1Aa is *CYP1A1*.

(2) Like hemoglobin, they are **hemoproteins.**

(3) They are widely distributed across species. Bacteria possess cytochrome P450s, and P450$_{cam}$ (involved in the metabolism of camphor) of *Pseudomonas putida* is the only P450 isoform whose crystal structure has been established.

(4) They are present in highest amount in liver. In liver and most other tissues, they are present mainly in the **membranes of the smooth endoplasmic reticulum,** which constitute part of the microsomal fraction when tissue is subjected to subcellular fractionation (Chapter 2). In hepatic microsomes, cytochrome P450s can comprise as much as 20% of the total protein. P450s are found in most tissues, though often in low amount compared with liver. In the adrenal, they are found in mitochondria as well as in the endoplasmic reticulum; the various hydroxylases present in that organ play an important role in cholesterol and steroid biosynthesis (Chapter 48). The mitochondrial cytochrome P450 system differs from the microsomal system in that it uses an NADPH-linked flavoprotein, **adrenodoxin reductase,** and a nonheme iron-sulfur protein, adrenodoxin. In addition, the specific P450 isoforms involved in steroid biosynthesis are generally much more restricted in their substrate specificity.

(5) There are at least six isoforms of cytochrome P450 present in the endoplasmic reticulum of human liver, each with wide and somewhat overlapping substrate specificities and acting on both xenobiotics and endogenous compounds. The genes for many isoforms of P450 (from both humans and animals such as the rat) have been isolated and studied in detail in recent years.

(6) NADPH, not NADH, is involved in the reaction mechanism of cytochrome P450. The enzyme that uses NADPH to yield the reduced cytochrome P450, shown in the left-hand side of the above equation, is called **NADPH-cytochrome P450 reductase.** Electrons are transferred from NADPH to NADPH-cytochrome P450 reductase and then to cytochrome P450. This leads to the reductive activation of molecular oxygen, and one atom of oxygen is subsequently inserted into the substrate. Cytochrome b_5, another hemoprotein found in the membranes of the smooth

endoplasmic reticulum (Chapter 13), may be involved as an electron donor in some cases.

(7) Lipids are also components of the cytochrome P450 system. The preferred lipid is **phosphatidylcholine,** which is the major lipid found in membranes of the endoplasmic reticulum.

(8) Most isoforms of cytochrome P450 are inducible. For instance, the administration of phenobarbital or of many other drugs causes hypertrophy of the smooth endoplasmic reticulum and a three- to fourfold increase in the amount of cytochrome P450 within 4–5 days. The mechanism of induction has been studied extensively and in most cases involves increased transcription of mRNA for cytochrome P450. However, certain cases of induction involve stabilization of mRNA, enzyme stabilization, or other mechanisms (eg, an effect on translation).

Induction of cytochrome P450 has important clinical implications, since it is a biochemical mechanism of drug interaction. A drug interaction has occurred when the effects of one drug are altered by prior, concurrent, or later administration of another. As an illustration, consider the situation when a patient is taking the anticoagulant warfarin to prevent blood clotting. This drug is metabolized by CYP2C9. Concomitantly, the patient is started on phenobarbital to treat a certain type of epilepsy, but the dose of warfarin is not changed. After 5 days or so, the level of CYP2C9 in the patient's liver will be elevated three- to fourfold. This in turn means that warfarin will be metabolized much more quickly than before, and its dosage will have become inadequate. Therefore, the dose must be increased if warfarin is to be therapeutically effective. To pursue this example further, a problem could arise later on if the phenobarbital is discontinued but the increased dosage of warfarin stays the same. The patient will be at risk of bleeding, since the high dose of warfarin will be even more active than before, because the level of CYP2C9 will decline once phenobarbital has been stopped.

Another example of enzyme induction involves CYP2E1, which is induced by consumption of ethanol. This is a matter for concern, because this P450 metabolizes certain widely used solvents and also components found in tobacco smoke, many of which are established carcinogens. Thus, if the activity of CYP2E1 is elevated by induction, this may increase the risk of carcinogenicity developing from exposure to such compounds.

(9) Certain isoforms of cytochrome P450 (eg, CYP1A1) are particularly involved in the metabolism of polycyclic aromatic hydrocarbons (PAHs) and related molecules; for this reason they were formerly called **aromatic hydrocarbon hydroxylases (AHH).** These enzymes are important in the metabolism of PAHs and in carcinogenesis produced by these agents. For example, in the lung it may be involved in the conversion of inactive PAHs (procarcinogens), inhaled by smoking, to active carcinogens

by hydroxylation reactions. Smokers have higher levels of this enzyme in some of their cells and tissues than do nonsmokers. Some reports have indicated that the activity of this enzyme may be elevated (induced) in the placenta of a woman who smokes, thus potentially altering the quantities of metabolites of PAHs (some of which could be harmful) to which the fetus is exposed.

(10) Recent findings have shown that certain cytochrome P450s exist in **polymorphic forms,** some of which exhibit low catalytic activity. These observations are one important explanation for the variations in drug responses noted among many patients. One P450 exhibiting polymorphism is CYP2D6, which is involved in the metabolism of debrisoquin (an antihypertensive drug; see Table 61–2) and sparteine (an antiarrhythmic and oxytocic drug). Certain polymorphisms of CYP2D6 cause poor metabolism of these and a variety of other drugs, so that they can accumulate in the body, resulting in untoward consequences.

Table 61–1 summarizes some principal features of cytochrome P450s.

CONJUGATION REACTIONS PREPARE XENOBIOTICS FOR EXCRETION IN PHASE TWO OF THEIR METABOLISM

In phase 1 reactions, xenobiotics are generally converted to more polar, hydroxylated derivatives. In phase 2 reactions, these derivatives are conjugated with molecules such as glucuronic acid, sulfate, or glutathione. This renders them even more water-soluble, and they are eventually excreted in the urine or bile.

Five Types of Phase 2 Reactions Are Described Here

A. Glucuronidation: The glucuronidation of bilirubin is discussed in Chapter 34; the reactions

Table 61–1. Some properties of cytochrome P450s.

- Involved in the metabolism of many xenobiotics and also endogenous compounds such as steroids.
- All are hemoproteins.
- Often exhibit broad substrate specificity.
- Catalyze reactions involving introduction of one atom of oxygen into the substrate and one into water.
- Their hydroxylated products are more water-soluble than their generally lipophilic substrates.
- Liver contains highest amounts, but found in most tissues.
- Located in the smooth endoplasmic reticulum or in mitochondria (steroidogenic enzymes).
- In some cases, their products are mutagenic or carcinogenic.
- Most have molecular mass of about 55 kDa.
- Many are inducible.
- Some exhibit polymorphisms, which can result in atypical drug metabolism.

whereby xenobiotics are glucuronidated are essentially similar. UDP-glucuronic acid is the glucuronyl donor, and a variety of glucuronosyltransferases, present in both the endoplasmic reticulum and cytosol, are the catalysts. Molecules such as 2-acetylaminofluorene (a carcinogen), aniline, benzoic acid, meprobamate (a tranquilizer), phenol, and many steroids are excreted as glucuronides. The glucuronide may be attached to oxygen, nitrogen, or sulfur groups of the substrates. Glucuronidation is probably the most frequent conjugation reaction.

B. Sulfation: Some alcohols, arylamines, and phenols are sulfated. The sulfate donor in these and other biologic sulfation reactions (eg, sulfation of steroids, glycosaminoglycans, glycolipids, and glycoproteins) is adenosine 3′-phosphate-5′-phosphosulfate (PAPS) (Chapter 26); this compound is called "active sulfate."

C. Conjugation With Glutathione: Glutathione (γ-glutamyl cysteinyl glycine) is a tripeptide consisting of glutamic acid, cysteine, and glycine (Figure 5–4). Glutathione is commonly abbreviated GSH (because of the sulfhydryl group of its cysteine) and is the business part of the molecule. A number of potentially toxic electrophilic xenobiotics (such as certain carcinogens) are conjugated to the nucleophilic GSH in reactions that can be represented as follows:

$$R + GSH \rightarrow R\text{---}S\text{---}G$$

where R = an electrophilic xenobiotic. The enzymes catalyzing these reactions are called **glutathione** *S*-transferases and are present in high amounts in liver cytosol and in lower amounts in other tissues. A variety of glutathione *S*-transferases are present in human tissue. They exhibit different substrate specificities and can be separated by electrophoretic and other techniques. If the potentially toxic xenobiotics were not conjugated to GSH, they would be free to combine covalently with DNA, RNA, or cell protein and could thus lead to serious cell damage. GSH is therefore an important defense mechanism against certain toxic compounds, such as some drugs and carcinogens. If the levels of GSH in a tissue such as liver are lowered (as can be achieved by the administration to rats of certain compounds that react with GSH), then that tissue can be shown to be more susceptible to injury by various chemicals that would normally be conjugated to GSH. Glutathione conjugates are subjected to further metabolism before excretion. The glutamyl and glycinyl groups belonging to glutathione are removed by specific enzymes, and an acetyl group (donated by acetyl-CoA) is added to the amino group of the remaining cysteinyl moiety. The resulting compound is a mercapturic acid, a conjugate of L-acetylcysteine, which is then excreted in the urine.

Glutathione has other important functions in human cells, apart from its role in xenobiotic metabolism.

1. It participates in the decomposition of potentially toxic hydrogen peroxide in the reaction catalyzed by glutathione peroxidase (Chapter 22).

2. It is an important intracellular reductant, helping to maintain essential SH groups of enzymes in their reduced state. This role is discussed in Chapter 22, and its involvement in the hemolytic anemia caused by deficiency of glucose-6-phosphate dehydrogenase is discussed in Chapters 22 and 60.

3. A metabolic cycle involving GSH as a carrier has been implicated in the transport of certain amino acids across membranes in the kidney. The first reaction of the cycle is shown below.

$$\text{Amino acid} + \text{GSH} \rightarrow \gamma\text{-Glutamyl amino acid} + \text{Cysteinylglycine}$$

This reaction helps transfer certain amino acids across the plasma membrane, the amino acid being subsequently hydrolyzed from its complex with GSH and the GSH being resynthesized from cysteinylglycine. The enzyme catalyzing the above reaction is γ-glutamyltransferase (GGT). It is present in the plasma membrane of renal tubular cells and in the endoplasmic reticulum of hepatocytes. The enzyme has diagnostic value because it is released into the blood from hepatic cells in various hepatobiliary diseases.

D. Other Reactions: The two most important are acetylation and methylation.

1. Acetylation–Acetylation is represented by

$$X + \text{Acetyl-CoA} \rightarrow \text{Acetyl-X} + \text{CoA}$$

where X represents a xenobiotic. As for other acetylation reactions, acetyl-CoA (active acetate) is the acetyl donor. These reactions are catalyzed by acetyltransferases present in the cytosol of various tissues, particularly liver. The drug isoniazid, used in the treatment of tuberculosis, is subject to acetylation. Polymorphic types of acetyltransferases exist, resulting in individuals who are classified as slow or fast acetylators, and influence the rate of clearance of drugs such as isoniazid from blood. Slow acetylators are more subject to certain toxic effects of isoniazid because the drug persists longer in these individuals.

2. Methylation–A few xenobiotics are subject to methylation by methyltransferases, employing *S*-adenosylmethionine (Figure 32–21) as the methyl donor.

THE ACTIVITIES OF XENOBIOTIC-METABOLIZING ENZYMES ARE AFFECTED BY AGE, SEX, AND OTHER FACTORS

Various factors affect the activities of the enzymes metabolizing xenobiotics. The activities of these enzymes may differ substantially among species. Thus, for example, the possible toxicity or carcinogenicity of xenobiotics cannot be extrapolated freely from one species to another. There are significant differences in enzyme activities among individuals, many of which appear to be due to genetic factors. The activities of some of these enzymes vary according to age and sex.

Intake of various xenobiotics such as phenobarbital, PCBs, or certain hydrocarbons can cause enzyme induction. It is thus important to know whether or not an individual has been exposed to these inducing agents in evaluating biochemical responses to xenobiotics. Metabolites of certain xenobiotics can inhibit or stimulate the activities of xenobiotic-metabolizing enzymes. Again, this can affect the doses of certain drugs that are administered to patients.

RESPONSES TO XENOBIOTICS INCLUDE PHARMACOLOGIC, TOXIC, IMMUNOLOGIC, AND CARCINOGENIC EFFECTS

Xenobiotics are metabolized in the body by the reactions described above. When the xenobiotic is a drug, phase 1 reactions may produce its active form or may diminish or terminate its action if it is pharmacologically active in the body without prior metabolism. The diverse effects produced by drugs comprise the area of study of pharmacology; here it is important to appreciate that drugs act primarily through biochemical mechanisms. Table 61–2 summarizes three important reactions to drugs that reflect

Table 61–2. Some important drug reactions due to mutant or polymorphic forms of enzymes or proteins.[1]

Enzyme or Protein Affected	Reaction or Consequence
Glucose-6-phosphate dehydrogenase (G6PD) [mutations]	Hemolytic anemia following ingestion of drugs such as primaquine
Ca^{2+} release channel (ryanodine receptor) in the sarcoplasmic reticulum [mutations]	Malignant hyperthermia following administration of certain anesthetics (eg, halothane)
CYP2D6 [polymorphisms]	Slow metabolism of certain drugs (eg, debrisoquin), resulting in their accumulation

[1]G6PD deficiency is discussed in Chapters 22 and 60 and malignant hyperthermia in Chapter 58. Many other examples of drug reactions based on polymorphism or mutation are available.

genetically determined differences in enzyme and protein structure among individuals—part of the field of study known as "pharmacogenetics."

Certain xenobiotics are very toxic even at low levels (eg, cyanide). On the other hand, there are few xenobiotics, including drugs, that do not exert some toxic effects if sufficient amounts are administered. The toxic effects of xenobiotics cover a wide spectrum. However, there are three general types of effects (Figure 61–1), which are mentioned briefly here because of their relationship to xenobiotic metabolism.

The first is cell injury (cytotoxicity), which can be severe enough to result in cell death. There are many mechanisms by which xenobiotics injure cells. The one considered here is covalent binding to cell macromolecules of reactive species of xenobiotics produced by metabolism. These macromolecular targets include DNA, RNA, and protein. If the macromolecule to which the reactive xenobiotic binds is essential for short-term cell survival, eg, a protein or enzyme involved in some critical cellular function such as oxidative phosphorylation or regulation of the permeability of the plasma membrane, then severe effects on cellular function could become evident quite rapidly.

Second, the reactive species of a xenobiotic may bind to a protein, altering its antigenicity. The xenobiotic is said to act as a **hapten,** ie, a small molecule that by itself does not stimulate antibody synthesis but will combine with antibody once formed. The resulting antibodies can then damage the cell by several immunologic mechanisms that grossly perturb normal cellular biochemical processes.

Third, reactions of activated species of chemical carcinogens with DNA are thought to be of great importance in chemical carcinogenesis (Chapter 62). Some chemicals (eg, benzo[α]pyrene) require activation by monooxygenases in the endoplasmic reticulum to become carcinogenic (they are thus called **indirect carcinogens**). The activities of the monooxygenases and of other xenobiotic-metabolizing enzymes present in the endoplasmic reticulum thus help to determine whether such compounds become carcinogenic or are "detoxified." Other chemicals (eg, various alkylating agents) can react directly (**direct carcinogens**) with DNA without undergoing intracellular chemical activation.

The enzyme **epoxide hydrolase** is of interest because it can exert a protective effect against certain carcinogens. The products of the action of certain monooxygenases on some procarcinogen substrates are **epoxides.** Epoxides are highly reactive and mutagenic or carcinogenic or both. Epoxide hydrolase—like cytochrome P450, also present in the membranes of the endoplasmic reticulum—acts on these compounds, converting them into much less reactive dihydrodiols. The reaction catalyzed by epoxide hydrolase can be represented as follows:

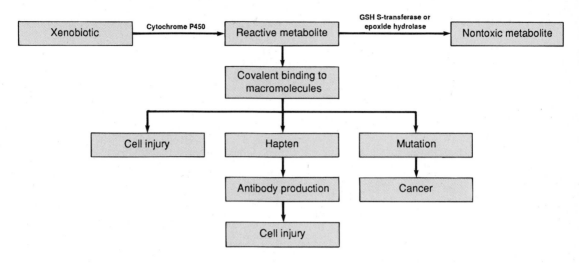

Figure 61–1. Simplified scheme showing how metabolism of a xenobiotic can result in cell injury, immunologic damage, or cancer. In this instance, the conversion of the xenobiotic to a reactive metabolite is catalyzed by a cytochrome P450, and the conversion of the reactive metabolite (eg, an epoxide) to a nontoxic metabolite is catalyzed by either a GSH S-transferase or epoxide hydrolase.

$$\underset{\text{Epoxide}}{\overset{\displaystyle -\overset{|}{\underset{\diagdown O \diagup}{C}}-\overset{|}{C}-}{}} + H_2O \rightarrow \underset{\text{Dihydrodiol}}{\overset{\displaystyle -\overset{|}{\underset{HO}{C}}-\overset{|}{\underset{OH}{C}}-}{}}$$

It is hoped that further knowledge of cytochrome P450s and other enzymes of xenobiotic metabolism will result in improved methods of assessing the safety of drugs, in helping to avoid undesirable drug interactions, and in helping to dispose of potentially toxic environmental pollutants. It is likely that DNA-based tests will be developed to help screen individuals for mutations that result in serious adverse reactions to some drugs.

SUMMARY

Xenobiotics are chemical compounds foreign to the body, such as drugs, food additives, and environmental pollutants; more than 200,000 have been identified. Xenobiotics are metabolized in two phases. The major reaction of phase 1 is hydroxylation catalyzed by a variety of monooxygenases, also known as cytochrome P450s. In phase 2, the hydroxylated species are conjugated with a variety of hydrophilic compounds such as glucuronic acid, sulfate, or glutathione. The combined operation of these two phases renders lipophilic compounds into water-soluble compounds that can be eliminated from the body.

Cytochrome P450s catalyze reactions that introduce one atom of oxygen derived from molecular oxygen into the substrate, yielding a hydroxylated product. NADPH and NADPH-cytochrome P450 reductase are involved in the complex reaction mechanism. All cytochrome P450s are hemoproteins and generally have a wide substrate specificity, acting on many exogenous and endogenous substrates. They represent the most versatile biocatalyst known. Members of 11 families of cytochrome P450 are found in human tissue. Cytochrome P450s are generally located in the endoplasmic reticulum of cells and are particularly enriched in liver. Many cytochrome P450s are inducible. This has important implications in phenomena such as drug interaction. Mitochondrial cytochrome P450s also exist and are involved in cholesterol and steroid biosynthesis. They use a non-heme iron-containing sulfur protein, adrenodoxin, not required by microsomal isoforms. Cytochrome P450s, because of their catalytic activities, play major roles in the reactions of cells to chemical compounds and in chemical carcinogenesis.

Phase 2 reactions are catalyzed by enzymes such as glucuronosyltransferases, sulfotransferases, and glutathione S-transferases, using UDP-glucuronic acid, PAPS (active sulfate), and glutathione, respectively, as donors. Glutathione not only plays an important role in phase 2 reactions but is also an intracellular reducing agent and is involved in the transport of certain amino acids into cells.

Xenobiotics can produce a variety of biologic effects, including pharmacologic responses, toxicity, immunologic reactions, and cancer.

REFERENCES

Coon MJ et al: Cytochrome P450: Progress and predictions. FASEB J 1992;6:669.

Gelboin HV: Cytochrome P450 and monoclonal antibodies. Pharmacol Rev 1993;45:413.

Guengerich FP: Characterization of human cytochrome P450 enzymes. FASEB J 1992;6:745.

Jakoby WB, Ziegler DM: The enzymes of detoxication. J Biol Chem 1990;265:20715.

Kalow W, Grant DM: Pharmacogenetics. In: *The Metabolic and Molecular Bases of Inherited Disease,* 7th ed. Scriver CR et al (editors). McGraw-Hill, 1995.

Katzung BG (editor): *Basic & Clinical Pharmacology,* 6th ed. Appleton & Lange, 1994.

Nebert DW et al: The P450 superfamily: Update on new sequences, gene mapping, and recommended nomenclature. DNA Cell Biol 1991;10:1.

Pacifici GM, Fracchia GN (editors): *Advances in Drug Metabolism in Man.* Office for Official Publications of the European Communities, 1995. (The various chapters contain up-to-date descriptions of the major enzymes involved in drug metabolism in human tissues.)

Porter TD, Coon MJ: Cytochrome P-450: Multiplicity of isoforms, substrates and catalytic and regulatory mechanisms. J Biol Chem 1991;266:13469.

Rushmore TH, Pickett CB: Glutathione *S*-transferases, structure, regulation, and therapeutic implications. J Biol Chem 1993;268:11475.

Cancer, Cancer Genes, & Growth Factors

62

Robert K. Murray, MD, PhD

INTRODUCTION

Cancer cells are characterized by three properties: (1) diminished or unrestrained control of growth; (2) invasion of local tissues; and (3) spread, or metastasis, to other parts of the body. Cells of benign tumors also show diminished control of growth but do not invade local tissue or spread to other parts of the body.

In this chapter, we discuss some biochemical aspects of cancer. The key issues are to explain in biochemical terms the **uncontrolled growth** of cancer cells and their ability to **invade** and **metastasize.** Certain genes controlling growth and interactions with other normal cells are apparently abnormal in structure or regulation in cancer cells. Information on how cell growth—both normal and pathologic—is controlled is limited, and knowledge of specific genes involved in growth regulation is even more meager. Little is known as yet about the biochemical basis of metastasis, so that coverage of this topic will be brief. At least some types of cancer (eg, certain leukemias) can be regarded as examples of abnormal differentiation. Again, astonishingly little is known about the molecular basis of differentiation. However, many workers in this area think that further research on **oncogenes, tumor suppressor genes, growth factors** and their receptors, DNA repair systems, and regulation of the cell cycle will provide insight into the nature of the disturbed control of growth, of differentiation (where applicable), and of cell-cell interaction exhibited by cancer cells. There has been a recent surge of interest in elucidating the molecular basis of genetic susceptibility to cancer. One example is isolation of the *BRCA1* gene, which increases susceptibility to breast and ovarian cancer. Thus, most of these topics will be discussed in some detail.

BIOMEDICAL IMPORTANCE

Cancer is the second most common cause of death in the USA after cardiovascular disease. Humans of all ages develop cancer, and a wide variety of organs are affected. The incidence of many cancers increases with age, so that as people live longer, more will develop the disease. Apart from individual suffering, the economic burden to society is immense.

PHYSICAL, CHEMICAL, AND BIOLOGIC AGENTS CAN CAUSE CANCER

Agents causing cancer fall into three broad groups: radiant energy, chemical compounds, and viruses. Three genetic causes of cancer are discussed later in this chapter.

Radiant Energy Can Be Carcinogenic

Ultraviolet rays, x-rays, and γ-rays are mutagenic and carcinogenic. These rays damage DNA in several ways. Ultraviolet radiation may cause pyrimidine dimers to form. Apurinic or apyrimidinic sites may form by elimination of corresponding bases. Single- and double-strand breaks or cross-linking of strands may occur. Damage to DNA is presumed to be the basic mechanism of carcinogenicity with radiant energy, but the details are unclear. Repair of DNA is discussed in Chapter 38. Apart from direct effects on DNA, x-rays and γ-rays cause **free radicals** to form in tissues. The resultant OH·, superoxide, and other radicals can interact with DNA and other macromolecules, leading to molecular damage and thereby probably contributing to carcinogenic effects of radiant energy.

Many Chemicals Are Carcinogenic

A wide variety of chemical compounds are carcinogenic (Table 62–1); the structures of three of the most widely studied are shown in Figure 62–1. Most of the compounds listed in Table 62–1 have been tested by administration to rodents or other animals.

Table 62–1. Some chemical carcinogens.

Class	Compound
Polycyclic aromatic hydrocarbons	Benzo[a]pyrene, dimethylbenzanthracene
Aromatic amines	2-Acetylaminofluorene, N-methyl-4-aminoazobenzene (MAB)
Nitrosamines	Dimethylnitrosamine, diethylnitrosamine
Various drugs	Alkylating agents (eg, cyclophosphamide), diethylstilbestrol
Naturally occurring compounds	Dactinomycin, aflatoxin B_1
Inorganic compounds	Arsenic, asbestos, beryllium, cadmium, chromium

However, many substances are associated with the development of cancer in humans. It is estimated that approximately 80% of human cancers are caused by environmental factors, principally chemicals. Exposure to such compounds can occur because of a person's occupation (eg, benzene, asbestos), diet (eg, aflatoxin B_1, which is produced by the mold *Aspergillus flavus* and sometimes found as a contaminant of peanuts and other foodstuffs), or life-style (eg, cigarette smoking) or in other ways (eg, certain therapeutic drugs can be carcinogenic). We shall present only a few important generalizations that have emerged from the study of chemical carcinogenesis.

A. Structure: Both organic and inorganic molecules may be carcinogenic (Table 62–1). The diversity of these compounds indicates that they do not possess one common structural feature that confers carcinogenicity.

B. Action: The organic carcinogens have been the most thoroughly studied. Some, such as mechlorethamine and β-propiolactone, have been found to interact directly with target molecules (direct carcinogens), but others require prior metabolism to become carcinogenic **(procarcinogens)** (Chapter 61). The process whereby one or more enzyme-catalyzed reactions convert procarcinogens to active carcinogens is called **metabolic activation.** Any intermediate compounds formed are **proximate carcinogens** (there may be one or more), and the fi-

nal compound that reacts with cellular components (eg, DNA) is the **ultimate carcinogen.** A possible sequence can be displayed as follows:

Procarcinogen → Proximate carcinogen → Ultimate carcinogen

The procarcinogen itself is not a chemically reactive species, whereas the ultimate carcinogen is often highly reactive. At least two reactions are required to convert the procarcinogen 2-acetylaminofluorene (2-AAF) to the ultimate carcinogen, the sulfate ester of N-hydroxy-AAF. An important generalization is that ultimate carcinogens are usually **electrophiles** (ie, molecules deficient in electrons), which readily attack nucleophilic (electron-rich) groups in DNA, RNA, and proteins.

C. Metabolism of Chemical Carcinogens: The metabolism of procarcinogens and other xenobiotics involves monooxygenases and transferases (Chapter 61). The enzymes responsible for metabolic activation of procarcinogens are principally species of **cytochrome P450,** located in the endoplasmic reticulum. These are the same enzymes that are involved in the metabolism of other xenobiotics, such as drugs and environmental pollutants (eg, PCBs). The activities of the enzymes metabolizing chemical carcinogens are affected by a number of factors, such as species, genetic considerations, age, or sex. The variations in activities of these enzymes help explain the often appreciable differences in the carcinogenicity of chemicals among different species and individuals of the same species. Many of the above points are discussed in more detail in Chapter 61.

D. Covalent Binding: When chemical carcinogens are administered to animals or placed in cultured cells, it can be shown (eg, by using radioactive carcinogens) that they or their derivatives generally bind covalently to cellular macromolecules, including DNA, RNA, and proteins. The chemical natures of the adducts formed by interaction of certain ultimate carcinogens with their target molecules have been determined. Most interest has focused on products formed with DNA. Carcinogens have been found to interact with the purine, pyrimidine, or phosphodiester groups of DNA. The most common site of attack is guanine, and the addition of various

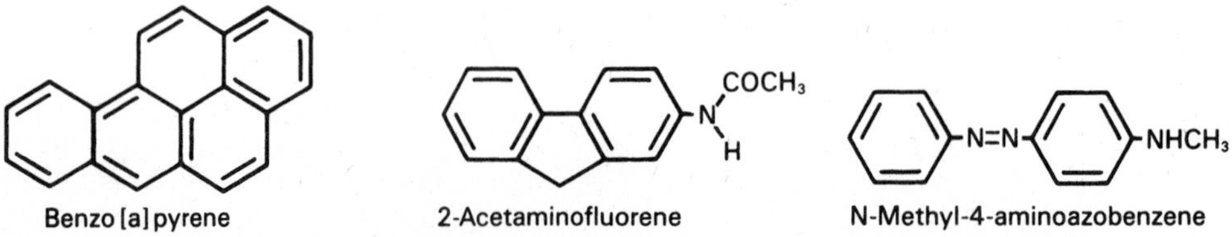

Figure 62–1. Structures of three important experimentally used chemical carcinogens.

carcinogens to the N_2, N_3, N_7, O_6, and O_8 atoms of this base has been observed.

E. Damage to DNA: The covalent interaction of direct carcinogens or ultimate carcinogens with DNA can result in several types of damage; this damage can be repaired, as described in Chapter 38. Despite the existence of repair systems, certain modifications of DNA by chemical carcinogens persist for relatively long periods of time. It is possible that these persistent unrepaired lesions are of special importance in generating mutations critical to carcinogenesis.

F. Mutagens: Most chemical carcinogens are mutagens. This has been demonstrated using the Ames assay (see below) and other tests. At a molecular level, transitions, transversions, and other types of mutation (Chapters 38 and 40) have been shown to occur following exposure of certain bacteria to ultimate carcinogens. It has been assumed that some types of cancer are due to mutations in somatic cells that affect key regulatory processes in these cells. Direct evidence of this has now been obtained (see the discussion of oncogenes and tumor suppressor genes below).

Since testing the carcinogenicity of chemicals in animals is slow and expensive, assays for screening the potential carcinogenicity of chemical compounds have been developed. Many are based on detection of the mutagenicity of chemical carcinogens. Such assays are more rapid and less expensive than detecting tumors in animals. None are ideal, since the ultimate test of a carcinogen is to show that it causes tumors in animals. However, one assay based on detecting mutagenicity, the **Ames assay,** has proved useful in screening for potential carcinogens. This assay uses a specially constructed strain of *Salmonella typhimurium* that has a mutation (His⁻) in a gene that codes for one of the enzymes involved in the synthesis of histidine. Thus, these particular salmonellae cannot synthesize histidine, which must be present in the medium for growth to occur. When a mutation caused by a carcinogen occurs at the site of the His⁻ mutation, the latter mutation can restore its reading sequence, converting it to His⁺. The progeny from bacteria containing such a reverse mutation can now synthesize histidine and thus grow in a medium lacking it. Such salmonellae can be detected as readily observable and quantifiable colonies growing on agar plates.

One problem with the use of bacteria in mutagenicity tests is that they do not contain the spectrum of monooxygenases found in higher animals. Thus, if a compound requires activation to become a mutagenic or carcinogenic species, this may not occur when bacteria are used. Ames circumvented this problem by incubating the agents to be tested in a postmitochondrial supernatant of rat liver (the S-9 fraction, which is the supernatant fraction after centrifuging a rat liver homogenate at 9000 g for a suit-

able period of time). The S-9 fraction contains fragments of endoplasmic reticulum and thus most of the various monooxygenases and other enzymes required to activate potential mutagens and carcinogens.

The Ames assay identifies approximately 90% of known carcinogens. It is becoming routine to test newly synthesized chemicals by this assay, particularly if they are to be introduced commercially or widely used in industry. Compounds giving a positive reaction should undergo further testing, including assessment of carcinogenicity in animals.

G. Initiation and Promotion: In certain organs such as skin and liver, carcinogenesis has at least two stages. The classic example is skin. Typically, identical areas of the skin of a group of mice are painted once with **benzo[a]pyrene.** If no other subsequent treatment is used, no skin tumors develop (Figure 62–2). However, if the application of benzo[a]pyrene is followed by several applications of croton oil, many tumors subsequently develop. Applications of croton oil alone (ie, no pretreatment with benzo[a]-pyrene) do not result in skin tumors. Many other variants of this basic protocol have been carried out, permitting the following conclusions: (1) The stage of carcinogenesis caused by application of benzo[a]-pyrene is called "initiation"; this stage appears to be rapid and irreversible. It is presumed to involve an irreversible modification of DNA, perhaps resulting in one or more mutations. Benzo[a]pyrene is thus called an "initiating agent." (2) The second, much slower (ie, months or years) stage of carcinogenesis, resulting from application of croton oil, is called "promo-

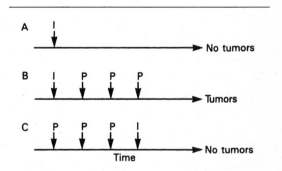

Figure 62–2. Diagrammatic representation of the stages of initiation and promotion of chemical carcinogenesis in skin. **A:** One application of the initiator (eg, benzo[a]pyrene) is made to the skin of a number of mice. **B:** The application of the initiator is followed by a number of applications of a promoter (eg, croton oil) at, for instance, weekly intervals. **C:** The promoter is applied first, and then the initiator is applied. Benign tumors of the skin (papillomas) may appear in about 100 days; malignant tumors (carcinomas) take about 1 year to appear. Many other informative variants of this protocol have been performed, all substantiating the basic concepts of initiation and promotion. (I, initiator; P, promoter.)

tion." Croton oil is thus a promoting agent, or **promoter.** Promoters are incapable of causing initiation. (3) Most carcinogens are capable of acting as both initiating and promoting agents.

A large number of compounds, including phenobarbital and saccharin, can act as promoters in various organs. The active agent of croton oil is a mixture of phorbol esters. The most active phorbol ester is 12-*O*-tetradecanoylphorbol-13-acetate (TPA), which has numerous effects. The most interesting finding has been that **protein kinase C** can act as a receptor for TPA. Stimulation of the activity of this enzyme by interaction with TPA may result in the phosphorylation of a number of membrane proteins, leading to effects on transport and other functions. This important result ties in the action of certain tumor promoters to the field of transmembrane signaling (see the discussion of growth factors, below). Many tumor promoters appear to act by causing alterations of gene expression, but the precise mechanisms by which promoters influence the initiated cell to become a tumor cell remain to be determined.

DNA Is the Critical Macromolecule in Carcinogenesis

This statement is supported by the following observations: (1) Cancer cells beget cancer cells—ie, the essential changes responsible for cancer are transmitted from mother to daughter cells. This is consistent with the behavior of DNA. (2) Both irradiation and chemical carcinogens damage DNA and are capable of causing mutations in DNA. (3) Many tumor cells exhibit abnormal chromosomes. (4) Transfection experiments (see below) indicate that purified DNA (oncogenes) from cancer cells can transform normal cells into (potential) cancer cells. (5) Genes that increase susceptibility to cancer have been isolated. However, **epigenetic factors** (eg, methylation of DNA) may also play a role in carcinogenesis.

Some DNA and RNA Viruses Are Carcinogenic

Oncogenic viruses contain either DNA or RNA as their genome (Table 62–2). Only a few important features of the major members of these two classes will be described here.

Polyomavirus and **SV40 viruses** have played an important role in the development of current ideas about viral oncogenesis, though they are not implicated in the causation of human tumors. They are both small (containing a genome of about 5 kb), and their circular genomes code for only about five or six proteins. Under certain circumstances, infection of appropriate cells with these viruses can result in malignant transformation. Specific viral proteins are known to be involved. In the case of SV40, these proteins (often called **antigens,** because they were detected by immunologic methods) are known as T ("large T") and t ("small t"), and in the case of poly-

Table 62–2. Some important tumor viruses.[1]

Class	Members
DNA viruses	
Papovavirus	Polyomavirus, SV40 virus, papillomavirus
Adenovirus	Adenoviruses 12, 18, and 31
Herpesvirus	Epstein-Barr virus
Hepadnavirus	Hepatitis B virus
RNA viruses	
Retrovirus type C	Murine sarcoma and leukemia viruses, avian sarcoma and leukemia viruses, human T cell leukemia viruses I and II
Retrovirus type B	Mouse mammary tumor virus

[1]The major viruses implicated in the causation of tumors in humans are Epstein-Barr virus (Burkitt's lymphoma, nasopharyngeal cancer, B cell lymphomas), hepatitis B virus (hepatocellular carcinoma), human papillomaviruses (a variety of tumors including anogenital cancers), and human T-cell leukemia-lymphoma virus type 1 (adult T cell leukemia). It has been estimated that virus-linked human cancers are responsible for 15% of total cancer incidence.

omavirus, they are known as T, mid-T, and t. (T refers to the fact that the first of these proteins was detected in a tumor.) How these proteins cause malignant transformation is still under investigation; the T antigens are known to bind tightly to DNA and cause alterations in gene expression. These proteins show cooperative effects, suggesting that alteration of more than one reaction or process is required for transformation.

Some types of adenovirus are known to cause transformation of certain animal cells. There is considerable interest in the Epstein-Barr virus, since it is associated with Burkitt's lymphoma and nasopharyngeal carcinoma in humans. Hepatitis B virus may be associated with some cases of liver cancer in humans.

Since much of the knowledge of oncogenes obtained in recent years has emerged from the study of RNA-containing tumor viruses (retroviruses; see below), the subsequent discussion of oncogenes contains frequent references to these viruses.

BOTH MORPHOLOGIC AND BIOCHEMICAL CHANGES OCCUR UPON MALIGNANT TRANSFORMATION

When cultured cells are infected with certain oncogenic viruses, they may undergo malignant transformation. The most important morphologic and biochemical changes occurring upon transformation are listed in Table 62–3. These changes affect cell shape, motility, adhesiveness to the culture dish, growth, and a number of biochemical processes. They reflect the primary processes that cause—and the secondary changes that result from—conversion

Table 62–3. Some changes shown by cultured cells which suggest that malignant transformation has occurred (eg, after infection by an oncogenic virus). However, the crucial test of malignancy is the ability of cells to grow into tumors in vivo.

- Alterations of morphology: Transformed cells often have a much rounder shape than control cells.
- Increased cell density (loss of contact inhibition of growth): Transformed cells often form multilayers, while control cells usually form a monolayer.
- Loss of anchorage dependence: Transformed cells can grow without attachment to the surface of the culture dish and will often grow in agar.
- Loss of contact inhibition of movement: Transformed cells grow over one another, while normal cells stop moving when they come into contact with each other.
- A variety of biochemical changes, including an increased rate of glycolysis, alterations of the cell surface (eg, changes in the composition of glycoproteins or glycosphingolipids), and secretion of certain proteases.
- Alterations of cytoskeletal structures, such as actin filaments.
- Diminished requirement for growth factors and, often, increased secretion of certain growth factors into the surrounding medium.

from the normal to the malignant state. The ability to equate transformation approximately with acquisition of malignant properties has been of tremendous importance in cancer research. However, acquisition by cells of the changes collectively known as transformation does not necessarily mean that such cells will display the same biologic properties as tumor cells in vivo; to be classified as genuine tumor cells, cells must yield tumors when injected into a suitable host animal.

ONCOGENES PLAY A CRUCIAL ROLE IN CARCINOGENESIS

Oncogenes are genes capable of causing cancer. Their discovery has had a major impact on research on the fundamental mechanisms involved in carcinogenesis. Oncogenes were first recognized as unique genes of tumor-causing viruses that are responsible for the process of transformation (viral oncogenes).

(1) Oncogenes of Rous Sarcoma Virus: Analyses of the oncogene of the Rous sarcoma virus and its product have been particularly revealing. The genome of this retrovirus contains four genes named *gag, pol, env,* and *src.* This can be shown schematically as follows:

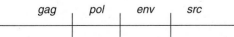

The *gag* gene codes for group-specific antigens of the virus, *pol* for the **reverse transcriptase** that characterizes retroviruses, and *env* for certain glycopro-

teins of the viral envelope. A protein-tyrosine kinase was shown to be the product of *src* (ie, the sarcoma-causing gene) that is responsible for transformation. This finding was of fundamental importance. It revealed a specific biochemical mechanism (ie, abnormal phosphorylation of a number of proteins) that could explain, at least in part, how a tumor virus could cause the pleiotropic (ie, diverse) effects of transformation. The critical cell proteins, whose abnormal phosphorylation presumably leads to transformation, are still to be defined. **Vinculin,** a protein found in focal adhesion plaques (structures involved in intercellular adhesion), is one candidate. The abnormal phosphorylation of vinculin in focal adhesion plaques could help explain the rounding-up of cells and their diminished adhesion to the substratum and to one another observed during transformation (Table 62–3). Certain glycolytic enzymes appear to be target proteins for the *src* protein-tyrosine kinase; this is in keeping with the observation that transformed cells often show increased rates of glycolysis. The product of *src* may also interact with the kinase catalyzing phosphorylation of **phosphatidylinositol** to phosphatidylinositol mono- and bisphosphate, increasing turnover of the polyphosphoinositide cycle. When phosphatidylinositol 4,5-bisphosphate is hydrolyzed by the action of phospholipase C, two second messengers are released: inositol triphosphate and diacylglycerol (Chapter 44). The first compound mediates release of Ca^{2+} from intracellular sites of storage (eg, the endoplasmic reticulum). Diacylglycerol stimulates the activity of the plasma membrane-bound protein kinase C, which in turn phosphorylates a number of proteins, some of which may be components of ion pumps. Specifically, it has been proposed that mild alkalinization of the cell, brought about by activation of an Na^+/H^+ antiport system (Chapter 43), could play a role in stimulating mitosis. Thus, the product of *src* may affect a large number of cellular processes by its ability to phosphorylate various target proteins and enzymes and by stimulating the pathway of synthesis of the polyphosphoinositides.

(2) Protein-Tyrosine Kinases in Normal and Transformed Cells: The observation that Rous sarcoma virus contained a protein-tyrosine kinase stimulated much research on the phosphorylation of tyrosine. It is now known that most if not all normal cells contain protein-tyrosine kinase activity. The amount of phosphotyrosine in most normal cells is low but is usually elevated in cells transformed by an oncogenic virus containing a protein-tyrosine kinase, though the amount is still relatively small (about 1% of the total phosphoamino acids [mainly phosphoserine, phosphothreonine, and phosphotyrosine] in such cells). Certain receptors (eg, for epidermal growth factor, insulin, and platelet-derived growth factor) found in both normal and transformed cells have protein-tyrosine activities that are stimulated upon interaction with their ligands (see the discussion of growth factors, below). Protein-tyrosine kinase activities thus play important roles in both normal and transformed cells.

(3) Oncogenes of Other Retroviruses: In addition to the oncogenes of Rous sarcoma virus, approximately 20 oncogenes of other retroviruses have been recognized. About half of the products of these viral oncogenes are **protein kinases,** mostly of the tyrosine type. Some viral oncogenes are listed in Table 62–4, along with their products. While some of those listed encode protein kinases, the remainder encode various other proteins with interesting biologic activities. The product of the *erb*-B gene of avian erythroblastosis virus is a truncated form of the receptor for epidermal growth factor, and that of the *sis* oncogene of simian sarcoma virus is a truncated B chain of platelet-derived growth factor. The product of the oncogene *(fms)* of one type of viral isolate of feline sarcoma virus is a **macrophage colony-stimulating factor.** On the other hand, the product of the *myc* oncogene, originally found in chicken myelocytoma viruses, is a DNA-binding protein, which may affect the control of mitosis. The product of the *ras* oncogene of murine sarcoma viruses binds GTP, has GTPase activity, and appears to be related to the G proteins that regulate the activity of the important plasma membrane enzyme, adenylyl cyclase (Chapter 44).

(4) Proto-Oncogenes: A key issue raised by the discovery of viral oncogenes relates to their origin. Use of nucleic acid hybridization (Chapter 42) revealed that normal cells contained DNA sequences similar—if not identical—to those of the viral oncogenes. Thus, the viruses apparently incorporated cellular genes into their genomes during their passages through cells. The retention of such genes in their genomes indicated that they must confer a selective advantage on the affected viruses, presumably related to the altered growth properties of transformed cells.

The cellular sequences were found to be conserved in a wide range of eukaryotic cells, suggesting that they were important components of normal cells. In addition, mRNA species and proteins derived from these normal sequences could be detected at various stages of their development or life cycles. The genes present in normal cells thus have been designated **proto-oncogenes,** and their products are believed to play important roles in normal differentiation and other cellular processes.

(5) Oncogenes From Tumor Cells: Experiments using DNA extracted from tumors have also provided evidence for the existence of oncogenes. The method used for detecting such cellular oncogenes is called **gene transfer** or **DNA transfection.** It depends on the fact that certain genes present in tumors can cause transformation of "normal" cultured cells. DNA is isolated from tumor cells and added to recipient cells, often a line of mouse fibroblasts known as NIH/3T3 cells. DNA isolated from tumor cells is precipitated with calcium phosphate (to facilitate endocytosis) and added to NIH/3T3 cells in tissue culture. The cells are observed microscopically over a period of 1–2 weeks for formation of foci of transformed cells. If transformation occurs, the NIH/3T3 cells change their morphology from flat to rounded cells that grow in characteristic foci. The procedure is repeated several times using DNA extracted from the transformed cells, thus reducing the amount of DNA not involved in transformation that was transfected and facilitating identification (eg, by the Southern blot technique, using a suitable probe [Chapter 42]) of the specific gene involved. A considerable number of cellular oncogenes have been recognized in this manner, with some related to the *ras* oncogene of murine sarcoma viruses. These cellular oncogenes either are identical to normal genes or show very small structural differences from their normal counterparts (see below). In the former case,

Table 62–4. Some oncogenes of retroviruses.[1]

Oncogene	Retrovirus	Origin	Oncogene Product	Subcellular Location
abl	Abelson murine leukemia virus	Mouse	Protein-tyrosine kinase	Plasma membrane
erb-B	Avian erythroblastosis virus	Chicken	Truncated EGF receptor	Plasma membrane
fes	Feline sarcoma virus	Cat	Protein-tyrosine kinase	Plasma membrane
fos	Murine sarcoma virus	Mouse	Transcription factor (AP-1); complexes with *jun*	Nucleus
jun	Avian sarcoma virus	Chicken	Transcription factor (AP-1); complexes with *fos*	Nucleus
myc	Myelocytoma virus 29	Chicken	DNA-binding protein	Nucleus
sis	Simian sarcoma virus	Monkey	Truncated PDGF (B chain)	Membranes ? Secreted
src	Rous sarcoma virus	Chicken	Protein-tyrosine kinase	Plasma membrane

EGF, epidermal growth factor; PDGF, platelet-derived growth factor.
[1]Modified and reproduced, with permission, from Franks LM, Teich NM (editors): *Introduction to the Cellular and Molecular Biology of Cancer.* Oxford Univ Press, 1986.

the regulation of their expression may be abnormal in cancer cells.

(6) Abbreviations for Cellular and Viral Oncogenes: The abbreviation c-*onc* (cellular oncogene, eg, c-*ras*) is used to designate an oncogene present in tumor cells. The species present in normal cells—ie, its proto-oncogene—can be conveniently referred to as the corresponding c-*onc* proto-oncogene (eg, the c-*ras* proto-oncogene). Similarly, a viral oncogene is designated v-*onc* (viral oncogene, eg, v-*ras*), with its proto-oncogene being referred to as a v-*onc* proto-oncogene (eg, the v-*ras* proto-oncogene).*

Proto-Oncogenes Are Activated to Oncogenes by Various Mechanisms

Five mechanisms will be discussed that alter the expression or structure of proto-oncogenes and thus participate in their becoming oncogenes. For the sake of convenience, the process whereby transcription of a gene is increased (from zero or a relatively low level) will be designated as **activation.** Familiarity with the mechanisms involved in activation is crucial for understanding contemporary thinking about carcinogenesis.

(1) Promoter Insertion: Certain retroviruses lack oncogenes (eg, avian leukemia viruses) but may cause cancer over a longer period of time—months rather than days—than those which do contain oncogenes. As for other retroviruses, when these particular viruses infect cells, a DNA copy (cDNA) of their RNA genome is synthesized by reverse transcriptase, and the cDNA is integrated into the host genome. The integrated double-stranded cDNA is called a "provirus." The cDNA copies of retroviruses are flanked at both ends by sequences named long terminal repeats, as are certain transposons ("jumping genes") found in bacteria and plants (Chapter 38).

---

*Where *ras* = a gene present in certain viruses that cause rat sarcomas.

The long terminal repeat sequences appear to be important in the mechanism of proviral integration, and they can act as promoters of transcription (Chapter 39). For example, following infection of chicken B lymphocytes by certain avian leukemia viruses, the proviruses become integrated near the *myc* gene. The *myc* gene is activated by an upstream, adjacent viral long terminal repeat acting as a promoter, resulting in transcription of both the corresponding *myc* mRNA and translation of its product in such cells (Figure 62–3). A B cell tumor ensues, though the precise role of the products of the *myc* gene in the overall process is not clear. Similar events occur following infection of various cells with other retroviruses.

(2) Enhancer Insertion: In some cases, the provirus is inserted downstream from the *myc* gene, or upstream from it but oriented in the reverse direction; nevertheless, the *myc* gene becomes activated (Figure 62–4). Such activation cannot be due to promoter insertion, since a promoter sequence must be upstream of the gene whose transcription it increases and the sequence must be in the correct 5' to 3' direction. Instead, enhancer sequences (Chapters 39 and 41) present in the long terminal repeat sequences of the retroviruses under consideration appear to be involved.

The above two mechanisms—promoter and enhancer insertion—commonly operate in viral carcinogenesis. They can be classified as examples of **insertional mutagenesis.** Proto-oncogenes other than *myc* may also be involved.

(3) Chromosomal Translocations: As mentioned earlier, many tumor cells exhibit chromosomal abnormalities. One type of chromosomal change seen in cancer cells is translocation. The basis of a translocation is that a piece of one chromosome is split off and then joined to another chromosome. If the second chromosome donates material to the first, the translocation is said to be "reciprocal." Characteristic translocations are found in a number of tumor cells. One important translocation is the **Philadelphia**

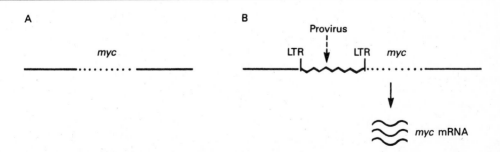

Figure 62–3. Schematic representation of how promoter insertion may activate a proto-oncogene. *A:* Normal chicken chromosome, showing an inactive *myc* gene. *B:* An avian leukemia virus has integrated in the chromosome in its proviral form, adjacent to the *myc* gene. Its right-hand long terminal repeat (LTR), containing a strong promoter, lies just upstream of the *myc* gene and activates that gene, resulting in transcription of *myc* mRNA. For simplicity, only one strand of DNA is depicted and other details have been omitted.

A

B

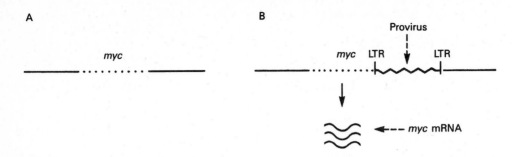

Figure 62–4. Schematic representation showing how enhancer insertion may activate a proto-oncogene. ***A:*** Normal chicken chromosome, showing an inactive *myc* gene. ***B:*** An avian leukemia virus has integrated in the chromosome in its proviral form, adjacent to the *myc* gene. However, in this instance, the site of integration is just downstream of the myc gene and it cannot act as a promoter (Figure 62–6). Instead, a certain proviral sequence acts as an enhancer element, leading to activation of the upstream *myc* gene and its transcription. For simplicity, only one strand of DNA is depicted and other details have been omitted.

chromosome, involving chromosomes 9 and 22 and occurring in chronic granulocytic leukemia.

Burkitt's lymphoma is a fast-growing cancer of human B lymphocytes. In certain cases, an example of a reciprocal translocation (Figure 62–5) is found that has illuminated the mechanisms of activation of potential cellular oncogenes. Chromosomes 8 and 14 are involved. The segment of chromosome 8 that breaks off and moves to chromosome 14 contains the *myc* gene. As shown in Figure 62–6, the transposition places the previously inactive *myc* gene under the influence of enhancer sequences in the genes coding for the heavy chains of immunoglobulins. This juxtaposition results in activation of transcription of the *myc* gene. Apparently, synthesis of greatly increased amounts of the DNA-binding protein coded for by the *myc* gene act to "drive" or "force" the cell toward becoming malignant, perhaps by an

effect on the regulation of mitosis. This mechanism is similar to enhancer insertion, except that chromosomal translocation (rather than integration of provirus) is responsible for placing the proto-oncogene (ie, *myc*) under the influence of an enhancer.

(4) Gene Amplification: Amplification of certain genes (Chapter 41) is found in a number of tumors. One method of bringing about gene amplification in tumors is by administration of the anticancer drug **methotrexate,** an inhibitor of the enzyme dihydrofolate reductase. Tumor cells can become resistant to the action of this drug. The basis of this phenomenon is that the gene for dihydrofolate reductase becomes amplified, resulting in an increase of the activity of the enzyme (up to 400-fold). The amplified genes, measuring up to 1000 kb or more in length, may be detected as homogeneously staining regions on a specific chromosome. Alternatively, they are de-

Figure 62–5. Schematic representation of the reciprocal translocation involved in Burkitt's lymphoma. The chromosomes involved are 8 and 14. A segment from the end of the q arm of chromosome 8 breaks off and moves to chromosome 14. The reverse process moves a small segment from the q arm of chromosome 14 to chromosome 8. The *myc* gene is contained in the small piece of chromosome 8 that was transferred to chromosome 14; it is thus placed next to genes transcribing the heavy chains of immunoglobulin molecules and itself becomes activated.

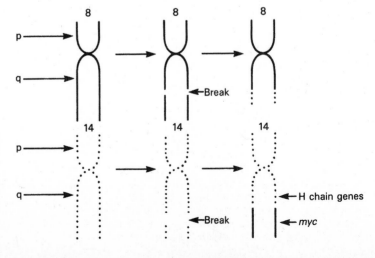

A   B

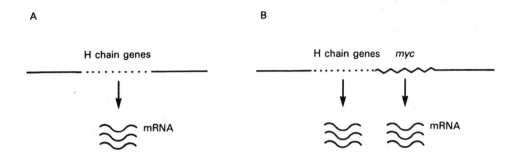

Figure 62–6. Schematic representation showing how the translocation involved in Burkitt's lymphoma may activate the *myc* proto-oncogene. ***A:*** A small segment of chromosome 14 prior to the translocation. The segment shown contains the genes encoding regions of heavy chains of immunoglobulins. ***B:*** Following the translocation, the previously inactive *myc* gene is placed under the influence of enhancer sequences in the genes encoding the heavy chains and is thus activated, resulting in transcription. For simplicity, only one strand of DNA is depicted and other details have been omitted.

tected as double-minute chromosomes, which are minichromosomes lacking centromeres. The precise relationship of homogeneously staining regions to double-minute chromosomes is under investigation. Certain cellular oncogenes can also be amplified in like manner and are thus activated. Increased amounts of the products of certain oncogenes (such as c-*ras*) produced by gene amplification may play a role in the progression of tumor cells to a more malignant state (see the discussion of the progression of tumors, below).

(5) Point Mutation: The v-*ras* oncogene was originally detected in certain murine retroviruses. Its product, a polypeptide of molecular mass 21 kDa (thus named p21), appears to be related to the G proteins that modulate the activity of adenylyl cyclase (see above and Chapter 44) and thus plays a key role in cellular responses to many hormones and drugs. Analyses by DNA sequencing of the c-*ras* proto-oncogene from normal human cells and of the c-*ras* oncogene from a cancer of the human bladder showed that they differed solely in one base, resulting in an amino acid substitution at position 12 of p21. This intriguing result has been confirmed by analyses of c-*ras* genes from other human tumors. In each case, the results were consistent; the gene isolated from the tumor exhibited only a single-point mutation, in comparison with the c-*ras* proto-oncogene from normal cells. The position of the mutation varied somewhat, so that a few other amino acid substitutions also were observed. These mutations in p21 appear to affect its conformation and to diminish its activity as a GTPase. The lower activity of GTPase could result in chronic stimulation of the activity of adenylyl cyclase, which normally is diminished when GDP is formed from GTP (Chapter 44). The resulting stimulation of the activity of adenylyl cyclase can result in a number of effects on cellular metabolism exerted by the increased amount of cAMP affecting the activities of

various cAMP-dependent protein kinases. These events may assist in tipping the balance of cellular metabolism toward a state favoring transformation or its maintenance.

General Comments on Activation of Oncogenes

Of the five mechanisms described above, the first four (promoter insertion, enhancer insertion, chromosome translocation, and gene amplification) involve an increase in amount of the product of an oncogene due to increased transcription but no alteration of the structure of the product of the oncogene. Thus, it appears that increased amounts of the product of an oncogene may be sufficient to push a cell toward becoming malignant. The fifth mechanism, a point mutation, involves a change in the structure of the product of the oncogene but not necessarily any change in its amount. This finding implies that the presence of a structurally abnormal key regulatory protein in a cell may also be sufficient to tip the scale toward malignancy.

When considering the role of oncogenes in cancer, it is important to bear in mind that activation of oncogenes is not the sole pathway to malignancy. As described below, a combination of their activation combined with inactivation of tumor suppressor genes is involved in colorectal and probably most cancers. In some cases, their activation may be only a secondary occurrence associated with transformation rather than a causal event. **Epigenetic changes** also may be involved in certain instances, since some chemicals that apparently do not alter DNA are known to be carcinogens. An important finding was that the activation of c-*ras* in rat mammary cancers induced by nitrosomethylurea was apparently due to a specific G → A transition type of mutation, demonstrating that oncogenes are involved in chemical carcinogenesis. Much more work is needed to examine the possible involvement of oncogenes in the phenomena of initiation, promotion, tumor progression, and metastasis.

Mechanisms of Action of Oncogenes

The following are general mechanisms by which the products of oncogenes may stimulate growth (Figure 62–7).

They may act on key intracellular pathways involved in growth control, uncoupling them from the need for an exogenous stimulus. Relevant examples (described above) are the product of *src* acting as a protein-tyrosine kinase, the product of *ras* acting to stimulate the activity of adenylyl cyclase, and the product of *myc* acting as a DNA-binding protein. Each of these could affect the control of mitosis, the first two by events involving phosphorylation of key regulatory proteins. A major deficiency in our knowledge of cell growth is that until recently remarkably little was known about molecular aspects of the regulation of mitosis, even in normal cells. Advances in knowledge of cyclin and cdc (cell division-cycle) genes (Chapter 38) are rapidly changing this situation. An important area of current research is analyzing interactions between the products of certain oncogenes (and of tumor suppressor genes) with cyclin and related proteins.

The products of oncogenes (eg, the *sis* oncogene) may also imitate the action of a polypeptide growth factor or imitate an occupied receptor for a growth factor (eg, the *erb*-B oncogene) (see below).

Other mechanisms whereby oncogenes stimulate growth are currently being defined.

POLYPEPTIDE GROWTH FACTORS ARE MITOGENIC

The study of growth factors is rapidly becoming of major interest. A variety of such factors have been isolated and partly characterized (Table 62–5). Until recently, only very small quantities of most growth factors were available for study. However, the genes for a number of growth factors have now been cloned, confirming their separate identities and making usable amounts available through recombinant DNA technology. The growth factors known to date affect many different types of cells, eg, cells from the blood, nervous system, mesenchymal tissues, and epithelial tissues. They exert a mitogenic response on their target cells, though special conditions may be necessary to demonstrate this, such as depriving cells in culture of serum so that they have become quiescent prior to exposure to a growth factor. **Platelet-derived growth factor** (PDGF), released from the α granules of platelets, probably plays a role in normal wound healing. Various growth factors appear to play key roles in regulating differentiation of stem cells to form various types of mature hematopoietic

cells. Growth inhibitory factors also exist (eg, transforming growth factor [TGFβ] exerts inhibitory effects on the growth of certain cells). Thus, chronic exposure to increased amounts of a growth factor or to decreased amounts of a growth inhibitory factor could alter the balance of cellular growth.

Growth Factors Act in an Endocrine, Paracrine, or Autocrine Manner (See also Chapter 44.)

The effects of growth factors may be **endocrine** in the sense that, like hormones, they may be synthesized elsewhere in the body and pass in the circulation to their target cells. They may be **paracrine,** ie, synthesized in certain cells and secreted there to affect neighboring cells, though the synthesizing cells are not themselves affected because they lack suitable receptors. And certain growth factors can affect the cells that synthesize them, a mode of action characterized as **autocrine.** For instance, a factor may be secreted and then attach to its cell of origin, provided that cell possesses appropriate receptors. Alternatively, if a certain amount of the factor is not secreted, its presence inside the cell may directly stimulate various processes.

Growth Factors Act on Mitosis Via Transmembrane Signal Transduction

Relatively little is known about how growth factors operate at the molecular level. Like polypeptide hormones (Chapter 44), they must transmit a message across the plasma membrane to the interior of the cell (transmembrane signal transduction). In the case of growth factors, the message will ultimately affect one or more processes involved in mitosis. Most growth factors have high-affinity protein receptors on the plasma membrane of target cells. The genes for the receptors for many growth factors have been cloned and models of the structures of the receptors constructed. They generally have short membrane-spanning segments and external and cytoplasmic domains of varying lengths. The ligands bind to the external domains. A number of receptors (eg, for EGF, insulin, and PDGF) have been found to exhibit protein-tyrosine kinase activities, reminiscent of the product of the v-*src* gene (see above). This kinase activity, located in the cytoplasmic domains, causes autophosphorylation of the receptor protein and also phosphorylates other target proteins. The receptor-ligand complexes are subjected to endocytosis in coated vesicles (see discussion of LDL receptors in Chapter 27); it is not yet clear whether the receptors recirculate to the cell surface. The precise events resulting in transmembrane signaling are still under investigation and differ among the various factors. The case of PDGF will be described as an example.

Phospholipase C is stimulated following exposure of cells to PDGF, resulting in hydrolysis of phos-

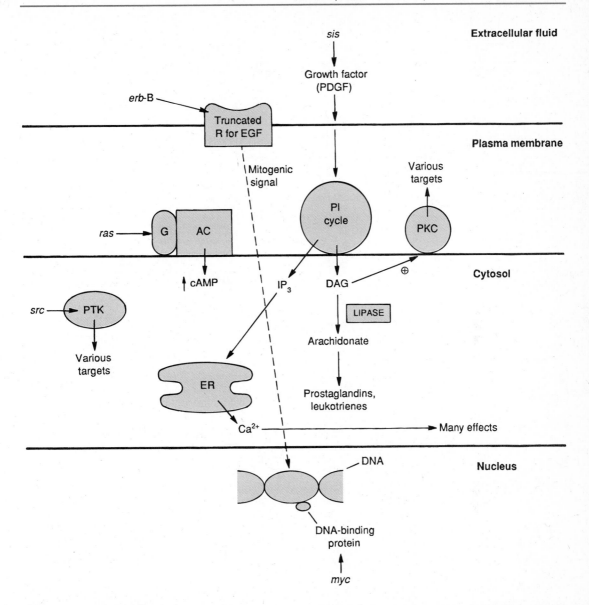

Figure 62–7. Schematic representation of mechanisms by which the products of certain oncogenes may alter cellular metabolism and thereby stimulate growth. The products of the following five oncogenes are shown: *sis* (encoding the B chain of PDGF), *erb*-B (encoding a truncated form of the receptor for EGF), *ras* (encoding a G type protein), *src* (encoding a protein-tyrosine kinase), and *myc* (encoding a DNA-binding protein). The product of *sis* stimulates the PI cycle, the product of *erb*-B stimulates mitosis, the product of *ras* stimulates adenylyl cyclase, the product of *src* phosphorylates various target proteins on tyrosine residues, and *myc* alters gene expression. Cyclic AMP can affect a number of cellular processes by activating cAMP-dependent protein kinases. Protein kinase C affects the activities of a number of target proteins. Ca^{2+} has numerous effects, as do prostaglandins and leukotrienes produced from arachidonate. (PDGF, platelet-derived growth factor; R, receptor; EGF, epidermal growth factor; G, G protein; AC, adenylyl cyclase; cAMP, cyclic AMP; PI, phosphatidylinositol; PKC, protein kinase C; PTK, protein-tyrosine kinase; IP_3, inositol triphosphate; DAG, diacylglycerol; ER, endoplasmic reticulum.)

Table 62–5. Some polypeptide growth factors.[1]

Growth Factor	Source	Function
Epidermal growth factor (EGF)	Mouse salivary gland	Stimulates growth of many epidermal and epithe-lial cells
Erythropoietin	Kidney, urine	Regulates development of early erythropoietic cells
Insulin-like growth factors I and II (IGF-I and IGF-II, also named somatomedins C and A)	Serum	Stimulate sulfate incorporation into cartilage, are mitogenic for chondrocytes, and exert insulinlike effects on many cells
Interleukin-1 (IL-1)	Conditioned media	Stimulates production of IL-2 by T cells
Interleukin-2 (IL-2)	Conditioned media	Stimulates growth of T cells
Nerve growth factor (NGF)	Mouse salivary gland	Tropic effect on sympathetic and certain sensory neurons
Platelet-derived growth factor (PDGF)	Platelets	Stimulates growth of mesenchymal and glial cells
Transforming growth factor (TGFα)	Conditioned media of transformed or tumor cells	Is similar to EGF
Transforming growth factor (TGFβ)	Kidney, platelets	Exerts both stimulatory and inhibitory effects on certain cells

[1]Modified and reproduced, with permission, from Franks LM, Teich NM (editors): *Introduction to the Cellular and Molecular Biology of Cancer.* Oxford Univ Press, 1986.

phatidylinositol 4,5-bisphosphate to form inositol triphosphate and diacylglycerol (see v-*src,* above, and Figure 44–6). These two second messengers can affect intracellular release of Ca^{2+} and stimulation of the activity of protein kinase C, respectively, thus affecting a large number of cellular reactions. The subsequent hydrolysis of diacylglycerol by phospholipase A_2, liberating arachidonic acid, can also result in the production of prostaglandins and leukotrienes, which themselves may exert many biologic activities (Chapter 25). Exposure of cells to PDGF can result in rapid (minutes to 1–2 hours) activation of certain cellular proto-oncogenes (eg, *myc* and *fos*). It seems likely that gene activation, whether of normal genes or proto-oncogenes, is involved in the action of most growth factors.

Growth Factors and Oncogenes Interact in Several Ways

The products of several oncogenes are either growth factors or parts of the receptors for growth factors (Table 62–4).

The B chain of PDGF contains 109 amino acids. It seems likely that the B chain is biologically active as a homodimer, without involvement of the A chain. The discovery that v-*sis* encodes 100 of the 109 amino acids of the B chain of PDGF revealed a direct relationship between oncogenes and growth factors. It also suggested that autocrine stimulation by PDGF—supplying a chronic mitogenic stimulus— could be an important factor in the mechanism of transformation of cells by v-*sis*. Indeed, many cultured tumor cells are known to secrete growth factors into their surrounding media and also to possess receptors for these molecules.

Sequence analysis of v-*erb*-B revealed that it encoded a truncated form of the receptor EGF, with much of the external domain of the receptor being deleted but the protein-tyrosine kinase activity being retained. It has been suggested that the abnormal form of the receptor for EGF encoded by v-*erb*-B may be continuously active when present in cells, simulating an occupied receptor. As in the case of autocrine stimulation by PDGF, this could result in a chronic mitogenic signal, "driving" cells toward the transformed state.

Transforming growth factor (TGFβ) was originally thought to be a positive growth factor, since it caused fibroblasts to behave as if they were transformed. TGFβ is now known to inhibit the growth of most cell types except fibroblasts. It will inhibit the growth of the monkey kidney cells that synthesize it. TGFβ may activate the *sis* gene in fibroblasts, perhaps explaining its growth-stimulating effect in these cells. How it produces its inhibitory effects on other cells remains to be established (see Mismatch Repair Genes, below). As we shall now see, there is other evidence that certain genes involved in neoplasia code for products that slow cell growth, and accordingly they have been designated tumor suppressor genes. Thus, the control of cell growth is complex, and both positive and negative regulatory factors are involved.

RB1, A TUMOR SUPPRESSOR GENE, IS INVOLVED IN THE GENESIS OF RETINOBLASTOMA

Genes other than oncogenes have been found to play a major role in the causation of at least some types of cancers. These are tumor suppressor genes, sometimes called recessive oncogenes or anti-oncogenes. They operate quite differently from oncogenes (Table 62–6) in that their inactivation (as opposed to activation) removes certain constraints on control of growth. An important model for understanding how such genes function is retinoblastoma. This is a malignant tumor of retinal neuroblasts, which are precursor cells of photoreceptor cells in the retina. In some cases, this tumor is inherited, whereas in others it does not appear to be hereditary. In 1971, Knudson suggested that development of retinoblastomas depended upon two mutations. He postulated that in hereditary cases of retinoblastoma, the first mutation was present in the germ cell line and the second occurred in retinoblasts. In sporadic (nonhereditary) cases, both mutations were thought to occur in retinoblasts. These ideas were verified by subsequent work. The gene involved in formation of retinoblastomas *(RB1)* has been isolated and sequenced, and its protein product, pRB, has also been partially characterized (Table 62–7). Analyses using RFLPs and Southern blots revealed that appropriate control specimens from individuals with hereditary retinoblastomas were heterozygous in the region of the *RB1* gene (reflecting one normal and one abnormal allele), whereas all tumor specimens examined were homozygous in this region. This phenomenon is called **loss of heterozygosity,** reflecting the fact that the tumor has both alleles mutated and is frequently found when analyzing tumor suppressor genes. pRB is a nuclear phosphoprotein whose phosphorylation status is apparently important with respect to its function. During G_0 or G_1, the protein is hypophosphorylated, whereas its phosphorylation increases in late G_1 and the early S phases, returning to a low level of phosphorylation in G_0–G_1. The hypophosphorylated form of pRB forms complexes with a number of viral proteins, such as the SV40 large T antigen; pRB is inactive in such complexes, abolish-

Table 62–7. Some properties of the *RB1* gene and its protein product.

- Gene is located on chromosome 13q14.
- Familial retinoblastomas have identical mutations in both alleles (loss of heterozygosity).
- Protein product (pRB) of molecular mass 110 kDa is expressed in many cells.
- pRB is a nuclear phosphoprotein whose phosphorylation is tightly regulated during the cell cycle.
- pRB binds certain viral proteins (eg, large T antigen of SV40), forming inactive complexes.
- pRB may regulate cell proliferation by binding to certain transcription factors (eg, E2F) that are active in S phase, thus slowing cell cycling.

ing its negative regulatory effect on cell division and helping to explain the increased multiplication shown by cells transformed by such viruses. In situations where oncogenic viruses are not involved, it appears to function by binding in G_0/G_1 to a set of proteins, some of which are transcription factors that are active in the S phase of the cycle and thus slowing cell cycling.

The main tumors in which inactivation of pRB is an important factor include small cell lung cancers, adenocarcinomas of the prostate, and tumors originating from retina, bone, and connective tissues.

THE *p53* TUMOR SUPPRESSOR GENE ACTS AS A GUARDIAN OF THE GENOME

Another extremely important tumor suppressor gene is *p53,* encoding a protein (designated p53) of molecular mass 53 kDA. Some of the principal features of this gene and its product are summarized in Table 62–8. The protein product, like pRB, is nuclear in location, subject to phosphorylation-dephosphorylation, and binds certain viral proteins. p53 appears to have at least three major effects: (1) It acts as a transcriptional activator, regulating certain genes involved in cell division. (2) It acts as a G_1 checkpoint control for DNA damage. If excess damage to DNA has occurred—eg, following ultraviolet irradiation—the activity of p53 increases, resulting in inhibition of cell division and allowing time for repair. If cell division were to proceed unchecked, the DNA damage would be replicated, introducing permanent muta-

Table 62–6. Some differences between oncogenes and tumor suppressor genes.[1]

Oncogenes	Tumor Suppressor Genes
Mutation in one of the two alleles sufficient for activity; act dominant to wild-type	Mutations in both alleles or a mutation in one followed by a loss of or reduction to homozygosity in the second
"Gain of function" of a protein that signals cell division	Loss of function of a protein
Mutation arises in somatic tissue, not inherited	Mutation present in germ cell (can be inherited) or somatic tissue
Some tissue preference	Strong tissue preference (eg, effect of *RB1* gene in the retina)

[1]Adapted from Levine AJ: The p53 tumor-suppressor gene. (Editorial.) N Engl J Med 1992;326:1350.

Table 62–8. Some properties of the *p53* gene and its protein product.

- Gene is located on short arm of chromosome 17.
- Product (p53) is nuclear phosphoprotein of 53 kDa.
- p53 binds specific sequences in dsDNA.
- p53 acts as a transcriptional regulator, presumably regulating genes that participate in cell division.
- p53 binds various viral proteins (eg, SV40 large T antigen), forming inactive oligomeric complexes.

tions into the genome. On the other hand, if p53 is inactivated, as in a large number of tumors (see below), then it could not fulfill this function, DNA damage would accumulate in the cell, and the cell would be genetically less stable. Thus, p53 has been proposed to act as a "guardian of the genome" or as a "molecular policeman." (3) A third function of p53—one that has attracted much experimental attention—is that it participates in the initiation of apoptosis (Gr "a falling off"). In most adult tissues, homeostasis is a balance between cell growth and cell death. Apoptosis can be defined as programmed cell death, controlled by specific gene-regulated events, and occurs in many physiologic (eg, embryonic growth and remodeling of tissues) and pathologic situations (eg, hormone deprivation and perhaps the cell death seen in Alzheimer's, Huntington's and Parkinson's diseases [Chapter 64]). Cells undergoing apoptosis shrink, their chromatin condenses, their DNA fragments, and their membranes exhibit blebs. Activation of a Ca^{2+}-Mg^{2+}-dependent endonuclease that cleaves internucleosomal DNA is one key event in apoptosis. A number of genes involved in apoptosis have been identified. Some of these are oncogenes, such as *myc* (stimulates) and *bcl-2* (which inhibits apoptosis). Another gene involved in apoptosis in certain cells, such as those of lymphocytic origin, is *APO-1* (or *Fas*), which encodes a receptor for a ligand that appears to be a member of the tumor necrosis factor family of growth factors. Interaction of the ligand with the receptor triggers apoptosis in susceptible cells. In the nematode *Caenorhabditis elegans,* a key pro-apoptotic gene is *ced-3,* encoding a putative cysteine protease, related to mammalian interleukin-1-β-converting enzyme (ICE). When *ced-3* was deleted in *C elegans,* apoptotic death of target cells was prevented. Hydrolases (eg, nucleases, proteases) are key players in apoptosis, no matter its cellular location, as they cause degradation of vital biomolecules.

How p53 participates in apoptosis is under examination. For instance, it could activate key genes involved in apoptosis, repress genes necessary for cell survival, or be a component of the biochemical series of events that result in apoptosis. The purpose of this function of p53 is that it hastens the death of potentially dangerous cells—eg, those damaged by ultraviolet irradiation—which have the potential to become malignant. In this role, p53 has been referred to as a "guardian of the tissues."

From another viewpoint, if apoptosis could be turned on in tumor cells (eg, by introducing a functioning *p53* gene into only them), this could have great therapeutic potential.

Mutations in *p53* Occur Frequently in Many Human Tumors

Because of many studies in the past few years, more is known about the occurrence of mutations in this gene in human cancers than about their occurrence in any other gene. Some of the major findings are summarized in Table 62–9. Only a few points will be commented upon further.

(1) It is clear that a wide variety of mutations in the *p53* gene occur in human cancer.

(2) CpG dinucleotides are hot spots for spontaneous mutations, reflecting the tendency of 5-methylcytosine residues in these dinucleotides to spontaneously deaminate.

(3) Aflatoxin B_1 is a potent hepatocarcinogen, and many individuals (in, for instance, China and Africa) are exposed to it. By in vitro mutagenesis experiments, it has been found to cause mainly G to T transversions. It is thus of great interest that liver cancers obtained from patients in areas where there is high exposure to this carcinogen were found to contain the same transversion.

(4) A similar transversion was found frequently in lung cancers, which are associated with smoking and exposure to chemicals such as benzo[a]pyrene. These two latter sets of findings and others of a similar kind indicate that determination of the specific types of mutations (transversions, transitions at CpG dinucleotides, etc) in genes from human tumors may help reveal their causes (eg, exposure to environmental chemicals or spontaneous mutations). This is an example of using molecular epidemiology to pinpoint the cause of cancer and contribute to its prevention.

Table 62–9. Summary of major features of *p53* mutations found in human tumors.[1]

- Mutations in the *p53* gene are the most common genetic alterations in human cancer and are frequent in colon, breast, and lung cancers.
- At least 100 different mutations have been detected in this gene.
- The mutations are usually found at highly conserved codons.
- The mutational spectrum differs among different cancers.
- Transitions occurring at CpG dinucleotides (a hot spot) are frequent in neoplasms of the colon, brain, and lymphoid tissue.
- Transversions are frequent in lung cancer and liver cancer and are found mainly at one specific site (G to T mutations at codon 249) in certain hepatocellular carcinomas, occurring in areas where exposure to aflatoxin B_1 and hepatitis B virus is frequent.

[1]Data from Hollstein M et al: p53 mutations in human cancers. Science 1991;253:49. Control tissues uninvolved by cancer did not exhibit mutations in the *p53* gene. The mutations studied occurred in somatic cells; mutations in the *p53* gene have been found in germ cells in subjects with the Li-Fraumeni cancer syndrome, a syndrome in which a high incidence of certain cancers is inherited. In a transition, a purine is substituted for a purine or a pyrimidine for a pyrimidine. In a transversion, a purine is substituted for a pyrimidine or vice versa.

A GENETIC MODEL FOR COLORECTAL CANCER SUGGESTS THAT MUTATIONS IN FIVE OR SIX GENES MAY BE NECESSARY FOR ITS DEVELOPMENT

Colorectal cancer is a common malignancy; approximately 150,000 cases occur in the USA annually. Along with cancer of the lung and breast, it constitutes about 40% of all human cancers. Most colorectal cancers appear to arise from benign tumors called **adenomas.** It is possible to obtain from patients specimens of normal colorectal tissue, of adenomas of varying sizes, of colorectal cancer, and of metastatic colorectal cancer; these can be compared with respect to parameters such as chromosomal karyotypes, oncogenes, tumor suppressor genes, and extent of methylation of DNA.

Fearon and Vogelstein have proposed a genetic model to explain the development of colorectal cancer (Figure 62–8). The major features are the following: (1) A number of steps are involved, in agreement with a body of knowledge indicating that cancer is a multistep process. (2) Mutation of tumor suppressor genes on chromosomes 5, 18, and 17 (the last being p53) is involved. Mutation and consequent activation of the *ras* oncogene on chromosome 12 are also involved. (3) Thus, at least four genes and probably five or six (or even more) are implicated. More tumor suppressor genes are affected than oncogenes. (4) The precise order of changes is not as important as the accumulation of changes. (5) Additional mutations are necessary to permit spread and metastasis. In other malignancies, the spectrum of oncogenes and tumor suppressor genes involved probably differs from that implicated in this model.

The above represents a stimulation for future work, but it is probably simplistic. Similar data are being obtained for other major tumors. It is hoped that these findings will be useful in terms of prevention, diagnosis, prediction of prognosis, and treatment. For instance, with reference to treatment, it may be possible to design agents that inactivate oncogenes (eg, antisense oligodeoxynucleotides), or it may be possible to replace—by gene therapy—mutant tumor suppressor genes by their wild-type counterparts.

MISMATCH REPAIR GENES ARE INVOLVED IN THE DEVELOPMENT OF HEREDITARY NONPOLYPOSIS COLON CANCER

The two major DNA repair systems in humans are the mismatch repair system and the nucleotide excision repair system. They have attracted considerable attention, because the first has been implicated in the development of hereditary nonpolyposis colon cancer (HNPCC) and the second in the causation of xeroderma pigmentosum (see Case no. 1, Chapter 65).

The mismatch repair system corrects errors that may occur while DNA is being copied, eg, insertion of an incorrect nucleotide. It is a relatively simple system, involving approximately six proteins. As in most repair systems (see Chapter 38), these proteins collectively recognize the abnormal site, cut it out, and replace it with the correct sequence. In *E coli*, products of three genes (*MutS, MutL,* and *MutH*) are involved in the initial stages of mismatch repair. The key components of this system are indicated in Figure 62–9.

In contrast, the nucleotide excision repair system removes larger areas of DNA damaged by chemicals or irradiation. It is more complex than the mismatch

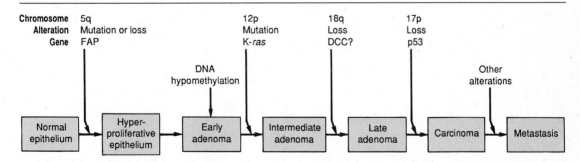

Figure 62–8. Representation of a genetic model for colorectal cancer. The tumor suppressor genes implicated are located on chromosomes 5, 18, and 17 (p53). That on chromosome 5 is mutated in patients with familial adenomatous polyposis and is designated FAP, whereas that on chromosome 18 is frequently deleted in colorectal cancers and is thus designated DCC. Hypomethylation of DNA has been shown to occur early in the development of colorectal cancer and may play a significant role in the genetic instability exhibited by tumors. The adenomas were classified according to size, the early adenomas being less than 1.0 cm, the intermediate and late both exceeding this size but the late also containing foci of cancer cells. (Reproduced, with permission, from Fearon ER, Vogelstein B: A genetic model for colorectal tumorigenesis. Cell 1990;61:759. Copyright © Cell Press.)

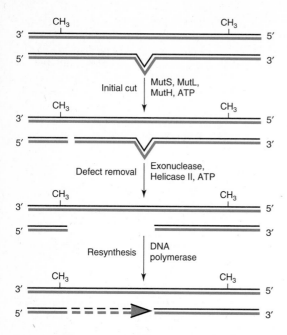

Figure 62–9. Simplified representation of mismatch repair in *E coli*. A DNA mismatch (indicated by the bend) can be repaired from the 5′ side, as shown here, or from the 3′ side. DNA ligase is needed to close the gap filled in by DNA polymerase (III). The methyl (CH₃) groups on the upper strand are attached to adenine in GATC sequences; their transient absence in the newly synthesized strand targets repair to this strand. Details of how the system operates in human cells are under investigation. (Reproduced, with permission, from Marx J: DNA repair comes into its own. Science 1994;266:728.)

repair system, involving at least 12 gene products (Chapter 65).

Hereditary nonpolyposis colon cancer (HNPCC) is a relatively common syndrome characterized by the early onset of cancer of the colon and tumors of certain other organs. It was observed that cells from this type of cancer showed microsatellite instability. Microsatellites are short sequences of DNA repeated many times. Compared with the DNA of normal tissue, the microsatellite sequences of cancer cells from patients with HNPCC varied in length, indicating that the tumors had either gained or lost sequences. This could occur if the two strands of DNA slipped relative to each other during replication. Depending on the direction of slippage, the new strand would be either shorter or longer than the parent strand. Thus, the new dsDNA will contain a small loop of unpaired DNA, which in normal tissues would be corrected by the mismatch repair system. The observed abnormalities suggested that the mismatch repair system was not functioning properly in these tumors and was thus probably altered. It was then found that HNPCC was linked to a gene on chromosome 2, which was

isolated and turned out to be the equivalent of *mutS*. The gene was thus designated *hMSH2*. This gene appears to account for approximately 60% of cases of HNPCC. Three more human genes were then isolated (*hMLH1*, *hPMS1*, and *hPMS2*), and all were found to be homologous to specific genes of *E coli* that are involved in mismatch repair. *hMLH1*, the equivalent of *MutL* in *E coli*, appears to account for most of the remaining 40% of cases of HNPCC. Tumor cells with microsatellite instability were unable to carry out mismatch repair, indicating that mutations had occurred in the genes (ie, *hMSH2* and *hMLH1*) involved in this process.

An important connection between microsatellite instability and lack of responsiveness to a particular growth factor has been established. Normally, TGFβ has an inhibitory effect on the growth of epithelial cells. This is mediated by TGFβ receptors, which are heteromers consisting of a mixture of type I and type II receptor (R) molecules. In cell lines derived from human colon cancers, which had high rates of microsatellite instability, mRNA transcripts for RII were found to be markedly reduced or absent. Confirming this, the same cell lines showed little or no binding of radiolabeled TGFβ, indicating that their TGFβ receptors were defective. A number of types of mutations were detected in the gene encoding RII. Thus, in this situation, defective repair (manifested by microsatellite instability) appears to be directly linked to development of mutations in a gene encoding a growth factor receptor, which if inactivated permits escape from normal growth control (Figure 62–10). Defective repair could also contribute to some of the events depicted in Figure 62–8.

It appears that defective mismatch repair may be a critical event in the development of at least certain types of tumors. Defects in mismatch repair have also been found in some nonhereditary colon cancers. It will be important to establish the prevalence of defects in mismatch repair in different cancers. Inability to repair damage may lead to activation or in-

Mutations in genes (eg *hMSH2*, *hMLH1*) encoding proteins of the mismatch repair system
↓
Defective mismatch repair system
↓
Unrepaired mutations causing microsatellite instability, affecting the RII gene, among others
↓
Defective TGFβ receptor
↓
Cells lose inhibitory growth response to TGFβ

Figure 62–10. Schematic diagram of a chain of events that could lead to loss of growth control of cells in hereditary nonpolyposis colon cancer (HNPCC).

activation of various genes in tumors and allow them to diverge rapidly from their normal parental genes. This could be important in the phenomenon of tumor progression (see below). The identification of these genes also allows tests to be performed, such as screening for abnormal *hMSH2* and *hMLH1* genes in high-risk families.

A NUMBER OF GENES THAT INCREASE SUSCEPTIBILITY TO CANCER HAVE BEEN ISOLATED

It is well known that certain types of cancer (eg, of the breast) show an increased familial incidence. Considerable progress has now been made in isolating genes that increase susceptibility to cancer. A number of these genes and the conditions with which they are associated are listed in Table 62–10; they include *RB1* and *p53*.

A recent triumph has been the isolation of *BRCA1*, a gene that influences susceptibility to early-onset breast and ovarian cancer. Genetic factors contribute to approximately 5% of all cases of breast cancer but to approximately 30% of all cases diagnosed before age 30. *BRCA1* specifies a protein of 1863 amino acids containing a zinc finger domain in its amino terminal region. Its function has not been established, but it could be a transcription factor. A variety of different types of germline mutations were found in the gene, presumably predisposing to development of cancer. Approximately 12 other genes associated with familial types of cancer have been mapped to specific chromosomal locations, and isolation of the genes should soon be forthcoming. One of these is a second gene associated with increased susceptibility to breast cancer (but not to ovarian cancer), *BRCA2*, mapped to chromosome 13q12–13. The identifica-

tion and further study of these genes (eg, to determine whether they act as tumor suppressor genes or by other mechanisms) should facilitate early diagnosis of different types of familial cancer and provide insights into mechanisms of development of cancer in various tissues.

THE MALIGNANCY OF TUMOR CELLS TENDS TO PROGRESS

Once a cell becomes a tumor cell, the composition and behavior of its progeny do not remain static. Instead, there is a tendency for malignancy to increase. This is manifested by increasingly abnormal karyotypes, increasing rates of growth, and an increasing tendency to invade and metastasize. The important phenomenon of progression appears to reflect a fundamental instability of the genome of tumor cells. It seems likely that mutations in DNA repair genes are involved in this phenomenon, creating a mutator phenotype. The activation of additional oncogenes may be involved. Cells with faster rates of growth have a selective advantage.

It is important to distinguish the biochemical profiles of cells that have just been transformed from those of fast-growing, highly malignant tumor cells. The former may show few differences from normal cells apart from key changes leading to cancer (eg, activation of one or more oncogenes), and those changes generally are associated with transformation (Table 62–3). Analyses of such cells can disclose the key biochemical alterations that result in transformation. The biochemical profile of highly malignant cells may be very different from that of normal cells. Many changes in enzyme profile and other biochemical parameters have occurred (Table 62–11), some of which are secondary to rapid growth rate and others

Table 62–10. Some cloned genes that increase susceptibility to cancer.[1,2]

	Gene	Chromosome	Clinical Expression
Familial adenomatous polyposis	*APC*	5q21	Characterized by the development of many early-onset adenomatous polyps, which are immediate precursors of colorectal cancers.
Familial breast and ovarian cancer	*BRCA1*	17q21	See text.
Li-Fraumeni syndrome	*p53*	17p13	A rare syndrome involving cancers at different sites, developing at an early age, and often exhibiting multiple primary tumors.
Neurofibromatosis type 1	*NF1*	17q11	Condition that varies from a few café au lait spots to development of thousands of neurofibromas.
Neurofibromatosis type 2	*NF2*	22q12	Major characteristic is development of bilateral acoustic neuromas.
Retinoblastoma	*RB1*	13q14	See text.
Wilms' tumor	*WT1*	11p13	A specific type of kidney tumor developing in childhood.

[1]Adapted from Knudson AG: All in the (cancer) family. Nature Genetics 1993;5:103.
[2]All of the conditions listed in the table are hereditary. Several other genes causing familial cancer have also been cloned (eg, for multiple endocrine neoplasia type 2A and for the von Hippel-Lindau syndrome).

Table 62–11. Biochemical changes often found in fast-growing tumor cells.

- Increased activity of ribonucleotide reductase.
- Increased synthesis of RNA and DNA.
- Decreased catabolism of pyrimidines.
- Increased rates of aerobic and anaerobic glycolysis.
- Alterations of isozyme profiles, often to a fetal pattern.
- Synthesis of fetal proteins (eg, carcinoembryonic antigen).
- Loss of differentiated biochemical functions (eg, diminished synthesis of specialized proteins).
- Inappropriate synthesis of certain growth factors and hormones.

probably due to chromosomal instability. These fast-growing cells tend to maximize the anabolic processes involved in growth (eg, DNA and RNA synthesis), cut down on catabolic functions (eg, catabolism of pyrimidines), and dispense with the differentiated functions shown by their normal ancestors. In other words, they are concentrating almost exclusively upon growth. They also show biochemical changes that reflect altered gene regulation, such as the synthesis of certain fetal proteins (some of which can be used as tumor markers; see below) and the inappropriate manufacture of growth factors or hormones. Analyses of such cells are unlikely to reveal the initial key events responsible for transformation, partly because these events are obscured by a myriad of changes secondary to progression. However, knowledge of the biochemical profile of such cells is extremely important in choosing chemotherapy, as it is exactly this type of cell that the oncologist is usually called upon to deal with.

DRUGS USED IN CANCER CHEMOTHERAPY ACT AT MANY BIOCHEMICAL SITES

Surgery, radiotherapy, and chemotherapeutic agents are the major modalities used to treat patients with cancer. The development of anticancer drugs is—like numerous other areas of medicine—in many ways an exercise in applied biochemistry. The basic problem is to make available drugs (natural products or synthetic) that kill cancer cells effectively but are not excessively toxic to normal cells. Table 62–12 lists seven major classes of compounds that have been widely used in the treatment of cancer. Insofar as unrestrained cell division is a feature that typifies many malignant tumors, many of these agents are used because they inhibit DNA synthesis. For this reason, they are also likely to damage normal tissues whose cells divide continuously (eg, the gut and bone marrow).

The **growth fraction** (percentage of the cells of a tumor that are constantly in cycle) is an important concept in cancer chemotherapy; tumors with a high growth fraction are usually more responsive to cancer chemotherapy than cells that are dormant in the G_0 phase of the cycle. It is useful to classify anticancer drugs as **cell cycle specific** (CCS) or **cell cycle nonspecific** (CCNS). The former (including methotrexate, fluorouracil, and cytarabine) act on proliferating cells, whereas the latter (eg, alkylating agents and cisplatin) act irrespective of the state of proliferation. CCS agents can be further subdivided into **phase-specific** agents (eg, agents acting on the S phase of the cycle [eg, cytarabine] or on the G_2-M phase [eg, bleomycin]). In all cancer therapy, it is vital to diagnose and treat the tumor early; otherwise, the tumor burden may be too great for therapy to be successful. During therapy, it is desirable to elimi-

Table 62–12. Some drugs used in cancer chemotherapy.

Class of Compound	Example	Site of Action	Treatment Use
Alkylating agents	Melphalan	Alkylates DNA and other molecules	Myeloma
Antimetabolites			
Purine antagonists	Mercaptopurine	Converted to a "fraudulent" nucleotide and inhibits purine synthesis	Acute myelocytic leukemia
Pyrimidine antagonists	Fluorouracil	Converted to a "fraudulent" nucleotide and inhibits thymidylate synthetase	Colorectal cancer
Folate antagonists	Methotrexate	Inhibits dihydrofolate reductase	Choriocarcinoma
Antitumor antibiotics	Doxorubicin	Intercalates in DNA and stabilizes the DNA-topoisomerase II complex	Hodgkin's disease
Other agents	Cisplatin	Causes strand breakage in DNA	Carcinoma of the lung
	Hydroxyurea	Inhibits ribonucleotide reductase	Chronic myelocytic leukemia
Plant compounds	Vinblastine	Binds tubulin and inhibits microtubule formation	Kaposi's sarcoma
Sex hormones	Estrogens	Block effects of androgens in prostatic tumors	Cancer of the prostate
Corticosteroids	Prednisone	Inhibits proliferation of lymphocytes	Myeloma

Many agents listed above are used along with other agents in combination chemotherapy and are not necessarily the current treatment of choice for the conditions indicated. Other classes of chemotherapeutic agents also exist, such as the biologic response modifiers (eg, interferons).

nate all clonogenic cells ("tumor stem cells"), since they have the potential for unlimited multiplication. In general, high-dose intermittent therapy is more likely to accomplish this than low-dose continuous therapy, because it exposes tumor cells to a higher level of the agent used. Combination chemotherapy (combining three or four agents) has proved successful in some cancers because the drugs may act synergistically, the onset of resistance may be delayed, and toxicity is often less.

P-GLYCOPROTEIN PLAYS AN IMPORTANT ROLE IN MULTIDRUG RESISTANCE IN CANCER CHEMOTHERAPY

A major problem in cancer chemotherapy is development of resistance to the agents used. While the drugs used are effective initially, often after a time (eg, several months) the tumor cells develop mechanisms that render the agents ineffective (acquired resistance).

Resistance may also be intrinsic; ie, a particular drug is *never* effective against the target tumor cells. For example, various bacteria are sensitive to penicillin because it inhibits specific biochemical reactions involved in the synthesis of their cell walls. Penicillin has no effect on eukaryotic cells because they do not synthesize cell walls; these cells thus lack the target site of action of penicillin. Other explanations for intrinsic resistance by cells include a high capacity for metabolic inactivation of the administered drug, usually mediated by cytochrome P450 species (Chapter 61).

The acquired resistance of cancer cells to chemotherapeutic agents is believed to reflect their high spontaneous mutation rates. Resistance may be not only to the drug but also to other structurally unrelated anticancer agents. For instance, if resistance to methotrexate develops, the tumor cells may also be found to be resistant to antitumor antibiotics such as doxorubicin and to plant compounds such as vincristine. This is known as **multidrug resistance** and is of great importance because its development is frequently the cause of failure of cancer chemotherapy, a common event during the treatment of cancer.

The molecular basis of multidrug resistance has been illuminated by the discovery that cancer cells which exhibit this phenomenon often have increased amounts of a protein designated **P-glycoprotein** (P = permeability) on their plasma membranes. This protein acts as an energy-dependent pump, ejecting a wide variety of anticancer drugs from cancer cells and thus diminishing their efficacy. Because of its therapeutic importance, P-glycoprotein has been the subject of much study, and considerable information has accumulated (Table 62–13). In cultured cells, gene amplification and point mutations (perhaps af-

Table 62–13. Some properties of P-glycoprotein.

- Phosphorylated glycoprotein of 170 kDa present in the plasma membrane, 1280 amino acids, 12 transmembrane regions, 2 ATP-binding folds; shows homology to bacterial hemolysin B.
- Acts as energy-dependent efflux pump expelling a variety of drugs and thus mediating MDR.
- Member of a multigene family (including *mdr*1 and *mdr*2).
- Widely distributed and highly conserved across species.
- Present in some normal cells of kidney, gut, etc.
- Amount increased in resistant cultured cells and in resistant tumor cells in vivo; increased amount in vivo correlates with poor prognosis.
- Can also be induced by other stimuli such as heat shock.
- Gene amplification and/or point mutations involved in MDR in cultured cells.
- When transfected into cells can confer MDR.
- Can be inhibited by chemosensitizers such as verapamil (a calcium blocker) or cyclosporine.

MDR, multidrug resistance.

fecting regulation) have been found in resistant cancer cells showing elevated amounts of P-glycoprotein; the mechanisms that regulate increase of P-glycoprotein in cancer cells in vivo have not yet been established. Much effort is being expended to develop inhibitors of P-glycoprotein; the use of drugs such as verapamil and cyclosporine (chemosensitizers) has shown some promise. P-glycoprotein must have physiologic functions, since it is present in normal organs such as kidney and gut; it may play a role in the excretion of potentially toxic compounds from cells of these organs.

A variety of other mechanisms have been shown to be involved in the resistance of cancer cells to certain classes of drugs (Table 62–14).

METASTASIS IS THE MOST DANGEROUS PROPERTY OF TUMOR CELLS

Metastasis is the spread of cancer cells from a primary site of origin to other tissues where they grow as secondary tumors, and it is the major problem presented by the disease. Metastasis is a complex phenomenon to analyze in humans, and knowledge of its biochemical basis is sparse. Because it reflects a failure in cell-cell interaction, much attention has naturally focused on comparisons of the biochemistry of the surfaces of normal and malignant cells. Many changes have been documented at the surfaces of malignant cells (Table 62–15), though not all are directly relevant to the problem of metastasis.

At present, considerable research is devoted to developing suitable animal model systems for the study of metastasis. Many studies also are being done to uncover the possible roles of certain proteases (eg, type 4 collagenase) and of certain glycoproteins and glycosphingolipids of the cell surface in the phenom-

Table 62–14. Some biochemical mechanisms of drug resistance found in cancer cells.[1]

General Mechanism	Drug(s)	Specific Example
Decreased uptake	Methotrexate	Mutant carrier system
Increased efflux	Certain anticancer drugs	Mediated by P-glycoprotein
Insufficient activation	Cyclophosphamide	Decrease of activating cytochrome P450 species
Increased inactivation of drug	Cytarabine	Increased activity of deaminases acting on cytosine
Drug sequestration	Cisplatin	Increase of one type of metallothionein[2]
Mutation in target enzyme	Methotrexate	Mutant DHFR[3]
Increase of target enzyme	Methotrexate	Amplification of gene for DHFR[3]
Rapid repair mechanisms	Certain methylating agents	Increase of specific repair enzyme (O^6-Alkylguanine-DNA-alkyl-transferase)

[1]Data from Hays JD, Wolf CR: Molecular mechanisms of drug resistance. Biochem J 1990;22:281.
[2]Metallothioneins are low-molecular-weight cytosolic metalloproteins, rich in cysteine, present in high amounts in liver and kidney, and which bind and thus detoxify various metals (eg, Zn, Cu, Hg, and Cd). They are inducible on exposure to such metals (see Chapter 59).
[3]DHFR, dihydrofolate reductase.

enon of metastasis. For instance, it is possible that changes in the oligosaccharide chains of cell glycoproteins (secondary to alterations of activities of specific glycoprotein glycosyltransferases) may be critical in permitting metastasis to occur. A search for alterations in proteins involved in cell-tissue and cell-cell interactions (eg, integrins, cadherins, and other cell adhesion molecules such as the neural cell adhesion molecules [N-CAMs]) is also an active area of research. Elucidation of the biochemical mechanisms involved in metastasis could provide a basis for the rational development of more effective anticancer therapies.

Another area related to metastasis concerns the blood supply of tumors. Folkman demonstrated that progressive tumor growth is angiogenesis-dependent. Tumor cells can secrete angiogenic growth factors, such as basic or acidic fibroblast growth factor (bFGF, aFGF), which promote the proliferation of endothelial cells and the formation of new capillaries. This raises the intriguing possibility of designing drugs that could kill tumors by specifically inhibiting the growth of their blood vessels without affecting their normal counterparts.

BIOCHEMICAL LABORATORY TESTS ARE ESSENTIAL IN THE MANAGEMENT OF PATIENTS WITH CANCER

Biochemical laboratory tests help in the management of patients with cancer. Many cancers are associated with the abnormal production of enzymes, proteins, and hormones (see the discussion of the progression of tumors, above), which can be measured in plasma or serum. These molecules are known as **tumor markers.** Measurement of some tumor markers is now an essential feature of management of some types of cancer (Table 62–16). Applications of tumor markers in diagnosis and management of cancer are listed in Table 62–17. Three major conclusions have emerged from the study of tumor markers (McIntire, 1984): (1) No single marker is useful for all types of cancer or for all patients with a given type of cancer. For this reason, the use of a battery of tumor markers is sometimes advantageous. (2) Markers are most often detected in advanced stages of cancer rather than early stages,

Table 62–15. Some changes that have been detected at the surfaces of malignant cells.[1]

- Alterations of permeability
- Alterations in transport properties
- Diminished adhesion
- Increased agglutinability by many lectins
- Alterations of the activities of a number of enzymes (eg, certain proteases)
- Alterations of surface charge
- Appearance of new antigens
- Loss of certain antigens
- Alterations of the oligosaccharide chains of glycoproteins
- Changes of glycolipid constituents

[1]Adapted from Robbins JC, Nicolson GL: Surfaces of normal and transformed cells. In: *Cancer: A Comprehensive Treatise.* Vol 4. Becker FF (editor). Plenum Press, 1975.

Table 62–16. Clinically useful tumor markers.[1]

Marker	Associated Cancer
Carcinoembryonic antigen (CEA)	Colon, lung, breast, pancreas
Alpha-fetoprotein (AFP)	Liver, germ cell
Human chorionic gonadotropin (hCG)	Trophoblast, germ cell
Calcitonin (CT)	Thyroid (medullary carcinoma)
Prostatic acid phosphatase (PAP)[2]	Prostate

[1]Adapted, with permission, from McIntire KR: Tumor markers: How useful are they? Hosp Pract (Dec) 1984;19:55.
[2]Measurement of prostate-specific antigen (PSA) is useful to monitor recurrence or persistence of prostatic cancer after surgery. Increase of PAP generally indicates local extension or metastases.

Table 62–17. Applications of tumor markers.[1]

Detection: Screening in asymptomatic persons.
Diagnosis: Differentiating malignant from benign conditions.
Monitoring: Predicting effect of therapy and detecting recurrent cancer.
Classification: Choosing therapy and predicting tumor behavior (prognosis).
Staging: Defining extent of disease.
Localization: Nuclear scanning of injected radioactive antibodies.
Therapy: Cytotoxic agents directed to marker-containing cells.

[1]Adapted, with permission, from McIntire KR: Tumor markers: How useful are they? Hosp Pract (Dec) 1984;19:55.

when they would be more helpful. (3) Of the uses of markers listed in Table 62–17, the most successful have been the monitoring of responses to therapy and the detection of early recurrence.

SUMMARY

Cancer (malignant) cells are characterized by loss of growth control, invasiveness, and metastasis. Benign tumor cells have lost growth control but do not metastasize. Cancer can be caused by physical, chemical, and biologic agents. These agents damage or alter DNA, so that cancer is truly a disease of the genome. Initiation and promotion are well-established phenomena in chemical carcinogenesis. Initiation probably damages DNA, but the mechanisms involved in promotion remain to be clarified. Phorbol esters are much-studied promoting agents; they activate protein kinase C, which exerts a variety of effects.

Much current interest in cancer is focused on the study of oncogenes and tumor suppressor genes. The concept of oncogenes arose from studies of oncogenic viruses. Normal cells contain potential precursors of oncogenes, designated proto-oncogenes. Activation of these genes to oncogenes is achieved by at least five mechanisms (promoter and enhancer insertion, chromosomal translocation, gene amplification,

and point mutation). Activated oncogenes influence cellular growth by perturbing normal cellular mechanisms of growth control, by acting as growth factors or receptors, and probably by other means as well.

Growth factors act in endocrine, paracrine, or autocrine manners. They affect cell division, and some may exert negative effects (eg, TGFβ). The products of some oncogenes (eg, that of *sis*) act as growth factors.

Tumor suppressor genes are now recognized as key players in the genesis of cancer. The effect of oncogenes on cell growth has been compared to putting one's foot on the accelerator of an automobile, whereas the action of tumor suppressor genes resembles taking one's foot off the brake. Important tumor suppressor genes are *RB1* and *p53*, both of which are nuclear phosphoproteins and probably affect the transcription of genes involved in regulating events in the cell cycle. An important genetic model for colorectal cancer invokes the interplay of tumor suppressor genes and the *ras* oncogene. Mutations in mismatch repair genes have been found to be associated with hereditary nonpolyposis colon cancer, and loss of responsiveness to the growth inhibitory effect of TGFβ appears important in the development of this type of tumor. A number of cancer susceptibility genes have been isolated; they include *RB1*, *p53*, and *BRCA1*. Tumor progression reflects an instability of the tumor genome, possibly due at least in part to defects in DNA repair systems and activation of additional oncogenes.

Tumor progression reflects an instability of the tumor genome. The phenomenon of metastasis is being investigated vigorously and may involve changes in surface glycoconjugates, altered activities of proteases, and alterations of cell adhesive molecules.

Agents used in cancer chemotherapy act at a variety of biochemical loci. Resistance to anticancer drugs is an important phenomenon; one mechanism of multidrug resistance involves P-glycoprotein. Certain enzymes and proteins are released by tumor cells into the blood and are used as tumor markers in diagnosis and for monitoring therapy and recurrence of disease.

REFERENCES

Cavenee WK, White RL: The genetic basis of cancer. Sci Am (Mar) 1995;272:72.

Fearon ER, Vogelstein B: A genetic model for colorectal tumorigenesis. Cell 1990;61:759.

Franks LM, Teich NM (editors): *Introduction to the Cellular and Molecular Biology of Cancer,* 2nd ed. Oxford Univ Press, 1991.

Gottesman MM, Pastan I: Biochemistry of multidrug resistance mediated by the multidrug transporter. Annu Rev Biochem 1993;62:385.

Hamel PA et al: Speculations on the roles of *RB1* in tissue-

specific differentiation, tumor initiation, and tumor progression. FASEB J 1993;7:846.

Hollstein M et al: p53 mutations in human cancers. Science 1991;253:49.

Knudson AG: All in the (cancer) family. Nature Genetics 1993;5:103.

Kohn EC, Liotta LA: Molecular insights into cancer invasion: Strategies for prevention and intervention. Cancer Res 1995;55:1856.

Levine AJ: Tumor suppressor genes. Sci Am Science and Medicine (Jan/Feb) 1995;2:29.

Loeb LA: Mutator phenotype may be required for multi-stage carcinogenesis. Cancer Res 1991;51:3075.

Markowitz S et al: Inactivation of the type II TGF-beta receptor in colon cancer cells with microsatellite instability. Science 1995;268:1336.

Marx J: DNA repair comes into its own. Science 1994;266: 728.

Marx J: Oncogenes reach a milestone. Science 1994;266: 1942.

Miki Y et al: A strong candidate for the breast and ovarian cancer susceptibility gene *BRCA1*. Science 1994;266:66.

Modrich P: Mismatch repair, genetic stability, and cancer. Science 1994;266:1959.

Sancar A: Mechanisms of DNA excision repair. Science 1994;266:1954.

Tannock IF, Hill RP (editors): *The Basic Science of Oncology,* 2nd ed. Pergamon, 1992.

Thompson CB: Apoptosis in the pathogenesis and treatment of disease. Science 1995;267:1456.

Weinberg RA: The retinoblastoma protein and cell cycle control. Cell 1995;81:323.

Biochemical & Genetic Bases of Disease

63

Robert K. Murray, MD, PhD

INTRODUCTION

In Chapter 1 it was stated that a knowledge of biochemistry is important for the understanding and maintenance of health and for the understanding and effective treatment of diseases. Put another way, a knowledge of biochemistry offers a rational framework, based on the experimental approach (as opposed to a received dogma), with which to view health and disease. Numerous examples of biochemical abnormalities resulting in loss of health and development of disease have been encountered throughout this text. However, these examples have necessarily appeared in a random fashion.

This chapter has the following objectives:

(1) To reemphasize the importance of defining the precise biochemical abnormalities that lead to loss of health or to the development of disease because, once these are understood, rational therapy may follow.

(2) To discuss briefly the causes of disease from a biochemical perspective and to emphasize that diseases result from changes of either the structures, function, or amounts of certain molecules (eg, DNA, proteins) or from perturbations of various biochemical reactions or processes.

(3) To present biochemical aspects of disease in general and briefly cover genetic diseases in particular.

Chapter 64 considers the biochemical basis of a number of neuropsychiatric disorders. Chapter 65 gives concrete examples of the value of biochemical knowledge in the clinical setting by presenting nine case histories, each covering one of the major causes of disease listed in Table 1–1.

ELUCIDATION OF THE BIOCHEMICAL BASES OF HEALTH AND DISEASE GENERALLY LEADS TO RATIONAL TREATMENT

Elucidation of the biochemical bases of health and disease generally leads to rational treatment of disease. It should be evident from reading this text that the maintenance of health is dependent on an adequate intake of water, calories, vitamins, and certain minerals, amino acids, and fatty acids. To select one of these groups of compounds, the recognition of the fundamental roles played by vitamins in metabolism and of the results of their deficiencies has made for rational treatment of vitamin deficiencies, eg, the treatment of scurvy, rickets, and beri-beri by administration of vitamin C, vitamin D, and thiamin, respectively. This illustrates the basic medical principle that it is generally necessary to know the cause and mechanisms involved in generating a disease in order to devise rational and effective treatment for it. At the present time, the causes of many important diseases, such as Alzheimer's disease, atherosclerosis, rheumatoid arthritis, and schizophrenia, are still unknown. Thus, the treatments of such conditions are generally symptomatic and empirical, not based on correcting the underlying abnormality and generally not very effective. The economic and human costs of such ignorance are immense.

However, even if the molecular nature of a disease has been shown, it may be impossible to institute appropriate treatment because of limitations—technologic or other—in the ability to correct the basic abnormality. Such is the case for sickle cell anemia, for which it is not yet possible to correct the underlying abnormality in DNA. The pace of research on gene therapy suggests that this statement may not apply for very long to this disease or to many other genetic disorders.

ALL DISEASES HAVE A BIOCHEMICAL BASIS

Table 1–1 lists the major causes of disease. Life as we know it depends upon biochemical reactions; if these cease, death results. Health depends on the regulated, harmonious functioning of the thousands of biochemical reactions and processes that occur in normal cells and that operate to maintain the constancy (with regard to pH, osmolality, concentration of electrolytes, etc) of the internal environment. Disease results from alterations of either the structures, eg, ultimately DNA in genetic diseases; or the amounts of certain biomolecules; or disturbances of important biochemical reactions or processes. These mechanisms, temporary or permanent, are induced by the causes listed in Table 1–1 and often lead to potentially severe changes of the internal environment, for which compensatory mechanisms can only operate for a finite period of time.

The above view of diseases is reductionist and, while useful conceptually, simplistic. Diseases are seen as abnormalities in the structure and function of cells, organs, and systems generated by biochemical mechanisms. However, patients experience illnesses, which reflect underlying disease processes but also reflect psychologic, cultural, and other factors. Physicians must treat the whole patient, taking into account social, psychologic, cultural, economic, and other factors, but always relying on a sound knowledge of biochemical, physiologic, and pathologic mechanisms.

SIX POINTS TO NOTE WHEN DISEASES ARE CONSIDERED FROM A BIOCHEMICAL STANDPOINT

(1) Many Diseases Are Determined Genetically: This topic is of such importance that it receives separate treatment below.

(2) All Classes of Biomolecules Found in Cells Are Affected in Structure, Function, or Amount in One or Another Disease: This point is illustrated with brief examples in Table 63–1. Biomolecules can be affected in a primary or a secondary manner; in genetic diseases, the primary defect resides in DNA, and the structures, functions, or amounts of the other biomolecules are affected secondarily.

(3) Biochemical Alterations That Cause Disease May Occur Rapidly or Slowly: Some diseases progress rapidly. For instance, death can occur within minutes or less after a massive coronary thrombosis, reflecting the fact that most tissues (brain and heart in this particular instance) are very sensitive

to lack of oxygen and fuel (eg, glucose for the brain). A vivid example of the reliance upon oxygen is the fact that cyanide (which inhibits cytochrome oxidase) kills within a few minutes. Massive loss of water and electrolytes in cholera can threaten life within hours. In general, rapid, large alterations of the amounts or distribution in the body of certain electrolytes (eg, K^+) become hazardous very quickly, at least in part because of the sensitivity of myocardial muscle to such changes. Severe alterations of pH can also only be tolerated for a short time. On the other hand, it may take years for the buildup of a biomolecule (eg, due to lack of a lysosomal enzyme responsible for its normal degradation) to affect organ function. An example is the relatively slow accumulation of sphingomyelin in liver and spleen that occurs in mild cases of Niemann-Pick disease. The above are examples of the time-honored clinical classification of diseases into acute and chronic categories.

(4) Diseases Can Be Caused by Deficiency or Excess of Certain Biomolecules: This statement is well illustrated by consideration of deficiency and excess of vitamin A (Chapter 53). Deficiency of this vitamin results in night blindness. On the other hand, excessive intake of vitamin A can result in acute or chronic states of toxicity. Similarly, deficiency of vitamin D results in rickets, but excess results in a potentially serious hypercalcemia. In thinking about nutritional deficiencies, it is useful to consider primary (poor diet) and secondary causes of deficiency. General causes of secondary deficiency include inadequate absorption, increased requirement, inadequate utilization, and increased excretion. Each of these four general causes can be brought about by a number of diseases or conditions.

(5) Almost Every Cell Organelle Has Been Involved in the Genesis of Various Diseases: This statement is illustrated in Table 63–2, which lists some of the organelles implicated in various diseases.

(6) Different Biochemical Mechanisms Can Produce Similar Pathologic, Clinical, and Laboratory Findings: The body has a limited number of ways of reacting to the causes of disease listed in Table 1–1. The most important of these are listed in

Table 63–1. Examples of the involvements of various biomolecules in diseases.

Biomolecule	Property Affected	Disease	Fundamental Cause
DNA	Structure	Sickle cell anemia	Mutation
RNA	Structure	Thalassemia (certain types)	Mutation leading to faulty splicing of mRNA
Protein	Structure/function	Sickle cell anemia	Mutation
Lipid (GM_2)	Amount ↑	Tay-Sachs	Mutation resulting in defective hexosaminidase A
Polysaccharide (glycogen)	Amount ↑	Glycogen storage disease	Mutation in gene for enzyme degrading glycogen (eg, phosphorylase)
GAG (dermatan and heparan sulfates)	Amount ↑	Hurler's syndrome	Mutation resulting in defective iduronidase
Electrolyte (Cl^-)	Amount in sweat ↑	Cystic fibrosis	Mutation in a membrane protein affecting transport of Cl^-
Water	Amount ↓	Cholera	Infection of small intestine by *Vibrio cholerae* leading to massive loss of water and electrolytes

Table 63–2. Involvements of the major intracellular organelles in various diseases.

Organelle	Disease(s)	Mechanism
Nucleus	Most genetic diseases	Mutations in DNA.
Mitochondrion	Several diseases including Leber's hereditary optic neuropathy and mitochondrial myopathies	Mutations in mitochondrial DNA affecting the structures of proteins (such as NADH dehydrogenase) that are encoded by the mitochondrial genome.
Endoplasmic reticulum	Chemical toxicities, eg, following intake of CCl_4	Enzymes in the endoplasmic reticulum such as cytochrome P450 activate various chemicals to potentially toxic species.
Golgi	I-cell disease	Decrease of GlcNAc phosphotransferase in the Golgi results in lysosomal enzymes being misdirected and secreted by affected cells.
Plasma membrane	Metastasis of cancer cells	Changes in the oligosaccharides of glycoproteins in this organelle are thought to be important in permitting metastasis.
Lysosome	Lysosomal storage diseases	Decreased activities (due to mutations) of the various hydrolases present in lysosomes result in accumulation of various biomolecules.
Peroxisome	Zellweger's (cerebrohepatorenal) syndrome and others	Decreased biogenesis of peroxisomes and decreased activity of certain peroxisomal enzymes, such as dihydroxyacetone phosphate acyltransferase.

Table 63–3. These processes, however, can be produced by a number of different stimuli. For example, many different bacteria and viruses can cause acute or chronic inflammation. Similarly, hepatomegaly can occasionally arise from accumulation of glucosylceramide, but much more commonly it is due to heart failure or metastases. Fibrosis of the liver (cirrhosis) can result from chronic intake of ethanol, excess of copper (Wilson's disease), excess of iron (primary hemochromatosis), deficiency of α_1-antitrypsin, and other causes. In addition, a variety of inborn errors of metabolism can lead to mental retardation, and many conditions can result in ketosis. Another example of different biochemical lesions producing a similar end point is when the local concentration of a compound exceeds its solubility point owing to excessive formation or decreased removal. This can result in its precipitation to form a calculus (stone). Calcium oxalate, magnesium ammonium phosphate, uric acid, and cystine may all form renal calculi, but they accumulate for different biochemical reasons. The generalization that emerges is that distinct biochemical causes can produce the same pathologic finding (eg, cirrhosis), clinical finding (eg, mental retardation), or laboratory finding (eg, ketosis). However, it is usually possible to distinguish among diseases that share some common findings by the history, physical examination, and appropriate laboratory tests.

THE MOLECULAR BASES OF MOST GENETIC DISEASES MAY BE REVEALED WITHIN THE NEXT DECADE

About 7000 diseases or disorders may have a genetic basis. Genetic diseases have been estimated to account for about 10% of hospitalized children, and many of the chronic diseases that afflict adults (eg, diabetes mellitus, atherosclerosis) have an important genetic component. The advent of recombinant DNA and methods for sequencing DNA have revolution-

Table 63–3. The major pathologic processes that occur in the human body.

Pathologic Process	One Cause	Example of Disease	Example of Biomolecules Involved
Inflammation, acute or chronic	Infections, bacterial or viral	Pneumonia	Mediators of inflammation (prostaglandins, leukotrienes)
Degenerations	Various chemicals	Fatty liver	Ethanol
Enlargement of an organ (eg, liver)	Accumulation of a compound	Gaucher's disease	Glucosylceramide
Atrophy (decrease in size) of an organ	Diminished blood supply	Atrophy of a kidney	Decrease of various nutrients supplied by the blood
Anemia	Lack of vitamin or mineral	Iron deficiency anemia	Iron
Neoplasia	Irradiation	Various leukemias	DNA damaged by irradiation
Cell death	Diminished blood supply	Myocardial infarction	Lack of oxygen
Fibrosis	Often follows cell death	Cirrhosis	Accumulation of collagen
Formation of calculus	High local concentration of a compound	Renal calculus in gout	Uric acid

ized genetics, particularly medical genetics. It has been predicted that through the use of these approaches, the molecular bases of the majority of genetic diseases will be elucidated within the next decade.

Chromosomal, Monogenic, and Multifactorial Disorders Make Up the Major Classes of Genetic Disease

Genetic diseases can be arranged into three major classes (Table 63–4): (1) chromosomal disorders, (2) monogenic (classic mendelian) disorders, and (3) multifactorial disorders which are the product of multiple genetic and environmental factors. The term polygenic denotes a disorder caused by multiple genetic factors independently of environmental influences. Two additional classes are somatic disorders, such as many types of cancer, in which mutations occur in somatic cells; and mitochondrial disorders due to mutations in the mt genome. These are considered in Chapters 62 and 64, respectively.

The **chromosomal disorders** will not be discussed here in any detail but include conditions in which there is an excess or loss of chromosomes, deletion of part of a chromosome, or a translocation. The best-known such condition is trisomy 21 (Down's syndrome). These disorders can be recognized by analysis of the karyotype (chromosomal pattern) of an individual. Chromosomal translocations are important in activating oncogenes (Chapter 62).

The **monogenic disorders** involve single mutant genes. They are classified as (1) autosomal dominant, (2) autosomal recessive, or (3) X-linked. The term "dominant" signifies that the mutation will be clinically evident even if only one chromosome is affected (heterozygous); the term "recessive" denotes that both chromosomes must be affected (homozygous). "X-linked" disorders are those in which the mutation is present on the X chromosome. Because females have two X chromosomes, they may be either heterozygous or homozygous for the affected gene. Thus, X-linked inheritance in females can be dominant or recessive. On the other hand, males have only one X chromosome, so that they will be affected if they inherit the mutant gene. Each of these three classes has its own characteristic pattern of inheritance. A textbook of medical genetics should be consulted for details.

Multifactorial disorders involve the action of a number of genes. The pattern of inheritance of these conditions does not conform to classic mendelian genetic principles. Less is known about this category of diseases, but it is assuming increasing importance because of the common adult diseases such as ischemic heart disease and hypertension that are members of this group. Efforts are now going forward to develop methods that will reveal the multiple genes which contribute to multifactorial disorders.

Mutations lie at the heart of most genetic disorders. They have already been discussed in Chapter 40 and will not be covered systematically here. However, a major thrust in molecular genetics is to determine their causes and also to develop sensitive, reliable methods for their detection. Knowledge of mutations has expanded greatly in recent years. For instance, CpG dinucleotides are known to be "hot spots" for mutations (see discussion of p53 in Chapter 62). A novel type of mutation (the trinucleotide repeat expansion) has been found to be the cause of a number of important chronic neurodegenerative disorders. Furthermore, mutations of the mitochondrial genome have been demonstrated in a number of other neurologic diseases. These two latter topics are discussed in the next chapter.

Genetic Diseases Produce Their Pathologic Consequences by Altering DNA, RNA, Proteins, and Cell Function

A mutation in a structural gene may affect the structure of the encoded protein, whether it be an enzyme or a noncatalytic protein. If an enzyme is affected, an **inborn error of metabolism** may result.

Table 63–4. Examples of each of the three major classes of genetic disease.

Class	Example	Comment
Chromosomal	Trisomy 21 (Down's syndrome)	Prevalence increases with maternal age.
	Chronic myelogenous leukemia	Presence of Philadelphia chromosome.
Monogenic	Familial hypercholesterolemia	Autosomal dominant; mutation in gene for LDL receptor.
	Huntington's chorea	Autosomal dominant; diagnostic probe now available.
	Cystic fibrosis	Autosomal recessive; majority of cases due to deletion of a Phe residue in a membrane protein regulating Cl^- transport (the CFTR protein.
	Sickle cell anemia	Autosomal recessive; mutation of Glu $\rightarrow$ Val at β^6 position of globin.
	Phenylketonuria	Autosomal recessive; mutation of gene for Phe hydroxylase.
	Duchenne muscular dystrophy	X-linked; affects synthesis of dystrophin.
	Hemophilia	X-linked; affects synthesis of factor VIII (AHG).
Multifactorial	Ischemic heart disease	Complex genetics; study of DNA polymorphisms for lipoproteins holds promise for resolving genetic susceptibility.
	Essential hypertension	A single-gene theory has its proponents.

The concept of an inborn error of metabolism was first proposed by the English clinician Sir Archibald Garrod in the early 1900s, based on his studies of alkaptonuria, albinism, cystinuria, and pentosuria. An inborn error of metabolism is a genetic disorder in which a specific enzyme is affected, producing a metabolic block, which may have pathologic consequences as explained below. The following depicts a metabolic block:

$$
\begin{array}{ccc}
 & \text{Increased X, Y} & \\
\text{E} & \uparrow & \text{*E} \\
\text{S} \rightarrow \text{P} & \text{Increased S} \longrightarrow\!\!\!\!\parallel\!\rightarrow \text{Decreased P} \\
\text{Normal} & \text{Block} &
\end{array}
$$

where *E = mutant enzyme and X, Y = alternative products of the metabolism of S. As shown, a block can have three results: (1) decreased formation of the product P, (2) accumulation of the substrate S behind the block, and (3) increased formation of metabolites (X, Y) of the substrate S, resulting from its accumulation. Any one of the these three results may have pathologic effects. In the case of phenylketonuria (Chapter 32), the mutant enzyme is phenylalanine hydroxylase, resulting in the following situation:

Increased phenylpyruvic acid

$$
\begin{array}{cc}
\uparrow & \text{*E} \\
\text{Increased phenylalanine} \longrightarrow\!\!\!\!\parallel\!\rightarrow \text{Decreased tyrosine}
\end{array}
$$

Thus, patients with PKU synthesize less tyrosine (they are often fair skinned, because tyrosine is used for the synthesis of melanin), have increased plasma levels of phenylalanine, and also exhibit increased amounts of phenylpyruvate and other metabolites of phenylalanine in their body fluids and urine. The precise cause of toxicity in phenylketonuria is not clear and may reflect decreased availability of tyrosine for protein and neurotransmitter synthesis in the brain and also inhibitory effects of the high level of phenylalanine on the transport of other amino acids into brain. The crucial point is that decreased formation of product or accumulation of substrate or other metabolites behind a block—as well as alterations of feedback regulation (eg, if an allosteric site on an enzyme is affected by mutation)—alter the flux through metabolic pathways, leading to pathologic effects.

It is also important to understand that some inborn errors of metabolism are essentially harmless. These are usually due to blocks in peripheral areas of metabolism, where neither diminished formation of product nor accumulation of its precursor perturbs the cell (eg, pentosuria). On the other hand, inborn errors of the tricarboxylic acid cycle are virtually unknown; because of its central importance in metabolism, any block in this cycle would have disastrous consequences for a cell and probably result in its death at a very early stage of development.

If a structural gene for a noncatalytic protein is affected by a mutation, in many cases a mutant protein is synthesized. Even a change of one amino acid, as in the case of hemoglobin S (Chapter 7), can have disastrous pathologic consequences. The mutant protein may not function properly (certain mutant hemoglobins), may aggregate (hemoglobin S), or may move very slowly through the cell (eg, α_1-antitrypsin).

Table 63–5 summarizes different levels at which the pathologic consequences of genetic diseases can be exerted. The levels shown are not mutually exclusive; for example, all genetic diseases depend upon changes in the structure of DNA.

Table 63–5. Levels at which the pathologic effects of genetic diseases are expressed. All genetic diseases are due to changes in DNA. However, it is useful to consider their pathologic consequences at the levels indicated. Some diseases are shown acting at two levels, and most of the defects in nonenzymatic proteins listed affect organ functions. Treatment of genetic diseases can also be aimed at the different levels shown.[1]

DNA
Altered nuclear DNA (various types of mutations). Altered mitochondrial DNA (various types of mutations).
RNA
Altered splicing (eg, certain cases of thalassemia).
Proteins
(a) Altered enzymes: decreased activity, absent, increased activity (rare).
(b) Altered nonenzymatic proteins:
  Transport: albumin (analbuminemia), Hb (HbS).
  Protective: γ-globulin (agammaglobulinemia), fibrinogen (afibrinogenemia).
  Structural: collagen (altered in various collagen diseases).
  Hormonal: thyroglobulin (deficient in certain cases of familial goiter).
  Contractile: dystrophin (decreased or absent in Duchenne muscular dystrophy).
  Receptor: LDL receptor (deficient or altered in familial hypercholesterolemia).
Cell and organ consequences
Deficiency of product: melanin (albinism).
Accumulation of toxic precursor: phenylalanine or various keto acids (PKU).
Disordered feedback regulation: of porphyrin synthesis (acute intermittent porphyria).
Altered membrane function: LDL receptor (familial hypercholesterolemia), altered renal transporter for cystine (cystinuria).
Altered compartmentalization: decreased GlcNAc phosphotransferase leads to misdirection (secretion) of lysosomal enzymes (I-cell disease).
Effect on cell or tissue architecture
  Cell shape: sickle cells (HbS)
  Altered organelle: decrease of peroxisomes (Zellweger's syndrome).
  Altered extracellular matrix: altered collagen (various collagen diseases).

[1]Adapted from Stanbury JB et al: *The Metabolic Basis of Inherited Disease,* 5th ed. McGraw-Hill, 1983.

Early Diagnosis of Certain Inborn Errors Is Imperative If Permanent Damage Is to Be Avoided

What should make one consider that a patient has an inborn error of metabolism? This is an important issue, since in some cases if treatment is not started immediately the infant will be permanently damaged (eg, phenylketonuria, galactosemia). A number of useful cues are listed in Table 63–6. The sources of material that can be analyzed and the main tests used in investigating patients or fetuses suspected of having genetic diseases are listed in Table 63–7. Chorionic villus sampling and amniocentesis apply only to investigation of the fetus.

RFLPs and Other Markers Have Revolutionized Genetic Linkage Studies and Enormously Facilitated the Isolation of Genes Causing Disease

Recombinant DNA technology (Chapter 42) has made possible the isolation and detailed analysis of many genes involved in disease. How has this been accomplished? Before tackling this question, it is important to have an understanding of the concept of genetic linkage.

In essence, genetic linkage reflects the proximity of genes in chromosomes. If two genes are located on different chromosomes, they will show independent assortment at meiosis and are not linked. However, if two genes are adjacent to each other, it is unlikely that they will be separated in meiosis, and they will be tightly linked. Genes separated from each other but located on the same chromosome will probably be inherited together unless recombination occurs during meiosis. The more distant they are from each other on the same chromosome, the greater is the chance of recombination occurring. The unit that measures genetic linkage is the morgan (M); this is the genetic length of a chromosome over which one recombination event occurs per meiosis. Two chromosomes are 1 centimorgan (cM) apart if they are separated 1% of the time during transmission from

Table 63–6. Cues to the diagnosis of an inborn error of metabolism.

- Family history of a genetic disease
- Positive screening test (eg, for PKU)
- Sick newborn with other conditions (infections, cardiovascular, etc) excluded
- History of poor physical and mental development and of failure to thrive
- Physical findings such as enlarged organs (eg, hepatomegaly in Gaucher's disease and Niemann-Pick disease)
- Unusual odor from breath or urine (eg, maple syrup urine disease)
- Low blood sugar (eg, glycogen storage disease)
- Low blood pH (eg, due to accumulation of organic acids in maple syrup urine disease)

Table 63–7. Major tests used in the diagnosis of genetic diseases.

Material
Plasma, red cells, white cells, fibroblasts, urine, organ biopsy, chorionic villus sample, amniotic cells

General Tests
Blood and urine glucose, blood pH, blood ammonia, amino acids in plasma and urine, detection of organic acids in urine, various color tests for miscellaneous metabolites

Specific Tests
Measurement of low amount of product or elevated amount of precursor (eg, phenylalanine in PKU)
Measurement of activity of candidate enzyme in red cells, white cells, or tissue biopsy
Electrophoresis (eg, for HbS)
Analyses by Southern blotting of restriction fragment length polymorphisms (RFLPs) and other features of DNA structure linked to or causing specific diseases (eg, HbS, Huntington's disease, Duchenne muscular dystrophy)
Identification of novel metabolite in urine or plasma by GLC-MS

parent to child. For the haploid genome (approximately 3×10^9 bp), 1 cM is equivalent to approximately 1 million bp.

Suppose that we are seeking a gene involved in the causation of a particular disease. If close by that gene a locus can be identified (eg, an RFLP or other marker; see below), then it can serve as a marker locus for studies on inheritance (by linkage analysis) and isolation of the disease gene. If the marker locus is adjacent to the gene in question, the two loci will show tight linkage when studied in members of families afflicted with the disease. Data obtained from studies of families can be analyzed to determine if the two loci are linked. The lod score is the logarithm to the base 10 of the odds in favor of linkage. A lod score of 3 or more (ie, odds of 1000:1 or more in favor) is generally accepted as proof of linkage, and a lod score of –2 (ie, odds of 100:1 against) indicate that the two loci are not linked. Important considerations in performing linkage analysis are that the family be informative for the loci involved (that one parent be heterozygous, thus providing DNA restriction digest fragments that will differ in migration and will thus be helpful diagnostically) and that the alignment of alleles (the phase) in the parents be known or calculable. Large families with histories extending over several generations also facilitate linkage studies enormously (see Huntington's disease, Chapter 64). Readers are advised to consult a textbook of genetics for wider reading on genetic linkage and genetic analysis, since this subject has been central to permitting the isolation of genes involved in disease.

Before the advent of recombinant DNA technology, relatively few markers were available to locate genes involved in disease. The finding and subsequent widespread use of restriction fragment length polymorphisms (RFLPs) changed this situation. However, RFLPs are being displaced by microsatellites (small repeat units of 2–6 bp that occur in tandem), which afford a much wider—and thus more informative—spectrum of polymorphisms than do RFLPs. Already, through the Human Genome Project (see below), hundreds of these marker loci have become available for each chromosome. This means that much shorter stretches of DNA have to be searched to find affected genes, greatly facilitating their isolation. Having markers that flank a gene on both sides is particularly useful, since it helps pinpoint the gene reliably (see Case No. 7 [cystic fibrosis], in Chapter 65).

Positional Candidate Cloning Is Becoming the Most Powerful Method of Isolating Disease Genes

Francis Collins, Director of the United States National Center for Human Genome Research, has recently summarized the approaches that are available for cloning genes causing particular diseases. They include the following.

(1) Functional approach: In this case, the gene is identified on the basis of the biochemical defect in the disease, without reference to its chromosomal location. For instance, when it was discovered that the phenotypic defect in HbS was replacement of glutamate by valine (Chapter 7), it became evident that the basic lesion would be a mutation in the gene encoding β-globin. This was subsequently confirmed by sequencing. It should be noted that implication of the β-globin gene in the causation of sickle cell anemia did not immediately facilitate its isolation, as methods for gene isolation were only to become available a decade or so later.

(2) Candidate gene approach: This term is often reserved for genes whose function, if lost by mutation, could explain the nature of the disease under consideration. For example, mutations in rhodopsin were considered as one of the causes of blindness due to retinitis pigmentosa; subsequent research showed that this was actually the case. Similarly, genes encoding dopamine receptors have been considered to be candidate genes for schizophrenia (Chapter 64), but this has not been confirmed. The implication of the p53 gene in the Li-Fraumeni syndrome (Chapter 62) is an example of the candidate gene approach.

(3) Positional cloning: In this case, no functional information about the gene product is available, and isolation is based solely on its chromosomal position. That information is acquired by linkage analysis of multiple affected families. At least 42 genes causing disease have been cloned in this manner. One classic case is isolation of the cystic fibrosis gene (Case No. 7 in Chapter 65), in which genetic linkage studies revealed two markers that flanked the gene, leaving an area of approximately 1.5 million bp that had to be surveyed for the presence of a gene encoding a protein, and which exhibited mutations only in individuals with cystic fibrosis. Another example is the isolation of the gene for Duchenne muscular dystrophy (DMD; Case No. 6, Chapter 65). In this case, as in a number of other isolations of disease genes, the presence of specific cytogenetic rearrangements was very important in helping to pinpoint an area on the X chromosome in which the gene was located.

(4) The positional candidate approach: This is rapidly becoming the method of choice. It depends on the rapidly increasing knowledge of the human genome afforded by the Human Genome Project. Basically, a chromosomal subregion is pinpointed by linkage studies and the subregion is surveyed (via a detailed reference map) to see what candidate genes are located there. This was the approach used in the isolation of the gene causing Marfan's syndrome (Chapter 57). The gene was mapped to chromosome 15q and almost simultaneously the protein fibrillin was mapped to this area. Very soon thereafter, mutations in the gene encoding fibrillin were detected in patients with Marfan's syndrome.

Partial cDNA sequences for many human genes have now been obtained. These are known as expressed sequence tags (ESTs). An international effort is under way to map approximately 50,000 of these on the human genome to at most 0.5 Mb (megabase) intervals, thus affording a relatively high-density transcript map. It has been estimated that more than half of the human transcripts will be placed on the human genome map by the end of 1996. The bottom line is that as more and more information on human gene transcripts obtained from the Human Genome Project is entered into appropriate databases, the positional candidate approach will tremendously speed up the detection and isolation of disease genes. This will fulfill one of the major goals of the project (see below). Methods for identifying the multiple genes involved in multifactorial diseases (see above) are also being developed.

Once a gene is assigned to a particular chromosome or chromosomal region, a variety of techniques are available with which to isolate it and sequence it (Table 63–8). (See also Chapter 42.)

The Human Genome Project Has Major Implications for the Study of Health and Disease

The Human Genome Project is an international cooperative effort that will provide a huge body of genetic information basic to the study of health and disease. The long-term goal is to sequence the entire human genome and the genomes of several other model organisms that have been basic to the study of genetics (eg, *Escherichia coli,* several yeasts, *Drosophila, Caenorhabditis elegans,* and the mouse). Details of the mapping and sequencing strategies involved will not be presented here; rather, it is our intention to give an overall impression of the present

Table 63–8. Principal methods used to identify and isolate genes involved in disease.

Procedure	Comments
Extensive linkage studies	Large families with defined pedigrees are desirable. Dominant genes are easier to recognize than recessive ones.
Use of probes to define marker loci	Probes identify specific RFLPs, microsatellites, etc; hundreds are now available, and it will soon be possible to saturate the entire genome with them. It is desirable to flank the gene on both sides, clearly delineating it.
Use of rodent or human somatic cell hybrids	Permits assignment of a gene to one specific chromosome.
Use of a specific chromosomal library	If a gene is assigned to one chromosome, use of a gene library constructed from it simplifies the search for it.
Fluorescence in situ hybridization (FISH)	Permits localization of a gene to one chromosomal band.
Use of pulsed-field gel electrophoresis to separate large fragments	Permits isolation of large DNA fragments obtained by use of restriction endonucleases (rare cutters) that result in very limited cutting of DNA chromosomes.
Chromosome walking	Involves repeated cloning of overlapping DNA segments; the procedure is laborious and can usually cover only 100–200 kb.
Chromosome jumping	By cutting DNA into relatively large fragments and circularizing it, one can move more quickly and cover greater lengths of DNA than with chromosomal walking.
Cloning via yeast artificial chromosomes (YACs), cosmids, phages, and plasmids	Permits isolation of fragments of varying lengths.
Detection of expression of mRNAs in tissues by Northern blotting using one or more fragments of the gene as a probe	The mRNA should be expressed in affected tissues.
Polmerase chain reaction (PCR)	Can be used to amplify fragments of the gene.
DNA sequencing	Establishes the highest-resolution physical map. Identifies open reading frame.

status of the Project, particularly as it relates to the human genome. The initial part of the project (1991–1995) had several short-term goals, some of which are listed in Table 63–9. Considerable progress has been made toward realizing many of these goals, only some of which will be described here.

To understand some of the goals of the Human Genome Project, it is important to appreciate the concept of a genetic map, a cytogenetic map, and a physical map (Figure 63–1). A genetic map (or linkage

Table 63–9. Some initial short-term goals of the Human Genome Project.[1]

- Establish a human genetic map with markers 2–5 cM apart; this will require 600–1500 markers.
- Establish a physical map of all 24 human chromosomes (22 autosomal, X and Y) with markers spaced at approximately 100,000 bp.
- Assemble overlapping sets (contigs) of cloned DNA or closely spaced unambiguously ordered markers.
- Perform a limited amount of DNA sequencing and develop new methods to reduce the cost of sequencing.
- Generate initial genetic maps of the genomes of the model organisms mentioned in this text.
- Develop informatics, research training, technology, and technology transfer and analyze ethical, legal, and social implications of the Project.

[1]A number of these goals have been met. Some scientists consider that enough mapping information and appropriate technology are available that the time is now ripe for very large-scale flat-out sequencing of the human genome, perhaps achievable by the year 2001.

map) represents the location of genetic markers within a stretch of DNA; RFLPs initially were used, but microsatellites have now replaced them. Such maps are more abstract than physical maps because they reflect inheritance patterns rather than physical distances. A cytogenetic map, particularly using fluorescence in situ hybridization (FISH; see Chapter 42), is also very useful in confirming the positions of genes on chromosomes. It provides a coarse map but does not have the resolution of the two other types of maps. A physical map consists of ordered landmarks at known distances from one another. The ultimate physical map will be the base sequence of each chromosome. Contig maps are constructed from contigs, which are sets of overlapping and ordered clones. Yeast accessory chromosomes (YACs) are used in large-scale mapping because they carry much larger fragments of DNA than other vectors, greatly facilitating progress. Sequence-tagged sites (STSs) are short sequences of DNA (100–1000 bp) that are amplified by the polymerase chain reaction (PCR); their special utility is that if the primers used for a set of STSs are known, then an individual investigator can access them from the laboratory, solely by performing appropriate PCRs.

Impressive progress has been made in constructing both a genetic map and a physical map of the human genome. For instance, in 1994 one large consortium described a comprehensive human linkage map generated from microsatellite markers. It consisted of 5840 loci of which 970 were uniquely ordered. The

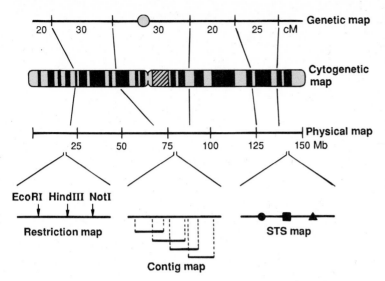

Figure 63–1. Comparison of genetic, cytogenetic, and physical maps of a chromosome. For the genetic map, the positions of several hypothetical genetic markers are shown along with the genetic distances in centimorgans between them. The circle shows the position of the centromere. For the cytogenetic map, the classical banding pattern of a hypothetical chromosome is shown. For the physical map, the approximate physical locations of the above genetic markers are shown, along with the relative physical distances in megabase pairs (Mb). Examples of a restriction map, a contig map, and an STS map are also shown., (Reproduced, with permission, from Green ED, Waterston RH: The Human Genome Project: Prospects and implications for clinical medicine. JAMA 1991;266:1966. Copyright © 1991 by the American Medical Association.)

map had markers at an average density of 0.7 cM, providing a high-density genetic map, one of the first goals of the Human Genome Project. Reference was made above to the use of expressed sequence tags in mapping. A number of experts now believe that enough information has been accumulated from the initial mapping studies to justify the commencement of large-scale high-throughput sequencing of the entire human genome. Part of this confidence is based on the rate of progress in sequencing the genome of *C elegans* (approximately one-fifth of its genome of 100 million bp sequenced by mid 1995) and other species, including certain bacteria (see below). One investigator has estimated that three laboratories specializing in large-scale sequencing could sequence 99% of the human genome in about five years with an accuracy of 99.9% and a cost of $20–25 million per laboratory per year.

The major medical implications of the Human Genome Project are that it will provide a tremendous body of background information which will be invaluable for elucidating the molecular bases and treatments of genetic diseases (Figure 63–2). In the opinion of Nobelist Watson, "it will provide a molecular explanation for the biochemistry of health and disease."

The Entire Genomes of *Haemophilus influenzae* and *Mycoplasma genitalium* Have Been Sequenced

The ramifications of the Human Genome Project are not of course confined to the human sphere. The technology that it has helped create is being used to

explore all biologic systems. In 1995, the entire genomic sequences of two bacteria, the first species to claim this honor, were reported by two teams, both led by Craig Venter.

H influenzae in its wild form can cause serious infections in humans. Its sequence (1.8 Mb) was ob-

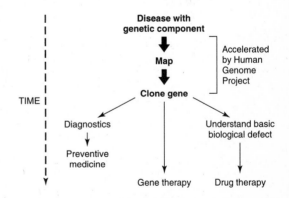

Figure 63–2. Progress in molecular medicine. The time needed to clone a disease has been rapidly shrinking, based upon maps and technologies emanating from the Human Genome Project. Improved diagnostic capabilities often result relatively quickly from gene discovery. In some instances, this allows powerful preventive medicine strategies to be initiated (colonoscopy for individuals at high risk for colon cancer, for example). The timetable for development of effective gene or drug therapies is much less predictable. (Reproduced, with permission, from Collins FS: Positional cloning moves from perditional to traditional. Nature Genetics 1995;9:347.)

Table 63–10. Summary of the gene content in *H influenzae* and *M genitalium* sorted by functional category. The number of genes in each functional category is listed for each organism. The number in parentheses indicates the percentage of the putatively identified genes devoted to each functional category. For the category of unassigned genes, the percentage of the genome indicated in parentheses represents the percentage of the total number of putative coding regions.[1]

Biologic Role	*H influenzae*[2]	*M genitalium*[3]
Amino acid biosynthesis	68 (6.8)	1 (0.3)
Biosynthesis of cofactors	54 (5.4)	5 (1.6)
Cell envelope	84 (8.3)	17 (5.3)
Cellular processes	53 (5.3)	21 (6.6)
Cell division	16	4
Cell killing	5	2
Chaperones	6	7
Detoxification	3	1
Protein secretion	15	6
Transformation	8	1
Central intermediary metabolism	30 (3)	6 (1.9)
Energy metabolism	112 (10.4)	31 (9.7)
Aerobic	4	3
Amino acids and amines	4	0
Anaerobic	24	0
ATP-proton force interconversion	9	8
Electron transport	9	0
Entner-Doudoroff	9	0
Fermentation	8	0
Gluconeogenesis	2	0
Glycolysis	10	10
Pentose phosphate pathway	3	2
Pyruvate dehydrogenase	4	4
Sugars	15	4
TCA cycle	11	0
Fatty acid and phospholipid metabolism	25 (2.5)	6 (1.9)
Purines, pyrimidines, nucleosides, and nucleotides	53 (5.3)	19 (6.0)
2′-Deoxyribonucleotide metabolism	8	3
Nucleotide and nucleoside interconversions	3	1
Purine ribonucleotide biosynthesis	18	3
Pyrimidine ribonucleotide biosynthesis	5	0
Salvage of nucleosides and nucleotides	13	10
Sugar-nucleotide biosynthesis and conversions	6	2
Regulatory functions	64 (6.3)	7 (2.2)
Replication	87 (8.6)	32 (10.0)
Degradation of DNA	8	1
DNA replication, restriction, modification, recombination, and repair	76	31
Transcription	27 (2.7)	12 (3.8)
Degradation of RNA	10	2
RNA synthesis and modification, DNA transcription	17	10
Translation	141 (14)	101 (31.8)
Transport and binding proteins	123 (12.2)	34 (10.7)
Amino acids and peptides	38	10
Anions	8	3
Carbohydrates	30	12
Cations	24	1
Other transporters	22	8
Other categories	93 (9.2)	27 (8.2)
Unassigned role	763 (43)	152 (32)
No database match	389	96
Match hypothetical proteins	347	56

[1]Modified and reproduced, with permission, from Fraser CM et al: The minimal gene complement of *Mycoplasma genitalium.* Science 1995;270:397.

[2]The genome of *H influenzae* contains 1,830,137 bp. It has 1743 predicted coding regions (genes), of which 1007 are assigned and 736 are unassigned.

[3]The genome of *M genitalium* has 580,070 bp. It has 470 predicted coding regions (genes), of which 318 are assigned and 152 are unassigned. It should be noted that many bacterial genes are polycistronic and generally lack introns, making the interpretation of sequence information much easier than in the case of eukaryotes. Sonication was used to produce the fragments that were sequenced; it breaks DNA randomly (unlike restriction enzymes), producing overlapping fragments, from which the total sequences were reconstructed. The approach used avoids the necessity for preliminary extensive mapping.

tained by sonicating (shotgunning) its DNA, sequencing the fragments, and reassembling the sequence based on overlaps. In contrast to the human genome, mapping was not required. A similar approach was employed on *Mycoplasma genitalium* (0.58 Mb, believed to contain the smallest genome of any self-replicating organism), an organism found in ciliated epithelial cells of primate genital and respiratory tracts. Amazingly, its sequence was obtained in a total of 8 weeks. Comparisons of the size of the two genomes and their numbers of genes are set forth in Table 63–10, along with identification of the biologic roles of many of the genes that were sequenced. Examination of the genes identified gives one a taste of the kind of information that will emerge when the human genome is sequenced. The approach used should be applicable to most bacteria. The information obtained will be particularly valuable in studies in the areas of comparative evolution and pathogenicity.

SUCCESSFUL TREATMENT IS AVAILABLE FOR SOME GENETIC DISEASES

In general, treatment of genetic disease employs one of the following strategies: (1) attempts to correct the metabolic consequences of a disease by administration of the missing product or by limiting the availability of substrate; (2) attempts to replace the absent enzyme or protein or to increase its activity; (3) attempts to remove excess of a stored compound; (4) attempts to correct the basic genetic abnormality by gene therapy (Table 63–11).

Some of these approaches are quite effective for certain disorders, eg, the dietary treatments of phenylketonuria and galactosemia, replacement therapy for hemophilia and agammaglobulinemia, and removal of iron by periodic bleeding in hemochromatosis. However, attempts at enzyme therapy have so far met with mostly limited success. Problems include good sources of human enzymes (placentas have been useful in some instances), targeting enzymes in sufficient amounts to the appropriate organ, and maintaining their activities in tissues, in which they may be rapidly degraded. Enzyme therapy is particularly difficult in the case of the brain, because administered enzymes must be made to cross the blood-brain barrier. Many attempts have been made to use liposomes to deliver enzymes (or DNA and other molecules) to target organs, but so far success has not been spectacular. Because of the tremendous progress in the field of recombinant DNA technology (Chapter 42), a great deal of thought and effort are being devoted to strategy (4), gene therapy.

Gene Therapy Has Been Shown in Clinical Trials to Be Feasible, But Its Efficacy Must Be Greatly Improved

It is important first to consider the major criteria that should be satisfied to permit the use of gene therapy in humans (Table 63–12). Only somatic gene therapy is permissible in humans at present, because germ cell gene therapy carries the risk of transmitting genetic alterations to offspring.

In theory, gene therapy can involve gene replacement, correction, or augmentation. In **replacement,** the mutant gene would be removed and replaced with a normal gene. In **correction,** only the mutated area of the affected gene would be corrected and the remainder left unchanged. Neither approach has yet

Table 63–11. Four major classes of treatment strategies for genetic diseases: (1) attempts to manage the metabolic consequences of the disease by administering the missing product or by limiting the availability of substrate; (2) attempts to replace the missing enzyme or protein or to increase its activity; (3) attempts to remove excess of a stored compound; (4) attempts to correct the basic genetic abnormality.[1]

Class of Treatment	Principle	Disease	Treatment or Comment
(1)	Replace missing product Limit substrate	Familial goiter Phenylketonuria	Administration of levothyroxine Diet low in phenylalanine
(2)	Replace mutant enzyme Replace missing protein	Gaucher's disease Hemophilia	Injections of β-glucosidase (alglucerase) Injections of factor VIII (AHG)
(3)	Increase activity of mutant enzyme by supplying large amounts of cofactor Increase activity of mutant enzyme by induction	Methylmalonic aciduria Crigler-Najjar syndrome	Injections of vitamin B_{12} Administration of phenobarbital
(4)	Replace diseased organ carrying defective gene by normal organ Introduce enzyme into somatic cells by gene therapy	Galactosemia[2] Many possible candidates	Liver transplantation Clinical trials already conducted on individuals with adenosine deaminase deficiency, cystic fibrosis, familial hypercholesterolemia, etc.

[1]Adapted from Stanbury JB et al: *The Metabolic Basic of Inherited Disease,* 5th ed. McGraw-Hill, 1983.
[2]A diet low in lactose is first-line treatment for infants with galactosemia who have been diagnosed early.

Table 63–12. Some criteria to be satisfied prior to initiation of gene therapy.[1]

- The prognosis of the disease should be accurately predictable and should be adverse.
- The gene should be isolated and its regulatory regions defined.
- The target cells should be identified, and safe methods of introducing the gene into them should be available.
- Evidence should be available (eg, from cultured cells and animal studies) that the gene functions adequately and produces no deleterious effects.

[1]Based on Weatherall DJ: *The New Genetics and Clinical Practice*, 3rd ed. Oxford Univ Press, 1991.

been developed. **Augmentation** involves introduction of foreign genetic material into a cell to compensate for the defective product of the mutant gene and is the sole type of gene therapy available at present. The foreign material can be introduced into affected cells by any of the methods listed in Table 63–13; for gene therapy in humans, the gene of interest is usually administered via a viral vector or via plasmid-liposome complexes. It should be noted that the cells which take up the gene will contain both the mutant and the exogenously derived gene. Moreover, if introduced via a retrovirus, the foreign gene integrates at random sites on chromosomes, may interrupt (by insertional mutagenesis) the expression of certain host cell genes, and is generally not subject to physiologic regulatory mechanisms. Various methods are being developed to target genes to specific sites in order to increase uptake and thus efficacy. Target cells have included bone marrow cells, fibroblasts, epithelial cells of the respiratory tract, hepatocytes, and various tumor cells.

The three major routes by which genes have been introduced into humans are via retroviruses, adenoviruses and plasmid-liposome complexes. The viruses are specially constructed to be replication-deficient by excision of appropriate genes. The gene of interest, part of an expression cassette containing it

Table 63–13. Some methods of introducing genes into cells for gene therapy.[1]

- Intranuclear injection
- Transfection (eg, with calcium phosphate)
- Electroporation (employs a pulsed electrical field)
- Retroviruses (specially constructed)
- Adenovirus (used for introduction of the cystic fibrosis gene into cells of the respiratory tract)
- Plasmid-liposome complexes
- Site-directed recombination (based on homologous recombination)

[1]The first three techniques listed are generally used on isolated or cultured cells. Most clinical trials to date have used modified retroviral or adenovirus vectors. Site-directed recombination, based on homologous recombination, is in experimental use; it is similar in principle to the technique used to produce gene knockouts (Chapter 42), except that functional genes are introduced.

and regulatory sequences, is ligated onto the viral genome by appropriate enzymes (Figure 63–3). Currently used retroviruses and adenoviruses can take inserts up to approximately 9 kb and 7.5 kb, respectively. In the case of retroviruses, the target cells must be actively growing, since cell division is required if the gene is to be integrated into the genome. For adenoviruses, this is not an important consideration, because their genome does not integrate with the host cell genome; however, this is also a disadvantage, as expression of the introduced gene gradually declines and additional treatments with the virus carrying it are required.

Administration of genetically engineered viruses containing the appropriate gene of interest can be performed ex vivo or in vivo. In the former case, cells are taken out of the body (eg, bone marrow cells or hepatocytes), treated with the vector in vitro, and then reintroduced into the body by an appropriate route. This is the method usually employed with retroviruses. For treatment in vivo, the vector can, for example, be introduced into the upper respiratory tract in an aerosolized form, as when a modified adenovirus containing the CFTR gene was administered to patients with cystic fibrosis.

A considerable number of clinical trials of gene therapy have now been completed or are still in progress. Diseases in which gene therapy have been used include adenosine deaminase deficiency (Case No. 8, Chapter 65); cystic fibrosis (Case No. 7, Chapter 65), familial hypercholesterolemia (Chapter 27), and various types of cancer. In the latter instance, many of the trials have introduced a marker gene into the cells to follow some aspect of the biologic behavior of the tumor cells rather than a therapeutic aim. Myoblast transfer has also been used in an attempt to treat Duchenne muscular dystrophy (Case No. 6, Chapter 65).

Crystal has summarized some conclusions regarding this novel therapy: (1) Gene therapy is feasible. Evidence has been obtained for expression of introduced genes, and at least transient improvements in clinical condition have occurred in a number of cases. (2) So far it has proved safe, although long-term follow-up studies will be important. Reactions to it have occurred, usually of an inflammatory or immune nature, but they appear to have been directed toward the vector or some other aspect of the method of administration rather than toward the introduced genetic material. (3) No genetic disease has been cured by this method. (4) A major problem has been efficacy, as the levels of expression of the desired gene product have often been low or transient. Inconsistency in results has also been a problem. Development of new vectors and other technologic advances should improve the situation. However, the overall approach appears rational and perhaps will eventually improve the quality of life of many individuals with disabling genetic disorders. Another dimension

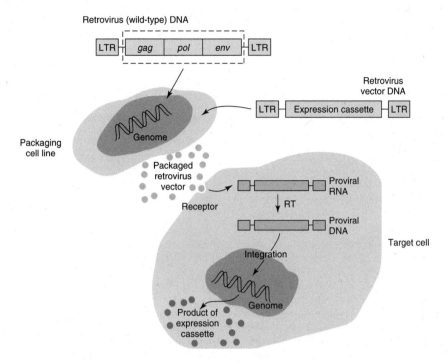

Figure 63–3. Retrovirus vector design, production, and gene transfer. Retroviruses are RNA viruses that replicate through a DNA intermediate. The retrovirus vectors administered to humans all use the Moloney murine leukemia virus as the base. The *gag, pol,* and *env* sequences are deleted from the virus, rendering it replication-deficient. The expression cassette is inserted, and the infectious replication deficient retrovirus is produced in a packaging cell line that contains the *gag, pol,* and *env* sequences that provide the proteins necessary to package the virus. The vector with its expression cassette enters the target cell via a specific receptor. In the cytoplasm, the reverse transcriptase (RT) carried by the vector converts the vector RNA into the proviral DNA that is randomly integrated into the target cell genome, where the expression cassette makes its product. (Reproduced, with permission, from Crystal RG: Transfer of genes to humans: Early lessons and obstacles to success. Science 1995;270:404.)

is added to this area by recent reports of the first attempts at in vivo fetal gene therapy in animals.

Antisense or Triplex Therapy May Be Useful in the Treatment of Certain Diseases

There has been considerable interest in synthesizing oligonucleotides (about 20 bases long) that will bind through base-pairing to a specific mRNA (an antisense compound) or to double-stranded DNA to form a triplex molecule. (There is evidence that antisense molecules exist in nature and are used to regulate the activities of some genes.) Obviously, sequence information on the target molecules must be available in order to synthesize specific oligonucleotides. In the first case mentioned above, translation of the specific mRNA molecules would be blocked, and in the second case transcription of a specific gene would be blocked. There is also some evidence that complexing of an antisense oligonucleotide with an mRNA molecule can make the latter more susceptible to digestion by nucleases. For convenience, here we shall refer to both types of compound referred to above as antisense compounds. Such oligonucleotides could act as drugs (genetic medicines), inhibiting the synthesis of the protein products of specific genes involved in various diseases or inhibiting the genes themselves. A number of clinical trials of these compounds are in progress. For instance, one such oligonucleotide appears to block the replication of HIV-1 virus by combining with its *gag* gene (similar to the *gag* gene of Rous sarcoma virus) (Chapter 62) and could thus prove useful in the treatment of AIDS. Other compounds are being investigated as treatments for autoimmune disorders or some types of cancer.

Some problems have been encountered in the development of clinically useful antisense compounds. Many oligonucleotides do not enter cells easily, and they may also be degraded inside cells by nucleases. To counter the latter problem, chemists have synthesized modified oligonucleotides in which a critical oxygen atom in each nucleotide is replaced by a sulfur atom (phosphorothioates). They also appear to have an undesirable capacity to bind to various proteins, thus producing unwanted side effects. A puzzling feature is that, at least in certain cases, control oligonucleotides (ie, not specific for the target mRNA or DNA molecules) have been found to produce effects similar to those observed with antisense compounds. This raises a question about whether these compounds are really acting as antisense molecules or are eliciting other responses. Despite these problems, it is hoped that antisense molecules will eventually prove of therapeutic use in a variety of diseases.

Molecular Medicine Is Advancing Rapidly

Table 63–14 lists some of the major advances made in the area of molecular medicine in the past decade or so, relating particularly to our understanding of disease; all have been dependent upon the development of recombinant DNA technology.

SUMMARY

Rational treatments for most diseases depend upon elucidating their biochemical or genetic causes (or both). However, despite precise knowledge of the cause, the most appropriate therapy for a disease may not be feasible at present, eg, the use of gene therapy for sickle cell anemia. From a biochemical viewpoint, the following points about diseases should be considered: (1) many have a genetic basis; (2) all classes of biomolecules found in cells may be affected in one or another disease; (3) biochemical disorders causing disease may occur rapidly or slowly; (4) excesses or deficiencies of molecules can cause disease; (5) almost every cellular organelle has been involved in the genesis of various diseases; and (6) different biochemical mechanisms can produce similar clinical findings.

About 7000 genetic disorders exist; the major categories are chromosomal, monogenic, and multifactorial disorders. Diseases due to somatic mutations (eg, many types of cancer) and to mutations in the mitochondrial genome also occur. These diseases usually affect cell function by producing mutant proteins. Mutant enzymes result in inborn errors of metabolism, which may be fatal or harmless. The consequences of inborn errors generally result from decreased formation of product or accumulation of substrate or metabolites derived from it. Early diagnosis may be imperative to prevent permanent damage; suitable laboratory tests are available to facilitate diagnosis.

In searching for genes that cause disease, the first step is to establish genetic linkage to a particular chromosome or chromosomal region. The availability of RFLPs and other new marker loci such as microsatellites has revolutionized this field. Various routes to isolating genes responsible for disease are discussed; they include functional, candidate gene, positional cloning, and positional candidate approaches. The Human Genome Project has already provided many marker loci and technologic advances that will accelerate discovery of the causes of most other genetic diseases. It is anticipated that full-scale sequencing of the human genome will begin very soon.

Various treatments for genetic diseases are available, including somatic gene therapy. The present status of this novel therapy is described.

Table 63–14. Some recent major achievements in, or relating to, molecular medicine.

- Major advances in mapping human genome, with large-scale sequencing now imminent
- Increased knowledge of homeobox genes and other genes relating to embryonic development and differentiation
- Development of gene knockouts, yielding animal models of disease and information on gene function
- Development of gene and antisense therapy
- Production of biotechnology-derived products (eg, erythropoietin, t-PA) based on recombinant DNA technology
- Development of various DNA-based methods for diagnosis and screening of disease
- Elucidation of the molecular bases of some major mendelian disorders (eg, Duchenne muscular dystrophy, cystic fibrosis, familial hypercholesterolemia)
- Isolation of oncogenes, tumor suppressor genes, and cancer susceptibility genes (eg, *BRCA1*) and proposal of a specific model for the development of colorectal cancer
- Insights into the molecular bases of a number of chronic neurologic diseases (eg, trinucleotide expansions)
- Discovery that mutations in mitochondrial genes can cause diseases

REFERENCES

Blau HM, Springer ML: Gene therapy: A novel form of drug delivery. N Engl J Med 1995;333:1204.

Brock DJH: *Molecular Genetics for the Clinician.* Cambridge Univ Press, 1993.

Cohen JS, Hogan ME: The new genetic medicines. Sci Am (Dec) 1994;271:76.

Collins FS: Positional cloning moves from perditional to traditional. Nature Genetics 1995;9:347.

Cotran RS, Kumar V, Robbins SL: *Robbins Pathologic Basis of Disease,* 5th ed. Saunders, 1994.

Coutelle C et al: The challenge of fetal gene therapy. Nature Medicine 1995;1:864.

Crystal RG: Transfer of genes to humans: Early lessons and obstacles to success. Science 1995;270:404.

Evens RP et al: Biotechnology and clinical medicine. (Two parts.) Hosp Phys (Jan and Feb) 1995;27, 26.

Fleischmann RD et al: Whole-genome random sequencing and assembly of *Haemophilus influenzae* Rd. Science 1995;269:496.

Fraser CM et al: The minimal gene complement of *Mycoplasma genitalium*. Science 1995;270:397.

Green ED, Cox DR, Myers RM: The Human Genome Project and its impact on the study of human disease. In: *The Metabolic and Molecular Bases of Inherited Disease,* 7th ed. Scriver CR et al (editors). McGraw-Hill, 1995.

Gura T: Antisense has growing pains. Science 1995;270: 575.

Hodgkin J, Plasterk RHA, Waterston RH: The nematode *Caenorhabditis elegans* and its genome. Science 1995; 270:410.

Jorde LB, Carey JC, White RL: *Medical Genetics.* Mosby, 1995.

Knoppers BM, Chadwick R: The Human Genome Project: Under an international ethical microscope. Science 1994;265:2035.

Lander ES, Schork NJ: Genetic dissection of complex traits. Science 1995;265:2037.

Leder P, Clayton DA, Rubenstein E (editors): *Scientific American Introduction to Molecular Medicine.* Scientific American, 1994.

Mulligan RC: The basic science of gene therapy. Science 1993;260:926.

Murray JC et al: A comprehensive human linkage map with centimorgan density. Cooperative Human Linkage Center (CHLC). Science 1994;265:2049.

Olson M: A time to sequence. Science 1995;270:394.

Scriver CR et al (editors): *The Metabolic and Molecular Bases of Inherited Disease,* 7th ed. McGraw-Hill, 1995.

Sedivy JM, Joyner A: *Gene Targeting.* Freeman, 1992.

Smithies O: Animal models of human genetic diseases. Trends Genet 1993;9:112.

Watson JD et al: *Recombinant DNA,* 2nd ed. Scientific American, 1992.

Weatherall DJ: *The New Genetics and Clinical Practice,* 3rd ed. Oxford Univ Press, 1991.

Wilkie AOM: The molecular basis of genetic dominance. J Med Genet 1994;31:89.

64

The Biochemical Basis of Some Neuropsychiatric Disorders

Robert K. Murray, MD, PhD

INTRODUCTION

This chapter does not provide systematic coverage of such topics as the mechanisms of nerve transmission, properties of neurotransmitters, properties of synapses, or details of ion channels in the brain. These subjects are well covered in many textbooks of neurophysiology, neurobiology, molecular biology, and neuropharmacology, some of which are listed in the references at the end of this chapter. We shall assume that the reader has a degree of familiarity with the basic concepts of neurophysiology and neuroanatomy. Instead, we shall discuss eight neuropsychiatric disorders (or groups of disorders)—myasthenia gravis, Huntington's disease, strokes, diseases due to mutations in mitochondrial DNA, the fragile X syndrome and other disorders due to triplet repeats, Parkinson's disease, Alzheimer's disease, and schizophrenia—which contemporary biochemical, cell biologic, and genetic approaches are illuminating and in some cases offering prospects for new therapies. Certain of the conditions are frequent, others are less frequent, but all offer valuable insights and perspectives.

BIOMEDICAL IMPORTANCE

The two major challenges of contemporary biology are determining the mechanisms involved in development and differentiation and unraveling the workings of the nervous system. The techniques made available by recombinant DNA technology are playing a major role in both efforts. Clinically, neuropsychiatric disorders are important, among other reasons because they are often chronic (eg, schizophrenia) and devastating (eg, strokes, Huntington's disease, Alzheimer's disease, and schizophrenia), since they can destroy intellectual functions. Research in the neurosciences will not only provide us with a deeper understanding of our own natures but will also provide rational bases for effective therapy of many extremely disabling conditions.

According to the National Foundation for Brain Research (USA), the direct costs of brain disorders

(psychiatric disease, neurologic disease, alcohol abuse, and substance abuse) exceed $400 billion annually. The costs of just one disorder—dementia (of which Alzheimer's disease is the most important example)—exceed the costs of cancer and coronary heart disease. The direct costs represented approximately one-seventh of total health care costs in the USA in 1991. The indirect costs (eg, lost wages, criminal justice system) were calculated to exceed the direct costs. For each $100 spent on the care of patients with Alzheimer's disease, the USA invested only 50 cents for research into the problem.

TO UNDERSTAND MYASTHENIA GRAVIS, IT IS NECESSARY TO UNDERSTAND EVENTS OCCURRING AT THE NEUROMUSCULAR JUNCTION

Myasthenia gravis is characterized by recurrent episodes of muscle weakness after exercise; muscles supplied by cranial nerves are most affected. It is improved by administration of drugs that inhibit acetylcholinesterase (see below), a fact that is used in diagnosis and treatment. In myasthenia gravis, for reasons that are not clear, the body forms autoantibodies against the acetylcholine receptor (AChR) of the neuromuscular junction; these antibodies cause destruction of the receptors and lead to marked reduction in their number. Thus, to understand this disorder, it is necessary to know the basic facts concerning neuromuscular transmission.

A diagram of the neuromuscular junction and of some of the events occurring there is shown in Figure 64–1. The junction consists of a single nerve terminal separated from the postsynaptic region by the synaptic cleft. The motor endplate is the specialized portion of the muscle membrane involved in the junction. **Junctional folds** are prominent; they contain a high density of AChRs in close proximity to the nerve terminal.

The overall process at the junction can be considered to take place in six steps (refer to the numbers in Figure 64–1).

794

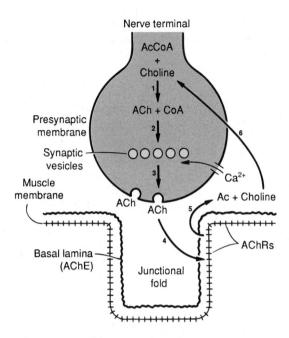

Figure 64–1. Schematic representation of some of the events at the neuromuscular junction. Part of a nerve terminal is shown lying in close apposition to a muscle endplate. The overall process whereby acetylcholine stimulates its receptor may be considered to occur in six steps (see text). (AcCoA, acetyl-CoA; ACh, acetylcholine; AChE, acetylcholinesterase; AChRs, acetylcholine receptors; Ac, acetate.)

(1) Synthesis of acetylcholine occurs in the cytosol of the nerve terminal, employing the enzyme **choline acetyltransferase,** which catalyzes the following reaction:

Acetyl-CoA + Choline→ Acetylcholine + CoA

(2) Acetylcholine is then incorporated into small membrane-bound particles called **synaptic vesicles** and stored therein; the assembly of synaptic vesicles resembles the events described for assembly of transport vesicle (Chapter 43), involving SNAREs, etc.

(3) Release of acetylcholine from these vesicles into the synaptic cleft is the next step. This occurs by **exocytosis,** which involves fusion of the vesicles with the presynaptic membranes. In the resting state, single quanta (about 10,000 molecules of the transmitter, probably corresponding to the contents of one synaptic vesicle) are released spontaneously, resulting in small miniature endplate potentials. When a nerve ending is depolarized by transmission of a nerve impulse, this process opens voltage-sensitive Ca^{2+} channels, permitting an influx of Ca^{2+} from the synaptic space into the nerve terminal. This Ca^{2+} plays an essential role in the exocytosis that releases

acetylcholine (contents of approximately 200 vesicles) into the synaptic space.

(4) The released acetylcholine diffuses rapidly across the synaptic cleft to its receptors in the junctional folds. When two molecules of acetylcholine bind to a receptor, it undergoes a conformational change, opening a channel in the receptor that permits a flux of cations across the membrane. The consequent entry of Na^+ results in depolarization of the muscle membrane, forming the **endplate potential.** This in turn depolarizes the adjacent muscle membrane, and action potentials are generated and transmitted along the fiber, resulting in contraction (Chapter 58).

(5) When the channel closes, the acetylcholine dissociates and is hydrolyzed by **acetylcholinesterase,** which catalyzes the following reaction:

Acetylcholine + H_2O → Acetate + Choline

This important enzyme is present in high amounts in the basal lamina of the synaptic space.

(6) Choline is recycled into the nerve terminal by an active transport mechanism, where it can be used again for synthesis of acetylcholine.

The Acetylcholine Receptor of the Neuromuscular Junction Is a Transmitter-Gated Ion Channel

The acetylcholine receptor of the neuromuscular junction has been studied intensively because of its key role in neuromuscular transmission. Some important properties of this receptor are listed in Table 64–1. As shown, it is a membrane glycoprotein consisting of five subunits, two of which (α) are identical. The cDNAs for the different subunits have been cloned and amino acid sequences derived for each. The fact that α-bungarotoxin binds tightly to the pro-

Table 64–1. Some properties of the acetylcholine receptor of the neuromuscular junction.[1]

- Is a nicotinic receptor (ie, nicotine is an agonist for the receptor).
- Is a membrane glycoprotein of molecular mass about 275 kDa.
- Contains five subunits, consisting of $\alpha_2\beta\gamma\delta$.[2]
- Only the α-subunit binds acetylcholine with high affinity.
- Two molecules of acetylcholine must bind to open the ion channel, which permits flow of both Na^+ and K^+; the receptor is thus a transmitter-gated ion channel.
- The snake venom α-bungarotoxin binds tightly to the α–subunit and can be used to label the receptor or as an affinity ligand to purify it.
- Autoantibodies to the receptor are implicated in the causation of myasthenia gravis.

[1]Adapted from Kandel ER, Schwartz JH, Jessell TM (editors): *Principles of Neural Science,* 3rd ed. Appleton & Lange, 1991.
[2]The subunit composition shown is that of the fetal enzyme. In the adult, the single γ chain is replaced by an ϵ chain. There is a different gene for each type of subunit.

tein has proved useful in many studies, such as measurement of the number of receptors in samples of normal and abnormal muscle (see below).

The disposition of the protein in its membrane is illustrated in Figure 64–2. As shown, all five subunits span the membrane and contribute to the ion channel. The channel is closed in the absence of acetylcholine. When two molecules of the transmitter bind to the receptor, one to each α subunit, the protein undergoes a conformational change that results in opening of the ion channel for approximately 1 ms. During this time, Na^+ flows in and K^+ flows out. As mentioned above, it is the entry of Na^+ that results in depolarization of the muscle membrane, generating the endplate potential. Because the presence or absence of acetylcholine itself results in opening

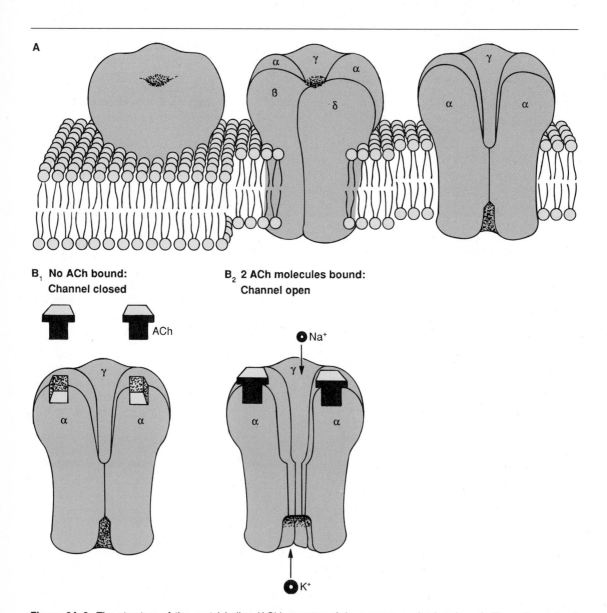

Figure 64–2. The structure of the acetylcholine (ACh) receptor of the neuromuscular junction. **A:** Three-dimensional model of the nicotinic acetylcholine-activated ion channel based on the model of Karlin and coworkers. The receptor-channel complex consists of several subunits. One molecule of acetylcholine binds to each α-subunit prior to channel opening. All subunits contribute to the pore. **B:** When two molecules of acetylcholine bind to the portions of the α-subunits exposed to the membrane surface, the receptor-channel changes conformation, opening a pore in the portions of the receptor embedded in the lipid bilayer. Both K^+ and Na^+ flow through the open channel down their individual electrochemical gradients. (Reproduced, with permission, from Kandel ER, Schwartz JH, Jessell TM (editors): *Principles of Neural Science,* 3rd ed. Appleton & Lange, 1991.)

and closing of the channel, the acetylcholine receptor is referred to as a "transmitter-gated" ion channel (in contrast to the two other types of gated channels— voltage-gated and mechanically gated channels).

In Myasthenia Gravis, Autoantibodies Damage Acetylcholine Receptors and Reduce Their Numbers

Electrophysiologic studies (eg, electromyography) indicate that the abnormality in myasthenia gravis occurs at the motor endplate and not at the presynaptic membrane. In fact, normal amounts of acetylcholine appear to be liberated in this condition. In contrast, the use of radiolabeled bungarotoxin to measure the number of acetylcholine receptors in samples of muscle shows that they are markedly reduced in patients with severe myasthenia gravis. A variety of findings suggest that a disturbance of the immune system is implicated in the pathogenesis of myasthenia gravis. For instance, thymic and lymphoid hyperplasia are frequently observed, and thymomas are not infrequent. About 10% of patients have some other autoimmune disease. Interestingly, in certain cases the fetus may be affected in utero when the mother has myasthenia gravis, suggesting transplacental transfer of some factor, such as antibodies. An illuminating finding was that injection of purified acetylcholine receptor (obtained from the electric organ [electroplax] of electric eels) into rabbits produced circulating antibodies to the receptor and also a disease resembling myasthenia gravis. Subsequently it was shown that autoantibodies to the receptor can be demonstrated in humans with myasthenia gravis, and are present in approximately 100% of patients with a severe form. The autoantibodies attach to the acetylcholine receptors in the neuromuscular junction and damage them, resulting in so-called focal lysis (Figure 64–3). The damaged receptors are subject to endocytosis, which accelerates their turnover and reduces their number. Proteins of the complement system (Chapter 59) play an important role in this type of cell injury. Why the au-

toantibodies are produced in the first place has not been adequately explained.

Inhibitors of Cholinesterase Increase the Amount of Acetylcholine at the Neuromuscular Junction, Affording Treatment for Myasthenia Gravis

Drugs that inhibit acetylcholinesterase, the enzyme involved in the degradation of acetylcholine, effectively increase concentration of the neurotransmitter at the motor endplate and prolong its action. Such agents are the mainstay of treatment of myasthenia gravis. There are two types of inhibitors of cholinesterase: reversible and irreversible. Both short-acting and relatively long-acting reversible inhibitors are useful in myasthenia gravis. The former have found most use in diagnosis. The diagnosis is suspected from a combination of the history, clinical findings, and certain laboratory tests (eg, electromyography, measurement of circulating autoantibodies). A diagnostic test for myasthenia gravis consists of administering an appropriate dose of the short-acting agent **edrophonium (Tensilon);** if the patient has myasthenia gravis, this maneuver should result in a rapid but short-lived improvement of muscle strength. **Pyridostigmine** and **neostigmine** are effective longer-acting agents that are widely used in the treatment of myasthenia gravis. Diisopropylfluorophosphate (DFP) is an example of an irreversible inhibitor of cholinesterase; it binds covalently to a serine residue in the active center of acetylcholinesterase (see discussion of serine proteases in Chapter 10). DFP is a toxic "nerve gas" and has also been used as an insecticide.

Suppression of the abnormal immune response involved in myasthenia gravis may also be needed. This commonly involves use of corticosteroids, but severe cases may require administration of cytotoxic immunosuppressants such as azathioprine. Thymectomy may be of benefit. Plasmapheresis can temporarily reduce the level of autoantibodies in the plasma.

Formation of autoantibodies to the ACh receptors present in the myoneural junction

↓

Damage to the receptors (focal lysis) by the auto-antibodies, resulting in cross-linking and endocytosis

↓

Marked reduction in number of receptors

↓

Clinical signs, particularly episodic weakness of muscles supplied by cranial nerves

Figure 64–3. Scheme of causation of myasthenia gravis. The reason for the initial formation of autoantibodies directed against the receptors has not been established.

HUNTINGTON'S DISEASE IS GENETICALLY TRANSMITTED

Huntington's disease, or Huntington's chorea, is characterized by brief involuntary movements (chorea) and progressive deterioration of higher neural functions. It usually begins between 35 and 50 years of age and progresses relentlessly, with a mean duration of approximately 15 years after onset. The disease is inherited in an autosomal dominant manner with complete penetrance; 50% of the children of an affected parent are at risk.

Neurons in the **corpus striatum** (caudate nucleus

and putamen) are most affected by Huntington's disease. Many die and are partly replaced by glial cells (gliosis). The cell death affects some subtypes of neurons more than others; interneurons tend not to be affected. Decreases are observed in the levels of some neurotransmitters (eg, γ-aminobutyric acid [GABA] and acetylcholine), of some enzymes involved in their synthesis (eg, glutamic decarboxylase and choline acetyltransferase), of some neuropeptides (eg, substance P, cholecystokinin, and metenkephalin), and of some receptors (eg, for dopamine, acetylcholine, and serotonin). Increases in the levels of other members of most of these four classes of molecules are also found, eg, of dopamine and norepinephrine, somatostatin, and neurotensin. These various changes are probably secondary, reflecting cell damage and cell death consequent upon the genetic lesion, but they are important in that they provide the neurochemical basis of the signs exhibited by patients.

The Gene for Huntington's Disease Has Been Isolated, and the Mutation Involved Is a Trinucleotide Repeat Expansion

In 1983, Gusella and coworkers described the occurrence of an RFLP on chromosome 4 that was closely linked to the gene responsible for Huntington's disease. This was one of the first studies to use RFLPs in linkage studies, and the successful outcome gave a tremendous boost to this type of effort. The subjects studied included a very large family (about 5000 individuals) around Lake Maracaibo in Venezuela, many of whose members were afflicted with Huntington's disease. The largeness of the family provided an excellent pedigree for analysis. The locus detected was very close to the end of the short arm of chromosome 4. Despite this promising start, various difficulties were encountered in precisely localizing the gene, and a further 10 years elapsed before it was isolated and sequenced. Some of the major findings regarding the gene and its gene product are listed in Table 64–2. The mutation involved was found to be a trinucleotide (CAG) repeat expansion (CAG encodes glutamine). Trinucleotide (triplet) repeat expansions were first reported in 1991, when they were found to be the basis of the fragile X syndrome. They were subsequently reported to be involved in a few other chronic neurodegenerative diseases (see below) prior to discovery of the Huntington disease gene. The gene encodes a protein (huntingtin) of 348 kDa, whose function has not yet been identified; thus, the role of the mutant protein in the pathologic mechanisms involved in Huntington's disease remains to be determined. For example, it is not yet clear whether the neuropathology found in Huntington's disease derives from a loss of function of huntingtin or whether some new function, initiating neuronal damage, is acquired by huntingtin.

Table 64–2. Some properties of the gene responsible for Huntington's disease.

- Located near the tip of chromosome 4 (4p16.3), close to the telomere.
- Approximately 210 kb in length.
- Encodes transcripts that are widely expressed inside and outside the central nervous system.
- Encodes a product (huntingtin) of approximate molecular mass 348 kDa, whose function is unknown.
- CAG repeats specifying glutamine are located at the 5′ end of the normal gene, within the putative protein-coding region.
- The mutation in the gene is an example of a trinucleotide repeat expansion.
- Genes from normal individuals had 11–24 CAG repeats; genes from patients with Huntington's disease had 42–86 repeats.
- Increasing length of trinucleotide repeat expansion appears to correlate with earlier onset and increasing severity.
- Mutation may involve a gain of function, although this has not been established; perhaps the polyglutamine stretch is involved in cross-linking to other proteins.
- Isolation of the gene greatly facilitates precise diagnosis but has not so far illuminated pathologic mechanisms.

The above findings have made it possible, using appropriate DNA probes, to tell members of affected families—if they wish to know—whether or not they will be afflicted with the condition (presymptomatic testing) and to offer prenatal diagnosis. At present, there is no specific treatment for Huntington's disease: genetic counseling, psychologic support, and symptomatic therapy are all that can be offered.

Excitotoxins May Cause Neuronal Death in Huntington's Disease via Their Actions on the NMDA Subtype of Glutamate Receptor

Huntington's disease is characterized by the death of certain neurons (see above). Without identification of the function (enzyme, receptor, etc) of huntingtin, the gene product involved in Huntington's disease, it is difficult to be certain how this occurs. One hypothesis is that it is due to certain chemicals (excitotoxins) that may be exogenous or endogenous (normal products of metabolism). To understand this concept more fully, some knowledge of **glutamate receptors** is necessary.

Glutamate and aspartate are **excitatory amino acids** in the brain; glutamate may be responsible for about 75% of the excitatory neurotransmission in the brain. Using selective agonists and antagonists, glutamate receptors have been subdivided into five classes: (1) NMDA (N-methyl-D-aspartate); (2) AMPA (α-amino-3-hydroxy-5-methyl-4-isoxazole propionic acid), formerly called quisqualate receptors; (3) kainate (a compound isolated from seaweed); (4) L-AP$_4$ (a synthetic agonist); and (5) metabotropic receptors. The first four are cation channels, whereas the metabotropic receptors are linked to intracellular production of diacylglycerol and inositol triphosphate by the polyphosphoinositide pathway (Chapter 44).

Of great interest is the finding that injection of **kainate** into the striatum of rats causes death of neurons but preserves glial cells and axons. It acts by causing release of excess glutamate. Indeed, incubation of neurons with relatively low concentrations of glutamate can cause cell death, which is dependent on the presence of Ca^{2+} in the medium. Injection of **quinolinate,** a metabolite of tryptophan that stimulates the NMDA receptor, also causes neuronal death but spares interneurons, just as Huntington's disease does. These compounds are thus called "excitotoxins." The NMDA receptor may be particularly involved in the cell damage and death observed in Huntington's disease, though the precise chain of events is not definable until the gene product is identified. The NMDA receptor (Figure 64–4) is complex in that it contains at least five distinct sites (a site that binds the transmitter glutamate, a regulatory site that binds glycine, a voltage-dependent Mg^{2+}-binding site, a site that binds phencyclidine, and a site that binds Zn^{2+}. It opens when it is occupied by an agonist, such as glutamate, and when open it allows the influx of Ca^{2+} and Na^{+} into the cell. If the NMDA receptor is chronically stimulated (eg, by excess glutamate secreted by neurons owing to depolarization of their plasma membranes), this allows accumulation of Ca^{2+}, which is toxic to cells when present in high levels and may result in cell death by mechanisms similar to those described for myocardial infarction.

(See Case No. 4 in Chapter 65.) In addition, Ca^{2+} influx is accompanied by an influx of Na^{+}, causing osmotic swelling and damage to cells. These findings may have therapeutic implications because it may be possible to partially inhibit NMDA receptors, using certain drugs such as dextrorphan (dextromethorphan), and thus reduce the numbers of dying cells.

The NMDA receptor is of interest in other respects. It is thought to play a central role in long-term potentiation, which results in an increase of synaptic efficacy in the hippocampus and other areas, and may be involved in memory and learning. Activation of this receptor has also been implicated in initiating seizures; in addition, repeated seizures can cause death of neurons, possibly involving excitotoxins.

A tentative scheme of some of the events involved in the causation of Huntington's disease is shown in Figure 64–5.

EXCITOTOXINS AND OTHER BIOCHEMICAL MECHANISMS ARE INVOLVED IN BRAIN DAMAGE DUE TO STROKE

In a stroke, localized brain damage occurs owing to diminished blood flow. The results can be devastating (eg, permanent loss of consciousness, paralysis, blindness, loss of speech), depending on the area of the brain affected and the size of the area affected. Strokes affect approximately 500,000 individuals an-

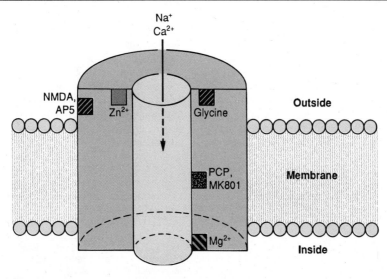

Figure 64–4. Schematic illustration of the NMDA receptor and the sites of action of different agents on the receptor. The NMDA receptor gates a cation channel that is permeable to Ca^{2+} and Na^{+} and is gated by Mg^{2+} in a voltage-dependent fashion; K^{+} is the counter-ion. The NMDA receptor channel is blocked by PCP and MK801, and the complex is regulated at two regulatory sites by glycine and Zn^{2+}. AP5 is a competitive antagonist at the NMDA site. (PCP, phencyclidine.) (Reproduced, with permission, from Cooper JR, Bloom FE, Roth RH: *The Biochemical Basis of Neuropharmacology,* 6th ed. Oxford Univ Press, 1991.)

Trinucleotide (CAG) expanded repeat mutation in Huntington's disease gene in the short arm of chromosome 4

↓

Alteration in the structure of its protein product (huntingtin), whose function has not been established

↓

Possible release of excitotoxins (eg, excess glutamate) causing excess stimulation of the NMDA receptor, intracellular accumulation of toxic amounts of Ca^{2+}, and death of striatal and other neurons

↓

Signs of Huntington's disease (chorea, progressive neurologic deterioration)

Figure 64–5. Tentative scheme of some of the events involved in the causation of Huntington's disease. It is not established whether the mutation results in a loss of function or gain of function of the affected protein product of the Huntington disease gene.

nually in the USA and are the third leading cause of death. As in the case of Alzheimer's disease, the economic costs involved in the medical care of patients with strokes are enormous.

In order to design appropriate therapies, it is necessary to understand the basic mechanisms involved in brain damage from stroke. The following are some important considerations. Most strokes are caused by thromboembolic events in the cerebral arteries. This reduces the supply of oxygen and glucose to the brain, both of which are vital to its metabolism and without which permanent cell damage occurs rapidly (< 1 hour). There are three subsequent stages in the development of brain damage due to a cerebral thrombosis.

(1) Induction: Ischemia causes depolarization of the neuronal membrane, leading to release of **glutamate.** As in the case of Huntington's disease, this overexcites the NMDA receptors on adjacent neurons, leading to abnormally large influxes of Ca^{2+} and Na^+ and resultant cell injury or death. In addition, glutamate stimulates the AMPA-kainate receptors (leading to additional influx of Na^+) and also the metabotropic receptors, causing intracellular release of inositol triphosphate and diacylglycerol.

(2) Amplification: Further buildup of intracellular Ca^{2+} occurs by the following mechanisms: (a) increased intracellular Na^+ activates Na^+-Ca^{2+} transporters; (b) voltage-gated Ca^{2+} channels are activated by depolarization; and (c) inositol triphosphate releases Ca^{2+} into the cytosol from within the endoplasmic reticulum. All three mechanisms favor buildup of intracellular levels of Ca^{2+}, which causes additional release of glutamate, further exciting neighboring neurons.

(3) Expression: The high level of intracellular Ca^{2+} activates Ca^{2+}-dependent nucleases, proteases, and phospholipases. Degradation of phospholipids can also lead to formation of **platelet-activating factor** (PAF) and release of arachidonic acid, whose metabolism can generate **eicosanoids;** both of these lipids can cause vasoconstriction, worsening the thrombosis. In addition, the metabolism of eicosanoids leads to formation of oxygen free radicals, which cause peroxidative damage of the lipids of neuronal membranes. Since glutamate plays a prominent role in the above series of events, the overall process has been described as the **glutamate cascade.** Many additional factors are also involved in ischemic cerebral damage.

Treatment of stroke is designed to limit the damage due to thrombosis. Experimental treatments are being developed to minimize the biochemical effects of the glutamate cascade (Table 64–3). As in the case of thromboses in the coronary arteries (Chapter 59), early thrombolytic treatment may restore blood flow. Several large clinical trials of t-PA are in progress. For stroke, t-PA may be more suitable than streptokinase because the latter is more likely to cause hemorrhage, which can be disastrous in the brain.

Prevention of stroke includes control of risk factors such as hypertension, diabetes mellitus, smoking, and hyperlipidemia. Low-dose aspirin administration appears to be effective in reducing the risk of stroke in patients who are experiencing transient ischemic attacks.

Table 64–3. Some experimental treatments under development to limit the glutamate cascade occurring during a stroke. (The three stages of the cascade are described in the text.)[1]

Stage of Cascade	Biochemical Problems	Approaches
Induction	Excess release of glutamate	Inhibit synthesis of glutamate by methionine sulfoximine. Lower body temperature if patient is feverish; this lowers brain metabolism and may inhibit release of glutamate.
	Activation of NMDA receptors	NMDA receptor antagonists, such as dextrorphan, CGS 19755, and MK-801.
Amplification	Influx of Ca^{2+} via voltage-gated Ca^{2+} channels	Dihydropyridine derivatives (eg, nimodipine) to block these channels.
Expression	Production of free radicals	Free radical inhibitors (eg, 21-aminosteroids).

[1]Adapted from Zivin JA, Choi DW: Stroke therapy. Sci Am (July) 1991;265:56.

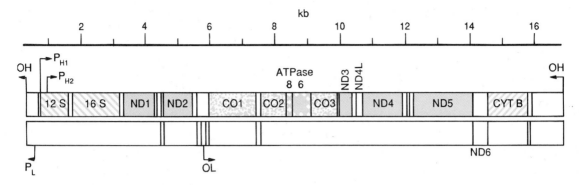

Figure 64–6. Maps of human mitochondrial genes. The maps represent the heavy (upper strand) and light (lower map) strands of linearized mitochondrial (mt) DNA, showing the genes for the subunits of NADH-coenzyme Q oxidoreductase (ND1 through ND6), cytochrome *c* oxidase (CO1 through CO3), cytochrome *b* (CYT B), and ATP synthase (ATPase 8 and 6) and for the 12S and 16S ribosomal mtRNAs. The transfer RNAs are denoted by small open boxes. The origin of heavy-strand (OH) and light-strand (OL) replication and the promoters for the initiation of heavy-strand (PH1 and PH2) and light-strand (PL) transcription are indicated by arrows. (Reproduced, with permission, from Moraes CT et al: Mitochondrial DNA deletions in progressive external ophthalmoplegia and Kearns-Sayre syndrome. N Engl J Med 1989;320:1293.)

MUTATIONS IN MITOCHONDRIAL DNA CAUSE SOME MYOPATHIES AND NEUROLOGIC DISEASES

Human mitochondria contain two to ten copies of a small circular double-stranded DNA molecule that makes up approximately 1% of total cellular DNA (Figure 64–6 and Table 64–4). An important feature of human mitochondrial (mt) DNA is that, because all mitochondria are contributed by the ovum during zygote formation, it is transmitted by maternal non-mendelian inheritance. Thus, in diseases resulting from mutations of mtDNA, in theory an affected mother would pass the disease to all her children but

Table 64–4. Some major features of the structure and function of human mitochondrial DNA.[1]

- Is circular, double-stranded, and contains a heavy (H) and a light (L) chain.
- Contains 16,569 bp, fully sequenced.
- Encodes 13 protein subunits of the respiratory chain (of a total of about 67):
  Seven subunits of NADH dehydrogenase (complex I)
  Cytochrome b of complex III
  Three subunits of cytochrome oxidase (complex IV)
  Two subunits of ATP synthase
- Encodes large (16s) and small (12s) mt ribosomal RNAs.
- Encodes 22 mt tRNA molecules.
- Genetic code differs slightly from the standard code:
  UGA (standard stop codon) is read as Trp.
  AGA and AGG (standard codons for Arg) are read as stop codons.
- Contains very few untranslated sequences.
- High mutation rate (five to ten times that of nuclear DNA).
- Comparisons of mtDNA sequences provide evidence about evolutionary origins of primates and other species.

[1]Adapted from Harding AE: Neurological disease and mitochondrial genes. Trends Neurol Sci 1991;14:132.

only her daughters would transmit the trait. However, in some cases (see below), deletions in mtDNA occur during oogenesis and are thus not inherited from the mother.

A number of diseases have now been shown to be due to mutations of mtDNA (Table 64–5). One group is composed of the mitochondrial myopathies, characterized by the presence in skeletal muscle of mitochondria of abnormal shape and size. These abnormal mitochondria typically result in muscle fibers appearing as **ragged red fibers** (detected in muscle biopsy by use of Gomori trichrome or other stains). The first four conditions listed in Table 64–5 are examples of mitochondrial myopathies. In ocular myopathy, the muscles responsible for eye movement are affected primarily, resulting in progressive external ophthalmoplegia, which also occurs, along with several other findings, in the Kearns-Sayre syndrome. Myoclonus epilepsy with ragged red fibers (MERRF syndrome) and mitochondrial encephalomyopathy with lactacidosis and stroke-like episodes (MELAS syndrome) can also be classified as mitochondrial encephalomyopathies, since the brain is usually severely affected. The two other conditions listed in Table 64–5 (Leber's hereditary optic neuropathy [LHON] and maternal retinitis pigmentosa are diseases resulting from mutations in mtDNA but are not mitochondrial myopathies since muscle is not necessarily affected.

Many patients with progressive external ophthalmoplegia (including those with Kearns-Sayre syndrome) have been found to exhibit deletions in mtDNA, and duplications have also been found in patients with Kearns-Sayre syndrome. The extent of the deletion often correlates with the severity of the progressive external ophthalmoplegia. The deletions

Table 64–5. Some mitochondrial myopathies and diseases resulting from mutations in mitochondrial genes.[1]

Disease	Ragged Red Fibers	Mutation[2]	Clinical Expression
Ocular myopathy	+	Deletions	Progressive external ophthalmoplegia generally presenting in childhood.
Kearn-Sayre syndrome (KSS)	+	Deletions, duplications	Progressive external ophthalmoplegia, pigmentary retinopathy, and cardiac conduction defects presenting in childhood.
Myoclonus epilepsy with ragged red fibers (MERFF)	+	Point mutation in lysine tRNA (position 8344)	Myoclonus, epilepsy, and ataxia presenting in childhood or adulthood.
Mitochondrial encephalopathy with lactic acidosis and stroke-like episodes (MELAS)	+	Point mutation in leucine tRNA (position 3423)	Intermittent vomiting, proximal limb weakness, and stroke-like episodes occurring in adolescence or later.
Leber's hereditary optic neuropathy (LHON)	–	Point mutation in NADH-ubiquinone reductase (position 11,778)	Acute or subacute loss of vision in men around the age of 20.
Maternal retinitis pigmentosa (RP)	–	Point mutation in subunit 6 of H^+ ATPase (position 8993)	Occurred in one family along with dementia, seizures, ataxia, and muscle weakness.

[1]Adapted from Harding AE: Neurological disease and mitochondrial genes. Trends Neurol Sci 1991;14:132.
[2]Other mutations may also occur in the conditions listed. The occurrence of mutations in tRNA species appears to correlate with the mitochondrial proliferation observed in certain of these diseases, whereas mutations in the genes encoding enzymes of the respiratory chain do not.

appear to occur during oogenesis and are thus not inherited from the mother. Moreover, they may not be present in all cells—eg, leukocytes may not be affected. On the other hand, patients with the MERRF syndrome, the MELAS syndrome, LHON, and familial maternally transmitted retinitis pigmentosa exhibit specific point mutations (Table 64–5). Such point mutations, unlike deletions, are usually transmitted from the mother. Laboratory tests of value in the diagnosis of diseases due to mutation of mtDNA are listed in Table 64–6.

Several additional points regarding diseases due to mutations of mtDNA merit comment.

(1) The DNA from mitochondria of patients with these diseases may be composed of a mixture of mutant and normal DNA (heteroplasmy) or of entirely mutant DNA (homoplasmy). The amount of mutant DNA generally correlates with severity of the disease.

(2) The majority of the mitochondrial polypeptides (about 54/67) are coded by nuclear genes; thus, mutations in these latter genes may also affect mitochondrial structure and function. In certain conditions (eg, LHON), interactions between nuclear and mt genes may be important in determining the phenotype.

Table 64–6. Laboratory tests of value in diagnosing diseases due to mutations in mtDNA.

- Presence of ragged red fibers in muscle biopsy (not in Leber's hereditary optic neuropathy nor in maternal retinitis pigmentosa).
- Assays of various enzymes of the respiratory chain.
- Analyses of mtDNA (eg, from muscle or in some cases leukocytes).

(3) How do mt mutations produce their specific tissue effects? For example, why is the optic nerve so specifically affected in LHON? Probably the answer lies in the underlying physiology of the tissue affected; presumably, the activity of the NADH-coenzyme Q reductase (which is affected in LHON) is particularly important for this tissue. Interestingly, chronic administration of sodium azide (which inhibits cytochrome *c* oxidase) to certain mammals produces a neuropathy similar to LHON.

(4) Abnormalities of mtDNA may be present in some individuals with Parkinson's disease (see below) and may also be involved in the aging process.

(5) Some individuals with diabetes mellitus have been found to have mutations in mtDNA, eg, in the $tRNA^{Leu}$ gene.

FRAGILE SITES AND VARIOUS CHRONIC NEURODEGENERATIVE DISEASES ARE DUE TO TRINUCLEOTIDE REPEAT EXPANSIONS

As discussed above, Huntington's disease is due to a type of mutation known as a trinucleotide repeat expansion, sometimes called a dynamic mutation. The first disease in which this type of mutation was found was fragile X syndrome. Its molecular basis, an expansion of a CGG repeat, was reported in 1991. Since that time, the number of diseases in which trinucleotide repeat expansions have been implicated has mushroomed to double figures; some of these conditions are listed in Table 64–7.

Two types of abnormalities, fragile sites on chro-

Table 64–7. Diseases exhibiting trinucleotide repeat expansions.[1,2]

Condition[3]	Repeat Sequence	Repeat Localization[3]	Repeat Number
FRAXA	CGG	5′ untranslated	Large (200–1000)
FRAXE	CGG/CCG	?	Large (200–1000)
FRAXF	CGG/CCG	?	Large (300–500)
FRA16AC	GG/CCG	?	Very large (1000–2000)
SBMA	CAG	ORF	Small (<100)
Huntington's disease	CAG	ORF	Small (<150)
SCA1	CAG	ORF	Small (<100)
DRPLA/HRS	CAG	ORF	Small (<100)
MJD	CAG	ORF	Small (<100)
Myotonic dystrophy	CTG/CAG	3′ untranslated	Very large (200–4000)

[1]Adapted from Willems PJ: Dynamic mutations hit double figures. Nature Genetics 1994;8:213.
[2]Genes involved in some of the above conditions have been isolated. Examples: *FMR-1* in the fragile X syndrome (FRAXA), the *Huntington* gene in Huntington's disease, the *myotonin kinase* gene in myotonic dystrophy, and the *androgen receptor* gene in SBMA. FRAXF and FRA-16A (on chromosome 16) are not apparently associated with pathologic consequences.
  Most of the above conditions exhibit greater or lesser degrees of anticipation. The parent in whom expansion occurs (eg, exclusively through the mother in fragile X syndrome, more often through the father in Huntington's disease) has not been indicated in the table. The majority of the conditions listed exhibit anticipation, though this is not found in X-linked spinal and bulbar muscular atrophy.
  The reader is referred to textbooks of internal medicine or neurology for discussions of the clinical findings in the various conditions listed.
[3]**Key:** FRAXA, fragile X syndrome; FRAXE, FRAXE syndrome; SBMA, X-linked spinobulbar muscular atrophy (Kennedy's disease); SCA1, spinocerebellar ataxia, type 1; DRPLA/HRS, dentatorubral pallidoluysian atrophy and its allelic variant, Haw River syndrome; MJD, Machado-Joseph disease; ORF, open reading frame.

mosomes and chronic neurodegenerative diseases, have been associated with trinucleotide repeat expansions. The mechanisms involved in trinucleotide expansion have not been established; postulated reasons include slippage of DNA polymerase on abnormal DNA secondary structures formed by the expansions.

CGG/CCG Expansions Lead to Fragile Sites on Chromosomes

CGG/CCG expansions lead to fragile sites on chromosomes, such as are found in the fragile X syndrome and in several other conditions (see notes to Table 64–7). The fragile X syndrome is an X-linked recessive disorder characterized by mild to moderately severe mental retardation and large head, ears, and testicles. Its prevalence is approximately 1:1250 males in North America, and it accounts for an appreciable proportion of males who are institutionalized for mental retardation. After Down's syndrome, it is the major cause of mental retardation associated with an identifiable chromosomal abnormality. Cytogenetically, it is detected by a constriction near the end (Xq27) of the long arm of the X chromosome, so that the tip of the long arm looks like a small knob attached to the rest of the chromosome by a thin stalk. The fragile site (named *FRAXA* in the fragile X syndrome) can be induced to appear in leukocyte cultures by depletion of folate or addition of methotrexate, both of which inhibit replication of DNA; however, the accuracy of testing in female carriers is less than 30%. Other fragile sites (*FRAXE* and *FRAXF*)

nearby *FRAXA* on the distal end of Xq have been identified, and *FRA16A* is located on chromosome 16. *FRAXE* is associated with mild mental retardation, but *FRAXF* and *FRA16A* may not be associated with any pathologic conditions. The cause of the fragility associated with CGG/CCG expansions has not been elucidated. However, the fragile sites so far discovered are located near a CpG island, and there appears to be a correlation among amplification of CGG/CCG repeats, hypermethylation, and expression of fragile sites. It is possible that methylation at CCG sites might stabilize regions of tetraplex DNA formed by CCG tracts, and this could lead to obstruction of DNA replication and of chromatin condensation, resulting in a fragile site. On the other hand, from studies on *FRA16A*, it has been suggested that methylation may be a consequence rather than a cause of the genesis of fragile sites, so that the mechanisms involved in fragility remain unsettled.

The genetics of fragile X syndrome are intriguing. A small increase in expansion appears to render the trinucleotide repeat unstable. This is called a premutation. A man who carries a premutation is a normal transmitting male, and all his daughters carry the premutation but are normal. Their sons are at risk that the premutation will expand, and if it reaches 200 repeats, this constitutes a full mutation. For women with full mutations, there is a 50% risk that each of their sons will be affected and that each daughter will inherit the full mutation. Approximately half of women with a full mutation are mildly retarded. Expansion of the trinucleotide repeat appears to suppress transcription of the gene *(FMR-1)* affected in the fragile X syndrome; the function of its protein product has not been identified. Screening of high-risk families and prenatal diagnosis are now possible using DNA testing.

Expansion of trinucleotide repeats affords an explanation of the phenomenon of anticipation. This is the tendency for some genetic diseases to show an earlier age of onset or more severe expression in more recent generations of a family. Most of the diseases listed in Table 64–7 show anticipation.

Trinucleotide Repeat Expansions Cause a Number of Chronic Neurodegenerative Diseases

As shown in Table 64–7, CAG repeat expansions are implicated in the cause of at least six chronic neurodegenerative disorders, with the distinct possibility of others yet to be discovered. Most of the conditions are of late onset, and cell death (?apoptosis) is prominent in all of them. The mutations occur in the open reading frames of the genes involved, with the expansions being smaller than those found in the fragile chromosome conditions. It is not established whether the mutations are associated with losses of function of the involved proteins or possibly with gains of new functions. The polyglutamine tracts encoded by the expansions are thought to be functional and could play a role in binding to DNA or could be involved in transglutamination, thus giving rise to protein aggregates that could cause neuronal damage. However, these suggestions are speculative, with little substantiating evidence. Another interesting possibility is that proteins with these abnormal polyglutamine tracts could somehow participate in apoptosis; cell death is a prominent feature of these chronic neurodegenerative diseases and of Alzheimer's disease (see below).

Myotonic dystrophy differs in several respects from the other conditions listed in Table 64–7. The mutation is in the 3′ untranslated end of the gene, which raises a question about how an untranslated sequence can generate neuropathology. The trinucleotide expansion is much larger than in the case of the other conditions, and anticipation is a very prominent feature.

THE SIGNS OF PARKINSON'S DISEASE REFLECT A DEFICIENCY OF DOPAMINE IN THE SUBSTANTIA NIGRA AND CORPUS STRIATUM

Parkinson's disease is characterized by tremor, bradykinesia (slowness and poverty of movement), rigidity, and postural instability. It rarely occurs before age 40. A genetic component has been considered for years, but not proved. Interest has increased recently, since it was found that there may be a defi-

ciency of the activity of complex I of the respiratory chain in the substantia nigra of some individuals with Parkinson's disease. It is estimated that 1% of people over age 50 have this condition and that over 1 million people in the USA suffer from it. Patients who exhibit these signs are said to exhibit **parkinsonism;** Parkinson's disease is just one cause of parkinsonism (see below).

The key pathologic characteristic of Parkinson's disease is degeneration of the pigmented cells in the substantia nigra. When normal, these cells synthesize and use dopamine as a neurotransmitter and thus are said to be dopaminergic. Dopaminergic neurons are found in a number of areas of the brain, including the nigrostriatal, mesolimbic, mesocortical, and tubero-hypophysial systems.

Figure 64–7 illustrates the overall process involved when dopamine acts as a neurotransmitter; as in the case of acetylcholine (Figure 64–1), it can be conveniently divided into six steps. A similar process is involved in the action of other monoamine neurotransmitters (eg, norepinephrine and epinephrine), though there are sometimes important differences in individuals steps.

(1) Synthesis of dopamine occurs from tyrosine, involving a number of enzymatic reactions. The reaction catalyzed by tyrosine hydroxylase is rate-limiting.

(2) Storage of dopamine in synaptic vesicles is the next step. Entry of dopamine is driven by a pH gradient established by a protein—present in the vesicular membrane—that pumps protons into the vesicle at the expense of ATP.

(3) Release of dopamine involves exocytosis. The mechanism is similar to that described above for release of acetylcholine.

(4) Binding of dopamine to its postsynaptic receptors then occurs. The amine reaches its receptors by diffusion across the synaptic cleft. As described later in this chapter, there are at least five classes of dopamine receptors. These produce their effector actions by affecting adenylyl cyclase positively or negatively or in at least one case by affecting another signaling system (ie, phospholipase C and the polyphosphoinositide cycle).

(5) Reuptake of dopamine occurs. This is achieved by a high-affinity transporter (which uses ATP) present in the presynaptic membrane. The recycled dopamine can again be incorporated into synaptic vesicles and reused as a transmitter.

(6) Degradation of dopamine can occur within the synaptic cleft or, following reuptake, within the presynaptic terminal. Monoamine oxidase B (MAO-B) is present in the outer membrane of mitochondria within the presynaptic terminal and is also present in the synaptic cleft. MAO-B and MAO-A are distinguished from each other by their preferences for different substrates and by their different susceptibilities to various inhibitors. Both enzymes can act on

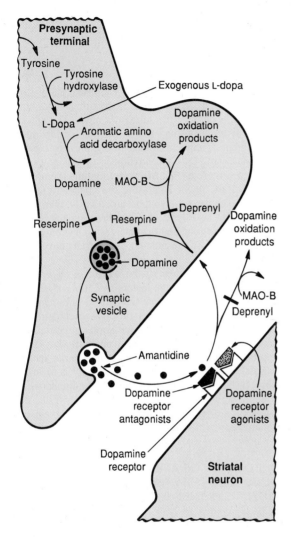

Figure 64–7. Schematic diagram illustrating the release of dopamine by a neuron in the substantia nigra and also showing the sites of action of drugs that ameliorate or induce parkinsonism. The six steps described in the text are illustrated but are not numbered in the figure. The sites of actions of drugs (L-dopa, deprenyl, amantadine, and dopamine receptor agonists [eg, bromocriptine]) that are used to treat parkinsonism are also shown; in general these drugs elevate local levels of dopamine, thus countering its low level in parkinsonism. Sites of action of certain drugs (reserpine and dopamine receptor antagonists, such as many neuroleptics) that induce parkinsonism by depleting dopamine or competing with it are also indicated. (Reproduced, with permission, from Sweeney PJ: New concepts in Parkinson's disease. Hosp Pract [Off Ed] Apr 1991;26:84.)

dopamine, forming 3,4-dihydroxyphenylacetaldehyde (DOPAC). The latter is converted to homovanillic acid by the action of catechol-O-methyltransferase (COMT) (Figure 49–3). This fact is of importance in relation to studies on schizophrenia (see below), since the level of homovanillic acid in the cerebrospinal fluid can be used to follow dopamine metabolism in the brain. **Deprenyl** acts to inhibit MAO-B, whereas clorgyline is a specific inhibitor of the type A enzyme. Figure 64–7 also shows the sites of action of a number of drugs that are used to treat parkinsonism and of several (see below) that cause parkinsonism as an undesirable side effect; all six steps set forth above are affected by various drugs.

The degenerative process referred to above causes a marked decrease in synthesis of dopamine and a resultant drop in its level in the substantia nigra and corpus striatum (caudate nucleus and putamen). The lowering of dopamine raises the ratio of acetylcholine listed to dopamine in cells of the nigrostriatal system because levels of acetylcholine are not so affected. This imbalance contributes to the various disorders of movement found in Parkinson's disease.

Why the dopaminergic cells in the substantia nigra are damaged in the first place is not clear. There may be a genetic susceptibility in some cases. In part, the cell damage may reflect the process of aging, since about 13% of the cells of the substantia nigra are lost per decade from the age of 25 onward. It is estimated that signs of Parkinson' disease appear when the level of dopamine in the nigrostriatal system has dropped by 80%. Viral infections can cause parkinsonism, but this appears to be a relatively minor cause nowadays. Exposure to high levels of Mn^{2+} has caused parkinsonism in miners. Drugs (eg, reserpine and the neuroleptics [antipsychotics]) are an important cause of parkinsonism, which can be reversible when therapy is stopped. As shown in Figure 64–7, reserpine depletes dopamine by inhibiting dopamine storage in presynaptic cells, and many neuroleptics block dopamine receptors. The amino acid derivative **β-N-methyl-amino-L-alanine** (BMAA), present in the seeds of the cycad tree, can cause parkinsonism and may account for the high incidence of this condition in Guam, where many inhabitants have used the seeds as part of their diet. BMAA is thus a neurotoxin. In recent years, particular interest has focused on **1-methyl-4-phenyl-1,2,3,6-tetrahydropyridine** (MPTP) as a cause of acute parkinsonism in intravenous drug abusers. This compound is a by-product in the synthesis of a synthetic heroin derivative and contaminates preparations of it synthesized illicitly. MPTP is converted by monoamine oxidase B to 1-methyl-4-phenyl-pyridinium ion (MPP⁺), which appears to be the actual neurotoxic agent. MPP⁺ is taken up via the dopamine uptake system of the dopaminergic cells of the substantia nigra and damages them. Interestingly, in view of its possible involvement in some cases of Parkinson's disease, this may inhibit complex I of the respiratory chain, resulting in a decline of the level of ATP. Because of its free radical nature, MPP^+ may act by causing oxidative damage (Chapter 60). In addition, dopamine itself may be oxidized to form potentially toxic radicals. The knowledge gained regarding these neurotoxins raises the issue as to how much neurotoxins may contribute to the genesis of Parkinson's disease.

Levodopa Crosses the Blood-Brain Barrier and Is Converted to Dopamine in the Brain, Thus Providing Replacement Therapy for Parkinson's Disease

When it became apparent in the 1960s that levels of dopamine were low in the nigrostriatal system, a new era began in the treatment of Parkinson's disease. There are three major approaches to treatment.

A. Anticholinergic Therapy: Anticholinergic therapy was the main therapy before the knowledge regarding the importance of low levels of dopamine was acquired and is still used in the treatment of certain patients.

B. Treatment With Dopamine Precursors: Dopamine itself does not cross the blood-brain barrier, so it cannot be used in treatment of Parkinson's disease. Administration of dopamine precursors, however, restores the levels of dopamine to more nearly normal. The drug of choice is **levodopa** (L-dopa; L-3,4-dihydroxyphenylalanine), which does cross the blood-brain barrier and is subsequently converted to dopamine by **dopa decarboxylase** (also called aromatic amino acid decarboxylase; Figure 64–7). However, most of the administered dopa is decarboxylated in the periphery, and only about 1% reaches the brain. Higher doses of levodopa are not the answer, because they cause severe nausea and vomiting. The solution has been to administer levodopa along with **carbidopa,** a compound that inhibits the activity of peripheral dopa decarboxylase but does not cross the blood-brain barrier and so does not inhibit the conversion of levodopa to dopamine in the brain. A meal high in protein provides amino acids that compete for uptake of levodopa by intestinal cells; eating a meal low in protein diminishes this effect. Levodopa has become the mainstay of the treatment of patients with Parkinson's disease since its introduction in the 1960s. In fact, the diagnosis of Parkinson's disease requires demonstration of a beneficial response to this drug, whereas many patients with parkinsonism do not show much response to it. However, after about 5 years, its effects wear off, and alternative therapy is then required.

C. Dopamine Receptor Agonists: These drugs simulate the effects of dopamine itself (Figure 64–7). Bromocriptine is useful in this respect.

Other Drugs Are Useful in the Treatment of Parkinson's Disease Because They Affect the Metabolism of Dopamine or Are Neuroprotective

Other treatments are available, based mainly on the knowledge of depletion of dopamine and how neurotoxins damage the substantia nigra. The antiviral agent **amantadine** induces release of dopamine from the presynaptic terminals of dopaminergic cells, thus increasing its concentration at the postsynaptic junction (Figure 64–7). **Mazindol** inhibits the uptake of dopamine, again increasing the concentration of dopamine at the postsynaptic junction. Deprenyl inhibits MAO-B (Figure 64–7), decreasing the degradation of dopamine and thus increasing its concentration. However, the results of clinical trials have been disappointing.

Surgical approaches to the treatment of Parkinson's disease include the injection of minced adrenal autografts into the caudate nucleus or putamen; the adrenal medulla synthesizes various catecholamines, including dopamine. Fetal tissue implants, containing embryonic nigral cells, are also under study. These approaches may be less effective than initial results suggested. Another approach under study is to inject fibroblasts that have been genetically engineered to contain high activities of tyrosine hydroxylase, the rate-limiting enzyme in the pathway of synthesis of dopamine.

DEPOSITION OF AMYLOID β PROTEIN OCCURS IN THE BRAINS OF INDIVIDUALS WITH ALZHEIMER'S DISEASE

Alzheimer's disease is an incurable neuropsychiatric condition in which progressive impairment of cognitive functions occurs, usually accompanied by affective and behavioral disturbances. Some 2 million people in the USA suffer from Alzheimer's disease, and its prevalence is likely to increase as more people live longer. Some cases have a familial (genetic) basis, but the majority appear to be sporadic. Alzheimer's disease is the commonest cause of **dementia,** which can be defined as a progressive decline in intellectual functions, due to an organic cause, that interferes substantially with an individual's activities. Alzheimer's disease places a tremendous burden on families and on the health care system, as, sooner or later, most patients cannot look after themselves. The economic cost of Alzheimer's disease in the USA is estimated to be approximately $40 billion annually. The usual age at onset is over 65 years, but it can occur in the early 40s; survival ranges from 2 to 20 years. Loss of memory is often the first sign. The disease usually progresses inexorably, and terminal patients are completely incapacitated.

The basic pathologic picture is of a degenerative process characterized by the loss of cells in certain areas of the brain (eg, the cortex and hippocampus). Apoptosis (Chapter 62) may be involved in the cell death in Alzheimer's disease and other neurodegenerative diseases discussed in this chapter. At the microscopic level, amyloid plaques surrounded by nerve cells containing neurofibrillary tangles (paired helical filaments formed from a hyperphosphorylated form of the microtubule associated protein, tau) are hallmarks. Deposits of amyloid are frequent in small blood vessels.

Intensive research is under way to determine the cause of Alzheimer's disease. Significant progress appears to have been made recently. Particular interest has focused on the presence of **amyloid β protein** (AβP), the major constituent of the plaques found in Alzheimer's disease. The term "amyloid" refers to a group of diverse extracellular protein deposits found in many different diseases. Amyloid proteins usually stain blue with iodine, like starch, which accounts for the name. The **amyloid cascade hypothesis** holds that deposition of AβP is the cause of the pathologic changes observed in the brains of victims of Alzheimer's disease and that other changes, such as neurofibrillary tangles and vascular alterations, are secondary. AβP is derived from a larger precursor protein named **amyloid precursor protein** (APP), whose gene is located on chromosome 21 close to the area affected in Down's syndrome (trisomy 21). Individuals with Down's syndrome who survive to age 50 often suffer from Alzheimer's disease.

APP is a transmembrane protein of about 770 amino acids. AβP is a peptide of 39–42 amino acids derived by proteolytic cleavage from the carboxyl terminal end of APP. It forms an insoluble extracellular deposit. APP can apparently be cleaved by at least two different proteinases. The first (designated **secretase**) produces a soluble fragment containing only part of AβP (Figure 64–8). The second, an endosomal-lysosomal protease, generates a fragment containing the entire sequence of AβP; it is speculated that this fragment precipitates extracellularly and leads to the neurofibrillary tangles and other histopathologic features of Alzheimer's disease.

As described below, mutations in certain genes have been found in some patients with Alzheimer's disease. One of these genes is that encoding APP. It is possible that mutations in this gene could lead to increased deposition of AβP in neurons.

A second part of the amyloid cascade hypothesis regarding the causation of Alzheimer's disease is that AβP or AβP-containing fragments are directly or indirectly neurotoxic. There is evidence that exposure of neurons to AβP can increase their intracellular concentration of Ca^{2+}. Some protein kinases, including that involved in phosphorylation of tau, are regulated by levels of Ca^{2+}. Thus, increase of Ca^{2+} may lead to hyperphosphorylation of tau and formation of

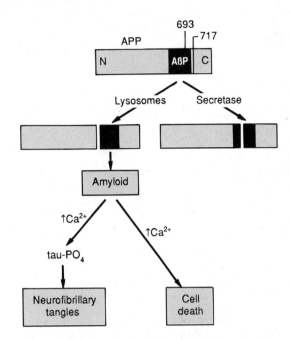

Figure 64–8. Diagram of the amyloid cascade hypothesis. Amyloid precursor protein (APP) can be processed by two different routes. The first involves secretase, which produces peptides that do not contain the entire AβP molecule, are soluble, and do not precipitate to produce amyloid. The second involves the endosomal-lysosomal proteases and produces amyloid β protein (AβP) or peptides containing it; these precipitate to form amyloid. It is speculated that this leads to neurofibrillary tangles and cell death. In this model, precipitation of amyloid precedes and leads to the formation of neurofibrillary tangles. (Reproduced, with permission, from Hardy JA, Higgins GA: Alzheimer's disease: The amyloid cascade hypothesis. Science 1992,256:184. Copyright © by the American Association for the Advancement of Science, 1992.)

the paired helical filaments present in the neurofibrillary tangles. Figure 64–9 shows a tentative scheme of the possible sequence of events in certain cases of Alzheimer's disease.

Genes Involved in Alzheimer's Disease Have Been Isolated

Certain cases of Alzheimer's disease are known to have a familial basis. In recent years, chromosomal loci carrying such genes have been pinpointed by genetic linkage analysis studies, followed by intensive efforts to isolate the genes responsible (Chapter 63). Genes involved in causing the disease display mutations which are not found in normal subjects. Some results of these studies are listed in Table 64–8. The finding of mutations in the gene on chromosome 21 encoding APP caused great excitement, as it seemed a likely candidate for involvement in most cases of Alzheimer's disease. However, this gene appears to

Mutation (in certain cases) of the gene for APP, located on chromosome 21

↓

Proteolysis of APP, resulting in abnormal fragments containing AβP

↓

Deposition of AβP, or fragments containing AβP, leading to increased levels of intracellular Ca^{2+} in neurons

↓

Increased levels of Ca^{2+} may activate the protein kinase phosphorylating the microtubule-associated tau protein, resulting in formation of the paired helical filaments seen in neurofibrillary tangles

↓

Deposition of AβP and hyperphosphorylated tau, over many years, resulting in plaque formation and neurofibrillary tangles

Figure 64–9. Tentative scheme of a possible sequence of events involved in the causation of certain cases of Alzheimer's disease. Mutations in the gene for APP are found in only a small number of cases. (APP, amyloid precursor protein; AβP, amyloid β protein.) (Based on Hardy JA, Higgins GA: Alzheimer's disease: The amyloid cascade hypothesis. Science 1992;256:184.)

be involved in only a very small number of cases of Alzheimer's disease. Recently, two genes have been isolated that are involved in certain early-onset familial cases of Alzheimer's disease. They appear to encode transmembrane proteins, which show homology to the SPE-4 protein of *Caenorhabditis elegans.* This protein appears to participate in the fibrous body-membrane organelle (FBMO) complex during spermatogenesis. The FBMO is a specialized organelle that may participate in the transport of soluble and membrane-bound proteins. This has led to speculations that the proteins (aptly named "presenilins") encoded by the newly discovered genes may be in-

Table 64–8. Genes involved in familial types of Alzheimer's disease.[1]

Gene	Type of Alzheimer's Disease	Chromosome	Protein
APP	Familial with amyloid precursor protein mutation	21	APP[2]
S182	Early onset, aggressive	14	Integral membrane protein
E5-1	Early onset	1	Integral membrane protein
APOE4	Adult onset	19	ApoE4

[1]The above conditions are inherited in an autosomal dominant manner. The two integral membrane proteins listed show extensive homology and have been referred to as presenilins. Their exact functions and their interactions (if any) with APP and apoE4 are under investigation.
[2]APP, amyloid precursor protein.

volved in transport or processing of APP. Another possibility is that the gene products could be receptors or ion channels. Whatever their precise nature, they are sure to open up new avenues of research.

An intriguing finding has been that *APOE4,* the gene encoding apolipoprotein E4 (apoE4) [Chapter 27] on chromosome 19, has been found to be a susceptibility gene for late-onset Alzheimer's disease. It advances the onset of this condition by as much as 20 years. How might this relate to the genesis of Alzheimer's disease? One possibility is as follows. Tau is the major microtubule-associated protein (Chapter 58). Studies have indicated that if apoE3 (or apoE2) is mixed with tau under appropriate conditions, a precipitate forms, whereas apoE4 does not form such a precipitate. The binding of apoE4 to tau is evidently much looser than that of the other isoforms. Thus, it has been suggested that when apoE4 is the isoform present, tau will be free to interact with another molecule of tau, leading to the formation of paired helical filaments which are precursors of neurofibrillary tangles. It has been suggested that if specific therapy could be devised to counteract the negative effects of apoE4, this could push back the onset of Alzheimer's disease in affected patients to as late as 90 years of age.

Other Factors Have Been Implicated in the Causation of Alzheimer's Disease

There are other hypotheses concerning the causation of Alzheimer's disease. For instance, it has been proposed that it is caused by infection with a slow virus, though no such virus has been isolated with any consistency. Another line of research has been stimulated by the finding that plaques from patients with Alzheimer's disease often contain increased amounts of aluminum. Thus, it has been proposed that one cause of Alzheimer's disease is ingestion of elevated amounts of aluminum. This brings up the issue whether increased uptake of aluminum into plaques is a primary or secondary event—ie, do elevated amounts of aluminum cause Alzheimer's disease in certain individuals, or do the increased amounts of aluminum reflect increased uptake brought about by cell damage caused by some other process? The answer is not clear. The brains of patients who have died from Alzheimer's disease often show a marked reduction in **choline acetyltransferase** and **acetylcholine.** Levels of other neurotransmitters are also frequently decreased. These alterations appear to be secondary changes due to cell damage and not primary events that initiate Alzheimer's disease.

A major hope is that research into the mechanisms involved in the causation of Alzheimer's disease will lead to the development of improved diagnostic tests and effective therapy. For instance, compounds that could inhibit formation of AβP or keep it in a soluble form might be of therapeutic value. At present, a definitive diagnosis of Alzheimer's disease can often only be made at autopsy, upon finding the characteristic plaques. One test that is in the developmental stage is based on measurement of the amount of APP in cerebrospinal fluid by the use of a specific monoclonal antibody. It is claimed that levels of APP are substantially (about two-thirds) lower in the cerebrospinal fluid of patients with Alzheimer's disease than in normal subjects.

No specific drug therapy for Alzheimer's disease is yet available, though it may eventually prove possible to alter the deposition of AβP in tissues by drugs. Brain-derived nerve growth factor has been shown to be deficient in certain areas of the brains of patients with Alzheimer's disease; its possible therapeutic benefit is being studied in animals with surgically induced neuronal degeneration.

To summarize, while amyloid plays a central role in Alzheimer's disease, it is not yet clear whether in the great majority of cases its deposition is a primary event or whether it occurs secondary to various other biochemical phenomena. The recent discovery of mutations in genes, at first glance unrelated to amyloid deposition, should open up new pathways of research that might lead to discovery of the mechanisms involved in this tragic disease.

GENETIC, NEURODEVELOPMENTAL, AND DOPAMINERGIC FACTORS MAY BE INVOLVED IN THE CAUSATION OF SCHIZOPHRENIA

The schizophrenic disorders are mental disorders, usually characterized by psychotic symptoms reflecting disturbances of thinking, feeling, and behavior. It is useful (see below) to distinguish between "positive symptoms" (eg, hallucinations, delusions, bizarre behavior) and "negative symptoms" (social withdrawal, emotional blunting, etc). There are a number of subtypes, eg, disorganized (hebephrenic), catatonic, and paranoid. For simplicity, we shall use the term schizophrenia to refer to the group of disorders.

Schizophrenia occurs throughout the world and afflicts approximately 1% of the population of North America. Usually its onset occurs before age 45, often in the teens, and it tends to run a chronic course. It is a major medical problem, often with devastating consequences for the victim and family members; even at the present time, when relatively effective drug therapy is available, schizophrenics occupy approximately 20% of all hospital beds.

The cause or causes of schizophrenia are unknown. Various psychologic, social, developmental, environmental, anatomic, genetic, biochemical, and other factors have been invoked. Here we shall consider why it has been so difficult to define a genetic basis for schizophrenia, indicate that there are structural abnormalities in the brains of schizophrenics that must be taken into account, and discuss the dopamine hypothesis of schizophrenia.

Genetic Linkage Studies in Schizophrenia Have Suffered From a Lack of Replication

A variety of approaches (eg, family histories, analyses of consanguinity, adoption studies, studies on mono- and dizygotic twins) have indicated a significant genetic contribution to schizophrenia. For instance, a child born to parents both of whom are schizophrenic has a 39% chance of developing the condition. The concordance among monozygotic twins is 47%. However, these studies have not indicated whether schizophrenia is a monogenic, polygenic, or multifactorial condition. If one of the latter two possibilities is the case, it will obviously make the search for a genetic basis more complex.

Attempts to define **linkage** of genes possibly involved in schizophrenia to a particular chromosome have produced conflicting results. Conflicting results have also been reported from linkage studies on other psychiatric disorders (eg, major depressive illness and manic depression). Some possible explanations are listed in Table 64–9. Because a disease occurs in a family does not necessarily mean that it has a genetic origin. For instance, poor psychologic interactions among family members could lead to abnormal behavior patterns. The best solution to these problems in linkage analysis is to obtain and study in depth large multigenerational families that share a common genetic defect and exhibit similar symptoms. In the case of schizophrenia, until clear linkage is established, it is likely that a number of candidate genes (eg, for dopamine receptors and for enzymes involved in catecholamine metabolism) will be selected and analyzed, looking for mutations that could play a role in the disorder.

Table 64–9. Some reasons for the difficulties in locating genes for schizophrenia. (Most of the points also apply to other psychiatric disorders, such as depression and manic depression.)[1]

- Schizophrenia may have multiple genetic causes (genetic heterogeneity).
- Large, multigenerational families are needed to increase efficiency of linkage analysis.
- Looseness of diagnostic criteria across various countries results in, for instance, misdiagnosis of relatives.
- Faulty statistical methodology (eg, misapplication of the Lod score) is applied.
- Postmortem samples of brain in various stages of preservation are used for biochemical analyses (eg, of dopamine).
- Prior treatment (eg, by neuroleptics) can alter the biochemical profile of brain samples analyzed.

[1]Based on Barnes DM: Troubles encountered in gene linkage land. Science 1989;243:313.

Structural Abnormalities That May Have a Developmental Basis Are Observed in the Brains of Schizophrenics

Structural abnormalities in the medial temporal lobe (parahippocampal gyrus, hippocampus, and amygdala) have been observed in the brains of many schizophrenics. These areas are involved in integrating and processing information from the association cortex. For instance, an altered orientation of hippocampal pyramidal cells has been observed that may reflect a defective pattern of neuronal migration. It may be that some genetic abnormality affects molecules involved in cell migration or cell adhesion, resulting in the observed findings. In turn, the altered cell orientation could produce secondary effects on various neurochemical parameters. In support of the finding that the brains of many schizophrenics show structural abnormalities is the fact that ventricular enlargement is frequently observed, though the difference from normal is often not significant enough to make this diagnostic.

Dopamine Has Been Implicated in Schizophrenia, But the Issue Is Not Yet Resolved

There has been no shortage of biochemical theories of schizophrenia; at various times, acetylcholine, γ-aminobutyric acid (GABA), norepinephrine, opiates, peptides, and other molecules have all been invoked in its causation. However, for the past 30 years, dopamine has attracted much of the attention. This occurred after the successful introduction of neuroleptic drugs (antipsychotics) in the early 1950s for the treatment of psychoses, including schizophrenia. It was then observed that many schizophrenics receiving neuroleptic treatment developed parkinsonism. This suggested that the neuroleptics were acting to lower levels of dopamine (see discussion of Parkinson's disease, above). This observation and other findings that support the involvement of dopamine in schizophrenia are summarized in Table 64–10. The original dopamine hypothesis of schizophrenia postulated that schizophrenia was a manifestation of hyperdopaminergia, as compared with Parkinson's disease, which can be considered as a state of hypodopaminergia.

Biochemical observations regarding the role of dopamine in schizophrenia have generally fallen into three categories.

A. Brain Dopamine Levels: A number of observers have reported variable increases in the amounts of dopamine in brain tissue samples of patients with schizophrenia.

B. Dopamine Metabolites: Measurements of dopamine metabolites in brain or body fluids have tended to stress **homovanillic acid,** the principal metabolite in humans. Elevations of this metabolite have been observed in the cerebrospinal fluid of pa-

Table 64–10. Observations that support the dopamine hypothesis of schizophrenia.[1]

- Neuroleptic drugs (antipsychotics) often induce parkinsonism, suggesting that they are acting to lower levels of dopamine.
- Neuroleptics appear to act primarily by reducing dopamine activity in mesolimbic dopamine neurons.
- A variety of other drugs (eg, levodopa, amphetamine) that act on dopamine metabolism (dopamine-mimetic) induce the signs and symptoms of schizophrenia.
- Prolonged treatment with neuroleptics leads to a fall in the levels of homovanillic acid in cerebrospinal fluid, generally correlating with a positive response to therapy.
- The antipsychotic potencies of most neuroleptics correlate with their binding to D2 receptors.
- Analyses of postmortem specimens and use of in vivo positron tomography indicate that the density of D2 receptors is increased in the brains of schizophrenics.
- Negative symptoms of schizophrenia may be associated with low dopamine activity in the prefrontal cortex.

[1]Based on Seeman P: Dopamine receptors and the dopamine hypothesis of schizophrenia. Synapse 1987;1:133; and on Davis KL et al: Dopamine in schizophrenia: A review and reconceptualization. Am J Psychiatr 1991;148:144.

tients with schizophrenia, the amount declining during response to drug therapy; but again, variability has been observed.

C. Dopamine Receptors: Measurements of dopamine (D) receptors have led to the most consistent result, ie, that the level of D2 receptors (see below) appears to be increased in the brains of schizophrenics and in particular of those who are drug-naive. Importantly, studies have shown that the antipsychotic potencies of most major neuroleptics correlate with their abilities to compete with dopamine in vitro for the D2 receptors. Reasons for variability in the results mentioned above include the use of postmortem tissues and the fact that most schizophrenics have been treated with neuroleptics, which in themselves (ie, independently of schizophrenia) may alter the levels of various receptors and enzymes in the brain.

The results obtained on D2 receptors have focused much interest on dopamine receptors. Largely owing to gene cloning, five classes of these receptors have been distinguished to date (Table 64–11). All appear to be transmembrane proteins; some at least are glycoproteins; and most appear to be coupled to G proteins. The D2, D3, and D4 receptors appear to resemble each other. Apart from the importance of the D2 receptor referred to above, the most interesting finding has been that the D4 receptor exhibits five polymorphic variants. This is the first receptor in the catecholamine family that has been found to display polymorphic variation in the human population. A key question is whether there may be a correlation between one of these variants and the susceptibility to schizophrenia or response to drug therapy. It is already known that the atypical neuroleptic **clozapine** (which does not cause parkinsonism or tardive dyski-

Table 64–11. Some properties of dopamine receptors. New receptors, isolated using appropriate probes and genomic or cDNA libraries, are being reported frequently.[1]

- At least five different major classes are recognized (D1–D5), with some subtypes formed by alternative splicing.
- They are membrane proteins, at least some of which are glycosylated.
- Most have seven transmembrane domains with cytoplasmic loops.
- Most appear to be coupled to G proteins.
- Some are positively coupled to adenylyl cyclase (eg, D1), at least one (D2) negatively.
- At least one subtype (of D1) appears to be coupled to phospholipase C.
- Some, at least, are regulated by phosphorylation.
- Drug affinities of most neuroleptics for the D2 receptor reflect their potencies in treating schizophrenia.
- The various receptors show different anatomic distributions.
- The D4 receptor binds the atypical neuroleptic clozapine with an affinity ten times higher than that of D2 sites.
- Five distinct subtypes of the D4 receptor have been recognized, it being the first member of the catecholamine family of receptors to show polymorphic variation in the human population.

[1]Based on Strange PG: Interesting times for dopamine receptors. Trends Neurol Sci 1991;14:43, and Iversen L: Which D4 do you have? Nature 1992;358:109.

nesia, serious side effects of most neuroleptics) exhibits a tenfold greater affinity for D4 receptors than for D2 receptors.

Other Theories of Causation of Schizophrenia

In view of some of the inconsistencies in the results mentioned above, the original dopamine hypothesis has been reformulated to postulate that there is abnormal, though not necessarily excessive, dopaminergic activity in schizophrenia. In some areas of the brain, there may be high dopaminergic activity, whereas in others there may be low activity. This viewpoint is supported by observations that there may be low dopamine activity in the prefrontal cortex of the brain of schizophrenics, perhaps correlating with the negative symptoms of the disorder. Other studies indicate that since prefrontal dopamine neurons may normally inhibit the activity of subcortical dopamine neurons, a lowering of dopamine in the former could lead to elevated dopaminergic activity in subcortical neurons. It should also be borne in mind that other neurotransmitters may also play a role in schizophrenia, either on their own or by interacting with dopaminergic systems.

Thus, despite numerous studies, it is not clear whether alterations of dopamine metabolism play a primary role in the causation of schizophrenia. It is possible that many of the changes in dopamine metabolism could be secondary (the smoke, not the fire) to other more basic phenomena. However, at the very least, detailed knowledge of dopamine receptors is leading to increasing refinement in the drug treatment of schizophrenia.

The foregoing discussion illustrates that the biochemistry of schizophrenia is complex. The demonstration of clear-cut mutations that play a causative role in schizophrenia would help illuminate the role of specific proteins (eg, receptors, enzymes) in the disease. A signpost pointing in this direction is the finding that susceptibility to schizophrenia in some individuals is associated with deletions on chromosome 22q11. This region has been implicated by genetic linkage analysis to contain lesions that increase susceptibility to schizophrenia. Another area of current research activity concerns trinucleotide repeat expansion mutations. Time will presumably tell whether they are involved in the genesis of schizophrenia or of other psychoses. In part, this issue has arisen because there is some evidence that the phenomenon of anticipation (see above) is apparently observed in certain families with a history of schizophrenia.

THE TECHNIQUES ARE AT HAND TO ESTABLISH THE MOLECULAR BASIS OF MANY NEUROPSYCHIATRIC DISORDERS

The use of recombinant DNA technology has made it feasible to establish the molecular basis of every disease that has a genetic origin. The work on cystic fibrosis (Case No. 7 in Chapter 65) demonstrated that it is possible to successfully search relatively large areas of DNA for abnormal genes. Thus, within this decade it may be possible to determine the molecular basis of most neuropsychiatric conditions that have a genetic basis. Significant advances have already been made regarding disorders not discussed above. For instance, mutations in the genes encoding rhodopsin, peripherin, and the β-subunit of rod phosphodiesterase (all proteins involved in vision) have been shown to be responsible for various forms of retinitis pigmentosa, a relatively common cause of blindness.

Gene therapy for neurologic disorders is beginning. It is proposed, for example, using the technique of MRI-guided stereotaxis, to inject mouse fibroblasts carrying a retroviral vector containing the gene for the thymidine kinase of the herpes simplex virus into inoperable **human brain tumors.** The retrovirus would then infect neighboring tumor cells, introducing the gene for thymidine kinase and thus rendering the recipient tumor cells susceptible to the drug ganciclovir; normal cells are not affected by this compound. This novel approach has been called "molecular surgery."

SUMMARY

The acetylcholine receptors of the myoneural junction are transmitter-gated ion channels. In myasthenia gravis, autoantibodies form to these receptors, destroying many of them. This is manifested by episodic attacks of muscle weakness. Drugs that inhibit cholinesterase are effective, but often the abnormal immune response must be suppressed by appropriate measures.

The gene for Huntington's disease is located on the short arm of chromosome 4. It encodes a protein (huntingtin) whose function is unknown. Normal subjects have 11–24 CAG trinucleotide repeats near the 5' end of the coding region of the gene, whereas individuals with Huntington's disease have 42–86 CAG repeats.

Studies of mechanisms of cell injury and death in both Huntington's disease and in strokes have shown that one major factor is excitation of the NMDA class of glutamate receptor. Elucidation of this and other biochemical mechanisms of injury to cells of the brain should lead to improved therapy for neuropsychiatric conditions in which cell damage is a feature.

Human mtDNA has been sequenced and is known to code for mt ribosomal RNAs, for mt transfer RNAs and also for 13 polypeptides of the respiratory chain. Deletions, duplications, and point mutations in mtDNA have been detected and shown to cause a variety of myopathies, neurologic disorders, and some cases of diabetes mellitus. Mitochondrial diseases are transmitted by maternal inheritance.

Novel mutations consisting of repeats involving GC-rich triplets are the cause of the fragile X syndrome, of spinal bulbar muscular atrophy, and of myotonic dystrophy. Use of techniques that detect such mutations will facilitate the diagnosis of these and other possibly related conditions.

Parkinson's disease is caused by degeneration of the cells of the substantia nigra, resulting in a deficiency of dopamine in the nigrostriatal system. Levodopa is effective therapy, at least in the early stages of the disease.

Amyloid plaques and neurofibrillary tangles are key features of Alzheimer's disease. The amyloid cascade hypothesis postulates that deposition of amyloid β protein, derived from amyloid precursor protein, is an initial event in the causation of the disease and that amyloid β protein, or fragments of it, are neurotoxic, leading to formation of the neurofibrillary tangles and other features. However, the possibility that deposition of amyloid β protein is secondary in most cases to other events has not been excluded. Several genes, including *APOE4,* have been implicated in the causation of Alzheimer's disease.

The introduction of neuroleptic drugs to treat psychoses led to evidence that implicated dopamine in the causation of schizophrenia. Much information sustains this view, but the dopamine hypothesis of schizophrenia remains unproved. Cloning of dopamine receptors has led to the recognition of five classes, with the D4 class showing a number of polymorphic variants. Intensive research is in progress in the attempt to find genetic causes of schizophrenia.

REFERENCES

Bajjalieh SM, Scheller RH: The biochemistry of neurotransmitter secretion. J Biol Chem 1995;270:1971.

Barnes DM: Troubles encountered in gene linkage land. Science 1989;243:313. [Discusses why published data for gene linkage studies of four different mental disorders have been difficult to replicate.]

Cooper JR, Bloom FE, Roth RH: *The Biochemical Basis of Neuro-pharmacology,* 6th ed. Oxford Univ Press, 1991.

Culver KW et al: In vivo gene transfer with retroviral-producer cells for treatment of experimental brain tumors. Science 1992;256:1550.

Drachman DB: Myasthenia gravis. N Engl J Med 1994; 330:1797.

Fisher M (editor): *Stroke Therapy.* Butterworth-Heinemann, 1995.

Ganong WF: *Review of Medical Physiology,* 17th ed. Appleton & Lange, 1995.

Horn JP: The heroic age of neurophysiology. Hosp Pract (July) 1992;27:65. [First of a series of six articles on the physiologic basis of neuromuscular transmission disorders.]

Housman D: Gain of glutamines, gain of function? Nature Genetics 1995;10:3.

The Huntington's Disease Collaborative Research Group: A novel gene containing a trinucleotide repeat that is expanded and unstable on Huntington's disease chromosomes. Cell 1993;72:971.

Hyman BT, Tanzi R: Molecular epidemiology of Alzheimer's disease. N Engl J Med 1995;333:1283.

Kandel ER, Schwartz JH, Jessell TM: *Principles of Neural Science,* 3rd ed. Elsevier, 1991.

Karayiorgou M et al: Schizophrenia susceptibility associated with interstitial deletions of chromosome 22q11. Proc Natl Acad Sci USA 1995;92:7612.

Levitan IB, Kaczmarek LK: *The Neuron: Cell and Molecular Biology.* Oxford Univ Press, 1991.

Lipton SA, Rosenberg PA: Excitatory amino acids as a final common pathway for neurologic disorders. N Engl J Med 1994;330:614.

Pearlman AL, Collins RC: *Neurobiology of Disease.* Oxford Univ Press, 1990.

Roses AD: Apolipoprotein E and Alzheimer disease. Sci Am Science and Medicine (Sept/Oct) 1995;16.

Schapira AHV: Nuclear and mitochondrial genetics in Parkinson's disease. J Med Genet 1995;32:411.

Sedvall G, Farde L: Chemical brain anatomy in schizophrenia. Lancet 1995;346:743.

Seeman P: Dopamine receptors and psychosis. Sci Am Science and Medicine (Sept/Oct) 1995;28.

Sherrington R et al: Cloning of a gene bearing missense mutations in early-onset Alzheimer's disease. Nature 1995;375:754.

Sisodia SS, Price DL: Role of the β-amyloid protein in Alzheimer's disease. FASEB J 1995;9:366.

Sudhof TC: The synaptic vesicle cycle: A cascade of protein-protein interactions. Nature 1995;375:645.

Sweeney PJ: New concepts in Parkinson's disease. Hosp Pract (April) 1991;26:84.

Wallace DC: Diseases of the mitochondrial genome. Annu Rev Biochem 1992;61:1175.

Willems PJ: Dynamic mutations hit double figures. Nature Genetics 1994;8:213.

Wright AF: New insights into genetic eye disease. Trends Genet 1992;8:85.

65

Biochemical Case Histories

Robert K. Murray, MD, PhD

INTRODUCTION

In this final chapter, nine case histories are presented and discussed. They illustrate the importance of a knowledge of biochemistry for the understanding of disease. The cases selected cover the causes of disease listed in Table 1–1, including two examples of diseases classified as genetic (Duchenne muscular dystrophy and cystic fibrosis). The classification used in Table 1–1 is artificial, since some diseases belong to more than one class; eg, xeroderma pigmentosum (here classified as physical) is an inherited disease. Also, genetic factors contribute to some diseases presented here that have not been classified as genetic (eg, myocardial infarction and diabetes mellitus type 1). The list of causes in Table 1–1 could have been extended, for instance, to include "neoplastic" and "psychologic." On the other hand, the causes of neoplasms can be considered under "physical" (cancer due to irradiation), "biologic" (in view of the role of oncogenic viruses in causing certain types of cancer), and "chemical" (approximately 80% of human tumors appear to have chemical causes). With regard to psychologic diseases, there has been great interest of late in the likelihood that at least certain cases of schizophrenia (Chapter 64) have a genetic basis. Most of the diseases described here are common, or relatively common, in a global sense. However, two (xeroderma pigmentosum and severe combined immunodeficiency disease due to deficiency of ADA) are relatively rare. They are included because they nicely illustrate the class of disease under which they are listed and also because they illustrate two biologic principles: the importance of DNA repair and of the immune system as protective mechanisms. In addition, ADA deficiency is the first disease for which gene therapy has been performed in humans.

It should be noted that the reference values cited in the cases below are not necessarily the same as those listed in the Appendix. This is because reference values from different laboratories can vary somewhat, in part due to different methodologies.

CASE NO. 1: XERODERMA PIGMENTOSUM

Classification
Physical; exposure to ultraviolet (UV) irradiation.

History and Physical Examination
A 12-year-old boy presented at a dermatology clinic with a skin tumor on his right cheek. He had always avoided exposure to sunlight because it made his skin blister. His skin had scattered areas of hyperpigmentation and other areas where it looked mildly atrophied. Because of the presence of a skin tumor at such a young age, the history of avoidance of sunlight, and the other milder skin lesions, the dermatologist made a tentative diagnosis of xeroderma pigmentosum (pigmented dry skin).

Laboratory Findings
Histologic examination of the excised tumor showed that it was a squamous carcinoma (a common type of skin cancer in older people but not in a boy of this age). A sample of skin was taken for the preparation of fibroblasts. A research laboratory in the hospital specialized in radiobiology and was set up to measure the amount of **thymine dimers** formed following exposure to UV light. The patient's fibroblasts and control fibroblasts were exposed to UV light, and cell samples were taken at 8-hourly intervals for a total of 32 hours postirradiation. Extracts of DNA were prepared and the numbers of dimers remaining at each time point indicated were determined. Whereas only 24% of the dimers formed persisted in DNA extracted from the normal cells at 32 hours, approximately 95% were found in the extract from the patient's cells at 32 hours. This confirmed the diagnosis of xeroderma pigmentosum.

Discussion
Xeroderma pigmentosum is a rare autosomal recessive condition in which the mechanisms for repair of DNA subsequent to damage by UV irradiation are defective. This arises because of mutations in the genes encoding components of the nucleotide excision pathway of repair. The major damage inflicted on DNA by UV irradiation is the formation of thymine dimers, where covalent bonds are formed between carbons 5

and 5 and 6 and 6 of adjacent intrachain thymine residues. Other types of damage can also occur. Detailed analyses of the processes involved in removal of pyrimidine dimers have been performed. In *E coli*, an endonucleolytic cleavage, catalyzed by a specific UV endonuclease, occurs on either side of the damage, releasing a 12- to 13-base oligonucleotide. The polymerization step involves DNA polymerase (α or δ, with β perhaps being involved also), and the final step is sealing by DNA ligase. This pathway is called nucleotide excision repair (NER). The NER pathway operates in humans, and its details are being elucidated. It appears to be generally similar to the pathway in *E coli*. The most notable difference is that a much larger oligonucleotide (about 30 bases) is excised in humans. A simplified scheme of the initial steps in the human pathway is shown in Figure 65–1.

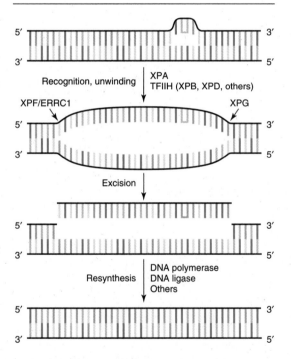

Figure 65–1. A simplified scheme of the nucleotide excision repair pathway involved in the repair of UV damage in skin. The site of damage is shown by the bulge (site of intrachain thymine dimers) in the upper strand of DNA. XPA recognizes and binds to the site of damage. XPB and XPD are helicases that cause unwinding of the double helix prior to excision. XPF and ERRC1 cut on the 5' side of the lesion, and XPG cuts on the 3' side. Thus, a stretch of about 30 bases is removed and is filled in by DNA polymerase and joined by DNA ligase. Mutations in genes encoding the various proteins involved in the process are the cause of xeroderma pigmentosum. If the proteins are nonfunctional, repair will be defective, resulting in mutations that can lead to cancer of the skin. (Reproduced, with permission, from Marx J: DNA repair comes into its own. Science 1994;266:728.)

At least seven genes (those encoding XPA through XPG; XPC is not shown in Figure 65–1, as its function is still under examination) have been implicated in NER is humans. Mutations in any one of these genes can cause xeroderma pigmentosum.

The involvement of these genes was shown in the following manner. It was observed that when cultured cells from individuals with xeroderma pigmentosum were cocultured with cells from other individuals with the condition, the defect in DNA repair could sometimes be corrected. This indicated that one set of cells was providing a normal gene product to the other, thus correcting the defect. In this manner, at least seven complementation groups, corresponding to the seven genes and their protein products mentioned above, have been recognized.

If UV damage is not repaired, mutations in DNA will result and may cause cancer. Patients with xeroderma pigmentosum often suffer from a variety of skin cancers from an early age. Figure 65–2 summarizes the mechanisms involved in the disease. It should be recalled that mutations in genes involved in another pathway of DNA repair, mismatch repair, have been implicated in the causation of another type of cancer, hereditary nonpolyposis colon cancer (Chapter 62). The parents of this boy were told that he would have to be watched throughout life for the development of new skin cancers. In addition, he was advised to continue to avoid sunlight and to use an appropriate sunscreen ointment. Although xeroderma pigmentosum is a rare condition, the existence of a variety of mechanisms for repairing DNA following exposure to different types of irradiation and to chemical damage is of great protective importance. Without their existence, life on this planet would be fraught with even more danger than it already is! For instance, it has been estimated that patients with xeroderma pigmentosum have a 1000-fold greater chance of developing skin cancer than do normal individuals.

Exposure to UV light causes formation of thymine dimers in skin cells (keratinocytes)
↓
Thymine dimers are not excised or not correctly repaired because of mutations in the genes encoding components of the nucleotide excision repair pathway
↓
Damage to DNA persists; replication leads to synthesis of DNA containing mutations
↓
Mutant DNA mediates carcinogenesis (perhaps by activation of oncogenes)
↓
Multiple skin tumors may result

Figure 65–2. Summary of mechanisms involved in the causation of xeroderma pigmentosum.

CASE NO. 2: KWASHIORKOR

Classification

Nutritional; deficiency of protein.

History and Physical Examination

A two-year old girl in a developing country was brought by her mother to the outpatient department of the local hospital. The mother had four children, the youngest of whom was 3 months of age and still being breast-fed. The father had broken his leg in an accident during the previous year and had not been able to work since. Family income was thus low, and they were not able to buy milk and meat on a regular basis. Their main subsistence food was a starchy gruel, high in carbohydrate and low in protein. The mother stated that the daughter had been eating poorly for the past month, had intermittent diarrhea for some time, had recently developed a cough and fever, and had become very irritable and apathetic.

On examination, the girl was found to be both underweight for her height and small for her age. She was pale, very weak, and drowsy. Her temperature was 40.5 °C. Her midarm circumference was well below normal, consistent with protein-energy malnutrition. Her skin was flaky and her hair dry and brittle. The abdomen was distended and the liver moderately enlarged. Generalized edema was evident. Rales were heard over the lower lobe of the left lung. The doctor on duty made the diagnoses of kwashiorkor, diarrhea, pneumonia, and possible bacteremia.

Laboratory Findings

Samples of blood were taken for analysis. Results were subsequently reported as hemoglobin 6.0 g/dL (normal for a 2-year-old: 11–14 g/dL), total serum protein 4.4 g/dL (normal: 6–8 g/dL), and albumin 2.0 g/dL (normal: 3.5–5.5 g/dL). Stool and blood samples were taken for culture; a gram-negative anaerobe was later reported in both. The white blood cell count was 18,000/μL. Chest x-rays revealed mottled opacities in the lower lobe of the left lung, consistent with acute bronchopneumonia.

Treatment

In many cases, it is undesirable to treat patients with mild to moderate protein-energy malnutrition in hospital, because this only increases the chance of infection. However, in view of the fever, weakness, drowsiness, and severe edema, this child was admitted. She was immediately started on a broad-spectrum antibiotic and intravenous dextrose-saline infusion. Tragically, her condition worsened and she died approximately 12 hours after admission. Autopsy revealed severe fatty liver and bronchopneumonia.

Discussion

Protein-energy malnutrition is the commonest nutritional disorder in many parts of the world. As many as 1 billion people suffer from various degrees of severity of protein-energy malnutrition. Kwashiorkor is one end of the spectrum in which the essential feature is protein deficiency, with relatively adequate energy intake. Marasmus is at the other end and is due to severe and prolonged restriction not only of protein but of all food. Massive loss of muscle and fat occurs in marasmus. Anorexia occurs in kwashiorkor, leading to further restriction of food intake. Intermediate forms are called marasmic kwashiorkor. The major differences between kwashiorkor and marasmus are set forth in Table 65–1. The hallmarks of kwashiorkor are hypoalbuminemia, edema, and fatty liver. The hypoalbuminemia reflects the inadequate supply of amino acids derived from protein, thus impairing the synthesis of albumin and other proteins (eg, transferrin) by the liver. The edema, at least in part, is due to low oncotic pressure in the plasma due to hypoalbuminemia (Chapter 59). Synthesis of plasma proteins by the liver is also decreased; this in turn impairs the export of triglycerides and other lipids from the liver, resulting in a fatty liver.

Kwashiorkor is the word used by members of the Ga tribe in Ghana to describe "the sickness the older gets when the next child is born." It follows weaning from breast milk and exposure to a diet low in protein and high in carbohydrate. The edema, poor skin and hair, and fatty liver seen in kwashiorkor are mainly due to protein deficiency. Deficiencies of vitamins and minerals are often found in patients with kwashiorkor. Hormones are important in the generation of protein-energy malnutrition. The exposure to a high intake of carbohydrate keeps levels of insulin high and levels of epinephrine and cortisol low in kwashiorkor, as opposed to marasmus. The combination of low insulin and high cortisol greatly favors catabolism of muscle; thus, muscle wasting is greater in marasmus than in kwashiorkor. In addition, because of the lower levels of epinephrine, fat is not mobilized to the same extent in kwashiorkor. The low dietary intake of protein in kwashiorkor results in decreased synthesis of plasma proteins, especially albumin and transferrin, and also decreased synthesis of hemoglobin. Impaired protein synthesis in the liver along with sufficient dietary carbohydrate to en-

Table 65–1. Differences between kwashiorkor and marasmus.

	Kwashiorkor	Marasmus
Edema	Present	Absent
Hypoalbuminemia	Present, may be severe	Mild
Fatty liver	Present	Absent
Levels of insulin	Maintained	Low
Levels of cortisol	Normal	High
Muscle wasting	Absent or mild	May be very severe
Body fat	Diminished	Absent

sure lipid synthesis lead to the accumulation of triacylglycerols in the liver (fatty liver). The immune system is impaired in protein-energy malnutrition, particularly T-cell function. Individuals are thus very susceptible to infections (eg, causing diarrhea), and infections worsen the situation by placing a higher metabolic demand on the body, eg, through fever. Figure 65–3 summarizes some of the mechanisms involved in the causation of kwashiorkor.

Kwashiorkor is entirely preventable if children are given a well-balanced diet containing adequate amounts of protein and of the essential amino acids.

CASE NO. 3: CHOLERA

Classification
Biologic, bacterial.

History and Physical Examination
A 21-year-old female student working in a developing country suddenly began to pass profuse watery stools almost continuously. She soon started to vomit, her general condition declined abruptly, and she was rushed to the local village hospital. On admission she was cyanotic, skin turgor was poor, blood pressure was 70/50 mm Hg (normal: 120/80 mm Hg), and her pulse was rapid and weak. The doctor on duty diagnosed cholera, took a stool sample for culture, and started treatment immediately.

Treatment
Treatment consisted of intravenous administration of a solution made up in the hospital containing 5 g NaCl, 4 g NaHCO$_3$, and 1 g KCl per liter of pyrogen-free distilled water. This solution was initially given rapidly (100 mL/kg) until the blood pressure was normal and pulse strong. She was also started on tetracycline. On the second day, she was able to take the oral rehydration solution recommended by the

World Health Organization (WHO) for the treatment of cholera, consisting of 20 g glucose, 3.5 g NaCl, 2.5 g NaHCO$_3$, and 1.5 g KCl per liter of drinking water. Solid food was reinstituted on the fourth day after admission. She continued to recover rapidly and was discharged 7 days after admission.

Discussion
Cholera is an important infectious disease endemic in certain Asian countries and other parts of the world. It is due to *Vibrio cholerae,* a bacterium that secretes a protein **enterotoxin.** The enterotoxin is made up of one A subunit (composed of one A$_1$ and one A$_2$ peptide joined by a disulfide link) and five B subunits and has a molecular mass of approximately 84 kDa. In the small intestine, the toxin attaches by means of the B subunits to the ganglioside G$_{M1}$ (Figure 16–19) present in the plasma membrane of mucosal cells; the A subunit then dissociates, and the A$_1$ peptide passes across to the inner aspect of the plasma membrane. It catalyzes the ADP-ribosylation (using NAD as donor) of the G$_s$ regulatory protein, which inhibits the GTPase activity and fixes it in its active form. Thus, **adenylyl cyclase** becomes chronically activated (Chapter 44). This results in an elevation of cAMP, which is thought to activate a protein kinase that phosphorylates one or more membrane proteins involved in active transport. The consequence of this chain of events is that absorption of NaCl into the intestinal cells by a neutral NaCl transport cosystem is inhibited and active secretion of Cl$^-$ is stimulated. These events lead to massive secretion of Na$^+$ and water into the lumen of the small intestine, producing the liquid stools characteristic of cholera. The histologic structure of the small intestine remains remarkably unaffected, despite the outpouring of not only Na$^+$ and water but also Cl$^-$, HCO$_3$, and K$^+$. It is the loss of these constituents that results in the marked fluid loss, low blood volume, acidosis, and K$^+$ depletion found in serious cases of

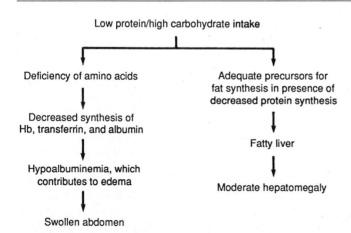

Figure 65–3. Summary of some of the mechanisms involved in the causation of kwashiorkor.

cholera that can prove fatal unless appropriate replacement therapy (as described above) is begun immediately. Figure 65–4 summarizes the mechanisms involved in the causation of the diarrhea of cholera. The recognition and easy availability of appropriate replacement fluids, such as oral rehydration solution, has led to tremendous improvement in the treatment of cholera. It should be emphasized that glucose is an essential component of oral rehydration solution; whereas the cholera toxin inhibits absorption of NaCl by intestinal cells, it does not inhibit glucose-facilitated Na^+ transport into these cells.

CASE NO. 4: MYOCARDIAL INFARCTION

Classification
Oxygen lack.

History and Physical Examination
A 46-year-old businessman was admitted to the emergency department of his local hospital complaining of severe retrosternal pain of 2 hours' duration. He had previously been admitted to hospital once for treatment of a small myocardial infarction, but despite this he continued to smoke heavily. His blood pressure was 150/90 mm Hg (his normal was 140/80 mm Hg), pulse was 60/min, and he was sweating profusely. There was no evidence of cardiac failure. Because of the admitting diagnosis of

---

Ingestion of *V cholerae*
↓
Release of enterotoxin (containing A_1, A_2, and B subunits) in small intestine
↓
Binding of enterotoxin to GM_1 ganglioside
↓
Release of A_1 subunit, which crosses the plasma membrane of intestinal mucosal cells
↓
A_1 subunit catalyzes ADP-ribosylation of the G_S protein of the plasma membrane with loss of its GTPase activity
↓
Chronic activation of adenylyl cyclase
↓
Level of cAMP is elevated in intestinal mucosal cells
↓
Activation of one or more cAMP-dependent protein kinases, resulting in alteration of transport systems of these cells by phosphorylation of certain transport proteins
↓
Massive losses of Na^+, Cl^-, HCO_3^-, K^+, and H_2O

Figure 65–4. Summary of mechanisms involved in the causation of the diarrhea of cholera.

---

myocardial infarction, he was given morphine to relieve his pain and apprehension and immediately transferred to a cardiac care unit, where continuous electrocardiographic monitoring was started at once.

Laboratory Findings
The initial ECG showed S–T segment elevation and other changes in certain leads, indicative of an acute anterior transmural left ventricular infarction. Blood was taken at 4 hours and subsequently at regular intervals for measurement of the MB isozyme of CK; at 4 hours, the activity of this isozyme was slightly elevated, and at 12 hours the elevation was fourfold. The plasma cholesterol was moderately elevated (6.5 mmol/L), and triacylglycerol levels were normal.

Treatment
The attending cardiologist, after reviewing all aspects of the case, decided to administer streptokinase by cardiac catheterization because of the diagnosis of anterior transmural myocardial infarction seen within 4 hours after the onset of symptoms. Chest pain began to disappear after 12 hours, and the patient felt increasingly comfortable. He was discharged from hospital 10 days later under the care of his family doctor with advice to start a cholesterol-lowering regimen (including reduction of intake of saturated fat and administration of a drug inhibiting HMG-CoA reductase) and to stop smoking.

Discussion
A transmural myocardial infarction is generally caused by an occlusive or near-occlusive thrombus lying in close proximity to an atherosclerotic plaque. Generally the diagnosis can be made from the clinical history, the electrocardiographic results, and serial measurement of the CK-MB isozyme. The overall aims of treatment are to prevent death from cardiac arrhythmias by administration of appropriate drugs and to limit the size of the infarction. In this case, the decision was made to limit the size of the infarct by intracoronary administration of streptokinase, which can dissolve the thrombus (Chapter 59). For long-term therapy, measures to reduce the plasma cholesterol were started by prescribing a drug that inhibits HMG-CoA reductase.

The causes of the atherosclerotic lesion in the coronary artery that led to the thrombus can be discussed only very briefly here; a textbook of pathology should be consulted for the details. Atherosclerosis is predisposed to by high levels of LDL, low levels of HDL, unknown polygenic factors, and a variety of risk factors such as hypertension, elevated cholesterol, and smoking; this patient had an elevated cholesterol and was a heavy smoker. The presence of oxidized LDL in atherosclerotic lesions appears to be particularly important in that it encourages recruitment of macrophages and stimulates the release of various growth factors. The intima of arteries is initially affected, and macrophages, plasma lipoproteins, glycosaminoglycans, collagen, and calcium accumulate in a lesion called a fatty streak. Platelets and fibrin can deposit on the luminal aspect of the blood vessel, and monoclonally derived smooth muscle cells present in the medial layer of the artery grow into the intimal lesion, attracted by growth factors released by macrophages and platelets (eg, platelet-derived growth factor). The overall lesion is now an intimal plaque. Inflammation and hemorrhage into the plaque can occur, leading to rupture of its surface and exposure of its underlying constituents to the blood. Platelets will adhere to the exposed collagen, and a thrombus is initiated (Chapter 59).

If the thrombus occludes 90% of the vessel wall, blood flow through the affected vessel may cease (total ischemia), and capillary hemoglobin will be very rapidly depleted of oxygen. The normal metabolism of the myocardium is aerobic, with most of its ATP being derived from oxidative phosphorylation. The anoxia secondary to total ischemia results in a switch of myocardial metabolism to anaerobic glycolysis, which generates only about one-tenth of the ATP produced by oxidative phosphorylation. Not only does this switch in metabolism occur, but the flow of substrates into the myocardium via the blood and the removal of metabolic products from it are also greatly reduced. This accumulation of intracellular metabolites increases the intracellular osmotic pressure, resulting in cell swelling, affecting the permeability of the plasma membrane. Thus, the affected myocardium exhibits depletion of ATP, accumulation of lactic acid, development of severe acidosis, and marked reduction in contractile force. Synthesis of macromolecules and of nucleotides ceases under these metabolic conditions. The accumulation of lactate and hydrogen ions inhibits glycolysis at the level of glyceraldehyde phosphate dehydrogenase. Oxidized NAD becomes deficient because it is not regenerated by the terminal electron transport chain, which does not function in the absence of oxygen. As the level of ATP drops owing to depression of oxidative phosphorylation, the level of ADP rises initially. However, ADP is converted to AMP by muscle adenylyl kinase, and the AMP is further degraded by ADA to adenosine. Adenosine itself is further converted to inosine and other products of purine catabolism. All of this markedly depletes the adenine nucleotide pool, which is a key component of normal cellular metabolism. It is known that the level of ATP in the canine myocardium drops to approximately 10% of control values after about 40 minutes of severe ischemia. The exhaustion of the adenine nucleotide pool coincides with the development of irreversible cellular damage but does not necessarily cause it. At present, it is not possible to state the precise metabolic changes that commit a cell irreversibly to dying. Various studies have implicated depletion of ATP, activation of intracellular phospholipases (which will produce damage to cell membranes), activation of proteases, and accumulation of intracellular Ca^{2+}. Some of the mechanisms involved in the causation of an acute myocardial infarction are summarized in Figure 65–5.

Generally speaking, the earlier attempts are made to reestablish perfusion through an area of ischemic myocardium, the better. By 6 hours, irreversible damage has probably been done, and some cells are probably irreversibly damaged by as little as 1 hour of complete ischemia. Thus, administration of streptokinase or t-PA must be prompt, and certainly within 12 hours. Recombinant human tissue plasminogen activator (t-PA) is at least as effective as streptokinase, if not more so; it is not antigenic and seldom causes profound hypotension, as streptokinase can. However, it is approximately ten times more expensive. Accordingly, many centers use streptokinase unless it is contraindicated by allergy, previous exposure, or profound hypotension.

The biochemical events that occur if reperfusion of an ischemic area of myocardium is established (eg, following administration of streptokinase) are also of great biochemical and clinical interest. Reperfusion can itself lead to cell death, a condition known

---

Development of atherosclerosis (usually polygenic)
↓
Formation of large thrombus in a coronary artery
↓
Deprivation of blood supply (ischemia) to myocardium
↓
Shift to anaerobic glycolysis → decreased synthesis of ATP, depletion of adenine nucleotide pool
↓
Increase of NADH due to inactive terminal electron transport chain; due to lack of oxygen
↓
Accumulation of lactic acid and other metabolites, causing increased intracellular osmolarity and altered membrane permeability
↓
Decrease of pH in heart muscle cells
↓
Increasingly inefficient contraction of heart muscle
↓
Cessation of contraction
↓
Activation of membrane phospholipases, degradation of proteins by proteases, influx of Ca^{2+}
↓
Death of affected area of heart muscle

Figure 65–5. Summary of mechanisms involved in the causation of an acute myocardial infarction. The arrows do not in all cases imply a strict causal connection.

as reperfusion injury. One mechanism whereby this can occur is that damage to the plasma membrane (eg, to various of its ion pumps) may have occurred during the period of ischemia, seriously altering its permeability properties. This will affect the membrane potential and will also permit a flood of compounds such as Ca^{2+} to enter from the plasma. High levels of intracellular Ca^{2+} can wreak havoc inside the cell by activating or inhibiting various enzymes in an unregulated manner. There is also considerable interest in the possibility that free radicals, eg, highly reactive partially reduced metabolites of oxygen such as superoxide ($O_2^{\bar{}}$) and hydroxyl radicals ($OH^{\bullet}$), may play a role in reperfusion injury. $OH^{\bullet}$ radicals are particularly reactive. They can be generated by myocardial cells or by circulating blood cells, such as polymorphonuclear leukocytes. They damage cells by causing lipid peroxidation, breakage of DNA strands, and oxidation of SH groups in proteins. Superoxide anion is generated by the transfer of a single electron to O_2,

$$O_2 + e^- \rightarrow O_2^{\bar{}}$$

in reactions catalyzed by cytochrome P450, xanthine oxidase, and the respiratory burst oxidase (NADPH oxidase) present in polymorphonuclear leukocytes. The enzyme superoxide dismutase catalyzes the following reaction:

$$O_2^{\bar{}} + O_2^{\bar{}} + 2H^+ \rightarrow H_2O_2 + O_2$$

It thus scavenges superoxide anions, converting them to the less reactive hydrogen peroxide. Experiments have been performed to determine whether administration of superoxide dismutase during reperfusion can protect the myocardium from reperfusion injury, but the results have been equivocal. The role, if any, of superoxide anions in reperfusion injury thus remains to be settled; however, at present there is considerable interest in the role of free radicals in many types of cell injury and disease.

CASE NO. 5: ACUTE INTOXICATION WITH ETHANOL

Classification
Chemical.

History and Physical Examination
A 52-year-old man was admitted to the emergency department in coma. Apparently, he had become increasingly depressed after the death of his wife 1 month previously. Before her death he had been a moderate drinker, but his consumption of alcohol had increased markedly over the last few weeks. He had also been eating poorly. His married daughter had dropped around to see him on Sunday morning and had found him unconscious on the living room couch. Two empty bottles of rye whisky were found on the living room table. On examination, he could not be roused, his breathing was deep and noisy, alcohol could be smelled on his breath, and his temperature was 35.5 °C (normal: 36.3–37.1 °C). The diagnosis on admission was coma due to excessive intake of alcohol.

Laboratory Findings
The pertinent laboratory results on blood were alcohol 500 mg/dL, glucose 2.7 mmol/L (normal: 3.6–6.1), lactate 8.0 mmol/L (normal: 0.5-2.2), and pH of 7.21 (normal: 7.35–7.45). These results were consistent with the admitting diagnosis, accompanied by a metabolic acidosis.

Treatment
Because of the very high level of blood alcohol and the coma, it was decided to start hemodialysis immediately. This directly eliminates the toxic ethanol from the body but is only required in very serious cases of ethanol toxicity. In this case, the level of blood alcohol fell rapidly and the patient regained consciousness later the same day. Intravenous glucose (5%) was administered after dialysis was stopped to counteract the hypoglycemia that this patient exhibited. The patient made a good recovery and was referred for psychiatric counseling.

Discussion
Excessive consumption of alcohol is a major health problem in most societies. The present case deals with the acute, toxic effects of a very large intake of ethanol. A related problem, which is not discussed here but which has many biochemical aspects, is the development of liver cirrhosis in individuals who maintain a high intake of ethanol (eg, 80 g of absolute ethanol daily, or about 1 quart of 80 proof whisky) for more than 10 years.

From a biochemical viewpoint, the major question concerning the present case is how does ethanol produce its diverse acute effects, including coma, lactacidosis, and hypoglycemia? The clinical viewpoint is how best to treat this condition.

The metabolism of ethanol was described in Chapter 27; it occurs mainly in the liver and involves two routes. The first and major route uses alcohol dehydrogenase and acetaldehyde dehydrogenase, converting ethanol via acetaldehyde to acetate, which is then converted to acetyl-CoA. Reduced $NADH^+ + H^+$ is produced in both of these reactions. The intracellular NADH/NAD ratio can thus be increased appreciably by ingestion of large amounts of ethanol. In turn, this can affect the K_{eq} of a number of important metabolic reactions that use these two cofactors. High levels of NADH favor formation of lactate from pyruvate, accounting for the lactic acidosis. This diminishes the concentration of pyruvate (required for the pyruvate carboxylase reaction) and thus inhibits gluconeogenesis. In severe cases, when liver glycogen is depleted and no longer available for glycogenolysis, hypoglycemia results. The second route involves a microsomal cytochrome P450 (microsomal ethanol oxidizing system), also producing acetaldehyde. Acetaldehyde is a highly reactive molecule and can form adducts with proteins, nucleic acids, and other molecules. It appears likely that its ability to react with various molecules is involved in the causation of the toxic effects of ethanol. Ethanol also appears to be able to interpolate into biologic membranes, expanding them and increasing their fluidity. When the membranes affected are excitable, this results in alterations of their action potentials, impairs active transport across them, and also affects neurotransmitter release. All of these depress cerebral function and, if severe enough, can produce coma and death from respiratory paralysis. Figure 65–6 summarizes the major mechanisms involved in the causation of toxicity by ethanol.

CASE NO. 6:
DUCHENNE MUSCULAR DYSTROPHY

Classification
Genetic.

History and Physical Examination
A 4-year-old boy was brought to a children's hospital clinic. His mother was concerned because she

had noticed that her son was walking awkwardly, fell over frequently, and had difficulty climbing stairs. There were no siblings, but the mother had a brother who died at age 19 of muscular dystrophy. The examining physician noted muscle weakness in both the pelvic and shoulder girdle. Modest enlargement of the calf muscles was noted. Because of the muscle weakness and its distribution, the pediatrician made a tentative diagnosis of Duchenne muscular dystrophy.

Laboratory and Other Findings
The activity in serum of creatine kinase was markedly elevated. Electromyographic findings were characteristic of muscular dystrophy, whereas nerve conduction studies were normal. A biopsy of calf muscle showed areas of muscle necrosis and some variation in the size of muscle fibers. Using a cDNA probe available for dystrophin, a Southern blot test revealed an absence of the fragment corresponding to the gene for this protein. Analysis of a small portion of the muscle biopsy by a Western blot also showed an absence of the protein dystrophin.

Discussion
The family history, the typical distribution of muscle weakness, the elevation of the activity in serum of CK, the electromyographic findings, the muscle biopsy, and the laboratory findings revealing abnormalities in the gene for dystrophin confirmed the pediatrician's provisional diagnosis of Duchenne muscular dystrophy. This is a severe X-linked degenerative disease of muscle. It has an incidence of ap-

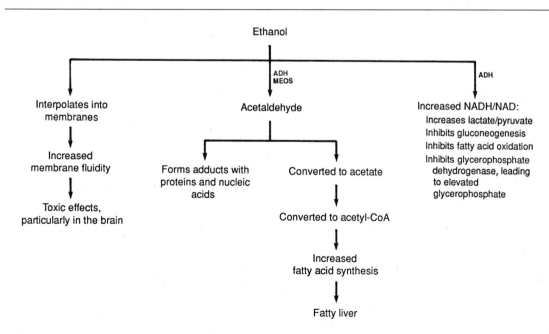

Figure 65–6. Summary of mechanisms involved in the causation of toxicity by ethanol. (ADH, alcohol dehydrogenase; MEOS, microsomal ethanol oxidizing system—a species of cytochrome P450.)

proximately 1:3500 live male births. It affects young boys, who first show loss of strength in their proximal muscles, leading to a waddling gait, difficulty in standing up, and eventually very severe weakness.

Various studies led to localization of the defect to the middle of the short arm of the X chromosome and to subsequent identification of a segment of DNA that was deleted in patients with Duchenne muscular dystrophy. Using the corresponding nondeleted segment from normal individuals, a cDNA was isolated derived by reverse transcription from a transcript (mRNA) of 14 kb that was expressed in fetal and adult skeletal muscle. This was cloned and the protein product was identified as dystrophin, a 400-kDa rod-shaped protein of approximately 3685 amino acids. Dystrophin was absent or markedly reduced in electrophoretic gels of muscle from patients with Duchenne muscular dystrophy and from mice with an X-linked muscular dystrophy (mdx). Antibodies against dystrophin were used to study its localization in muscle; it appears to be located in the sarcolemma (plasma membrane) of normal muscle and was absent or markedly deficient in the sarcolemma of patients with Duchenne muscular dystrophy. A less severe reduction in the amount of dystrophin, or a reduction in its size, is the cause of Becker's muscular dystrophy, a milder type of muscular dystrophy. Figure 65–7 summarizes the mechanisms involved in the causation of Duchenne muscular dystrophy.

Dystrophin appears to have four domains, two of which are similar to domains present in a α-actinin and one to a domain in spectrin, a protein of the cytoskeleton of the red blood cell. The function of dystrophin may be to link the cytoskeleton of the muscle cell, through a transmembrane complex of proteins and glycoproteins, to laminin. Thus, deficiency of dystrophin may disrupt this linkage, leading to increased osmotic fragility of dystrophic muscle or permitting excessive influx of Ca^{2+}. Since the extracellular matrix serves as a matrix for repair, such a function would help explain why dystrophic

muscle is more prone to injury and less able to repair itself. The gene coding for dystrophin is one of the largest human genes recognized to date (2300 kb, 79 exons), which helps explain the observation that approximately one-third of cases of Duchenne muscular dystrophy are new mutations. Attempts are being made to produce dystrophin by recombinant DNA technology and perhaps eventually administer it to patients. Assay of dystrophin cannot be used in prenatal diagnosis because the protein is expressed only by muscle cells and not by amniotic cells or samples of chorionic villus. However, the availability of cDNA probes for dystrophin facilitates prenatal diagnosis of Duchenne muscular dystrophy by chorionic villus sampling or amniocentesis. The classification of the various types of muscular dystrophy will also be facilitated by determining in which types dystrophin is affected and in which types it does not appear to play a role. Certain types of muscular dystrophy have been shown to be due to mutations in the genes encoding various members of the transmembrane complex referred to above. The demonstration of lack of dystrophin as a cause of Duchenne muscular dystrophy has been one of the major accomplishments of the application of the new molecular biology to human disease.

Treatment

At present, no specific therapy for Duchenne muscular dystrophy exists. Treatment in this case was thus essentially symptomatic. The boy was encouraged to exercise, and regular attendance at a specialized muscular dystrophy clinic was started, so that complications could be treated as they arose. The mother was advised to seek genetic counseling.

Myoblast transfer is being investigated in several centers as a treatment for Duchenne muscular dystrophy. The rationale is that donor myoblasts injected into muscles of affected individuals can fuse with host muscle fibers and donate their nuclei, leading to transcription and synthesis of normal dystrophin. Results of a clinical trial of myoblast transfer in 12 boys were reported in 1995. Little or no persistence of the injected myoblasts in 11 of the 12 boys was noted, and no improvement in the strength of the muscle that received the injection was noted over the 6-month period of the trial. It was concluded that human muscle is a less suitable site for myoblast transfer than is murine muscle (in which preclinical trials were performed). Furthermore, immune responses to the injected myoblasts (obtained from healthy fathers and brothers of the boys) may have interfered with stable engraftment of the myoblasts. Much more research is required if this method is to prove useful. Clinical trials on other types of gene therapy for Duchenne muscular dystrophy are imminent.

Deletion of part of the structural gene for dystrophin, located on the X chromosome

↓

Diminished synthesis of the mRNA for dystrophin

↓

Low levels or absence of dystrophin

↓

Muscle contraction/relaxation affected; precise mechanisms not elucidated

↓

Progressive, usually fatal muscular weakness

Figure 65–7. Summary of mechanisms involved in the causation of Duchenne muscular dystrophy.

CASE NO. 7: CYSTIC FIBROSIS

Classification
Genetic.

History and Physical Examination

A 1-year-old girl, an only child of Caucasian racial background, was brought to the clinic at the Hospital for Sick Children by her mother. She had been feverish for the past 24 hours and was coughing frequently. The mother stated that her daughter had experienced three attacks of "bronchitis" since birth, each of which had been treated with antibiotics by their family physician. The mother had also noted that her daughter had been passing somewhat bulky, foul-smelling stools for the past several months and was not gaining weight as expected. In view of the history of pulmonary and gastrointestinal problems, the attending pediatrician suspected that his patient might have cystic fibrosis, though no family history of this condition was elicited.

Laboratory Findings

Chest x-rays showed signs consistent with bronchopneumonia. Fecal fat was increased. A quantitative pilocarpine iontophoresis sweat test was performed, and the serum Cl⁻ was 70 meq/L (> 60 meq/L is a positive test); the test was repeated a few days later with similar results.

Treatment

The child was given an appropriate antibiotic and referred to the cystic fibrosis clinic for further care. A comprehensive program was instituted to look after all aspects of her health, including psychosocial considerations. She was started on a pancreatic enzyme preparation (given with each meal) and placed on a high-calorie diet supplemented with multivitamins and vitamin E. Postural drainage was begun for the thick pulmonary secretions. Subsequent infections were treated promptly with appropriate antibiotics. At age 6 years, she had grown normally and had been relatively free of infection for a year—attending school and making satisfactory progress.

Discussion

Cystic fibrosis is the most common serious genetic disease among whites in North America. It affects approximately 1:2500 individuals and is inherited as an autosomal recessive disease; about one person in 25 is a carrier. It is a disease of the exocrine glands, with the gastrointestinal and respiratory tracts being most affected. A diagnostic hallmark is the presence of high amounts of NaCl in sweat, reflecting an underlying abnormality in exocrine gland function (see below). For reasons that are not fully understood, the pancreatic ducts and the ducts of certain other exocrine glands become filled with viscous mucus, which leads to obstruction. This mucus is also pres-

ent in the bronchioles, leading to their obstruction; this favors the growth of certain bacteria (eg, *Staphylococcus aureus* and *Pseudomonas aeruginosa*) that cause recurrent bronchopulmonary infections, eventually seriously compromising lung function. In turn, the pulmonary disease can lead to right ventricular hypertrophy and possible heart failure. Patients may die of pulmonary infection or heart failure. More patients have been living into their 20s and later, as the condition is diagnosed early and appropriate comprehensive therapy started; intensive research on various therapies is ongoing, and the picture is likely to become brighter.

The classic clinical presentation is that of a young child with a history of recurrent pulmonary infection and signs of exocrine pancreatic insufficiency (eg, fatty, bulky stools). The disease is, however, clinically heterogeneous, which at least partly reflects heterogeneity at the molecular level (see below). Approximately 15% of patients have sufficient residual pancreatic function to be classified as "pancreatic-sufficient"; these patients do better than the remainder, who are "pancreatic-insufficient." A family history of cystic fibrosis may be obtained if other siblings have been affected. Sometimes problems due to lack of pancreatic secretions can be present at birth, the infant presenting with intestinal obstruction due to very thick meconium (meconium ileus). Other patients, less severely affected, may not be diagnosed until they are in their teens. Cystic fibrosis also affects the genital tract, and most males and the majority of females are infertile. The most useful laboratory test is the quantitative pilocarpine iontophoresis sweat test.

In 1989, results of a collaborative program between Canadian and American scientists revealed the nature of the genetic lesion in the majority of sufferers from cystic fibrosis. The gene for cystic fibrosis was the first to be cloned based solely on its position determined by linkage analysis (positional cloning) and constituted an enormous amount of painstaking labor and a tremendous triumph for "reverse genetics." This success has shown that, at least in theory, the molecular basis of any genetic disease can be revealed using similar approaches. Earlier studies had shown the gene to be located on chromosome 7, based on linkage to an RFLP. Additional RFLPs were found that flanked the cystic fibrosis gene, encompassing an area of approximately 1.5 million bp. A combination of "chromosome jumping" and "chromosome walking" permitted cloning of some 500 kb of DNA, approximately 250 kb of which turned out to be the cystic fibrosis gene. The gene was recognized because part of it was used to probe a cDNA library made from mRNAs of sweat gland duct cells; these cells would be expected to manifest the defect in salt transport exhibited by patients with cystic fibrosis. A positive clone was identified and the cDNA sequenced. This cDNA was approximately 6000 bp in length and contained an open reading frame, consistent with encoding a protein. Extensive studies in which cDNA sequences from normal and affected individuals were compared disclosed an intriguing mutation, consisting of the deletion of three bases resulting in the deletion of phenylalanine residue number 508 (ΔF508) from the protein product of the gene. This mutation has been found only in patients with cystic fibrosis and never in a normal subject. The protein product of the cystic fibrosis gene encodes an integral membrane protein of approximately 170 kDa. It has been named the CFTR (cystic fibrosis transmembrane conductance regulator) protein. Table 65–2 summarizes the main methods used to identify the cystic fibrosis gene, and some characteristics of the gene and the CFTR protein are listed in Table 65–3. The protein consists of two similar halves, each containing six transmembrane regions and a nucleotide (ATP)-binding fold (NBF). The two halves of the molecule are joined by a regulatory domain. F508 is located in NBF1 (Figure 65–8). The protein shows similarities in structure to certain other proteins that use ATP to transport molecules across cell membranes (eg, P-glycoprotein, involved in resistance to cancer chemotherapy; see Chapter 62). CFTR protein is a cAMP-activated chloride channel; thus, abnormal function of this channel is consistent with and explains the characteristically high chloride content of sweat referred to above. Deletion of ΔF508 is usually associated with the PI phenotype. Exactly how mutations in this channel cause the elevation of sweat chloride and the viscous mucus that is so troublesome in cystic fibrosis remains to be determined; however, Figure 65–9 summarizes one sequence of events that could explain the causation of airway problems in individuals with cystic fibrosis. It has been observed that certain mutations can greatly slow the intracellular transport of CFTR to the cell surface during its biogenesis, so that in some individuals with cystic fibrosis the reduced function of CFTR may represent more a decrease in its amount in the plasma membrane than an alteration of its function.

Table 65–2. Summary of major strategies used in detecting the gene for cystic fibrosis.

- From study of a large number of families with cystic fibrosis, assignment of the gene to chromosome 7 was made by demonstrating linkage to several RFLPs on that chromosome.
- Further narrowing to a smaller region of chromosome 7 was accomplished by use of additional RFLPs.
- Chromosome jumping and chromosome walking were used to isolate clones.
- The affected region was sequenced by looking for mutations in DNA from patients that were not present in DNA from normal individuals, for exons expressed as mRNAs in tissues affected by cystic fibrosis (eg, pancreas and lung), for sequences conserved across species, and for an open reading frame (indicating an expressed protein).

Table 65–3. Some characteristics of the gene for the CFTR protein and of the protein itself.

- About 250,000-bp gene on chromosome 7.
- 25 exons.
- mRNA of 6129 bp.
- Transmembrane protein of 1480 amino acids.
- CFTR protein contains two NBFs and one regulatory domain.
- Commonest mutation in cystic fibrosis is deletion of ΔF508 present in the first NBF.
- CFTR protein is a cAMP-responsive chloride channel.
- Shows homology to other proteins that use ATP to effect transport across membranes (eg, P-glycoprotein).

Key: CFTR, cystic fibrosis transmembrane regulatory protein; F, phenylalanine; NBF, nucleotide-binding fold.

Although deletion of ΔF508 is the most frequent mutation, many other mutations (now more than 500) have also been detected, including small deletions, insertions, and missense and nonsense mutations. Variations in the frequency of specific mutations occur among different populations. In North America, approximately 70% of cystic fibrosis carriers have the mutation resulting in deletion of ΔF508. Identification of a specific mutation in cystic fibrosis was expected to lead quickly to identification of carriers by use of a specific oligodeoxynucleotide probe, synthesized to detect the mutation. However, the fact that 30% of cases are composed of at least 500 different mutations has slowed progress. Universal screening will probably not proceed—even if it proves to be economically feasible—until it is certain that at least 95% of the mutations can be detected by use of a battery of probes; otherwise, there would be too many false negatives and parents could not be assured that their children would not be affected. How-

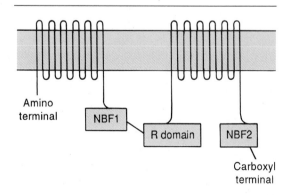

Figure 65–8. Diagram of the structure of the CFTR protein (not to scale). The protein contains 12 transmembrane segments (probably helical), two nucleotide-binding folds or domains (NBF1 and NBF2), and one regulatory (R) domain. NBF1 and NBF2 probably bind ATP and couple its hydrolysis to transport of Cl⁻. Phe 508, the major locus of mutations in cystic fibrosis, is located in NBF1.

Mutations in DNA
↓
Alterations in the structure and function of the CFTR protein
↓
Decreased output of chloride from epithelial cells; uptake of Na$^+$ is also affected, and is excessive, leading secondarily to increased uptake of water
↓
Increased viscosity of mucus in the airway
↓
Impaired mucociliary clearance, bacterial colonization, and recurrent infections

Figure 65–9. Summary of possible mechanisms involved in the airways of individuals with cystic fibrosis who have pulmonary pathology. In individuals of Caucasian origin, approximately 70% of the mutations occur at one locus, resulting in deletion of ΔF508 from the CFTR protein; about 500 other mutations have been identified. Similar electrolyte events affect the mucus in the pancreatic ductules.

ever, probes designed to detect deletion of ΔF508 and a limited number of other mutations are currently being used in appropriate instances to confirm the presence of cystic fibrosis in infants, to detect carriers, and in prenatal diagnosis using chorionic villus sampling.

Because of the stimulus provided by the research described above, intense effort is ongoing to develop new therapies for cystic fibrosis. Already it has been shown that introduction of cDNAs encoding the CFTR protein can apparently correct defects in chloride transport in cultured cells from patients with cystic fibrosis. An adenovirus encoding the human CFTR protein has been introduced into the respiratory tract of rats and shown to direct synthesis of the CFTR protein; a somewhat similar study has been completed in humans. Results of a clinical trial were reported in 1994 in which a recombinant adenovirus vector *(AdCFTR)* containing the normal human *CFTR* cDNA protein was administered to the nasal and bronchial epithelium of four individuals with cystic fibrosis. The vector was shown to express the *CFTR* cDNA in the respiratory tract epithelium. At high doses, a transient systemic and pulmonary syndrome was observed, perhaps mediated by interleukin-6.

Follow-up at 6–12 months revealed no long-term adverse effects. It should be noted that this study did not attempt to correct the cystic fibrosis phenotype but rather to demonstrate the feasibility of the approach. It is possible that immune responses may limit the use of adenovirus vectors, and it remains to be determined whether the level of expression of CFTR will be high enough and sustained enough to be of benefit to individuals with cystic fibrosis.

Use of an aerosolized recombinant preparation of human DNase that digests the DNA of microorganisms present in the respiratory tract has proved useful. The possibility that specific drugs can interact with and improve the function of the mutated CFTR protein is being actively explored.

CASE NO. 8: SEVERE COMBINED IMMUNODEFICIENCY DISEASE (SCID) DUE TO DEFICIENCY OF ADENOSINE DEAMINASE (ADA)

Classification
Immunologic, genetic.

History and Physical Examination
A little girl aged 11 months was brought by her parents to a children's hospital. She had had a number of attacks of pneumonia and thrush (oral infection usually due to *Candida albicans*) since birth. The major findings of a thorough workup were very low levels of circulating lymphocytes (severe lymphopenia) and low levels of circulating immunoglobulins. The examining pediatrician suspected SCID.

Laboratory Findings
Analysis of a sample of red blood cells revealed a very low activity of ADA and also a very high level (about 50 times normal) of dATP. This confirmed the diagnosis of SCID due to deficiency of ADA, the enzyme that converts adenosine to inosine (Chapter 36):

$$\text{Adenosine} \rightarrow \text{Inosine} + NH_4^+$$

Treatment
Appropriate antibiotic treatment was commenced, and the child was also given periodic injections of immune globulin. In addition, she was started on weekly intramuscular injections of bovine ADA conjugated to polyethylene glycol. Bovine ADA is relatively nonimmunogenic, and conjugation to polyethylene glycol prolongs its half-life; it has been shown to be beneficial in the treatment of ADA deficiency. Her parents were informed that bone marrow transplantation was the most useful therapy, but the treatment was declined. However, despite initial improvement following administration of polyethylene glycol-conjugated ADA, the patient's condition began to worsen. In view of recent reports of successes with ADA gene therapy, this treatment was offered, and the parents consented. Lymphocytes and mononuclear cells were isolated from her blood using a Ficoll gradient. They were then grown in the presence of interleukin-2 (to stimulate cell division) and infected with a modified retrovirus containing inserts encoding ADA and also an enzyme governing neomycin resistance (NeoR gene, acting as a marker to show that gene transfer had been achieved). The

autologous gene-treated cells were then injected intravenously. The child received similar injections once a month over the next year, and in addition continued to receive polyethylene glycol-conjugated ADA. Measurements of the activity of ADA revealed a sustained elevation (about 20% of normal) of the enzyme in circulating lymphocytes after 6 months of treatment; analyses using the PCR technique with NeoR probes revealed that approximately the same percentage of lymphocytes contained inserted genetic material.

Discussion

Deficiency of the activity of ADA is inherited as an autosomal recessive condition. Most of the mutations in the gene for ADA so far detected have been base substitutions, though deletions have also been detected. These mutations result in diminished activity or stability of ADA. The block of activity of ADA results in accumulation of adenosine, which in turn results in accumulation of deoxyadenosine and dATP. Elevated levels of the latter are toxic, particularly to lymphocytes, which normally exhibit a high activity of ADA. Thus, lymphocytes are injured or killed, resulting in impairment of both cell and humoral immunity.

Adenosine deaminase deficiency has become important because it is the first disease to be treated by somatic cell gene therapy. Several patients have been treated by protocols similar to that described above. One reason for selecting ADA deficiency as a suitable condition for somatic gene therapy was that cells that express the gene for ADA would have a selective advantage for growth over uncorrected cells. Results of a clinical trial on two children with ADA deficiency and SCID were published in 1995. The trial was begun in 1990, using retroviral-mediated transfer of the ADA gene into their T cells. The number of blood T cells normalized, as did many cellular and humoral immune responses. Gene treatment ended after 2 years, but integrated vector and ADA gene expression persisted. It was concluded that gene therapy can be a safe and effective addition to the treatment of this condition. A simplified scheme of the events involved in the causation of ADA deficiency is shown in Figure 65–10.

CASE NO. 9: DIABETES MELLITUS TYPE I WITH KETOACIDOSIS

Classification

Endocrine deficiency, lack of insulin.

History and Physical Examination

A 14-year-old girl was admitted to a children's hospital in coma. Her mother stated that the girl had been in good health until approximately 2 weeks previously, when she developed a sore throat and mod-

Mutations in the gene for ADA located on chromosome 20

↓

Deficient activity of ADA

↓

Increased levels of adenosine, deoxyadenosine, and dATP

↓

Toxicity to lymphocytes

↓

Decreased synthesis of immunoglobulins and impaired cell-mediated immunity

Figure 65–10. Summary of probable events in the causation of SCID due to ADA deficiency.

erate fever. She subsequently lost her appetite and generally did not feel well. Several days before admission she began to complain of undue thirst and also started to get up several times during the night to urinate. Her family doctor was out of town, and the mother felt reluctant to contact another physician. However, on the day of admission the girl had started to vomit, had become drowsy and difficult to arouse, and accordingly had been brought to the emergency department. On examination she was dehydrated, her skin was cold, she was breathing in a deep sighing manner (Kussmaul's respiration), and her breath had a fruity odor. Her blood pressure was 90/60 and her pulse rate was 115/min. She could not be aroused. A diagnosis of severe type I insulin-dependent diabetes mellitus with resulting ketoacidosis and coma was made by the intern on duty.

Laboratory Findings

Plasma (normal levels in parentheses)
Glucose, 35 mmol/L (3.6–6.1 mmol/L)
β-Hydroxybutyrate, 13.0 mmol/L (< 0.25 mmol/L)
Acetoacetate, 2.8 mmol/L (< 0.2 mmol/L)
Bicarbonate, 5 mmol/L (24–28 mmol/L)
Urea, 12 mmol/L (2.9–8.9 mmol/L)
Arterial blood H$^+$, 89 nmol/L [pH 7.05] (44.7–45.5 nmol/L)[pH 7.35–7.45]
Potassium, 5.8 mmol/L (3.5–5.0 mmol/L)
Creatinine, 160 μmol/L (60–132 μmol/L)
Urine
Glucose, ++++
Ketone bodies, ++++

The above confirmed the admitting diagnosis.

Treatment

The most important measures in treatment of diabetic ketoacidosis are intravenous administration of insulin and saline fluids. This patient was given intravenous insulin (10 units/h) added to 0.9% NaCl. Glu-

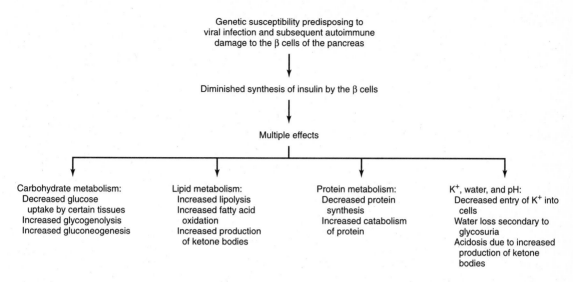

Figure 65–11. Summary of mechanisms involved in the causation of the ketoacidosis of diabetes mellitus type I.

cose was withheld until the level of plasma glucose fell below 250 mg/dL. KCl was also administered cautiously, with plasma K^+ levels monitored every hour initially. Continual monitoring of K^+ levels is extremely important in the management of diabetic ketoacidosis because inadequate management of K^+ balance is the main cause of death. Bicarbonate is not needed routinely but may be required if acidosis is very severe.

Discussion

The precise cause of type I insulin-dependent diabetes mellitus has not been elucidated. A likely chain of events is the following. Patients with this type of diabetes have a genetic susceptibility, which may predispose to a viral infection. The infection and consequent inflammatory reaction apparently alter the antigenicity of the surface of the pancreatic B cells and set up an autoimmune reaction, involving both cytotoxic antibodies and T lymphocytes. This leads to widespread destruction of B cells, resulting in type I diabetes mellitus. Perhaps the sore throat this patient had several weeks before admission reflected the initiating viral infection.

The marked hyperglycemia, glucosuria, ketonemia, and ketonuria confirmed the diagnosis of dia-

betic ketoacidosis. The low pH indicated a severe acidosis due to the greatly increased production of acetoacetic acid and β-hydroxybutyric acid. The low levels of bicarbonate and Pco_2 confirmed the presence of a metabolic acidosis with partial respiratory compensation (by hyperventilation). The values of urea and creatinine indicated some renal impairment (due to diminished renal perfusion because of low blood volume secondary to dehydration), dehydration, and increased degradation of protein. A high plasma level of potassium is often found in diabetic ketoacidosis owing to a lowered uptake of potassium by cells in the absence of insulin.

Thus, the clinical picture in diabetic ketoacidosis reflects the abnormalities in the metabolism of carbohydrate, lipid, and protein that occur when plasma levels of insulin are sharply reduced. These abnormalities and the causes of the disturbances of K^+, water, and H^+ metabolism found in diabetic ketoacidosis are summarized in Figure 65–11. The increased osmolality of plasma due to hyperglycemia also contributes to the development of coma in diabetic ketoacidosis. It should be apparent that the rational management of a patient with diabetic ketoacidosis depends on thorough familiarity with the actions of insulin.

REFERENCES

Blaese RM et al: T lymphocyte-directed gene therapy for ADA⁻ SCID: Initial results after 4 years. Science 1995;270:475.

Cleaver JE, Kraemer KH: Xeroderma pigmentosum and Cockayne syndrome. In: *The Metabolic and Molecular*

Bases of Inherited Disease, 7th ed. Scriver CR et al (editors). McGraw-Hill, 1995.

Collins FS: Cystic fibrosis: Molecular biology and therapeutic implications. Science 1992;256:774.

Crystal RG et al: Administration of an adenovirus containing the human *CFTR* cDNA to the respiratory tract of individuals with cystic fibrosis. Nat Genet 1994;8:42.

Foster DW: Diabetes mellitus. In: *Harrison's Principles of Internal Medicine,* 13th ed. Isselbacher KJ et al (editors). McGraw-Hill, 1994.

Fuster VR et al: The pathogenesis of coronary artery disease and the acute coronary syndrome. (Two parts.) N Engl J Med 1992;326:245, 319.

Greenough WB III: Cholera. In: *Cecil Textbook of Medicine,* 19th ed. Wyngaarden JB, Smith LH Jr, Bennett JC (editors). Saunders, 1992.

Halliwell B, Gutteridge JMC: *Free Radicals in Biology and Medicine,* 2nd ed. Clarendon Press, 1989.

Halperin ML, Rolleston FS: *Clinical Detective Stories: A Problem-Based Approach to Clinical Cases in Energy and Acid-Base Metabolism.* Portland Press, 1993.

Knowles MR et al: A controlled study of adenoviral-vector-mediated gene transfer in the nasal epithelium of patients with cystic fibrosis. N Engl J Med 1995;333:823.

Leiden KJM: Gene therapy: Promises, pitfalls and prognosis. N Engl J Med 1995;333:871.

Lieber CS: Medical disorders of alcoholism. Seminars in Medicine of the Beth Israel Hospital, Boston: N Engl J Med 1995;333:1058.

Marx J: DNA repair comes into its own. Science 1994; 333:1058.

Mendell JT et al: Myoblast transfer in the treatment of Duchenne's muscular dystrophy. N Engl J Med 1995; 333:832.

Rommens JR et al: Identification of the cystic fibrosis gene: Chromosome walking and jumping. Science 1989; 245:1059.

Sancar A: Mechanisms of DNA excision repair. Science 1994;266:1954.

Waterlow JC: *Protein Energy Metabolism.* Arnold, 1992.

Worton R: Muscular dystrophies: Diseases of the dystrophin-glycoprotein complex. Science 1995;270:755.

Worton RG, Brooke MH: The X-linked muscular dystrophies. In: *The Metabolic and Molecular Bases of Inherited Disease,* 7th ed. Scriver CR et al (editors). McGraw-Hill, 1995.

Appendix*

*Modified and reproduced, with permission, from Krupp MA, Schroeder SA, Tierney LM Jr (editors): *Current Medical Diagnosis & Treatment 1987.* Appleton & Lange, 1987.

CHEMICAL CONSTITUENTS OF BLOOD & BODY FLUIDS

Validity of Numerical Values in Reporting Laboratory Results

The value reported from a clinical laboratory after determination of the concentration or amount of a substance in a specimen represents the best value obtainable with the method, reagents, instruments, and technical personnel involved in obtaining and processing the material.

Accuracy is the degree of agreement of the determination with the "true" value (eg, the known concentration in a control sample). **Precision** denotes the reproducibility of the analysis and is expressed in terms of variation among several determinations on the same sample. **Reliability** is a measure of the congruence of accuracy and precision.

Precision is not absolute but is subject to variation inherent in the complexity of the method, the stability of reagents, the accuracy of the primary standard, the sophistication of the equipment, and the skill of the technical personnel. Each laboratory should maintain data on precision (reproducibility) that can be expressed statistically in terms of the standard deviation from the mean value obtained by repeated analyses of the same sample. For example, the precision in determination of cholesterol in serum in a good laboratory may be the mean value ± 5 mg/dL. The 95% confidence limits are ± 2 SD, or ± 10 mg/dL. Thus, any value reported is "accurate" within a range of 20 mg/dL. Therefore, the reported value 200 mg/dL means that the true value lies between 190 and 210 mg/dL. For the determination of serum potassium with a variance of 1 SD of ± 0.1 mmol/L, values ± 0.2 mmol could be obtained on the same specimen. A report of 5.5 could represent at best the range 5.3–5.7 mmol/L. That is, the two results—5.3 mmol/L and 5.7 mmol/L—might be obtained on analysis of the same sample and still be within the limits of precision of the test.

Physicians should obtain from the laboratory the values for the variation of a given determination as a basis for deciding whether one reported value represents a change from another on the same patient.

Interpretation of Laboratory Tests

Normal values are those that fall within 2 SD of the mean value for the normal population. This normal range encompasses 95% of the population. Many factors may affect values and influence the normal range; by the same token, various factors may produce values that are normal under the prevailing conditions but outside the 95% limits determined under other circumstances. These factors include age; race; sex; environment; posture; diurnal and other cyclic variations; fasting or postprandial state, foods eaten; drugs; and level of exercise.

Normal or reference values vary with the method employed, the laboratory, and conditions of collection and preservation of specimens. The normal values established by individual laboratories should be clearly expressed to ensure proper interpretation.

Interpretation of laboratory results must always be related to the condition of the patient. A low value may be the result of deficit or of dilution of the substance measured, eg, low serum sodium. Deviation from normal may be associated with a specific disease or with some drug consumed by the subject—eg, elevated serum uric acid levels may occur in patients with gout or may be due to treatment with chlorothiazides or with antineoplastic agents.

Values may be influenced by the method of collection of the specimen. Inaccurate collection of a 24-hour urine specimen, variations in concentration of the randomly collected urine specimen, hemolysis in a blood sample, addition of an inappropriate anticoagulant, and contaminated glassware or other apparatus are examples of causes of erroneous results.

Note: Whenever an unusual or abnormal result is obtained, all possible sources of error must be considered before responding with therapy based on the laboratory report. Laboratory medicine is a specialty, and experts in the field should be consulted whenever results are unusual or in doubt.

Effect of Meals and Posture on Concentration of Substances in Blood

A. Meals: The usual normal values for blood tests have been determined by assay of "fasting" specimens collected after 8–12 hours of abstinence from food. With few exceptions, water is usually permitted as desired.

Few routine tests are altered from usual fasting values if blood is drawn 3–4 hours after breakfast. When blood is drawn 3–4 hours after lunch, values are more likely to vary from those of the true fasting state. Valid measurement of triacylglycerol (triglyceride) in serum or plasma requires abstinence from food for 10–14 hours.

B. Posture: Plasma volume measured in a person who has been supine for several hours is 12–15% greater than in a person who has been up and about or standing for an hour or so. It follows that measurements performed on blood obtained after the subject has been lying down for an hour or more will yield lower values than when blood has been obtained after the same subject has been upright. An intermediate change apparently occurs with sitting.

A tourniquet applied for 1 minute instead of 3 minutes produced the following changes in reported values: total protein, +5%; iron, +6.7%; cholesterol, +5%; AST (SGOT), +9.3%; and bilirubin, +8.4%. Decreases were observed for potassium, –6%; and creatinine, –2.3%.

Validity of Laboratory Tests*

The clinical value of a test is related to its specificity and sensitivity and the incidence of the disease in the population tested.

Sensitivity means percentage of positive results in patients with the disease. The test for phenylketonuria is highly sensitive: a positive result is obtained in all who have the disease (100% sensitivity). The carcinoembryonic antigen (CEA) test has low sensitivity: only 72% of those with carcinoma of the colon test positive when the disease is extensive, and only 20% are positive with early disease. Lower sensitivity occurs in the early stages of many diseases— in contrast to the higher sensitivity in well-established disease.

Specificity means percentage of negative results

among people who do not have the disease. The test for phenylketonuria is highly specific: 99.9% of normal individuals give a negative result. In contrast, the CEA test for carcinoma of the colon has a variable specificity: about 3% of nonsmoking individuals give a false-positive result (97% specificity), whereas 20% of smokers give a false-positive result (80% specificity). The overlap of serum thyroxine levels between hyperthyroid patients and those taking oral contraceptives or those who are pregnant is an example of a change in specificity from that prevailing in a different set of individuals.

The **predictive value** of a positive test defines the percentage of positive results that are true positives. This is related fundamentally to the incidence of the disease. In a group of patients on a urology service, the incidence of renal disease is higher than in the general population, and the serum creatinine level will have a higher predictive value in that group than for the general population.

Formulas for definitions:

$$\text{Sensitivity} = \frac{\text{True positive}}{\text{True positive + false negative}} \times 100$$

$$\text{Specificity} = \frac{\text{True negative}}{\text{True negative + false positive}} \times 100$$

$$\text{Predictive value} = \frac{\text{True positive}}{\text{True positive + false positive}} \times 100$$

Before ordering a test, attempt to determine whether test sensitivity, specificity, and predictive value are adequate to provide useful information. To be useful, the result should influence diagnosis, prognosis, or therapy; lead to a better understanding of the disease process; and benefit the patient.

NORMAL LABORATORY VALUES (Blood [B], Plasma [P], Serum [S], Urine [U])

HEMATOLOGY

Bleeding time: Ivy method, 1–7 minutes (60–420 seconds). Template method, 3–9 minutes (180–540 seconds).

Cellular measurements of red cells: Average diameter = 7.3 μm (5.5–8.8 μm). Mean corpuscular volume (MCV): Men, 80–94 fL; women, 81–99 fL (by Coulter counter). Mean corpuscular hemoglobin (MCH): 27–32 pg. Mean corpuscular hemoglobin concentration (MCHC):32–36 g/dL red blood cells (32–36%).

---

*This section is an abridged version of an article by Krieg AF, Gambino R, Galen RS: Why are clinical laboratory tests performed? When are they valid? JAMA 1975;233:76. Reprinted from the *Journal of the American Medical Association*. Copyright © 1975 by American Medical Association. See also Galen RS, Gambino SR: *Beyond Normality: The Predictive Value and Efficiency of Medical Diagnosis*. Wiley, 1975.

Clot retraction: Begins in 1–3 hours; complete in 6–24 hours. No clot lysis in 24 hours.

Fibrinogen split products: Negative > 1:4 dilution.

Fragility of red cells: Begins at 0.45–0.38% NaCl; complete at 0.36–0.3% NaCl.

Hematocrit (PCV): Men, 40–52% (0.4–0.52); women, 37–47% (0.37–0.47).

Hemoglobin: [B] Men, 14–18 g/dL (2.09–2.79 mmol/L as Hb tetramer); women, 12–16 g/dL (1.86–2.48 mmol/L). [S] 2–3 mg/dL.

Partial thromboplastin time: Activated, 25–37 seconds.

Platelets: 150,000–400,000/μL (0.15–0.4 × 10^{12}/L).

Prothrombin: [P] 11–14.5 seconds. International Normalized Ratio (INR): [P] 2.0–3.0.

Red blood count (RBC): Men, 4.5–6.2 million/μL (4.5–6.2 × 10^{12}/L); women, 4–5.5 million/μL (4–5.5 × 10^{12}/L).

Reticulocytes: 0.2–2% of red cells.

Sedimentation rate: Less than 20 mm/h (Westergren); 0–10 mm/h (Wintrobe).

White blood count (WBC) and differential: 5000–10,000/μL (5–10 × 10^9/L).

Myelocytes	0 %
Juvenile neutrophils	0 %
Band neutrophils	0–5 %
Segmented neutrophils	40–60%
Lymphocytes	20–40%
Eosinophils	1–3 %
Basophils	0–1 %
Monocytes	4–8 %
Lymphocytes: Total, 1500–4000/μL	
B cell	5–25%
T cell	60–88%
Suppressor	10–43%
Helper	32–66%
H:S	> 1

CLINICAL CHEMISTRY

Acetone and acetoacetate: [S] 0.3–2 mg/dL (3–20 mg/L).

Adrenal hormones and metabolites:
Aldosterone: [U] 2–26 μg/24 h (5.5–72 nmol/d). Values vary with sodium and potassium intake.
Catecholamines: [U] Total, < 100 μg/24 h. Epinephrine, < 10 μg/24 h (< 100 nmol/d); norepinephrine, < 100 μg/24 h (< 590 nmol/d).
Cortisol, free: [U] 20–100 μg/24 h (0.55–2.76 μmol/d).
11,17-Hydroxycorticoids: [U] Men, 4–12 mg/24 h; women, 4–8 mg/24 h.
17-Ketosteroids: [U] Under 8 years, 0–2 mg/24 h; adolescents, 2–20 mg/24 h. (1 mg = 3.5 μmol.)
Metanephrine: [U] < 1.3 mg/24 h (< 6.6 μmol/d) or < 2.2 μg/mg creatinine.
Vanillylmandelic acid (VMA): [U] Up to 7 mg/24 h (< 35 μmol/d).

Aminotransferases:
Aspartate aminotransferase (AST; SGOT): 0–41 IU/L at 37 °C.
Alanine aminotransferase (ALT; SGPT): 0–45 IU/L at 37 °C.

Ammonia: [P] (as NH_3) 10–80 μg/dL (5–50 μmol/L).

Amylase: [S] 80–180 units/dL (Somogyi).

α_1-Antitrypsin: [S] > 180 mg/dL.

Ascorbic acid: [P] 0.4–1.5 mg/dL (23–85 μmol/L).

Bicarbonate: [S] 24–28 meq/L (24–28 mmol/L).

Bilirubin: [S] Total, 0.2–1.2 mg/dL (3.5–20.5 μmol/L). Direct (conjugated), 0.1–0.4 mg/dL (< 7 μmol/L). Indirect, 0.2–0.7 mg/dL (< 12 μmol/L).

Calcium: [S] 8.5–10.3 mg/dL (2.1–2.6 mmol/L). Values vary with albumin concentration.

Calcium, ionized: [S] 4.25–5.25 mg/dL; 2.1–2.6 meq/L (1.05–1.3 mmol/L).

β-Carotene: [S, fasting] 50–300 μg/dL (0.9–5.58 μmol/L).

Ceruloplasmin: [S] 25–43 mg/dL (1.7–2.9 μmol/L).

Chloride: [S or P] 96–106 meq/L (96–106 mmol/L).

Cholesterol: [S or P] 150–220 mg/dL (3.9–5.72 mmol/L). (See Lipid fractions, below.)

Cholesteryl esters: [S] 65–75% of total cholesterol.

CO_2 content: [S or P] 24–29 meq/L (24–29 mmol/L).

Complement: [S] C3 (β_{1C}), 90–250 mg/dL. C4 (β_{1E}), 10–60 mg/dL. Total (CH$_{50}$), 75–160 mg/dL.

Copper: [S or P] 100–200 μg/dL (16–31 μmol/L).

Cortisol: [P] 8:00 AM, 5–25 μg/dL (138–690 nmol/L); 8:00 PM < 10 μg/dL (275 nmol/L).

Creatine kinase (CK): [S] 10–50 IU/L at 30 °C.

Creatine kinase isoenzymes: BB, 0%; MB, 0–3%; MM, 97–100%.

Creatinine: [S or P] 0.7–1.5 mg/dL (62–132 μmol/L).

Cyanocobalamin: [S] 200 pg/mL (148 pmol/L).

Epinephrine: [P] Supine, < 100 pg/mL (< 550 pmol/L).

Fecal fat: < 30% dry weight.

Ferritin: [S] Adult women, 20–120 ng/mL; men, 30–300 ng/mL. Child to 15 years, 7–140 ng/mL.

Folic acid: [S] 2–20 ng/mL (4.5–45 nmol/L). [RBC] > 140 ng/mL (> 318 nmol/L).

Glucose: [S or P] 65–110 mg/dL (3.6–6.1 mmol/L).

γ-Glutamyl transpeptidase: [S] < 30 units/L at 30 °C.

Haptoglobin: [S] 40–170 mg of hemoglobin-binding capacity.

Iron: [S] 50–175 µg/dL (9–31.3 µmol/L).

Iron-binding capacity: [S] Total, 250–410 µg/dL (44.7–73.4 µmol/L). Percent saturation, 20–55%.

Lactate: [B, special handling] Venous, 4–16 mg/dL (0.44–1.8 mmol/L).

Lactate dehydrogenase (LDH): [S] 55–140 IU/L at 30 °C; SMA, 100–225 IU/L at 37 °C; SMAC, 60–200 IU/L at 37 °C.

Lead: [U] < 80 µg/24 h (< 0.4 µmol/d).

Lipase: [S] < 150 units/L.

Lipid fractions: [S or P] Desirable levels: HDL cholesterol, > 40 mg/dL; LDL cholesterol, < 180 mg/dL; VLDL cholesterol, < 40 mg/dL. (To convert to mmol/L, multiply by 0.026.)

Lipids, total: [S] 450–1000 mg/dL (4.5–10 g/L).

Magnesium: [S or P] 1.8–3 mg/dL (0.75–1.25 mmol/L).

Norepinephrine: [P] Supine, < 500 pg/mL (< 3 nmol/L).

Osmolality: [S] 280–296 mosm/kg water (280–296 mmol/kg water).

Oxygen:
Capacity: [B] 16–24 vol%. Values vary with hemoglobin concentration.
Arterial content: [B] 15–23 vol%. Values vary with hemoglobin concentration.
Arterial % saturation: 94–100% of capacity.
Arterial PO_2 (PaO_2): 80–100 mm Hg (10.67–13.33 kPa) (sea level). Values vary with age.

$PaCO_2$: [B, arterial] 35–45 mm Hg (4.7–6 kPa).

pH: [B, arterial] 7.35–7.45 (H^+ 44.7–45.5 nmol/L).

Phosphatase, acid: [S] 1–5 units (King-Armstrong), 0.1–0.63 units (Bessey-Lowry).

Phosphatase, alkaline: [S] Adults, 5–13 units (King-Armstrong); 0.8–2.3 (Bessey-Lowry); SMA, 30–85 IU/L at 37 °C; SMAC, 30–115 IU/L at 37 °C.

Phospholipid: [S] 145–200 mg/dL (1.45–2 g/dL).

Phosphorus, inorganic: [S, fasting] 3–4.5 mg/dL (1–1.5 mmol/L).

Porphyrins:
Delta-aminolevulinic acid: [U] 1.5–7.5 mg/24 h (11–57 µmol/d).

Coproporphyrin: [U] < 230 µg/24 h (< 350 nmol/d).
Uroporphyrin: [U] < 50 µg/24 h (< 60 nmol/d).
Porphobilinogen: [U] < 2 mg/24 h (< 8.8 µmol/d).

Potassium: [S or P] 3.5–5 meq/L (3.5–5 mmol/L).

Protein:
Total: [S] 6–8 g/dL (60–80 g/L).
Albumin: [S] 3.5–5.5 g/dL (35–55 g/L).
Globulin: [S] 2–3.6 g/dL (20–36 g/L).
Fibrinogen: [P] 0.2–0.6 g/dL (2–6 g/L).

Pyruvate: [B] 0.6–1 mg/dL (70–114 µmol/L).

Serotonin: [B] 5–20 µg/dL (0.2–1.14 µmol/L).

Sodium: [S or P] 136–145 meq/L (136–145 mmol/L).

Specific gravity: [B] 1.056 (varies with hemoglobin and protein concentration). [S] 1.0254–1.0288 (varies with protein concentration).

Sulfate: [S or P] As sulfur, 0.5–1.5 mg/dL (156–468 µmol/L).

Transferrin: [S] 200–400 mg/dL (23–45 µmol/L).

Triglycerides: [S] < 165 mg/dL (1.9 mmol/L). (See Lipid fractions, above.)

Urea nitrogen: [S or P] 8–25 mg/dL (2.9–8.9 mmol/L). Do not use anticoagulant containing ammonium oxalate.

Uric acid: [S or P] Men, 3–9 mg/dL (0.18–0.54 mmol/L); women, 2.5–7.5 mg/dL (0.15–0.45 mmol/L).

Urobilinogen: [U] 0–2.5 mg/24 h (70–470 µmol/d).

Urobilinogen, fecal: 40–280 mg/24 h (70–470 µmol/d).

Vitamin A: [S] 15–60 µg/dL (0.53–2.1 µmol/L).

Vitamin B$_{12}$: [S] > 200 pg/mL (> 148 pmol/L).

Vitamin D: [S] Cholecalciferol (D$_3$): 25-Hydroxycholecalciferol, 8–55 ng/mL (19.4–137 nmol/L); 1,25-dihydroxycholecalciferol, 26–65 pg/mL (62–155 pmol/L); 24,25-dihydroxycholecalciferol, 1–5 ng/mL (2.4–12 nmol/L).

Volume, blood (Evans blue dye method): Adults, 2990–6980 mL. Women, 46.3–85.5 mL/kg; men, 66.2–97.7 mL/kg.

Zinc: [S] 50–150 µg/dL (7.65–22.95 µmol/L).

HORMONES, SERUM OR PLASMA

Adrenal:
Aldosterone: [P] Supine, normal salt intake, 2–9 ng/dL (56–250 pmol/L); increased when upright.
Cortisol: [S] 8:00 AM, 5–20 µg/dL (0.14–0.55 µmol/L); 8:00 PM, < 10 µg/dL (0.28 µmol/L).

Deoxycortisol: [S] After metyrapone, > 7 µg/dL (> 0.2 µmol/L).

Dopamine: [P] < 135 pg/mL.

Epinephrine: [P] < 0.1 ng/mL (< 0.55 nmol/L).

Norepinephrine: [P] < 0.5 µg/L (< 3 nmol/L).

Gonad:

Testosterone, free: [S] Men, 10–30 ng/dL; women, 0.3–2 ng/dL. (1 ng/dL = 0.035 nmol/L.)

Testosterone, total: [S] Prepubertal, < 100 ng/dL; adult men, 300–1000 ng/dL; adult women, 20–80 ng/dL; luteal phase, up to 120 ng/dL.

Estradiol (E_2): [S, special handling] Men, 12–34 pg/mL; women, menstrual cycle 1–10 days, 24–68 pg/mL; 11–20 days, 50–300 pg/mL; 21–30 days, 73–149 pg/mL (by RIA). (1 pg/mL = 3.6 pmol/L.)

Progesterone: [S] Follicular phase, 0.2–1.5 mg/mL; luteal phase, 6–32 ng/mL; pregnancy, > 24 ng/mL; men, < 1 ng/mL = 3.2 nmol/L

Islets:

Insulin: [S] 4–25 µU/mL (29–181 pmol/L).

C-peptide: [S] 0.9–4.2 ng/mL.

Glucagon: [S, fasting] 20–100 pg/mL.

Kidney:

Renin activity: [P, special handling] Normal sodium intake: Supine, 1–3 ng/mL/h; standing, 3–6 ng/mL/h. Sodium depleted: Supine, 2–6 ng/mL/h; standing, 3–20 ng/mL/h.

Parathyroid: Parathyroid hormone levels vary with method and antibody. Correlate with serum calcium.

Pituitary:

Growth hormone (GH): [S] Adults, 1–10 ng/mL (46–465 pmol/L) (by RIA).

Thyroid-stimulating hormone (TSH): [S] < 10 µU/mL.

Follicle-stimulating hormone (FSH): [S] Prepubertal, 2–12 mIU/mL; adult men, 1–15 mIU/mL; adult women, 1–30 mIU/mL; castrate or postmenopausal, 30–200 mIU/mL (by RIA).

Luteinizing hormone (LH): [S] Prepubertal, 2–12 mIU/mL; adult men, 1–15 mIU/mL; adult women, < 30 mIU/mL; castrate or postmenopausal, > 30 mIU/mL.

Corticotropin (ACTH): [P] 8:00–10:00 AM, up to 100 pg/mL (22 pmol/L).

Prolactin: [S] 1–25 ng/mL (0.4–10 nmol/L).

Somatomedin C: [P] 0.4–2 U/mL.

Antidiuretic hormone (ADH; vasopressin): [P] Serum osmolality 285 mosm/kg, 0–2 pg/mL; > 290 mosm/kg, 2–12+ pg/mL.

Placenta:

Estriol (E_3): [S] Men and nonpregnant women, < 0.2 µg/dL (< 7 nmol/L) (by RIA).

Chorionic gonadotropin: [S] Beta subunit: Men, < 9 mIU/mL; pregnant women after implantation, > 10 mIU/mL.

Stomach:

Gastrin: [S, special handling] Up to 100 pg/mL (47 pmol/L). Elevated, > 200 pg/mL.

Pepsinogen I: [S] 25–100 ng/mL.

Thyroid:

Thyroxine, free (FT_4): [S] 0.8–2.4 ng/dL (10–30 pmol/L).

Thyroxine, total (TT_4): [S] 5–12 µg/dL (65–156 nmol/L) (by RIA).

Thyroxine-binding globulin capacity: [S] 12–28 µg T_4/dL (150–360 nmol T_4/dL).

Triiodothyronine (T_3): [S] 80–220 ng/dL (1.2–3.3 nmol/L).

Reverse triiodothyronine (rT_3): [S] 30–80 ng/dL (0.45–1.2 nmol/L).

Triiodothyronine uptake (RT_3U): [S] 25–36%; as TBG assessment (RT_3U ratio), 0.85–1.15.

Calcitonin: [S] < 100 pg/mL (< 29.2 pmol/L).

CEREBROSPINAL FLUID VALUES

Appearance: Clear and colorless.

Cells: Adults, 0–5 mononuclears/µL; infants, 0–20 mononuclears/µL.

Glucose: 50–85 mg/dL (2.8–4.7 mmol/L). (Draw serum glucose at same time.)

Pressure (reclining): Newborns, 30–90 mm H_2O; children, 50–100 mm H_2O; adults, 70–200 mm H_2O (avg = 125 mm H_2O).

Proteins: Total, 20–45 mg/dL (200–450 mg/L) in lumbar cerebrospinal fluid. IgG, 2–6 mg/dL (0.02–0.06 g/L).

Specific gravity: 1.003–1.008.

RENAL FUNCTION TESTS

_p_-Aminohippurate (PAH) clearance (RPF): Men, 560–830 mL/min; women, 490–700 mL/min.

Creatinine clearance, endogenous (GFR): Approximates inulin clearance (see below).

Filtration fraction (FF): Men, 17–21%; women, 17–23%. (FF = GFR/RPF.)

Fractional excretion of sodium:

$$FENa^+ = \frac{(U/P)Na^+}{(U/P)Cr} \times 100$$

Inulin clearance (GFR): Men, 110–150 mL/min; women, 105–132 mL/min (corrected to 1.73 m^2 surface area).

Maximal glucose reabsorptive capacity (Tm_G): Men, 300–450 mg/min; women, 250–350 mg/min.

Maximal PAH excretory capacity (Tm_{PAH}): 80–90 mg/min.

Osmolality: On normal diet and fluid intake: Range 500–850 mosm/kg water. Achievable range, normal kidney: Dilution 40–80 mosm; concentration (dehydration) up to 1400 mosm/kg water (at least three to four times plasma osmolality).

Specific gravity of urine: 1.003–1.030.

ABBREVIATIONS

A (Å)	Angstrom unit(s) (10^{-10} m, 0.1 nm)	**CLIP**	Corticotropin-like intermediate lobe peptide
AA	Amino acid	**CMP**	Cytidine monophosphate; 5′-phosphoribosyl cytosine
α-AA	α-Amino acid		
ABP	α-Androgen-binding protein	**CoA-SH**	Free (uncombined) coenzyme A. A pantothenic acid-containing nucleotide that functions in the metabolism of fatty acids, ketone bodies, acetate, and amino acids
ACAT	Acyl-CoA:cholesterol acyltransferase		
ACE	Angiotensin-converting enzyme		
ACh	Acetylcholine		
AChR	Acetylcholine receptor		
ACP	Acyl carrier protein		O
ACTH	Adrenocorticotropic hormone, adrenocorticotropin, corticotropin		‖
		CoA-S·C·CH₃	Acetyl-CoA, "activated acetate." The form in which acetate is "activated" by combination with coenzyme A for participation in various reactions
Acyl-CoA	An acyl derivative of coenzyme A (eg, butyryl-CoA)		
ADH	Alcohol dehydrogenase		
ADH	Antidiuretic hormone (vasopressin)	**COMT**	Catechol-O-methyltransferase
ADP	Adenosine diphosphate	**CPK**	Creatine phosphokinase
Ala	Alanine	**CRH**	Corticotropin-releasing hormone
ALA	Aminolevulinic acid	**CRP**	C-reactive protein
ALT	Alanine aminotransferase	**CT**	Calcitonin
AMP	Adenosine monophosphate	**CTP**	Cytidine triphosphate
ANF	Atrial natriuretic factor	**Cys**	Cysteine
APTT	Activated partial thromboplastin time	**D-**	Dextrorotatory
		D₂ (vitamin)	Ergocalciferol
Arg	Arginine	**D₃ (vitamin)**	Cholecalciferol
Asn	Asparagine	**1,25(OH)₂D₃**	1,25-Dihydroxycholecalciferol
Asp	Aspartic acid	**dA**	Deoxyadenosine
AST	Aspartate aminotransferase	**DAG**	Diacylglycerol
ATP	Adenosine triphosphate	**dATP**	Deoxyadenosine triphosphate
BAL	Dimercaprol (British antilewisite)	**DBH**	Dopamine β-hydroxylase
BLI	Bombesin-like immunoreactivity	**dC**	Deoxycytosine
BPG	Bisphosphoglycerate (diphosphoglycerate)	**dG**	Deoxyguanosine
		DHEA	Dehydroepiandrosterone
cAMP	3′,5′-Cyclic adenosine monophosphate; cyclic AMP	**DHT**	Dihydrotestosterone
		DIT	Diidotyrosine
CBG	Corticosteroid-binding globulin	**DNA**	Deoxyribonucleic acid
CBP	Calcium-binding protein	**DNP**	Dinitrophenol
CCCP	m-Chlorocarbonyl cyanide phenylhydrazone	**DOC**	Deoxycorticosterone
		Dopa	3,4-Dihydroxyphenylalanine
CCK	Cholecystokinin	**DPG**	Diphosphoglycerate (bisphosphoglycerate)
CDP	Cytidine diphosphate		
Cer	Ceramide	**dpm**	Disintegrations per minute
cGMP	3′,5′-Guanosine monophosphate; cyclic GMP	**dT**	Deoxythymidine
		dTMP	Deoxythymidine 5′-monophosphate
CK	Creatine kinase	**dUMP**	Deoxyuridine 5′-monophosphate

E	Enzyme (also Enz)
E$_2$	Estradiol
E.C.	Enzyme commission number (IUB) system
ECF	Extracellular fluid
ECM	Extracellular matrix
EDTA	Ethylenediaminetetraacetic acid. A reagent used to chelate divalent metals
EGF	Epidermal growth factor
Enz	Enzyme (also E)
Eq	Equivalent
eu	Enzyme unit
FAD	Flavin adenine dinucleotide (oxidized)
FADH$_2$	Flavin adenine dinucleotide (reduced)
FDA	Food & Drug Administration
FFA	Free fatty acids
FGF	Fibroblast growth factor
Figlu	Formiminoglutamic acid
FMN	Flavin mononucleotide
FP	Flavoprotein
FSF	Fibrin-stabilizing factor
FSH	Follicle-stimulating hormone
FSHRH	Follicle-stimulating hormone-releasing hormone
g	Gram(s)
g	Gravity
GAG	Glycosaminoglycan
Gal	Galactose
GalNAc	N-Acetylgalactosamine
GAP	GnRH-associated peptide
GDP	Guanosine diphosphate
GFR	Glomerular filtration rate
GH	Growth hormone
GHRH	Growth hormone-releasing hormone
GHRIH	Growth hormone releasing-inhibiting hormone (somatostatin)
GIP	Gastric inhibitory polypeptide
GLC	Gas-liquid chromatography
Glc	Glucose
GlcNAc	N-Acetylglucosamine
GlcUA	Glucuronic acid
Gln	Glutamine
Glu	Glutamic acid
Gly	Glycine
GMP	Guanosine monophosphate
GnRH	Gonadotropin-releasing hormone
GRE	Glucocorticoid response element
GTP	Guanosine triphosphate
Hb	Hemoglobin
hCG	Human chorionic gonadotropin
hCS	Human chorionic somatomammotropin
HDL	High-density lipoprotein
H$_2$folate	Dihydrofolate

H$_4$ folate	Tetrahydrofolate
His	Histidine
HMG-CoA	3-Hydroxy-3-methylglutaryl-CoA
Hyl	Hydroxylysine
Hyp	4-Hydroxyproline
ICD	Isocitric dehydrogenase
IDL	Intermediate-density lipoprotein
IDP	Inosine diphosphate
IF	Initiation factor (for protein synthesis)
IGF	Insulin-like growth factor
IL	Interleukin
Ile	Isoleucine
IMP	Inosine monophosphate; hypoxanthine ribonucleotide
INH	Isonicotinic acid hydrazide (isoniazid)
IP$_3$	Inositol triphosphate
ITP	Inosine triphosphate
IU	International unit(s)
IUB	International Union of Biochemistry
α-KA	α-Keto acid
kcal	Kilocalorie (calorie)
α-KG	α-Ketoglutarate
kJ	Kilojoule
K$_m$	Substrate concentration producing half-maximal velocity (Michaelis constant)
L-	Levorotatory
LCAT	Lecithin:cholesterol acyltransferase
LDH	Lactate dehydrogenase
LDL	Low-density lipoprotein
Leu	Leucine
LH	Luteinizing hormone
LHRH	Luteinizing hormone-releasing hormone
LPH	Lipotropin
LTH	Luteotropic hormone
Lys	Lysine
M	Molar
MAO	Monoamine oxidase
MCH	Mean corpuscular hemoglobin
MCHC	Mean corpuscular hemoglobin concentration
MCV	Mean corpuscular volume
Met	Methionine
MIF	Müllerian-inhibiting factor
MIT	Monoiodotyrosine
mol	Mole(s)
M$_r$	Molecular weight
MRF	Melanocyte-releasing factor
MRH	Melanocyte-releasing hormone
MRIH	Melanocyte release-inhibiting hormone
mRNA	Messenger RNA
MSA	Multiplication-stimulating activity
MSH	Melanocyte-stimulating hormone

MW	Molecular weight		**rRNA**	Ribosomal RNA
NAD⁺	Nicotinamide adenine dinucleotide (oxidized)		**S (Sf) units**	Svedberg units of flotation
			SDA	Specific dynamic action
NADH	Nicotinamide adenine dinucleotide (reduced)		**SDS**	Sodium dodecyl sulfate
			Ser	Serine
NADP⁺	Nicotinamide adenine dinucleotide phosphate (oxidized)		**SGOT**	Serum glutamic-oxaloacetic transaminase
NADPH	Nicotinamide adenine dinucleotide phosphate (reduced)		**SGPT**	Serum glutamic-pyruvic transaminase
NDP	(Any) nucleoside diphosphate		**SH**	Sulfhydryl
NeuAc	*N*-Acetylneuraminic acid		**SHBG**	Sex hormone-binding globulin
NGF	Nerve growth factor		**SLR**	*Streptococcus lactis* R
NTP	(Any) nucleoside triphosphate		**snRNA**	Small nuclear RNA
OA	Oxaloacetic acid		**SRIH**	Somatostatin (growth hormone release-inhibiting hormone)
OD	Optical density			
3b-OHSD	3β-Hydroxysteroid dehydrogenase		**STP**	Standard temperature and pressure (273° absolute, 760 mm Hg)
P_i	Inorganic phosphate (orthophosphate)			
PCR	Polymerase chain reaction		T_3	Triiodothyronine
PCV	Packed cell volume		T_4	Tetraiodothyronine; thyroxine
PDGF	Platelet-derived growth factor		**TBG**	Thyroxine-binding globulin
PEPCK	Phosphoenolpyruvate carboxykinase		**TBPA**	Thyroxine-binding prealbumin
			TEBG	Testosterone-estrogen-binding globulin
PFK	Phosphofructokinase			
PG	Prostaglandin		**TG**	Triacylglycerol (formerly called triglyceride)
Phe	Phenylalanine			
P_i	Inorganic phosphate		**TGF**	Transforming growth factor
PIH	Prolactin release-inhibiting hormone		**Thr**	Threonine
			TLC	Thin-layer chromatography
PL	Placental lactogen		Tm_{Ca}	Tubular maximum for calcium
PL	Pyridoxal		Tm_G	Tubular maximum for glucose
PLP	Pyridoxal phosphate		**TMP**	Thymidine monophosphate
PNMT	Phenylethanolamine-*N*-methyl transferase		**t-PA**	Tissue plasminogen activator
			TRH	Thyrotropin-releasing hormone
PNPA	*p*-Nitrophenylate anion		**Tris**	Tris(hydroxymethyl)aminomethane, a buffer
POMC	Pro-opiomelanocortin			
PP	Pancreatic polypeptide		**tRNA**	Transfer RNA (see also sRNA)
PP_i	Inorganic pyrophosphate		**Trp**	Tryptophan
PRH	Prolactin-releasing hormone		**TSH**	Thyroid-stimulating hormone; thyrotropin
PRIH	Prolactin release-inhibiting hormone			
			TTP	Thymidine triphosphate
PRL	Prolactin		**Tyr**	Tyrosine
Pro	Proline		**UDP**	Uridine diphosphate
PRPP	5-Phosphoribosyl-1-pyrophosphate		**UDPGal**	Uridine diphosphate galactose
			UDPGlc	Uridine diphosphate glucose
PT	Prothrombin time		**UDPGlcUA**	Uridine diphosphoglucuronic acid
PTA	Plasma thromboplastin antecedent		**UDPGluc**	Uridine diphosphoglucuronic acid
			UMP	Uridine monophosphate; uridine-5′-phosphate; uridylic acid
PTC	Plasma thromboplastin component			
PTH	Parathyroid hormone		**UTP**	Uridine triphosphate
PTT	Partial thromboplastin time		V_{max}	Maximal velocity
RBC	Red blood cell		**Val**	Valine
RDA	Recommended daily allowance		**VHDL**	Very high density lipoprotein
RE	Retinol equivalents		**VIP**	Vasoactive intestinal polypeptide
RFLP	Restriction fragment length polymorphism		**VLDL**	Very low density lipoprotein
			VMA	Vanillylmandelic acid
			VNTR	Variable number of tandem repeats
RNA	Ribonucleic acid			
RQ	Respiratory quotient		**vol%**	Volumes percent

Index

Liver, *(cont.)*
  vitamin D and, 544
  vitamin K cycle, 623, 623*f*
Liver disease
  ammonia intoxication and, 644
  α1-antiproteinase deficiency and, 714–715
Locus control region (LCR), 460
Long interspersed repeat sequence (LINE), 402
Loop, 398
Looped domain, 399
Lovastatin, serum cholesterol and, 281
Low-density lipoprotein (LDL), 254
  metabolism, 260
Low-density lipoprotein (LDL) receptor
  regulation, 276–277
  structure, 591*f*
Lung, extensibility, elastin and, 670–671
Lung surfactant, 245
Lung surfactant deficiency, respiratory distress syndrome and, 251
Luteal phase, 574–575
Luteinizing hormone (LH), 527–528
  testicular steroidogenesis and, 569–570
Luteotropic hormone, 527
Lyase, 548
Lymphatic vessel, nutrient absorption and, 641–643
Lymphocyte homing, selectins and, 661–663
Lymphoma, Burkitt's, 764
  myc proto-oncogene and, 765*f*
  reciprocal translocation and, 764*f*
Lysine, symbol and structural formula, 25*t*
L-Lysine, catabolism, 323*f*
Lysogenic pathway, 451
Lysolecithin, 151, 249
  structure, 152*f*
Lysophospholipase, 248
Lysophospholipid, 151
Lysosomal enzyme, I-cell disease and, 664
Lysosomal hydrolase, genetic deficiencies, 664–665, 665*t*
Lysosome, 657
  marker and major functions, 9*t*
Lysozyme, cleft region, catalytic site, 80*f*
Lytic infection, 451
Lytic pathway, 451
D-Lyxose, physiologic importance, 139*t*

Macrophage colony-stimulating factor, viral oncogenes and, 762
α2-Macroglobulin, 715–716
Macromineral, essential, 630*t*
  nutritional requirements, 626*t*
Magnesium, essential macromineral, characteristics, 630*t*
Magnetic resonance imaging (MRI), protein structure and, 48
Major groove, 388, 450

Malabsorption
  disturbances due to, 646*t*
  fructose-sorbitol, 645
  monosaccharide, 645
Malabsorption syndrome, 635
Malaria, 4
Malate dehydrogenase, 172
  citric acid cycle and, 173
Malate shuttle, 132, 133*f*
Male reproductive system, pathophysiology, hormonal defects and, 571
Maleylacetoacetate *cis,trans* isomerase, 317
Maleylacetoacetate hydrolase, 317
Malic enzyme, 219
  inducer, repressor, activator, inhibitor, 198*t*
Malignant cell, surface changes, 777*t*
Malignant hyperthermia, 686
  calcium-release channel and, 694
  scheme of causation, 694*f*
Malignant transformation, changes occurring, 760–761, 762*t*
Malnutrition, protein-energy, 628
Malonate, respiratory chain and, 128
Malonate semialdehyde, 332
Malonyl-CoA
  biosynthesis, 217*f*
  fatty acid synthesis and, 216
Malonyl transacylase, 219
Maltose, 140–141
  source and clinical significance, 142*t*
Mammalian genome, redundancy, 400–402
Mammary gland, development, 577–578
Mammotropin, 327
Manganese, essential micromineral, characteristics, 631*t*
Mannosamine, 212
Mannose, glycoprotein, 650*t*
D-Mannose, physiologic importance, 139*t*
Mannose-binding protein, 665
Mannosidosis, 664–665, 665*t*
Maple syrup urine disease, 329
Marasmus, 109, 628
  kwashiorkor compared, 816*t*
Marble bone disease, 682
Marfan syndrome, 671
  causation of major signs, sequence of events, 673*f*
Mass spectrometry, 10
  glycoproteins and, 649
  peptide sequencing and, 38
Maternal retinitis pigmentosa, mutation and clinical expression, 802*t*
Maxam and Gilbert method, 473
Mazindol, Parkinson's disease therapy and, 806
Meal, concentration of substances in blood and, 830
Medicine
  preventive, 3
  relationship with biochemistry, 2
Megaloblastic anemia, 609, 610–611
  cause, 734*t*

Melanin, 340
Melanocyte-stimulating hormone (MSH), 530
Melanogenesis, 530
MELAS, 134
  mutation and clinical expression, 802*t*
Melatonin
  biosynthesis and metabolism, 338*f*
  serotonin forming, 337–338
Melphalan, site of action and use, 774*t*
Membrane, 483–507
  abnormalities, disorders associated, 506–507, 506*t*
  amphipathic, 485
  artificial models, 489
  assembly, 490–499, 499*t*
  asymmetry, 487
  enzymatic markers, 487*t*
  erythrocyte, 738–741, 739*t*
    principal proteins, 739*t*
    protein interactions, 740*f*
  fluid mosaic model, 488*f*
  fluid phospholipid matrix, 489–490
  glucose transport, 503–504, 504*f*
  lipid bilayers, 485–487
  nerve impulse transmission, 503
  plasma
    functions, 501–506
    macromolecule transport, 504–506
  protein insertion, 488*f*
  ratio of protein to lipid, 484*f*
  specialized functions, 499–500
  transport systems, 501*f*
  vesicle fusion, 491*f*
Membrane fatty acid-binding protein, 257
Membrane transport, insulin and, 587
Menadione, 622
Menaquinone-7, 622
Menkes' disease, 670*t*
  copper-binding P-type ATPase gene and, 713
  Wilson's disease compared, 714*t*
Menopause, 578
Menstrual cycle
  hormonal/physiologic changes during, 575*f*
  phases, 574–575
Menstruation, 574
Mercaptolactate-cysteine disulfiduria, 314
Mercaptopurine, 366
  site of action and use, 774*t*
  structure, 367*f*
3-Mercaptopyruvate-cysteine disulfiduria, defect, 316*t*
Mesobilirubinogen, structure, 354*f*
Messenger RNA. *See* mRNA
Metabolic acidosis, 304, 586
Metabolic activation, 758
Metabolic alkalosis, 304
Metabolic disorder, 307–308
  phenylalanine catabolism and, 320–325
Metabolic pathway, 11, 159–167
  categories, 158*f*
  regulation, 288, 289*t*
  strategies for studying, 12*f*

L A N G E
medical books

Available at your local health science bookstore
or by calling
Appleton & Lange toll-free
1-800-423-1359

A smart investment
in your medical career

Basic Science Textbooks

chemistry
amination & Board Review
cavage & King
5, ISBN 0-8385-0661-5, A0661-7
lor Atlas of Basic Histology
man
3, ISBN 0-8385-0445-0, A0445-5
6 First Aid for the USMLE Step 1
tudent-to-Student Guide
ushan, et al.
6, ISBN 0-8385-2597-0, A2597-1
etz, Melnick, & Adelberg's
dical Microbiology, 20/e
oks, Butel, & Ornston
5, ISBN 0-8385-6243-4, A6243-8
nual for Human Dissection
otographs with Clinical Applica-
ns
las
4, ISBN 0-8385-6133-0, A6133-1
ncise Pathology, 2/e
andrasoma & Taylor
5, ISBN 0-8385-1229-1, A1229-2
roduction to Clinical Psychiatry
in
6, ISBN 0-8385-4333-2, A4333-9
dical Biostatistics & Epidemiology
amination & Board Review
ex-Sorlie
5, ISBN 0-8385-6219-1, A6219-8

**Fundamentals of Medical Cell Biology
and Histology**
Fuller
1996, ISBN 0-8385-1384-0, A1384-5
Review of Medical Physiology, 17/e
Ganong
1995, ISBN 0-8385-8431-4, A8431-7
**First Aid for the USMLE Step 2
A Student-to-Student Guide**
Go, Curet-Salim, & Fullerton
1996, ISBN 0-8385-2591-1, A2591-4
Medical Epidemiology, 2/e
*Greenberg, Daniels, Flanders, Eley, & Bor-
ing*
1996, ISBN 0-8385-6206-X, A6206-5
Basic Histology, 8/e
Junqueira, Carneiro, & Kelley
1995, ISBN 0-8385-0567-8, A0567-6
Basic & Clinical Pharmacology, 6/e
Katzung
1995, ISBN 0-8385-0619-4, A0619-5
Pharmacology
Examination & Board Review, 4/e
Katzung & Trevor
1995, ISBN 0-8385-8067-X, A8067-9
First Aid for the Match
Le, Bhushan, & Amin
1996, ISBN 0-8385-2596-2, A2596-3
Medical Microbiology & Immunology
Examination & Board Review, 4/e
Levinson & Jawetz
1996, ISBN 0-8385-6225-6, A6225-5
Clinical Anatomy
Lindner
1989, ISBN 0-8385-1259-3, A1259-9

**Pathophysiology of Disease
An Introduction to Clinical Medicine**
McPhee, Lingappa, Ganong, & Lange
1995, ISBN 0-8385-7815-2, A7815-2
Harper's Biochemistry, 24/e
Murray, Granner, Mayes, & Rodwell
1996, ISBN 0-8385-3611-5, A3611-9
Pathology
Examination & Board Review
Newland
1995, ISBN 0-8385-7719-9, A7719-6
Basic Histology
Examination & Board Review, 3/e
Paulsen
1996, ISBN 0-8385-2282-3, A2282-0
Basic & Clinical Immunology, 8/e
Stites, Terr, & Parslow
1994, ISBN 0-8385-0561-9, A0561-9
Correlative Neuroanatomy, 22/e
Waxman & deGroot
1995, ISBN 0-8385-1091-4, A1091-6

Clinical Science Textbooks

Clinical Neurology, 3/e
Aminoff, Greenberg, & Simon
1996, ISBN 0-8385-1383-2, A1383-7
**Understanding Health Policy:
A Clinical Approach**
Bodenheimer & Grumbach
1995, ISBN 0-8385-3678-6, A3678-8

(more on reverse)

Clinical Cardiology, 6/e
Cheitlin, Sokolow, & McIlroy
1993, ISBN 0-8385-1093-0, A1093-2
Fluid & Electrolytes
Physiology & Pathophysiology
Cogan
1991, ISBN 0-8385-2546-6, A2546-8
Basic & Clinical Biostatistics, 2/e
Dawson & Trapp
1994, ISBN 0-8385-0542-2, A0542-9
Basic Gynecology and Obstetrics
Gant & Cunningham
1993, ISBN 0-8385-9633-9, A9633-7
Review of General Psychiatry, 4/e
Goldman
1995, ISBN 0-8385-8421-7, A8421-8
Principles of Clinical
Electrocardiography, 13/e
Goldschlager & Goldman
1990, ISBN 0-8385-7951-5, A7951-5
Basic & Clinical Endocrinology, 4/e
Greenspan & Baxter
1994, ISBN 0-8385-0560-0, A0560-1
Occupational Medicine
LaDou
1990, ISBN 0-8385-7207-3, A7207-2
Primary Care of Women
Lemcke, Pattison, Marshall, & Cowley
1995, ISBN 0-8385-9813-7, A9813-5
Clinical Anesthesiology, 2/e
Morgan & Mikhail
1996, ISBN 0-8385-1381-6, A1381-1
Dermatology
Orkin, Maibach, & Dahl
1991, ISBN 0-8385-1288-7, A1288-8
Rudolph's Fundamentals of
Pediatrics
Rudolph & Kamei
1994, ISBN 0-8385-8233-8, A8233-7
Genetics in Primary Care and Clinical
Medicine
Seashore & Wappner
1995, ISBN 0-8385-3128-8, A3128-4
Smith's General Urology, 14/e
Tanagho & McAninch
1995, ISBN 0-8385-8612-0, A8612-2
Clinical Oncology
Weiss
1993, ISBN 0-8385-1325-5, A1325-8
General Ophthalmology, 14/e
Vaughan, Asbury, & Riordan-Eva
1995, ISBN 0-8385-3127-X, A3127-6

CURRENT Clinical References

CURRENT Critical Care Diagnosis &
Treatment, 2/e
Bongard & Sue
1996, ISBN 0-8385-1454-5, A1454-6
CURRENT Diagnosis & Treatment in Car-
diology
Crawford
1995, ISBN 0-8385-1444-8, A1444-7

CURRENT Diagnosis & Treatment in
Vascular Surgery
Dean, Yao, & Brewster
1995, ISBN 0-8385-1351-4, A1351-4
CURRENT Obstetric & Gynecologic
Diagnosis & Treatment, 9/e
DeCherney & Pernoll
1996, ISBN 0-8385-1401-4, A1401-7
CURRENT Diagnosis & Treatment in
Gastroenterology
Grendell, McQuaid, & Friedman
1996, ISBN 0-8385-1448-0, A1448-8
CURRENT Pediatric Diagnosis &
Treatment, 12/e
Hay, Groothuis, Hayward, & Levin
1995, ISBN 0-8385-1446-4, A1446-2
CURRENT Emergency Diagnosis &
Treatment, 5/e
Saunders & Ho
1996, ISBN 0-8385-1450-2, A1450-4
CURRENT Diagnosis & Treatment in
Orthopedics
Skinner
1995, ISBN 0-8385-1009-4, A1009-8
CURRENT Medical Diagnosis &
Treatment 1996
Tierney, McPhee, & Papadakis
1996, ISBN 0-8385-1465-0, A1465-2
CURRENT Surgical Diagnosis &
Treatment, 11/e
Way
1996, ISBN 0-8385-1456-1, A1456-1

LANGE Clinical Manuals

Dermatology
Diagnosis and Therapy
Bondi, Jegasothy, & Lazarus
1991, ISBN 0-8385-1274-7, A1274-8
Practical Oncology
Cameron
1994, ISBN 0-8385-1326-3, A1326-6
Office & Bedside Procedures
Chesnutt, Dewar, Locksley, & Tureen
1992, ISBN 0-8385-1095-7, A1095-7
Psychiatry
Diagnosis & Therapy 2/e
Flaherty, Davis, & Janicak
1993, ISBN 0-8385-1267-4, A1267-2
Neonatology
**Management, Procedures, On-Call
Problems, Diseases and Drugs, 3/e**
Gomella, Cunningham, Eyal, & Zenk
1995, ISBN 0-8385-1331-X, A1331-6
Practical Gynecology
Jacobs & Gast
1994, ISBN 0-8385-1336-0, A1336-5
Drug Therapy, 2/e
Katzung
1991, ISBN 0-8385-1312-3, A1312-6

Ambulatory Medicine
The Primary Care of Families
Mengel & Schwiebert
1993, ISBN 0-8385-1294-1, A1294-6
Poisoning & Drug Overdose, 2/e
Olson
1994, ISBN 0-8385-1108-2, A1108-8
Internal Medicine
Diagnosis and Therapy, 3/e
Stein
1993, ISBN 0-8385-1112-0, A1112-0
Surgery
Diagnosis & Therapy
Stillman
1989, ISBN 0-8385-1283-6, A1283-9
Medical Perioperative Management
Wolfsthal
1989, ISBN 0-8385-1298-4, A1298-7

LANGE Handbooks

Handbook of Gynecology &
Obstetrics
Brown & Crombleholme
1992, ISBN 0-8385-3608-5, A3608-5
HIV/AIDS Primary Care Handbook
Carmichael, Carmichael, & Fischl
1996, ISBN 0-8385-3777-9, A3777-8
Pocket Guide to Diagnostic Tests
Detmer, McPhee, Nicoll, & Chou
1992, ISBN 0-8385-8020-3, A8020-8
Handbook of Poisoning
**Prevention, Diagnosis & Treatment,
12/e**
Dreisbach & Robertson
1987, ISBN 0-8385-3643-3, A3643-2
Handbook of Clinical Endocrinology,
2/e
Fitzgerald
1992, ISBN 0-8385-3615-8, A3615-0
Clinician's Pocket Reference, 7/e
Gomella
1993, ISBN 0-8385-1222-4, A1222-7
Surgery on Call, 2/e
Gomella & Lefor
1996, ISBN 0-8385-8746-1, A8746-8
Internal Medicine On Call, 2/e
Haist & Robbins
1996, ISBN 0-8385-4056-2, A4056-6
Obstetrics & Gynecology On Call
Horowitz & Gomella
1993, ISBN 0-8385-7174-3, A7174-4
Pocket Guide to Commonly
Prescribed Drugs, 2/e
Levine
1996, ISBN 0-8385-8099-8, A8099-2
Handbook of Pediatrics, 17/e
Merenstein, Kaplan, & Rosenberg
1994, ISBN 0-8385-3657-3, A3657-2

 Appleton & Lange • P.O. Box 120041 • Stamford, CT • 06912-0041 • 1-800-423-1359